CLEMENTE · ANATOMY

ANATOMY

A Regional Atlas of the Human Body

Carmine D. Clemente, Ph.D.

Professor of Anatomy
University of California at Los Angeles School of Medicine;
Professor of Surgery (Anatomy)
Charles R. Drew Postgraduate Medical School
Los Angeles, California

Third Edition

Lea & Febiger · Philadelphia–London

Urban & Schwarzenberg · Baltimore–Munich

The pictures on the end leaves of the present volume show reproductions of the work of "Andreas Vesalii de corporis humani fabrica libri septem", which appeared in Basel in 1542. Andreas Vesalius (* Brussels 1514, † 1564) urged and practiced the dissection of human bodies and may be considered to be the founder of Modern Anatomy.

Editor's address:

Carmine D. Clemente, Ph. D., Department of Anatomy, UCLA School of Medicine, Los Angeles, California 90024 USA

Publisher's addresses:

Lea & Febiger
200 Chester Field Parkway
Malvern, Pennsylvania 19355-9725
U.S.A.
(215) 251-2230

Urban & Schwarzenberg
Landwehrstraße 61
D-8000 München 2
Germany

The illustrations of this atlas were originally published in

Sobotta, Atlas der Anatomie des Menschen, edited by H. Ferner and J. Staubesand, Urban & Schwarzenberg,

with the exception of figures 16, 18, 38, 39, 178, 183, 315, 334, 337, 365, 384, 405, 407, 506, 517, 584, 595, 596, 607, 611, 616, 617, 742, 749, 752, 780, 782, 792, 804 (from Benninghoff/Goerttler, Lehrbuch der Anatomie des Menschen, edited by H. Ferner and J. Staubesand, Urban & Schwarzenberg); 6, 36, 73, 96, 97, 113, 133, 150, 200, 271, 298, 304, 311, 342, 400, 401, 457 (from Lothar Wicke, Atlas of Radiologic Anatomy, 3rd. Ed., Urban & Schwarzenberg); 13, 14, 34, 63, 64, 120, 132, 136, 170, 176, 188, 208, 209, 212, 214, 219, 220, 226, 227, 230, 239, 240, 243–246, 253, 257, 258, 296, 299, 300, 320, 331, 332, 388, 393, 399, 444, 446, 448, 459, 486–491, 515, 516, 521, 574, 587, 588, 592, 593, 615, 634, 635, 640, 642, 654–657, 678, 686, 712–721, 723–725, 733, 735, 737, 739, 740, 746, 747, 779, 787, 789, 793, 801 (from Eduard Pernkopf, Atlas of Topographical and Applied Human Anatomy, 2nd Ed., edited by H. Ferner, Urban & Schwarzenberg).
Figures 17, 119, 232, 502, 504 were drawn for this atlas by Jill Penkhus.

The first edition of this atlas was distributed in North America by Lea & Febiger, Philadelphia

10 9 8 7 6 5 4

Library of Congress Cataloging-in-Publication Data

Clemente, Carmine D.
Anatomy, a regional atlas of the human body.

Includes index.
1. Anatomy, Surgical and topographical--Atlases.
I. Title. [DNLM: 1. Anatomy, Regional--atlases.
QS 17 C626a]
QM531.C57 1986 611'.0022'2 85-11157
ISBN 0-8067-0323-7

Printed in Germany by Neue Stalling, Oldenburg

ISBN 0-8067-0323-7

PREFACE TO THE THIRD EDITION

Once again my primary objective in the preparation of this edition has been to facilitate the task of the learning of anatomy by students in the medical, dental and other health professions. Because of the popularity and acceptance of the two previous editions, it has been possible in this edition to expand the coverage of certain regions of the body that were treated more thinly in the 1975 and 1981 editions. I extend my appreciation to the publisher for allowing me this freedom. It is sometimes said that if a book used in the teaching of a major professional field ever achieves a third edition, there is a possibility that it has finally reached maturity. I hope that this is the case with this atlas. There are many changes in this book and a large number of these were suggested by students and by my professional colleagues. I have the good fortune of being able to teach anatomy to first-year medical students at UCLA as well as to surgical residents at the Martin Luther King, Jr. Memorial Hospital here in Los Angeles. This consistent exposure to students who are studying human anatomy for the first time and others who are reviewing anatomy during their residency training period helps to keep me informed of student needs and how certain body regions could be more completely or more clearly demonstrated in the atlas.

There are 96 new figures included in this edition while six figures from the 2nd edition have been eliminated. This net increase of 90 figures represents a 13% enlargement of the atlas over the previous edition. Of the new figures, a total of 34 (35%) were added to the sections on the thorax, abdomen and pelvis, 37 (39%) were added to the limbs and 25 (26%) were added to the sections on neck and head. More specifically, for the thorax some of the figures added depict the bony thoracic cage, the position of the heart within the chest, the coronary arteries, the sympathetic trunks and the fetal heart. Four new cross sections at various levels of the abdomen along with additional figures on the anterior abdominal wall, the female inguinal region, lymphatics of the stomach, the appendix and its blood supply, the posterior abdominal wall, and the female perineum have been included. In the sections on the upper and lower extremities there is an additional coverage of the shoulder and elbow joints and especially the knee joint. Other figures illustrating the surface anatomy and cutaneous innervation of the limbs, as well as arteriograms and other X-rays have also been included. The section on neck and head has been enhanced by 10 new figures on the eye, 6 on the oral cavity, and 4 on the ear. Five other figures on the neck and dura mater have also been added.

As in previous editions the figures in this atlas are derived principally from the Johannes Sobotta collection and from the Eduard Pernkopf collection. The latter were initially published in Professor Pernkopf's multivolume classic textbook *Topographische Anatomie des Menschen* and then were later incorporated in atlas format. Additionally, a number of other figures have come from the Benninghoff-Goerttler textbooks and nearly all of the radiographs have come from the *Atlas of Radiologic Anatomy* (3rd Edition) by Dr. Lothar Wicke of Vienna. I wish to extend my sincere thanks to Professors Helmut Ferner of Vienna and Jochen Staubesand of Freiburg, who edit the German editions of the Sobotta and Pernkopf atlases and the Benninghoff-Goerttler books, and to Dr. Wicke for use of his elegant radiographs.

One of the more significant features of this edition is the fact that over 225 of the figures used in this book are precise repaintings of the older illustrations used in former editions. Further, a number of other figures which were black and white illustrations in previous editions now are presented in full color. This new art is, for the most part, brighter and more vivid in color than the old, and my many thanks go to the several artists who produced these new works because their creations strictly adhered to the older illustrations and resulted in meticulously perfect reproductions. These were necessary for the continued publications of the atlas, and I wish to thank the publisher for the significant expenditure involved.

I wish to thank the many students who have written to me for their kind comments and for offering suggestions of how this atlas might be improved. I both encourage and welcome users of this edition to do the same. My appreciation also goes to Dr. Caroline Belz Caloyeras who, with dedication, edited and typed the manuscript for this edition and helped prepare the index. At Urban & Schwarzenberg I wish to thank Mr. Michael Urban and Wulf J. Dietrich in Munich and Mr. Braxton Mitchell and Norman W. Och in Baltimore for their continuing support in this effort to create an atlas that best meets the needs of students. Finally and most importantly, I want to thank my wife, Julie, for her patience and for her compassionate and generous spirit in support of my writing efforts.

CARMINE D. CLEMENTE

Los Angeles, California, December, 1986

FROM THE PREFACE TO THE FIRST EDITION

Twenty-five years ago, while a student at the University of Pennsylvania, I marvelled at the clarity, completeness, and boldness of the anatomical illustrations of the original German editions of Professor Johannes Sobotta's Atlas and their excellent three-volume, English counterparts, the recent editions of which were authored by the late Professor Frank H. J. Figge. It is a matter of record that before World War II these atlases were the most popular ones consulted by American medical students. In the United States, with the advent of other anatomical atlases, the shortening of courses of anatomy in the medical schools, and the increase in publishing cost, the excellent but larger editions of the Sobotta atlases have become virtually unknown to a full generation of students. During the past twenty years of teaching Gross Anatomy at the University of California at Los Angeles I have found only a handful of students who are familiar with the beautiful and still unexcelled Sobotta illustration.

This volume introduces several departures from the former Sobotta atlases. It is the first English edition that presents the Sobotta plates in a regional sequence – the pectoral region and upper extremity, the thorax, the abdomen, the pelvis and perineum, the lower extremity, the back, vertebral column and spinal cord, and finally the neck and head. This sequence is consistent with that followed in many courses presented in the United States and Canada and one which should be useful to students in other countries.

Several illustrations never before published are presented in this Atlas. These have been drawn by Jill Penkhus, who for a number of years has been the resident medical artist in the Anatomy Department at the UCLA School of Medicine.

Many have contributed to bringing this Atlas to fruition. I wish to thank Dr. David S. Maxwell, Professor and Vice Chairman for Gross Anatomy and my colleague at UCLA, for his encouragement and suggestions. I also with to express my appreciation to Caroline Belz and Louise Campbell, who spent many hours proofreading and typing the original text. I especially wish to thank Mary Mansor for constructing the index – a most laborious task. I am grateful to Barbara Robins for her assistance in typing some of the early parts of the manuscript, and above all, to her sister Julie, who is my wife and who makes all of my efforts worthwhile through her encouragement and devotion.

Los Angeles, California, January 1975 CARMINE D. CLEMENTE

FROM THE PREFACE TO THE SECOND EDITION

Seldom does a professor experience more gratification than when students at home and throughout the world express kind words regarding a teaching resource which he had offered tentatively and, initially, with some trepidation. After publication of the first edition of this altlas, I received nearly two hundred letters from medical and dental students, students of kinesiology, physical therapy and nursing as well as residents in the surgical specialties. Not one failed to comment in some manner on the beauty of the reproduced Sobotta figures (for which the publishers should be commended), and most letters made excellent suggestions of items to be included in future editions, while others uncovered typographical errors or mistaken facts. I am most grateful because this feedback information allowed me to alter labels and correct misspellings in time for the subsequent reprintings of the first edition and to add important transition figures in this edition.

The wide distribution and adoption of this atlas has allowed an opportunity for its revision. I have observed the use of the first edition in the classroom and in the gross anatomy laboratory setting for over five years, and this has given me further direction in assembling this revision. The addition of 130 figures and the elimination of only 10 published in the first edition represents an increase of 120 figures, or slightly over 20 percent. These have not been distributed equally among the seven parts of the atlas. Nearly one-third of the new figures were added to the section on neck and head, another one-third to the sections on thorax and abdomen, and the remaining one-third to the extremities, the pelvis and the back.

Some special emphasis has been placed in this edition on the inclusion of figures illustrating radiographic views of different regions, lymphatic drainage, the anatomy of the newborn, more material on the oral cavity, as well as additional cross sections of the limbs.

I am most grateful to Dr. Caroline Belz who contributed superb editorial assistance in the preparation of the text for this edition and to Mrs. Joyce Fried who spent many hours helping to revise the index. From Urban & Schwarzenberg, I wish to thank Mr. Michael Urban, Dr. Richard Degkwitz and Mr. Klaus Gullath in Munich and Mr. Braxton Mitchell in Baltimore for their conceptual suggestions and for the practical solution of innumerable problems. And, as always, to my wife, Julie, I owe a continuing debt of gratitude for her selflessness and for her true understanding of the academic way of life.

Los Angeles, California, January 1981 CARMINE D. CLEMENTE

CONTENTS

PART I PECTORAL REGION AND UPPER EXTREMITY

A. Body Regions and Dermatomes FIGS. 1– 2
B. Mammary Gland and a typical spinal nerve FIGS. 3– 10
C. Anterior thoracic region and pectoral muscles FIGS. 11– 16
D. Axilla: Muscles and vessels; the brachial plexus FIGS. 17– 20
E. Shoulder Region: Muscles, vessels, nerves and bones FIGS. 21– 28
F. Shoulder joint FIGS. 29– 36
G. Upper extremity: Surface anatomy, superficial nerves and vessels and the deep fascia FIGS. 37– 48
H. Brachial region: Anterior and posterior FIGS. 49– 60
I. Upper extremity and cross sections of the brachium FIGS. 61– 64
J. Antebrachial region FIGS. 65– 79
K. Bones of the forearm, the elbow joint and cross sections of the antebrachium FIGS. 80–101
L. Dorsum of the hand FIGS. 102–105
M. Palmar aspect of the hand FIGS. 106–117
N. Skeleton of the wrist and hand; cross sections of the hand and fingers FIGS. 118–128
O. Joints of the wrist and fingers FIGS. 129–135

PART II THE THORAX

A. Thoracic cage: Clavicle, sternum, ribs and intercostal structures FIGS. 136–151
B. Pleura, bronchial tree and lungs FIGS. 152–169
C. Position of the heart and its valves in the thoracic cavity FIGS. 170–174
D. Newborn heart; fetal and adult circulations FIGS. 175–183
E. Mediastinum viewed from the right and left sides FIGS. 184–185
F. Middle mediastinum: Pericardium and heart FIGS. 186–213
G. Median sagittal and frontal sections of the thorax FIGS. 214–215
H. Posterior mediastinum: Organs, vessels, nerves, thoracic duct and diaphragm FIGS. 216–230

PART III THE ABDOMEN

A. Anterior abdominal wall inguinal region: Muscles, vessels and nerves FIGS. 231–254
B. Anterior abdominal wall and visceria of the newborn child FIGS. 255–256
C. Abdominal cavity. Organ relations and peritoneum FIGS. 257–264
D. Abdominal cavity: Celiac axis, stomach and spleen FIGS. 265–278
E. Abdominal cavity: Liver, biliary system and gall bladder FIGS. 279–284
F. Abdominal cavity: Duodenum and pancreas FIGS. 285–293
G. Abdominal cavity: Mesenteric vessels, nerves, lymphatics; the small and large intestine, rectum FIGS. 294–316
H. Posterior abdominal wall: Kidney, ilopsoas region, lumbar plexus, lumbar sympathetic trunk, lumbar lymphatic channels and lumbar vessels FIGS. 317–342

PART IV THE PERINEUM AND PELVIS

A. Male perineum: Muscles, vessels and nerves; penis FIGS. 343–352
B. Male pelvis: Bladder, seminal vesicle, prostate; muscles and vessels; urethra, male urogenital diaphragm FIGS. 353–366
C. Female perineum: External genitalia; female urogenital diaphragm; muscles, vessels and nerves FIGS. 367–371
D. Female pelvis: Muscles, uterus, ovary, blood vessels and lymphatics FIGS. 372–387
E. Osteology and ligaments of the pelvis FIGS. 388–398
F. Fetus *in utero;* X-ray of pelvis FIGS. 399–401

PART V THE LOWER EXTREMITY

A. Vessels, dermatomes and superficial nerves; surface anatomy and lymph nodes — FIGS. 402–415
B. Anterior thigh: Muscles, vessels and nerves — FIGS. 416–423
C. Gluteal region and posterior thigh — FIGS. 424–430
D. Femur and hip joint — FIGS. 431–439
E. Knee: Popliteal fossa and knee joint — FIGS. 440–459
F. Leg: Superficial nerves and vessels; anterior and lateral compartments — FIGS. 460–467
G. Dorsum of the foot — FIGS. 468–471
H. Leg: Posterior compartment; tibia and fibula — FIGS. 472–485
I. Cross sections of thigh and leg — FIGS. 486–491
J. Plantar aspect of the foot; bones of the foot — FIGS. 492–505
K. Tendons at the ankle; joints of the foot — FIGS. 506–514
L. Cross sections of the foot — FIGS. 515–516

PART VI THE BACK, VERTEBRAL COLUMN AND SPINAL CORD

A. Cutaneous nerves, dermatomes, regions of the back — FIGS. 517–521
B. Muscles of the back: Superficial, intermediate and deep — FIGS. 522–526
C. Posterior neck and suboccipital region — FIGS. 527–529
D. Atlas, axis, cervical joints, cervical vertebrae — FIGS. 530–543
E. Vertebral column, thoracic vertebrae and costovertebral joints — FIGS. 544–554
F. Lumbar vertebrae, sacrum and coccyx — FIGS. 555–562
G. Spinal cord, spinal roots, blood supply — FIGS. 563–573

PART VII THE NECK AND HEAD

A. Surface anatomy of the neck and the platysma muscle; vessels and nerves of the anterior and posterior triangles of the neck — FIGS. 574–582
B. The strap muscles, the great blood vessels and lymphatic channels of the neck; the thyroid gland — FIGS. 583–593
C. Prevertebral region: Muscles and vertebral artery — FIGS. 594–596
D. Submandibular region: Suprahyoid muscles and submandibular gland — FIGS. 597–600
E. Anterior skull, anterior face and muscles of facial expression — FIGS. 601–605
F. Lateral face, parotid gland, muscles of mastication, the lateral skull and the temporomandibular joint — FIGS. 606–617
G. Blood vessels, nerves and lymphatics of the head, including the superficial and deep face and the infratemporal fossa — FIGS. 618–625
H. Calvaria, diploic veins, dura mater and meningeal vessels — FIGS. 626–636
I. Dural sinuses, cranial cavity, inferior surface of the brain, internal carotid artery, base of the skull — FIGS. 637–647
J. The orbit and eye: Muscles, vessels, nerves and eyeball — FIGS. 648–681
K. External nose, nasal cavity, nasal sinuses and nasopharynx — FIGS. 682–695
L. Mandible, maxilla and teeth — FIGS. 696–721
M. Oral cavity, palate, tongue and oropharynx — FIGS. 722–747
N. The pharyngeal constrictor muscles; pharyngeal vessels and nerves; laryngopharynx — FIGS. 748–755
O. The larynx, laryngeal cartilages, muscles and vessels — FIGS. 756–772
P. The external, middle and internal ear; temporal bone and facial canal; the cochlea and semicircular canals — FIGS. 773–805

PART VIII INDEX — Pages 417–439

PART I: PECTORAL REGION AND UPPER EXTREMITY

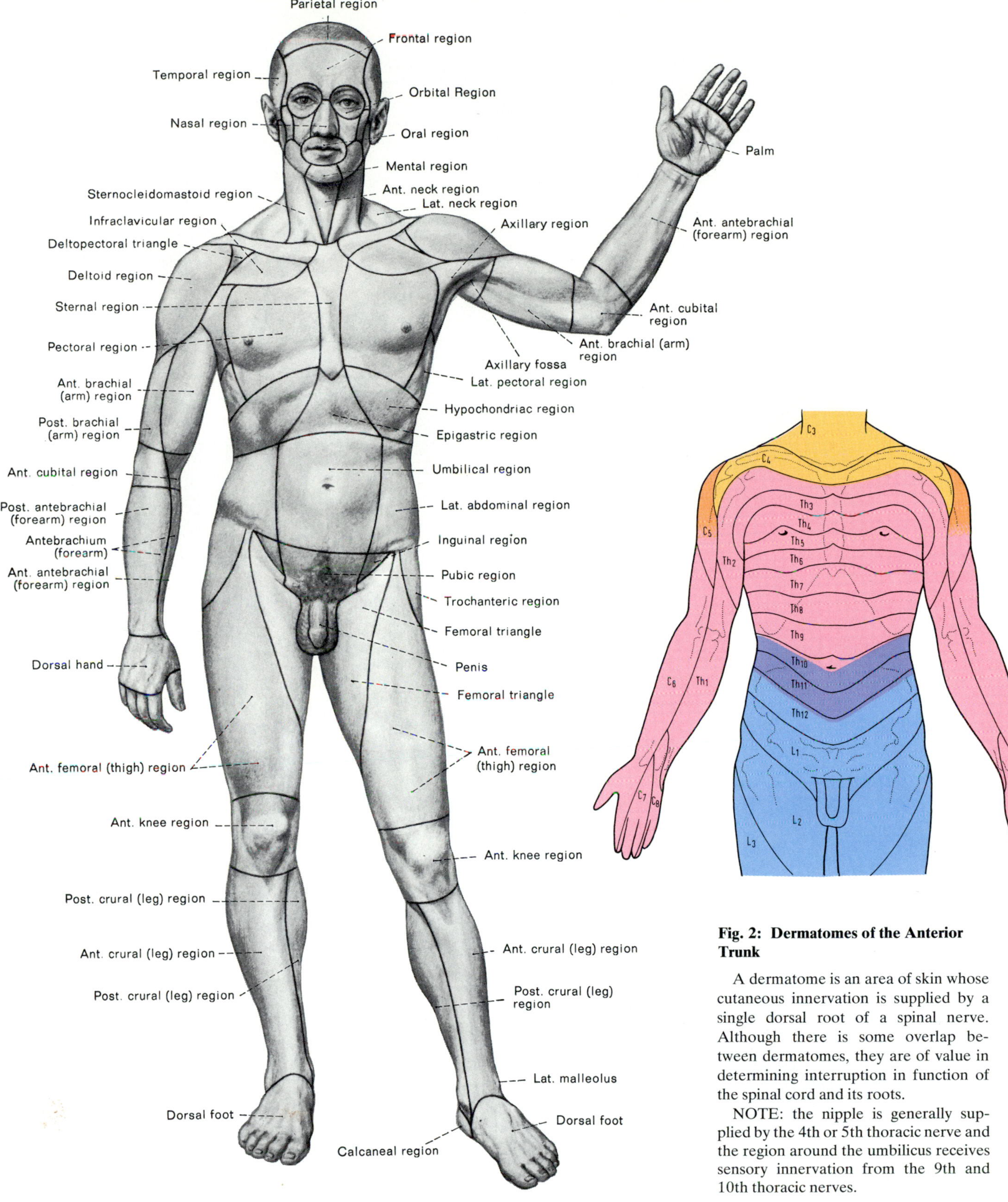

Parietal region
Frontal region
Temporal region
Orbital Region
Nasal region
Oral region
Mental region
Palm
Sternocleidomastoid region
Ant. neck region
Lat. neck region
Infraclavicular region
Axillary region
Ant. antebrachial (forearm) region
Deltopectoral triangle
Deltoid region
Sternal region
Ant. cubital region
Pectoral region
Ant. brachial (arm) region
Axillary fossa
Ant. brachial (arm) region
Lat. pectoral region
Hypochondriac region
Post. brachial (arm) region
Epigastric region
Ant. cubital region
Umbilical region
Post. antebrachial (forearm) region
Lat. abdominal region
Antebrachium (forearm)
Inguinal region
Ant. antebrachial (forearm) region
Pubic region
Trochanteric region
Femoral triangle
Dorsal hand
Penis
Femoral triangle
Ant. femoral (thigh) region
Ant. femoral (thigh) region
Ant. knee region
Ant. knee region
Post. crural (leg) region
Ant. crural (leg) region
Ant. crural (leg) region
Post. crural (leg) region
Post. crural (leg) region
Lat. malleolus
Dorsal foot
Dorsal foot
Calcaneal region
C3
C4
Th3
Th4
C5
Th5
Th2
Th6
Th7
Th8
Th9
Th10
C6
Th1
Th11
Th12
L1
C7
C8
L2
L3

Fig. 2: Dermatomes of the Anterior Trunk

A dermatome is an area of skin whose cutaneous innervation is supplied by a single dorsal root of a spinal nerve. Although there is some overlap between dermatomes, they are of value in determining interruption in function of the spinal cord and its roots.

NOTE: the nipple is generally supplied by the 4th or 5th thoracic nerve and the region around the umbilicus receives sensory innervation from the 9th and 10th thoracic nerves.

Fig. 1: The Regions of the Body: Anterior View

Every surface area and region of the body has been identified by a specific name in order to describe more precisely the location of anatomical structures. Note that regions are named after underlying or adjacent bones (sternal, parietal, frontal, temporal, infraclavicular, femoral) while other regions are named for underlying muscles (sternocleidomastoid, deltoid, pectoral). Still other regions are named after specialized anatomical structures (umbilical, oral, nasal).

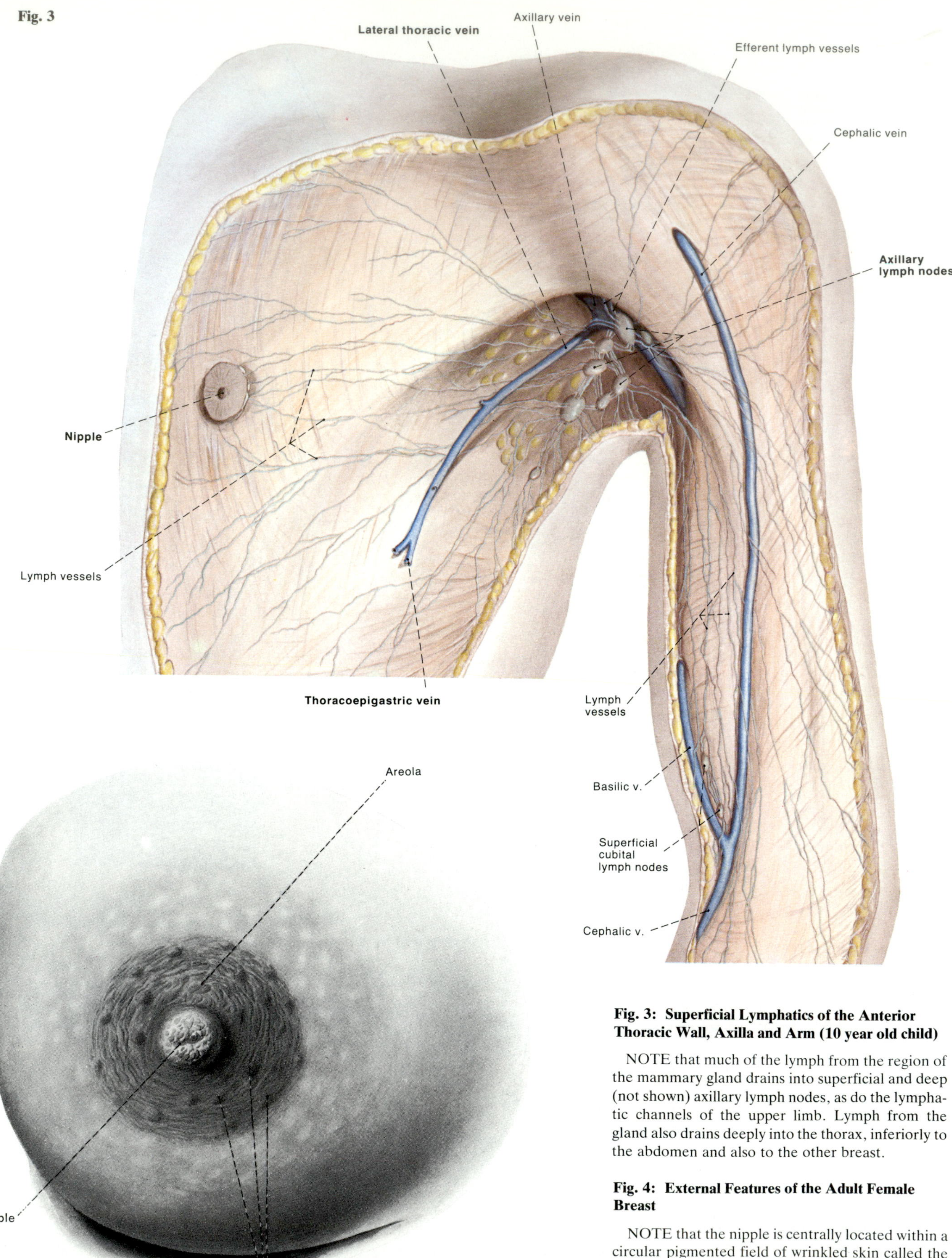
Fig. 3
Axillary vein
Lateral thoracic vein
Efferent lymph vessels
Cephalic vein
Axillary lymph nodes
Nipple
Lymph vessels
Thoracoepigastric vein
Lymph vessels
Basilic v.
Superficial cubital lymph nodes
Cephalic v.
Areola
Nipple
Fig. 4
Areolar glands *(Montgomery)*

Fig. 3: Superficial Lymphatics of the Anterior Thoracic Wall, Axilla and Arm (10 year old child)

NOTE that much of the lymph from the region of the mammary gland drains into superficial and deep (not shown) axillary lymph nodes, as do the lymphatic channels of the upper limb. Lymph from the gland also drains deeply into the thorax, inferiorly to the abdomen and also to the other breast.

Fig. 4: External Features of the Adult Female Breast

NOTE that the nipple is centrally located within a circular pigmented field of wrinkled skin called the areola. Areolar glands which lubricate the nipple during nursing may be observed as small surface elevations in the areola.

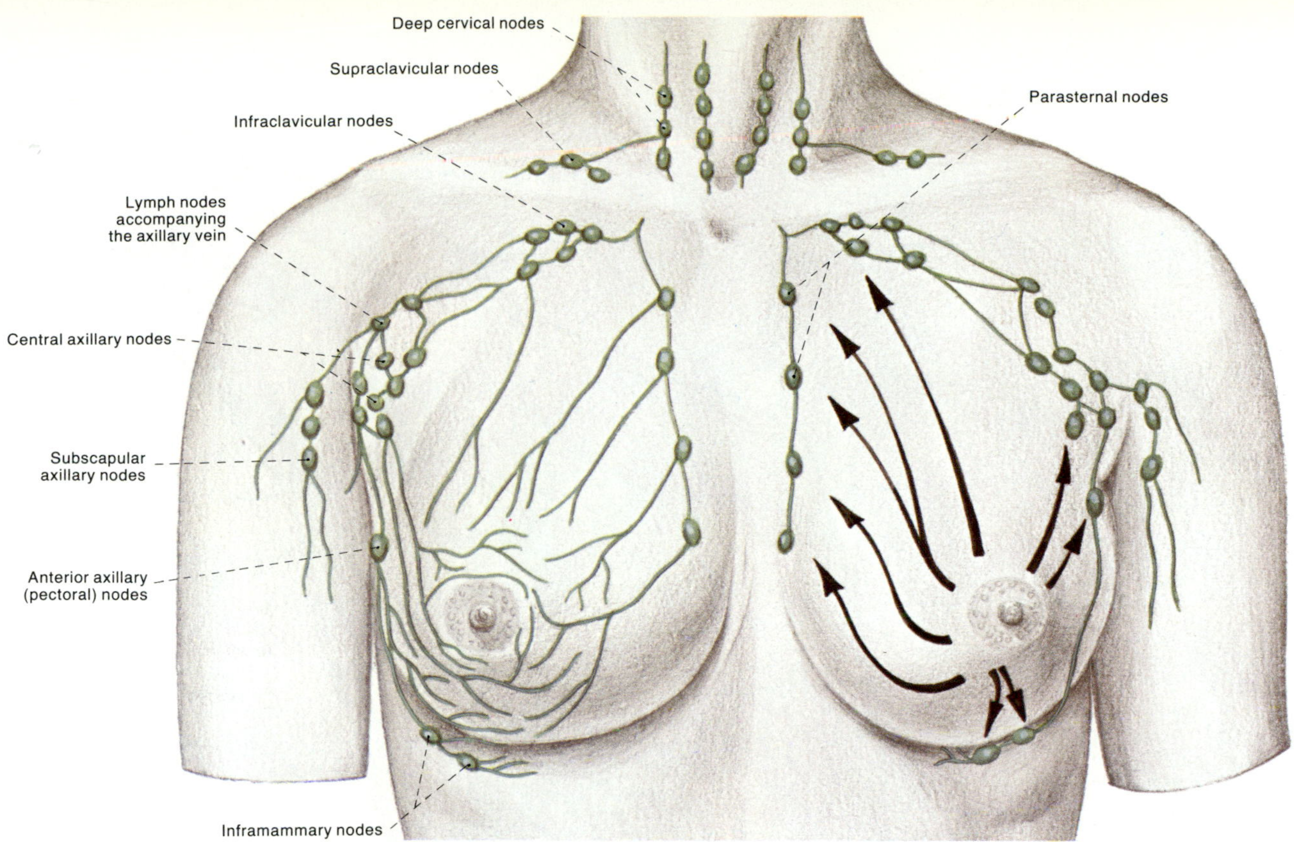
Deep cervical nodes
Supraclavicular nodes
Infraclavicular nodes
Parasternal nodes
Lymph nodes
accompanying
the axillary vein
Central axillary nodes
Subscapular
axillary nodes
Anterior axillary
(pectoral) nodes
Inframammary nodes

Fig. 5: Lymphatic Drainage from the Adult Female Breast

NOTE: 1) the extremely numerous lymph vessels of the breast originate in the connective tissue and laciferous ducts of the organ and communicate with the subareolar plexus deep to and around the nipple.

2) nearly three-quarters of the lymph from the breast passes laterally and superiorly to pectoral, subscapular, central axillary and infraclavicular nodes, as well as to nodes that course along the axillary vein. Most of the remaining lymph passes medially to parasternal nodes that course with the internal thoracic vessels. Lymph may also drain inferiorly along the anterior thoracic wall and communicate with upper abdominal nodes; some lymph may also drain to the opposite breast.

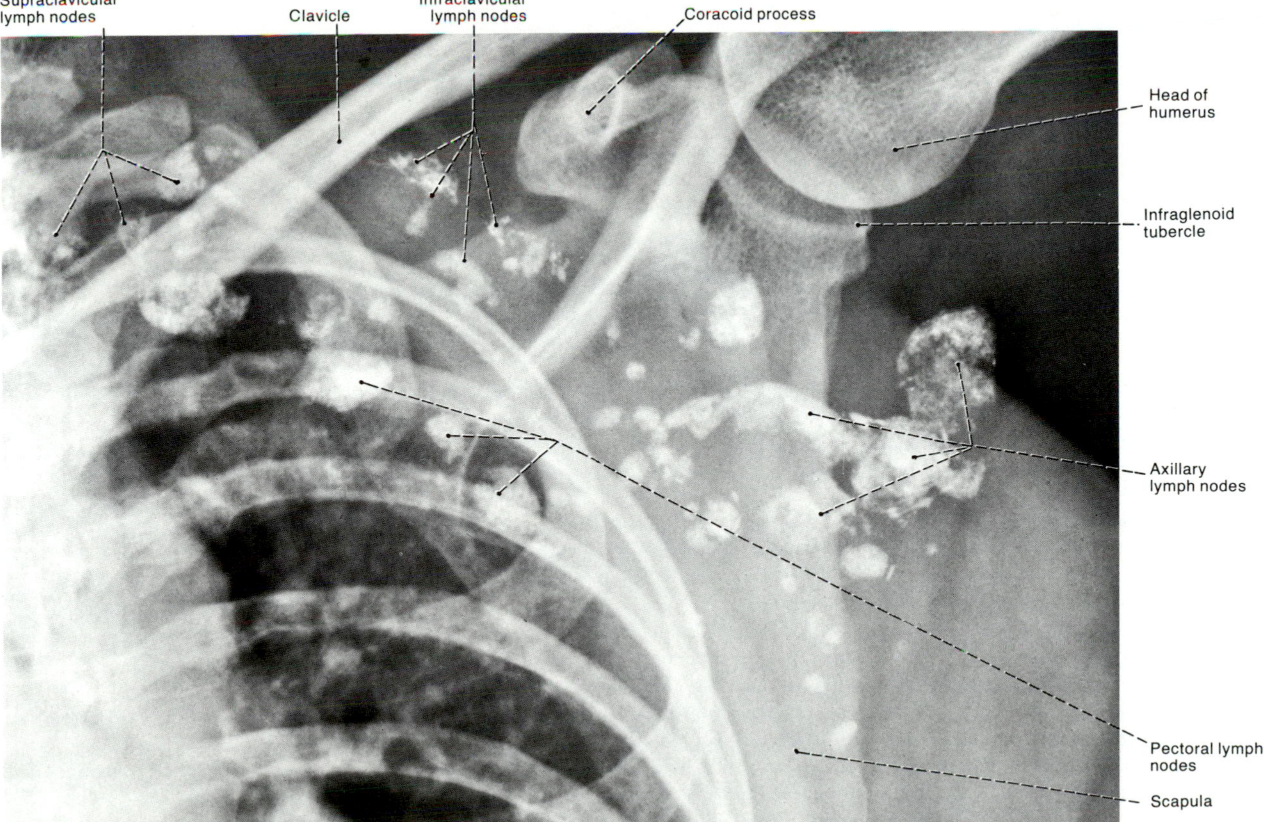
Supraclavicular
lymph nodes
Clavicle
Infraclavicular
lymph nodes
Coracoid process
Head of
humerus
Infraglenoid
tubercle
Axillary
lymph nodes
Pectoral lymph
nodes
Scapula

Fig. 6: Lymphangiogram of the Pectoral and Axillary Lymph Nodes

(From: Wicke, L., *Atlas of Radiologic Anatomy,* 3rd Edition, Urban & Schwarzenberg, Baltimore, 1982).

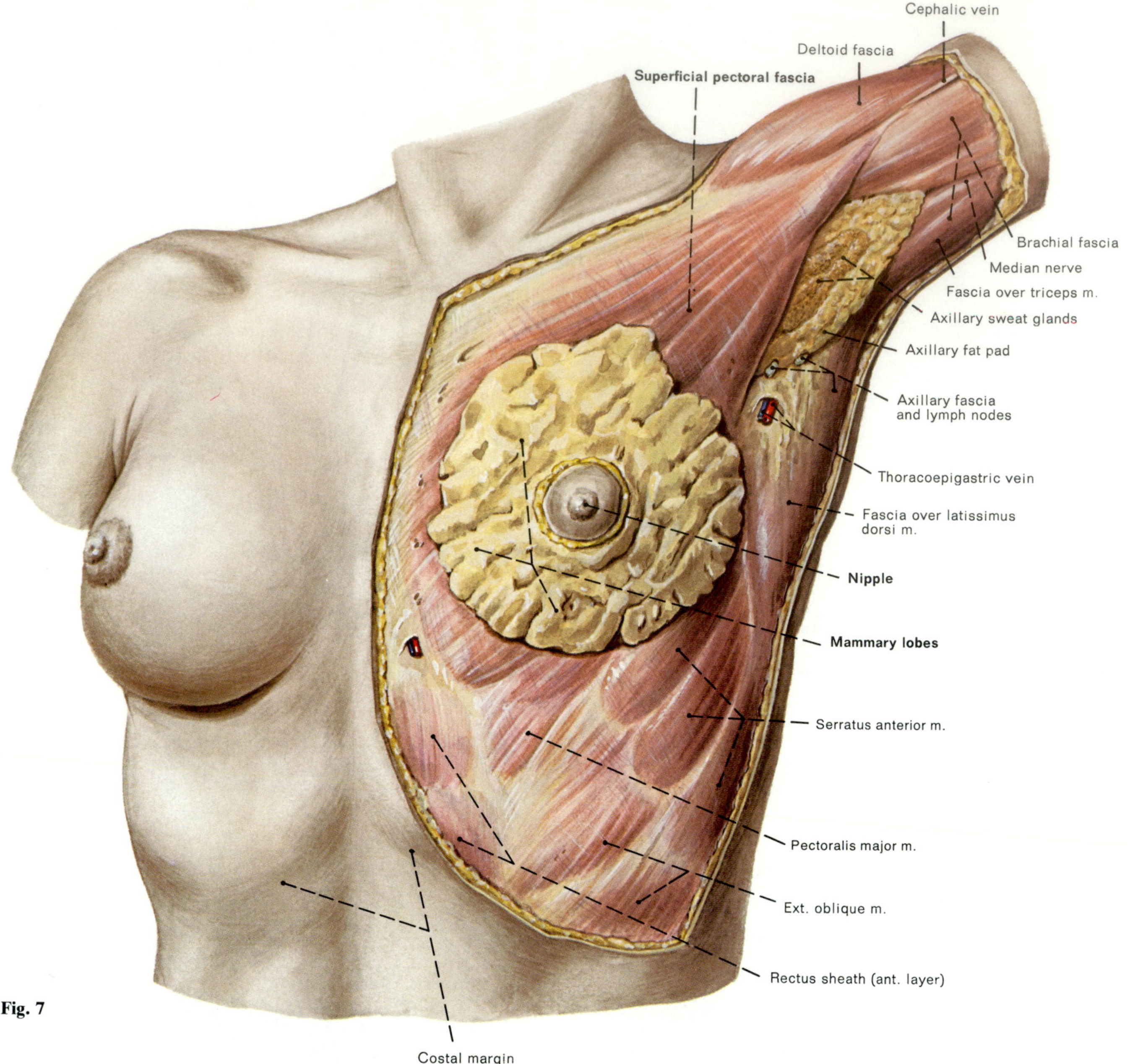

Cephalic vein
Deltoid fascia
Superficial pectoral fascia
Brachial fascia
Median nerve
Fascia over triceps m.
Axillary sweat glands
Axillary fat pad
Axillary fascia and lymph nodes
Thoracoepigastric vein
Fascia over latissimus dorsi m.
Nipple
Mammary lobes
Serratus anterior m.
Pectoralis major m.
Ext. oblique m.
Rectus sheath (ant. layer)
Costal margin

Fig. 7

Fig. 7: Anterior Pectoral Dissection (Adult Female)

NOTE: 1) the lobular nature of the mammary gland extending toward the axilla and its location anterior to the pectoralis major muscle.

2) the superficial axillary lymph and sudoriferous (sweat) glands.

Fig. 8: Sagittal Section through Mammary Gland of Gravid Female

NOTE: 1) the radial arrangement of the lobes of glandular tissue. These lobes are comprised of smaller lobules and are separated from one another by fat and the supporting connective tissue.

2) the lactiferous duct system. Each of the 15 to 20 lobes has its own duct which opens by means of a small orifice onto the nipple.

3) that the mammary gland is separated from the pectoralis major muscle by the pectoral fascia and that connective tissue strands (suspensory ligaments of Cooper) within the gland extend toward this fascia.

Fig. 9: Right Mammary Gland: Dissection of the Nipple

A circular piece of skin has been removed in this dissection. With the incised margin of the skin around the nipple retracted, the lactiferous ducts can be observed perforating onto the surface and arranged circumferentially around the nipple.

Fig. 10: Two Typical Spinal Nerves: Their Origin, Branches and Connections to the Sympathetic Trunk

NOTE: 1) each spinal nerve attaches to the spinal cord by two roots: an afferent or sensory dorsal root and an efferent or motor ventral root. Each dorsal root contains a spinal ganglion comprised of afferent neuron cell bodies.

2) the two spinal roots join to form the spinal nerve, which in turn divides into a dorsal ramus coursing posteriorly and a ventral ramus coursing anteriorly. During their course, these rami divide further to innervate the body segment with both sensory and motor fibers.

3) the spinal nerve communicates with the sympathetic trunk carrying preganglionic sympathetic fibers to the trunk (white ramus) and postganglionic fibers from the trunk (gray ramus).

Fig. 7

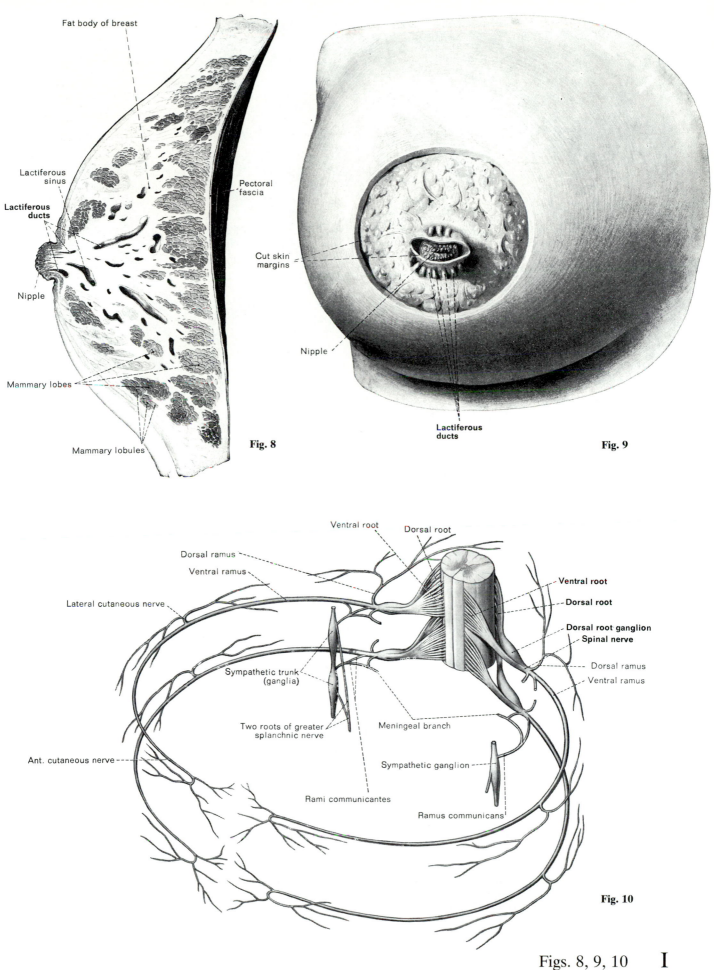

Fat body of breast
Lactiferous sinus
Pectoral fascia
Lactiferous ducts
Nipple
Mammary lobes
Mammary lobules

Fig. 8

Cut skin margins
Nipple
Lactiferous ducts

Fig. 9

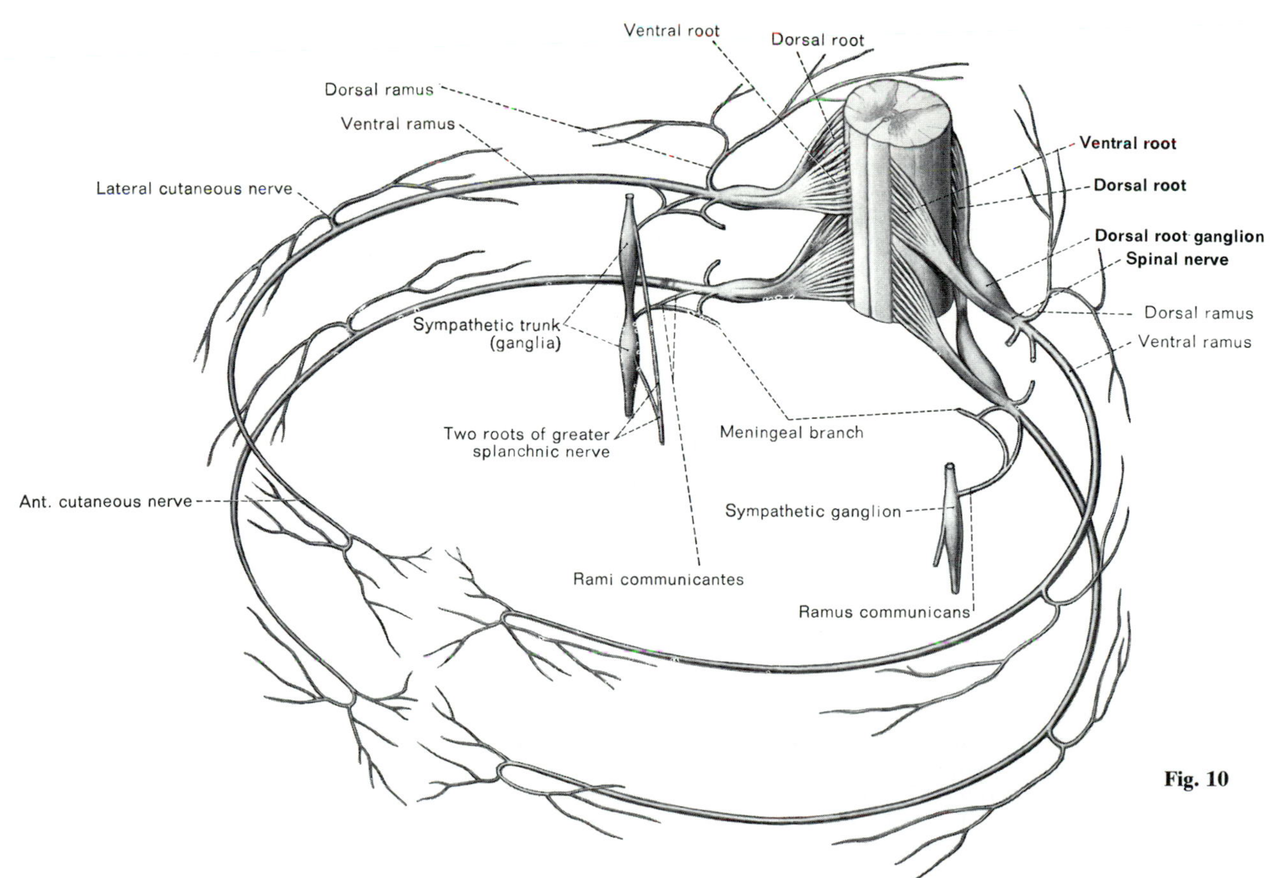

Ventral root
Dorsal root
Dorsal ramus
Ventral ramus
Lateral cutaneous nerve
Ventral root
Dorsal root
Dorsal root ganglion
Spinal nerve
Dorsal ramus
Ventral ramus
Sympathetic trunk (ganglia)
Two roots of greater splanchnic nerve
Meningeal branch
Ant. cutaneous nerve
Sympathetic ganglion
Rami communicantes
Ramus communicans

Fig. 10

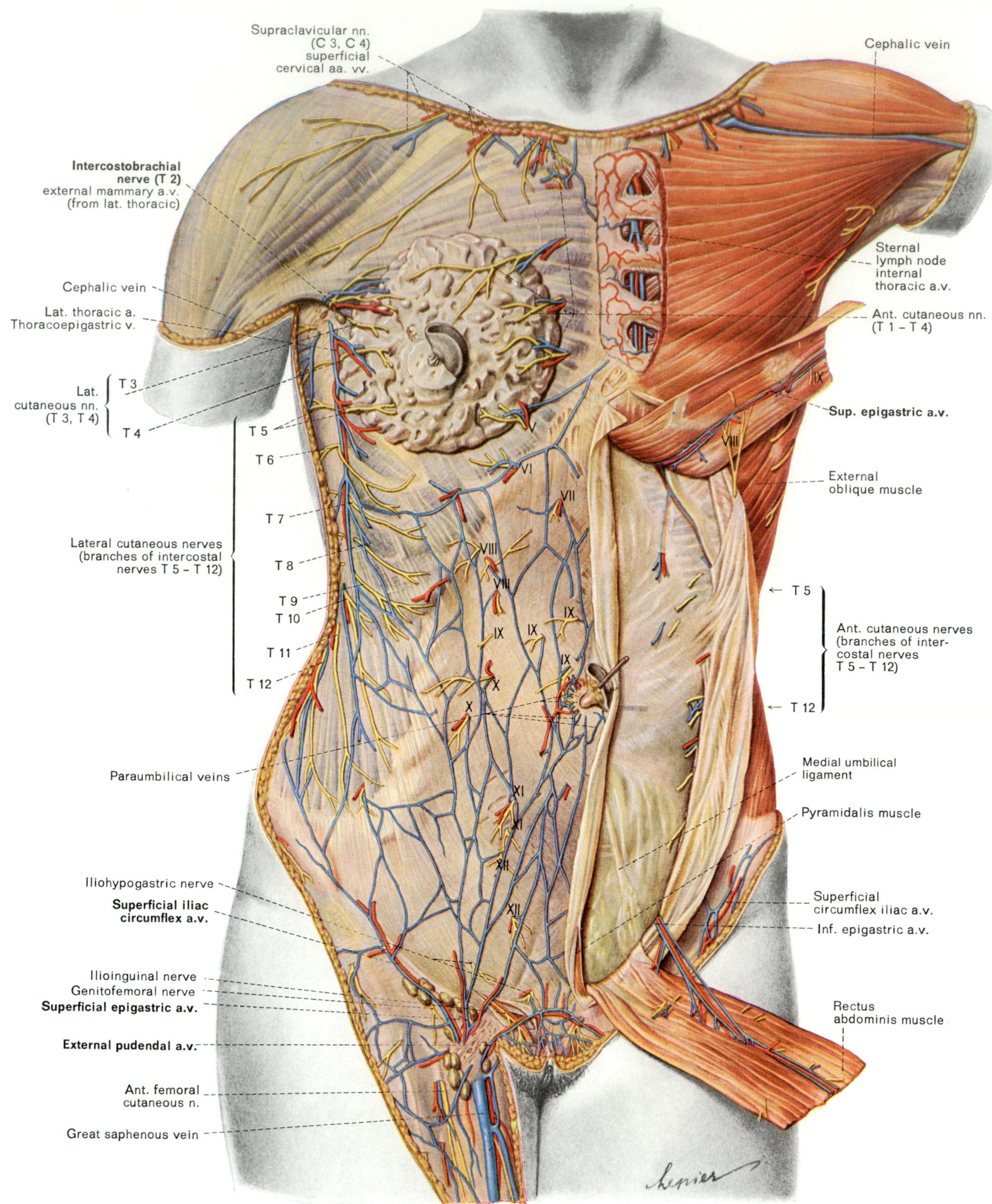
Supraclavicular nn.
(C 3, C 4)
superficial
cervical aa. vv.
Cephalic vein
Intercostobrachial
nerve (T 2)
external mammary a.v.
(from lat. thoracic)
Sternal
lymph node
internal
thoracic a.v.
Cephalic vein
Lat. thoracic a.
Thoracoepigastric v.
Ant. cutaneous nn.
(T 1 - T 4)
Lat.
cutaneous nn.
(T 3, T 4)
T 3
T 4
T 5
Sup. epigastric a.v.
T 6
External
oblique muscle
T 7
Lateral cutaneous nerves
(branches of intercostal
nerves T 5 - T 12)
T 8
T 9
T 10
T 11
T 12
T 5
Ant. cutaneous nerves
(branches of inter-
costal nerves
T 5 - T 12)
T 12
Paraumbilical veins
Medial umbilical
ligament
Pyramidalis muscle
Iliohypogastric nerve
Superficial iliac
circumflex a.v.
Superficial
circumflex iliac a.v.
Inf. epigastric a.v.
Ilioinguinal nerve
Genitofemoral nerve
Superficial epigastric a.v.
Rectus
abdominis muscle
External pudendal a.v.
Ant. femoral
cutaneous n.
Great saphenous vein

Fig. 11: Superficial Vessels and Nerves of the Ventral Trunk: Pectoral Region and Anterior Abdominal Wall

OBSERVE: 1) the cutaneous innervation of the anterior trunk which is derived from a) the supraclavicular nerves (C_3, C_4); b) the intercostal nerves (anterior cutaneous T 1 – T 12; lateral cutaneous T 2 – T 12); c) iliohypogastric and ilioinguinal nerves (L 1).

2) the thoracoepigastric venous anastomosis between the thoracoepigastric and lateral thoracic veins superiorly and the superficial circumflex iliac and superficial epigastric veins inferiorly.

3) the mammary gland: its innervation (T2–T6) and its blood supply (branches of internal thoracic artery, lateral thoracic artery). Additionally, the mammary gland may receive small branches from the intercostal arteries which may enter the deep surface of the gland.

4) the nipple at the level of T 4 and the umbilicus at the level of T 10.

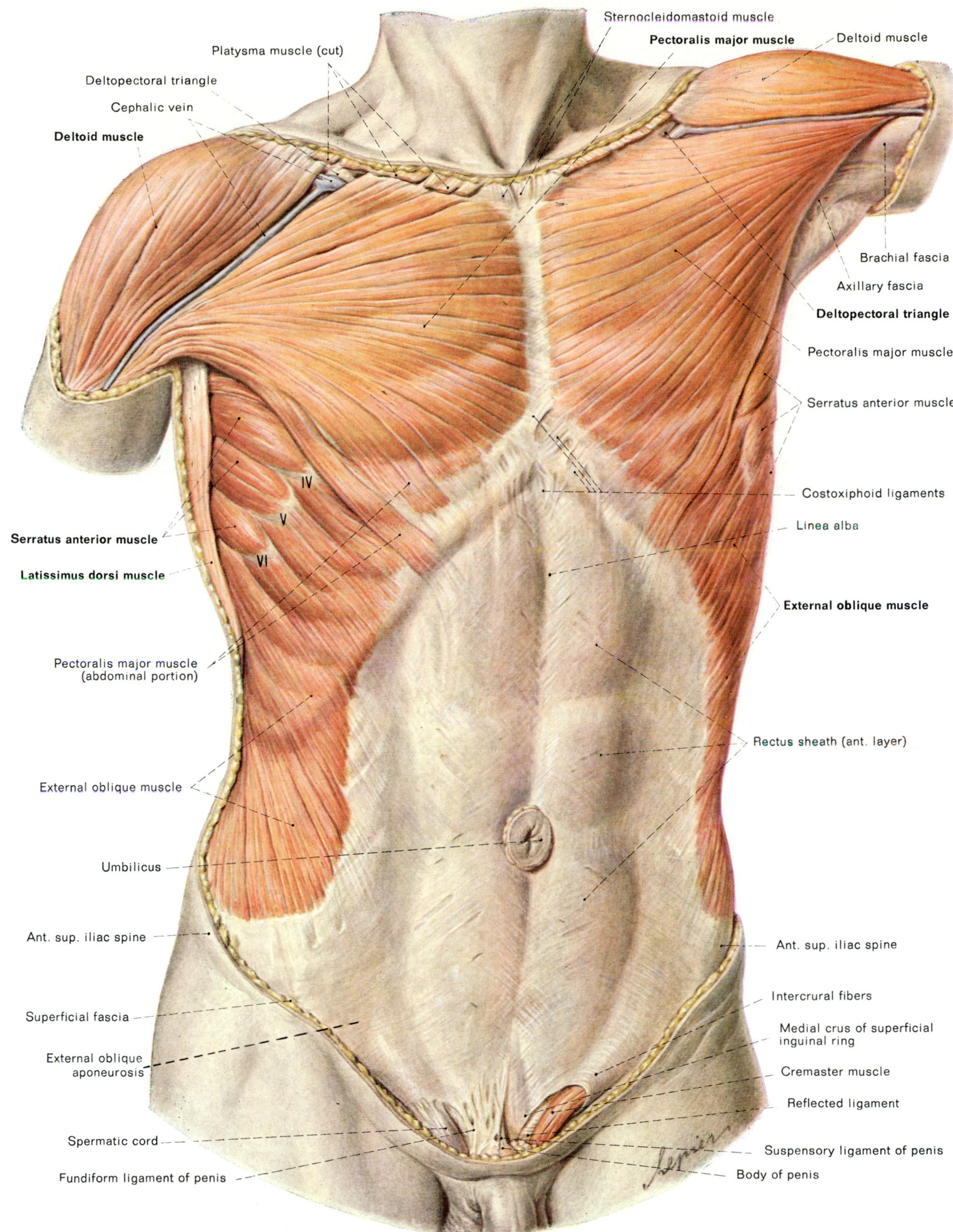

Sternocleidomastoid muscle
Pectoralis major muscle
Deltoid muscle
Platysma muscle (cut)
Deltopectoral triangle
Cephalic vein
Deltoid muscle
Brachial fascia
Axillary fascia
Deltopectoral triangle
Pectoralis major muscle
Serratus anterior muscle
IV
V
VI
Costoxiphoid ligaments
Linea alba
Serratus anterior muscle
Latissimus dorsi muscle
External oblique muscle
Pectoralis major muscle
(abdominal portion)
Rectus sheath (ant. layer)
External oblique muscle
Umbilicus
Ant. sup. iliac spine
Ant. sup. iliac spine
Intercrural fibers
Superficial fascia
Medial crus of superficial
inguinal ring
External oblique
aponeurosis
Cremaster muscle
Reflected ligament
Spermatic cord
Suspensory ligament of penis
Fundiform ligament of penis
Body of penis

Fig. 12: The Superficial Thoracic and Abdominal Muscle

OBSERVE that the pectoralis major arises from the medial half of the clavicle, the costal margin of the sternum, the 2nd to 6th ribs and the upper part of the aponeurosis of the external oblique. Also note the deltopectoral triangle and the course of the cephalic vein as it empties into the axillary vein.

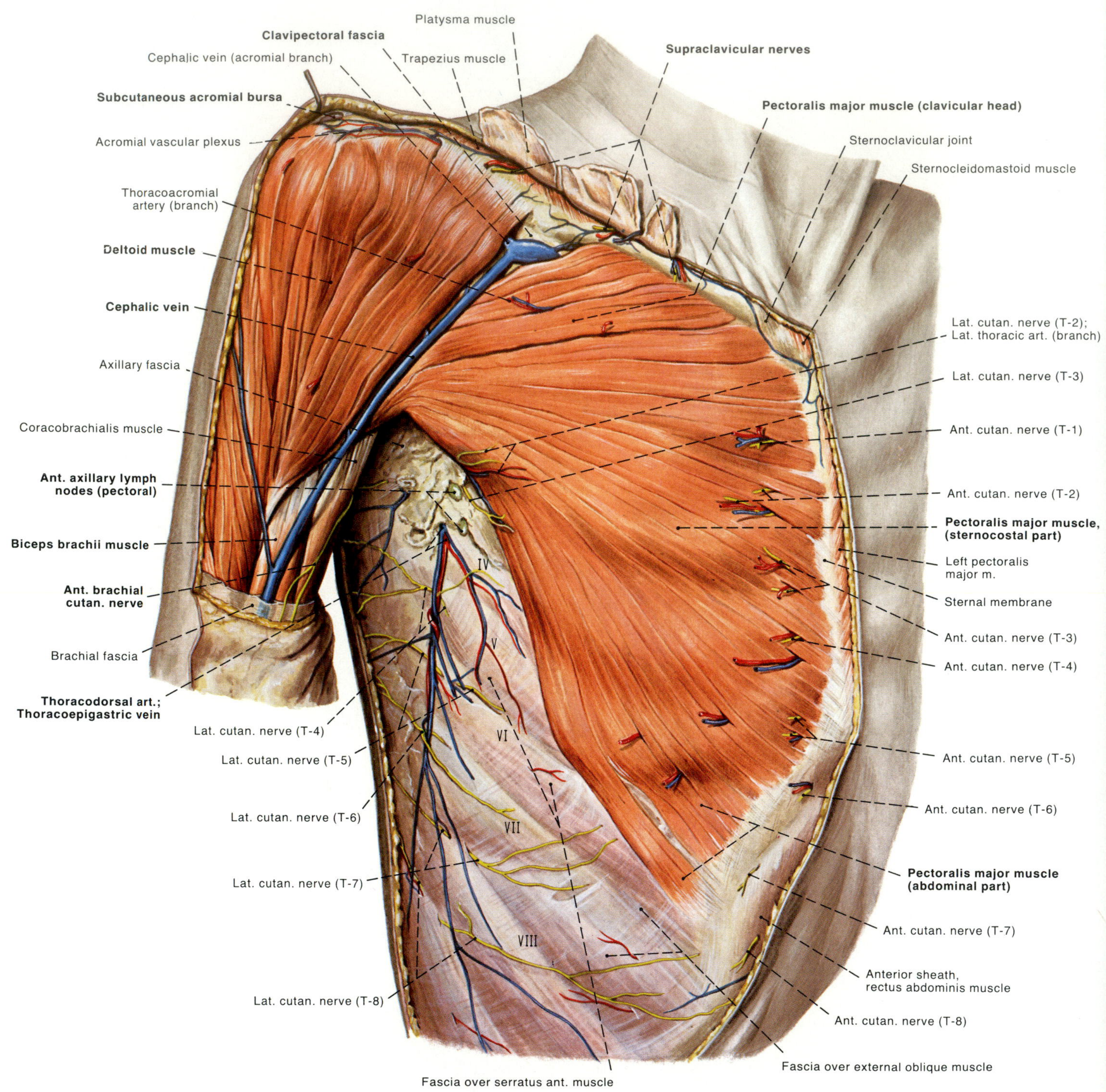

Platysma muscle
Clavipectoral fascia
Cephalic vein (acromial branch)
Trapezius muscle
Supraclavicular nerves
Subcutaneous acromial bursa
Pectoralis major muscle (clavicular head)
Acromial vascular plexus
Sternoclavicular joint
Sternocleidomastoid muscle
Thoracoacromial artery (branch)
Deltoid muscle
Cephalic vein
Lat. cutan. nerve (T-2); Lat. thoracic art. (branch)
Axillary fascia
Lat. cutan. nerve (T-3)
Coracobrachialis muscle
Ant. cutan. nerve (T-1)
Ant. axillary lymph nodes (pectoral)
Ant. cutan. nerve (T-2)
Pectoralis major muscle, (sternocostal part)
Biceps brachii muscle
Left pectoralis major m.
Ant. brachial cutan. nerve
Sternal membrane
IV
Ant. cutan. nerve (T-3)
Brachial fascia
V
Ant. cutan. nerve (T-4)
Thoracodorsal art.; Thoracoepigastric vein
Lat. cutan. nerve (T-4)
VI
Lat. cutan. nerve (T-5)
Ant. cutan. nerve (T-5)
Lat. cutan. nerve (T-6)
Ant. cutan. nerve (T-6)
VII
Lat. cutan. nerve (T-7)
Pectoralis major muscle (abdominal part)
Ant. cutan. nerve (T-7)
VIII
Anterior sheath, rectus abdominis muscle
Lat. cutan. nerve (T-8)
Ant. cutan. nerve (T-8)
Fascia over external oblique muscle
Fascia over serratus ant. muscle

Fig. 13: The Pectoralis Major and Deltoid Muscles, Anterior View

NOTE: 1) segmentally arranged lateral and anterior cutaneous nerves which branch from the intercostal nerves. These penetrate through the intercostal spaces approximately in the mid-axillary line (lateral cutaneous nerves) and, more anteriorly, along the lateral border of the sternum through the substance of the pectoralis major muscle (anterior cutaneous nerves).

2) the 4th to the 8th ribs are numbered sequentially with Roman numerals.

3) the thoracoepigastric vein and the thoracodorsal artery piercing the deep fascia and coursing along the mid-axillary line.

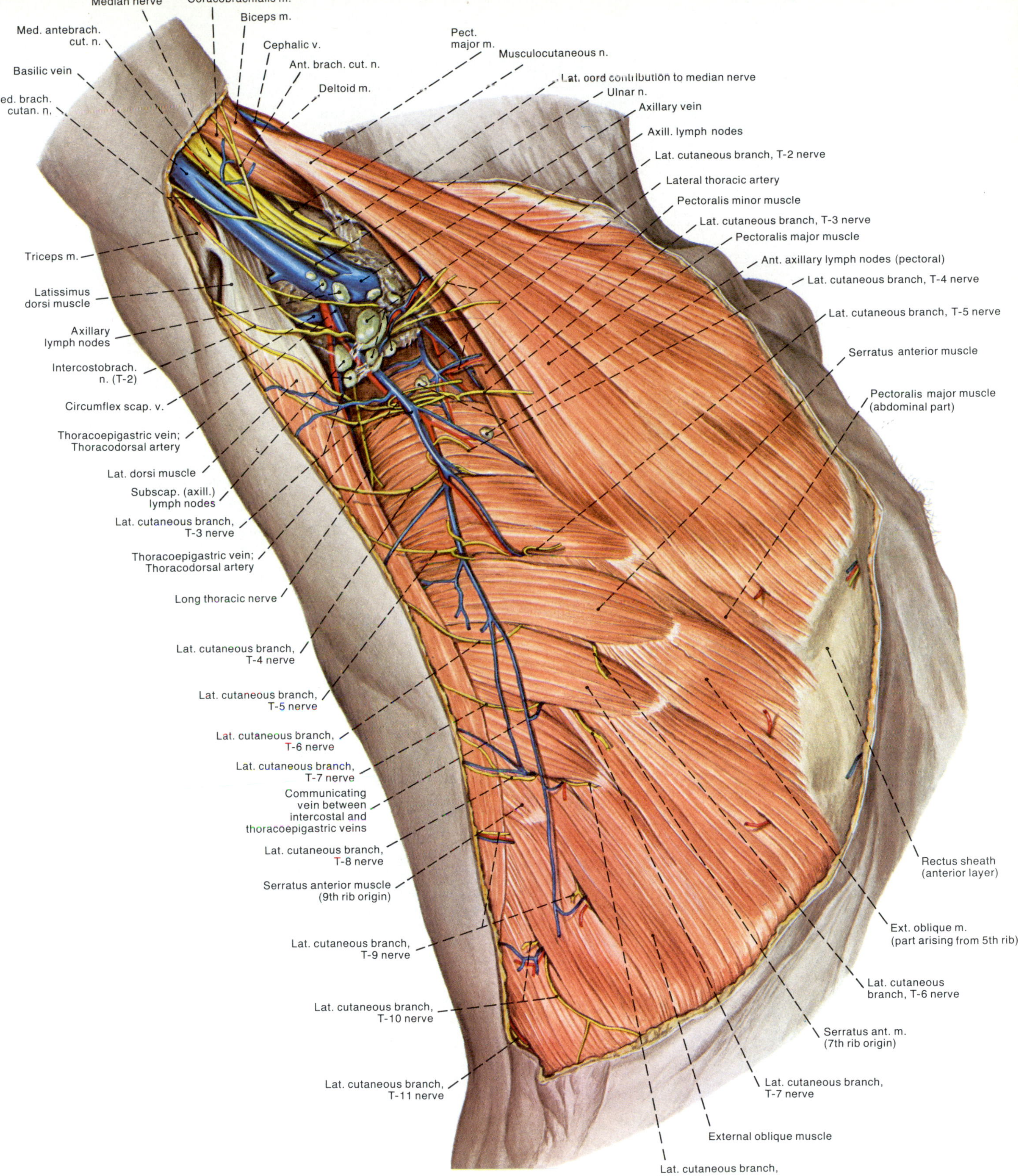

Median nerve
Coracobrachialis m.
Biceps m.
Med. antebrach. cut. n.
Cephalic v.
Ant. brach. cut. n.
Basilic vein
Deltoid m.
Med. brach. cutan. n.
Pect. major m.
Musculocutaneous n.
Lat. cord contribution to median nerve
Ulnar n.
Axillary vein
Axill. lymph nodes
Lat. cutaneous branch, T-2 nerve
Lateral thoracic artery
Pectoralis minor muscle
Lat. cutaneous branch, T-3 nerve
Pectoralis major muscle
Triceps m.
Ant. axillary lymph nodes (pectoral)
Lat. cutaneous branch, T-4 nerve
Latissimus dorsi muscle
Lat. cutaneous branch, T-5 nerve
Axillary lymph nodes
Serratus anterior muscle
Intercostobrach. n. (T-2)
Pectoralis major muscle (abdominal part)
Circumflex scap. v.
Thoracoepigastric vein; Thoracodorsal artery
Lat. dorsi muscle
Subscap. (axill.) lymph nodes
Lat. cutaneous branch, T-3 nerve
Thoracoepigastric vein; Thoracodorsal artery
Long thoracic nerve
Lat. cutaneous branch, T-4 nerve
Lat. cutaneous branch, T-5 nerve
Lat. cutaneous branch, T-6 nerve
Lat. cutaneous branch, T-7 nerve
Communicating vein between intercostal and thoracoepigastric veins
Lat. cutaneous branch, T-8 nerve
Rectus sheath (anterior layer)
Serratus anterior muscle (9th rib origin)
Ext. oblique m. (part arising from 5th rib)
Lat. cutaneous branch, T-9 nerve
Lat. cutaneous branch, T-6 nerve
Lat. cutaneous branch, T-10 nerve
Serratus ant. m. (7th rib origin)
Lat. cutaneous branch, T-11 nerve
Lat. cutaneous branch, T-7 nerve
External oblique muscle
Lat. cutaneous branch, T-8 nerve

Fig. 14: Lateral Aspect of the Upper Right Thoracic Wall and the Superficial Axillary Structures

NOTE: 1) the serratus anterior is a fleshy sheet of muscle that arises as separate muscular digitations from the upper eight of nine ribs. The lower four of these slips interdigitate with bundles of the external oblique muscle which also arise from the ribs.

2) the partially exposed superficial vessels and certain peripheral nerves from the brachial plexus. These become visible after the axillary space is entered by removing the skin of the armpit and the subjacent axillary fascia. The axillary sheath surrounding the axillary artery and the cords of the brachial plexus are still intact.

3) the intermediate and central axillary lymph nodes within the axillary space, as well as a few of the pectoral nodes that lie along the inferolateral border of the pectoralis major and the lateral surface of the serratus anterior.

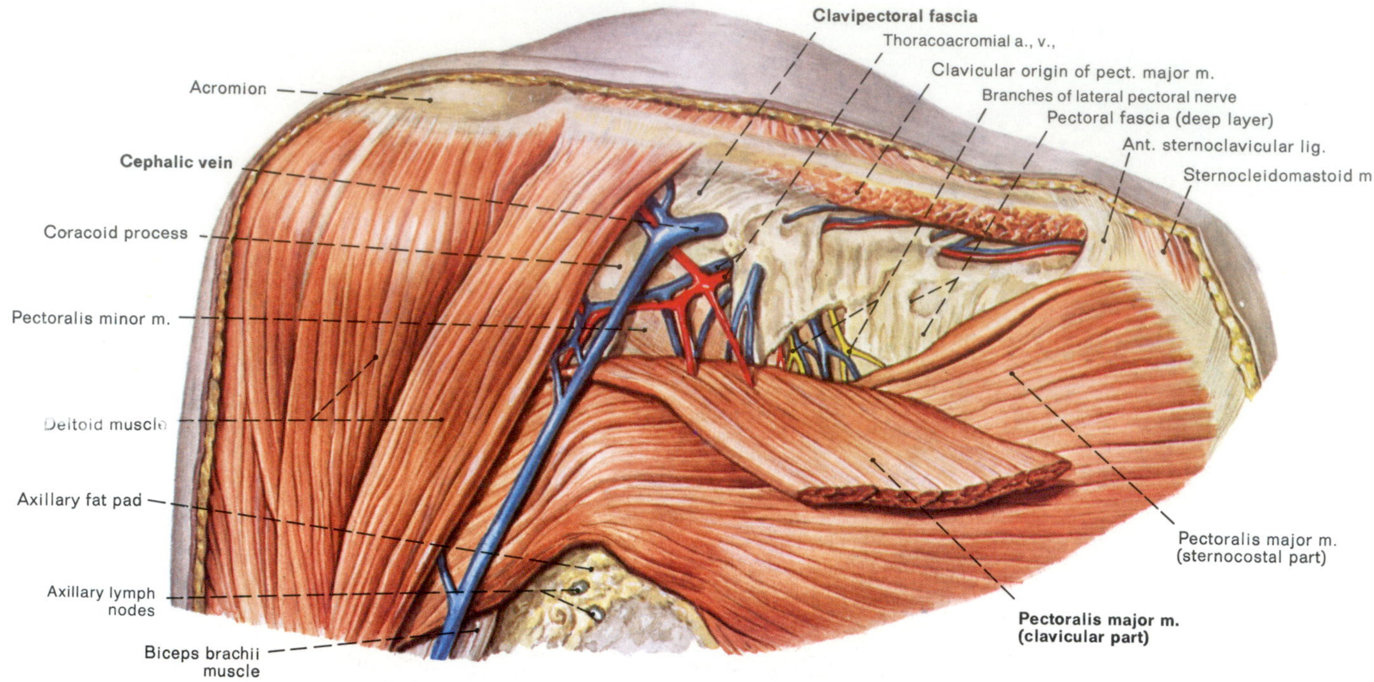

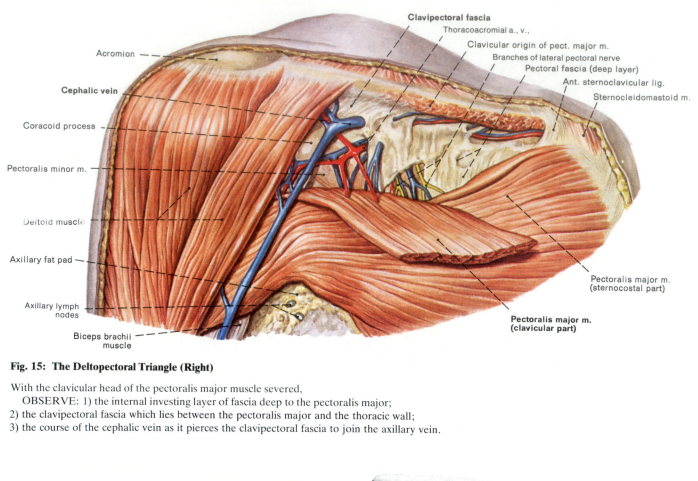

Clavipectoral fascia
Thoracoacromial a., v.,
Clavicular origin of pect. major m.
Branches of lateral pectoral nerve
Pectoral fascia (deep layer)
Ant. sternoclavicular lig.
Sternocleidomastoid m.
Acromion
Cephalic vein
Coracoid process
Pectoralis minor m.
Deltoid muscle
Axillary fat pad
Axillary lymph nodes
Biceps brachii muscle
Pectoralis major m. (sternocostal part)
Pectoralis major m. (clavicular part)

Fig. 15: The Deltopectoral Triangle (Right)

With the clavicular head of the pectoralis major muscle severed,

OBSERVE: 1) the internal investing layer of fascia deep to the pectoralis major;
2) the clavipectoral fascia which lies between the pectoralis major and the thoracic wall;
3) the course of the cephalic vein as it pierces the clavipectoral fascia to join the axillary vein.

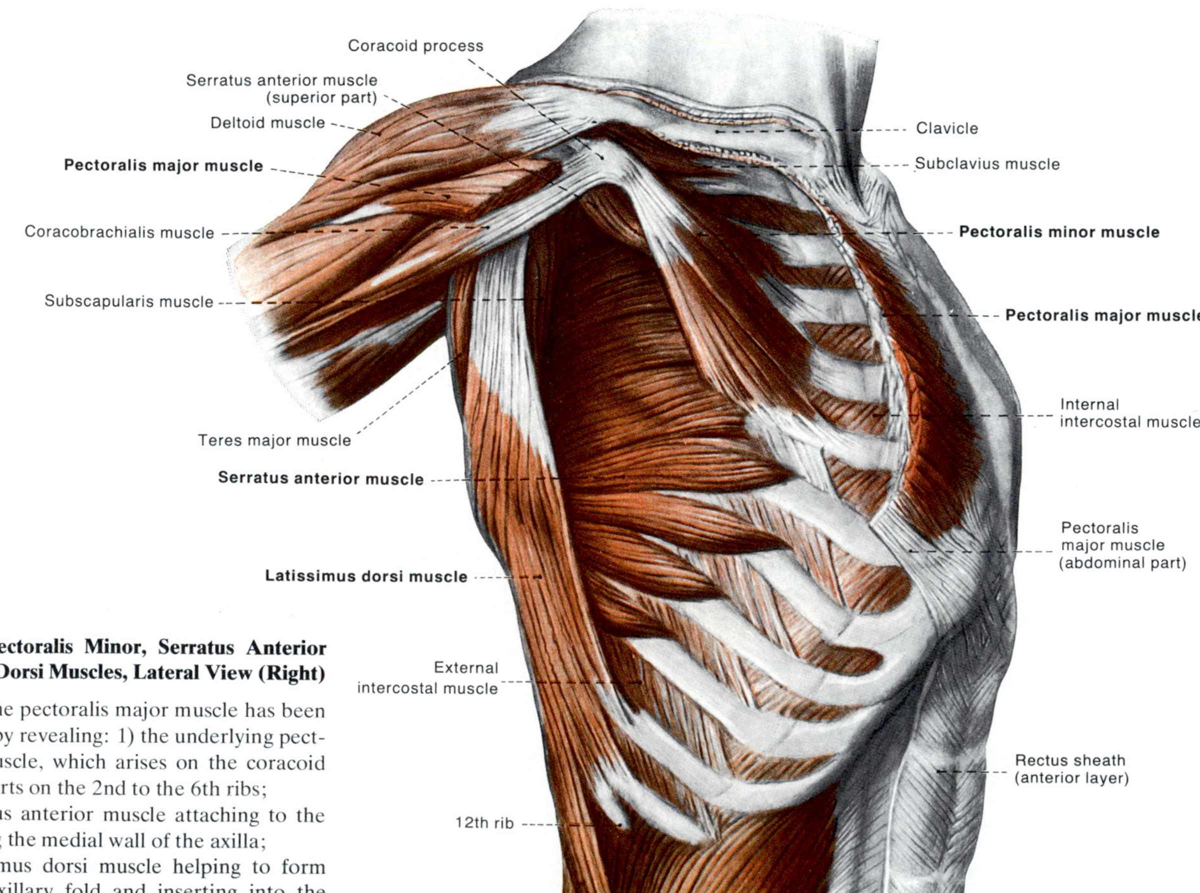

Coracoid process
Serratus anterior muscle (superior part)
Deltoid muscle
Pectoralis major muscle
Coracobrachialis muscle
Subscapularis muscle
Teres major muscle
Serratus anterior muscle
Latissimus dorsi muscle
External intercostal muscle
12th rib
Clavicle
Subclavius muscle
Pectoralis minor muscle
Pectoralis major muscle
Internal intercostal muscle
Pectoralis major muscle (abdominal part)
Rectus sheath (anterior layer)

Fig. 16: The Pectoralis Minor, Serratus Anterior and Latissimus Dorsi Muscles, Lateral View (Right)

NOTE that the pectoralis major muscle has been reflected, thereby revealing: 1) the underlying pectoralis minor muscle, which arises on the coracoid process and inserts on the 2nd to the 6th ribs;

2) the serratus anterior muscle attaching to the ribs and forming the medial wall of the axilla;

3) the latissimus dorsi muscle helping to form the posterior axillary fold and inserting into the humerus.

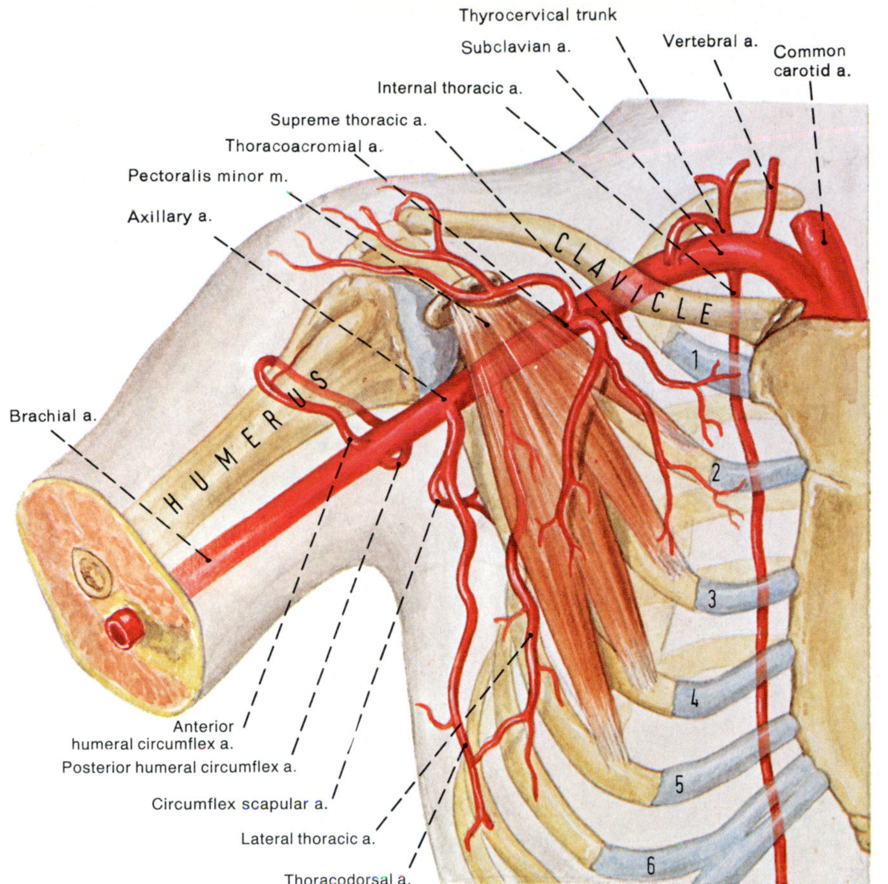
Thyrocervical trunk
Subclavian a.
Vertebral a.
Common carotid a.
Internal thoracic a.
Supreme thoracic a.
Thoracoacromial a.
Pectoralis minor m.
Axillary a.
CLAVICLE
HUMERUS
Brachial a.
1
2
3
4
5
6
Anterior humeral circumflex a.
Posterior humeral circumflex a.
Circumflex scapular a.
Lateral thoracic a.
Thoracodorsal a.

Fig. 17: The Branches of the Axillary Artery

NOTE: 1) as the subclavian artery passes beneath the clavicle, it becomes the axillary artery. The axillary artery becomes the brachial artery in the upper arm (at the level of the tendon of the teres major muscle). In the axilla the pectoralis minor muscle crosses anterior to the axillary artery, thereby, for descriptive purposes, dividing the vessel into three parts.

2) from the 1st part of the axillary artery (medial to the pectoralis minor and lateral to the clavicle) branches one vessel, the supreme thoracic artery. From the 2nd part (beneath the muscle) branch two vessels, the thoracoacromial artery and the lateral thoracic. From the 3rd part of the axillary (lateral to the pectoralis minor muscle) are derived three branches, the subscapular artery and the anterior and posterior humeral circumflex arteries.

3) that the subscapular artery is the largest branch from the axillary artery. It descends from the main trunk and, 4 to 5 cm from its origin, it usually divides into the thoracodorsal and circumflex scapular arteries (shown but not labelled). The thoracodorsal artery descends along the lateral aspect of the thorax to supply the latissimus dorsi muscle and other structures. The circumflex scapular artery traverses the triangular space and courses around the lateral border of the scapula to enter the infraspinatus fossa.

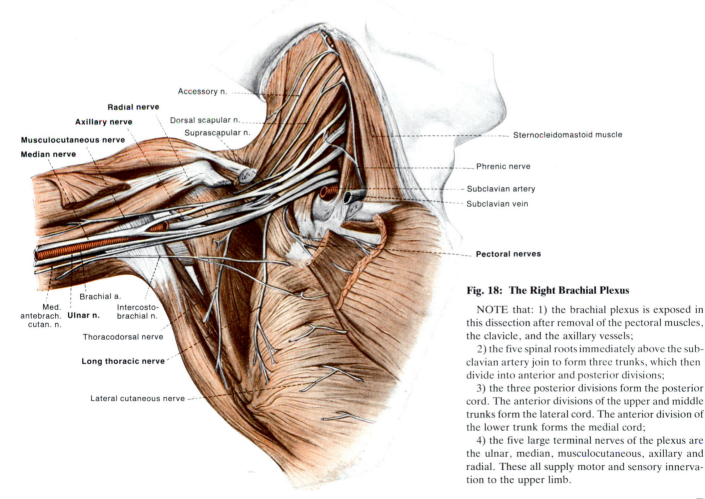

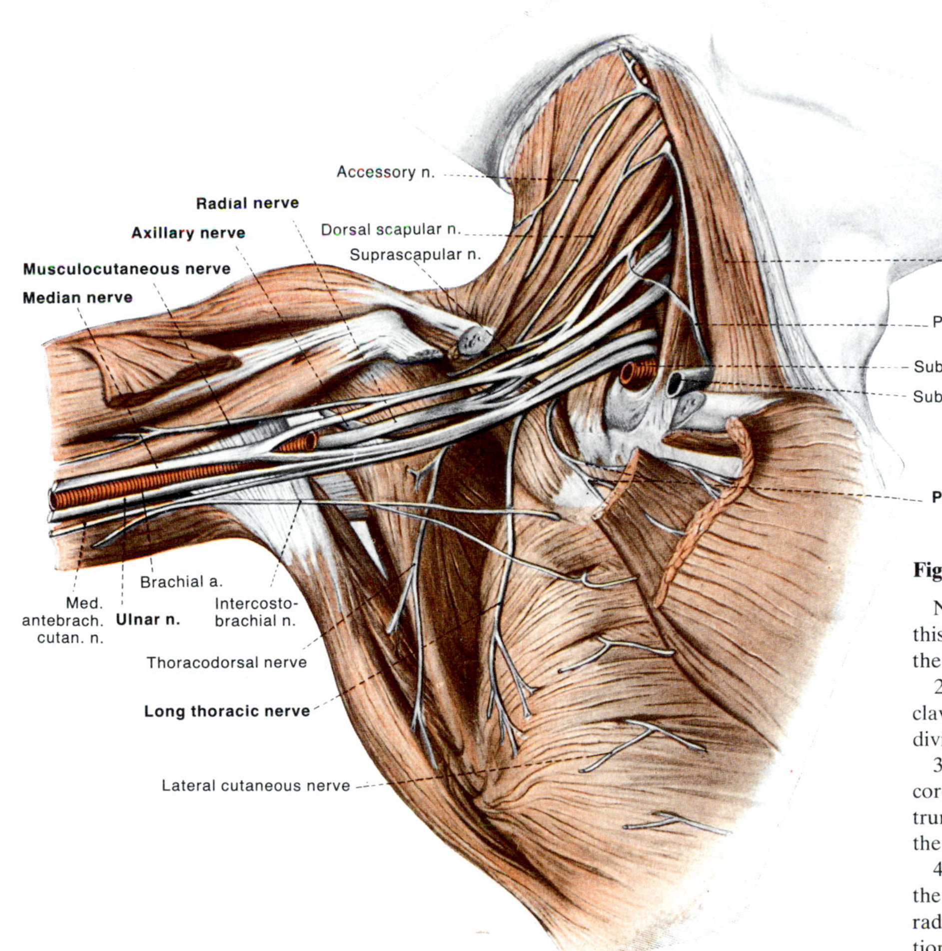
Accessory n.
Radial nerve
Axillary nerve
Dorsal scapular n.
Suprascapular n.
Musculocutaneous nerve
Median nerve
Sternocleidomastoid muscle
Phrenic nerve
Subclavian artery
Subclavian vein
Pectoral nerves
Brachial a.
Med. antebrach. cutan. n.
Ulnar n.
Intercosto-brachial n.
Thoracodorsal nerve
Long thoracic nerve
Lateral cutaneous nerve

Fig. 18: The Right Brachial Plexus

NOTE that: 1) the brachial plexus is exposed in this dissection after removal of the pectoral muscles, the clavicle, and the axillary vessels;

2) the five spinal roots immediately above the subclavian artery join to form three trunks, which then divide into anterior and posterior divisions;

3) the three posterior divisions form the posterior cord. The anterior divisions of the upper and middle trunks form the lateral cord. The anterior division of the lower trunk forms the medial cord;

4) the five large terminal nerves of the plexus are the ulnar, median, musculocutaneous, axillary and radial. These all supply motor and sensory innervation to the upper limb.

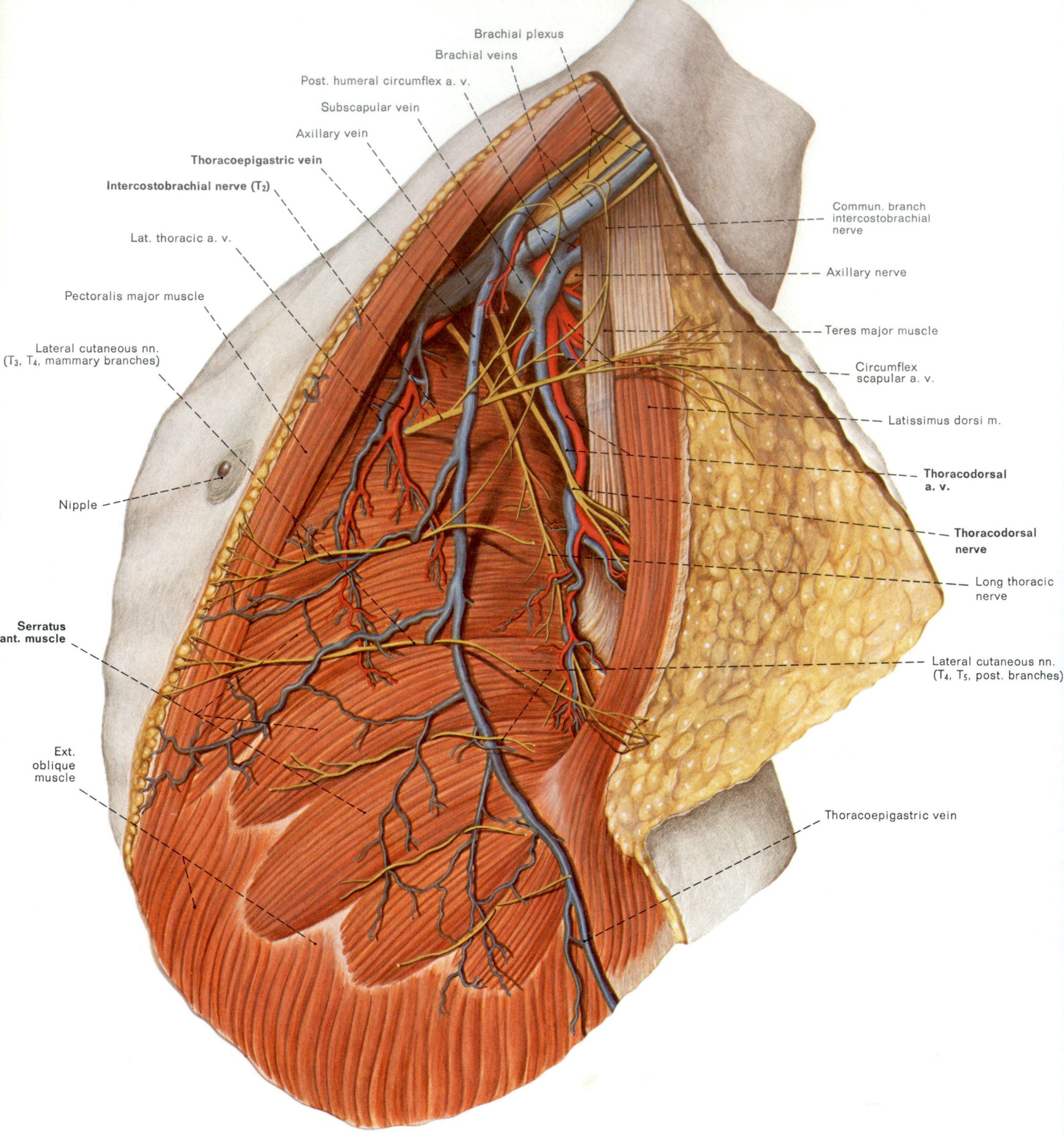
Brachial plexus
Brachial veins
Post. humeral circumflex a. v.
Subscapular vein
Axillary vein
Thoracoepigastric vein
Intercostobrachial nerve (T_2)
Lat. thoracic a. v.
Pectoralis major muscle
Lateral cutaneous nn.
(T_3, T_4, mammary branches)
Nipple
Serratus ant. muscle
Ext. oblique muscle
Commun. branch intercostobrachial nerve
Axillary nerve
Teres major muscle
Circumflex scapular a. v.
Latissimus dorsi m.
Thoracodorsal a. v.
Thoracodorsal nerve
Long thoracic nerve
Lateral cutaneous nn.
(T_4, T_5, post. branches)
Thoracoepigastric vein

Fig. 19: The Axilla: Superficial Vessels and Nerves (left)

OBSERVE: 1) that the boundaries of the axilla are, a) anteriorly, the pectoralis major muscle; b) posteriorly, the subscapularis, teres major and latissimus dorsi muscles; c) medially, the serratus anterior muscle covering the ribs, and; d) laterally, the bicipital groove of the humerus.

2) that the inferior portion of the serratus anterior muscle arises from the lower ribs as fleshy interdigitations with the external oblique muscle.

3) the serratus anterior is innervated by the long thoracic nerve (C 5, 6, 7) and the latissimus dorsi is innervated by the thoracodorsal nerve (C 5, 6, 7).

4) the axillary vein lies medial to the axillary artery and the cords of the brachial plexus.

5) the descending course of the thoracoepigastric vein and the lateral thoracic artery and vein.

Fig. 19

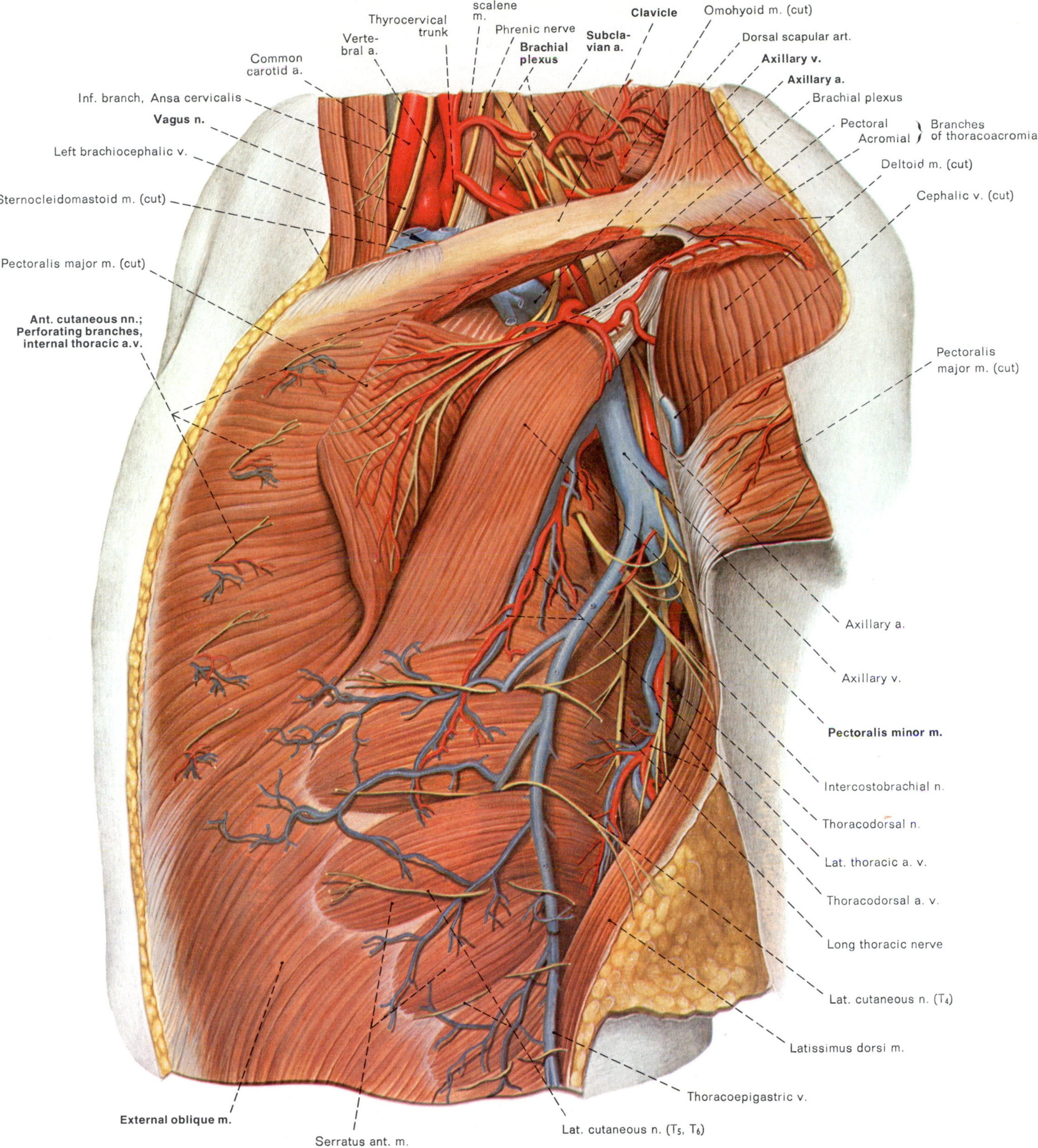

Ant. scalene m.
Thyrocervical trunk
Verte-bral a.
Phrenic nerve
Brachial plexus
Subcla-vian a.
Clavicle
Omohyoid m. (cut)
Dorsal scapular art.
Axillary v.
Axillary a.
Brachial plexus
Pectoral
Acromial
Branches of thoracoacromial a.
Deltoid m. (cut)
Cephalic v. (cut)
Common carotid a.
Inf. branch, Ansa cervicalis
Vagus n.
Left brachiocephalic v.
Sternocleidomastoid m. (cut)
Pectoralis major m. (cut)
Ant. cutaneous nn.; Perforating branches, internal thoracic a.v.
Pectoralis major m. (cut)
Axillary a.
Axillary v.
Pectoralis minor m.
Intercostobrachial n.
Thoracodorsal n.
Lat. thoracic a. v.
Thoracodorsal a. v.
Long thoracic nerve
Lat. cutaneous n. (T4)
Latissimus dorsi m.
Thoracoepigastric v.
External oblique m.
Serratus ant. m.
Lat. cutaneous n. (T5, T6)

Fig. 20: The Axilla (left): Deep Vessels and Nerves

OBSERVE: 1) that the subclavian artery becomes the axillary artery as it passes beneath the clavicle;

2) that the pectoralis minor muscle is helpful in describing the underlying axillary artery in its course through the axilla, since the three parts of the axillary artery are medial, beneath and lateral to the pectoralis minor;

3) that the axillary vein courses medial to the axillary artery and it receives tributaries not only from the upper extremity but from the thorax as well;

4) that the axillary artery is surrounded by the three cords of the brachial plexus;

5) that the thoracoacromial artery divides into pectoral, acromial, deltoid and small clavicular branches (the latter are not shown in the figure);

6) that the intercostobrachial (T_2) nerve pierces the thoracic cage through the 2nd intercostal space in its course toward the axilla and arm.

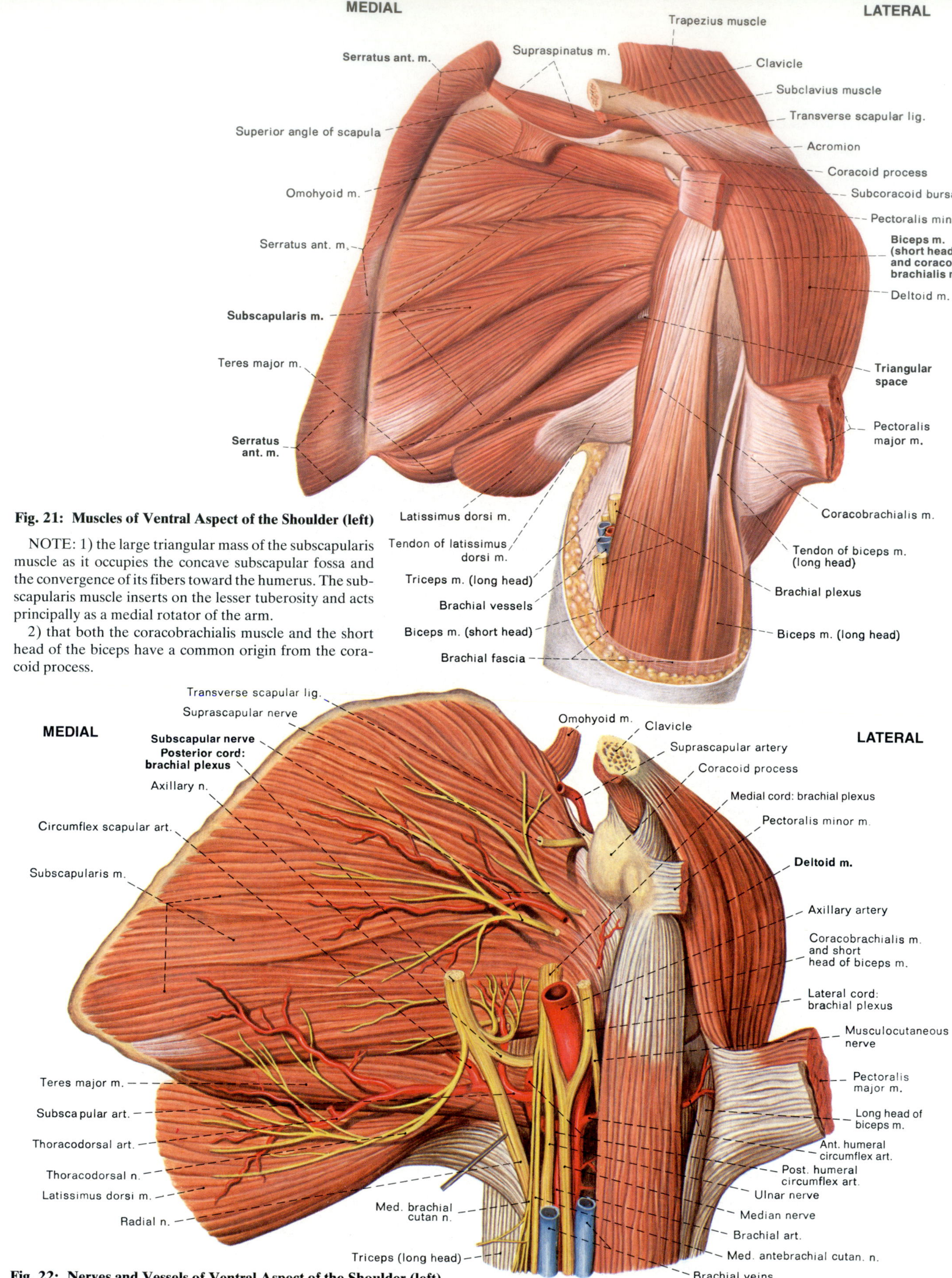
MEDIAL
LATERAL
Trapezius muscle
Supraspinatus m.
Serratus ant. m.
Clavicle
Subclavius muscle
Transverse scapular lig.
Superior angle of scapula
Acromion
Coracoid process
Omohyoid m.
Subcoracoid bursa
Pectoralis minor m.
Serratus ant. m.
Biceps m. (short head) and coraco-brachialis m.
Deltoid m.
Subscapularis m.
Teres major m.
Triangular space
Pectoralis major m.
Serratus ant. m.
Coracobrachialis m.
Latissimus dorsi m.
Tendon of latissimus dorsi m.
Tendon of biceps m. (long head)
Triceps m. (long head)
Brachial vessels
Brachial plexus
Biceps m. (short head)
Biceps m. (long head)
Brachial fascia

Fig. 21: Muscles of Ventral Aspect of the Shoulder (left)

NOTE: 1) the large triangular mass of the subscapularis muscle as it occupies the concave subscapular fossa and the convergence of its fibers toward the humerus. The subscapularis muscle inserts on the lesser tuberosity and acts principally as a medial rotator of the arm.

2) that both the coracobrachialis muscle and the short head of the biceps have a common origin from the coracoid process.

Transverse scapular lig.
Suprascapular nerve
MEDIAL
Subscapular nerve
Posterior cord: brachial plexus
Axillary n.
Omohyoid m.
Clavicle
Suprascapular artery
LATERAL
Coracoid process
Medial cord: brachial plexus
Circumflex scapular art.
Pectoralis minor m.
Subscapularis m.
Deltoid m.
Axillary artery
Coracobrachialis m. and short head of biceps m.
Lateral cord: brachial plexus
Musculocutaneous nerve
Teres major m.
Pectoralis major m.
Subscapular art.
Long head of biceps m.
Thoracodorsal art.
Ant. humeral circumflex art.
Thoracodorsal n.
Post. humeral circumflex art.
Latissimus dorsi m.
Ulnar nerve
Med. brachial cutan n.
Median nerve
Radial n.
Brachial art.
Med. antebrachial cutan. n.
Triceps (long head)
Brachial veins

Fig. 22: Nerves and Vessels of Ventral Aspect of the Shoulder (left)

NOTE: the relationship of the medial, lateral and posterior cords of the brachial plexus to the axillary artery. The posterior cord has been pulled medially from behind the plexus in this dissection.

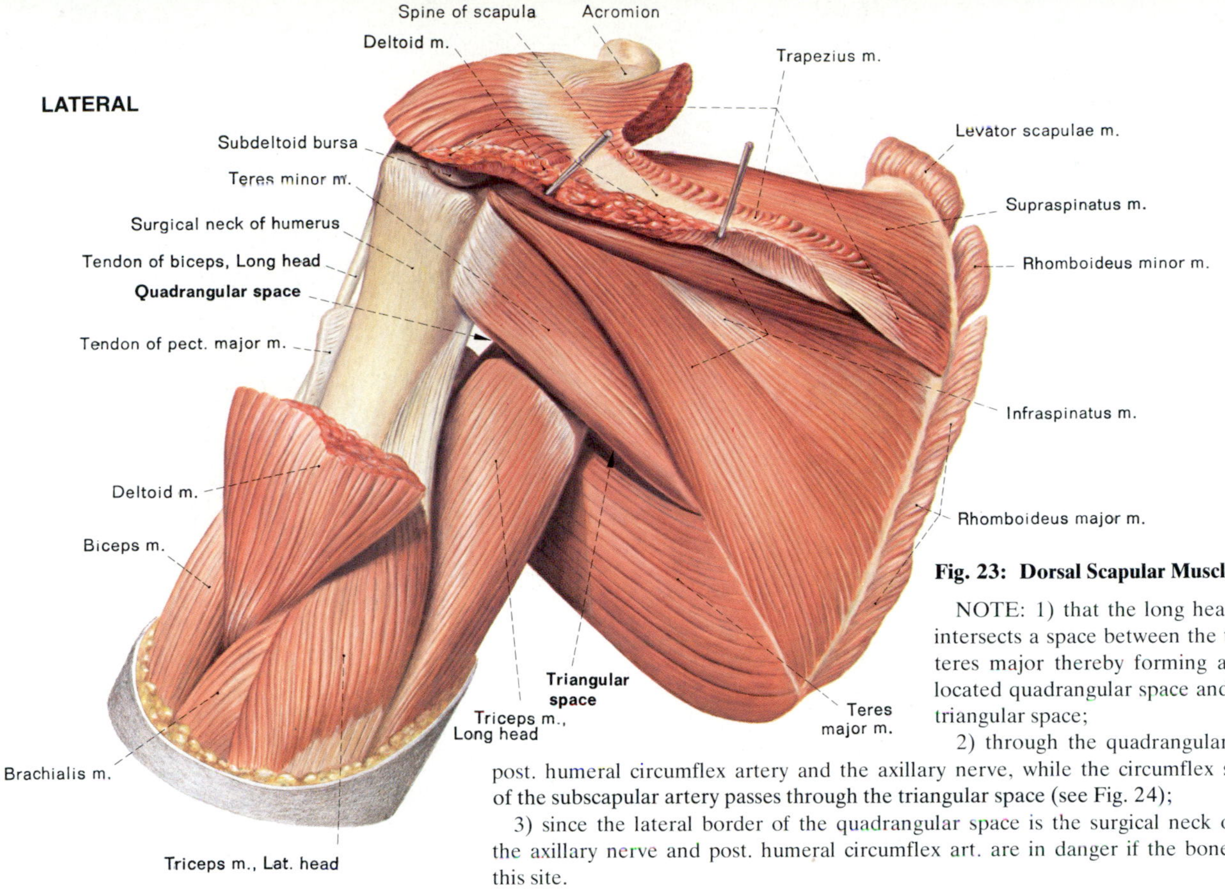

Spine of scapula
Acromion
Deltoid m.
Trapezius m.
LATERAL
MEDIAL
Levator scapulae m.
Subdeltoid bursa
Teres minor m.
Supraspinatus m.
Surgical neck of humerus
Tendon of biceps, Long head
Rhomboideus minor m.
Quadrangular space
Tendon of pect. major m.
Infraspinatus m.
Deltoid m.
Rhomboideus major m.
Biceps m.
Triangular space
Triceps m., Long head
Teres major m.
Brachialis m.
Triceps m., Lat. head

Fig. 23: Dorsal Scapular Muscles (left)

NOTE: 1) that the long head of the triceps intersects a space between the teres minor and teres major thereby forming a more laterally located quadrangular space and a more medial triangular space;

2) through the quadrangular space pass the post. humeral circumflex artery and the axillary nerve, while the circumflex scapular branch of the subscapular artery passes through the triangular space (see Fig. 24);

3) since the lateral border of the quadrangular space is the surgical neck of the humerus, the axillary nerve and post. humeral circumflex art. are in danger if the bone is fractured at this site.

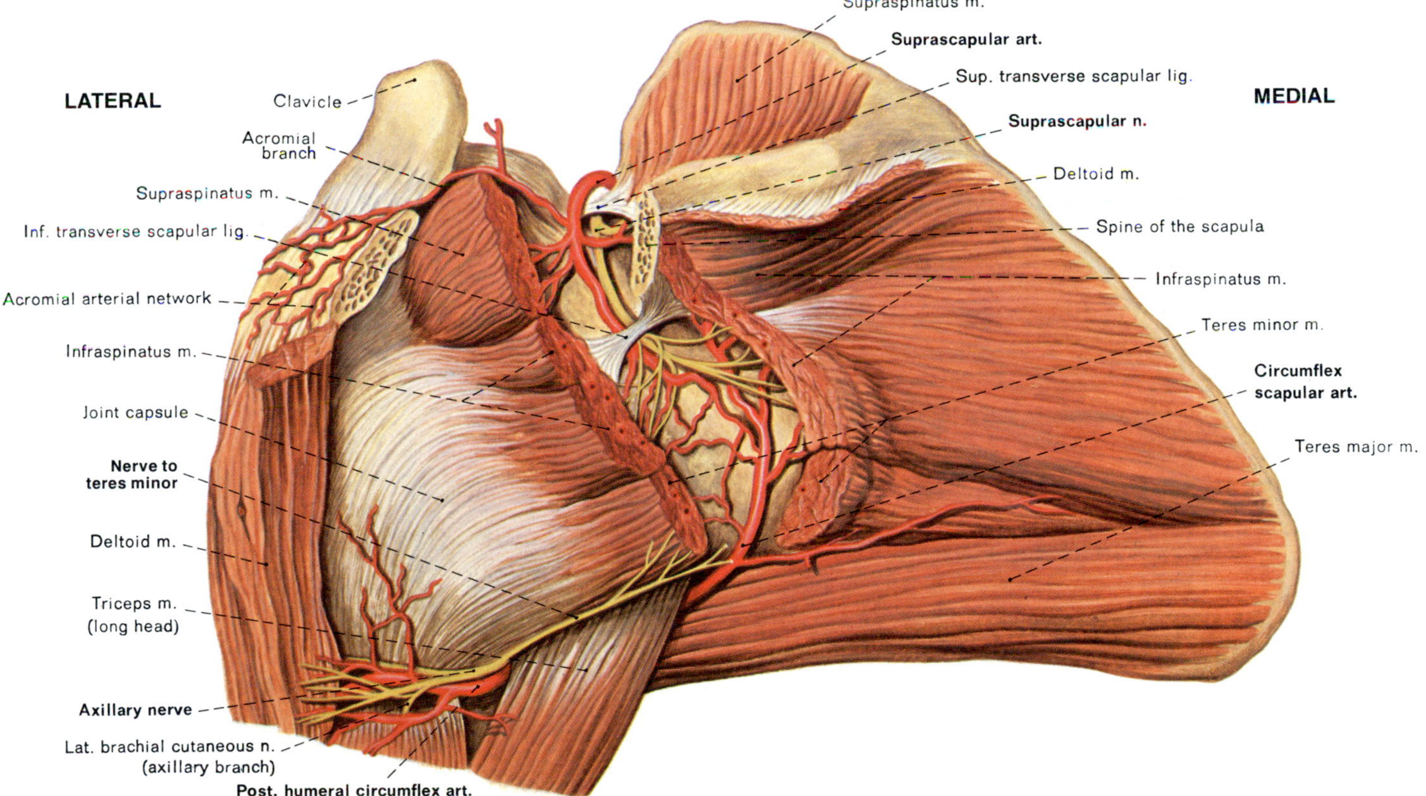
Supraspinatus m.
Suprascapular art.
Sup. transverse scapular lig.
LATERAL
MEDIAL
Clavicle
Suprascapular n.
Acromial branch
Deltoid m.
Supraspinatus m.
Spine of the scapula
Inf. transverse scapular lig.
Infraspinatus m.
Acromial arterial network
Teres minor m.
Infraspinatus m.
Circumflex scapular art.
Joint capsule
Teres major m.
Nerve to teres minor
Deltoid m.
Triceps m. (long head)
Axillary nerve
Lat. brachial cutaneous n. (axillary branch)
Post. humeral circumflex art.

Fig. 24: Nerves and Vessels of Dorsal Scapular Region (left)

NOTE: 1) that the sup. transverse scapular lig. bridges across the scapular notch and the suprascapular nerve passes beneath the ligament while the suprascapular artery usually passes above it;

2) that the axillary nerve supplies four structures: the deltoid muscle, the teres minor muscle, the capsule of the shoulder joint and the skin over the shoulder joint;

3) that the axillary nerve and post. humeral circumflex artery achieve the dorsal aspect of the shoulder through the quadrangular space, while the circumflex scapular artery reaches the infraspinatus fossa through the triangular space.

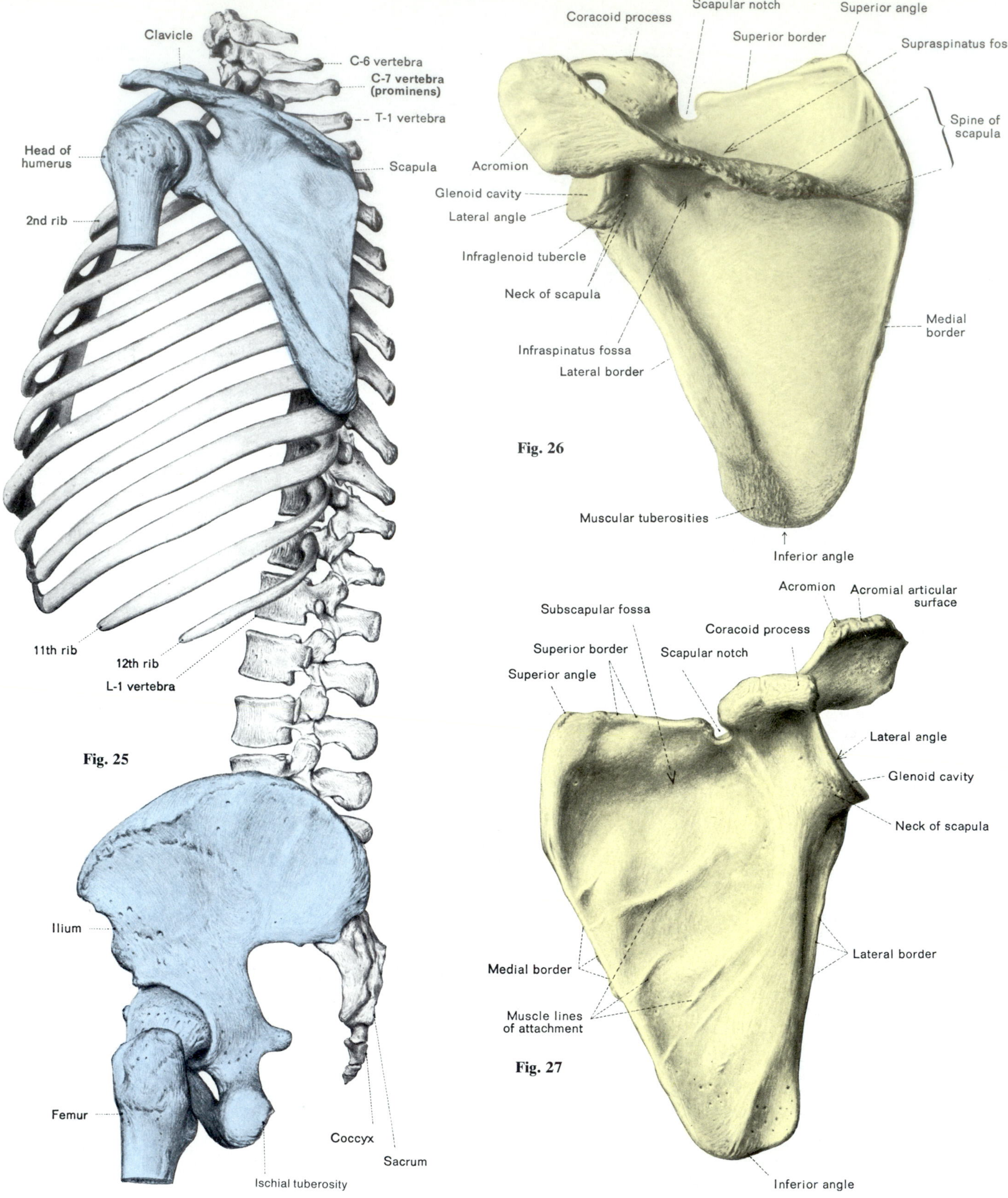
Clavicle
C-6 vertebra
C-7 vertebra (prominens)
T-1 vertebra
Head of humerus
Scapula
2nd rib
11th rib
12th rib
L-1 vertebra
Fig. 25
Ilium
Femur
Coccyx
Sacrum
Ischial tuberosity
Coracoid process
Scapular notch
Superior angle
Superior border
Supraspinatus fossa
Spine of scapula
Acromion
Glenoid cavity
Lateral angle
Infraglenoid tubercle
Neck of scapula
Medial border
Infraspinatus fossa
Lateral border
Fig. 26
Muscular tuberosities
Inferior angle
Acromion
Acromial articular surface
Subscapular fossa
Coracoid process
Superior border
Scapular notch
Superior angle
Lateral angle
Glenoid cavity
Neck of scapula
Lateral border
Medial border
Muscle lines of attachment
Fig. 27
Inferior angle

Fig. 25: Skeleton of Trunk with Scapula and Pelvis

NOTE that the flat triangular shaped scapula articulates with the head of the humerus and the clavicle. It is also attached to the rib cage by muscles.

Fig. 26: The Left Scapula (Dorsal Surface)

NOTE that the socket for the head of the humerus is formed by the glenoid cavity and is further enlarged by the coracoid and acromial processes along with their related ligaments. The spine of the scapula separates the dorsal surface into supraspinatus and infraspinatus fossae.

Fig. 27: The Left Scapula (Ventral Surface)

NOTE that much of the ventral surface is a concave fossa within which lies the subscapularis muscle.

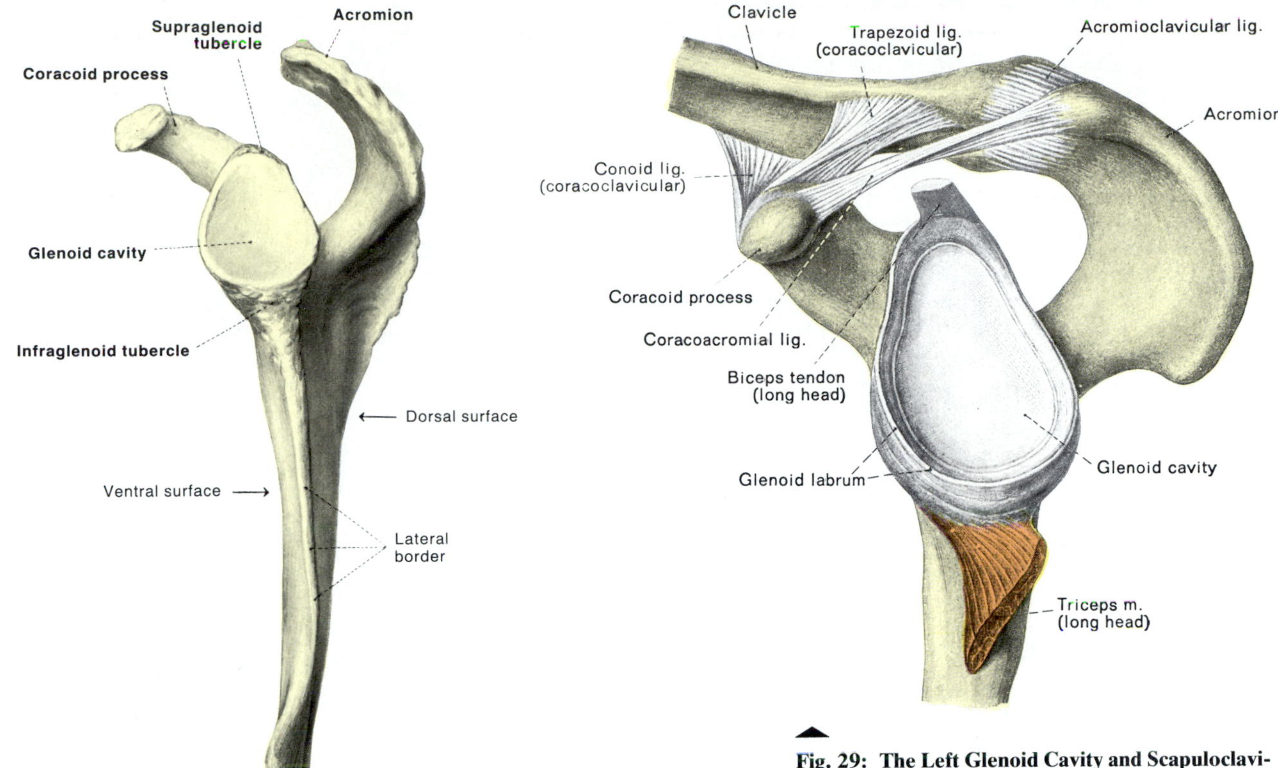

Acromion
Supraglenoid tubercle
Coracoid process
Glenoid cavity
Infraglenoid tubercle
Dorsal surface
Ventral surface
Lateral border
Clavicle
Trapezoid lig. (coracoclavicular)
Acromioclavicular lig.
Acromion
Conoid lig. (coracoclavicular)
Coracoid process
Coracoacromial lig.
Biceps tendon (long head)
Glenoid labrum
Glenoid cavity
Triceps m. (long head)

Fig. 28: The Left Scapula, Lateral View

NOTE: 1) the supraglenoid and infraglenoid tubercles from which arise the long heads of the biceps and triceps muscles, respectively;

2) the anteriorly projecting coracoid process to which are attached the pectoralis minor, short head of the biceps, and the coracobrachialis muscles.

Fig. 29: The Left Glenoid Cavity and Scapuloclavicular Joint (Lateral View)

NOTE: 1) exposure of the glenoid cavity was achieved by removal of the articular capsule at the glenoid labrum;

2) the attachment of the tendon of the long head of the biceps muscle at the supraglenoid tubercle and that of the long head of the triceps at the infraglenoid tubercle have been left intact;

3) the shallowness of the glenoid cavity is slightly deepened (4 to 6 mm.) by the glenoid labrum;

4) the protection afforded to the shoulder joint superiorly by the acromion, coracoid process and clavicle and their ligamentous attachments and by the tendon of the long head of the biceps.

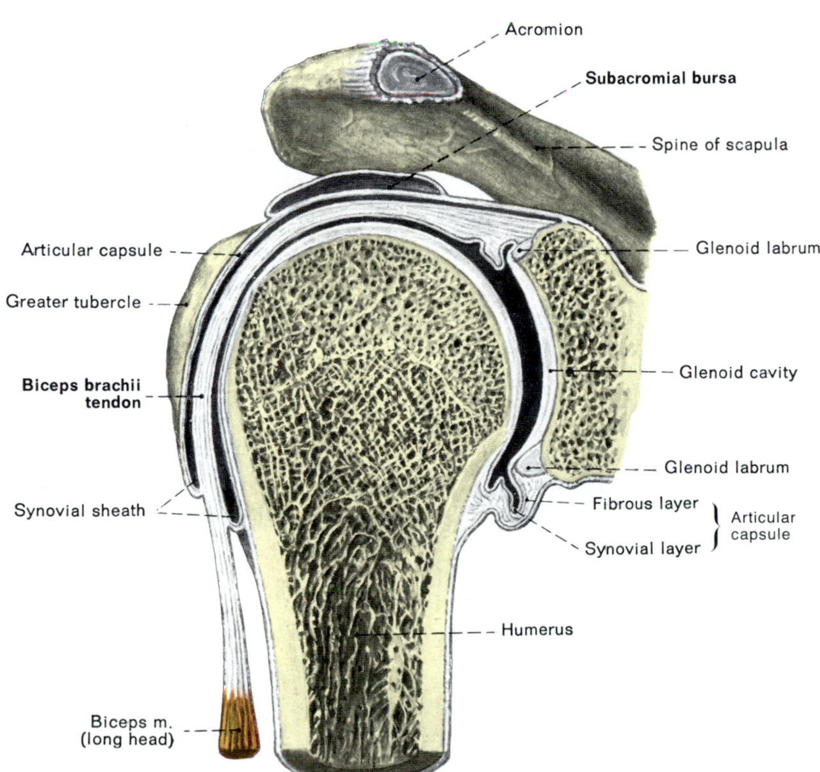

Acromion
Subacromial bursa
Spine of scapula
Articular capsule
Glenoid labrum
Greater tubercle
Biceps brachii tendon
Glenoid cavity
Glenoid labrum
Synovial sheath
Fibrous layer
Synovial layer
Articular capsule
Humerus
Biceps m. (long head)

Fig. 30: Frontal Section through Right Shoulder Joint

NOTE: 1) that the tendon of the long head of the biceps arises from the supraglenoid tuberosity and is at once enclosed by a reflection of the synovial sheath. Thus, although the tendon passes through the joint, it is not contained within the synovial cavity of the joint;

2) the capsule of the joint is composed of a dense fibrous outer layer and a thin synovial inner layer. It is this thin inner layer which reflects itself around the biceps tendon in its course through the joint;

3) a bursa is a sac lined with a synovial-like membrane containing a small amount of fluid. Bursae are found at sites subjected to friction and normally do not communicate with the joint capsule. In the shoulder, separate bursae are found between the capsule and the subscapularis, infraspinatus and deltoid tendons as well as other muscles, and between the capsule and the coracoid and acromial processes (subacromial bursa).

Fig. 31: Left Shoulder Joint and Acromioclavicular Joint (Anterior View)

NOTE: 1) that the clavicle is attached by ligaments to both the acromion (acromioclavicular lig) and the coracoid process (coracoclavicular lig.) of the scapula. The acromion and coracoid process are themselves connected by the coracoacromial ligament;

2) neither the acromion nor the clavicle articulates directly with the humerus, whereas the coracoid process and the glenoid labrum afford attachment of the scapula to the humerus;

3) the acromion, coracoid process and clavicle assist in the protection of the shoulder joint from above. Thus, the joint is weakest inferiorly and anteriorly, the directions in which most dislocations occur;

4) the glenohumeral ligaments are thickened bands which tend to strenghten somewhat the capsule of the joint anteriorly;

5) the position of the long tendon of the biceps traversing the articular cavity to its point of attachment on the supraglenoid tubercle.

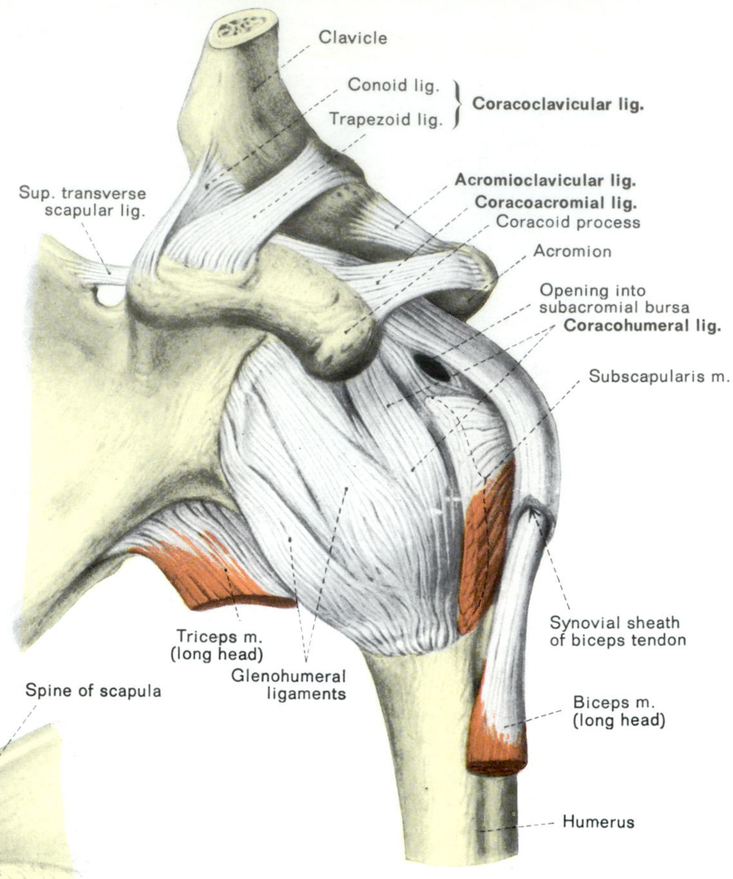
Clavicle
Conoid lig.
Trapezoid lig.
Coracoclavicular lig.
Sup. transverse scapular lig.
Acromioclavicular lig.
Coracoacromial lig.
Coracoid process
Acromion
Opening into subacromial bursa
Coracohumeral lig.
Subscapularis m.
Synovial sheath of biceps tendon
Triceps m. (long head)
Glenohumeral ligaments
Biceps m. (long head)
Humerus

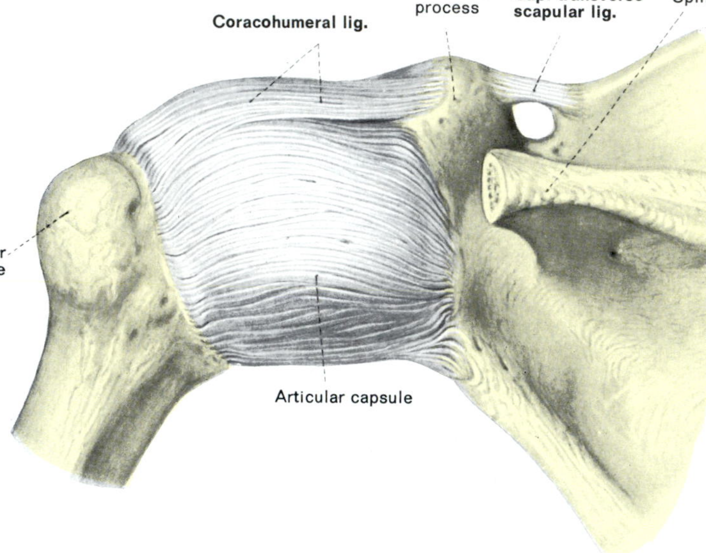
Coracohumeral lig.
Coracoid process
Sup. transverse scapular lig.
Spine of scapula
Greater tubercle
Articular capsule

Fig. 32: Capsule of Left Shoulder Joint (Posterior View)

NOTE: 1) the articular capsule completely surrounds the joint, being attached beyond the glenoid cavity on the scapula above and to the anatomical neck of the humerus below;

2) the superior part of the capsule is further strengthened by the coracohumeral ligament.

Fig. 33: Left Shoulder Joint (Posterior View)

NOTE: 1) the shoulder joint is a freely moving ball and socket joint. The capsule of the joint is not drawn tightly between the humeral head and scapula but attached loosely over these bony structures;

2) the tendons of the supraspinatus, infraspinatus and teres minor blend superiorly and posteriorly with the capsule of the joint. These muscles along with the subscapularis anteriorly form a muscular encasement lending some support in the maintenance of the head of the humerus in its socket;

3) the close relationship of the long head of the triceps to the capsule of joint. When the arm is abducted the triceps is drawn even closer to the capsule to help prevent dislocation.

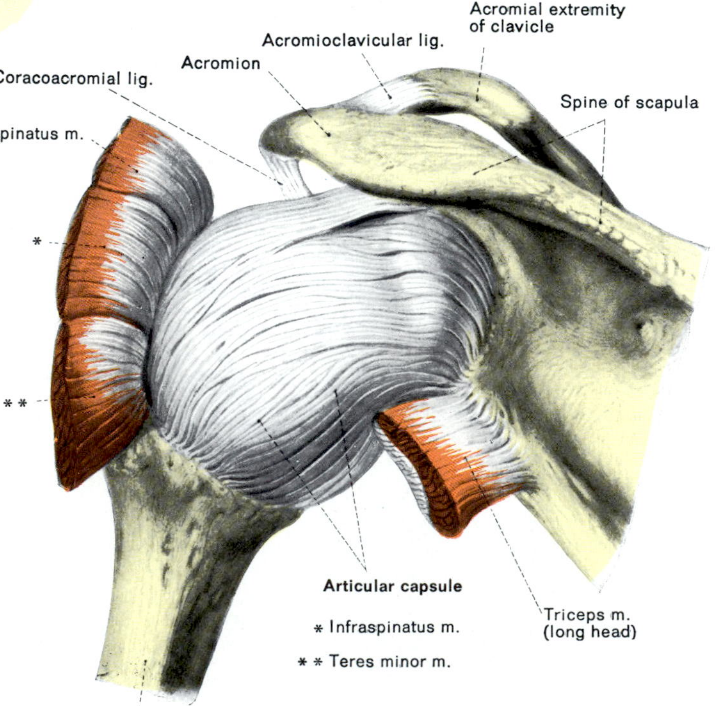
Acromial extremity of clavicle
Acromioclavicular lig.
Acromion
Coracoacromial lig.
Spine of scapula
Supraspinatus m.
*
**
Articular capsule
* Infraspinatus m.
** Teres minor m.
Triceps m. (long head)
Body of humerus

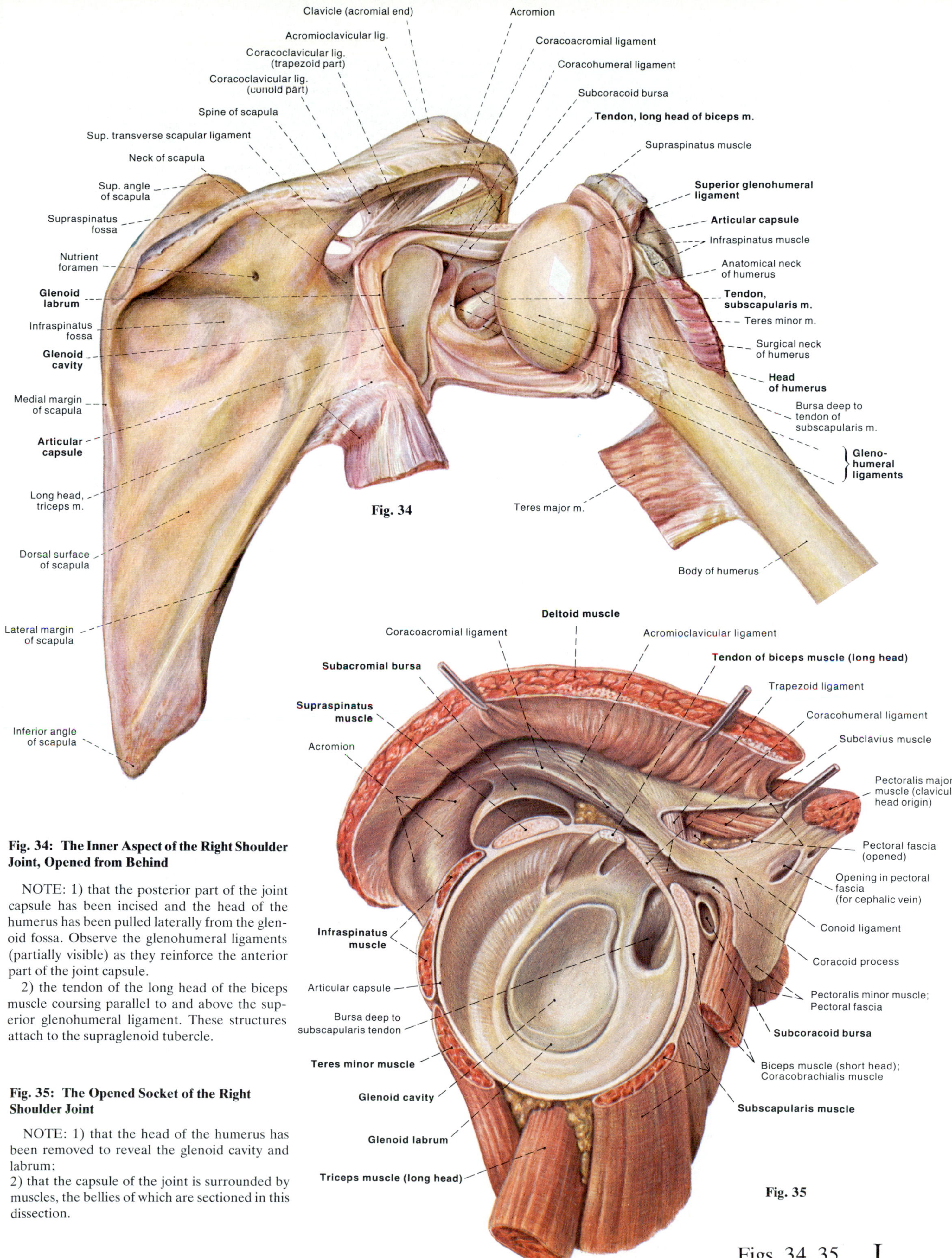
Clavicle (acromial end)
Acromion
Acromioclavicular lig.
Coracoacromial ligament
Coracoclavicular lig. (trapezoid part)
Coracohumeral ligament
Coracoclavicular lig. (conoid part)
Subcoracoid bursa
Spine of scapula
Tendon, long head of biceps m.
Sup. transverse scapular ligament
Supraspinatus muscle
Neck of scapula
Sup. angle of scapula
Superior glenohumeral ligament
Supraspinatus fossa
Articular capsule
Infraspinatus muscle
Nutrient foramen
Anatomical neck of humerus
Glenoid labrum
Tendon, subscapularis m.
Teres minor m.
Infraspinatus fossa
Surgical neck of humerus
Glenoid cavity
Head of humerus
Medial margin of scapula
Bursa deep to tendon of subscapularis m.
Articular capsule
Gleno-humeral ligaments
Long head, triceps m.
Fig. 34
Teres major m.
Dorsal surface of scapula
Body of humerus
Lateral margin of scapula
Inferior angle of scapula
Deltoid muscle
Coracoacromial ligament
Acromioclavicular ligament
Subacromial bursa
Tendon of biceps muscle (long head)
Trapezoid ligament
Supraspinatus muscle
Coracohumeral ligament
Acromion
Subclavius muscle
Pectoralis major muscle (clavicular head origin)
Pectoral fascia (opened)
Opening in pectoral fascia (for cephalic vein)
Infraspinatus muscle
Conoid ligament
Coracoid process
Articular capsule
Pectoralis minor muscle; Pectoral fascia
Bursa deep to subscapularis tendon
Subcoracoid bursa
Teres minor muscle
Biceps muscle (short head); Coracobrachialis muscle
Glenoid cavity
Subscapularis muscle
Glenoid labrum
Triceps muscle (long head)
Fig. 35

Fig. 34: The Inner Aspect of the Right Shoulder Joint, Opened from Behind

NOTE: 1) that the posterior part of the joint capsule has been incised and the head of the humerus has been pulled laterally from the glenoid fossa. Observe the glenohumeral ligaments (partially visible) as they reinforce the anterior part of the joint capsule.

2) the tendon of the long head of the biceps muscle coursing parallel to and above the superior glenohumeral ligament. These structures attach to the supraglenoid tubercle.

Fig. 35: The Opened Socket of the Right Shoulder Joint

NOTE: 1) that the head of the humerus has been removed to reveal the glenoid cavity and labrum;
2) that the capsule of the joint is surrounded by muscles, the bellies of which are sectioned in this dissection.

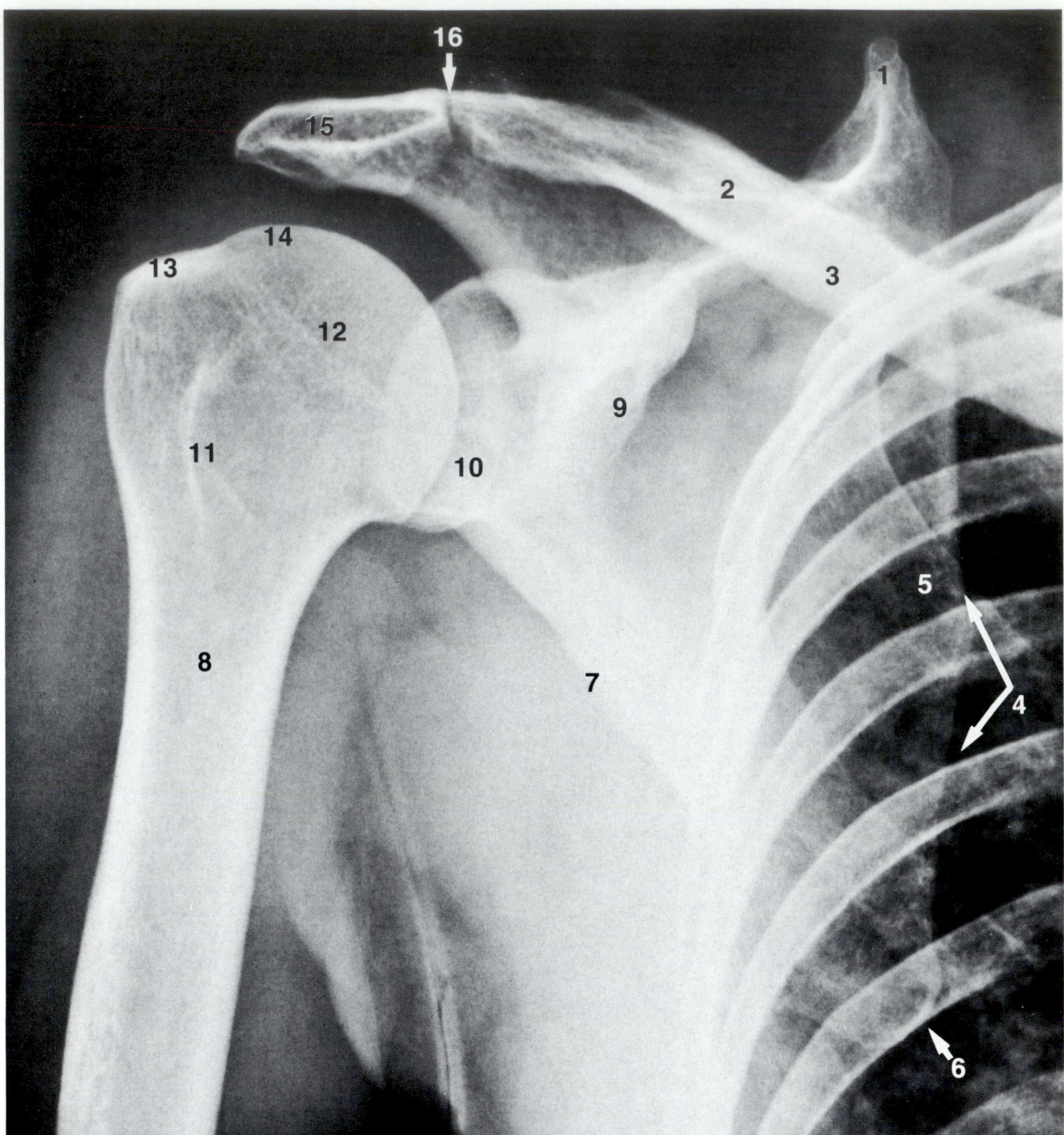
16
1
15
2
14
13
3
12
9
11
10
5
8
7
4
6

Fig. 36: Radiograph of the Right Shoulder Region

1. Superior angle of scapula
2. Spine of scapula
3. Clavicle
4. Medial margin of scapula
5. Second rib
6. Inferior angle of scapula
7. Lateral margin of scapula
8. Surgical neck of humerus
9. Coracoid process
10. Glenoid cavity
11. Lesser tubercle
12. Anatomical neck of humerus
13. Greater tubercle
14. Head of humerus
15. Acromion
16. Acromioclavicular joint

NOTE: 1) the clavicle, scapula and humerus are importantly involved in the radiographic anatomy of the shoulder region. The acromioclavicular joint is a planar joint formed by the apposition of the lateral end of the clavicle and the medial border of the acromion.

2) the glenohumeral joint, more commonly called the shoulder joint, is remarkably loose and provides a free range of movement. Observe the wide separation between the humeral head and glenoid cavity, and that only a small part of the humeral head is in contact with the cavity at any time.

3) that inferior *dislocations* of the head of the humerus are common because of minimal protection to this part of the joint. A *shoulder separation* results from the dislocation of the acromion beneath the lateral end of the clavicle, usually because of a strong blow to the lateral aspect of the joint.

4) the smooth surface of the head of the humerus and that it is almost hemispherical in shape. Covered with hyalin cartilage, the humeral head is slightly constricted at the anatomical neck, which forms a circumferential line separating the articular part superomedially from the greater and lesser tubercles inferolaterally.

5) that below the tubercles, the humerus shows another constriction that is called the surgical neck. Fractures of the humerus frequently occur at this site. (From Wicke, L. *Atlas of Radiologic Anatomy*, 3rd Edition, Urban & Schwarzenberg, Baltimore, 1982).

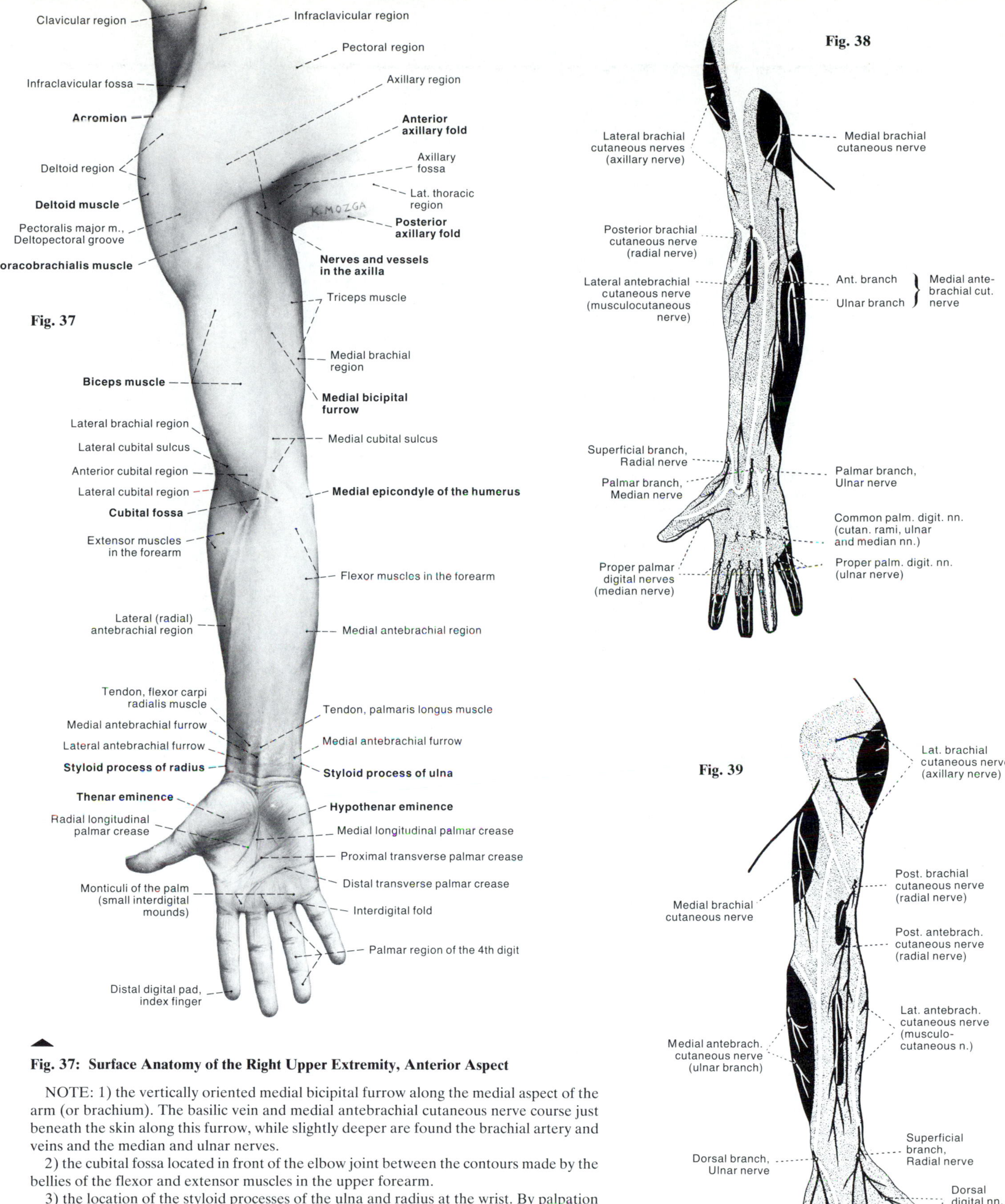

Fig. 37
Clavicular region
Infraclavicular region
Pectoral region
Infraclavicular fossa
Axillary region
Acromion
Anterior axillary fold
Deltoid region
Axillary fossa
Lat. thoracic region
Deltoid muscle
Posterior axillary fold
Pectoralis major m., Deltopectoral groove
Coracobrachialis muscle
Nerves and vessels in the axilla
Triceps muscle
Medial brachial region
Biceps muscle
Medial bicipital furrow
Lateral brachial region
Medial cubital sulcus
Lateral cubital sulcus
Anterior cubital region
Lateral cubital region
Medial epicondyle of the humerus
Cubital fossa
Extensor muscles in the forearm
Flexor muscles in the forearm
Lateral (radial) antebrachial region
Medial antebrachial region
Tendon, flexor carpi radialis muscle
Tendon, palmaris longus muscle
Medial antebrachial furrow
Lateral antebrachial furrow
Medial antebrachial furrow
Styloid process of radius
Styloid process of ulna
Thenar eminence
Hypothenar eminence
Radial longitudinal palmar crease
Medial longitudinal palmar crease
Proximal transverse palmar crease
Distal transverse palmar crease
Monticuli of the palm (small interdigital mounds)
Interdigital fold
Palmar region of the 4th digit
Distal digital pad, index finger
Fig. 38
Lateral brachial cutaneous nerves (axillary nerve)
Medial brachial cutaneous nerve
Posterior brachial cutaneous nerve (radial nerve)
Lateral antebrachial cutaneous nerve (musculocutaneous nerve)
Ant. branch
Ulnar branch
Medial antebrachial cut. nerve
Superficial branch, Radial nerve
Palmar branch, Median nerve
Palmar branch, Ulnar nerve
Common palm. digit. nn. (cutan. rami, ulnar and median nn.)
Proper palmar digital nerves (median nerve)
Proper palm. digit. nn. (ulnar nerve)
Fig. 39
Lat. brachial cutaneous nerves (axillary nerve)
Post. brachial cutaneous nerve (radial nerve)
Medial brachial cutaneous nerve
Post. antebrach. cutaneous nerve (radial nerve)
Lat. antebrach. cutaneous nerve (musculo-cutaneous n.)
Medial antebrach. cutaneous nerve (ulnar branch)
Superficial branch, Radial nerve
Dorsal branch, Ulnar nerve
Dorsal digital nn. (radial n.)
Dorsal digital nerves (ulnar n.)
Proper palmar digital nerves (median n.)
Proper palmar digital nn. (ulnar n.)

Fig. 37: Surface Anatomy of the Right Upper Extremity, Anterior Aspect

NOTE: 1) the vertically oriented medial bicipital furrow along the medial aspect of the arm (or brachium). The basilic vein and medial antebrachial cutaneous nerve course just beneath the skin along this furrow, while slightly deeper are found the brachial artery and veins and the median and ulnar nerves.

2) the cubital fossa located in front of the elbow joint between the contours made by the bellies of the flexor and extensor muscles in the upper forearm.

3) the location of the styloid processes of the ulna and radius at the wrist. By palpation determine that the radial styloid process extends about 1 cm lower than that of the ulna.

Fig. 38: The Distribution of Cutaneous Nerves, Anterior Aspect of the Right Upper Extremity

Fig. 39: The Distribution of Cutaneous Nerves, Posterior Aspect of the Right Upper Extremity

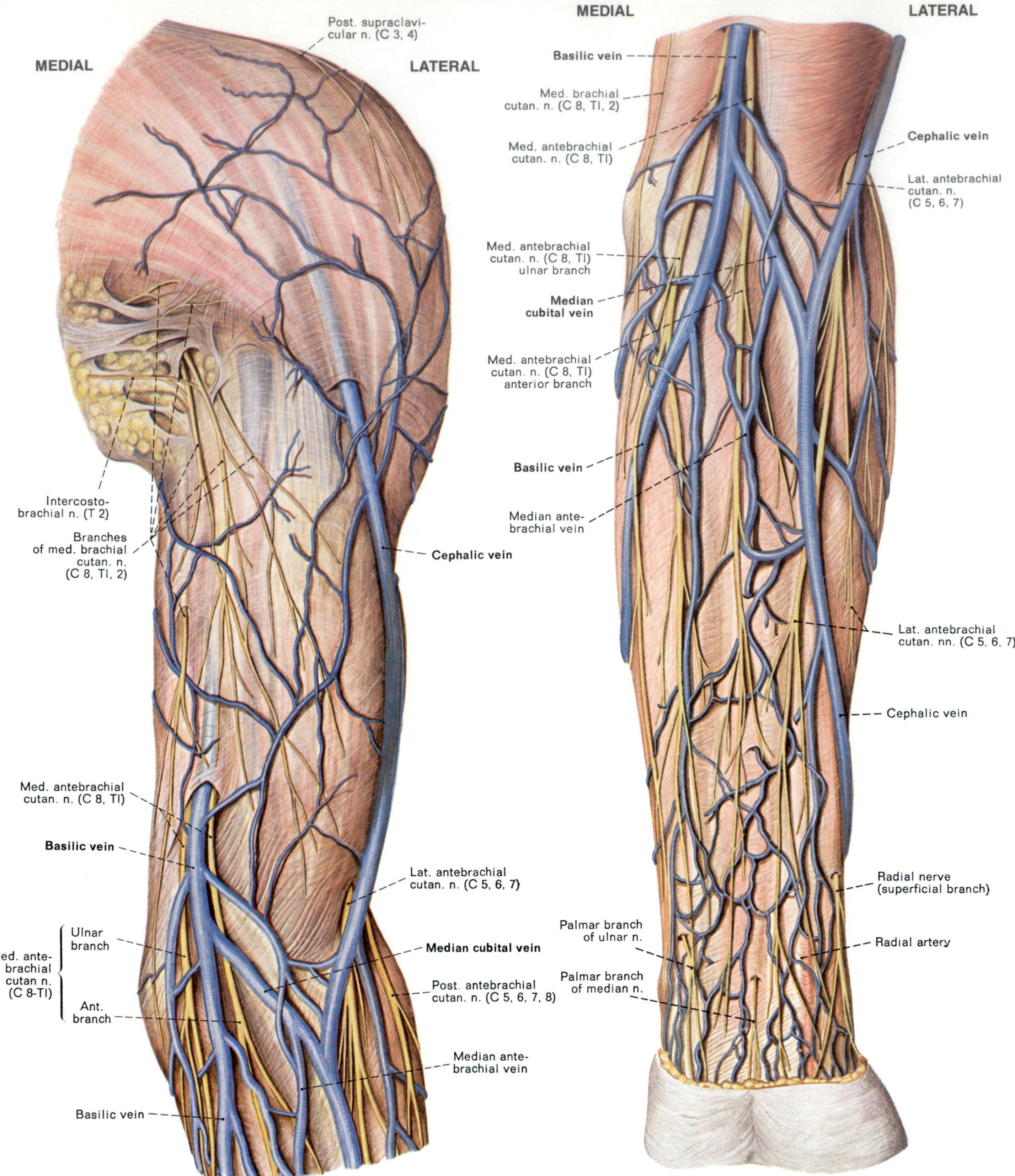

MEDIAL
LATERAL
Post. supraclavicular n. (C 3, 4)
Intercostobrachial n. (T 2)
Branches of med. brachial cutan. n. (C 8, TI, 2)
Cephalic vein
Med. antebrachial cutan. n. (C 8, TI)
Basilic vein
Lat. antebrachial cutan. n. (C 5, 6, 7)
Med. antebrachial cutan n. (C 8-TI)
Ulnar branch
Ant. branch
Median cubital vein
Post. antebrachial cutan. n. (C 5, 6, 7, 8)
Median antebrachial vein
Basilic vein
MEDIAL
LATERAL
Basilic vein
Med. brachial cutan. n. (C 8, TI, 2)
Med. antebrachial cutan. n. (C 8, TI)
Cephalic vein
Lat. antebrachial cutan. n. (C 5, 6, 7)
Med. antebrachial cutan. n. (C 8, TI) ulnar branch
Median cubital vein
Med. antebrachial cutan. n. (C 8, TI) anterior branch
Basilic vein
Median antebrachial vein
Lat. antebrachial cutan. nn. (C 5, 6, 7)
Cephalic vein
Radial nerve (superficial branch)
Palmar branch of ulnar n.
Radial artery
Palmar branch of median n.

Fig. 40: Arm; Superficial Veins and Cutaneous Nerves of Left Upper Limb (Anterior Surface)

NOTE: 1) the basilic vein ascends on the medial (ulnar) aspect of the arm, pierces the deep fascia and at the lower border of the teres major joins the brachial vein to form the axillary vein. The cephalic vein ascends laterally in the arm in its course toward the axillary vein.

2) the principal sensory nerves of the anterior arm are the medial and lateral brachial cutaneous nerves and the intercostobrachial nerve.

Fig. 41: Forearm; Superficial Veins and Cutaneous Nerves of Left Upper Limb (Anterior Surface)

NOTE: 1) the median cubital vein, interconnecting the cephalic and the basilic veins in the cubital fossa.

2) the main sensory nerves of the anterior forearm are the medial antebrachial cutaneous nerve (derived from the medial cord of the brachial plexus) and the lateral antebrachial cutaneous nerve, which is the continuation of the musculocutaneous nerve.

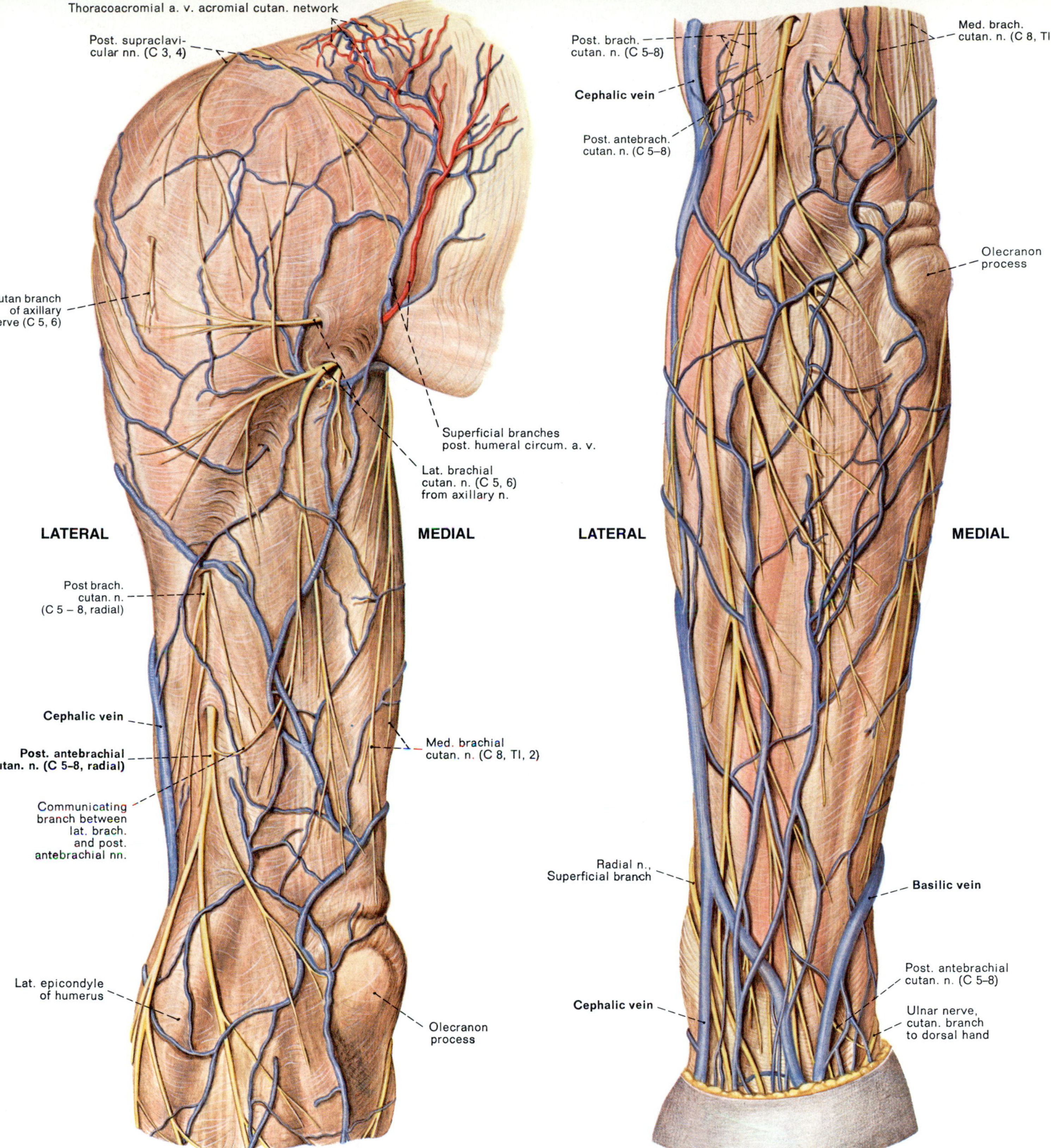
Thoracoacromial a. v. acromial cutan. network
Post. supraclavicular nn. (C 3, 4)
Cutan branch of axillary nerve (C 5, 6)
Superficial branches post. humeral circum. a. v.
Lat. brachial cutan. n. (C 5, 6) from axillary n.
LATERAL
MEDIAL
Post brach. cutan. n. (C 5 – 8, radial)
Cephalic vein
Post. antebrachial cutan. n. (C 5-8, radial)
Med. brachial cutan. n. (C 8, TI, 2)
Communicating branch between lat. brach. and post. antebrachial nn.
Lat. epicondyle of humerus
Olecranon process
Post. brach. cutan. n. (C 5-8)
Cephalic vein
Post. antebrach. cutan. n. (C 5-8)
Med. brach. cutan. n. (C 8, TI, 2)
Olecranon process
LATERAL
MEDIAL
Radial n., Superficial branch
Basilic vein
Post. antebrachial cutan. n. (C 5-8)
Cephalic vein
Ulnar nerve, cutan. branch to dorsal hand

Fig. 42: Arm; Superficial Veins and Cutaneous Nerves of Left Upper Limb (Posterior Surface)

NOTE: 1) that the posterior surface of the arm receives cutaneous innervation from branches of the radial (posterior brachial cutaneous n.) and axillary (lateral brachial cutaneous n.) nerves, both of which are derived from the posterior cord of the brachial plexus.

2) the posterior antebrachial cutaneous nerve (from the radial n.) perforates the lateral head of the triceps about five centimeters above the elbow. It pierces the superficial fascia on the posterolateral surface of the arm, but sends cutaneous branches to the dorsal surface of the forearm, as well as a communicating branch to cutaneous rami of the axillary nerve.

Fig. 43: Forearm; Superficial Veins and Cutaneous Nerves of the Left Upper Limb (Posterior Surface)

NOTE: 1) branches of the radial nerve (posterior antebrachial cutaneous and superficial radial) contribute the principal cutaneous innervation to the posterior aspect of the forearm.

2) at the wrist the dorsal branch of the ulnar nerve passes backward onto the dorsal surface of the wrist and hand.

3) the basilic (ulnar side, medial) and cephalic (radial side, lateral) veins commence on the dorsum of the hand and wrist and ascend in the forearm.

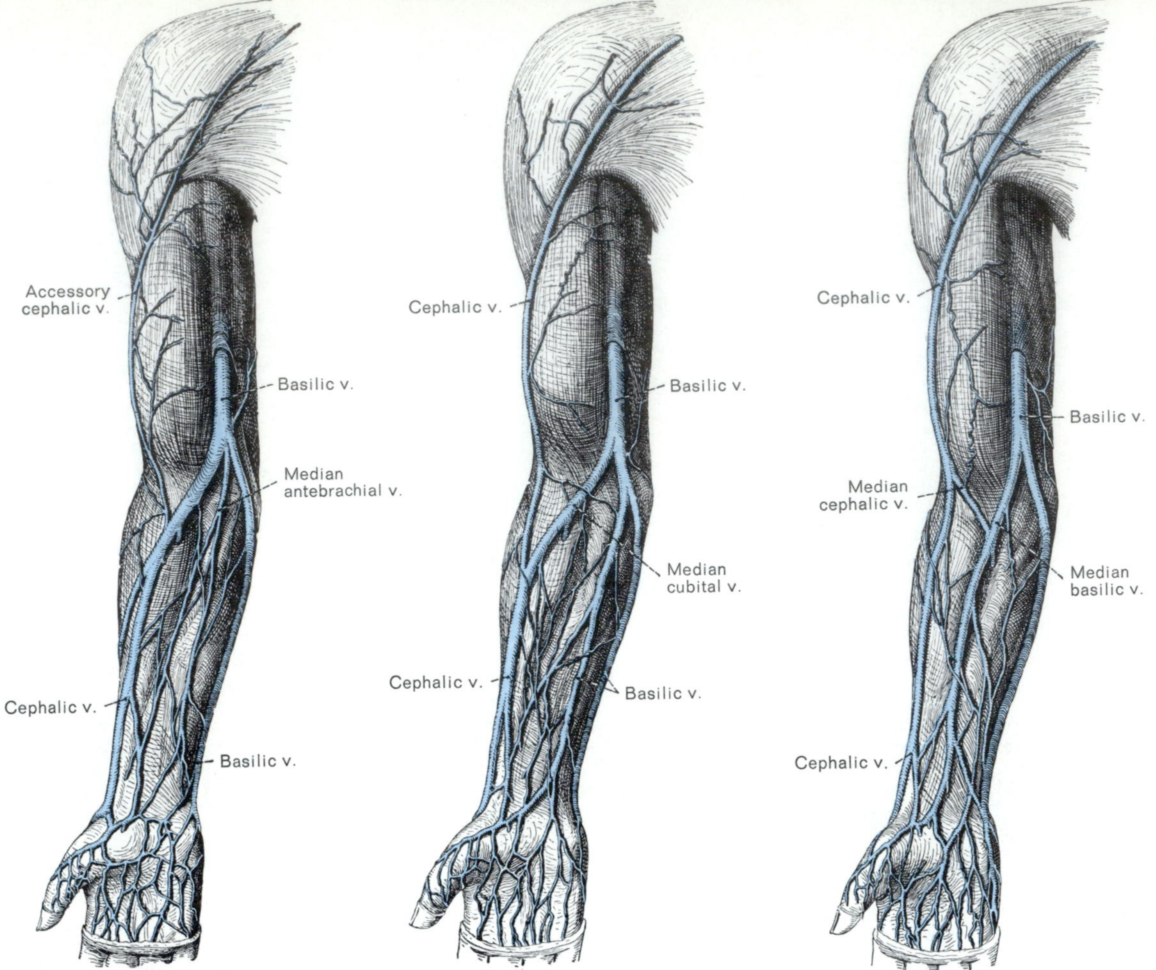
Accessory cephalic v.
Basilic v.
Median antebrachial v.
Cephalic v.
Basilic v.
Cephalic v.
Basilic v.
Median cubital v.
Cephalic v.
Basilic v.
Cephalic v.
Basilic v.
Median cephalic v.
Median basilic v.
Cephalic v.

Fig. 44: Variations in the Venous Pattern of the Upper Extremity

NOTE that the superficial veins of the upper extremity are quite variable and yet are of significance clinically. The median cubital vein in the cubital fossa is especially used for the withdrawal of blood and for procedures necessitating the injection of fluids into the vascular system. In this regard, care must be taken not to injure the median nerve nor puncture the brachial artery, which both lie deep to the median cubital vein and the subjacent bicipital aponeurosis on which the vein rests.

Fig. 45: Dermatomes of the Upper Limb

NOTE: the dermatomes of the upper limb are supplied by the 5th cervical to the 1st thoracic segments of the spinal cord. The boundary between the 5th cervical and the 1st thoracic dermatomes ventrally is called the ventral axial line of the upper limb. The 4th cervical dermatome lies in the neck. Commencing with the 5th cervical dermatome and proceeding radially around the upper limb, the dermatomes can be followed sequentially to the 1st thoracic and thus, the ventral axial line.

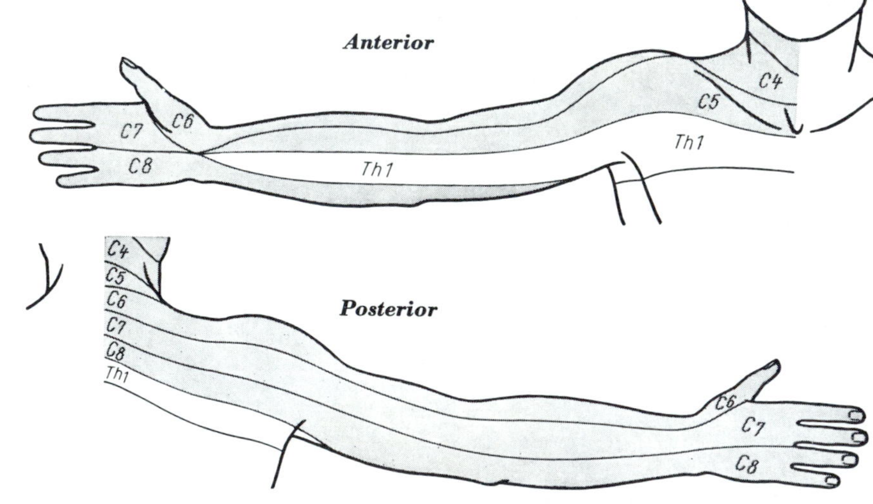
Anterior
C4
C5
C6
C7
C8
Th1
Th1
Posterior
C4
C5
C6
C7
C8
Th1
C6
C7
C8

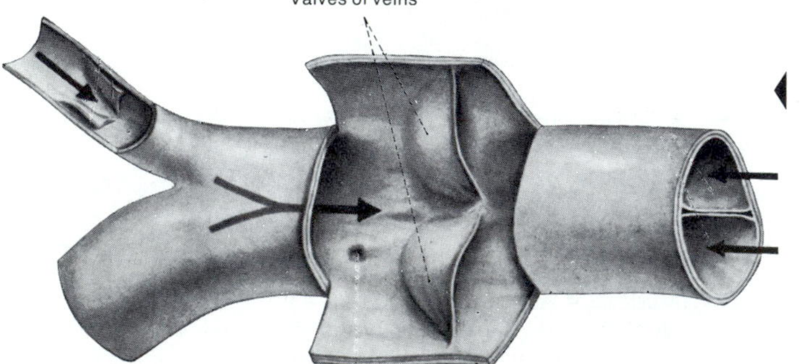
Valves of veins

Fig. 46: Valves Visible in an Opened Segment of Vein

NOTE: 1) venous valves are formed by semilunar folds of the tunica intima strengthened by connective tissue containing a thick network of elastic fibers.

2) arrows on left (pointing to the right) indicate the normal direction of blood flow, i.e. toward the heart, in which case the valves open easily. Arrows on the right (pointing to the left) indicate the reversed direction of blood flow, whereby regurgitated blood quickly becomes trapped in the expanded sinus, causing the valves to close and thus to interrupt the backflow. Incompetent valves result in the formation of varicosed veins.

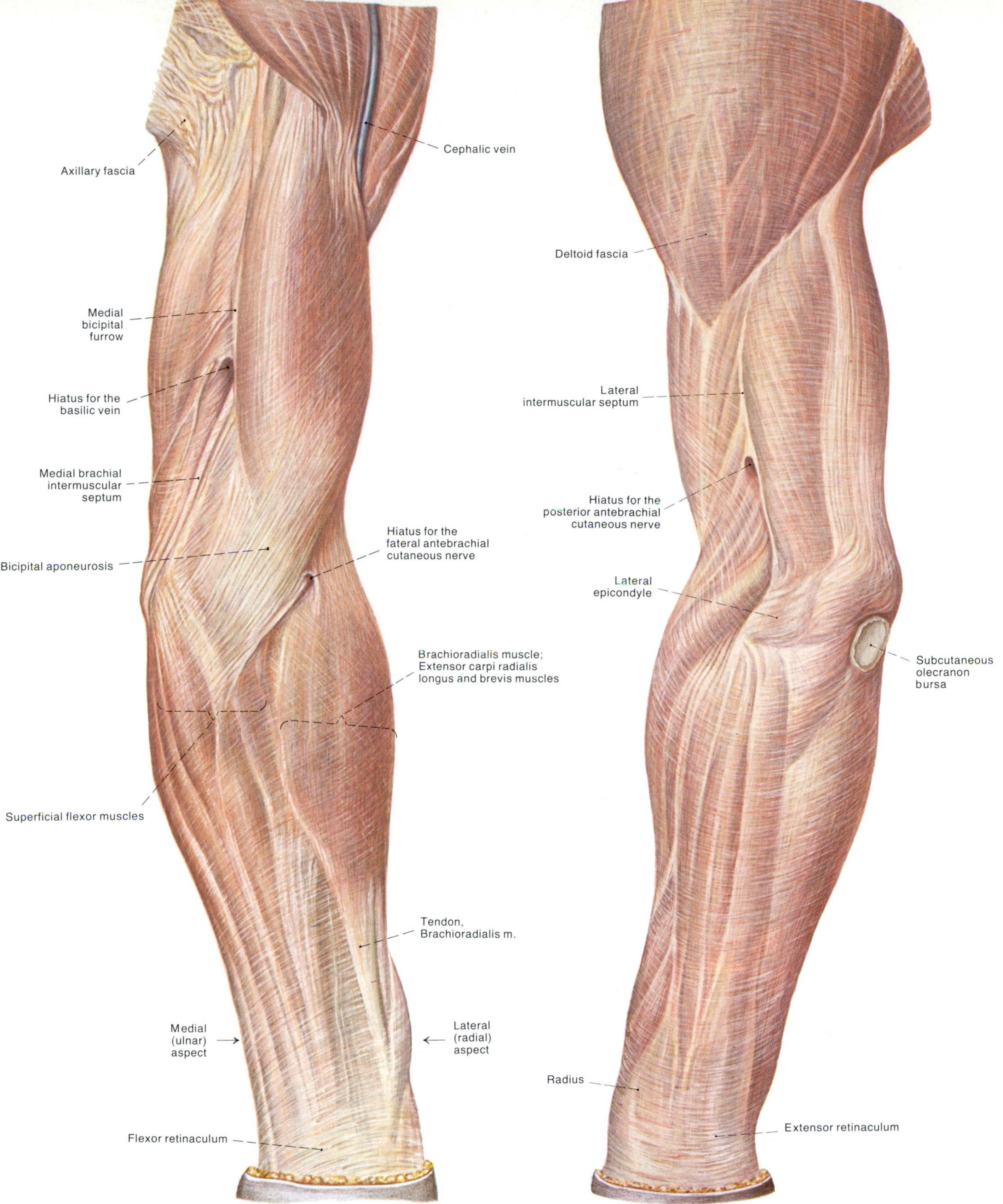
Axillary fascia
Cephalic vein
Medial bicipital furrow
Hiatus for the basilic vein
Medial brachial intermuscular septum
Bicipital aponeurosis
Hiatus for the fateral antebrachial cutaneous nerve
Brachioradialis muscle; Extensor carpi radialis longus and brevis muscles
Superficial flexor muscles
Tendon, Brachioradialis m.
Medial (ulnar) aspect
Lateral (radial) aspect
Flexor retinaculum
Deltoid fascia
Lateral intermuscular septum
Hiatus for the posterior antebrachial cutaneous nerve
Lateral epicondyle
Subcutaneous olecranon bursa
Radius
Extensor retinaculum

Fig. 47: The Deep Fascia Overlying the Brachial and Antebrachial Muscles of the Left Upper Extremity, Anterior View

NOTE: 1) the medial bicipital furrow that separates the flexor from the extensor compartments of the arm, and the hiatuses in the deep fascia through which course the basilic vein and the lateral antebrachial cutaneous nerve.

2) the bicipital aponeurosis. This is an extension medially from the tendon of insertion of the biceps brachii muscle that courses over the brachial artery to fuse with the deep fascia of the forearm.

Fig. 48: The Deep Fascia Overlying the Brachial and Antebrachial Muscles of the Left Upper Extremity, Posterior View

NOTE: 1) the subcutaneous olecranon bursa that lubricates the posterolateral aspect of the olecranon between the sites of attachment of the triceps and anconeus muscles.

2) the hiatus in the lateral intermuscular septum through which the posterior antebrachial cutaneous nerve penetrates the deep fascia to reach the superficial fascia and skin of the posterior forearm.

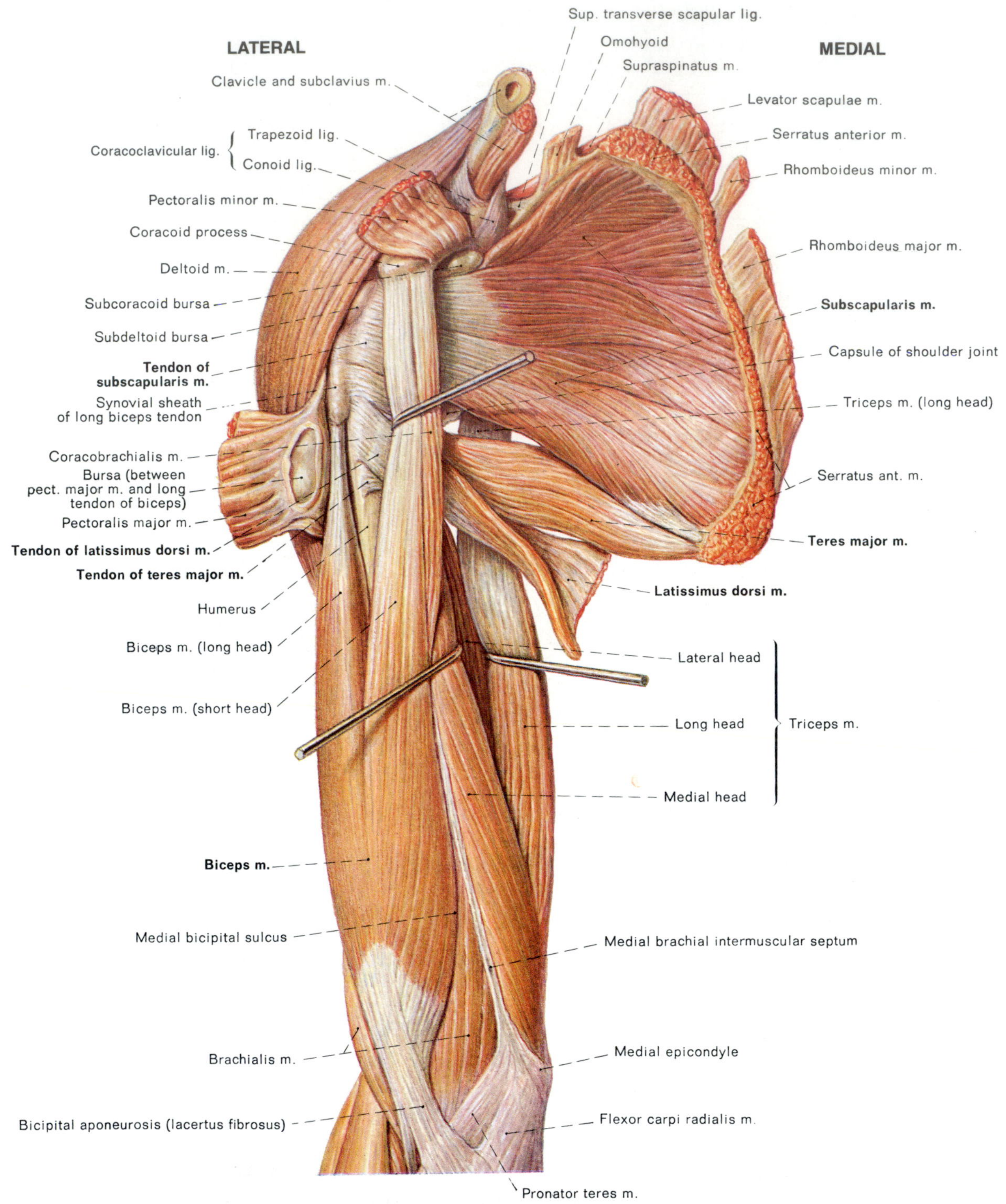

LATERAL
MEDIAL
Sup. transverse scapular lig.
Omohyoid
Supraspinatus m.
Clavicle and subclavius m.
Levator scapulae m.
Coracoclavicular lig.
Trapezoid lig.
Conoid lig.
Serratus anterior m.
Rhomboideus minor m.
Pectoralis minor m.
Coracoid process
Rhomboideus major m.
Deltoid m.
Subcoracoid bursa
Subscapularis m.
Subdeltoid bursa
Capsule of shoulder joint
Tendon of subscapularis m.
Synovial sheath of long biceps tendon
Triceps m. (long head)
Coracobrachialis m.
Bursa (between pect. major m. and long tendon of biceps)
Serratus ant. m.
Pectoralis major m.
Teres major m.
Tendon of latissimus dorsi m.
Tendon of teres major m.
Latissimus dorsi m.
Humerus
Biceps m. (long head)
Lateral head
Biceps m. (short head)
Long head
Triceps m.
Medial head
Biceps m.
Medial bicipital sulcus
Medial brachial intermuscular septum
Brachialis m.
Medial epicondyle
Bicipital aponeurosis (lacertus fibrosus)
Flexor carpi radialis m.
Pronator teres m.

Fig. 49: Muscles of the Right Shoulder and Arm (Anterior View)

NOTE: 1) the insertion of the subscapularis muscle on the lesser tubercle of the humerus. Distal to this, from medial to lateral, insert the teres major, latissimus dorsi and pectoralis major muscles.

2) attaching to the coracoid process are the pectoralis minor m., coracobrachialis m. and the short head of the biceps m.

3) the insertion of the pectoralis major m. and the long tendon of the biceps muscle are frequently separated by a bursa.

4) in the arm, the flexor compartment (biceps, coracobrachialis and brachialis) is separated from the extensor compartment (triceps) by an intermuscular septum of deep fascia.

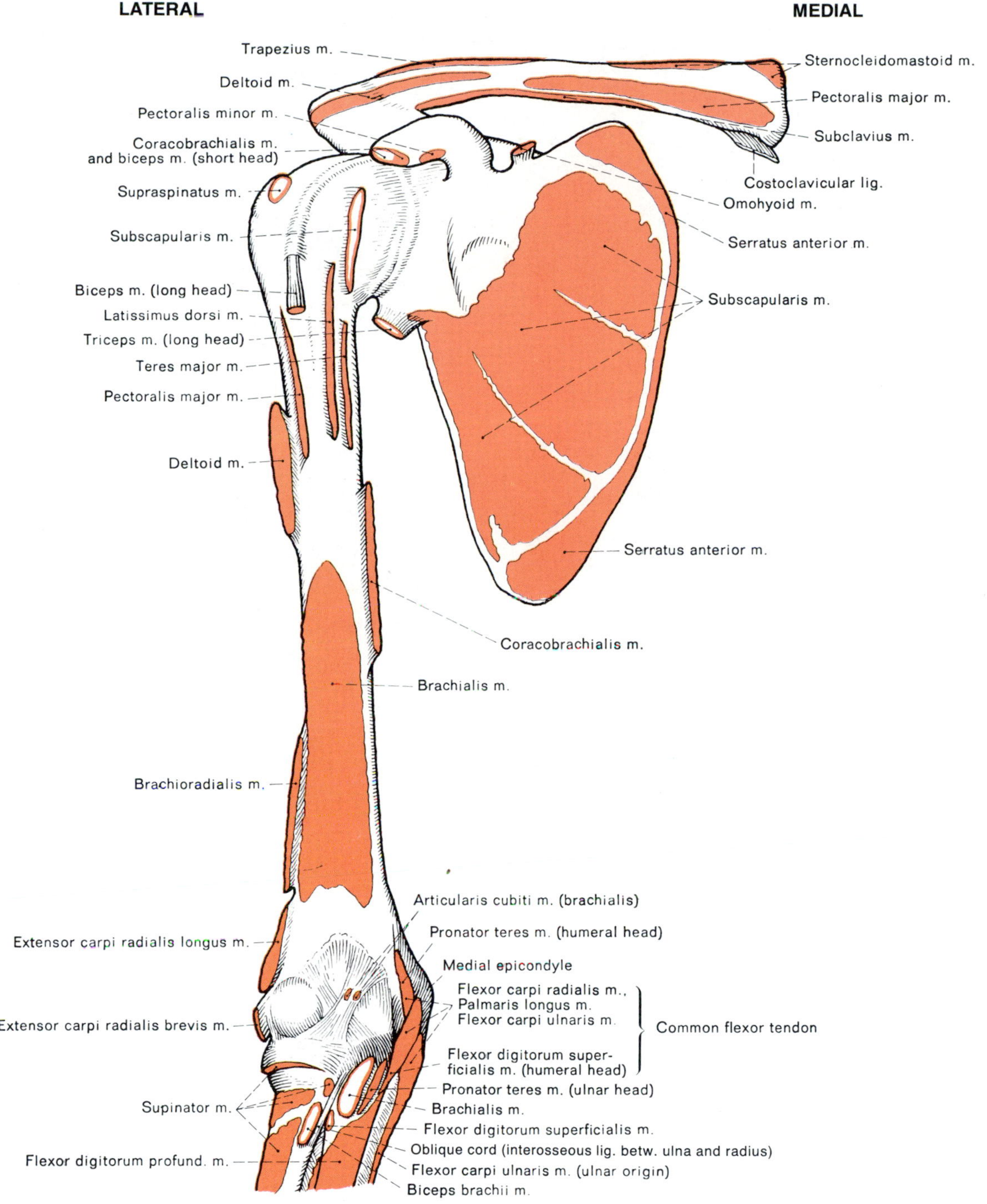
LATERAL
MEDIAL
Trapezius m.
Deltoid m.
Pectoralis minor m.
Coracobrachialis m. and biceps m. (short head)
Supraspinatus m.
Subscapularis m.
Biceps m. (long head)
Latissimus dorsi m.
Triceps m. (long head)
Teres major m.
Pectoralis major m.
Deltoid m.
Sternocleidomastoid m.
Pectoralis major m.
Subclavius m.
Costoclavicular lig.
Omohyoid m.
Serratus anterior m.
Subscapularis m.
Serratus anterior m.
Coracobrachialis m.
Brachialis m.
Brachioradialis m.
Articularis cubiti m. (brachialis)
Pronator teres m. (humeral head)
Medial epicondyle
Flexor carpi radialis m.,
Palmaris longus m.
Flexor carpi ulnaris m.
Common flexor tendon
Flexor digitorum superficialis m. (humeral head)
Pronator teres m. (ulnar head)
Brachialis m.
Flexor digitorum superficialis m.
Oblique cord (interosseous lig. betw. ulna and radius)
Flexor carpi ulnaris m. (ulnar origin)
Biceps brachii m.
Extensor carpi radialis longus m.
Extensor carpi radialis brevis m.
Supinator m.
Flexor digitorum profund. m.

Fig. 50: Anterior View of Bones of the Upper Limb (Including Proximal End of Radius and Ulna) Showing Attachments of Muscles

NOTE: 1) the broad *origin* of the subscapularis in the subscapular fossa of the scapula and its *insertion* on the lesser tubercle of the humerus proximal to the insertions of the latissimus dorsi and teres major muscles. The subscapularis is an adductor and medial rotator of the arm.

2) the brachialis muscle *arises* from the distal three-fifths of the anterior surface of the humerus and *inserts* on the coronoid process of the ulna. This muscle is the strongest flexor of the forearm.

3) the short head of the biceps m. *arises* with the coracobrachialis m. from the coracoid process, while the long head *arises* from the supraglenoid tubercle of the scapula.

4) the coracobrachialis *inserts* onto the shaft of the humerus near its middle, while the biceps inserts onto the tuberosity of the radius and onto the deep fascia of the forearm by way of the bicipital aponeurosis.

5) the coracobrachialis flexes and adducts the arm at the shoulder joint, while the biceps flexes and supinates the forearm, with the long head assisting in flexion of the arm at the shoulder joint.

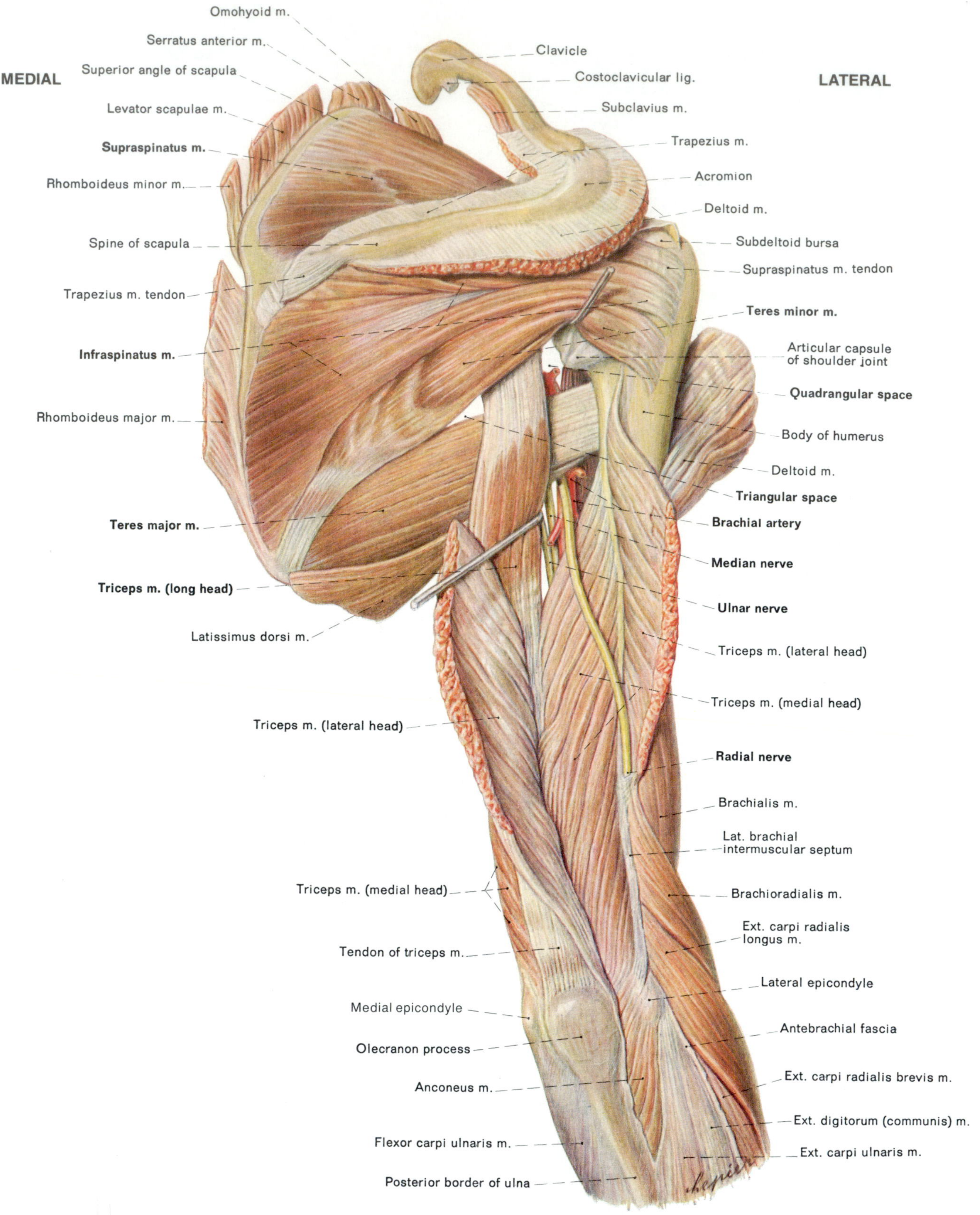

MEDIAL
LATERAL
Omohyoid m.
Serratus anterior m.
Superior angle of scapula
Levator scapulae m.
Supraspinatus m.
Rhomboideus minor m.
Spine of scapula
Trapezius m. tendon
Infraspinatus m.
Rhomboideus major m.
Teres major m.
Triceps m. (long head)
Latissimus dorsi m.
Triceps m. (lateral head)
Triceps m. (medial head)
Tendon of triceps m.
Medial epicondyle
Olecranon process
Anconeus m.
Flexor carpi ulnaris m.
Posterior border of ulna
Clavicle
Costoclavicular lig.
Subclavius m.
Trapezius m.
Acromion
Deltoid m.
Subdeltoid bursa
Supraspinatus m. tendon
Teres minor m.
Articular capsule of shoulder joint
Quadrangular space
Body of humerus
Deltoid m.
Triangular space
Brachial artery
Median nerve
Ulnar nerve
Triceps m. (lateral head)
Triceps m. (medial head)
Radial nerve
Brachialis m.
Lat. brachial intermuscular septum
Brachioradialis m.
Ext. carpi radialis longus m.
Lateral epicondyle
Antebrachial fascia
Ext. carpi radialis brevis m.
Ext. digitorum (communis) m.
Ext. carpi ulnaris m.

Fig. 51: Muscles of the Shoulder and Deep Arm (Posterior View)

NOTE: 1) that with the deltoid muscle and the lateral head of the triceps muscle severed, the course of the radial nerve in the upper arm is revealed.

2) the sequential insertions of the supraspinatus, infraspinatus and teres minor muscles on the greater tubercle of the humerus.

3) the boundaries of the quadrangular and triangular spaces.

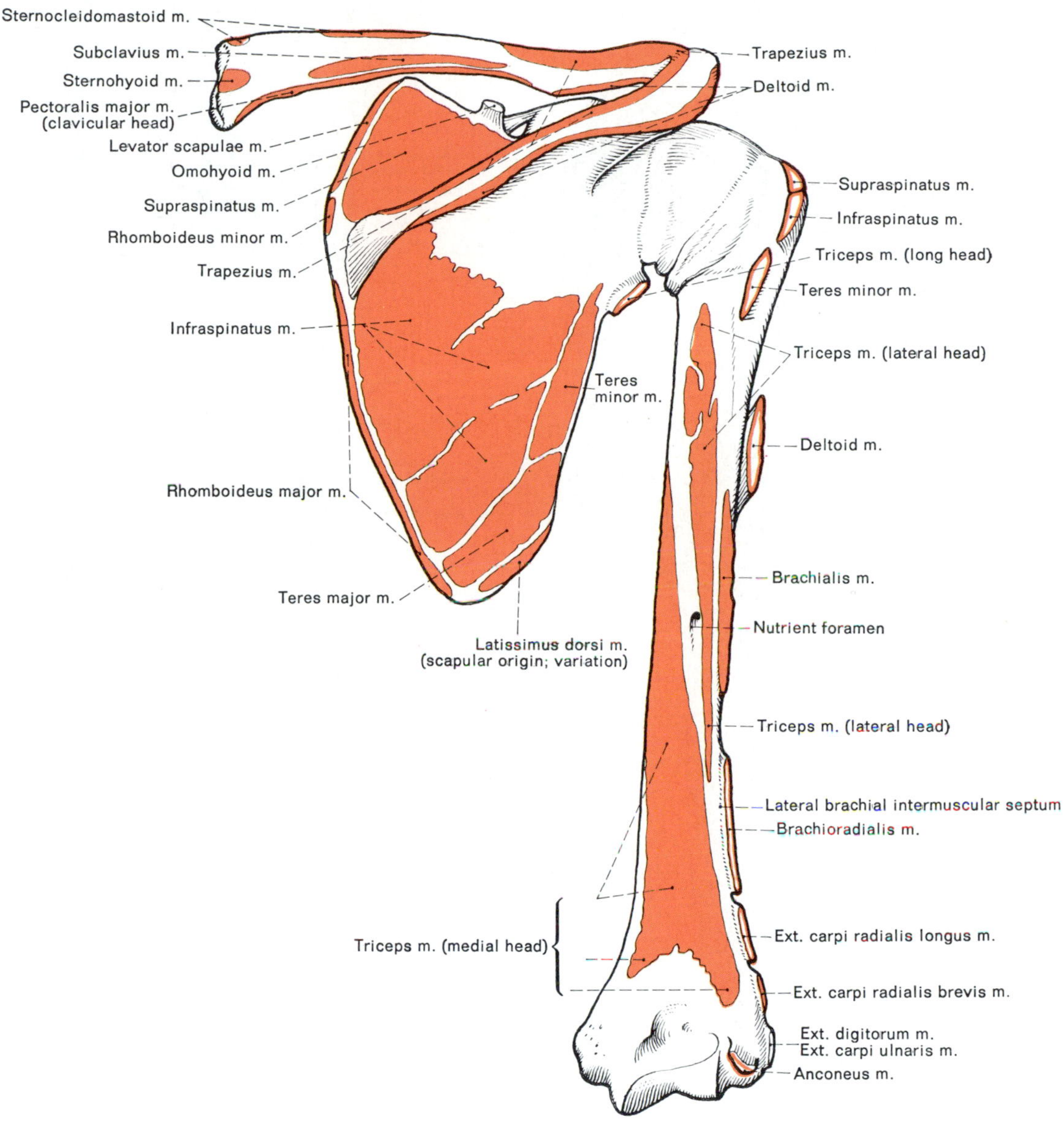
MEDIAL
LATERAL
Sternocleidomastoid m.
Subclavius m.
Sternohyoid m.
Pectoralis major m.
(clavicular head)
Levator scapulae m.
Omohyoid m.
Supraspinatus m.
Rhomboideus minor m.
Trapezius m.
Infraspinatus m.
Rhomboideus major m.
Teres major m.
Latissimus dorsi m.
(scapular origin; variation)
Teres
minor m.
Trapezius m.
Deltoid m.
Supraspinatus m.
Infraspinatus m.
Triceps m. (long head)
Teres minor m.
Triceps m. (lateral head)
Deltoid m.
Brachialis m.
Nutrient foramen
Triceps m. (lateral head)
Lateral brachial intermuscular septum
Brachioradialis m.
Triceps m. (medial head)
Ext. carpi radialis longus m.
Ext. carpi radialis brevis m.
Ext. digitorum m.
Ext. carpi ulnaris m.
Anconeus m.

Fig. 52: Posterior View of the Clavicle, Scapula and Humerus Showing Muscle Attachments

NOTE: 1) that dorsally the vertebral border of the scapula is attached to the trunk by the levator scapulae and the rhomboideus major and minor muscles, whereas on the ventral scapular surface (see Fig. 50) the serratus anterior attaches along the vertebral border.

2) the supraspinatus muscle arising from the medial two-thirds of the supraspinatus fossa.

3) the origins of the teres major and teres minor muscles along the axillary border of the scapula, and the broad origin of the infraspinatus muscle from the infraspinatus fossa.

4) that onto the spine of the scapula and extending to the lateral third of the clavicle are attached the trapezius and deltoid muscles.

5) most of the posterior surface of the humerus affords origin to the medial and lateral heads of the triceps muscle, while the long head arises from the infraglenoid tubercle of the scapula.

6) the alignment of the tendinous insertions of the supraspinatus, infraspinatus, and teres minor muscles which, along with the tendon of the subscapularis (see Figs. 49 and 50), from the so-called "rotator cuff" of the shoulder joint. This musculotendinous cuff significantly strengthens the joint, but is also the site of inflammatory reactions resulting in pain brought on by trauma or aging.

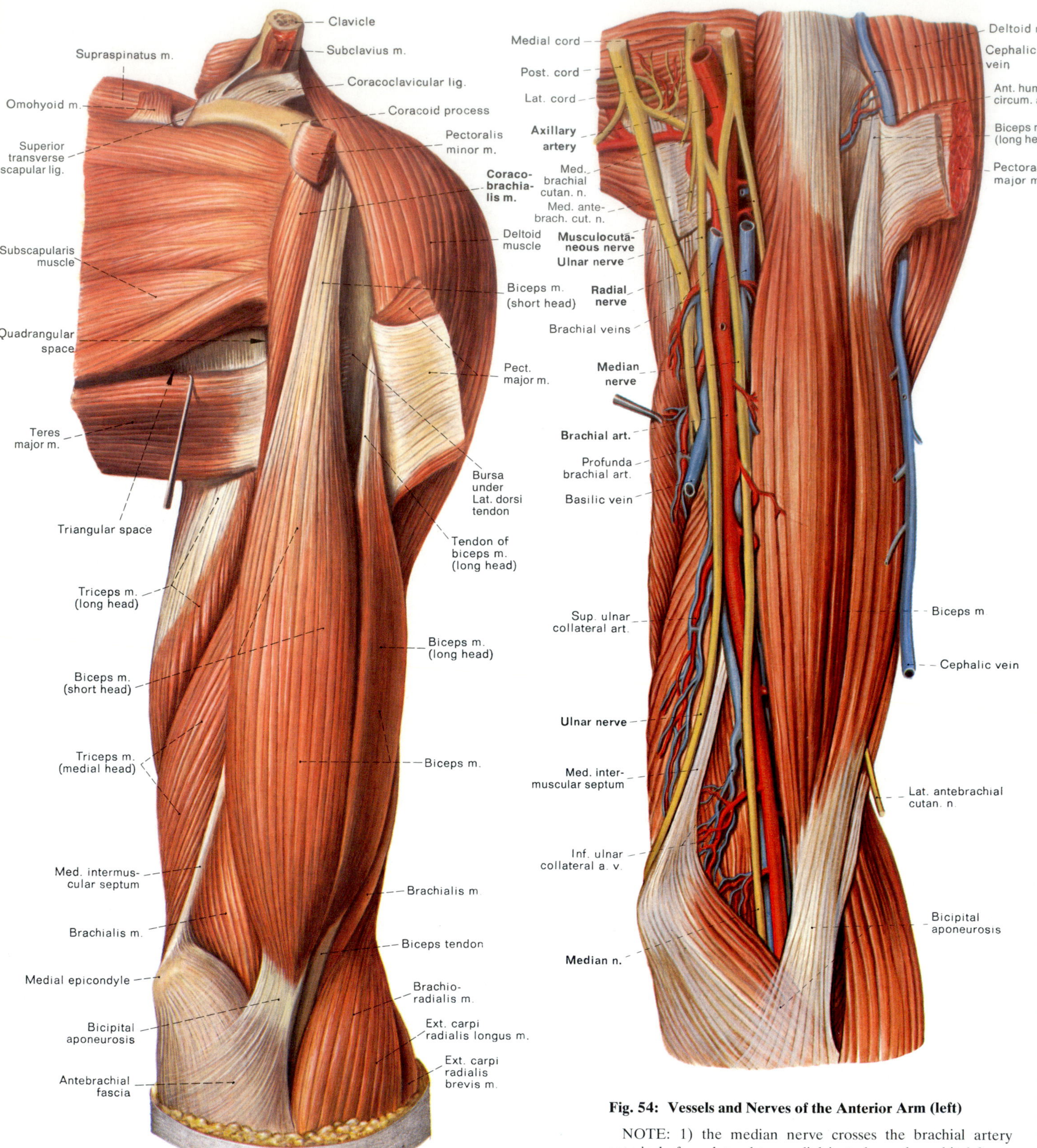

Clavicle
Subclavius m.
Supraspinatus m.
Coracoclavicular lig.
Omohyoid m.
Coracoid process
Superior transverse scapular lig.
Pectoralis minor m.
Coraco-brachia-lis m.
Subscapularis muscle
Deltoid muscle
Biceps m. (short head)
Quadrangular space
Pect. major m.
Teres major m.
Bursa under Lat. dorsi tendon
Triangular space
Tendon of biceps m. (long head)
Triceps m. (long head)
Biceps m. (long head)
Biceps m. (short head)
Triceps m. (medial head)
Biceps m.
Med. intermus-cular septum
Brachialis m.
Brachialis m.
Biceps tendon
Medial epicondyle
Brachio-radialis m.
Bicipital aponeurosis
Ext. carpi radialis longus m.
Antebrachial fascia
Ext. carpi radialis brevis m.
Medial cord
Deltoid m.
Cephalic vein
Post. cord
Ant. hum. circum. a.
Lat. cord
Axillary artery
Biceps m. (long head)
Med. brachial cutan. n.
Pectoralis major m.
Med. ante-brach. cut. n.
Musculocuta-neous nerve
Ulnar nerve
Radial nerve
Brachial veins
Median nerve
Brachial art.
Profunda brachial art.
Basilic vein
Biceps m.
Sup. ulnar collateral art.
Cephalic vein
Ulnar nerve
Med. inter-muscular septum
Lat. antebrachial cutan. n.
Inf. ulnar collateral a. v.
Bicipital aponeurosis
Median n.

Fig. 53: Superficial View of Muscles on the Anterior Aspect of the Left Arm

NOTE: that the biceps muscle extends across both the shoulder and elbow joints, but that the coracobrachialis muscle extends only across the shoulder joint.

Fig. 54: Vessels and Nerves of the Anterior Arm (left)

NOTE: 1) the median nerve crosses the brachial artery anteriorly from lateral to medial just above the cubital fossa.

2) neither the ulnar nor median nerve gives off branches in the arm.

3) the origin of the median nerve from both the medial and lateral cords of the brachial plexus. Observe that also arising from the medial cord are the medial brachial and medial antebrachial cutaneous nerves, the former coursing to the brachium (or arm) and arising more proximally from the medial cord than the latter, which courses to the antebrachium (or forearm).

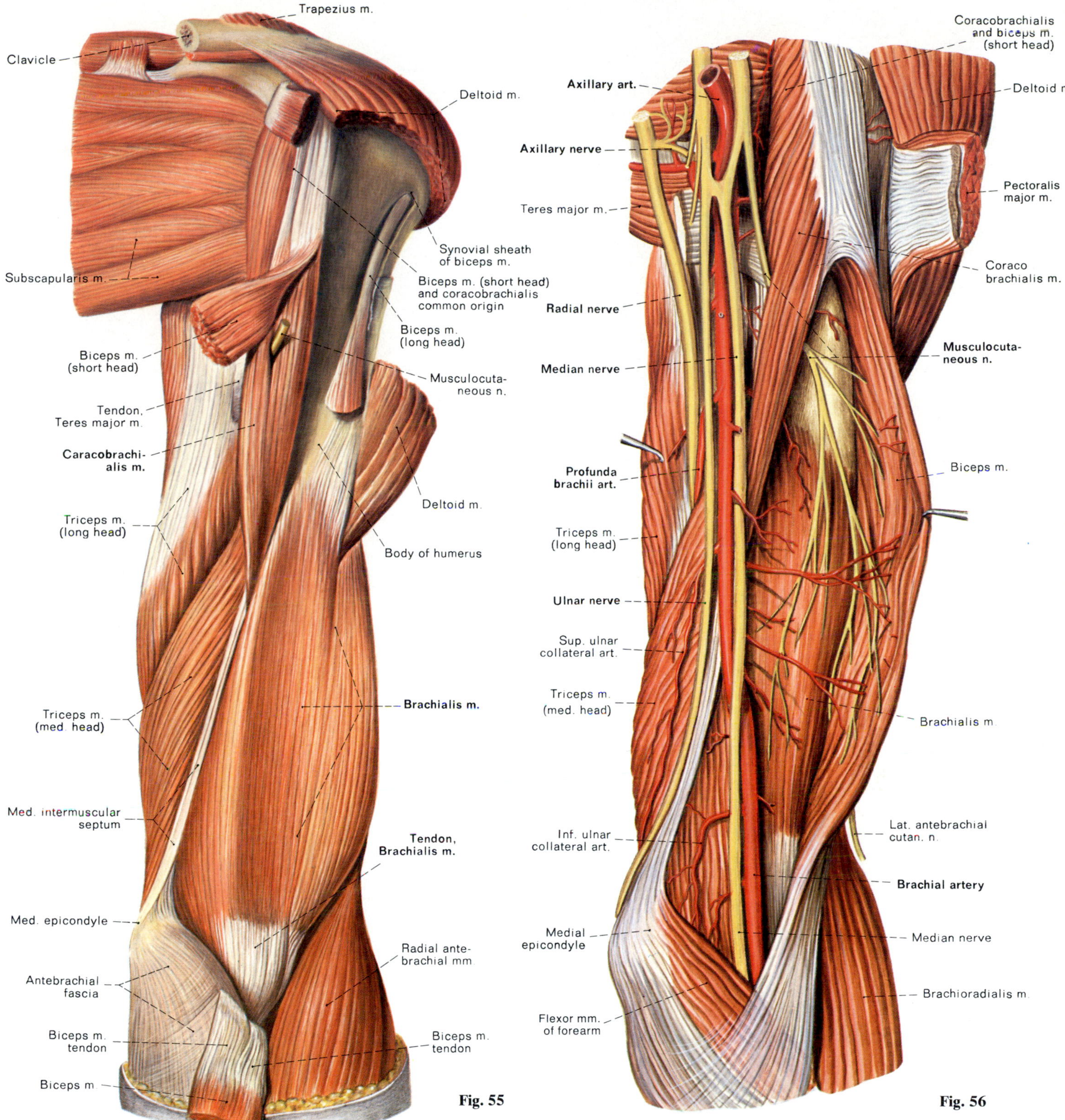

Trapezius m.
Clavicle
Deltoid m.
Synovial sheath of biceps m.
Biceps m. (short head) and coracobrachialis common origin
Subscapularis m.
Biceps m. (long head)
Biceps m. (short head)
Musculocutaneous n.
Tendon, Teres major m.
Caracobrachialis m.
Deltoid m.
Triceps m. (long head)
Body of humerus
Brachialis m.
Triceps m. (med. head)
Med. intermuscular septum
Tendon, Brachialis m.
Med. epicondyle
Radial antebrachial mm.
Antebrachial fascia
Biceps m. tendon
Biceps m. tendon
Biceps m.
Fig. 55
Coracobrachialis and biceps m. (short head)
Axillary art.
Deltoid m.
Axillary nerve
Pectoralis major m.
Teres major m.
Coraco brachialis m.
Radial nerve
Musculocutaneous n.
Median nerve
Profunda brachii art.
Biceps m.
Triceps m. (long head)
Ulnar nerve
Sup. ulnar collateral art.
Triceps m. (med. head)
Brachialis m.
Inf. ulnar collateral art.
Lat. antebrachial cutan. n.
Brachial artery
Medial epicondyle
Median nerve
Brachioradialis m.
Flexor mm. of forearm
Fig. 56

Fig. 55: Deep View of Muscles on the Anterior Aspect of the Left Arm

NOTE: In this dissection both the long head and short head of the biceps brachii muscle have been severed and reflected in order to reveal the underlying brachialis muscle. The coracobrachialis muscle has been left intact.

Fig. 56: The Nerves and Arteries of the Anterior Arm (left)

NOTE: 1) the short head of the biceps muscle has been pulled aside to reveal the musculocutaneous nerve which supplies the coracobrachialis, biceps and brachialis muscles. This nerve continues into the forearm as the lateral antebrachial cutaneous nerve.

2) the superficial course of the brachial artery in the arm. Its branches include the profunda brachii artery and the superior and inferior ulnar collateral arteries in addition to its muscular branches.

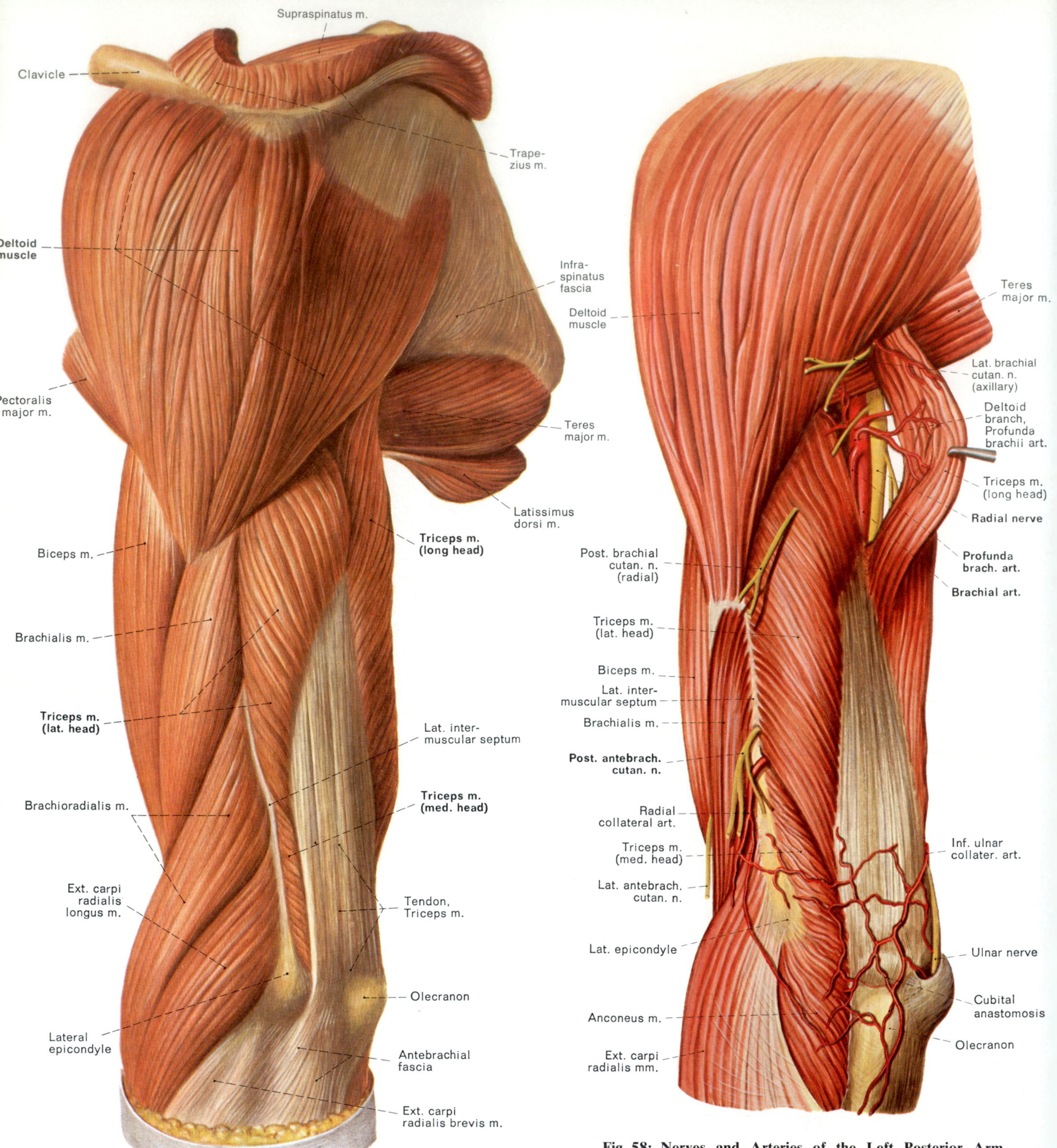

Supraspinatus m.
Clavicle
Trape-
zius m.
Deltoid
muscle
Infra-
spinatus
fascia
Pectoralis
major m.
Teres
major m.
Latissimus
dorsi m.
Biceps m.
Triceps m.
(long head)
Brachialis m.
Triceps m.
(lat. head)
Lat. inter-
muscular septum
Brachioradialis m.
Triceps m.
(med. head)
Ext. carpi
radialis
longus m.
Tendon,
Triceps m.
Olecranon
Lateral
epicondyle
Antebrachial
fascia
Ext. carpi
radialis brevis m.
Deltoid
muscle
Teres
major m.
Lat. brachial
cutan. n.
(axillary)
Deltoid
branch,
Profunda
brachii art.
Triceps m.
(long head)
Radial nerve
Post. brachial
cutan. n.
(radial)
Profunda
brach. art.
Brachial art.
Triceps m.
(lat. head)
Biceps m.
Lat. inter-
muscular septum
Brachialis m.
Post. antebrach.
cutan. n.
Radial
collateral art.
Triceps m.
(med. head)
Inf. ulnar
collater. art.
Lat. antebrach.
cutan. n.
Lat. epicondyle
Ulnar nerve
Cubital
anastomosis
Anconeus m.
Olecranon
Ext. carpi
radialis mm.

Fig. 57: The Muscles of the Arm (Lateral View)

NOTE: 1) the deltoid muscle acting as a whole abducts the arm. The clavicular portion flexes and medially rotates the arm, while the scapular portion extends and laterally rotates the arm.

2) the lateral intermuscular septum separates the anterior muscular compartment from the posterior muscular compartment.

Fig. 58: Nerves and Arteries of the Left Posterior Arm (Superficial Branches)

NOTE: 1) the origin of the profunda brachii artery from the brachial artery and its relationship to the radial nerve. The long head of the triceps has been pulled medially.

2) the relationship of the ulnar nerve to the olecranon process and the vascular anastomosis around the elbow.

3) the site of attachment of the deltoid muscle on the humerus, and the relationship of this attachment to the uppermost fibers of the brachialis muscle, the lateral intermuscular septum, and the lateral head of the triceps muscle (see also Fig. 50).

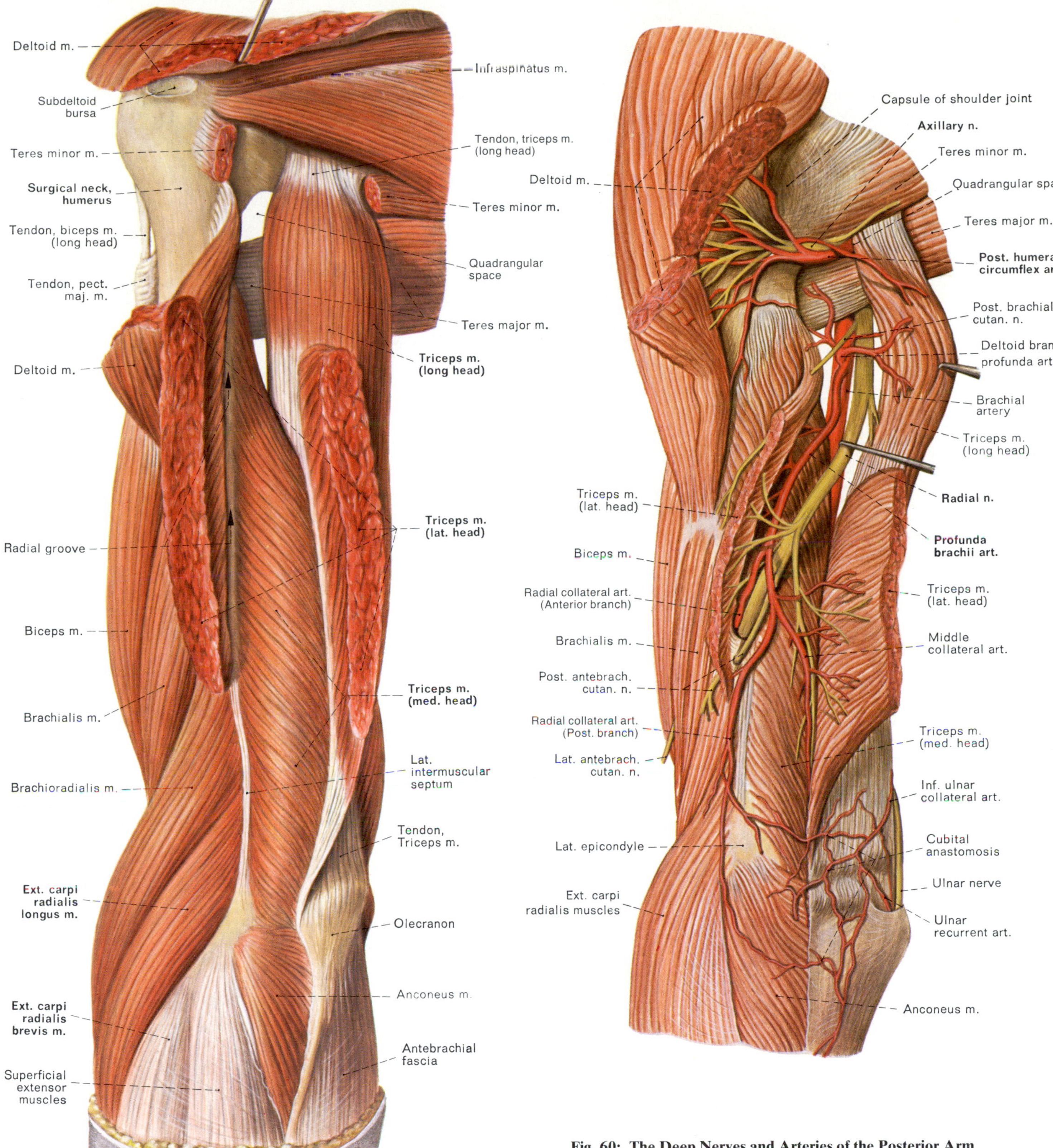

Deltoid m.
Infraspinatus m.
Subdeltoid bursa
Teres minor m.
Tendon, triceps m. (long head)
Surgical neck, humerus
Teres minor m.
Tendon, biceps m. (long head)
Quadrangular space
Tendon, pect. maj. m.
Teres major m.
Deltoid m.
Triceps m. (long head)
Triceps m. (lat. head)
Radial groove
Biceps m.
Triceps m. (med. head)
Brachialis m.
Lat. intermuscular septum
Brachioradialis m.
Tendon, Triceps m.
Ext. carpi radialis longus m.
Olecranon
Anconeus m.
Ext. carpi radialis brevis m.
Antebrachial fascia
Superficial extensor muscles
Capsule of shoulder joint
Axillary n.
Teres minor m.
Deltoid m.
Quadrangular space
Teres major m.
Post. humeral circumflex art.
Post. brachial cutan. n.
Deltoid branch, profunda art.
Brachial artery
Triceps m. (long head)
Triceps m. (lat. head)
Radial n.
Profunda brachii art.
Biceps m.
Triceps m. (lat. head)
Radial collateral art. (Anterior branch)
Brachialis m.
Middle collateral art.
Post. antebrach. cutan. n.
Radial collateral art. (Post. branch)
Triceps m. (med. head)
Lat. antebrach. cutan. n.
Inf. ulnar collateral art.
Cubital anastomosis
Lat. epicondyle
Ulnar nerve
Ext. carpi radialis muscles
Ulnar recurrent art.
Anconeus m.

Fig. 59: Deep Muscles of the Arm and Shoulder (Posterolateral View)

NOTE: in this dissection much of the deltoid and teres minor muscles was removed, and the lateral head of the triceps muscle was transected and reflected. Observe the radial groove between the medial and lateral heads of the triceps.

Fig. 60: The Deep Nerves and Arteries of the Posterior Arm

NOTE: 1) the course of the axillary nerve and posterior humeral circumflex artery through the quadrangular space to achieve the deltoid muscle and dorsal shoulder region.

2) the course of the radial nerve and profunda brachii artery along the musculospiral groove to the posterior brachial region. The groove lies along the body of the humerus between the origins of the lateral and medial heads of the triceps muscle.

3) the common insertion of the triceps muscle onto the olecranon process of the ulna.

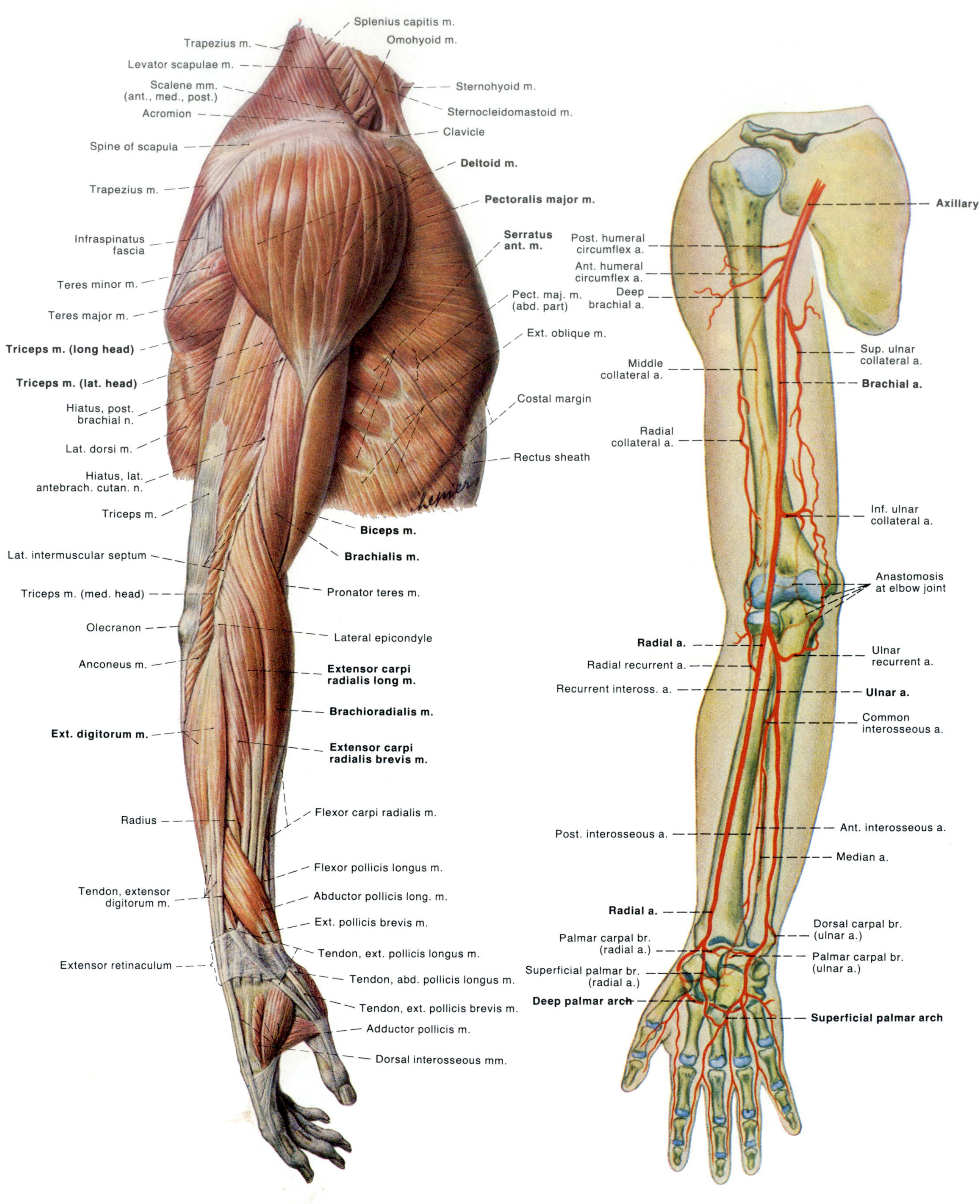
Splenius capitis m.
Trapezius m.
Omohyoid m.
Levator scapulae m.
Scalene mm.
(ant., med., post.)
Sternohyoid m.
Acromion
Sternocleidomastoid m.
Clavicle
Spine of scapula
Deltoid m.
Trapezius m.
Pectoralis major m.
Infraspinatus
fascia
Serratus
ant. m.
Teres minor m.
Pect. maj. m.
(abd. part)
Teres major m.
Ext. oblique m.
Triceps m. (long head)
Triceps m. (lat. head)
Costal margin
Hiatus, post.
brachial n.
Lat. dorsi m.
Rectus sheath
Hiatus, lat.
antebrach. cutan. n.
Triceps m.
Biceps m.
Lat. intermuscular septum
Brachialis m.
Triceps m. (med. head)
Pronator teres m.
Olecranon
Lateral epicondyle
Anconeus m.
Extensor carpi
radialis long m.
Brachioradialis m.
Ext. digitorum m.
Extensor carpi
radialis brevis m.
Radius
Flexor carpi radialis m.
Flexor pollicis longus m.
Tendon, extensor
digitorum m.
Abductor pollicis long. m.
Ext. pollicis brevis m.
Tendon, ext. pollicis longus m.
Extensor retinaculum
Tendon, abd. pollicis longus m.
Tendon, ext. pollicis brevis m.
Adductor pollicis m.
Dorsal interosseous mm.
Axillary a.
Post. humeral
circumflex a.
Ant. humeral
circumflex a.
Deep
brachial a.
Sup. ulnar
collateral a.
Middle
collateral a.
Brachial a.
Radial
collateral a.
Inf. ulnar
collateral a.
Anastomosis
at elbow joint
Radial a.
Ulnar
recurrent a.
Radial recurrent a.
Recurrent inteross. a.
Ulnar a.
Common
interosseous a.
Post. interosseous a.
Ant. interosseous a.
Median a.
Radial a.
Dorsal carpal br.
(ulnar a.)
Palmar carpal br.
(radial a.)
Palmar carpal br.
(ulnar a.)
Superficial palmar br.
(radial a.)
Deep palmar arch
Superficial palmar arch

Fig. 61: Muscles of the Thorax and Right Upper Extremity, Lateral View

Fig. 62: Arteries of the Upper Extremity (Schematic Representation)

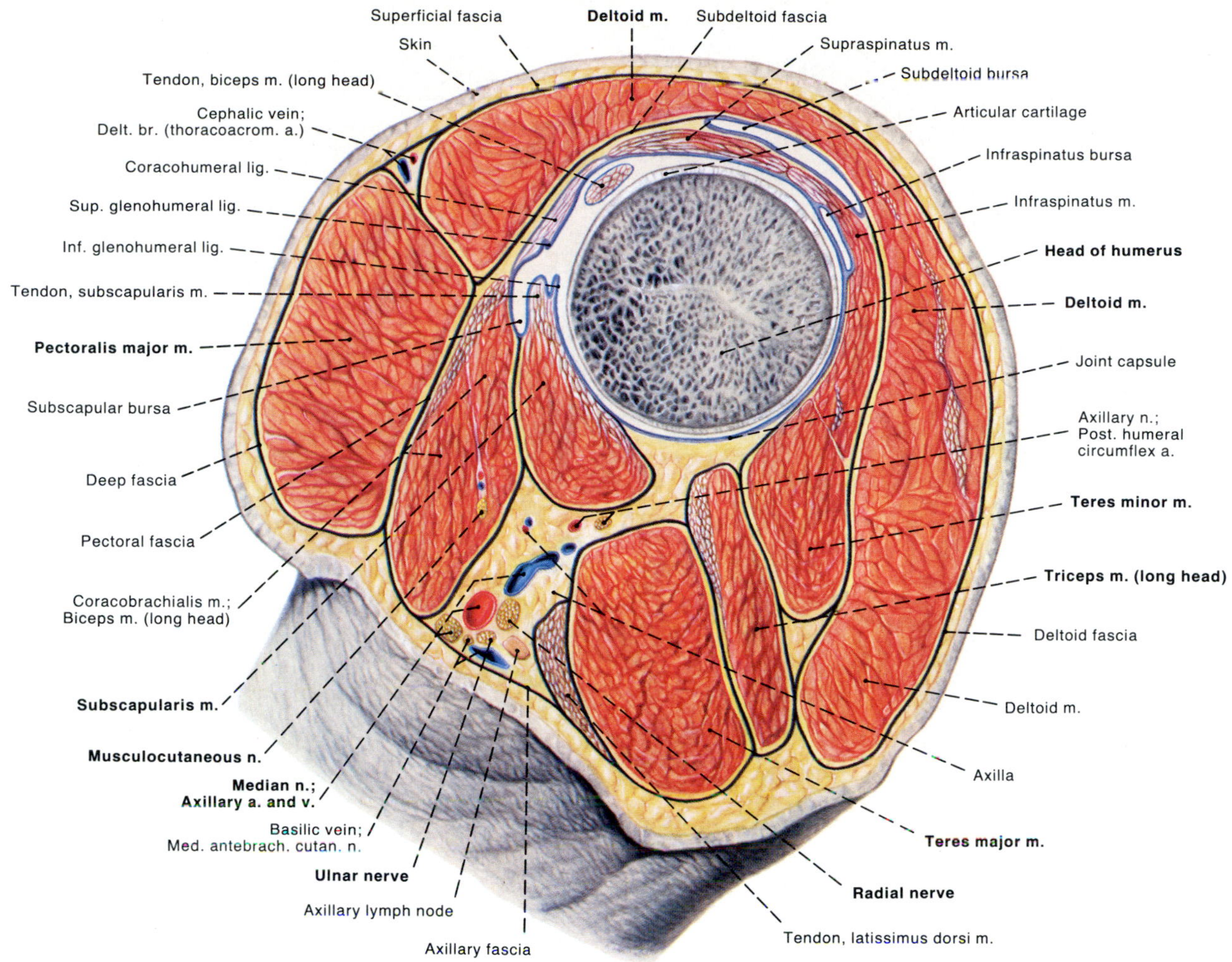

Superficial fascia
Deltoid m.
Subdeltoid fascia
Skin
Supraspinatus m.
Subdeltoid bursa
Tendon, biceps m. (long head)
Articular cartilage
Cephalic vein;
Delt. br. (thoracoacrom. a.)
Infraspinatus bursa
Coracohumeral lig.
Infraspinatus m.
Sup. glenohumeral lig.
Inf. glenohumeral lig.
Head of humerus
Tendon, subscapularis m.
Deltoid m.
Pectoralis major m.
Joint capsule
Subscapular bursa
Axillary n.;
Post. humeral circumflex a.
Deep fascia
Teres minor m.
Pectoral fascia
Triceps m. (long head)
Coracobrachialis m.;
Biceps m. (long head)
Deltoid fascia
Deltoid m.
Subscapularis m.
Axilla
Musculocutaneous n.
Median n.;
Axillary a. and v.
Basilic vein;
Med. antebrach. cutan. n.
Teres major m.
Ulnar nerve
Radial nerve
Axillary lymph node
Tendon, latissimus dorsi m.
Axillary fascia

Fig. 63: Cross Section of the Right Upper Extremity Through the Head of the Humerus, Viewed from Above

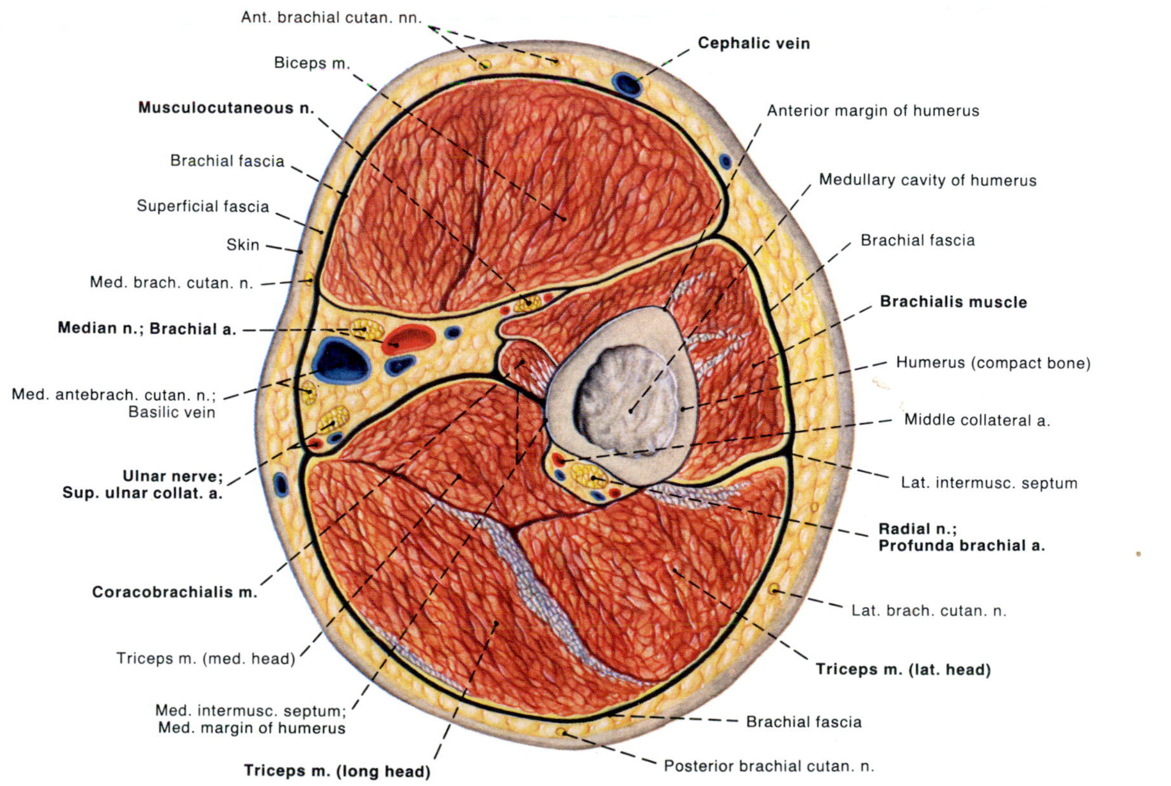
Ant. brachial cutan. nn.
Biceps m.
Cephalic vein
Musculocutaneous n.
Anterior margin of humerus
Brachial fascia
Medullary cavity of humerus
Superficial fascia
Skin
Brachial fascia
Med. brach. cutan. n.
Brachialis muscle
Median n.; Brachial a.
Humerus (compact bone)
Med. antebrach. cutan. n.;
Basilic vein
Middle collateral a.
Ulnar nerve;
Sup. ulnar collat. a.
Lat. intermusc. septum
Radial n.;
Profunda brachial a.
Coracobrachialis m.
Lat. brach. cutan. n.
Triceps m. (med. head)
Triceps m. (lat. head)
Med. intermusc. septum;
Med. margin of humerus
Brachial fascia
Triceps m. (long head)
Posterior brachial cutan. n.

Fig. 64: Cross Section of the Right Upper Extremity Through the Middle of the Humerus, Viewed from Above

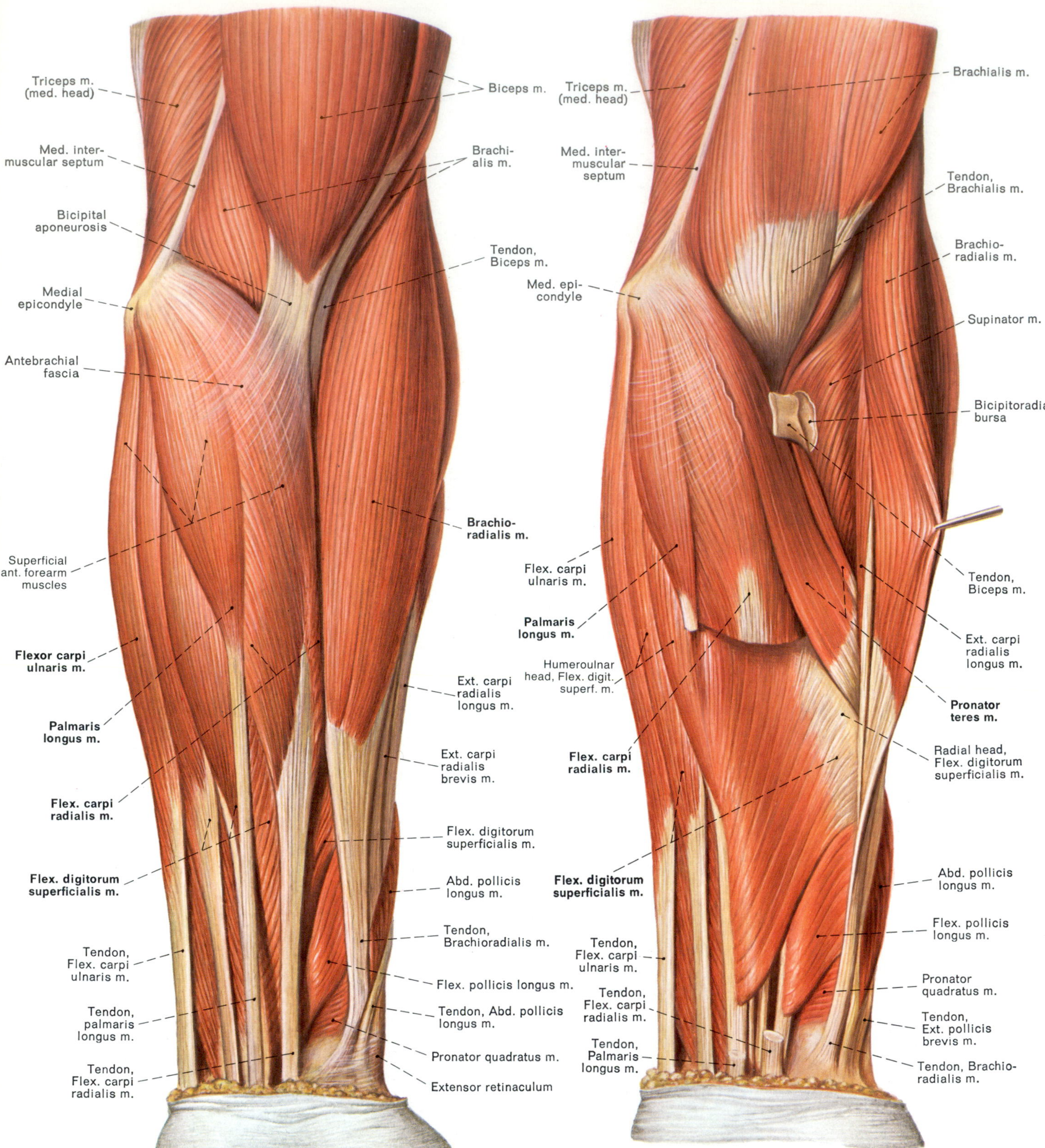

Triceps m.
(med. head)
Biceps m.
Med. inter-
muscular septum
Brachi-
alis m.
Bicipital
aponeurosis
Tendon,
Biceps m.
Medial
epicondyle
Antebrachial
fascia
Brachio-
radialis m.
Superficial
ant. forearm
muscles
Flexor carpi
ulnaris m.
Ext. carpi
radialis
longus m.
Palmaris
longus m.
Ext. carpi
radialis
brevis m.
Flex. carpi
radialis m.
Flex. digitorum
superficialis m.
Flex. digitorum
superficialis m.
Abd. pollicis
longus m.
Tendon,
Brachioradialis m.
Tendon,
Flex. carpi
ulnaris m.
Flex. pollicis longus m.
Tendon,
palmaris
longus m.
Tendon, Abd. pollicis
longus m.
Pronator quadratus m.
Tendon,
Flex. carpi
radialis m.
Extensor retinaculum
Triceps m.
(med. head)
Brachialis m.
Med. inter-
muscular
septum
Tendon,
Brachialis m.
Brachio-
radialis m.
Med. epi-
condyle
Supinator m.
Bicipitoradial
bursa
Flex. carpi
ulnaris m.
Tendon,
Biceps m.
Palmaris
longus m.
Ext. carpi
radialis
longus m.
Humeroulnar
head, Flex. digit.
superf. m.
Pronator
teres m.
Radial head,
Flex. digitorum
superficialis m.
Flex. carpi
radialis m.
Flex. digitorum
superficialis m.
Abd. pollicis
longus m.
Flex. pollicis
longus m.
Tendon,
Flex. carpi
ulnaris m.
Pronator
quadratus m.
Tendon,
Flex. carpi
radialis m.
Tendon,
Ext. pollicis
brevis m.
Tendon,
Palmaris
longus m.
Tendon, Brachio-
radialis m.

Fig. 65: The Left Anterior Forearm Muscles, Superficial Group

NOTE: 1) that the brachioradialis m. is studied with the posterior forearm muscles instead of the anterior muscles.

2) that the anterior forearm muscles arise from the medial epicondyle of the humerus and include the pronator teres (not labelled), flexor carpi radialis, palmaris longus and flexor carpi ulnaris. Beneath these is the flexor digitorum superficialis.

Fig. 66: The Flexor Digitorum Superficialis Muscle and Related Muscles (left)

NOTE: 1) the palmaris longus, flexor carpi radialis and tendon of the biceps have been cut to reveal the flexor digitorum superficialis and pronator teres.

2) the triangular cubital fossa is bounded medially by the superficial flexors and laterally by the extensors. Its floor is the brachialis muscle.

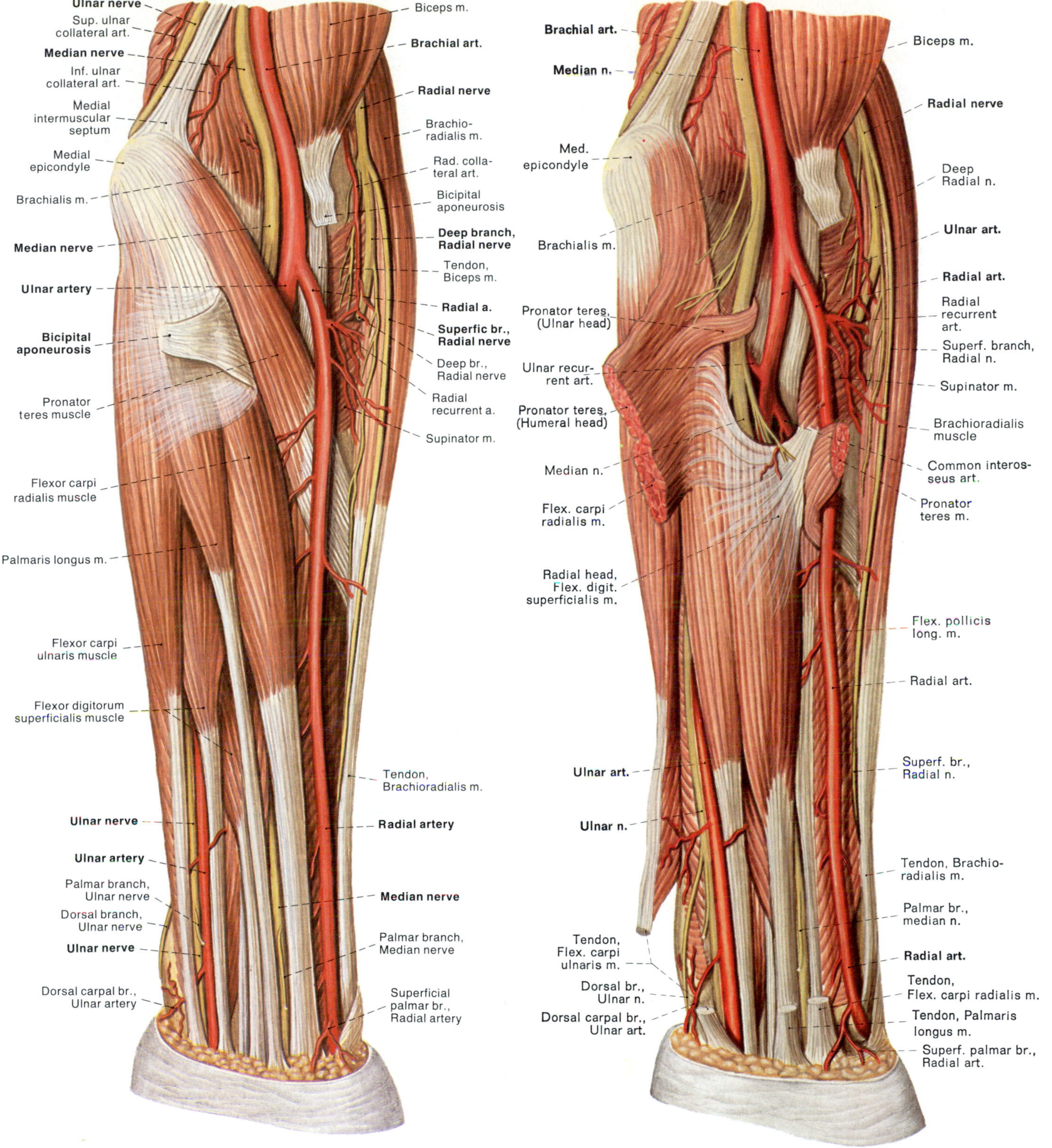

Ulnar nerve
Sup. ulnar collateral art.
Median nerve
Inf. ulnar collateral art.
Medial intermuscular septum
Medial epicondyle
Brachialis m.
Median nerve
Ulnar artery
Bicipital aponeurosis
Pronator teres muscle
Flexor carpi radialis muscle
Palmaris longus m.
Flexor carpi ulnaris muscle
Flexor digitorum superficialis muscle
Ulnar nerve
Ulnar artery
Palmar branch, Ulnar nerve
Dorsal branch, Ulnar nerve
Ulnar nerve
Dorsal carpal br., Ulnar artery
Biceps m.
Brachial art.
Radial nerve
Brachio-radialis m.
Rad. colla-teral art.
Bicipital aponeurosis
Deep branch, Radial nerve
Tendon, Biceps m.
Radial a.
Superfic br., Radial nerve
Deep br., Radial nerve
Radial recurrent a.
Supinator m.
Tendon, Brachioradialis m.
Radial artery
Median nerve
Palmar branch, Median nerve
Superficial palmar br., Radial artery
Brachial art.
Median n.
Med. epicondyle
Brachialis m.
Pronator teres, (Ulnar head)
Ulnar recur-rent art.
Pronator teres, (Humeral head)
Median n.
Flex. carpi radialis m.
Radial head, Flex. digit. superficialis m.
Ulnar art.
Ulnar n.
Tendon, Flex. carpi ulnaris m.
Dorsal br., Ulnar n.
Dorsal carpal br., Ulnar art.
Biceps m.
Radial nerve
Deep Radial n.
Ulnar art.
Radial art.
Radial recurrent art.
Superf. branch, Radial n.
Supinator m.
Brachioradialis muscle
Common interos-seus art.
Pronator teres m.
Flex. pollicis long. m.
Radial art.
Superf. br., Radial n.
Tendon, Brachio-radialis m.
Palmar br., median n.
Radial art.
Tendon, Flex. carpi radialis m.
Tendon, Palmaris longus m.
Superf. palmar br., Radial art.

Fig. 67: Nerves and Arteries, Anterior Aspect of the Left Forearm (Superficial View)

NOTE: 1) the bicipital aponeurosis has been reflected to reveal the underlying median nerve, brachial artery, and tendon of insertion of the biceps brachii muscle.

2) the brachioradialis muscle has been pulled laterally (toward the radial side) in order to expose the course of the radial artery and the division of the radial nerve into its superficial and deep branches.

3) that the radial artery, as it descends in the forearm, courses anterior to the tendon of the biceps brachii muscle, the supinator muscle, the tendon of insertion of the pronator teres, and the belly of the flexor pollicis longus (the latter not labelled in this figure, but see Fig. 68).

Fig. 68: Nerves and Arteries, Anterior Aspect of Left Forearm

NOTE: 1) the pronator teres and flexor carpi radialis muscles are reflected just below the cubital fossa to reveal the origins of the ulnar and radial arteries.

2) at the wrist, the flexor carpi ulnaris muscle is severed to expose the ulnar nerve and artery.

3) that the median nerve lies deep to the flexor digitorum superficialis throughout much of its course in the forearm, but just above the wrist it usually becomes visible between the tendons, with the flexor pollicis longus and flexor carpi radialis on its radial side and the tendons of the palmaris longus and flexor digitorum superficialis on its ulnar side.

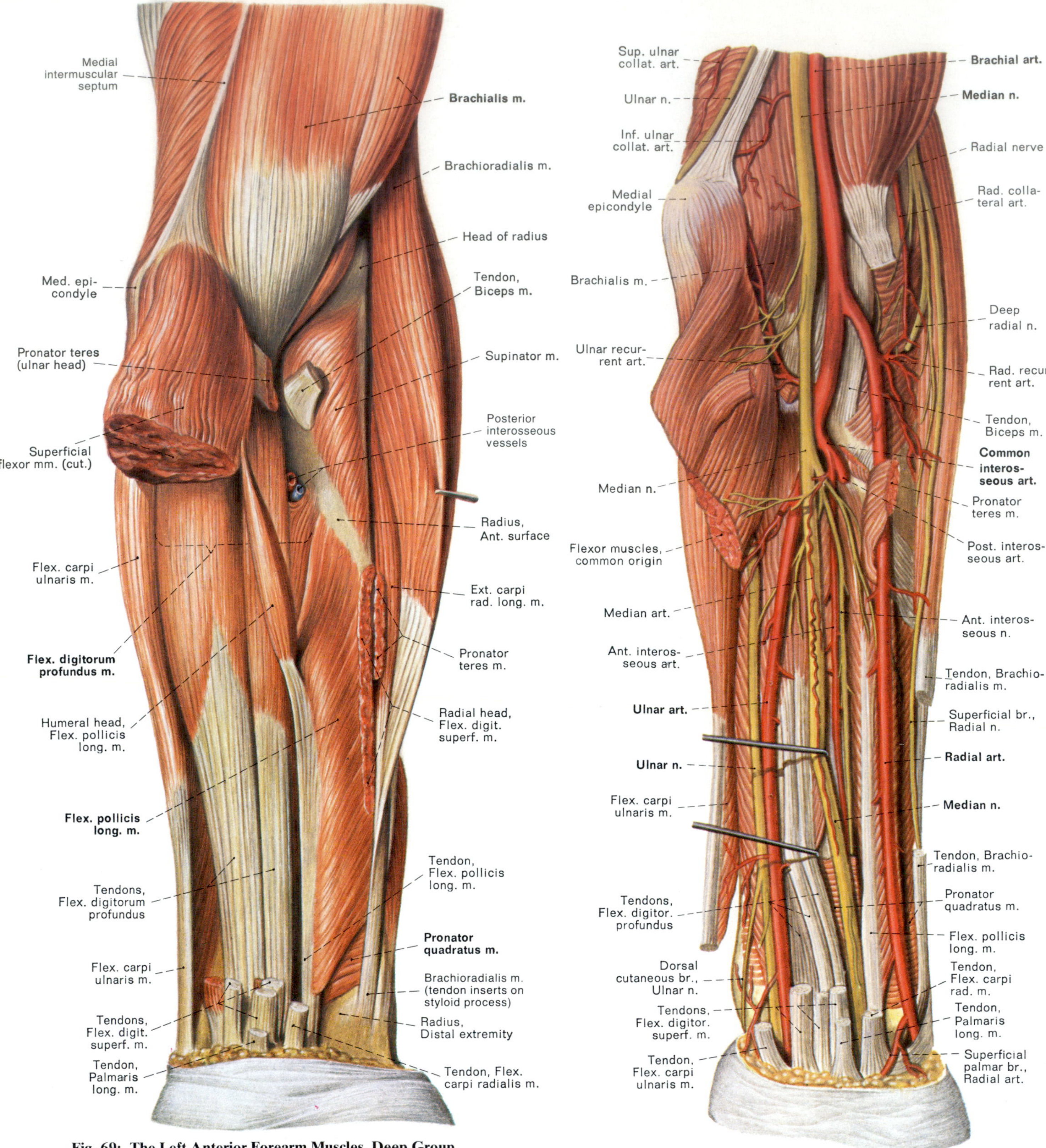

Medial intermuscular septum
Brachialis m.
Brachioradialis m.
Head of radius
Med. epi-condyle
Tendon, Biceps m.
Pronator teres (ulnar head)
Supinator m.
Posterior interosseous vessels
Superficial flexor mm. (cut.)
Radius, Ant. surface
Flex. carpi ulnaris m.
Ext. carpi rad. long. m.
Flex. digitorum profundus m.
Pronator teres m.
Humeral head, Flex. pollicis long. m.
Radial head, Flex. digit. superf. m.
Flex. pollicis long. m.
Tendon, Flex. pollicis long. m.
Tendons, Flex. digitorum profundus
Pronator quadratus m.
Flex. carpi ulnaris m.
Brachioradialis m. (tendon inserts on styloid process)
Tendons, Flex. digit. superf. m.
Radius, Distal extremity
Tendon, Palmaris long. m.
Tendon, Flex. carpi radialis m.
Sup. ulnar collat. art.
Brachial art.
Ulnar n.
Median n.
Inf. ulnar collat. art.
Radial nerve
Medial epicondyle
Rad. colla-teral art.
Brachialis m.
Deep radial n.
Ulnar recur-rent art.
Rad. recur-rent art.
Tendon, Biceps m.
Common interos-seous art.
Median n.
Pronator teres m.
Flexor muscles, common origin
Post. interos-seous art.
Median art.
Ant. interos-seous n.
Ant. interos-seous art.
Tendon, Brachio-radialis m.
Ulnar art.
Superficial br., Radial n.
Ulnar n.
Radial art.
Flex. carpi ulnaris m.
Median n.
Tendon, Brachio-radialis m.
Tendons, Flex. digitor. profundus
Pronator quadratus m.
Flex. pollicis long. m.
Dorsal cutaneous br., Ulnar n.
Tendon, Flex. carpi rad. m.
Tendons, Flex. digitor. superf. m.
Tendon, Palmaris long. m.
Tendon, Flex. carpi ulnaris m.
Superficial palmar br., Radial art.

Fig. 69: The Left Anterior Forearm Muscles, Deep Group

NOTE: 1) the superficial anterior forearm muscles have been removed to reveal the three muscles of the deep group. These include the flexor digitorum profundus, the flexor pollicis longus and the pronator quadratus.

2) the pronator quadratus is a small quadrangular muscle situated at the distal end of the forearm beneath the tendons of the flexor digitorum profundus and flexor pollicis longus. It is only partially shown in this dissection and can better be seen in Fig. 114.

3) the tendons of the flexor digitorum profundus to the ring and little fingers and the middle and index fingers appear to be fused at the wrist, as if they were two structures rather than four.

Fig. 70: Nerves and Arteries of Left Anterior Forearm (Deep Dissection)

NOTE the division of the brachial artery into the radial and ulnar. The common interosseous artery branches from the ulnar artery and divides immediately into anterior and posterior interosseous arteries. Observe the courses of the median and ulnar nerves.

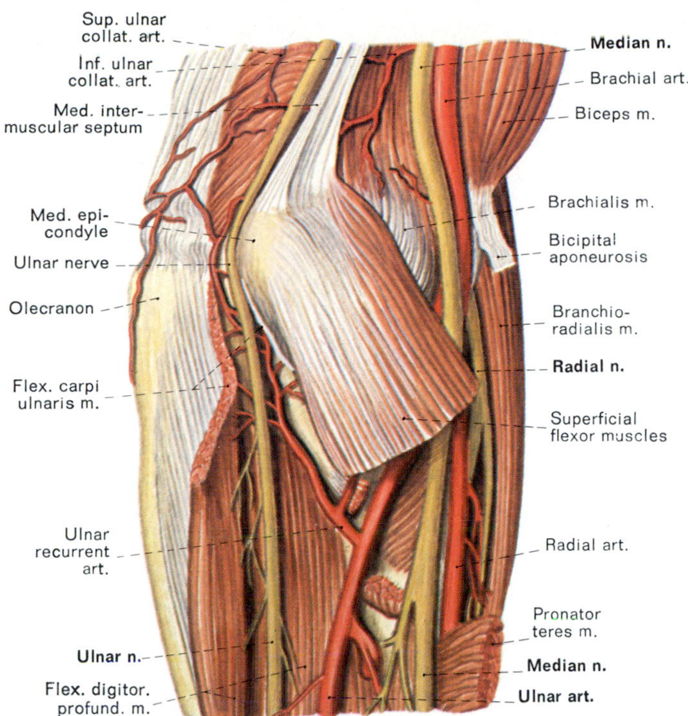
Sup. ulnar collat. art.
Inf. ulnar collat. art.
Med. inter-muscular septum
Med. epi-condyle
Ulnar nerve
Olecranon
Flex. carpi ulnaris m.
Ulnar recurrent art.
Ulnar n.
Flex. digitor. profund. m.
Median n.
Brachial art.
Biceps m.
Brachialis m.
Bicipital aponeurosis
Branchio-radialis m.
Radial n.
Superficial flexor muscles
Radial art.
Pronator teres m.
Median n.
Ulnar art.

Fig. 71: Nerves and Arteries at the Elbow (Medial View)

NOTE: the ulnar nerve enters the forearm directly behind the medial epicondyle.

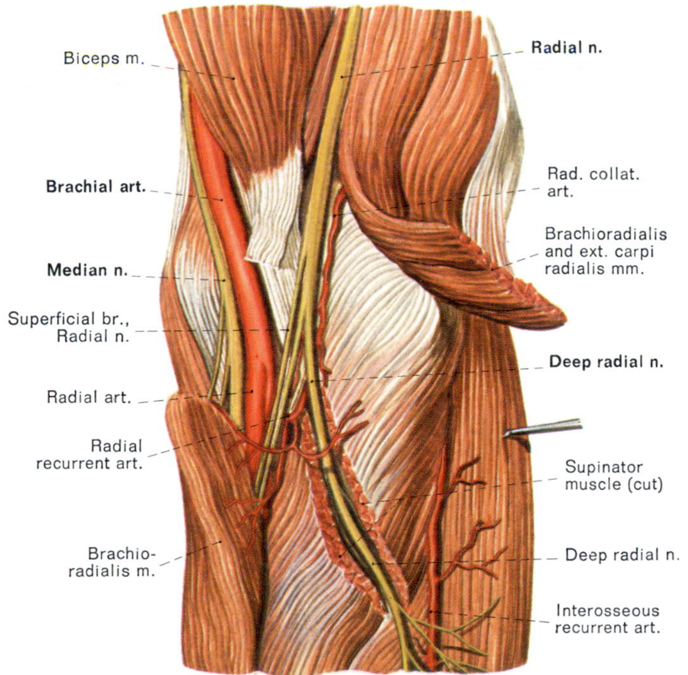
Biceps m.
Brachial art.
Median n.
Superficial br., Radial n.
Radial art.
Radial recurrent art.
Brachio-radialis m.
Radial n.
Rad. collat. art.
Brachioradialis and ext. carpi radialis mm.
Deep radial n.
Supinator muscle (cut)
Deep radial n.
Interosseous recurrent art.

Fig. 72: Nerves and Arteries at the Elbow (Lateral View)

NOTE: the deep radial nerve passes into the forearm in front of the lateral part of the elbow joint. It then courses dorsally through the supinator muscle to supply the posterior forearm muscles.

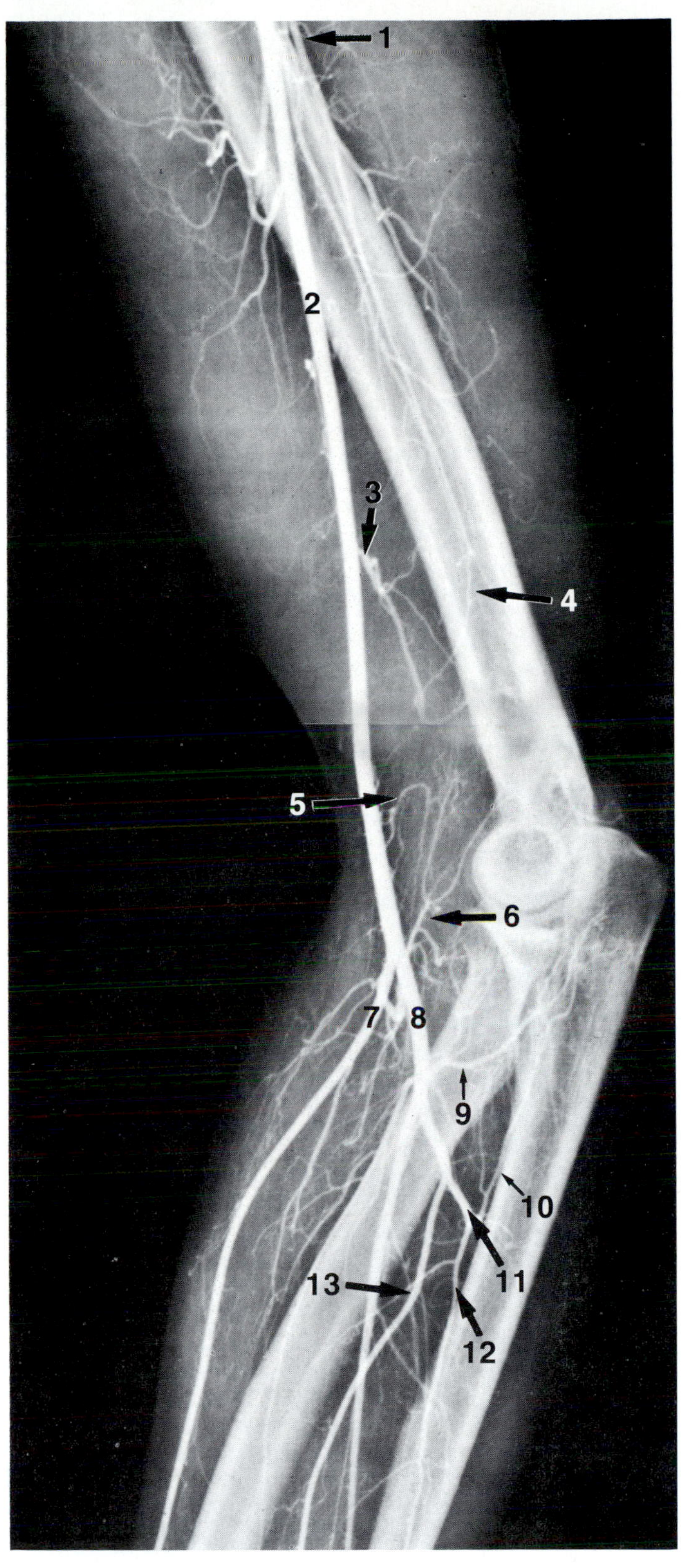
1
2
3
4
5
6
7
8
9
10
11
12
13

Fig. 73: A Brachial Arteriogram Showing the Origins of the Vessels that Supply the Elbow and Forearm

1. Profunda brachii artery
2. Brachial artery
3. Superior ulnar collateral artery
4. Radial collateral artery
5. Inferior ulnar collateral artery
6. Radial recurrent artery
7. Radial artery
8. Ulnar artery
9. Ulnar recurrent artery
10. Recurrent interosseous artery
11. Common interosseous artery
12. Posterior interosseous artery
13. Anterior interosseous artery

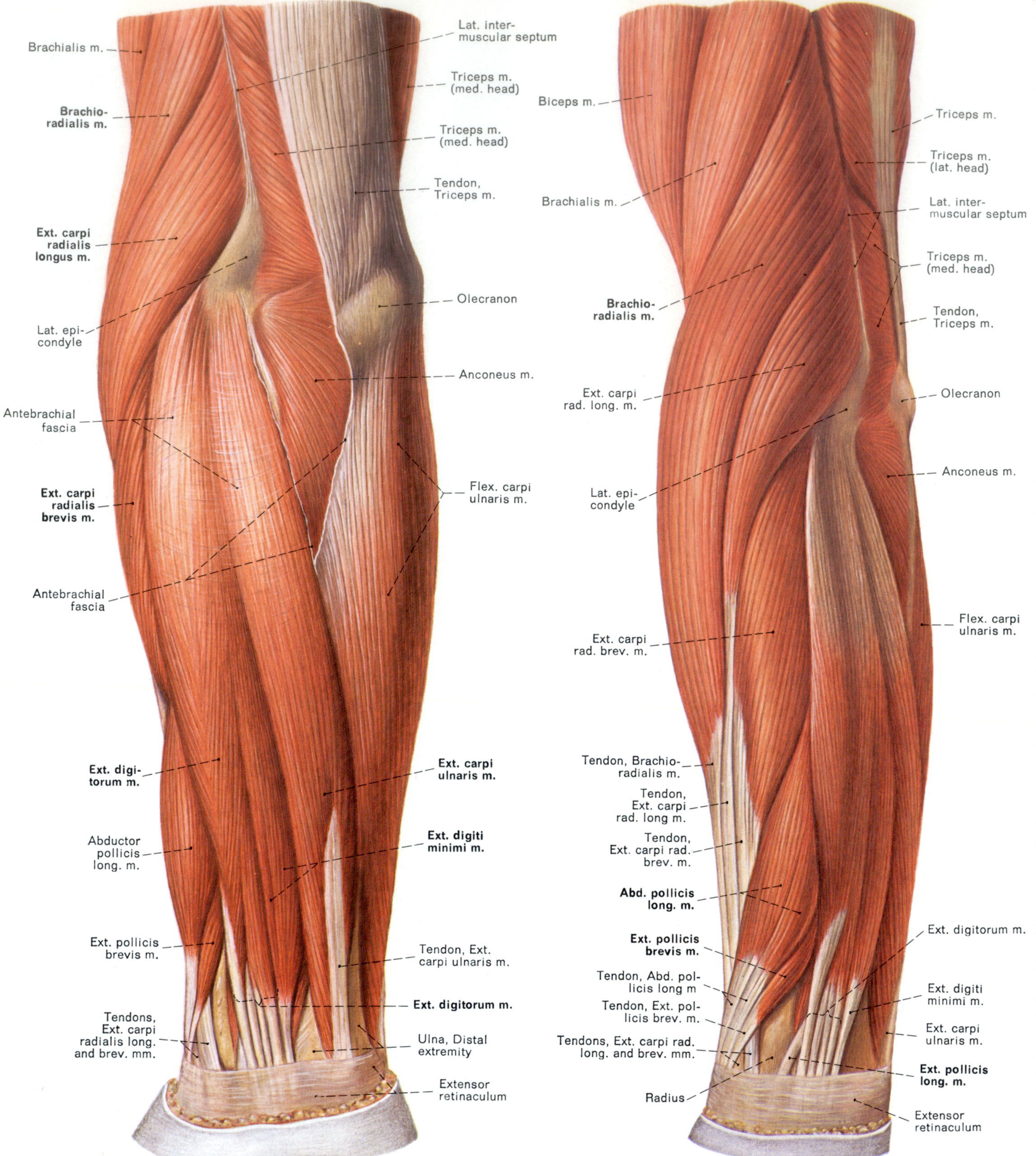
Brachialis m.
Brachio-
radialis m.
Ext. carpi
radialis
longus m.
Lat. epi-
condyle
Antebrachial
fascia
Ext. carpi
radialis
brevis m.
Antebrachial
fascia
Ext. digi-
torum m.
Abductor
pollicis
long. m.
Ext. pollicis
brevis m.
Tendons,
Ext. carpi
radialis long.
and brev. mm.
Lat. inter-
muscular septum
Triceps m.
(med. head)
Triceps m.
(med. head)
Tendon,
Triceps m.
Olecranon
Anconeus m.
Flex. carpi
ulnaris m.
Ext. carpi
ulnaris m.
Ext. digiti
minimi m.
Tendon, Ext.
carpi ulnaris m.
Ext. digitorum m.
Ulna, Distal
extremity
Extensor
retinaculum
Biceps m.
Brachialis m.
Brachio-
radialis m.
Ext. carpi
rad. long. m.
Lat. epi-
condyle
Ext. carpi
rad. brev. m.
Tendon, Brachio-
radialis m.
Tendon,
Ext. carpi
rad. long m.
Tendon,
Ext. carpi rad.
brev. m.
Abd. pollicis
long. m.
Ext. pollicis
brevis m.
Tendon, Abd. pol-
licis long m
Tendon, Ext. pol-
licis brev. m.
Tendons, Ext. carpi rad.
long. and brev. mm.
Radius
Triceps m.
Triceps m.
(lat. head)
Lat. inter-
muscular septum
Triceps m.
(med. head)
Tendon,
Triceps m.
Olecranon
Anconeus m.
Flex. carpi
ulnaris m.
Ext. digitorum m.
Ext. digiti
minimi m.
Ext. carpi
ulnaris m.
Ext. pollicis
long. m.
Extensor
retinaculum

Fig. 74: Posterior Muscles of the Left Forearm, Superficial Group

NOTE that most of the superficial extensor muscles arise from the lateral epicondyle of the humerus. These are the extensor carpi radialis brevis, the extensor digitorum, the extensor digiti minimi and the extensor carpi ulnaris. The brachioradialis and extensor carpi radialis longus arise from the supracondylar ridge.

Fig. 75: Posterior Muscles (Superficial) of the Left Forearm, Lateral View

NOTE: 1) the superficial location of the brachioradialis muscle.

2) three muscles of the thumb: extensor pollicis longus, extensor pollicis brevis and abductor pollicis longus.

3) the closely investing extensor retinaculum under which the extensor tendons pass into the dorsum of the hand.

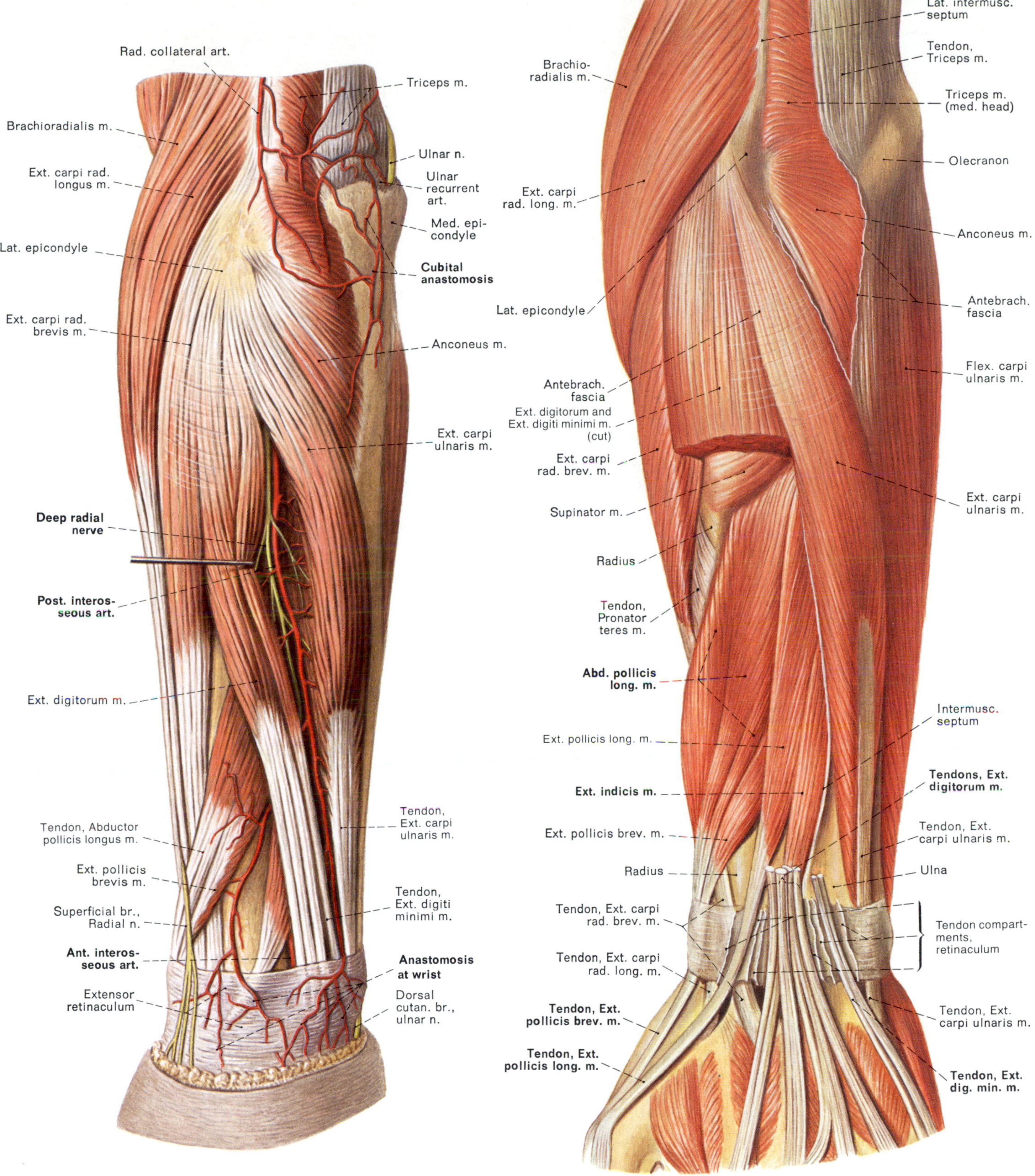
Rad. collateral art.
Triceps m.
Brachioradialis m.
Ulnar n.
Ulnar recurrent art.
Ext. carpi rad. longus m.
Med. epicondyle
Lat. epicondyle
Cubital anastomosis
Ext. carpi rad. brevis m.
Anconeus m.
Ext. carpi ulnaris m.
Deep radial nerve
Post. interosseous art.
Ext. digitorum m.
Tendon, Ext. carpi ulnaris m.
Tendon, Abductor pollicis longus m.
Ext. pollicis brevis m.
Tendon, Ext. digiti minimi m.
Superficial br., Radial n.
Ant. interosseous art.
Anastomosis at wrist
Extensor retinaculum
Dorsal cutan. br., ulnar n.
Lat. intermusc. septum
Tendon, Triceps m.
Brachioradialis m.
Triceps m. (med. head)
Olecranon
Ext. carpi rad. long. m.
Anconeus m.
Antebrach. fascia
Lat. epicondyle
Flex. carpi ulnaris m.
Antebrach. fascia
Ext. digitorum and Ext. digiti minimi m. (cut)
Ext. carpi rad. brev. m.
Ext. carpi ulnaris m.
Supinator m.
Radius
Tendon, Pronator teres m.
Abd. pollicis long. m.
Intermusc. septum
Ext. pollicis long. m.
Tendons, Ext. digitorum m.
Ext. indicis m.
Tendon, Ext. carpi ulnaris m.
Ext. pollicis brev. m.
Radius
Ulna
Tendon, Ext. carpi rad. brev. m.
Tendon compartments, retinaculum
Tendon, Ext. carpi rad. long. m.
Tendon, Ext. pollicis brev. m.
Tendon, Ext. carpi ulnaris m.
Tendon, Ext. pollicis long. m.
Tendon, Ext. dig. min. m.

Fig. 76: Nerves and Arteries of the Left Posterior Forearm

NOTE: 1) the extensor digiti minimi and extensor digitorum have been separated from the extensor carpi ulnaris to expose the deep radial nerve and posterior interosseous artery coursing inferiorly in the posterior forearm.

2) the anastomoses at the elbow and wrist.

Fig. 77: The Thumb Muscles of the Left Posterior Forearm

NOTE: 1) that the three thumb muscles (abductor pollicis longus and extensors pollicis brevis and longus) are exposed when the extensor digitorum and extensor digiti minimi muscles are partially removed. Observe also the extensor indicis muscle coursing to the index finger.

2) the tendon compartments formed by the extensor retinaculum at the wrist.

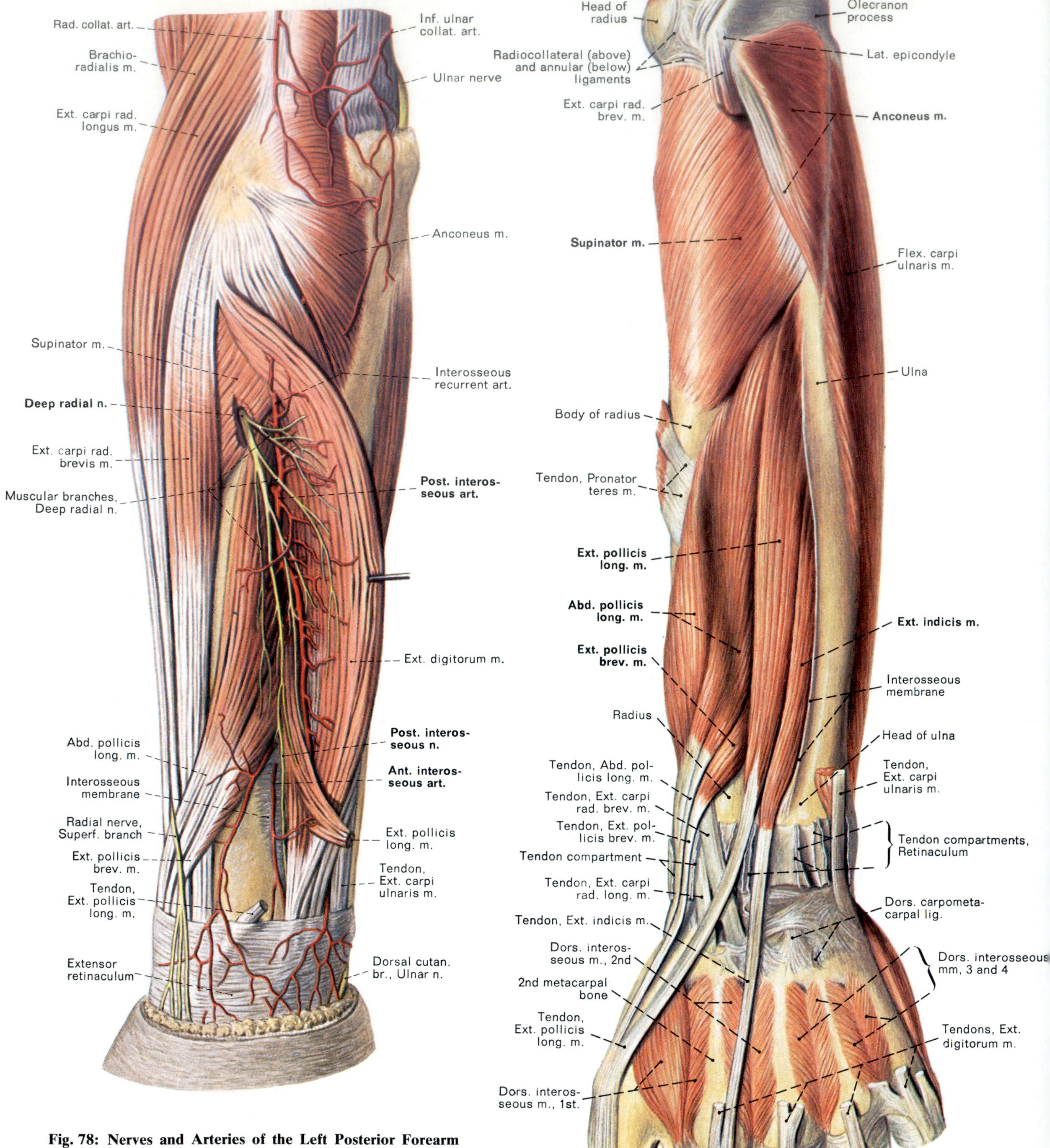
Rad. collat. art.
Brachio-radialis m.
Ext. carpi rad. longus m.
Inf. ulnar collat. art.
Ulnar nerve
Anconeus m.
Supinator m.
Deep radial n.
Ext. carpi rad. brevis m.
Muscular branches, Deep radial n.
Interosseous recurrent art.
Post. interos-seous art.
Ext. digitorum m.
Abd. pollicis long. m.
Interosseous membrane
Radial nerve, Superf. branch
Ext. pollicis brev. m.
Tendon, Ext. pollicis long. m.
Extensor retinaculum
Post. interos-seous n.
Ant. interos-seous art.
Ext. pollicis long. m.
Tendon, Ext. carpi ulnaris m.
Dorsal cutan. br., Ulnar n.
Head of radius
Radiocollateral (above) and annular (below) ligaments
Ext. carpi rad. brev. m.
Olecranon process
Lat. epicondyle
Anconeus m.
Supinator m.
Flex. carpi ulnaris m.
Ulna
Body of radius
Tendon, Pronator teres m.
Ext. pollicis long. m.
Abd. pollicis long. m.
Ext. indicis m.
Ext. pollicis brev. m.
Interosseous membrane
Radius
Head of ulna
Tendon, Abd. pollicis long. m.
Tendon, Ext. carpi rad. brev. m.
Tendon, Ext. pollicis brev. m.
Tendon compartment
Tendon, Ext. carpi rad. long. m.
Tendon, Ext. indicis m.
Dors. interosseous m., 2nd
2nd metacarpal bone
Tendon, Ext. pollicis long. m.
Dors. interosseous m., 1st.
Tendon, Ext. carpi ulnaris m.
Tendon compartments, Retinaculum
Dors. carpometacarpal lig.
Dors. interosseous mm, 3 and 4
Tendons, Ext. digitorum m.

Fig. 78: Nerves and Arteries of the Left Posterior Forearm (Deep Dissection)

NOTE: 1) the extensor digitorum muscle is separated from the extensor carpi radialis brevis and pulled medially to reveal the posterior interosseous artery and deep radial nerve.

2) the emergence of the deep radial nerve to the posterior forearm through the supinator muscle.

3) the posterior interosseous nerve is a continuation inferiorly of the deep radial nerve and may be seen coursing deep to the extensor pollicis longus muscle. This muscle has been cut at the wrist in this dissection.

Fig. 79: The Left Posterior Forearm Muscles, Deep Group

NOTE: 1) all of the superficial posterior forearm muscles have been removed except the anconeus. The five deep muscles are the supinator, abductor pollicis longus, extensors pollicis brevis and longus and extensor indicis.

2) the supinator is a broad muscle, arising from the lateral epicondyle of the humerus and from the ridge of the ulna. It courses obliquely to insert around the upper third of the radius.

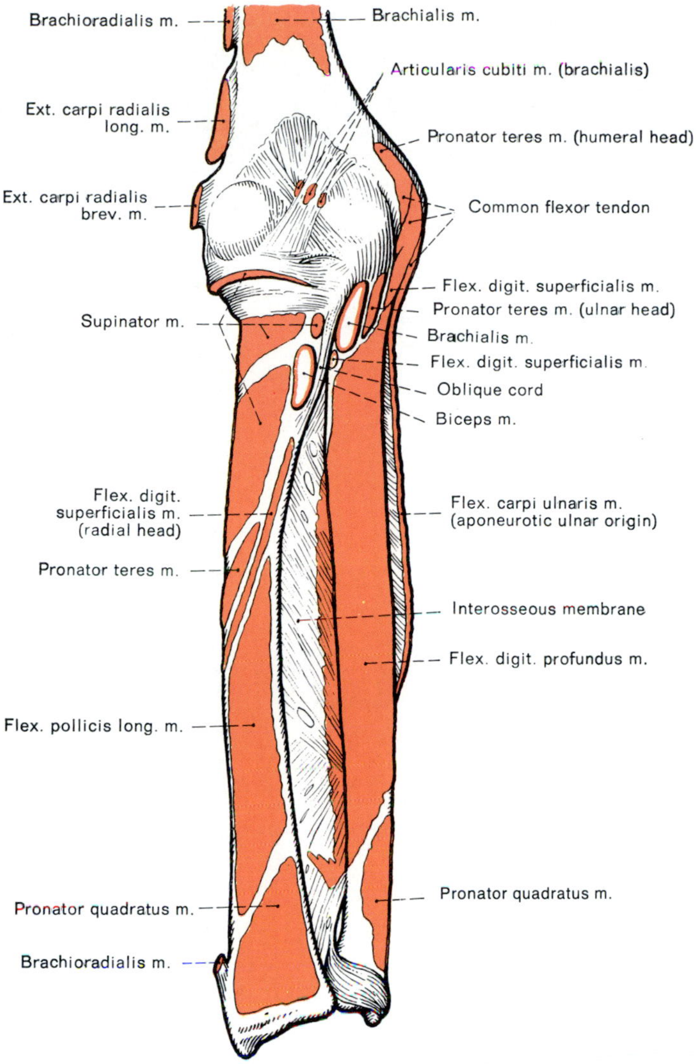

Brachioradialis m.
Brachialis m.
Articularis cubiti m. (brachialis)
Ext. carpi radialis long. m.
Pronator teres m. (humeral head)
Ext. carpi radialis brev. m.
Common flexor tendon
Flex. digit. superficialis m.
Pronator teres m. (ulnar head)
Supinator m.
Brachialis m.
Flex. digit. superficialis m.
Oblique cord
Biceps m.
Flex. digit. superficialis m. (radial head)
Flex. carpi ulnaris m. (aponeurotic ulnar origin)
Pronator teres m.
Interosseous membrane
Flex. digit. profundus m.
Flex. pollicis long. m.
Pronator quadratus m.
Pronator quadratus m.
Brachioradialis m.

Fig. 80: Bones of the Right Forearm Showing Attachments of Muscles (Anterior View)

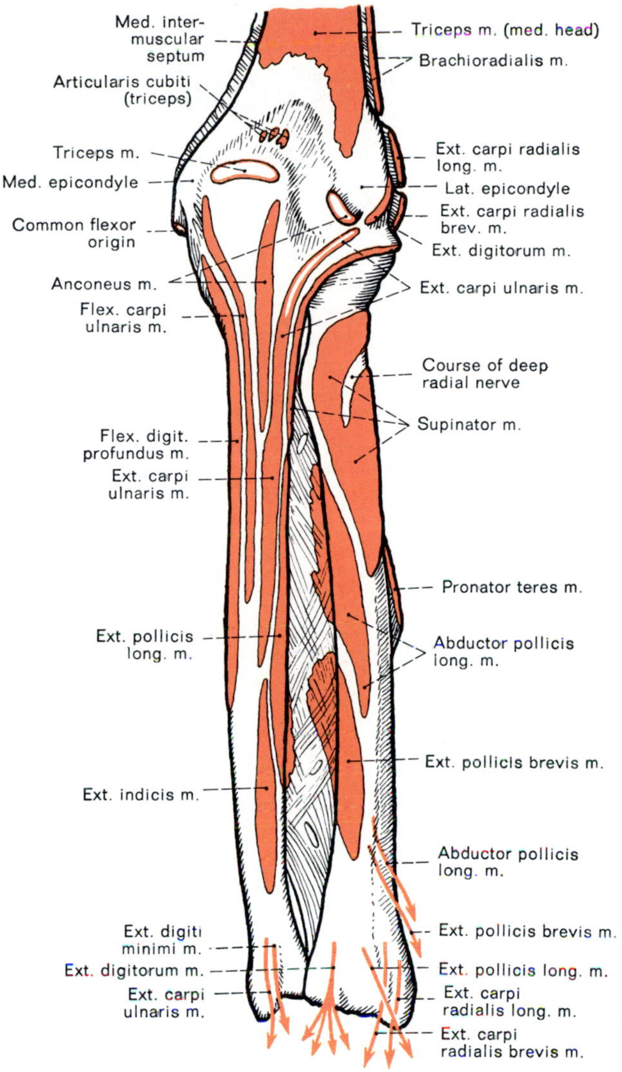

Med. inter-muscular septum
Triceps m. (med. head)
Brachioradialis m.
Articularis cubiti (triceps)
Triceps m.
Ext. carpi radialis long. m.
Med. epicondyle
Lat. epicondyle
Ext. carpi radialis brev. m.
Common flexor origin
Ext. digitorum m.
Anconeus m.
Ext. carpi ulnaris m.
Flex. carpi ulnaris m.
Course of deep radial nerve
Supinator m.
Flex. digit. profundus m.
Ext. carpi ulnaris m.
Pronator teres m.
Ext. pollicis long. m.
Abductor pollicis long. m.
Ext. pollicis brevis m.
Ext. indicis m.
Abductor pollicis long. m.
Ext. digiti minimi m.
Ext. pollicis brevis m.
Ext. digitorum m.
Ext. pollicis long. m.
Ext. carpi ulnaris m.
Ext. carpi radialis long. m.
Ext. carpi radialis brevis m.

Fig. 81: Bones of the Right Forearm Showing the Attachments of Muscles (Posterior View)

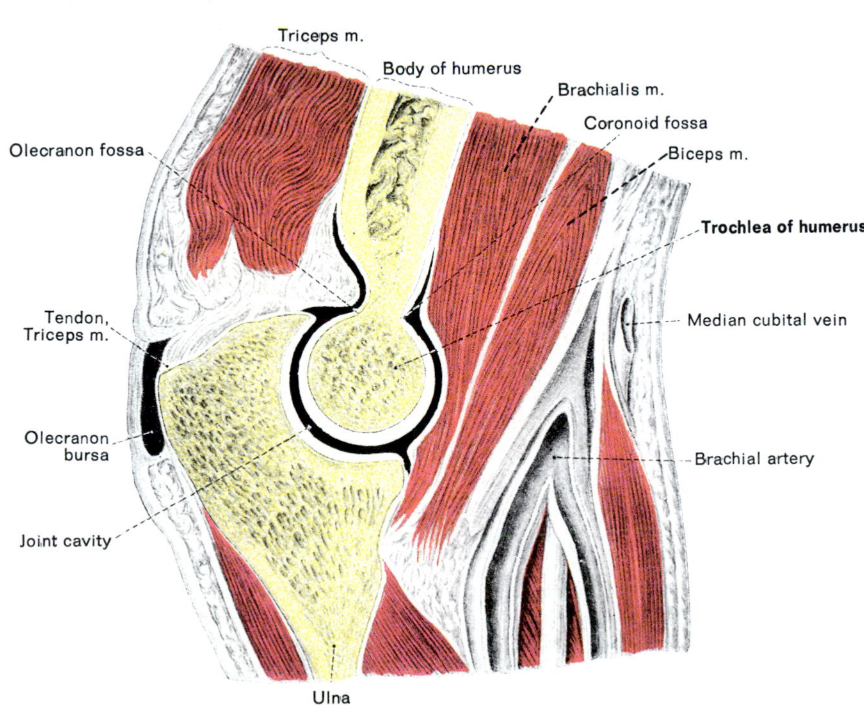
Triceps m.
Body of humerus
Brachialis m.
Coronoid fossa
Olecranon fossa
Biceps m.
Trochlea of humerus
Tendon, Triceps m.
Median cubital vein
Olecranon bursa
Brachial artery
Joint cavity
Ulna

Fig. 82: Sagittal Section of the Left Elbow Joint

NOTE: 1) the trochlea of the humerus articulates with the trochlear notch of the ulna to form a ginglymus or hinge joint. The adaptation of these two articular surfaces is such that only flexion and extension can take place and not lateral displacement.

2) the posterior aspect of the olecranon process is separated from the skin by a subcutaneous bursa, and the insertion of the triceps on the olecranon.

3) that the brachialis muscle lies immediately adjacent to the joint capsule anteriorly, and courses to insert on the roughened depression on the anterior surface of the coronoid process and onto the tuberosity of the ulna.

4) fractures of the distal extremity of the humerus most frequently occur from falls on the outstretched hand because the resulting force is transmitted through the bones of the forearm to the brachium, while fracture of the olecranon generally results from direct trauma to the bone by a fall on the point of the elbow. Posterior dislocation of the ulna and attached radius is the most common dislocation at the elbow joint, again frequently from falls on the outstretched and abducted hand.

Figs. 83 and 84: The Left Humerus

NOTE: 1) the humerus consists of a body and two extremities. The head of the humerus is shaped as a hemisphere and articulates with the scapula at the glenoid cavity.

2) the anatomical neck is a constricted zone just distal to the head of the humerus, and the surgical neck, where fractures frequently occur, lies just below the two tubercles.

3) the greater and lesser tubercles are roughened prominences which allow the insertion of muscles: the supraspinatus, infraspinatus, and the teres minor on the greater tubercle and the subscapularis on the lesser.

4) within the intertubercular sulcus passes the tendon of the long head of the biceps.

5) adjacent to the radial groove courses the radial nerve, which is therefore endangered by fractures of the humerus. If the radial nerve does become injured by a fracture, a clinical condition called *wrist drop* develops because this nerve supplies all the extensors of the wrist and fingers.

6) the distal extremity affords articulation with the radius and ulna; the rounded capitulum (Fig. 84) articulates with the head of the radius, while the grooved trochlea fits into the trochlear notch of the ulna.

7) the radial fossa above the capitulum (Fig. 84) within which the radial head is received upon flexion of the forearm. Additionally, note the coronoid fossa which receives the coronoid process of the ulna.

8) the olecranon fossa (Fig. 83) for the olecranon process of the ulna. Finally, observe the medial and lateral epicondyles which allow attachment of muscles.

Greater tubercle
Head
Anatomical neck
Surgical neck
Body of humerus
Radial (spiral) groove
Posterior surface
Olecranon fossa
Lateral epicondyle
Medial epicondyle
Trochlea
Ulnar n. sulcus

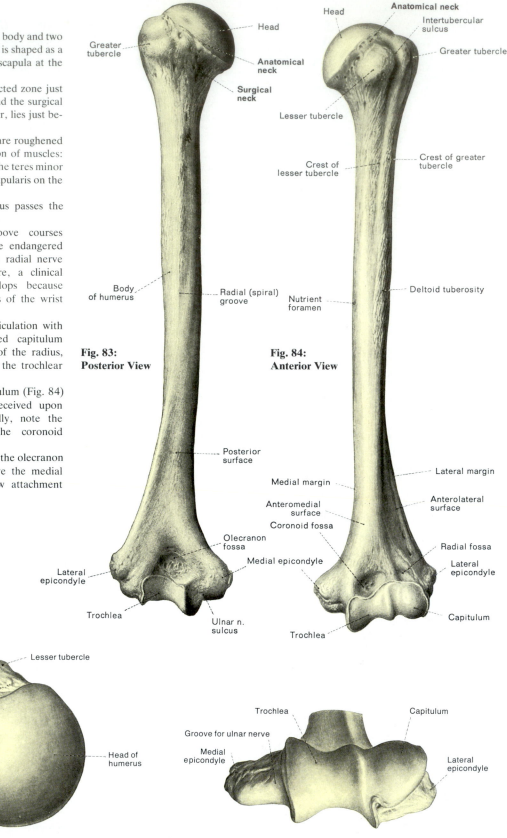

Fig. 83: Posterior View

Head
Anatomical neck
Intertubercular sulcus
Greater tubercle
Lesser tubercle
Crest of lesser tubercle
Crest of greater tubercle
Deltoid tuberosity
Nutrient foramen
Lateral margin
Medial margin
Anteromedial surface
Anterolateral surface
Coronoid fossa
Radial fossa
Lateral epicondyle
Medial epicondyle
Capitulum
Trochlea

Fig. 84: Anterior View

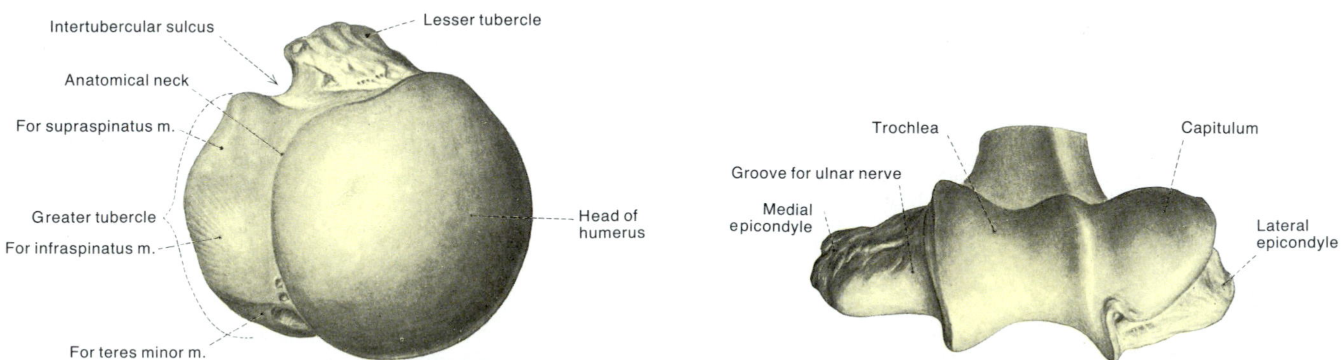
Intertubercular sulcus
Lesser tubercle
Anatomical neck
For supraspinatus m.
Greater tubercle
For infraspinatus m.
Head of humerus
For teres minor m.
Trochlea
Capitulum
Groove for ulnar nerve
Medial epicondyle
Lateral epicondyle

Fig. 85: The Head of the Left Humerus, Viewed from Above

NOTE the sequential sites of attachment for the supraspinatus muscle, infraspinatus muscle, and teres minor muscle on the greater tubercle. The humeral head is coated with hyaline cartilage, which is more abundant in the central part and becomes thinner nearer the circumference.

Fig. 86: The Distal Extremity of the Left Humerus, Viewed from Below

NOTE that the medial epicondyle is larger than the lateral epicondyle and affords attachment for the flexor muscles in the anterior forearm compartment. The ulnar nerve courses along a groove on the dorsal surface of the medial epicondyle and enters the forearm between the two heads of the flexor carpi ulnaris.

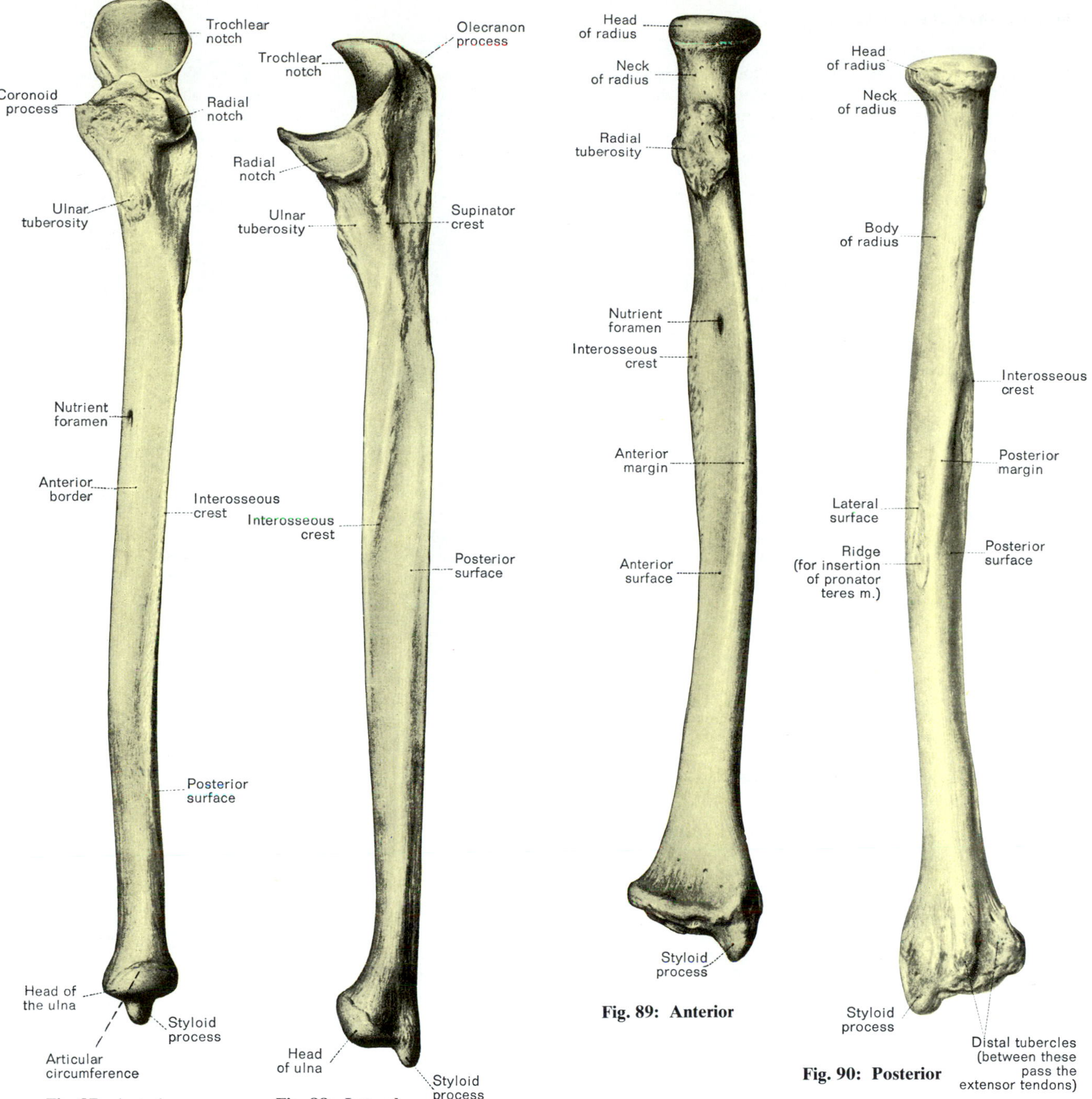
Trochlear notch
Coronoid process
Radial notch
Ulnar tuberosity
Nutrient foramen
Anterior border
Interosseous crest
Posterior surface
Head of the ulna
Styloid process
Articular circumference
Olecranon process
Trochlear notch
Radial notch
Ulnar tuberosity
Supinator crest
Interosseous crest
Posterior surface
Head of ulna
Styloid process
Head of radius
Neck of radius
Radial tuberosity
Nutrient foramen
Interosseous crest
Anterior margin
Anterior surface
Styloid process
Head of radius
Neck of radius
Body of radius
Interosseous crest
Posterior margin
Lateral surface
Ridge (for insertion of pronator teres m.)
Posterior surface
Styloid process
Distal tubercles (between these pass the extensor tendons)

Fig. 87: Anterior **Fig. 88: Lateral** **Fig. 89: Anterior** **Fig. 90: Posterior**

The Left Ulna (Figs. 87 and 88)

NOTE: 1) the ulna is the medial bone of the forearm. It presents a superior extremity, a body or shaft and an inferior extremity.

2) the superior extremity is marked by two processes, the olecranon and coronoid processes and two concave cavities, the radial notch for articulation with the radius, and the trochlear notch which serves for articulation with the trochlea of the humerus. The brachialis muscle inserts on the tuberosity of the ulna.

3) the tapering body of the ulna affords attachment of the interosseous membrane. The distal extremity is marked by the head of the ulna laterally, and the styloid process posteromedially. The head of the ulna is attached to an articular disc which, in turn, articulates with the triquetral bone. Onto the styloid process of the ulna is attached the ulnar collateral ligament of the wrist joint.

The Left Radius (Figs. 89 and 90)

NOTE: 1) the radius is situated lateral to the ulna in the forearm. It has a body and two extremities. The proximal extremity articulates with both the humerus and ulna. The larger distal extremity articulates inferiorly with the carpal bones (scaphoid, lunate and triquetrum) and medially with the ulna.

2) the proximal extremity is marked by a cylindrical head which articulates with both the capitulum of the humerus and the radial notch of the ulna. Just beneath the neck on the anteromedial aspect is found the radial tuberosity on which is inserted the biceps tendon.

3) the interosseous membrane is attached along the interosseous crest of the shaft.

4) the styloid process distally gives attachment to the brachioradialis muscle and the radial collateral ligament of the radiocarpal joint.

Fig. 91: The Left Elbow Joint, Anterior View

NOTE: 1) the elbow joint is a hinge (ginglymus) joint in which the trochlear notch of the ulna receives the trochlea of the humerus, and the shallow fovea on the head of radius articulates with the capitulum of the humerus.

2) the entire joint is encased by an articular capsule which tends to be loose to allow flexion and extension of the forearm. The capsule is thickened medially by the ulnar collateral ligament and laterally by the radial collateral ligament.

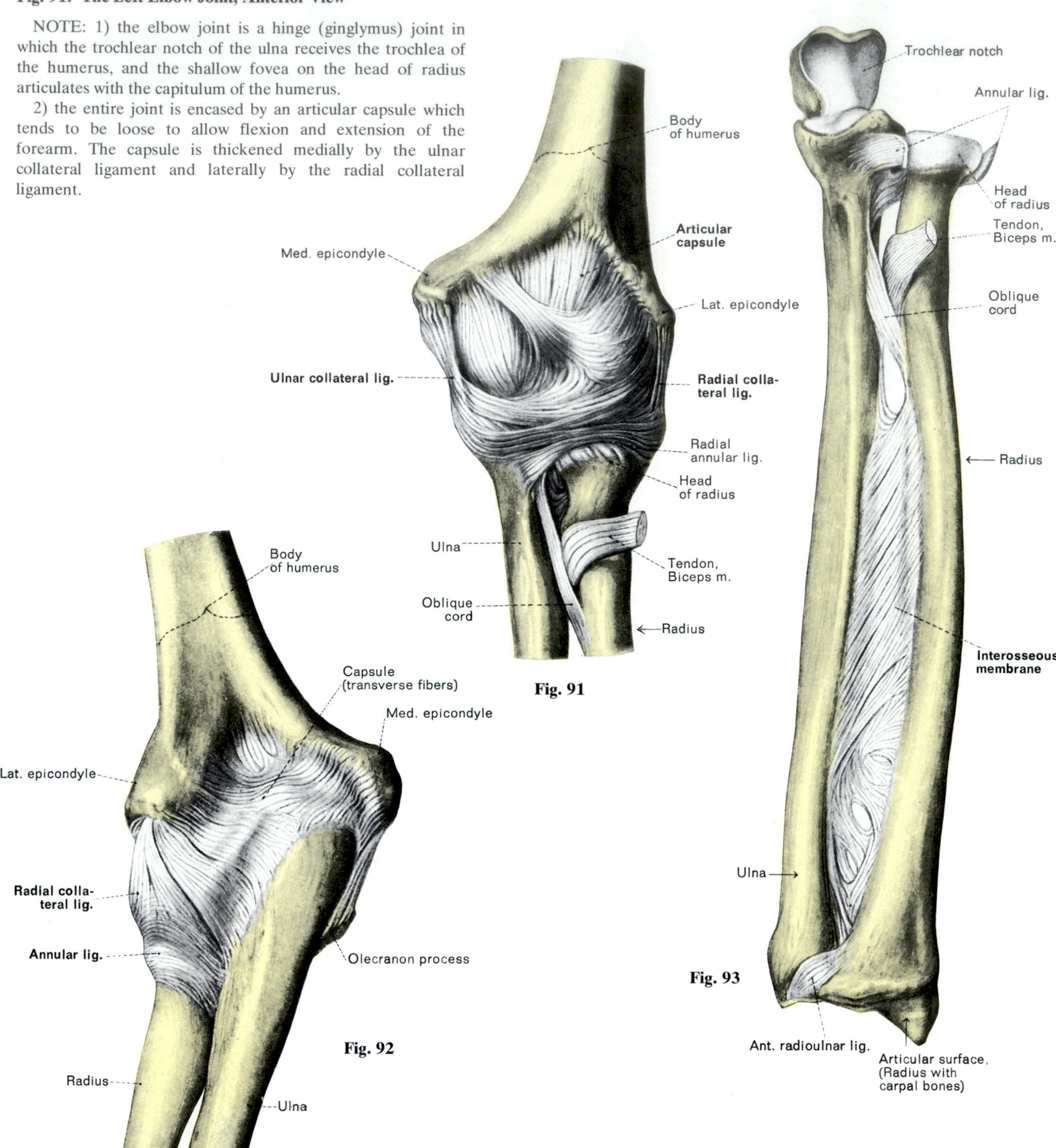
Body of humerus
Med. epicondyle
Articular capsule
Lat. epicondyle
Ulnar collateral lig.
Radial collateral lig.
Radial annular lig.
Head of radius
Ulna
Tendon, Biceps m.
Oblique cord
Radius
Fig. 91
Trochlear notch
Annular lig.
Head of radius
Tendon, Biceps m.
Oblique cord
Radius
Interosseous membrane
Ulna
Fig. 93
Ant. radioulnar lig.
Articular surface, (Radius with carpal bones)
Body of humerus
Capsule (transverse fibers)
Med. epicondyle
Lat. epicondyle
Radial collateral lig.
Annular lig.
Olecranon process
Fig. 92
Radius
Ulna

Fig. 92: The Left Elbow Joint, Posterolateral View

NOTE: 1) the fan-shaped form of the radial collateral ligament. It attaches superiorly to the lateral epicondyle and blends inferiorly with the capsule of the joint.

2) the superior portion of the radial annular ligament also blends with the articular capsule.

3) the transverse fibers of the articular capsule forming a band between the medial and lateral epicondyles and bridging the olecranon fossa.

Fig. 93: Radioulnar Joints, Anterior View (left)

NOTE: 1) articulations between the radius and ulna occur proximally, along the shafts of the two bones and distally.

2) the proximal joint is a pivot (trochoid) type joint and consists of the head of the radius which rotates within the radial notch of the ulna. This joint is protected by the lower part of the capsule of the elbow joint and by an underlying annular ligament attached at both ends to the ulna and forming a circular band around the head of the radius.

3) the broad, fibrous interosseous membrane extends obliquely between the bones, while distally the head of the ulna articulates with the ulnar notch of the radius.

Fig. 94: Roentgenogram of the Elbow Joint of 5½ Year Old Boy

NOTE: 1) the shaft of a long bone is called the *diaphysis* while a center of ossification, distinct from the shaft and usually at the extremity of a long bone, is known as an *epiphysis*.

2) the epiphysis of the head of the radius is as yet not formed in the 5½-year old, while ossification has commenced in the humeral capitulum.

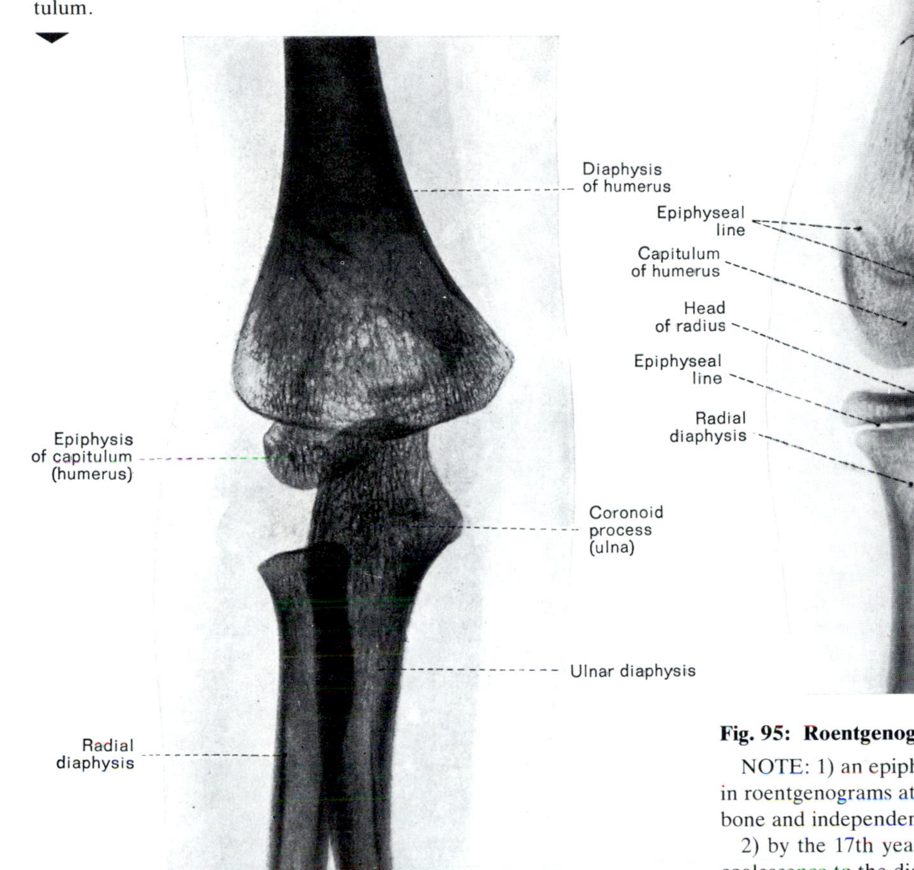

Diaphysis of humerus
Epiphysis of capitulum (humerus)
Coronoid process (ulna)
Ulnar diaphysis
Radial diaphysis

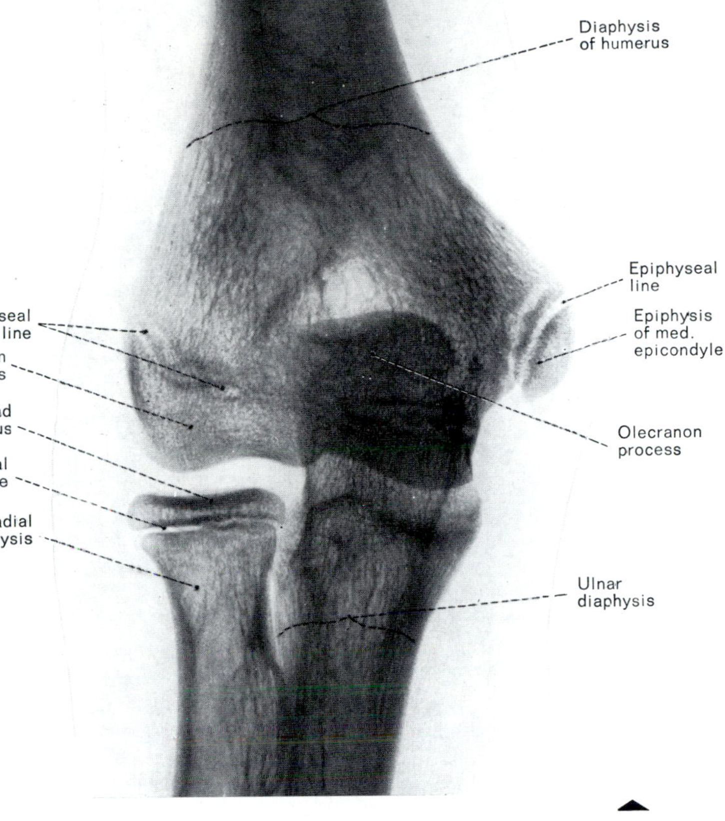

Diaphysis of humerus
Epiphyseal line
Epiphysis of med. epicondyle
Epiphyseal line
Capitulum of humerus
Head of radius
Epiphyseal line
Radial diaphysis
Olecranon process
Ulnar diaphysis

Fig. 95: Roentgenogram of the Elbow Joint of 17 Year Old Male

NOTE: 1) an epiphyseal line (containing cartilage) can be identified in roentgenograms at the plane of junction between the main part of a bone and independently ossifying centers, the epiphyses.

2) by the 17th year ossification of the epiphyseal centers and their coalescence to the diaphyses of the bones of the elbow joint are nearly complete.

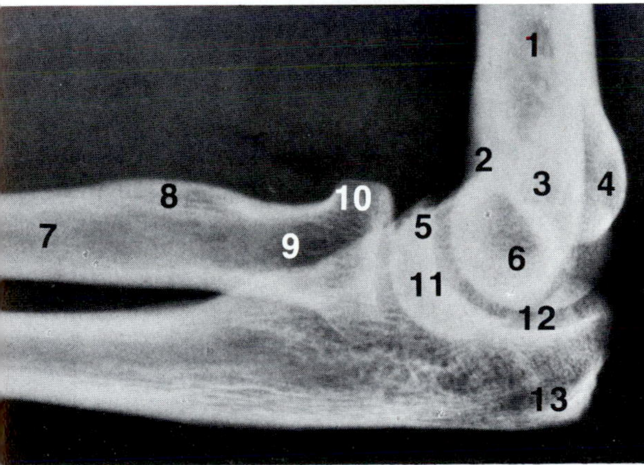

1
2
3
4
5
6
7
8
9
10
11
12
13

Fig 96: Roentgenogram of the Left Elbow Joint in an Adult, Lateral Projection

NOTE the following bony structures:

1. Body of humerus
2. Radial fossa
3. Olecranon fossa
4. Medial epicondyle
5. Coronoid process of ulna
6. Trochlea of humerus
7. Body of radius
8. Radial tuberosity
9. Neck of radius
10. Head of radius
11. Capitulum of humerus
12. Trochlear notch
13. Olecranon

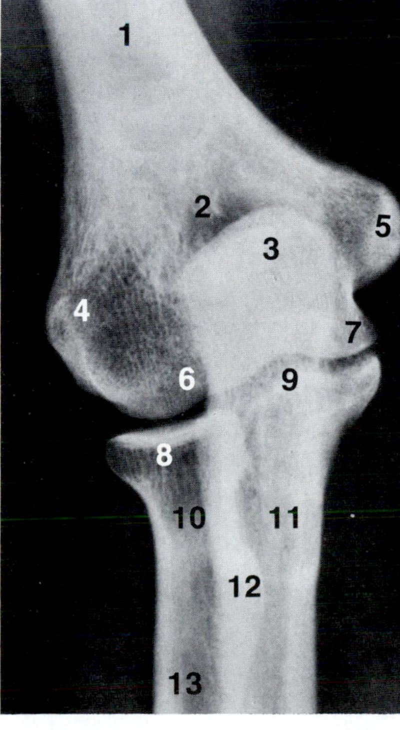

1
2
3
4
5
6
7
8
9
10
11
12
13

Fig. 97: Roentgenogram of the Right Elbow Joint in an Adult (A–P Projection)

NOTE the following bony structures:

1. Body of humerus
2. Olecranon fossa
3. Olecranon
4. Lateral epicondyle
5. Medial epicondyle
6. Capitulum of humerus
7. Trochlea of humerus
8. Head of radius
9. Coronoid process of ulna
10. Neck of radius
11. Ulna
12. Radial tuberosity
13. Body of radius

(Figures 96 and 97 come from Wicke, L. *Atlas of Radiologic Anatomy,* 3rd Edition, Urban & Schwarzenberg, Baltimore, 1982).

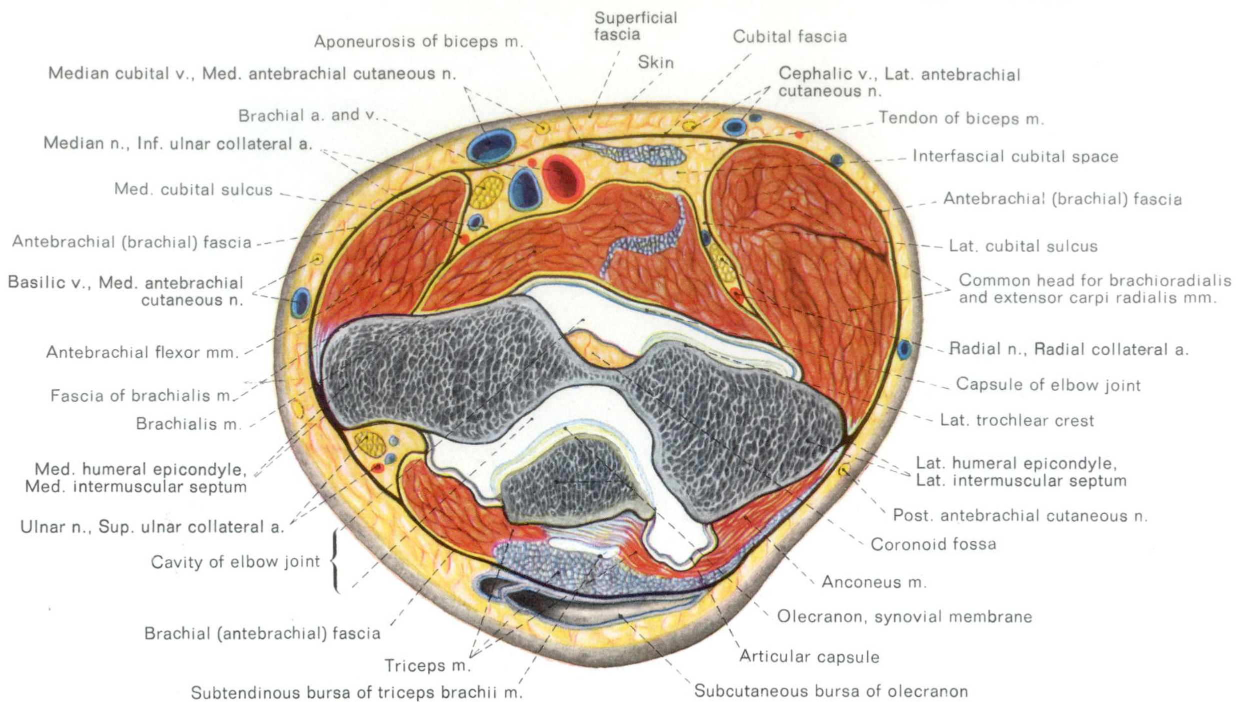
Superficial fascia
Aponeurosis of biceps m.
Cubital fascia
Skin
Median cubital v., Med. antebrachial cutaneous n.
Cephalic v., Lat. antebrachial cutaneous n.
Brachial a. and v.
Tendon of biceps m.
Median n., Inf. ulnar collateral a.
Interfascial cubital space
Med. cubital sulcus
Antebrachial (brachial) fascia
Antebrachial (brachial) fascia
Lat. cubital sulcus
Basilic v., Med. antebrachial cutaneous n.
Common head for brachioradialis and extensor carpi radialis mm.
Antebrachial flexor mm.
Radial n., Radial collateral a.
Capsule of elbow joint
Fascia of brachialis m.
Lat. trochlear crest
Brachialis m.
Med. humeral epicondyle, Med. intermuscular septum
Lat. humeral epicondyle, Lat. intermuscular septum
Post. antebrachial cutaneous n.
Ulnar n., Sup. ulnar collateral a.
Coronoid fossa
Cavity of elbow joint
Anconeus m.
Olecranon, synovial membrane
Brachial (antebrachial) fascia
Articular capsule
Triceps m.
Subtendinous bursa of triceps brachii m.
Subcutaneous bursa of olecranon

Fig. 98: Cross Section Through the Right Upper Extremity at the Level of the Elbow Joint

NOTE: 1) the position of the ulnar nerve and superior ulnar collateral artery behind the medial epicondyle of the humerus and medial to the olecranon of the ulna (see also Figs. 58, 60 and 76).

2) that the median nerve lies to the ulnar (medial) side of the brachial vein and artery in the cubital fossa (see also Figs. 54, 56 and 70), and that all three of these structures lie deep to the cubital fascia and median cubital vein (see also Figs. 40 and 41).

3) that at this level the radial nerve and radial collateral artery lie between the common origins of the brachioradialis and extensors carpi radialis longus and brevis muscles *and* the deeply located brachialis muscle, which is about to insert on the ulna (see also Figs. 67 and 68).

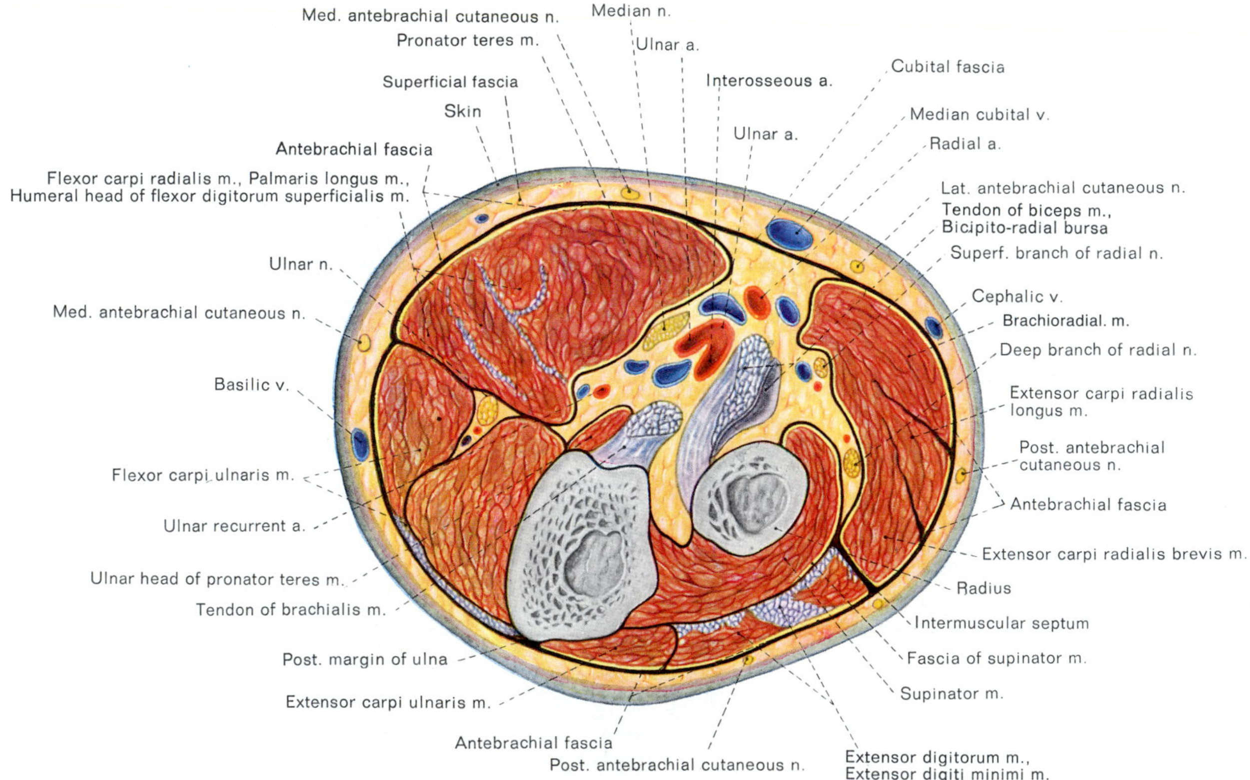
Med. antebrachial cutaneous n.
Median n.
Pronator teres m.
Ulnar a.
Cubital fascia
Superficial fascia
Interosseous a.
Skin
Median cubital v.
Ulnar a.
Antebrachial fascia
Radial a.
Flexor carpi radialis m., Palmaris longus m., Humeral head of flexor digitorum superficialis m.
Lat. antebrachial cutaneous n.
Tendon of biceps m., Bicipito-radial bursa
Superf. branch of radial n.
Ulnar n.
Cephalic v.
Med. antebrachial cutaneous n.
Brachioradial. m.
Deep branch of radial n.
Basilic v.
Extensor carpi radialis longus m.
Post. antebrachial cutaneous n.
Flexor carpi ulnaris m.
Antebrachial fascia
Ulnar recurrent a.
Extensor carpi radialis brevis m.
Ulnar head of pronator teres m.
Radius
Tendon of brachialis m.
Intermuscular septum
Post. margin of ulna
Fascia of supinator m.
Supinator m.
Extensor carpi ulnaris m.
Antebrachial fascia
Post. antebrachial cutaneous n.
Extensor digitorum m., Extensor digiti minimi m.

Fig. 99: Cross Section Through the Proximal Third of the Right Forearm

NOTE: 1) that this section is cut at the level at which the common interosseous artery branches from the ulnar artery (see also Fig. 70), and at the sites at which the tendons of insertion of the biceps brachii and brachialis muscles attach to the radius and ulna, respectively (see also Fig. 80).

2) that at this level the radial nerve has already divided into its superficial and deep branches (see also Fig. 67 and 68) and that the ulnar nerve lies under cover of the flexor carpi ulnaris (see also Fig. 70).

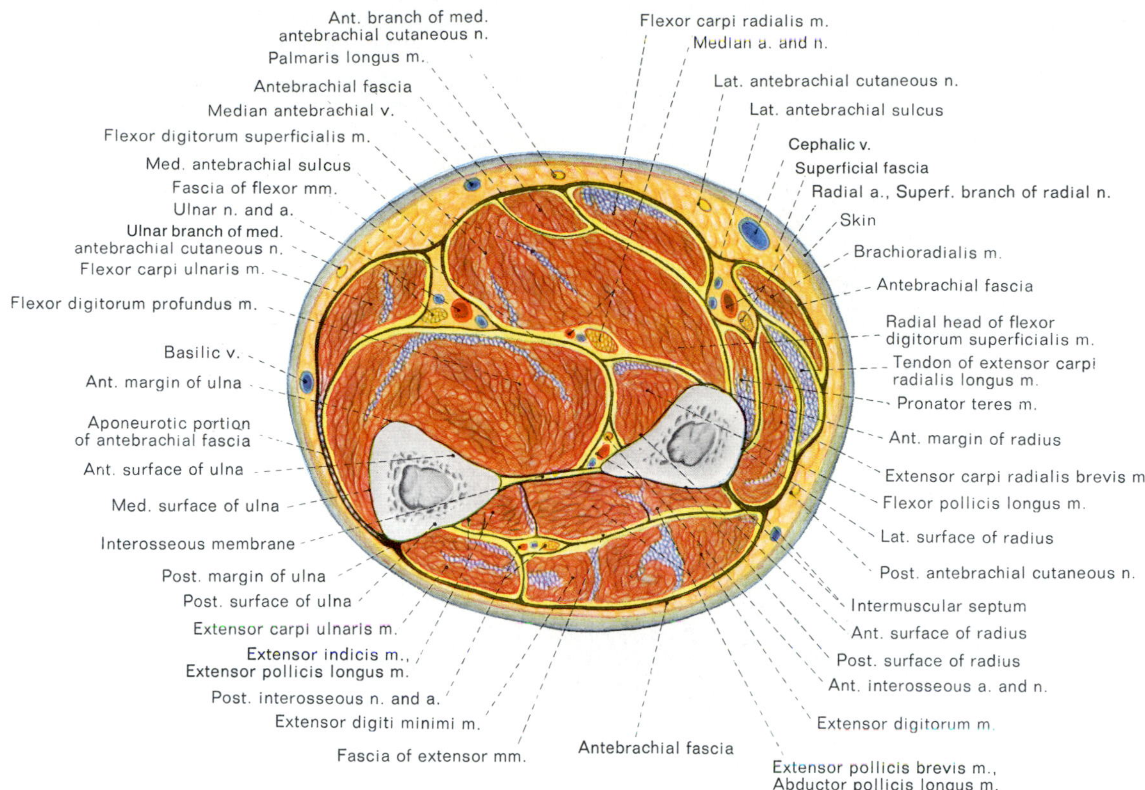
Ant. branch of med. antebrachial cutaneous n.
Palmaris longus m.
Antebrachial fascia
Median antebrachial v.
Flexor digitorum superficialis m.
Med. antebrachial sulcus
Fascia of flexor mm.
Ulnar n. and a.
Ulnar branch of med. antebrachial cutaneous n.
Flexor carpi ulnaris m.
Flexor digitorum profundus m.
Basilic v.
Ant. margin of ulna
Aponeurotic portion of antebrachial fascia
Ant. surface of ulna
Med. surface of ulna
Interosseous membrane
Post. margin of ulna
Post. surface of ulna
Extensor carpi ulnaris m.
Extensor indicis m., Extensor pollicis longus m.
Post. interosseous n. and a.
Extensor digiti minimi m.
Fascia of extensor mm.
Antebrachial fascia
Flexor carpi radialis m.
Median a. and n.
Lat. antebrachial cutaneous n.
Lat. antebrachial sulcus
Cephalic v.
Superficial fascia
Radial a., Superf. branch of radial n.
Skin
Brachioradialis m.
Antebrachial fascia
Radial head of flexor digitorum superficialis m.
Tendon of extensor carpi radialis longus m.
Pronator teres m.
Ant. margin of radius
Extensor carpi radialis brevis m.
Flexor pollicis longus m.
Lat. surface of radius
Post. antebrachial cutaneous n.
Intermuscular septum
Ant. surface of radius
Post. surface of radius
Ant. interosseous a. and n.
Extensor digitorum m.
Extensor pollicis brevis m., Abductor pollicis longus m.

Fig. 100: Cross Section Through the Middle Third of the Right Forearm

NOTE: 1) that in the middle of the forearm, the ulna, radius, interosseous membrane, and intermuscular septa quite clearly delineate the posterior compartment, extending posteriorly and laterally, from the anterior compartment (located anteriorly and medially).

2) the median nerve coursing down the forearm deep to the flexor digitorum superficialis and anterior to the flexor digitorum profundus and flexor pollicis longus (see also Figs. 68–70).

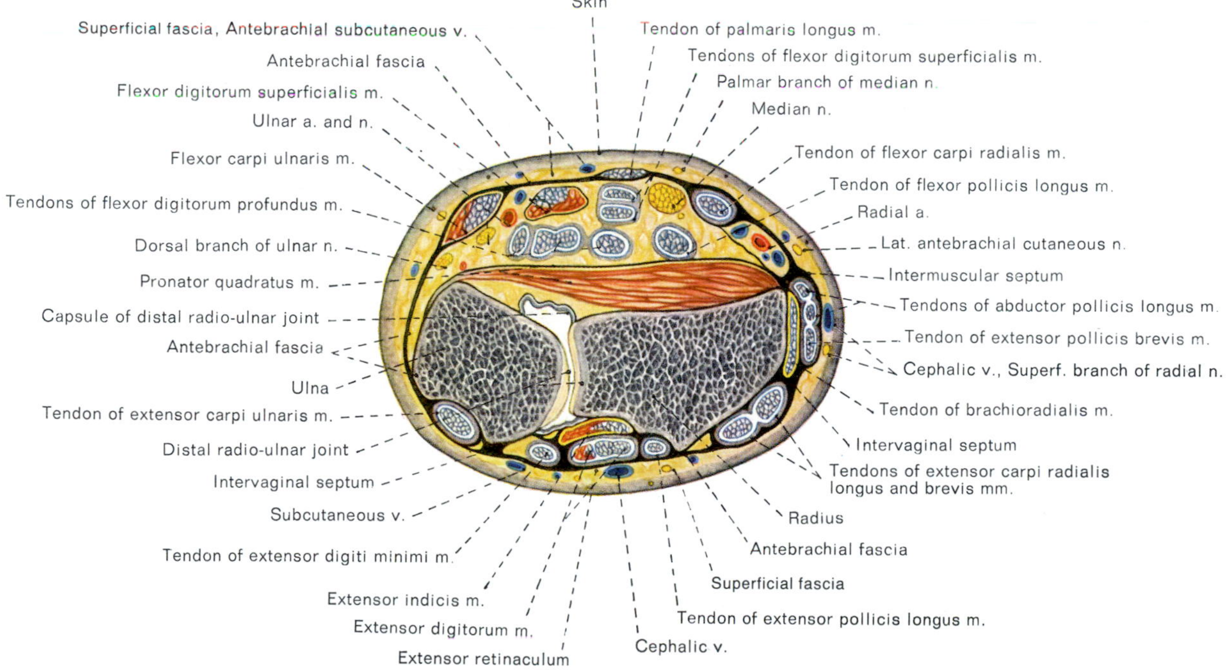
Skin
Superficial fascia, Antebrachial subcutaneous v.
Antebrachial fascia
Flexor digitorum superficialis m.
Ulnar a. and n.
Flexor carpi ulnaris m.
Tendons of flexor digitorum profundus m.
Dorsal branch of ulnar n.
Pronator quadratus m.
Capsule of distal radio-ulnar joint
Antebrachial fascia
Ulna
Tendon of extensor carpi ulnaris m.
Distal radio-ulnar joint
Intervaginal septum
Subcutaneous v.
Tendon of extensor digiti minimi m.
Extensor indicis m.
Extensor digitorum m.
Extensor retinaculum
Tendon of palmaris longus m.
Tendons of flexor digitorum superficialis m.
Palmar branch of median n.
Median n.
Tendon of flexor carpi radialis m.
Tendon of flexor pollicis longus m.
Radial a.
Lat. antebrachial cutaneous n.
Intermuscular septum
Tendons of abductor pollicis longus m.
Tendon of extensor pollicis brevis m.
Cephalic v., Superf. branch of radial n.
Tendon of brachioradialis m.
Intervaginal septum
Tendons of extensor carpi radialis longus and brevis mm.
Radius
Antebrachial fascia
Superficial fascia
Tendon of extensor pollicis longus m.
Cephalic v.

Fig. 101: Cross Section Through the Distal Third of the Right Forearm

NOTE: 1) that at this level the ulnar nerve has already given off its dorsal branch, which supplies the ulnar part of the dorsum of the hand and the dorsal surfaces of the little finger and half of the ring finger (see also Fig. 102).

2) the extreme vulnerability of the ulnar nerve and artery, median nerve, radial artery, as well as all of the flexor tendons on the anterior aspect of the wrist (see also Fig. 68), and the extensor tendons and cephalic vein dorsally.

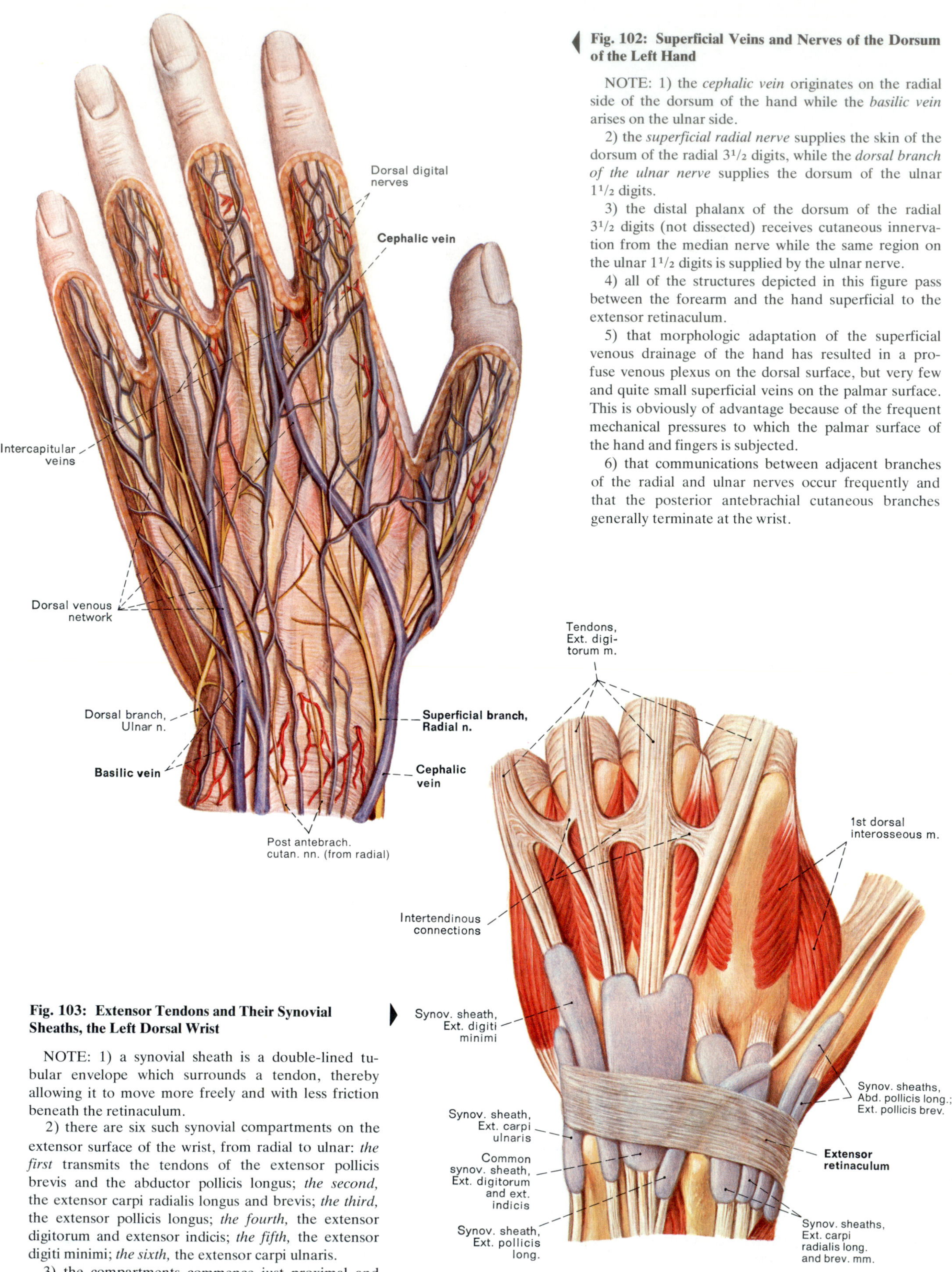

Dorsal digital nerves
Cephalic vein
Intercapitular veins
Dorsal venous network
Dorsal branch, Ulnar n.
Basilic vein
Superficial branch, Radial n.
Cephalic vein
Post antebrach. cutan. nn. (from radial)
Tendons, Ext. digitorum m.
1st dorsal interosseous m.
Intertendinous connections
Synov. sheath, Ext. digiti minimi
Synov. sheaths, Abd. pollicis long.; Ext. pollicis brev.
Synov. sheath, Ext. carpi ulnaris
Extensor retinaculum
Common synov. sheath, Ext. digitorum and ext. indicis
Synov. sheath, Ext. pollicis long.
Synov. sheaths, Ext. carpi radialis long. and brev. mm.

Fig. 102: Superficial Veins and Nerves of the Dorsum of the Left Hand

NOTE: 1) the *cephalic vein* originates on the radial side of the dorsum of the hand while the *basilic vein* arises on the ulnar side.

2) the *superficial radial nerve* supplies the skin of the dorsum of the radial $3^1/_2$ digits, while the *dorsal branch of the ulnar nerve* supplies the dorsum of the ulnar $1^1/_2$ digits.

3) the distal phalanx of the dorsum of the radial $3^1/_2$ digits (not dissected) receives cutaneous innervation from the median nerve while the same region on the ulnar $1^1/_2$ digits is supplied by the ulnar nerve.

4) all of the structures depicted in this figure pass between the forearm and the hand superficial to the extensor retinaculum.

5) that morphologic adaptation of the superficial venous drainage of the hand has resulted in a profuse venous plexus on the dorsal surface, but very few and quite small superficial veins on the palmar surface. This is obviously of advantage because of the frequent mechanical pressures to which the palmar surface of the hand and fingers is subjected.

6) that communications between adjacent branches of the radial and ulnar nerves occur frequently and that the posterior antebrachial cutaneous branches generally terminate at the wrist.

Fig. 103: Extensor Tendons and Their Synovial Sheaths, the Left Dorsal Wrist

NOTE: 1) a synovial sheath is a double-lined tubular envelope which surrounds a tendon, thereby allowing it to move more freely and with less friction beneath the retinaculum.

2) there are six such synovial compartments on the extensor surface of the wrist, from radial to ulnar: *the first* transmits the tendons of the extensor pollicis brevis and the abductor pollicis longus; *the second,* the extensor carpi radialis longus and brevis; *the third,* the extensor pollicis longus; *the fourth,* the extensor digitorum and extensor indicis; *the fifth,* the extensor digiti minimi; *the sixth,* the extensor carpi ulnaris.

3) the compartments commence just proximal and terminate somewhat distal to the extensor retinaculum.

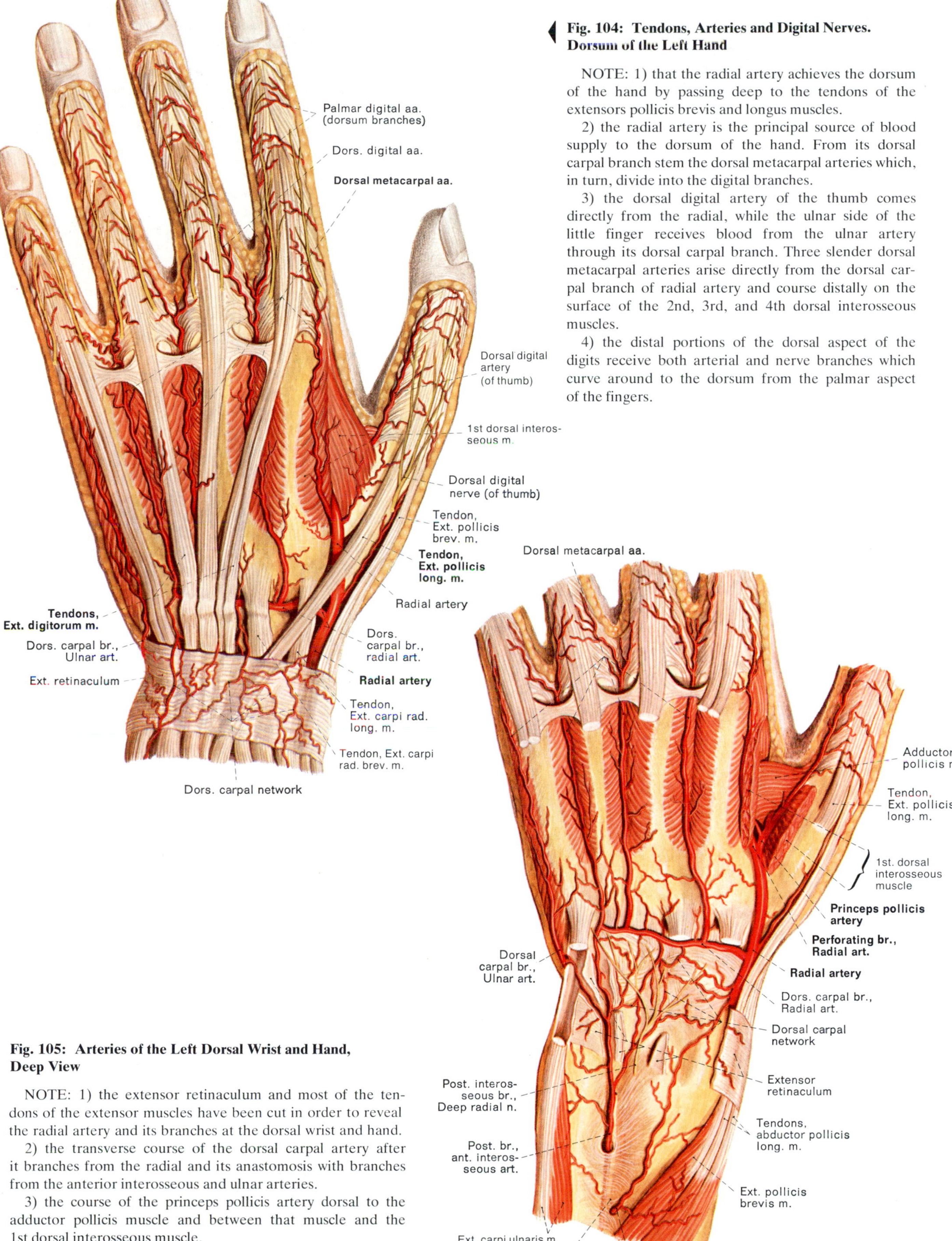
Palmar digital aa.
(dorsum branches)
Dors. digital aa.
Dorsal metacarpal aa.
Dorsal digital
artery
(of thumb)
1st dorsal interos-
seous m.
Dorsal digital
nerve (of thumb)
Tendon,
Ext. pollicis
brev. m.
Tendon,
Ext. pollicis
long. m.
Radial artery
Tendons,
Ext. digitorum m.
Dors. carpal br.,
Ulnar art.
Ext. retinaculum
Dors.
carpal br.,
radial art.
Radial artery
Tendon,
Ext. carpi rad.
long. m.
Tendon, Ext. carpi
rad. brev. m.
Dors. carpal network
Dorsal metacarpal aa.
Adductor
pollicis m.
Tendon,
Ext. pollicis
long. m.
1st. dorsal
interosseous
muscle
Princeps pollicis
artery
Perforating br.,
Radial art.
Radial artery
Dors. carpal br.,
Radial art.
Dorsal carpal
network
Extensor
retinaculum
Tendons,
abductor pollicis
long. m.
Ext. pollicis
brevis m.
Dorsal
carpal br.,
Ulnar art.
Post. interos-
seous br.,
Deep radial n.
Post. br.,
ant. interos-
seous art.
Ext. carpi ulnaris m.
Interosseous
membrane
Perforating br.,
Ant. interosseous art.

Fig. 104: Tendons, Arteries and Digital Nerves. Dorsum of the Left Hand

NOTE: 1) that the radial artery achieves the dorsum of the hand by passing deep to the tendons of the extensors pollicis brevis and longus muscles.

2) the radial artery is the principal source of blood supply to the dorsum of the hand. From its dorsal carpal branch stem the dorsal metacarpal arteries which, in turn, divide into the digital branches.

3) the dorsal digital artery of the thumb comes directly from the radial, while the ulnar side of the little finger receives blood from the ulnar artery through its dorsal carpal branch. Three slender dorsal metacarpal arteries arise directly from the dorsal carpal branch of radial artery and course distally on the surface of the 2nd, 3rd, and 4th dorsal interosseous muscles.

4) the distal portions of the dorsal aspect of the digits receive both arterial and nerve branches which curve around to the dorsum from the palmar aspect of the fingers.

Fig. 105: Arteries of the Left Dorsal Wrist and Hand, Deep View

NOTE: 1) the extensor retinaculum and most of the tendons of the extensor muscles have been cut in order to reveal the radial artery and its branches at the dorsal wrist and hand.

2) the transverse course of the dorsal carpal artery after it branches from the radial and its anastomosis with branches from the anterior interosseous and ulnar arteries.

3) the course of the princeps pollicis artery dorsal to the adductor pollicis muscle and between that muscle and the 1st dorsal interosseous muscle.

4) the manner in which the dorsal metacarpal arteries divide to form dorsal digital vessels.

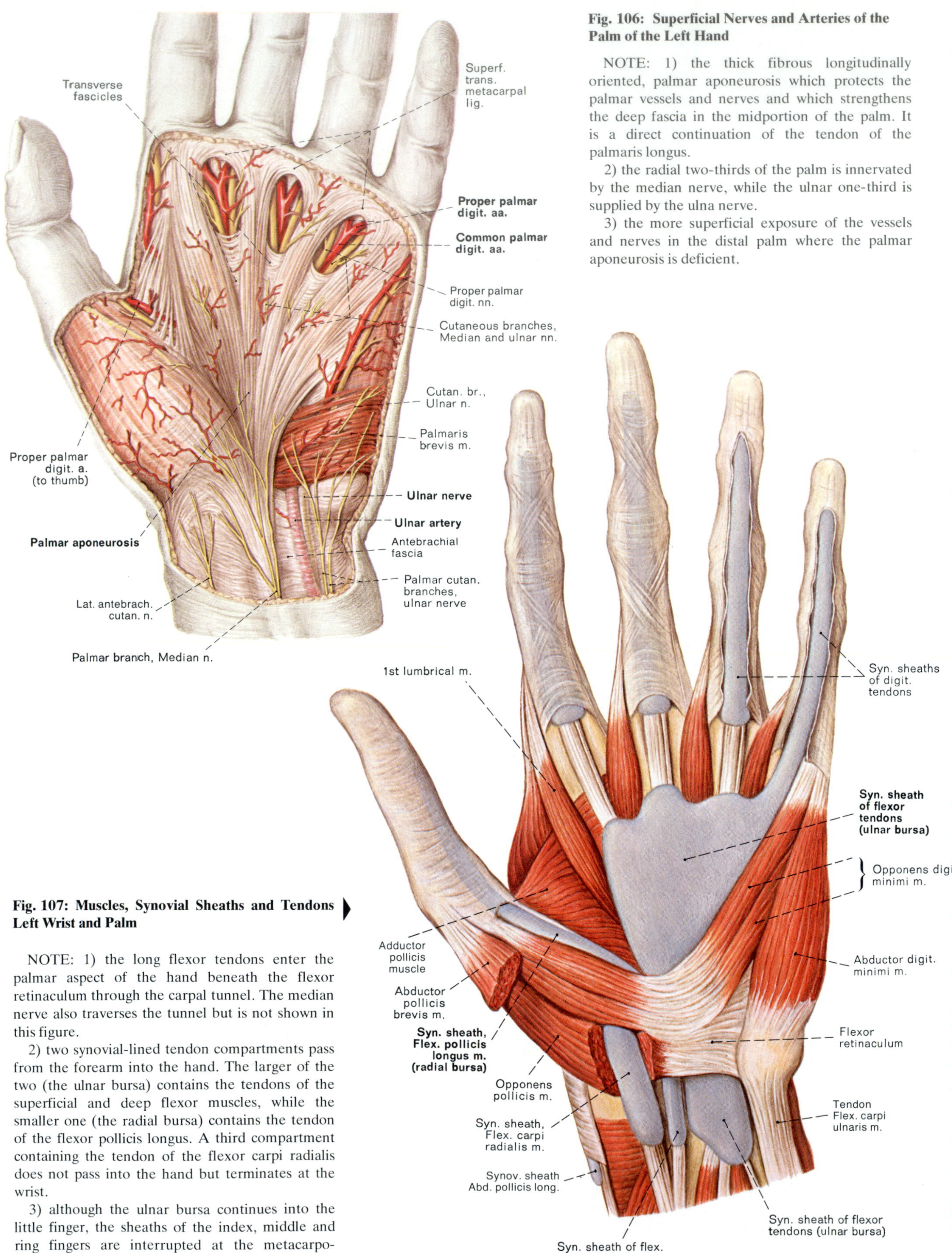

Transverse fascicles
Superf. trans. metacarpal lig.
Proper palmar digit. aa.
Common palmar digit. aa.
Proper palmar digit. nn.
Cutaneous branches, Median and ulnar nn.
Cutan. br., Ulnar n.
Palmaris brevis m.
Proper palmar digit. a. (to thumb)
Ulnar nerve
Ulnar artery
Palmar aponeurosis
Antebrachial fascia
Palmar cutan. branches, ulnar nerve
Lat. antebrach. cutan. n.
Palmar branch, Median n.
1st lumbrical m.
Syn. sheaths of digit. tendons
Syn. sheath of flexor tendons (ulnar bursa)
Opponens digit. minimi m.
Adductor pollicis muscle
Abductor digit. minimi m.
Abductor pollicis brevis m.
Syn. sheath, Flex. pollicis longus m. (radial bursa)
Flexor retinaculum
Opponens pollicis m.
Tendon Flex. carpi ulnaris m.
Syn. sheath, Flex. carpi radialis m.
Synov. sheath Abd. pollicis long.
Syn. sheath of flexor tendons (ulnar bursa)
Syn. sheath of flex. pollicis longus (radial bursa)

Fig. 106: Superficial Nerves and Arteries of the Palm of the Left Hand

NOTE: 1) the thick fibrous longitudinally oriented, palmar aponeurosis which protects the palmar vessels and nerves and which strengthens the deep fascia in the midportion of the palm. It is a direct continuation of the tendon of the palmaris longus.

2) the radial two-thirds of the palm is innervated by the median nerve, while the ulnar one-third is supplied by the ulna nerve.

3) the more superficial exposure of the vessels and nerves in the distal palm where the palmar aponeurosis is deficient.

Fig. 107: Muscles, Synovial Sheaths and Tendons Left Wrist and Palm

NOTE: 1) the long flexor tendons enter the palmar aspect of the hand beneath the flexor retinaculum through the carpal tunnel. The median nerve also traverses the tunnel but is not shown in this figure.

2) two synovial-lined tendon compartments pass from the forearm into the hand. The larger of the two (the ulnar bursa) contains the tendons of the superficial and deep flexor muscles, while the smaller one (the radial bursa) contains the tendon of the flexor pollicis longus. A third compartment containing the tendon of the flexor carpi radialis does not pass into the hand but terminates at the wrist.

3) although the ulnar bursa continues into the little finger, the sheaths of the index, middle and ring fingers are interrupted at the metacarpophalangeal joint.

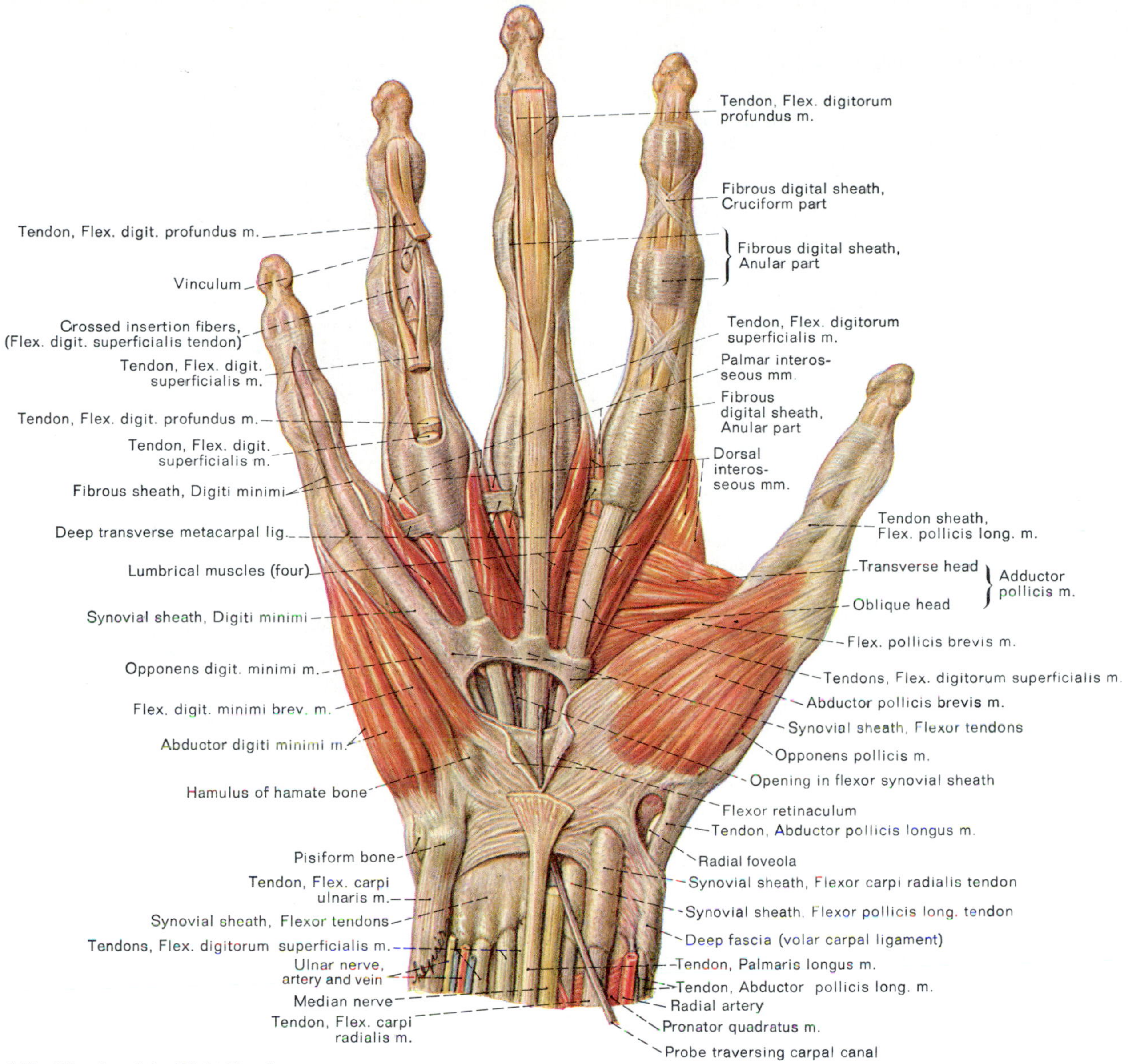

Tendon, Flex. digitorum profundus m.
Fibrous digital sheath, Cruciform part
Fibrous digital sheath, Anular part
Tendon, Flex. digitorum superficialis m.
Palmar interosseous mm.
Fibrous digital sheath, Anular part
Dorsal interosseous mm.
Tendon sheath, Flex. pollicis long. m.
Transverse head
Oblique head
Adductor pollicis m.
Flex. pollicis brevis m.
Tendons, Flex. digitorum superficialis m.
Abductor pollicis brevis m.
Synovial sheath, Flexor tendons
Opponens pollicis m.
Opening in flexor synovial sheath
Flexor retinaculum
Tendon, Abductor pollicis longus m.
Radial foveola
Synovial sheath, Flexor carpi radialis tendon
Synovial sheath, Flexor pollicis long. tendon
Deep fascia (volar carpal ligament)
Tendon, Palmaris longus m.
Tendon, Abductor pollicis long. m.
Radial artery
Pronator quadratus m.
Probe traversing carpal canal
Tendon, Flex. digit. profundus m.
Vinculum
Crossed insertion fibers, (Flex. digit. superficialis tendon)
Tendon, Flex. digit. superficialis m.
Tendon, Flex. digit. profundus m.
Tendon, Flex. digit. superficialis m.
Fibrous sheath, Digiti minimi
Deep transverse metacarpal lig.
Lumbrical muscles (four)
Synovial sheath, Digiti minimi
Opponens digit. minimi m.
Flex. digit. minimi brev. m.
Abductor digiti minimi m.
Hamulus of hamate bone
Pisiform bone
Tendon, Flex. carpi ulnaris m.
Synovial sheath, Flexor tendons
Tendons, Flex. digitorum superficialis m.
Ulnar nerve, artery and vein
Median nerve
Tendon, Flex. carpi radialis m.

Fig. 108: Muscles of the Right Hand

NOTE: 1) how the flexor digit. profundus achieves its insertion on the distal phalanx. The tendon of the flexor digit. superficialis divides into two slips, allowing the corresponding deep flexor tendon to pass.

2) that along the fingers, the tendons are encased in a synovial sheath and then they are bound by both crossed and transverse (cruciate and annular) fibrous sheaths.

3) the muscles of the thenar eminence: abductor pollicis brevis, flexor pollicis brevis and the underlying opponens pollicis muscle.

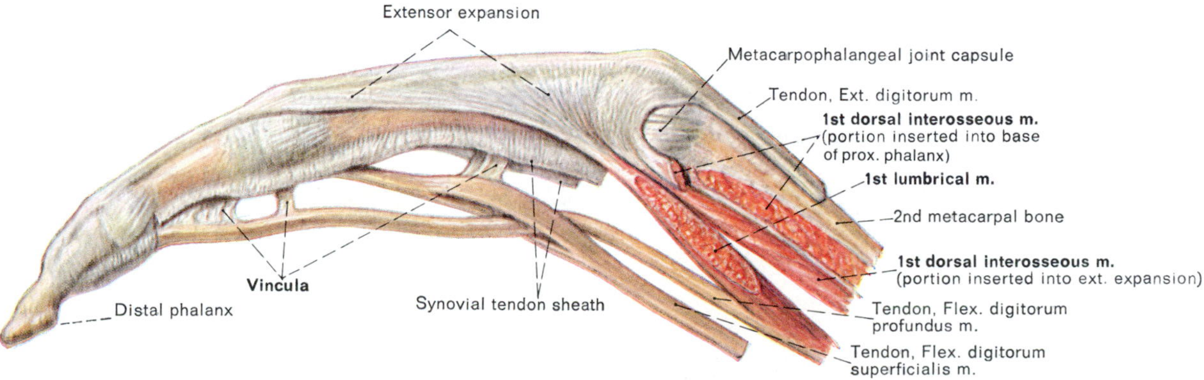
Extensor expansion
Metacarpophalangeal joint capsule
Tendon, Ext. digitorum m.
1st dorsal interosseous m.
(portion inserted into base of prox. phalanx)
1st lumbrical m.
2nd metacarpal bone
1st dorsal interosseous m.
(portion inserted into ext. expansion)
Tendon, Flex. digitorum profundus m.
Tendon, Flex. digitorum superficialis m.
Synovial tendon sheath
Vincula
Distal phalanx

Fig. 109: Tendon Insertions, Index Finger of Right Hand (Radial Side)

NOTE: 1) the flexor digit. superficialis inserts on the middle phalanx while the flexor digit. profundus inserts on the distal phalanx.

2) the dorsal interosseous and lumbrical muscles join the diverging extensor fibers in the formation of the extensor expansion on the dorsum of the finger.

3) the vincula are remnants of mesotendons and attach both superficial and deep flexor tendons to the digital sheath.

Fig. 110: Nerves and Arteries of the Left Palm, Superficial Palmar Arch

NOTE: 1) the median nerve enters the palm beneath the flexor retinaculum and supplies the muscles of the thenar eminence: abductor pollicis brevis, opponens pollicis and flexor pollicis brevis (superficial head). Additionally, it supplies the two most lateral lumbrical muscles and the palmar surface of the lateral three and one-half fingers.

2) the superficial location of the small but important "recurrent" branch of the median nerve which supplies several of the thenar muscles. Its location, just below the deep fascia on the thenar eminence, makes it vulnerable to injury.

3) the ulnar nerve enters the palm superficial to the flexor retinaculum, supplies the medial one and one-half fingers and all the remaining musculature of the hand: three hypothenar muscles, seven interosseous muscles, two medial lumbrical muscles, the adductor pollicis and the flexor pollicis brevis (deep head).

4) the superficial palmar arch is derived principally from the ulnar artery. It crosses the palm to the radial side superficial to the nerves and tendons and is joined by a palmar branch of the radial artery. Three or four common palmar digital arteries arise from the arch, proceed distally and divide into proper palmar digital arteries which course along the fingers with corresponding digital nerves.

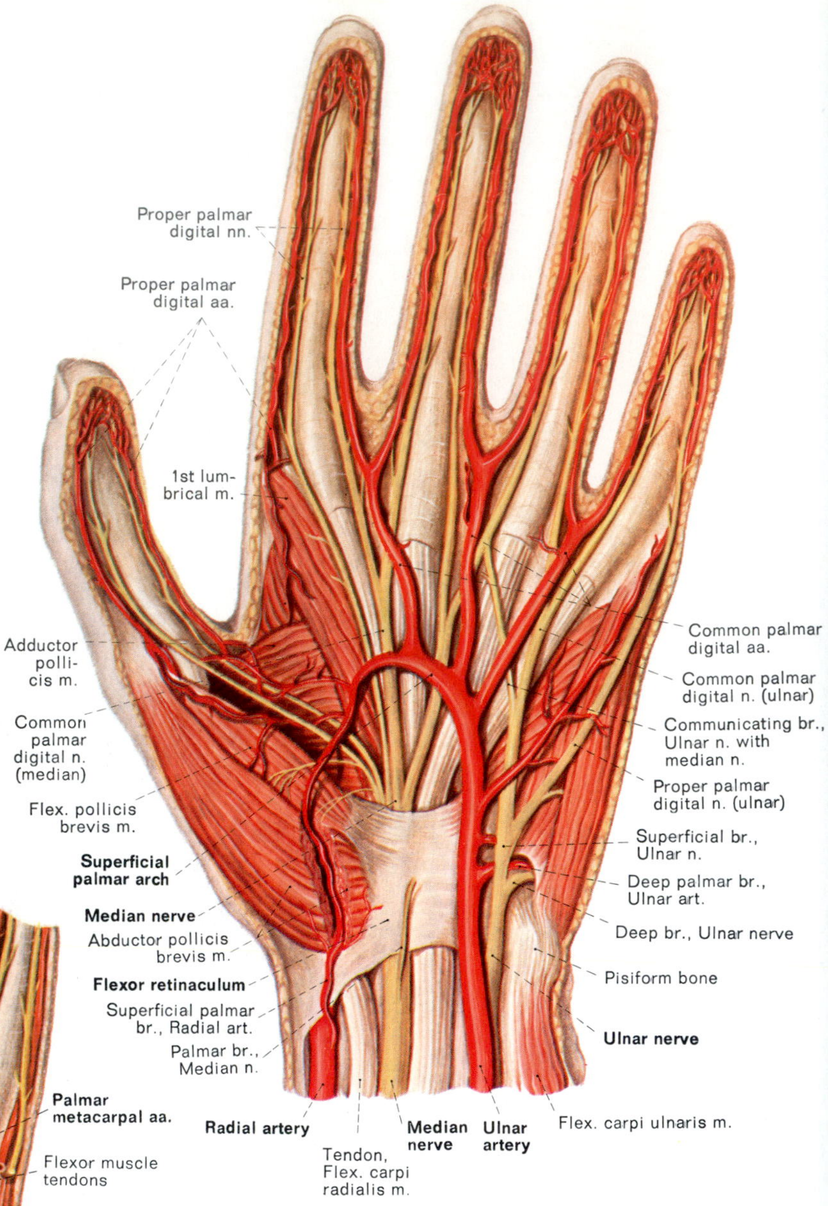

Proper palmar digital nn.
Proper palmar digital aa.
1st lumbrical m.
Adductor pollicis m.
Common palmar digital n. (median)
Flex. pollicis brevis m.
Superficial palmar arch
Median nerve
Abductor pollicis brevis m.
Flexor retinaculum
Superficial palmar br., Radial art.
Palmar br., Median n.
Radial artery
Tendon, Flex. carpi radialis m.
Median nerve
Ulnar artery
Flex. carpi ulnaris m.
Common palmar digital aa.
Common palmar digital n. (ulnar)
Communicating br., Ulnar n. with median n.
Proper palmar digital n. (ulnar)
Superficial br., Ulnar n.
Deep palmar br., Ulnar art.
Deep br., Ulnar nerve
Pisiform bone
Ulnar nerve

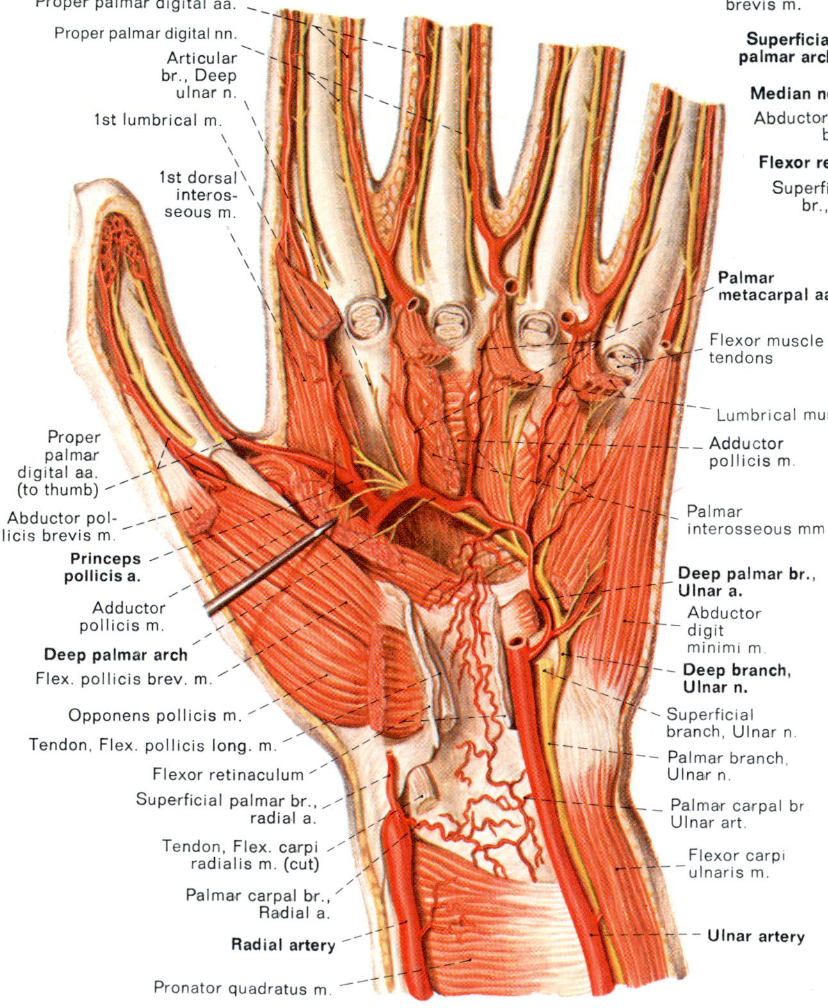

Proper palmar digital aa.
Proper palmar digital nn.
Articular br., Deep ulnar n.
1st lumbrical m.
1st dorsal interosseous m.
Proper palmar digital aa. (to thumb)
Abductor pollicis brevis m.
Princeps pollicis a.
Adductor pollicis m.
Deep palmar arch
Flex. pollicis brev. m.
Opponens pollicis m.
Tendon, Flex. pollicis long. m.
Flexor retinaculum
Superficial palmar br., radial a.
Tendon, Flex. carpi radialis m. (cut)
Palmar carpal br., Radial a.
Radial artery
Pronator quadratus m.
Palmar metacarpal aa.
Flexor muscle tendons
Lumbrical muscles
Adductor pollicis m.
Palmar interosseous mm
Deep palmar br., Ulnar a.
Abductor digit minimi m.
Deep branch, Ulnar n.
Superficial branch, Ulnar n.
Palmar branch, Ulnar n.
Palmar carpal br. Ulnar art.
Flexor carpi ulnaris m.
Ulnar artery

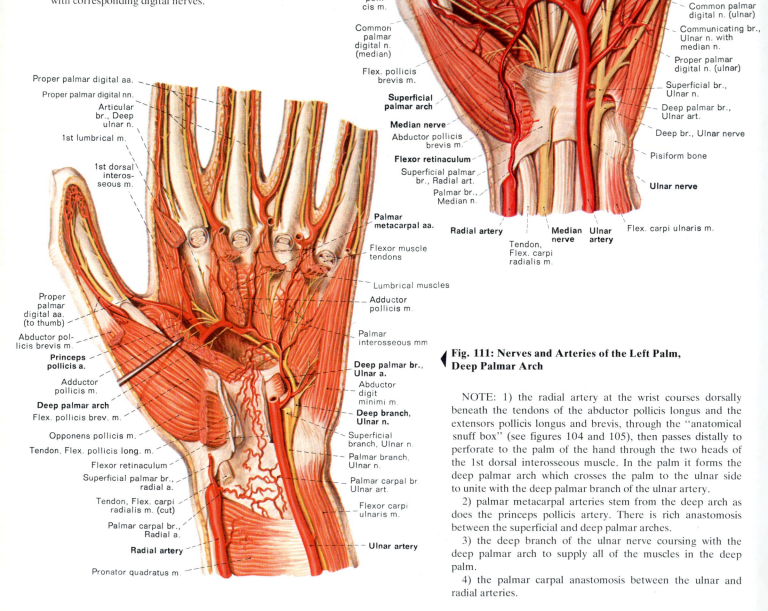

Fig. 111: Nerves and Arteries of the Left Palm, Deep Palmar Arch

NOTE: 1) the radial artery at the wrist courses dorsally beneath the tendons of the abductor pollicis longus and the extensors pollicis longus and brevis, through the "anatomical snuff box" (see figures 104 and 105), then passes distally to perforate to the palm of the hand through the two heads of the 1st dorsal interosseous muscle. In the palm it forms the deep palmar arch which crosses the palm to the ulnar side to unite with the deep palmar branch of the ulnar artery.

2) palmar metacarpal arteries stem from the deep arch as does the princeps pollicis artery. There is rich anastomosis between the superficial and deep palmar arches.

3) the deep branch of the ulnar nerve coursing with the deep palmar arch to supply all of the muscles in the deep palm.

4) the palmar carpal anastomosis between the ulnar and radial arteries.

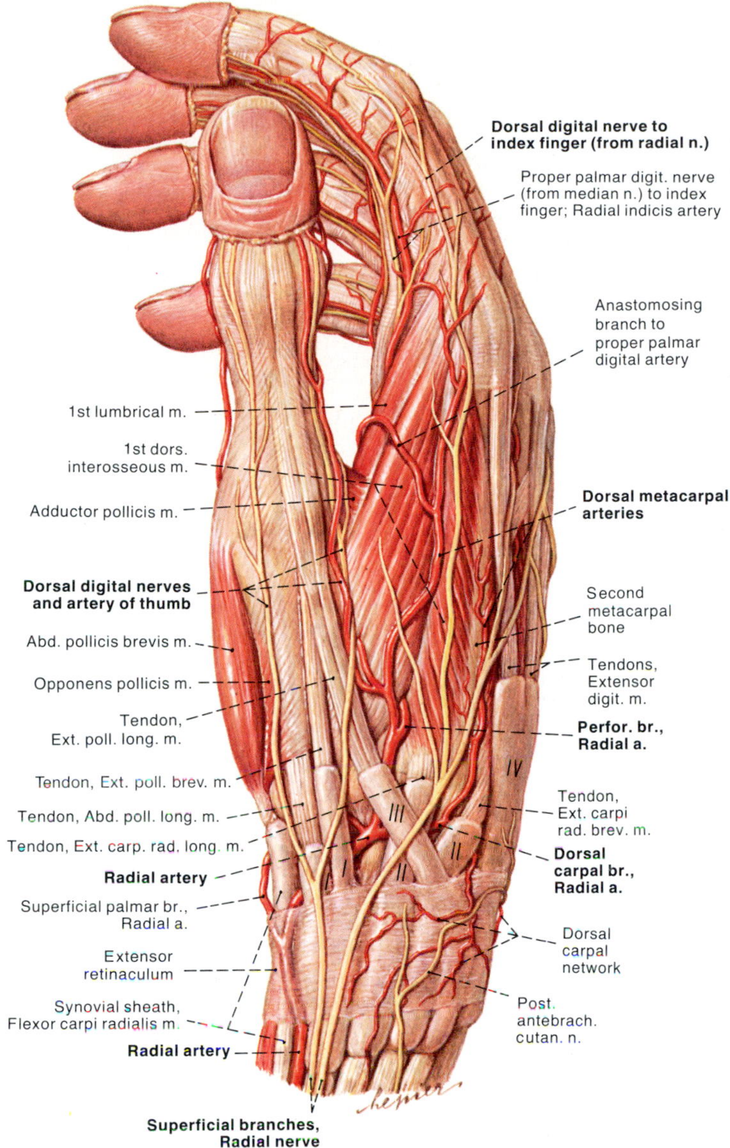

Dorsal digital nerve to index finger (from radial n.)
Proper palmar digit. nerve (from median n.) to index finger; Radial indicis artery
Anastomosing branch to proper palmar digital artery
1st lumbrical m.
1st dors. interosseous m.
Adductor pollicis m.
Dorsal metacarpal arteries
Dorsal digital nerves and artery of thumb
Second metacarpal bone
Abd. pollicis brevis m.
Tendons, Extensor digit. m.
Opponens pollicis m.
Tendon, Ext. poll. long. m.
Perfor. br., Radial a.
IV
Tendon, Ext. poll. brev. m.
Tendon, Ext. carpi rad. brev. m.
III
Tendon, Abd. poll. long. m.
Tendon, Ext. carp. rad. long. m.
II
Dorsal carpal br., Radial a.
I
Radial artery
Superficial palmar br., Radial a.
Dorsal carpal network
Extensor retinaculum
Post. antebrach. cutan. n.
Synovial sheath, Flexor carpi radialis m.
Radial artery
Superficial branches, Radial nerve

Fig. 112: Superficial Nerves, Arteries and Tendons on the Radial Aspect of the Right Hand

I–IV = Synovial tendon sheaths
I = Abductor pollicis longus and extensor pollicis brevis tendon sheaths
II = Extensor carpi radialis longus and brevis tendon sheaths
III = Extensor pollicis longus tendon sheath
IV = Extensor digitorum and extensor indicis tendon sheath

NOTE: 1) that only the skin and superficial fascia have been removed in the dissection and the cutaneous nerves and superficial arteries to the thumb, radial side of the index finger, and dorsum of the hand have been preserved.

2) the superficial branches of the radial nerve to the hand (see also Figs. 41, 76 and 102). These supply the dorsum of the thumb, nearly to the tip, as well as the lateral (radial) half of the dorsum of the hand. Additionally, the radial nerve supplies the proximal part of the dorsum of the index, middle, and lateral half of the ring fingers, as far as the proximal interphalangeal joint, since the median nerve sends branches around the digits to supply the more distal parts of the fingers.

3) the distribution of the radial artery to the thumb and dorsum of the hand (see also Figs. 104 and 105). Observe the dorsal digital branch to the thumb, the radial indicis branch to the index finger, the perforating branch that penetrates between the two heads of the first dorsal interosseous muscle and the dorsal carpal branch from which the dorsal metacarpal arteries arise.

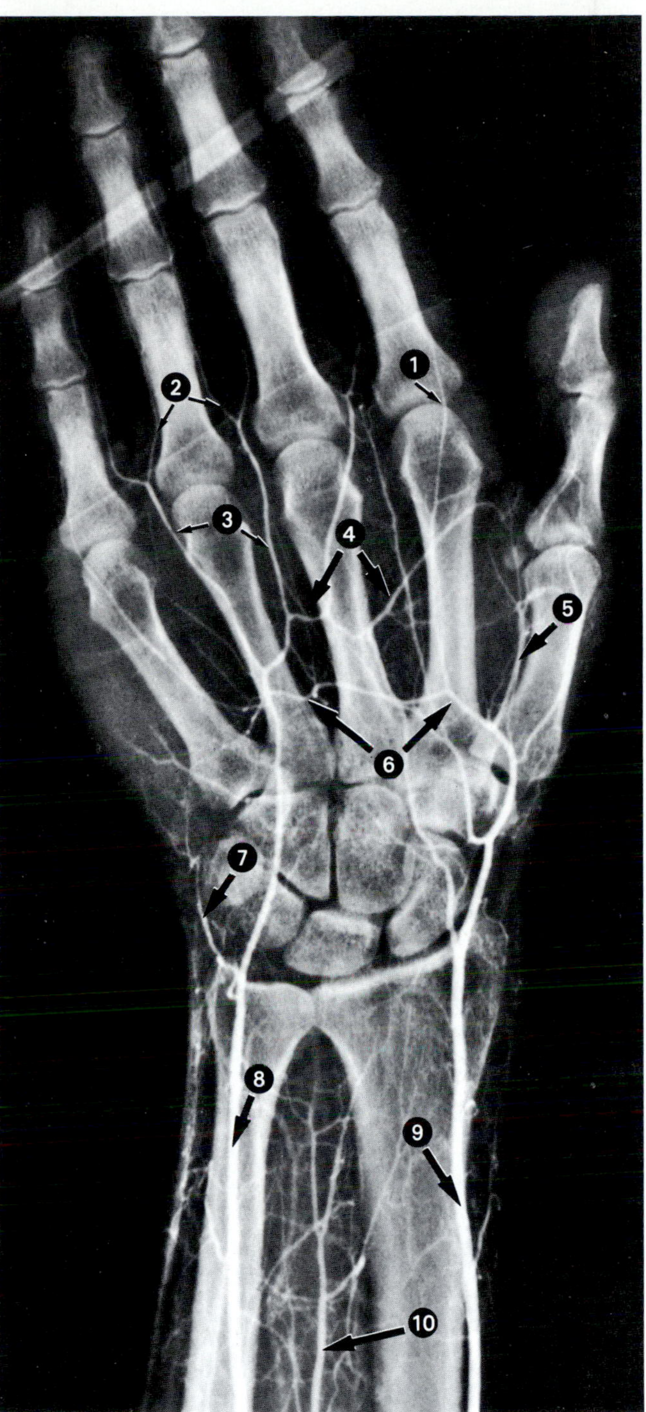

1
2
3
4
5
6
7
8
9
10

Fig. 113: An Arteriogram of the Hand

1. Radial indicis artery
2. Proper palmar digital arteries
3. Common palmar digital arteries
4. Superficial palmar arch
5. Princeps pollicis artery
6. Deep palmar arch
7. Deep palmar branch of ulnar artery
8. Ulnar artery
9. Radial artery
10. Anterior interosseous artery

NOTE: 1) that the radial artery (9), after crossing the palm to form the deep palmar arch (6), anastomoses with the ulnar artery (8) near the base of the 4th matacarpal bone. Frequently, this anastomosis occurs more medially near the base of the 5th metacarpal bone.

2) that the ulnar artery (8) enters the hand medial to the pisiform bone and hamulus of the hamate bone and then descends to about the mid-metacarpal level before turning radially to form the superficial palmar arch (4) (From Wicke, L., *Atlas of Radiologic Anatomy,* 3rd Edition, Urban & Schwarzenberg, Baltimore, 1982).

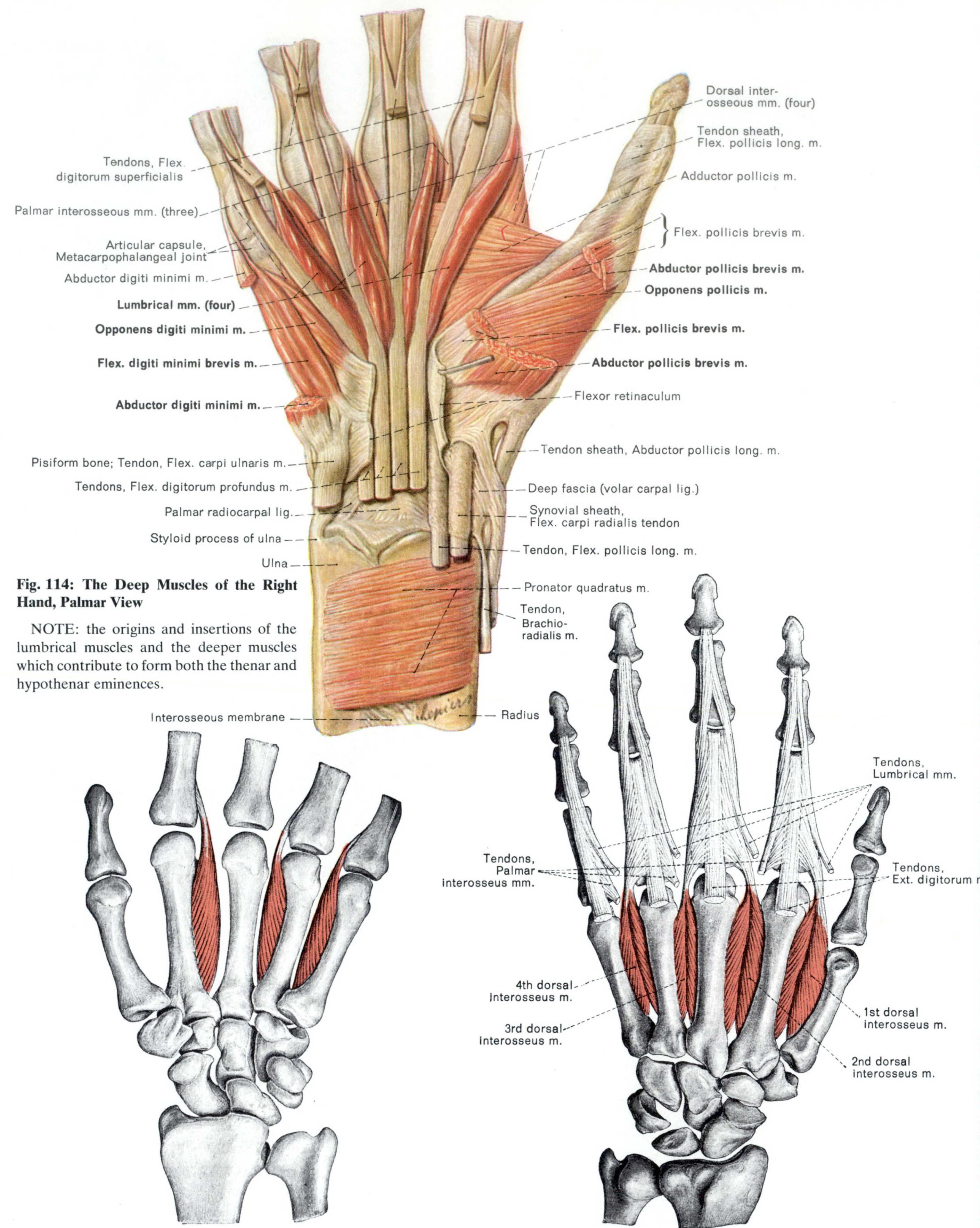

Dorsal inter-
osseous mm. (four)
Tendon sheath,
Flex. pollicis long. m.
Adductor pollicis m.
Flex. pollicis brevis m.
Abductor pollicis brevis m.
Opponens pollicis m.
Flex. pollicis brevis m.
Abductor pollicis brevis m.
Flexor retinaculum
Tendon sheath, Abductor pollicis long. m.
Deep fascia (volar carpal lig.)
Synovial sheath,
Flex. carpi radialis tendon
Tendon, Flex. pollicis long. m.
Pronator quadratus m.
Tendon,
Brachio-
radialis m.
Radius
Tendons, Flex.
digitorum superficialis
Palmar interosseous mm. (three)
Articular capsule,
Metacarpophalangeal joint
Abductor digiti minimi m.
Lumbrical mm. (four)
Opponens digiti minimi m.
Flex. digiti minimi brevis m.
Abductor digiti minimi m.
Pisiform bone; Tendon, Flex. carpi ulnaris m.
Tendons, Flex. digitorum profundus m.
Palmar radiocarpal lig.
Styloid process of ulna
Ulna
Interosseous membrane
Tendons,
Lumbrical mm.
Tendons,
Palmar
interosseus mm.
Tendons,
Ext. digitorum m
4th dorsal
interosseus m.
3rd dorsal
interosseus m.
1st dorsal
interosseus m.
2nd dorsal
interosseus m.

Fig. 114: The Deep Muscles of the Right Hand, Palmar View

NOTE: the origins and insertions of the lumbrical muscles and the deeper muscles which contribute to form both the thenar and hypothenar eminences.

Fig. 115: The Three Palmar Interosseus Muscles (left, Palmar View)

Fig. 116: The Four Dorsal Interosseus Muscles (left, Dorsal View)

NOTE: whereas the three palmar interosseus muscles (figure 115) are adductors of the fingers, the four dorsal interosseus muscles (figure 116) are abductors. All of the interossei flex the metacarpophalangeal joint and extend the interphalangeal joints, and they are all supplied by the ulnar nerve.

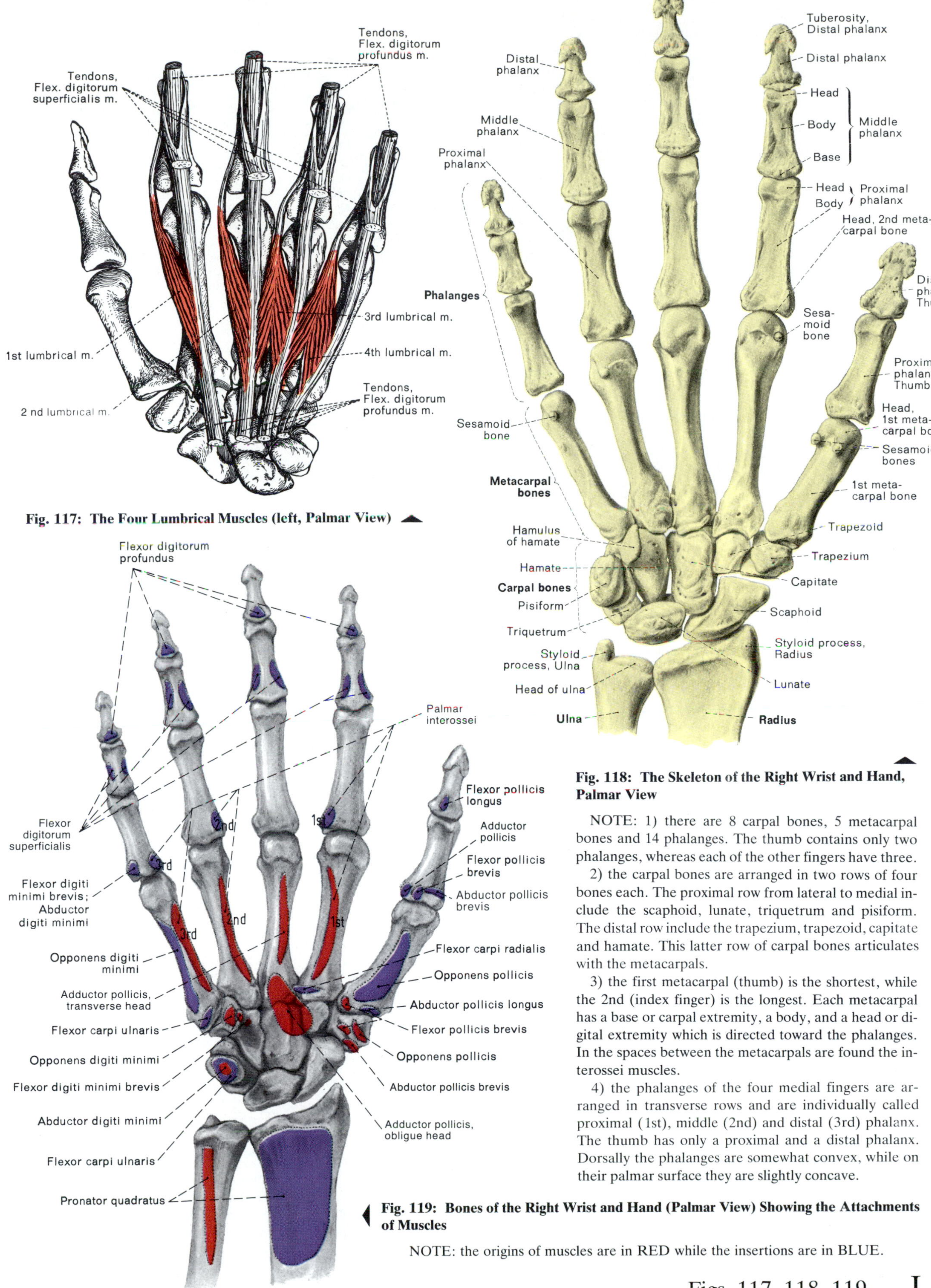

Tendons, Flex. digitorum profundus m.
Tendons, Flex. digitorum superficialis m.
3rd lumbrical m.
4th lumbrical m.
1st lumbrical m.
2 nd lumbrical m.
Tendons, Flex. digitorum profundus m.
Tuberosity, Distal phalanx
Distal phalanx
Head
Body
Base
Middle phalanx
Distal phalanx
Middle phalanx
Proximal phalanx
Head
Body
Proximal phalanx
Head, 2nd metacarpal bone
Distal phalanx, Thumb
Phalanges
Sesamoid bone
Proximal phalanx, Thumb
Sesamoid bone
Head, 1st metacarpal bone
Sesamoid bones
Metacarpal bones
1st metacarpal bone
Hamulus of hamate
Trapezoid
Hamate
Trapezium
Carpal bones
Capitate
Pisiform
Scaphoid
Triquetrum
Styloid process, Ulna
Styloid process, Radius
Head of ulna
Lunate
Ulna
Radius
Flexor digitorum profundus
Palmar interossei
Flexor pollicis longus
Flexor digitorum superficialis
Adductor pollicis
Flexor pollicis brevis
2nd
1st
3rd
Abductor pollicis brevis
Flexor digiti minimi brevis; Abductor digiti minimi
2nd
1st
3rd
Flexor carpi radialis
Opponens digiti minimi
Opponens pollicis
Adductor pollicis, transverse head
Abductor pollicis longus
Flexor carpi ulnaris
Flexor pollicis brevis
Opponens digiti minimi
Opponens pollicis
Flexor digiti minimi brevis
Abductor pollicis brevis
Abductor digiti minimi
Adductor pollicis, obligue head
Flexor carpi ulnaris
Pronator quadratus

Fig. 117: The Four Lumbrical Muscles (left, Palmar View)

Fig. 118: The Skeleton of the Right Wrist and Hand, Palmar View

NOTE: 1) there are 8 carpal bones, 5 metacarpal bones and 14 phalanges. The thumb contains only two phalanges, whereas each of the other fingers have three.

2) the carpal bones are arranged in two rows of four bones each. The proximal row from lateral to medial include the scaphoid, lunate, triquetrum and pisiform. The distal row include the trapezium, trapezoid, capitate and hamate. This latter row of carpal bones articulates with the metacarpals.

3) the first metacarpal (thumb) is the shortest, while the 2nd (index finger) is the longest. Each metacarpal has a base or carpal extremity, a body, and a head or digital extremity which is directed toward the phalanges. In the spaces between the metacarpals are found the interossei muscles.

4) the phalanges of the four medial fingers are arranged in transverse rows and are individually called proximal (1st), middle (2nd) and distal (3rd) phalanx. The thumb has only a proximal and a distal phalanx. Dorsally the phalanges are somewhat convex, while on their palmar surface they are slightly concave.

Fig. 119: Bones of the Right Wrist and Hand (Palmar View) Showing the Attachments of Muscles

NOTE: the origins of muscles are in RED while the insertions are in BLUE.

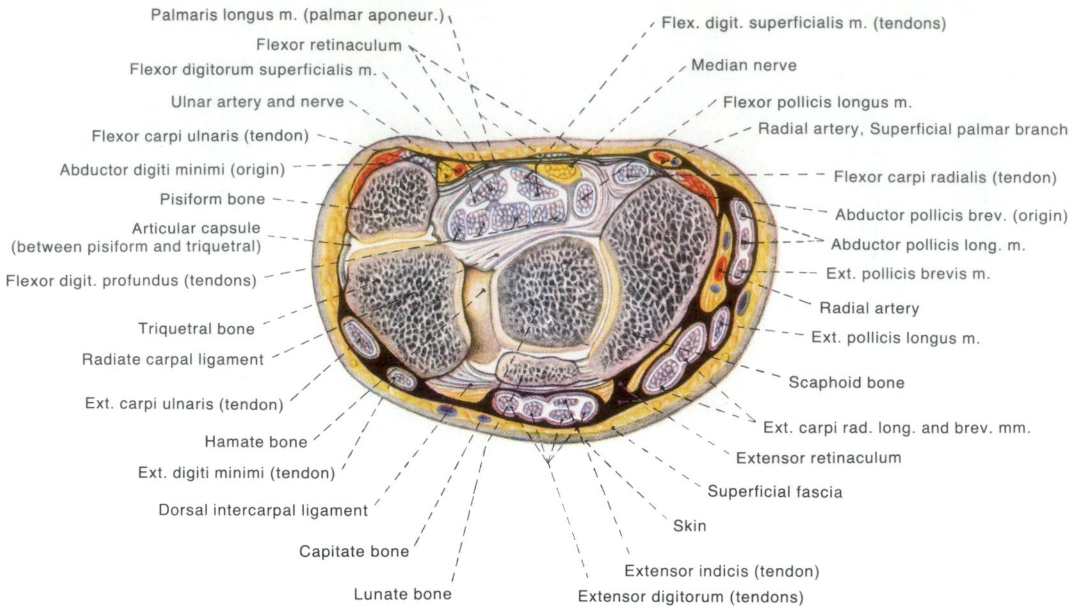

Palmaris longus m. (palmar aponeur.)
Flexor retinaculum
Flexor digitorum superficialis m.
Ulnar artery and nerve
Flexor carpi ulnaris (tendon)
Abductor digiti minimi (origin)
Pisiform bone
Articular capsule (between pisiform and triquetral)
Flexor digit. profundus (tendons)
Triquetral bone
Radiate carpal ligament
Ext. carpi ulnaris (tendon)
Hamate bone
Ext. digiti minimi (tendon)
Dorsal intercarpal ligament
Capitate bone
Lunate bone
Flex. digit. superficialis m. (tendons)
Median nerve
Flexor pollicis longus m.
Radial artery, Superficial palmar branch
Flexor carpi radialis (tendon)
Abductor pollicis brev. (origin)
Abductor pollicis long. m.
Ext. pollicis brevis m.
Radial artery
Ext. pollicis longus m.
Scaphoid bone
Ext. carpi rad. long. and brev. mm.
Extensor retinaculum
Superficial fascia
Skin
Extensor indicis (tendon)
Extensor digitorum (tendons)

Fig. 120: Cross Section Through the Hand at the Level of the First Row of Carpal Bones

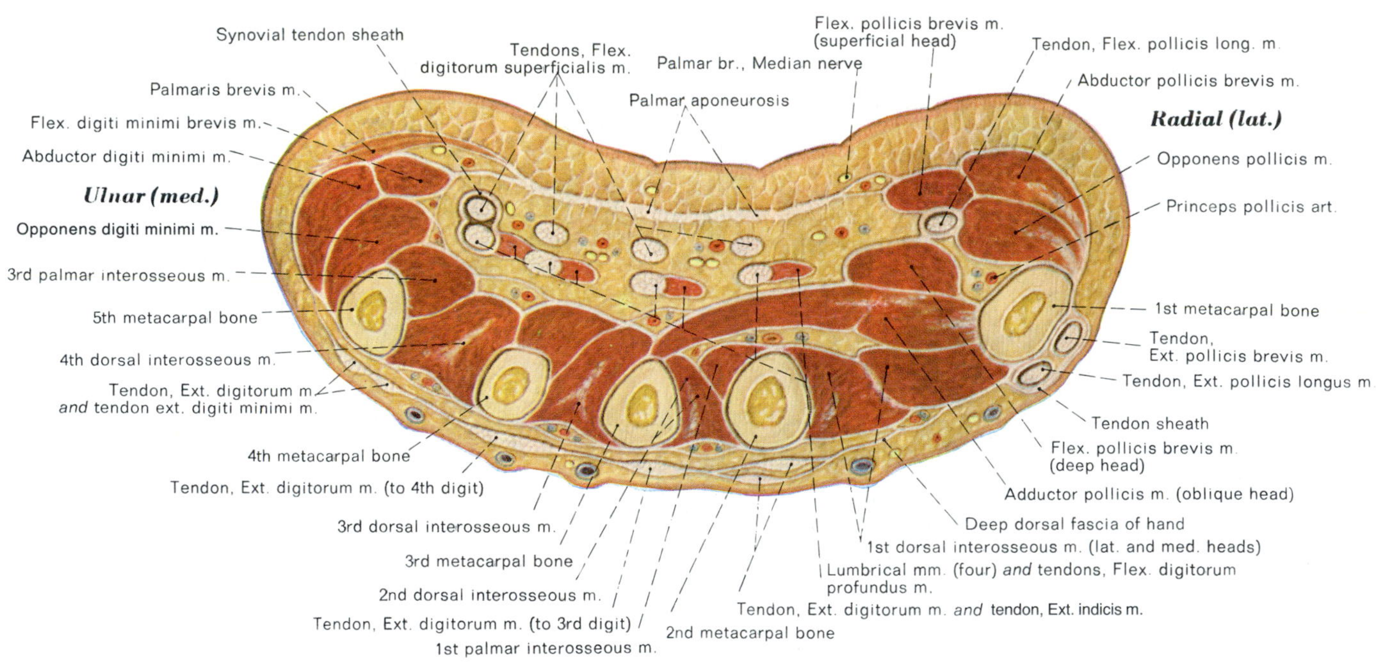

Synovial tendon sheath
Tendons, Flex. digitorum superficialis m.
Palmar br., Median nerve
Flex. pollicis brevis m. (superficial head)
Tendon, Flex. pollicis long. m.
Abductor pollicis brevis m.
Radial (lat.)
Palmaris brevis m.
Palmar aponeurosis
Flex. digiti minimi brevis m.
Abductor digiti minimi m.
Ulnar (med.)
Opponens digiti minimi m.
3rd palmar interosseous m.
5th metacarpal bone
4th dorsal interosseous m.
Tendon, Ext. digitorum m. and tendon ext. digiti minimi m.
4th metacarpal bone
Tendon, Ext. digitorum m. (to 4th digit)
3rd dorsal interosseous m.
3rd metacarpal bone
2nd dorsal interosseous m.
Tendon, Ext. digitorum m. (to 3rd digit)
1st palmar interosseous m.
2nd metacarpal bone
Opponens pollicis m.
Princeps pollicis art.
1st metacarpal bone
Tendon, Ext. pollicis brevis m.
Tendon, Ext. pollicis longus m.
Tendon sheath
Flex. pollicis brevis m. (deep head)
Adductor pollicis m. (oblique head)
Deep dorsal fascia of hand
1st dorsal interosseous m. (lat. and med. heads)
Lumbrical mm. (four) and tendons, Flex. digitorum profundus m.
Tendon, Ext. digitorum m. and tendon, Ext. indicis m.

Fig. 121: Cross Section of the Right Hand Through the Metacarpal Bones

IDENTIFY a) the four dorsal interossei which act as abductors of the fingers and which fill the intervals between the metacarpal bones, b) the three palmar interossei which act as adductors of the fingers, c) the thenar and hypothenar muscles and d) the tendons and lumbrical muscles in the palmar compartment.

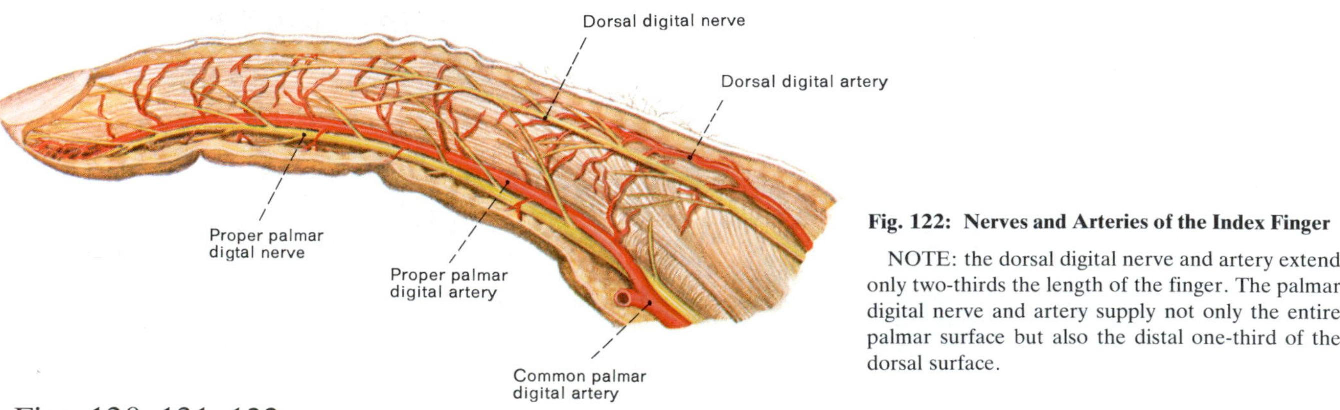

Dorsal digital nerve
Dorsal digital artery
Proper palmar digtal nerve
Proper palmar digital artery
Common palmar digital artery

Fig. 122: Nerves and Arteries of the Index Finger

NOTE: the dorsal digital nerve and artery extend only two-thirds the length of the finger. The palmar digital nerve and artery supply not only the entire palmar surface but also the distal one-third of the dorsal surface.

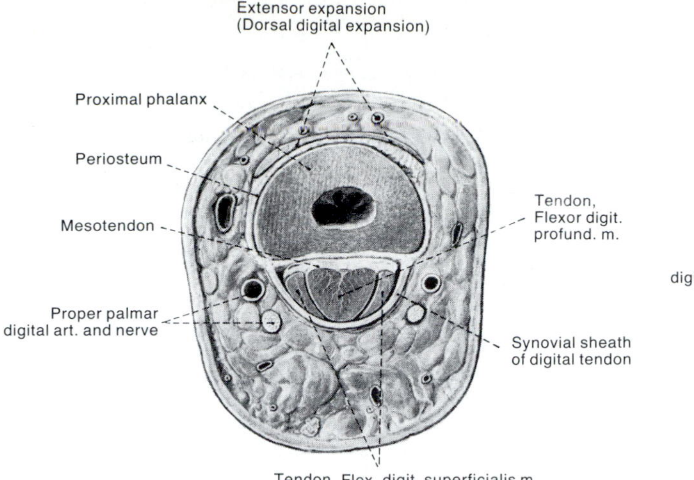
Extensor expansion
(Dorsal digital expansion)
Proximal phalanx
Periosteum
Mesotendon
Proper palmar
digital art. and nerve
Tendon,
Flexor digit.
profund. m.
Synovial sheath
of digital tendon
Tendon, Flex. digit. superficialis m.

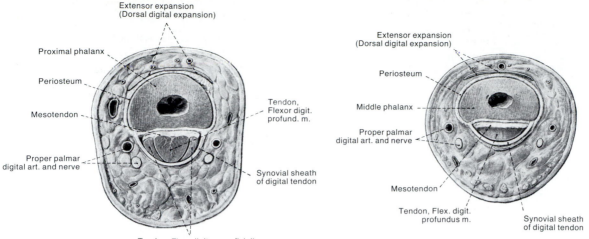

Fig. 123: Cross Section of the Middle Finger Through the Proximal Phalanx

NOTE: 1) the extensor expansion (also called dorsal digital expansion or extensor hood) over the dorsal aspect of the proximal phalanx and part of the middle phalanx. Into this triangular-shaped aponeurosis blend the digital tendon of the extensor digitorum, as well as the tendons of insertion of the adjacent interossous muscles and those of the lumbrical muscles (see also Fig. 109)

2) the synovial sheath on the palmar aspect of the phalanx and the disposition of the superficial and deep flexor tendons of the digit.

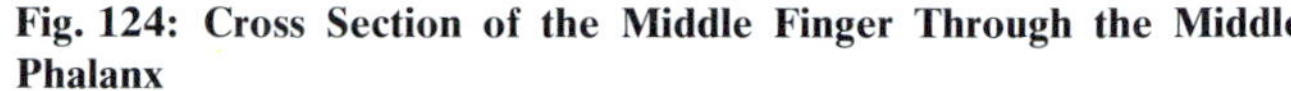
Extensor expansion
(Dorsal digital expansion)
Periosteum
Middle phalanx
Proper palmar
digital art. and nerve
Mesotendon
Tendon, Flex. digit.
profundus m.
Synovial sheath
of digital tendon

Fig. 124: Cross Section of the Middle Finger Through the Middle Phalanx

NOTE: 1) the location of the proper palmar digital arteries and nerves within the subcutaneous fibrous tissue on the lateral aspects of the deep flexor tendon (see also Figs. 110, 122 and 123). Knowing the location of these structures is important both for the application of local anesthesia to the digit and for the cessation of severe bleeding (hemostasis).

2) that the tendon of only the flexor digitorum profundus is present on the palmar aspect of the middle phalanx (see also Figs. 108, 109).

Free margin

Body of nail
(unguis)

Cuticle
(eponychium)

Lunula

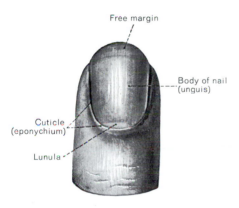

Fig. 125: Finger Nail, Normal Position (Dorsal View)

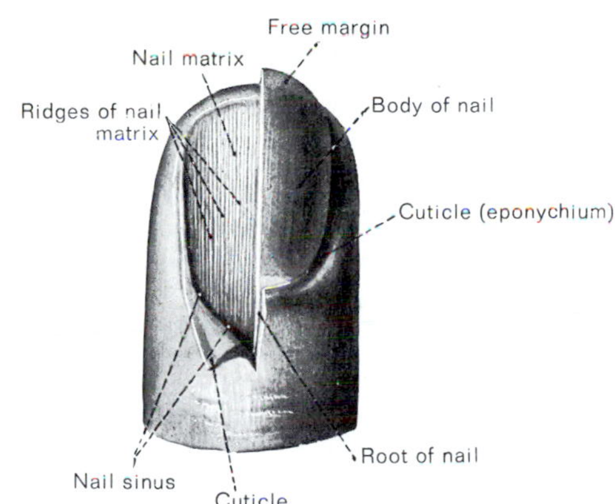
Free margin
Nail matrix
Ridges of nail
matrix
Body of nail
Cuticle (eponychium)
Root of nail
Nail sinus
Cuticle

Fig. 126: Left Half of Finger Nail Bed Exposed

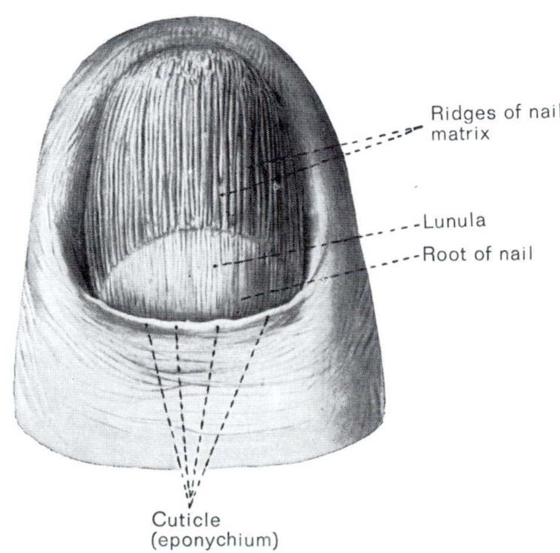
Ridges of nail
matrix
Lunula
Root of nail
Cuticle
(eponychium)

Fig. 127: Nail Bed of Thumb after Removal of Nail

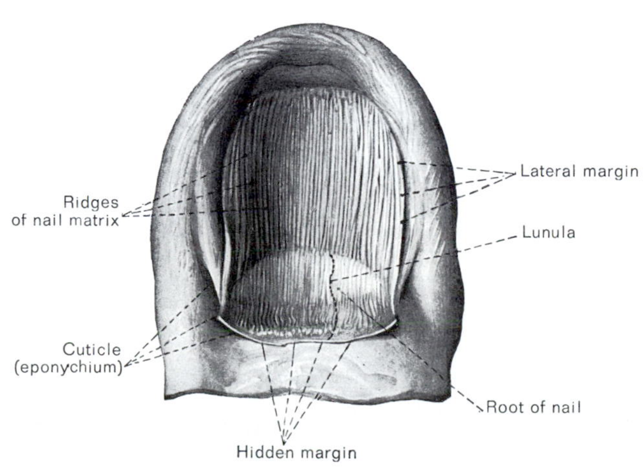
Lateral margin
Ridges
of nail matrix
Lunula
Cuticle
(eponychium)
Root of nail
Hidden margin

Fig. 128: Nail Bed of Thumb and Reflection of Cuticle

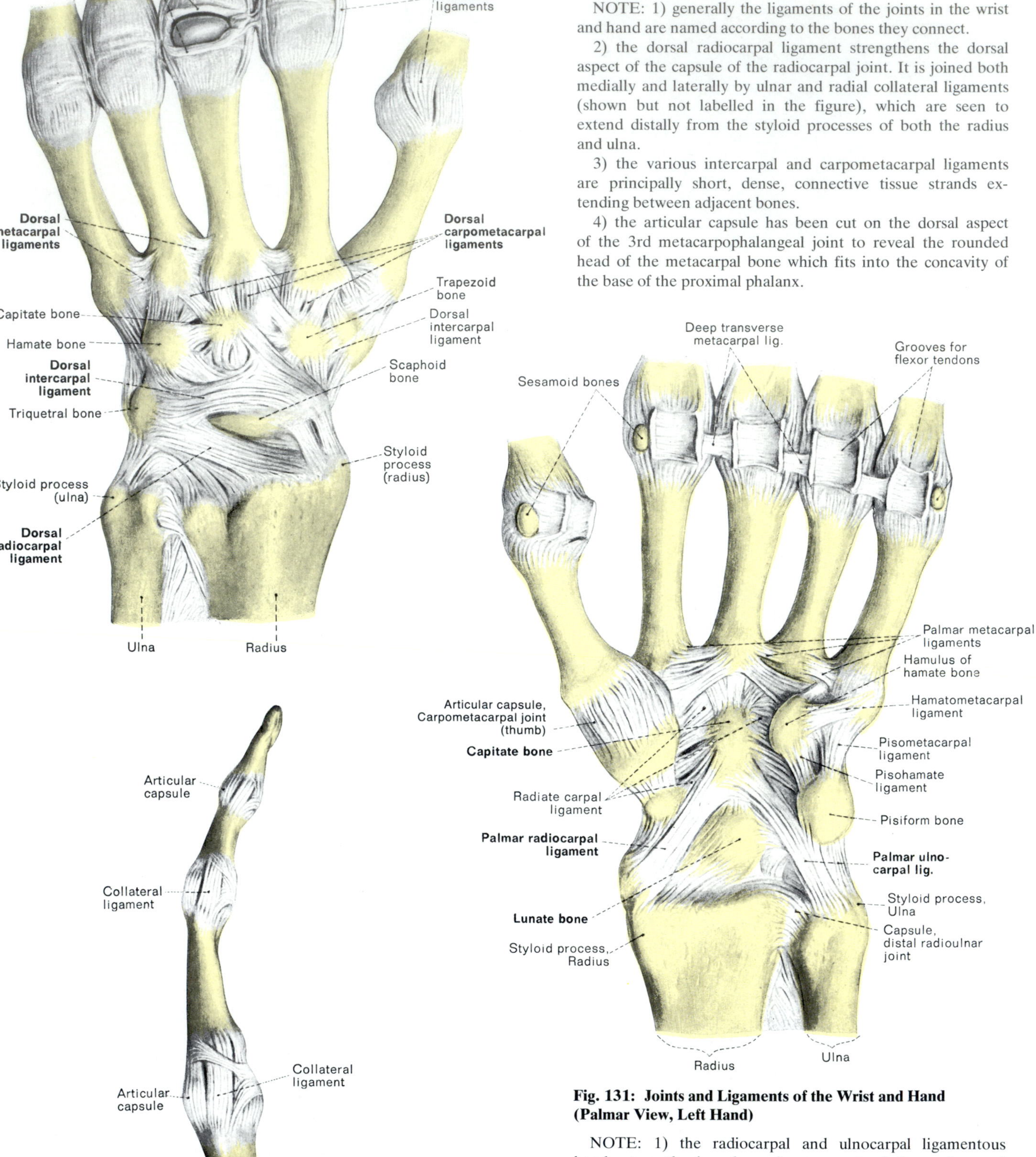

Articular capsule,
3rd metacarpophalangeal joint
Collateral
ligaments
Dorsal
metacarpal
ligaments
Dorsal
carpometacarpal
ligaments
Trapezoid
bone
Capitate bone
Dorsal
intercarpal
ligament
Hamate bone
Dorsal
intercarpal
ligament
Scaphoid
bone
Triquetral bone
Styloid
process
(radius)
Styloid process
(ulna)
Dorsal
radiocarpal
ligament
Ulna
Radius
Deep transverse
metacarpal lig.
Grooves for
flexor tendons
Sesamoid bones
Palmar metacarpal
ligaments
Hamulus of
hamate bone
Articular capsule,
Carpometacarpal joint
(thumb)
Hamatometacarpal
ligament
Capitate bone
Pisometacarpal
ligament
Pisohamate
ligament
Radiate carpal
ligament
Pisiform bone
Palmar radiocarpal
ligament
Palmar ulno-
carpal lig.
Styloid process,
Ulna
Lunate bone
Capsule,
distal radioulnar
joint
Styloid process,
Radius
Radius
Ulna
Articular
capsule
Collateral
ligament
Collateral
ligament
Articular
capsule

Fig. 129: Joints and Ligaments of the Wrist and Hand (Dorsal View, Left Hand)

NOTE: 1) generally the ligaments of the joints in the wrist and hand are named according to the bones they connect.

2) the dorsal radiocarpal ligament strengthens the dorsal aspect of the capsule of the radiocarpal joint. It is joined both medially and laterally by ulnar and radial collateral ligaments (shown but not labelled in the figure), which are seen to extend distally from the styloid processes of both the radius and ulna.

3) the various intercarpal and carpometacarpal ligaments are principally short, dense, connective tissue strands extending between adjacent bones.

4) the articular capsule has been cut on the dorsal aspect of the 3rd metacarpophalangeal joint to reveal the rounded head of the metacarpal bone which fits into the concavity of the base of the proximal phalanx.

Fig. 130: Joints and Ligaments of the Middle Finger

NOTE: the articular capsules of the metacarpophalangeal and interphalangeal joints are strengthened by longitudinally oriented collateral ligaments.

Fig. 131: Joints and Ligaments of the Wrist and Hand (Palmar View, Left Hand)

NOTE: 1) the radiocarpal and ulnocarpal ligamentous bands strengthening the palmar aspect of the radiocarpal joint.

2) identify several strong ligaments in the palmar hand including the pisohamate, pisometacarpal and the radiate ligament surrounding the capitate bone.

3) the bases of the metacarpal bones are joined by the palmar metacarpal ligaments, while the more distal heads of these bones are interconnected by the deep transverse metacarpal ligament.

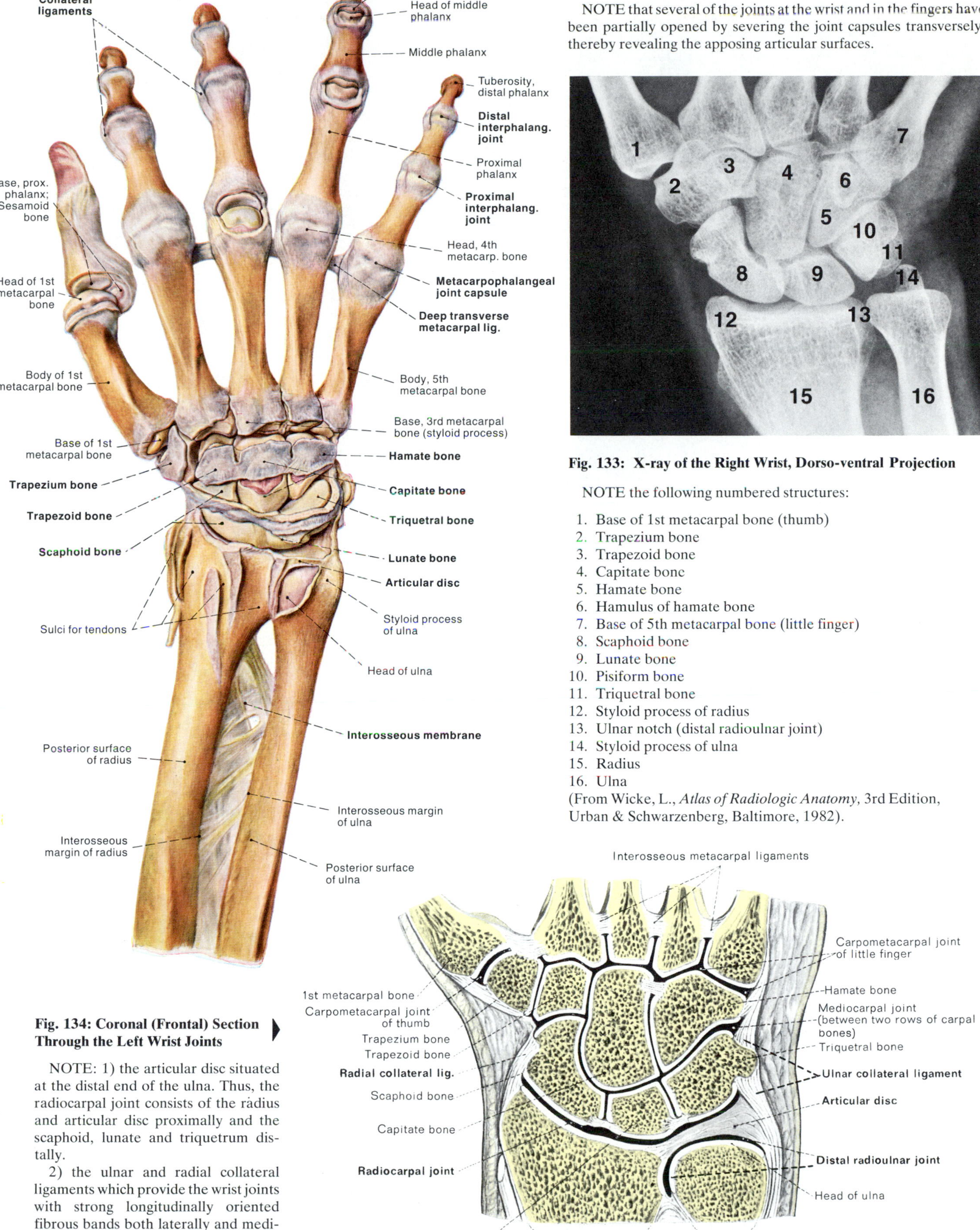

Base of distal phalanx
Collateral ligaments
Head of middle phalanx
Middle phalanx
Tuberosity, distal phalanx
Distal interphalang. joint
Proximal phalanx
Base, prox. phalanx; Sesamoid bone
Proximal interphalang. joint
Head, 4th metacarp. bone
Head of 1st metacarpal bone
Metacarpophalangeal joint capsule
Deep transverse metacarpal lig.
Body of 1st metacarpal bone
Body, 5th metacarpal bone
Base, 3rd metacarpal bone (styloid process)
Base of 1st metacarpal bone
Hamate bone
Trapezium bone
Capitate bone
Trapezoid bone
Triquetral bone
Scaphoid bone
Lunate bone
Articular disc
Styloid process of ulna
Sulci for tendons
Head of ulna
Interosseous membrane
Posterior surface of radius
Interosseous margin of ulna
Interosseous margin of radius
Posterior surface of ulna
1
2
3
4
5
6
7
8
9
10
11
12
13
14
15
16
Interosseous metacarpal ligaments
Carpometacarpal joint of little finger
Hamate bone
1st metacarpal bone
Carpometacarpal joint of thumb
Mediocarpal joint (between two rows of carpal bones)
Trapezium bone
Trapezoid bone
Triquetral bone
Radial collateral lig.
Ulnar collateral ligament
Scaphoid bone
Articular disc
Capitate bone
Distal radioulnar joint
Radiocarpal joint
Head of ulna
Lunate bone
Radius
Ulna

Fig. 132: Joints of the Right Wrist and Fingers, Viewed from the Dorsal Aspect

NOTE that several of the joints at the wrist and in the fingers have been partially opened by severing the joint capsules transversely, thereby revealing the apposing articular surfaces.

Fig. 133: X-ray of the Right Wrist, Dorso-ventral Projection

NOTE the following numbered structures:

1. Base of 1st metacarpal bone (thumb)
2. Trapezium bone
3. Trapezoid bone
4. Capitate bone
5. Hamate bone
6. Hamulus of hamate bone
7. Base of 5th metacarpal bone (little finger)
8. Scaphoid bone
9. Lunate bone
10. Pisiform bone
11. Triquetral bone
12. Styloid process of radius
13. Ulnar notch (distal radioulnar joint)
14. Styloid process of ulna
15. Radius
16. Ulna

(From Wicke, L., *Atlas of Radiologic Anatomy*, 3rd Edition, Urban & Schwarzenberg, Baltimore, 1982).

Fig. 134: Coronal (Frontal) Section Through the Left Wrist Joints

NOTE: 1) the articular disc situated at the distal end of the ulna. Thus, the radiocarpal joint consists of the radius and articular disc proximally and the scaphoid, lunate and triquetrum distally.

2) the ulnar and radial collateral ligaments which provide the wrist joints with strong longitudinally oriented fibrous bands both laterally and medially.

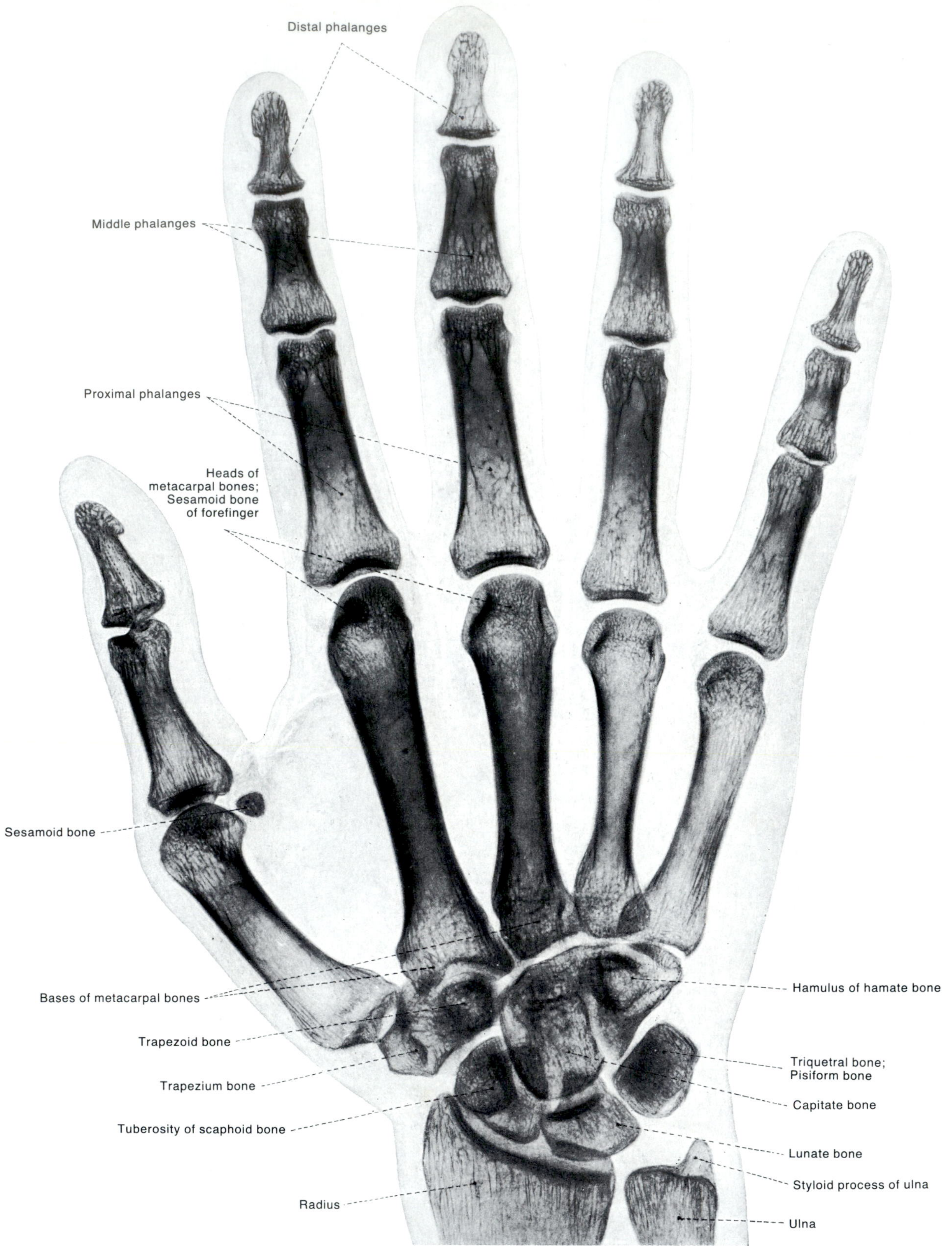

Distal phalanges
Middle phalanges
Proximal phalanges
Heads of metacarpal bones; Sesamoid bone of forefinger
Sesamoid bone
Bases of metacarpal bones
Trapezoid bone
Trapezium bone
Tuberosity of scaphoid bone
Radius
Hamulus of hamate bone
Triquetral bone; Pisiform bone
Capitate bone
Lunate bone
Styloid process of ulna
Ulna

Fig. 135: X-ray on Adult Left Hand

NOTE: 1) the expanded distal surface of the radius and its articulation inferolaterally with the scaphoid bone and inferomedially with the lunate. The distal extremity of the ulna is lengthened by the styloid process and has an interposed fibrocartilaginous articular disc between it, the radius, and the triquetral bone;

2) the shadow of the pisiform overlies that of the triquetrum. Thus, the proximal row of carpal bones includes the scaphoid, lunate, and triquetrum (with its superimposed pisiform);

3) the articulation of the scaphoid with the trapezium and trapezoid distally and the lunate and capitate laterally;

4) the size and central location of the capitate bone within the wrist. In addition to the capitate, the distal row of carpal bones includes the trapezium, trapezoid, and hamate.

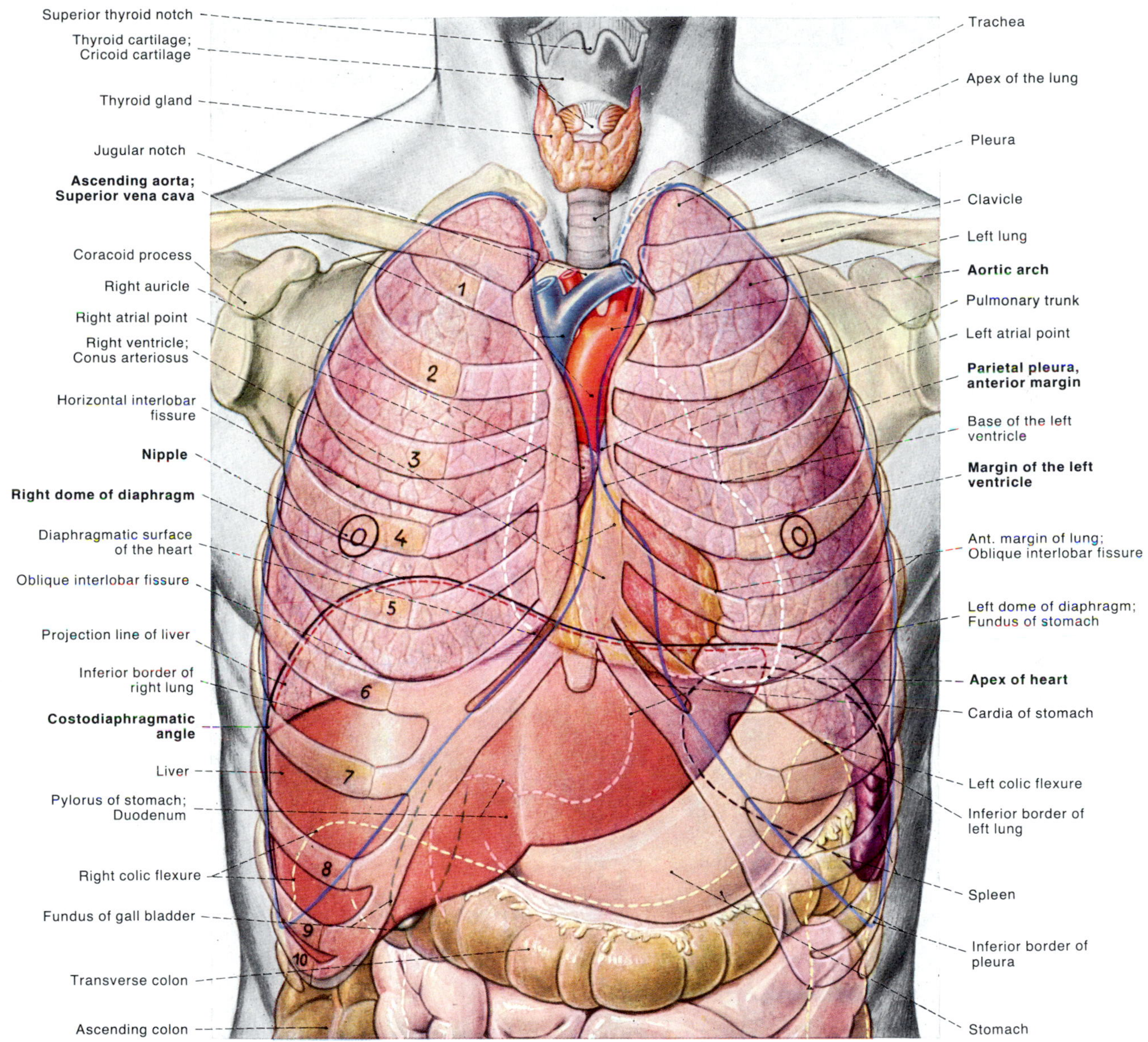
Superior thyroid notch
Thyroid cartilage; Cricoid cartilage
Thyroid gland
Jugular notch
Ascending aorta; Superior vena cava
Coracoid process
Right auricle
Right atrial point
Right ventricle; Conus arteriosus
Horizontal interlobar fissure
Nipple
Right dome of diaphragm
Diaphragmatic surface of the heart
Oblique interlobar fissure
Projection line of liver
Inferior border of right lung
Costodiaphragmatic angle
Liver
Pylorus of stomach; Duodenum
Right colic flexure
Fundus of gall bladder
Transverse colon
Ascending colon
Trachea
Apex of the lung
Pleura
Clavicle
Left lung
Aortic arch
Pulmonary trunk
Left atrial point
Parietal pleura, anterior margin
Base of the left ventricle
Margin of the left ventricle
Ant. margin of lung; Oblique interlobar fissure
Left dome of diaphragm; Fundus of stomach
Apex of heart
Cardia of stomach
Left colic flexure
Inferior border of left lung
Spleen
Inferior border of pleura
Stomach
1
2
3
4
5
6
7
8
9
10

Fig. 136: Thoracic and Upper Abdominal Viscera Viewed from the Anterior Aspect

NOTE: 1) the outline of the heart and great vessels (white broken line) deep to the anteromedial portion of the lungs;

2) that the liver, lying below the diaphragm, extends superiorly as high as the fourth interspace on the right and to the fifth interspace on the left (red broken line);

3) the superficial position of the superior vena cava and ascending aorta just deep to the manubrium of the sternum in the superior mediastinum;

4) that an upper triangular region containing the great vessels and a lower triangular region overlying the heart (area of superficial cardiac dullness) are not covered by pleura.

5) the reflections of the pleura (blue lines) over both lungs. Observe that the anterior margins of the lung and pleura on the left side are indented to form the cardiac notch (see also Fig. 153 and 173).

6) the position of the nipple over the 4th rib (or 4th intercostal space) in the male and in the young female. Observe also the apex of the heart, oriented toward the left and deep to the 5th intercostal space (i.e. between the 5th and 6th ribs; see also Fig. 173)

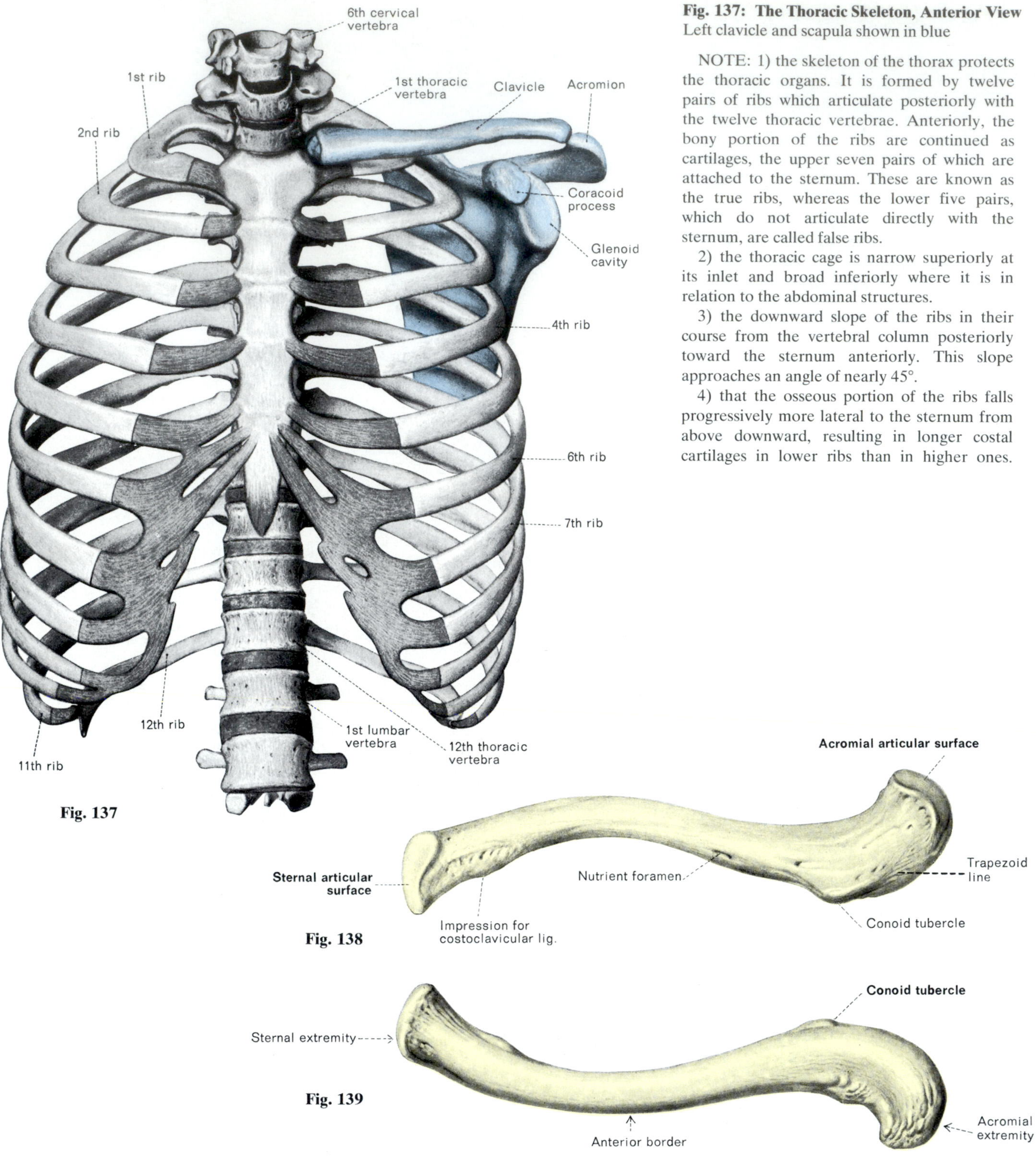
6th cervical vertebra
1st rib
1st thoracic vertebra
Clavicle
Acromion
2nd rib
Coracoid process
Glenoid cavity
4th rib
6th rib
7th rib
12th rib
1st lumbar vertebra
12th thoracic vertebra
11th rib
Acromial articular surface
Trapezoid line
Sternal articular surface
Nutrient foramen
Conoid tubercle
Impression for costoclavicular lig.
Conoid tubercle
Sternal extremity
Anterior border
Acromial extremity

Fig. 137

Fig. 138

Fig. 139

Fig. 137: The Thoracic Skeleton, Anterior View
Left clavicle and scapula shown in blue

NOTE: 1) the skeleton of the thorax protects the thoracic organs. It is formed by twelve pairs of ribs which articulate posteriorly with the twelve thoracic vertebrae. Anteriorly, the bony portion of the ribs are continued as cartilages, the upper seven pairs of which are attached to the sternum. These are known as the true ribs, whereas the lower five pairs, which do not articulate directly with the sternum, are called false ribs.

2) the thoracic cage is narrow superiorly at its inlet and broad inferiorly where it is in relation to the abdominal structures.

3) the downward slope of the ribs in their course from the vertebral column posteriorly toward the sternum anteriorly. This slope approaches an angle of nearly 45°.

4) that the osseous portion of the ribs falls progressively more lateral to the sternum from above downward, resulting in longer costal cartilages in lower ribs than in higher ones.

Fig. 138: The Left Clavicle, Inferior View

NOTE: 1) the clavicle is a double-curved bone which articulates medially with the sternum just above the first rib and laterally with the acromion of the scapula.

2) the inferior surface has roughened areas for the attachment of the costoclavicular ligament medially and the conoid and trapezoid fascicles of the coracoclavicular ligament laterally. The subclavius muscle attaches along the middle third of the inferior surface.

Fig. 139: The Left Clavicle, Superior View

NOTE: 1) the superior surface of the clavicle affords attachment of the pectoralis major and sternocleidomastoid muscles medially and the deltoid and trapezius muscles laterally.

2) the acromioclavicular articulation associates the clavicle with all the movements of the scapula, while the sternal articulation of the clavicle is a more secure and less movable joint.

Fig. 140: The Thoracic Skeleton, Posterior View
Left clavicle and scapula shown in blue

NOTE: 1) posteriorly, the skeleton of the thoracic cage consists of the twelve thoracic vertebrae and the posterior parts of the twelve pairs of ribs. Observe the mode of articulation of the ribs to the vertebral column. On an articulated skeleton, confirm that the extremity on the head of typical ribs possesses two articular facets separated by a crest (see Fig. 143: 8th rib). These two facets articulate with the bodies of two adjacent vertebrae, while the crest is attached to the intervertebral disc. The lower of these two facets articulates with the vertebra that numerically corresponds with the rib, and the more superior of the two facets articulates with the adjacent vertebra above.

2) that the scapula affords bony protection posteriorly to the upper lateral aspect of the thoracic cage, but the more medial portions of the intercostal spaces are protected only by soft tissues and, therefore, are vulnerable, as they are in the mid-axillary line and anteriorly. Recall that the medial (vertebral) border of the scapula is attached to the spinal column and thoracic cage by muscles and connective tissue only, and its articulations are to the clavicle and humerus, thus forming the pectoral girdle.

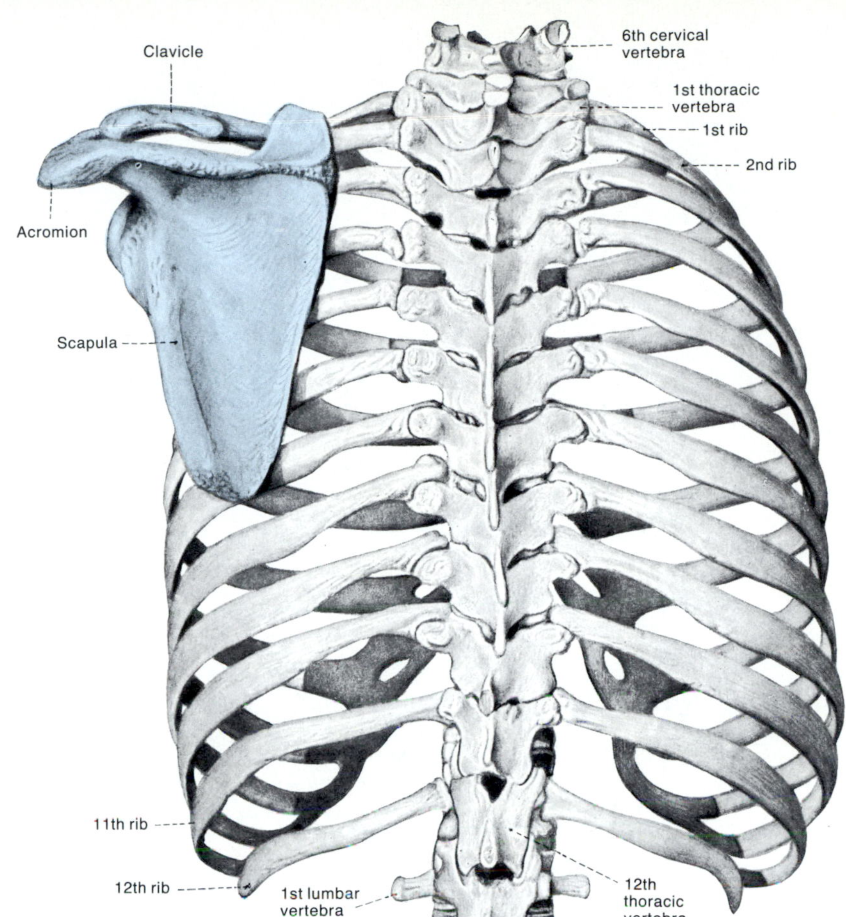
Clavicle
6th cervical vertebra
1st thoracic vertebra
1st rib
2nd rib
Acromion
Scapula
11th rib
12th rib
1st lumbar vertebra
12th thoracic vertebra

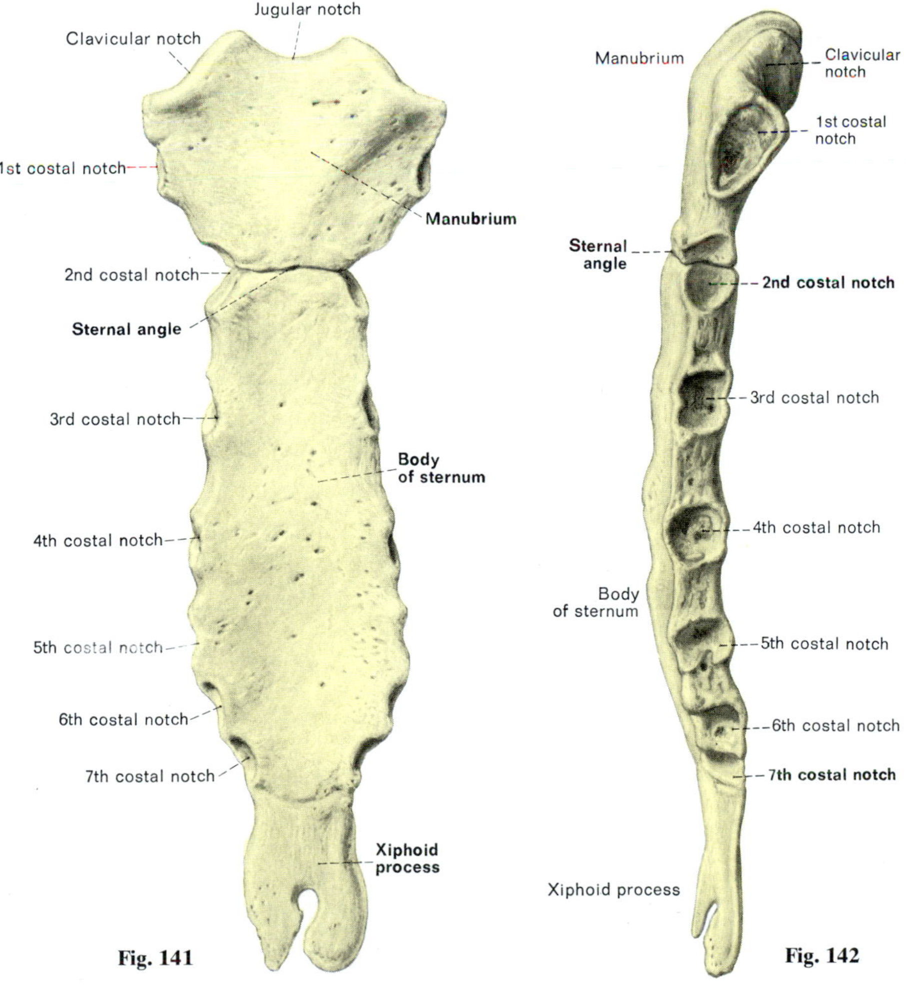
Jugular notch
Clavicular notch
1st costal notch
Manubrium
2nd costal notch
Sternal angle
3rd costal notch
Body of sternum
4th costal notch
5th costal notch
6th costal notch
7th costal notch
Xiphoid process
Manubrium
Clavicular notch
1st costal notch
Sternal angle
2nd costal notch
3rd costal notch
4th costal notch
Body of sternum
5th costal notch
6th costal notch
7th costal notch
Xiphoid process

Fig. 141

Fig. 142

Fig. 141: The Sternum, Anterior View

NOTE: 1) the sternum consists of three parts: the manubrium, the body and the xiphoid process and forms the middle portion of the anterior wall of the thorax.

2) the manubrium articulates with the body of the sternum at somewhat of an angle called the sternal angle. The xiphoid process is thin and usually cartilaginous.

3) the concave jugular notch which marks the superior border of the manubrium. Lateral to this are the two clavicular notches, each of which receives the medial end of the respective clavicle.

4) the entire sternum measures between 15 and 20 cm (6 to 7 inches). Its anterior surface is roughened and affords attachment to the sternocleidomastoid and pectoralis major muscles.

Fig. 142: The Sternum, Lateral View

NOTE: 1) the clavicle and the 1st rib articulate with the manubrium. The 2nd rib articulates at the sternal angle where the sternal manubrium and body join. The 3rd to the 6th ribs articulate with the body of the sternum, while the 7th joins the sternum inferiorly at the junction of the xiphoid process.

2) A line projected posteriorly through the sternal angle would meet the vertebral column at the 4th thoracic vertebral level, while the xiphisternal junction lies at vertebral level T-9.

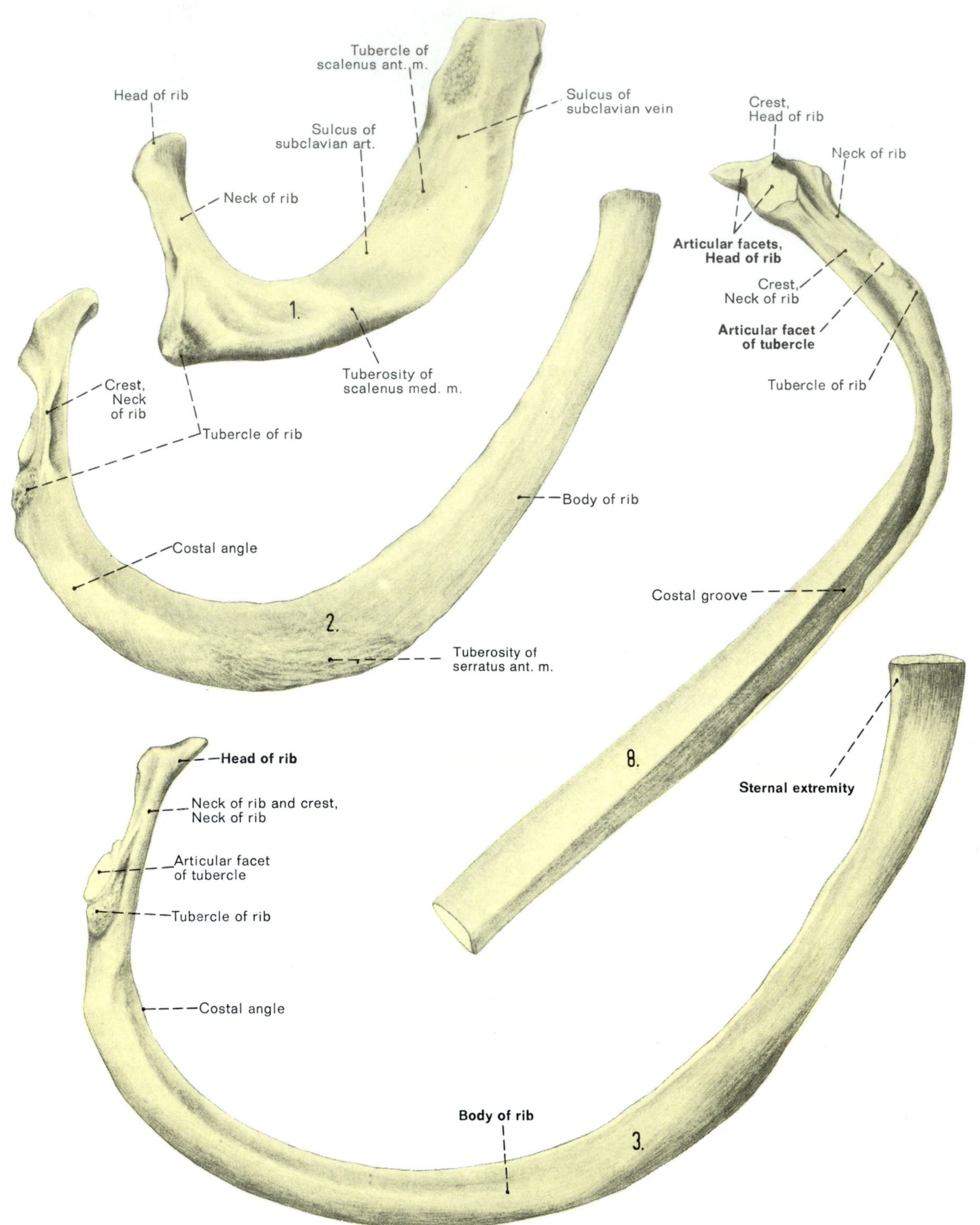
Tubercle of scalenus ant. m.
Head of rib
Sulcus of subclavian vein
Sulcus of subclavian art.
Neck of rib
1.
Tuberosity of scalenus med. m.
Crest, Neck of rib
Tubercle of rib
Costal angle
Body of rib
2.
Tuberosity of serratus ant. m.
Crest, Head of rib
Neck of rib
Articular facets, Head of rib
Crest, Neck of rib
Articular facet of tubercle
Tubercle of rib
Costal groove
8.
Sternal extremity
Head of rib
Neck of rib and crest, Neck of rib
Articular facet of tubercle
Tubercle of rib
Costal angle
Body of rib
3.

Fig. 143: The 1st, 2nd, 3rd and 8th Right Ribs

NOTE: 1) the superior surfaces of the first, second and third ribs are illustrated in this figure, while the inferior surface of the eighth rib is shown.

2) each rib has a vertebral extremity directed posteriorly, and a sternal extremity directed anteriorly. The body of the rib is the shaft which stretches between these two extremities.

3) the vertebral extremity is marked by a head, a neck and a tubercle. The head contains two facets for articulation with the bodies of the thoracic vertebrae, while the tubercle consists of a non-articular roughened elevation and an articular facet for connection with the transverse process of the thoracic vertebrae.

4) the 1st, 2nd, 10th, 11th and 12th ribs present somewhat different structural characteristics from the 3rd through the 9th ribs. The 1st rib is the most curved of all the ribs and has only a single articular facet on the head of the rib. The 10th, 11th and 12th ribs also have only a single facet on the rib head. The 2nd rib is shaped similar to the 1st rib but is longer.

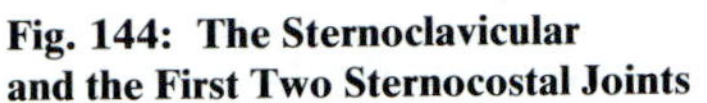

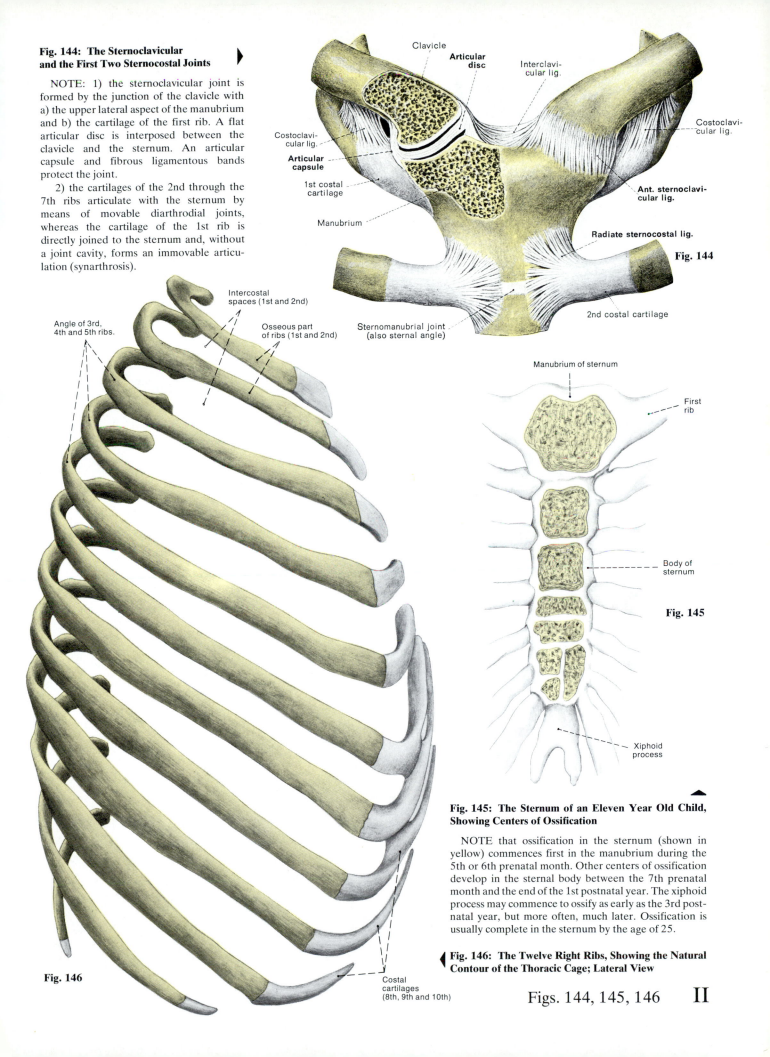

Fig. 144: The Sternoclavicular and the First Two Sternocostal Joints

NOTE: 1) the sternoclavicular joint is formed by the junction of the clavicle with a) the upper lateral aspect of the manubrium and b) the cartilage of the first rib. A flat articular disc is interposed between the clavicle and the sternum. An articular capsule and fibrous ligamentous bands protect the joint.

2) the cartilages of the 2nd through the 7th ribs articulate with the sternum by means of movable diarthrodial joints, whereas the cartilage of the 1st rib is directly joined to the sternum and, without a joint cavity, forms an immovable articulation (synarthrosis).

Clavicle
Articular disc
Interclavicular lig.
Costoclavicular lig.
Costoclavicular lig.
Articular capsule
1st costal cartilage
Manubrium
Ant. sternoclavicular lig.
Radiate sternocostal lig.
2nd costal cartilage
Sternomanubrial joint (also sternal angle)

Fig. 144

Intercostal spaces (1st and 2nd)
Angle of 3rd, 4th and 5th ribs.
Osseous part of ribs (1st and 2nd)
Costal cartilages (8th, 9th and 10th)

Fig. 146

Manubrium of sternum
First rib
Body of sternum
Xiphoid process

Fig. 145

Fig. 145: The Sternum of an Eleven Year Old Child, Showing Centers of Ossification

NOTE that ossification in the sternum (shown in yellow) commences first in the manubrium during the 5th or 6th prenatal month. Other centers of ossification develop in the sternal body between the 7th prenatal month and the end of the 1st postnatal year. The xiphoid process may commence to ossify as early as the 3rd postnatal year, but more often, much later. Ossification is usually complete in the sternum by the age of 25.

Fig. 146: The Twelve Right Ribs, Showing the Natural Contour of the Thoracic Cage; Lateral View

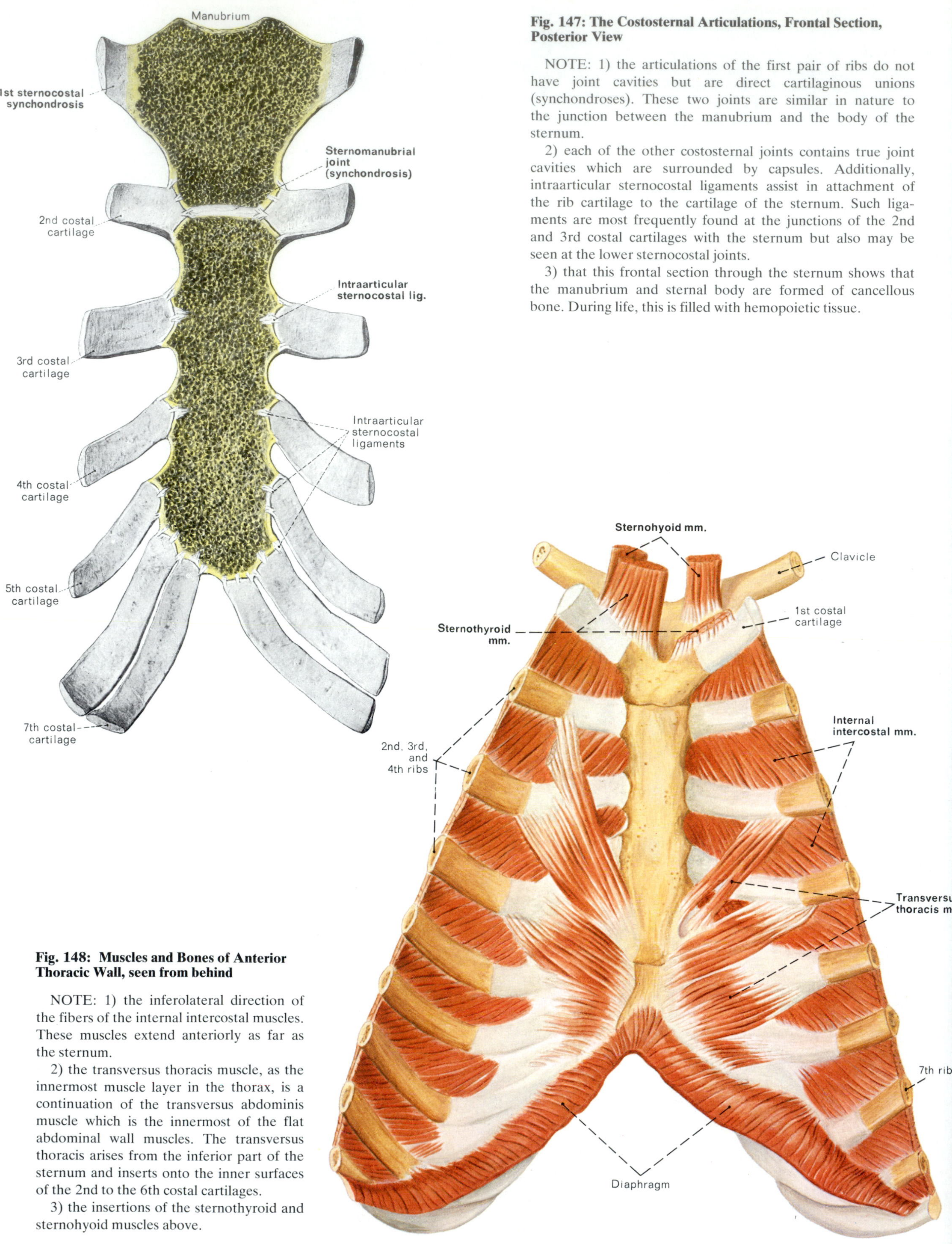
Manubrium
1st sternocostal synchondrosis
Sternomanubrial joint (synchondrosis)
2nd costal cartilage
Intraarticular sternocostal lig.
3rd costal cartilage
Intraarticular sternocostal ligaments
4th costal cartilage
5th costal cartilage
7th costal cartilage
Sternohyoid mm.
Clavicle
1st costal cartilage
Sternothyroid mm.
Internal intercostal mm.
2nd, 3rd, and 4th ribs
Transversu thoracis m.
7th rib
Diaphragm

Fig. 147: The Costosternal Articulations, Frontal Section, Posterior View

NOTE: 1) the articulations of the first pair of ribs do not have joint cavities but are direct cartilaginous unions (synchondroses). These two joints are similar in nature to the junction between the manubrium and the body of the sternum.

2) each of the other costosternal joints contains true joint cavities which are surrounded by capsules. Additionally, intraarticular sternocostal ligaments assist in attachment of the rib cartilage to the cartilage of the sternum. Such ligaments are most frequently found at the junctions of the 2nd and 3rd costal cartilages with the sternum but also may be seen at the lower sternocostal joints.

3) that this frontal section through the sternum shows that the manubrium and sternal body are formed of cancellous bone. During life, this is filled with hemopoietic tissue.

Fig. 148: Muscles and Bones of Anterior Thoracic Wall, seen from behind

NOTE: 1) the inferolateral direction of the fibers of the internal intercostal muscles. These muscles extend anteriorly as far as the sternum.

2) the transversus thoracis muscle, as the innermost muscle layer in the thorax, is a continuation of the transversus abdominis muscle which is the innermost of the flat abdominal wall muscles. The transversus thoracis arises from the inferior part of the sternum and inserts onto the inner surfaces of the 2nd to the 6th costal cartilages.

3) the insertions of the sternothyroid and sternohyoid muscles above.

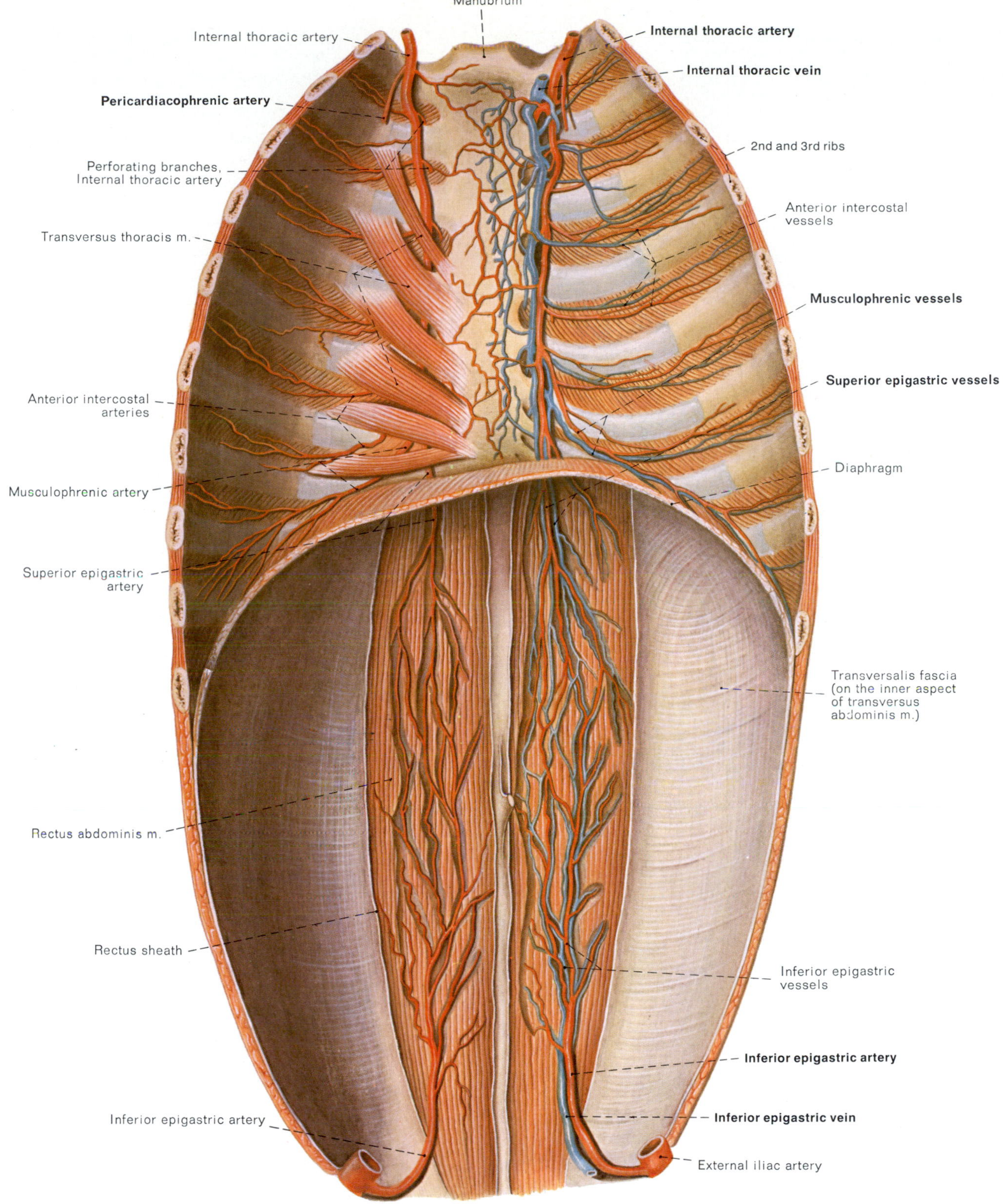
Manubrium
Internal thoracic artery
Internal thoracic artery
Internal thoracic vein
Pericardiacophrenic artery
2nd and 3rd ribs
Perforating branches, Internal thoracic artery
Anterior intercostal vessels
Transversus thoracis m.
Musculophrenic vessels
Superior epigastric vessels
Anterior intercostal arteries
Diaphragm
Musculophrenic artery
Superior epigastric artery
Transversalis fascia (on the inner aspect of transversus abdominis m.)
Rectus abdominis m.
Rectus sheath
Inferior epigastric vessels
Inferior epigastric artery
Inferior epigastric artery
Inferior epigastric vein
External iliac artery

Fig. 149: Blood Vessels of the Thoracic and Abdominal Wall, viewed from the inside

NOTE: 1) the principal vessels dissected include the internal thoracic and inferior epigastric arteries and veins and their terminal branches.

2) the internal thoracic artery branches from the subclavian artery and descends behind the costal cartilages on the inner aspect of the anterior thoracic wall parallel to the lateral margin of the sternum. In its course the internal thoracic artery gives rise to the pericardiacophrenic artery, small anterior mediastinal vessels to the thymus and bronchial structures, perforating branches to the chest wall, anterior intercostal branches which anastomose with the intercostal arteries, and finally it terminates as the musculophrenic and superior epigastric arteries.

3) the superior epigastric artery anastomoses with the inferior epigastric artery, which is a branch of the external iliac artery and which ascends in the abdominal wall from below. The anastomosis occurs in the substance of the rectus abdominis muscle.

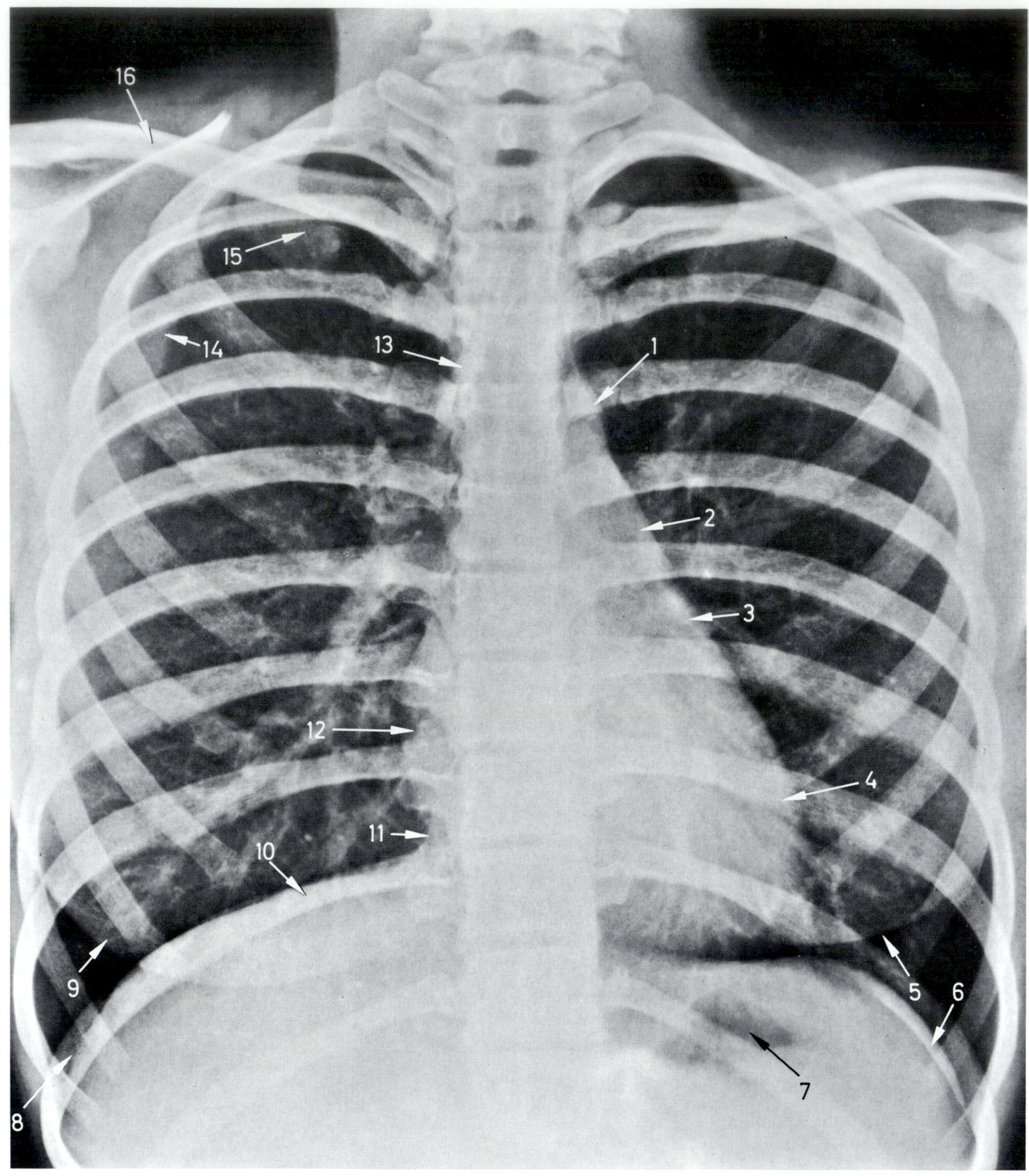
16
15
14
13
1
2
3
12
4
11
10
9
5
6
7
8

Fig. 150: Posterior-Anterior Radiograph of the Thorax Showing the Heart and Lungs

NOTE the contour of the heart and great vessels (arch of aorta 1, pulmonary trunk 2, inferior vena cava 11, and superior vena cava, 13) and the relationship of these structures to the vertebral column. Observe that the left margin of the heart is formed by the left auricle (3) and left ventricle (4) and that it slopes toward the apex which usually lies about 9 centimeters to the left of the midsternal line, deep to the 5th intercostal space. The right margin of the heart (12) projects as a curved line slightly to the right of the vertebral column (and sternum). Also observe that the heart rests on the diaphragm (6) and note the obvious contours of the left (5) and right (9) breasts. (From Wicke, L. *Atlas of Radiographic Anatomy,* 3rd Edition, Urban & Schwarzenberg, Baltimore, 1982).

1. Arch of aorta
2. Pulmonary trunk
3. Left auricle
4. Left ventricle
5. Contour of left breast
6. Diaphragm
7. Air in fundus of stomach
8. Costodiaphragmatic recess
9. Contour of right breast
10. Diaphragm
11. Inferior vena cava
12. Right atrium
13. Superior vena cava
14. Medial border of scapula
15. First rib
16. Right clavicle

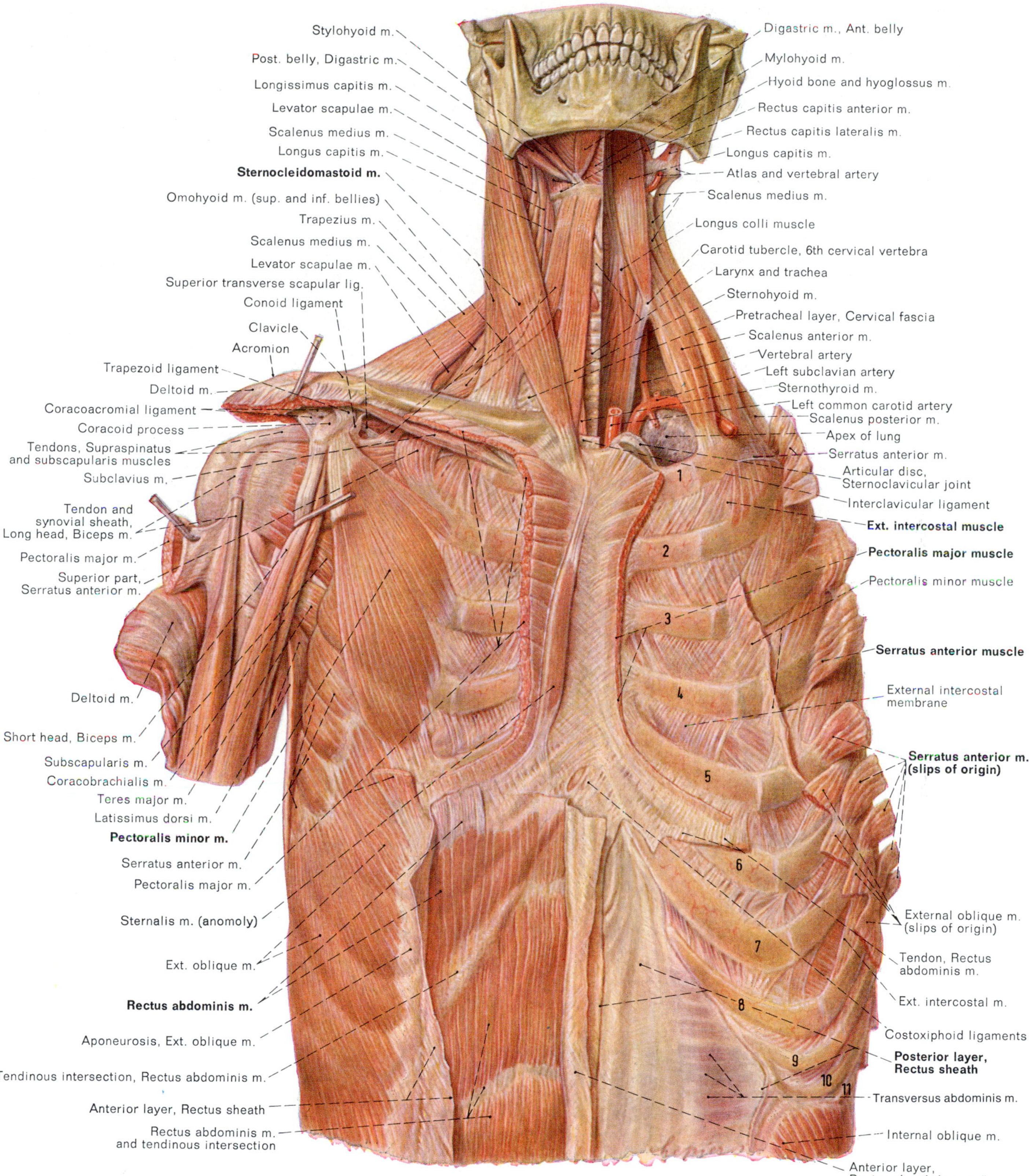
Stylohyoid m.
Post. belly, Digastric m.
Longissimus capitis m.
Levator scapulae m.
Scalenus medius m.
Longus capitis m.
Sternocleidomastoid m.
Omohyoid m. (sup. and inf. bellies)
Trapezius m.
Scalenus medius m.
Levator scapulae m.
Superior transverse scapular lig.
Conoid ligament
Clavicle
Acromion
Trapezoid ligament
Deltoid m.
Coracoacromial ligament
Coracoid process
Tendons, Supraspinatus and subscapularis muscles
Subclavius m.
Tendon and synovial sheath, Long head, Biceps m.
Pectoralis major m.
Superior part, Serratus anterior m.
Deltoid m.
Short head, Biceps m.
Subscapularis m.
Coracobrachialis m.
Teres major m.
Latissimus dorsi m.
Pectoralis minor m.
Serratus anterior m.
Pectoralis major m.
Sternalis m. (anomoly)
Ext. oblique m.
Rectus abdominis m.
Aponeurosis, Ext. oblique m.
Tendinous intersection, Rectus abdominis m.
Anterior layer, Rectus sheath
Rectus abdominis m. and tendinous intersection
Digastric m., Ant. belly
Mylohyoid m.
Hyoid bone and hyoglossus m.
Rectus capitis anterior m.
Rectus capitis lateralis m.
Longus capitis m.
Atlas and vertebral artery
Scalenus medius m.
Longus colli muscle
Carotid tubercle, 6th cervical vertebra
Larynx and trachea
Sternohyoid m.
Pretracheal layer, Cervical fascia
Scalenus anterior m.
Vertebral artery
Left subclavian artery
Sternothyroid m.
Left common carotid artery
Scalenus posterior m.
Apex of lung
Serratus anterior m.
Articular disc, Sternoclavicular joint
Interclavicular ligament
Ext. intercostal muscle
Pectoralis major muscle
Pectoralis minor muscle
Serratus anterior muscle
External intercostal membrane
Serratus anterior m. (slips of origin)
External oblique m. (slips of origin)
Tendon, Rectus abdominis m.
Ext. intercostal m.
Costoxiphoid ligaments
Posterior layer, Rectus sheath
Transversus abdominis m.
Internal oblique m.
Anterior layer, Rectus sheath (opened)
1
2
3
4
5
6
7
8
9
10
11

Fig. 151: Anterior Cervical, Thoracic and Abdominal Musculature

NOTE: 1) on the right side the muscles of the shoulder and upper arm are demonstrated following the removal of the pectoralis major muscle. The anterior layer of the rectus sheath has been opened.

2) on the left side the upper limb has been removed, as have all of the more superficial trunk and cervical musculature, revealing the thoracic cage and the deeper soft tissues.

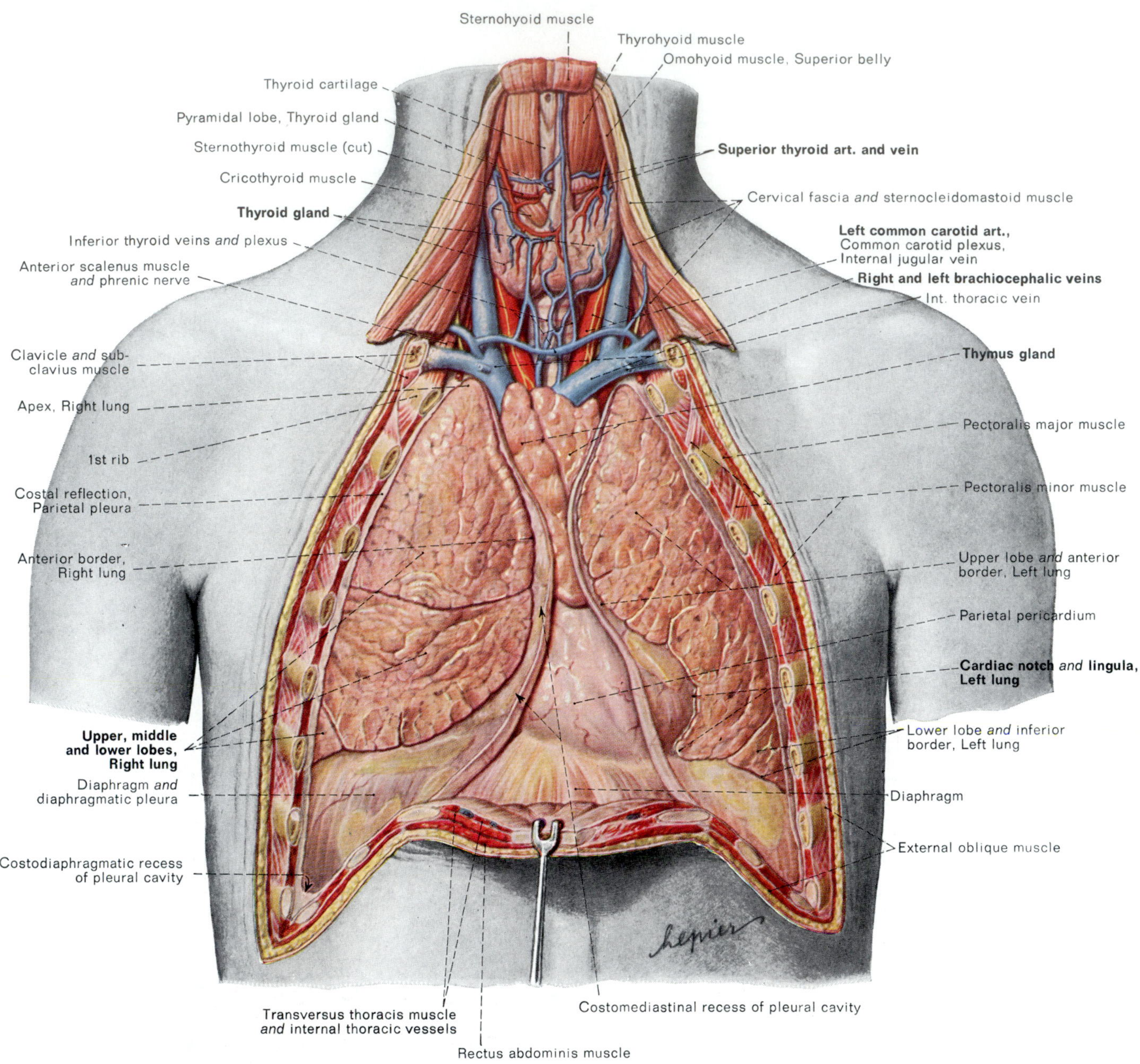

Sternohyoid muscle
Thyrohyoid muscle
Omohyoid muscle, Superior belly
Thyroid cartilage
Pyramidal lobe, Thyroid gland
Sternothyroid muscle (cut)
Superior thyroid art. and vein
Cricothyroid muscle
Cervical fascia and sternocleidomastoid muscle
Thyroid gland
Left common carotid art., Common carotid plexus, Internal jugular vein
Inferior thyroid veins and plexus
Anterior scalenus muscle and phrenic nerve
Right and left brachiocephalic veins
Int. thoracic vein
Clavicle and subclavius muscle
Thymus gland
Apex, Right lung
Pectoralis major muscle
1st rib
Costal reflection, Parietal pleura
Pectoralis minor muscle
Anterior border, Right lung
Upper lobe and anterior border, Left lung
Parietal pericardium
Cardiac notch and lingula, Left lung
Upper, middle and lower lobes, Right lung
Lower lobe and inferior border, Left lung
Diaphragm and diaphragmatic pleura
Diaphragm
External oblique muscle
Costodiaphragmatic recess of pleural cavity
Transversus thoracis muscle and internal thoracic vessels
Costomediastinal recess of pleural cavity
Rectus abdominis muscle
Lepier

Fig. 152: The Thoracic Viscera and the Root of the Neck, Anterior Exposure

NOTE: 1) the anterior thoracic wall has been removed along with the medial portion of both clavicles to reveal the normal position of the heart, lungs, thymus and thyroid gland. The great vessels at the superior aperture to the thorax are also exposed.

2) the parietal pleura, likewise, has been removed arteriorly. The thymus is situated between the two lungs superiorly, whereas inferiorly is found the bare area of the heart. With the heart's apex directed more toward the left, a deficiency can be observed along the border of the left lung. This is called the cardiac notch.

3) the basal surface of both lungs and the inferior aspect of the heart rest on the diaphragm. Superiorly, the apex of each lung extends slightly above the level of the first rib.

4) at the root of the neck, the common carotid artery and internal jugular vein lie lateral and somewhat posterior to the thyroid gland.

5) the rather transverse course in the superior mediastinum of the left brachiocephalic vein in contrast to the nearly vertical course of the right brachiocephalic vein. On either side each of these vessels is formed by a junction of the internal jugular and subclavian veins. The two brachiocephalic veins join (deep to the thymus) to form the superior vena cava.

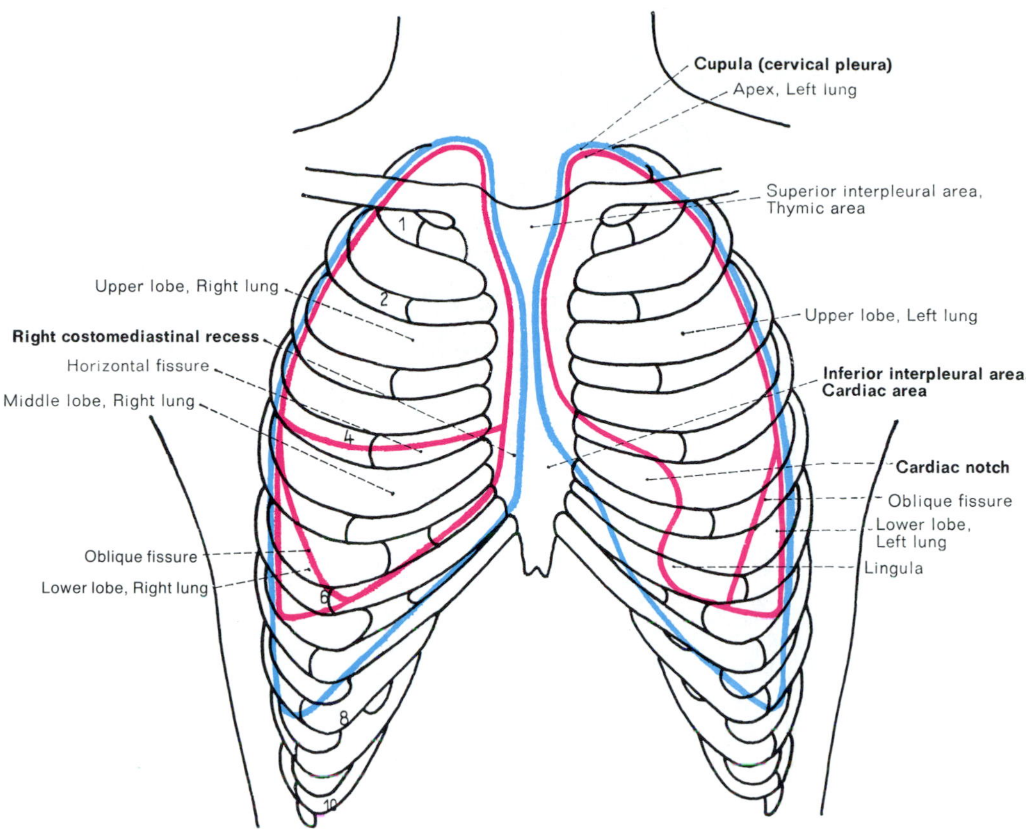

Cupula (cervical pleura)
Apex, Left lung
Superior interpleural area,
Thymic area
Upper lobe, Right lung
Right costomediastinal recess
Horizontal fissure
Middle lobe, Right lung
Upper lobe, Left lung
Inferior interpleural area,
Cardiac area
Cardiac notch
Oblique fissure
Lower lobe,
Left lung
Lingula
Oblique fissure
Lower lobe, Right lung
1
2
4
6
8
10

Fig. 153 and 154:
Pleural Reflections (blue) and Lungs (red) Projected onto Thoracic Wall

Fig. 153: Anterior View

NOTE: 1) each lung is invested by two layers of pleural membrane which are continuous with each other at the hilum of the lung, thereby forming an invaginated sac. The parietal layer of pleura (represented in blue) is the outermost of the two layers and lines the inner surface of the thoracic wall and the superior surface of the diaphragm. The visceral, innermost layer of pleura closely invests and adheres to the surfaces of the lungs (represented in red).

2) the potential space between the two pleural layers is called the pleural cavity and contains only a small amount of serous fluid in the healthy individual, but may contain considerable fluid and blood in pathological conditions.

3) although the parietal pleura is a continuous sheet for each lung, portions of it are described in relation to their adjacent surfaces. Thus, lining the inner surface of the ribs is the costal pleura, while the diaphragmatic and mediastinal pleurae are applied onto the surfaces of the diaphragm and mediastinal structures. Superiorly, the apex of each lung extends above the clavicle into the root of the neck. This is covered by the cupula, or cervical pleura.

4) because of the curvature of the diaphragm, a narrow recess is formed around its periphery into which the surface of the lung (visceral pleura) does not extend. This potential space, lying between reflections of the costal and diaphragmatic pleurae, is called the costodiaphragmatic recess and is of clinical importance, since it may be punctured and drained without damage to lung tissue.

5) similarly, the costomediastinal recess is another pleural space which is situated anterior to the heart, and which is formed at that site by the reflections of the costal and mediastinal pleurae.

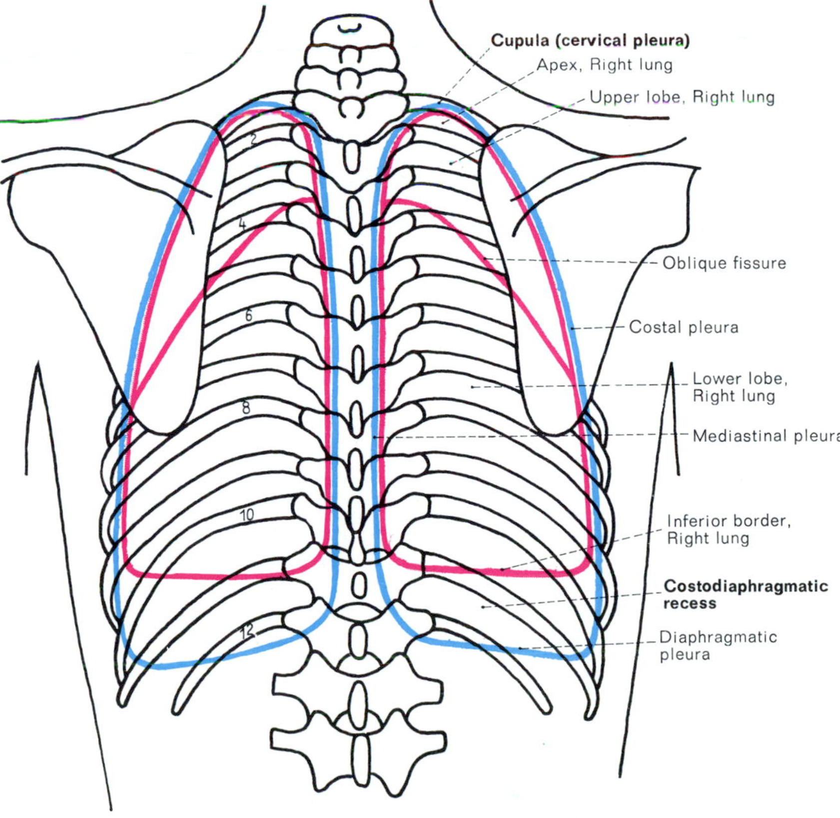

Cupula (cervical pleura)
Apex, Right lung
Upper lobe, Right lung
Oblique fissure
Costal pleura
Lower lobe,
Right lung
Mediastinal pleura
Inferior border,
Right lung
Costodiaphragmatic
recess
Diaphragmatic
pleura
2
4
6
8
10
12

Fig. 154: Posterior View

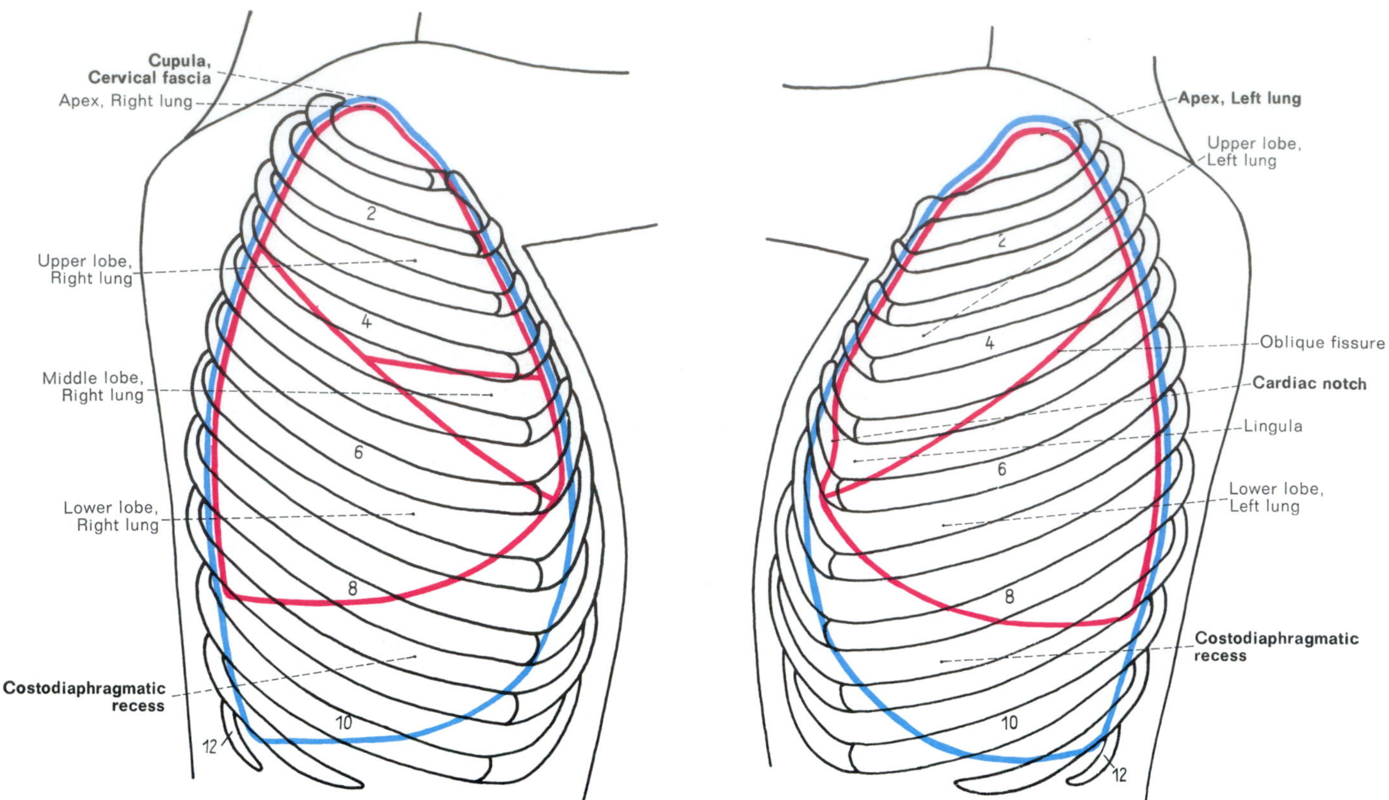
Cupula,
Cervical fascia
Apex, Right lung
Upper lobe,
Right lung
Middle lobe,
Right lung
Lower lobe,
Right lung
Costodiaphragmatic
recess
12
2
4
6
8
10
Apex, Left lung
Upper lobe,
Left lung
Oblique fissure
Cardiac notch
Lingula
Lower lobe,
Left lung
Costodiaphragmatic
recess
2
4
6
8
10
12

Fig. 155: Right Lateral View **Fig. 156: Left Lateral View**
Pleural Reflections (blue) and Lungs (red) Projected onto Thoracic Wall

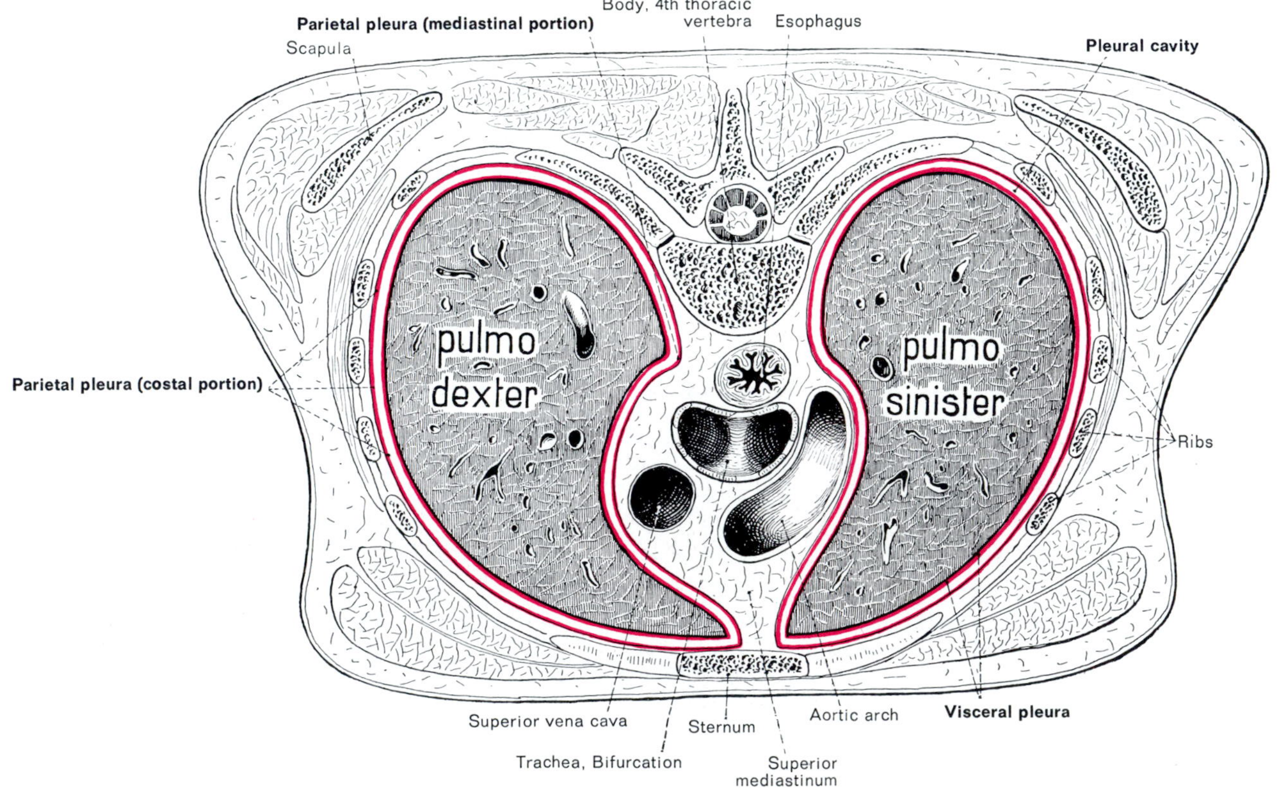
Body, 4th thoracic
vertebra
Parietal pleura (mediastinal portion)
Esophagus
Scapula
Pleural cavity
pulmo
dexter
pulmo
sinister
Parietal pleura (costal portion)
Ribs
Superior vena cava
Sternum
Aortic arch
Visceral pleura
Trachea, Bifurcation
Superior
mediastinum

Fig. 157: Cross Section of Thorax at the Level of the Tracheal Bifurcation and the 4th Thoracic Vertebra (Pleura is shown in red)

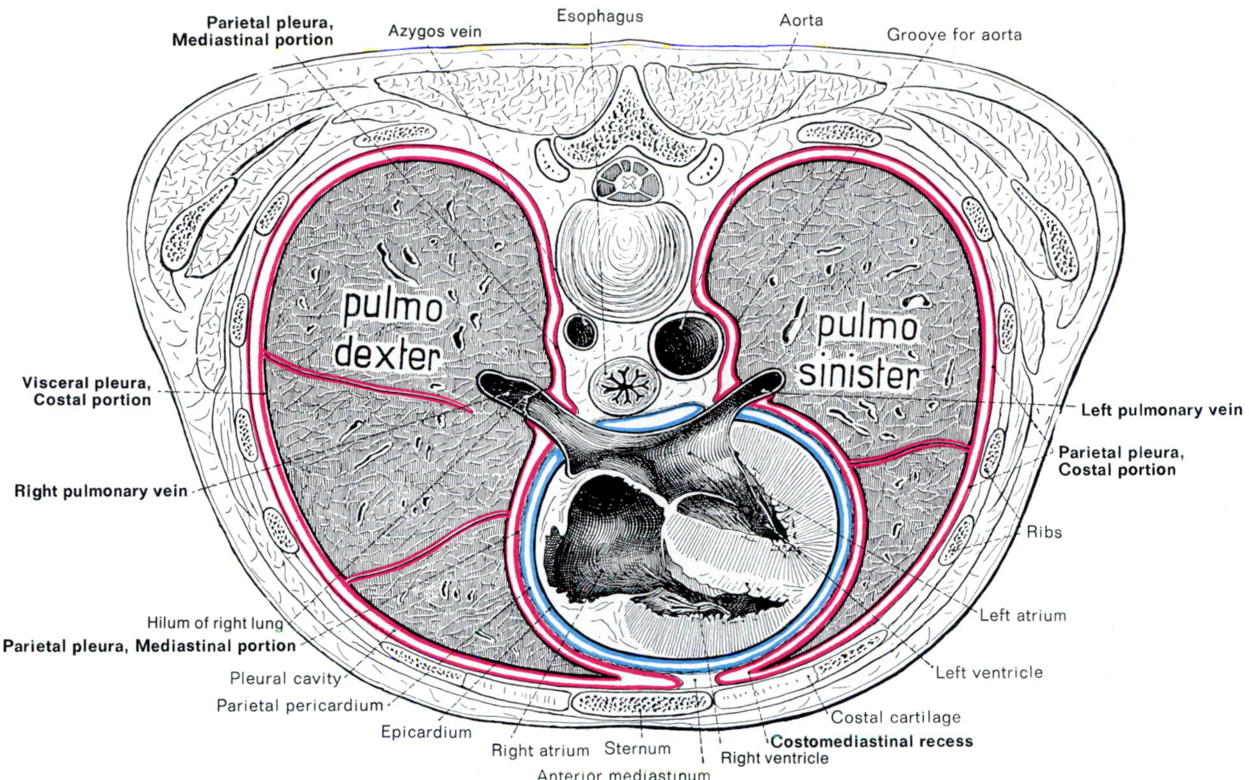

Parietal pleura,
Mediastinal portion
Azygos vein
Esophagus
Aorta
Groove for aorta
pulmo
dexter
pulmo
sinister
Visceral pleura,
Costal portion
Left pulmonary vein
Parietal pleura,
Costal portion
Right pulmonary vein
Ribs
Hilum of right lung
Parietal pleura, Mediastinal portion
Left atrium
Left ventricle
Pleural cavity
Parietal pericardium
Costal cartilage
Epicardium
Costomediastinal recess
Right atrium
Sternum
Right ventricle
Anterior mediastinum

Fig. 158: Cross Section of Thorax Through the Hilum of the Lung at the Level of the Pulmonary Vein (Pleura in red, Pericardium in blue)

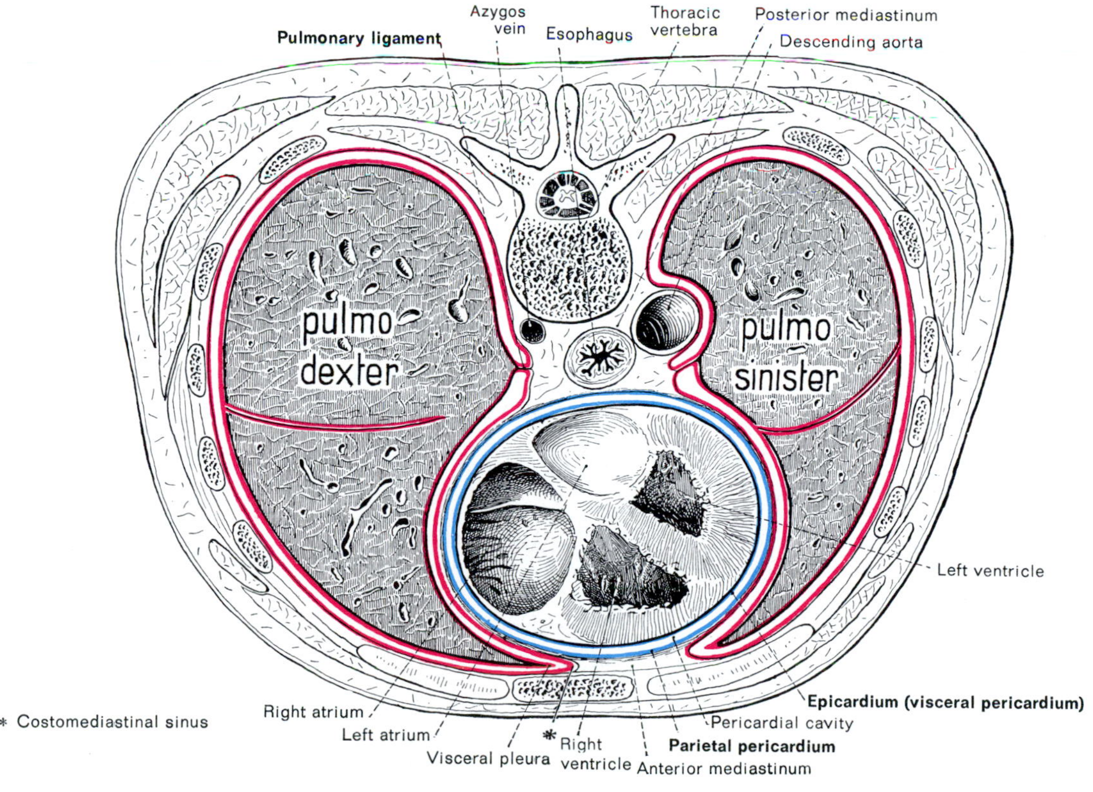

Pulmonary ligament
Azygos vein
Esophagus
Thoracic vertebra
Posterior mediastinum
Descending aorta
pulmo
dexter
pulmo
sinister
Left ventricle
Epicardium (visceral pericardium)
* Costomediastinal sinus
Right atrium
Pericardial cavity
Left atrium
* Right ventricle
Parietal pericardium
Visceral pleura
Anterior mediastinum

Fig. 159: Cross Section of the Thorax Inferior to the Hilum at the Level of the Pulmonary Ligament (Pleura in red, Pericardium in blue)

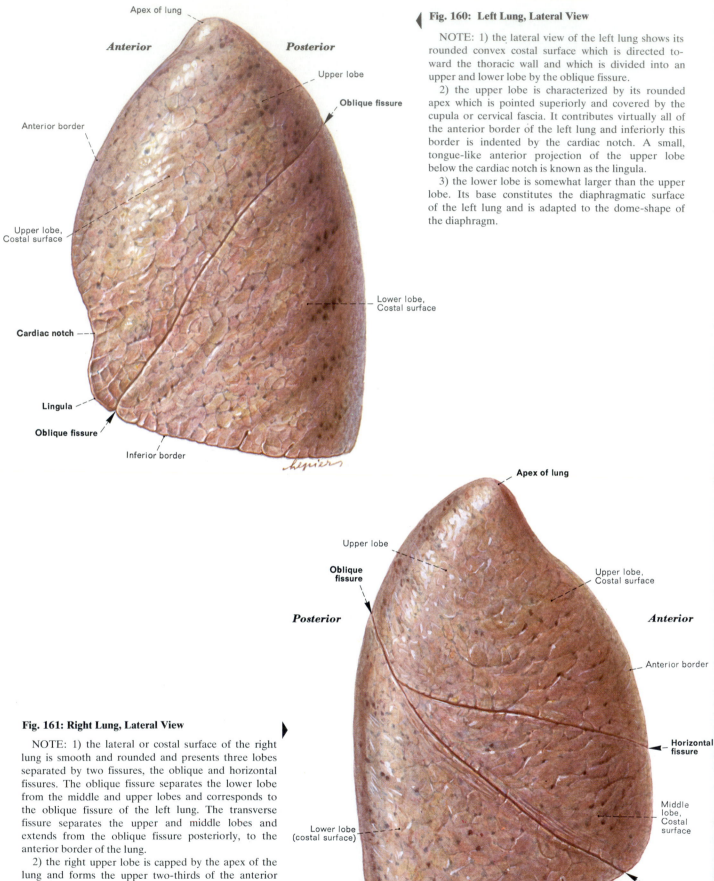

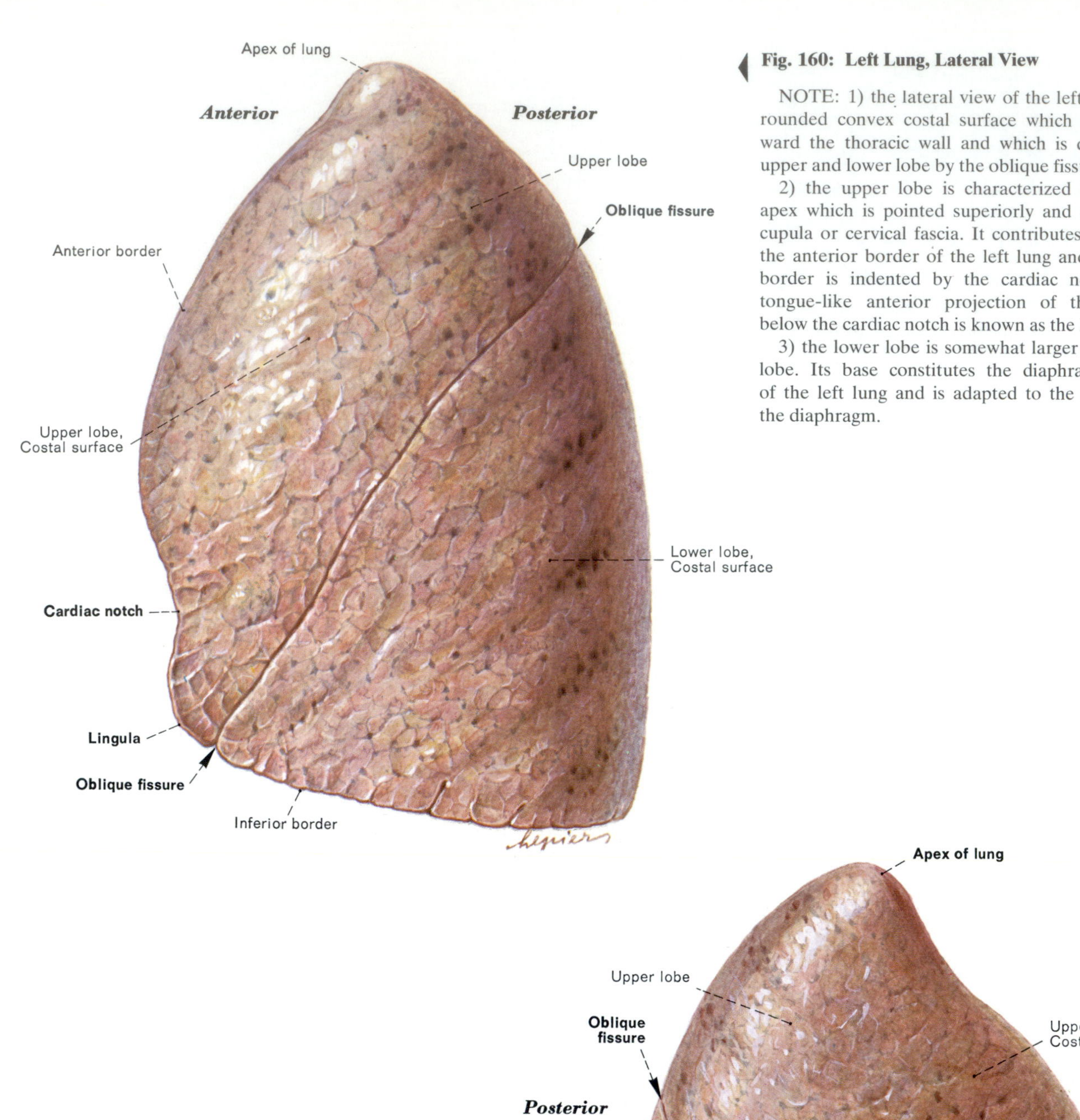
Apex of lung
Anterior
Posterior
Upper lobe
Oblique fissure
Anterior border
Upper lobe, Costal surface
Lower lobe, Costal surface
Cardiac notch
Lingula
Oblique fissure
Inferior border

Fig. 160: Left Lung, Lateral View

NOTE: 1) the lateral view of the left lung shows its rounded convex costal surface which is directed toward the thoracic wall and which is divided into an upper and lower lobe by the oblique fissure.

2) the upper lobe is characterized by its rounded apex which is pointed superiorly and covered by the cupula or cervical fascia. It contributes virtually all of the anterior border of the left lung and inferiorly this border is indented by the cardiac notch. A small, tongue-like anterior projection of the upper lobe below the cardiac notch is known as the lingula.

3) the lower lobe is somewhat larger than the upper lobe. Its base constitutes the diaphragmatic surface of the left lung and is adapted to the dome-shape of the diaphragm.

Apex of lung
Upper lobe
Oblique fissure
Upper lobe, Costal surface
Posterior
Anterior
Anterior border
Horizontal fissure
Middle lobe, Costal surface
Lower lobe (costal surface)
Oblique fissure
Inferior border
Lower lobe (base of lung)

Fig. 161: Right Lung, Lateral View

NOTE: 1) the lateral or costal surface of the right lung is smooth and rounded and presents three lobes separated by two fissures, the oblique and horizontal fissures. The oblique fissure separates the lower lobe from the middle and upper lobes and corresponds to the oblique fissure of the left lung. The transverse fissure separates the upper and middle lobes and extends from the oblique fissure posteriorly, to the anterior border of the lung.

2) the right upper lobe is capped by the apex of the lung and forms the upper two-thirds of the anterior border. The middle lobe forms the inferior third of the anterior border, while the lower lobe (as in the left lung) constitutes the entire inferior border and diaphragmatic surface.

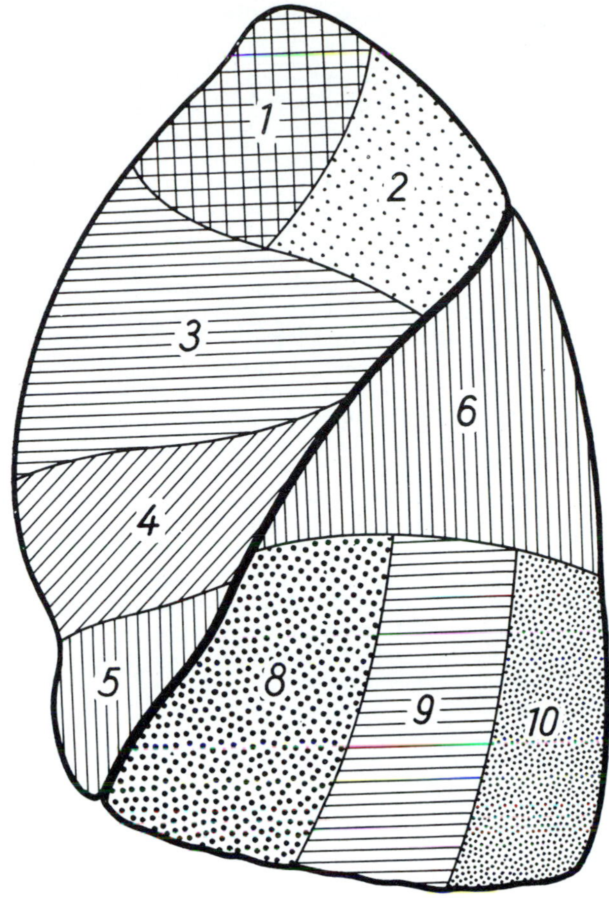

1
2
3
4
5
6
8
9
10

Fig. 162: Left Lung, Bronchopulmonary Segments, Lateral View

NOTE: 1) bronchopulmonary segments are anatomical subdivisions of the lung, each of which is supplied by its own segmental tertiary bronchus and artery, and drained by intersegmental veins.

2) the trachea divides into two primary bronchi, each of which supplies an entire lung. Each primary bronchus divides into secondary or lobar bronchi. There are two lobar bronchi on the left and three on the right, each supplying a single lobe. The secondary bronchi divide into the segmental or tertiary bronchi, which are distributed to the bronchopulmonary segments. Usual descriptions of the bronchopulmonary segments enumerate 8 to 10 segments in the left lung.

3) the bronchopulmonary segments of the left lung are numbered and named as follows:

Upper lobe
1 Apical } Frequently considered as a single segment
2 Posterior }
3 Anterior
4 Superior } Lingular
5 Inferior }

Lower lobe
6 Superior
7 Medial basal } Usually considered as a single segment;
8 Anterior basal } Medial basal cannot be seen from lateral view.
9 Lateral basal
10 Posterior basal

4) in the left lower lobe the medial basal bronchus arises separate from the anterior basal in only about 13 % of humans studied. Thus, in most instances the medial basal and anterior basal segments combine as an anteromedial basal segment.

Fig. 163: Right Lung, Bronchopulmonary Segments, Lateral View

NOTE: 1) the concept of subdividing the lungs into functional bronchopulmonary segments allows the surgeon to determine whether segments of lung might be resected in operations in preference to entire lobes.

2) although minor variations exist in the division of the bronchial tree, a significant consistency has become recognized in the bronchopulmonary segmentation. The nomenclature utilized here was offered by Jackson and Huber in 1943 (Dis. of Chest **9:** 319–326) and has now become generally accepted because it is the simplest and most straightforward of the many suggested.

3) the bronchopulmonary segments of the right lung are numbered and named as follows:

Upper lobe
1 Apical
2 Posterior
3 Anterior

Middle lobe
4 Lateral
5 Medial

Lower lobe
6 Superior
7 Medial basal (cannot be seen from lateral view)
8 Anterior basal
9 Lateral basal
10 Posterior basal

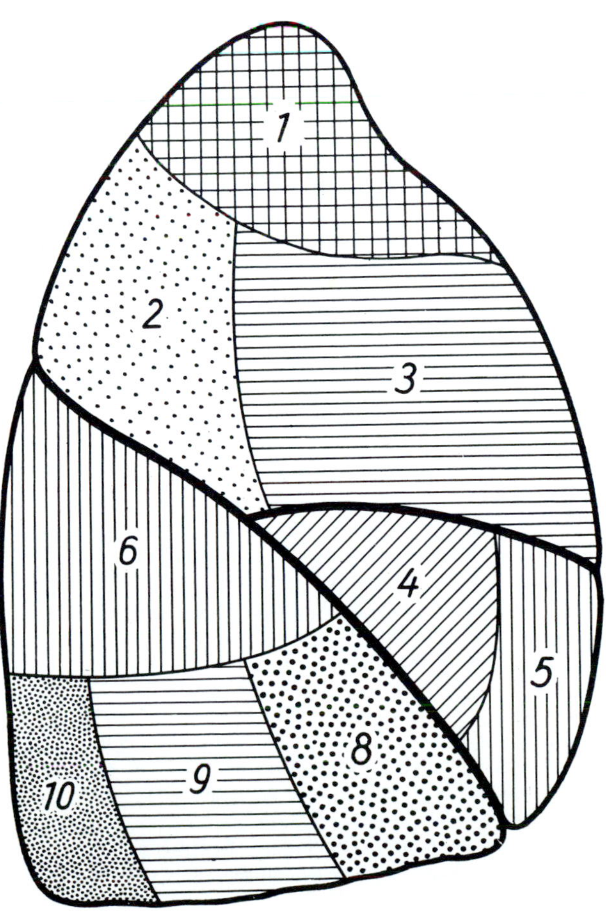

1
2
3
4
5
6
8
9
10

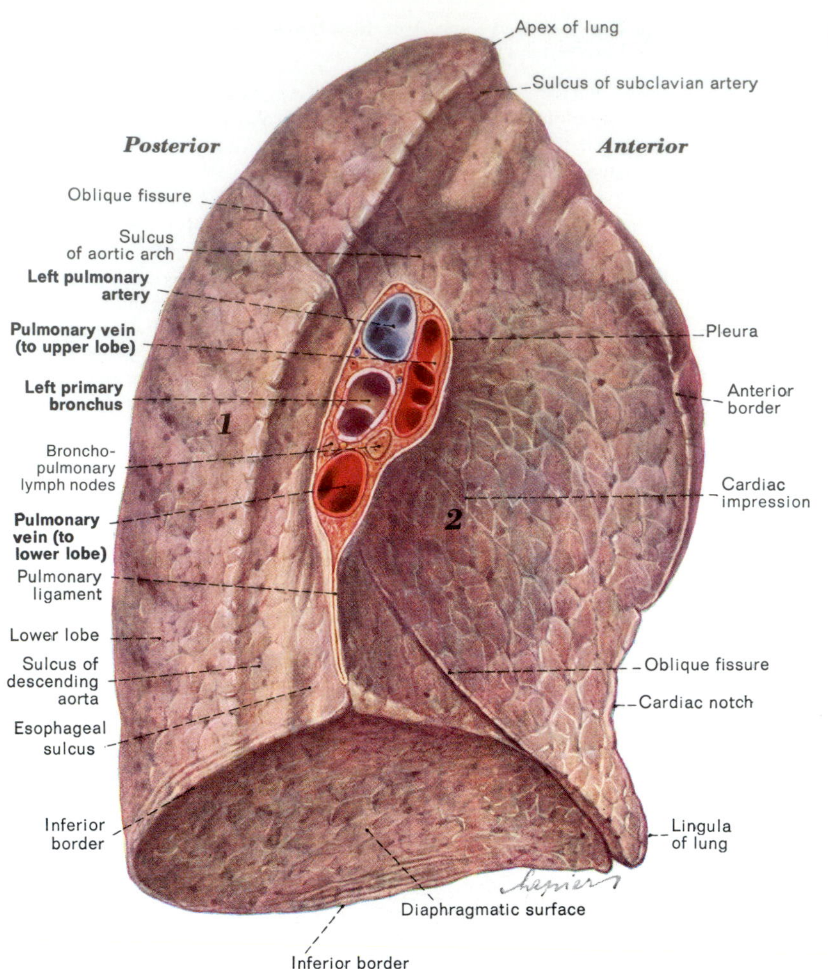
Apex of lung
Sulcus of subclavian artery
Posterior
Anterior
Oblique fissure
Sulcus of aortic arch
Left pulmonary artery
Pulmonary vein (to upper lobe)
Left primary bronchus
1
Broncho-pulmonary lymph nodes
Pulmonary vein (to lower lobe)
Pulmonary ligament
Lower lobe
Sulcus of descending aorta
Esophageal sulcus
Inferior border
Pleura
Anterior border
Cardiac impression
2
Oblique fissure
Cardiac notch
Lingula of lung
Diaphragmatic surface
Inferior border

Fig. 164: Left Lung, Mediastinal and Diaphragmatic Surfaces

NOTE: 1) the pulmonary arteries are shown in blue since their blood contains less oxygen than that in the pulmonary veins which are shown in red.

2) the concave diaphragmatic surface of the left lung is shaped to cover most of the convex dome of the diaphragm which is completely covered by diaphragmatic pleura. However, the lung does not completely fill the peripheral rim of the diaphragm, thereby forming the costodiaphragmatic recess.

3) the mediastinal (or medial) surface of the left lung is also concave and presents the contours of the adjacent organs in the mediastinum. The large anterior concavity is the cardiac impression. Also found are grooves for the aortic arch and the descending aorta as well as the subclavian artery superiorly and the esophagus inferiorly.

4) the structures which form the root of the left lung at the hilum include the left pulmonary artery, found most superior and below which is found the left bronchus. The left pulmonary veins lie anterior and inferior to the artery and bronchus. The oblique fissure extends across the mediastinal surface from the costal surface to the diaphragm, completely dividing the lung into its two lobes.

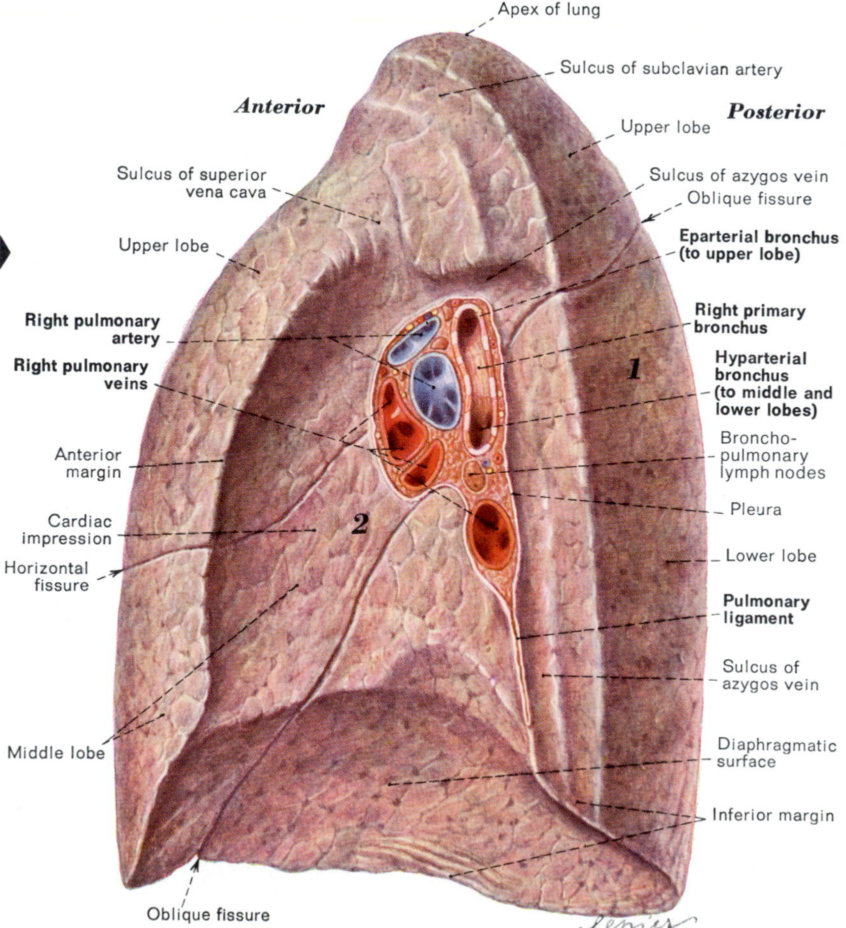
Apex of lung
Sulcus of subclavian artery
Anterior
Posterior
Upper lobe
Sulcus of superior vena cava
Sulcus of azygos vein
Oblique fissure
Upper lobe
Eparterial bronchus (to upper lobe)
Right pulmonary artery
Right primary bronchus
Right pulmonary veins
1
Hyparterial bronchus (to middle and lower lobes)
Anterior margin
Broncho-pulmonary lymph nodes
Pleura
Cardiac impression
2
Horizontal fissure
Lower lobe
Pulmonary ligament
Sulcus of azygos vein
Middle lobe
Diaphragmatic surface
Inferior margin
Oblique fissure

Fig. 165: Right Lung, Mediastinal and Diaphragmatic Surfaces

NOTE: 1) the diaphragmatic surface of the right lung, similar to that on the left, is shaped to the contour of the diaphragm, while the mediastinal surface superiorly shows grooves for the superior vena cava and subclavian artery. Just above the hilum of the right lung is the arched sulcus for the azygos vein, and this is continued inferiorly behind the root of the lung. The cardiac impression on the right lung is somewhat more shallow than on the left.

2) since the right bronchus frequently branches before the right pulmonary artery it is not unusual for the most superior structure at the root of the right lung to be the bronchus to the upper lobe (eparterial bronchus). The pulmonary artery lies anterior to the bronchus, while the pulmonary veins are located anterior and inferior to these structures.

3) the hilum of the lung is ensheathed by parietal pleura, the layers of which come into contact inferiorly to form the pulmonary ligament. It extends from the inferior border of the hilum to a point just above the diaphragm.

4) the numbers 1 and 2 on this figure and on Fig. 164 refer to the costal and mediastinal portions of the medial surface of the lung.

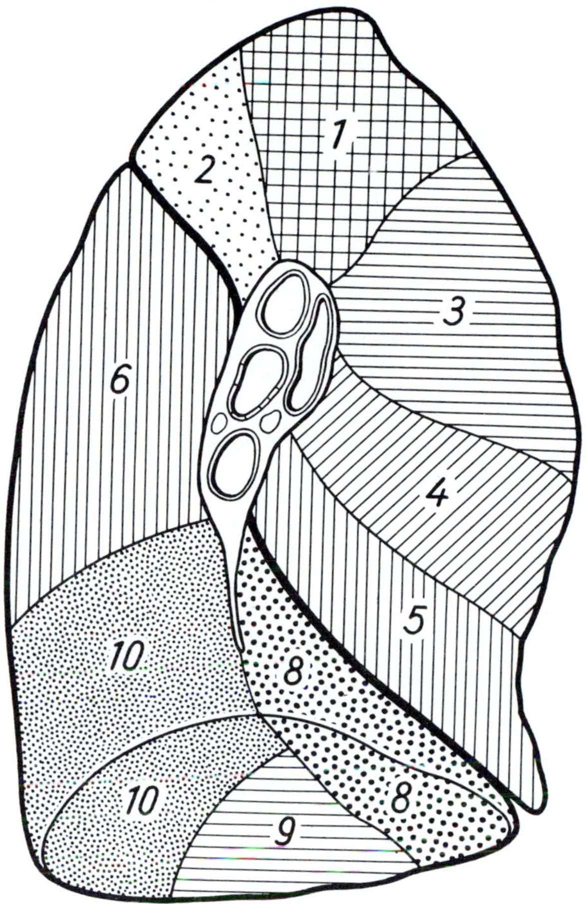
1
2
3
4
5
6
8
10
10
8
9

Fig. 166: Left Lung, Bronchopulmonary Segments, Medial View

NOTE: the bronchopulmonary segments of the left lung have been identified as follows:

Upper lobe

1 Apical } Frequently considered as one segment
2 Posterior
3 Anterior
4 Superior } Lingular
5 Inferior

Lower lobe

6 Superior
7 Medial basal*
8 Anterior basal*
9 Lateral basal
10 Posterior basal

* the medial basal and anterior basal segments were at one time frequently considered as a single bronchopulmonary segment. Today, however, they have been recognized as separate segments in a majority of left lungs. Therefore, on this figure that portion of segment 8 just inferior to the oblique fissure should be marked 7 and identified as medial basal.

Fig. 167: Right Lung, Bronchopulmonary Segments, Medial View

NOTE: the bronchopulmonary segments of the right lung have been identified as follows:

Upper lobe

1 Apical
2 Posterior
3 Anterior

Middle lobe

4 Lateral (not seen from this view)
5 Medial

Lower lobe

6 Superior
7 Medial basal
8 Anterior basal
9 Lateral basal
10 Posterior basal

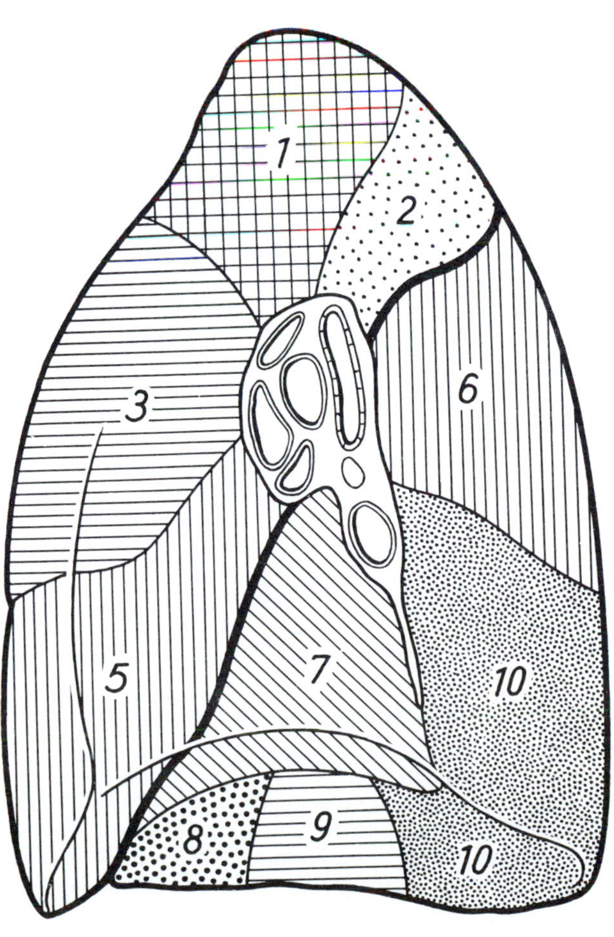
1
2
3
6
5
7
10
8
9
10

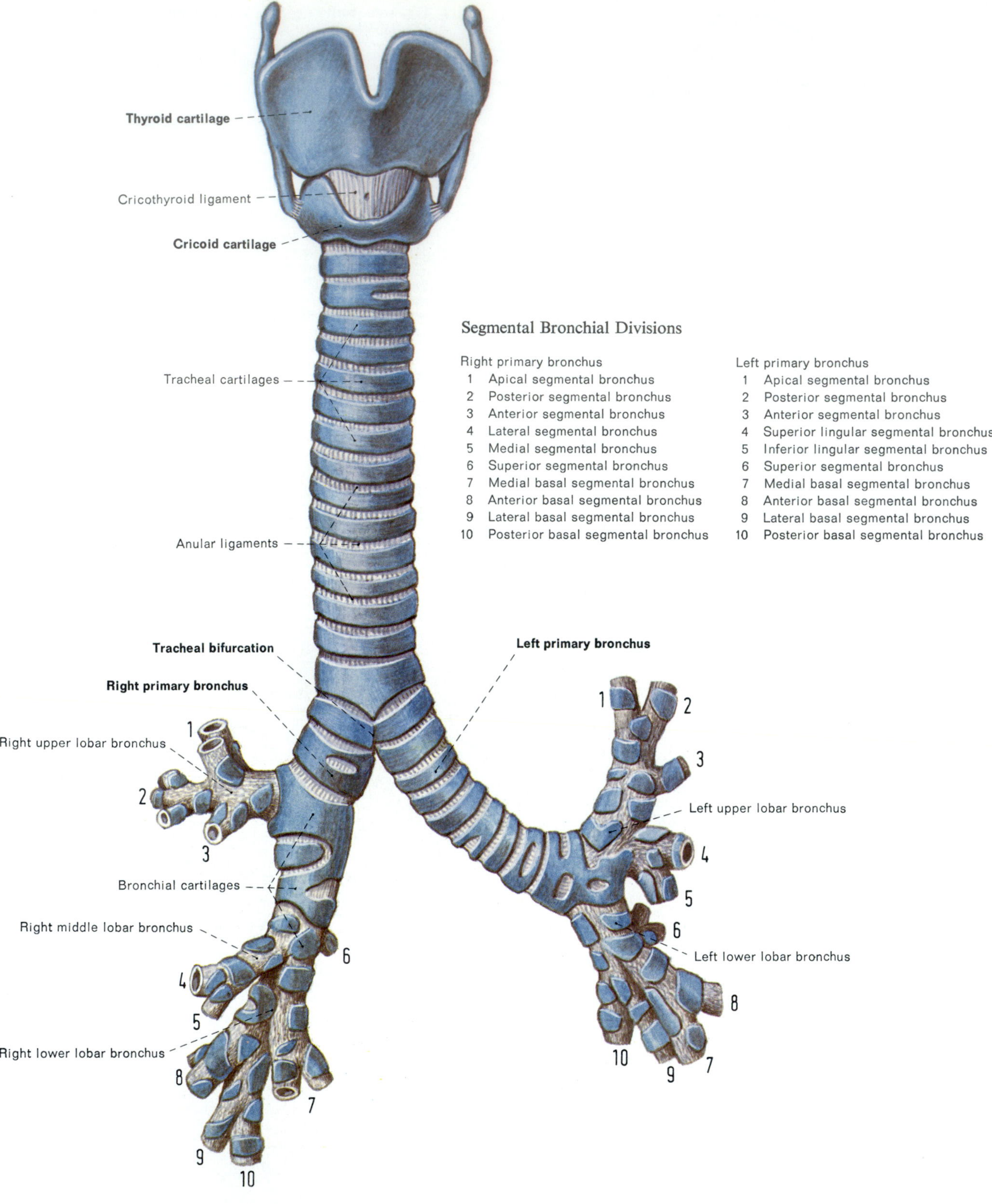

Thyroid cartilage
Cricothyroid ligament
Cricoid cartilage
Tracheal cartilages
Anular ligaments
Tracheal bifurcation
Right primary bronchus
Right upper lobar bronchus
Bronchial cartilages
Right middle lobar bronchus
Right lower lobar bronchus
Left primary bronchus
Left upper lobar bronchus
Left lower lobar bronchus
1
2
3
4
5
6
7
8
9
10
Segmental Bronchial Divisions
Right primary bronchus
1 Apical segmental bronchus
2 Posterior segmental bronchus
3 Anterior segmental bronchus
4 Lateral segmental bronchus
5 Medial segmental bronchus
6 Superior segmental bronchus
7 Medial basal segmental bronchus
8 Anterior basal segmental bronchus
9 Lateral basal segmental bronchus
10 Posterior basal segmental bronchus
Left primary bronchus
1 Apical segmental bronchus
2 Posterior segmental bronchus
3 Anterior segmental bronchus
4 Superior lingular segmental bronchus
5 Inferior lingular segmental bronchus
6 Superior segmental bronchus
7 Medial basal segmental bronchus
8 Anterior basal segmental bronchus
9 Lateral basal segmental bronchus
10 Posterior basal segmental bronchus

Fig. 168: Anterior Aspect of Larynx, Trachea, and Bronchi

NOTE: 1) the trachea bifurcates into two principal (primary) bronchi. These then divide into lobar (secondary) bronchi which in turn give rise to segmental (tertiary) bronchi.

2) the larynx is located in the anterior aspect of the neck, and its thyroid and cricoid cartilages can be felt through the skin.

3) the thyroid cartilage, projected posteriorly, lies at the level of the 4th and 5th cervical vertebrae, while the cricoid cartilage is at the 6th cervical level. The trachea commences at the lower end of the cricoid and extends slightly more than four inches before bifurcating into the two primary bronchi at the level of T-4. Two inches of trachea lie above the suprasternal notch in the neck, while about two inches of trachea are intrathoracic above its bifurcation.

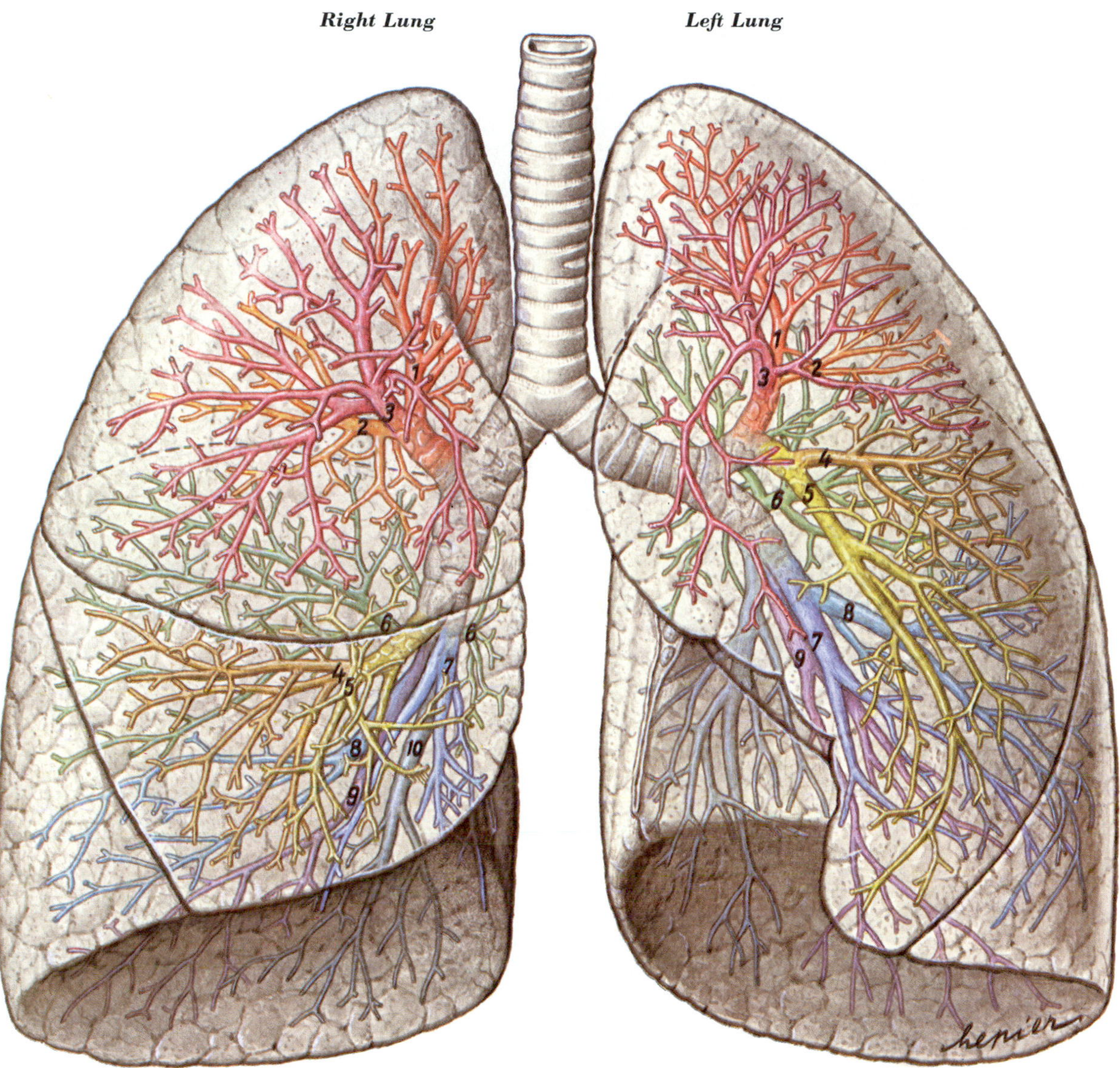
Right Lung
Left Lung

Fig. 169: Diagram of the Bronchial Tree and its Lobar and Bronchopulmonary Divisions, Anterior View

NOTE: 1) as the trachea divides, the left primary bronchus diverges at a more abrupt angle than the right primary bronchus to reach their respective lungs. Thus, the left bronchus is directed more transversely and the right bronchus more inferiorly.

2) on the right side the upper lobar bronchus branches from the primary bronchus almost immediately, even above the pulmonary artery (eparterial), while the bronchus directed toward the middle and lower lobes branches below the position of the main stem of the pulmonary artery (hyparterial).

3) on the left side the initial lobar bronchus, branching from the primary bronchus, is directed upward and lateralward to the upper lobe segments and its lingular segments. The remaining lobar bronchus is directed inferiorly and soon divides into the segmental bronchi of the lower lobe.

4) the segmental bronchi numbered above are as follows:

Right lung:

1	Apical	6	Superior
2	Posterior	7	Medial basal
3	Anterior	8	Anterior basal
4	Lateral	9	Lateral basal
5	Medial	10	Posterior basal

Left lung:

1	Apical	6	Superior
2	Posterior	7	Medial basal
3	Anterior	8	Anterior basal
4	Superior lingular	9	Lateral basal
5	Inferior lingular	10	Posterior basal

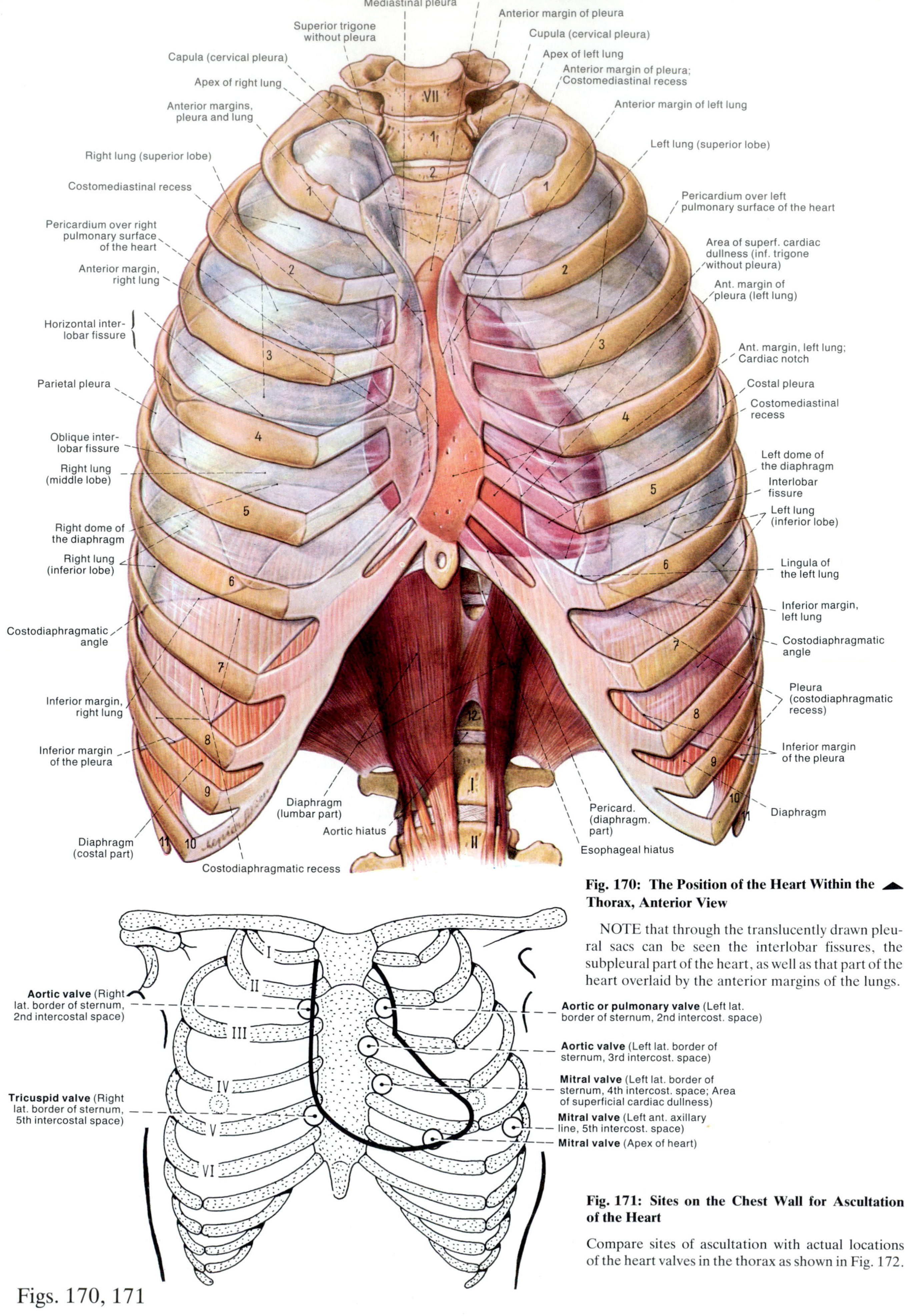

Mediastinal pleura
Aortic recess of pericardium
Anterior margin of pleura
Superior trigone without pleura
Cupula (cervical pleura)
Capula (cervical pleura)
Apex of left lung
Apex of right lung
Anterior margin of pleura; Costomediastinal recess
Anterior margins, pleura and lung
Anterior margin of left lung
VII
Left lung (superior lobe)
Right lung (superior lobe)
Costomediastinal recess
Pericardium over left pulmonary surface of the heart
Pericardium over right pulmonary surface of the heart
Area of superf. cardiac dullness (inf. trigone without pleura)
Anterior margin, right lung
Ant. margin of pleura (left lung)
Horizontal interlobar fissure
Ant. margin, left lung; Cardiac notch
Parietal pleura
Costal pleura
Costomediastinal recess
Oblique interlobar fissure
Left dome of the diaphragm
Right lung (middle lobe)
Interlobar fissure
Left lung (inferior lobe)
Right dome of the diaphragm
Right lung (inferior lobe)
Lingula of the left lung
Inferior margin, left lung
Costodiaphragmatic angle
Costodiaphragmatic angle
Pleura (costodiaphragmatic recess)
Inferior margin, right lung
Inferior margin of the pleura
Inferior margin of the pleura
Diaphragm
Diaphragm (lumbar part)
Pericard. (diaphragm. part)
Aortic hiatus
Diaphragm (costal part)
Esophageal hiatus
Costodiaphragmatic recess
1
2
3
4
5
6
7
8
9
10
11
12
I
II
III
IV
V
VI
Aortic valve (Right lat. border of sternum, 2nd intercostal space)
Aortic or pulmonary valve (Left lat. border of sternum, 2nd intercost. space)
Aortic valve (Left lat. border of sternum, 3rd intercost. space)
Mitral valve (Left lat. border of sternum, 4th intercost. space; Area of superficial cardiac dullness)
Tricuspid valve (Right lat. border of sternum, 5th intercostal space)
Mitral valve (Left ant. axillary line, 5th intercost. space)
Mitral valve (Apex of heart)

Fig. 170: The Position of the Heart Within the Thorax, Anterior View ▲

NOTE that through the translucently drawn pleural sacs can be seen the interlobar fissures, the subpleural part of the heart, as well as that part of the heart overlaid by the anterior margins of the lungs.

Fig. 171: Sites on the Chest Wall for Ascultation of the Heart

Compare sites of ascultation with actual locations of the heart valves in the thorax as shown in Fig. 172.

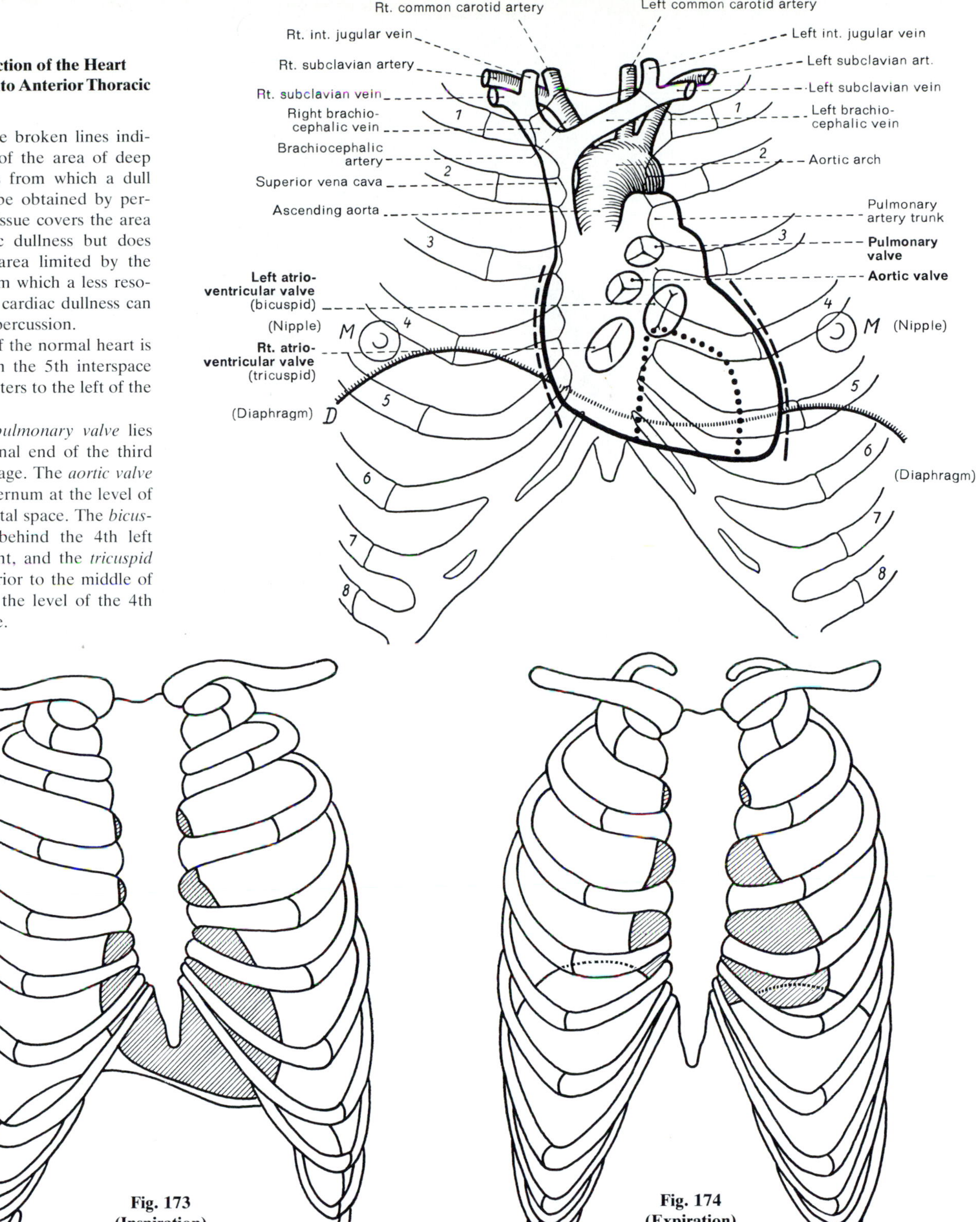

Rt. common carotid artery
Left common carotid artery
Rt. int. jugular vein
Left int. jugular vein
Rt. subclavian artery
Left subclavian art.
Rt. subclavian vein
Left subclavian vein
Right brachio-cephalic vein
Left brachio-cephalic vein
Brachiocephalic artery
Aortic arch
Superior vena cava
Ascending aorta
Pulmonary artery trunk
Pulmonary valve
Aortic valve
Left atrio-ventricular valve (bicuspid)
(Nipple)
M
Rt. atrio-ventricular valve (tricuspid)
(Diaphragm)
D
1
2
3
4
5
6
7
8

Fig. 172: Projection of the Heart and its Valves onto Anterior Thoracic Wall

NOTE: 1) the broken lines indicate the limits of the area of deep cardiac dullness from which a dull resonance can be obtained by percussion. Lung tissue covers the area of deep cardiac dullness but does not cover the area limited by the dotted lines from which a less resonant superficial cardiac dullness can be obtained by percussion.

2) the apex of the normal heart is usually found in the 5th interspace about 9 centimeters to the left of the midline.

3) that the *pulmonary valve* lies behind the sternal end of the third left costal cartilage. The *aortic valve* is behind the sternum at the level of the 3rd intercostal space. The *bicuspid valve* lies behind the 4th left sternocostal joint, and the *tricuspid valve* lies posterior to the middle of the sternum at the level of the 4th intercostal space.

Fig. 173 (Inspiration)

Fig. 174 (Expiration)

Fig. 173 and 174: Positions of the Heart During Full Inspiration (Fig. 173) and During Full Expiration (Fig. 174)

NOTE: 1) during full inspiration (Fig. 173), the thorax is enlarged by a lowering of the diaphragm due to contraction of its muscle fibers and by elevation and expansion of the thoracic cage (ribs and sternum). The thorax expands antero-posteriorly, transversely and vertically and, therefore, during inspiration the shape of the heart becomes more oblong (i.e. its transverse diameter is decreased), and its apex and diaphragmatic surface are lowered. Inspiration must be accompanied by relaxation of the muscles of the anterior abdominal wall, allowing protrusion of the abdomen and a lowering of the abdominal viscera.

2) full expiration (Fig. 174) occurs because of an elevation of the diaphragm which results from relaxation of its muscle fibers and because the ribs and sternum are lowered. Thus, the capacity of the thoracic cage is diminished, causing an elevation in the diaphragmatic surface of the heart, as well as the apex of the heart, with a concomitant increase in the heart's transverse diameter. Expiration is accompanied by contraction of the muscles of the anterior abdominal wall and an elevation of abdominal viscera, which push the relaxed diaphragm upward.

3) an excellent account of the movements of the thoracic cage during respiration is presented by Professor R. J. Last: *Anatomy, Regional and Applied,* 7th edition, Churchill-Livingstone, 1984, pages 219 – 223).

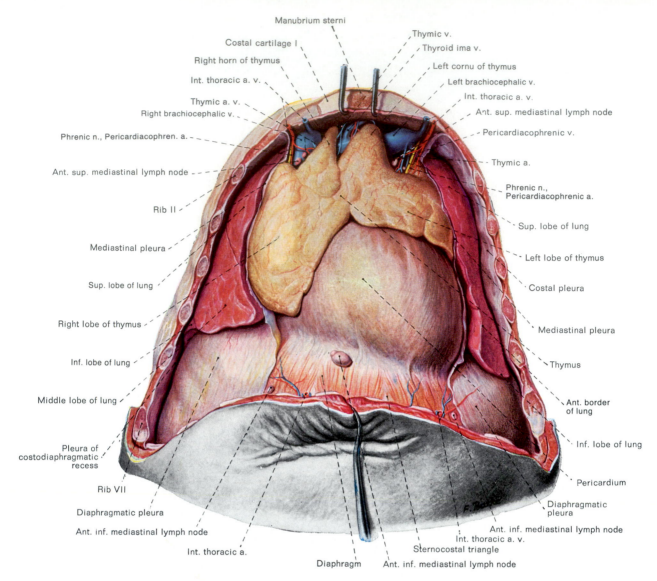

Manubrium sterni
Costal cartilage I
Right horn of thymus
Int. thoracic a. v.
Thymic a. v.
Right brachiocephalic v.
Phrenic n., Pericardiacophren. a.
Ant. sup. mediastinal lymph node
Rib II
Mediastinal pleura
Sup. lobe of lung
Right lobe of thymus
Inf. lobe of lung
Middle lobe of lung
Pleura of costodiaphragmatic recess
Rib VII
Diaphragmatic pleura
Ant. inf. mediastinal lymph node
Int. thoracic a.
Diaphragm
Ant. inf. mediastinal lymph node
Sternocostal triangle
Int. thoracic a. v.
Ant. inf. mediastinal lymph node
Diaphragmatic pleura
Pericardium
Inf. lobe of lung
Ant. border of lung
Thymus
Mediastinal pleura
Costal pleura
Left lobe of thymus
Sup. lobe of lung
Phrenic n., Pericardiacophrenic a.
Thymic a.
Pericardiacophrenic v.
Ant. sup. mediastinal lymph node
Int. thoracic a. v.
Left brachiocephalic v.
Left cornu of thymus
Thyroid ima v.
Thymic v.

Fig. 175: Anterior View of the Pericardial Sac and Thymus in the Newborn Child

NOTE that the great vessels of the superior mediastinum and the parietal pericardium over the base of the heart are covered by the thymus gland. The thymus also extends into the root of the neck in the newborn child.

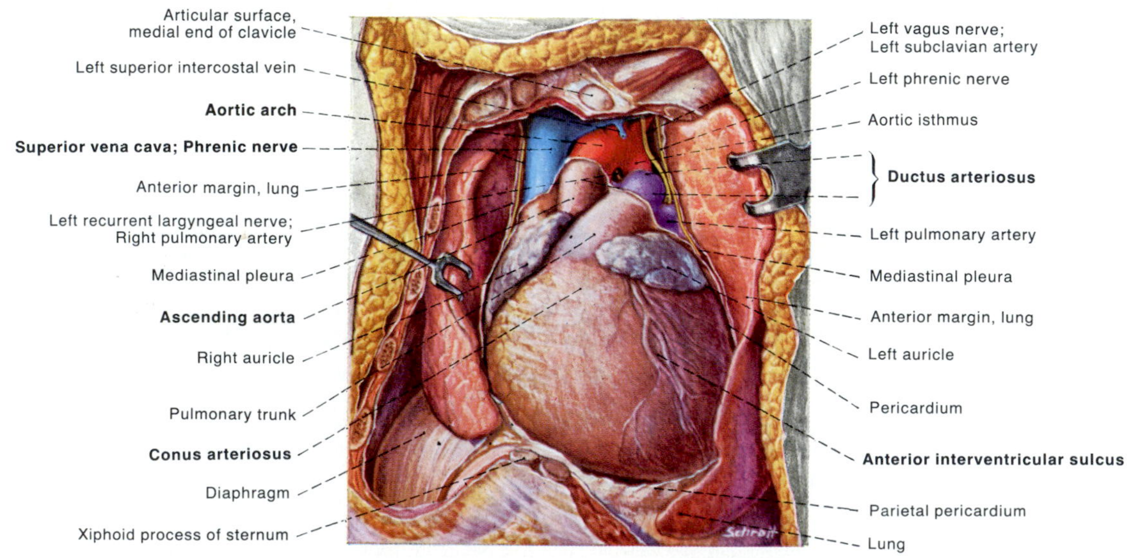

Articular surface, medial end of clavicle
Left superior intercostal vein
Aortic arch
Superior vena cava; Phrenic nerve
Anterior margin, lung
Left recurrent largyngeal nerve; Right pulmonary artery
Mediastinal pleura
Ascending aorta
Right auricle
Pulmonary trunk
Conus arteriosus
Diaphragm
Xiphoid process of sternum
Left vagus nerve; Left subclavian artery
Left phrenic nerve
Aortic isthmus
Ductus arteriosus
Left pulmonary artery
Mediastinal pleura
Anterior margin, lung
Left auricle
Pericardium
Anterior interventricular sulcus
Parietal pericardium
Lung

Fig. 176: Anterior View of the Heart and Great Vessels of a Newborn Child After Removal of the Thymus

NOTE that the ductus arteriosus is still an enlarged structure immediately after birth.

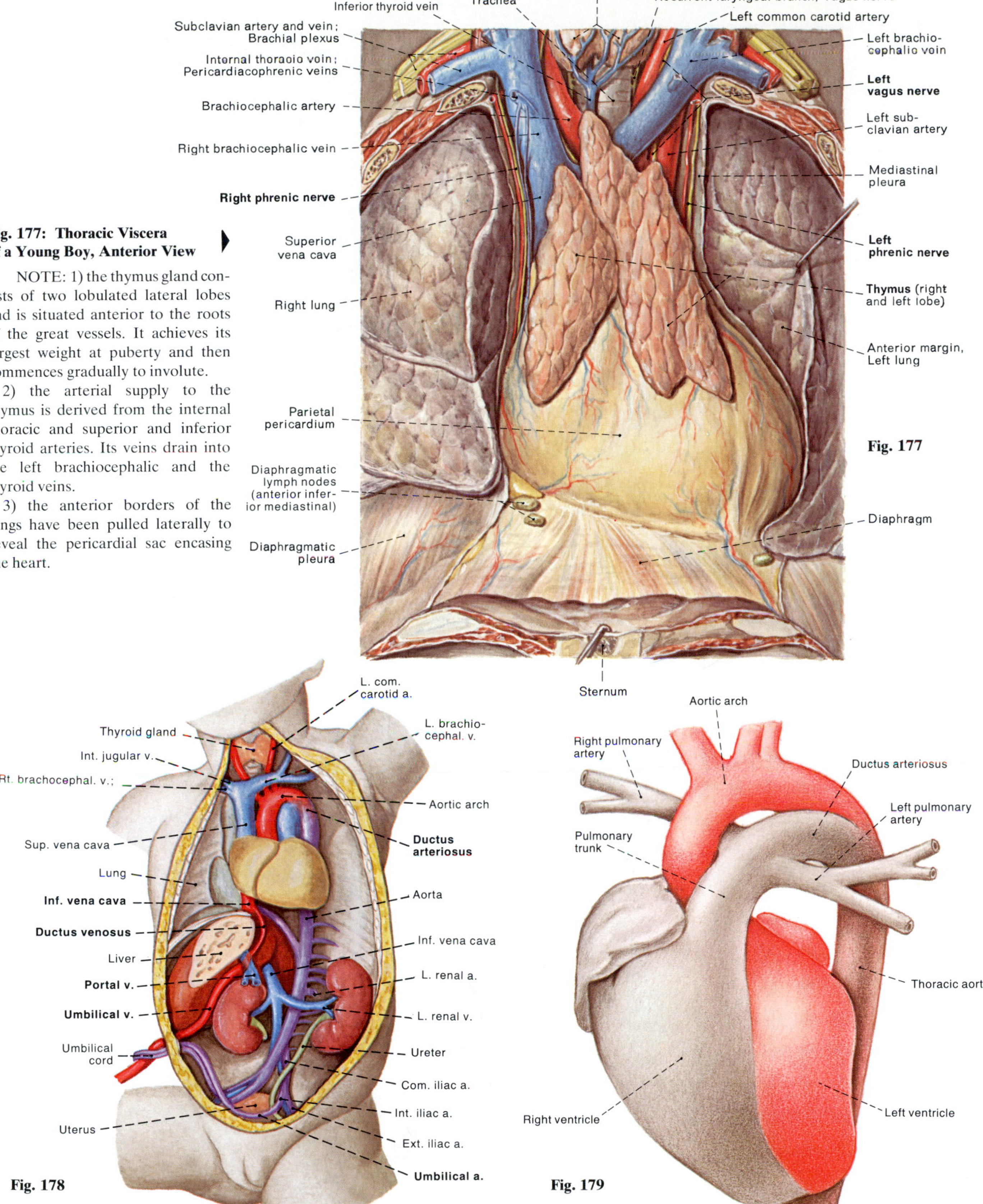
Thyroid gland
Trachea
Inferior thyroid vein
Recurrent laryngeal branch, Vagus nerve
Left common carotid artery
Subclavian artery and vein; Brachial plexus
Internal thoracic vein; Pericardiacophrenic veins
Left brachio-cephalic vein
Left vagus nerve
Brachiocephalic artery
Left sub-clavian artery
Right brachiocephalic vein
Mediastinal pleura
Right phrenic nerve
Left phrenic nerve
Superior vena cava
Thymus (right and left lobe)
Right lung
Anterior margin, Left lung
Parietal pericardium
Fig. 177
Diaphragmatic lymph nodes (anterior inferior mediastinal)
Diaphragm
Diaphragmatic pleura
Sternum
L. com. carotid a.
L. brachio-cephal. v.
Thyroid gland
Int. jugular v.
Rt. brachocephal. v.;
Aortic arch
Ductus arteriosus
Sup. vena cava
Lung
Aorta
Inf. vena cava
Ductus venosus
Inf. vena cava
Liver
L. renal a.
Portal v.
Umbilical v.
L. renal v.
Umbilical cord
Ureter
Com. iliac a.
Int. iliac a.
Uterus
Ext. iliac a.
Umbilical a.
Fig. 178
Aortic arch
Right pulmonary artery
Ductus arteriosus
Left pulmonary artery
Pulmonary trunk
Thoracic aorta
Right ventricle
Left ventricle
Fig. 179

Fig. 177: Thoracic Viscera of a Young Boy, Anterior View

NOTE: 1) the thymus gland consists of two lobulated lateral lobes and is situated anterior to the roots of the great vessels. It achieves its largest weight at puberty and then commences gradually to involute.

2) the arterial supply to the thymus is derived from the internal thoracic and superior and inferior thyroid arteries. Its veins drain into the left brachiocephalic and the thyroid veins.

3) the anterior borders of the lungs have been pulled laterally to reveal the pericardial sac encasing the heart.

Fig. 178: Fetal Circulation in the Newborn Female

NOTE that the shades of color depict the oxygen level in the blood. Observe the umbilical vein, ductus arteriosus and umbilical artery – all structures peculiar to fetal circulation.

Fig. 179: The Ductus Arteriosus in the Neonatal Heart

NOTE that the blood in the thoracic aorta is a mixture, part of it coming from the right heart by way of the ductus arteriosus and part of it coming from the left heart through the aorta.

Fig. 180: The Opened Right Atrium in a 310 mm Human Fetus ▶

NOTE that in the fetus blood pressure is greater in the right atrium than in the left and foramen ovale II is covered by the septum primum on the left side of septum secundum. This allows blood to pass easily from right to left. After birth, pressure is greater in the left atrium, which results in closure of the foramen.

Fig. 181: Circulation in the Fetus

NOTE: 1) that some oxygenated blood from the umbilical vein flows through the foramen ovale to the left ventricle and out the aorta to the head and limbs. A considerable amount of oxygenated blood also goes through the right ventricle, pulmonary trunk, and then through the ductus arteriosus to the aorta.

2) the shunt formed by the ductus venosus to bypass the liver.

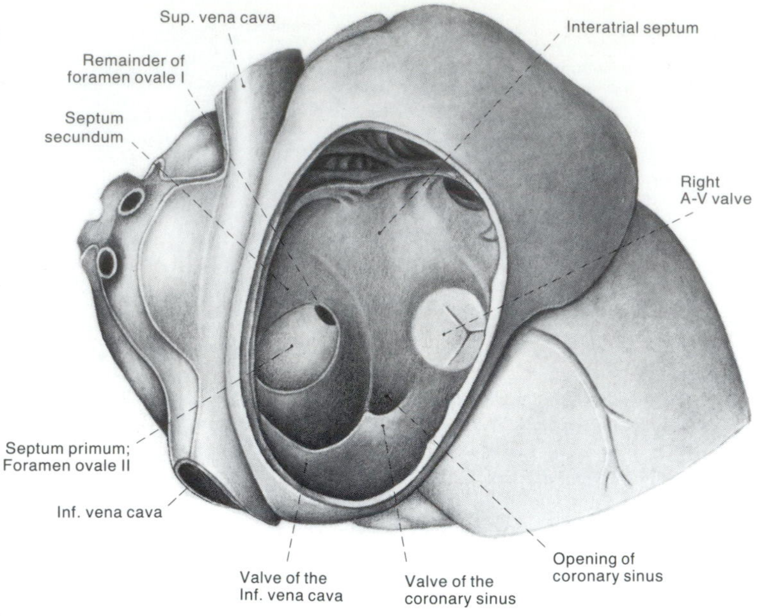

▼

Sup. vena cava
Interatrial septum
Remainder of foramen ovale I
Septum secundum
Right A-V valve
Septum primum; Foramen ovale II
Inf. vena cava
Valve of the Inf. vena cava
Valve of the coronary sinus
Opening of coronary sinus

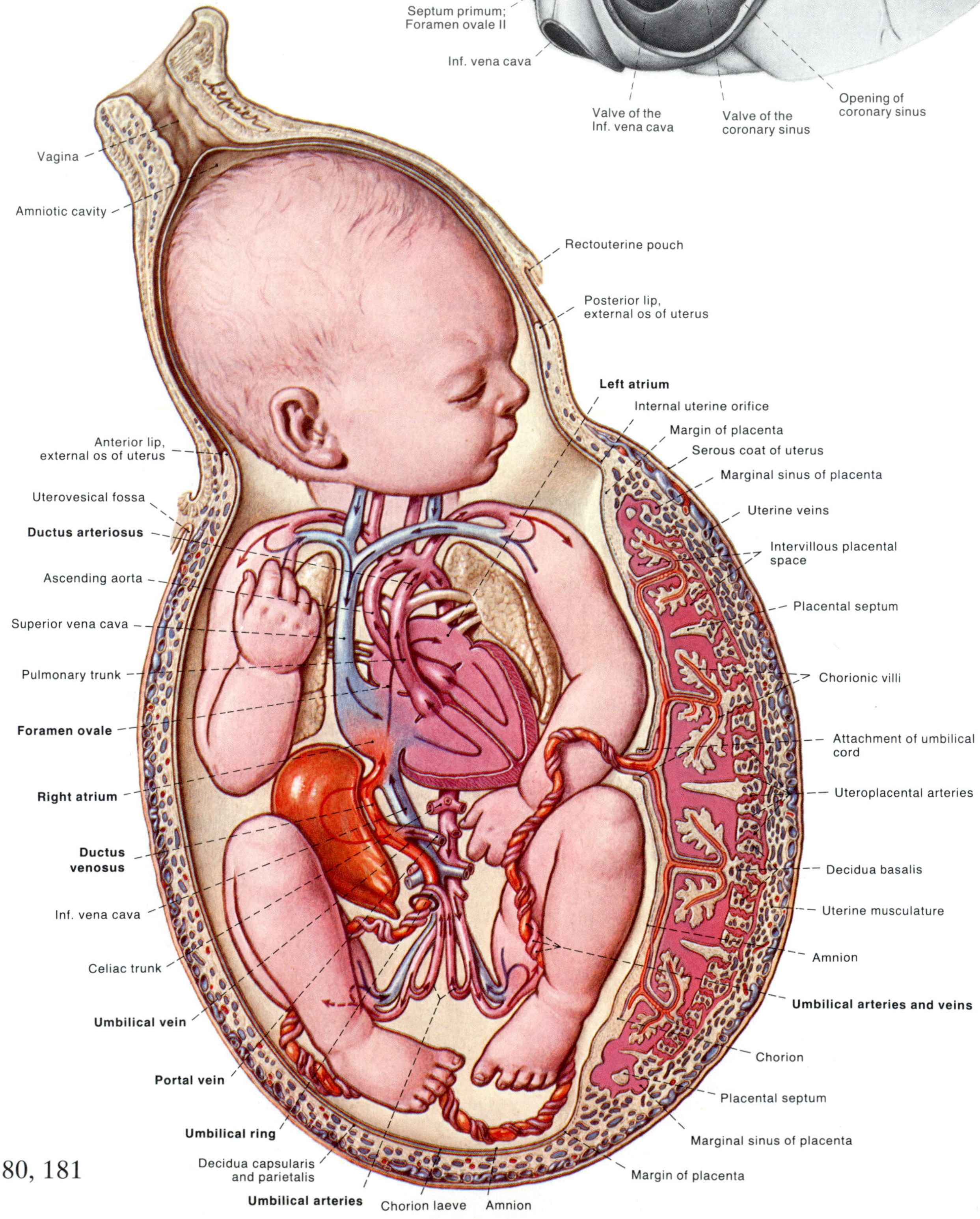

Vagina
Amniotic cavity
Rectouterine pouch
Posterior lip, external os of uterus
Left atrium
Internal uterine orifice
Margin of placenta
Serous coat of uterus
Marginal sinus of placenta
Anterior lip, external os of uterus
Uterovesical fossa
Uterine veins
Ductus arteriosus
Intervillous placental space
Ascending aorta
Placental septum
Superior vena cava
Pulmonary trunk
Chorionic villi
Foramen ovale
Attachment of umbilical cord
Right atrium
Uteroplacental arteries
Ductus venosus
Decidua basalis
Inf. vena cava
Uterine musculature
Amnion
Celiac trunk
Umbilical arteries and veins
Umbilical vein
Chorion
Portal vein
Placental septum
Umbilical ring
Marginal sinus of placenta
Decidua capsularis and parietalis
Margin of placenta
Umbilical arteries
Chorion laeve
Amnion

Figs. 180, 181

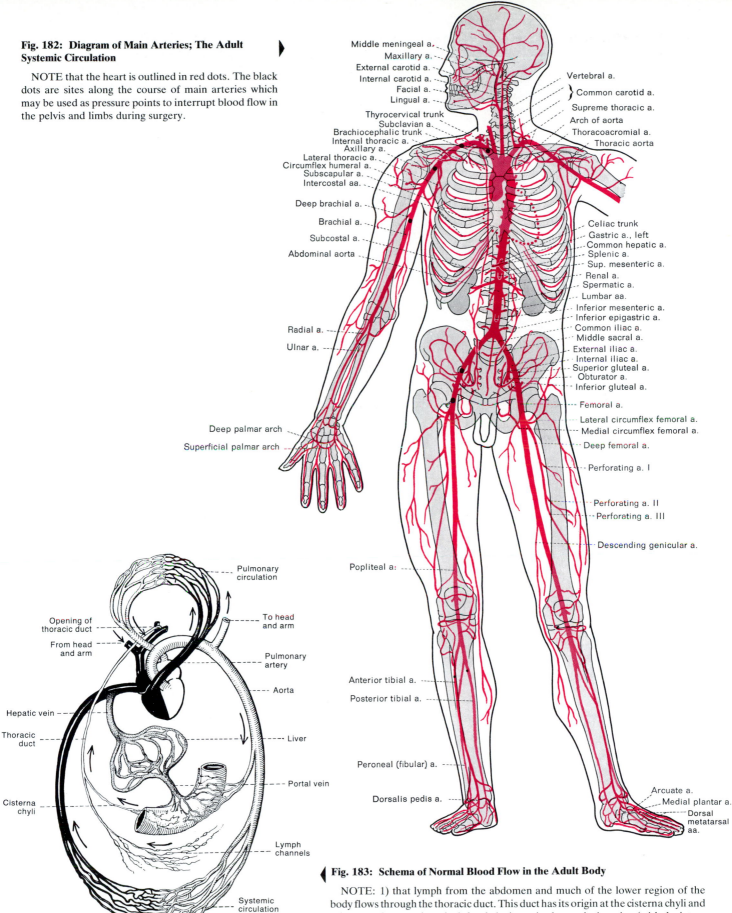

Fig. 182: Diagram of Main Arteries; The Adult Systemic Circulation

NOTE that the heart is outlined in red dots. The black dots are sites along the course of main arteries which may be used as pressure points to interrupt blood flow in the pelvis and limbs during surgery.

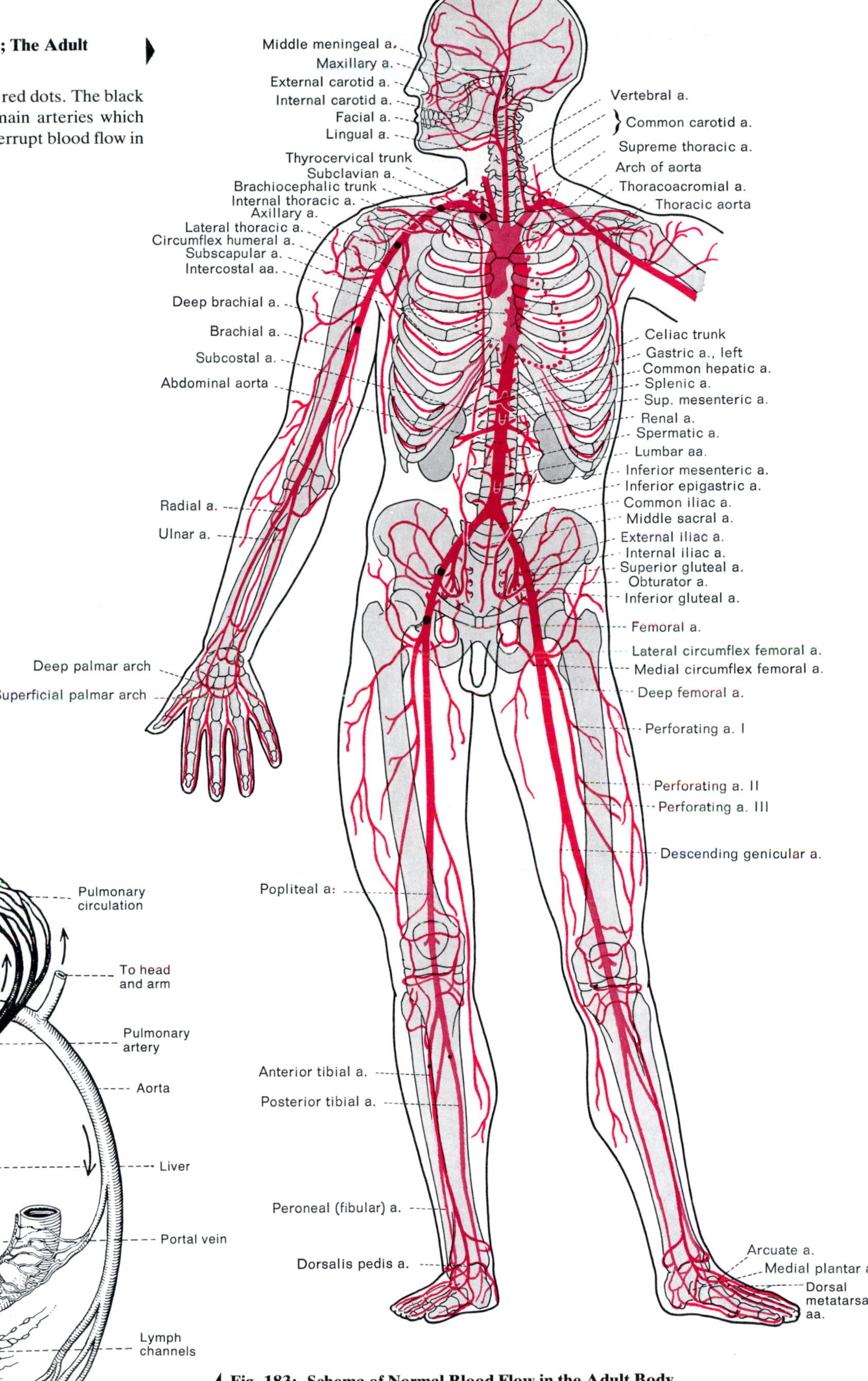

Middle meningeal a.
Maxillary a.
External carotid a.
Internal carotid a.
Facial a.
Lingual a.
Thyrocervical trunk
Subclavian a.
Brachiocephalic trunk
Internal thoracic a.
Axillary a.
Lateral thoracic a.
Circumflex humeral a.
Subscapular a.
Intercostal aa.
Deep brachial a.
Brachial a.
Subcostal a.
Abdominal aorta
Radial a.
Ulnar a.
Deep palmar arch
Superficial palmar arch
Popliteal a.
Anterior tibial a.
Posterior tibial a.
Peroneal (fibular) a.
Dorsalis pedis a.
Vertebral a.
Common carotid a.
Supreme thoracic a.
Arch of aorta
Thoracoacromial a.
Thoracic aorta
Celiac trunk
Gastric a., left
Common hepatic a.
Splenic a.
Sup. mesenteric a.
Renal a.
Spermatic a.
Lumbar aa.
Inferior mesenteric a.
Inferior epigastric a.
Common iliac a.
Middle sacral a.
External iliac a.
Internal iliac a.
Superior gluteal a.
Obturator a.
Inferior gluteal a.
Femoral a.
Lateral circumflex femoral a.
Medial circumflex femoral a.
Deep femoral a.
Perforating a. I
Perforating a. II
Perforating a. III
Descending genicular a.
Arcuate a.
Medial plantar a.
Dorsal metatarsal aa.

Pulmonary circulation
Opening of thoracic duct
To head and arm
From head and arm
Pulmonary artery
Aorta
Hepatic vein
Thoracic duct
Liver
Portal vein
Cisterna chyli
Lymph channels
Systemic circulation

Fig. 183: Schema of Normal Blood Flow in the Adult Body

NOTE: 1) that lymph from the abdomen and much of the lower region of the body flows through the thoracic duct. This duct has its origin at the cisterna chyli and subsequently opens into the left subclavian vein close to its junction (with the internal jugular vein) into the left brachiocephalic vein. In this manner circulating lymph in the tissues is returned to the general circulation.

2) the systemic, pulmonary and portal circulatory channels.

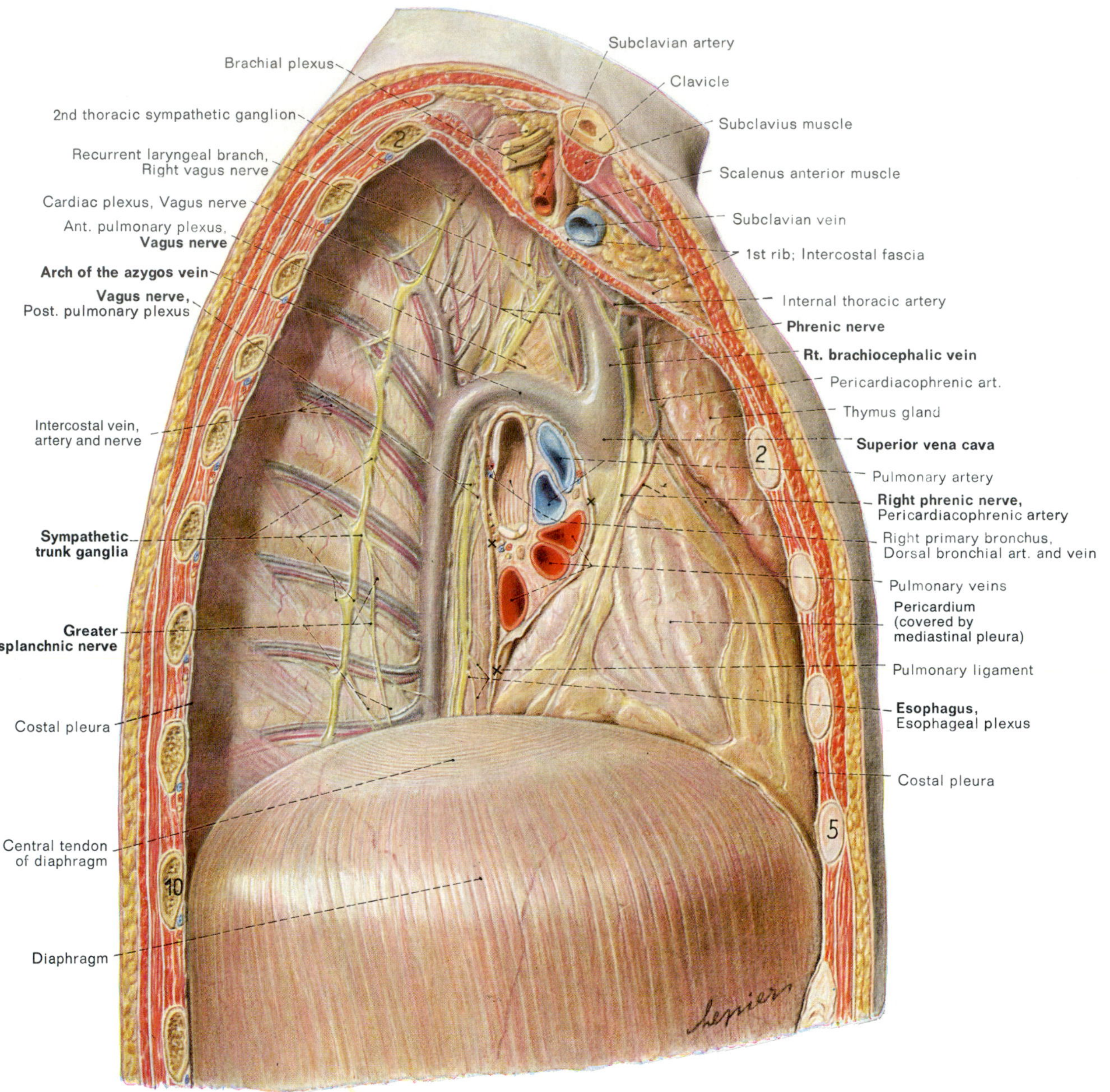

Subclavian artery
Brachial plexus
Clavicle
2nd thoracic sympathetic ganglion
Subclavius muscle
Recurrent laryngeal branch,
Right vagus nerve
Scalenus anterior muscle
Cardiac plexus, Vagus nerve
Subclavian vein
Ant. pulmonary plexus,
Vagus nerve
1st rib; Intercostal fascia
Arch of the azygos vein
Internal thoracic artery
Vagus nerve,
Post. pulmonary plexus
Phrenic nerve
Rt. brachiocephalic vein
Pericardiacophrenic art.
Thymus gland
Intercostal vein,
artery and nerve
Superior vena cava
Pulmonary artery
Right phrenic nerve,
Pericardiacophrenic artery
Sympathetic trunk ganglia
Right primary bronchus,
Dorsal bronchial art. and vein
Pulmonary veins
Pericardium
(covered by
mediastinal pleura)
Greater splanchnic nerve
Pulmonary ligament
Esophagus,
Esophageal plexus
Costal pleura
Costal pleura
Central tendon
of diaphragm
Diaphragm

Fig. 184: The Right Side of the Mediastinum

NOTE: 1) with the right lung removed and the structures at its hilum transected, the organs of the mediastinum are exposed and their right lateral surface viewed.

2) the right side of the heart covered by the pericardium and the course of the phrenic nerve and pericardiacophrenic vessels.

3) the ascending course of the azygos vein, its arch and its junction with the superior vena cava.

4) that the right vagus nerve descends in the thorax behind the root of the right lung to form the posterior pulmonary plexus. It then helps form the esophageal plexus and leaves the thorax on the posterior aspect of the esophagus.

5) the dome of the diaphragm on the right side, taking the rounded form of the underlying liver. The inferior (diaphragmatic) surface of the heart rests on the diaphragm.

6) the position of the thoracic sympathetic chain of ganglia coursing longitudinally along the inner surface of the thoracic wall. Observe the greater splanchnic nerve.

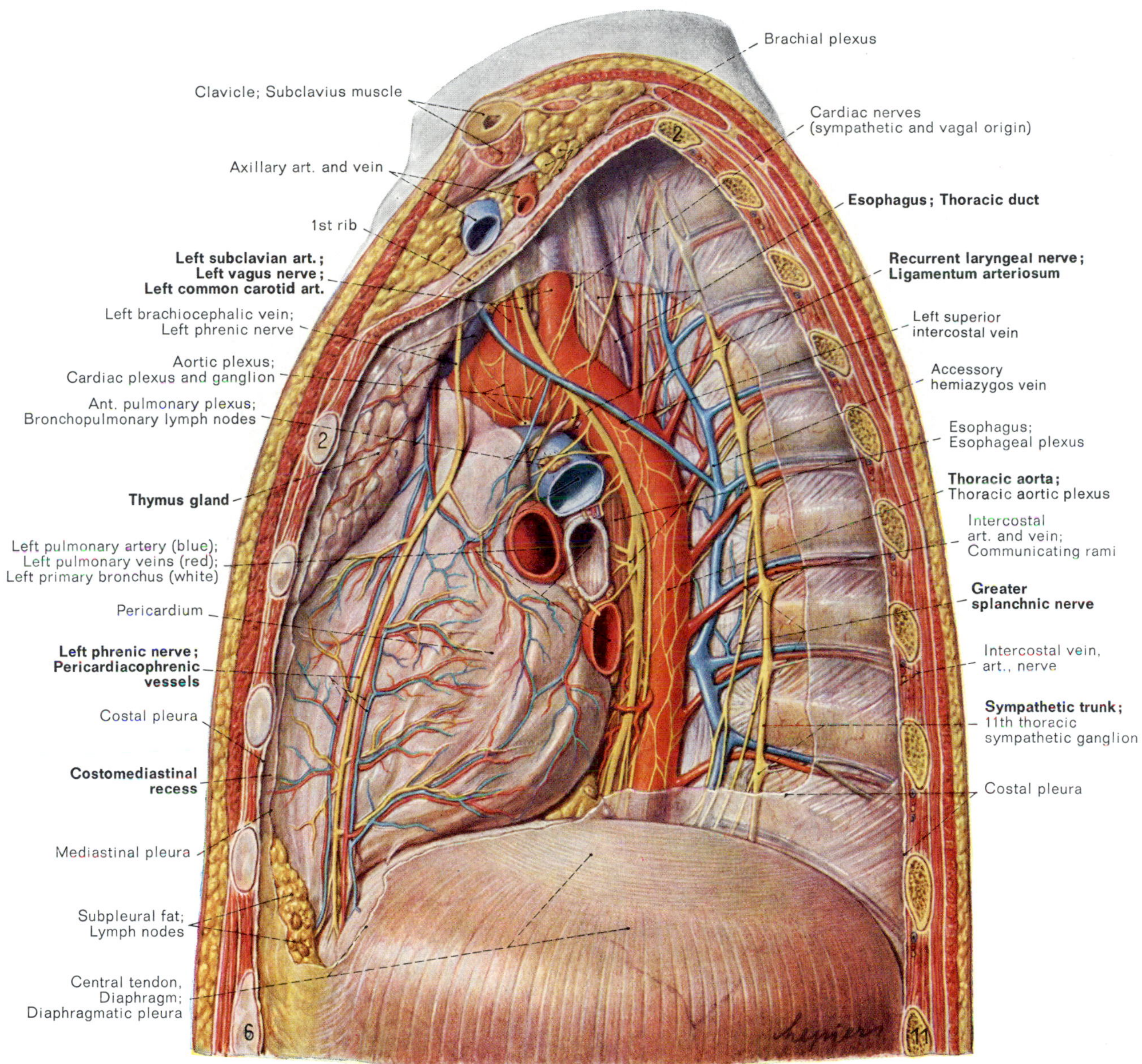
Brachial plexus
Clavicle; Subclavius muscle
Cardiac nerves
(sympathetic and vagal origin)
Axillary art. and vein
Esophagus; Thoracic duct
1st rib
Left subclavian art.;
Left vagus nerve;
Left common carotid art.
Recurrent laryngeal nerve;
Ligamentum arteriosum
Left brachiocephalic vein;
Left phrenic nerve
Left superior
intercostal vein
Aortic plexus;
Cardiac plexus and ganglion
Accessory
hemiazygos vein
Ant. pulmonary plexus;
Bronchopulmonary lymph nodes
Esophagus;
Esophageal plexus
Thymus gland
Thoracic aorta;
Thoracic aortic plexus
Left pulmonary artery (blue);
Left pulmonary veins (red);
Left primary bronchus (white)
Intercostal
art. and vein;
Communicating rami
Pericardium
Greater
splanchnic nerve
Left phrenic nerve;
Pericardiacophrenic
vessels
Intercostal vein,
art., nerve
Costal pleura
Sympathetic trunk;
11th thoracic
sympathetic ganglion
Costomediastinal
recess
Costal pleura
Mediastinal pleura
Subpleural fat;
Lymph nodes
Central tendon,
Diaphragm;
Diaphragmatic pleura
2
2
6
11

Fig. 185: The Left Side of the Mediastinum

NOTE: 1) with the left lung removed along with most of the mediastinal pleura, the structures of the mediastinum are observed from their left side.

2) the left phrenic nerve and pericardiacophrenic vessels coursing to the diaphragm along the pericardium covering the left side of the heart.

3) the aorta ascends about two inches before it arches posteriorly and to the left of the vertebral column. The descending thoracic aorta commences at about the level of the 4th thoracic vertebra and as it descends, it comes to lie directly anterior to the vertebral column. The intercostal arteries branch directly from the thoracic aorta.

4) the left vagus nerve lies lateral to the aortic arch and gives off its recurrent laryngeal branch which passes inferior to the ligamentum arteriosum. The left vagus then continues to descend, contributes to the esophageal plexus, and enters the abdomen on the anterior aspect of the esophagus.

5) the position of the thymus gland anterior to the root of the great vessels at their attachments to the heart.

6) that the typical intercostal artery and vein course along the inferior border of their respective rib. Because the superior border of the ribs is free of vessels and nerves, it is a safer site for injection or drainage of the thorax.

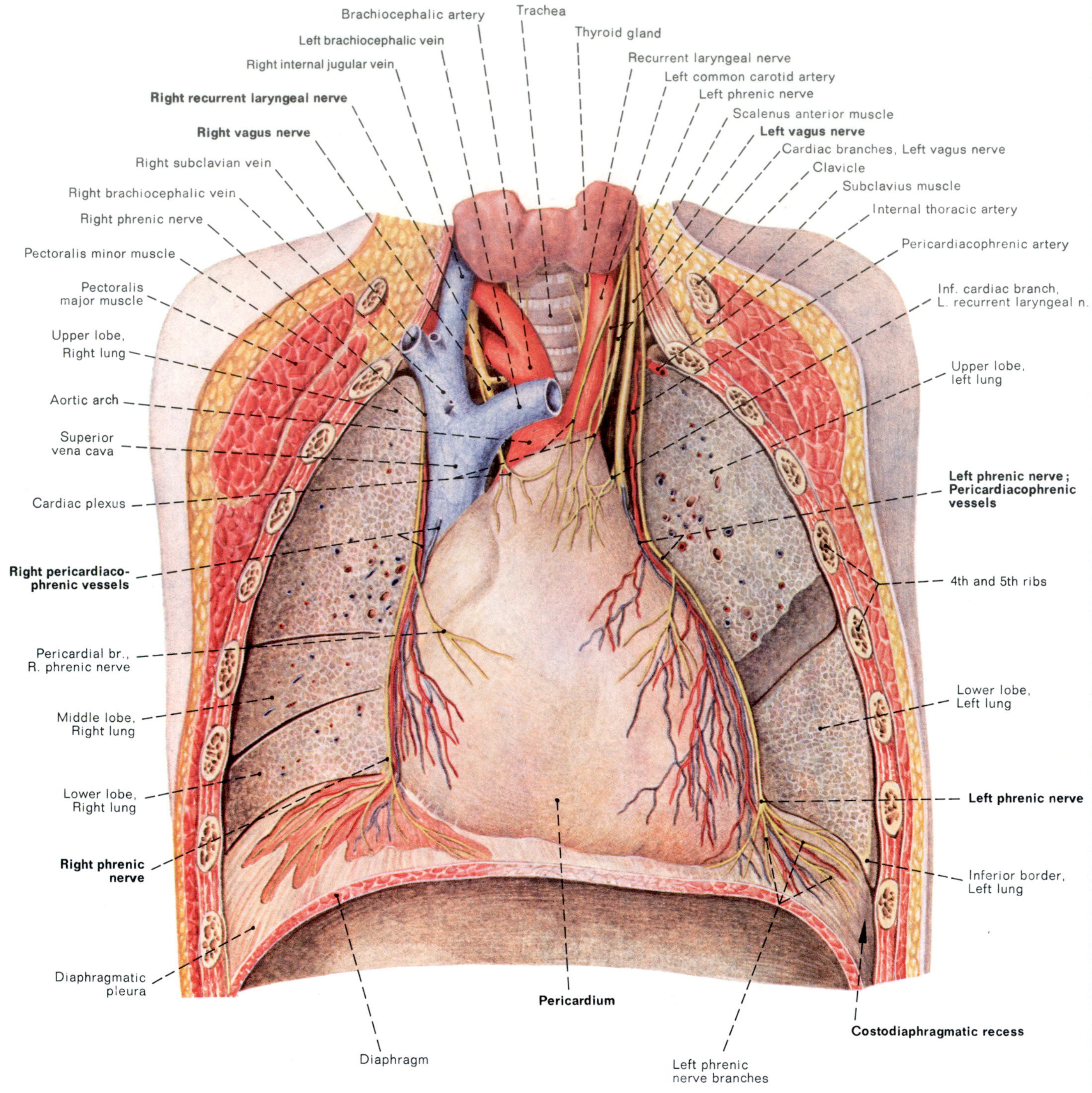
Brachiocephalic artery
Trachea
Thyroid gland
Left brachiocephalic vein
Right internal jugular vein
Recurrent laryngeal nerve
Left common carotid artery
Right recurrent laryngeal nerve
Left phrenic nerve
Scalenus anterior muscle
Right vagus nerve
Left vagus nerve
Cardiac branches, Left vagus nerve
Right subclavian vein
Clavicle
Right brachiocephalic vein
Subclavius muscle
Right phrenic nerve
Internal thoracic artery
Pectoralis minor muscle
Pericardiacophrenic artery
Pectoralis major muscle
Inf. cardiac branch, L. recurrent laryngeal n.
Upper lobe, Right lung
Upper lobe, left lung
Aortic arch
Superior vena cava
Left phrenic nerve; Pericardiacophrenic vessels
Cardiac plexus
Right pericardiaco-phrenic vessels
4th and 5th ribs
Pericardial br., R. phrenic nerve
Lower lobe, Left lung
Middle lobe, Right lung
Lower lobe, Right lung
Left phrenic nerve
Right phrenic nerve
Inferior border, Left lung
Diaphragmatic pleura
Pericardium
Costodiaphragmatic recess
Diaphragm
Left phrenic nerve branches

Fig. 186: The Adult Heart, Pericardium and Superior Mediastinal Structures, Anterior View

NOTE: 1) in this frontal section through the thorax, the anterior thoracic wall and the anterior aspect of the lungs and diaphragm have been removed, leaving the pericardium, its contents and its associated vessels and nerves intact. The courses of the vagus nerves and their branches in the superior mediastinum are also demonstrated.

2) the phrenic nerves originate in the neck (C3, 4, 5) and descend almost vertically to innervate the diaphragm. In the superior mediastinum they join the pericardiacophrenic vessels, travel along the lateral surfaces of the pericardium and are distributed principally to the diaphragm, sending some sensory fibers to the pericardium as well.

3) the pericardium is formed by an outer fibrous layer which is lined by an inner serous sac. As the heart develops it invaginates into the inner serous sac, thereby being covered by a visceral layer of serous pericardium (epicardium) and a parietal layer of serous pericardium. The outer fibrous pericardium has only a parietal layer.

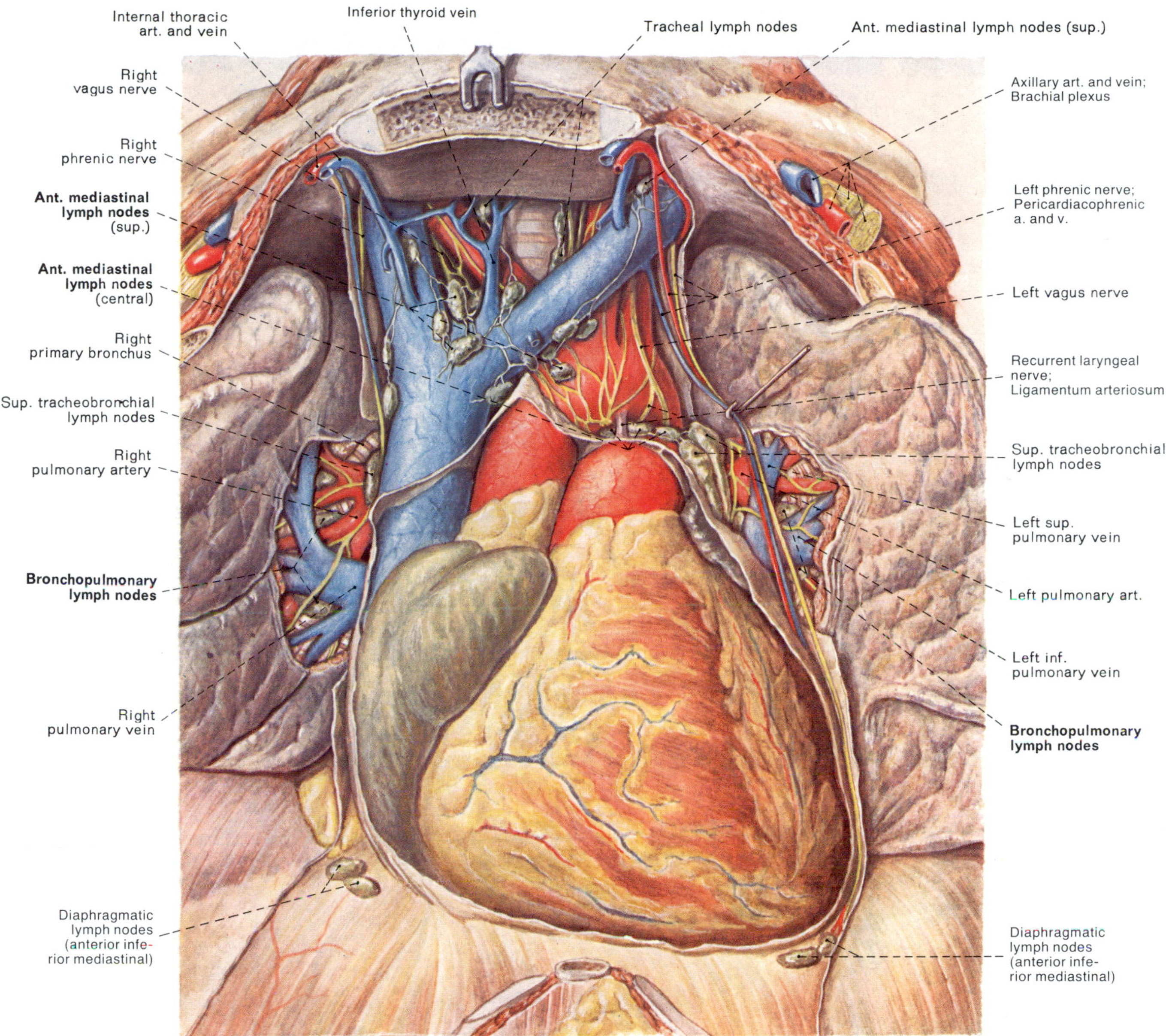

Internal thoracic art. and vein
Inferior thyroid vein
Tracheal lymph nodes
Ant. mediastinal lymph nodes (sup.)
Right vagus nerve
Axillary art. and vein; Brachial plexus
Right phrenic nerve
Ant. mediastinal lymph nodes (sup.)
Left phrenic nerve; Pericardiacophrenic a. and v.
Ant. mediastinal lymph nodes (central)
Left vagus nerve
Right primary bronchus
Recurrent laryngeal nerve; Ligamentum arteriosum
Sup. tracheobronchial lymph nodes
Sup. tracheobronchial lymph nodes
Right pulmonary artery
Left sup. pulmonary vein
Bronchopulmonary lymph nodes
Left pulmonary art.
Left inf. pulmonary vein
Right pulmonary vein
Bronchopulmonary lymph nodes
Diaphragmatic lymph nodes (anterior inferior mediastinal)
Diaphragmatic lymph nodes (anterior inferior mediastinal)

Fig. 187: Lymphatics of the Thorax, Anterior Aspect

NOTE: 1) in this dissection the anterior thoracic wall was removed along with the ventral portion of the pericardium. Further, the anterior borders of the lungs have been pulled laterally to reveal the lymphatic channels at the roots of the lungs. The thymus has also been removed and the manubrium reflected superiorly to expose the organs at the thoracic inlet and their associated lymphatics.

2) the lymph nodes in the anterior aspect of the thoracic cavity might be divided into those associated with the thoracic cage (parietal) and those associated with the organs (visceral). Probably all of the nodes indicated in this figure are visceral nodes.

3) situated ventrally are the *anterior mediastinal nodes* which include a superior group lying ventral to the brachiocephalic veins along their course in the superior mediastinum and at their junction to form the superior vena cava. A more centrally located group lies ventral to the arch of the aorta. Inferiorly, anterior diaphragmatic nodes are sometimes also classified as part of the anterior mediastinal nodes.

4) large numbers of lymph nodes are associated with the trachea, the bronchi and the other structures at the root of the lung. These nodes have been aptly named tracheal, tracheobronchial, bronchopulmonary and pulmonary.

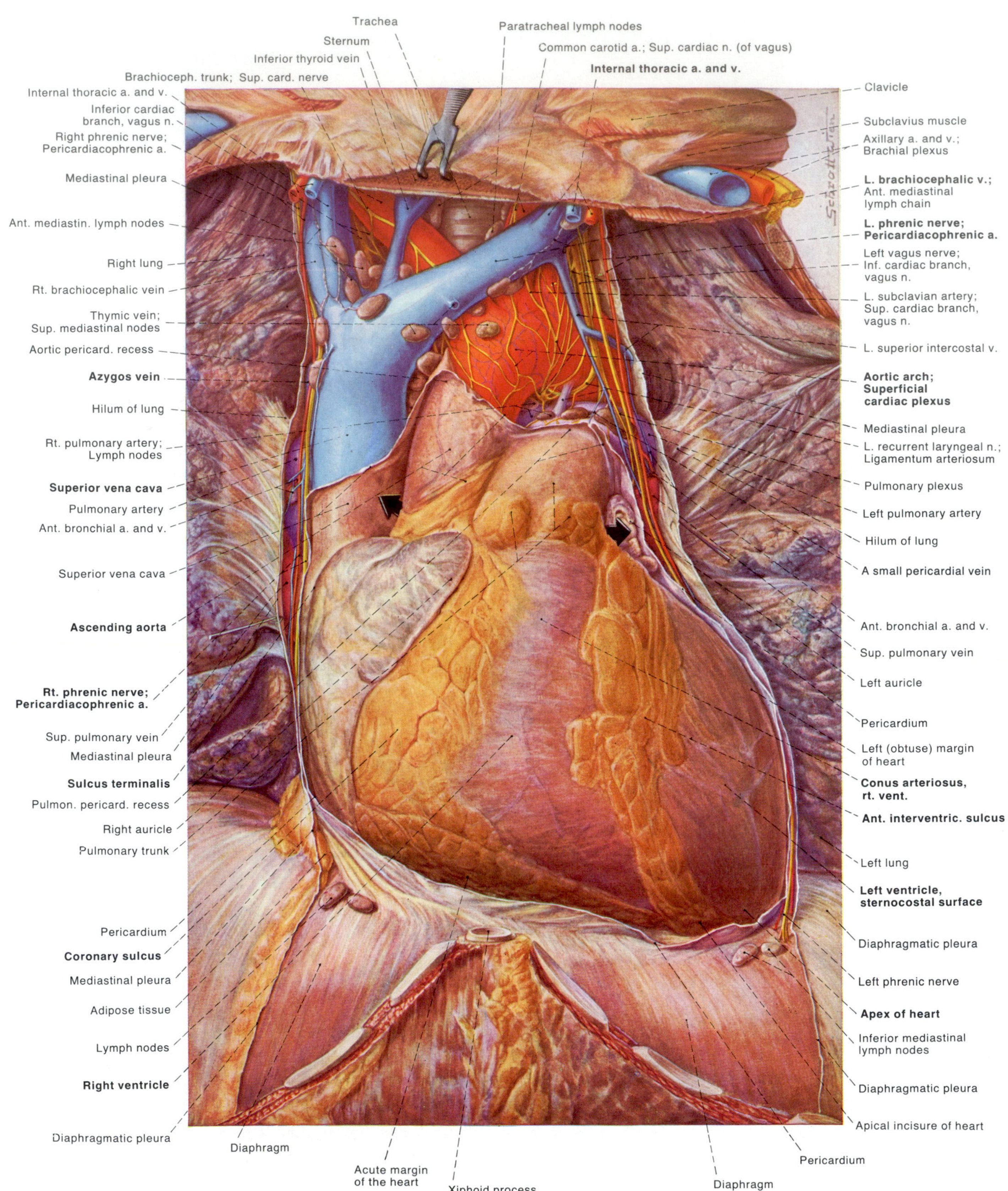
Trachea
Sternum
Inferior thyroid vein
Brachioceph. trunk; Sup. card. nerve
Internal thoracic a. and v.
Inferior cardiac branch, vagus n.
Right phrenic nerve; Pericardiacophrenic a.
Mediastinal pleura
Ant. mediastin. lymph nodes
Right lung
Rt. brachiocephalic vein
Thymic vein; Sup. mediastinal nodes
Aortic pericard. recess
Azygos vein
Hilum of lung
Rt. pulmonary artery; Lymph nodes
Superior vena cava
Pulmonary artery
Ant. bronchial a. and v.
Superior vena cava
Ascending aorta
Rt. phrenic nerve; Pericardiacophrenic a.
Sup. pulmonary vein
Mediastinal pleura
Sulcus terminalis
Pulmon. pericard. recess
Right auricle
Pulmonary trunk
Pericardium
Coronary sulcus
Mediastinal pleura
Adipose tissue
Lymph nodes
Right ventricle
Diaphragmatic pleura
Diaphragm
Acute margin of the heart
Xiphoid process
Paratracheal lymph nodes
Common carotid a.; Sup. cardiac n. (of vagus)
Internal thoracic a. and v.
Clavicle
Subclavius muscle
Axillary a. and v.; Brachial plexus
L. brachiocephalic v.; Ant. mediastinal lymph chain
L. phrenic nerve; Pericardiacophrenic a.
Left vagus nerve; Inf. cardiac branch, vagus n.
L. subclavian artery; Sup. cardiac branch, vagus n.
L. superior intercostal v.
Aortic arch; Superficial cardiac plexus
Mediastinal pleura
L. recurrent laryngeal n.; Ligamentum arteriosum
Pulmonary plexus
Left pulmonary artery
Hilum of lung
A small pericardial vein
Ant. bronchial a. and v.
Sup. pulmonary vein
Left auricle
Pericardium
Left (obtuse) margin of heart
Conus arteriosus, rt. vent.
Ant. interventric. sulcus
Left lung
Left ventricle, sternocostal surface
Diaphragmatic pleura
Left phrenic nerve
Apex of heart
Inferior mediastinal lymph nodes
Diaphragmatic pleura
Apical incisure of heart
Pericardium
Diaphragm

Fig. 188: The Heart and Great Vessels, Anterior View

NOTE that the anterior portion of the pericardium has been removed along with the remnants of the thymus to reveal the heart in its normal position within the middle mediastinum. The arrow is in the transverse pericardial sinus.

Fig. 188

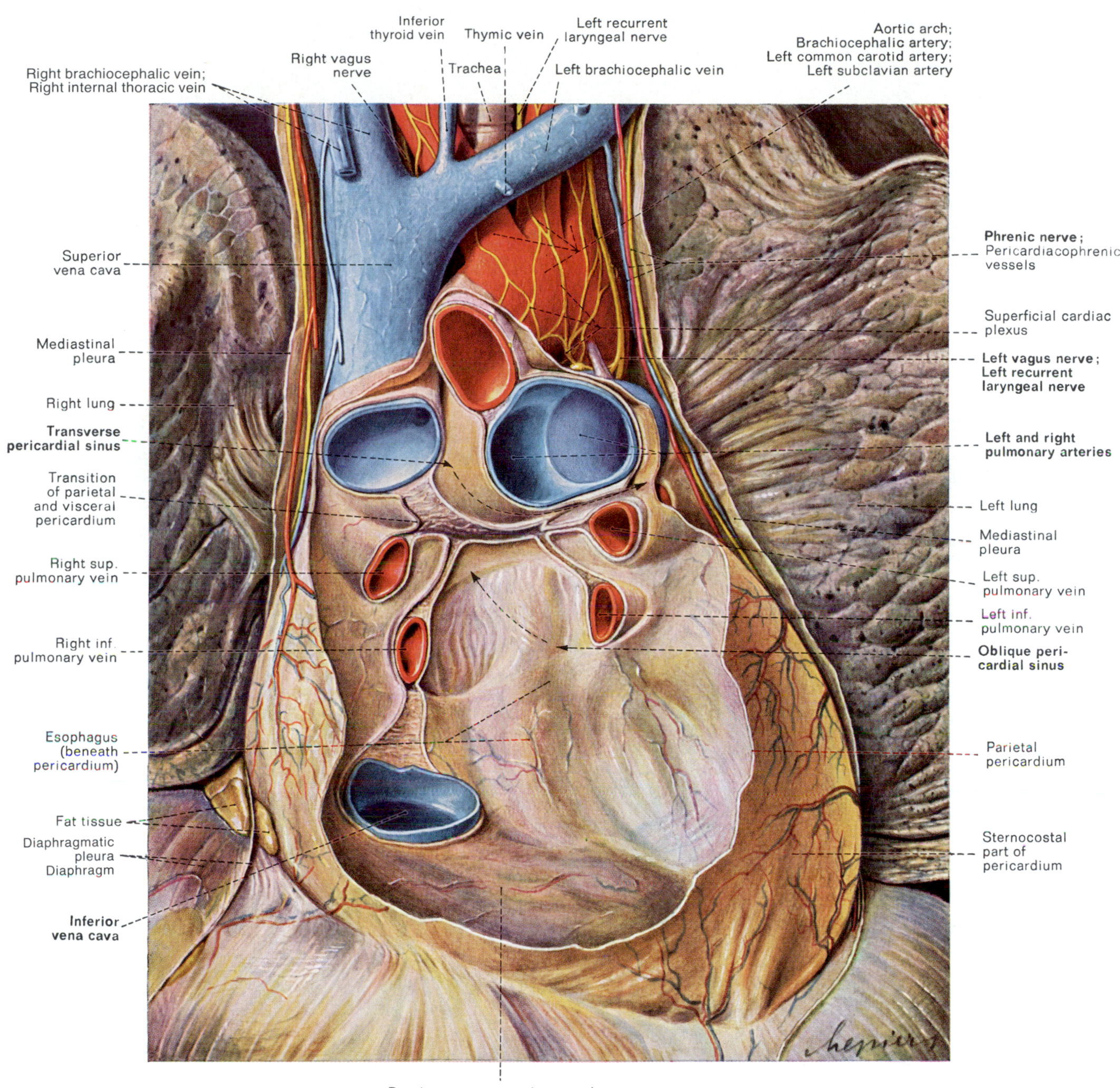

Inferior thyroid vein
Thymic vein
Left recurrent laryngeal nerve
Aortic arch; Brachiocephalic artery; Left common carotid artery; Left subclavian artery
Right vagus nerve
Trachea
Left brachiocephalic vein
Right brachiocephalic vein; Right internal thoracic vein
Superior vena cava
Phrenic nerve; Pericardiacophrenic vessels
Superficial cardiac plexus
Mediastinal pleura
Left vagus nerve; Left recurrent laryngeal nerve
Right lung
Transverse pericardial sinus
Left and right pulmonary arteries
Transition of parietal and visceral pericardium
Left lung
Mediastinal pleura
Right sup. pulmonary vein
Left sup. pulmonary vein
Left inf. pulmonary vein
Right inf. pulmonary vein
Oblique pericardial sinus
Esophagus (beneath pericardium)
Parietal pericardium
Fat tissue
Diaphragmatic pleura
Diaphragm
Sternocostal part of pericardium
Inferior vena cava
Diaphragmatic part of pericardium

Fig. 189: Interior of the Pericardium, Anterior View

NOTE: 1) the pericardium has been opened anteriorly and the heart has been severed from the great vessels and removed. Eight vessels have been cut: the superior and inferior venae cavae, the four pulmonary veins, the pulmonary artery and the aorta.

2) the oblique pericardial sinus is located in the central portion of this posterior wall and is bounded by the pericardial reflections over the pulmonary veins and the venae cavae (venous mesocardium). With the heart in place, the oblique pericardial sinus may be palpated by inserting several fingers behind the heart and probing superiorly until the blind pouch of the sinus is felt.

3) the transverse pericardial sinus lies behind the pericardial reflection surrounding the aorta and pulmonary artery (arterial mesocardium). It may be located by probing from right to left with the index finger immediately posterior to the pulmonary trunk.

4) the site of bifurcation of the pulmonary trunk beneath the arch of the aorta and the course of the left recurrent laryngeal nerve beneath the ligamentum arteriosum (not labelled).

Fig. 190: The Coronary Vessels, Anterior View

NOTE: 1) both the left and right coronary arteries arise from the ascending aorta. The *left coronary* is directed toward the left and soon divides into a *descending anterior interventricular branch* which courses toward the apex and a circumflex branch which passes posteriorly as far as the posterior interventricular sulcus.

2) the right coronary artery is directed toward the right, passing to the posterior aspect of the heart within the coronary sulcus. In its course, branches from the right coronary supply the anterior surface of the right side (anterior cardiac artery). Its largest branch is the posterior interventricular artery which courses toward the apex on the posterior or diaphragmatic surface of the heart.

3) the principal veins of the heart drain into the coronary sinus which in turn flows into the right atrium. The distribution and course of the veins is generally similar to the arteries.

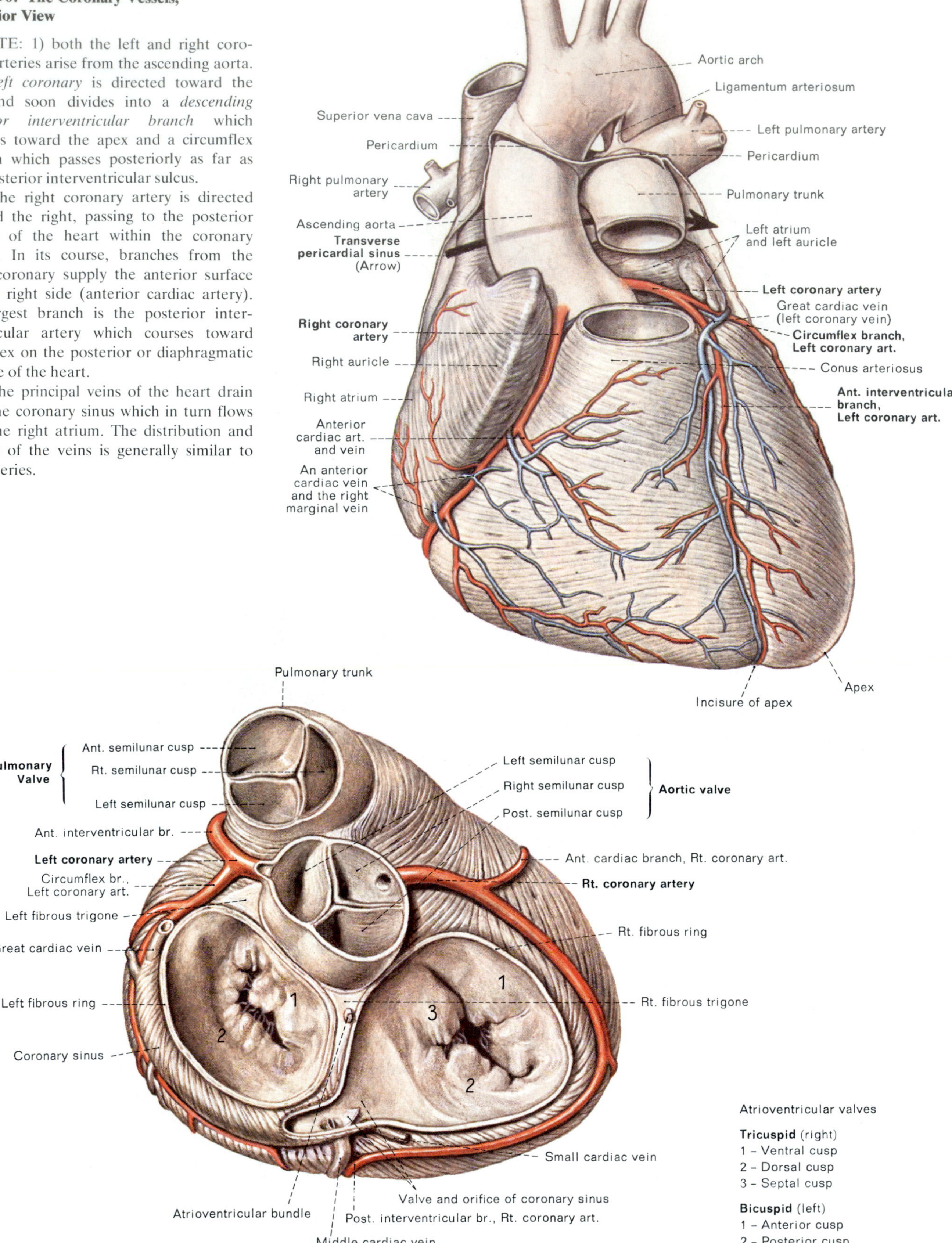
Aortic arch
Ligamentum arteriosum
Superior vena cava
Left pulmonary artery
Pericardium
Pericardium
Right pulmonary artery
Pulmonary trunk
Ascending aorta
Transverse pericardial sinus (Arrow)
Left atrium and left auricle
Left coronary artery
Great cardiac vein (left coronary vein)
Right coronary artery
Circumflex branch, Left coronary art.
Right auricle
Conus arteriosus
Right atrium
Ant. interventricular branch, Left coronary art.
Anterior cardiac art. and vein
An anterior cardiac vein and the right marginal vein
Apex
Incisure of apex
Pulmonary trunk
Pulmonary Valve
Ant. semilunar cusp
Rt. semilunar cusp
Left semilunar cusp
Left semilunar cusp
Right semilunar cusp
Post. semilunar cusp
Aortic valve
Ant. interventricular br.
Left coronary artery
Ant. cardiac branch, Rt. coronary art.
Rt. coronary artery
Circumflex br., Left coronary art.
Left fibrous trigone
Rt. fibrous ring
Great cardiac vein
Left fibrous ring
Rt. fibrous trigone
Coronary sinus
1
2
1
3
2
Atrioventricular valves
Tricuspid (right)
1 - Ventral cusp
2 - Dorsal cusp
3 - Septal cusp
Small cardiac vein
Valve and orifice of coronary sinus
Atrioventricular bundle
Post. interventricular br., Rt. coronary art.
Middle cardiac vein
Bicuspid (left)
1 - Anterior cusp
2 - Posterior cusp

Fig. 191: The Valves of the Heart and the Origin of the Coronary Vessels, Superior View

NOTE that the left coronary artery arises from the aortic sinus behind the left semilunar cusp and the right coronary artery stems from the aortic sinus behind the right semilunar cusp.

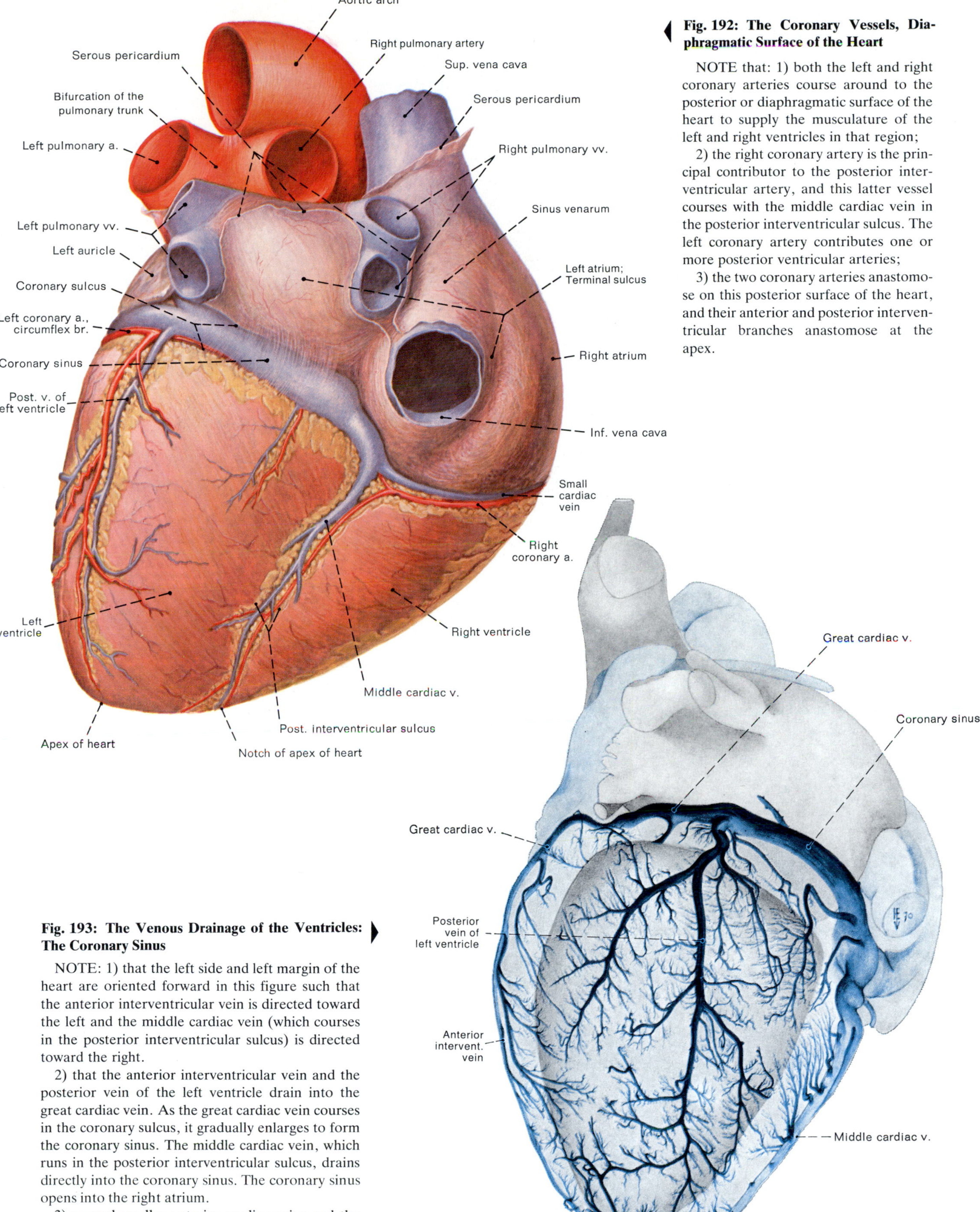
Aortic arch
Serous pericardium
Right pulmonary artery
Sup. vena cava
Bifurcation of the
pulmonary trunk
Serous pericardium
Left pulmonary a.
Right pulmonary vv.
Sinus venarum
Left pulmonary vv.
Left auricle
Left atrium;
Terminal sulcus
Coronary sulcus
Left coronary a.,
circumflex br.
Right atrium
Coronary sinus
Post. v. of
left ventricle
Inf. vena cava
Small
cardiac
vein
Right
coronary a.
Left
ventricle
Right ventricle
Middle cardiac v.
Post. interventricular sulcus
Apex of heart
Notch of apex of heart
Great cardiac v.
Coronary sinus
Great cardiac v.
Posterior
vein of
left ventricle
Anterior
intervent.
vein
Middle cardiac v.

Fig. 192: The Coronary Vessels, Diaphragmatic Surface of the Heart

NOTE that: 1) both the left and right coronary arteries course around to the posterior or diaphragmatic surface of the heart to supply the musculature of the left and right ventricles in that region;

2) the right coronary artery is the principal contributor to the posterior interventricular artery, and this latter vessel courses with the middle cardiac vein in the posterior interventricular sulcus. The left coronary artery contributes one or more posterior ventricular arteries;

3) the two coronary arteries anastomose on this posterior surface of the heart, and their anterior and posterior interventricular branches anastomose at the apex.

Fig. 193: The Venous Drainage of the Ventricles: The Coronary Sinus

NOTE: 1) that the left side and left margin of the heart are oriented forward in this figure such that the anterior interventricular vein is directed toward the left and the middle cardiac vein (which courses in the posterior interventricular sulcus) is directed toward the right.

2) that the anterior interventricular vein and the posterior vein of the left ventricle drain into the great cardiac vein. As the great cardiac vein courses in the coronary sulcus, it gradually enlarges to form the coronary sinus. The middle cardiac vein, which runs in the posterior interventricular sulcus, drains directly into the coronary sinus. The coronary sinus opens into the right atrium.

3) several smaller anterior cardiac veins and the right marginal vein cannot be seen from this view (see Figure 190).

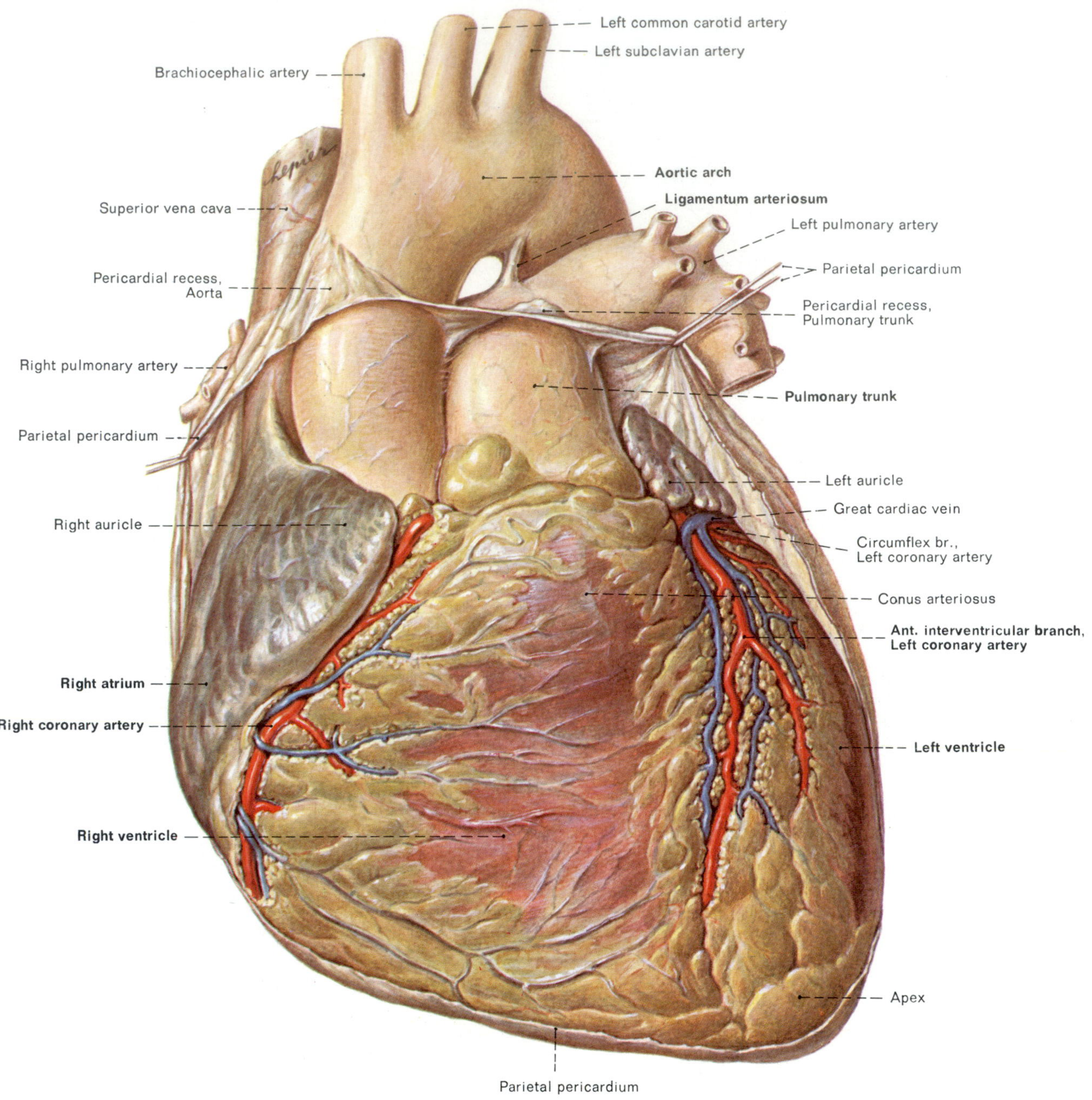
Left common carotid artery
Left subclavian artery
Brachiocephalic artery
Aortic arch
Ligamentum arteriosum
Superior vena cava
Left pulmonary artery
Parietal pericardium
Pericardial recess, Aorta
Pericardial recess, Pulmonary trunk
Right pulmonary artery
Pulmonary trunk
Parietal pericardium
Left auricle
Great cardiac vein
Right auricle
Circumflex br., Left coronary artery
Conus arteriosus
Ant. interventricular branch, Left coronary artery
Right atrium
Right coronary artery
Left ventricle
Right ventricle
Apex
Parietal pericardium

Fig. 194: Ventral View of the Heart and Great Vessels

NOTE: 1) the heart is a muscular organ with its apex pointed inferiorly, toward the left and slightly anteriorly. The base of the heart is opposite to the apex and is, therefore, directed superiorly and toward the right. The great vessels attach to the heart at its base, and the pericardium is reflected over these vessels at their origin.

2) the anterior surface of the heart is its sternocostal surface. The auricular portion of the right atrium and especially the right ventricle is seen from this anterior view; also a small part of the left ventricle is visible on the left side.

3) the pulmonary trunk originates from the right ventricle. To its right and slightly behind can be seen the aorta which arises from the left ventricle. The superior vena cava can be seen opening into the upper aspect of the right atrium.

4) the ligamentum arteriosum. This fibrous structure, attaching the left pulmonary artery to the arch of the aorta, is the postnatal remnant of the fetal ductus arteriosus which, before birth, acted as a shunt diverting some of the blood directed for the lungs back into the aorta for general systemic distribution.

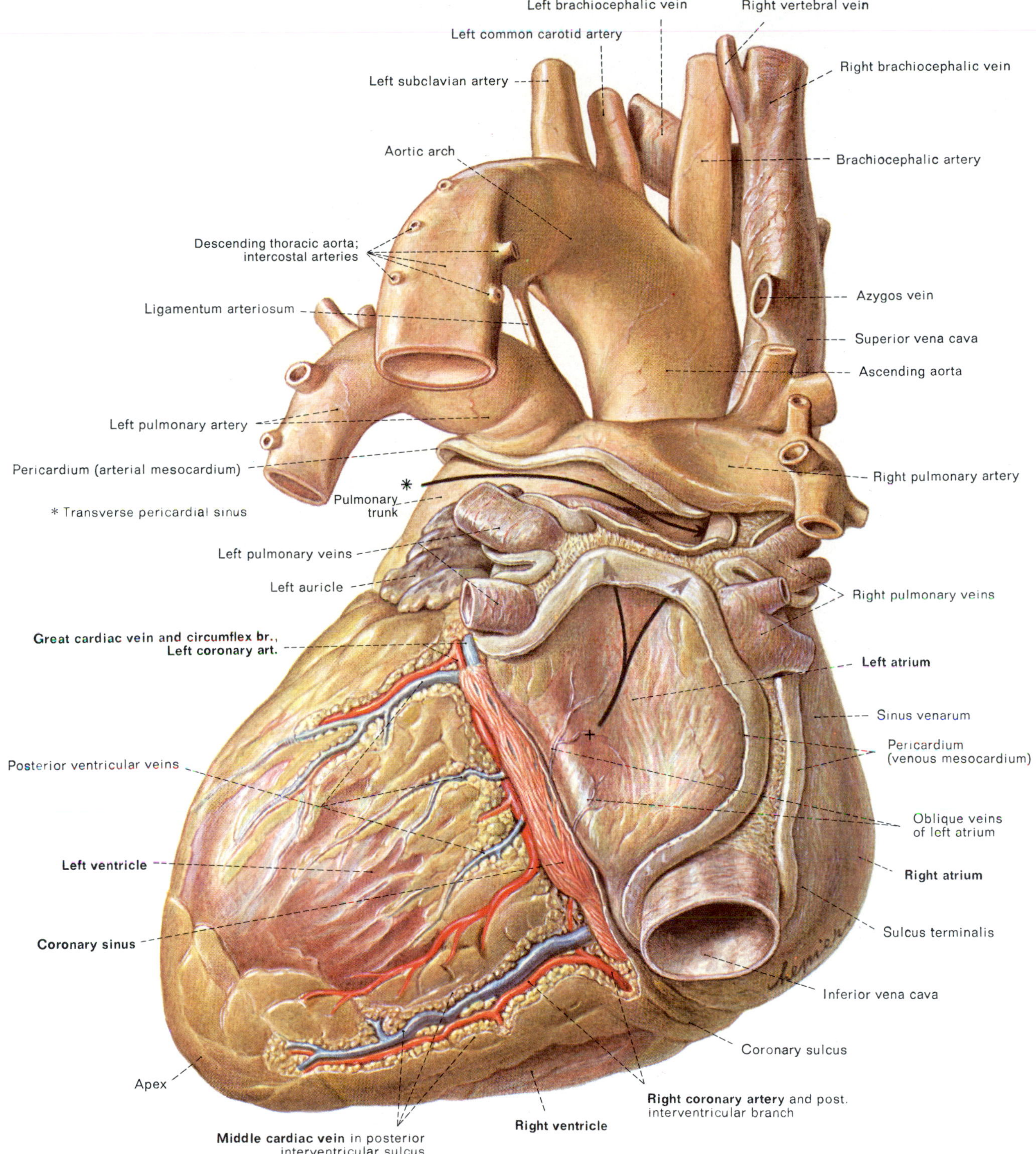
Left brachiocephalic vein
Right vertebral vein
Left common carotid artery
Right brachiocephalic vein
Left subclavian artery
Aortic arch
Brachiocephalic artery
Descending thoracic aorta; intercostal arteries
Azygos vein
Ligamentum arteriosum
Superior vena cava
Ascending aorta
Left pulmonary artery
Pericardium (arterial mesocardium)
Right pulmonary artery
* Transverse pericardial sinus
Pulmonary trunk
Left pulmonary veins
Left auricle
Right pulmonary veins
Great cardiac vein and circumflex br., Left coronary art.
Left atrium
Sinus venarum
Pericardium (venous mesocardium)
Posterior ventricular veins
Oblique veins of left atrium
Left ventricle
Right atrium
Coronary sinus
Sulcus terminalis
Inferior vena cava
Coronary sulcus
Apex
Right coronary artery and post. interventricular branch
Right ventricle
Middle cardiac vein in posterior interventricular sulcus

Fig. 195: Posterior View of the Heart and Great Vessels

NOTE: 1) the two pericardial sinuses. The long transverse arrow indicates the *transverse pericardial sinus* which lies between the arterial mesocardium and the venous mesocardium. The double arrows lie in the *oblique pericardial sinus,* the boundaries of which are actually demarcated by the pericardial reflections around the pulmonary veins.

2) the transverse sinus can be identified by placing your index finger behind the pulmonary artery and aorta with the heart *in situ.* The oblique sinus is open inferiorly, while superiorly it forms a closed sac. This sinus can also be felt by cupping your fingers behind the heart and pushing superiorly.

3) the coronary sinus separates the posterior surface regions of the atria (above and to the right) and the ventricles (below and to the left). The posterior atrial surface principally consists of the left atrium, into which flow the pulmonary veins, although below and to the right can be seen the right atrium and its inferior vena cava. The posterior atrial surface lies anterior to the vertebral column. The posterior ventricular surface is formed principally by the left ventricle and this surface lies over the diaphragm.

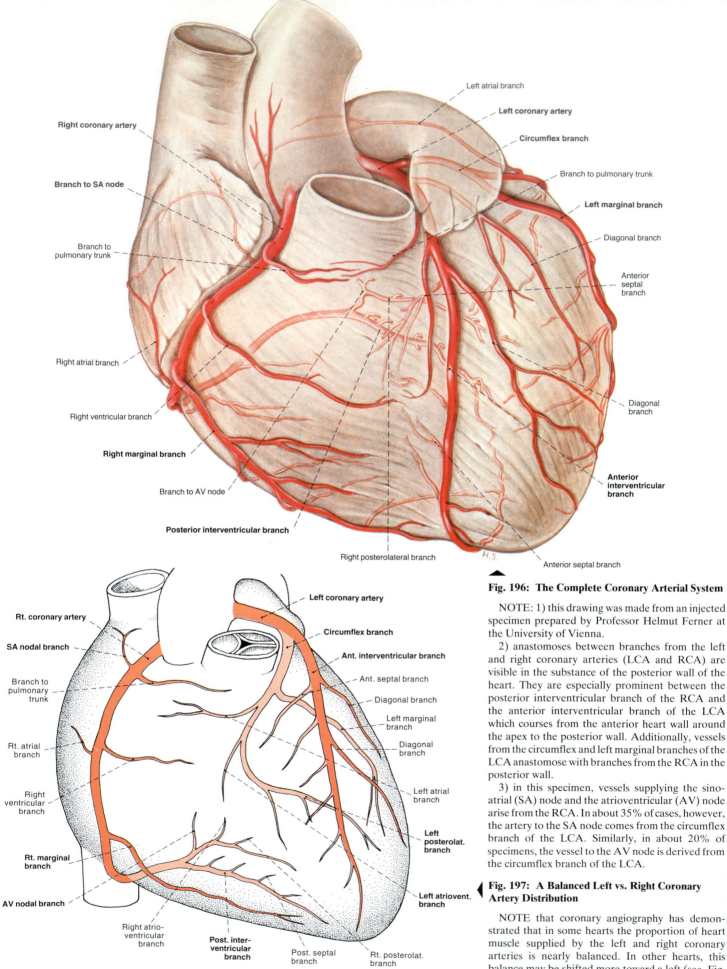

Left atrial branch
Left coronary artery
Circumflex branch
Right coronary artery
Branch to pulmonary trunk
Branch to SA node
Left marginal branch
Diagonal branch
Branch to pulmonary trunk
Anterior septal branch
Right atrial branch
Diagonal branch
Right ventricular branch
Right marginal branch
Anterior interventricular branch
Branch to AV node
Posterior interventricular branch
Right posterolateral branch
H.S.
Anterior septal branch

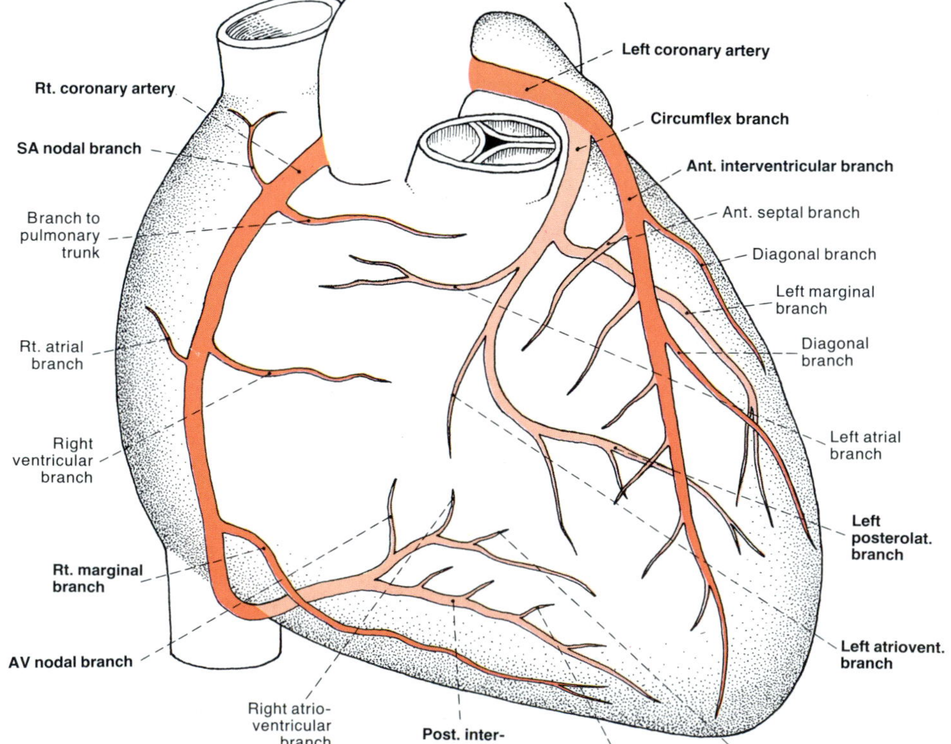

Left coronary artery
Rt. coronary artery
Circumflex branch
SA nodal branch
Ant. interventricular branch
Branch to pulmonary trunk
Ant. septal branch
Diagonal branch
Left marginal branch
Rt. atrial branch
Diagonal branch
Left atrial branch
Right ventricular branch
Left posterolat. branch
Rt. marginal branch
Left atriovent. branch
AV nodal branch
Right atrio-ventricular branch
Post. inter-ventricular branch
Post. septal branch
Rt. posterolat. branch

Fig. 196: The Complete Coronary Arterial System

NOTE: 1) this drawing was made from an injected specimen prepared by Professor Helmut Ferner at the University of Vienna.

2) anastomoses between branches from the left and right coronary arteries (LCA and RCA) are visible in the substance of the posterior wall of the heart. They are especially prominent between the posterior interventricular branch of the RCA and the anterior interventricular branch of the LCA which courses from the anterior heart wall around the apex to the posterior wall. Additionally, vessels from the circumflex and left marginal branches of the LCA anastomose with branches from the RCA in the posterior wall.

3) in this specimen, vessels supplying the sinoatrial (SA) node and the atrioventricular (AV) node arise from the RCA. In about 35% of cases, however, the artery to the SA node comes from the circumflex branch of the LCA. Similarly, in about 20% of specimens, the vessel to the AV node is derived from the circumflex branch of the LCA.

Fig. 197: A Balanced Left vs. Right Coronary Artery Distribution

NOTE that coronary angiography has demonstrated that in some hearts the proportion of heart muscle supplied by the left and right coronary arteries is nearly balanced. In other hearts, this balance may be shifted more toward a left (see. Fig. 198) or right (see Fig. 199) coronary artery dominance.

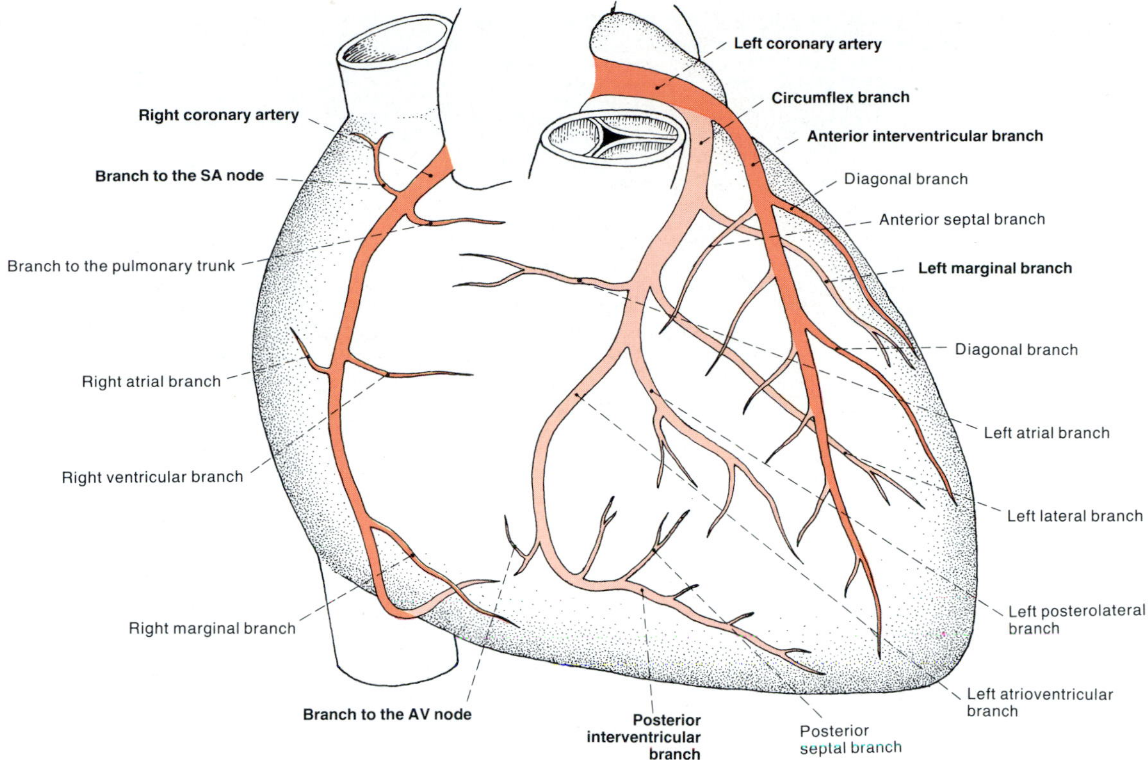

Left coronary artery
Circumflex branch
Anterior interventricular branch
Right coronary artery
Branch to the SA node
Diagonal branch
Anterior septal branch
Branch to the pulmonary trunk
Left marginal branch
Diagonal branch
Right atrial branch
Left atrial branch
Right ventricular branch
Left lateral branch
Left posterolateral branch
Right marginal branch
Left atrioventricular branch
Branch to the AV node
Posterior interventricular branch
Posterior septal branch

Fig. 198: A Dominant Distribution to the Heart Wall from the Left Coronary Artery

Observe that in hearts containing a dominant left coronary artery, there is very little contribution of blood from the right coronary artery to the posterior wall of the left ventricle. In these cases the posterior interventricular artery arises from the left coronary artery as a continuation of the enlarged circumflex branch.

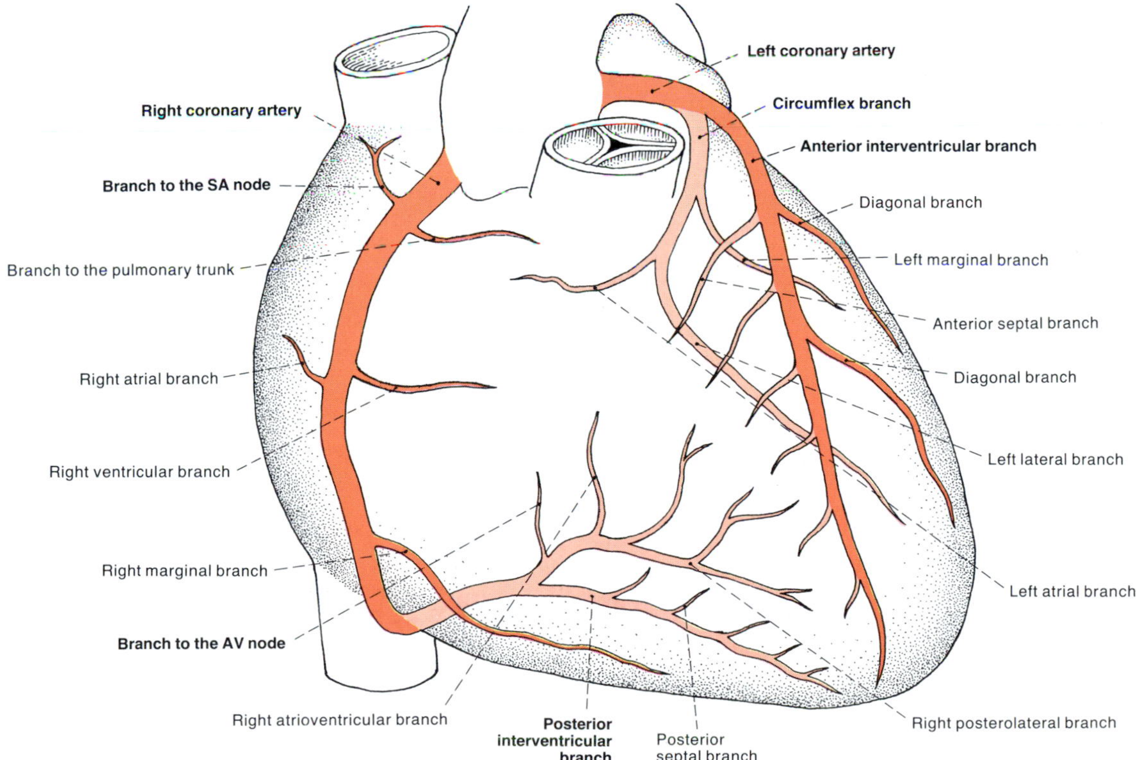

Left coronary artery
Circumflex branch
Right coronary artery
Anterior interventricular branch
Branch to the SA node
Diagonal branch
Branch to the pulmonary trunk
Left marginal branch
Anterior septal branch
Right atrial branch
Diagonal branch
Right ventricular branch
Left lateral branch
Right marginal branch
Left atrial branch
Branch to the AV node
Right atrioventricular branch
Posterior interventricular branch
Posterior septal branch
Right posterolateral branch

Fig. 199: A Dominant Distribution to the Heart Wall from the Right Coronary Artery

NOTE that in hearts containing a dominant right coronary artery, the posterior wall of the left ventricle receives a larger share of its blood from the right coronary artery when compared with a more balanced coronary distribution (see Fig. 197). In these instances the posterior interventricular artery arises from the right coronary, and the circumflex and marginal branches of the left coronary are relatively smaller. (Drawings in Figures 197, 198 and 199 come from data published by M. Kaltenbach and F. Spahn of Frankfurt, West-Germany: Z. Kardiol., **64**: 193–202, 1975.)

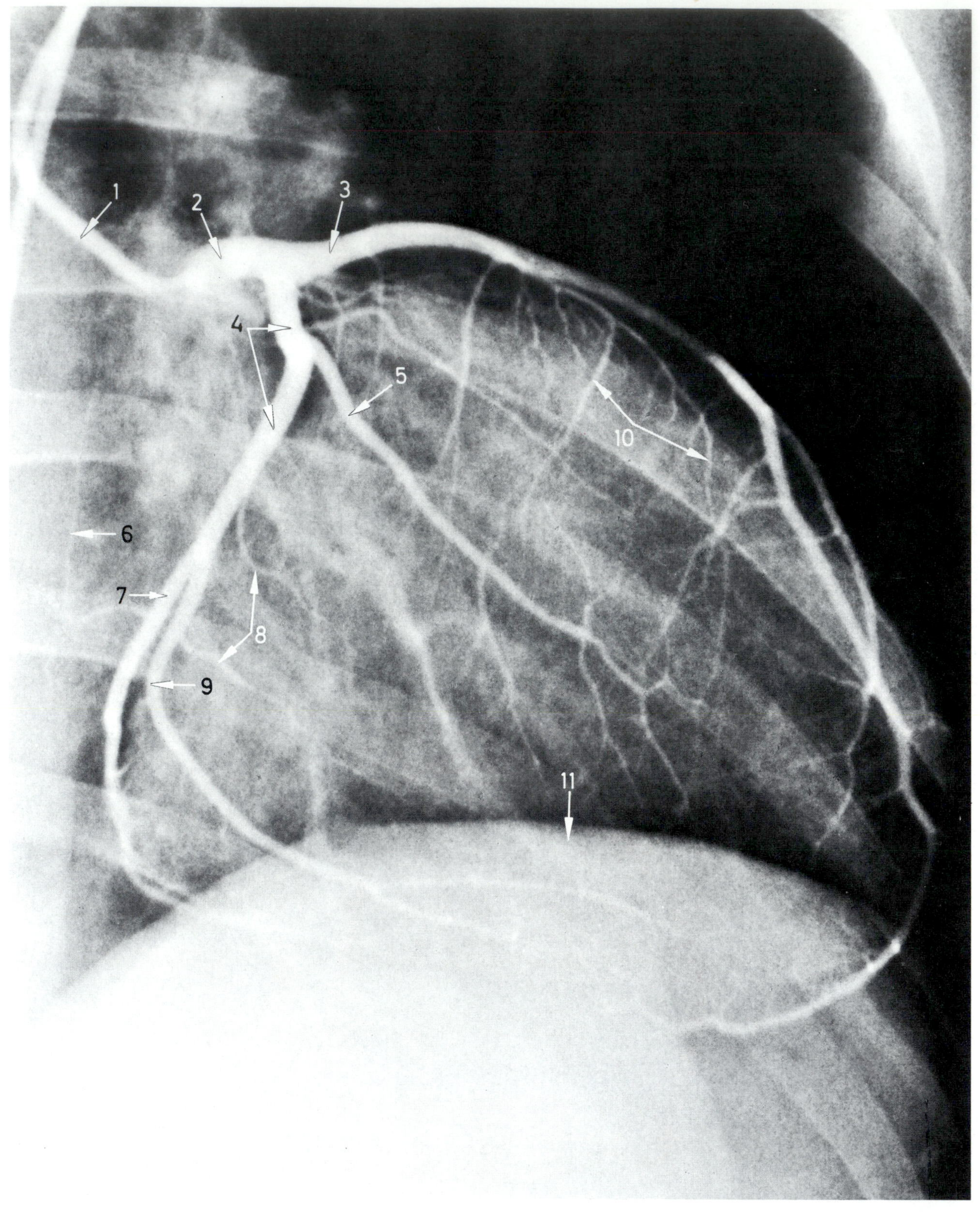
1
2
3
4
5
6
7
8
9
10
11

Fig. 200: Left Coronary Arteriogram

NOTE that this arteriogram of the left coronary artery is viewed from a right antero-oblique direction. (From Wicke, L. *Atlas of Radiographic Anatomy,* 3rd Edition, Urban & Schwarzenberg, Baltimore, 1983).

1. Catheter
2. Left coronary artery
3. Anterior interventricular branch
4. Circumflex branch
5. Left marginal branch of circumflex
6. Posterior atrial branch
7. Left posterolateral branch of circumflex
8. Posterior ventricular branches
9. Posterior interventricular branch
10. Septal branches
11. Diaphragm

Fig. 200

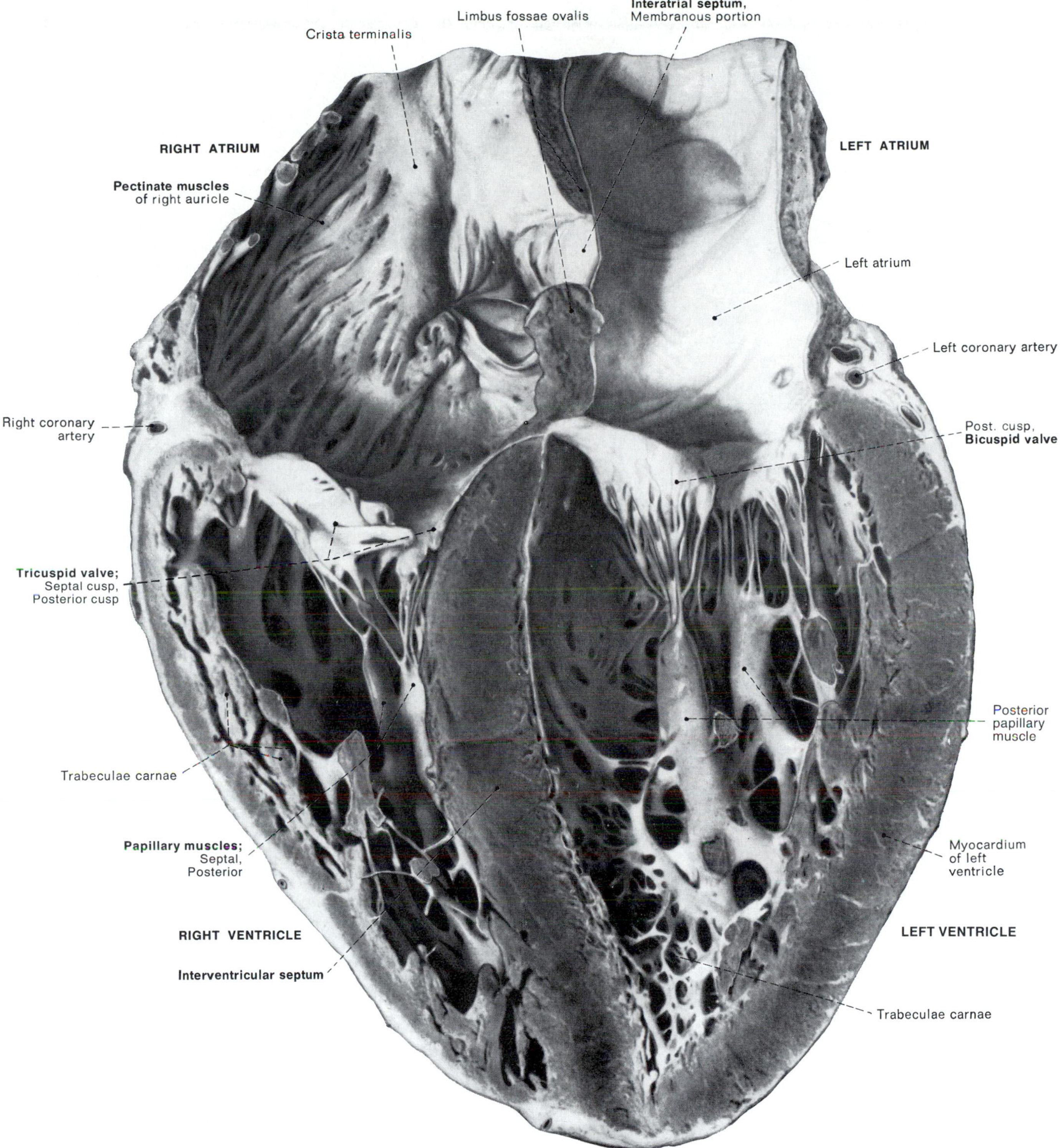
Interatrial septum,
Membranous portion
Limbus fossae ovalis
Crista terminalis
RIGHT ATRIUM
Pectinate muscles
of right auricle
LEFT ATRIUM
Left atrium
Left coronary artery
Right coronary
artery
Post. cusp,
Bicuspid valve
Tricuspid valve;
Septal cusp,
Posterior cusp
Posterior
papillary
muscle
Trabeculae carnae
Papillary muscles;
Septal,
Posterior
Myocardium
of left
ventricle
RIGHT VENTRICLE
LEFT VENTRICLE
Interventricular septum
Trabeculae carnae

Fig. 201: Frontal Section Through the Heart, Showing Dorsal Half of Heart

NOTE: 1) the human heart is a four-chambered muscular organ consisting of an atrium and a ventricle on each side. The walls of the ventricles are thicker than those of the atria. The atrial chambers are separated by an interatrial septum, and this is continuous with the interventricular septum dividing the two ventricles. Blood passes simultaneously from the two atria into their respective ventricles through the atrioventricular valves.

2) on the right side the atrioventricular valve (AV valve) consists of three cusps and is called the tricuspid valve. On the left side the AV valve has two cusps and is called the bicuspid or mitral valve.

3) the inner surfaces of the atria are relatively smooth, whereas muscular projections, the papillary muscles, protrude from the inner walls of the ventricles to attach to the cusps of the AV valves by way of fibrous, thin cords, the chordae tendineae. (These cords are shown but not labelled.)

4) other elevated muscular bundles on the inner heart wall do not attach to the valves. In the ventricles these are called the trabeculae carnae, while in the right auricle they are named the pectinate muscles.

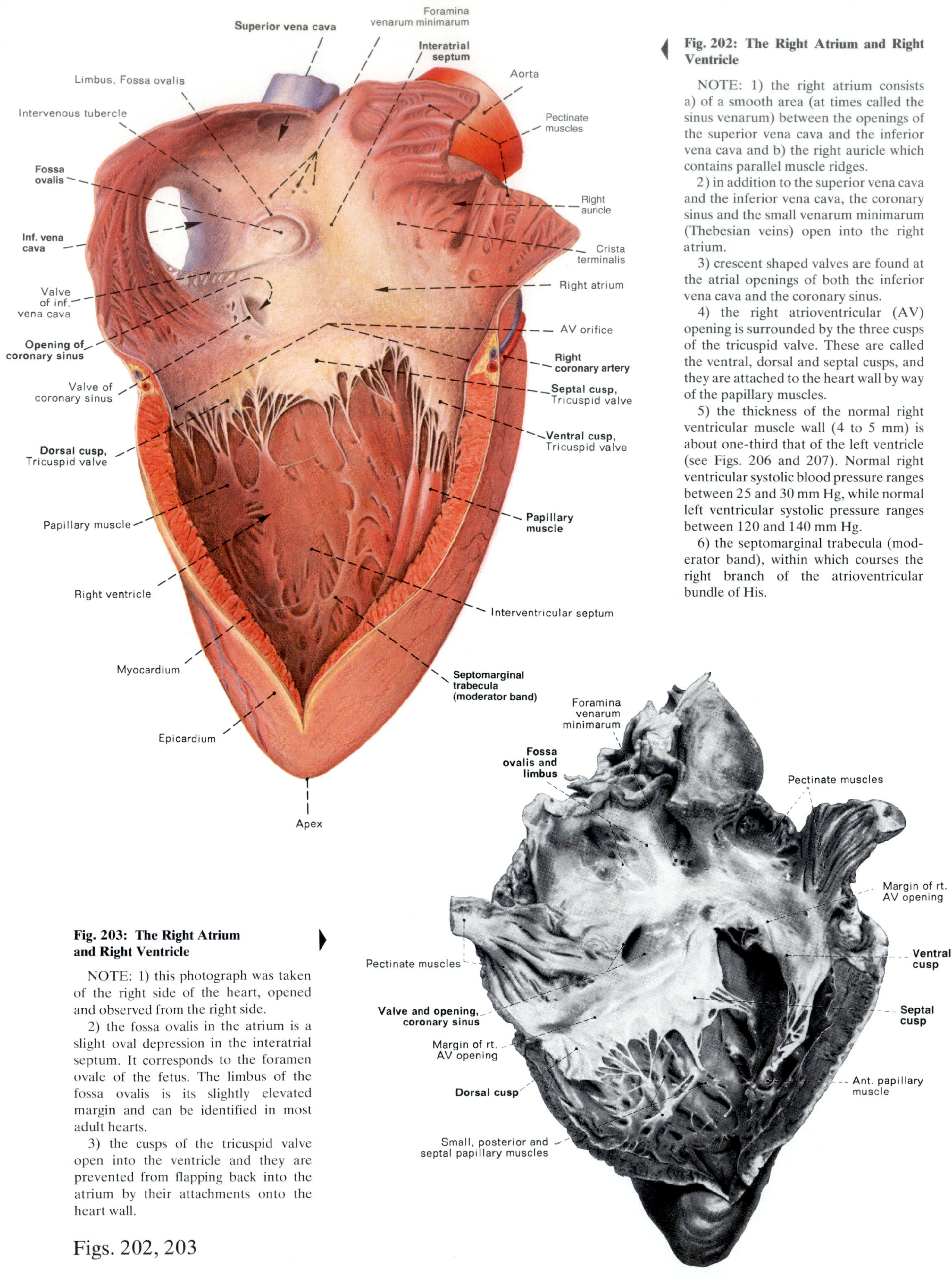

Superior vena cava
Foramina venarum minimarum
Interatrial septum
Aorta
Limbus, Fossa ovalis
Intervenous tubercle
Pectinate muscles
Fossa ovalis
Right auricle
Inf. vena cava
Crista terminalis
Valve of inf. vena cava
Right atrium
AV orifice
Opening of coronary sinus
Right coronary artery
Valve of coronary sinus
Septal cusp, Tricuspid valve
Ventral cusp, Tricuspid valve
Dorsal cusp, Tricuspid valve
Papillary muscle
Papillary muscle
Right ventricle
Interventricular septum
Myocardium
Septomarginal trabecula (moderator band)
Epicardium
Apex
Foramina venarum minimarum
Fossa ovalis and limbus
Pectinate muscles
Margin of rt. AV opening
Ventral cusp
Pectinate muscles
Septal cusp
Valve and opening, coronary sinus
Margin of rt. AV opening
Ant. papillary muscle
Dorsal cusp
Small, posterior and septal papillary muscles

Fig. 202: The Right Atrium and Right Ventricle

NOTE: 1) the right atrium consists a) of a smooth area (at times called the sinus venarum) between the openings of the superior vena cava and the inferior vena cava and b) the right auricle which contains parallel muscle ridges.

2) in addition to the superior vena cava and the inferior vena cava, the coronary sinus and the small venarum minimarum (Thebesian veins) open into the right atrium.

3) crescent shaped valves are found at the atrial openings of both the inferior vena cava and the coronary sinus.

4) the right atrioventricular (AV) opening is surrounded by the three cusps of the tricuspid valve. These are called the ventral, dorsal and septal cusps, and they are attached to the heart wall by way of the papillary muscles.

5) the thickness of the normal right ventricular muscle wall (4 to 5 mm) is about one-third that of the left ventricle (see Figs. 206 and 207). Normal right ventricular systolic blood pressure ranges between 25 and 30 mm Hg, while normal left ventricular systolic pressure ranges between 120 and 140 mm Hg.

6) the septomarginal trabecula (moderator band), within which courses the right branch of the atrioventricular bundle of His.

Fig. 203: The Right Atrium and Right Ventricle

NOTE: 1) this photograph was taken of the right side of the heart, opened and observed from the right side.

2) the fossa ovalis in the atrium is a slight oval depression in the interatrial septum. It corresponds to the foramen ovale of the fetus. The limbus of the fossa ovalis is its slightly elevated margin and can be identified in most adult hearts.

3) the cusps of the tricuspid valve open into the ventricle and they are prevented from flapping back into the atrium by their attachments onto the heart wall.

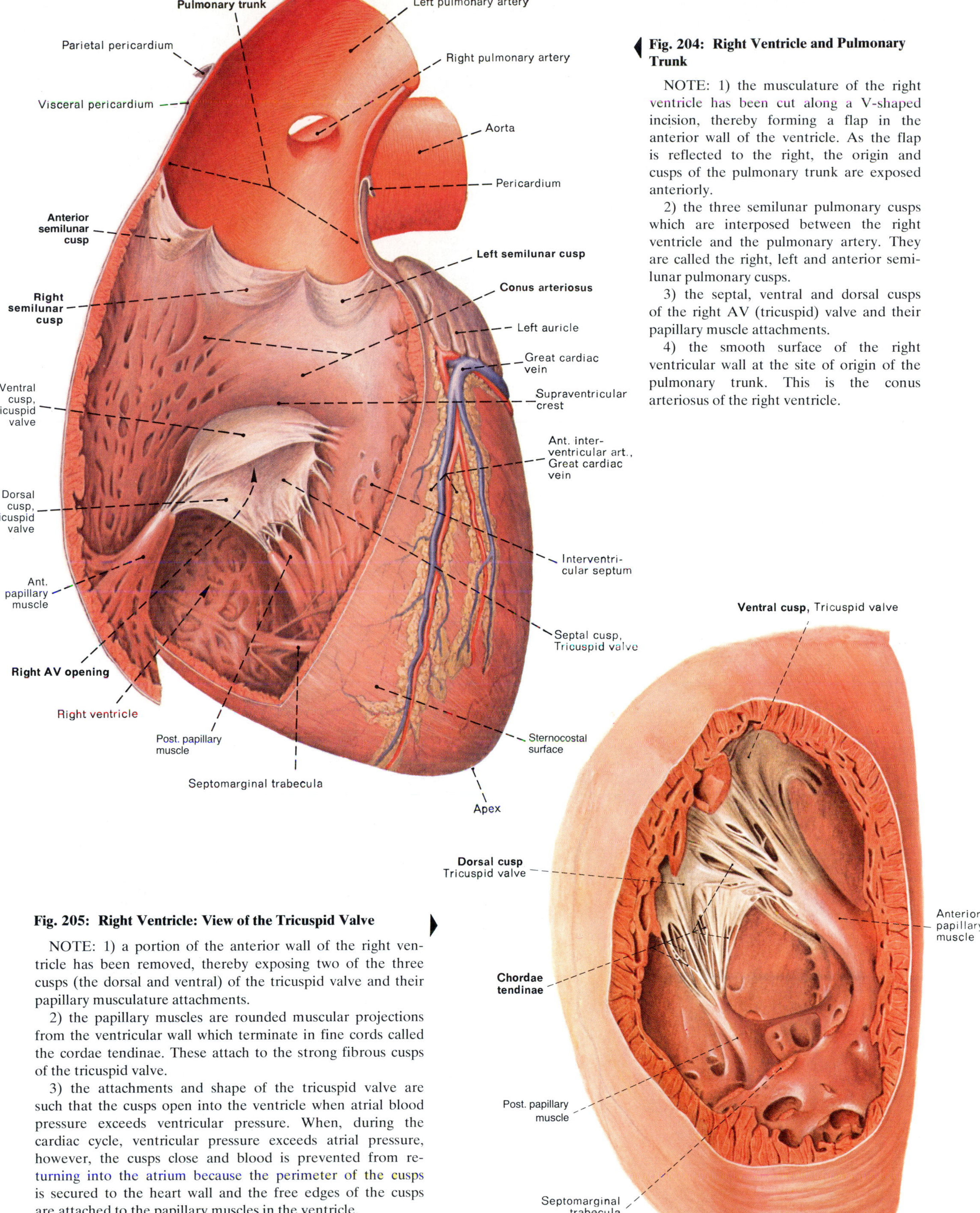

Pulmonary trunk
Left pulmonary artery
Parietal pericardium
Right pulmonary artery
Visceral pericardium
Aorta
Pericardium
Anterior semilunar cusp
Left semilunar cusp
Right semilunar cusp
Conus arteriosus
Left auricle
Great cardiac vein
Ventral cusp, ricuspid valve
Supraventricular crest
Ant. interventricular art., Great cardiac vein
Dorsal cusp, ricuspid valve
Interventricular septum
Ant. papillary muscle
Septal cusp, Tricuspid valve
Right AV opening
Right ventricle
Sternocostal surface
Post. papillary muscle
Septomarginal trabecula
Apex
Ventral cusp, Tricuspid valve
Dorsal cusp Tricuspid valve
Anterior papillary muscle
Chordae tendinae
Post. papillary muscle
Septomarginal trabecula

Fig. 204: Right Ventricle and Pulmonary Trunk

NOTE: 1) the musculature of the right ventricle has been cut along a V-shaped incision, thereby forming a flap in the anterior wall of the ventricle. As the flap is reflected to the right, the origin and cusps of the pulmonary trunk are exposed anteriorly.

2) the three semilunar pulmonary cusps which are interposed between the right ventricle and the pulmonary artery. They are called the right, left and anterior semilunar pulmonary cusps.

3) the septal, ventral and dorsal cusps of the right AV (tricuspid) valve and their papillary muscle attachments.

4) the smooth surface of the right ventricular wall at the site of origin of the pulmonary trunk. This is the conus arteriosus of the right ventricle.

Fig. 205: Right Ventricle: View of the Tricuspid Valve

NOTE: 1) a portion of the anterior wall of the right ventricle has been removed, thereby exposing two of the three cusps (the dorsal and ventral) of the tricuspid valve and their papillary musculature attachments.

2) the papillary muscles are rounded muscular projections from the ventricular wall which terminate in fine cords called the cordae tendinae. These attach to the strong fibrous cusps of the tricuspid valve.

3) the attachments and shape of the tricuspid valve are such that the cusps open into the ventricle when atrial blood pressure exceeds ventricular pressure. When, during the cardiac cycle, ventricular pressure exceeds atrial pressure, however, the cusps close and blood is prevented from returning into the atrium because the perimeter of the cusps is secured to the heart wall and the free edges of the cusps are attached to the papillary muscles in the ventricle.

4) that the septomarginal trabecula, or moderator band, containing the right branch of the atrioventricular bundle has been severed close to the right ventricular wall.

Fig. 206: The Left Atrium and Left Ventricle

NOTE: 1) in this specimen a longitudinal section has been made through the left side of the heart, thereby exposing the smooth-walled left atrium above and the thickened, muscular-walled left ventricle below. The left atrium opens into the left ventricle through the left atrioventricular (AV) orifice at which is located the left AV valve. This valve is also called the mitral or bicuspid valve.

2) the mitral valve consists of two cusps, an anterior cusp and a posterior cusp, and these are attached to the left ventricular wall by means of papillary muscles in a manner similar to that seen on the right side of the heart. Observe the chordae tendineae interposed between the cusps and the papillary muscles.

3) the interatrial septum is marked by the valve of the foramen ovale (falx septi), which represents the remnant of the septum primum during the development of the interatrial septum.

4) the left atrium receives the four pulmonary veins (two from each lung) while the left ventricle leads into (indicated by arrow) the aorta.

5) the thick muscular layer of the heart, called the myocardium, is lined on its inner surface by a layer of endothelium, consistent and continuous with the inner lining of blood vessels that enter and leave the heart. The outer covering, called the epicardium, is a serous membrane which serves as the visceral layer of pericardium.

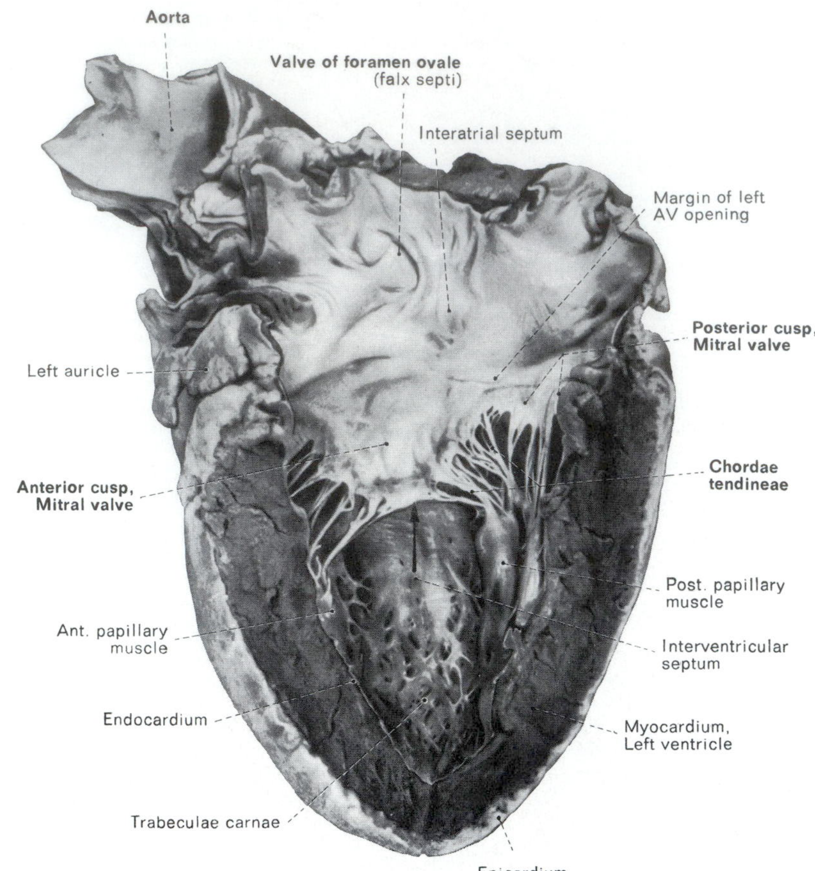

Aorta
Valve of foramen ovale
(falx septi)
Interatrial septum
Margin of left
AV opening
Posterior cusp,
Mitral valve
Left auricle
Chordae
tendineae
Anterior cusp,
Mitral valve
Post. papillary
muscle
Ant. papillary
muscle
Interventricular
septum
Endocardium
Myocardium,
Left ventricle
Trabeculae carnae
Epicardium

Fig. 207: The Left Ventricle and Ascending Aorta

NOTE: 1) the heart has been cut longitudinally in such a manner that the left ventricular cavity is exposed along with the origin of the ascending aorta.

2) the aortic opening is guarded by three semilunar cusps. These are named the posterior, left and right semilunar aortic cusps. Behind each cusp a small *cul de sac* is formed by the cusp and the wall of the aorta. These small dilated pockets are called the aortic sinuses (sinuses of Valsalva) and from the aortic sinuses behind the left and right semilunar aortic cusps, the left and right coronary arteries arise.

3) during ventricular contraction the blood pressure in the left ventricle is elevated over that in the aorta, thereby causing the aortic valve to open and blood to pass into the aorta. Soon, however, the aortic pressure exceeds ventricular pressure and blood then tends to rush back into the ventricle. In the normal heart, the aortic sinuses trap the regurgitating blood and thereby force the aortic cusps to close.

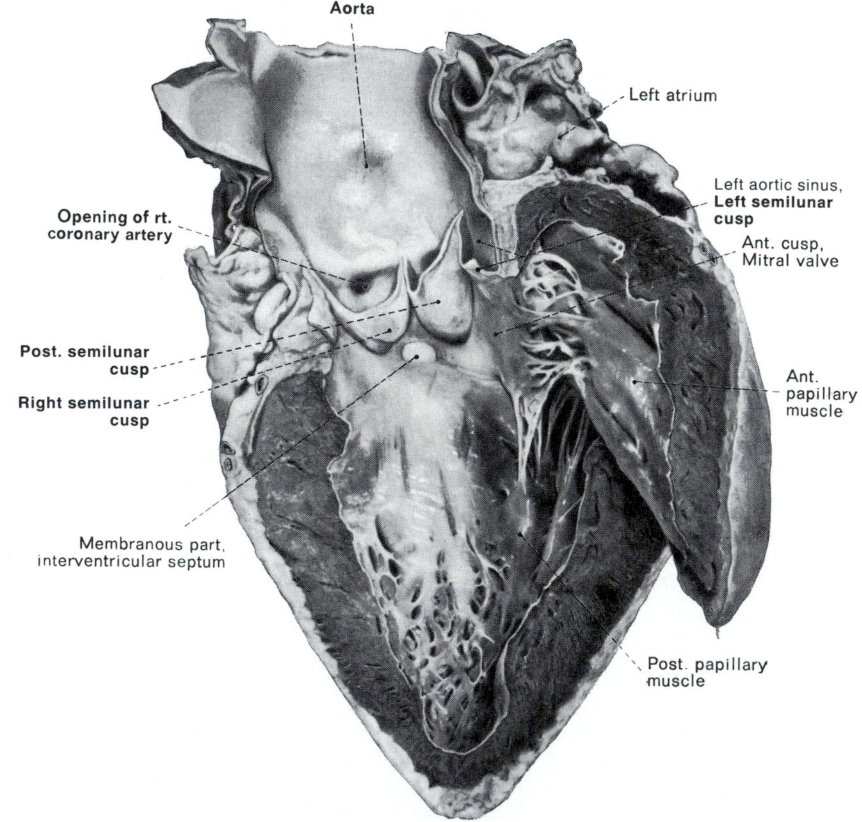

Aorta
Left atrium
Left aortic sinus,
Left semilunar
cusp
Opening of rt.
coronary artery
Ant. cusp,
Mitral valve
Post. semilunar
cusp
Ant.
papillary
muscle
Right semilunar
cusp
Membranous part,
interventricular septum
Post. papillary
muscle

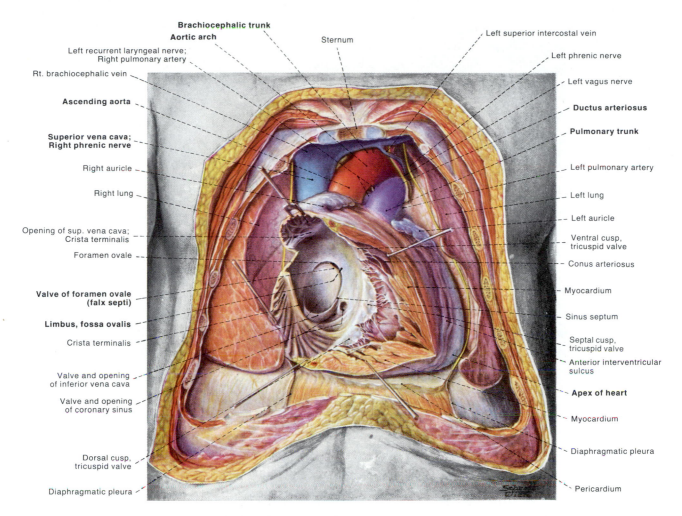

Brachiocephalic trunk
Aortic arch
Sternum
Left superior intercostal vein
Left recurrent laryngeal nerve;
Right pulmonary artery
Left phrenic nerve
Rt. brachiocephalic vein
Left vagus nerve
Ascending aorta
Ductus arteriosus
Superior vena cava;
Right phrenic nerve
Pulmonary trunk
Right auricle
Left pulmonary artery
Right lung
Left lung
Left auricle
Opening of sup. vena cava;
Crista terminalis
Ventral cusp,
tricuspid valve
Foramen ovale
Conus arteriosus
Myocardium
Valve of foramen ovale
(falx septi)
Sinus septum
Limbus, fossa ovalis
Crista terminalis
Septal cusp,
tricuspid valve
Anterior interventricular
sulcus
Valve and opening
of inferior vena cava
Apex of heart
Valve and opening
of coronary sinus
Myocardium
Diaphragmatic pleura
Dorsal cusp,
tricuspid valve
Diaphragmatic pleura
Pericardium

Fig. 208: Heart of Newborn Child: View of Right Atrium and Ventricle, Fossa Ovalis and Tricuspid Valve

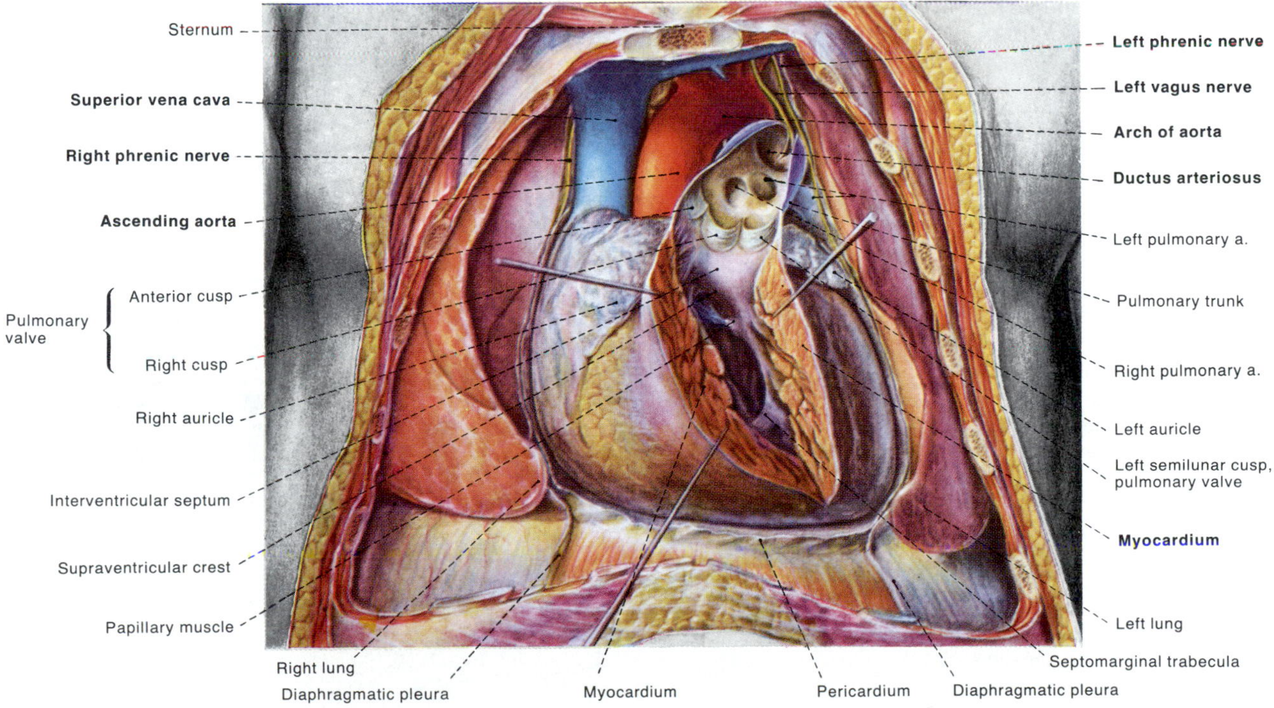
Sternum
Left phrenic nerve
Superior vena cava
Left vagus nerve
Arch of aorta
Right phrenic nerve
Ductus arteriosus
Ascending aorta
Left pulmonary a.
Pulmonary
valve
Anterior cusp
Pulmonary trunk
Right cusp
Right pulmonary a.
Right auricle
Left auricle
Left semilunar cusp,
pulmonary valve
Interventricular septum
Myocardium
Supraventricular crest
Papillary muscle
Left lung
Right lung
Septomarginal trabecula
Diaphragmatic pleura
Myocardium
Pericardium
Diaphragmatic pleura

Fig. 209: Heart of Newborn Child: View of Pulmonary Trunk and Opened Right Ventricle

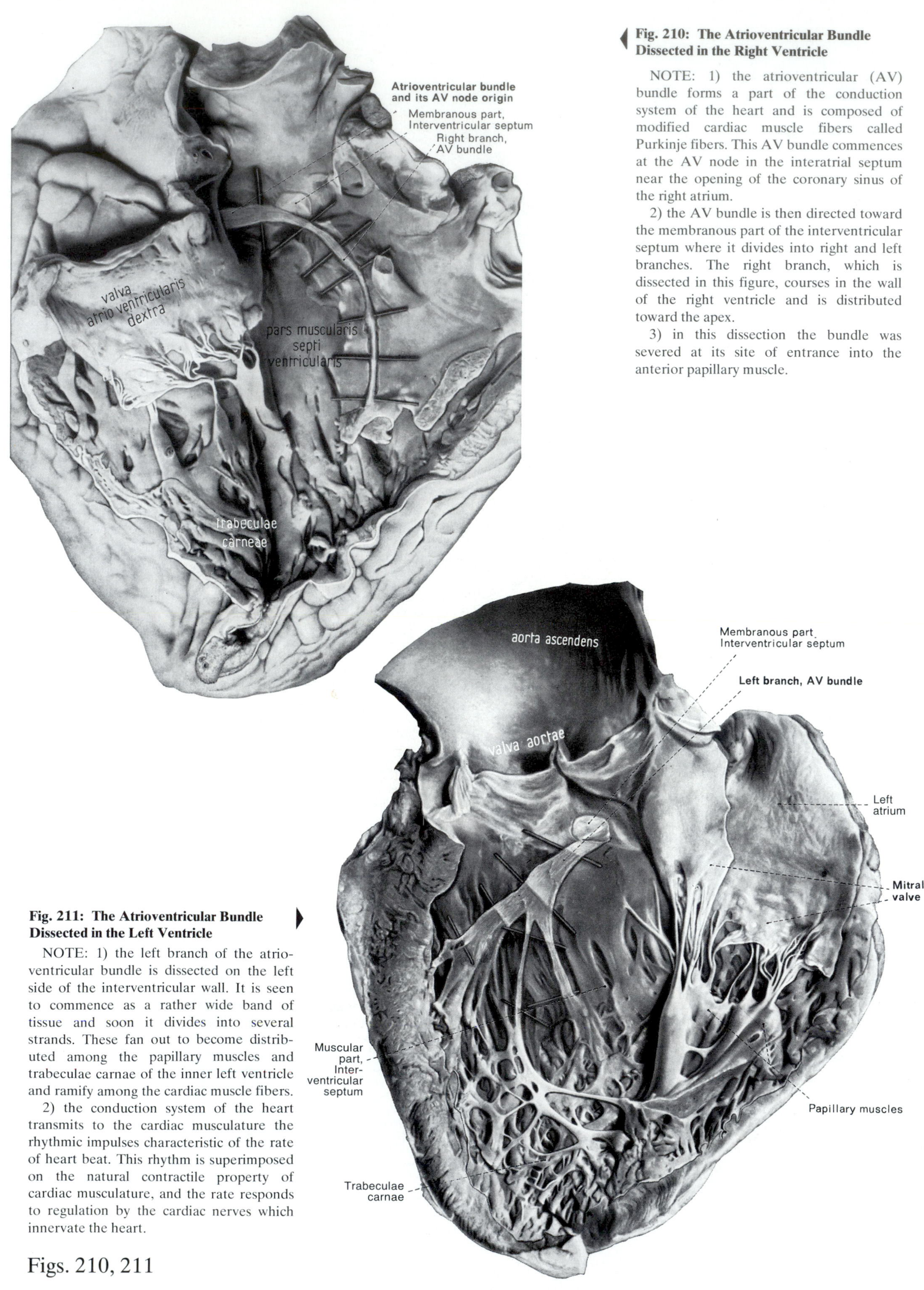
Atrioventricular bundle
and its AV node origin
Membranous part,
Interventricular septum
Right branch,
AV bundle
valva
atrio ventricularis
dextra
pars muscularis
septi
ventricularis
trabeculae
carneae
aorta ascendens
valva aortae
Membranous part
Interventricular septum
Left branch, AV bundle
Left
atrium
Mitral
valve
Muscular
part,
Inter-
ventricular
septum
Papillary muscles
Trabeculae
carnae

Fig. 210: The Atrioventricular Bundle Dissected in the Right Ventricle

NOTE: 1) the atrioventricular (AV) bundle forms a part of the conduction system of the heart and is composed of modified cardiac muscle fibers called Purkinje fibers. This AV bundle commences at the AV node in the interatrial septum near the opening of the coronary sinus of the right atrium.

2) the AV bundle is then directed toward the membranous part of the interventricular septum where it divides into right and left branches. The right branch, which is dissected in this figure, courses in the wall of the right ventricle and is distributed toward the apex.

3) in this dissection the bundle was severed at its site of entrance into the anterior papillary muscle.

Fig. 211: The Atrioventricular Bundle Dissected in the Left Ventricle

NOTE: 1) the left branch of the atrioventricular bundle is dissected on the left side of the interventricular wall. It is seen to commence as a rather wide band of tissue and soon it divides into several strands. These fan out to become distributed among the papillary muscles and trabeculae carnae of the inner left ventricle and ramify among the cardiac muscle fibers.

2) the conduction system of the heart transmits to the cardiac musculature the rhythmic impulses characteristic of the rate of heart beat. This rhythm is superimposed on the natural contractile property of cardiac musculature, and the rate responds to regulation by the cardiac nerves which innervate the heart.

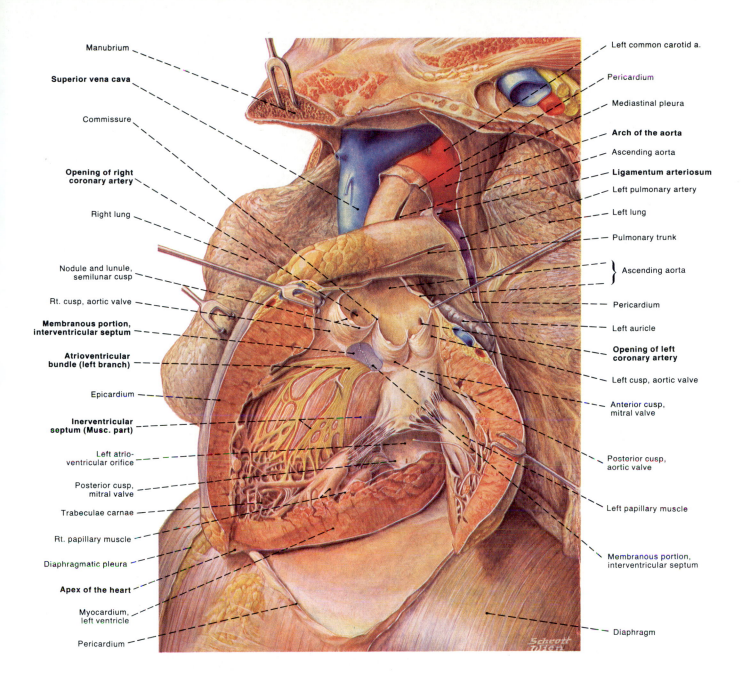

Manubrium
Superior vena cava
Commissure
Opening of right coronary artery
Right lung
Nodule and lunule, semilunar cusp
Rt. cusp, aortic valve
Membranous portion, interventricular septum
Atrioventricular bundle (left branch)
Epicardium
Ineventricular septum (Musc. part)
Left atrio-ventricular orifice
Posterior cusp, mitral valve
Trabeculae carnae
Rt. papillary muscle
Diaphragmatic pleura
Apex of the heart
Myocardium, left ventricle
Pericardium
Left common carotid a.
Pericardium
Mediastinal pleura
Arch of the aorta
Ascending aorta
Ligamentum arteriosum
Left pulmonary artery
Left lung
Pulmonary trunk
Ascending aorta
Pericardium
Left auricle
Opening of left coronary artery
Left cusp, aortic valve
Anterior cusp, mitral valve
Posterior cusp, aortic valve
Left papillary muscle
Membranous portion, interventricular septum
Diaphragm

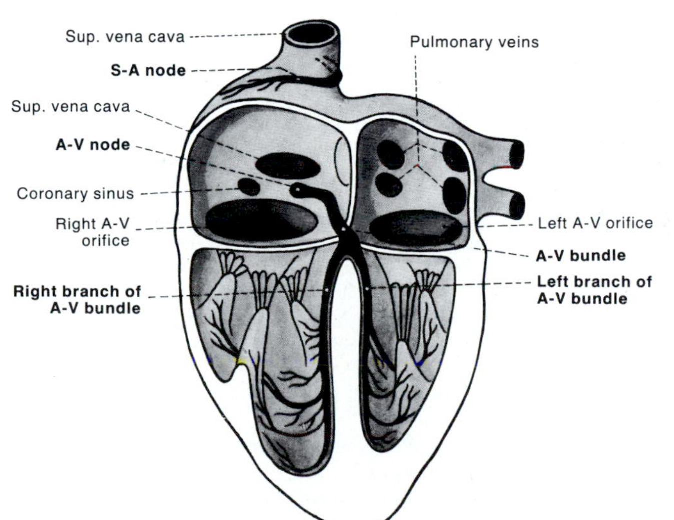

Sup. vena cava
S-A node
Sup. vena cava
A-V node
Coronary sinus
Right A-V orifice
Right branch of A-V bundle
Pulmonary veins
Left A-V orifice
A-V bundle
Left branch of A-V bundle

Fig. 212: Lateral View of the Atrioventricular Bundle (Left Branch)

NOTE: 1) that in this dissection the left branch of the A-V bundle has been dissected (and shown in yellow) in its course along the left side of the interventricular septum. It is observed from within the opened left ventricle;

2) the anterior and posterior cusps and the associated papillary muscles of the mitral valve, and the right, posterior and left cusps of the aortic valve.

Fig. 213: Diagram of the Conduction System of the Heart

NOTE that the sinoatrial node (S-A node) is located in the wall of the right atrium at the junction of the superior vena cava with the right atrium. The atrioventricular node (A-V node) lies in the septal wall of the right atrium near the opening of the coronary sinus.

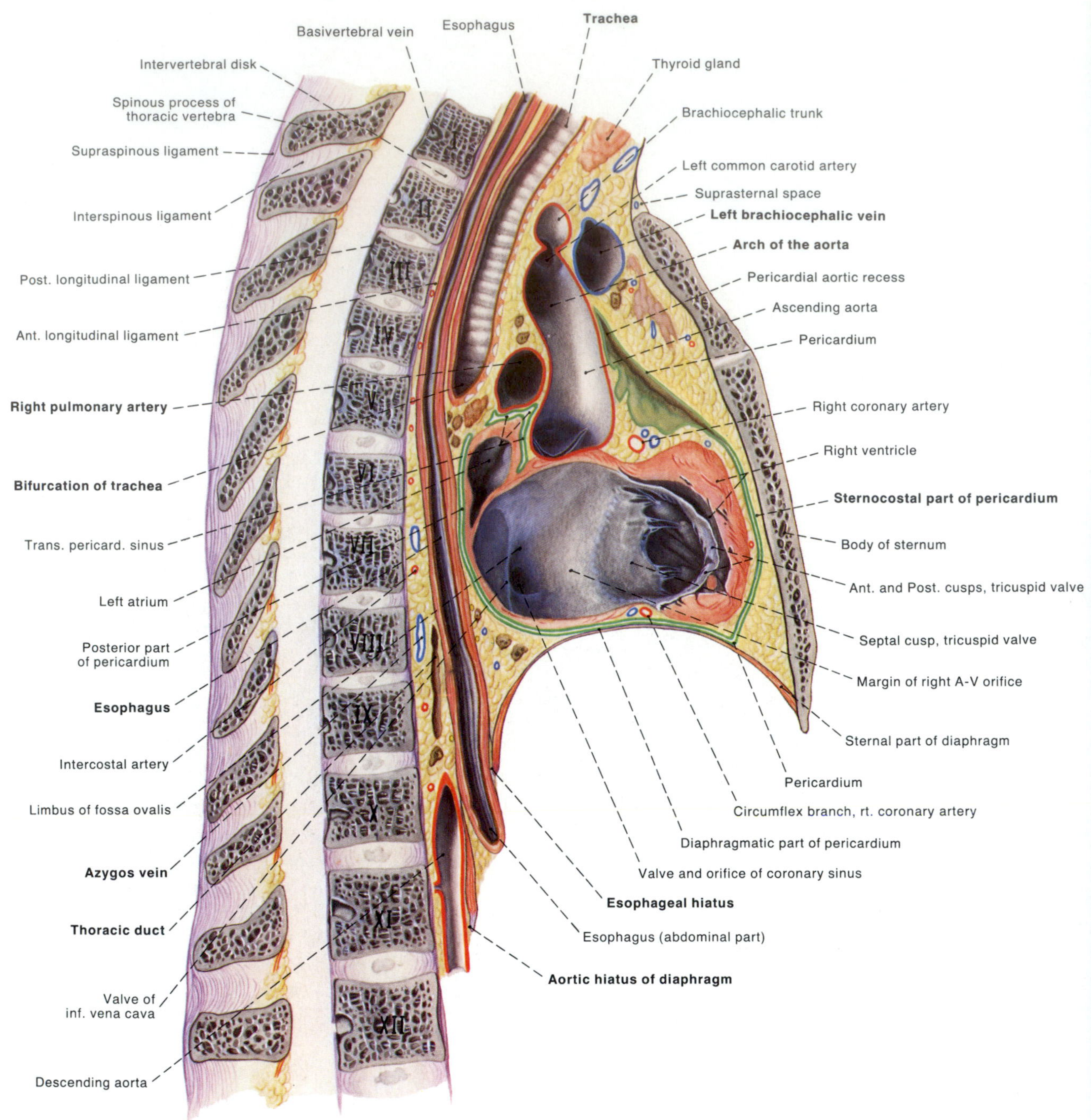

Basivertebral vein
Esophagus
Trachea
Intervertebral disk
Thyroid gland
Spinous process of thoracic vertebra
Brachiocephalic trunk
Supraspinous ligament
Left common carotid artery
Suprasternal space
Interspinous ligament
Left brachiocephalic vein
Arch of the aorta
Post. longitudinal ligament
Pericardial aortic recess
Ascending aorta
Ant. longitudinal ligament
Pericardium
Right pulmonary artery
Right coronary artery
Right ventricle
Bifurcation of trachea
Sternocostal part of pericardium
Body of sternum
Trans. pericard. sinus
Ant. and Post. cusps, tricuspid valve
Left atrium
Septal cusp, tricuspid valve
Posterior part of pericardium
Margin of right A-V orifice
Esophagus
Sternal part of diaphragm
Intercostal artery
Pericardium
Limbus of fossa ovalis
Circumflex branch, rt. coronary artery
Diaphragmatic part of pericardium
Azygos vein
Valve and orifice of coronary sinus
Esophageal hiatus
Thoracic duct
Esophagus (abdominal part)
Aortic hiatus of diaphragm
Valve of inf. vena cava
Descending aorta
I
II
III
IV
V
VI
VII
VIII
IX
X
XI
XII

Fig. 214: A Median Sagittal Section of the Thorax and Vertebral Column Viewed from the Right Side

NOTE that: 1) the bodies of the thoracic vertebrae have been numbered from I to XII;

2) a line drawn directly back from the manubriosternal joint (not labelled) to the spinal column crosses the lower end of the body of the 4th thoracic vertebra, whereas the superior border of the sternum projects back to the 2nd thoracic vertebra, and the lowest point of the xiphoid process lies at the level of the 9th thoracic vertebra;

3) the trachea bifurcates at the level of the upper border of the 5th thoracic vertebra. The esophageal hiatus lies at the upper border of the 10th thoracic vertebra, while somewhat lower, the aortic hiatus is seen at the 11th or between the 11th and 12th thoracic vertebra;

4) because the diaphragmatic opening for the inferior vena cava is to the right of the midline, it cannot be seen in this section; however, the vena caval foramen lies at a level between the 8th and 9th thoracic vertebrae.

5) a median sagittal section of the thorax goes through the right atrium and right ventricle. This is because only one-third of the normal heart lies to the right of the midsternal line, while two-thirds lies to the left of the midline.

6) the transverse pericardial sinus posterior to the aorta. This sinus also courses across the midline posterior to the pulmonary trunk, but in this figure the pulmonary trunk has already bifurcated and, thus, the sinus is shown below the right pulmonary artery.

7) in its course to the right atrium, the left brachiocephalic vein crosses the midline anterior to the arch of the aorta, at the site at which the brachiocephalic trunk arises from the aortic arch (see also Fig. 186, 188).

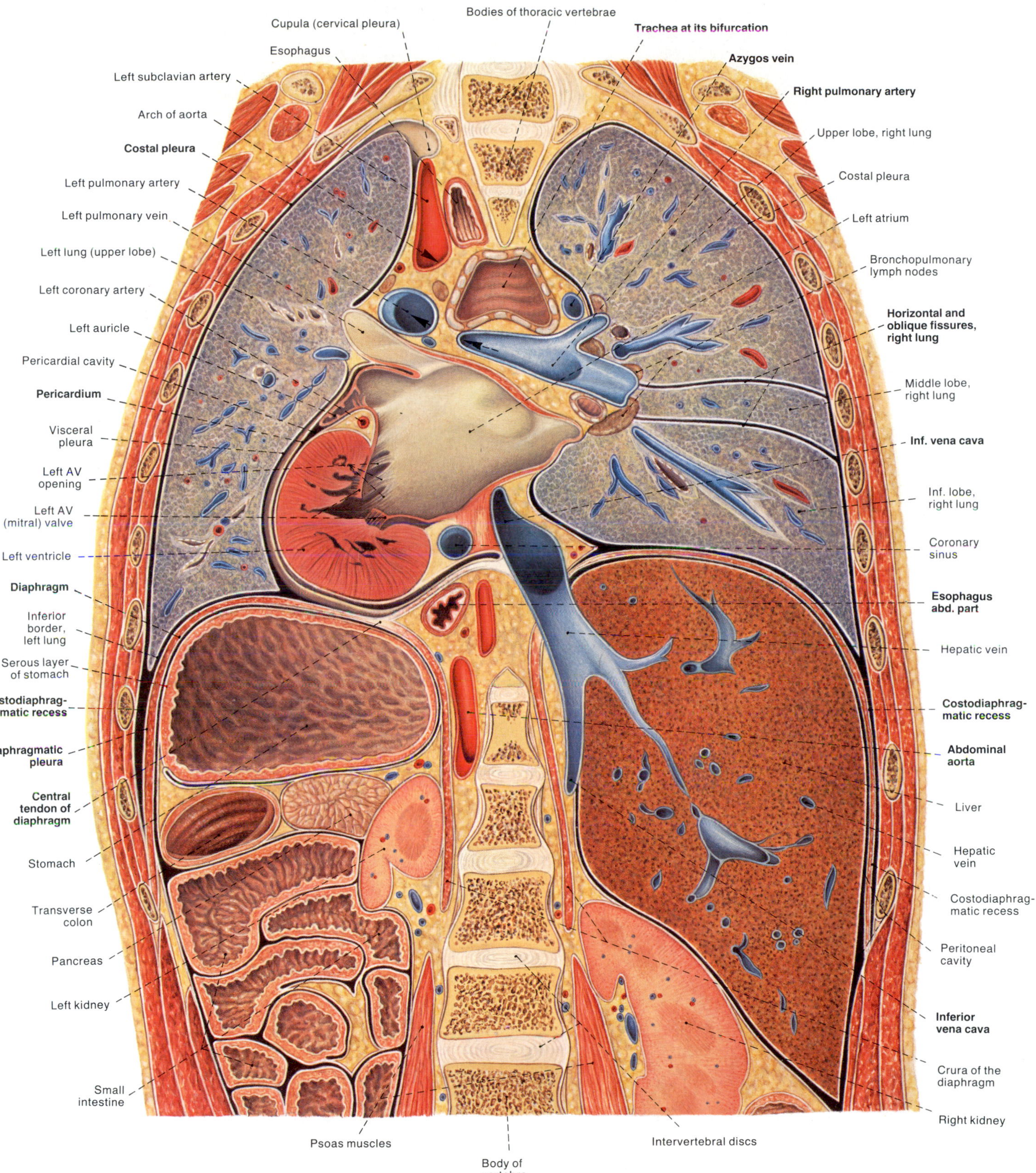

Cupula (cervical pleura)
Bodies of thoracic vertebrae
Trachea at its bifurcation
Esophagus
Azygos vein
Left subclavian artery
Right pulmonary artery
Arch of aorta
Upper lobe, right lung
Costal pleura
Costal pleura
Left pulmonary artery
Left atrium
Left pulmonary vein
Bronchopulmonary lymph nodes
Left lung (upper lobe)
Left coronary artery
Horizontal and oblique fissures, right lung
Left auricle
Pericardial cavity
Middle lobe, right lung
Pericardium
Visceral pleura
Inf. vena cava
Left AV opening
Inf. lobe, right lung
Left AV (mitral) valve
Coronary sinus
Left ventricle
Diaphragm
Esophagus abd. part
Inferior border, left lung
Hepatic vein
Serous layer of stomach
Costodiaphragmatic recess
Costodiaphragmatic recess
Diaphragmatic pleura
Abdominal aorta
Central tendon of diaphragm
Liver
Stomach
Hepatic vein
Transverse colon
Costodiaphragmatic recess
Peritoneal cavity
Pancreas
Left kidney
Inferior vena cava
Crura of the diaphragm
Small intestine
Right kidney
Psoas muscles
Intervertebral discs
Body of vertebra

Fig. 215: A Frontal Section of the Thorax and Abdomen from Behind (Dorsal View)

NOTE: 1) that from this dorsal view, the right side of the specimen is on the reader's right. Observe also that the pulmonary arteries and their branches are indicated in blue, as are veins (such as the hepatic veins) that also carry blood with low levels of oxygen saturation.

2) that the A-P plane of this frontal section in the thorax lies in front of the descending aorta but at the plane of etrance into the thorax of the inferior vena cava through the vena caval foramen. Observe that the esophagus is seen only in the superior mediastinum and at its entrance into the stomach just below the diaphragm, while the trachea has been cut at its point of bifurcation.

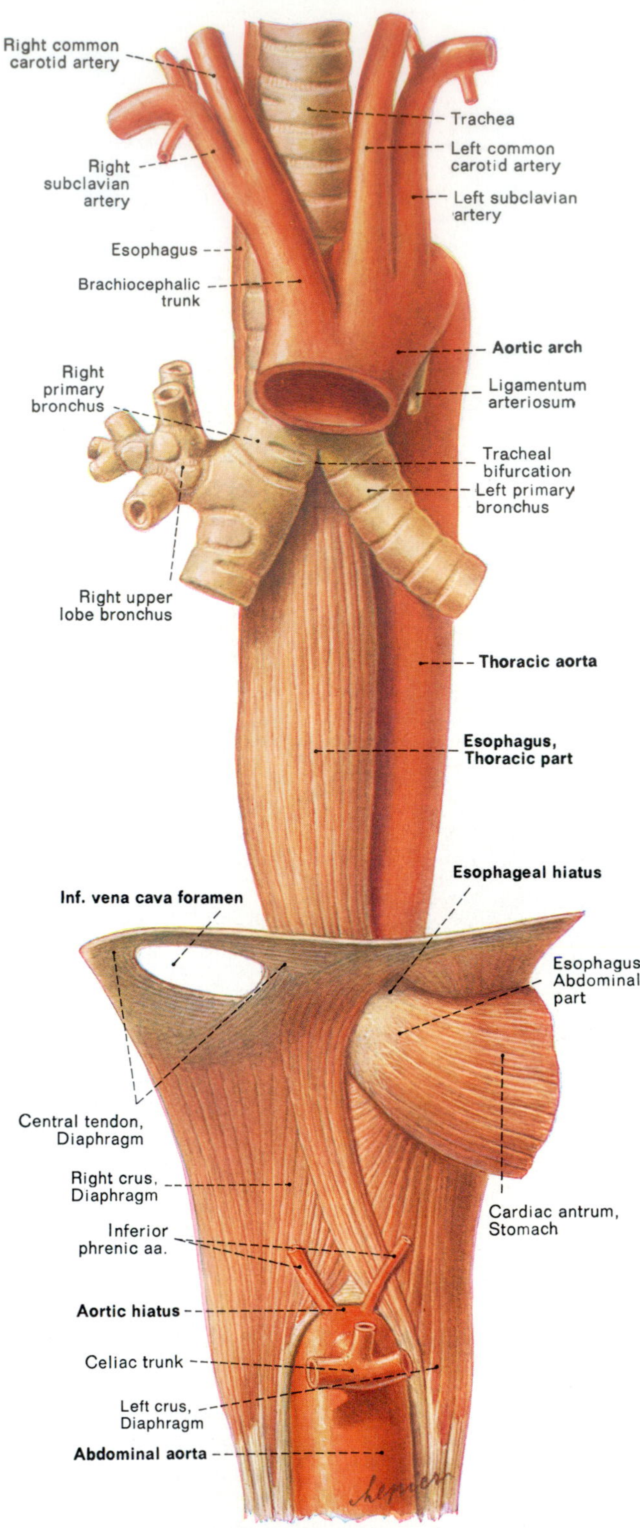
Right common carotid artery
Trachea
Left common carotid artery
Right subclavian artery
Left subclavian artery
Esophagus
Brachiocephalic trunk
Aortic arch
Right primary bronchus
Ligamentum arteriosum
Tracheal bifurcation
Left primary bronchus
Right upper lobe bronchus
Thoracic aorta
Esophagus, Thoracic part
Esophageal hiatus
Inf. vena cava foramen
Esophagus, Abdominal part
Central tendon, Diaphragm
Right crus, Diaphragm
Cardiac antrum, Stomach
Inferior phrenic aa.
Aortic hiatus
Celiac trunk
Left crus, Diaphragm
Abdominal aorta

Fig. 216: The Aorta and Lower Esophagus at the Tracheal Bifurcation and Diaphragm

NOTE: 1) at the level of the bifurcation of the trachea (T-5), the esophagus lies between the trachea and the thoracic aorta. The esophagus then descends into the thorax with the aorta somewhat to its left. In the lower thorax the esophagus bends to the left and thus crosses over the aorta anteriorly from right to left.

2) the esophagus enters the abdomen through the esophageal hiatus of the diaphragm, while the aorta passes through the aortic hiatus.

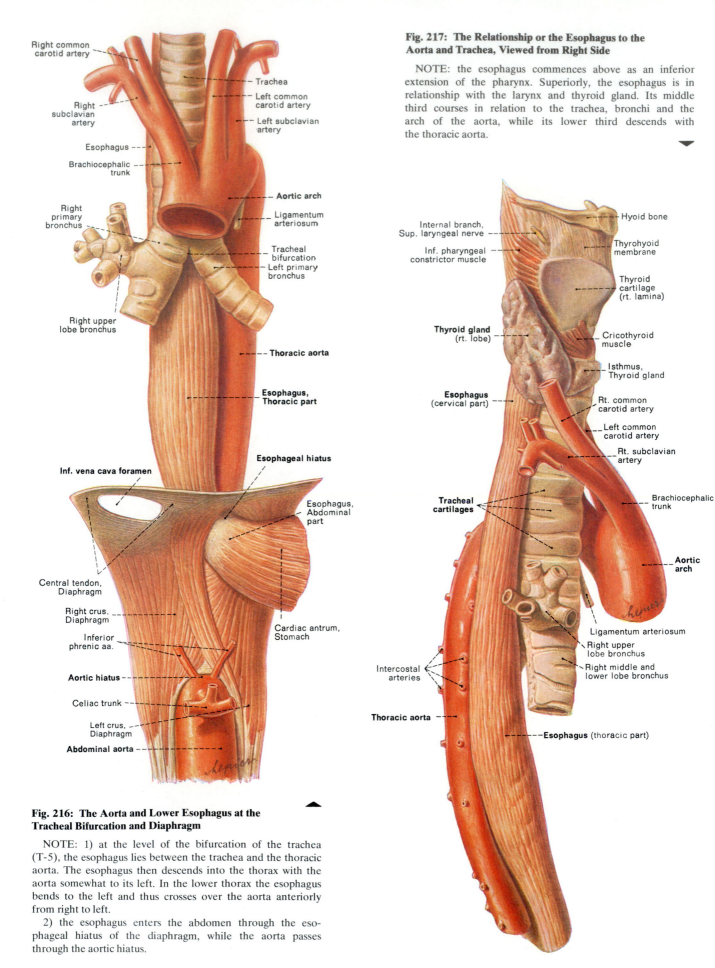

Fig. 217: The Relationship or the Esophagus to the Aorta and Trachea, Viewed from Right Side

NOTE: the esophagus commences above as an inferior extension of the pharynx. Superiorly, the esophagus is in relationship with the larynx and thyroid gland. Its middle third courses in relation to the trachea, bronchi and the arch of the aorta, while its lower third descends with the thoracic aorta.

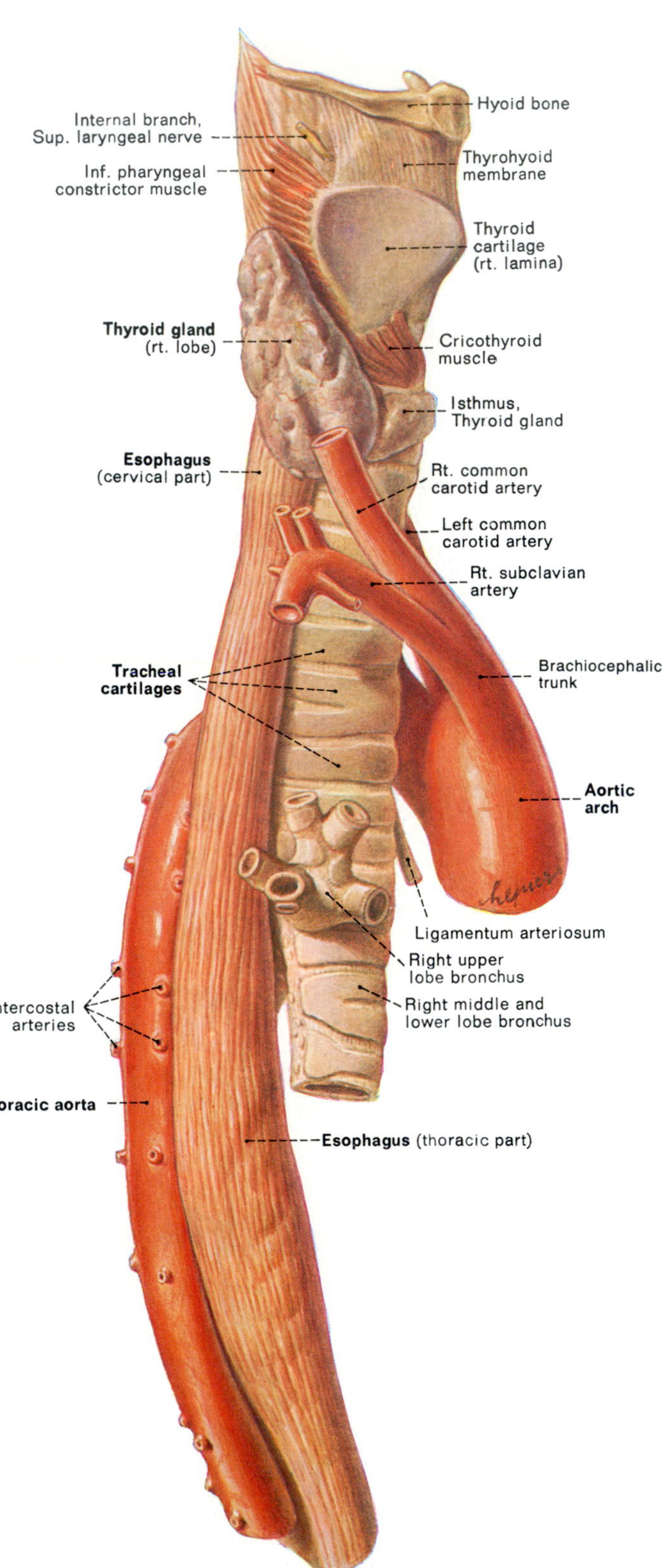
Hyoid bone
Internal branch, Sup. laryngeal nerve
Thyrohyoid membrane
Inf. pharyngeal constrictor muscle
Thyroid cartilage (rt. lamina)
Thyroid gland (rt. lobe)
Cricothyroid muscle
Isthmus, Thyroid gland
Esophagus (cervical part)
Rt. common carotid artery
Left common carotid artery
Rt. subclavian artery
Tracheal cartilages
Brachiocephalic trunk
Aortic arch
Ligamentum arteriosum
Right upper lobe bronchus
Intercostal arteries
Right middle and lower lobe bronchus
Thoracic aorta
Esophagus (thoracic part)

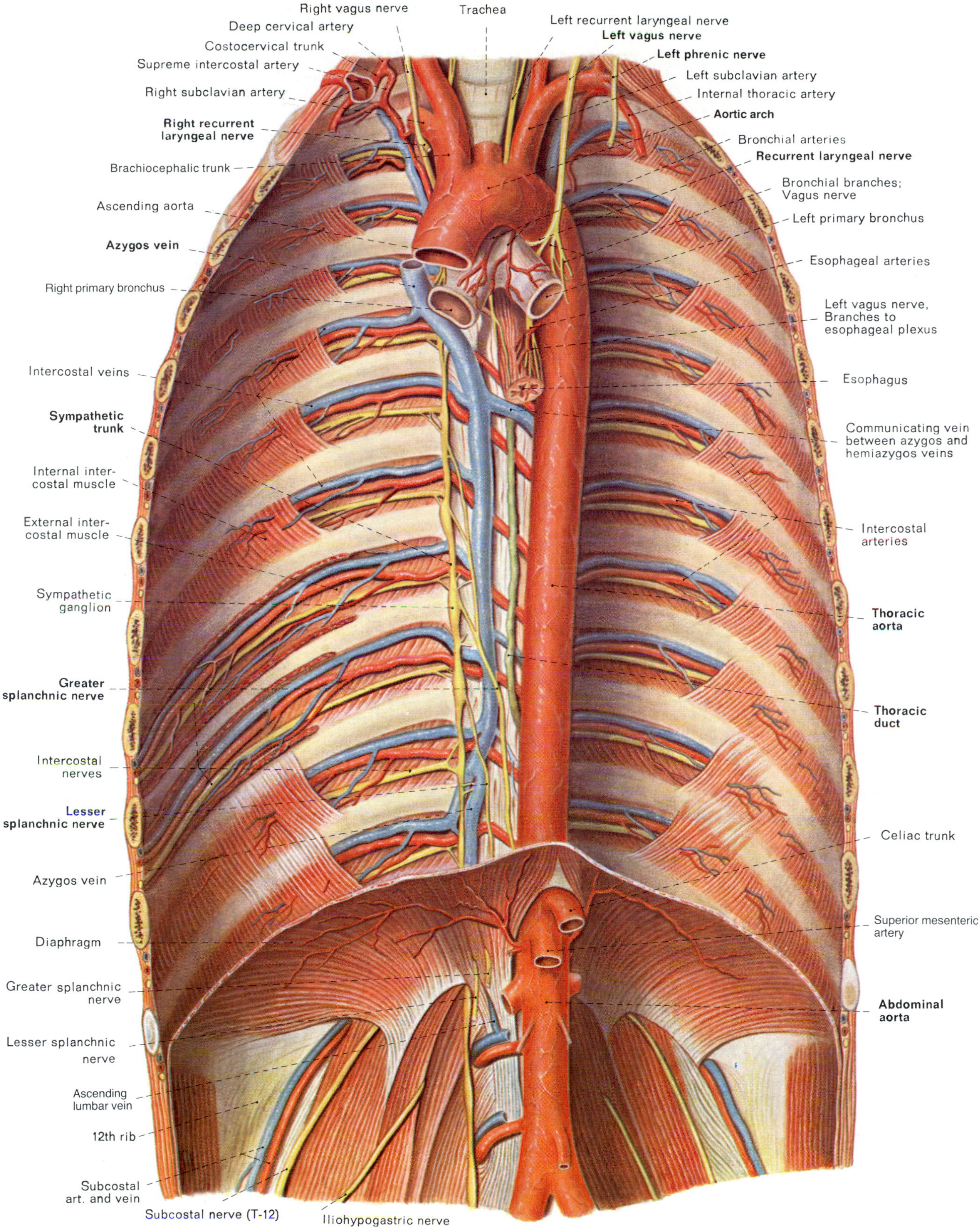

Right vagus nerve
Trachea
Deep cervical artery
Left recurrent laryngeal nerve
Costocervical trunk
Left vagus nerve
Supreme intercostal artery
Left phrenic nerve
Left subclavian artery
Right subclavian artery
Internal thoracic artery
Aortic arch
Right recurrent laryngeal nerve
Bronchial arteries
Recurrent laryngeal nerve
Brachiocephalic trunk
Bronchial branches; Vagus nerve
Ascending aorta
Left primary bronchus
Azygos vein
Esophageal arteries
Right primary bronchus
Left vagus nerve, Branches to esophageal plexus
Intercostal veins
Esophagus
Sympathetic trunk
Communicating vein between azygos and hemiazygos veins
Internal intercostal muscle
External intercostal muscle
Intercostal arteries
Sympathetic ganglion
Thoracic aorta
Greater splanchnic nerve
Thoracic duct
Intercostal nerves
Lesser splanchnic nerve
Celiac trunk
Azygos vein
Superior mesenteric artery
Diaphragm
Greater splanchnic nerve
Abdominal aorta
Lesser splanchnic nerve
Ascending lumbar vein
12th rib
Subcostal art. and vein
Subcostal nerve (T-12)
Iliohypogastric nerve

Fig. 218: Vessels and Nerves of the Dorsal Thoracic Wall

NOTE that the aorta ascends from the left ventricle, arches to the left behind the left pulmonary hilum, and descends through most of the thorax, just to the left side of the vertebral column. In this course through the posterior mediastinum, the aorta gradually shifts toward the midline, which it achieves by the time it traverses the aortic hiatus to enter the abdomen.

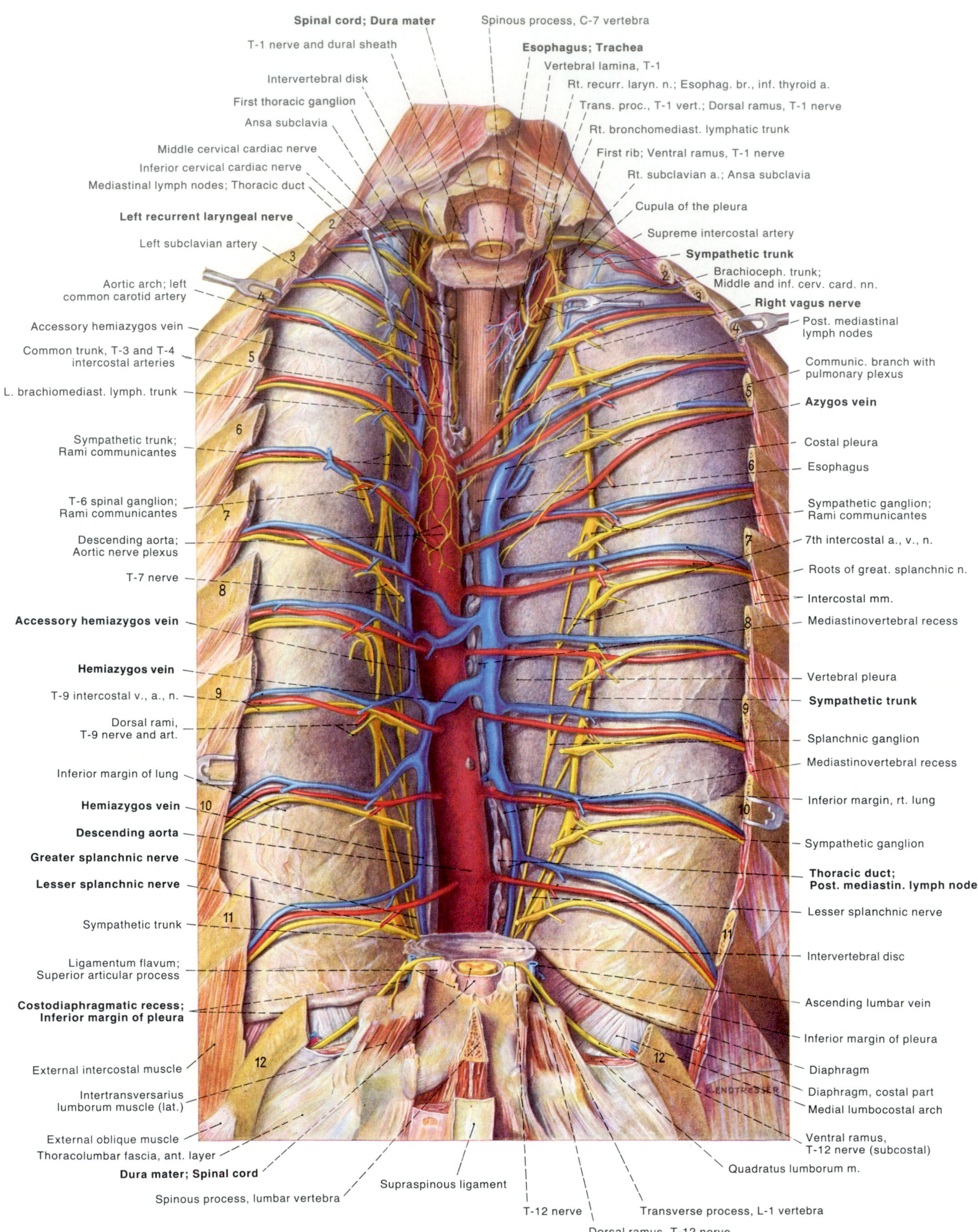

Spinal cord; Dura mater
Spinous process, C-7 vertebra
T-1 nerve and dural sheath
Esophagus; Trachea
Vertebral lamina, T-1
Intervertebral disk
Rt. recurr. laryn. n.; Esophag. br., inf. thyroid a.
First thoracic ganglion
Trans. proc., T-1 vert.; Dorsal ramus, T-1 nerve
Ansa subclavia
Rt. bronchomediast. lymphatic trunk
Middle cervical cardiac nerve
First rib; Ventral ramus, T-1 nerve
Inferior cervical cardiac nerve
Rt. subclavian a.; Ansa subclavia
Mediastinal lymph nodes; Thoracic duct
Cupula of the pleura
Left recurrent laryngeal nerve
Supreme intercostal artery
Left subclavian artery
Sympathetic trunk
Brachioceph. trunk;
Middle and inf. cerv. card. nn.
Aortic arch; left
common carotid artery
Right vagus nerve
Accessory hemiazygos vein
Post. mediastinal
lymph nodes
Common trunk, T-3 and T-4
intercostal arteries
Communic. branch with
pulmonary plexus
L. brachiomediast. lymph. trunk
Azygos vein
Sympathetic trunk;
Rami communicantes
Costal pleura
Esophagus
T-6 spinal ganglion;
Rami communicantes
Sympathetic ganglion;
Rami communicantes
Descending aorta;
Aortic nerve plexus
7th intercostal a., v., n.
T-7 nerve
Roots of great. splanchnic n.
Intercostal mm.
Accessory hemiazygos vein
Mediastinovertebral recess
Hemiazygos vein
Vertebral pleura
T-9 intercostal v., a., n.
Sympathetic trunk
Dorsal rami,
T-9 nerve and art.
Splanchnic ganglion
Mediastinovertebral recess
Inferior margin of lung
Inferior margin, rt. lung
Hemiazygos vein
Descending aorta
Sympathetic ganglion
Greater splanchnic nerve
Thoracic duct;
Post. mediastin. lymph nodes
Lesser splanchnic nerve
Lesser splanchnic nerve
Sympathetic trunk
Intervertebral disc
Ligamentum flavum;
Superior articular process
Ascending lumbar vein
Costodiaphragmatic recess;
Inferior margin of pleura
Inferior margin of pleura
Diaphragm
External intercostal muscle
Diaphragm, costal part
Intertransversarius
lumborum muscle (lat.)
Medial lumbocostal arch
External oblique muscle
Ventral ramus,
T-12 nerve (subcostal)
Thoracolumbar fascia, ant. layer
Quadratus lumborum m.
Dura mater; Spinal cord
Supraspinous ligament
Spinous process, lumbar vertebra
T-12 nerve
Transverse process, L-1 vertebra
Dorsal ramus, T-12 nerve
K. ENDTRESSER

Fig. 219: Dorsal View of the Mediastinum and Lungs with the Thoracic Vertebral Column Removed

NOTE the azygos, hemiazygos and accessory azygos venous pattern that flows in the mediastinum but drains the intercostal spaces. Observe also the relationship of the thoracic sympathetic chain and its ganglia to the intercostal nerves.

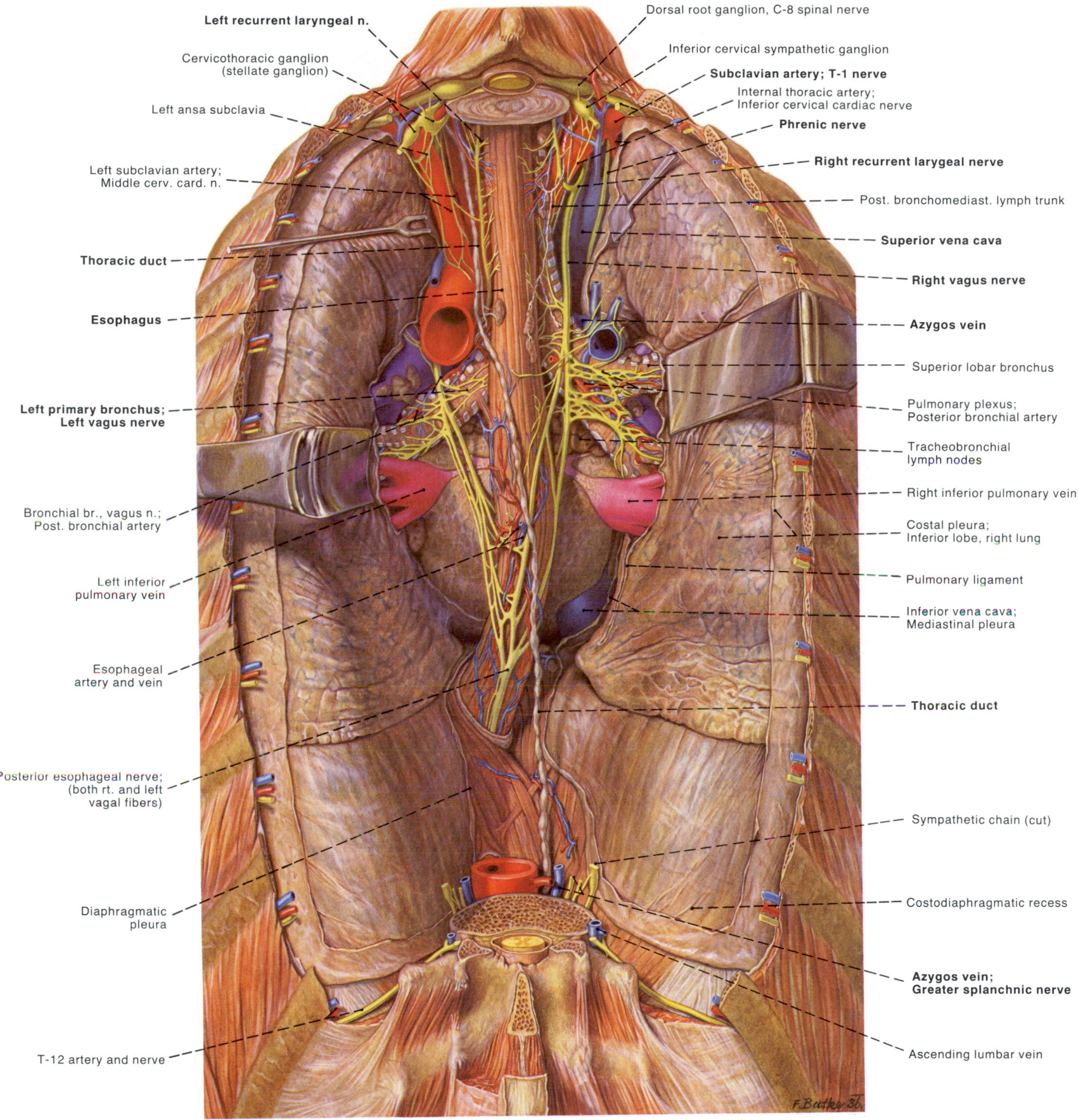

Left recurrent laryngeal n.
Cervicothoracic ganglion (stellate ganglion)
Left ansa subclavia
Left subclavian artery; Middle cerv. card. n.
Thoracic duct
Esophagus
Left primary bronchus; Left vagus nerve
Bronchial br., vagus n.; Post. bronchial artery
Left inferior pulmonary vein
Esophageal artery and vein
Posterior esophageal nerve; (both rt. and left vagal fibers)
Diaphragmatic pleura
T-12 artery and nerve
Dorsal root ganglion, C-8 spinal nerve
Inferior cervical sympathetic ganglion
Subclavian artery; T-1 nerve
Internal thoracic artery; Inferior cervical cardiac nerve
Phrenic nerve
Right recurrent larygeal nerve
Post. bronchomediast. lymph trunk
Superior vena cava
Right vagus nerve
Azygos vein
Superior lobar bronchus
Pulmonary plexus; Posterior bronchial artery
Tracheobronchial lymph nodes
Right inferior pulmonary vein
Costal pleura; Inferior lobe, right lung
Pulmonary ligament
Inferior vena cava; Mediastinal pleura
Thoracic duct
Sympathetic chain (cut)
Costodiaphragmatic recess
Azygos vein; Greater splanchnic nerve
Ascending lumbar vein
F. Batke 36

Fig. 220: Dorsal View of the Mediastinum after Removal of the Vertebral Column and Retraction of the Lungs

NOTE: 1) that in addition to the vertebral column, the descending thoracic aorta has been removed in order to reveal the mediastinal course of the thoracic duct and esophagus. Observe the gradual right to left course of the thoracic duct as it ascends in the mediastinum;

2) that the esophagus descends behind the trachea in the superior mediastinum, but lies directly behind the pericardium below the bifurcation of the trachea;

3) the autonomic plexus of nerves, lymph nodes and bronchial vessels which characterize the hilum of each lung;

4) the course of the right vagus nerve, and observe that the right recurrent laryngeal nerve arches behind the right subclavian artery in order to achieve the neck in its ascent to the larynx.

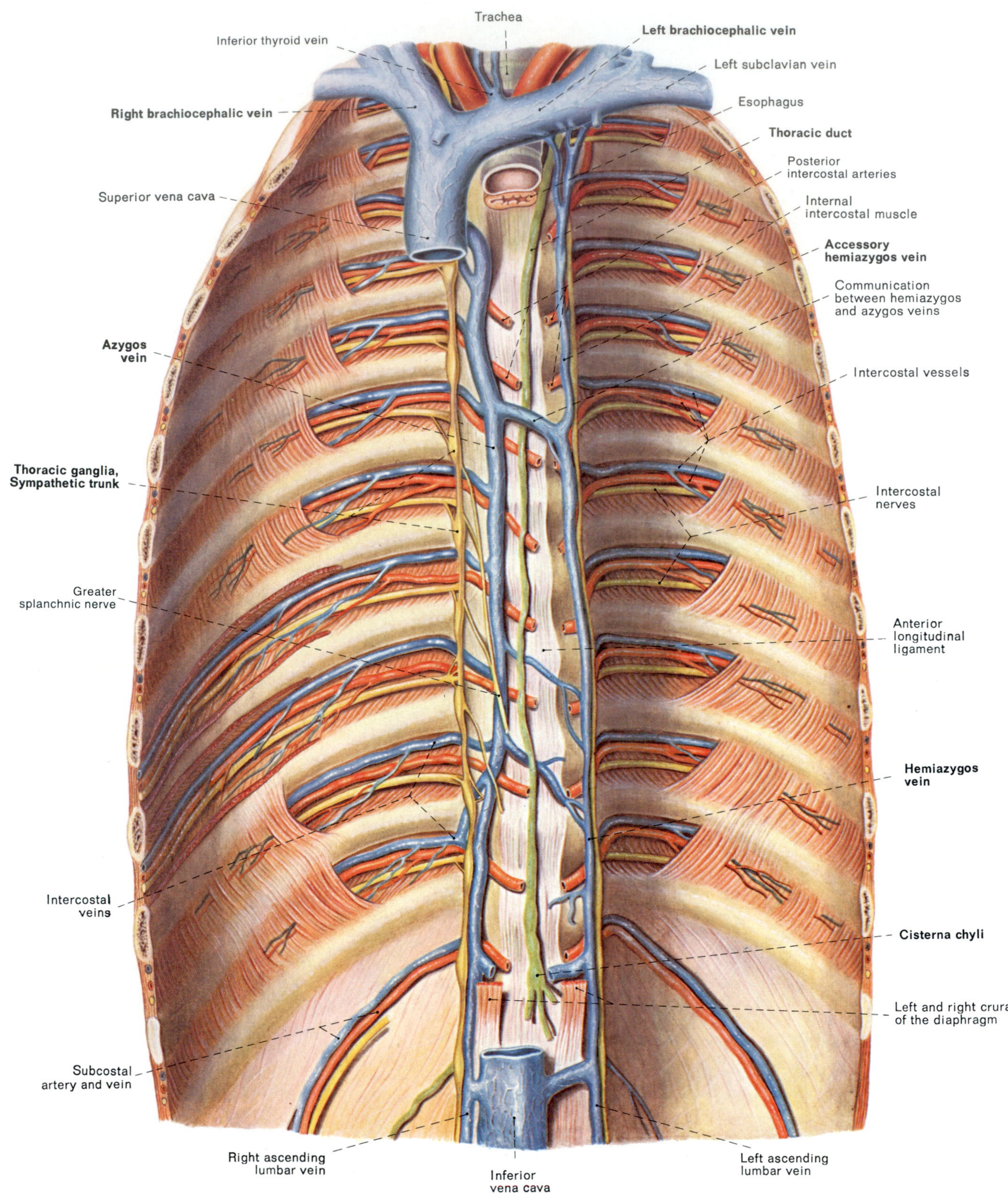
Trachea
Inferior thyroid vein
Left brachiocephalic vein
Left subclavian vein
Right brachiocephalic vein
Esophagus
Thoracic duct
Posterior intercostal arteries
Superior vena cava
Internal intercostal muscle
Accessory hemiazygos vein
Communication between hemiazygos and azygos veins
Azygos vein
Intercostal vessels
Thoracic ganglia, Sympathetic trunk
Intercostal nerves
Greater splanchnic nerve
Anterior longitudinal ligament
Hemiazygos vein
Intercostal veins
Cisterna chyli
Left and right crura of the diaphragm
Subcostal artery and vein
Right ascending lumbar vein
Inferior vena cava
Left ascending lumbar vein

Fig. 221: The Azygos System of Veins, the Thoracic Duct and Other Posterior Thoracic Wall Structures

NOTE: 1) with all the organs of the thorax and mediastinum removed or cut, the hemiazygos and accessory hemiazygos veins to the left of the vertebral column are seen communicating across the midline with the larger azygos vein. This latter vessel also ascends in the thorax to flow into the superior vena cava.

2) the thoracic duct as it arises from the cisterna chyli at the 1st lumbar level.

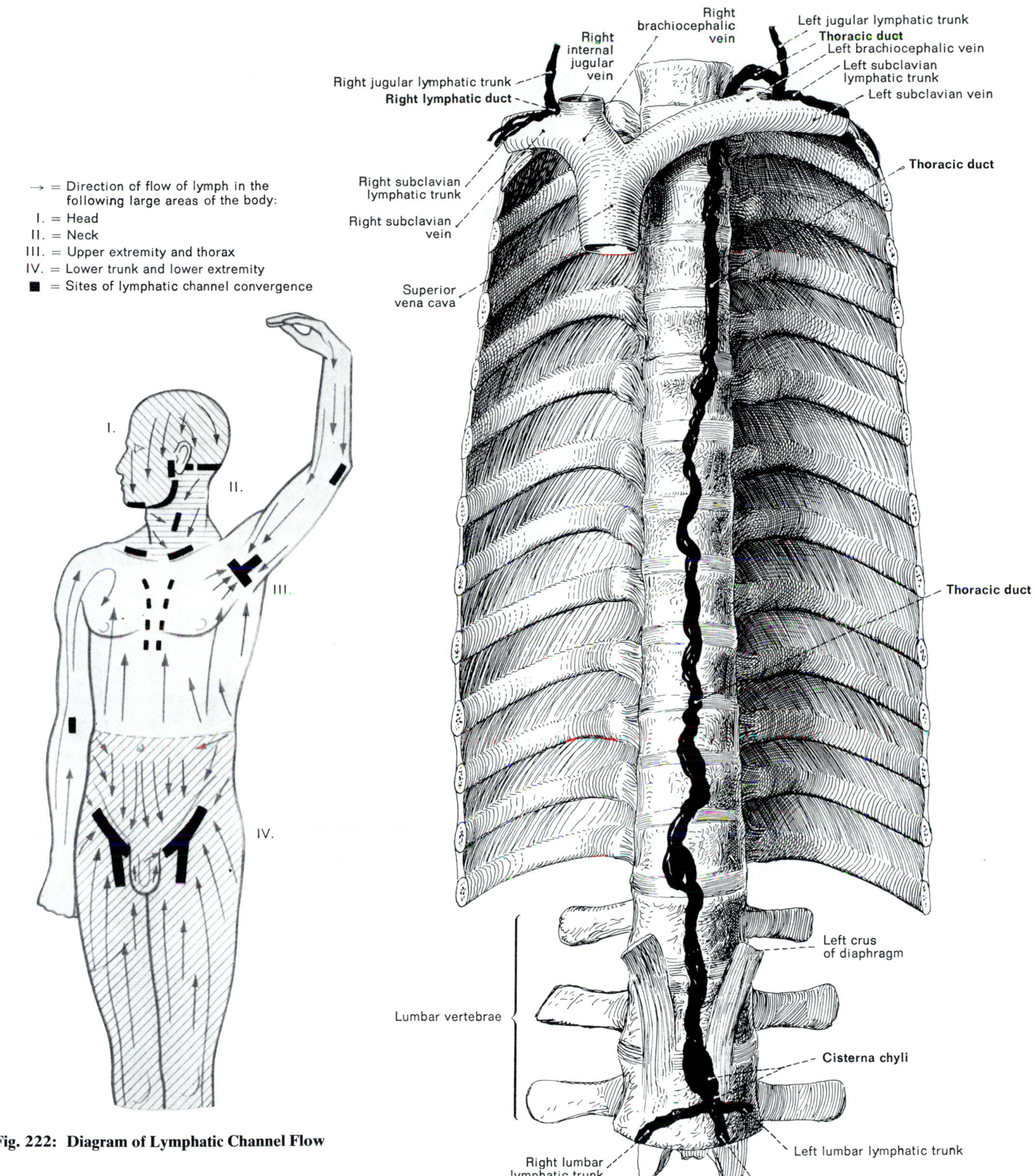
→ = Direction of flow of lymph in the following large areas of the body:
I. = Head
II. = Neck
III. = Upper extremity and thorax
IV. = Lower trunk and lower extremity
■ = Sites of lymphatic channel convergence
I.
II.
III.
IV.
Right brachiocephalic vein
Right internal jugular vein
Right jugular lymphatic trunk
Right lymphatic duct
Left jugular lymphatic trunk
Thoracic duct
Left brachiocephalic vein
Left subclavian lymphatic trunk
Left subclavian vein
Right subclavian lymphatic trunk
Right subclavian vein
Superior vena cava
Thoracic duct
Thoracic duct
Left crus of diaphragm
Lumbar vertebrae
Cisterna chyli
Left lumbar lymphatic trunk
Right lumbar lymphatic trunk
Intestinal lymphatic trunk

Fig. 222: Diagram of Lymphatic Channel Flow

Fig. 223: The Thoracic Duct: Its Origin and Course

NOTE: 1) the thoracic duct collects the lymph from most of the body tissues and transmits it back into the blood stream. It originates in the abdomen anterior to the 2nd lumbar vertebra at the cisterna chyli. The duct then ascends into the thorax through the aortic hiatus of the diaphragm slightly to the right of the midline. Within the posterior mediastinum of the thorax and still coursing just ventral to the vertebral column, it gradually crosses the midline to the left. The duct then ascends into the root of the neck on the left side and opens into the left subclavian vein near the junction of the right internal jugular vein.

2) the right lymphatic duct receives lymph from the right side of the head, neck and trunk and from the right upper extremity. It empties into the right subclavian vein.

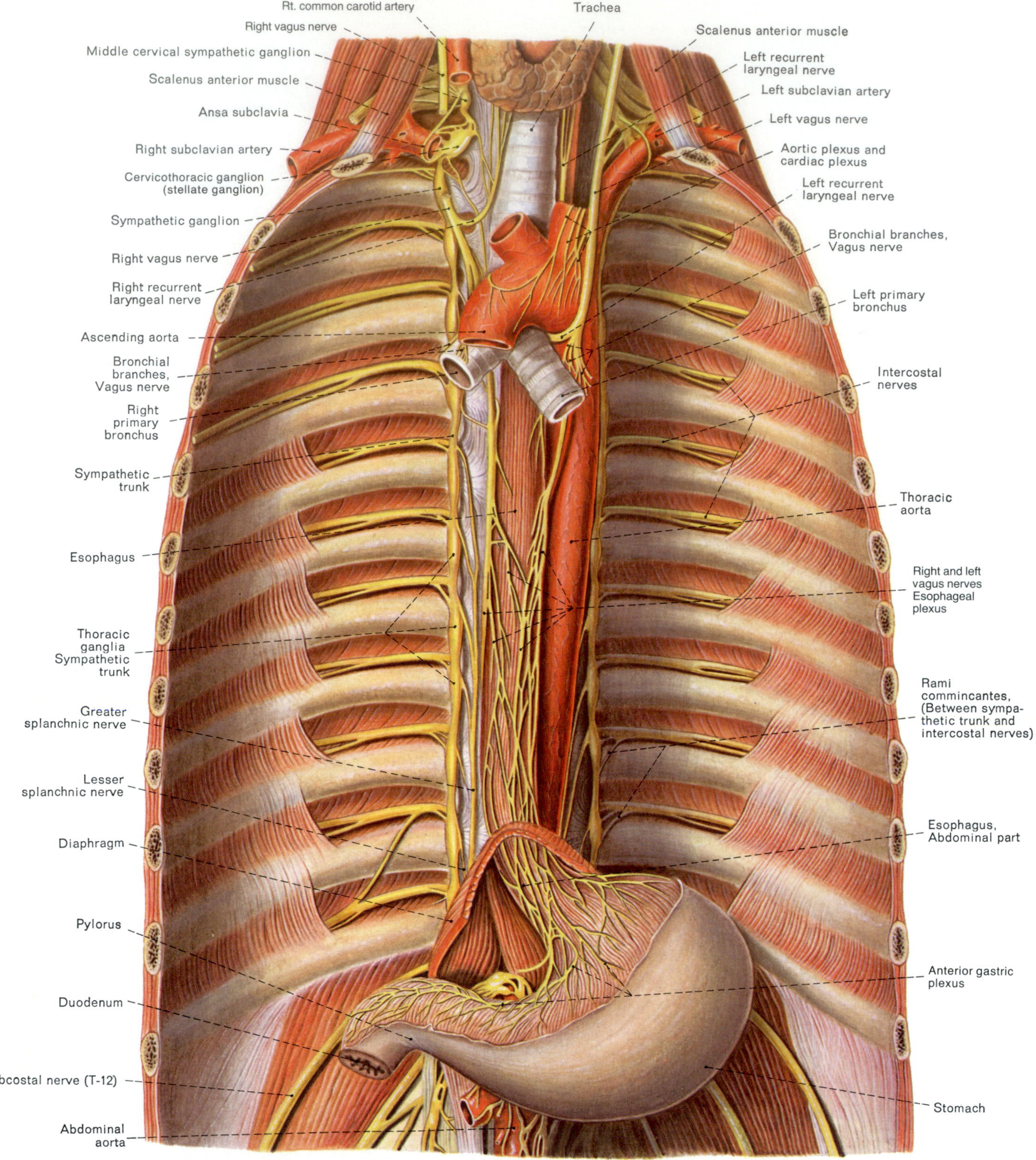

Rt. common carotid artery
Trachea
Right vagus nerve
Scalenus anterior muscle
Middle cervical sympathetic ganglion
Left recurrent laryngeal nerve
Scalenus anterior muscle
Left subclavian artery
Ansa subclavia
Left vagus nerve
Right subclavian artery
Aortic plexus and cardiac plexus
Cervicothoracic ganglion (stellate ganglion)
Left recurrent laryngeal nerve
Sympathetic ganglion
Bronchial branches, Vagus nerve
Right vagus nerve
Right recurrent laryngeal nerve
Left primary bronchus
Ascending aorta
Bronchial branches, Vagus nerve
Intercostal nerves
Right primary bronchus
Sympathetic trunk
Thoracic aorta
Esophagus
Right and left vagus nerves Esophageal plexus
Thoracic ganglia Sympathetic trunk
Rami commincantes, (Between sympa-thetic trunk and intercostal nerves)
Greater splanchnic nerve
Lesser splanchnic nerve
Esophagus, Abdominal part
Diaphragm
Pylorus
Anterior gastric plexus
Duodenum
Subcostal nerve (T-12)
Stomach
Abdominal aorta

Fig. 224: The Symphathetic Trunks and Vagus Nerves in the Thorax and Upper Abdomen

NOTE: 1) that the ganglionated sympathetic trunks on each side lie lateral to the bodies of the thoracic vertebrae and they are continued into the cervical region above the ribs and into the lumbar region below the diaphragm. Observe that each ganglion is connected to the intercostal nerves by means of rami communicantes. *White rami* communicantes mostly carry preganglionic symphatetic neurons to the ganglia, while the *gray rami* communicantes carry postganglionic fibers back to the spinal nerves. (This figure does not specifically label the white and gray rami.)

2) the course of the vagus nerves in the lower thorax. Below the aortic arch these nerves send branches to the bronchi and then descend to form much of the esophageal plexus. Below the diaphragm most of the fibers of the left vagus contribute to the formation of the anterior gastric nerve plexus, while most of the fibers of the right vagus form the posterior gastric nerve plexus.

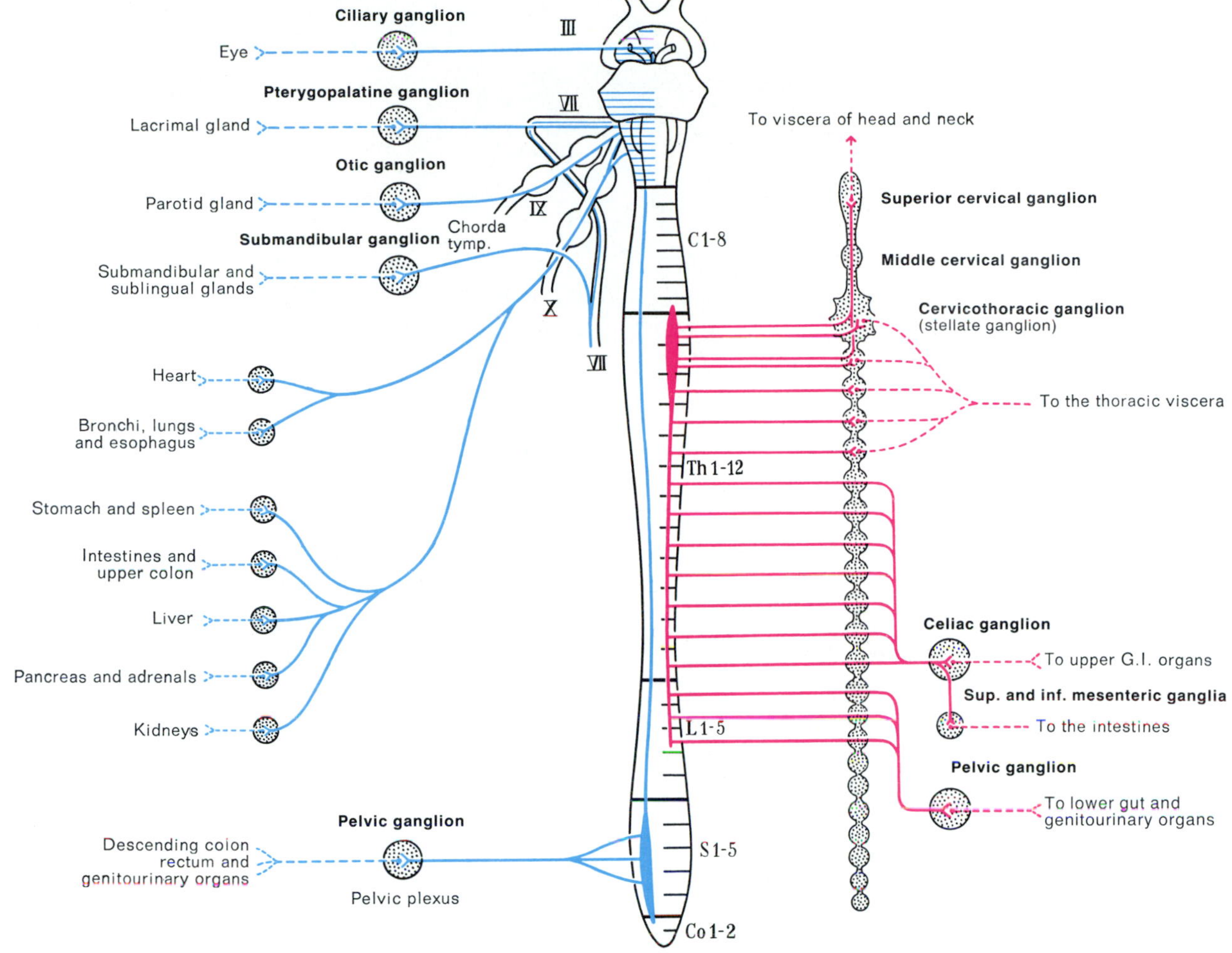

Ciliary ganglion
Eye
III
Pterygopalatine ganglion
Lacrimal gland
VII
To viscera of head and neck
Otic ganglion
Parotid gland
IX
Superior cervical ganglion
Chorda tymp.
Submandibular ganglion
C1-8
Submandibular and sublingual glands
Middle cervical ganglion
X
Cervicothoracic ganglion
(stellate ganglion)
VII
Heart
To the thoracic viscera
Bronchi, lungs and esophagus
Th 1-12
Stomach and spleen
Intestines and upper colon
Liver
Celiac ganglion
Pancreas and adrenals
To upper G.I. organs
Sup. and inf. mesenteric ganglia
Kidneys
L1-5
To the intestines
Pelvic ganglion
Pelvic ganglion
To lower gut and genitourinary organs
Descending colon rectum and genitourinary organs
S1-5
Pelvic plexus
Co1-2

Fig. 225: Diagram of the Autonomic Nervous System. Blue = parasympathetic; red = sympathetic; solid lines = presynaptic neurons; broken lines = postsynaptic neurons.

NOTE: 1) the autonomic nervous system, by definition, is a two motor neuron system with the neuron cell bodies of the *presynaptic neurons* (solid lines) somewhere within the central nervous system, and the cell bodies of the *postsynaptic neurons* (broken lines) located in ganglia distributed peripherally in the body.

2) the autonomic nervous system is comprised of the nerve fibers which supply all the glands and blood vessels of the body including the heart. In so doing, all the smooth and cardiac muscle tissues (sometimes called involuntary muscles) are thereby innervated.

3) the autonomic nervous system is composed of two major divisions called the parasympathetic (in blue) and sympathetic (in red) divisions. The autonomic regulation of visceral function is, therefore, a dualistic control, i.e., most organs receive postganglionic fibers of both parasympathetic and sympathetic source.

4) the *parasympathetic division* is sometimes called a craniosacral outflow because the preganglionic cell bodies of this division lie in the brainstem and in the sacral segments of the spinal cord. Parasympathetic preganglionic fibers are found in four cranial nerves, III (oculomotor), VII (facial), IX (glossopharyngeal) and X (vagus) and in the 2nd, 3rd and 4th sacral nerves.

5) these *pre*ganglionic parasympathetic fibers then synapse with *post*ganglionic parasympathetic cell bodies in peripheral ganglia. From these ganglia the *post*ganglionic nerve fibers innervate the various organs.

6) the *sympathetic division* is sometimes called the thoracolumbar outflow because the *pre*ganglionic sympathetic neuron cell bodies are located in the lateral horn of the spinal cord between the 1st thoracic spinal segment and the 2nd or 3rd lumbar spinal segment, (i.e., from T-1 to L-3).

7) these *pre*ganglionic fibers emerge from the cord with their corresponding spinal roots and communicate with the sympathetic trunk and its ganglia where some *pre*synaptic sympathetic fibers synapse with *post*ganglionic sympathetic neurons. Other presynaptic fibers (especially those of the upper thoracic segments) ascend in the sympathetic chain and synapse with *post*ganglionic neurons in the cervicothoracic, middle and superior cervical ganglia. *Post*ganglionic fibers from these latter ganglia are then distributed to the viscera of the head and neck. Still other *pre*synaptic sympathetic fibers do not synapse in the sympathetic chain of ganglia at all, but collect to form the splanchnic nerves. These nerves course to the collateral sympathetic ganglia (celiac, superior and inferior mesenteric and pelvic ganglia) where synapse with the *post*ganglionic neuron occurs. The *post*ganglionic neurons of the sympathetic division then course to the viscera to supply sympathetic innervation.

8) the functions of the parasympathetic and sympathetic divisions of the autonomic nervous system are antagonistic to each other. The parasympathetic division constricts the pupil, decelerates the heart, lowers blood pressure, relaxes the sphincters of the gut and contracts the longitudinal musculature of the hollow organs. It is the division which is active during periods of calm and tranquility and aids in digestion and absorption. In contrast, the *sympathetic division* dilates the pupil, accelerates the heart, increases blood pressure, contracts the sphincters of the gut and relaxes the longitudinal musculature of hollow organs. It is active when the organism is challenged. It prepares for fight and flight and generally comes to the individual's defense during periods of stress and adversity.

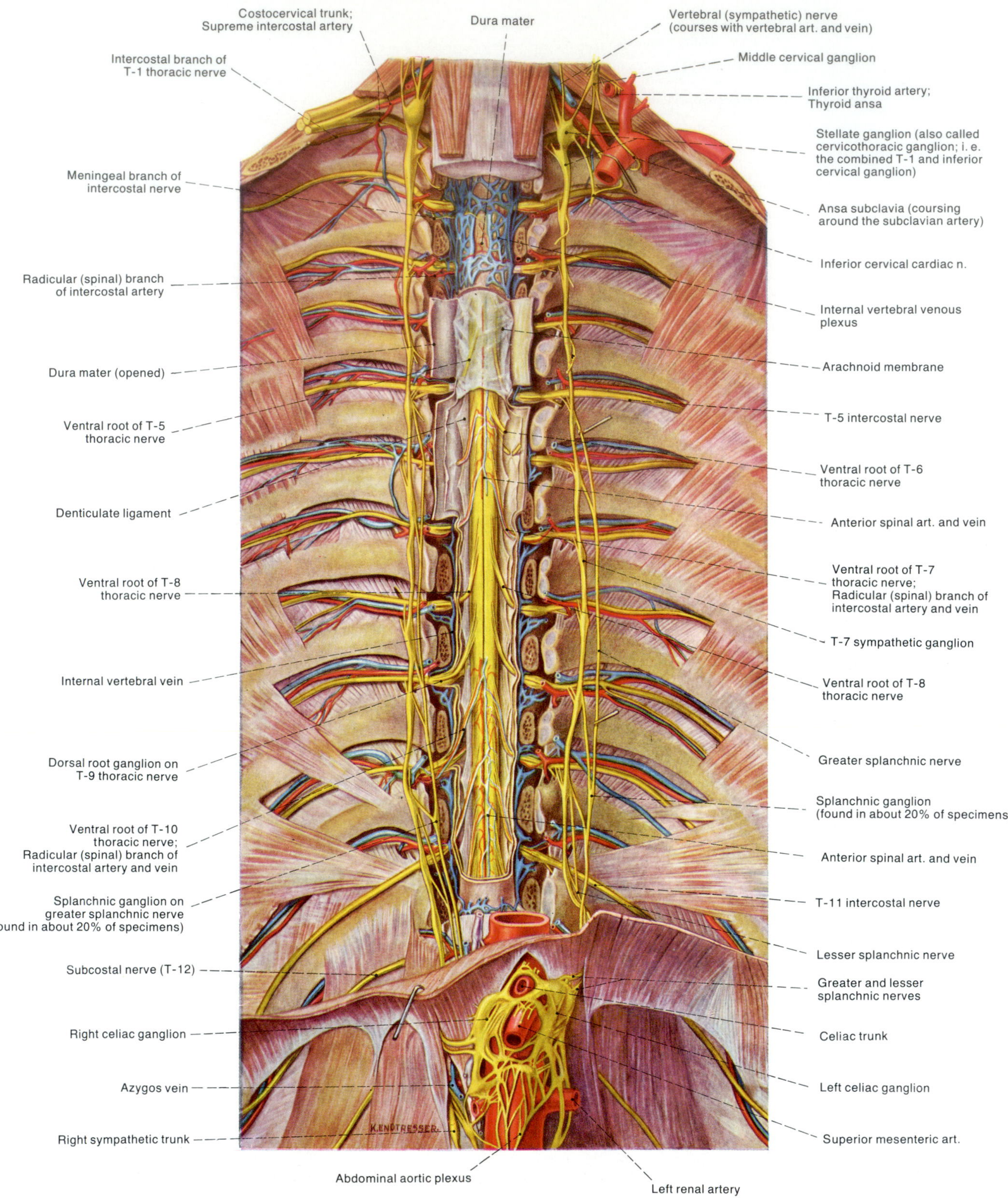

Costocervical trunk;
Supreme intercostal artery
Dura mater
Vertebral (sympathetic) nerve
(courses with vertebral art. and vein)
Intercostal branch of
T-1 thoracic nerve
Middle cervical ganglion
Inferior thyroid artery;
Thyroid ansa
Stellate ganglion (also called
cervicothoracic ganglion; i. e.
the combined T-1 and inferior
cervical ganglion)
Meningeal branch of
intercostal nerve
Ansa subclavia (coursing
around the subclavian artery)
Inferior cervical cardiac n.
Radicular (spinal) branch
of intercostal artery
Internal vertebral venous
plexus
Dura mater (opened)
Arachnoid membrane
T-5 intercostal nerve
Ventral root of T-5
thoracic nerve
Ventral root of T-6
thoracic nerve
Denticulate ligament
Anterior spinal art. and vein
Ventral root of T-7
thoracic nerve;
Radicular (spinal) branch of
intercostal artery and vein
Ventral root of T-8
thoracic nerve
T-7 sympathetic ganglion
Internal vertebral vein
Ventral root of T-8
thoracic nerve
Greater splanchnic nerve
Dorsal root ganglion on
T-9 thoracic nerve
Splanchnic ganglion
(found in about 20% of specimens)
Ventral root of T-10
thoracic nerve;
Radicular (spinal) branch of
intercostal artery and vein
Anterior spinal art. and vein
Splanchnic ganglion on
greater splanchnic nerve
(found in about 20% of specimens)
T-11 intercostal nerve
Lesser splanchnic nerve
Subcostal nerve (T-12)
Greater and lesser
splanchnic nerves
Right celiac ganglion
Celiac trunk
Azygos vein
Left celiac ganglion
Right sympathetic trunk
K. ENDTRESSER
Superior mesenteric art.
Abdominal aortic plexus
Left renal artery

Fig. 226: The Sympathetic Trunks and Intercostal Nerves and Vessels in the Posterior Part of the Thoracic Wall, Anterior View

NOTE: 1) the two sympathetic trunks (located in the posterior mediastinum of the thorax), their ganglia and the rami communicantes that interconnect the ganglia with the segmentally coursing intercostal nerves.

2) that the bodies of the lower 7 or 8 thoracic vertebrae have been removed in order to reveal, in this anterior view, the ventral surface of the spinal cord and the anterior spinal roots that help to form the thoracic nerves.

3) the **greater splanchnic nerve** which is formed from thoracic sympathetic ganglia T-5 to T-9 or T-10 and the **lesser splanchnic nerve** which is formed from sympathetic ganglia T-9 or T-10 and T-11. These nerves descend parallel to the sympathetic trunks and enter the abdomen by perforating the crura of the diaphragm. Most of the fibers of the greater splanchnic nerves terminate on the postganglionic neurons in the *celiac ganglia;* most fibers in the lesser splanchnic nerve synapse in the *aorticorenal ganglia* (shown near the renal arteries but not labelled), while the lowest splanchnic nerve, arising on each side from the T-12 sympathetic ganglion (not shown) ends in the renal autonomic plexus.

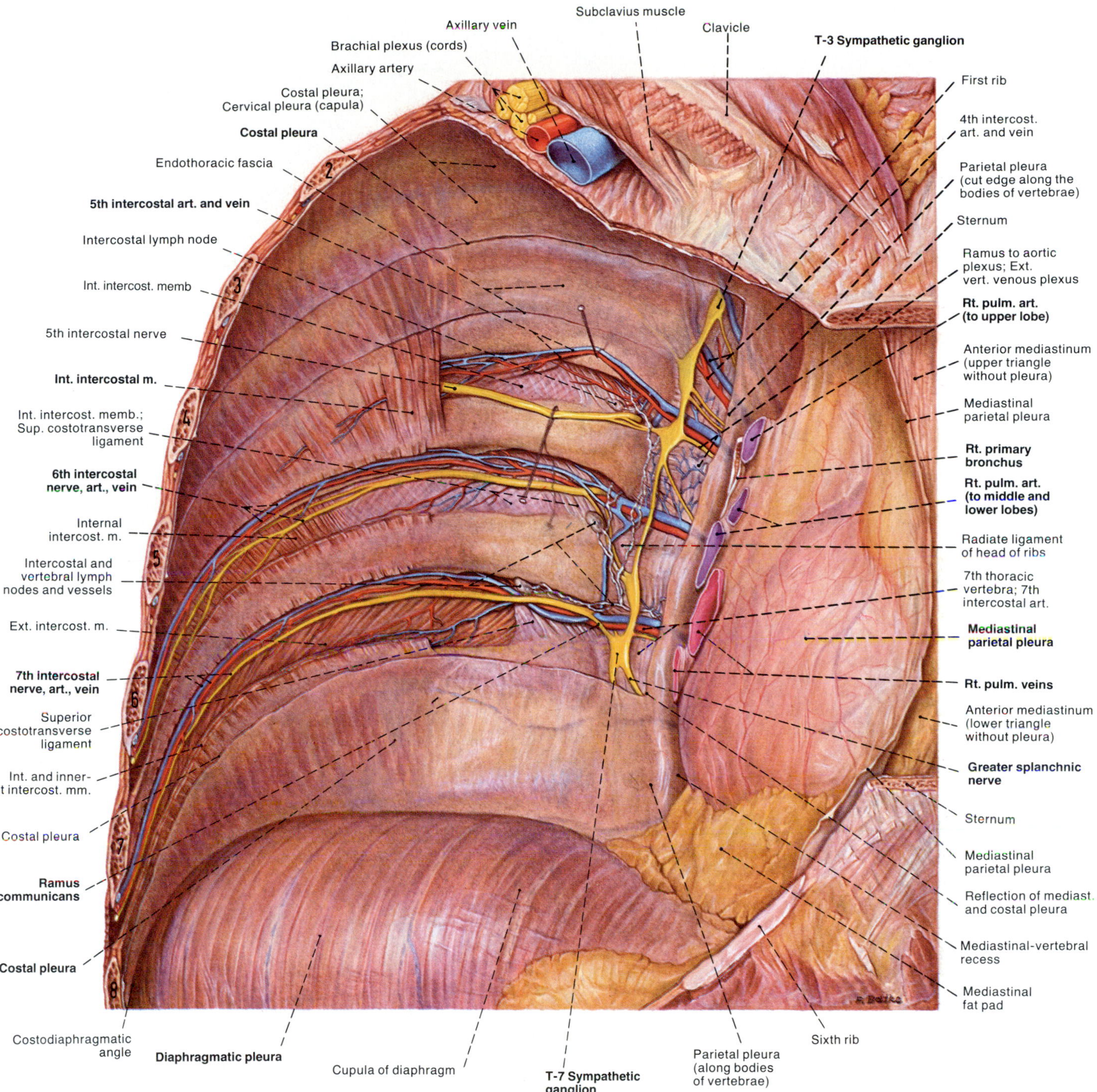
Axillary vein
Brachial plexus (cords)
Axillary artery
Costal pleura; Cervical pleura (capula)
Costal pleura
Endothoracic fascia
5th intercostal art. and vein
Intercostal lymph node
Int. intercost. memb
5th intercostal nerve
Int. intercostal m.
Int. intercost. memb.; Sup. costotransverse ligament
6th intercostal nerve, art., vein
Internal intercost. m.
Intercostal and vertebral lymph nodes and vessels
Ext. intercost. m.
7th intercostal nerve, art., vein
Superior costotransverse ligament
Int. and inner-most intercost. mm.
Costal pleura
Ramus communicans
Costal pleura
Costodiaphragmatic angle
Diaphragmatic pleura
Cupula of diaphragm
T-7 Sympathetic ganglion
Parietal pleura (along bodies of vertebrae)
Sixth rib
Subclavius muscle
Clavicle
T-3 Sympathetic ganglion
First rib
4th intercost. art. and vein
Parietal pleura (cut edge along the bodies of vertebrae)
Sternum
Ramus to aortic plexus; Ext. vert. venous plexus
Rt. pulm. art. (to upper lobe)
Anterior mediastinum (upper triangle without pleura)
Mediastinal parietal pleura
Rt. primary bronchus
Rt. pulm. art. (to middle and lower lobes)
Radiate ligament of head of ribs
7th thoracic vertebra; 7th intercostal art.
Mediastinal parietal pleura
Rt. pulm. veins
Anterior mediastinum (lower triangle without pleura)
Greater splanchnic nerve
Sternum
Mediastinal parietal pleura
Reflection of mediast. and costal pleura
Mediastinal-vertebral recess
Mediastinal fat pad
2
3
4
5
6
7
8

Fig. 227: The Right Posterior Thoracic Wall, the Sympathetic Trunk and the Intercostal Nerves and Vessels, Anterior View

NOTE: 1) that the right lung has been removed from the thoracic cavity, but clearly visible are the cut structures at the right pulmonary hilum; these include the right primary bronchus, the right pulmonary artery (severed distally enough to show its three branches to the lobes of the right lung) and the pulmonary veins.

2) that the costal and mediastinal reflections of parietal pleura have been stripped away to reveal the 5th, 6th and 7th intercostal spaces and the intercostal vessels and nerves that course within them around the thoracic wall. Observe that the intercostal vessels and nerves lie just below the ribs and that usually the nerve is slightly more inferior to the artery and vein within the intercostal space. This close relationship of the intercostal vessels and nerve to the lower border of the ribs in the posterior and lateral chest wall explains why it is safer to guide a needle through the intercostal space along the superior border of the rib when entrance into the thoracic cavity is necessary.

3) the sympathetic trunk and the 3rd and 8th sympathetic ganglia. Observe also that, in addition to the rami communicantes, the sympathetic trunk or ganglia contribute sympathetic fibers to the aortic plexus (seen coursing toward the left from this right sympathetic trunk), as well as to the splanchic nerves.

4) the intercostal and paravertebral lymph nodes and lymphatic channels.

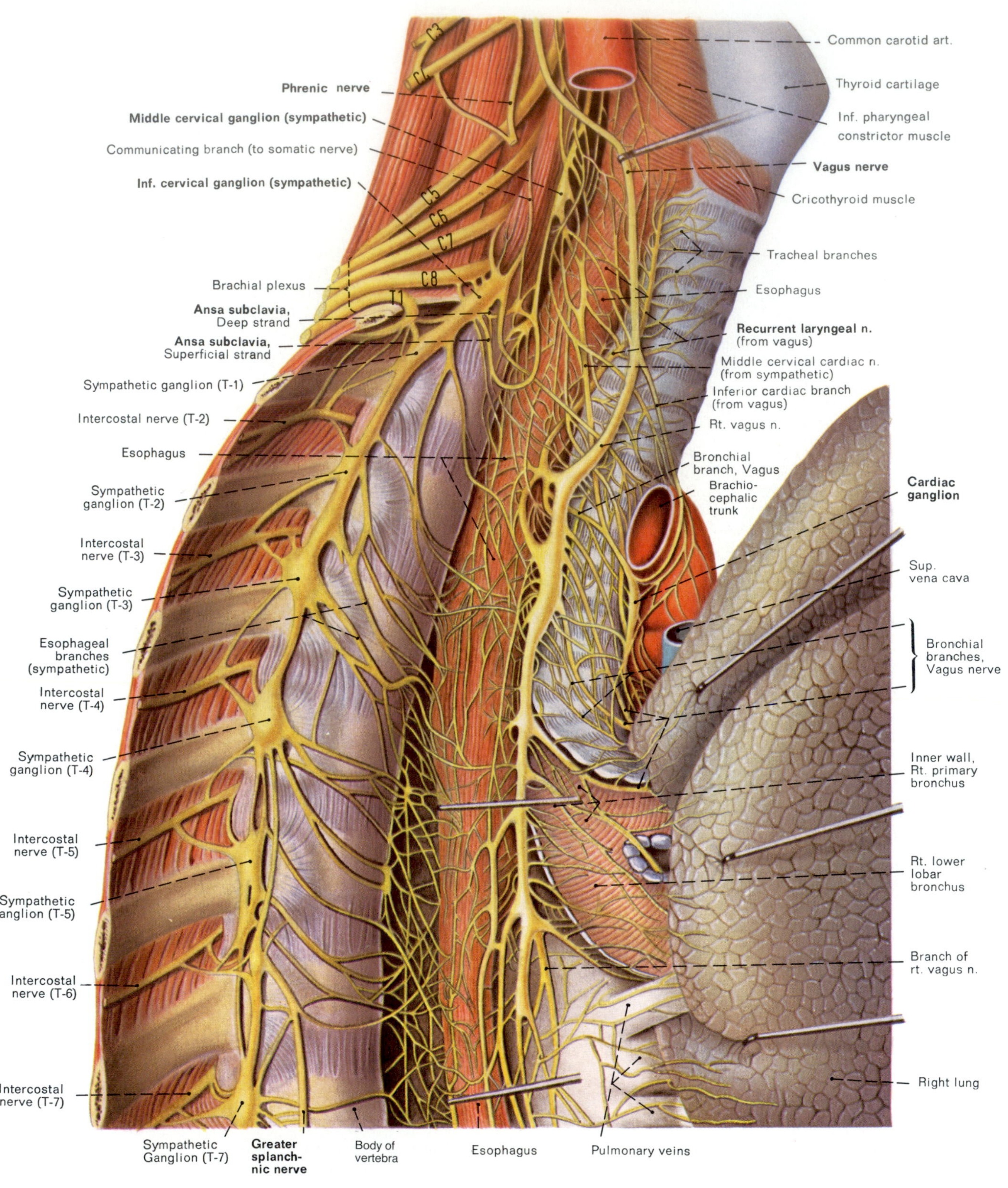
C3
C4
C5
C6
C7
C8
T1
Common carotid art.
Thyroid cartilage
Phrenic nerve
Middle cervical ganglion (sympathetic)
Inf. pharyngeal constrictor muscle
Communicating branch (to somatic nerve)
Vagus nerve
Inf. cervical ganglion (sympathetic)
Cricothyroid muscle
Tracheal branches
Brachial plexus
Esophagus
Ansa subclavia, Deep strand
Recurrent laryngeal n. (from vagus)
Ansa subclavia, Superficial strand
Middle cervical cardiac n. (from sympathetic)
Sympathetic ganglion (T-1)
Inferior cardiac branch (from vagus)
Intercostal nerve (T-2)
Rt. vagus n.
Esophagus
Bronchial branch, Vagus
Sympathetic ganglion (T-2)
Brachio-cephalic trunk
Cardiac ganglion
Intercostal nerve (T-3)
Sup. vena cava
Sympathetic ganglion (T-3)
Esophageal branches (sympathetic)
Bronchial branches, Vagus nerve
Intercostal nerve (T-4)
Sympathetic ganglion (T-4)
Inner wall, Rt. primary bronchus
Intercostal nerve (T-5)
Rt. lower lobar bronchus
Sympathetic ganglion (T-5)
Branch of rt. vagus n.
Intercostal nerve (T-6)
Intercostal nerve (T-7)
Right lung
Sympathetic Ganglion (T-7)
Greater splanch-nic nerve
Body of vertebra
Esophagus
Pulmonary veins

Fig. 228: The Cervical and Upper Thoracic Distribution of Autonomic Nerves

NOTE: the organs of the posterior mediastinum are viewed from the right side by pulling the lungs forward and removing certain of the organs. Observe the two major descending nerve trunks and their associated complexes: the right vagus nerve situated more anteriorly and the right sympathetic trunk descending more posteriorly in the thorax adjacent to the costovertebral joints.

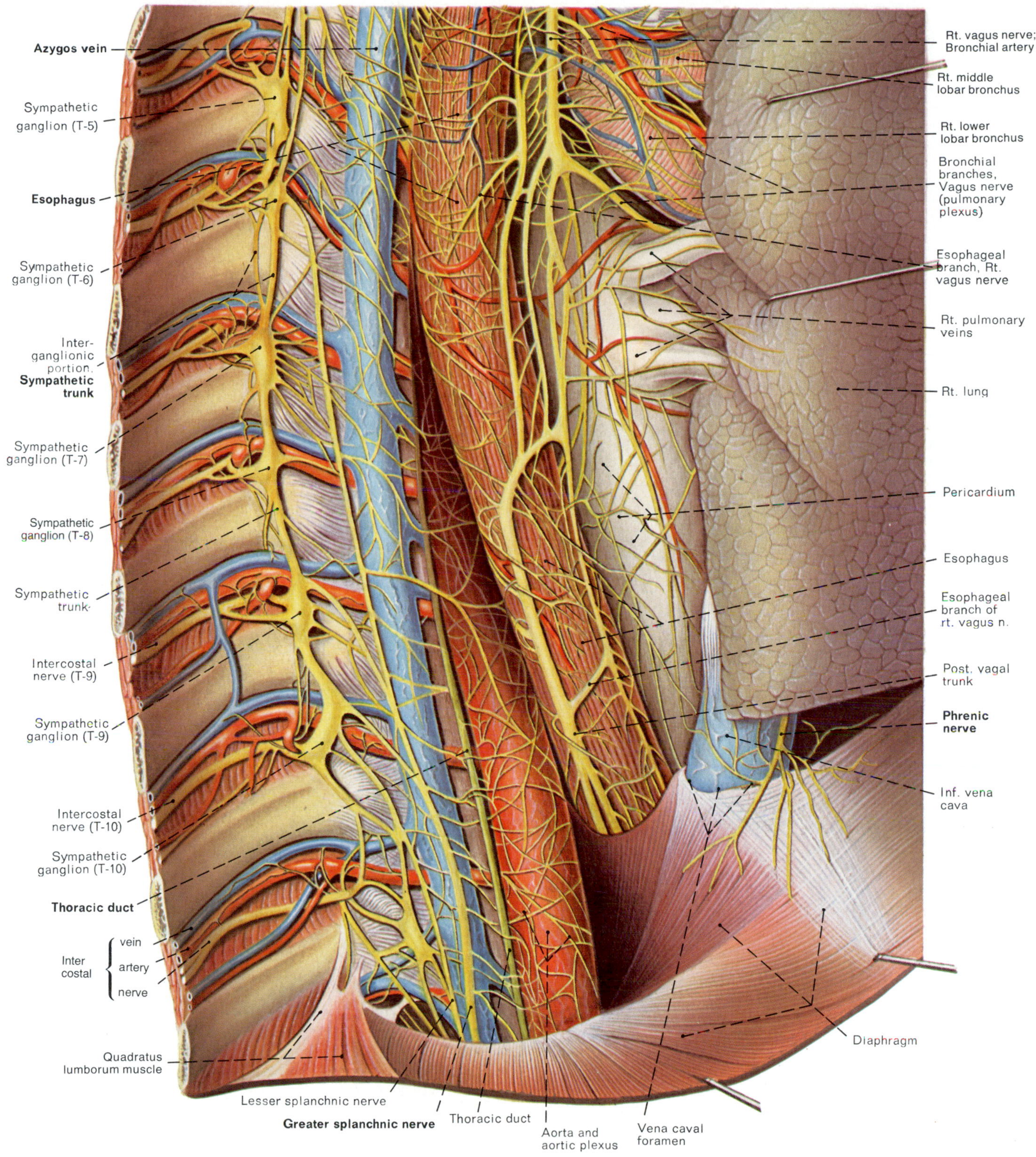

Azygos vein
Sympathetic ganglion (T-5)
Esophagus
Sympathetic ganglion (T-6)
Inter-ganglionic portion, Sympathetic trunk
Sympathetic ganglion (T-7)
Sympathetic ganglion (T-8)
Sympathetic trunk
Intercostal nerve (T-9)
Sympathetic ganglion (T-9)
Intercostal nerve (T-10)
Sympathetic ganglion (T-10)
Thoracic duct
Inter costal
vein
artery
nerve
Quadratus lumborum muscle
Lesser splanchnic nerve
Greater splanchnic nerve
Thoracic duct
Aorta and aortic plexus
Vena caval foramen
Diaphragm
Inf. vena cava
Phrenic nerve
Post. vagal trunk
Esophageal branch of rt. vagus n.
Esophagus
Pericardium
Rt. lung
Rt. pulmonary veins
Esophageal branch, Rt. vagus nerve
Bronchial branches, Vagus nerve (pulmonary plexus)
Rt. lower lobar bronchus
Rt. middle lobar bronchus
Rt. vagus nerve; Bronchial artery

Fig. 229: Lower Thoracic Portion of Autonomic Nervous System

NOTE: 1) the formation of the greater and lesser splanchnic nerves. The greater splanchnic nerve is derived from preganglionic sympathetic fibers which emerge from sympathetic ganglia T-6 to T-9 or T-10, whereas the lesser splanchnic nerve is derived from ganglia T-10 and T-11.

2) the right vagus nerve after contributing parasympathetic fibers to the esophageal plexus becomes the posterior vagal trunk dorsal to the esophagus as that organ passes through the diaphragm.

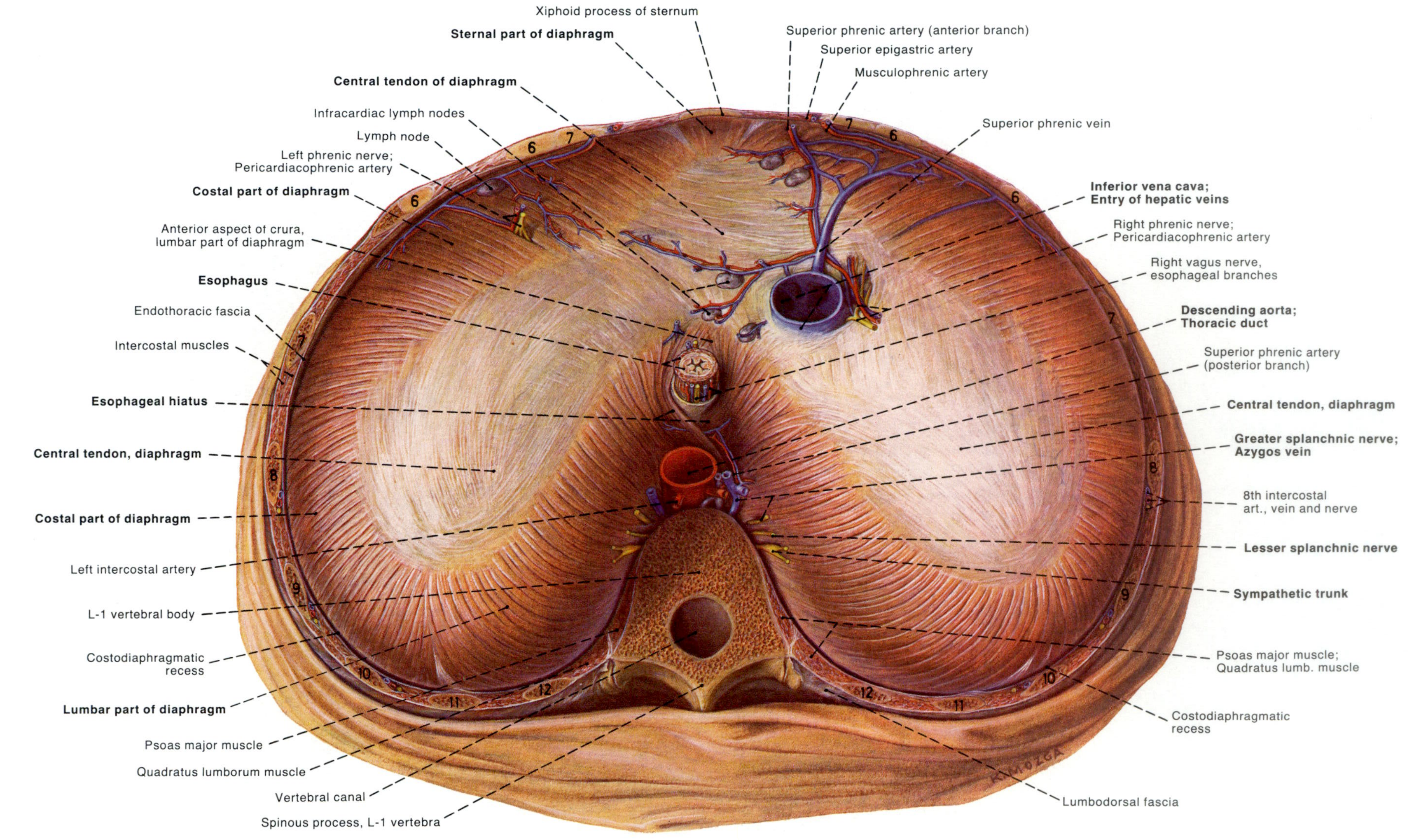
Xiphoid process of sternum
Sternal part of diaphragm
Central tendon of diaphragm
Infracardiac lymph nodes
Lymph node
Left phrenic nerve;
Pericardiacophrenic artery
Costal part of diaphragm
Anterior aspect of crura,
lumbar part of diaphragm
Esophagus
Endothoracic fascia
Intercostal muscles
Esophageal hiatus
Central tendon, diaphragm
Costal part of diaphragm
Left intercostal artery
L-1 vertebral body
Costodiaphragmatic
recess
Lumbar part of diaphragm
Psoas major muscle
Quadratus lumborum muscle
Vertebral canal
Spinous process, L-1 vertebra
Superior phrenic artery (anterior branch)
Superior epigastric artery
Musculophrenic artery
Superior phrenic vein
Inferior vena cava;
Entry of hepatic veins
Right phrenic nerve;
Pericardiacophrenic artery
Right vagus nerve,
esophageal branches
Descending aorta;
Thoracic duct
Superior phrenic artery
(posterior branch)
Central tendon, diaphragm
Greater splanchnic nerve;
Azygos vein
8th intercostal
art., vein and nerve
Lesser splanchnic nerve
Sympathetic trunk
Psoas major muscle;
Quadratus lumb. muscle
Costodiaphragmatic
recess
Lumbodorsal fascia
6
7
6
7
6
6
7
8
9
10
11
12
8
9
10
11
12

Fig. 230: The Diaphragm: Its Openings, Blood Vessels and Nerves Viewed From Above

NOTE: 1) that the thoracic cage has been servered transversely in a plane extending from the xiphoid process anteriorly to the body of the first lumbar vertebra posteriorly. The diaphragmatic pleurae and the pericardium have been stripped from the superior (thoracic) surface of the diaphragm;

2) the relative location of the vena caval orifice and the esophageal and aortic hiatuses. The *vena caval opening* lies to the right of the midline, is more anterior and higher (8th to 9th thoracic vertebra) than the other two. The *aortic hiatus* lies in the midline and is the most posterior and the most inferior (L-1) of these large apertures. The *esophageal hiatus*, also in the midline, lies anterior to the aortic hiatus at about the level of the 10th thoracic vertebra.

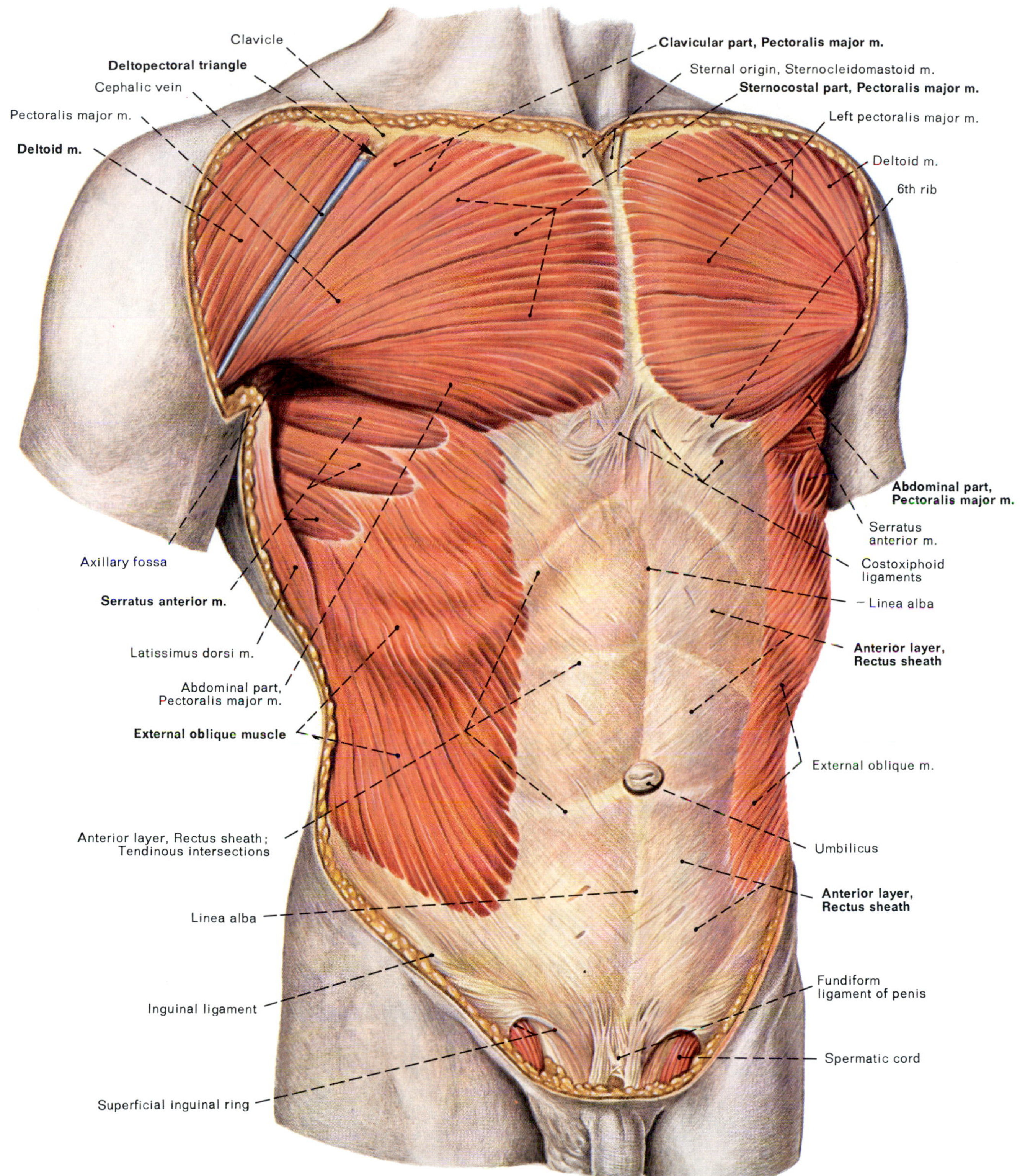
Clavicle
Deltopectoral triangle
Cephalic vein
Pectoralis major m.
Deltoid m.
Clavicular part, Pectoralis major m.
Sternal origin, Sternocleidomastoid m.
Sternocostal part, Pectoralis major m.
Left pectoralis major m.
Deltoid m.
6th rib
Abdominal part, Pectoralis major m.
Serratus anterior m.
Costoxiphoid ligaments
Linea alba
Anterior layer, Rectus sheath
External oblique m.
Umbilicus
Anterior layer, Rectus sheath
Fundiform ligament of penis
Spermatic cord
Axillary fossa
Serratus anterior m.
Latissimus dorsi m.
Abdominal part, Pectoralis major m.
External oblique muscle
Anterior layer, Rectus sheath; Tendinous intersections
Linea alba
Inguinal ligament
Superficial inguinal ring

Fig. 231: Superficial Musculature of the Anterior Abdominal and Thoracic Wall

NOTE: 1) the first layer of abdominal musculature consists principally of the external oblique muscle and its broad, flat aponeurosis which extends medially to the midline (forming the anterior layer of the sheath of the rectus abdominis muscle) and inferiorly as the inguinal ligament.

2) the external oblique arises by means of 7 or 8 fleshy slips from the outer surfaces of the lower ribs (ribs 5 to 12), thereby interdigitating with the fleshy origin of the serratus anterior muscle.

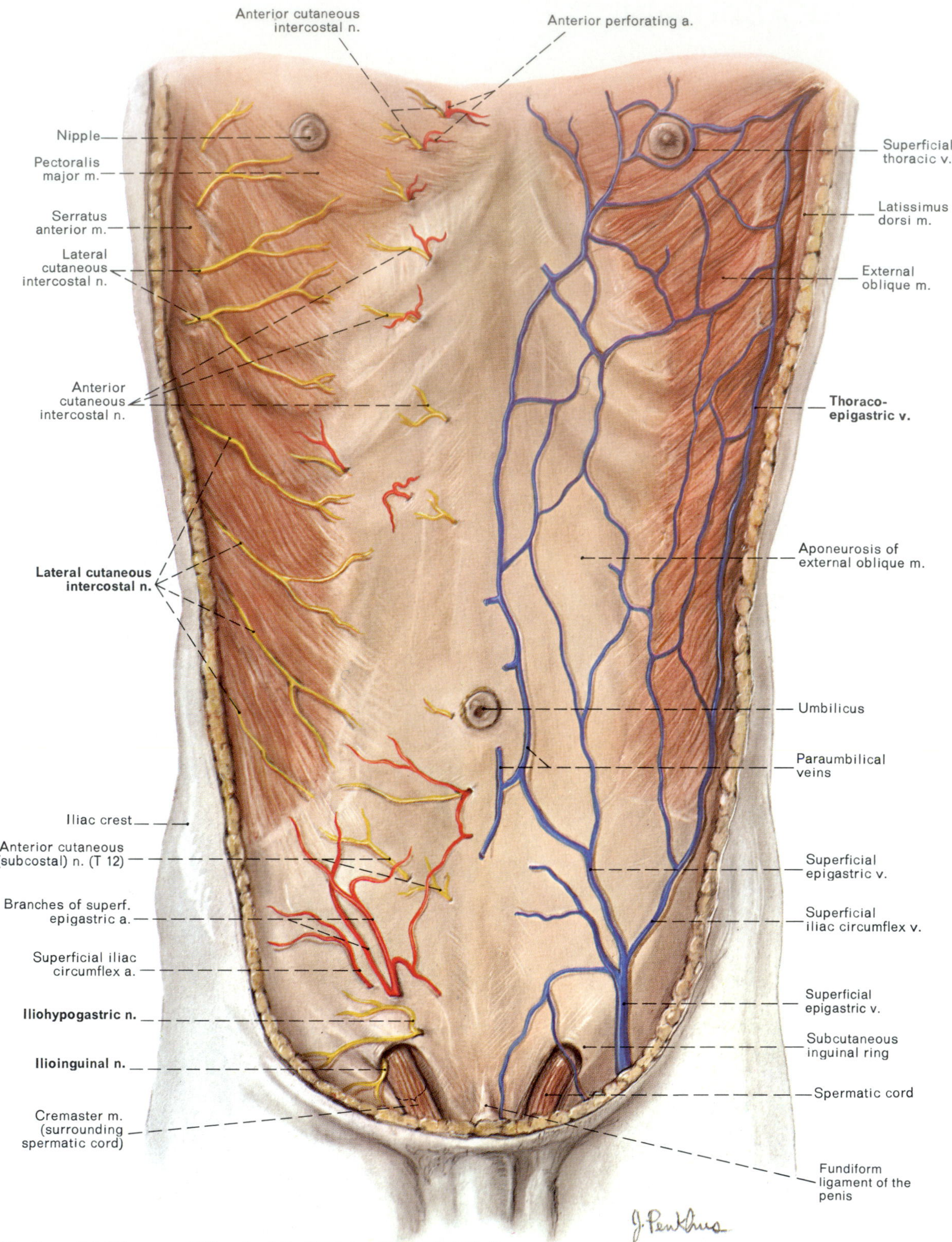

Anterior cutaneous intercostal n.
Anterior perforating a.
Nipple
Pectoralis major m.
Serratus anterior m.
Lateral cutaneous intercostal n.
Anterior cutaneous intercostal n.
Lateral cutaneous intercostal n.
Iliac crest
Anterior cutaneous (subcostal) n. (T 12)
Branches of superf. epigastric a.
Superficial iliac circumflex a.
Iliohypogastric n.
Ilioinguinal n.
Cremaster m. (surrounding spermatic cord)
Superficial thoracic v.v.
Latissimus dorsi m.
External oblique m.
Thoraco-epigastric v.
Aponeurosis of external oblique m.
Umbilicus
Paraumbilical veins
Superficial epigastric v.
Superficial iliac circumflex v.
Superficial epigastric v.
Subcutaneous inguinal ring
Spermatic cord
Fundiform ligament of the penis
J. Perkins

Fig. 232: Superficial Nerves and Vessels of the Anterior Abdominal Wall

NOTE: 1) the distribution of the superficial vessels and cutaneous nerves is demonstrated upon the removal of the skin and superficial fatty layers over the lower thoracic and anterior abdominal wall.

2) the thoracic intercostal nerves supply the abdominal surface with lateral and anterior cutaneous branches.

3) in the inguinal region, the ilioinguinal and iliohypogastric branches of the 1st lumbar nerve become superficial in the region of the superficial inguinal ring.

4) the branches of superficial epigastric artery (which arises from the femoral artery) as they ascend in the inguinal region toward the umbilicus. Note also the superficial branches of the intercostal arteries.

5) the thoracoepigastric vein which serves as a means of communication between the femoral vein and the axillary vein. In cases of obstruction of the portal vein, these superficial veins become greatly enlarged forming varicose veins over the abdominal wall (sometimes this condition is called *caput medusae*).

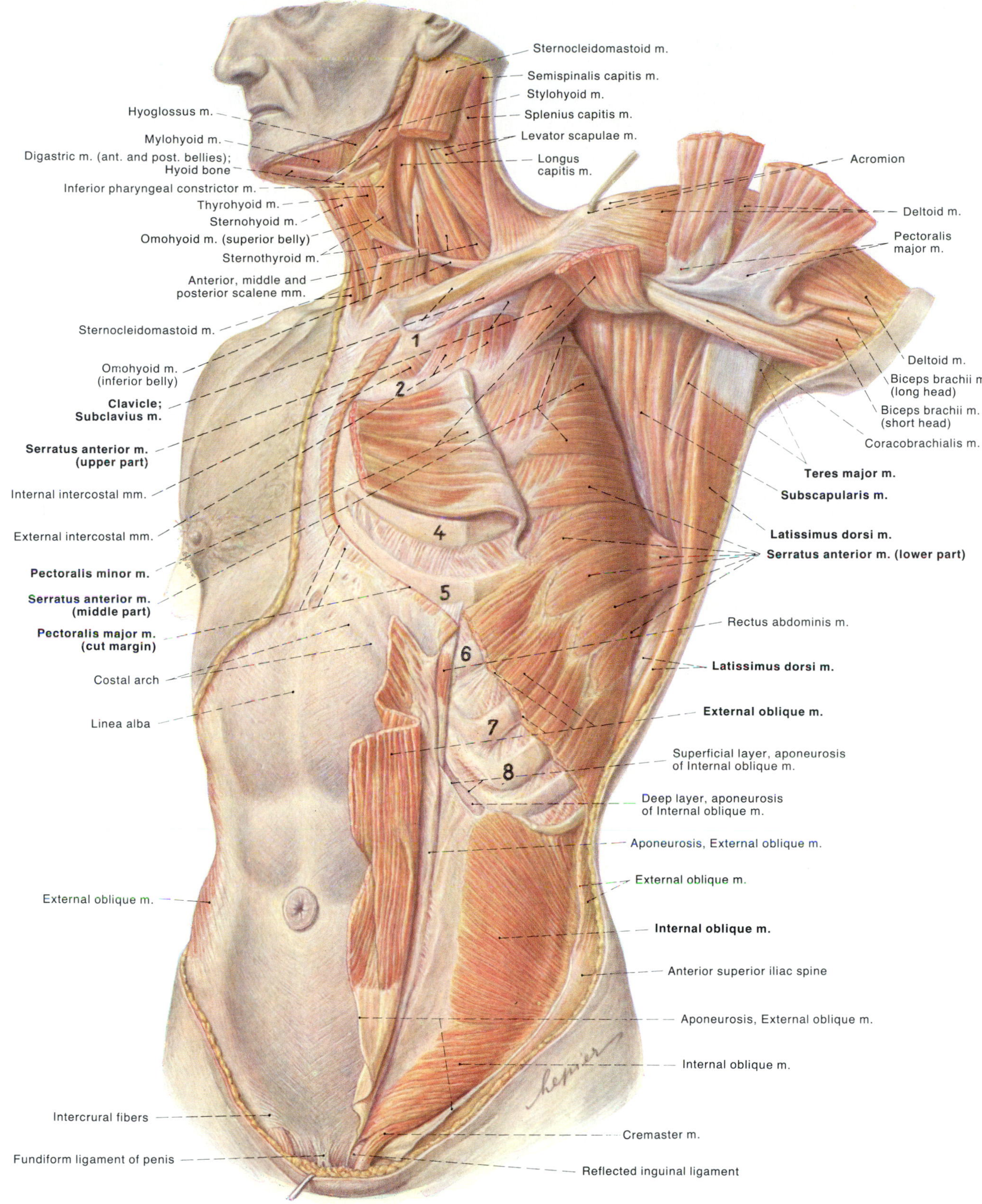
Sternocleidomastoid m.
Semispinalis capitis m.
Stylohyoid m.
Splenius capitis m.
Levator scapulae m.
Longus capitis m.
Acromion
Deltoid m.
Pectoralis major m.
Hyoglossus m.
Mylohyoid m.
Digastric m. (ant. and post. bellies);
Hyoid bone
Inferior pharyngeal constrictor m.
Thyrohyoid m.
Sternohyoid m.
Omohyoid m. (superior belly)
Sternothyroid m.
Anterior, middle and posterior scalene mm.
Sternocleidomastoid m.
Omohyoid m. (inferior belly)
Clavicle;
Subclavius m.
Serratus anterior m. (upper part)
Internal intercostal mm.
External intercostal mm.
Pectoralis minor m.
Serratus anterior m. (middle part)
Pectoralis major m. (cut margin)
Costal arch
Linea alba
External oblique m.
Intercrural fibers
Fundiform ligament of penis
1
2
4
5
6
7
8
Deltoid m.
Biceps brachii m. (long head)
Biceps brachii m. (short head)
Coracobrachialis m.
Teres major m.
Subscapularis m.
Latissimus dorsi m.
Serratus anterior m. (lower part)
Rectus abdominis m.
Latissimus dorsi m.
External oblique m.
Superficial layer, aponeurosis of Internal oblique m.
Deep layer, aponeurosis of Internal oblique m.
Aponeurosis, External oblique m.
External oblique m.
Internal oblique m.
Anterior superior iliac spine
Aponeurosis, External oblique m.
Internal oblique m.
Cremaster m.
Reflected inguinal ligament

Fig. 233: The Deeper Layers of the Musculature of the Trunk, Axilla and Neck

NOTE that: 1) the pectoralis major and minor muscles have been reflected to reveal the underlying digitations of the serratus anterior muscle as these strands attach to the ribs forming the thoracic wall. Observe that the external oblique muscle covers a relatively large portion of the infero-lateral thoracic wall as well as stretching across the anterior abdominal wall as its external muscular layer.

2) the external oblique muscle and the lower lateral part of its aponeurosis have been severed in a semicircular manner near their origin in order to reveal the underlying internal oblique muscle. There are certain generalities about the attachments of the external and internal oblique muscles which also apply to the deepest of the anterior abdominal wall muscles, the transversus abdominis muscle: a) they all attach to the last six ribs (the external oblique frequently extending one or two ribs higher), b) they all are attached to both the iliac crest and the inguinal ligament, and c) they all insert into the linea alba.

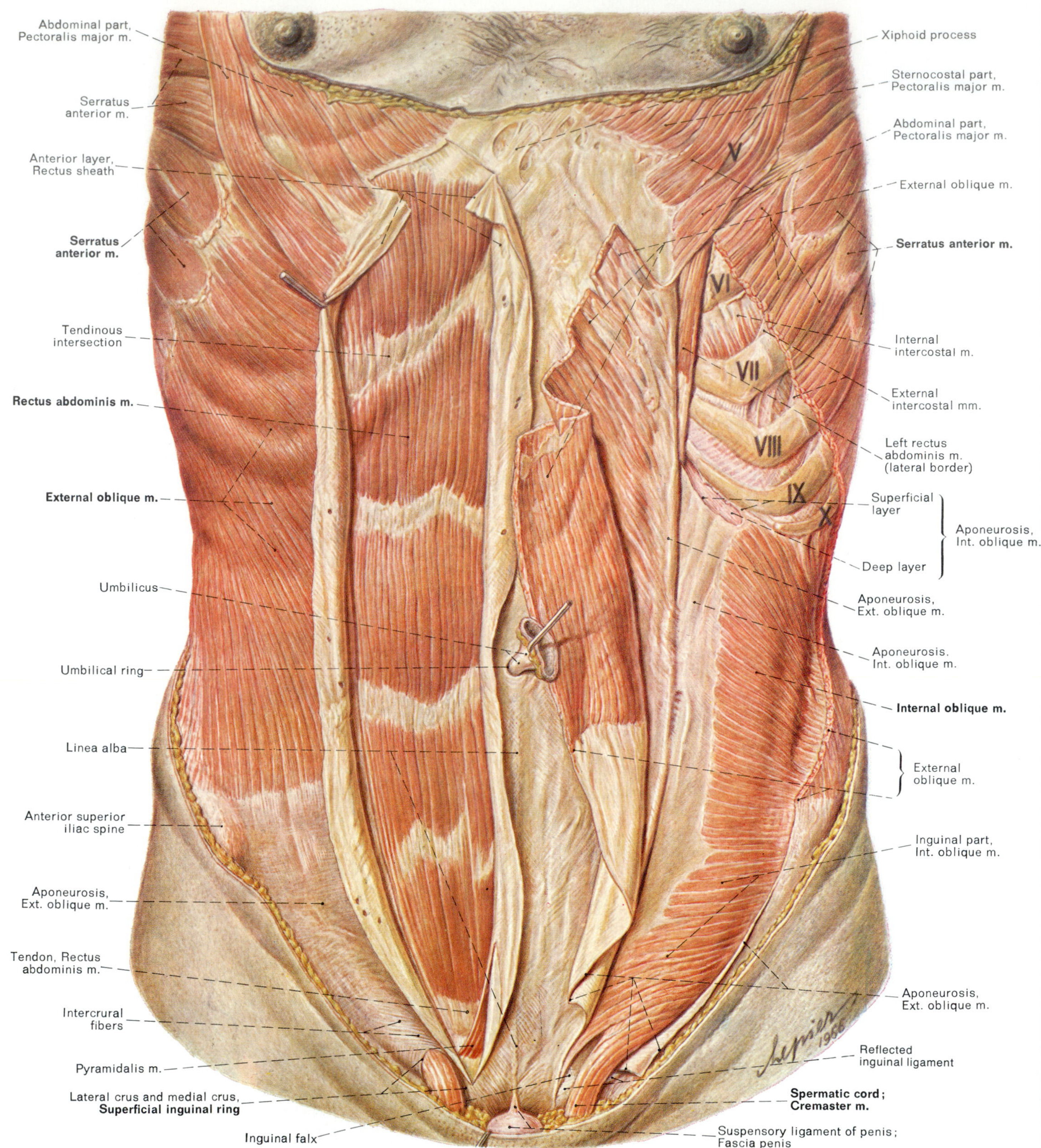

Abdominal part, Pectoralis major m.
Serratus anterior m.
Anterior layer, Rectus sheath
Serratus anterior m.
Tendinous intersection
Rectus abdominis m.
External oblique m.
Umbilicus
Umbilical ring
Linea alba
Anterior superior iliac spine
Aponeurosis, Ext. oblique m.
Tendon, Rectus abdominis m.
Intercrural fibers
Pyramidalis m.
Lateral crus and medial crus, Superficial inguinal ring
Inguinal falx
Xiphoid process
Sternocostal part, Pectoralis major m.
Abdominal part, Pectoralis major m.
External oblique m.
Serratus anterior m.
Internal intercostal m.
External intercostal mm.
Left rectus abdominis m. (lateral border)
Superficial layer
Deep layer
Aponeurosis, Int. oblique m.
Aponeurosis, Ext. oblique m.
Aponeurosis. Int. oblique m.
Internal oblique m.
External oblique m.
Inguinal part, Int. oblique m.
Aponeurosis, Ext. oblique m.
Reflected inguinal ligament
Spermatic cord; Cremaster m.
Suspensory ligament of penis; Fascia penis
V
VI
VII
VIII
IX
X

Fig. 234: Anterior Abdominal Wall: Rectus Abdominis and Internal Oblique Muscles

NOTE: 1) the specimen's right rectus sheath has been opened (reader's left) to reveal the right rectus abdominis muscle which is marked by transversely oriented tendinous intersections. On the specimen's left side, the external oblique muscle has been severed to reveal the second muscular layer, the internal oblique muscle. This also reveals ribs 6 through 10.

2) the muscle fibers of the external oblique course inferomedially (or in the same direction as you would put your hands in your side pockets), whereas most of the fibers of the internal oblique course in the opposite direction.

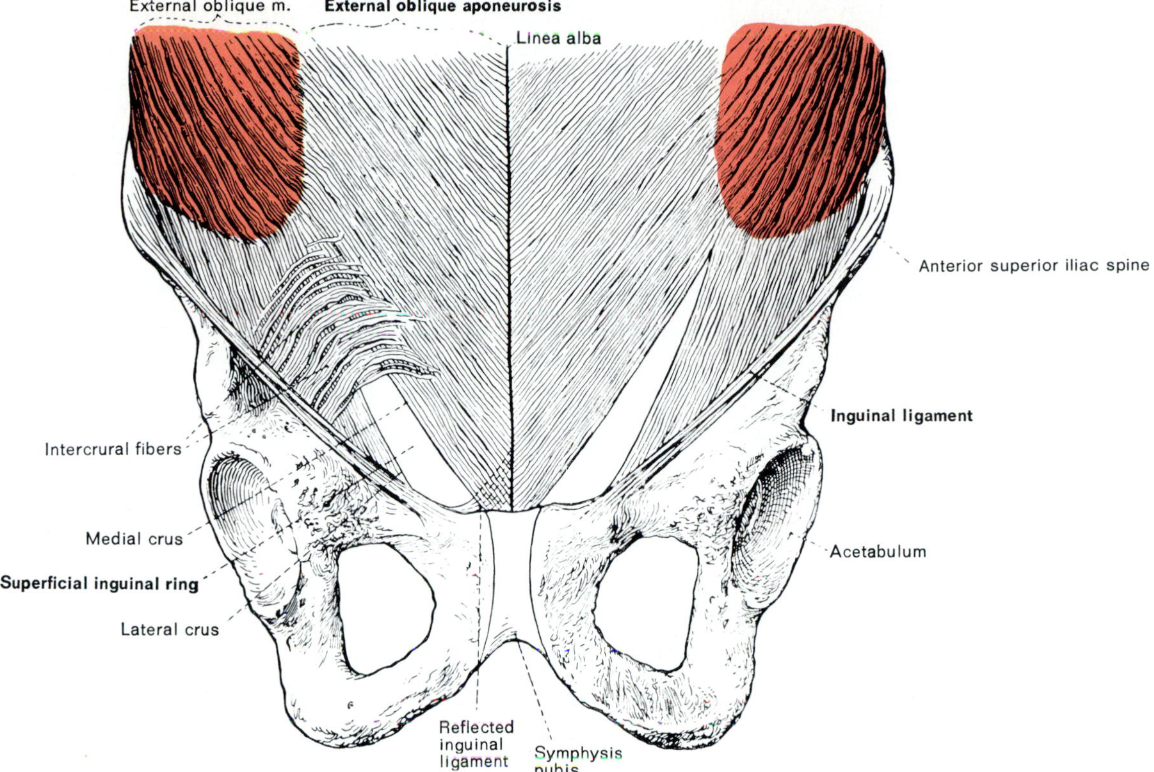

External oblique m.
External oblique aponeurosis
Linea alba
Anterior superior iliac spine
Inguinal ligament
Intercrural fibers
Medial crus
Superficial inguinal ring
Acetabulum
Lateral crus
Reflected inguinal ligament
Symphysis pubis

Fig. 235: The Aponeurosis of the External Oblique Muscle

NOTE: 1) the superficial inguinal ring is a triangular slit-like opening in the aponeurosis of the external oblique muscle. Observe how the intercrural fibers would strengthen the lateral aspect of the superficial ring by extending between the medial crus and lateral crus.

2) the inguinal ligament which extends between the anterior superior iliac spine and the pubic tubercle. This ligament is formed by the lowermost fibers of the external oblique aponeurosis and lends support to the inferior portion of the anterior abdominal wall.

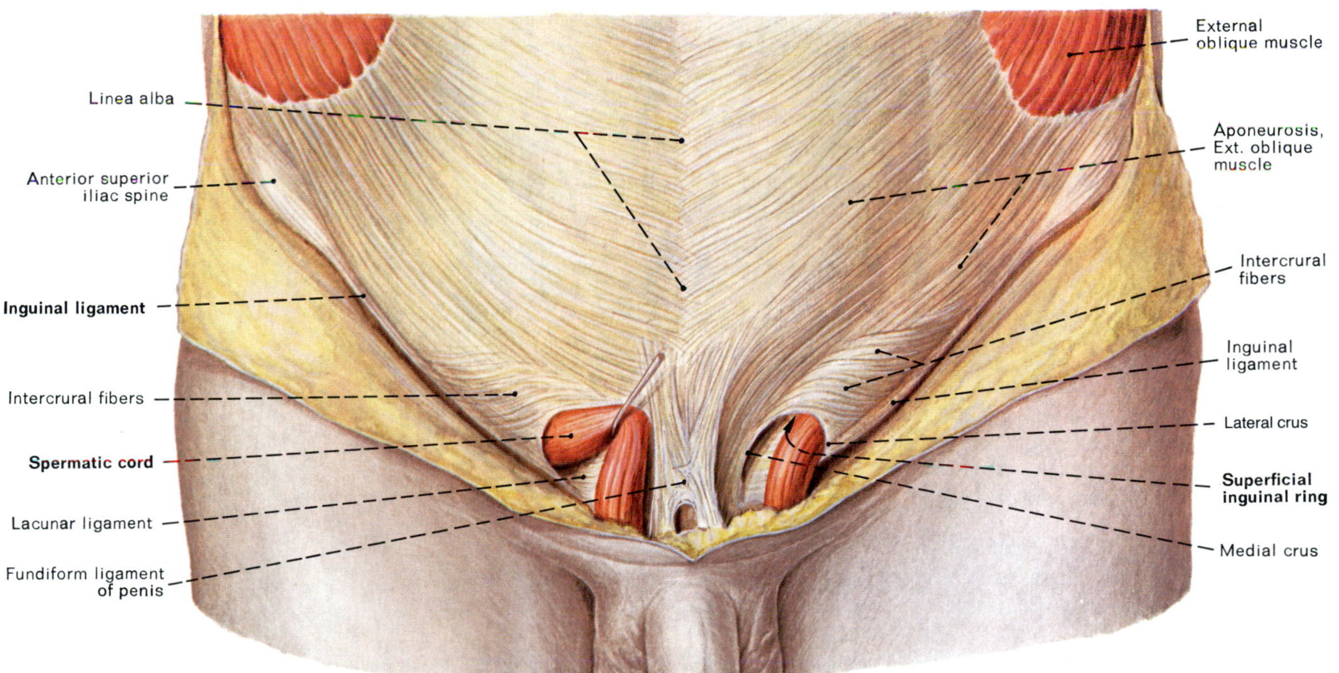

External oblique muscle
Linea alba
Aponeurosis, Ext. oblique muscle
Anterior superior iliac spine
Intercrural fibers
Inguinal ligament
Inguinal ligament
Intercrural fibers
Lateral crus
Spermatic cord
Superficial inguinal ring
Lacunar ligament
Medial crus
Fundiform ligament of penis

Fig. 236: The Superficial Inguinal Ring and Spermatic Cord

NOTE: 1) the superficial inguinal ring transmits the spermatic cord in the male and the round ligament of the uterus in the female. In this dissection, the right spermatic cord has been lifted in order to show the lateral crus of the inguinal ring.

2) the tendinous fibers of the aponeurosis are continuous with the fleshy fibers of the external oblique. They are directed inferiorly and medially and either decussate or insert into the linea alba.

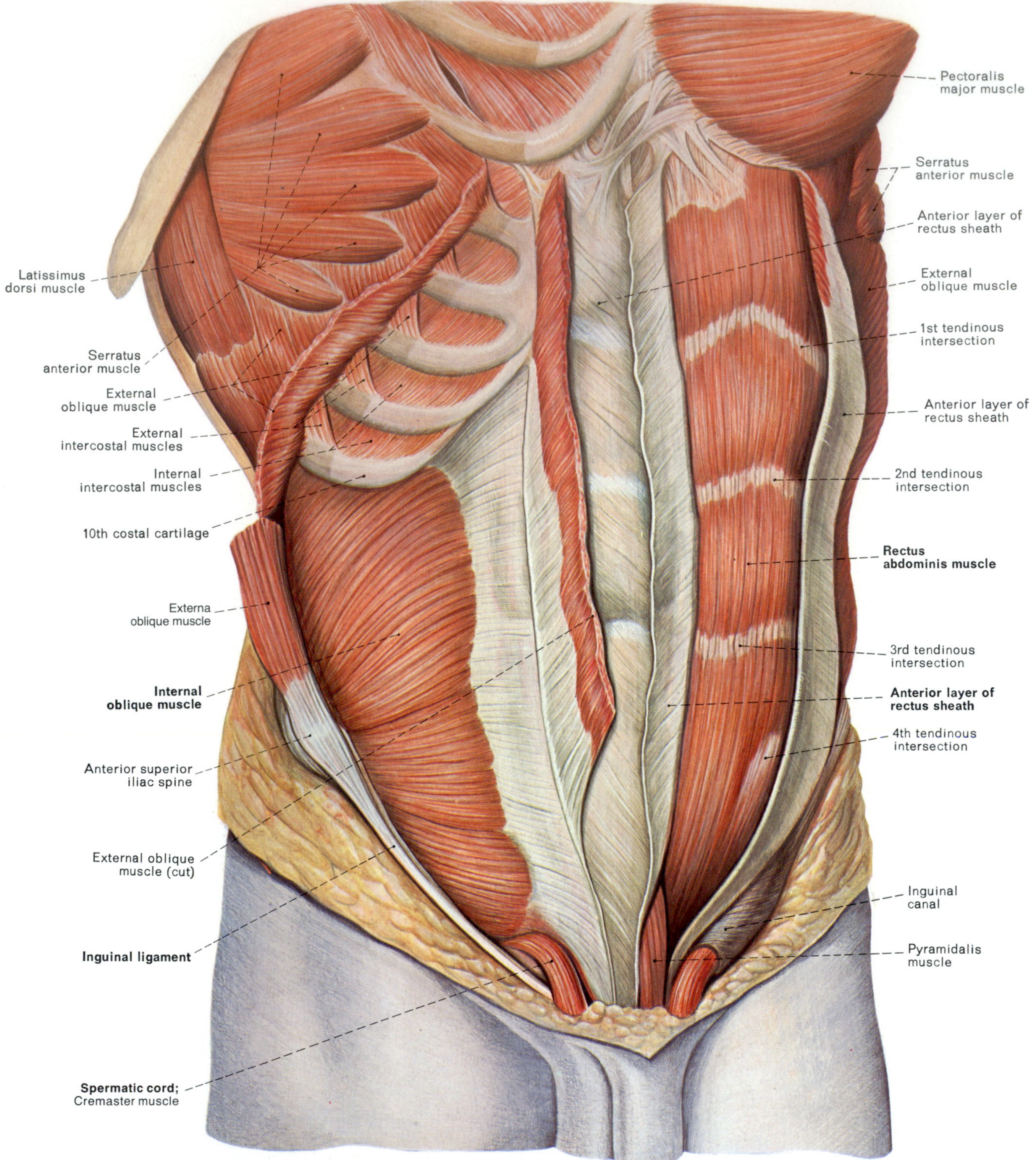

Pectoralis major muscle
Serratus anterior muscle
Anterior layer of rectus sheath
External oblique muscle
1st tendinous intersection
Anterior layer of rectus sheath
2nd tendinous intersection
Rectus abdominis muscle
3rd tendinous intersection
Anterior layer of rectus sheath
4th tendinous intersection
Inguinal canal
Pyramidalis muscle
Latissimus dorsi muscle
Serratus anterior muscle
External oblique muscle
External intercostal muscles
Internal intercostal muscles
10th costal cartilage
Externa oblique muscle
Internal oblique muscle
Anterior superior iliac spine
External oblique muscle (cut)
Inguinal ligament
Spermatic cord; Cremaster muscle

Fig. 237: Middle Layer of Abdominal Musculature: Internal Oblique Muscle

NOTE: 1) on the right side, the external oblique muscle has been severed and reflected to expose the right internal oblique muscle. On the left side, the anterior layer of the rectus sheath has been incised longitudinally to expose the left rectus abdominis muscle with its tendinous intersections and the small pyramidalis muscle.

2) the muscle fibers of the internal oblique muscle arise from the inguinal ligament, the iliac crest and the lumbar aponeurosis. They insert into the lower ribs above, and into an aponeurosis which contributes to the formation of the rectus sheath medially. Inferiorly, the aponeurosis of the internal oblique along with the aponeurosis of the transversus abdominis muscle forms the conjoint tendon, which is also known as the inguinal falx (shown, but not labelled). The conjoint tendon inserts into the pubic crest along with the lower end of the rectus sheath, thereby helping to provide strength to the inherently weak inguinal-pubic region.

3) the cremaster muscle. This muscular covering over the outer surface of the spermatic cord in the male represents an extension of the internal oblique (also possibly the transversus abdominis). It originates on the inguinal ligament and, after its fibers loop around the spermatic cord, inserts onto the pubis. Upon contraction this muscle lifts the testis within the scrotum toward the subcutaneous inguinal ring.

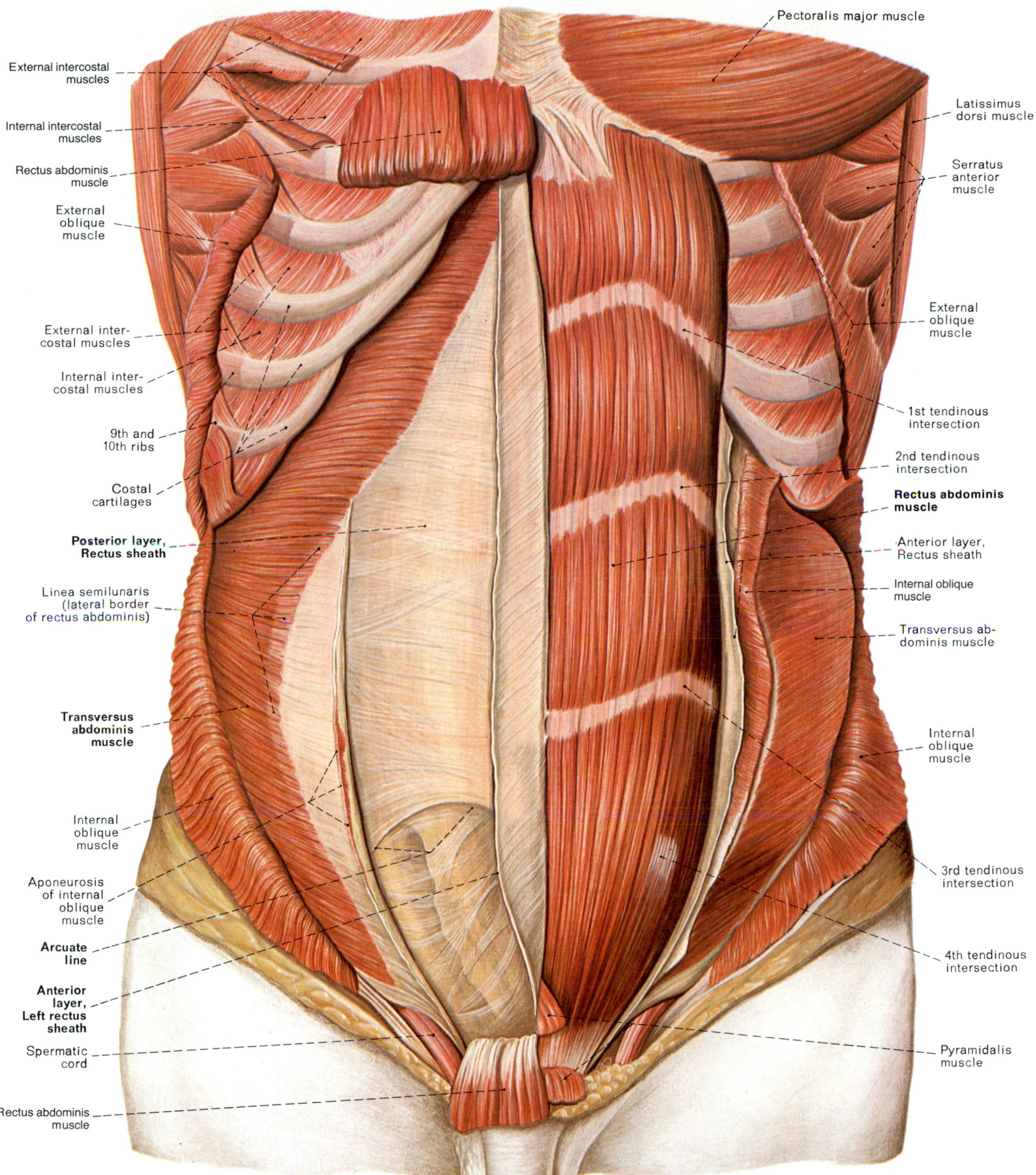

Pectoralis major muscle
External intercostal muscles
Internal intercostal muscles
Rectus abdominis muscle
External oblique muscle
External inter- costal muscles
Internal inter- costal muscles
9th and 10th ribs
Costal cartilages
Posterior layer, Rectus sheath
Linea semilunaris (lateral border of rectus abdominis)
Transversus abdominis muscle
Internal oblique muscle
Aponeurosis of internal oblique muscle
Arcuate line
Anterior layer, Left rectus sheath
Spermatic cord
Rectus abdominis muscle
Latissimus dorsi muscle
Serratus anterior muscle
External oblique muscle
1st tendinous intersection
2nd tendinous intersection
Rectus abdominis muscle
Anterior layer, Rectus sheath
Internal oblique muscle
Transversus ab- dominis muscle
Internal oblique muscle
3rd tendinous intersection
4th tendinous intersection
Pyramidalis muscle

Fig. 238: Deep Layer of Abdominal Musculature: The Transversus Abdominis Muscle

NOTE: 1) on the right, the external and internal oblique and rectus abdominis muscles have been resected demonstrating the transversus abdominis muscle. On the left, the rectus abdominis muscle remains intact but the small pyramidalis muscle was severed to show the insertion of the left rectus.

2) on the right, the posterior layer of the rectus sheath is exposed, while on the left, the anterior layer of the sheath has been opened to demonstrate the muscle. Below the arcuate line, the posterior sheath of the rectus abdominis is wanting, and the overlying rectus muscle lies directly anterior to the transversalis fascia.

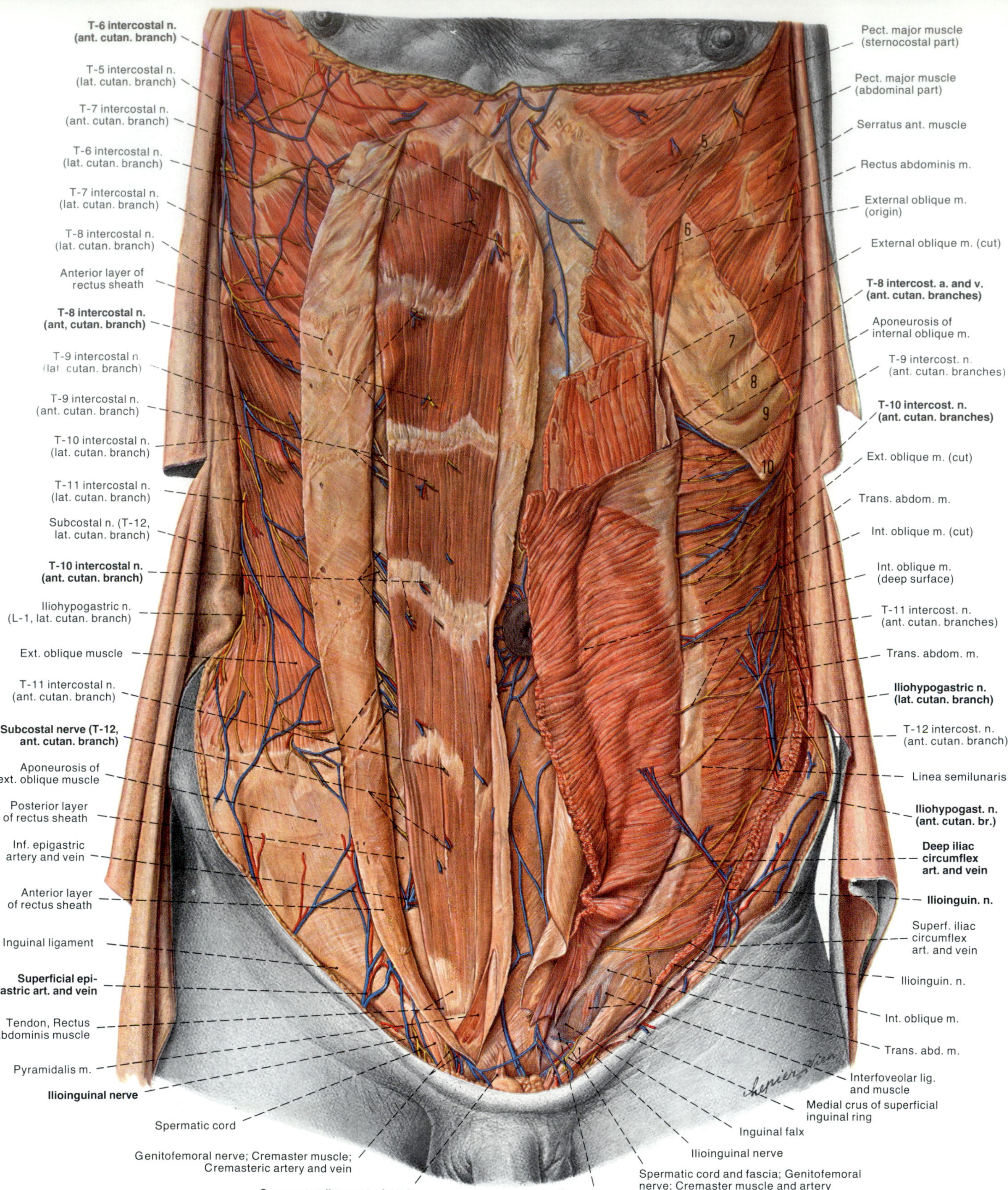

T-6 intercostal n.
(ant. cutan. branch)
T-5 intercostal n.
(lat. cutan. branch)
T-7 intercostal n.
(ant. cutan. branch)
T-6 intercostal n.
(lat. cutan. branch)
T-7 intercostal n.
(lat. cutan. branch)
T-8 intercostal n.
(lat. cutan. branch)
Anterior layer of
rectus sheath
T-8 intercostal n.
(ant, cutan. branch)
T-9 intercostal n.
(lat. cutan. branch)
T-9 intercostal n.
(ant. cutan. branch)
T-10 intercostal n.
(lat. cutan. branch)
T-11 intercostal n.
(lat. cutan. branch)
Subcostal n. (T-12,
lat. cutan. branch)
T-10 intercostal n.
(ant. cutan. branch)
Iliohypogastric n.
(L-1, lat. cutan. branch)
Ext. oblique muscle
T-11 intercostal n.
(ant. cutan. branch)
Subcostal nerve (T-12,
ant. cutan. branch)
Aponeurosis of
ext. oblique muscle
Posterior layer
of rectus sheath
Inf. epigastric
artery and vein
Anterior layer
of rectus sheath
Inguinal ligament
Superficial epi-
gastric art. and vein
Tendon, Rectus
abdominis muscle
Pyramidalis m.
Ilioinguinal nerve
Spermatic cord
Genitofemoral nerve; Cremaster muscle;
Cremasteric artery and vein
Suspensory ligament of penis
Reflected inguinal ligament
Spermatic cord and fascia; Genitofemoral
nerve; Cremaster muscle and artery
Ilioinguinal nerve
Inguinal falx
Medial crus of superficial
inguinal ring
Interfoveolar lig.
and muscle
Trans. abd. m.
Int. oblique m.
Ilioinguin. n.
Superf. iliac
circumflex
art. and vein
Ilioinguin. n.
Deep iliac
circumflex
art. and vein
Iliohypogast. n.
(ant. cutan. br.)
Linea semilunaris
T-12 intercost. n.
(ant. cutan. branch)
Iliohypogastric n.
(lat. cutan. branch)
Trans. abdom. m.
T-11 intercost. n.
(ant. cutan. branches)
Int. oblique m.
(deep surface)
Int. oblique m. (cut)
Trans. abdom. m.
Ext. oblique m. (cut)
T-10 intercost. n.
(ant. cutan. branches)
T-9 intercost. n.
(ant. cutan. branches)
Aponeurosis of
internal oblique m.
T-8 intercost. a. and v.
(ant. cutan. branches)
External oblique m. (cut)
External oblique m.
(origin)
Rectus abdominis m.
Serratus ant. muscle
Pect. major muscle
(abdominal part)
Pect. major muscle
(sternocostal part)
5
6
7
8
9
10

Fig. 239: Vessels and Nerves of the Anterior Abdominal Wall: The Lateral Cutaneous and Anterior Cutaneous Branches

NOTE: 1) that on the right side the *lateral cutaneous branches* of the lower six thoracic and first lumbar nerves and vessels are shown coursing in an inferomedial direction superficial to the external oblique muscle. On this side also the anterior lamina of the rectus sheath has been opened. The *anterior cutaneous branches* of these nerves supply the rectus abdominis muscle by penetrating the lateral border of the rectus sheath.

2) that on the left side (reader's right) the external and internal oblique muscles have been cut to show more clearly the anterior cutaneous branches in their segmental course around the abdominal wall on the surface of the transversus abdominis muscle. These vessels and nerves supply the three layers of flat abdominal muscles and then penetrate their aponeuroses to gain access to the rectus muscle. Note also that the L-1 nerve divides into the iliohypogastric and ilioinguinal nerves just above the iliac crest.

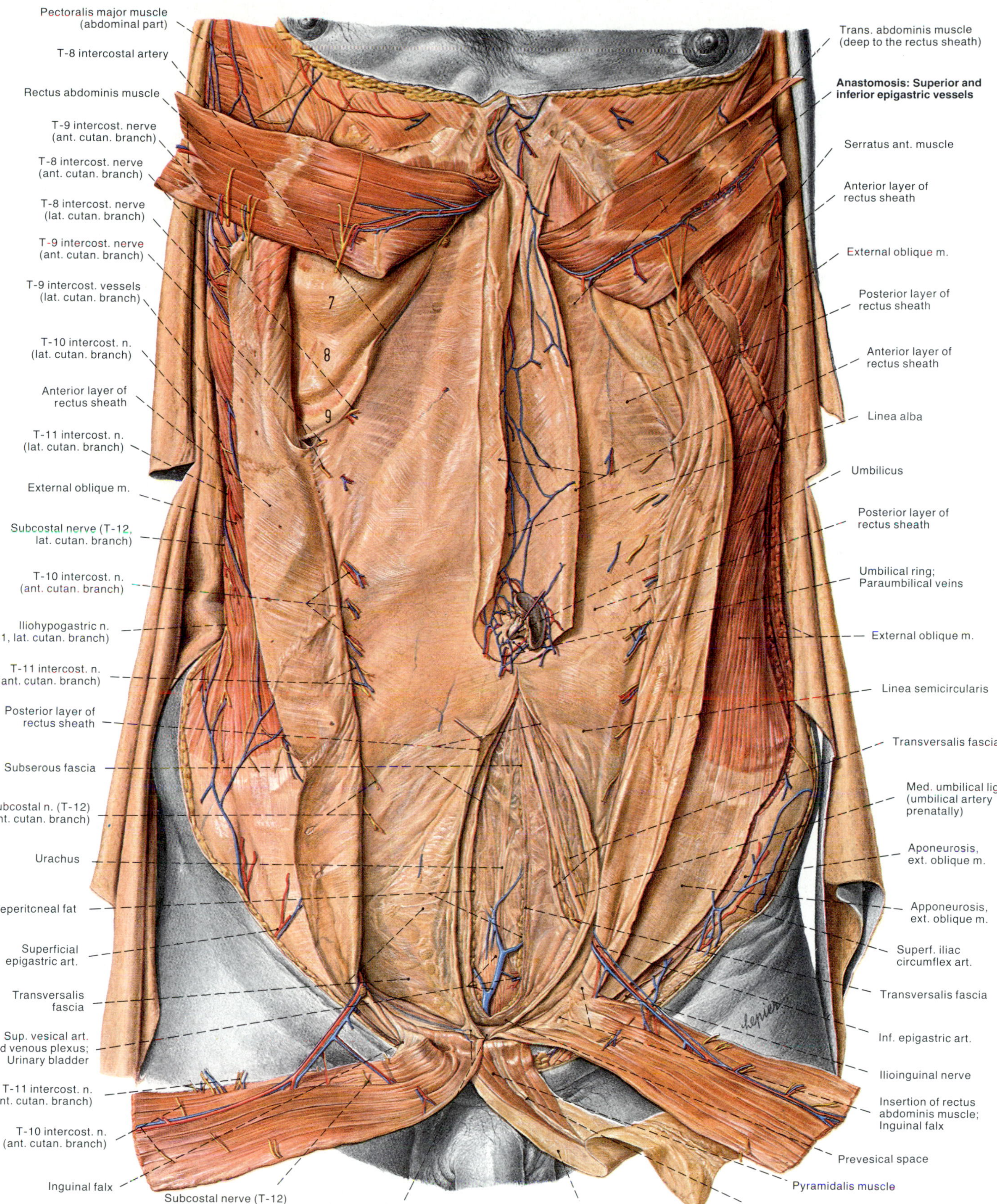
Pectoralis major muscle (abdominal part)
T-8 intercostal artery
Rectus abdominis muscle
T-9 intercost. nerve (ant. cutan. branch)
T-8 intercost. nerve (ant. cutan. branch)
T-8 intercost. nerve (lat. cutan. branch)
T-9 intercost. nerve (ant. cutan. branch)
T-9 intercost. vessels (lat. cutan. branch)
T-10 intercost. n. (lat. cutan. branch)
Anterior layer of rectus sheath
T-11 intercost. n. (lat. cutan. branch)
External oblique m.
Subcostal nerve (T-12, lat. cutan. branch)
T-10 intercost. n. (ant. cutan. branch)
Iliohypogastric n. (L-1, lat. cutan. branch)
T-11 intercost. n. (ant. cutan. branch)
Posterior layer of rectus sheath
Subserous fascia
Subcostal n. (T-12) (ant. cutan. branch)
Urachus
Preperitcneal fat
Superficial epigastric art.
Transversalis fascia
Sup. vesical art. and venous plexus; Urinary bladder
T-11 intercost. n. (ant. cutan. branch)
T-10 intercost. n. (ant. cutan. branch)
Inguinal falx
Subcostal nerve (T-12)
Adminiculum of linea alba; Symphysis pubis
Anterior layer of rectus sheath
Linea alba
Pyramidalis muscle
Prevesical space
Insertion of rectus abdominis muscle; Inguinal falx
Ilioinguinal nerve
Inf. epigastric art.
Transversalis fascia
Superf. iliac circumflex art.
Apponeurosis, ext. oblique m.
Aponeurosis, ext. oblique m.
Med. umbilical lig. (umbilical artery prenatally)
Transversalis fascia
Linea semicircularis
External oblique m.
Umbilical ring; Paraumbilical veins
Posterior layer of rectus sheath
Umbilicus
Linea alba
Anterior layer of rectus sheath
Posterior layer of rectus sheath
External oblique m.
Anterior layer of rectus sheath
Serratus ant. muscle
Anastomosis: Superior and inferior epigastric vessels
Trans. abdominis muscle (deep to the rectus sheath)
7
8
9

Fig. 240: Vessels and Nerves of the Anterior Abdominal Wall: The Epigastric Anastomosis

NOTE: 1) that the anterior layer of the rectus sheath has been opened on both sides and the two rectus abdominis muscles have been cut.
2) the descending course of the *superior epigastric artery and veins* (from the internal thoracic) and the ascending course of the *inferior epigastric artery and veins* (from the external iliac) found between the posterior layer of the rectus sheath and the rectus abdominis muscle on each side.

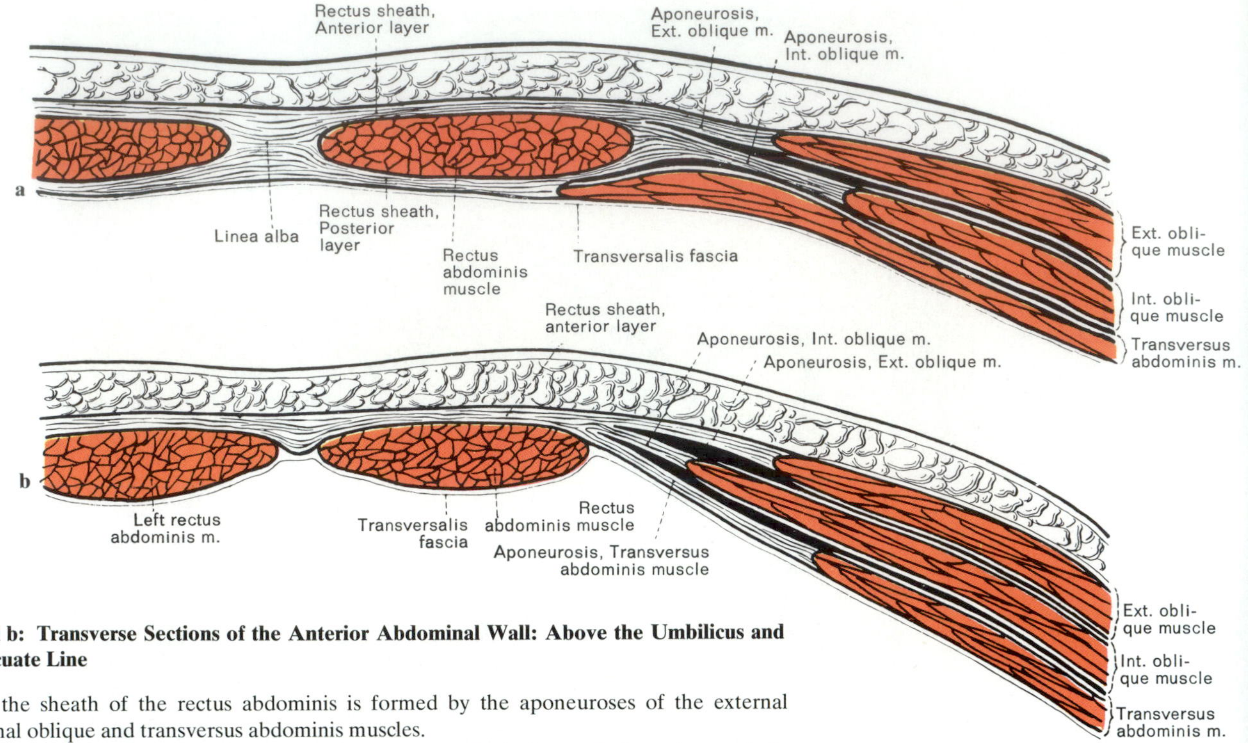
Rectus sheath,
Anterior layer
Aponeurosis,
Ext. oblique m.
Aponeurosis,
Int. oblique m.
a
Linea alba
Rectus sheath,
Posterior
layer
Rectus
abdominis
muscle
Transversalis fascia
Ext. obli-
que muscle
Int. obli-
que muscle
Transversus
abdominis m.
Rectus sheath,
anterior layer
Aponeurosis, Int. oblique m.
Aponeurosis, Ext. oblique m.
b
Left rectus
abdominis m.
Transversalis
fascia
Rectus
abdominis muscle
Aponeurosis, Transversus
abdominis muscle
Ext. obli-
que muscle
Int. obli-
que muscle
Transversus
abdominis m.

Fig. 241 a and b: Transverse Sections of the Anterior Abdominal Wall: Above the Umbilicus and Below the Arcuate Line

NOTE: 1) the sheath of the rectus abdominis is formed by the aponeuroses of the external oblique, internal oblique and transversus abdominis muscles.

2) the upper two-thirds of the sheath encloses the rectus muscle both anteriorly and posteriorly (a). To accomplish this, the internal oblique aponeurosis splits. Part of this aponeurosis joins the aponeurosis of the external oblique to form the anterior layer, while the other portion joins the aponeurosis of the transversus abdominis to form the posterior layer.

3) the lower one-third of the sheath (b), below the arcuate line, is deficient posteriorly, since the aponeuroses of all three muscles pass anterior to the rectus abdominis muscle.

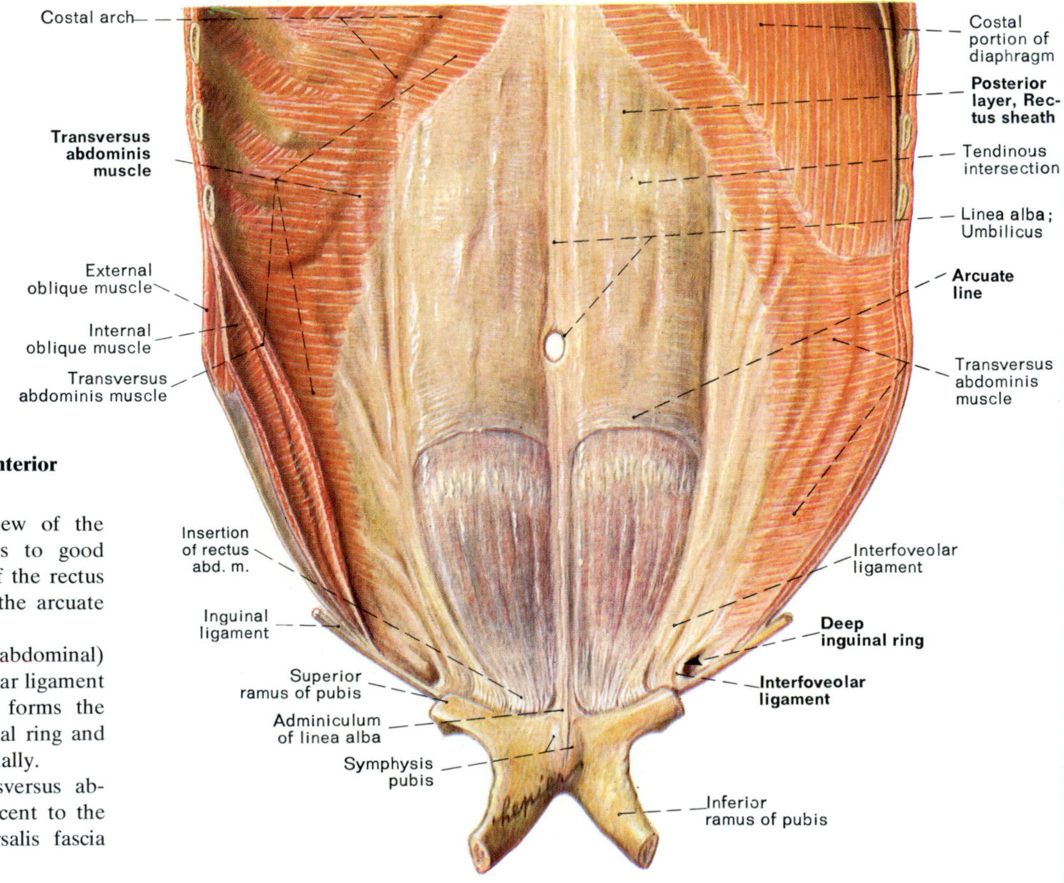
Costal arch
Costal
portion of
diaphragm
Posterior
layer, Rec-
tus sheath
Transversus
abdominis
muscle
Tendinous
intersection
Linea alba;
Umbilicus
External
oblique muscle
Arcuate
line
Internal
oblique muscle
Transversus
abdominis muscle
Transversus
abdominis
muscle
Insertion
of rectus
abd. m.
Interfoveolar
ligament
Inguinal
ligament
Deep
inguinal ring
Superior
ramus of pubis
Interfoveolar
ligament
Adminiculum
of linea alba
Symphysis
pubis
Inferior
ramus of pubis

Fig. 242: Inner Aspect of the Anterior Abdominal Wall

NOTE: 1) this posterior view of the anterior abdominal wall shows to good advantage the posterior layer of the rectus sheath and the relationship of the arcuate line to the umbilicus.

2) the opening of the deep (abdominal) inguinal ring and the interfoveolar ligament which is a fibrous band that forms the medial edge of the deep inguinal ring and then courses superiorly and medially.

3) the breadth of the transversus abdominis muscle which lies adjacent to the next inner layer, the transversalis fascia (not shown).

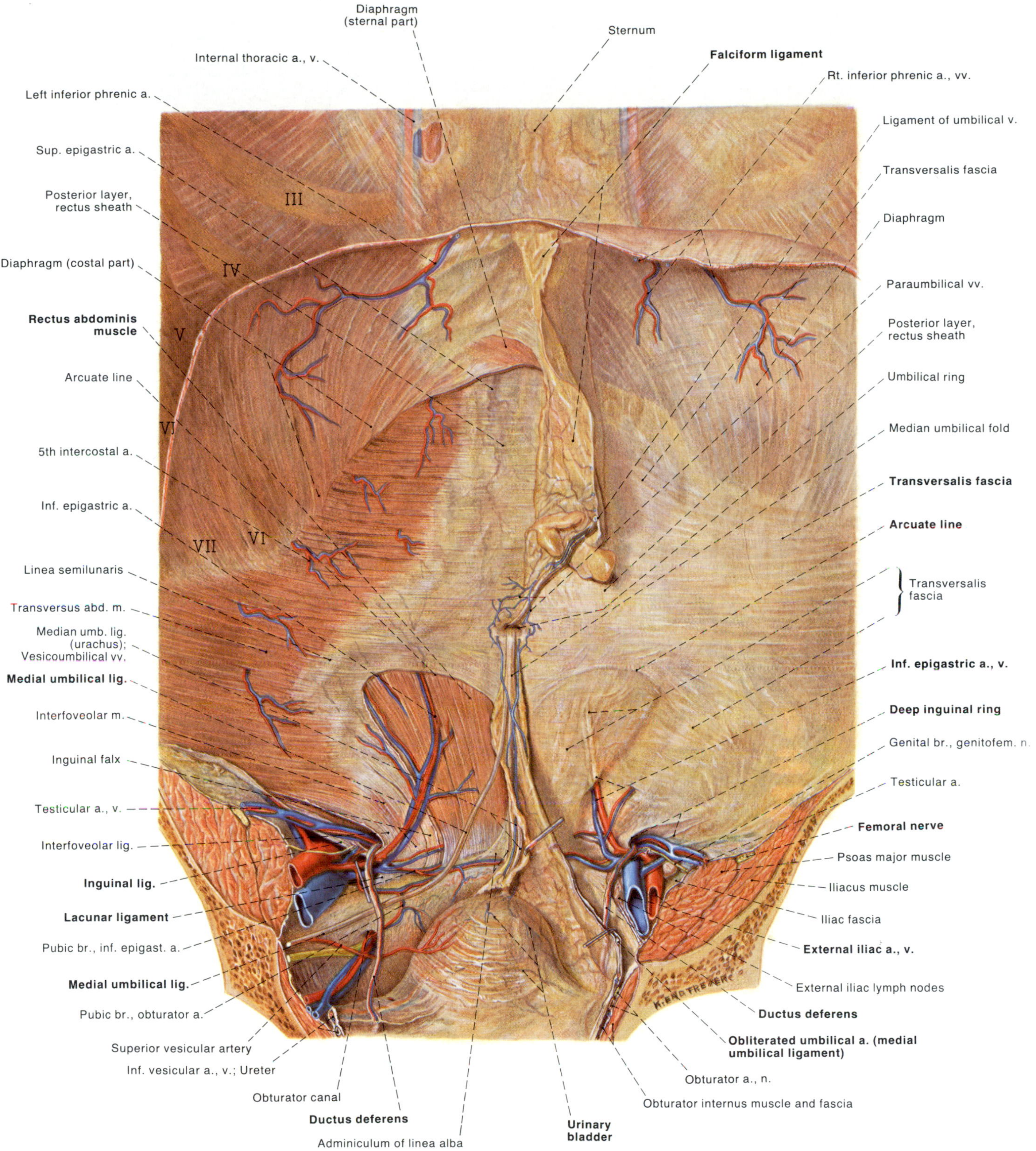

Diaphragm (sternal part)
Sternum
Internal thoracic a., v.
Falciform ligament
Rt. inferior phrenic a., vv.
Left inferior phrenic a.
Ligament of umbilical v.
Sup. epigastric a.
Transversalis fascia
Posterior layer, rectus sheath
III
Diaphragm
Diaphragm (costal part)
IV
Paraumbilical vv.
Rectus abdominis muscle
V
Posterior layer, rectus sheath
Arcuate line
Umbilical ring
VI
Median umbilical fold
5th intercostal a.
Transversalis fascia
Inf. epigastric a.
VII
VI
Arcuate line
Linea semilunaris
Transversalis fascia
Transversus abd. m.
Median umb. lig. (urachus); Vesicoumbilical vv.
Inf. epigastric a., v.
Medial umbilical lig.
Deep inguinal ring
Interfoveolar m.
Genital br., genitofem. n.
Inguinal falx
Testicular a.
Testicular a., v.
Femoral nerve
Interfoveolar lig.
Psoas major muscle
Inguinal lig.
Iliacus muscle
Lacunar ligament
Iliac fascia
Pubic br., inf. epigast. a.
External iliac a., v.
Medial umbilical lig.
External iliac lymph nodes
Pubic br., obturator a.
Ductus deferens
Superior vesicular artery
Obliterated umbilical a. (medial umbilical ligament)
Inf. vesicular a., v.; Ureter
Obturator a., n.
Obturator canal
Obturator internus muscle and fascia
Ductus deferens
Urinary bladder
Adminiculum of linea alba

Fig. 243: The Vessels and Umbilical Ligaments on the Inner Aspect of the Anterior Abdominal Wall

NOTE: 1) the course of the *median umbilical ligament* (remnant of the urachus) and the two *medial umbilical ligaments* (remnants of the umbilical arteries) from their origins in the pelvis to the umbilicus. When covered with peritoneum, these structures are called umbilical folds;

2) that the two *lateral umbilical folds* represent peritoneal reflections over the inferior epigastric vessels;

3) the abdominal inguinal rings through each of which course the ductus deferens and the testicular artery and vein;

4) the passage of the external iliac artery and vein beneath the inguinal ligament on each side, thereby achieving the anterior aspect of the thigh, where the vessels become the femoral artery and vein.

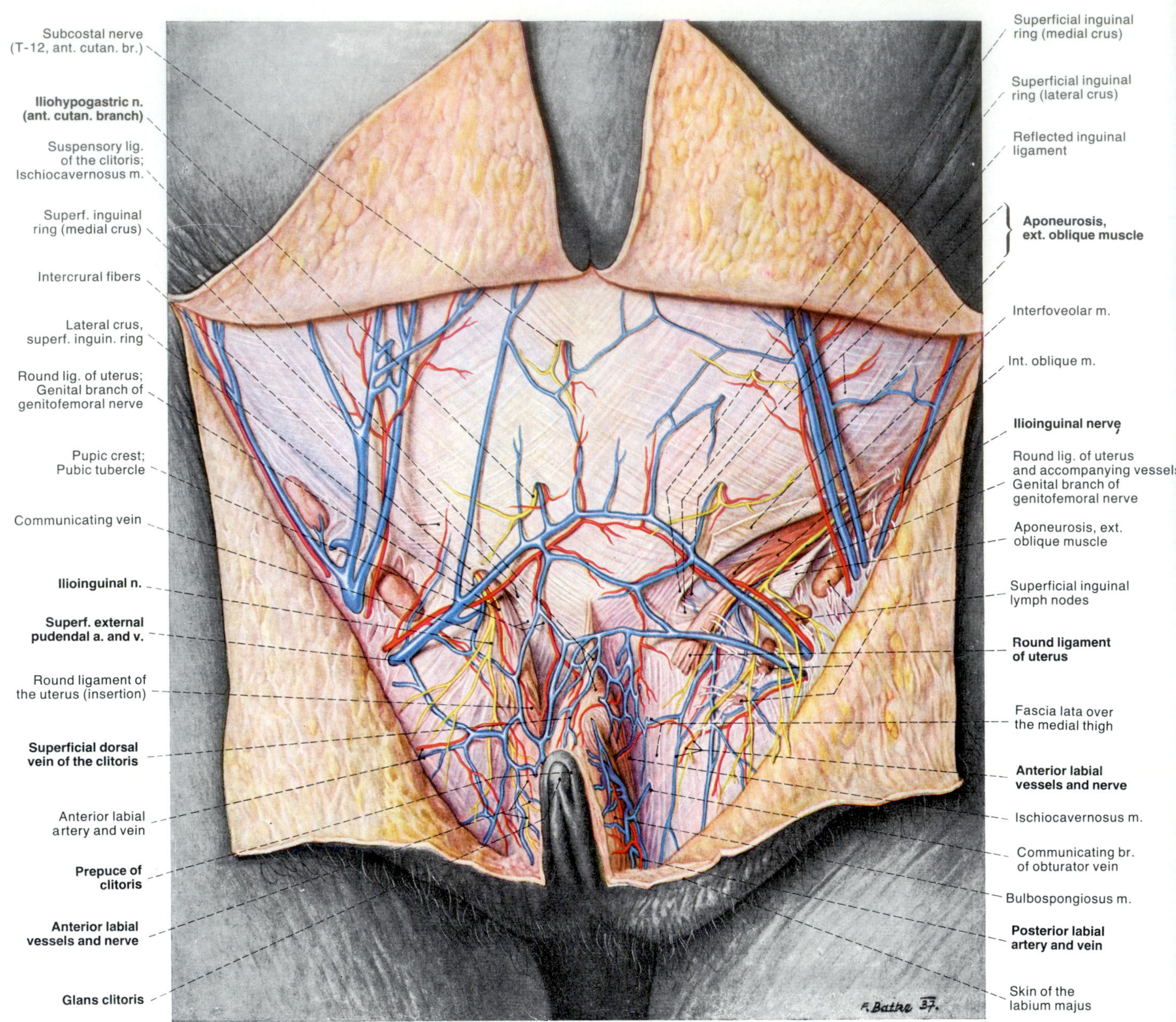

Subcostal nerve (T-12, ant. cutan. br.)
Iliohypogastric n. (ant. cutan. branch)
Suspensory lig. of the clitoris; Ischiocavernosus m.
Superf. inguinal ring (medial crus)
Intercrural fibers
Lateral crus, superf. inguin. ring
Round lig. of uterus; Genital branch of genitofemoral nerve
Pupic crest; Pubic tubercle
Communicating vein
Ilioinguinal n.
Superf. external pudendal a. and v.
Round ligament of the uterus (insertion)
Superficial dorsal vein of the clitoris
Anterior labial artery and vein
Prepuce of clitoris
Anterior labial vessels and nerve
Glans clitoris
Superficial inguinal ring (medial crus)
Superficial inguinal ring (lateral crus)
Reflected inguinal ligament
Aponeurosis, ext. oblique muscle
Interfoveolar m.
Int. oblique m.
Ilioinguinal nerve
Round lig. of uterus and accompanying vessels
Genital branch of genitofemoral nerve
Aponeurosis, ext. oblique muscle
Superficial inguinal lymph nodes
Round ligament of uterus
Fascia lata over the medial thigh
Anterior labial vessels and nerve
Ischiocavernosus m.
Communicating br. of obturator vein
Bulbospongiosus m.
Posterior labial artery and vein
Skin of the labium majus

Fig. 244: The Inguinal Region of the Anterior Abdominal Wall in the Female (1): Superficial Inguinal Ring and Superficial Vessels and Nerves

NOTE: 1) that the skin and superficial fascia have been reflected upward and laterally from the inguinal region in this female body, thereby exposing the aponeurosis of the external oblique muscle, the superficial inguinal ring, the superficial vessels and nerves of the lower anterior abdominal wall, a few superficial inguinal lymph nodes and the muscles and vessels of the clitoris.

2) that the superficial inguinal ring is an opening in the aponeurosis of the external oblique muscle. On the specimen's left (reader's right), the ring has been opened to reveal the lower course of the round ligament of the uterus and the ilioinguinal nerve and on the specimen's right, the genital branch of the genitofemoral nerve (L-1, L-2). Observe also the iliohypogastric nerve (branch of L-1) as it emerges through the aponeurosis to become a sensory nerve after supplying motor fibers to the underlying musculature.

3) that vessels supplying and draining this region include the superficial external pudendal, superficial iliac circumflex, and superficial epigastric arteries and veins (see labels, Fig. 245). All three of the arteries arise from the femoral artery, while their accompanying veins all drain into the great saphenous vein just before this latter vessel joins the femoral vein in the upper anteromedial aspect of the thigh.

4) that the *superficial external pudendal artery* courses medially across the round ligament (spermatic cord) near the superficial inguinal ring to supply the superficial structures of the lower abdomen and the labium majus (scrotum); the *superficial iliac circumflex artery* penetrates the fascia lata and courses laterally, immediately below the inguinal ligament, toward the anterior superior iliac spine; the *superficial epigastric artery* penetrates the cribriform fascia and ascends over the inguinal ligament within the superficial fascia of the lower abdominal wall between its (the superficial fascia's) superficial (Camper's) and deep (Scarpa's) layers. These two arteries are labelled on Fig. 245).

5) the surface anatomy of the clitoris, its prepuce, and its suspensory ligament (see also Figs. 367, 368 and 370). Observe the superficial dorsal vein of the clitoris; this vessel may drain into either the right or left superficial external pudendal veins, and note that it lies superficial to the (subfascial) dorsal arteries and vein of the clitoris (see Fig. 245).

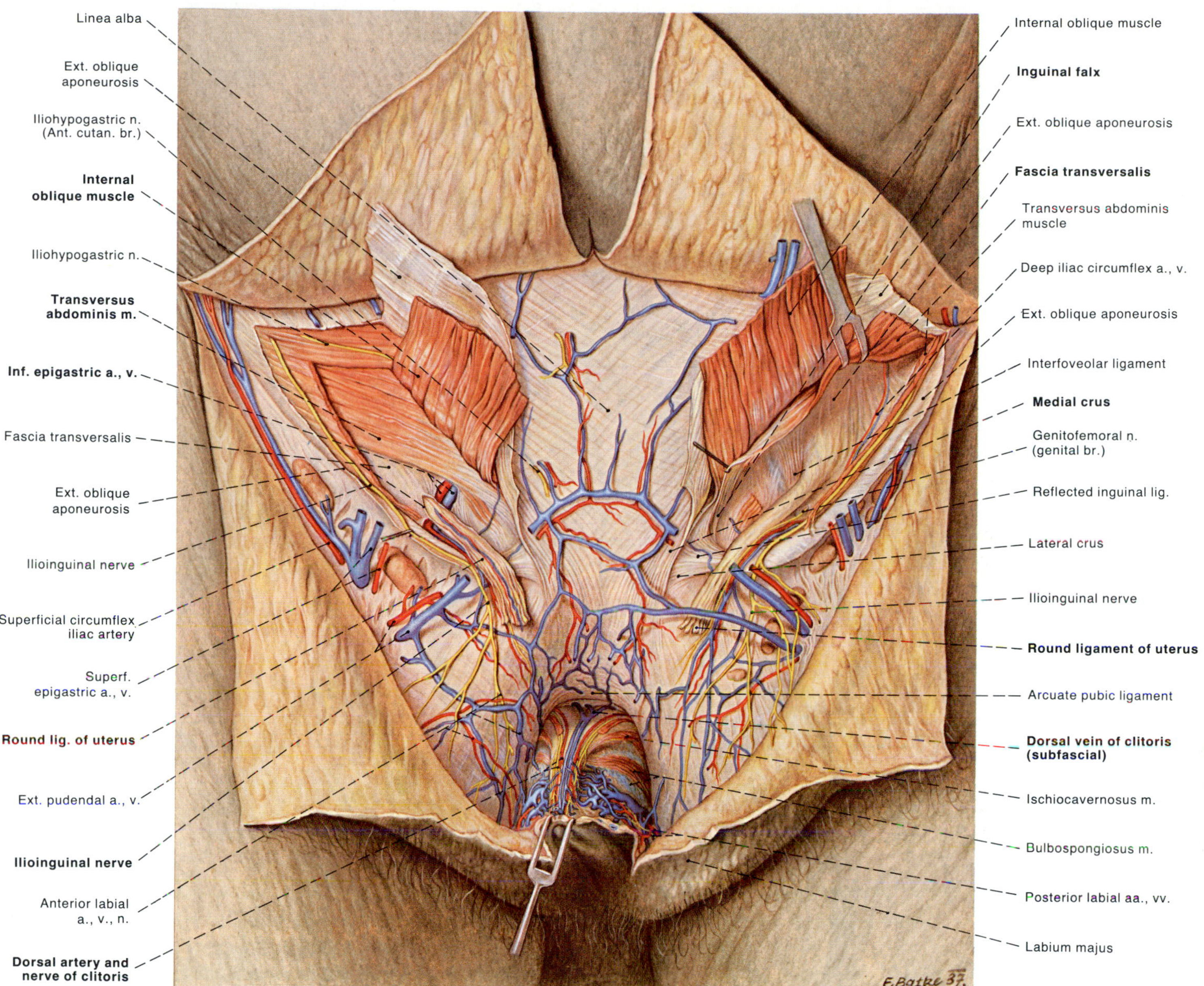

Linea alba
Ext. oblique aponeurosis
Iliohypogastric n. (Ant. cutan. br.)
Internal oblique muscle
Iliohypogastric n.
Transversus abdominis m.
Inf. epigastric a., v.
Fascia transversalis
Ext. oblique aponeurosis
Ilioinguinal nerve
Superficial circumflex iliac artery
Superf. epigastric a., v.
Round lig. of uterus
Ext. pudendal a., v.
Ilioinguinal nerve
Anterior labial a., v., n.
Dorsal artery and nerve of clitoris
Internal oblique muscle
Inguinal falx
Ext. oblique aponeurosis
Fascia transversalis
Transversus abdominis muscle
Deep iliac circumflex a., v.
Ext. oblique aponeurosis
Interfoveolar ligament
Medial crus
Genitofemoral n. (genital br.)
Reflected inguinal lig.
Lateral crus
Ilioinguinal nerve
Round ligament of uterus
Arcuate pubic ligament
Dorsal vein of clitoris (subfascial)
Ischiocavernosus m.
Bulbospongiosus m.
Posterior labial aa., vv.
Labium majus
F. Batke 37.

Fig. 245: The Inguinal Region of the Anterior Abdominal Wall in the Female (2): The Opened Inguinal Canal

NOTE: 1) that in this anterior view, the left (reader's right) inguinal canal has been completely opened by severing the overlying transversus abdominis and internal oblique muscles, as well as the lower lateral portion of the aponeurosis of the external oblique muscle;

2) that the female inguinal canal contains the round ligament of the uterus and the artery and vein which supply that ligament. The ilioinguinal branch of the L-1 spinal nerve, along with the genital branch of the genitofemoral nerve (L-1 and L-2), both emerge from the superficial inguinal ring with the round ligament;

3) that although the genital branch of the genitofemoral nerve enters the inguinal canal with the round ligament at the deep inguinal ring, the ilioinguinal nerve does not. It joins with the contents of the inguinal canal just deep to the aponeurosis of the external oblique;

4) that the ilioinguinal nerve and the genital branch of the genitofemoral nerve, in the female, both supply sensory innervation to the inguinal region and to the labia majora;

5) that the round ligament, when leaving the superficial inguinal ring, splits into a number of fibrous strands which then become enmeshed in the subcutaneous folds of the labium majus on each side. The labia majora are the homologues of the scrotal sacs in the male;

6) the inferior epigastric artery and vein coursing just medial to the deep inguinal ring, as they do in the male;

7) that the principal superficial vessels of the inguinal region are the superficial iliac circumflex, superficial epigastric, and the external pudendal arteries and veins.

8) the dorsal nerve and artery of the clitoris and the (subfascial) dorsal vein of the clitoris. The dorsal nerve of the clitoris comes from the posterior labial branch of the perineal nerve (from the pudendal nerve). The dorsal artery of the clitoris is one of the terminal branches of the internal pudendal artery, while the (subfascial) dorsal vein of the clitoris drains into the vesical plexus on the anterior aspect of the bladder within the pelvis.

Subcostal nerve (T-12, ant. cutan. br.)
Iliohypogastric n.
Suspensory ligament of penis
Aponeurosis, external oblique m.
Spermatic cord
Superficial iliac circumflex a., v.
Superficial epigastric a., v.
Ilioinguinal n.; Ext. pudendal a., v.
External pudendal artery and vein
Deep fascia of penis
Superficial dorsal vein of penis
Anterior scrotal nerve
External spermatic fascia
Coronal sulcus of glans penis
Prepuce
Glans penis

Medial crus, superf. inguinal ring
Aponeurosis, external oblique m.
Collateral and intercrural fibers
Cremasteric a., v., m.; Genital branch, genitofemoral n.
Superf. circumflex iliac a., v.
Ilioinguinal nerve
Superficial inguinal lymph nodes
Reflected inguinal ligament
Superficial epigastric a., v.
External pudendal a., v.
Fascia lata
Cremaster m.
Ischiocavernosus m.
Anterior scrotal vessels and nerve
Posterior scrotal artery and vein
Internal spermatic fascia
Dorsal vessels and nerve of the penis

F. Batke 1937

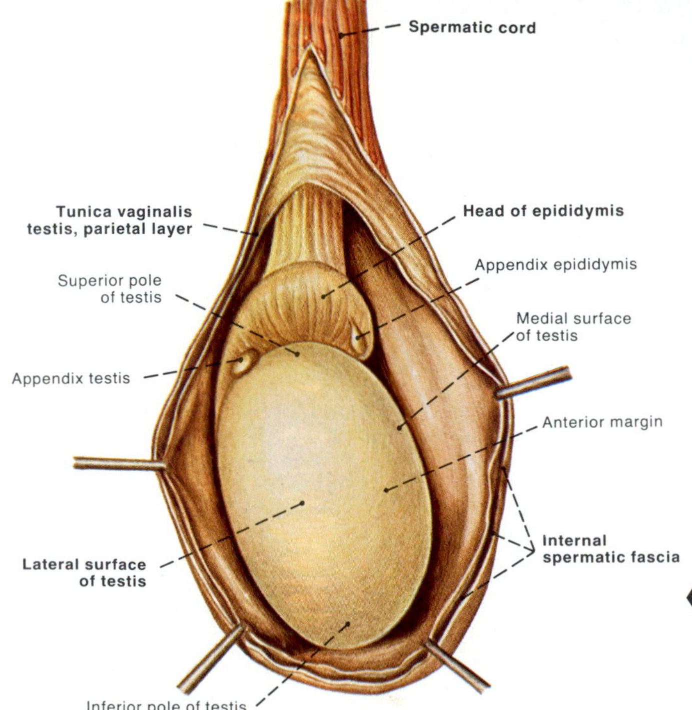

Spermatic cord
Tunica vaginalis testis, parietal layer
Head of epididymis
Superior pole of testis
Appendix epididymis
Medial surface of testis
Appendix testis
Anterior margin
Internal spermatic fascia
Lateral surface of testis
Inferior pole of testis

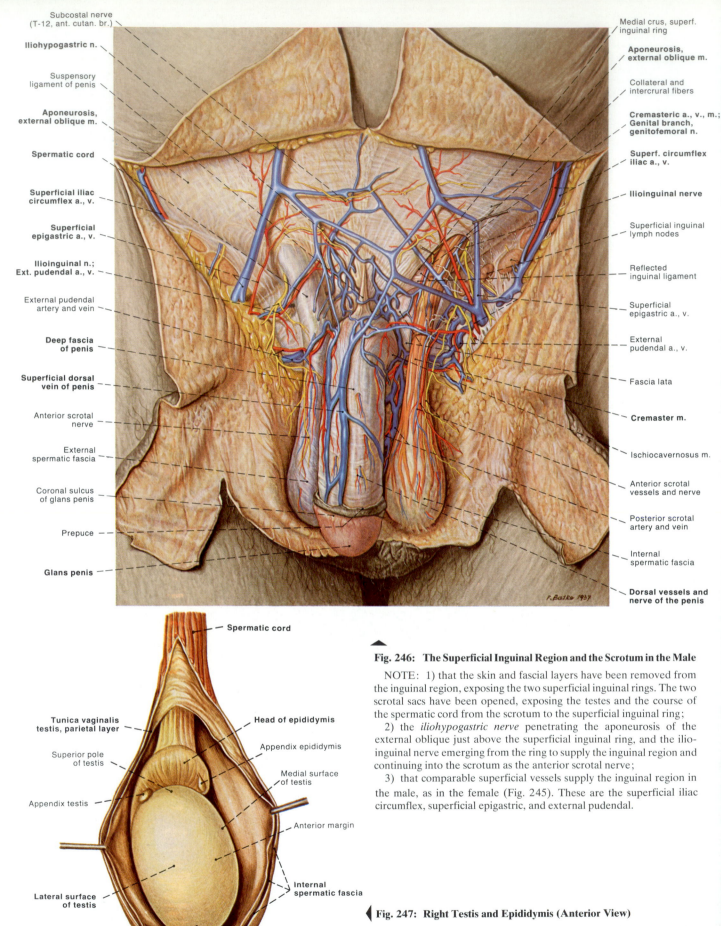

Fig. 246: The Superficial Inguinal Region and the Scrotum in the Male

NOTE: 1) that the skin and fascial layers have been removed from the inguinal region, exposing the two superficial inguinal rings. The two scrotal sacs have been opened, exposing the testes and the course of the spermatic cord from the scrotum to the superficial inguinal ring;

2) the *iliohypogastric nerve* penetrating the aponeurosis of the external oblique just above the superficial inguinal ring, and the ilio-inguinal nerve emerging from the ring to supply the inguinal region and continuing into the scrotum as the anterior scrotal nerve;

3) that comparable superficial vessels supply the inguinal region in the male, as in the female (Fig. 245). These are the superficial iliac circumflex, superficial epigastric, and external pudendal.

Fig. 247: Right Testis and Epididymis (Anterior View)

NOTE that the testis is suspended by its efferent duct system which consists of the head, body and tail of the epididymis, and that this eventually leads to the ductus deferens (see Fig. 250).

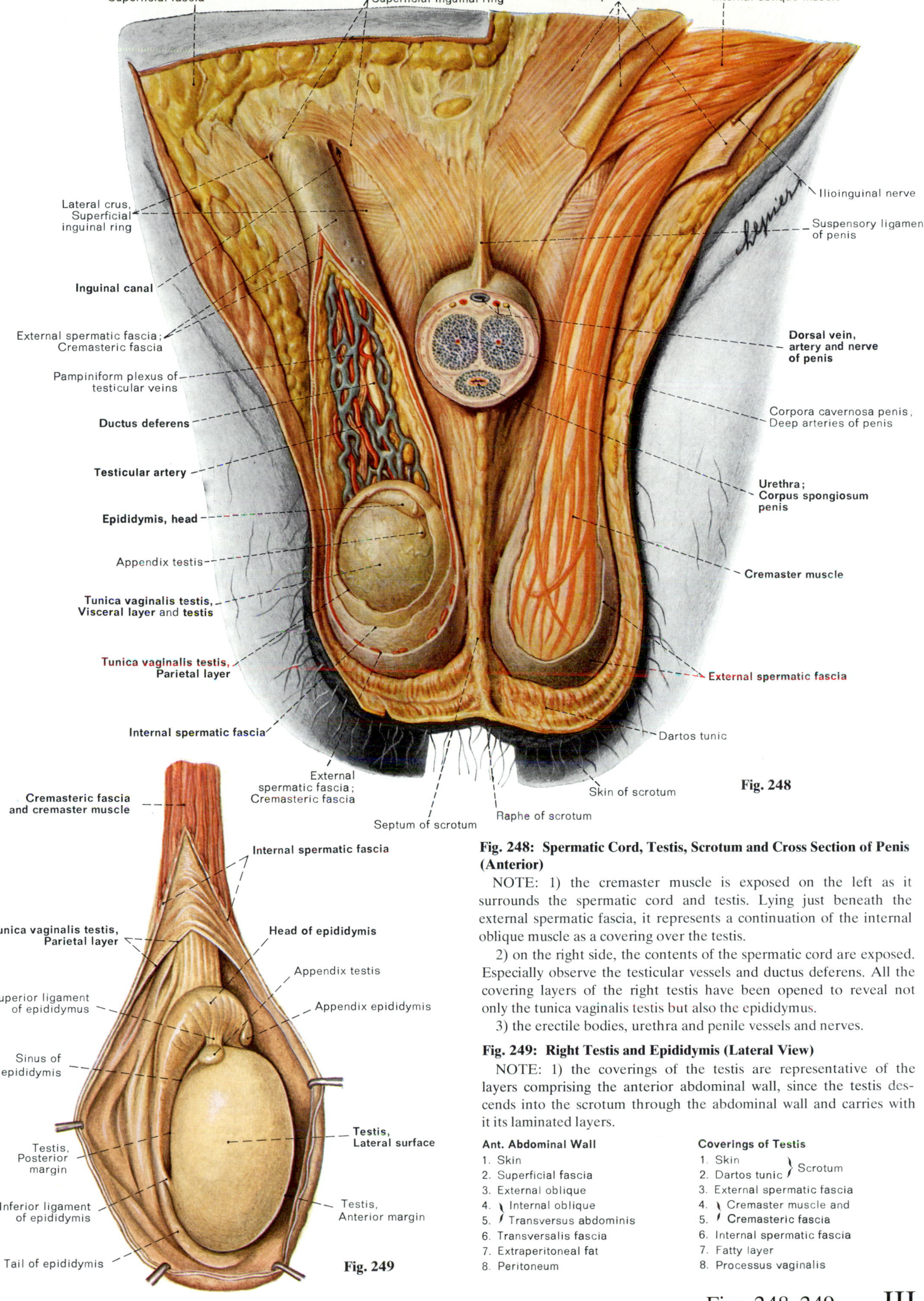
Superficial fascia
Cranial margin and medial crus,
Superficial inguinal ring
Aponeurosis,
Ext. oblique muscle
Internal oblique muscle
Ilioinguinal nerve
Suspensory ligament
of penis
Lateral crus,
Superficial
inguinal ring
Inguinal canal
External spermatic fascia;
Cremasteric fascia
Pampiniform plexus of
testicular veins
Ductus deferens
Testicular artery
Epididymis, head
Appendix testis
Tunica vaginalis testis,
Visceral layer and testis
Tunica vaginalis testis,
Parietal layer
Internal spermatic fascia
Dorsal vein,
artery and nerve
of penis
Corpora cavernosa penis;
Deep arteries of penis
Urethra;
Corpus spongiosum
penis
Cremaster muscle
External spermatic fascia
Dartos tunic
Skin of scrotum
External
spermatic fascia;
Cremasteric fascia
Septum of scrotum
Raphe of scrotum
Fig. 248
Cremasteric fascia
and cremaster muscle
Internal spermatic fascia
Tunica vaginalis testis,
Parietal layer
Head of epididymis
Appendix testis
Superior ligament
of epididymus
Appendix epididymis
Sinus of
epididymis
Testis,
Lateral surface
Testis,
Posterior
margin
Inferior ligament
of epididymis
Testis,
Anterior margin
Tail of epididymis
Fig. 249

Fig. 248: Spermatic Cord, Testis, Scrotum and Cross Section of Penis (Anterior)

NOTE: 1) the cremaster muscle is exposed on the left as it surrounds the spermatic cord and testis. Lying just beneath the external spermatic fascia, it represents a continuation of the internal oblique muscle as a covering over the testis.

2) on the right side, the contents of the spermatic cord are exposed. Especially observe the testicular vessels and ductus deferens. All the covering layers of the right testis have been opened to reveal not only the tunica vaginalis testis but also the epididymus.

3) the erectile bodies, urethra and penile vessels and nerves.

Fig. 249: Right Testis and Epididymis (Lateral View)

NOTE: 1) the coverings of the testis are representative of the layers comprising the anterior abdominal wall, since the testis descends into the scrotum through the abdominal wall and carries with it its laminated layers.

Ant. Abdominal Wall	Coverings of Testis
1. Skin	1. Skin } Scrotum
2. Superficial fascia	2. Dartos tunic } Scrotum
3. External oblique	3. External spermatic fascia
4. Internal oblique	4. Cremaster muscle and
5. Transversus abdominis	5. Cremasteric fascia
6. Transversalis fascia	6. Internal spermatic fascia
7. Extraperitoneal fat	7. Fatty layer
8. Peritoneum	8. Processus vaginalis

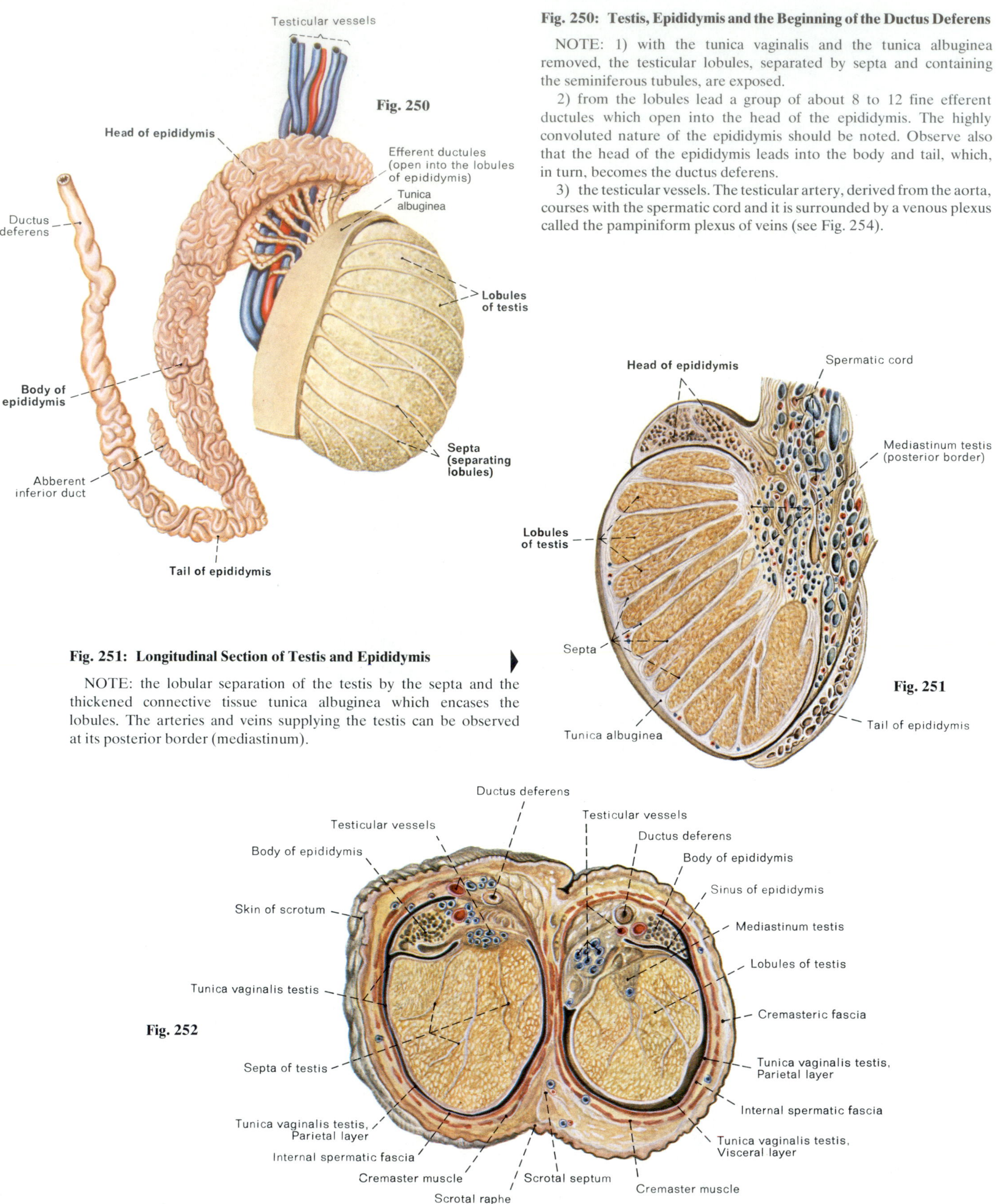

Testicular vessels
Fig. 250
Head of epididymis
Efferent ductules (open into the lobules of epididymis)
Tunica albuginea
Ductus deferens
Lobules of testis
Body of epididymis
Septa (separating lobules)
Abberent inferior duct
Tail of epididymis
Head of epididymis
Spermatic cord
Mediastinum testis (posterior border)
Lobules of testis
Septa
Fig. 251
Tunica albuginea
Tail of epididymis
Ductus deferens
Testicular vessels
Testicular vessels
Ductus deferens
Body of epididymis
Body of epididymis
Sinus of epididymis
Skin of scrotum
Mediastinum testis
Lobules of testis
Tunica vaginalis testis
Cremasteric fascia
Fig. 252
Septa of testis
Tunica vaginalis testis, Parietal layer
Tunica vaginalis testis, Parietal layer
Internal spermatic fascia
Internal spermatic fascia
Tunica vaginalis testis, Visceral layer
Cremaster muscle
Scrotal septum
Cremaster muscle
Scrotal raphe

Fig. 250: Testis, Epididymis and the Beginning of the Ductus Deferens

NOTE: 1) with the tunica vaginalis and the tunica albuginea removed, the testicular lobules, separated by septa and containing the seminiferous tubules, are exposed.

2) from the lobules lead a group of about 8 to 12 fine efferent ductules which open into the head of the epididymis. The highly convoluted nature of the epididymis should be noted. Observe also that the head of the epididymis leads into the body and tail, which, in turn, becomes the ductus deferens.

3) the testicular vessels. The testicular artery, derived from the aorta, courses with the spermatic cord and it is surrounded by a venous plexus called the pampiniform plexus of veins (see Fig. 254).

Fig. 251: Longitudinal Section of Testis and Epididymis

NOTE: the lobular separation of the testis by the septa and the thickened connective tissue tunica albuginea which encases the lobules. The arteries and veins supplying the testis can be observed at its posterior border (mediastinum).

Fig. 252: Cross Section of Testis and Scrotum

NOTE: the scrotum is divided by the median raphe and septum into two lateral compartments, each surrounding an ovoid-shaped testis. The two scrotal compartments do not communicate.

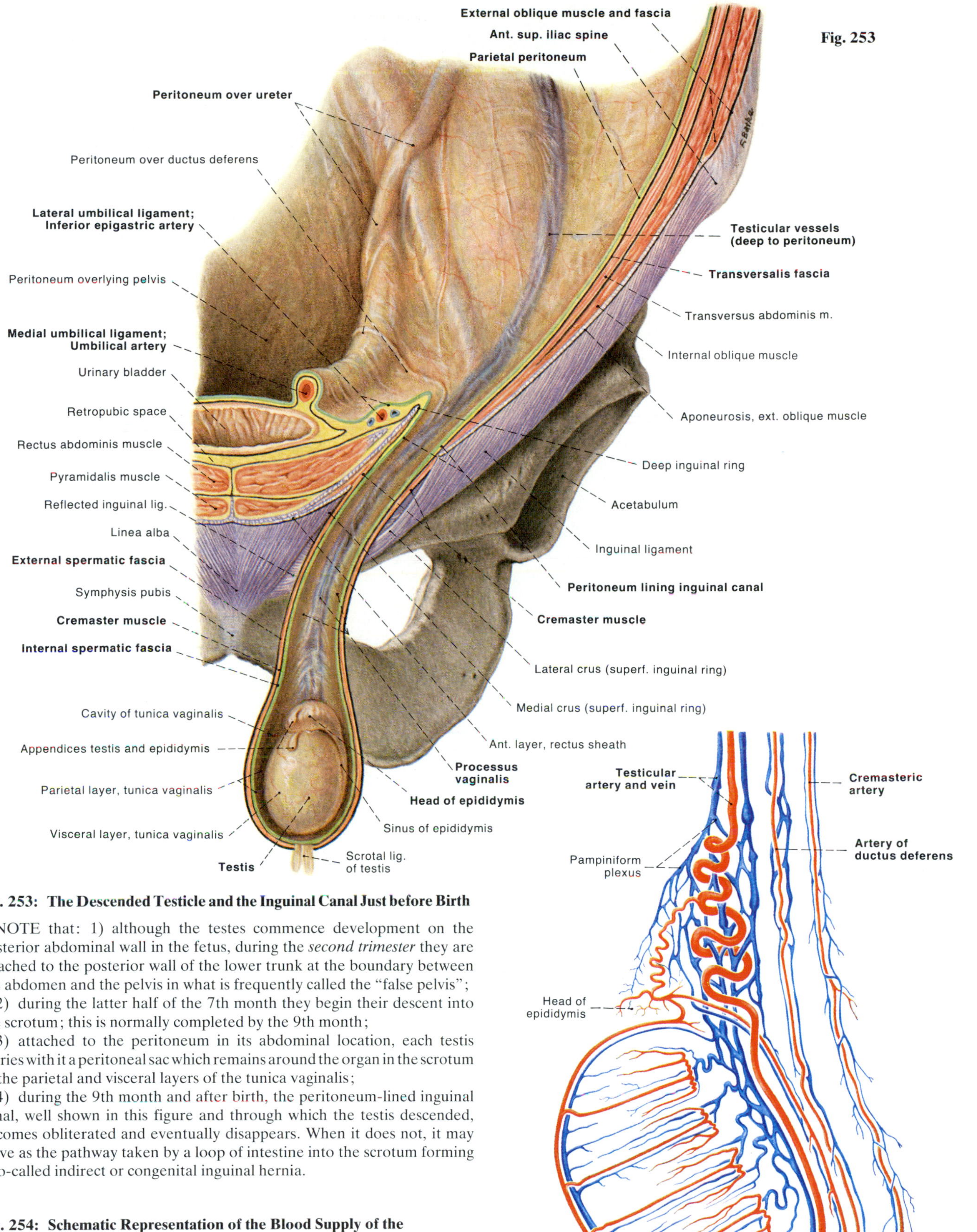

Fig. 253
External oblique muscle and fascia
Ant. sup. iliac spine
Parietal peritoneum
Peritoneum over ureter
Peritoneum over ductus deferens
Lateral umbilical ligament;
Inferior epigastric artery
Peritoneum overlying pelvis
Medial umbilical ligament;
Umbilical artery
Urinary bladder
Retropubic space
Rectus abdominis muscle
Pyramidalis muscle
Reflected inguinal lig.
Linea alba
External spermatic fascia
Symphysis pubis
Cremaster muscle
Internal spermatic fascia
Cavity of tunica vaginalis
Appendices testis and epididymis
Parietal layer, tunica vaginalis
Visceral layer, tunica vaginalis
Testis
Scrotal lig. of testis
Sinus of epididymis
Head of epididymis
Processus vaginalis
Ant. layer, rectus sheath
Medial crus (superf. inguinal ring)
Lateral crus (superf. inguinal ring)
Cremaster muscle
Peritoneum lining inguinal canal
Inguinal ligament
Acetabulum
Deep inguinal ring
Aponeurosis, ext. oblique muscle
Internal oblique muscle
Transversus abdominis m.
Transversalis fascia
Testicular vessels (deep to peritoneum)
Testicular artery and vein
Cremasteric artery
Artery of ductus deferens
Pampiniform plexus
Head of epididymis
Tail of epididymis
Fig. 254

Fig. 253: The Descended Testicle and the Inguinal Canal Just before Birth

NOTE that: 1) although the testes commence development on the posterior abdominal wall in the fetus, during the *second trimester* they are attached to the posterior wall of the lower trunk at the boundary between the abdomen and the pelvis in what is frequently called the "false pelvis";

2) during the latter half of the 7th month they begin their descent into the scrotum; this is normally completed by the 9th month;

3) attached to the peritoneum in its abdominal location, each testis carries with it a peritoneal sac which remains around the organ in the scrotum as the parietal and visceral layers of the tunica vaginalis;

4) during the 9th month and after birth, the peritoneum-lined inguinal canal, well shown in this figure and through which the testis descended, becomes obliterated and eventually disappears. When it does not, it may serve as the pathway taken by a loop of intestine into the scrotum forming a so-called indirect or congenital inguinal hernia.

Fig. 254: Schematic Representation of the Blood Supply of the Testis and Epididymis

NOTE that the testis and epididymis are served by the testicular artery (from the aorta), the artery of the ductus deferens (usually from the superior vesicle artery) and the cremasteric artery (from the inferior epigastric artery). The pampiniform plexus of veins drains into the testicular vein, which on the left side flows into the left renal vein and on the right side opens into the inferior vena cava.

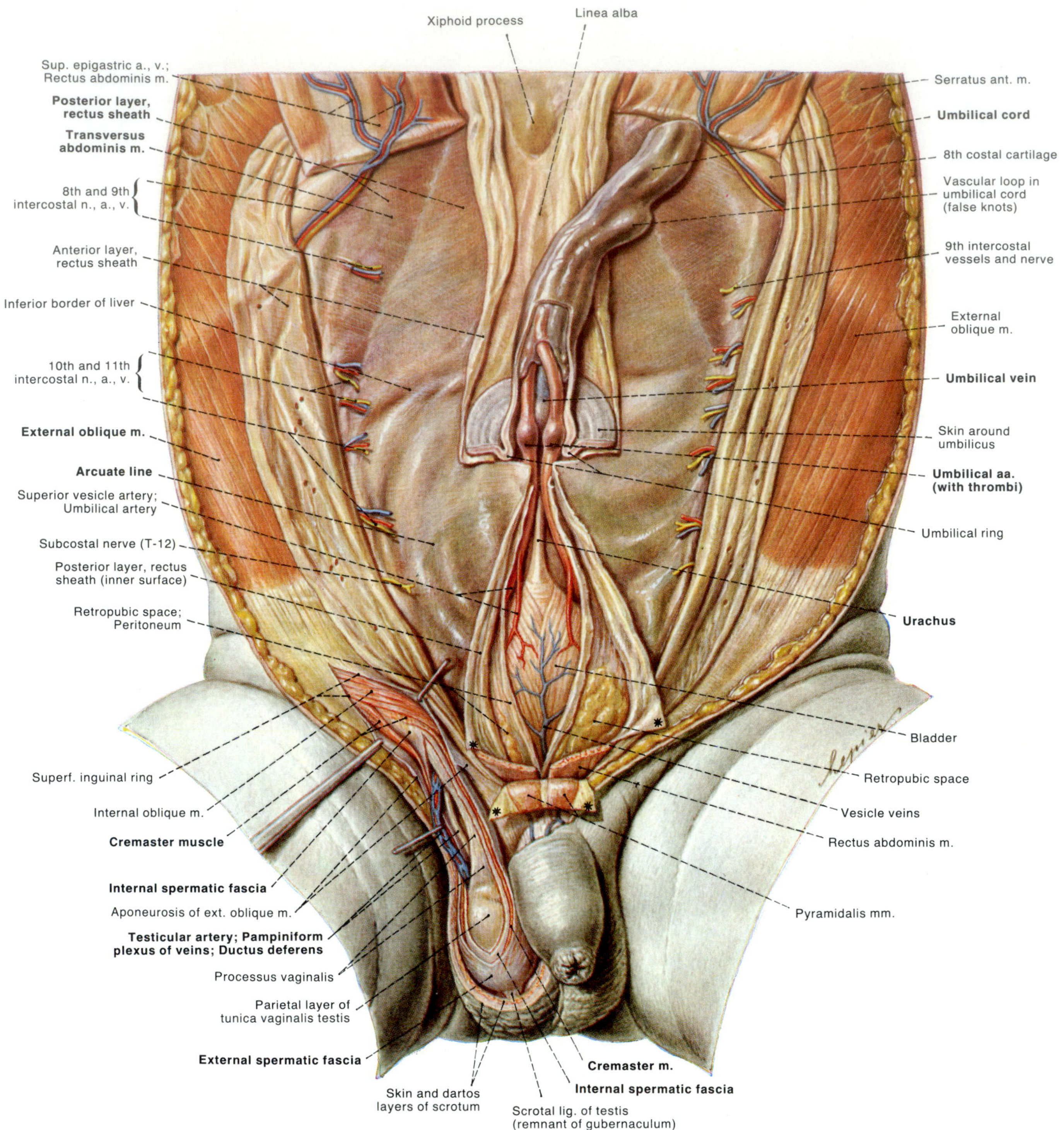
Xiphoid process
Linea alba
Sup. epigastric a., v.; Rectus abdominis m.
Posterior layer, rectus sheath
Transversus abdominis m.
8th and 9th intercostal n., a., v.
Anterior layer, rectus sheath
Inferior border of liver
10th and 11th intercostal n., a., v.
External oblique m.
Arcuate line
Superior vesicle artery; Umbilical artery
Subcostal nerve (T-12)
Posterior layer, rectus sheath (inner surface)
Retropubic space; Peritoneum
Superf. inguinal ring
Internal oblique m.
Cremaster muscle
Internal spermatic fascia
Aponeurosis of ext. oblique m.
Testicular artery; Pampiniform plexus of veins; Ductus deferens
Processus vaginalis
Parietal layer of tunica vaginalis testis
External spermatic fascia
Skin and dartos layers of scrotum
Scrotal lig. of testis (remnant of gubernaculum)
Internal spermatic fascia
Cremaster m.
Serratus ant. m.
Umbilical cord
8th costal cartilage
Vascular loop in umbilical cord (false knots)
9th intercostal vessels and nerve
External oblique m.
Umbilical vein
Skin around umbilicus
Umbilical aa. (with thrombi)
Umbilical ring
Urachus
Bladder
Retropubic space
Vesicle veins
Rectus abdominis m.
Pyramidalis mm.

Fig. 255: A Deep Dissection of the Anterior Abdominal Wall and the Umbilical Region in the Newborn

NOTE: 1) that the anterior layer of the rectus sheath has been severed and reflected laterally on each side. The two rectus abdominis muscles have been severed near the symphysis pubis and reflected superiorly (almost out of view) in order to reveal the posterior layer of the rectus sheath. Observe the arcuate line.

2) that an incision has been made in the linea alba between the umbilical ring and the symphysis pubis exposing the apex of the bladder, the urachus, the paired umbilical arteries and the single umbilical vein.

3) that the anterior aspect of the right spermatic cord and right scrotal sac have been opened to uncover the ductus deferens and the testis, the latter structure surrounded by the tunica vaginalis testis.

4) the severed umbilical cord which is usually between one and two centimeters in diameter and about 50 centimeters (20 inches) long. It contains the two umbilical arteries and the umbilical vein which are surrounded by a mucoid form of connective tissue called Wharton's jelly. Frequently the umbilical vessels form harmless loops in the cord called "false knots" (seen as bulges in this figure). More rarely looping of the cord may be of some functional significance and such "true knots" may alter the circulation to and from the fetus, causing vascular obstruction.

5) the bulges in the umbilical arteries. These are *in situ* blood clots, called thrombi, which occlude the arteries, but which are probably post-mortem phenomena in this instance.

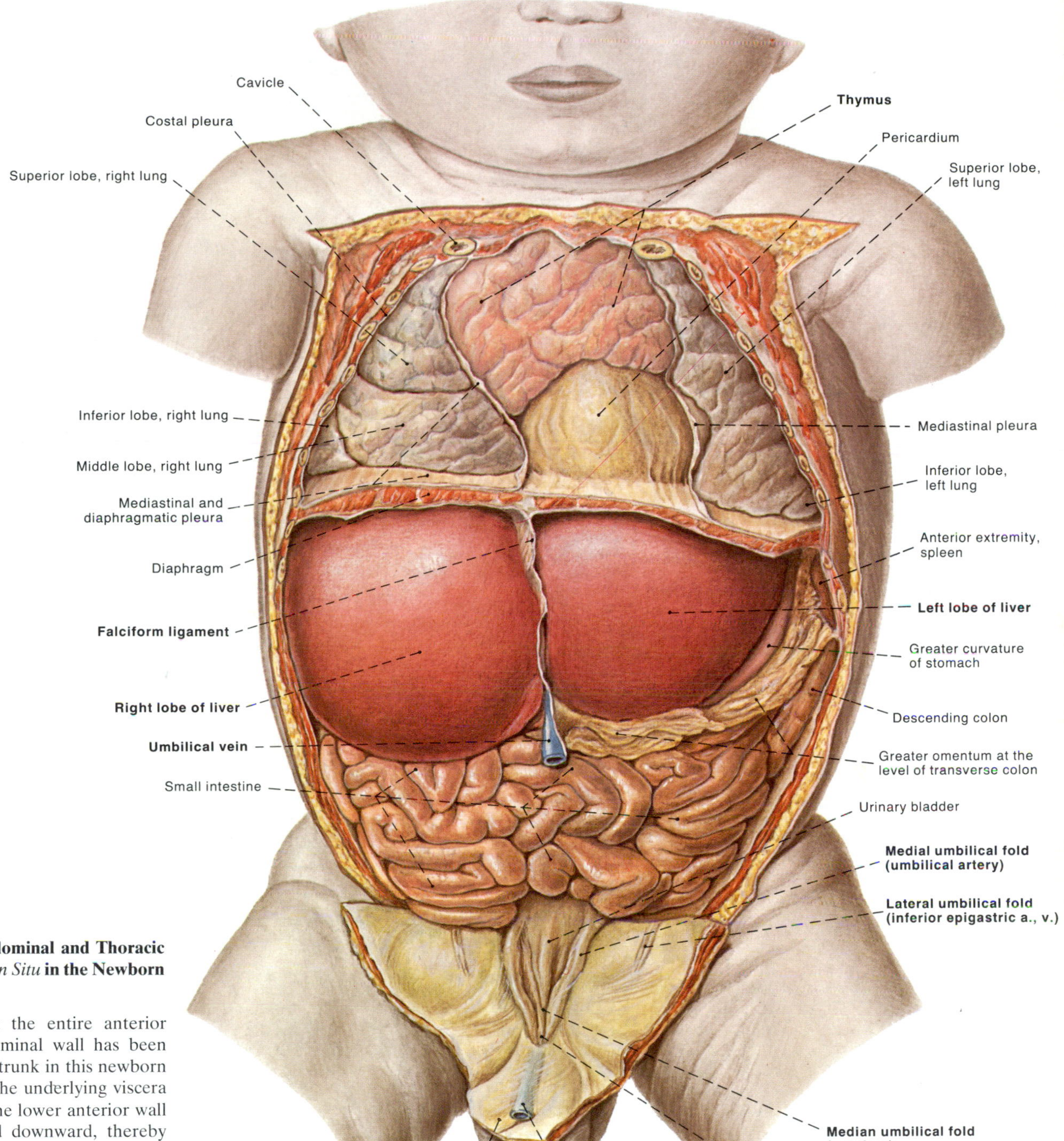

Cavicle
Costal pleura
Superior lobe, right lung
Thymus
Pericardium
Superior lobe, left lung
Inferior lobe, right lung
Middle lobe, right lung
Mediastinal and diaphragmatic pleura
Diaphragm
Falciform ligament
Right lobe of liver
Umbilical vein
Small intestine
Mediastinal pleura
Inferior lobe, left lung
Anterior extremity, spleen
Left lobe of liver
Greater curvature of stomach
Descending colon
Greater omentum at the level of transverse colon
Urinary bladder
Medial umbilical fold (umbilical artery)
Lateral umbilical fold (inferior epigastric a., v.)
Median umbilical fold (urachus)
Umbilical ring
Umbilical vein
Falciform ligament

Fig. 256: The Abdominal and Thoracic Viscera Observed *In Situ* in the Newborn Child

NOTE: 1) that the entire anterior thoracic and abdominal wall has been removed from the trunk in this newborn child, uncovering the underlying viscera in both cavities. The lower anterior wall has been reflected downward, thereby exposing the umbilical ligaments.

2) that the average newborn infant weighs about 3300 gr (7 lbs) and measures about 50 cm (20 inches) from the top of the head to the sole of the foot, and the umbilicus is located about 1.5 cm below the mid-point of this crown-heel length.

3) that the transverse diameter of the abdominal cavity in the newborn is greatest above the umbilicus, principally due to the inordinate proportion of the abdomen occupied by the liver. Observe that a greater area (than in the adult) of the anterior surface of both lobes lies immediately deep to the abdominal musculature and that the ribs (cut away in this figure) afford less protection to the upper abdomen.

4) that the average weight of the liver in the newborn is about 120 gr, and it constitutes 4 percent of the body weight at birth. In the adult, the liver weighs 12 to 13 times its weight at birth, but accounts for only 2.5 to 3.5 percent of the total body weight.

5) that most of the anterior surface of the stomach lies deep to the left lobe of the liver, allowing only a small portion of the greater curvature to be visible. The left lobe of the liver almost reaches the spleen.

6) that the loops of intestine form an oval-shaped mass, the greatest diameter of which is transverse in contrast to the adult in which it is vertical.

7) the broad-based truncated shape of the thoracic cavity and the large size of the thymus which weighs about 10 gr at birth (it accounts for 0.42 percent of body weight at birth, compared to 0.03 to 0.05 percent in the adult).

8) that the facts mentioned in 2, 3, 4, 6, and 7 are taken from: Crelin, Edmund S., *Functional Anatomy of the Newborn,* Yale University Press, New Haven, 1973. This is an excellent and short monograph (87 pages) which would be of benefit for any medical student to read.

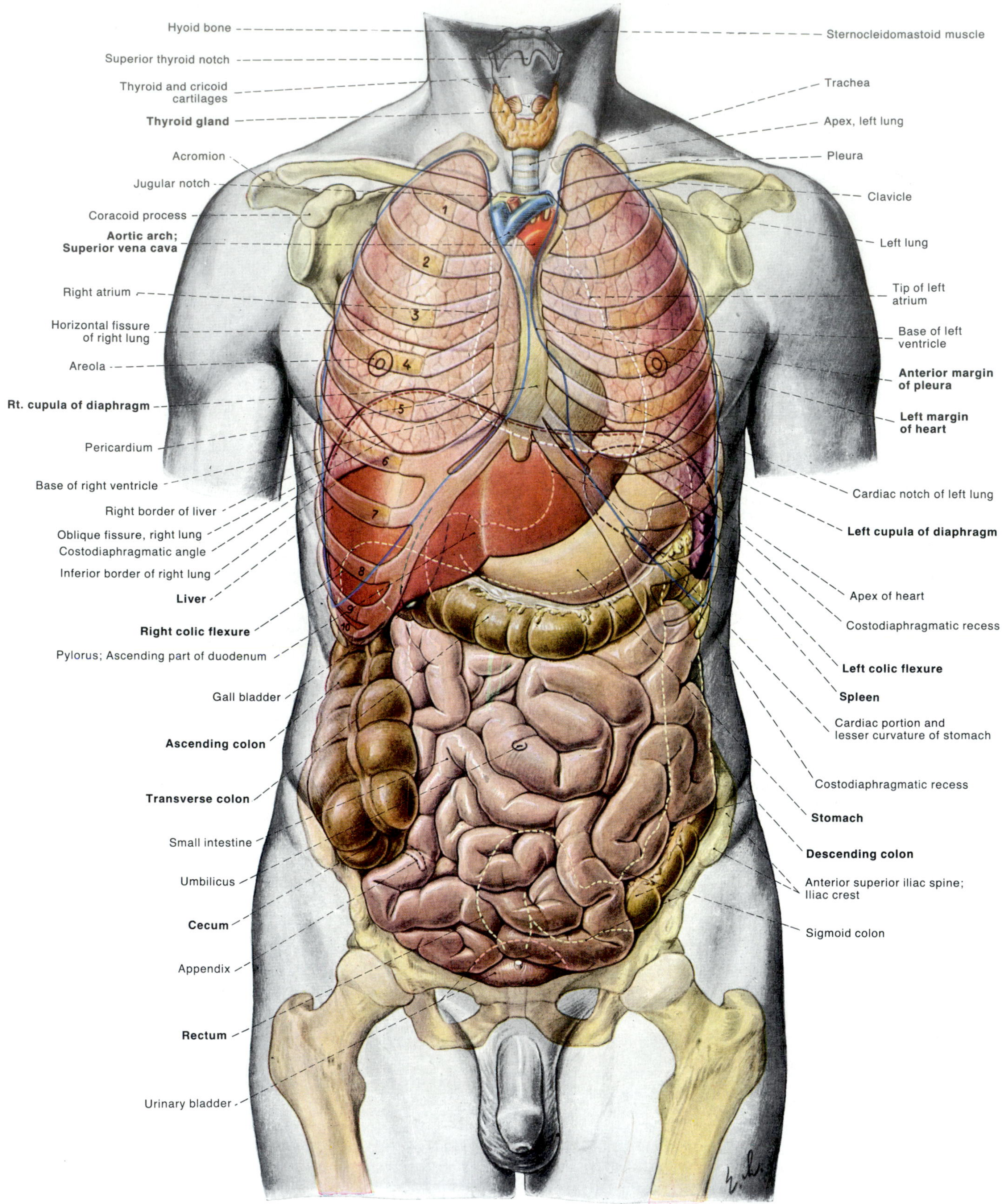

Hyoid bone
Superior thyroid notch
Thyroid and cricoid cartilages
Thyroid gland
Acromion
Jugular notch
Coracoid process
Aortic arch; Superior vena cava
Right atrium
Horizontal fissure of right lung
Areola
Rt. cupula of diaphragm
Pericardium
Base of right ventricle
Right border of liver
Oblique fissure, right lung
Costodiaphragmatic angle
Inferior border of right lung
Liver
Right colic flexure
Pylorus; Ascending part of duodenum
Gall bladder
Ascending colon
Transverse colon
Small intestine
Umbilicus
Cecum
Appendix
Rectum
Urinary bladder
Sternocleidomastoid muscle
Trachea
Apex, left lung
Pleura
Clavicle
Left lung
Tip of left atrium
Base of left ventricle
Anterior margin of pleura
Left margin of heart
Cardiac notch of left lung
Left cupula of diaphragm
Apex of heart
Costodiaphragmatic recess
Left colic flexure
Spleen
Cardiac portion and lesser curvature of stomach
Costodiaphragmatic recess
Stomach
Descending colon
Anterior superior iliac spine; Iliac crest
Sigmoid colon
1
2
3
4
5
6
7
8
9
10

Fig. 257: Frontal View of Thoracic and Abdominal Viscera

NOTE that the surface projections of the heart, stomach, gall bladder, spleen, transverse colon, descending and sigmoid colon, rectum, and urinary bladder are indicated by white broken outlines. The limits of the pleura are shown as solid blue lines and the spleen as a purple broken line. The gall bladder is shown as a broken blue line.

Fig. 257

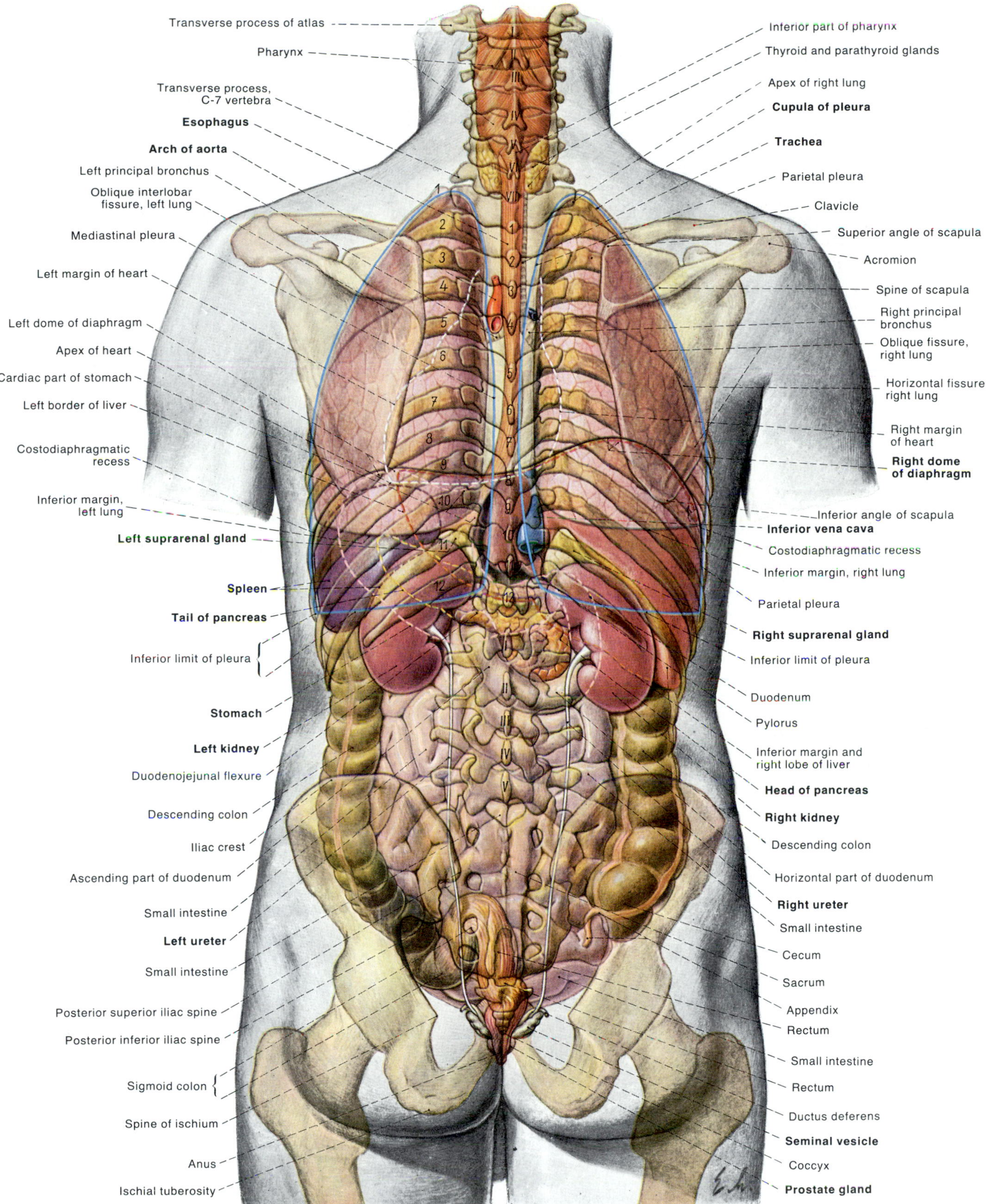

Transverse process of atlas
Pharynx
Transverse process, C-7 vertebra
Esophagus
Arch of aorta
Left principal bronchus
Oblique interlobar fissure, left lung
Mediastinal pleura
Left margin of heart
Left dome of diaphragm
Apex of heart
Cardiac part of stomach
Left border of liver
Costodiaphragmatic recess
Inferior margin, left lung
Left suprarenal gland
Spleen
Tail of pancreas
Inferior limit of pleura
Stomach
Left kidney
Duodenojejunal flexure
Descending colon
Iliac crest
Ascending part of duodenum
Small intestine
Left ureter
Small intestine
Posterior superior iliac spine
Posterior inferior iliac spine
Sigmoid colon
Spine of ischium
Anus
Ischial tuberosity
Inferior part of pharynx
Thyroid and parathyroid glands
Apex of right lung
Cupula of pleura
Trachea
Parietal pleura
Clavicle
Superior angle of scapula
Acromion
Spine of scapula
Right principal bronchus
Oblique fissure, right lung
Horizontal fissure, right lung
Right margin of heart
Right dome of diaphragm
Inferior angle of scapula
Inferior vena cava
Costodiaphragmatic recess
Inferior margin, right lung
Parietal pleura
Right suprarenal gland
Inferior limit of pleura
Duodenum
Pylorus
Inferior margin and right lobe of liver
Head of pancreas
Right kidney
Descending colon
Horizontal part of duodenum
Right ureter
Small intestine
Cecum
Sacrum
Appendix
Rectum
Small intestine
Rectum
Ductus deferens
Seminal vesicle
Coccyx
Prostate gland

Fig. 258: Posterior View of Thoracic and Abdominal Viscera

NOTE that the surface projections of the heart, stomach and duodenum are shown as white broken lines, the body and tail of the pancreas as yellow broken lines, the liver as brown broken lines, the superior pole of the spleen as a purple broken line, and the limits of the pleura as solid blue lines.

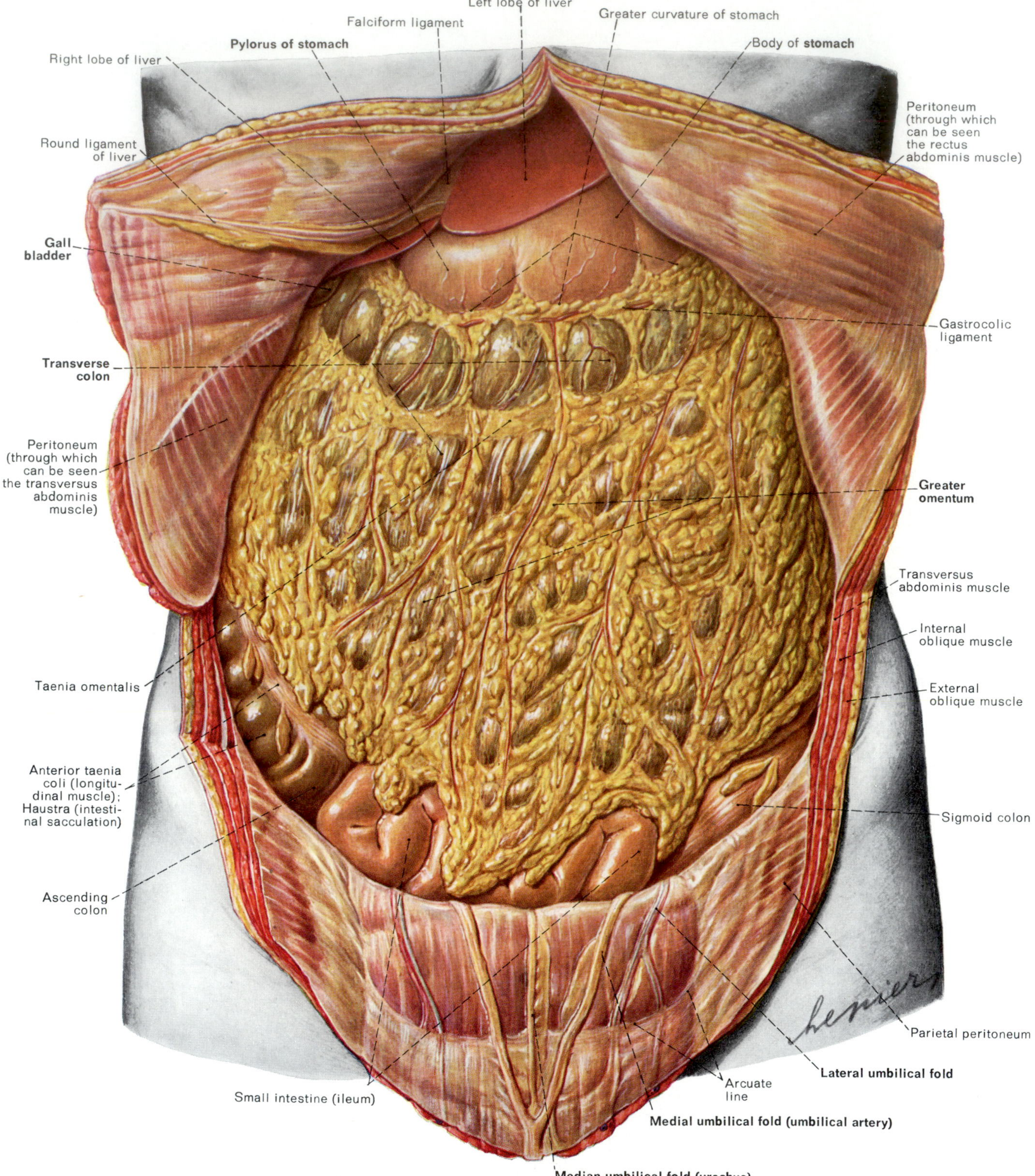

Left lobe of liver
Falciform ligament
Greater curvature of stomach
Pylorus of stomach
Body of stomach
Right lobe of liver
Peritoneum (through which can be seen the rectus abdominis muscle)
Round ligament of liver
Gall bladder
Gastrocolic ligament
Transverse colon
Peritoneum (through which can be seen the transversus abdominis muscle)
Greater omentum
Transversus abdominis muscle
Internal oblique muscle
External oblique muscle
Taenia omentalis
Anterior taenia coli (longitudinal muscle); Haustra (intestinal sacculation)
Sigmoid colon
Ascending colon
Parietal peritoneum
Lateral umbilical fold
Arcuate line
Small intestine (ileum)
Medial umbilical fold (umbilical artery)
Median umbilical fold (urachus)

Fig. 259: The Abdominal Cavity (1), the Viscera Left Intact

NOTE: 1) the greater omentum, which attaches along the greater curvature of the stomach, covers most of the intestines like an apron, and extends inferiorly almost as far as the pelvis.

2) the falciform ligament and round ligament of the liver (ligamentum teres). The falciform ligament is a remnant of the ventral mesogastrium. It extends between the liver and the anterior wall and separates the left and right lobes of the liver. The round ligament is the remains of the obliterated umbilical vein.

3) on the inner surface of the anterior wall identify

a) *the median umbilical fold* = the remains of the urachus, which in the fetus extends between the bladder and the umbilicus.

b) *the medial umbilical folds* = the obliterated umbilical arteries, which, before birth, coursed from the common iliac arteries to the umbilicus.

c) *the lateral umbilical folds* = which represent a reflection of peritoneum over the inferior epigastric vessels.

Fig. 259

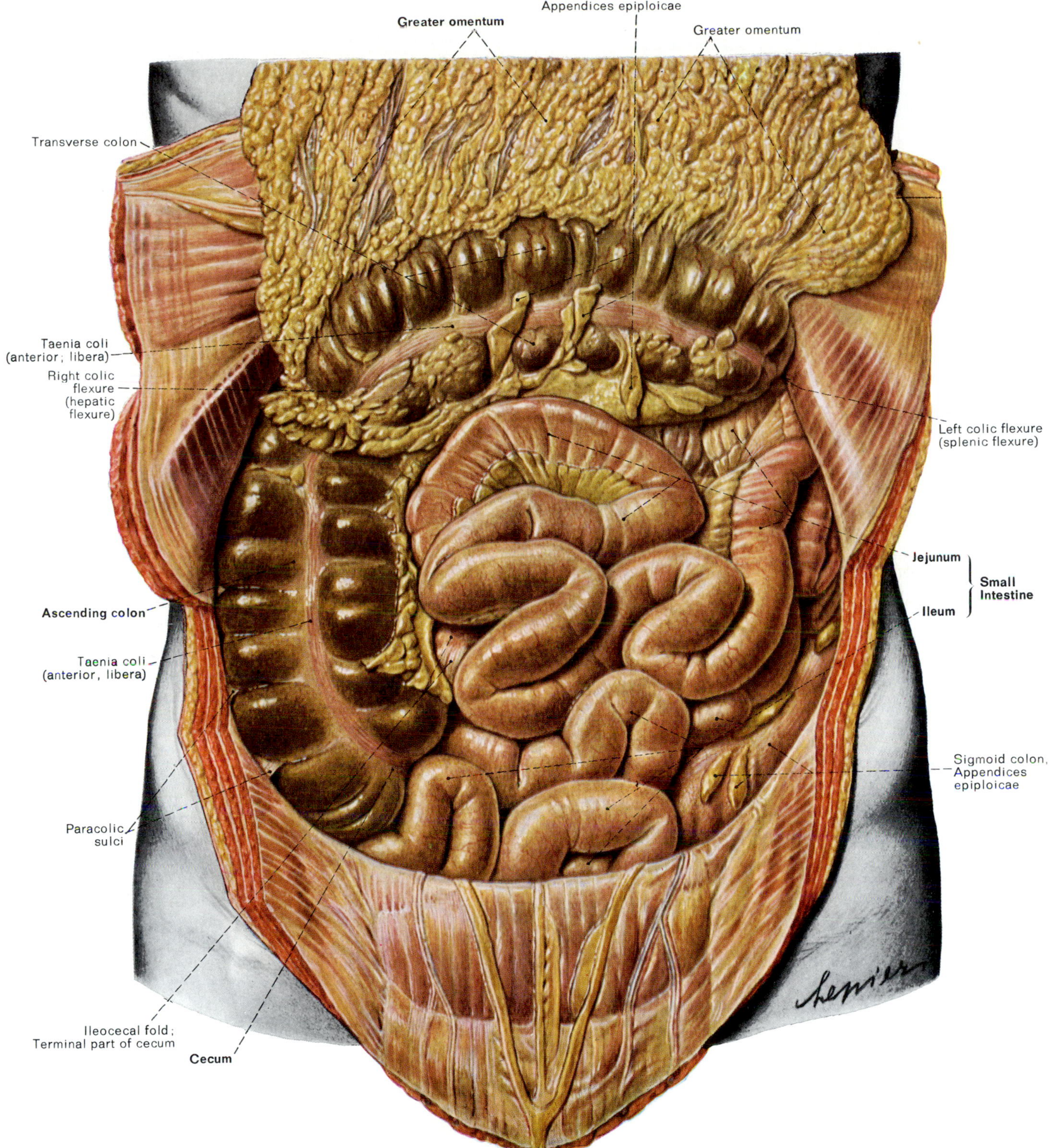
Appendices epiploicae
Greater omentum
Greater omentum
Transverse colon
Taenia coli
(anterior; libera)
Right colic
flexure
(hepatic
flexure)
Left colic flexure
(splenic flexure)
Jejunum
Small
Intestine
Ileum
Ascending colon
Taenia coli
(anterior, libera)
Sigmoid colon,
Appendices
epiploicae
Paracolic
sulci
Ileocecal fold;
Terminal part of cecum
Cecum

Fig. 260: The Abdominal Cavity (2), the Ascending Colon and the Transverse Colon and its Mesocolon

NOTE: 1) with the greater omentum reflected superiorly, the transverse colon comes into view as it crosses the abdominal cavity from right to left, in continuity with the ascending colon on the right and the descending colon (not shown in this figure; see Fig. 303 and 304) on the left. Observe the longitudinal muscles (taeniae) along the outer surface of the colon. Since these muscles are shorter than the other coats of the large intestine, they cause sacculations which are called haustrae.

2) that small, smooth irregular fatty masses called appendices epiploicae are suspended from the large intestine, thereby assisting in its identification.

3) below the mesocolon (inframesocolic) can be seen the small intestine which consists of three portions, the duodenum (see Figs. 285–287), jejunum and ileum. The outer walls of the small intestine are smooth and glistening and are not sacculated.

4) the small intestine measures about 22 feet in length, commencing at the pyloric end of the stomach and terminating at the ileocecal junction, which marks the commencement of the large intestine.

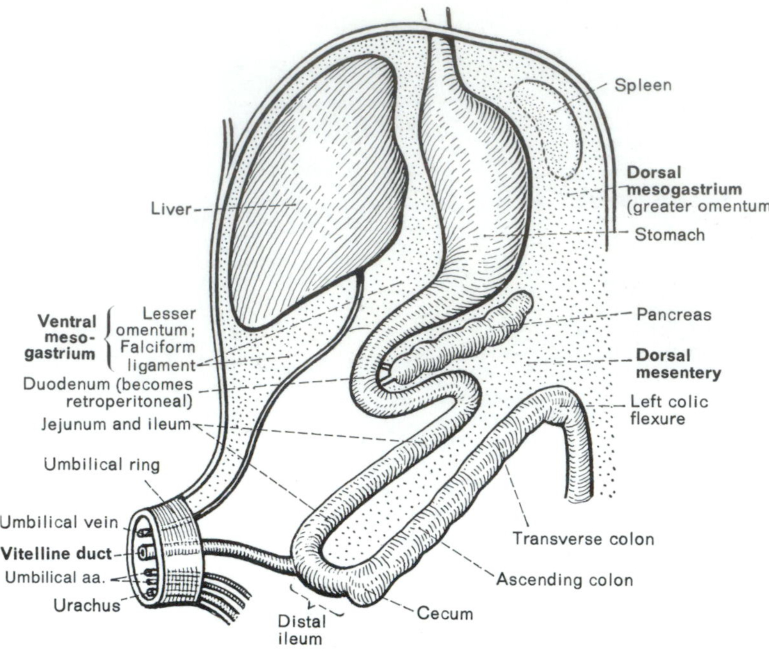
Spleen
Dorsal mesogastrium (greater omentum)
Liver
Stomach
Ventral meso-gastrium
Lesser omentum; Falciform ligament
Pancreas
Dorsal mesentery
Duodenum (becomes retroperitoneal)
Left colic flexure
Jejunum and ileum
Umbilical ring
Umbilical vein
Vitelline duct
Umbilical aa.
Urachus
Transverse colon
Ascending colon
Distal ileum
Cecum

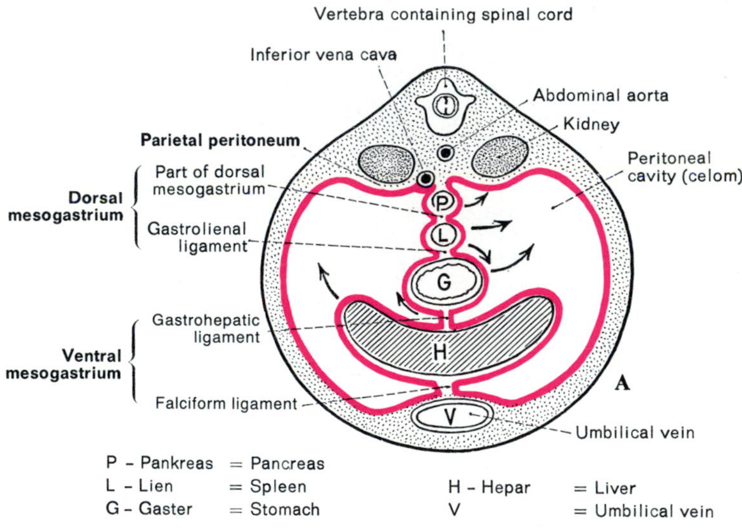
Vertebra containing spinal cord
Inferior vena cava
Abdominal aorta
Kidney
Parietal peritoneum
Peritoneal cavity (celom)
Dorsal mesogastrium
Part of dorsal mesogastrium
Gastrolienal ligament
P
L
G
H
Ventral mesogastrium
Gastrohepatic ligament
Falciform ligament
V
Umbilical vein
A
P - Pankreas = Pancreas
L - Lien = Spleen
G - Gaster = Stomach
H - Hepar = Liver
V = Umbilical vein

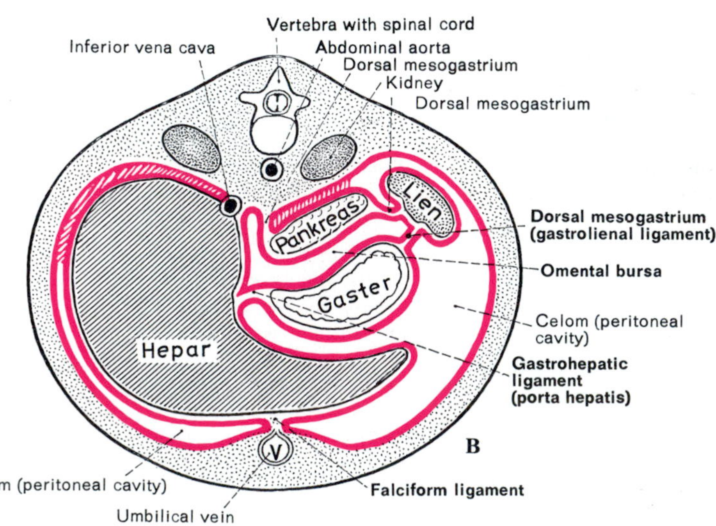
Vertebra with spinal cord
Inferior vena cava
Abdominal aorta
Dorsal mesogastrium
Kidney
Dorsal mesogastrium
Lien
Pankreas
Dorsal mesogastrium (gastrolienal ligament)
Omental bursa
Gaster
Celom (peritoneal cavity)
Hepar
Gastrohepatic ligament (porta hepatis)
V
B
Celom (peritoneal cavity)
Falciform ligament
Umbilical vein

Figs. 261, 262 A, 262 B

Fig. 261: The Developing Gastrointestinal Organs and their Mesenteries

NOTE: 1) as the primitive gastrointestinal tube develops within the abdominal celom, it is suspended to the body wall by primitive peritoneal reflections, both ventrally and dorsally. The early peritoneal attachments to the expanding stomach are called the ventral mesogastrium and dorsal mesogastrium, while the dorsal mesentery develops on the posterior aspect of the primitive small and large intestine.

2) the embryonic liver develops into the ventral mesogastrium, thereby dividing this ventral peritoneal attachment into:

a) a portion between the anterior body wall and the liver which eventually becomes the falciform ligament, and

b) a portion between the liver and the stomach which becomes the lesser omentum.

3) on the dorsal aspect:

a) the pancreas develops in relation to the primitive duodenum, both of which lose their mesenteries during gut rotation to become retroperitoneal;

b) the dorsal mesogastrium, attaching along the greater curvature of the stomach and rotating with the stomach, becomes the greater omentum. This eventually encases the transverse colon;

c) the dorsal mesentery remains attached to the small intestine, while the ascending and descending colon become displaced to the right and left side respectively, becoming adherent to the posterior body wall;

d) the sigmoid colon usually retains its mesentery while that of the rectum becomes obliterated.

4) near the cecal end of the small intestine the developing G.I. canal communicates with the vitelline duct. Before birth, this duct usually becomes resorbed; when it persists (3% of cases), it results in a diverticulum of the ileum called Meckel's diverticulum.

Cross Sectional Diagram of Development of Mesogastria
Fig. 262 A: Early Stage (about six weeks)

NOTE: 1) the primitive peritoneal reflections are indicated in red. The arrows show the direction of growth and, therefore, of movement by the various organs to achieve the positions shown in Fig. 262 B.

2) at this early stage, the peritoneum completely surrounds the organs in the upper abdominal region (visceral peritoneum) and attaches peripherally to the body wall (parietal peritoneum). Attaching along the posterior border of the stomach, the dorsal mesogastrium then surrounds the spleen and pancreas. Anterior to the stomach, the liver becomes interposed between the stomach and the anterior body wall. This forms the gastrohepatic ligament (also called lesser omentum) between the lesser curvature of the stomach and the liver, and the falciform ligament between the liver and the anterior body wall.

Cross Sectional Diagram of Development of Mesogastria
Fig. 262 B: Late Fetal Stage

NOTE: 1) with the rotation of the organs (in the direction of the arrows in Fig. 262 A), the liver grows into the celomic cavity toward the right and contacts the inferior vena cava, while the stomach rotates such that its dorsal mesogastrium (greater curvature) is shifted to the left. The pancreas and spleen still retain their position posterior to the stomach.

2) the reflection of dorsal mesogastrium between the stomach and spleen becomes established as the gastrolienal ligament while one layer of mesogastrium surrounding the pancreas (and duodenum) fuses to the posterior body wall. This latter development fixates these two organs with a layer of peritoneum on their anterior surface, causing them to become retroperitoneal. The omental bursa also develops posterior to the stomach and anterior to the pancreas.

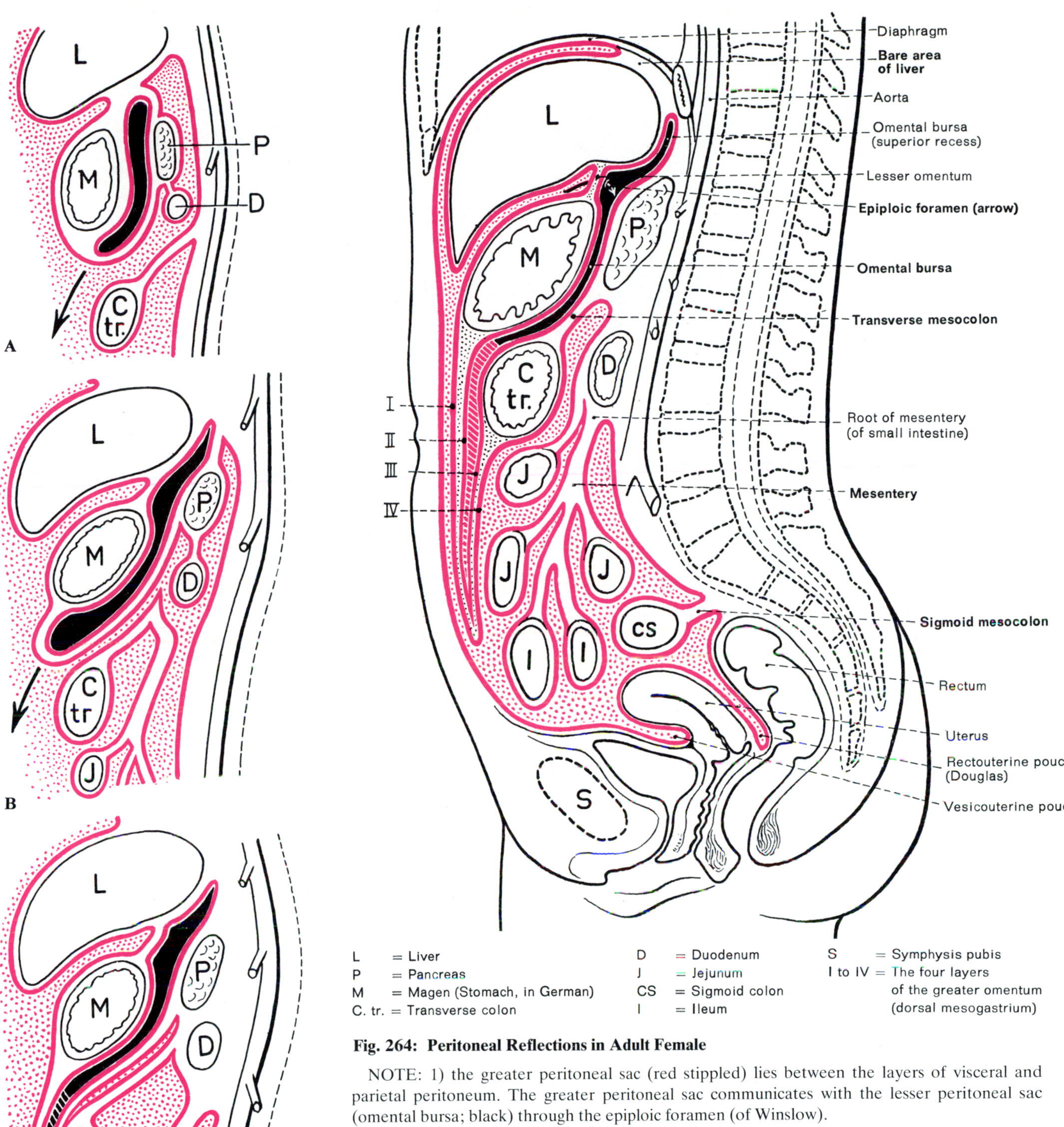

L
P
D
M
C tr.
A
B
C
J
Diaphragm
Bare area of liver
Aorta
Omental bursa (superior recess)
Lesser omentum
Epiploic foramen (arrow)
Omental bursa
Transverse mesocolon
Root of mesentery (of small intestine)
Mesentery
Sigmoid mesocolon
Rectum
Uterus
Rectouterine pouch (Douglas)
Vesicouterine pouch
I
II
III
IV
CS
S
L = Liver
P = Pancreas
M = Magen (Stomach, in German)
C. tr. = Transverse colon
D = Duodenum
J = Jejunum
CS = Sigmoid colon
I = Ileum
S = Symphysis pubis
I to IV = The four layers of the greater omentum (dorsal mesogastrium)

Fig. 264: Peritoneal Reflections in Adult Female

NOTE: 1) the greater peritoneal sac (red stippled) lies between the layers of visceral and parietal peritoneum. The greater peritoneal sac communicates with the lesser peritoneal sac (omental bursa; black) through the epiploic foramen (of Winslow).

2) dorsally, the roots of three distinct peritoneal mesenteries can be observed: a) the transverse mesocolon, b) the mesentery surrounding the small intestine, and c) the sigmoid mesocolon.

3) behind the stomach and transverse colon, observe the retroperitoneal pancreas and duodenum. Note also that a portion of the liver is not surrounded by peritoneum (bare area of the liver) and lies adjacent to the diaphragm.

Fig. 263 A, B, & C: Stages in the Development of the Omental Bursa (Sagittal Diagrams)

NOTE: 1) at four weeks the dorsal border of the stomach (Magen, M) grows faster than the ventral border assisting in rotation of the stomach on its long axis. The greater curvature and its dorsal mesogastrium becomes directed to the left, while the lesser curvature and the ventral mesogastrium is directed to the right.

2) by eight weeks (Fig. 263 A) the omental bursa (black) forms behind the stomach between the two leaves of dorsal mesogastrium. The pancreas and duodenum are still surrounded by dorsal mesentery. As gut rotation continues the dorsal mesogastrium extends inferiorly (Fig. 263 B, arrow) to form the greater omentum which becomes a double reflection (4 leaves) of the dorsal mesogastrium "trapping" the cavity of the omental bursa between the 2nd and 3rd leaves.

3) continued development (Fig. 263 C) results in a further descent of the greater omentum over the abdominal viscera and a fusion (cross-hatched) of the 2nd and 3rd leaves inferiorly.

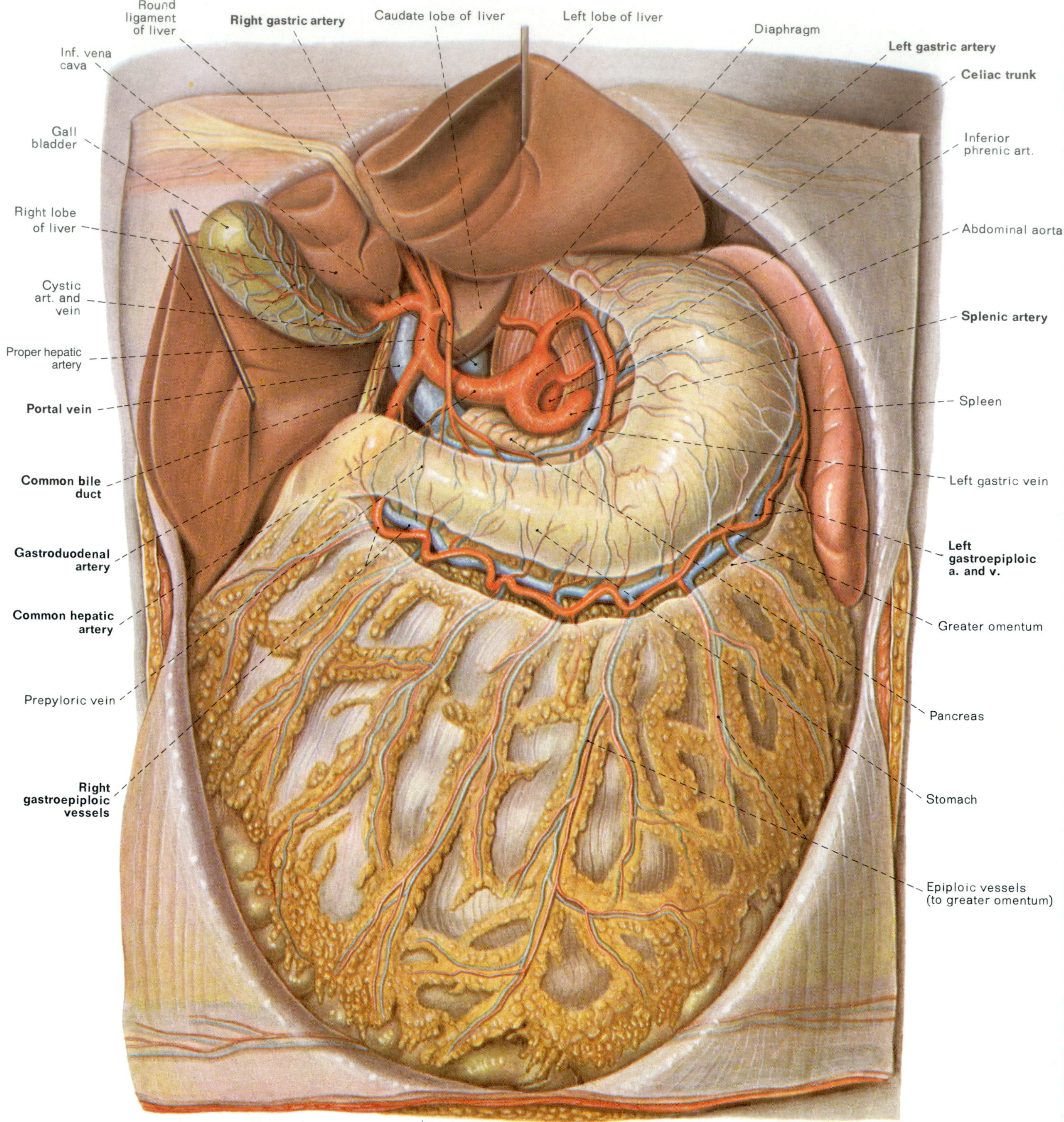
Round ligament of liver
Right gastric artery
Caudate lobe of liver
Left lobe of liver
Diaphragm
Left gastric artery
Celiac trunk
Inf. vena cava
Gall bladder
Inferior phrenic art.
Right lobe of liver
Abdominal aorta
Cystic art. and vein
Splenic artery
Proper hepatic artery
Portal vein
Spleen
Common bile duct
Left gastric vein
Gastroduodenal artery
Left gastroepiploic a. and v.
Common hepatic artery
Greater omentum
Prepyloric vein
Pancreas
Right gastroepiploic vessels
Stomach
Epiploic vessels (to greater omentum)

Fig. 265: The Abdominal Cavity (3): The Celiac Trunk and its Branches

NOTE: 1) the right and left lobes of the liver have been elevated and the lesser omentum has been removed between the lesser curvature of the stomach and the liver to reveal the celiac trunk and its branches and three major structures at the porta hepatis, the hepatic artery, portal vein and common bile duct.

2) the celiac trunk lies anterior to the 12th thoracic vertebra and almost immediately divides into the left gastric artery, and the hepatic and splenic arteries:

a) the left gastric artery courses along the lesser curvature of the stomach and anastomoses with the right gastric branch of the hepatic artery;

b) the hepatic artery courses to the right and gives off the gastroduodenal artery before dividing to enter the lobes of the liver;

c) the splenic artery courses to the left toward the hilum of the spleen;

d) the gastroduodenal artery gives rise to the right gastroepiploic artery which follows along the greater curvature of the stomach to anastomose with the left gastroepiploic branch of the splenic artery.

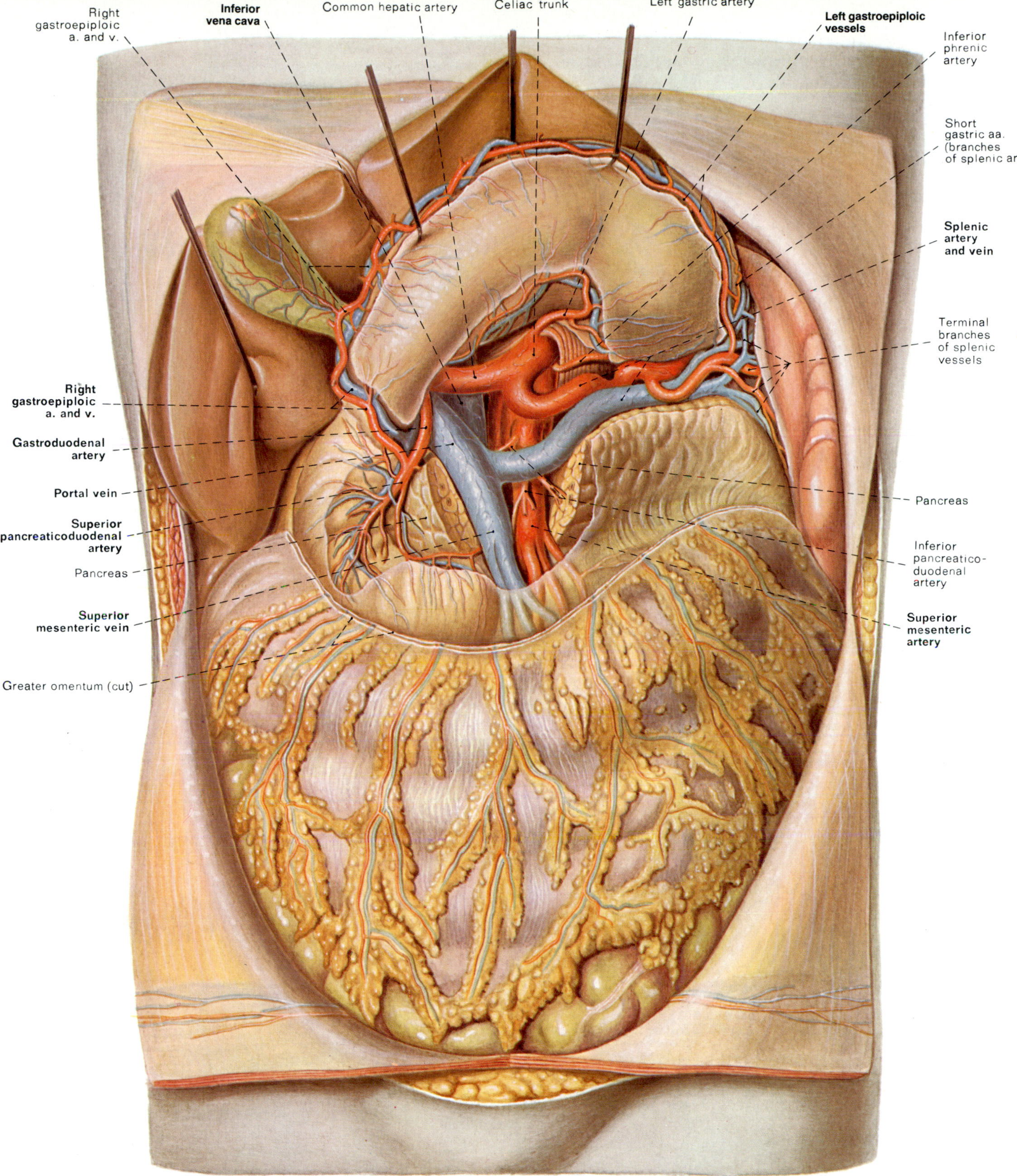

Right gastroepiploic a. and v.
Inferior vena cava
Common hepatic artery
Celiac trunk
Left gastric artery
Left gastroepiploic vessels
Inferior phrenic artery
Short gastric aa. (branches of splenic art.)
Splenic artery and vein
Terminal branches of splenic vessels
Right gastroepiploic a. and v.
Gastroduodenal artery
Portal vein
Superior pancreaticoduodenal artery
Pancreas
Superior mesenteric vein
Greater omentum (cut)
Pancreas
Inferior pancreatico-duodenal artery
Superior mesenteric artery

Fig. 266: The Abdominal Cavity (4): The Splenic Vessels and Formation of the Portal Vein

NOTE: 1) the attachment of the greater omentum has been cut along the greater curvature of the stomach. The stomach has been lifted to expose its posterior surface and the underlying pancreas, duodenum and blood vessels. A portion of the body of the pancreas has been removed to reveal the formation of the portal vein by the junction of the splenic and superior mesenteric veins.

2) the splenic artery in its tortuous course across the left upper abdomen to the hilum of the spleen. Observe also how the gastroduodenal artery lies posterior to the pyloric end of the stomach and divides into the right gastroepiploic artery and the superior pancreaticoduodenal artery. The origin of the left gastroepiploic artery from the splenic is also visible.

3) the root of the superior mesenteric artery as it branches from the abdominal aorta just below the celiac trunk.

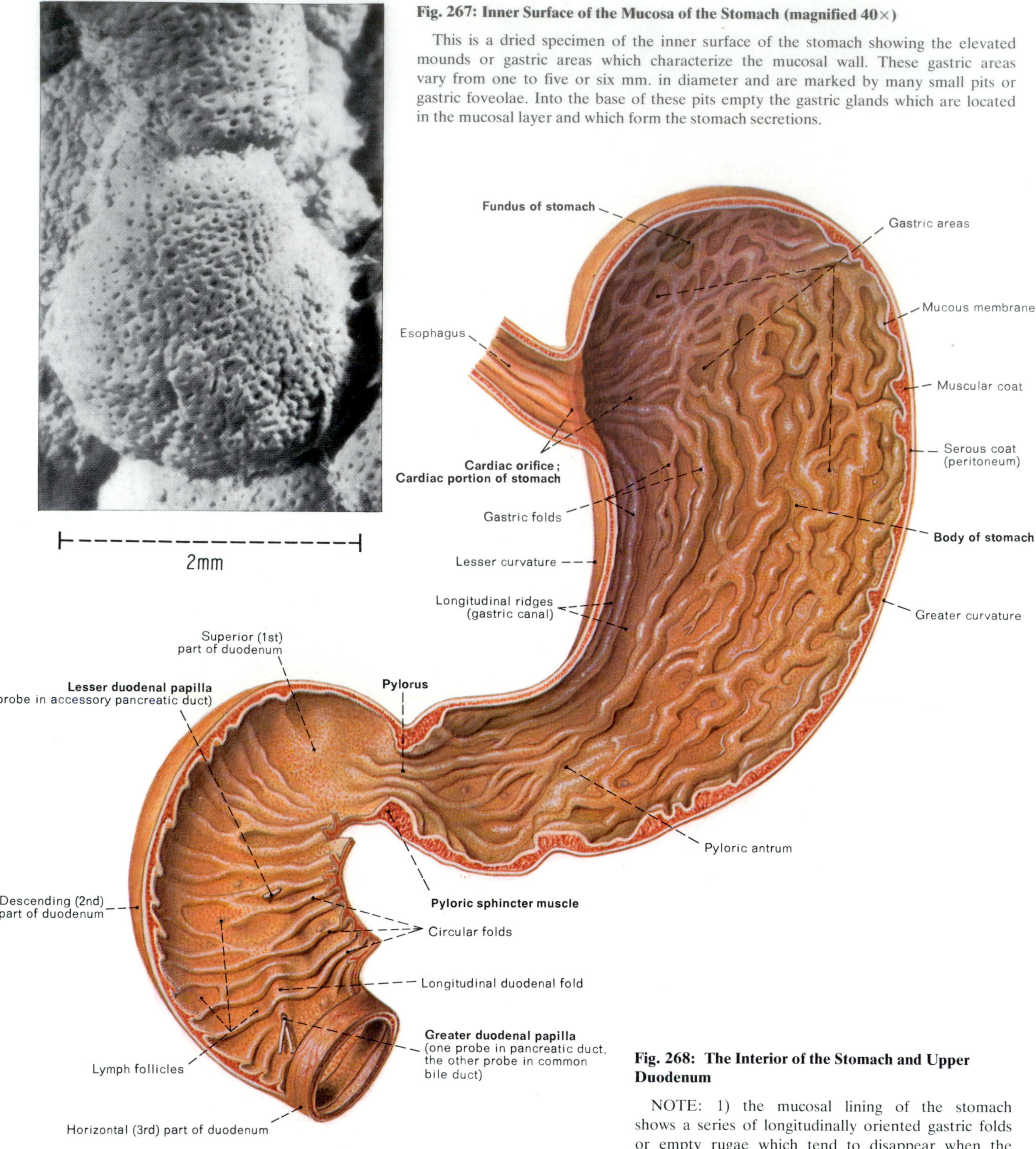
2mm
Fundus of stomach
Gastric areas
Mucous membrane
Esophagus
Muscular coat
Serous coat (peritoneum)
Cardiac orifice; Cardiac portion of stomach
Gastric folds
Body of stomach
Lesser curvature
Longitudinal ridges (gastric canal)
Greater curvature
Superior (1st) part of duodenum
Lesser duodenal papilla (probe in accessory pancreatic duct)
Pylorus
Pyloric antrum
Descending (2nd) part of duodenum
Pyloric sphincter muscle
Circular folds
Longitudinal duodenal fold
Greater duodenal papilla (one probe in pancreatic duct, the other probe in common bile duct)
Lymph follicles
Horizontal (3rd) part of duodenum

Fig. 267: Inner Surface of the Mucosa of the Stomach (magnified 40×)

This is a dried specimen of the inner surface of the stomach showing the elevated mounds or gastric areas which characterize the mucosal wall. These gastric areas vary from one to five or six mm. in diameter and are marked by many small pits or gastric foveolae. Into the base of these pits empty the gastric glands which are located in the mucosal layer and which form the stomach secretions.

Fig. 268: The Interior of the Stomach and Upper Duodenum

NOTE: 1) the mucosal lining of the stomach shows a series of longitudinally oriented gastric folds or empty rugae which tend to disappear when the stomach is full and distended. These folds are more regular along the lesser curvature and form the grooved gastric canal. The concept that food travels along this canal (magenstrasse) is not correct.

2) the surface of the first portion of the duodenum (superior) is smooth, whereas the circular ridges characteristic of the small intestine can be seen to commence in the second or descending portion of the duodenum.

3) the pyloric junction of the stomach with the duodenum. A circular muscle, the pyloric sphincter, guards this junction. It diminishes significantly in size the lumen of the gastrointestinal tract at this point. The pylorus is to the right of midline at the level of the 1st lumbar vertebra.

4) the openings in the wall of the duodenum. The greater duodenal papilla serves as the site of the openings of both the common bile duct and the main pancreatic duct. The accessory pancreatic duct opens two centimeters more proximally through the lesser duodenal papilla.

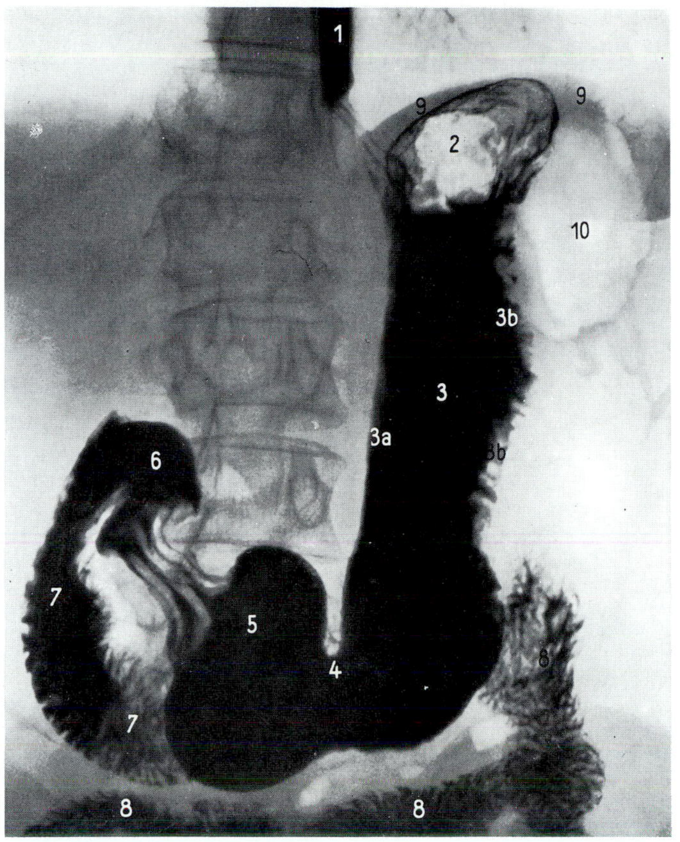

1
9
9
2
10
3b
3
3a
3b
6
7
5
4
8
7
8
8

1. Esophagus
2. Stomach fundus (air bubble)
3. Body of stomach
   3a Lesser curvature
   3b Greater curvature
4. Peristaltic constriction at angular notch
5. Pyloric antrum (expanded)
6. Bulb of superior duodenum (1st part)
7. Descending duodenum (2nd part)
8. Jejunum
9. Left dome of diaphragm
10. Gas in left colic flexure

Fig. 269: X-Ray of Lower Esophagus, Stomach, Duodenum, and Proximal Jejunum

NOTE: this is a normal "J-shaped" or "fish-hook" stomach. Observe that the cardiac and pyloric ends of the stomach are more securely attached to the posterior body wall, whereas the body and pyloric part are more mobile. Frequently in the upright position, the greater curvature hangs as low as the brim of the pelvis.

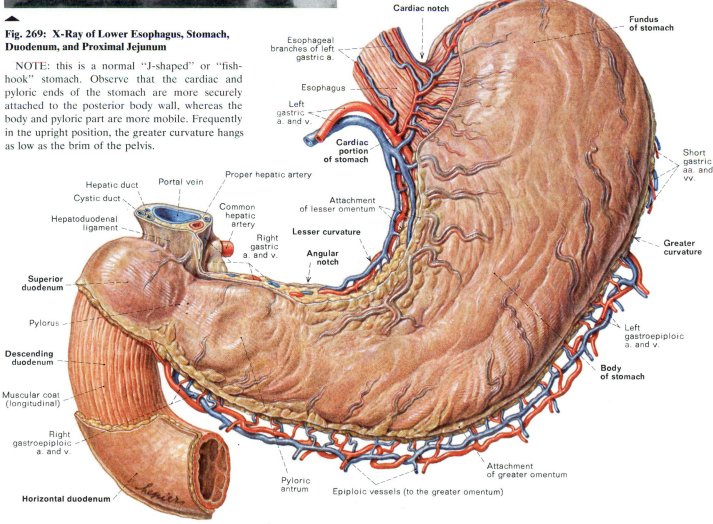

Fig. 270: Anterior View of Stomach

NOTE: 1) the stomach is a dilated muscular and mucosal sac situated in the gastrointestinal tract between the esophagus (cardiac end) and duodenum (pyloric end). It consists of an upper dome-shaped portion called the *fundus,* a middle portion, *the body,* and a tapering caudal part referred to as the *pyloric portion.*

2) although the shape of the stomach varies with the amount of its contents, it presents two curvatures. The greater curvature is directed toward the left and to it is attached the greater omentum. The greater curvature forms an acute angle with the esophagus called the cardiac notch. The lesser curvature constitutes the right border of the stomach and along this edge is attached the lesser omentum. Its lower third bends toward the duodenum beyond the angular notch.

3) the blood vessels supplying the stomach include the a) left and right gastric arteries along the lesser curvature, b) the left and right gastroepiploic arteries along the greater curvature, c) the short gastric branches of the splenic artery. Observe the esophageal branches of the left gastric artery supplying the cardiac end of the stomach.

Cardiac notch
Fundus of stomach
Esophageal branches of left gastric a.
Esophagus
Left gastric a. and v.
Cardiac portion of stomach
Short gastric aa. and vv.
Proper hepatic artery
Hepatic duct
Portal vein
Cystic duct
Common hepatic artery
Attachment of lesser omentum
Hepatoduodenal ligament
Lesser curvature
Right gastric a. and v.
Angular notch
Greater curvature
Superior duodenum
Pylorus
Left gastroepiploic a. and v.
Descending duodenum
Body of stomach
Muscular coat (longitudinal)
Right gastroepiploic a. and v.
Attachment of greater omentum
Pyloric antrum
Epiploic vessels (to the greater omentum)
Horizontal duodenum

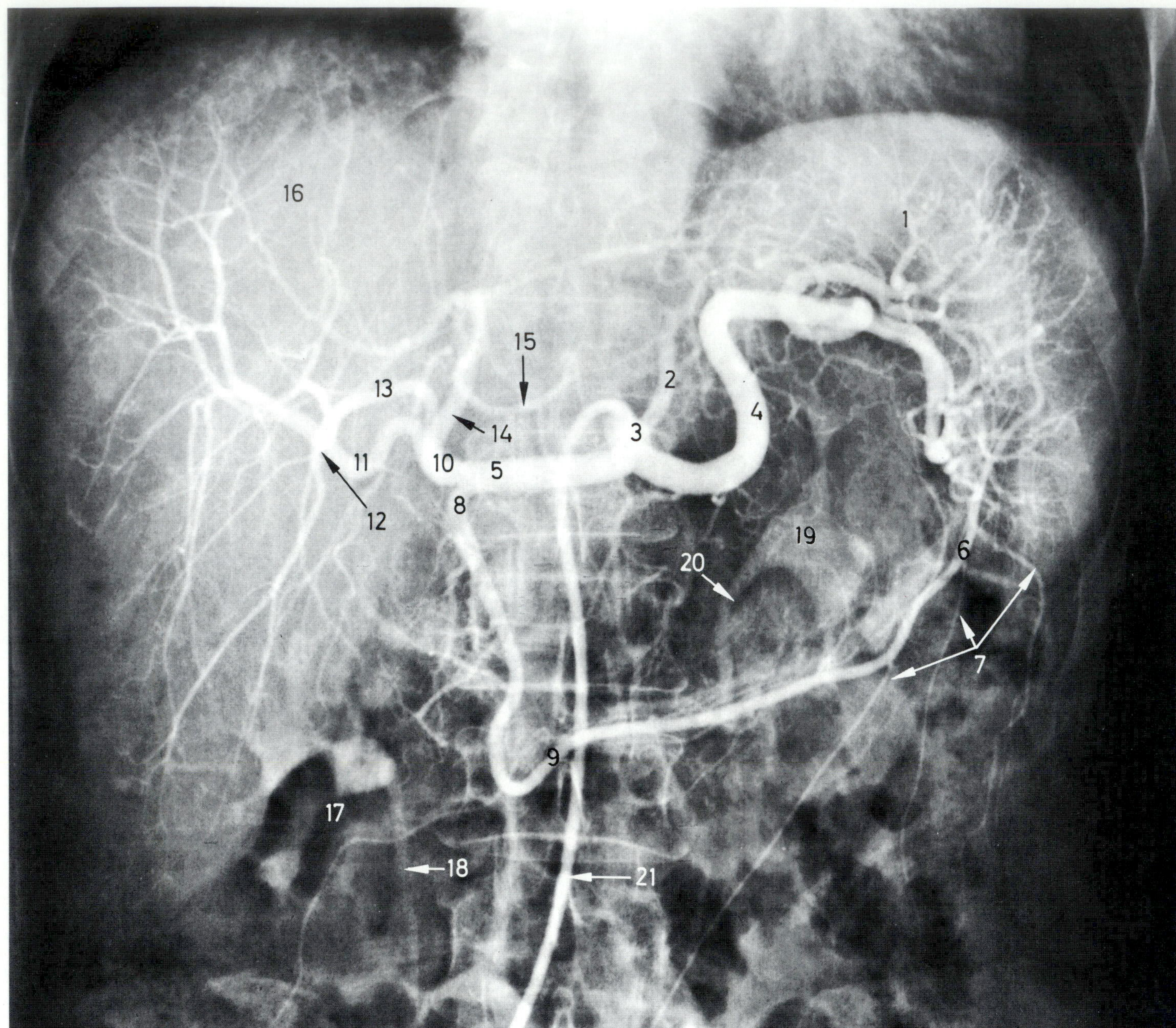

16
1
15
2
13
14
4
3
11
10
5
12
8
19
6
20
7
9
17
18
21

Fig. 271: A Celiac Trunk Arteriogram

NOTE: 1) that this figure is a negative print from an X-ray taken of the upper abdomen following the injection of a contrast medium through a catheter (*21,* arrow), which was introduced into the abdominal aorta and directed upward to the point where the celiac trunk (*3*) branches from the aorta. The principal subbranches of the celiac trunk, therefore, can be visualized.

2) the three primary vessels arising from the celiac trunk (*3*) are the *left gastric artery* (*2*), the *splenic artery* (*4*) which is directed in a tortuous pattern toward the spleen (*1*), and the *common hepatic artery* (*5*) which courses almost directly to the right.

3) that from one of the inferior hilar branches of the splenic artery arises the left gastroepiploic artery (*6*), which then gives origin to epiploic arteries (*7*, arrows) which descend to supply the greater omentum. The left gastroepiploic (*6*) courses along the greater curvature of the stomach to anastomose with the right gastroepiploic artery (*9*). This latter vessel arises from the gastroduodenal artery (*8*) which, in turn, courses inferiorly as a major vessel derived from the common hepatic artery (*5*).

4) that beyond the origin of the gastroduodenal artery (*8*) the common hepatic artery (*5*) is called the proper hepatic artery (*10*).

5) that the proper hepatic artery (*10*) soon divides into the *right hepatic artery* (*11*), from which branches the cystic artery (*12*) which supplies the gall bladder, the *middle hepatic artery* (*13*), and the *left hepatic artery* (*14*, arrow). The hepatic vessels, of course, supply the liver (*16*).

6) that from the right hepatic artery (*14*) in this individual branches the right gastric artery (*15*, arrow). Just as frequently the right gastric artery is found arising from the common hepatic artery (*5*). The right gastric artery courses around the lesser curvature of the stomach to anastomose with the left gastric (*2*). The latter vessel is one of the original branches of the celiac trunk.

7) that inferiorly can be seen the right renal pelvis (*17*) and the right ureter (*18*, arrow). Observe that because of the liver, these structures on the right side are significantly lower than the left renal pelvis (*19*) and the origin of the left ureter (*20*, arrow).

8) that this figure comes from Wicke, L., *Atlas of Radiologic Anatomy,* 3rd Edition, Urban & Schwarzenberg, Baltimore, 1982.

Fig. 271

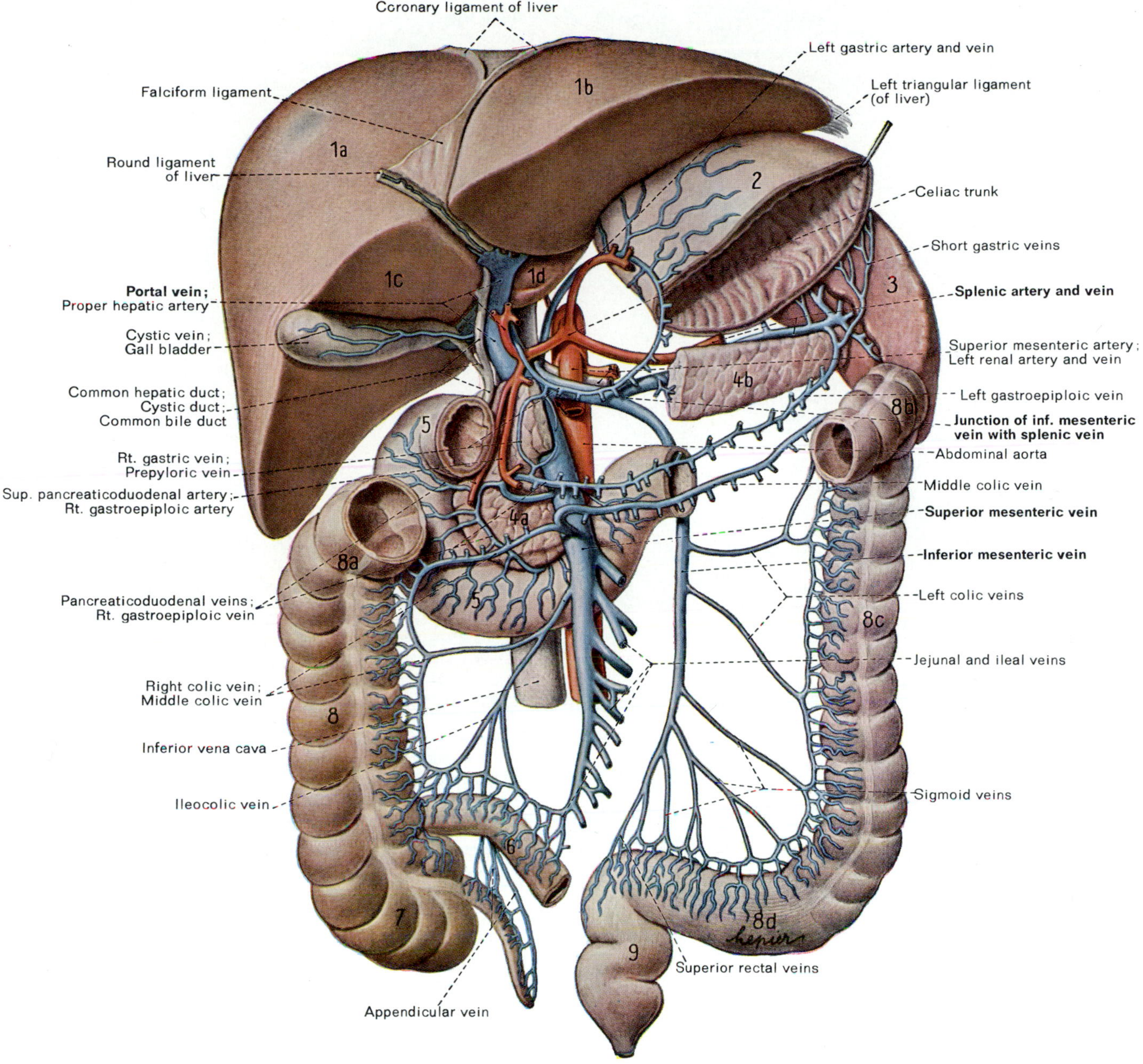
Coronary ligament of liver
Left gastric artery and vein
Falciform ligament
Left triangular ligament (of liver)
Round ligament of liver
Celiac trunk
Short gastric veins
Portal vein; Proper hepatic artery
Splenic artery and vein
Cystic vein; Gall bladder
Superior mesenteric artery; Left renal artery and vein
Common hepatic duct; Cystic duct; Common bile duct
Left gastroepiploic vein
Junction of inf. mesenteric vein with splenic vein
Rt. gastric vein; Prepyloric vein
Abdominal aorta
Sup. pancreaticoduodenal artery; Rt. gastroepiploic artery
Middle colic vein
Superior mesenteric vein
Inferior mesenteric vein
Pancreaticoduodenal veins; Rt. gastroepiploic vein
Left colic veins
Jejunal and ileal veins
Right colic vein; Middle colic vein
Inferior vena cava
Sigmoid veins
Ileocolic vein
Superior rectal veins
Appendicular vein
1a
1b
1c
1d
2
3
4a
4b
5
6
7
8
8a
8b
8c
8d
9

Fig. 272: Abdominal Portal System of Veins

1a Right lobe of liver
1b Left lobe of liver
1c Quadrate lobe of liver
1d Caudate lobe of liver
2 Stomach
3 Spleen
4a Head of pancreas
4b Tail of pancreas
5 Duodenum
6 Ileum
7 Cecum
8 Ascending colon
8a Right colic flexure
8b Left colic flexure
8c Descending colon
8d Sigmoid colon
9 Rectum
↑ = Junction of sup. mesenteric v. and splenic vein to form portal vein

NOTE: 1) the abdominal portal system of veins drains venous blood from the gastrointestinal tract, the gall bladder, pancreas and spleen through the liver via the large portal vein. This is done in order to subject this venous blood to the various functions of the liver before it is returned to the general systemic circulation by way of the hepatic veins into the inferior vena cava.

2) the portal vein is formed by the union of the superior mesenteric vein and the splenic vein. The inferior mesenteric vein drains into the splenic vein. At the esophageal end of the stomach (esophageal veins) and at the distal end of the rectum (inferior rectal veins), the portal system of veins anastomoses with the systemic veins. Certain disease states which may cause a reduction of blood flow through the liver result in greater use of these anastomotic channels in the return of blood in the portal system.

3) the functions of the liver are numerous, varied and vital to life. The liver secretes bile which is then stored in the gall bladder and released when food appears in the duodenum. Bile aids in the digestion and absorption of fats. The liver converts glucose to glycogen, stores the glycogen and then reconverts it to glucose again when needed. Further, the liver is involved in the synthesis of Vitamin A, heparin, prothrombin, fibrinogen and other substances. It functions also in detoxification of substances in the blood. It is involved in the breakdown of hemoglobin and stores both iron and copper.

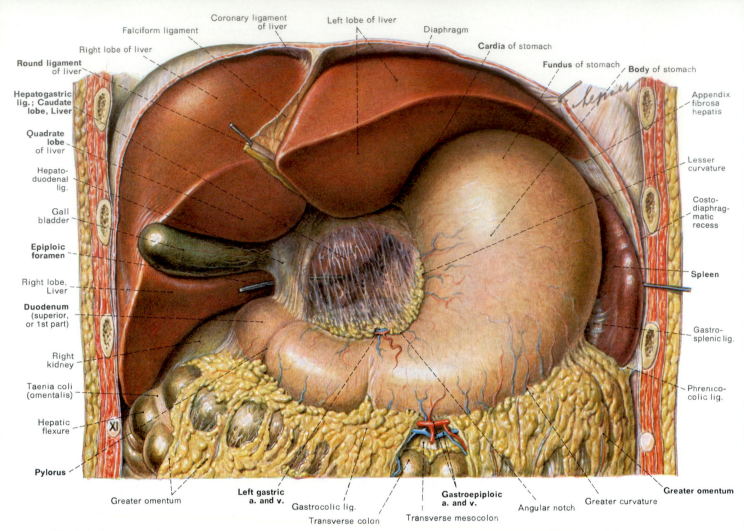

Coronary ligament of liver
Left lobe of liver
Falciform ligament
Diaphragm
Right lobe of liver
Cardia of stomach
Round ligament of liver
Fundus of stomach
Body of stomach
Hepatogastric lig.; Caudate lobe, Liver
Appendix fibrosa hepatis
Quadrate lobe of liver
Lesser curvature
Hepato-duodenal lig.
Costo-diaphrag-matic recess
Gall bladder
Epiploic foramen
Spleen
Right lobe, Liver
Duodenum (superior, or 1st part)
Gastro-splenic lig.
Right kidney
Taenia coli (omentalis)
Phrenico-colic lig.
Hepatic flexure
XI
Pylorus
Greater omentum
Greater omentum
Left gastric a. and v.
Gastrocolic lig.
Transverse colon
Gastroepiploic a. and v.
Transverse mesocolon
Angular notch
Greater curvature

Fig. 273: The Lesser Omentum, Stomach, Liver and Spleen

NOTE: 1) with the liver elevated, a probe has been inserted through the epiploic foramen into the vestibule of the omental bursa. By way of this opening, the greater peritoneal sac communicates with the lesser peritoneal sac. Observe that the lesser omentum consists of the hepatogastric and hepatoduodenal ligaments.

2) the epiploic foramen is situated just caudal to the liver and readily admits two fingers. It is bound superiorly by the caudate lobe of the liver, inferiorly by the superior or 1st part of the duodenum, posteriorly by the inferior vena cava, and anteriorly by the lesser omentum which ensheathes the structures of the porta hepatis (hepatic artery, portal vein and bile ducts).

3) the greater omentum extends along the greater curvature from the spleen to the duodenum. The gall bladder is situated between the right and quadrate lobes of the liver and projects just beyond the inferior border of the liver, thereby coming into contact directly with the anterior abdominal wall at this site.

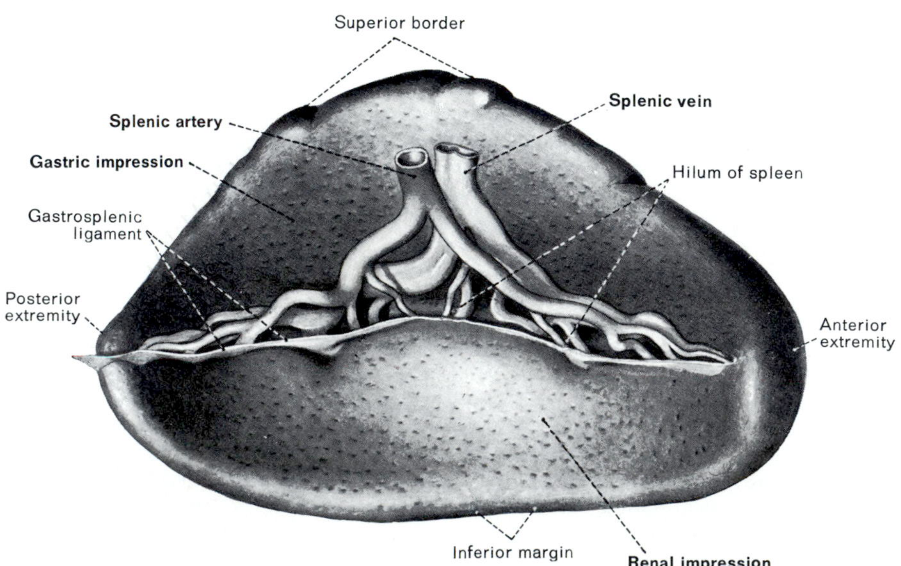
Superior border
Splenic vein
Splenic artery
Gastric impression
Hilum of spleen
Gastrosplenic ligament
Posterior extremity
Anterior extremity
Inferior margin
Renal impression

Fig. 274: The Spleen, Visceral Surface

NOTE that the spleen is situated in the left hypochondriac region between the fundus of the stomach and the diaphragm. Its visceral surface shows the contours of the organs related to it. A gastric impression and a renal impression conform to the shapes of the stomach and left kidney. Additionally, the left colic flexure, the tail of the pancreas and the left adrenal gland, which overlies the left kidney, are related to this visceral surface.

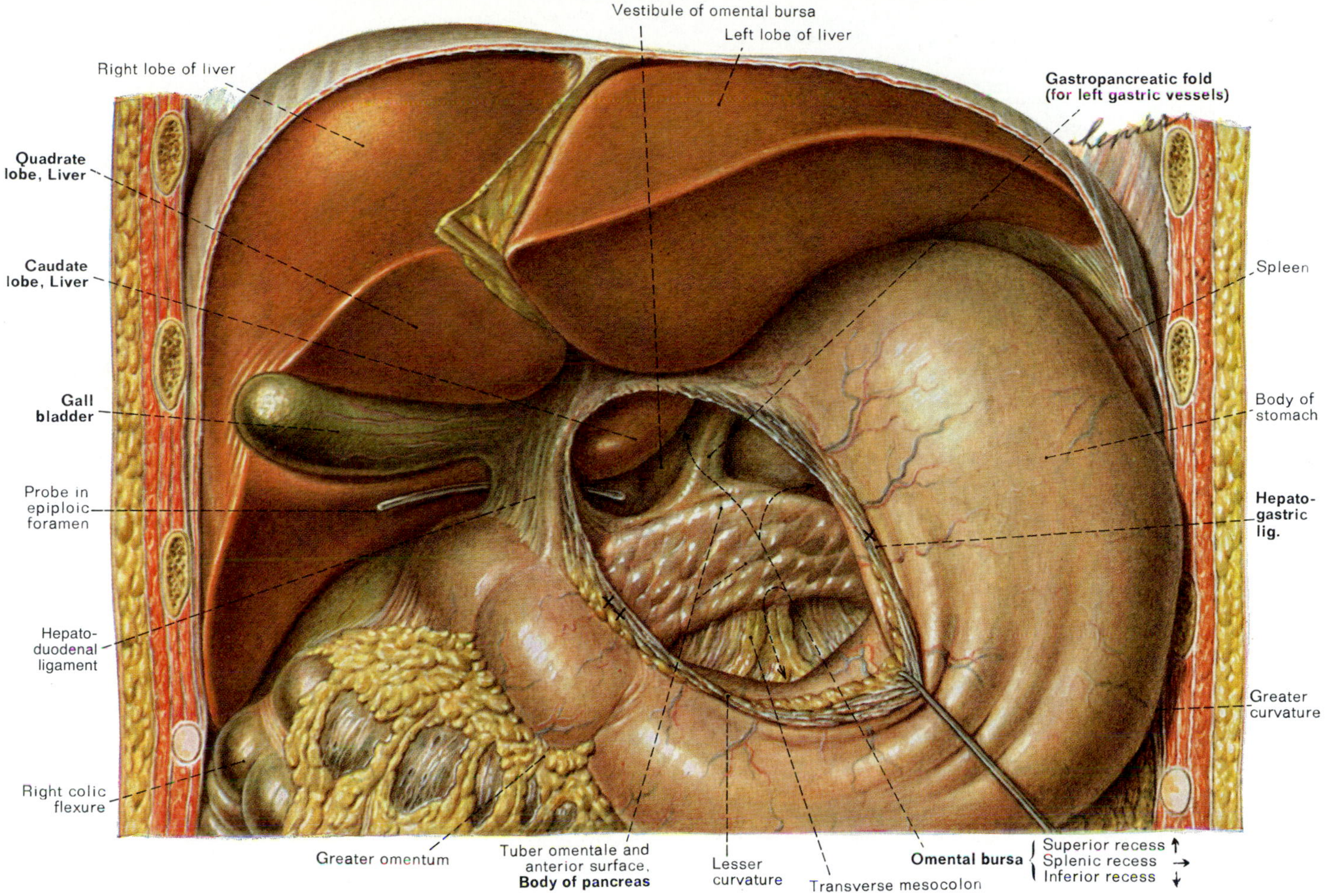

Vestibule of omental bursa
Left lobe of liver
Right lobe of liver
Gastropancreatic fold (for left gastric vessels)
Quadrate lobe, Liver
Caudate lobe, Liver
Spleen
Gall bladder
Body of stomach
Probe in epiploic foramen
Hepato-gastric lig.
Hepato-duodenal ligament
Greater curvature
Right colic flexure
Greater omentum
Tuber omentale and anterior surface, Body of pancreas
Lesser curvature
Transverse mesocolon
Omental bursa { Superior recess ↑ Splenic recess → Inferior recess ↓

Fig. 275: The Omental Bursa, Caudate Lobe of Liver and Body of Pancreas

NOTE: 1) the liver has been elevated and the lesser curvature of the stomach has been pulled down and to the left in order to enlarge the exposure obtained by opening the omental bursa through the hepatogastric ligament (indicated by x and xx). The superior, splenic and inferior recesses of this bursa have been indicated by the arrows.

2) the portion of the omental bursa adjacent to the epiploic foramen is called the vestibule. Observe the gastropancreatic fold which crosses the dorsal wall of the bursa. This fold is formed by a reflection of peritoneum covering the left gastric artery which courses (from its origin on the celiac trunk) to the left of the superior recess of the omental bursa to achieve its destination, the lesser curvature of the stomach.

3) exposure of the omental bursa in this manner reveals the caudate lobe of the liver which can be seen situated on the dorsal surface of the liver's right lobe as well as the anterior surface of the body of the pancreas coursing transversely behind the stomach.

4) the left lobe of the liver overlies the lesser curvature, the fundus and part of the body of the stomach. While the caudate lobe is situated to the right of the esophagus (not visible in this figure), the quadrate lobe, which lies between the fossa of the gall bladder and the round ligament, comes into contact with the pylorus and the first part of the duodenum.

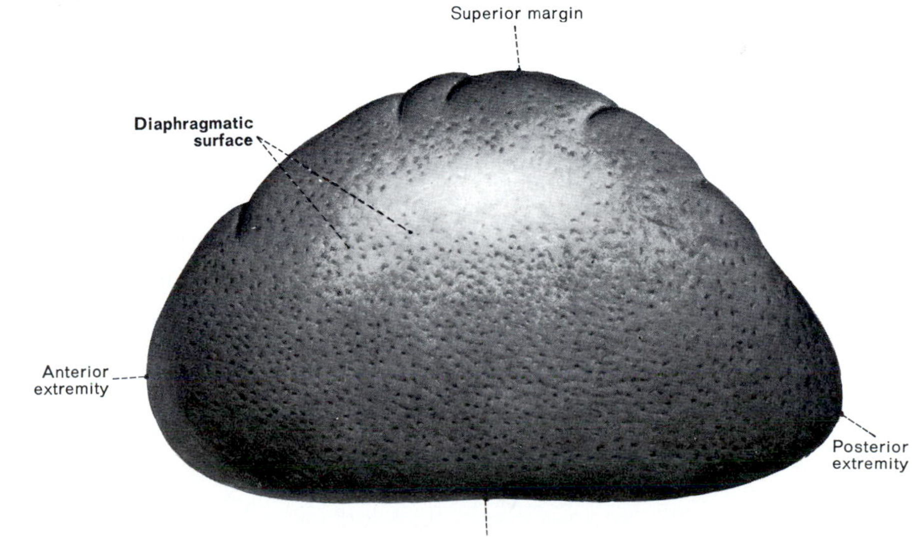

Superior margin
Diaphragmatic surface
Anterior extremity
Posterior extremity
Inferior margin

Fig. 276: The Spleen, Diaphragmatic Surface

NOTE: the diaphragmatic surface of the spleen is directed posterolaterally. It is smooth and convex and conforms to the concave abdominal surface of the adjacent diaphragm. Although the normal adult spleen may vary considerably in size from 100 grams to 400 grams, its proximity to the 9th, 10th and 11th ribs of the left side makes it vulnerable to costal fractures in this region.

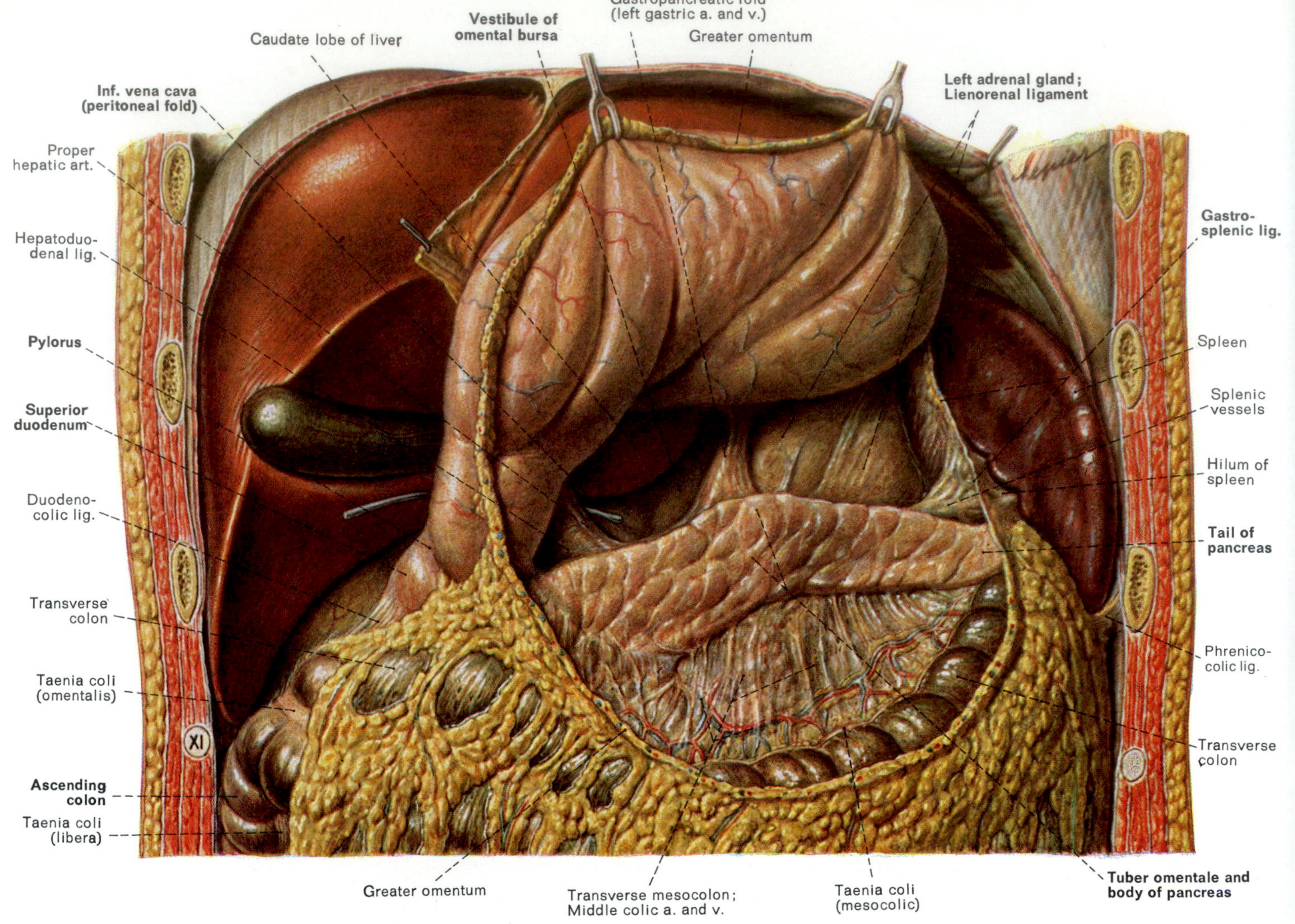
Gastropancreatic fold
(left gastric a. and v.)
Vestibule of
omental bursa
Caudate lobe of liver
Greater omentum
Inf. vena cava
(peritoneal fold)
Left adrenal gland;
Lienorenal ligament
Proper
hepatic art.
Gastro-
splenic lig.
Hepatoduo-
denal lig.
Pylorus
Spleen
Superior
duodenum
Splenic
vessels
Hilum of
spleen
Duodeno-
colic lig.
Tail of
pancreas
Transverse
colon
Phrenico-
colic lig.
Taenia coli
(omentalis)
XI
Transverse
colon
Ascending
colon
Taenia coli
(libera)
Greater omentum
Transverse mesocolon;
Middle colic a. and v.
Taenia coli
(mesocolic)
Tuber omentale and
body of pancreas

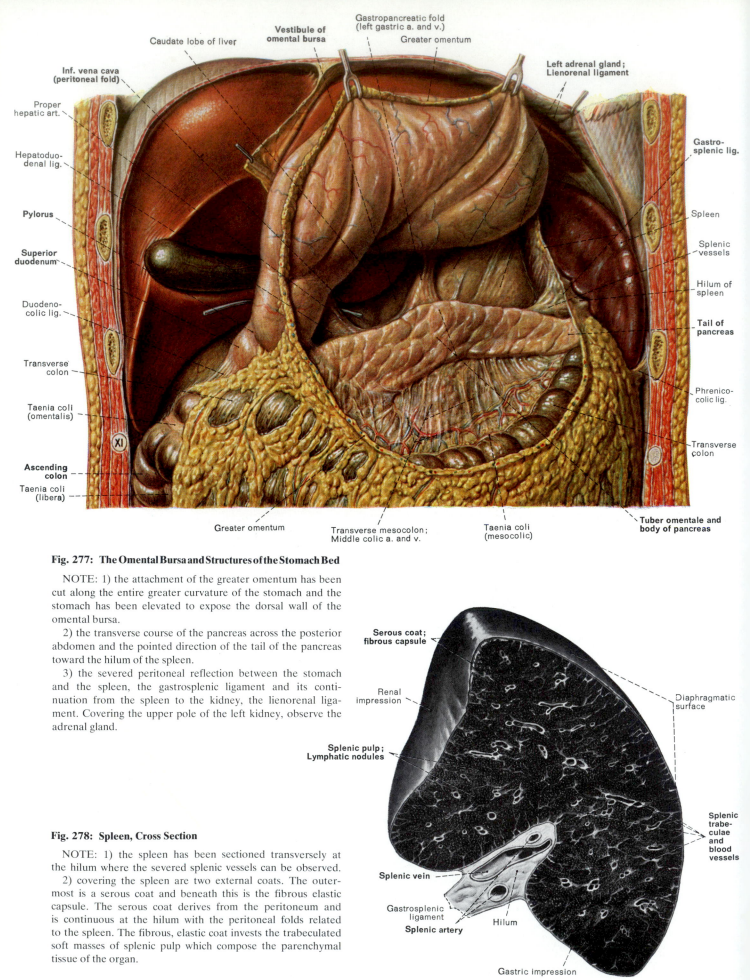

Fig. 277: The Omental Bursa and Structures of the Stomach Bed

NOTE: 1) the attachment of the greater omentum has been cut along the entire greater curvature of the stomach and the stomach has been elevated to expose the dorsal wall of the omental bursa.

2) the transverse course of the pancreas across the posterior abdomen and the pointed direction of the tail of the pancreas toward the hilum of the spleen.

3) the severed peritoneal reflection between the stomach and the spleen, the gastrosplenic ligament and its continuation from the spleen to the kidney, the lienorenal ligament. Covering the upper pole of the left kidney, observe the adrenal gland.

Serous coat;
fibrous capsule
Renal
impression
Diaphragmatic
surface
Splenic pulp;
Lymphatic nodules
Splenic
trabe-
culae
and
blood
vessels
Splenic vein
Gastrosplenic
ligament
Splenic artery
Hilum
Gastric impression

Fig. 278: Spleen, Cross Section

NOTE: 1) the spleen has been sectioned transversely at the hilum where the severed splenic vessels can be observed.

2) covering the spleen are two external coats. The outermost is a serous coat and beneath this is the fibrous elastic capsule. The serous coat derives from the peritoneum and is continuous at the hilum with the peritoneal folds related to the spleen. The fibrous, elastic coat invests the trabeculated soft masses of splenic pulp which compose the parenchymal tissue of the organ.

Left gastric lymph nodes
Left lobe of liver
Stomach
Hepatic lymph nodes
Right lobe of liver
Right gastric lymph nodes
Pyloric lymph nodes
Left gastroepiploic lymph nodes
Pancreas
Right gastroepiploic lymph nodes
Greater omentum

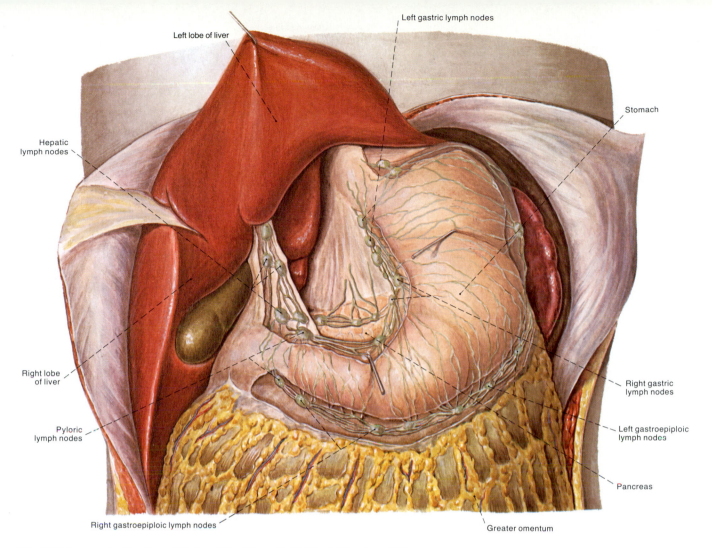

Fig. 279: Lymphatic Vessels and Nodes of the Stomach, the Porta Hepatis and Pancreas

NOTE that lymphatic vessels of the stomach follow the course of the right and left gastroepiploic vessels along the greater curvature and the right and left gastric vessels along the lesser curvature. Those in the porta hepatis course along the branches of the hepatic artery and, behind the stomach, nodes that drain the pancreas follow the course of the splenic artery. The efferent vessels from these nodes pass toward the preaortic nodes surrounding the celiac axis.

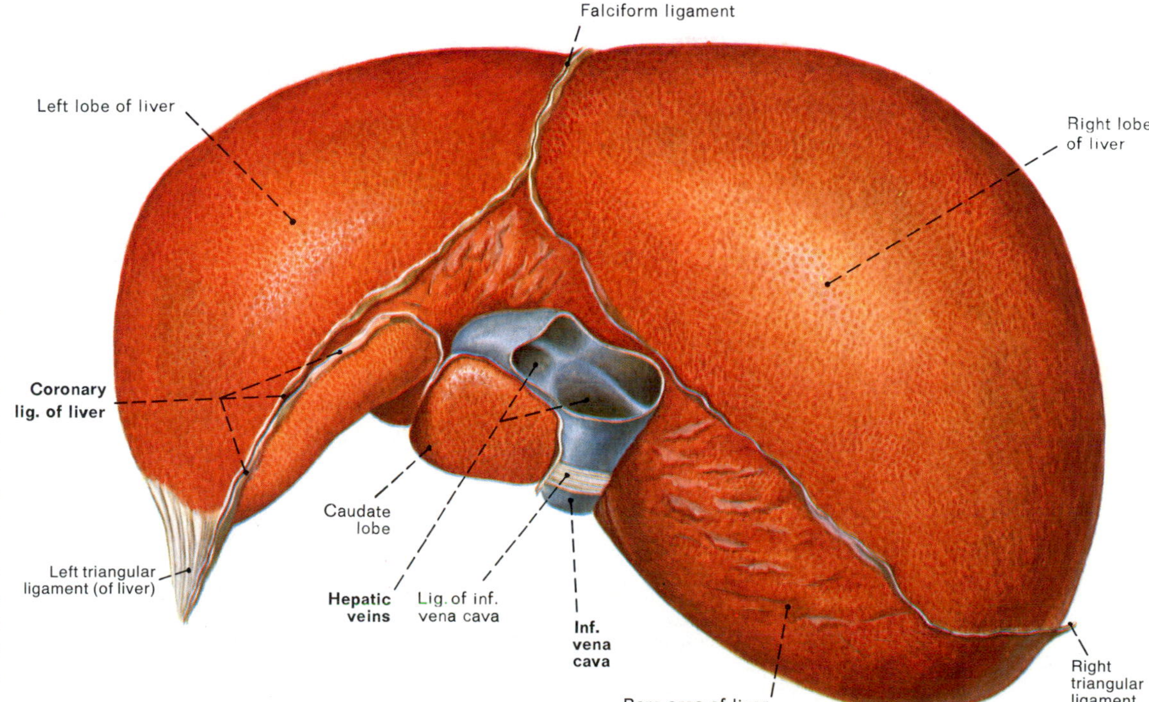

Falciform ligament
Left lobe of liver
Right lobe of liver
Coronary lig. of liver
Caudate lobe
Left triangular ligament (of liver)
Hepatic veins
Lig. of inf. vena cava
Inf. vena cava
Bare area of liver
Right triangular ligament

Fig. 280: Dorsocranial View of Liver

NOTE: 1) the visceral peritoneum closely adheres to the surface of the liver and is called the coronary ligament. Between the two leaves of the coronary ligament a portion of the liver is devoid of peritoneum. This is called the bare area of the liver and it is in direct contact with the diaphragm.

2) the hepatic veins as they converge superiorly from the liver lobes to empty into the inferior vena cava; this latter vessel lies in its sulcus on the posterior surface of the liver.

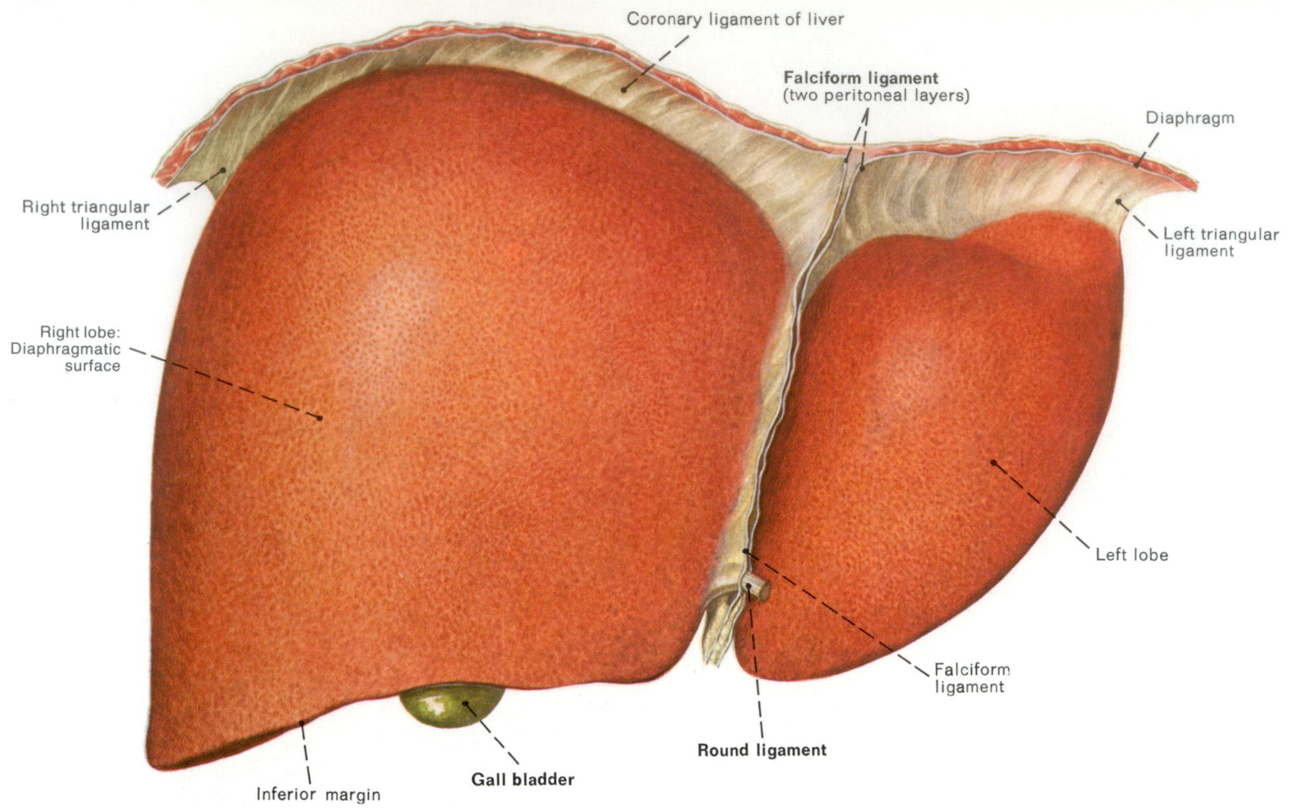

Coronary ligament of liver
Falciform ligament
(two peritoneal layers)
Diaphragm
Right triangular ligament
Left triangular ligament
Right lobe: Diaphragmatic surface
Left lobe
Falciform ligament
Round ligament
Gall bladder
Inferior margin

Fig. 281: Anterior Surface of the Liver (with Diaphragmatic Attachment)

NOTE: 1) the falciform ligament (derived from the ventral mesogastrium) separates the large right from the smaller left lobe of the liver. It contains the fibrous cord called the round ligament of the liver which is the resultant structure from the obliteration of the umbilical vein.

2) the fundus of the gall bladder extending below the sharply angled hepatic inferior margin.

Fig. 282: Posterior (Visceral) Surface of Liver and the Gall Bladder

NOTE: 1) the impressions made by the esophagus and stomach on the left lobe of the liver and the right kidney, right suprarenal gland, duodenum and transverse colon on the right lobe.

2) the sulcus formed by the inferior vena cava, which subdivides the caudate lobe from the right lobe. The gall bladder along with the portal vein, hepatic artery and common bile duct bound the quadrate lobe.

3) the continuity of the round ligament (umbilical vein) with the ligamentum venosum (ductus venosus).

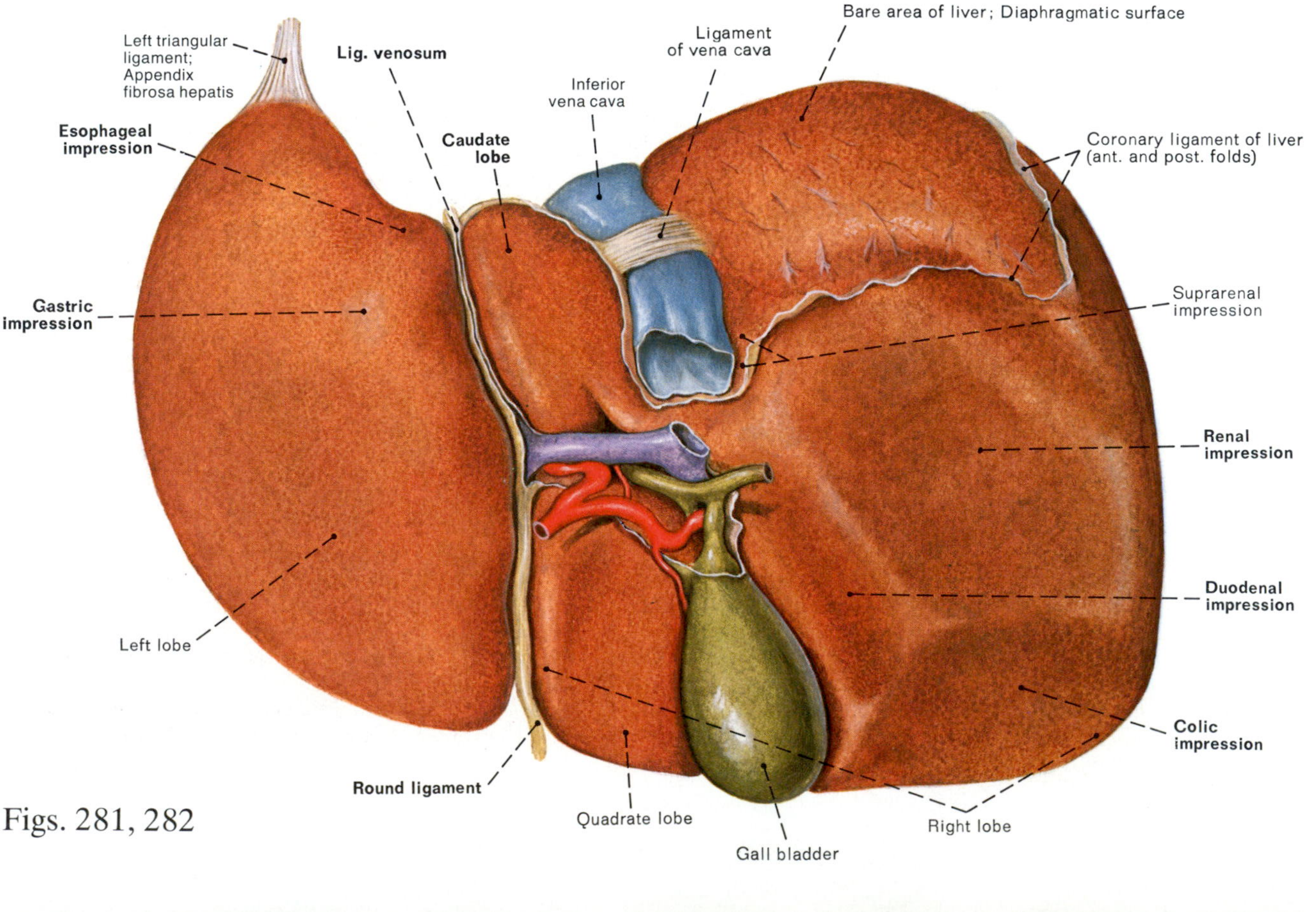

Bare area of liver; Diaphragmatic surface
Left triangular ligament; Appendix fibrosa hepatis
Lig. venosum
Ligament of vena cava
Inferior vena cava
Caudate lobe
Esophageal impression
Coronary ligament of liver (ant. and post. folds)
Gastric impression
Suprarenal impression
Renal impression
Duodenal impression
Left lobe
Colic impression
Round ligament
Quadrate lobe
Right lobe
Gall bladder

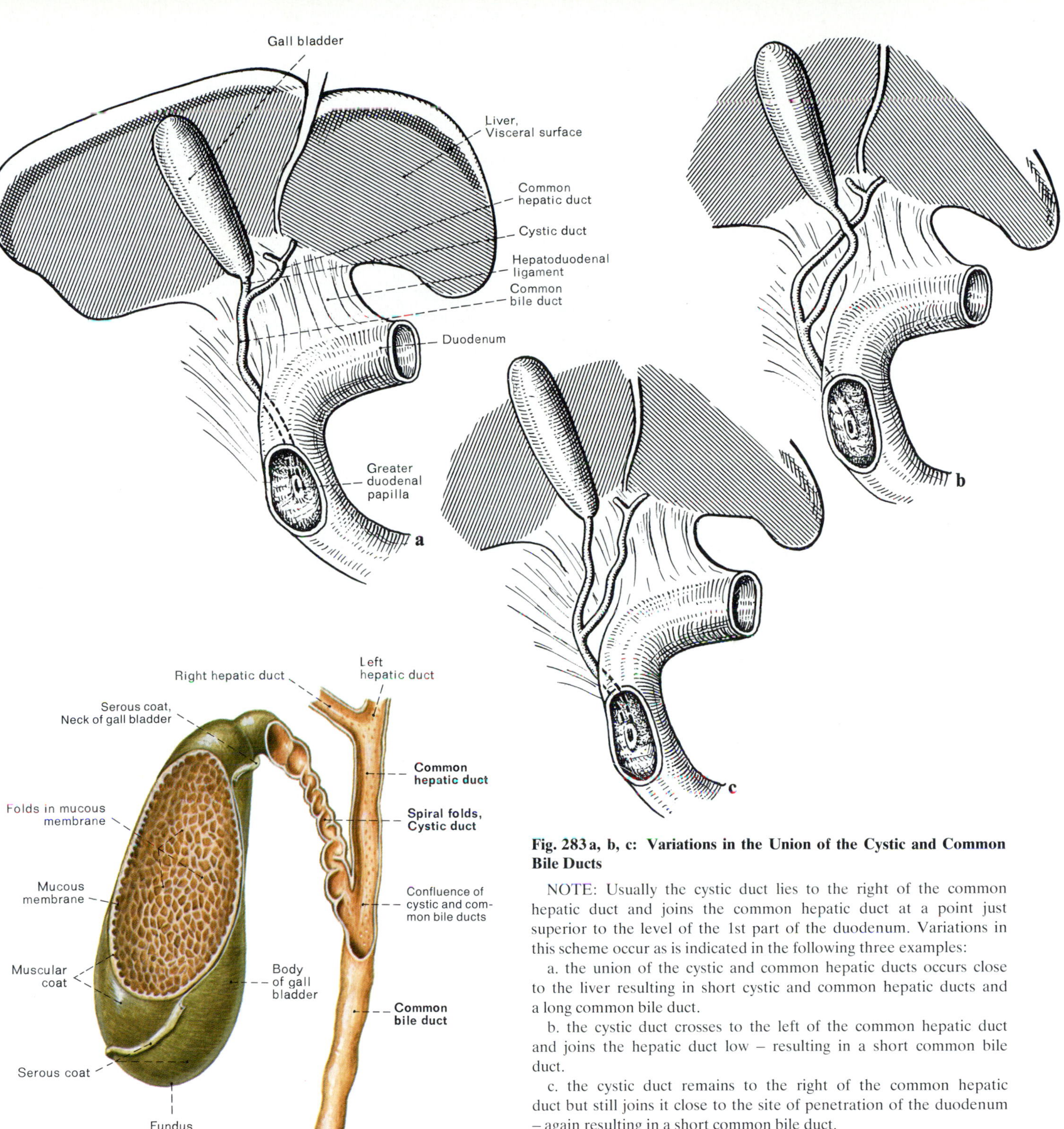

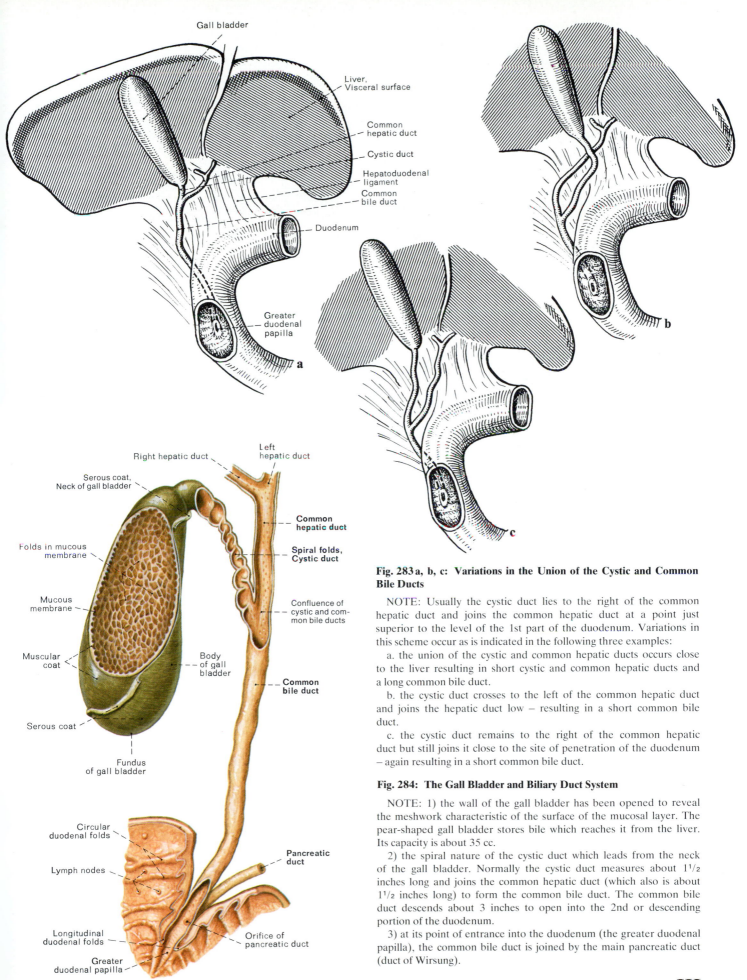

Gall bladder
Liver,
Visceral surface
Common
hepatic duct
Cystic duct
Hepatoduodenal
ligament
Common
bile duct
Duodenum
Greater
duodenal
papilla
a
b
c
Right hepatic duct
Left
hepatic duct
Serous coat,
Neck of gall bladder
Common
hepatic duct
Folds in mucous
membrane
Spiral folds,
Cystic duct
Mucous
membrane
Confluence of
cystic and com-
mon bile ducts
Muscular
coat
Body
of gall
bladder
Common
bile duct
Serous coat
Fundus
of gall bladder
Circular
duodenal folds
Pancreatic
duct
Lymph nodes
Longitudinal
duodenal folds
Orifice of
pancreatic duct
Greater
duodenal papilla

Fig. 283 a, b, c: Variations in the Union of the Cystic and Common Bile Ducts

NOTE: Usually the cystic duct lies to the right of the common hepatic duct and joins the common hepatic duct at a point just superior to the level of the 1st part of the duodenum. Variations in this scheme occur as is indicated in the following three examples:

a. the union of the cystic and common hepatic ducts occurs close to the liver resulting in short cystic and common hepatic ducts and a long common bile duct.

b. the cystic duct crosses to the left of the common hepatic duct and joins the hepatic duct low – resulting in a short common bile duct.

c. the cystic duct remains to the right of the common hepatic duct but still joins it close to the site of penetration of the duodenum – again resulting in a short common bile duct.

Fig. 284: The Gall Bladder and Biliary Duct System

NOTE: 1) the wall of the gall bladder has been opened to reveal the meshwork characteristic of the surface of the mucosal layer. The pear-shaped gall bladder stores bile which reaches it from the liver. Its capacity is about 35 cc.

2) the spiral nature of the cystic duct which leads from the neck of the gall bladder. Normally the cystic duct measures about 1½ inches long and joins the common hepatic duct (which also is about 1½ inches long) to form the common bile duct. The common bile duct descends about 3 inches to open into the 2nd or descending portion of the duodenum.

3) at its point of entrance into the duodenum (the greater duodenal papilla), the common bile duct is joined by the main pancreatic duct (duct of Wirsung).

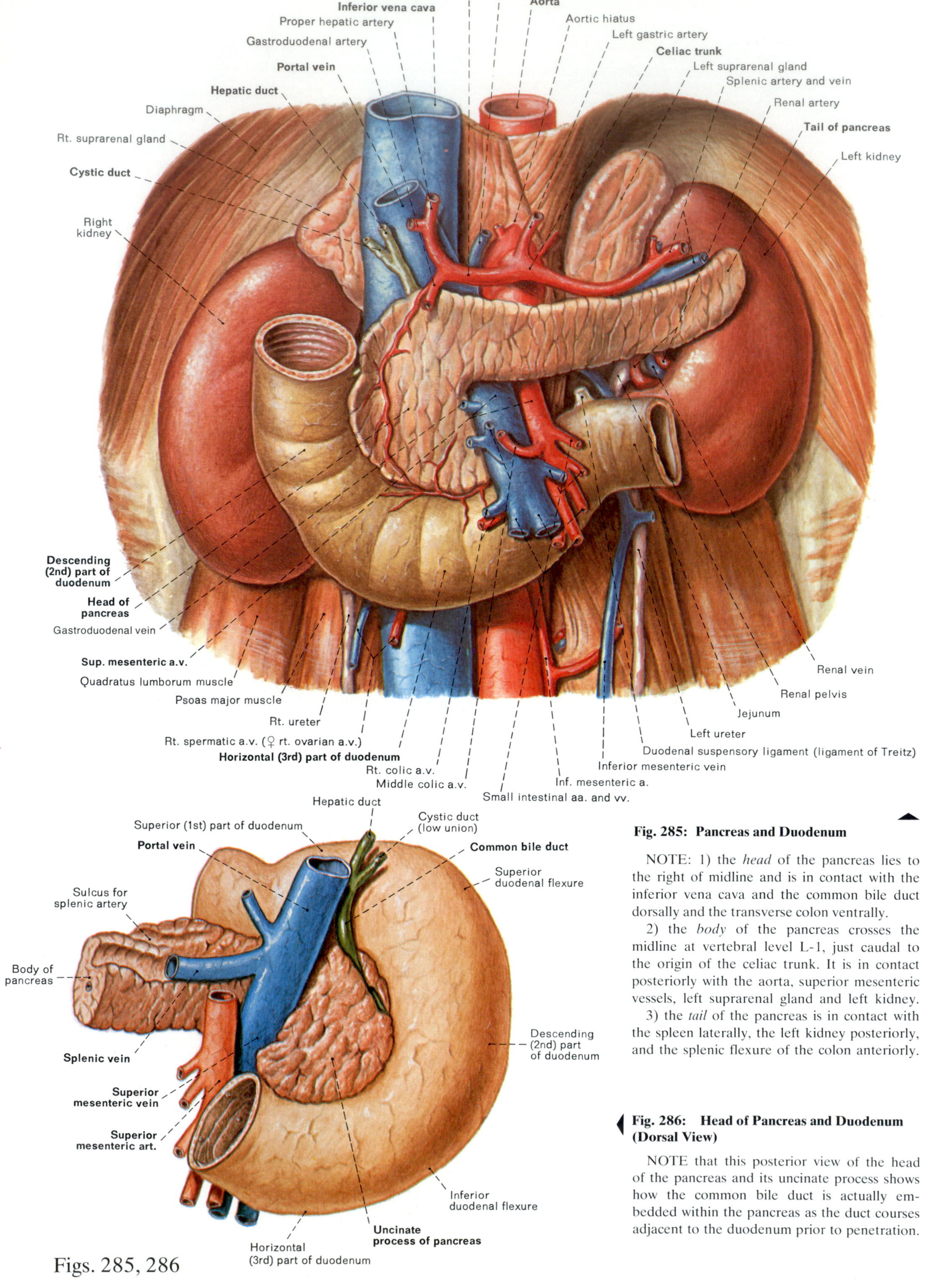

Common hepatic artery
Gastric (coronary) vein
Inferior vena cava
Aorta
Proper hepatic artery
Aortic hiatus
Gastroduodenal artery
Left gastric artery
Celiac trunk
Portal vein
Left suprarenal gland
Splenic artery and vein
Hepatic duct
Diaphragm
Renal artery
Tail of pancreas
Rt. suprarenal gland
Left kidney
Cystic duct
Right kidney
Descending (2nd) part of duodenum
Head of pancreas
Gastroduodenal vein
Sup. mesenteric a.v.
Quadratus lumborum muscle
Psoas major muscle
Rt. ureter
Rt. spermatic a.v. (♀ rt. ovarian a.v.)
Horizontal (3rd) part of duodenum
Rt. colic a.v.
Middle colic a.v.
Small intestinal aa. and vv.
Inf. mesenteric a.
Inferior mesenteric vein
Duodenal suspensory ligament (ligament of Treitz)
Left ureter
Jejunum
Renal pelvis
Renal vein
Hepatic duct
Superior (1st) part of duodenum
Cystic duct (low union)
Portal vein
Common bile duct
Superior duodenal flexure
Sulcus for splenic artery
Body of pancreas
Splenic vein
Descending (2nd) part of duodenum
Superior mesenteric vein
Superior mesenteric art.
Inferior duodenal flexure
Uncinate process of pancreas
Horizontal (3rd) part of duodenum

Fig. 285: Pancreas and Duodenum

NOTE: 1) the *head* of the pancreas lies to the right of midline and is in contact with the inferior vena cava and the common bile duct dorsally and the transverse colon ventrally.

2) the *body* of the pancreas crosses the midline at vertebral level L-1, just caudal to the origin of the celiac trunk. It is in contact posteriorly with the aorta, superior mesenteric vessels, left suprarenal gland and left kidney.

3) the *tail* of the pancreas is in contact with the spleen laterally, the left kidney posteriorly, and the splenic flexure of the colon anteriorly.

Fig. 286: Head of Pancreas and Duodenum (Dorsal View)

NOTE that this posterior view of the head of the pancreas and its uncinate process shows how the common bile duct is actually embedded within the pancreas as the duct courses adjacent to the duodenum prior to penetration.

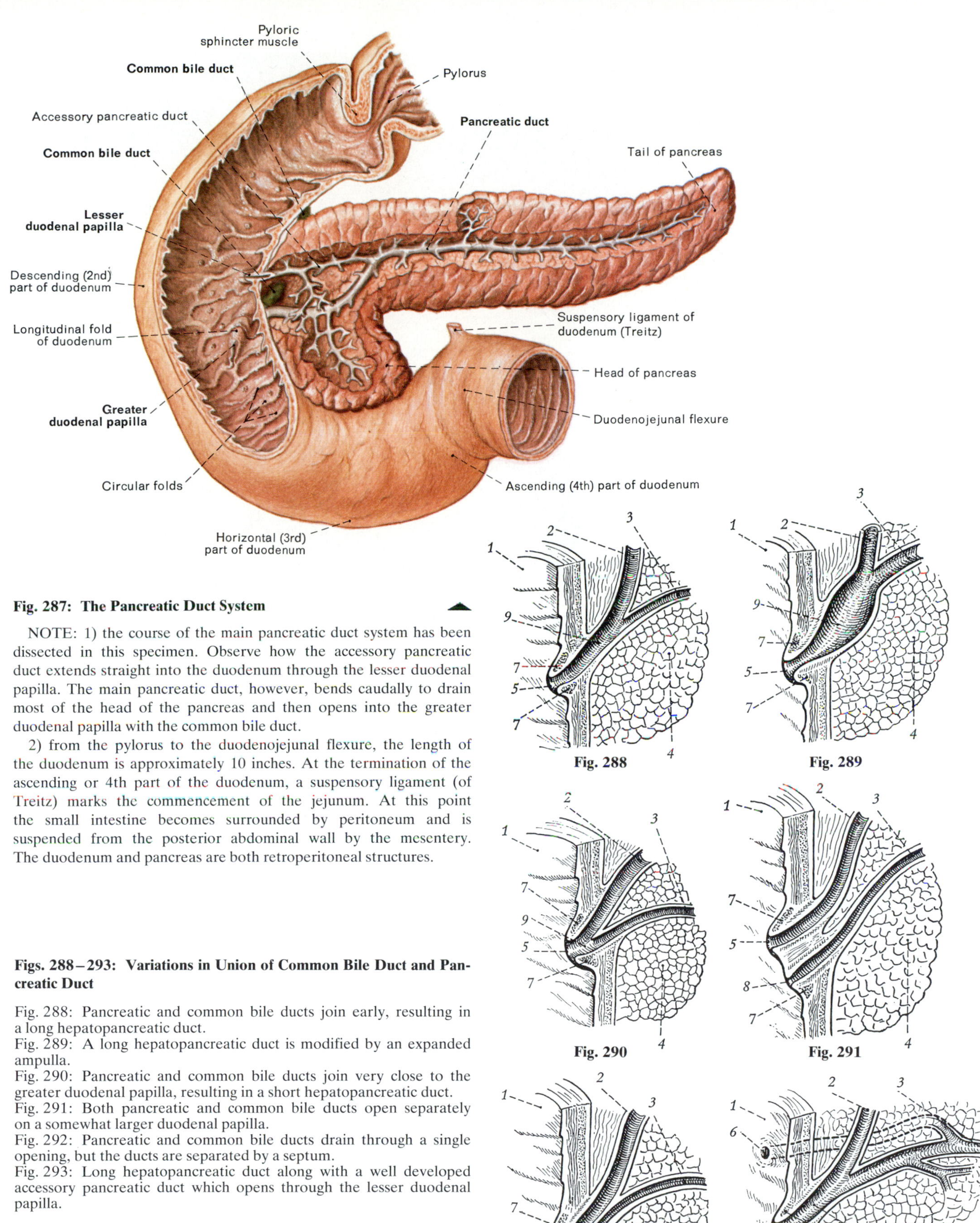

Pyloric sphincter muscle
Common bile duct
Pylorus
Accessory pancreatic duct
Pancreatic duct
Common bile duct
Tail of pancreas
Lesser duodenal papilla
Descending (2nd) part of duodenum
Longitudinal fold of duodenum
Suspensory ligament of duodenum (Treitz)
Head of pancreas
Greater duodenal papilla
Duodenojejunal flexure
Circular folds
Ascending (4th) part of duodenum
Horizontal (3rd) part of duodenum

Fig. 287: The Pancreatic Duct System

NOTE: 1) the course of the main pancreatic duct system has been dissected in this specimen. Observe how the accessory pancreatic duct extends straight into the duodenum through the lesser duodenal papilla. The main pancreatic duct, however, bends caudally to drain most of the head of the pancreas and then opens into the greater duodenal papilla with the common bile duct.

2) from the pylorus to the duodenojejunal flexure, the length of the duodenum is approximately 10 inches. At the termination of the ascending or 4th part of the duodenum, a suspensory ligament (of Treitz) marks the commencement of the jejunum. At this point the small intestine becomes surrounded by peritoneum and is suspended from the posterior abdominal wall by the mesentery. The duodenum and pancreas are both retroperitoneal structures.

Fig. 288 Fig. 289

Fig. 290 Fig. 291

Fig. 292 Fig. 293

Figs. 288–293: Variations in Union of Common Bile Duct and Pancreatic Duct

Fig. 288: Pancreatic and common bile ducts join early, resulting in a long hepatopancreatic duct.
Fig. 289: A long hepatopancreatic duct is modified by an expanded ampulla.
Fig. 290: Pancreatic and common bile ducts join very close to the greater duodenal papilla, resulting in a short hepatopancreatic duct.
Fig. 291: Both pancreatic and common bile ducts open separately on a somewhat larger duodenal papilla.
Fig. 292: Pancreatic and common bile ducts drain through a single opening, but the ducts are separated by a septum.
Fig. 293: Long hepatopancreatic duct along with a well developed accessory pancreatic duct which opens through the lesser duodenal papilla.

1 Duodenum
2 Common bile duct
3 Pancreatic duct (Wirsung)
4 Pancreas
5 Greater duodenal papilla
6 Accessory pancreatic duct (Santorini)
7 Sphincter (Oddi) at duodenal papilla
8 Pancreatic duct separate opening
9 Hepatopancreatic duct

Fig. 294: X-Ray of the Upper Gastrointestinal Tract.

NOTE: 1) that this figure is a positive print made from an X-ray film and, thus, the contrast medium appears black instead of the conventional white.

2) the sites of junction of the tubular organs comprising the upper gastrointestinal tract. The esophageal orifice at the cardiac end of the stomach (*1 a*) is closed. Observe that this is at the T-11 vertebral level.

3) that the pylorus of the stomach (not numbered) is the narrowed region leading into the superior part of the duodenum (*4*). The superior (*4*), descending (*5 a*), horizontal (*5 b*) and ascending (*5 c*) parts of the duodenum are easily recognizable, as is the duodenojejunal junction (*5 d*).

4) that the contrast medium outlines the internal circular folds or plicae of the jejunum (*6*) and also identifies a peristaltic contraction wave (*7 a, 7 b*).

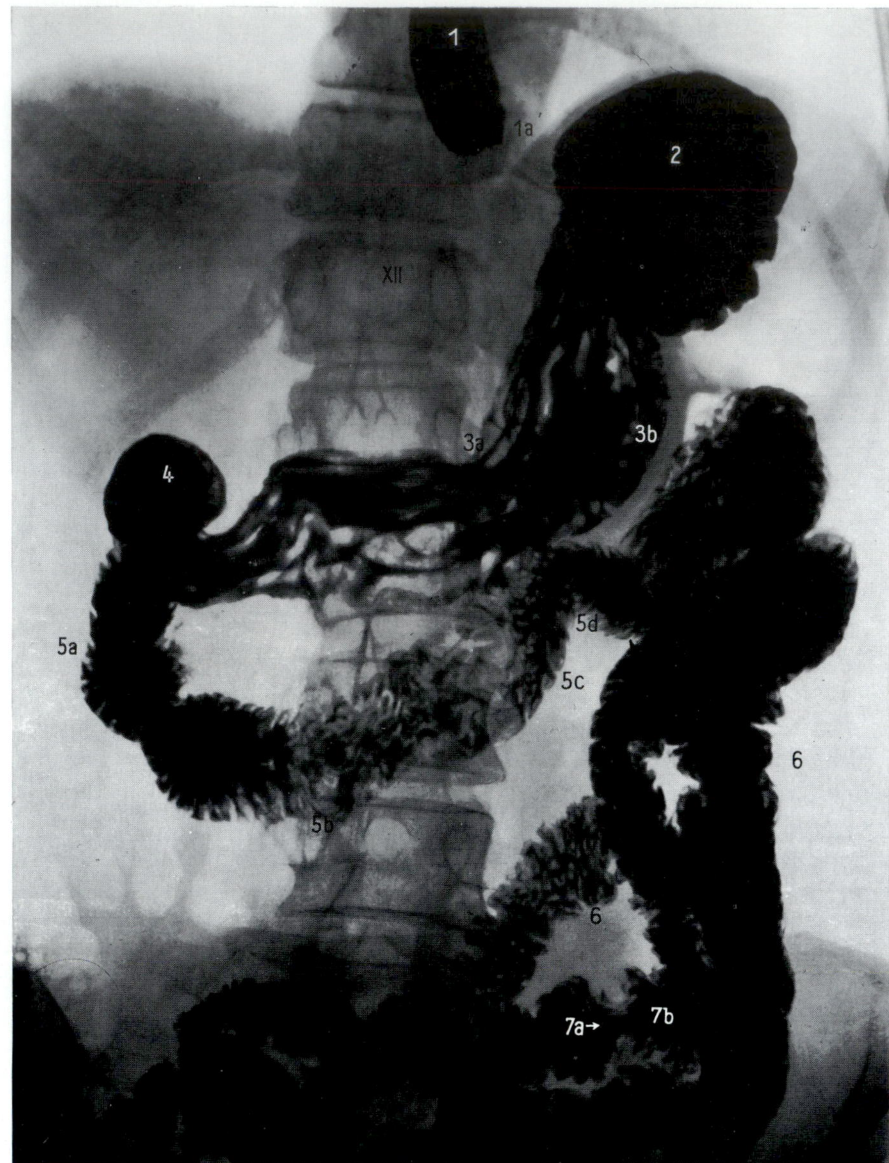
1
1a
2
XII
3a
3b
4
5a
5d
5c
6
5b
6
7a→
7b

1 Esophagus. 1a. Cardia closed, contrast medium in creases of mucous membrane
2 Fundus of stomach (filled)
3a Lesser curvature of stomach with longitudinal mucosal folds
3b Greater curvature with gastric folds and areas
4 Duodenal bulb, superior part of duodenum
5a Descending part of duodenum
5b Horizontal part of duodenum
5c Ascending part of duodenum
5d Duodenojejunal flexure
6 Jejunum with circular plicae
7 Peristaltic contraction wave from 7a to 7b

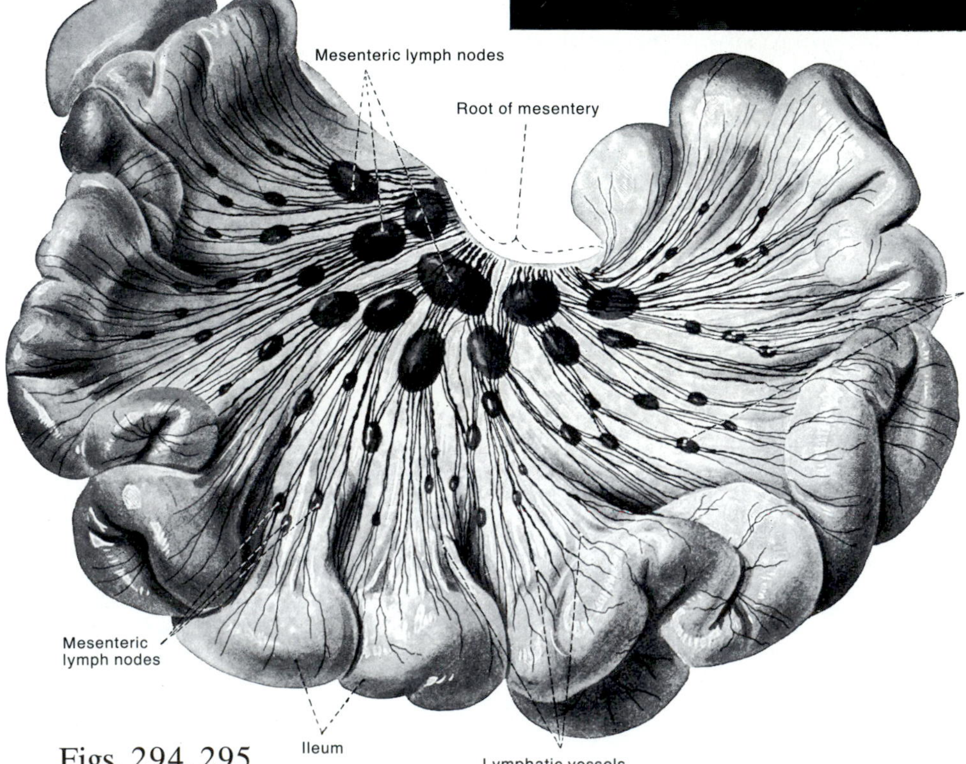
Mesenteric lymph nodes
Root of mesentery
Mesenteric lymph nodes
Mesenteric lymph nodes
Ileum
Lymphatic vessels

Fig. 295: Lymphatic Vessels and Nodes of the Ileum.

NOTE that this loop of ileum, injected with India ink, demonstrates the radial orientation of the lymphatic vessels from the organ to the root of the mesentery. Along the paths of these vessels are found the mesenteric lymph nodes.

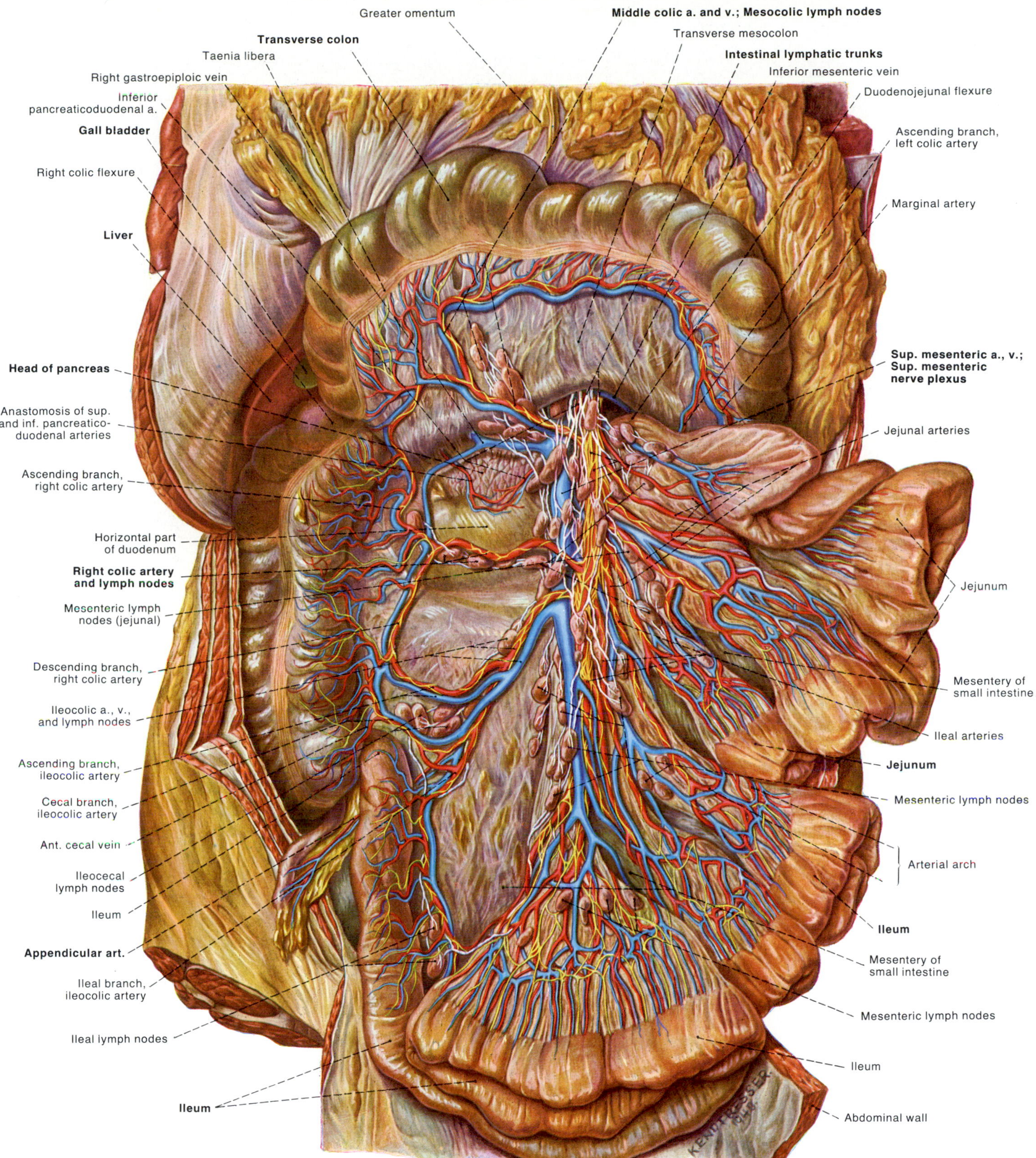

Greater omentum
Middle colic a. and v.; Mesocolic lymph nodes
Transverse mesocolon
Transverse colon
Intestinal lymphatic trunks
Taenia libera
Inferior mesenteric vein
Right gastroepiploic vein
Inferior pancreaticoduodenal a.
Duodenojejunal flexure
Gall bladder
Ascending branch, left colic artery
Right colic flexure
Marginal artery
Liver
Head of pancreas
Sup. mesenteric a., v.; Sup. mesenteric nerve plexus
Anastomosis of sup. and inf. pancreatico-duodenal arteries
Jejunal arteries
Ascending branch, right colic artery
Horizontal part of duodenum
Right colic artery and lymph nodes
Jejunum
Mesenteric lymph nodes (jejunal)
Descending branch, right colic artery
Mesentery of small intestine
Ileocolic a., v., and lymph nodes
Ileal arteries
Ascending branch, ileocolic artery
Jejunum
Cecal branch, ileocolic artery
Mesenteric lymph nodes
Ant. cecal vein
Arterial arch
Ileocecal lymph nodes
Ileum
Ileum
Appendicular art.
Mesentery of small intestine
Ileal branch, ileocolic artery
Mesenteric lymph nodes
Ileal lymph nodes
Ileum
Ileum
Abdominal wall

Fig. 296: Abdominal Cavity (5); Lymph Nodes, Vessels and Nerves Serving the Jejunum, Ileum, Ascending and Transverse Colon

NOTE that: 1) the transverse colon has been lifted to reveal its mesocolon along with the retroperitoneal head of the pancreas and duodenum. The loops of jejunum and ileum have been reflected to the left to show the root of the mesentery with its complex network of lymphatics, blood vessels and autonomic nerves.

2) chains of mesenteric lymph nodes follow the ileal and jejunal branches of the superior mesenteric artery and vein, as well as the ileocolic, right colic and middle colic branches of the same vessel.

3) the lymphatics from both the small and large intestine drain centrally and superiorly, eventually to interconnect with pre-aortic chains of nodes.

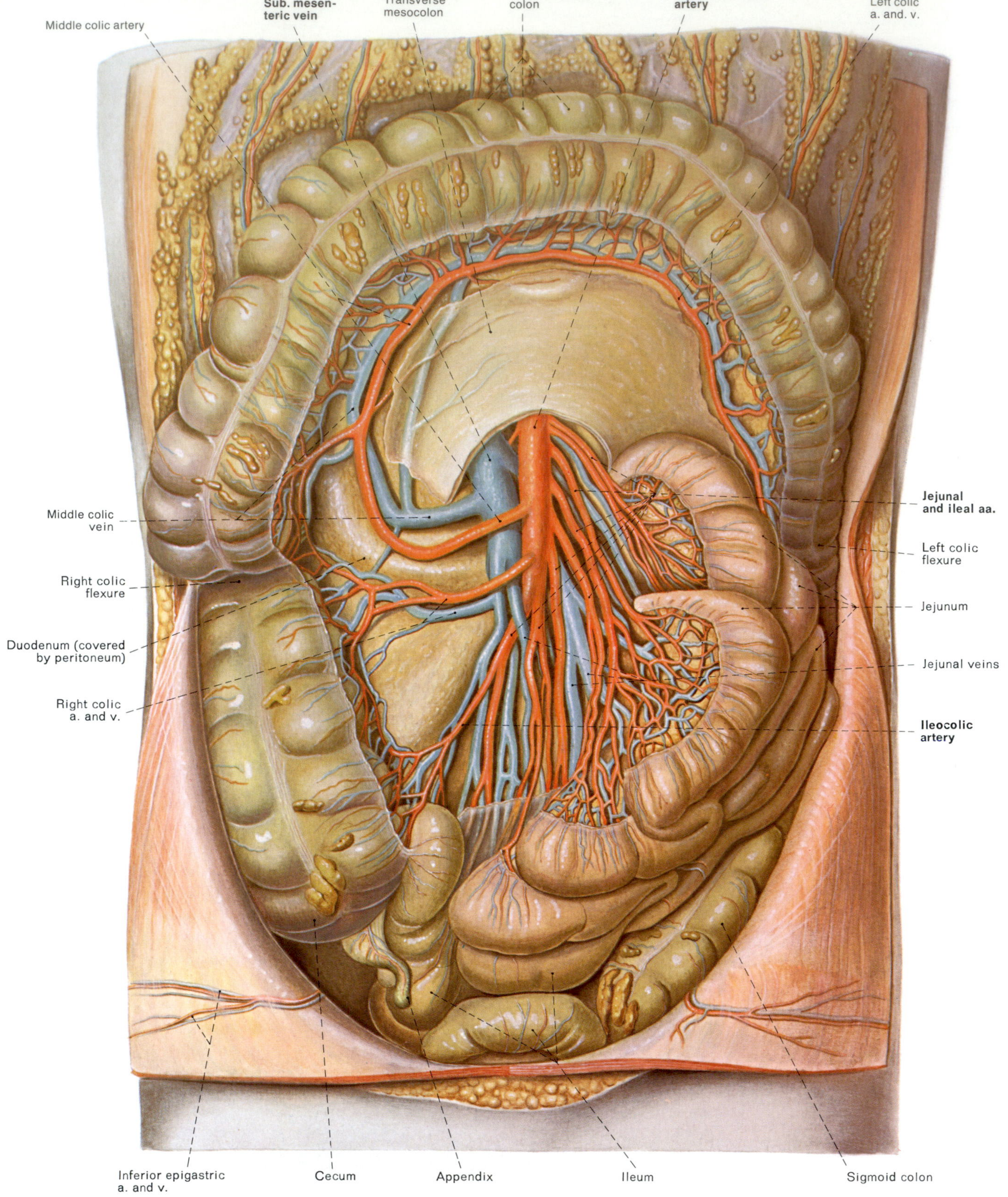
Middle colic artery
Sub. mesenteric vein
Transverse mesocolon
Transverse colon
Sub. mesenteric artery
Left colic a. and v.
Middle colic vein
Right colic flexure
Duodenum (covered by peritoneum)
Right colic a. and v.
Jejunal and ileal aa.
Left colic flexure
Jejunum
Jejunal veins
Ileocolic artery
Inferior epigastric a. and v.
Cecum
Appendix
Ileum
Sigmoid colon

Fig. 297: Abdominal Cavity (6): Superior Mesenteric Vessels and Branches

NOTE: 1) the transverse colon has been turned upward and the small intestine was pushed to the left. The peritoneal attachment along the coils of small intestine and colon has been dissected to reveal the branches of the superior mesenteric vessels which supply the small intestine, ascending colon and transverse colon.

2) the intestinal arteries (jejunal and ileal) branch from the left side of the superior mesenteric artery. There are about 12 of these vessels and they are distributed to the jejunum and ileum.

3) branching from the right side of the superior mesenteric artery are the ileocolic (to the ileocecal region), right colic (to the ascending colon) and middle colic (transverse colon) arteries. Observe the elaborate branching of these vessels and the rich anastomoses which exist among them.

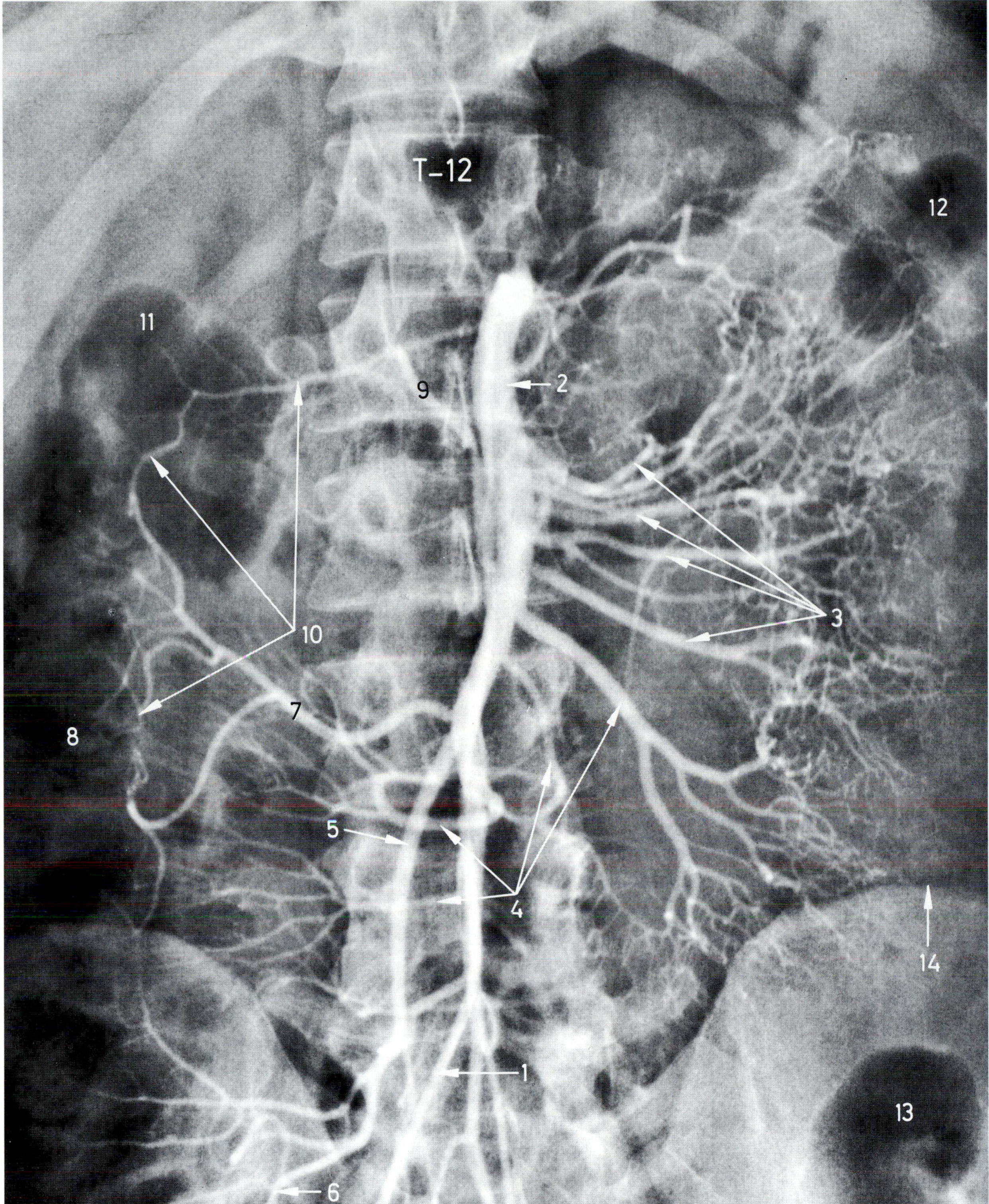
T-12
12
11
9
2
3
10
7
8
5
4
14
1
13
6

Fig. 298: Superior Mesenteric Arteriogram

NOTE: 1) that a catheter (*1*) has been inserted into the common iliac artery and through the abdominal aorta to the point of branching of the superior mesenteric artery (*2*). Contrast medium was injected in order to allow visualization of the principal branches of that vessel. The original radiograph is shown as a negative print.

2) the jejunal (*3*) and ileal (*4*) arteries branching as a sequence of vessels (about fifteen in number) which supply all of the small intestine beyond the duodenum.

3) the ileocolic artery (*5*) and its appendicular branch (*6*), the right colic (*7*), and middle colic (*9*) arteries which supply the cecum, ascending colon (*8*) and transverse colon. Observe that anastomoses among these vessels close to the colon contribute to the formation of the marginal artery (*10*).

4) that other structures can be identified for orientation. These are the right colic flexure (*11*), the left colic flexure (*12*), the sigmoid colon (*13*), the body of the T-12 vertebra and the iliac crest (*14*).

5) that this figure comes from Wicke, L., *Atlas of Radiologic Anatomy,* 3rd Edition, Urban & Schwarzenberg, Baltimore, 1982.

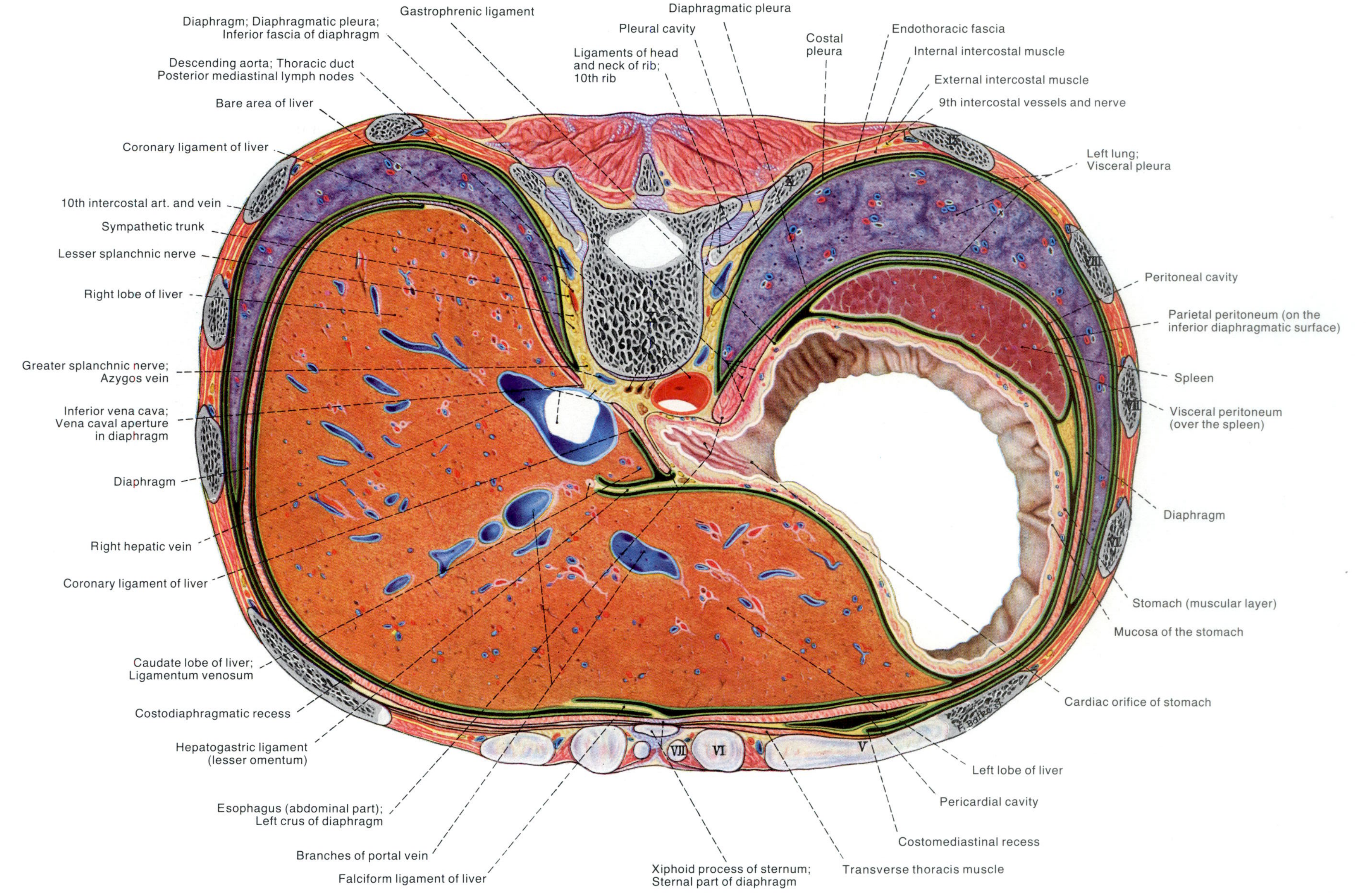

Diaphragm; Diaphragmatic pleura; Inferior fascia of diaphragm
Gastrophrenic ligament
Diaphragmatic pleura
Pleural cavity
Ligaments of head and neck of rib; 10th rib
Costal pleura
Endothoracic fascia
Internal intercostal muscle
External intercostal muscle
9th intercostal vessels and nerve
Descending aorta; Thoracic duct Posterior mediastinal lymph nodes
Bare area of liver
Coronary ligament of liver
Left lung; Visceral pleura
10th intercostal art. and vein
Sympathetic trunk
Lesser splanchnic nerve
Right lobe of liver
Peritoneal cavity
Parietal peritoneum (on the inferior diaphragmatic surface)
Greater splanchnic nerve; Azygos vein
Spleen
Inferior vena cava; Vena caval aperture in diaphragm
Visceral peritoneum (over the spleen)
Diaphragm
Diaphragm
Right hepatic vein
Coronary ligament of liver
Stomach (muscular layer)
Mucosa of the stomach
Caudate lobe of liver; Ligamentum venosum
Costodiaphragmatic recess
Cardiac orifice of stomach
Hepatogastric ligament (lesser omentum)
Left lobe of liver
Esophagus (abdominal part); Left crus of diaphragm
Pericardial cavity
Costomediastinal recess
Branches of portal vein
Falciform ligament of liver
Xiphoid process of sternum; Sternal part of diaphragm
Transverse thoracis muscle
VIII
VII
VI
V

Fig. 299: Transverse Section Through the Upper Abdomen at the Level of the Tenth Thoracic Vertebra

NOTE: 1) that a cross-section through the body of the 10th thoracic vertebra also cuts through the lower part of the thoracic cavity on each side and, therefore, includes the inferior portions of the lower lobes of both lungs (seen in purple). Observe the visceral pleura, closely adherent to the lungs, and the costal and diaphragmatic reflections of parietal pleura. Between the parietal and visceral layers of pleura is the pleural cavity, which extends inferolaterally as the costodiaphragmatic recess (labelled on the specimen's right).

2) that the liver, stomach, and spleen occupy the largest part of the abdominal cavity at this vertebral level. Observe that the esophageal aperture in the right crus of the diaphragm is located at T-10. Through this aperture course the vagal nerve trunks, as well as some esophageal branches of the left gastric vessels.

3) that the descending thoracic aorta, still in the posterior mediastinum, is slightly to the left of the midline and ventral to the T-10 vertebral body. Adjacent to it are found some lymph nodes and the thoracic duct. To the right are found the inferior vena cava, the lower part of the vena caval aperture, and the part of the posterior surface of the liver not covered by peritoneum and called the bare area of the liver. (Note that sections of the 5th to 10th ribs are labelled V to X.)

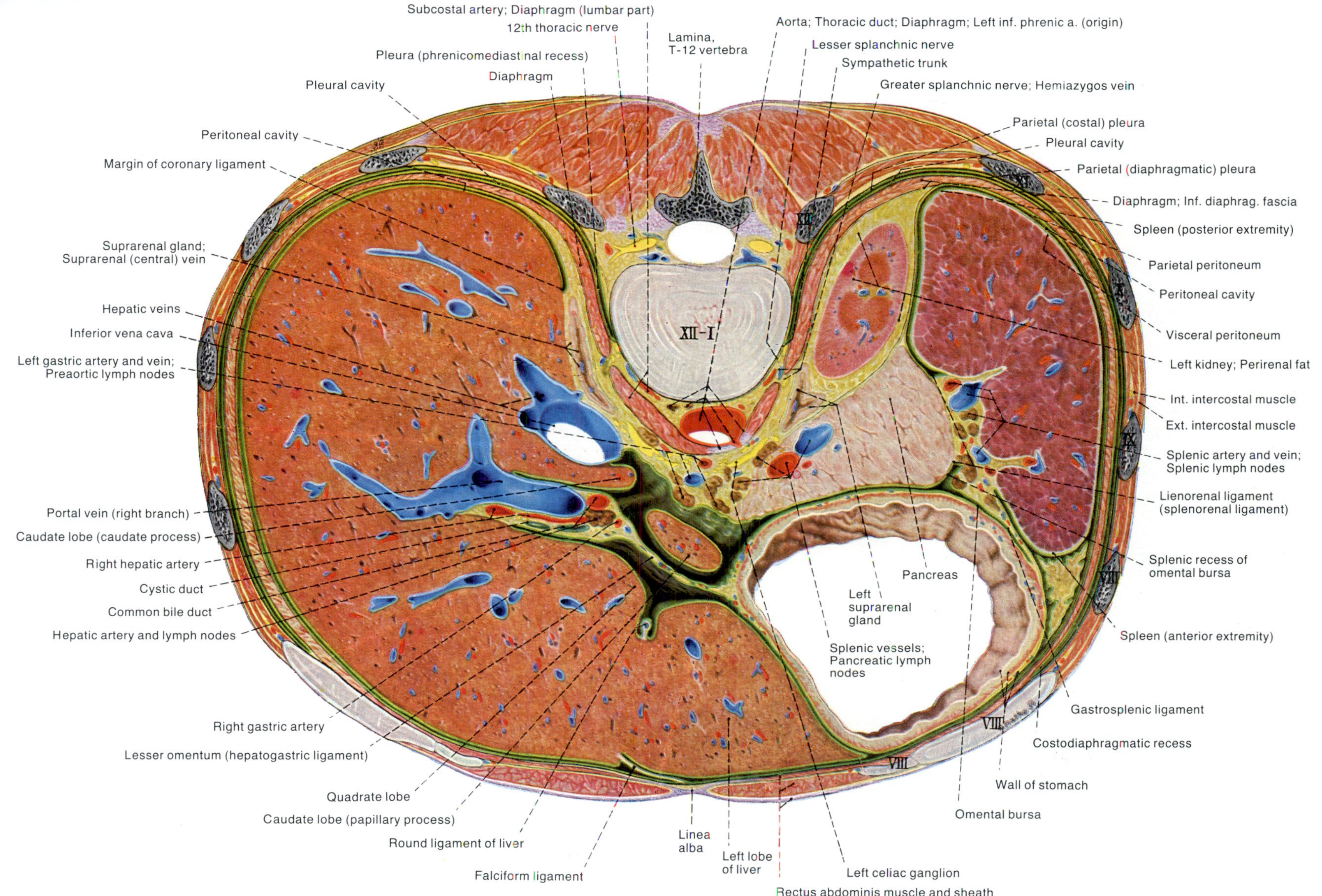
Subcostal artery; Diaphragm (lumbar part)
12th thoracic nerve
Pleura (phrenicomediastinal recess)
Diaphragm
Pleural cavity
Peritoneal cavity
Margin of coronary ligament
Suprarenal gland; Suprarenal (central) vein
Hepatic veins
Inferior vena cava
Left gastric artery and vein; Preaortic lymph nodes
Portal vein (right branch)
Caudate lobe (caudate process)
Right hepatic artery
Cystic duct
Common bile duct
Hepatic artery and lymph nodes
Right gastric artery
Lesser omentum (hepatogastric ligament)
Quadrate lobe
Caudate lobe (papillary process)
Round ligament of liver
Falciform ligament
Linea alba
Left lobe of liver
Rectus abdominis muscle and sheath
Left celiac ganglion
Lamina, T-12 vertebra
XII–I
Aorta; Thoracic duct; Diaphragm; Left inf. phrenic a. (origin)
Lesser splanchnic nerve
Sympathetic trunk
Greater splanchnic nerve; Hemiazygos vein
Parietal (costal) pleura
Pleural cavity
Parietal (diaphragmatic) pleura
Diaphragm; Inf. diaphrag. fascia
Spleen (posterior extremity)
Parietal peritoneum
Peritoneal cavity
Visceral peritoneum
Left kidney; Perirenal fat
Int. intercostal muscle
Ext. intercostal muscle
Splenic artery and vein; Splenic lymph nodes
Lienorenal ligament (splenorenal ligament)
Splenic recess of omental bursa
Spleen (anterior extremity)
Pancreas
Left suprarenal gland
Splenic vessels; Pancreatic lymph nodes
Gastrosplenic ligament
Costodiaphragmatic recess
Wall of stomach
Omental bursa
XI
XII
X
IX
VIII

Fig. 300: Transverse Section Through the Upper Abdomen at the Level of the Intervertebral Disc Between the Twelfth Thoracic and First Lumbar Vertebrae

NOTE: 1) that a cross-section of the abdomen between T-12 and L-1 goes through the hilum of the liver, the body of the stomach, the pancreas, the spleen and its hilar vessels, the upper pole of the left kidney, and the two suprarenal glands;

2) the quadrate and caudate lobes of the liver (see also Fig. 282). The *quadrate lobe,* located on the posteroinferior hepatic surface, is limited on the left by a fissure within which is found the round ligament of the liver and posteriorly by the hilar structures of the porta hepatis. The *caudate lobe,* located more superiorly on the posterior hepatic surface, is represented here by its caudate and papillary processes;

3) the hepatic artery, portal vein, the cystic duct, and the common bile duct. Also observe the inferior vena cava (into which flow the hepatic veins), the aorta, and the left gastric and splenic vessels;

4) the relationship of the pancreas to the stomach, spleen, left kidney, left suprarenal gland, the left crus of the diaphragm, and the aorta. Observe the many lymph nodes anterior to the aorta, at the hilum of the spleen, and at the porta hepatis. (Note that sections of the 8th to 12th ribs are labelled VIII to XII.)

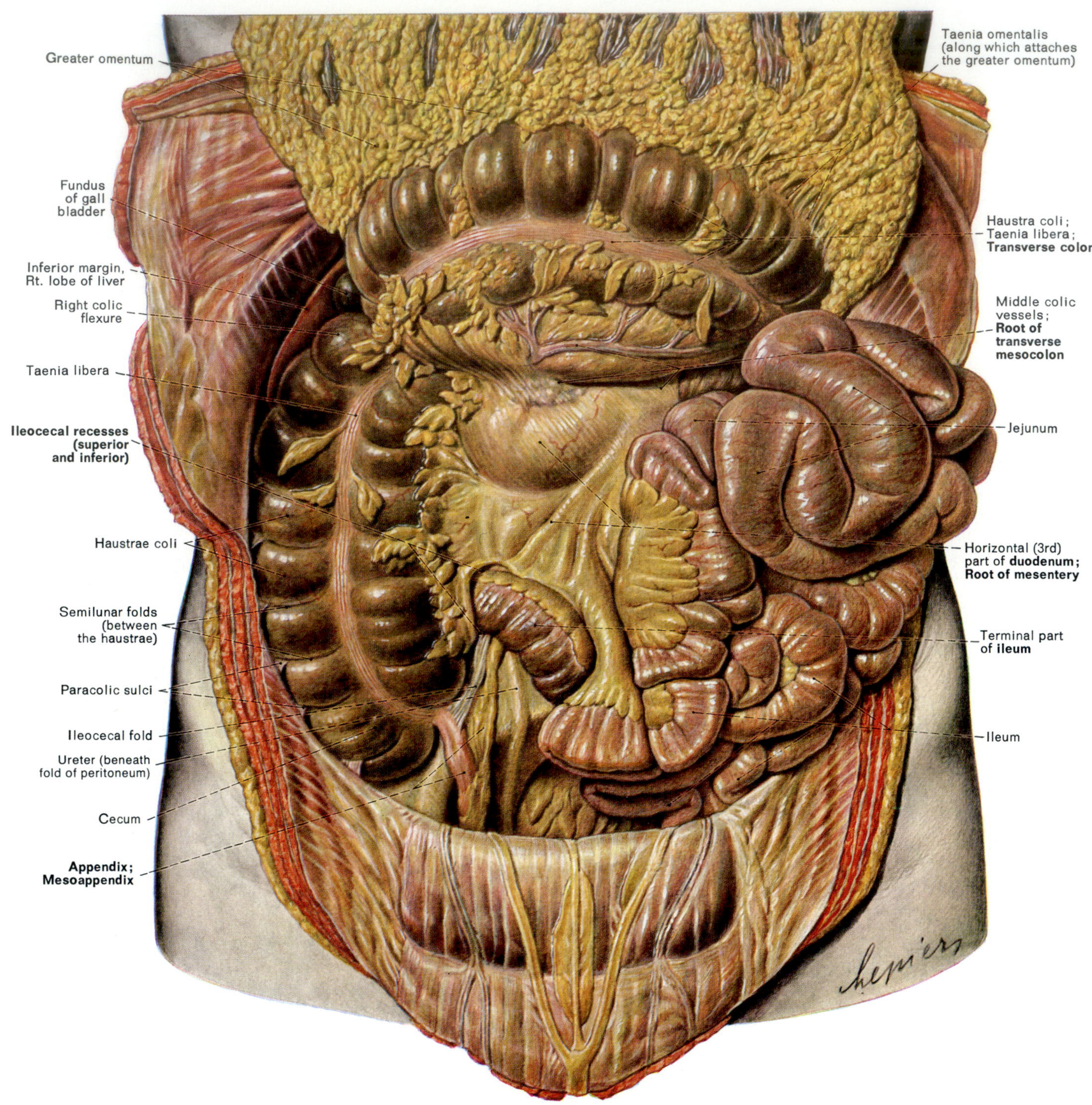

Greater omentum
Taenia omentalis (along which attaches the greater omentum)
Fundus of gall bladder
Haustra coli; Taenia libera; **Transverse colon**
Inferior margin, Rt. lobe of liver
Right colic flexure
Middle colic vessels; **Root of transverse mesocolon**
Taenia libera
Ileocecal recesses (superior and inferior)
Jejunum
Haustrae coli
Horizontal (3rd) part of **duodenum; Root of mesentery**
Semilunar folds (between the haustrae)
Terminal part of **ileum**
Paracolic sulci
Ileocecal fold
Ileum
Ureter (beneath fold of peritoneum)
Cecum
Appendix; Mesoappendix

Fig. 301: Abdominal Cavity (7): The Ascending and Transverse Colon

NOTE: 1) the greater omentum has been reflected superiorly and the jejunum and ileum have been pulled to the left in order to expose the root of the mesentery of the small intestine on the right side.

2) the horizontal (3rd) part of the duodenum which is retroperitoneal and which is covered by the smooth and glistening peritoneum. Observe also the distal portion of the ileum at its junction with the cecum. At this ileocecal junction, identify the ileocecal fold, the appendix and the mesoappendix. The appendix may extend cranially behind the cecum, toward the left and behind the ileum, or as demonstrated here, inferiorly over the pelvic brim.

3) the transverse colon and the small intestine beyond the duodenal junction are more mobile than most other organs because they are attached to the transverse mesocolon and the mesentery.

4) the retroperitoneal position of the right ureter as it descends over the pelvic brim on its course toward the bladder in the pelvis.

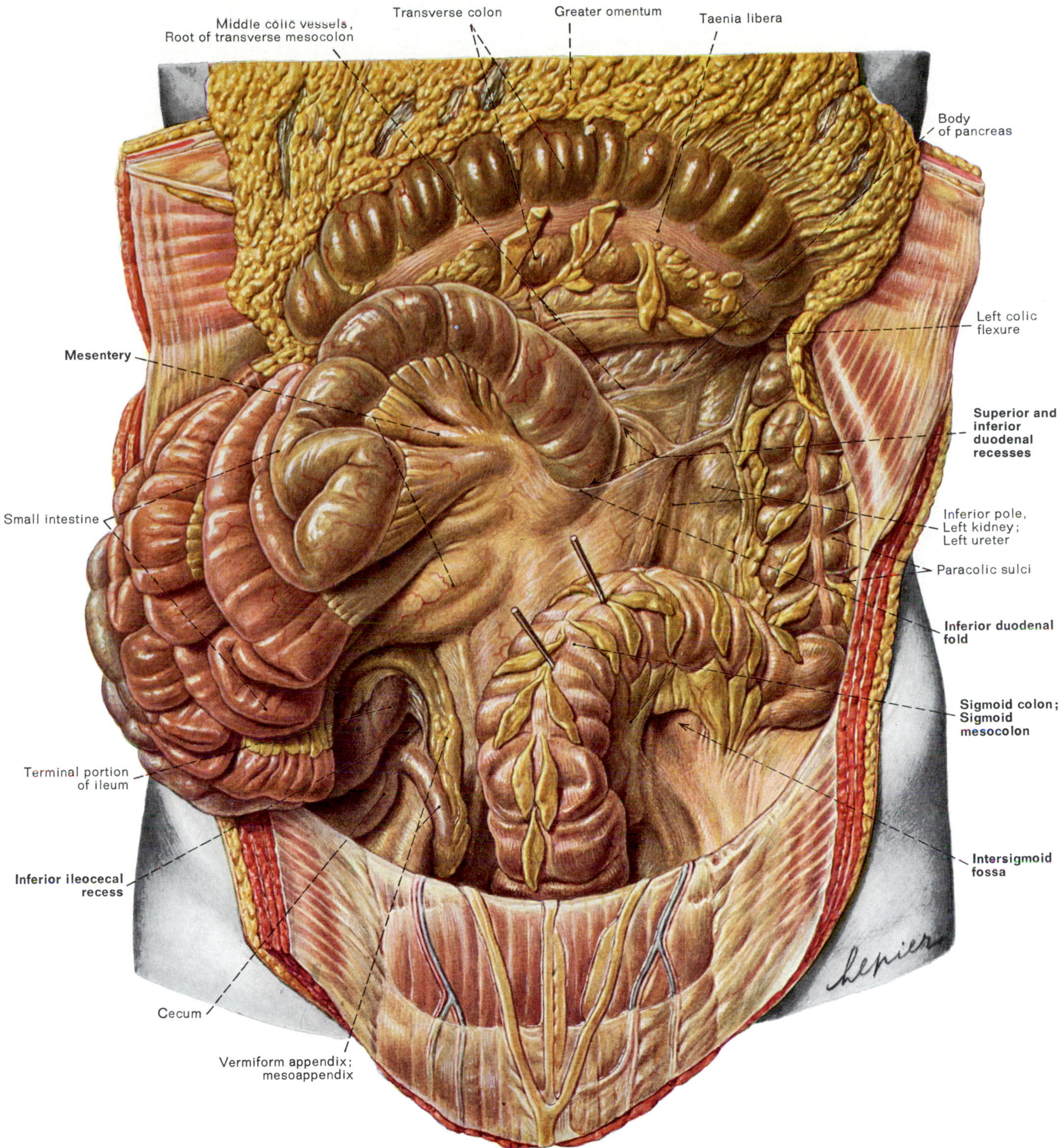

Middle colic vessels,
Root of transverse mesocolon
Transverse colon
Greater omentum
Taenia libera
Body
of pancreas
Left colic
flexure
Mesentery
Superior and
inferior
duodenal
recesses
Small intestine
Inferior pole,
Left kidney;
Left ureter
Paracolic sulci
Inferior duodenal
fold
Sigmoid colon;
Sigmoid
mesocolon
Terminal portion
of ileum
Intersigmoid
fossa
Inferior ileocecal
recess
Cecum
Vermiform appendix;
mesoappendix

Fig. 302: Abdominal Cavity (8): The Descending and Sigmoid Colon and the Duodenojejunal Junction

NOTE: 1) the transverse colon and greater omentum have been reflected superiorly and the jejunum and ileum have been pulled to the right to reveal the duodenojejunal junction and the descending and sigmoid colon.

2) as the ascending (4th) part of the duodenum becomes jejunum, and the small intestine acquires a mesentery at that site, there frequently are found duodenal fossae or recesses situated in relation to the junction. Look for the inferior and superior duodenal recesses (present in over 50% of the cases). These are of importance because they represent possible sites of intestinal herniae within the abdomen.

3) the mobility of the sigmoid colon because of its mesocolic attachment, in contrast to the descending colon which is more fixed to the posterior wall of the abdomen. Observe the intersigmoid fossa located behind the sigmoid mesocolon and between it and the peritoneum reflected over the external iliac vessels.

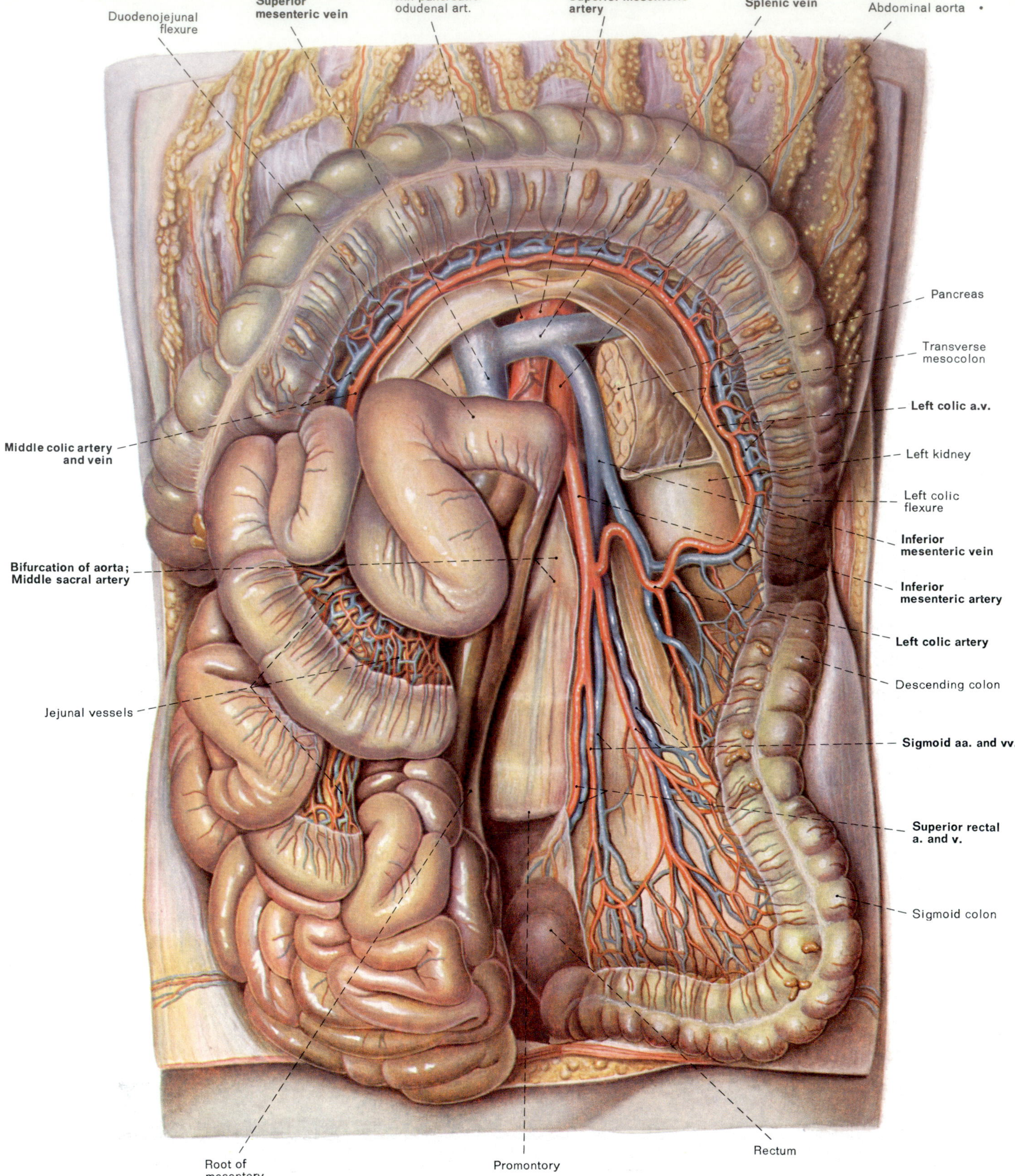
Duodenojejunal flexure
Superior mesenteric vein
Inf. pancreatic-odudenal art.
Superior mesenteric artery
Splenic vein
Abdominal aorta
Pancreas
Transverse mesocolon
Left colic a.v.
Middle colic artery and vein
Left kidney
Left colic flexure
Inferior mesenteric vein
Bifurcation of aorta; Middle sacral artery
Inferior mesenteric artery
Left colic artery
Descending colon
Jejunal vessels
Sigmoid aa. and vv.
Superior rectal a. and v.
Sigmoid colon
Rectum
Root of mesentery
Promontory

Fig. 303: Abdominal Cavity (9): The Inferior Mesenteric Vessels and Branches

NOTE: 1) the small intestine has been pushed to the right side of the abdomen to reveal the origin of the inferior mesenteric artery from the aorta, and the drainage of the inferior mesenteric vein into the splenic vein. Also a portion of the body of the pancreas and transverse mesocolon was removed to show these relationships more clearly.

2) the inferior mesenteric artery supplies the descending colon via the left colic artery. Observe the anastomosis around the margin of the large bowel between the left colic artery and the middle colic artery. Somewhat lower, the inferior mesenteric gives rise to the sigmoid arteries and the superior rectal vessels.

3) the bifurcation of the abdominal aorta retroperitoneally and the middle sacral artery descending in the midline.

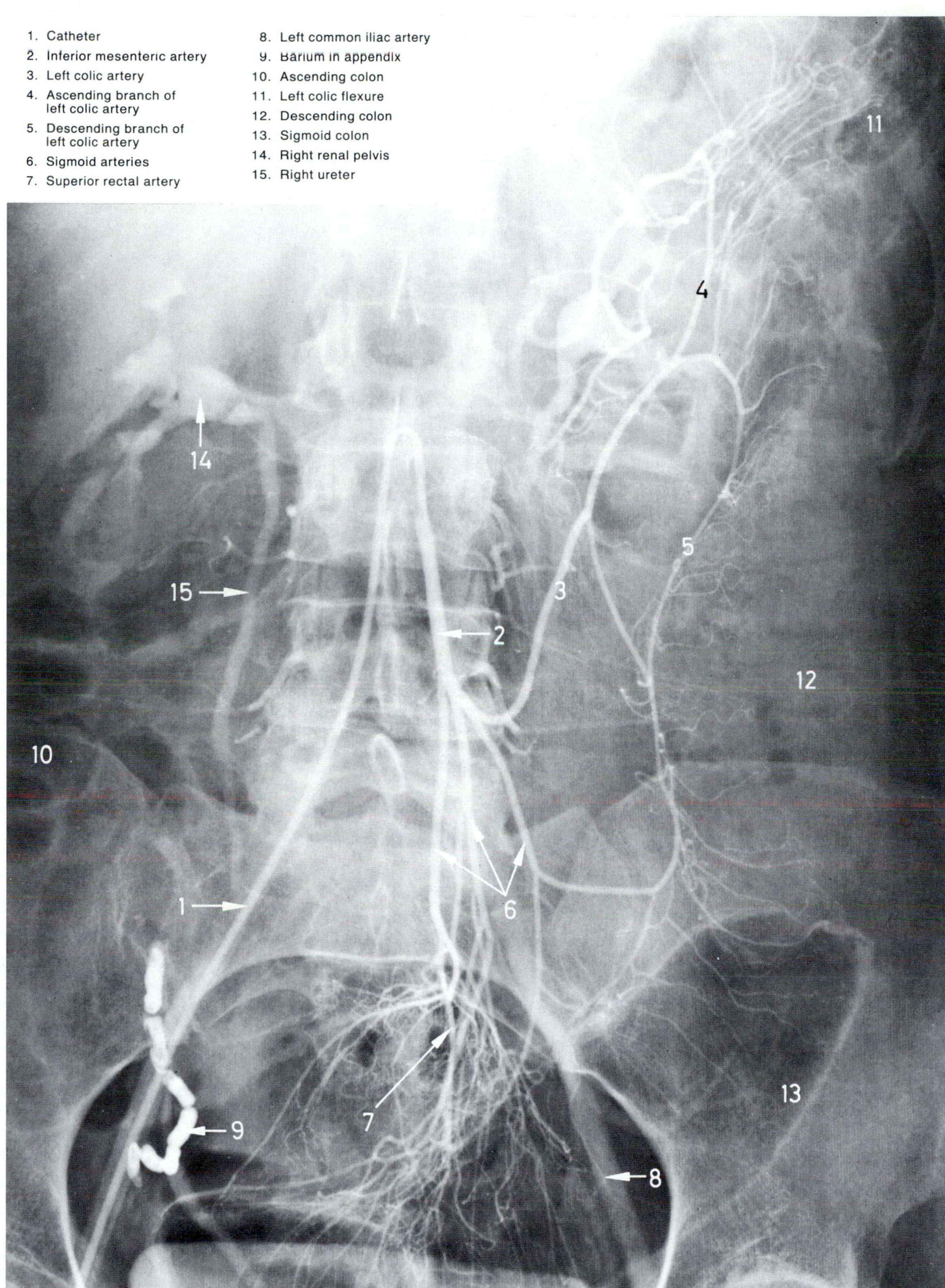

1. Catheter
2. Inferior mesenteric artery
3. Left colic artery
4. Ascending branch of left colic artery
5. Descending branch of left colic artery
6. Sigmoid arteries
7. Superior rectal artery
8. Left common iliac artery
9. Barium in appendix
10. Ascending colon
11. Left colic flexure
12. Descending colon
13. Sigmoid colon
14. Right renal pelvis
15. Right ureter
11
4
14
5
15
3
2
12
10
6
1
13
9
7
8

Fig. 304: An Inferior Mesenteric Arteriogram

NOTE that: 1) a catheter *(1)* has been inserted through the right internal iliac artery and directed superiorly into the abdominal aorta to the point of origin of the inferior mesenteric artery (*2*). A contrast medium was then injected into that artery in order to demonstrate its field of distribution;

2) the branches of the inferior mesenteric artery (*2*) shown in this arteriogram are quite normal. The left colic artery (*3*) shows both an ascending (*4*) and a descending branch (*5*);

3) several sigmoid arteries (*6*) supply the sigmoid colon (*13*), and these anastomose above with branches of the left colic artery (*3*) and below with the superior rectal artery (*7*).

4) that this figure comes from Wicke, L., *Atlas of Radiologic Anatomy*, 3rd Edition, Urban and Schwarzenberg, Baltimore, 1982.

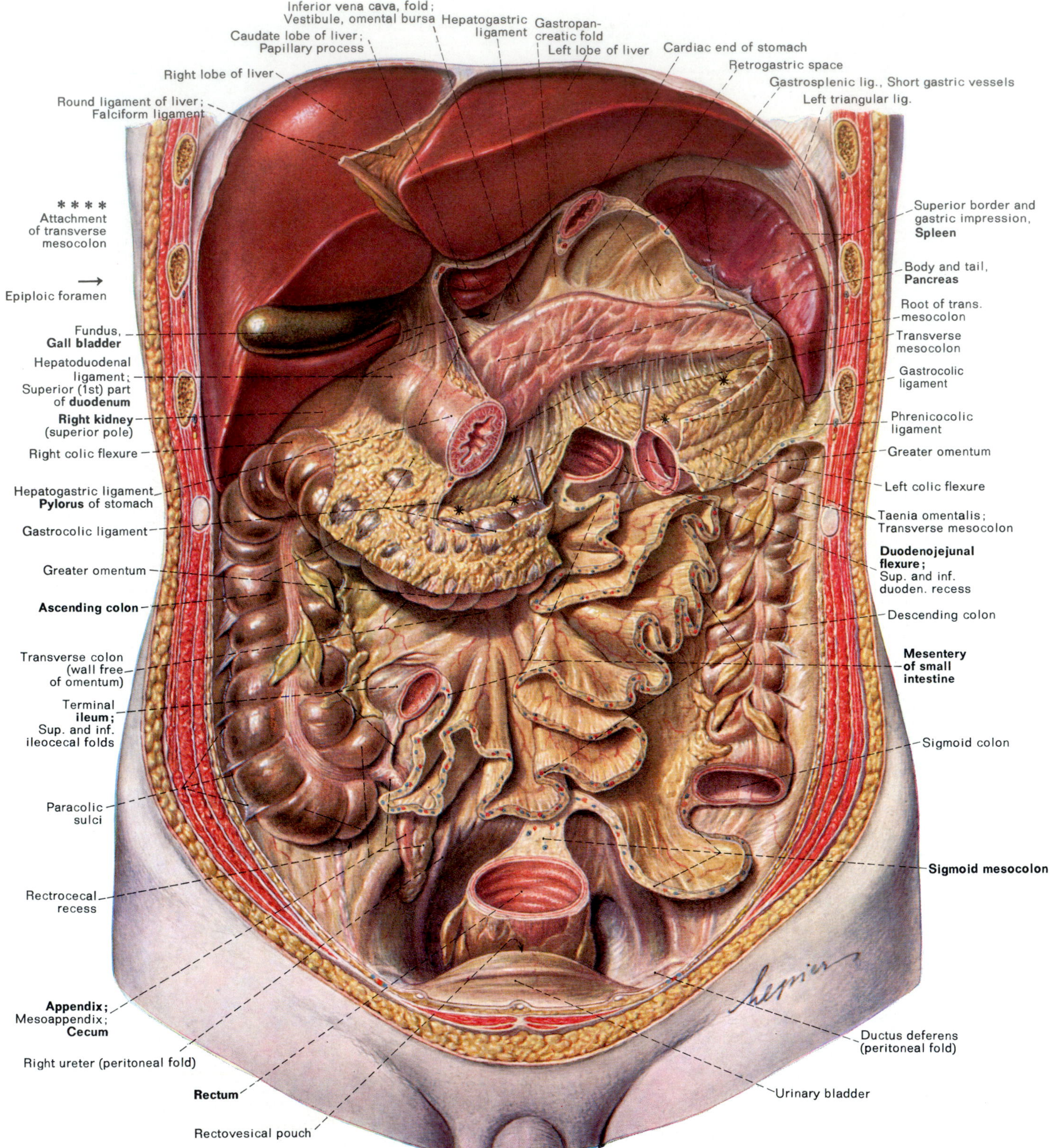
Inferior vena cava, fold;
Vestibule, omental bursa
Hepatogastric ligament
Gastropancreatic fold
Caudate lobe of liver;
Papillary process
Left lobe of liver
Cardiac end of stomach
Retrogastric space
Gastrosplenic lig., Short gastric vessels
Left triangular lig.
Right lobe of liver
Round ligament of liver;
Falciform ligament
* * * *
Attachment of transverse mesocolon
Epiploic foramen
Fundus, Gall bladder
Hepatoduodenal ligament;
Superior (1st) part of duodenum
Right kidney (superior pole)
Right colic flexure
Hepatogastric ligament
Pylorus of stomach
Gastrocolic ligament
Greater omentum
Ascending colon
Transverse colon (wall free of omentum)
Terminal ileum;
Sup. and inf. ileocecal folds
Paracolic sulci
Rectrocecal recess
Appendix;
Mesoappendix;
Cecum
Right ureter (peritoneal fold)
Rectum
Rectovesical pouch
Superior border and gastric impression, Spleen
Body and tail, Pancreas
Root of trans. mesocolon
Transverse mesocolon
Gastrocolic ligament
Phrenicocolic ligament
Greater omentum
Left colic flexure
Taenia omentalis;
Transverse mesocolon
Duodenojejunal flexure;
Sup. and inf. duoden. recess
Descending colon
Mesentery of small intestine
Sigmoid colon
Sigmoid mesocolon
Ductus deferens (peritoneal fold)
Urinary bladder

Fig. 305: Abdominal Cavity (10): The Large Intestine and the Mesenteries

NOTE: 1) the stomach was cut just proximal to the pylorus and the small intestine was severed at the duodenojejunal junction and at the distal ileum and then removed by cutting the mesentery. A portion of the transverse colon was removed along with the greater omentum, and the sigmoid colon was resected to reveal its mesocolon.

2) the mesentery of the small intestine extends obliquely across the posterior abdominal wall from the duodenojejunal flexure to the ilocecal junction. In this distance of about 6 to 7 inches, the mesentery is thrown into many folds to accommodate all the loops of jejunum and ileum.

3) the ascending and descending colon is fused to the posterior abdominal wall, while the transverse and sigmoid colon are suspended by their respective mesocolons.

4) that the vessels and nerves supplying the small intestine course *between* the layers of the mesentery to achieve the organ.

Fig. 306: The Ileocecal Junction

NOTE: 1) the terminal ileum, cecum and lower ascending colon have been opened anteriorly to reveal the ileocecal junction and the opening of the vermiform appendix.

2) the leaves of the ileocecal valve have been separated. Observe that this valve is formed by two reflected folds of the wall of the large intestine. These folds unite and then project further around the large intestine as the frenulum.

3) the orifice of the appendix opens into the cecum, although the position and the direction of the appendix are quite variable. Observe the semilunar folds, the sacculations (haustrae) and longitudinal muscle bands (taeniae) which characterize the large intestine.

Fig. 307: A Segment of Transverse Colon

NOTE: 1) a cut has been made along the taenia libera at the right end of this segment of transverse colon and the wall of the colon has been opened to reveal its inner surface. The greater omentum attaches along the taenia omentalis, the transverse mesocolon attaches along the taenia mesocolica, while the taenia libera is free from such attachments.

2) the appendices epiploicae. These small sac-like structures are appendages characteristic of the large intestine. They contain fat and their function is unknown.

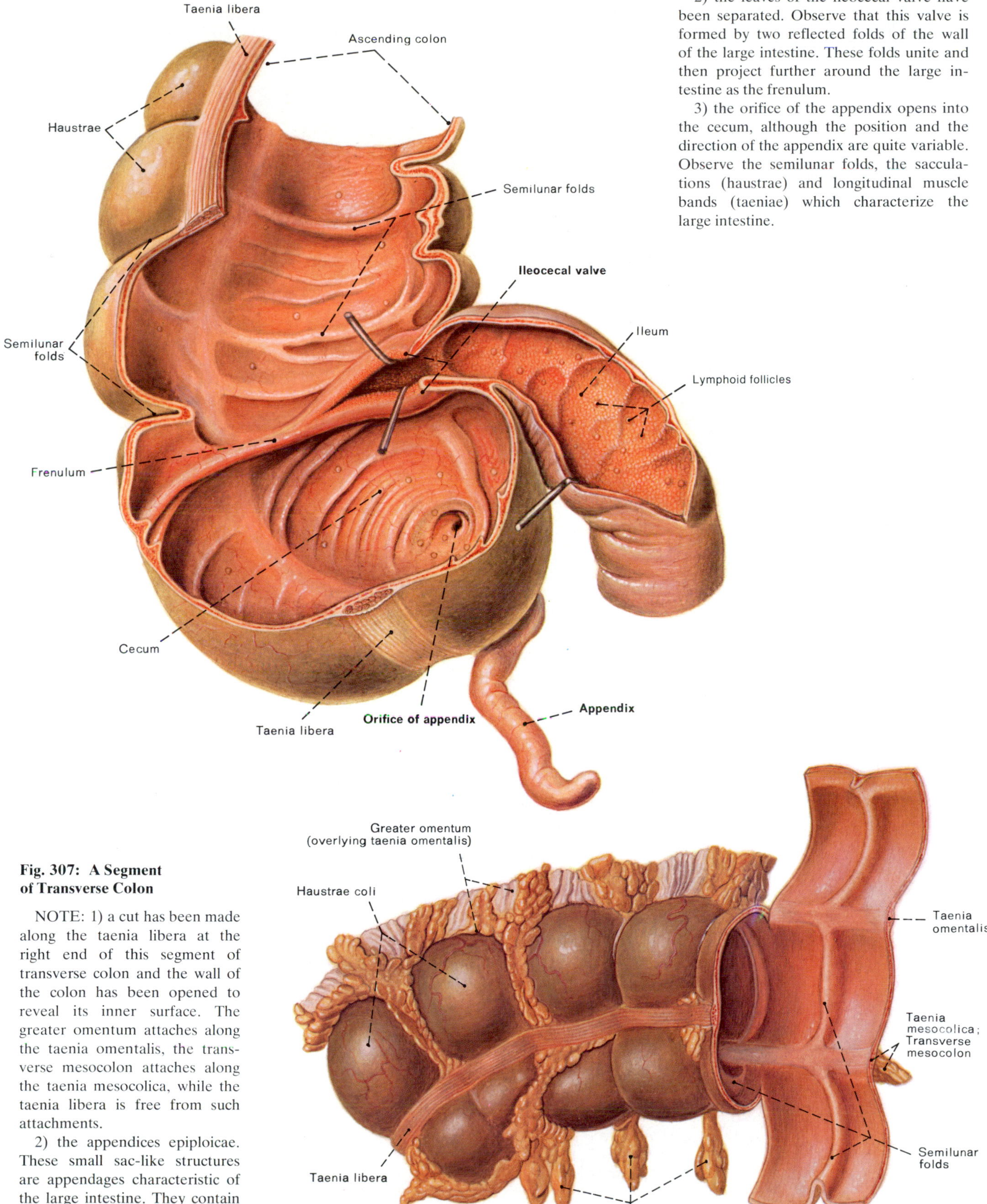

Taenia libera
Ascending colon
Haustrae
Semilunar folds
Ileocecal valve
Ileum
Semilunar folds
Lymphoid follicles
Frenulum
Cecum
Taenia libera
Orifice of appendix
Appendix
Greater omentum (overlying taenia omentalis)
Haustrae coli
Taenia omentalis
Taenia mesocolica; Transverse mesocolon
Semilunar folds
Taenia libera
Appendices epiploicae

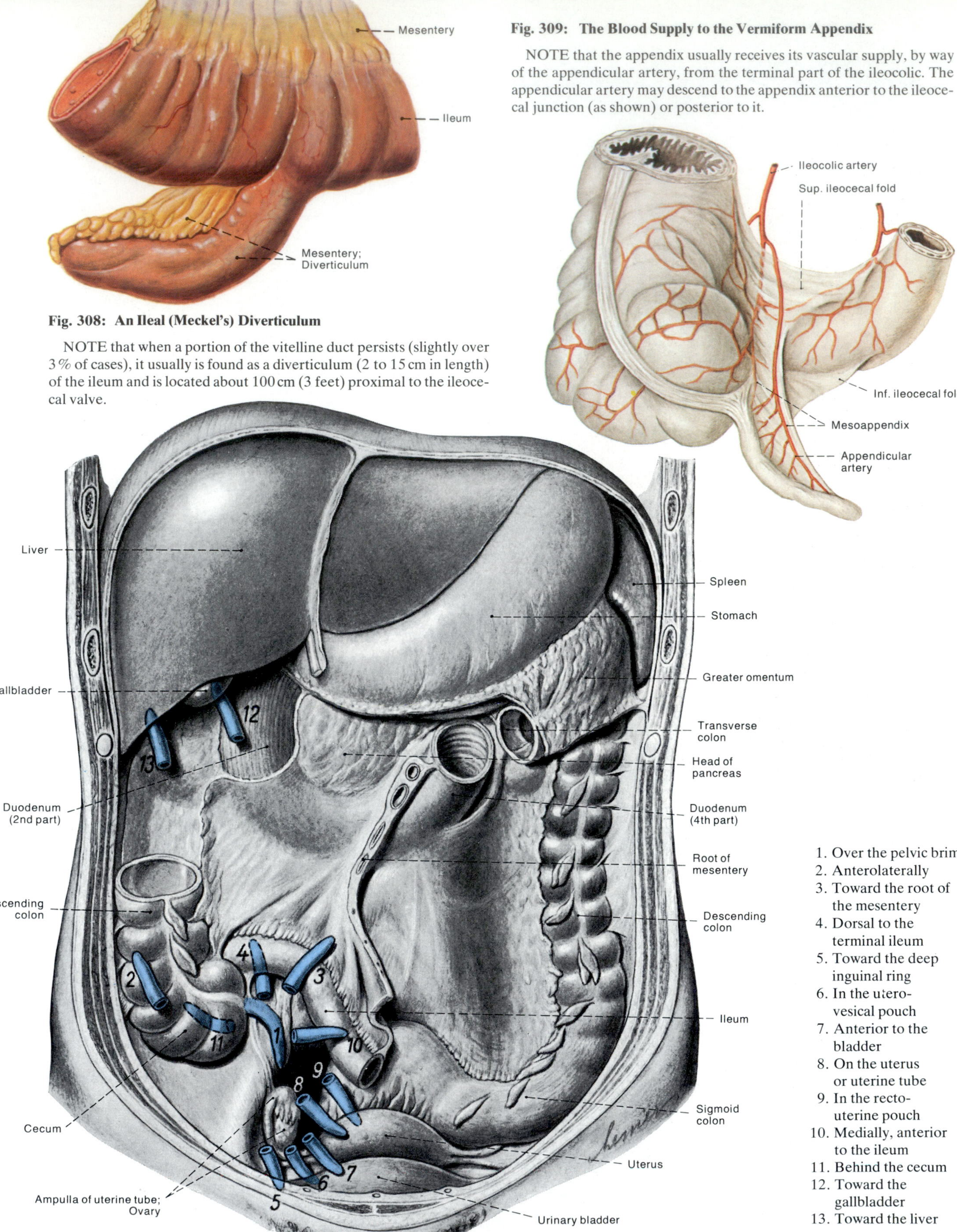

Mesentery
Ileum
Mesentery;
Diverticulum
Ileocolic artery
Sup. ileocecal fold
Inf. ileocecal fold
Mesoappendix
Appendicular
artery
Liver
Gallbladder
Duodenum
(2nd part)
Ascending
colon
Cecum
Ampulla of uterine tube;
Ovary
Spleen
Stomach
Greater omentum
Transverse
colon
Head of
pancreas
Duodenum
(4th part)
Root of
mesentery
Descending
colon
Ileum
Sigmoid
colon
Uterus
Urinary bladder
1. Over the pelvic brim
2. Anterolaterally
3. Toward the root of the mesentery
4. Dorsal to the terminal ileum
5. Toward the deep inguinal ring
6. In the utero-vesical pouch
7. Anterior to the bladder
8. On the uterus or uterine tube
9. In the recto-uterine pouch
10. Medially, anterior to the ileum
11. Behind the cecum
12. Toward the gallbladder
13. Toward the liver

Fig. 308: An Ileal (Meckel's) Diverticulum

NOTE that when a portion of the vitelline duct persists (slightly over 3% of cases), it usually is found as a diverticulum (2 to 15 cm in length) of the ileum and is located about 100 cm (3 feet) proximal to the ileocecal valve.

Fig. 309: The Blood Supply to the Vermiform Appendix

NOTE that the appendix usually receives its vascular supply, by way of the appendicular artery, from the terminal part of the ileocolic. The appendicular artery may descend to the appendix anterior to the ileocecal junction (as shown) or posterior to it.

Fig. 310: Variations in the Location of the Appendix

NOTE that at least 13 different locations (shown in blue) for the appendix have been described, the most common being 11, 1, 3 (and 10), and 4.

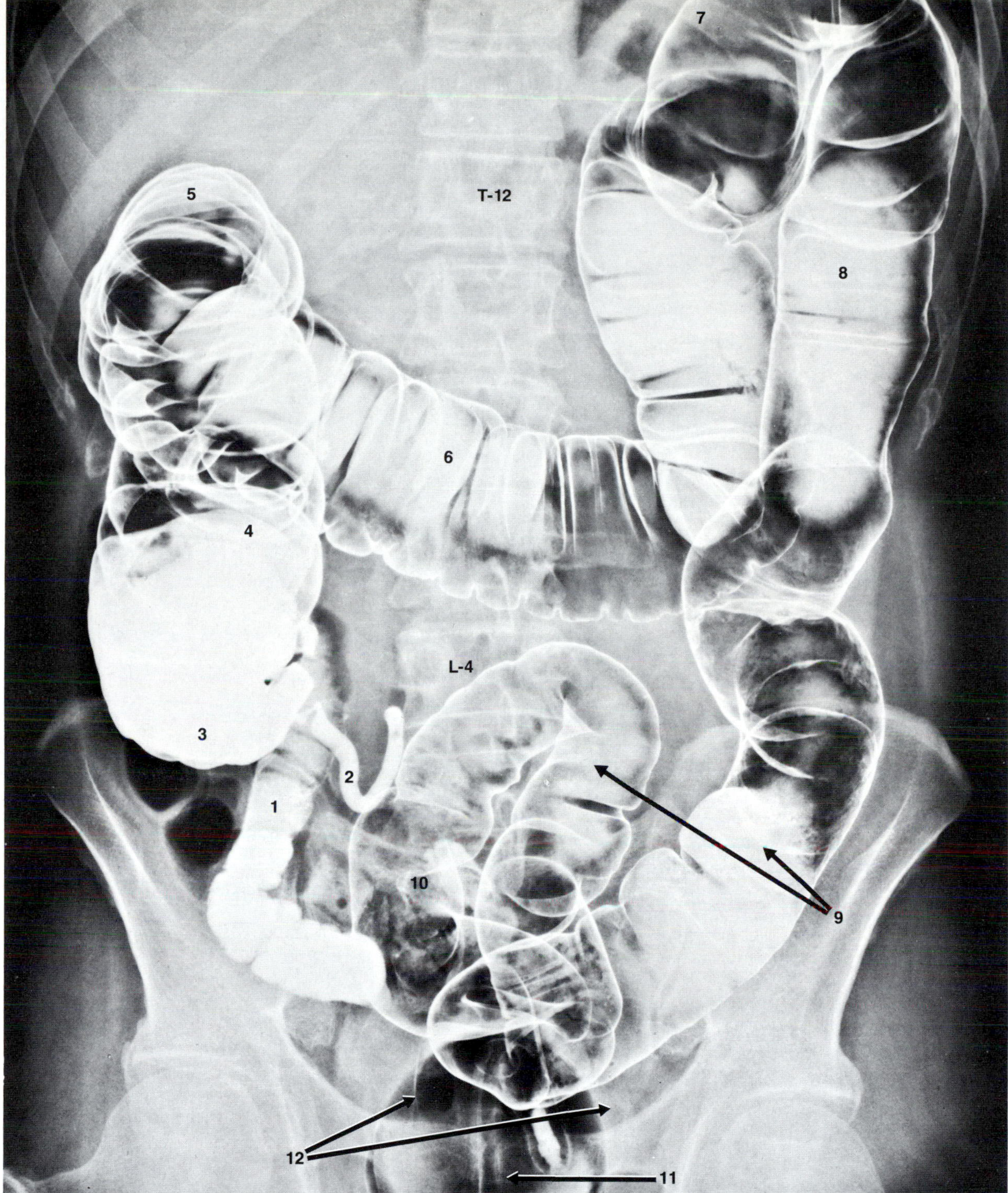

7
5
T-12
8
6
4
L-4
3
2
1
10
9
12
11

Fig. 311: The Radiographic Anatomy of the Large Intestine (Double Contrast)

NOTE: 1) that an excellent radiographic visualization may be achieved when a thin barium sulfate mixture is administered as an enema. The mixture is then expelled and the colon is insufflated with air (barium-air double contrast method).

2) that the large intestine is about 150 cm (5 feet) in length, and it extends from the ileocecal junction to the anus. Its diameter varies from 3.75 to 7.6 cm (1.5 to 3 inches) and, normally, is of greatest width at its commencement at the cecum, from whence it gradually narrows. Near the distal rectum it dilates once again as the rectal ampulla, and this leads to the anal canal.

3) the cecum (3) is usually located in the iliac fossa of the lower right abdominal quadrant, but because it is enveloped by peritoneum, its position may be variable. Observe that the cecum is a cul-de-sac that opens superiorly into the ascending colon (4). Its junction with the terminal ileum (1) occurs most often on the medial or posterior cecal surface. The appendix (2) extends from the cecum about 2 cm below the ileocecal opening.

4) that the right colic (or hepatic) flexure (5) is oriented inferiorly and anteriorly and continues to the left to become the transverse colon (6). This

5) that the transverse colon, suspended by its mesentery, crosses the abdomen. It ascends in the left upper quadrant toward the left colic (splenic) flexure (7), where it continues inferiorly as the descending colon (8).

6) that the descending colon becomes the sigmoid colon (9) at the inlet to the lesser pelvis. Surrounded by its mesentery, the sigmoid colon leads into the rectum (10), which then descends into the true pelvis.

7) the positions of the bodies of the 12th thoracic (T-12) and 4th lumbar (L-4) vertebrae, the symphysis pubis (11), and an air-filled balloon (12). (From Wicke, L., *Atlas of Radiologic Anatomy*, 3rd Edition, Urban and Schwarzenberg, Baltimore, 1982.)

Fig. 312: The Inner Surface of the Rectum and Anal Canal

NOTE: 1) at about the level of the 3rd sacral vertebra, the sigmoid colon becomes the rectum. Measuring about 5 inches in length, the rectum then becomes the anal canal which is the terminal $1^1/_2$ inches of the intestinal tract. The rectum is dilated near its junction with the anal canal, giving rise to the rectal ampulla.

2) the internal mucosa of the rectum is thrown into transverse folds of which there are usually three in number. They are also known as the valves of Houston.

3) below the rectal ampulla, the mucosa of the anal canal shows a series of longitudinal folds called the anal colums (columns of Morgagni). Each of these folds usually possesses an artery and vein, and between the anal columns are small fossae called the anal sinuses. The veins in this region are dilated and tortuous and can become varicosed, giving rise to a condition called hemorrhoids or piles.

4) distal to the anal column is an abrupt zone of epithelial transition. The stratified squamous of the distalmost part of the anal canal becomes the simple columnar of the rectum. This transition line can be identified in a living person and is referred to as Hilton's line.

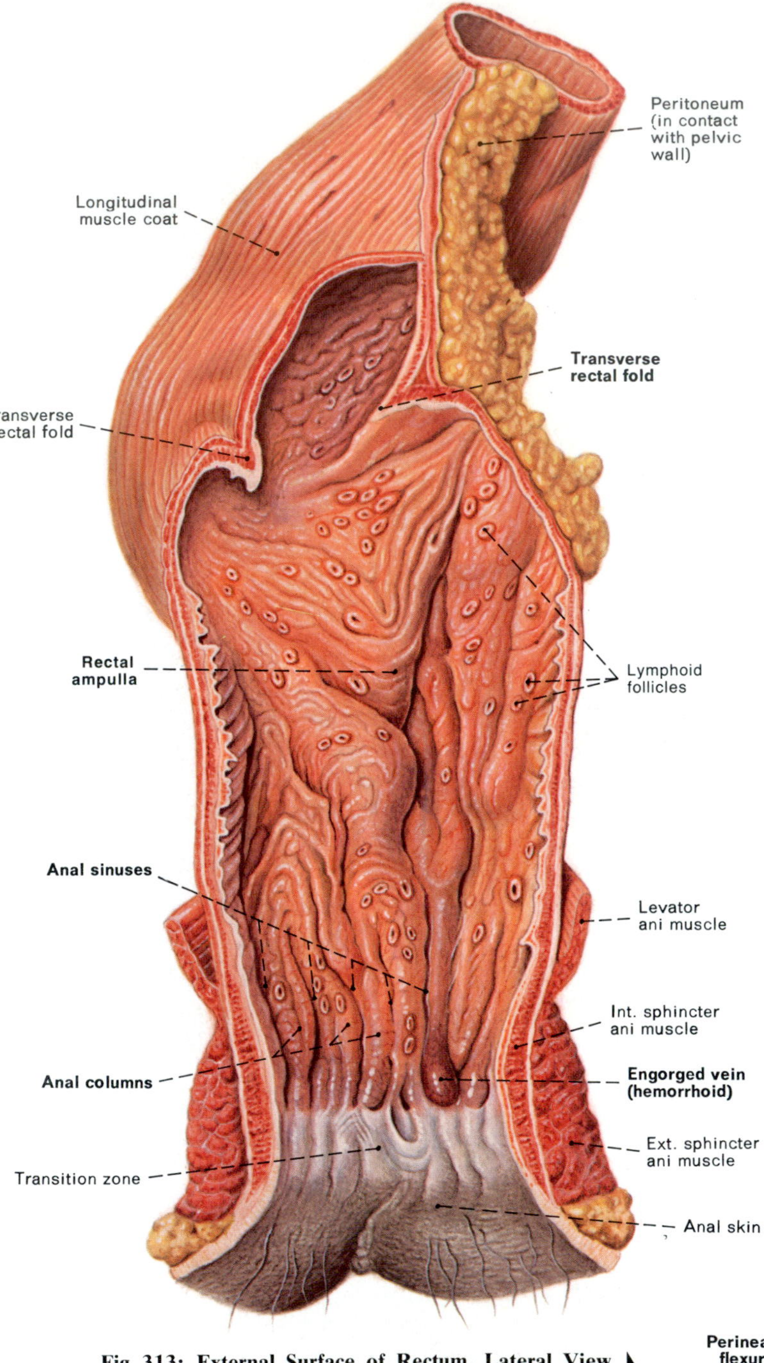

Peritoneum (in contact with pelvic wall)
Longitudinal muscle coat
Transverse rectal fold
Transverse rectal fold
Rectal ampulla
Lymphoid follicles
Anal sinuses
Levator ani muscle
Int. sphincter ani muscle
Anal columns
Engorged vein (hemorrhoid)
Ext. sphincter ani muscle
Transition zone
Anal skin

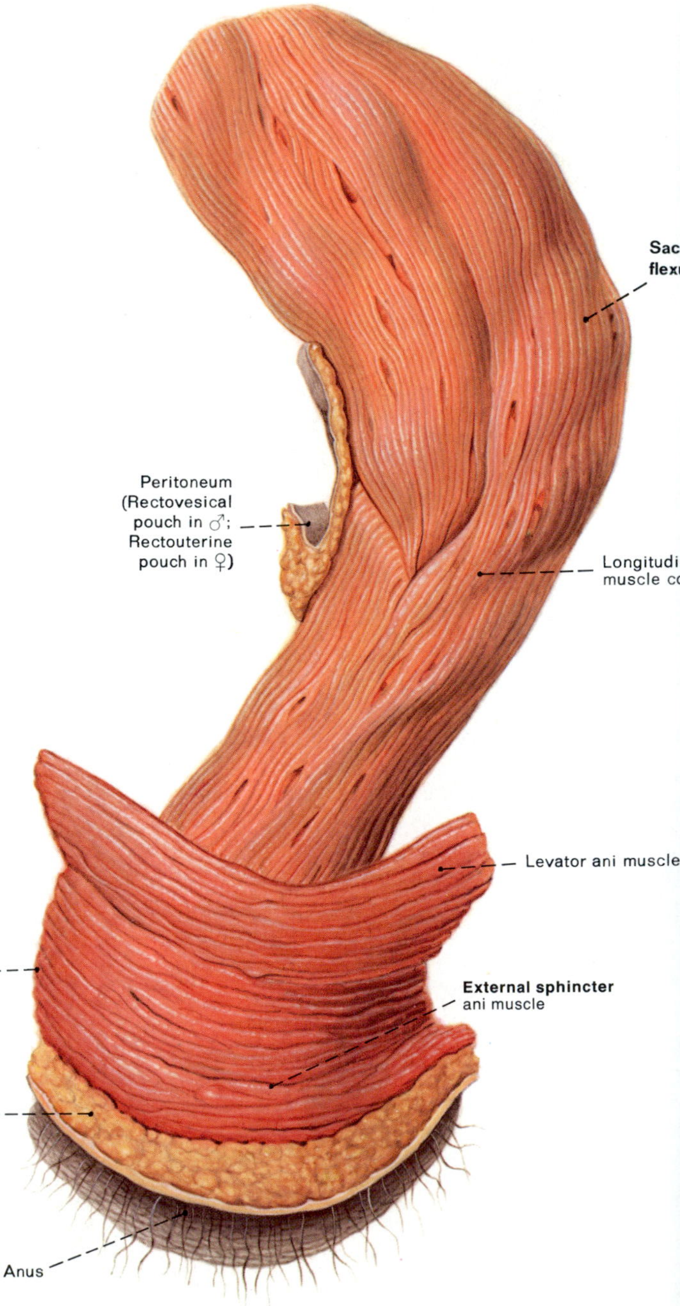

Sacr flexu
Peritoneum (Rectovesical pouch in ♂; Rectouterine pouch in ♀)
Longitudir muscle co
Levator ani muscle
Perineal flexure
External sphincter ani muscle
Superficial fascia
Anus

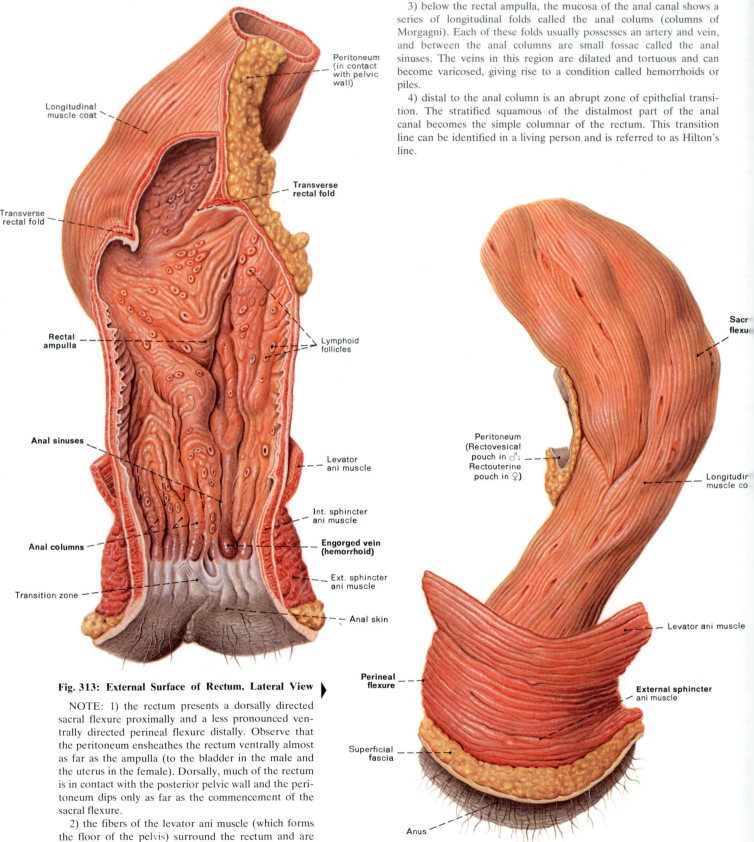

Fig. 313: External Surface of Rectum, Lateral View ▶

NOTE: 1) the rectum presents a dorsally directed sacral flexure proximally and a less pronounced ventrally directed perineal flexure distally. Observe that the peritoneum ensheathes the rectum ventrally almost as far as the ampulla (to the bladder in the male and the uterus in the female). Dorsally, much of the rectum is in contact with the posterior pelvic wall and the peritoneum dips only as far as the commencement of the sacral flexure.

2) the fibers of the levator ani muscle (which forms the floor of the pelvis) surround the rectum and are

(con't next page)

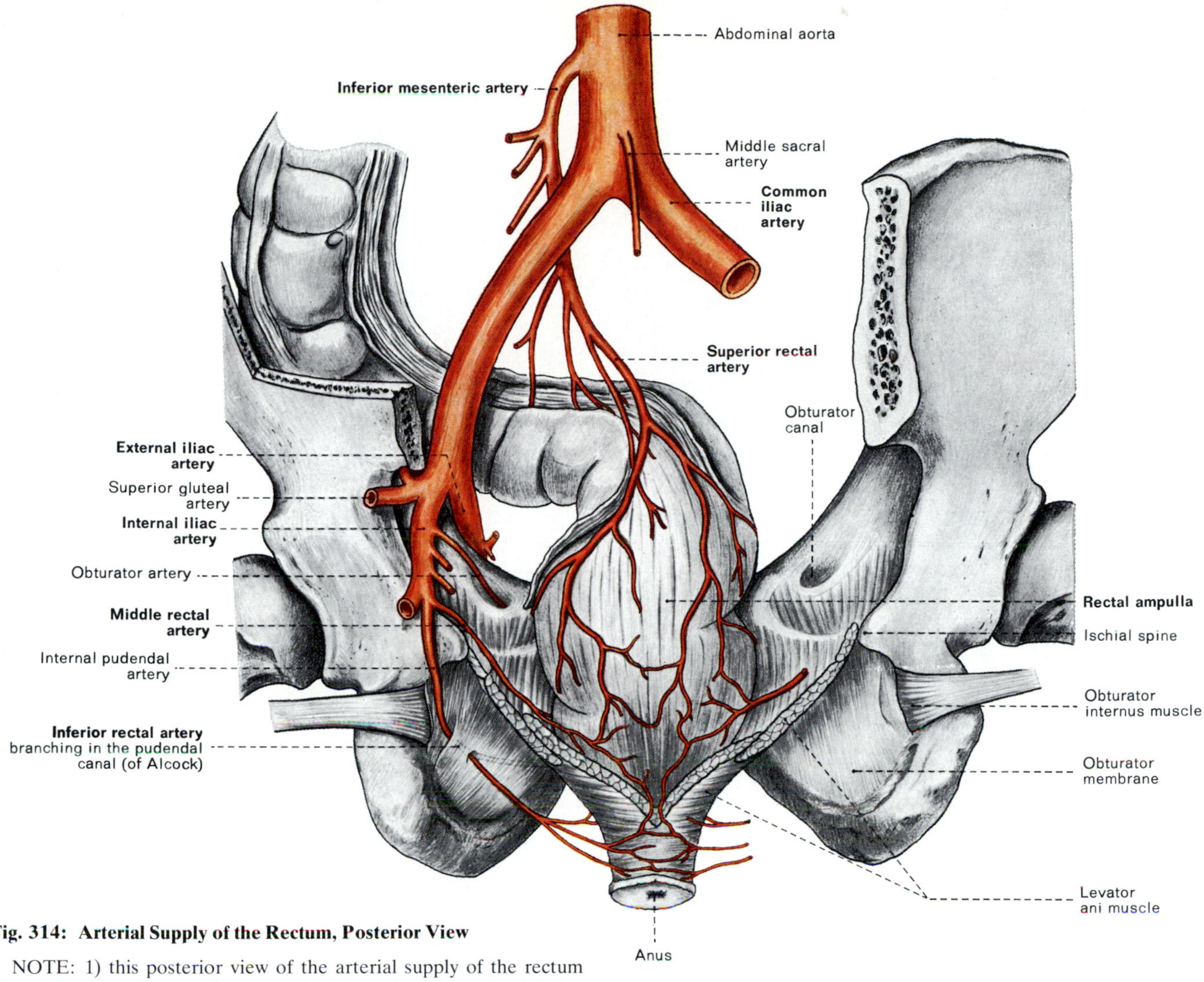

Abdominal aorta
Inferior mesenteric artery
Middle sacral artery
Common iliac artery
Superior rectal artery
Obturator canal
External iliac artery
Superior gluteal artery
Internal iliac artery
Obturator artery
Rectal ampulla
Ischial spine
Middle rectal artery
Internal pudendal artery
Obturator internus muscle
Inferior rectal artery branching in the pudendal canal (of Alcock)
Obturator membrane
Levator ani muscle
Anus

Fig. 314: Arterial Supply of the Rectum, Posterior View

NOTE: 1) this posterior view of the arterial supply of the rectum shows the superior rectal artery which branches from the inferior mesenteric artery, and distributes to the rectum as far as the ampulla. The middle rectal artery coming off the internal iliac artery and the inferior rectal artery branching from the internal pudendal artery supply the more caudal parts of the rectum.

2) from the bifurcation of the aorta, the middle sacral artery descends down the midline of the sacrum. The common iliac arteries branch into external and internal iliac vessels. From the internal iliac (hypogastric), many of the pelvic organs derive their blood supply.

3) the internal pudendal artery coursing within the pudendal canal (of Alcock). The inferior rectal artery branches from the internal pudendal within this canal.

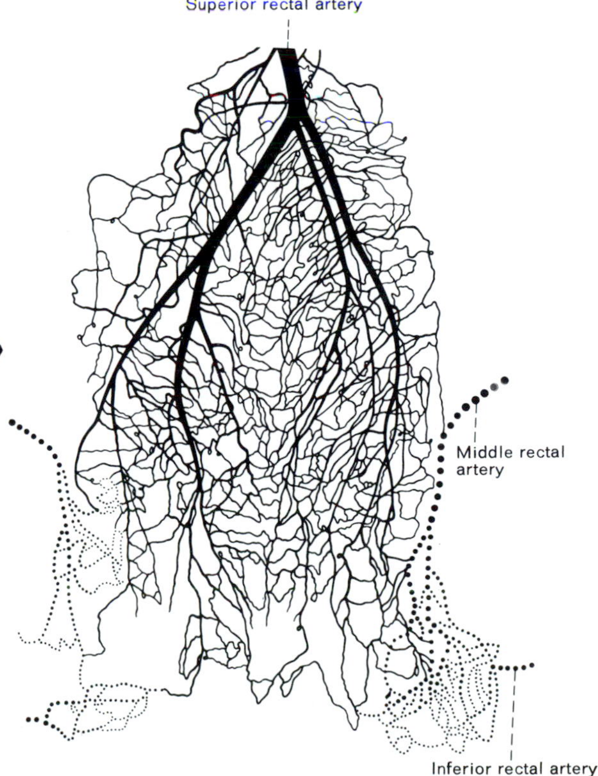
Superior rectal artery
Middle rectal artery
Inferior rectal artery

Fig. 315: Distribution Pattern of the Rectal Arteries

This figure shows the distribution of the superior, middle and inferior rectal arteries as they supply the rectum and the anal canal. Observe that the region of distribution of the superior rectal artery is much greater than either the middle or inferior rectal arteries. Note the rich anastomoses among these three vessels.

---

continued distally as the external sphincter ani muscle. The internal sphincter ani muscle (seen in Figure 312) is composed of smooth muscle and really represents a thickening of the inner circular muscle layer in the wall of the rectum. Oberve also the outer longitudinal (smooth) muscle layer.

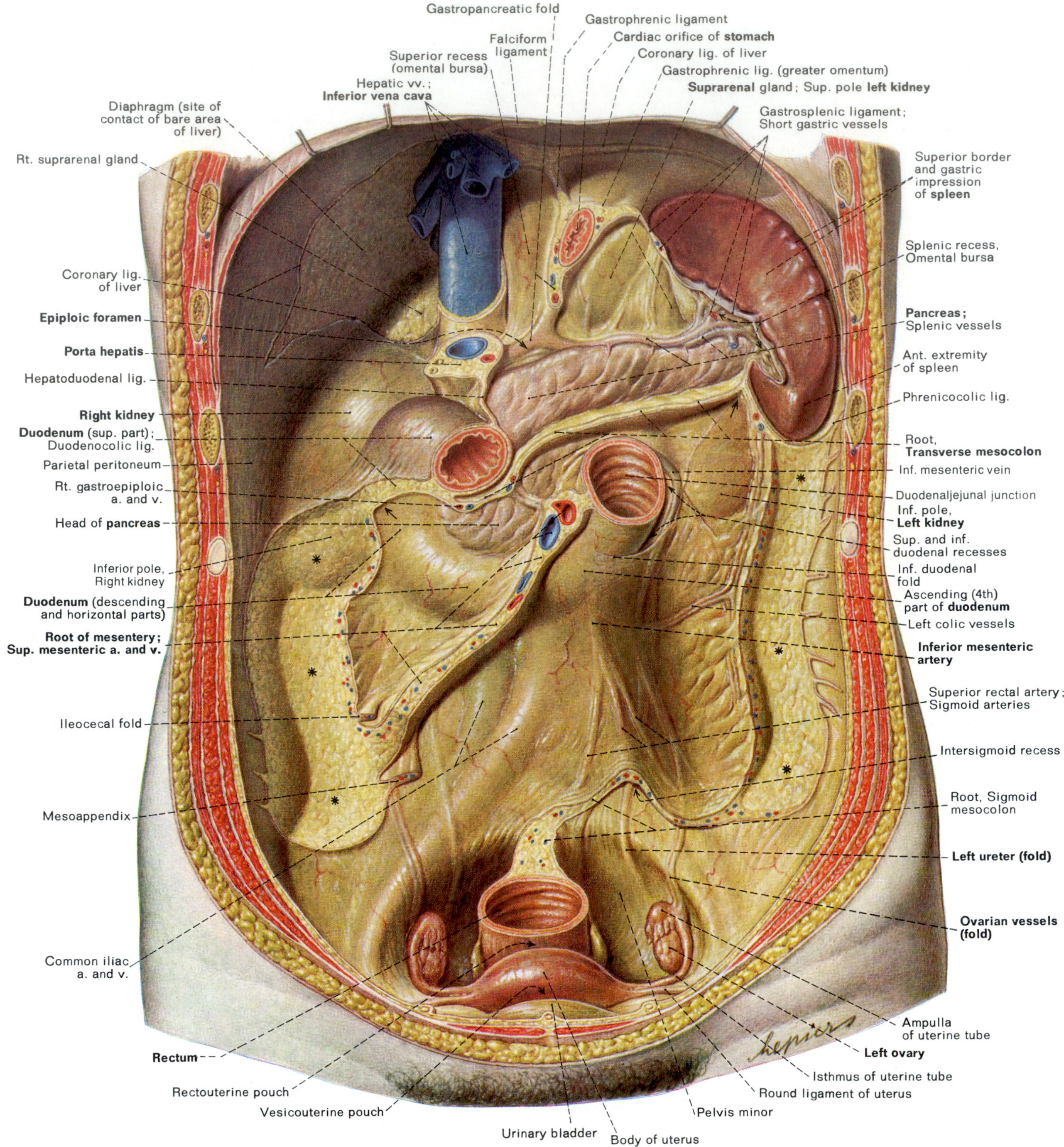

Gastropancreatic fold
Gastrophrenic ligament
Falciform ligament
Cardiac orifice of stomach
Superior recess (omental bursa)
Coronary lig. of liver
Gastrophrenic lig. (greater omentum)
Hepatic vv.; Inferior vena cava
Suprarenal gland; Sup. pole left kidney
Diaphragm (site of contact of bare area of liver)
Gastrosplenic ligament; Short gastric vessels
Rt. suprarenal gland
Superior border and gastric impression of spleen
Splenic recess, Omental bursa
Coronary lig. of liver
Epiploic foramen
Pancreas; Splenic vessels
Porta hepatis
Ant. extremity of spleen
Hepatoduodenal lig.
Phrenicocolic lig.
Right kidney
Duodenum (sup. part); Duodenocolic lig.
Root, Transverse mesocolon
Parietal peritoneum
Inf. mesenteric vein
Rt. gastroepiploic a. and v.
Duodenaljejunal junction
Head of pancreas
Inf. pole, Left kidney
Sup. and inf. duodenal recesses
Inferior pole, Right kidney
Inf. duodenal fold
Duodenum (descending and horizontal parts)
Ascending (4th) part of duodenum
Left colic vessels
Root of mesentery; Sup. mesenteric a. and v.
Inferior mesenteric artery
Superior rectal artery; Sigmoid arteries
Ileocecal fold
Intersigmoid recess
Root, Sigmoid mesocolon
Mesoappendix
Left ureter (fold)
Ovarian vessels (fold)
Common iliac a. and v.
Ampulla of uterine tube
Left ovary
Rectum
Isthmus of uterine tube
Rectouterine pouch
Round ligament of uterus
Vesicouterine pouch
Pelvis minor
Urinary bladder
Body of uterus

Fig. 316: The Abdominal Cavity (11): Posterior Abdominal Peritoneum (Female)

NOTE: 1) the stomach and intestine (except duodenum and rectum) have been removed and their associated mesenteries have been cut close to their roots on the posterior abdominal wall. The liver and its attachments and the gall bladder were removed, leaving intact the spleen and the retroperitoneal organs (duodenum, pancreas, adrenal glands, kidneys and ureters, aorta and inferior vena cava).

2) ascending and descending portions of the large intestine are fused to the posterior abdominal wall (sites marked by *) with a peritoneal layer covering their anterior surfaces. Observe the small black arrows at the right and left colic flexures.

3) the course of the ureters and ovarian vessels as they descend over the pelvic brim. Notice the ovary, uterine tubes and uterus located in the pelvis and their relationship to the rectum and bladder.

Fig. 316

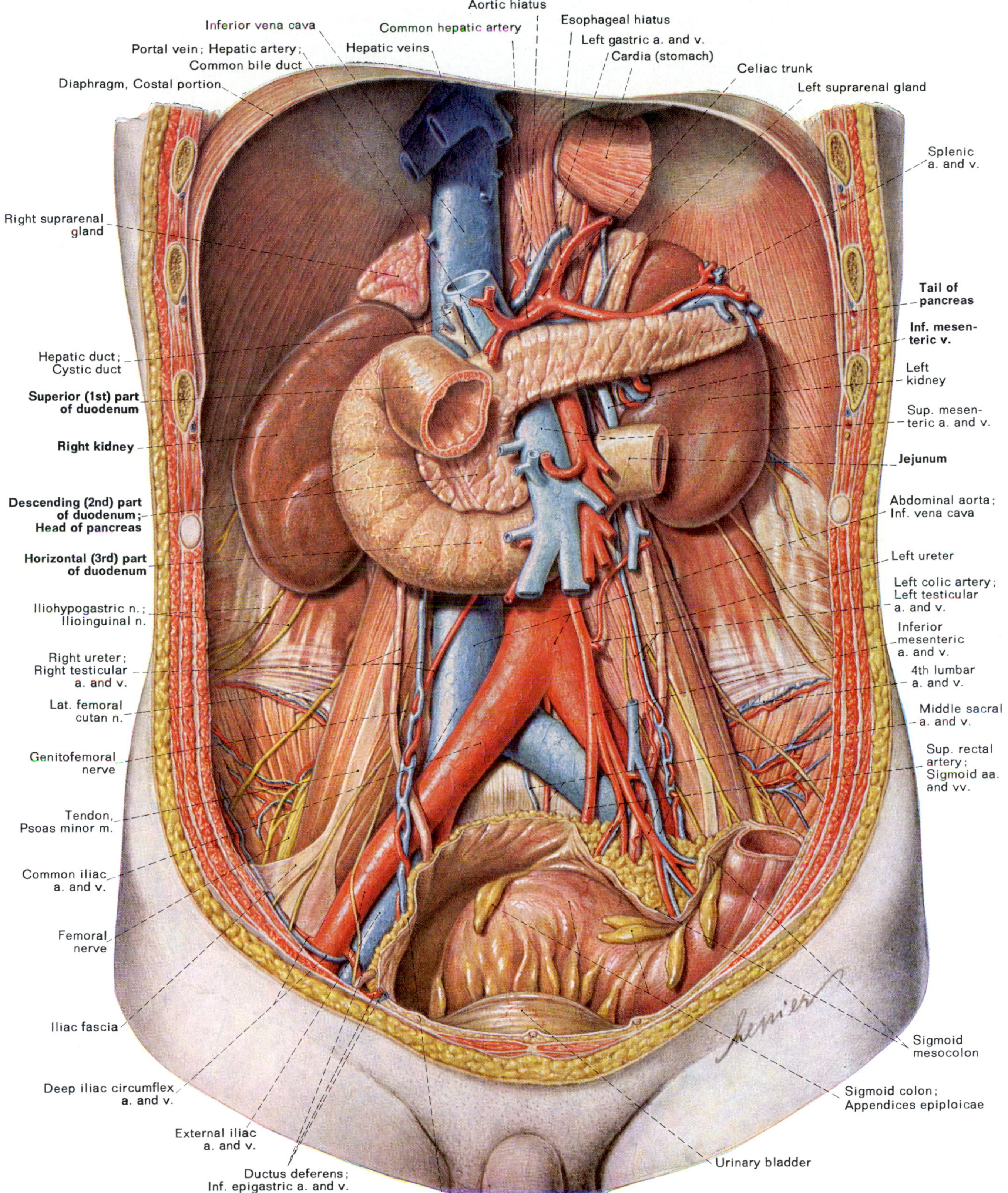
Aortic hiatus
Inferior vena cava
Common hepatic artery
Esophageal hiatus
Left gastric a. and v.
Portal vein; Hepatic artery;
Common bile duct
Hepatic veins
Cardia (stomach)
Celiac trunk
Diaphragm, Costal portion
Left suprarenal gland
Splenic a. and v.
Right suprarenal gland
Tail of pancreas
Inf. mesenteric v.
Hepatic duct; Cystic duct
Left kidney
Superior (1st) part of duodenum
Sup. mesenteric a. and v.
Right kidney
Jejunum
Descending (2nd) part of duodenum; Head of pancreas
Abdominal aorta; Inf. vena cava
Horizontal (3rd) part of duodenum
Left ureter
Left colic artery; Left testicular a. and v.
Iliohypogastric n.; Ilioinguinal n.
Inferior mesenteric a. and v.
Right ureter; Right testicular a. and v.
4th lumbar a. and v.
Lat. femoral cutan n.
Middle sacral a. and v.
Genitofemoral nerve
Sup. rectal artery; Sigmoid aa. and vv.
Tendon, Psoas minor m.
Common iliac a. and v.
Femoral nerve
Iliac fascia
Sigmoid mesocolon
Deep iliac circumflex a. and v.
Sigmoid colon; Appendices epiploicae
External iliac a. and v.
Urinary bladder
Ductus deferens; Inf. epigastric a. and v.
Peritoneum

Fig. 317: The Abdominal Cavity (12): The Retroperitoneal Organs (Male)

NOTE: 1) that the curvature of the duodenum lies ventral to the hilum of the right kidney, while the duodenojejunal junction is situated ventral to the lower medial border of the left kidney. Observe that the right kidney is slightly lower than the left.

2) the head of the pancreas lies ventral to the inferior vena cava and within the curve of the duodenum. A small extension of the head of the pancreas, the uncinate process, projects posterior to the superior mesenteric vessels. Upon crossing the midline at the level of the 1st lumbar vertebra, the posterior surface of the body and tail of the pancreas is in contact with the middle third of the left kidney.

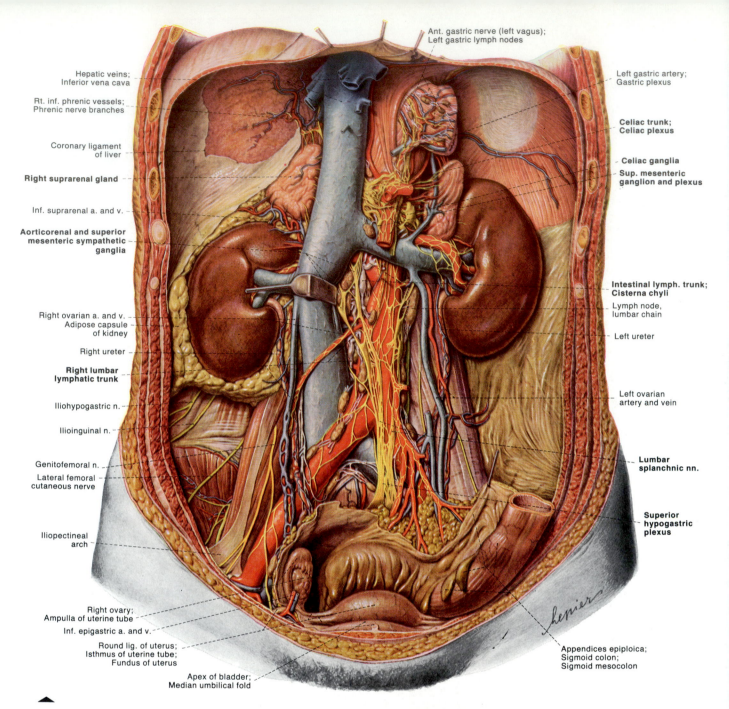

Ant. gastric nerve (left vagus); Left gastric lymph nodes
Hepatic veins; Inferior vena cava
Left gastric artery; Gastric plexus
Rt. inf. phrenic vessels; Phrenic nerve branches
Celiac trunk; Celiac plexus
Coronary ligament of liver
Celiac ganglia
Right suprarenal gland
Sup. mesenteric ganglion and plexus
Inf. suprarenal a. and v.
Aorticorenal and superior mesenteric sympathetic ganglia
Intestinal lymph. trunk; Cisterna chyli
Lymph node, lumbar chain
Right ovarian a. and v.
Adipose capsule of kidney
Left ureter
Right ureter
Right lumbar lymphatic trunk
Left ovarian artery and vein
Iliohypogastric n.
Ilioinguinal n.
Lumbar splanchnic nn.
Genitofemoral n.
Lateral femoral cutaneous nerve
Superior hypogastric plexus
Iliopectineal arch
Right ovary; Ampulla of uterine tube
Inf. epigastric a. and v.
Round lig. of uterus; Isthmus of uterine tube; Fundus of uterus
Appendices epiploica; Sigmoid colon; Sigmoid mesocolon
Apex of bladder; Median umbilical fold

Fig. 318: Abdominal Cavity (13): Lymphatics and Nerves of the Posterior Abdominal Wall

NOTE: 1) that lymphatic channels draining pelvic and posterior abdominal wall structures course along iliac nodes to right and left chains of lumbar nodes which parallel the aorta. These lumbar trunks are joined by intestinal trunks draining the G. I. tract in helping to form the cysterna chyli;

2) the sympathetic ganglia and plexuses which are found in relation to the abdominal aorta and its branches.

Fig. 319: Posterior Abdominal Wall in 5 Month Fetus

NOTE the lobulated appearance of the fetal kidneys. Even at birth this lobulation is obvious, but as maturation proceeds the surface slowly becomes smooth. Observe the large suprarenal glands and ureters, and the location of the testes.

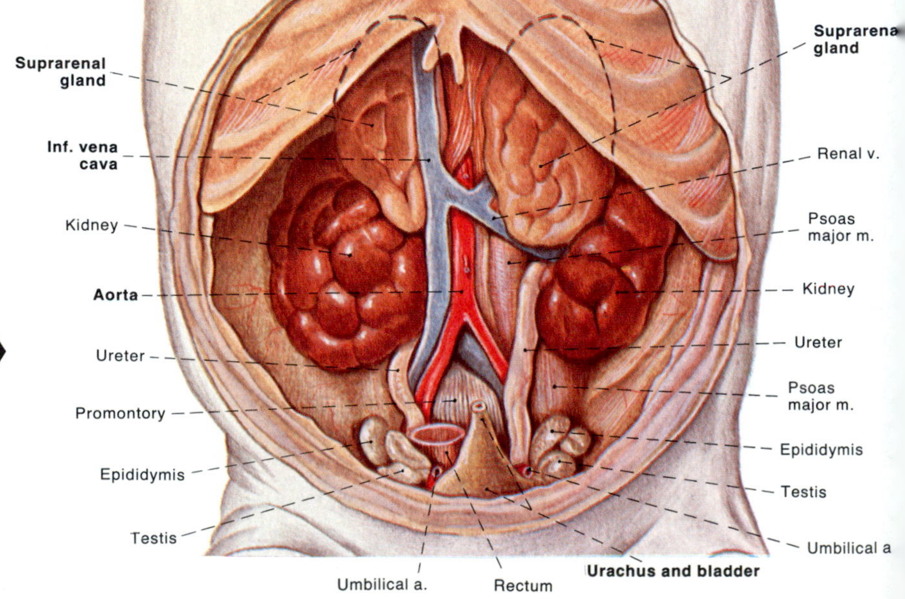

Suprarenal gland
Suprarena gland
Inf. vena cava
Renal v.
Kidney
Psoas major m.
Aorta
Kidney
Ureter
Ureter
Psoas major m.
Promontory
Epididymis
Epididymis
Testis
Testis
Umbilical a
Umbilical a.
Rectum
Urachus and bladder

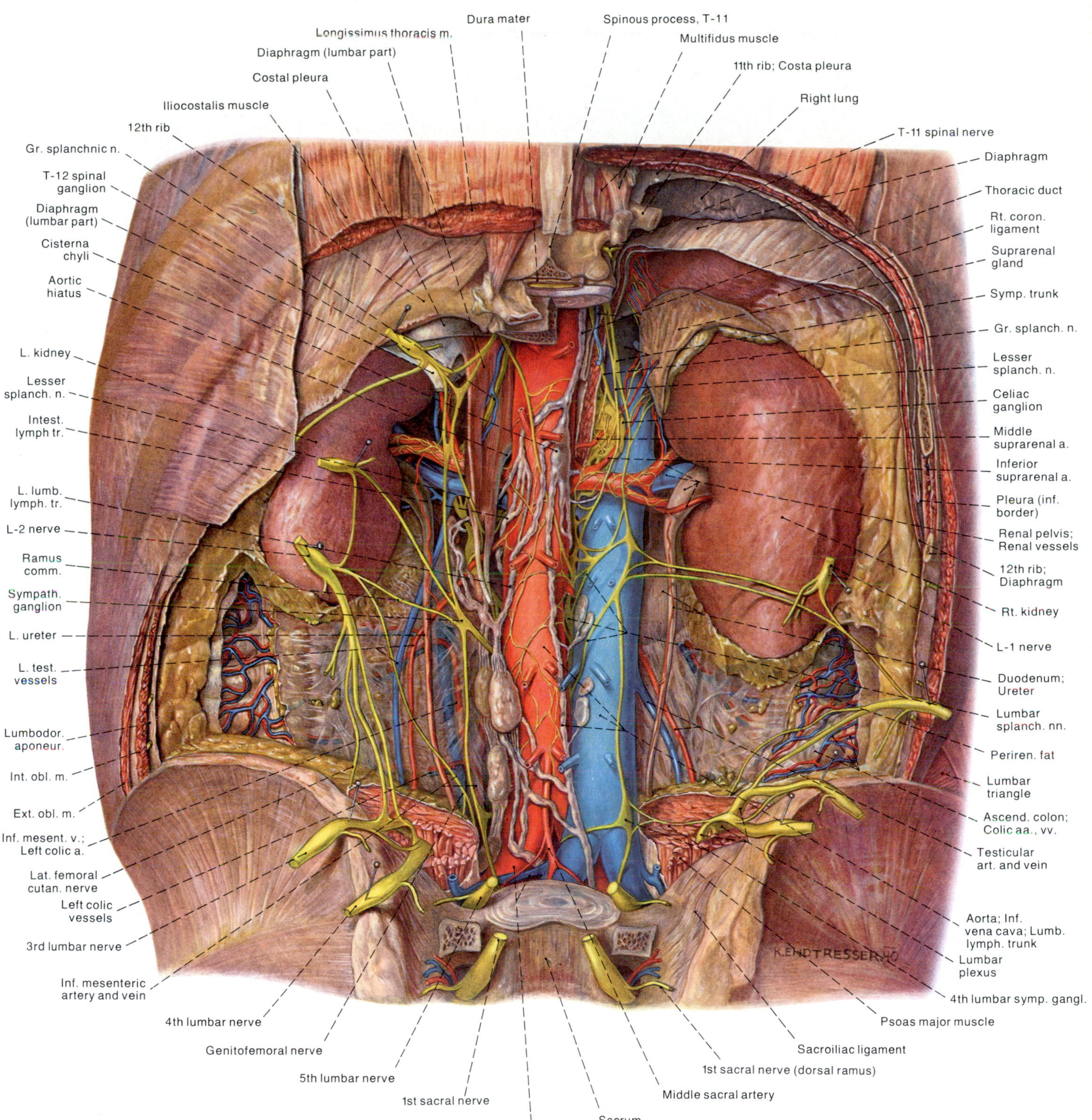

Dura mater
Spinous process, T-11
Longissimus thoracis m.
Multifidus muscle
Diaphragm (lumbar part)
11th rib; Costa pleura
Costal pleura
Right lung
Iliocostalis muscle
12th rib
T-11 spinal nerve
Gr. splanchnic n.
Diaphragm
T-12 spinal ganglion
Thoracic duct
Diaphragm (lumbar part)
Rt. coron. ligament
Cisterna chyli
Suprarenal gland
Aortic hiatus
Symp. trunk
Gr. splanch. n.
L. kidney
Lesser splanch. n.
Lesser splanch. n.
Celiac ganglion
Intest. lymph tr.
Middle suprarenal a.
Inferior suprarenal a.
L. lumb. lymph. tr.
Pleura (inf. border)
L-2 nerve
Renal pelvis; Renal vessels
Ramus comm.
12th rib; Diaphragm
Sympath. ganglion
Rt. kidney
L. ureter
L-1 nerve
L. test. vessels
Duodenum; Ureter
Lumbar splanch. nn.
Lumbodor. aponeur.
Periren. fat
Int. obl. m.
Lumbar triangle
Ext. obl. m.
Ascend. colon; Colic aa., vv.
Inf. mesent. v.; Left colic a.
Testicular art. and vein
Lat. femoral cutan. nerve
Left colic vessels
Aorta; Inf. vena cava; Lumb. lymph. trunk
3rd lumbar nerve
Lumbar plexus
Inf. mesenteric artery and vein
4th lumbar symp. gangl.
4th lumbar nerve
Psoas major muscle
Genitofemoral nerve
Sacroiliac ligament
5th lumbar nerve
1st sacral nerve (dorsal ramus)
1st sacral nerve
Middle sacral artery
Sacrum
Common iliac artery; Left iliolumbar vein
K. ENDTRESSER '70

Fig. 320: Abdominal Cavity (14): Posterior Abdominal Retroperitoneal Structures in the Lumbar Region, Viewed from Behind

NOTE: 1) that in this posterior view of the lumbar retroperitoneal space, the vertebral column (vertebrae T-12 to L-5) has been removed, as have the other soft tissue structures that form the posterior abdominal wall. On the right side the 11th and 12th ribs have also been resected in order to visualize the upper pole of the right kidney and the superimposed right suprarenal gland.

2) the inferior vena cava ascending to the right side of the abdominal aorta. Observe that the cut lumbar arteries and veins (unlabelled) join these great vessels at each segmental level in the lumbar region. Also note the lumbar lymphatic trunks ascending toward the cisterna chyli, which is located to the side of the aorta at about the L-1 level near the origin of the right renal artery. Closeby, observe the position of the celiac ganglion.

3) that the renal arteries send ascending branches to the suprarenal glands and descending branches to the ureters.

4) that the lumbar spinal nerves, having been severed from the spinal cord when the vertebral column was removed, are reflected to each side and pinned to surrounding structures. Some of their connections, however, to the lumbar sympathetic ganglia and trunks have been left intact.

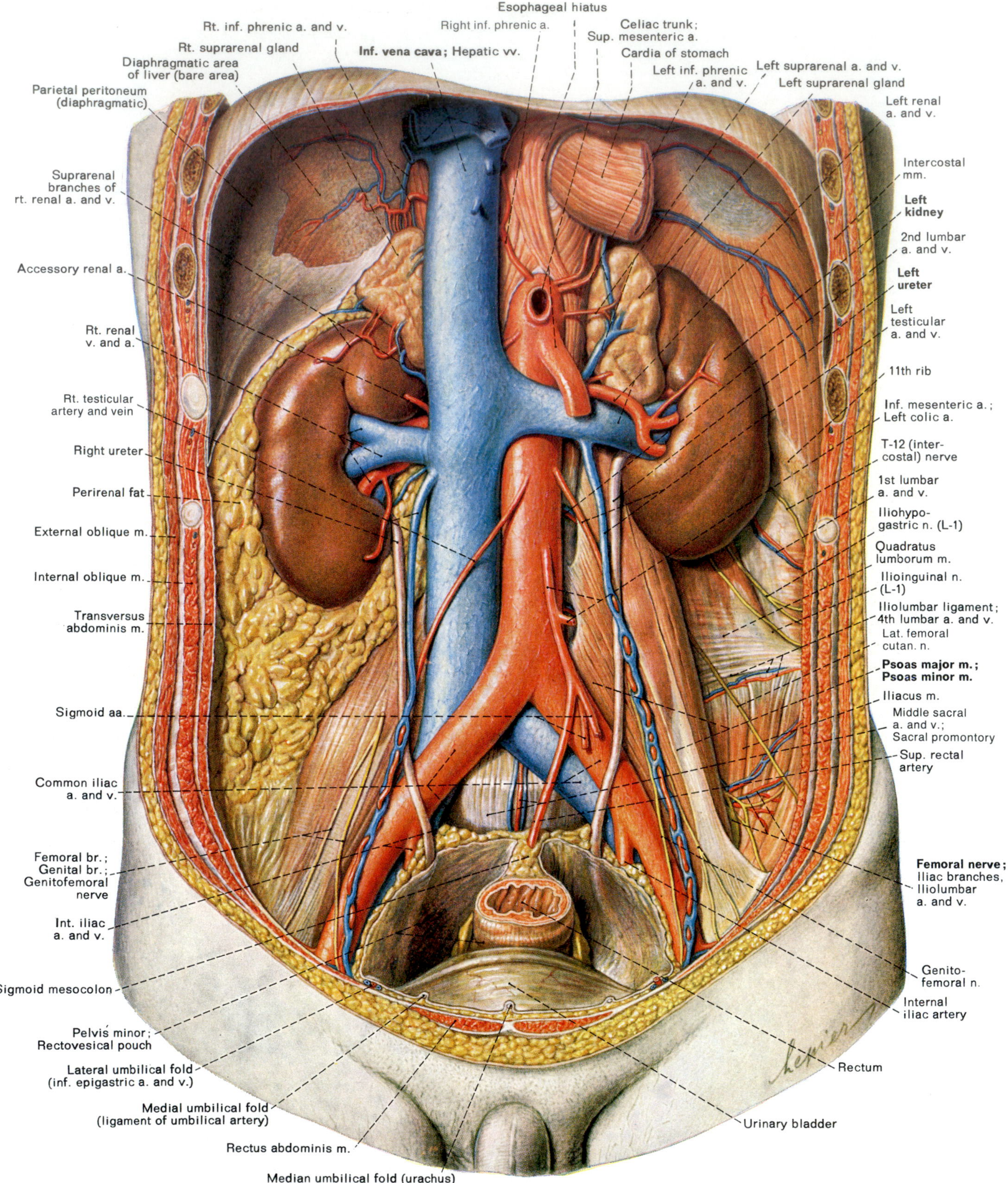
Esophageal hiatus
Rt. inf. phrenic a. and v.
Right inf. phrenic a.
Celiac trunk;
Sup. mesenteric a.
Rt. suprarenal gland
Inf. vena cava; Hepatic vv.
Cardia of stomach
Diaphragmatic area
of liver (bare area)
Left inf. phrenic
a. and v.
Left suprarenal a. and v.
Left suprarenal gland
Parietal peritoneum
(diaphragmatic)
Left renal
a. and v.
Intercostal
mm.
Suprarenal
branches of
rt. renal a. and v.
Left
kidney
2nd lumbar
a. and v.
Accessory renal a.
Left
ureter
Left
testicular
a. and v.
Rt. renal
v. and a.
11th rib
Rt. testicular
artery and vein
Inf. mesenteric a.;
Left colic a.
Right ureter
T-12 (inter-
costal) nerve
Perirenal fat
1st lumbar
a. and v.
External oblique m.
Iliohypo-
gastric n. (L-1)
Internal oblique m.
Quadratus
lumborum m.
Ilioinguinal n.
(L-1)
Transversus
abdominis m.
Iliolumbar ligament;
4th lumbar a. and v.
Lat. femoral
cutan. n.
Psoas major m.;
Psoas minor m.
Iliacus m.
Sigmoid aa.
Middle sacral
a. and v.;
Sacral promontory
Sup. rectal
artery
Common iliac
a. and v.
Femoral br.;
Genital br.;
Genitofemoral
nerve
Femoral nerve;
Iliac branches,
Iliolumbar
a. and v.
Int. iliac
a. and v.
Genito-
femoral n.
Sigmoid mesocolon
Internal
iliac artery
Pelvis minor;
Rectovesical pouch
Rectum
Lateral umbilical fold
(inf. epigastric a. and v.)
Medial umbilical fold
(ligament of umbilical artery)
Urinary bladder
Rectus abdominis m.
Median umbilical fold (urachus)

Fig. 321: The Abdominal Cavity (15): The Posterior Abdominal Wall

NOTE: 1) the kidneys, ureters, suprarenal glands, and the great vessels and their branches. The kidneys extend between vertebral levels T-12 and L-3. The ureters commence on the posterior aspect of the hilum of the kidneys, course inferiorly over the pelvic brim near the bifurcation of the common iliac arteries, eventually to terminate in the bladder.

2) the suprarenal glands as they cap the upper pole of each kidney. These are highly vascular and vital glands of internal secretion (endocrine).

3) the aorta which enters the abdomen through the aortic hiatus (T-12). The inferior vena cava lies to the right of the vertebral column. It is formed by the confluence of the two common iliac veins (at about L-5) and passes through the caval opening of the diaphragm.

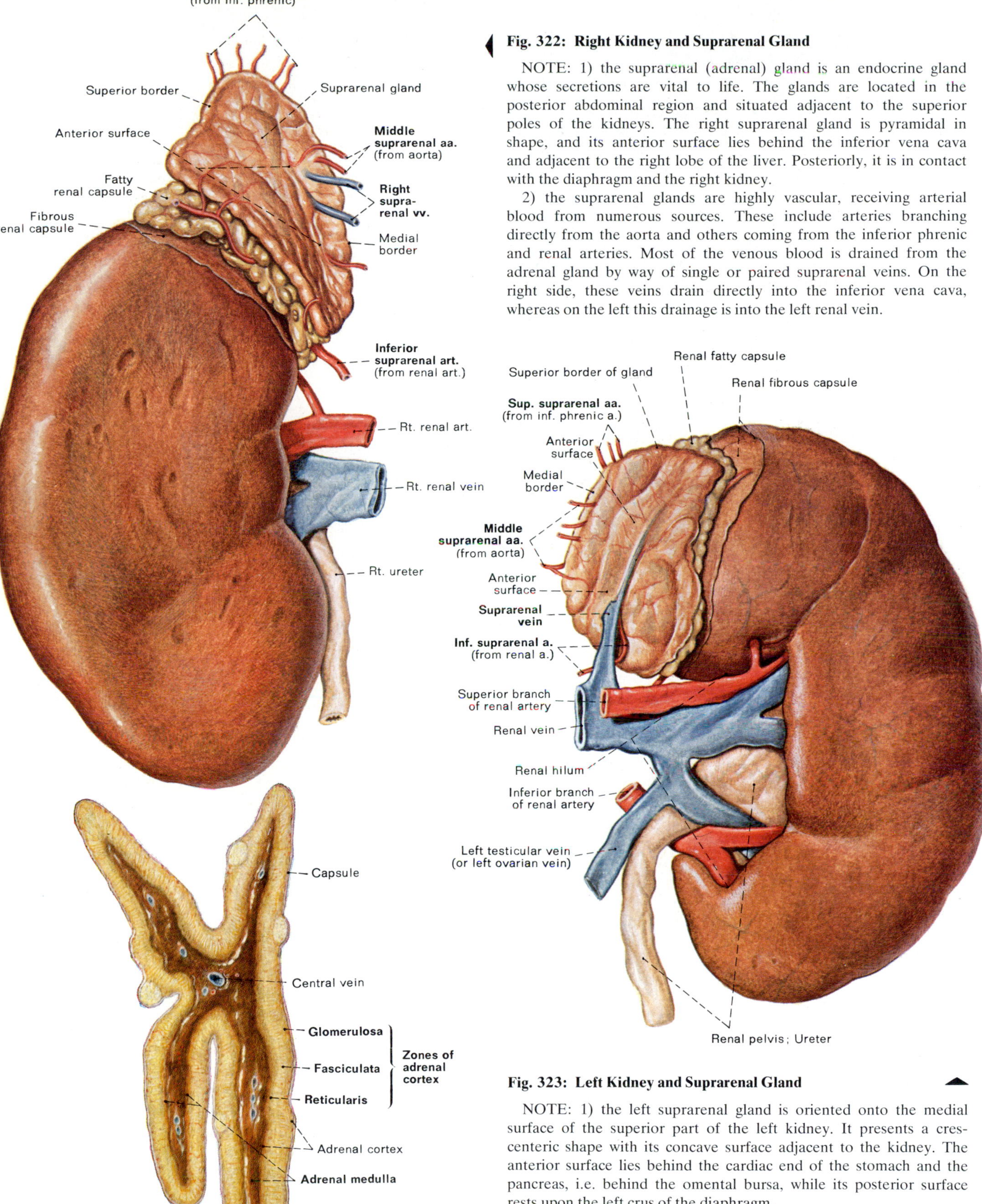

Superior suprarenal aa.
(from inf. phrenic)
Superior border
Suprarenal gland
Anterior surface
Middle
suprarenal aa.
(from aorta)
Fatty
renal capsule
Right
supra-
renal vv.
Fibrous
renal capsule
Medial
border
Inferior
suprarenal art.
(from renal art.)
Rt. renal art.
Rt. renal vein
Rt. ureter
Renal fatty capsule
Superior border of gland
Renal fibrous capsule
Sup. suprarenal aa.
(from inf. phrenic a.)
Anterior
surface
Medial
border
Middle
suprarenal aa.
(from aorta)
Anterior
surface
Suprarenal
vein
Inf. suprarenal a.
(from renal a.)
Superior branch
of renal artery
Renal vein
Renal hilum
Inferior branch
of renal artery
Left testicular vein
(or left ovarian vein)
Renal pelvis; Ureter
Capsule
Central vein
Glomerulosa
Fasciculata
Reticularis
Zones of
adrenal
cortex
Adrenal cortex
Adrenal medulla

Fig. 322: Right Kidney and Suprarenal Gland

NOTE: 1) the suprarenal (adrenal) gland is an endocrine gland whose secretions are vital to life. The glands are located in the posterior abdominal region and situated adjacent to the superior poles of the kidneys. The right suprarenal gland is pyramidal in shape, and its anterior surface lies behind the inferior vena cava and adjacent to the right lobe of the liver. Posteriorly, it is in contact with the diaphragm and the right kidney.

2) the suprarenal glands are highly vascular, receiving arterial blood from numerous sources. These include arteries branching directly from the aorta and others coming from the inferior phrenic and renal arteries. Most of the venous blood is drained from the adrenal gland by way of single or paired suprarenal veins. On the right side, these veins drain directly into the inferior vena cava, whereas on the left this drainage is into the left renal vein.

Fig. 323: Left Kidney and Suprarenal Gland

NOTE: 1) the left suprarenal gland is oriented onto the medial surface of the superior part of the left kidney. It presents a crescenteric shape with its concave surface adjacent to the kidney. The anterior surface lies behind the cardiac end of the stomach and the pancreas, i.e. behind the omental bursa, while its posterior surface rests upon the left crus of the diaphragm.

2) the combined weight of both suprarenal glands averages about 10 grams. The glands are each surrounded by an investing fibrous capsule, around which is a variable amount of areolar tissue.

Fig. 324: Section Through Suprarenal Gland

NOTE that the suprarenal gland is composed of an inner medulla and outer cortex. The cells of the medulla secrete epinephrine while the adrenal cortex secretes adrenal corticosteroids.

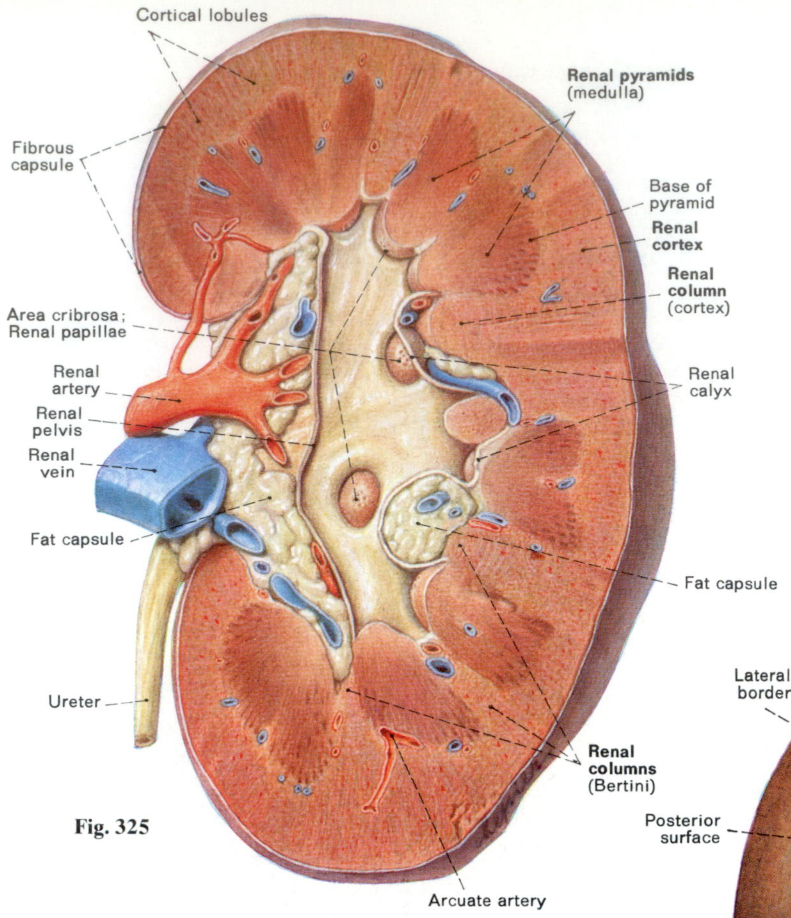

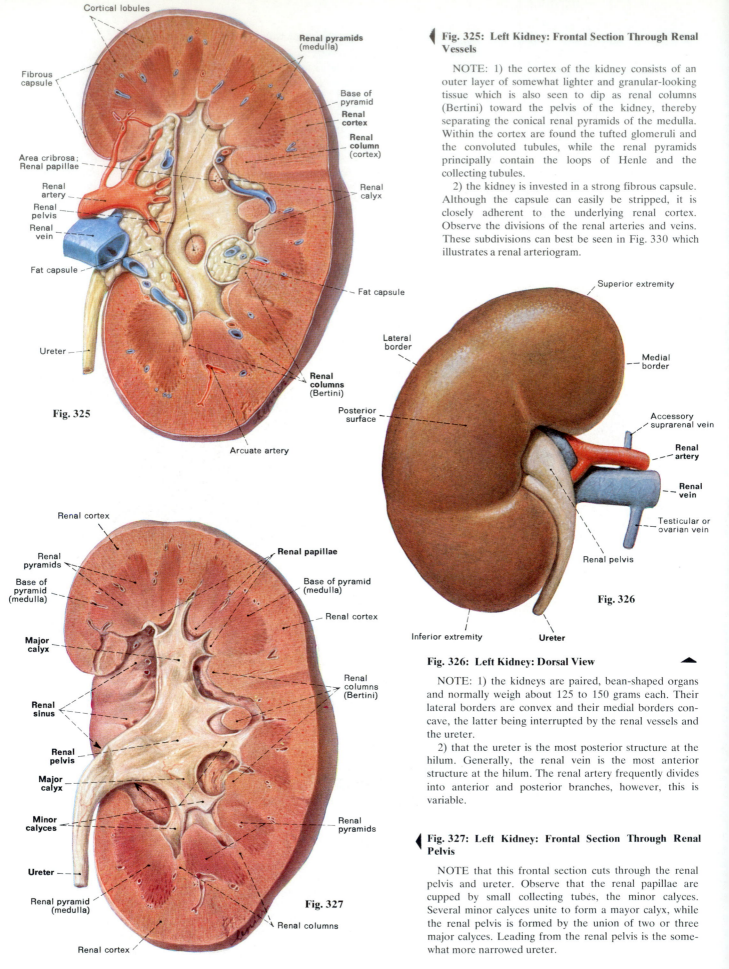

Cortical lobules
Renal pyramids (medulla)
Fibrous capsule
Base of pyramid
Renal cortex
Renal column (cortex)
Area cribrosa; Renal papillae
Renal artery
Renal calyx
Renal pelvis
Renal vein
Fat capsule
Fat capsule
Ureter
Renal columns (Bertini)
Arcuate artery

Fig. 325

Fig. 325: Left Kidney: Frontal Section Through Renal Vessels

NOTE: 1) the cortex of the kidney consists of an outer layer of somewhat lighter and granular-looking tissue which is also seen to dip as renal columns (Bertini) toward the pelvis of the kidney, thereby separating the conical renal pyramids of the medulla. Within the cortex are found the tufted glomeruli and the convoluted tubules, while the renal pyramids principally contain the loops of Henle and the collecting tubules.

2) the kidney is invested in a strong fibrous capsule. Although the capsule can easily be stripped, it is closely adherent to the underlying renal cortex. Observe the divisions of the renal arteries and veins. These subdivisions can best be seen in Fig. 330 which illustrates a renal arteriogram.

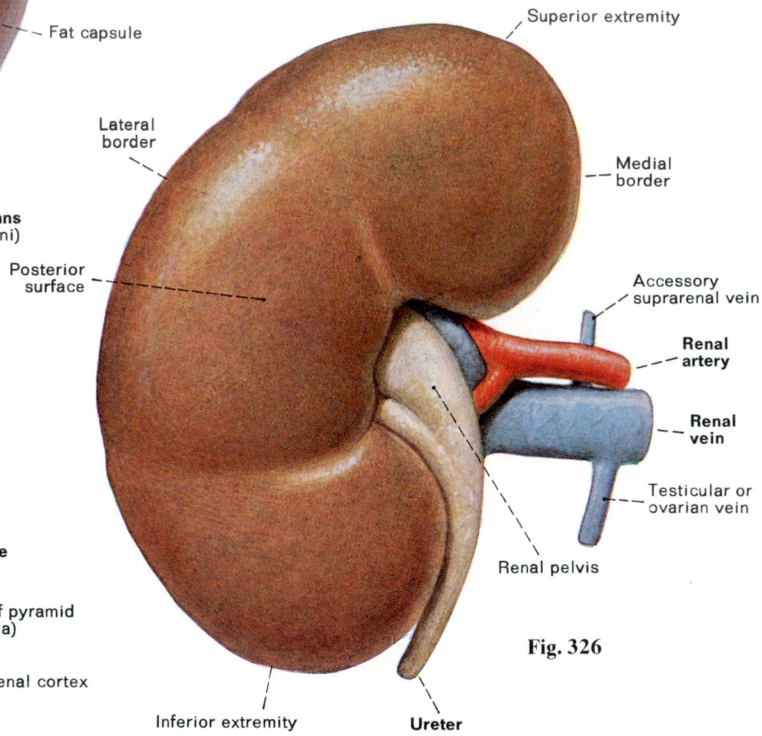

Superior extremity
Lateral border
Medial border
Posterior surface
Accessory suprarenal vein
Renal artery
Renal vein
Testicular or ovarian vein
Renal pelvis
Inferior extremity
Ureter

Fig. 326

Fig. 326: Left Kidney: Dorsal View

NOTE: 1) the kidneys are paired, bean-shaped organs and normally weigh about 125 to 150 grams each. Their lateral borders are convex and their medial borders concave, the latter being interrupted by the renal vessels and the ureter.

2) that the ureter is the most posterior structure at the hilum. Generally, the renal vein is the most anterior structure at the hilum. The renal artery frequently divides into anterior and posterior branches, however, this is variable.

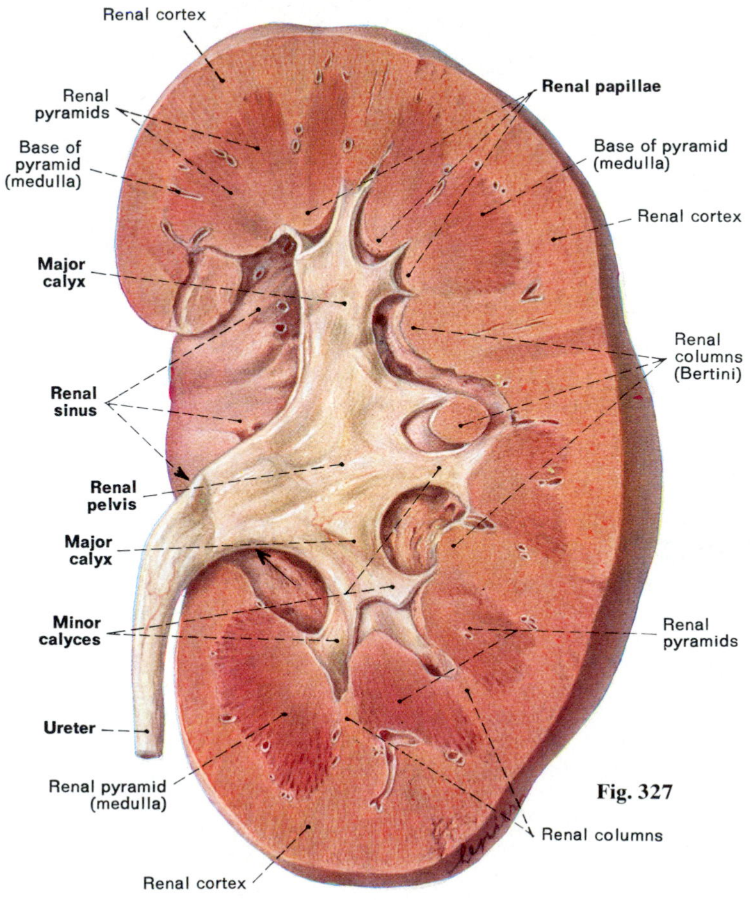

Renal cortex
Renal pyramids
Base of pyramid (medulla)
Renal papillae
Base of pyramid (medulla)
Renal cortex
Major calyx
Renal columns (Bertini)
Renal sinus
Renal pelvis
Major calyx
Minor calyces
Renal pyramids
Ureter
Renal pyramid (medulla)
Renal columns
Renal cortex

Fig. 327

Fig. 327: Left Kidney: Frontal Section Through Renal Pelvis

NOTE that this frontal section cuts through the renal pelvis and ureter. Observe that the renal papillae are cupped by small collecting tubés, the minor calyces. Several minor calyces unite to form a mayor calyx, while the renal pelvis is formed by the union of two or three major calyces. Leading from the renal pelvis is the somewhat more narrowed ureter.

Fig. 328: Retrograde Pyelogram

NOTE: 1) a radioopaque substance has been introduced into each ureter and forced in a retrograde manner into the renal pelvis, major calyces and minor calyces of each side. Observe that into the minor calyces project the renal papillae (5), resulting in radiolucent invaginations into the radio-opaque minor calyces.

2) the shadow of the superior extremity of the left kidney extending to the top of the body of the T-12 vertebra, while the right kidney is somewhat more inferior.

3) the lateral margins of the psoas major muscles. The ureters course toward the pelvis along the anterior surfaces of these muscles.

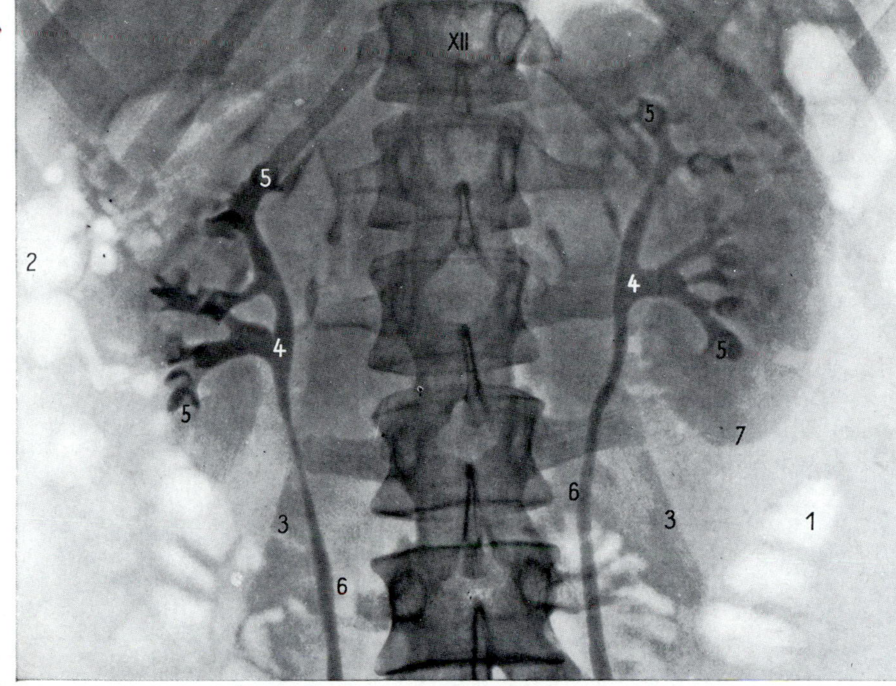

1 Descending colon
2 Ascending colon
3 Psoas major muscle (lateral border)
4 Renal pelvis
5 Renal papilla
6 Ureter
7 Inferior pole of left kidney
XII 12th thoracic vertebra

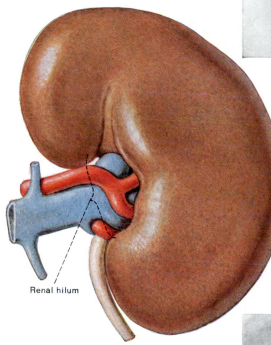

Fig. 329: Left Kidney: Ventral View

NOTE: 1) although the stem of the renal artery initially lies dorsal to the stem of the renal vein at the hilum, the renal artery divides into anterior and posterior branches. The anterior branch generally enters the kidney ventral to the renal vein, as is shown in this figure.

2) the anterior surfaces of the kidneys are covered by peritoneum and are in relationship to more ventrally located abdominal organs. In contrast, their posterior surfaces lie adjacent to the musculature of the posterior abdominal wall and the diaphragm.

Renal hilum

Fig. 330: Arteriogram of Right Renal Artery and its Branches

NOTE: 1) an arterial catheter (K) has been inserted into the femoral artery, passed through the abdominal aorta and into the right renal artery. Observe the division of the renal artery successively into interlobar arteries (4).

2) as the interlobar arteries reach the junction of the renal cortex and medulla, they arch over the bases of the pyramids forming the arcuate arteries (not numbered in this figure). From the arcuate arteries branch a series of interlobular arteries which extend through the afferent arterioles which enter the renal glomeruli.

3) the stomach (1) and superior (1st) part of the duodenum (2) which are filled with air.

1 Stomach
2 Superior (1st) part of duodenum
3 Right renal artery
4 Interlobar arteries
5 Interlobular arteries
K Catheter
XII 12th thoracic vertebra

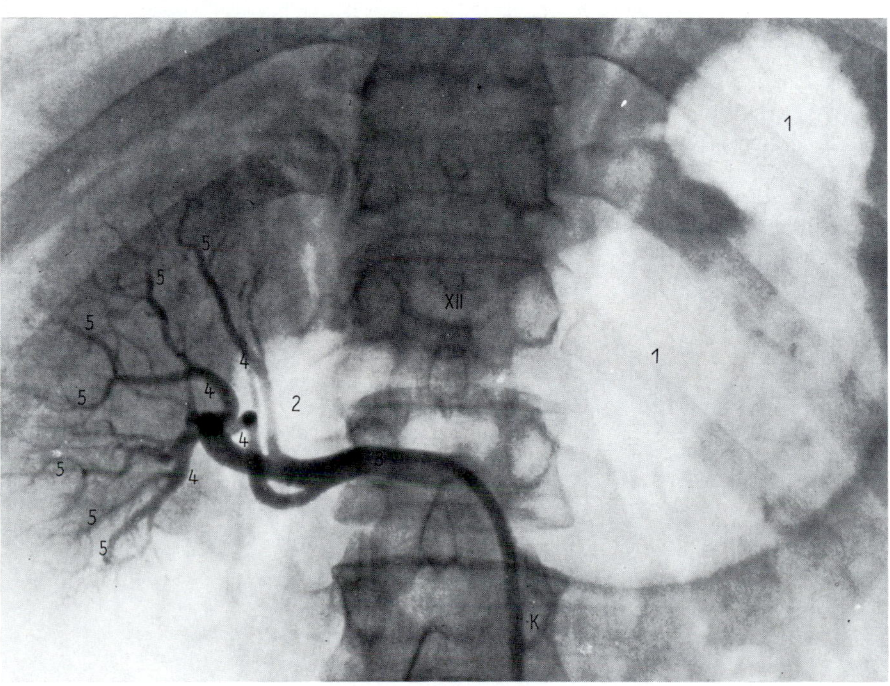

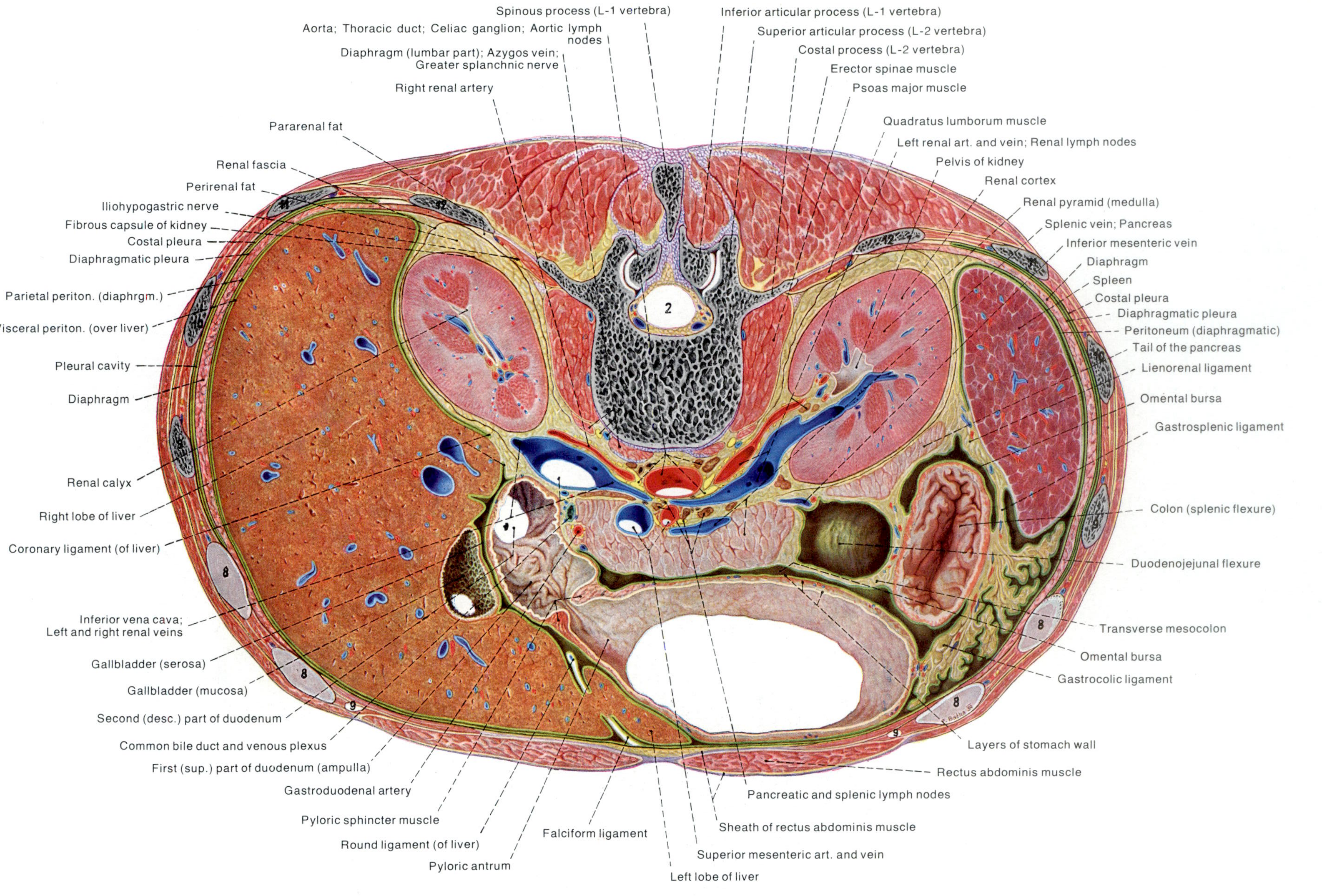
Spinous process (L-1 vertebra)
Aorta; Thoracic duct; Celiac ganglion; Aortic lymph nodes
Diaphragm (lumbar part); Azygos vein; Greater splanchnic nerve
Right renal artery
Pararenal fat
Renal fascia
Perirenal fat
Iliohypogastric nerve
Fibrous capsule of kidney
Costal pleura
Diaphragmatic pleura
Parietal periton. (diaphrgm.)
Visceral periton. (over liver)
Pleural cavity
Diaphragm
Renal calyx
Right lobe of liver
Coronary ligament (of liver)
Inferior vena cava; Left and right renal veins
Gallbladder (serosa)
Gallbladder (mucosa)
Second (desc.) part of duodenum
Common bile duct and venous plexus
First (sup.) part of duodenum (ampulla)
Gastroduodenal artery
Pyloric sphincter muscle
Round ligament (of liver)
Pyloric antrum
Falciform ligament
Left lobe of liver
Superior mesenteric art. and vein
Sheath of rectus abdominis muscle
Pancreatic and splenic lymph nodes
Rectus abdominis muscle
Layers of stomach wall
Gastrocolic ligament
Omental bursa
Transverse mesocolon
Duodenojejunal flexure
Colon (splenic flexure)
Gastrosplenic ligament
Omental bursa
Lienorenal ligament
Tail of the pancreas
Peritoneum (diaphragmatic)
Diaphragmatic pleura
Costal pleura
Spleen
Diaphragm
Inferior mesenteric vein
Splenic vein; Pancreas
Renal pyramid (medulla)
Renal cortex
Pelvis of kidney
Left renal art. and vein; Renal lymph nodes
Quadratus lumborum muscle
Psoas major muscle
Erector spinae muscle
Costal process (L-2 vertebra)
Superior articular process (L-2 vertebra)
Inferior articular process (L-1 vertebra)

Fig. 331: Transverse Section Through the Abdomen at the Level of the Body of the Second Lumbar Vertebra

NOTE: 1) that through a transverse section at this level are found the pylorus of the stomach and the first (or superior) part of the duodenum, from which can be seen continuing inferiorly the second (or descending) part of the duodenum. Observe also the gallbladder, the pancreas (dorsal to which are seen the superior mesenteric vessels), the duodenal-jejunal flexure of the small intestine and the left colic (or splenic) flexure of the large intestine.

2) that the left kidney (reader's right) is sectioned across the lower part of the left renal hilum, whereas the hilum of the right kidney is found below the level of this transection and shows well in Figure 332. Observe the long course of the left renal vein to the inferior vena cava, and that the latter vessel is located retroperitoneally behind the pancreas and duodenum and to the right side of the aorta.

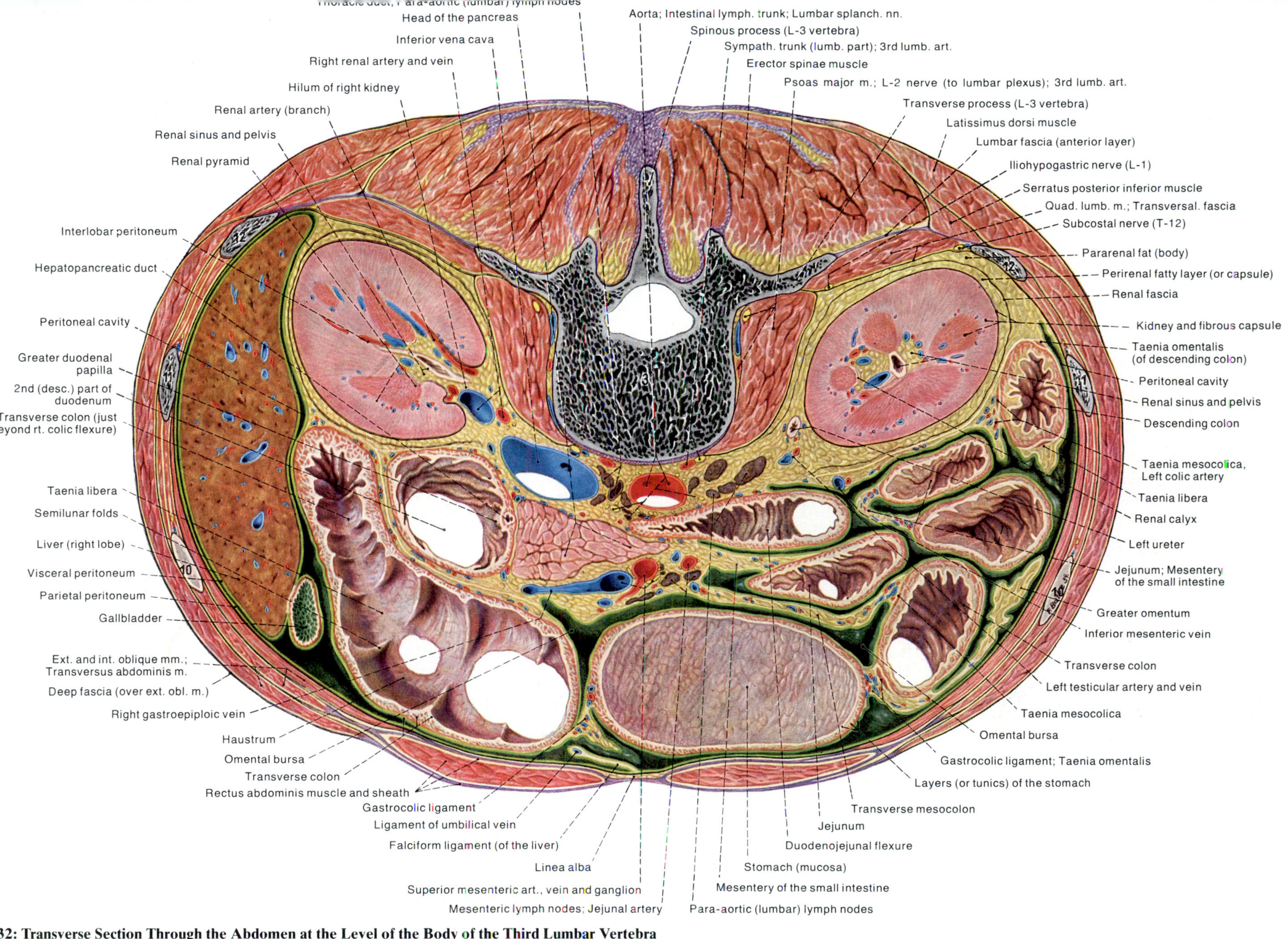

Thoracic duct; Para-aortic (lumbar) lymph nodes
Head of the pancreas
Inferior vena cava
Right renal artery and vein
Hilum of right kidney
Renal artery (branch)
Renal sinus and pelvis
Renal pyramid
Interlobar peritoneum
Hepatopancreatic duct
Peritoneal cavity
Greater duodenal papilla
2nd (desc.) part of duodenum
Transverse colon (just beyond rt. colic flexure)
Taenia libera
Semilunar folds
Liver (right lobe)
Visceral peritoneum
Parietal peritoneum
Gallbladder
Ext. and int. oblique mm.; Transversus abdominis m.
Deep fascia (over ext. obl. m.)
Right gastroepiploic vein
Haustrum
Omental bursa
Transverse colon
Rectus abdominis muscle and sheath
Gastrocolic ligament
Ligament of umbilical vein
Falciform ligament (of the liver)
Linea alba
Superior mesenteric art., vein and ganglion
Mesenteric lymph nodes; Jejunal artery
Para-aortic (lumbar) lymph nodes
Mesentery of the small intestine
Stomach (mucosa)
Duodenojejunal flexure
Jejunum
Transverse mesocolon
Layers (or tunics) of the stomach
Gastrocolic ligament; Taenia omentalis
Omental bursa
Taenia mesocolica
Left testicular artery and vein
Transverse colon
Inferior mesenteric vein
Greater omentum
Jejunum; Mesentery of the small intestine
Left ureter
Renal calyx
Taenia libera
Taenia mesocolica, Left colic artery
Descending colon
Renal sinus and pelvis
Peritoneal cavity
Taenia omentalis (of descending colon)
Kidney and fibrous capsule
Renal fascia
Perirenal fatty layer (or capsule)
Pararenal fat (body)
Subcostal nerve (T-12)
Quad. lumb. m.; Transversal. fascia
Serratus posterior inferior muscle
Iliohypogastric nerve (L-1)
Lumbar fascia (anterior layer)
Latissimus dorsi muscle
Transverse process (L-3 vertebra)
Psoas major m.; L-2 nerve (to lumbar plexus); 3rd lumb. art.
Erector spinae muscle
Sympath. trunk (lumb. part); 3rd lumb. art.
Spinous process (L-3 vertebra)
Aorta; Intestinal lymph. trunk; Lumbar splanch. nn.

Fig. 332: Transverse Section Through the Abdomen at the Level of the Body of the Third Lumbar Vertebra

NOTE: 1) the hilum of the right kidney, the descending (2nd) part of the duodenum with its greater duodenal papilla and sphincter of Oddi, the head of the pancreas just to the left of the descending part of the duodenum, and the transverse colon just beyond the right (hepatic) colic flexure. Also observe that the superior mesenteric vessels are anterior to the horizontal (3rd) part of the duodenum.

2) that the kidney is closely invested by a fibrous capsule. External to this capsule is a *perirenal* fatty layer, which is invested by the *renal fascia*. External to the renal fascia is the *pararenal fat* (body), which lies adjacent to the muscular structures of the posterior abdominal wall.

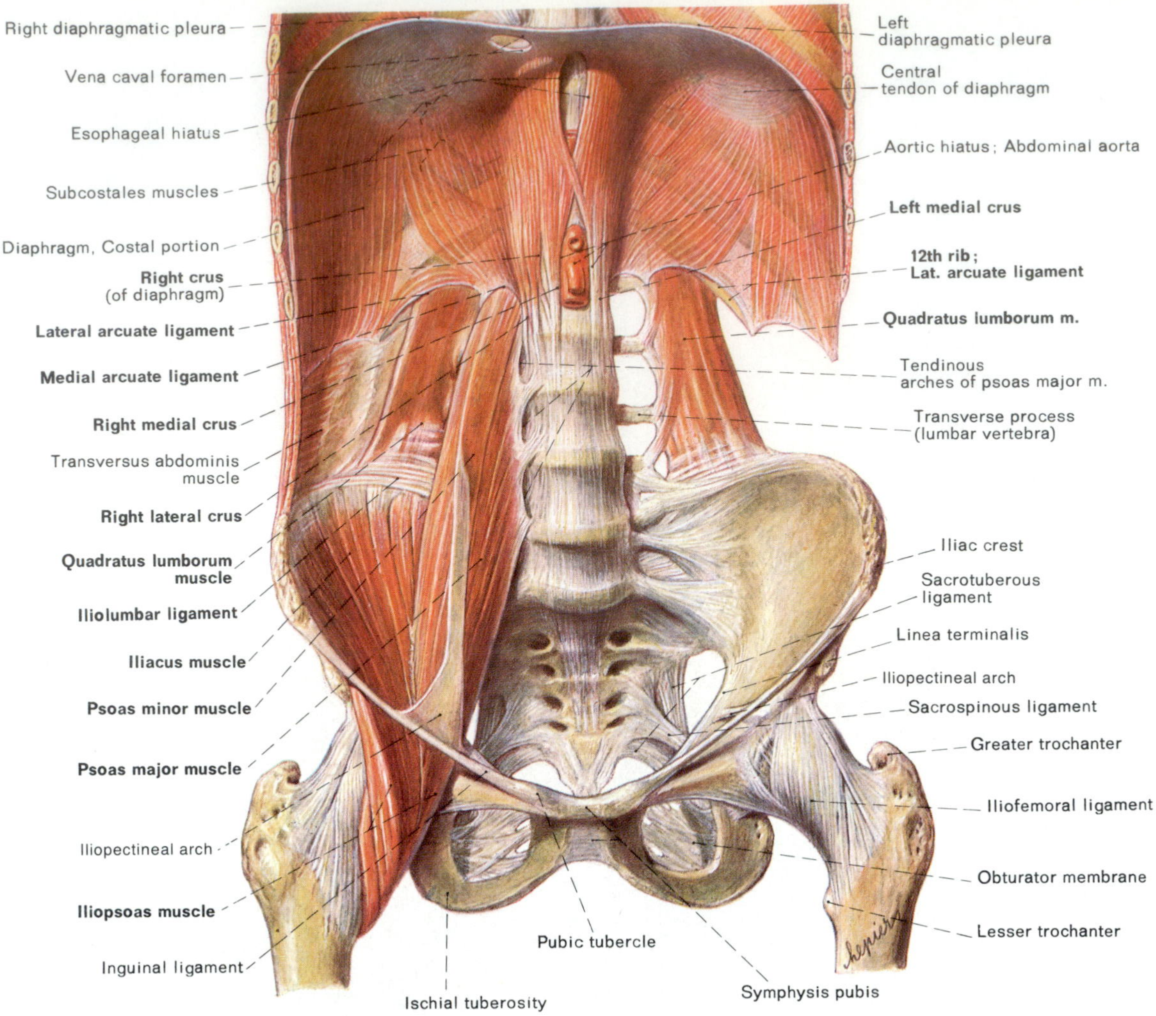

Right diaphragmatic pleura
Vena caval foramen
Esophageal hiatus
Subcostales muscles
Diaphragm, Costal portion
Right crus
(of diaphragm)
Lateral arcuate ligament
Medial arcuate ligament
Right medial crus
Transversus abdominis
muscle
Right lateral crus
Quadratus lumborum
muscle
Iliolumbar ligament
Iliacus muscle
Psoas minor muscle
Psoas major muscle
Iliopectineal arch
Iliopsoas muscle
Inguinal ligament
Ischial tuberosity
Pubic tubercle
Symphysis pubis
Left
diaphragmatic pleura
Central
tendon of diaphragm
Aortic hiatus; Abdominal aorta
Left medial crus
12th rib;
Lat. arcuate ligament
Quadratus lumborum m.
Tendinous
arches of psoas major m.
Transverse process
(lumbar vertebra)
Iliac crest
Sacrotuberous
ligament
Linea terminalis
Iliopectineal arch
Sacrospinous ligament
Greater trochanter
Iliofemoral ligament
Obturator membrane
Lesser trochanter

Fig. 333: The Diaphragm and Posterior Abdominal Wall Muscles

NOTE: 1) the posterior attachments of the diaphragm include a) the right and left crura which arise from the anterior and lateral aspects of the bodies of the upper 3 or 4 lumbar vertebrae, b) the right and left medial arcuate ligaments which are thickenings in the psoas fascia, and c) the lateral arcuate ligaments along the 12th rib overlying the quadratus lumborum muscle.

2) the psoas major and minor muscles descending from the lumbar vertebrae to join the iliacus in the lateral wall of the pelvis to insert onto the lesser trochanter of the femur. The iliopsoas is the most powerful flexor of the thigh at the hip joint.

3) the rectangular quadratus lumborum intervening between the 12th rib, the transverse processes of the lumbar vertebrae and the posteromedial iliac crest.

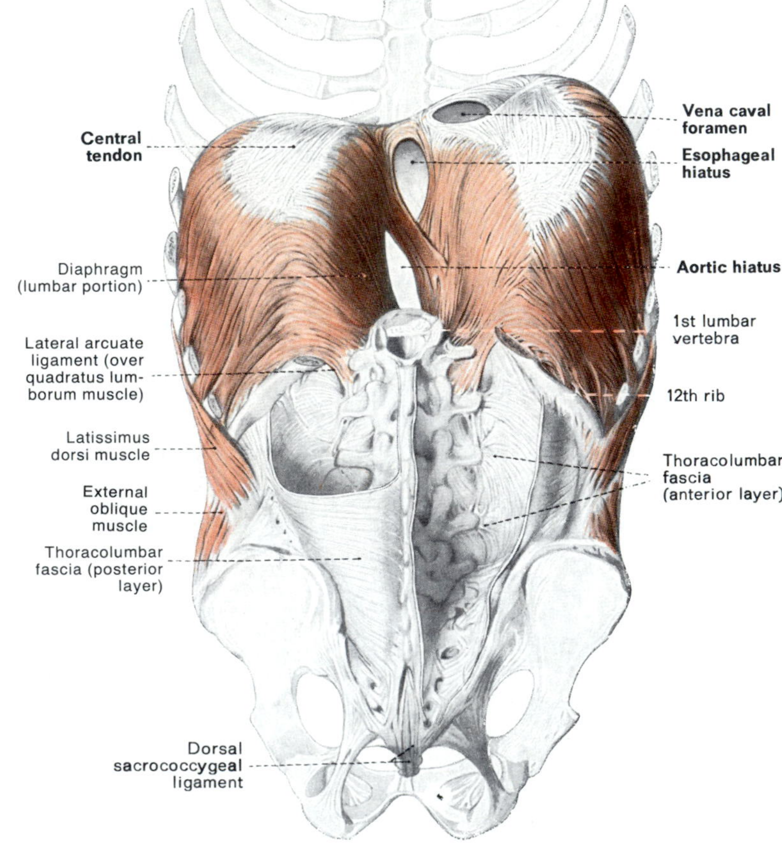

Central
tendon
Diaphragm
(lumbar portion)
Lateral arcuate
ligament (over
quadratus lum-
borum muscle)
Latissimus
dorsi muscle
External
oblique
muscle
Thoracolumbar
fascia (posterior
layer)
Dorsal
sacrococcygeal
ligament
Vena caval
foramen
Esophageal
hiatus
Aortic hiatus
1st lumbar
vertebra
12th rib
Thoracolumbar
fascia
(anterior layer)

Fig. 334: Posterior View of Diaphragm

NOTE: 1) the posterior half of the bony thorax (ribs and vertebral column) has been removed to reveal the diaphragm from behind. Observe that the caval opening is to the right of midline and more superior to those transmitting the esophagus and aorta.

2) that from their origin around the thoracic outlet, the muscular fibers of the diaphragm converge to insert into a central tendon. Observe the dome-shape of the diaphragm on each side and that the right dome extends more superiorly into the thoracic cavity because of the large right lobe of the liver.

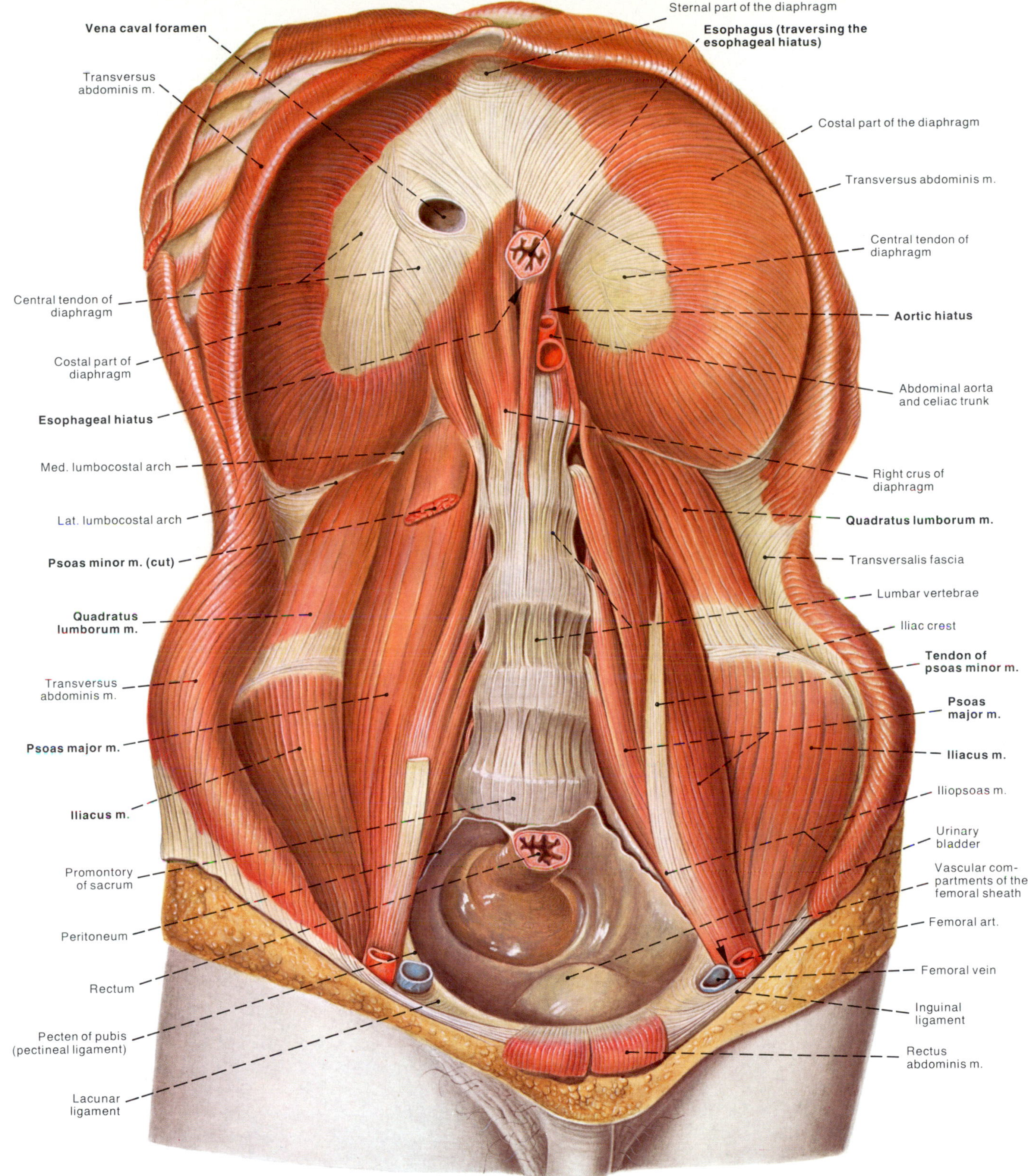

Sternal part of the diaphragm
Vena caval foramen
Esophagus (traversing the esophageal hiatus)
Transversus abdominis m.
Costal part of the diaphragm
Transversus abdominis m.
Central tendon of diaphragm
Central tendon of diaphragm
Aortic hiatus
Costal part of diaphragm
Abdominal aorta and celiac trunk
Esophageal hiatus
Med. lumbocostal arch
Right crus of diaphragm
Lat. lumbocostal arch
Quadratus lumborum m.
Psoas minor m. (cut)
Transversalis fascia
Lumbar vertebrae
Quadratus lumborum m.
Iliac crest
Tendon of psoas minor m.
Transversus abdominis m.
Psoas major m.
Psoas major m.
Iliacus m.
Iliopsoas m.
Iliacus m.
Urinary bladder
Promontory of sacrum
Vascular compartments of the femoral sheath
Peritoneum
Femoral art.
Femoral vein
Rectum
Inguinal ligament
Pecten of pubis (pectineal ligament)
Rectus abdominis m.
Lacunar ligament

Fig. 335: The Psoas Minor, Psoas Major, Iliacus, and Quadratus Lumborum Muscles and the Diaphragm

NOTE: 1) the *psoas minor muscle* lies anterior to the psoas major. It arises from the lateral aspect of the bodies of the T-12 and L-1 vertebrae and its insertion merges with the lower part of the psoas major above the inguinal ligament. It is innervated by the L-1 nerve.

2) the *psoas major muscle* arises from the transverse processes of the 5 lumbar vertebrae, as well as from the lateral aspect of the bodies of the 12th thoracic and all 5 lumbar vertebrae. It descends deep to the inguinal ligament and is joined by the iliacus muscle (see Fig. 333). The psoas major is supplied by the first 3 lumbar nerves, and through it course the lumbar nerves in the formation of the lumbar plexus (see Fig. 336).

3) the *iliacus muscle* arises from the inner aspect of the pelvis within the concavity called the iliac fossa, and from the lateral aspect of the sacrum. It converges with the psoas muscles deep to the inguinal ligament. The iliacus is supplied by branches of the femoral nerve.

4) the *quadratus lumborum muscle* arises from the medial part of the 12th rib and from the transverse processes of the upper 4 lumbar vertebrae. It inserts into the iliolumbar ligament and the iliac crest and is innervated by T-12 and the upper 3 or 4 lumbar nerves.

5) the relative locations of the vena caval, esophageal, and aortic hiatuses in the diaphragm.

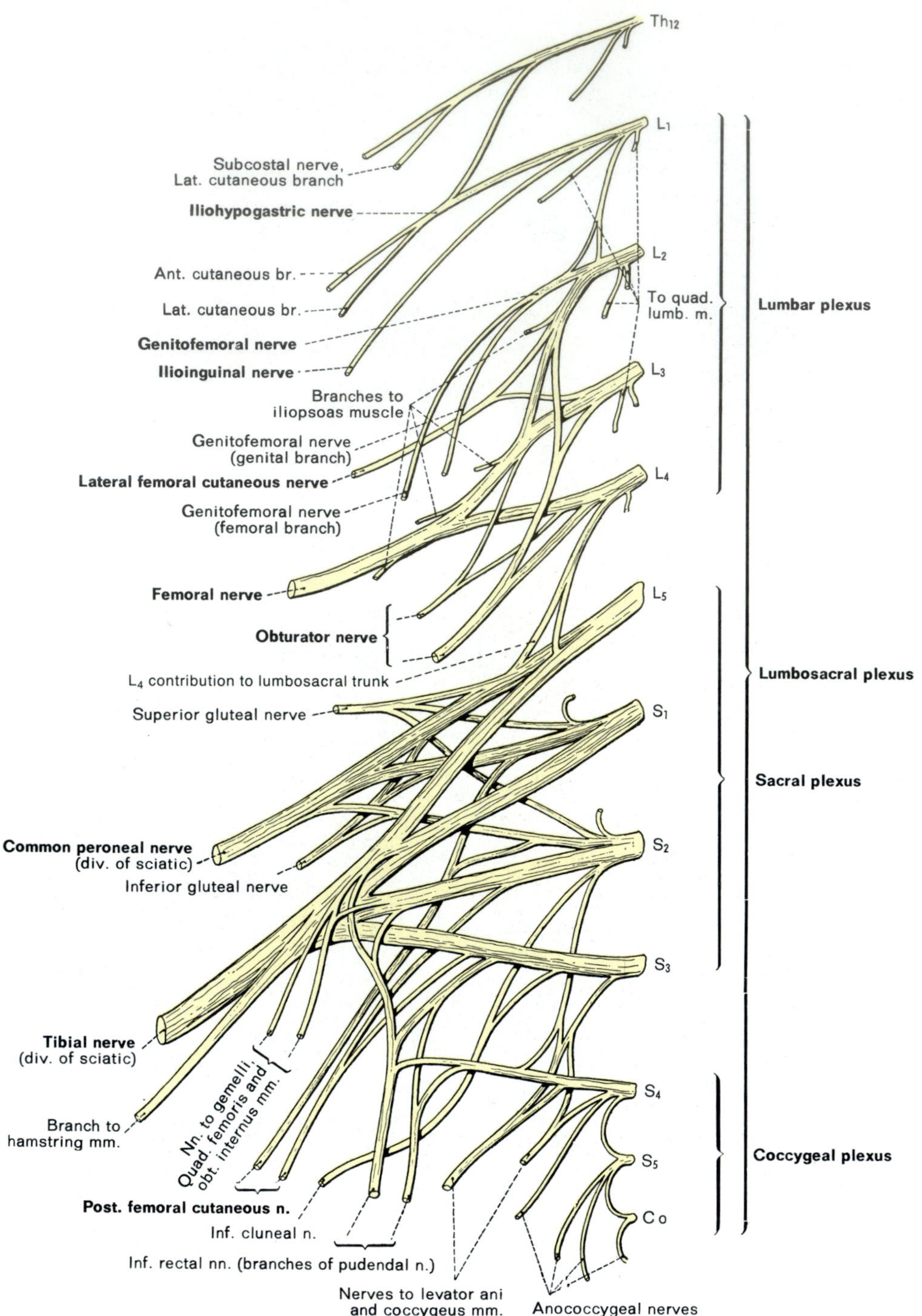

Th12
L1
Subcostal nerve,
Lat. cutaneous branch
Iliohypogastric nerve
Ant. cutaneous br.
L2
Lat. cutaneous br.
To quad. lumb. m.
Lumbar plexus
Genitofemoral nerve
Ilioinguinal nerve
L3
Branches to iliopsoas muscle
Genitofemoral nerve (genital branch)
Lateral femoral cutaneous nerve
L4
Genitofemoral nerve (femoral branch)
Femoral nerve
L5
Obturator nerve
L4 contribution to lumbosacral trunk
Lumbosacral plexus
Superior gluteal nerve
S1
Sacral plexus
Common peroneal nerve (div. of sciatic)
S2
Inferior gluteal nerve
S3
Tibial nerve (div. of sciatic)
Nn. to gemelli, Quad. femoris and obt. internus mm.
S4
Branch to hamstring mm.
S5
Coccygeal plexus
Post. femoral cutaneous n.
Co
Inf. cluneal n.
Inf. rectal nn. (branches of pudendal n.)
Nerves to levator ani and coccygeus mm.
Anococcygeal nerves

Fig. 336: Diagram of the Lumbar, Sacral, Pudendal and Coccygeal Nerve Plexuses

NOTE: 1) the 12th thoracic (subcostal) and 1st lumbar nerves are distributed principally to the lower abdominal wall. The anterior rami of the remaining segmental nerves of the lumbosacral plexus innervate the muscles and skin of the pelvis, perineum and lower extremity. L_1 divides into the iliohypogastric and ilioinguinal nerves.

2) L_2, L_3 and L_4 are the principal segments forming the lumbar plexus. L_1 contributes some fibers to the genitofemoral nerve. The main peripheral nerves derived from these segments include:

a) the genitofemoral nerve (L_1, L_2) c) the obturator nerve (L_2, L_3, L_4)
b) the lateral femoral cutaneous nerve (L_2, L_3) d) femoral nerve (L_2, L_3, L_4)

3) L_5, S_1, S_2 and S_3 with some contribution from L_4 form the sacral plexus. The principal peripheral nerves derived from these segments include:

a) the superior gluteal nerve (L_4, L_5, S_1)
b) the inferior gluteal nerve (L_5, S_1, S_2)
c) the sciatic nerve { common peroneal nerve (L_4, L_5, S_1, S_2); tibial nerve (L_4, L_5, S_1, S_2, S_3) }
d) the posterior femoral cutaneous nerve (S_1, S_2, S_3)

4) S_2, S_3 and S_4 also contribute to the pudendal plexus while S_4, S_5 and the coccygeal nerve form fine nerve filaments which innervate the skin of the anococcygeal region. The pudendal nerve is the main sensory and motor nerve of the perineum. Other sensory (inferior cluneal) and motor (nerve to levator ani and coccygeus muscles) branches of the pudendal plexus also come from sacral segments.

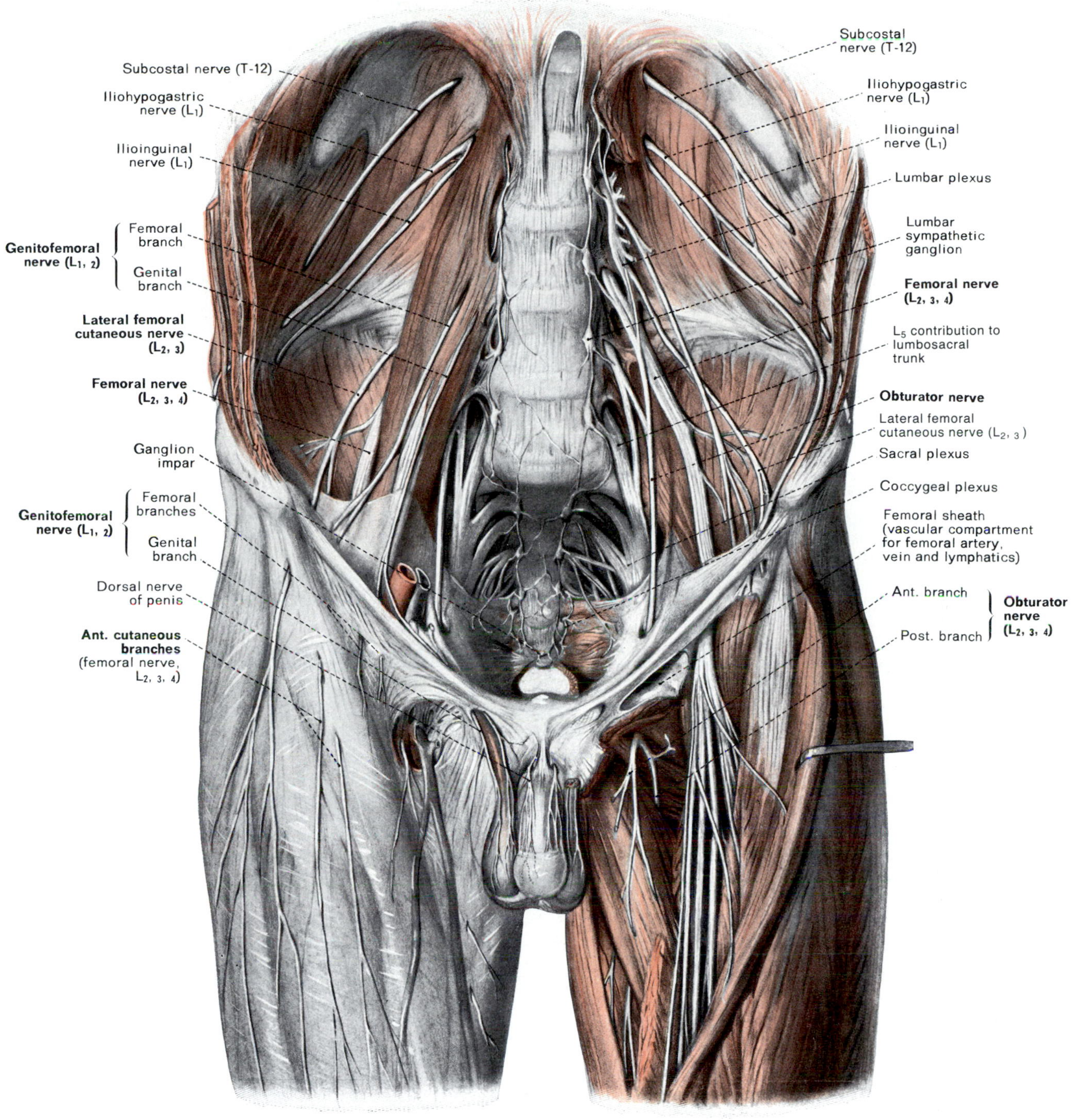
Subcostal nerve (T-12)
Iliohypogastric nerve (L_1)
Ilioinguinal nerve (L_1)
Genitofemoral nerve ($L_{1,2}$)
Femoral branch
Genital branch
Lateral femoral cutaneous nerve ($L_{2,3}$)
Femoral nerve ($L_{2,3,4}$)
Ganglion impar
Genitofemoral nerve ($L_{1,2}$)
Femoral branches
Genital branch
Dorsal nerve of penis
Ant. cutaneous branches (femoral nerve, $L_{2,3,4}$)
Subcostal nerve (T-12)
Iliohypogastric nerve (L_1)
Ilioinguinal nerve (L_1)
Lumbar plexus
Lumbar sympathetic ganglion
Femoral nerve ($L_{2,3,4}$)
L_5 contribution to lumbosacral trunk
Obturator nerve
Lateral femoral cutaneous nerve ($L_{2,3}$)
Sacral plexus
Coccygeal plexus
Femoral sheath (vascular compartment for femoral artery, vein and lymphatics)
Ant. branch
Post. branch
Obturator nerve ($L_{2,3,4}$)

Fig. 337: The Lumbosacral Plexus: Posterior Abdominal Wall and Anterior Thigh

NOTE: 1) on the left side, the psoas major and minor muscles have been removed in order to reveal the plexus of lumbar nerves more completely. As can be seen on the right side, these nerves emerge from the spinal cord and descend along the posterior abdominal wall within the substance of the psoas muscles. Observe that the 12th thoracic nerve (subcostal nerve) courses around the abdominal wall below the 12th rib.

2) the 1st lumbar nerve divides into an iliohypogastric and an ilioinguinal branch. The ilioinguinal nerve descends obliquely toward the iliac crest, penetrates through the transversus and internal oblique muscles to join the spermatic cord, becoming cutaneous at the superficial inguinal ring.

3) the genitofemoral nerve can be found coursing superficially on the surface of the psoas major muscle. It divides into a genital branch (which innervates the cremaster muscle and the skin of the scrotum) and a femoral branch (which is sensory to the upper anterior thigh.)

4) the femoral and obturator nerves. These nerves are derived from L_2, L_3 and L_4 and descend to innervate the anterior and medial groups of femoral muscles, respectively. The femoral nerve enters the thigh beneath the inguinal ligament, dividing into both motor and sensory branches, whereas the obturator nerve courses more medially through the obturator foramen to become the principal nerve of the adductor muscle group.

5) L_4 and L_5 nerve roots (lumbosacral trunk) join with the upper three sacral nerves to form the sacral plexus. From most of this plexus is derived the large sciatic nerve which leaves the pelvis through the greater sciatic foramen to achieve the gluteal region and the posterior aspect of the lower limb.

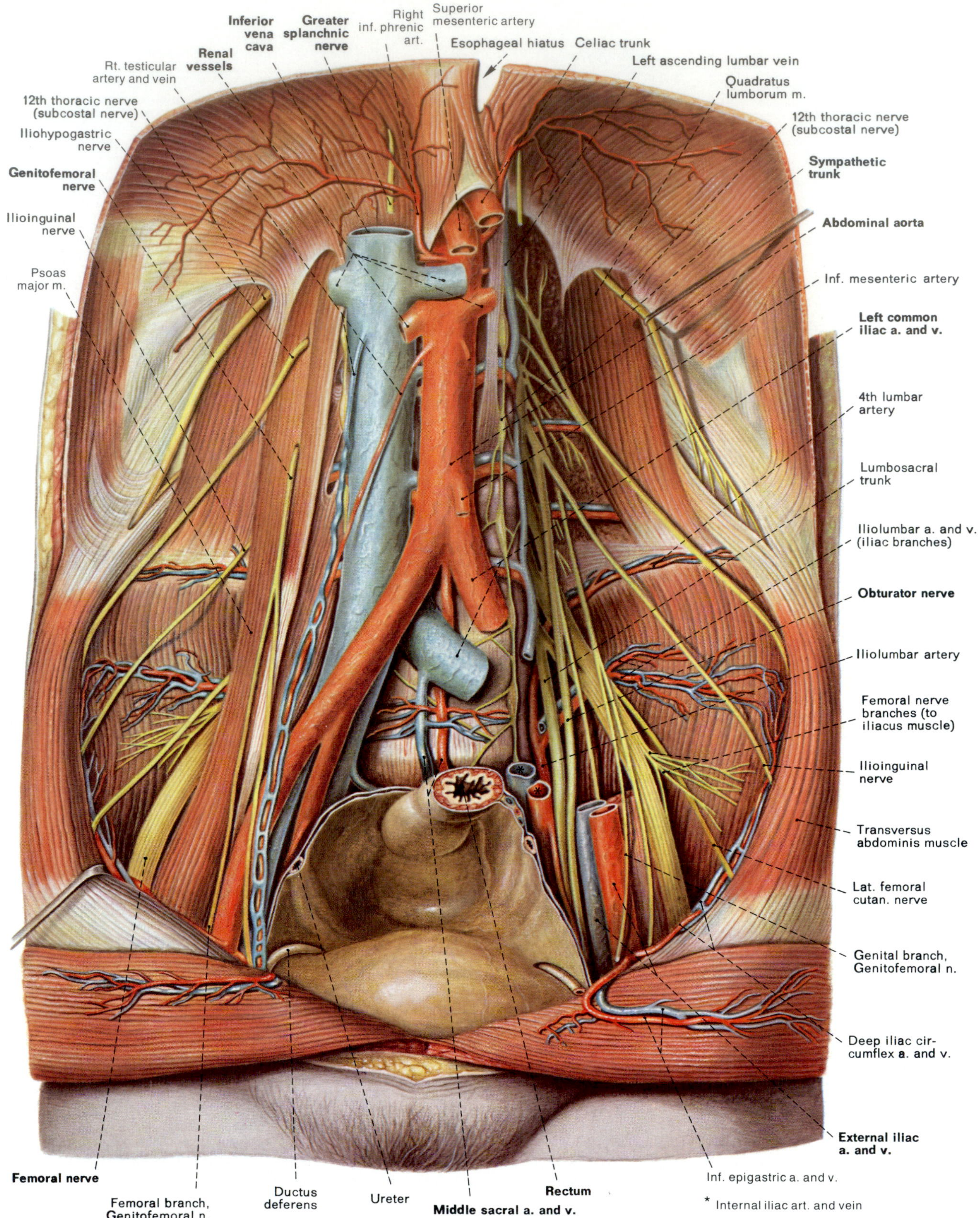

Inferior vena cava
Greater splanchnic nerve
Right inf. phrenic art.
Superior mesenteric artery
Esophageal hiatus
Celiac trunk
Renal vessels
Rt. testicular artery and vein
Left ascending lumbar vein
Quadratus lumborum m.
12th thoracic nerve (subcostal nerve)
12th thoracic nerve (subcostal nerve)
Iliohypogastric nerve
Sympathetic trunk
Genitofemoral nerve
Ilioinguinal nerve
Abdominal aorta
Psoas major m.
Inf. mesenteric artery
Left common iliac a. and v.
4th lumbar artery
Lumbosacral trunk
Iliolumbar a. and v. (iliac branches)
Obturator nerve
Iliolumbar artery
Femoral nerve branches (to iliacus muscle)
Ilioinguinal nerve
Transversus abdominis muscle
Lat. femoral cutan. nerve
Genital branch, Genitofemoral n.
Deep iliac circumflex a. and v.
External iliac a. and v.
Femoral nerve
Femoral branch, Genitofemoral n.
Ductus deferens
Ureter
Middle sacral a. and v.
Rectum
Inf. epigastric a. and v.
* Internal iliac art. and vein

Fig. 338: The Posterior Abdominal Vessels and Nerves

NOTE: 1) the abdominal viscera have been removed as well as the psoas muscles on the left side. Observe the greater splanchnic nerves as they pierce the diaphragmatic crura to enter the abdomen. Also note the inferior phrenic arteries nearby. Identify the abdominal sympathetic chain of ganglia situated anterolaterally on the vertebral column.

2) the testicular arteries arising from the aorta just below the renal arteries. Inferiorly, the testicular artery and vein join the ductus deferens to enter the abdominal inguinal ring just lateral to the inferior epigastric vessels. Observe the middle sacral vessels descending into the pelvis in the midline. The middle sacral artery originates from the aorta at its bifurcation, whereas the vein usually drains into the left common iliac vein.

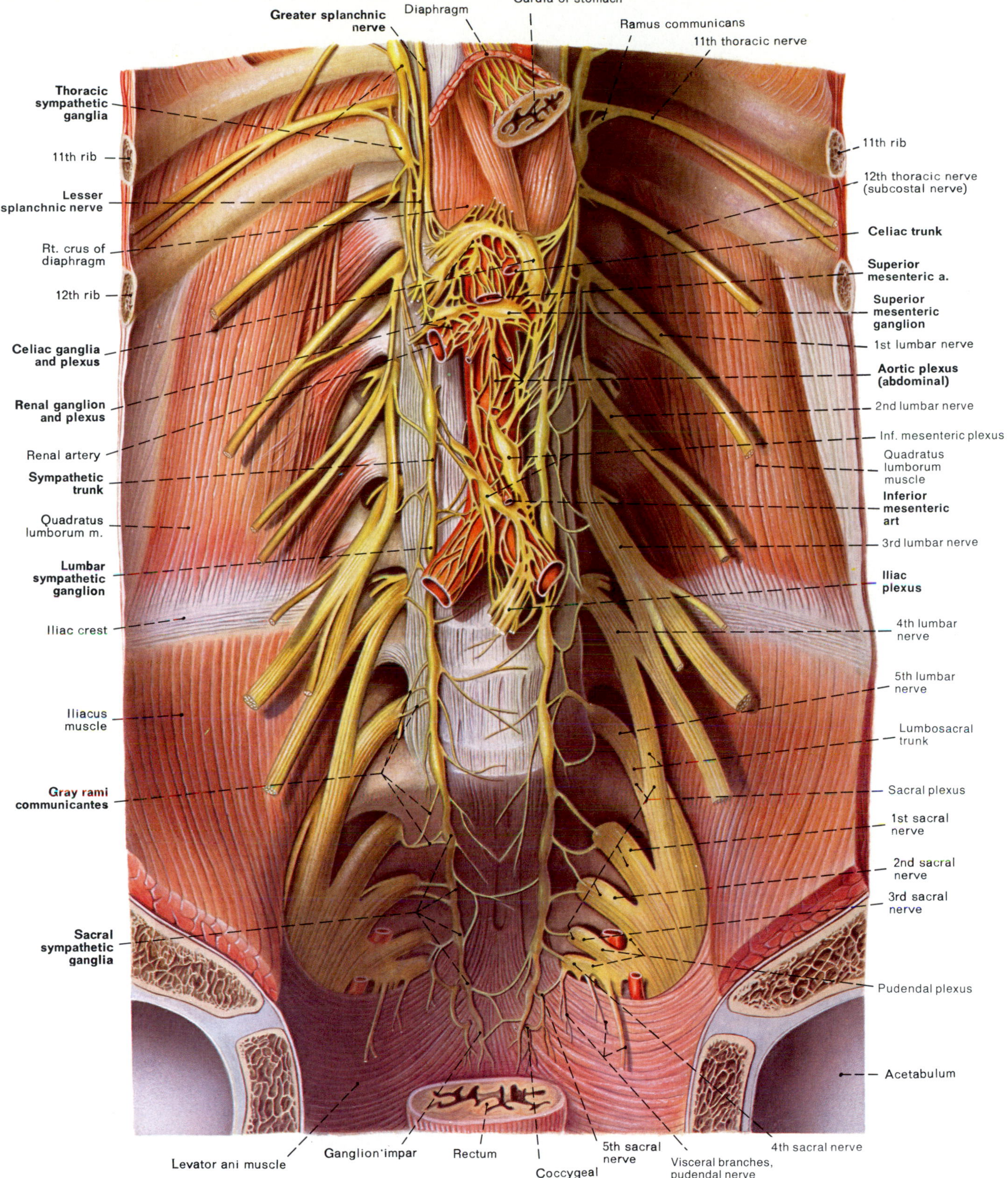
Greater splanchnic nerve
Diaphragm
Cardia of stomach
Ramus communicans
11th thoracic nerve
Thoracic sympathetic ganglia
11th rib
Lesser splanchnic nerve
Rt. crus of diaphragm
12th rib
Celiac ganglia and plexus
Renal ganglion and plexus
Renal artery
Sympathetic trunk
Quadratus lumborum m.
Lumbar sympathetic ganglion
Iliac crest
Iliacus muscle
Gray rami communicantes
Sacral sympathetic ganglia
11th rib
12th thoracic nerve (subcostal nerve)
Celiac trunk
Superior mesenteric a.
Superior mesenteric ganglion
1st lumbar nerve
Aortic plexus (abdominal)
2nd lumbar nerve
Inf. mesenteric plexus
Quadratus lumborum muscle
Inferior mesenteric art
3rd lumbar nerve
Iliac plexus
4th lumbar nerve
5th lumbar nerve
Lumbosacral trunk
Sacral plexus
1st sacral nerve
2nd sacral nerve
3rd sacral nerve
Pudendal plexus
Acetabulum
Levator ani muscle
Ganglion impar
Rectum
Coccygeal nerve
5th sacral nerve
Visceral branches, pudendal nerve
4th sacral nerve

Fig. 339: Lumbar Sympathetic Trunk and Abdominal Autonomic Ganglia

NOTE: 1) in addition to the two sympathetic chains of ganglia descending from the thorax through the abdomen and into the pelvis, the plexuses and their associated ganglia which overlie the major arteries branching from the aorta. Thus, the celiac plexuses along with the superior mesenteric, inferior mesenteric, renal, aortic and iliac plexuses form a dense network of autonomic fibers from which many of the abdominal and pelvic viscera receive sympathetic innervation.

2) the splanchnic nerves (containing preganglionic sympathetic fibers) as they join the upper abdominal ganglia where many of their fibers synapse with postganglionic sympathetic neurons.

3) that below L-2 only gray rami communicantes connect the ganglia of the sympathetic chain to the segmental nerves, since at these lower levels the rami consist only of postganglionic sympathetic fibers.

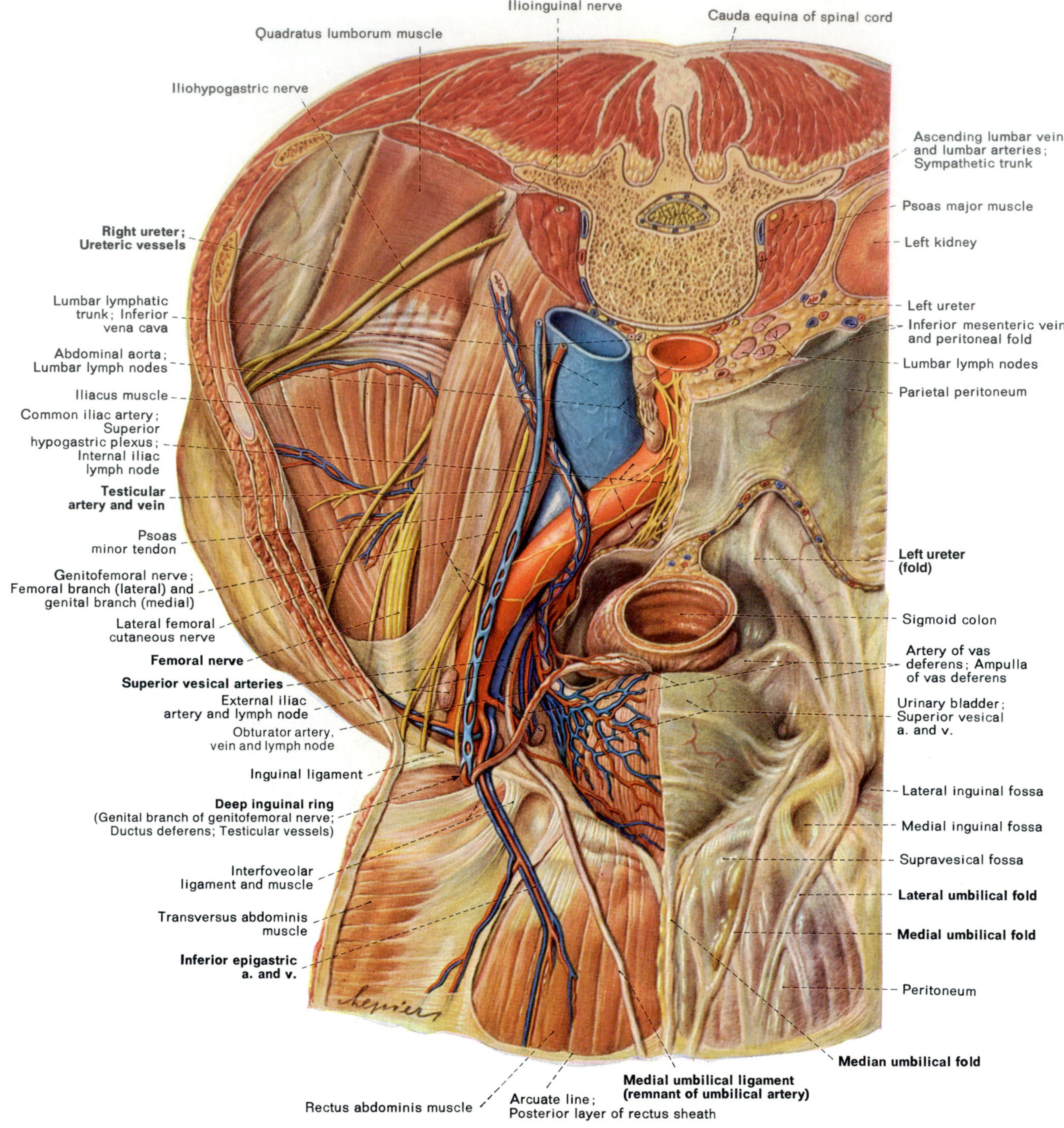

Ilioinguinal nerve
Cauda equina of spinal cord
Quadratus lumborum muscle
Iliohypogastric nerve
Ascending lumbar vein and lumbar arteries; Sympathetic trunk
Psoas major muscle
Right ureter; Ureteric vessels
Left kidney
Lumbar lymphatic trunk; Inferior vena cava
Left ureter
Inferior mesenteric vein and peritoneal fold
Abdominal aorta; Lumbar lymph nodes
Lumbar lymph nodes
Iliacus muscle
Parietal peritoneum
Common iliac artery; Superior hypogastric plexus; Internal iliac lymph node
Testicular artery and vein
Psoas minor tendon
Left ureter (fold)
Genitofemoral nerve; Femoral branch (lateral) and genital branch (medial)
Lateral femoral cutaneous nerve
Sigmoid colon
Femoral nerve
Artery of vas deferens; Ampulla of vas deferens
Superior vesical arteries
External iliac artery and lymph node
Urinary bladder; Superior vesical a. and v.
Obturator artery, vein and lymph node
Inguinal ligament
Lateral inguinal fossa
Deep inguinal ring
(Genital branch of genitofemoral nerve; Ductus deferens; Testicular vessels)
Medial inguinal fossa
Supravesical fossa
Interfoveolar ligament and muscle
Lateral umbilical fold
Transversus abdominis muscle
Medial umbilical fold
Inferior epigastric a. and v.
Peritoneum
Median umbilical fold
Medial umbilical ligament (remnant of umbilical artery)
Rectus abdominis muscle
Arcuate line; Posterior layer of rectus sheath

Fig. 340: Vessels and Nerves of the Inferior Abdomen and Pelvis

NOTE: 1) the anterior abdominal wall has been incised vertically on both sides and reflected inferiorly, thereby exposing its inner (posterior) surface. The body has been transected through the lumbar region at the caudal end of the 3rd lumbar vertebra. On the right side, the peritoneum investing the abdominal cavity has been stripped away, whereas on the left it has been left intact.

2) this dissection reveals the intact male pelvic viscera viewed from above. Observe the course of the ureters as they cross the pelvic brim anterior to the common iliac arteries. Within the pelvis, the ureter curves medially toward the bladder. At the lateral angle of the bladder, the ductus deferens on its path to the seminal vesicle courses ventral to the ureter.

3) the convergence of the testicular vessels, ductus deferens and genital branch of the genitofemoral nerve (innervation of cremaster muscle) at the abdominal inguinal ring to form the spermatic cord of the inguinal canal.

4) the origins within the pelvis of the umbilical ligaments of the anterior abdominal wall: a) the median umbilical fold extending from the bladder to the umbilicus (urachus), b) the medial umbilical fold formed by a peritoneal reflection over the obliterated umbilical artery, and c) the lateral umbilical fold formed by peritoneum covering the inferior epigastric vessels.

5) the superior hypogastric plexus or presacral nerve lying anterior to the bifurcation of the aorta and descending behind the peritoneum into the pelvis.

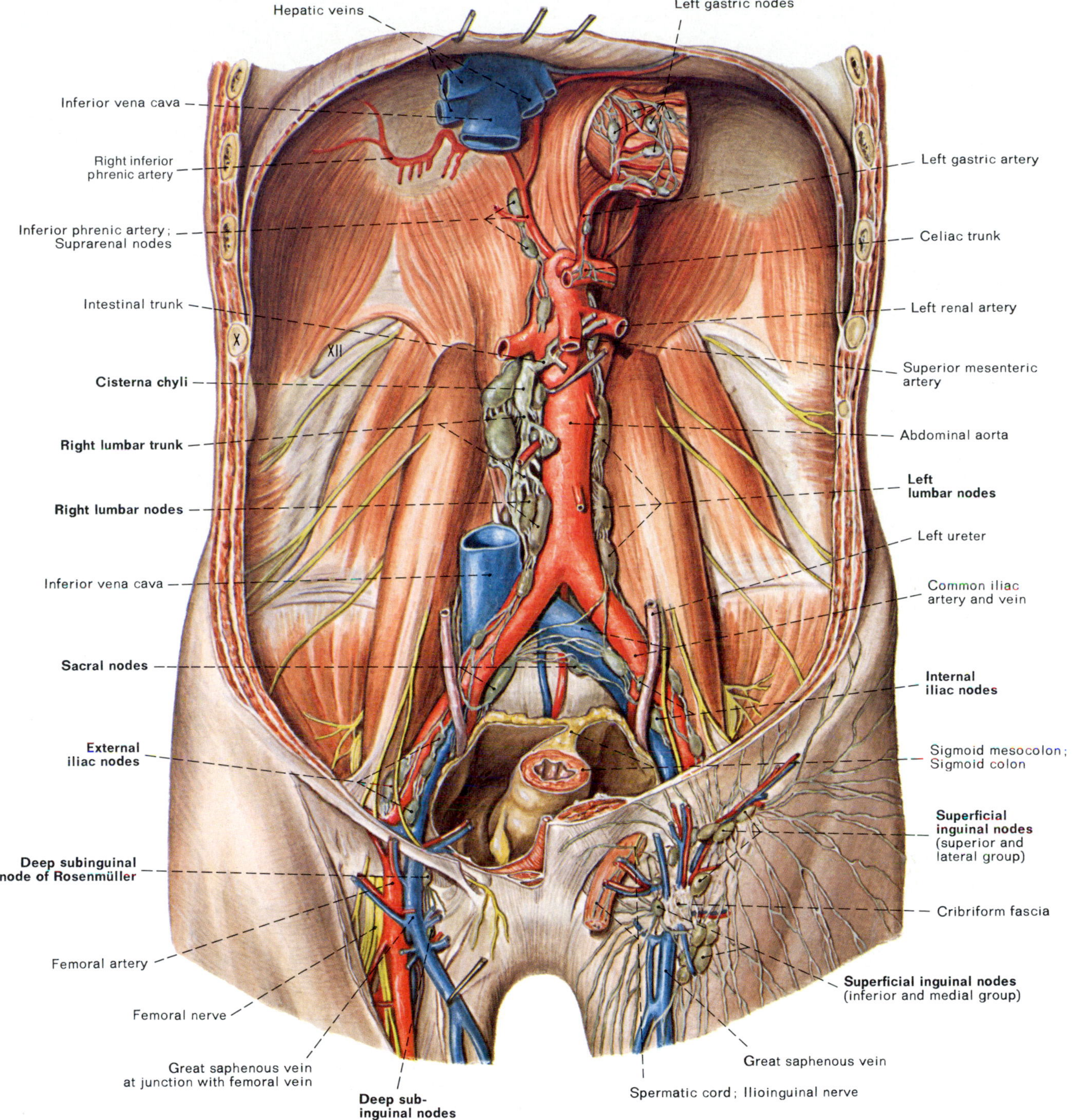

Hepatic veins
Left gastric nodes
Inferior vena cava
Right inferior phrenic artery
Left gastric artery
Inferior phrenic artery; Suprarenal nodes
Celiac trunk
Intestinal trunk
Left renal artery
X
XII
Superior mesenteric artery
Cisterna chyli
Abdominal aorta
Right lumbar trunk
Left lumbar nodes
Right lumbar nodes
Left ureter
Inferior vena cava
Common iliac artery and vein
Sacral nodes
Internal iliac nodes
External iliac nodes
Sigmoid mesocolon; Sigmoid colon
Superficial inguinal nodes (superior and lateral group)
Deep subinguinal node of Rosenmüller
Cribriform fascia
Femoral artery
Superficial inguinal nodes (inferior and medial group)
Femoral nerve
Great saphenous vein
Great saphenous vein at junction with femoral vein
Deep sub-inguinal nodes
Spermatic cord; Ilioinguinal nerve

Fig. 341: The Inguinal, Pelvic and Lumbar (Aortic) Lymph Nodes

NOTE: 1) chains of lymph nodes from the inguinal region to the diaphragm lie along the paths of the major blood vessels. In the inguinal region, the superficial inguinal lymph nodes lie just distal to the inguinal ligament and in the subcutaneous superficial fascia. They range from 10 to 20 in number and receive drainage from the genitalia, perineum, gluteal region and the anterior abdominal wall. More deeply, the subinguinal nodes receive drainage from the lower extremity. One of the deep subinguinal nodes (node of Rosenmüller or Cloquet) lies in the femoral ring.

2) within the pelvis and abdomen, visceral lymph nodes lie close to the organs which they drain. These visceral nodes then channel lymph through chains of parietal nodes which generally are located along the paths of the major arteries and veins. Thus, external iliac, internal iliac (hypogastric) and common iliac nodes are located in proximity to these vessels in the pelvis.

3) along the posterior abdominal wall are found right and left lumbar chains which course along the sides of the abdominal aorta. Other groups of nodes, the preaortic, are arranged along the roots of the major unpaired branches of the aorta, forming the celiac and superior and inferior mesenteric node chains.

4) at about the level of the 2nd lumbar vertebra there is a confluence of lymph channels which forms a dilated sac, the cisterna chyli. This is located somewhat posterior and to the right of the aorta and marks the commencement of the thoracic duct.

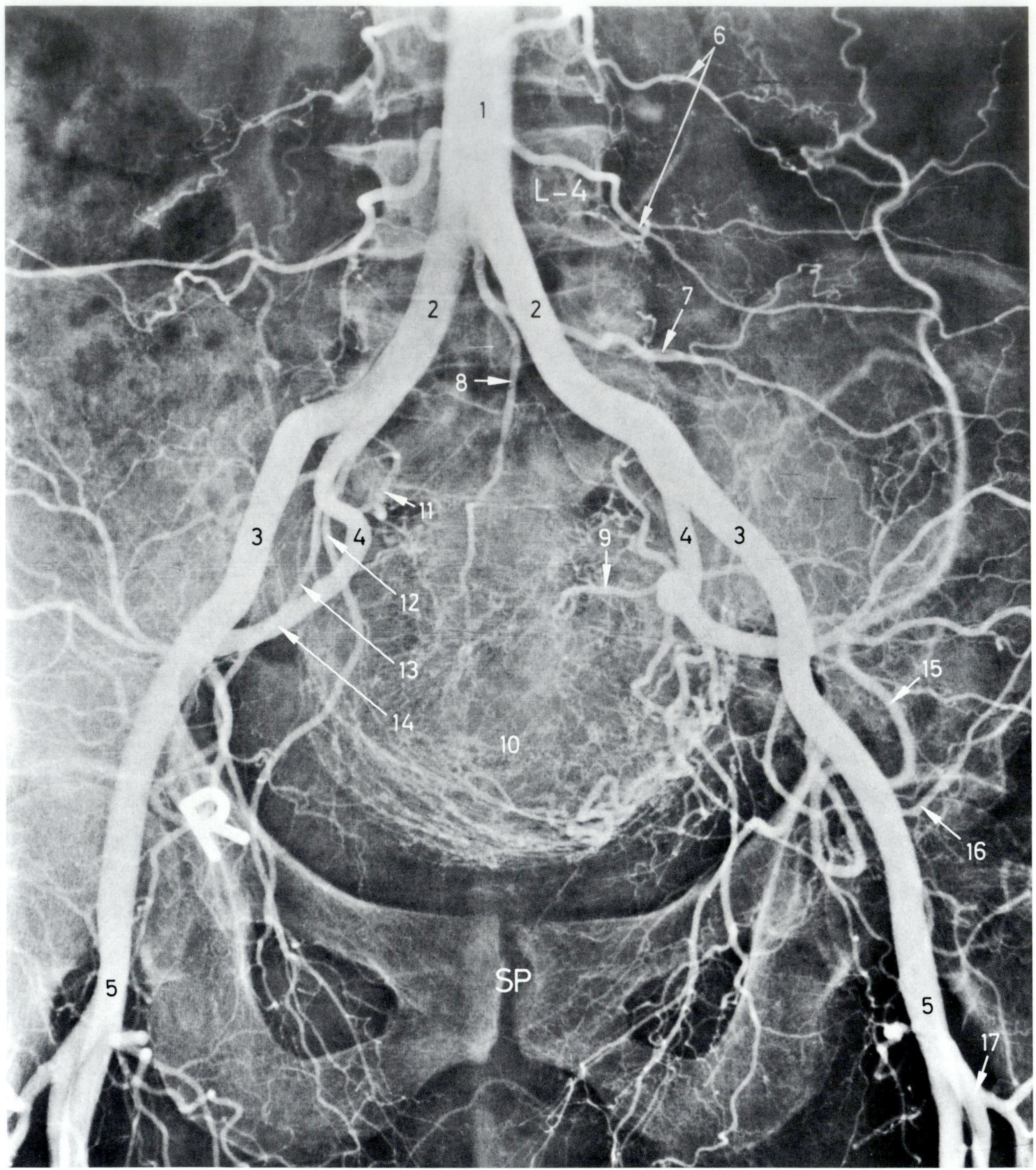

1
6
L-4
7
2
2
8
11
3
4
9
4
3
12
13
14
15
10
16
R
SP
5
5
17

Fig. 342: An Iliac Arteriogram in a Female

NOTE that the bifurcation of the aorta (*1*) into the two common iliac arteries (*2*) occurs at the lower border of the body of the L 4 vertebra. The common iliac vessels branch into external (*3*) and internal (*4*) iliac arteries. The internal iliac artery (*4*) on each side serves a number of branches to the pelvis, perineum, and gluteal region, while the external iliac artery (*3*), after giving off the inferior epigastric (*15*) and deep circumflex iliac (*16*) arteries, becomes the femoral artery below the inguinal ligament. (From Wicke, L., *Atlas of Radiologic Anatomy*, 3rd Edition, Urban and Schwarzenberg, Baltimore, 1982.)

1. Abdominal aorta
2. Common iliac artery
3. External iliac artery
4. Internal iliac artery
5. Femoral artery
6. Lumbar arteries
7. Iliolumbar artery
8. Median sacral artery
9. Uterine artery
10. Uterus
11. Lateral sacral artery
12. Obturator artery
13. Internal pudendal artery
14. Superior gluteal artery
15. Inferior epigastric artery
16. Deep circumflex iliac artery
17. Deep femoral artery

L 4 - 4 th lumbar vertebra
SP - Symphysis pubis

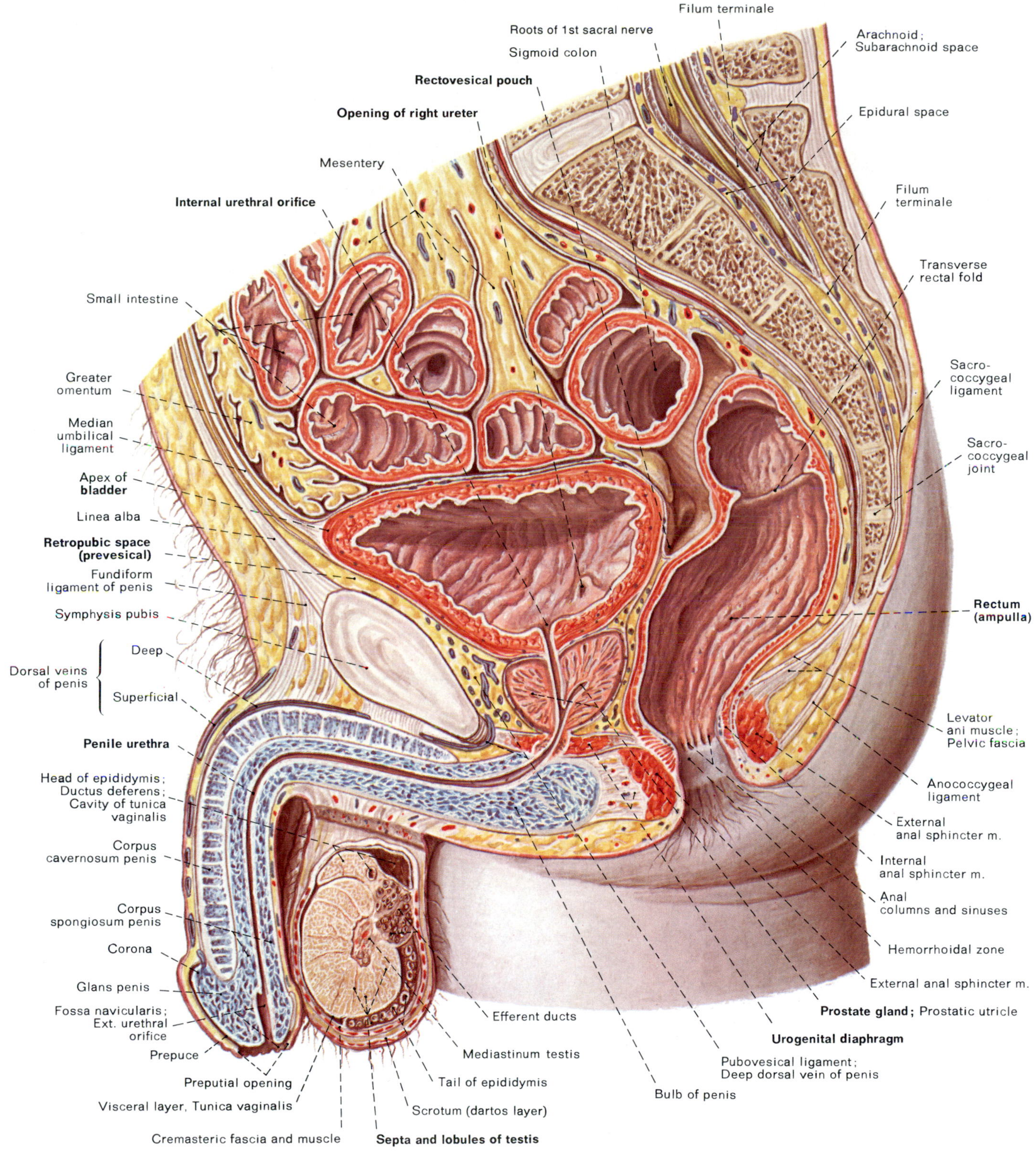
Filum terminale
Roots of 1st sacral nerve
Sigmoid colon
Rectovesical pouch
Opening of right ureter
Mesentery
Internal urethral orifice
Arachnoid; Subarachnoid space
Epidural space
Filum terminale
Transverse rectal fold
Small intestine
Greater omentum
Median umbilical ligament
Apex of bladder
Linea alba
Retropubic space (prevesical)
Fundiform ligament of penis
Symphysis pubis
Dorsal veins of penis
Deep
Superficial
Penile urethra
Head of epididymis; Ductus deferens; Cavity of tunica vaginalis
Corpus cavernosum penis
Corpus spongiosum penis
Corona
Glans penis
Fossa navicularis; Ext. urethral orifice
Prepuce
Preputial opening
Visceral layer, Tunica vaginalis
Cremasteric fascia and muscle
Septa and lobules of testis
Scrotum (dartos layer)
Tail of epididymis
Mediastinum testis
Efferent ducts
Bulb of penis
Pubovesical ligament; Deep dorsal vein of penis
Urogenital diaphragm
Prostate gland; Prostatic utricle
External anal sphincter m.
Hemorrhoidal zone
Anal columns and sinuses
Internal anal sphincter m.
External anal sphincter m.
Anococcygeal ligament
Levator ani muscle; Pelvic fascia
Rectum (ampulla)
Sacro-coccygeal joint
Sacro-coccygeal ligament

Fig. 343: Median Sagittal Section of the Male Pelvis and Perineum Showing the Pelvic Viscera and the External Genitalia

Anus
Anococcygeal ligament
Coccyx
Levator ani muscle
External anal sphincter muscle
Gluteal fascia
Gluteus maximus muscle
Ischiorectal fossa
Obturator fascia;
Pudendal canal (Alcock)
Sacrotuberous ligament
Obturator fascia;
Foramina for pudendal vessels and nn.
Femoral fascia
Ischial tuberosity
Foramina for perineal vessels and nn.
Adductor magnus muscle
Inferior fascia,
Urogenital diaphragm
Superficial transverse perineal muscle
Deep transverse perineal muscle
Gracilis muscle
Bulbourethral gland (Cowper)
Ischiocavernosus muscle
Central tendinous point of perineum
Bulbospongiosus muscle
Corpus spongiosum penis
Dartos layer of scrotum
Raphé of scrotum

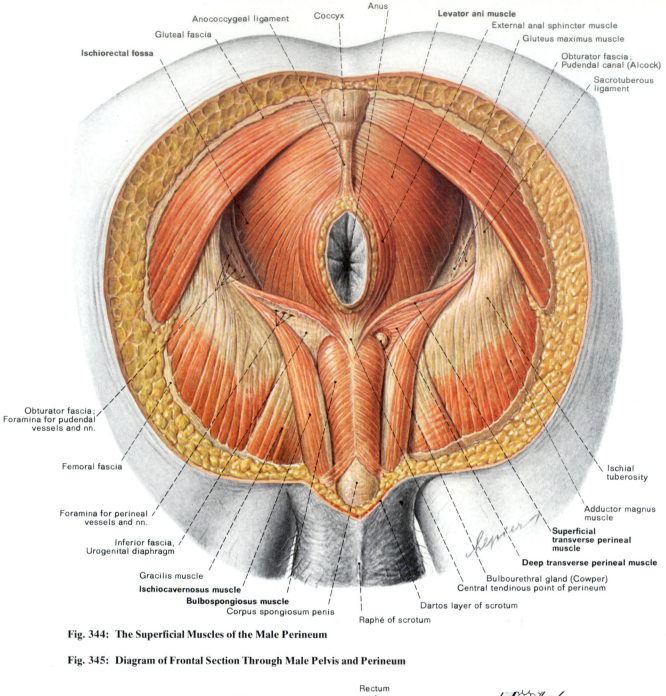

Fig. 344: The Superficial Muscles of the Male Perineum

Fig. 345: Diagram of Frontal Section Through Male Pelvis and Perineum

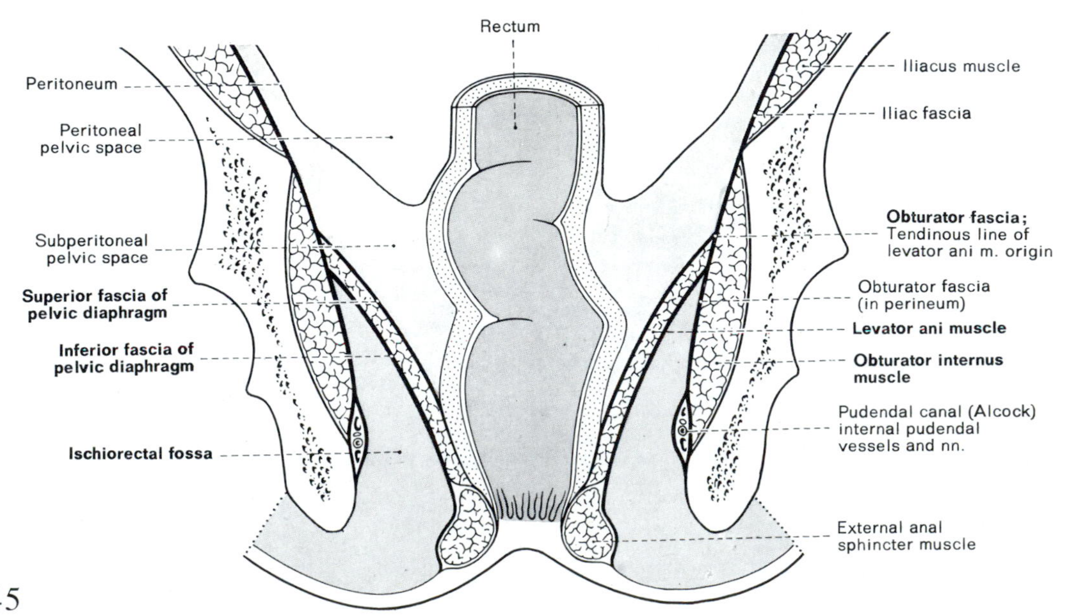
Rectum
Peritoneum
Iliacus muscle
Iliac fascia
Peritoneal pelvic space
Subperitoneal pelvic space
Obturator fascia;
Tendinous line of levator ani m. origin
Obturator fascia (in perineum)
Superior fascia of pelvic diaphragm
Levator ani muscle
Inferior fascia of pelvic diaphragm
Obturator internus muscle
Pudendal canal (Alcock) internal pudendal vessels and nn.
Ischiorectal fossa
External anal sphincter muscle

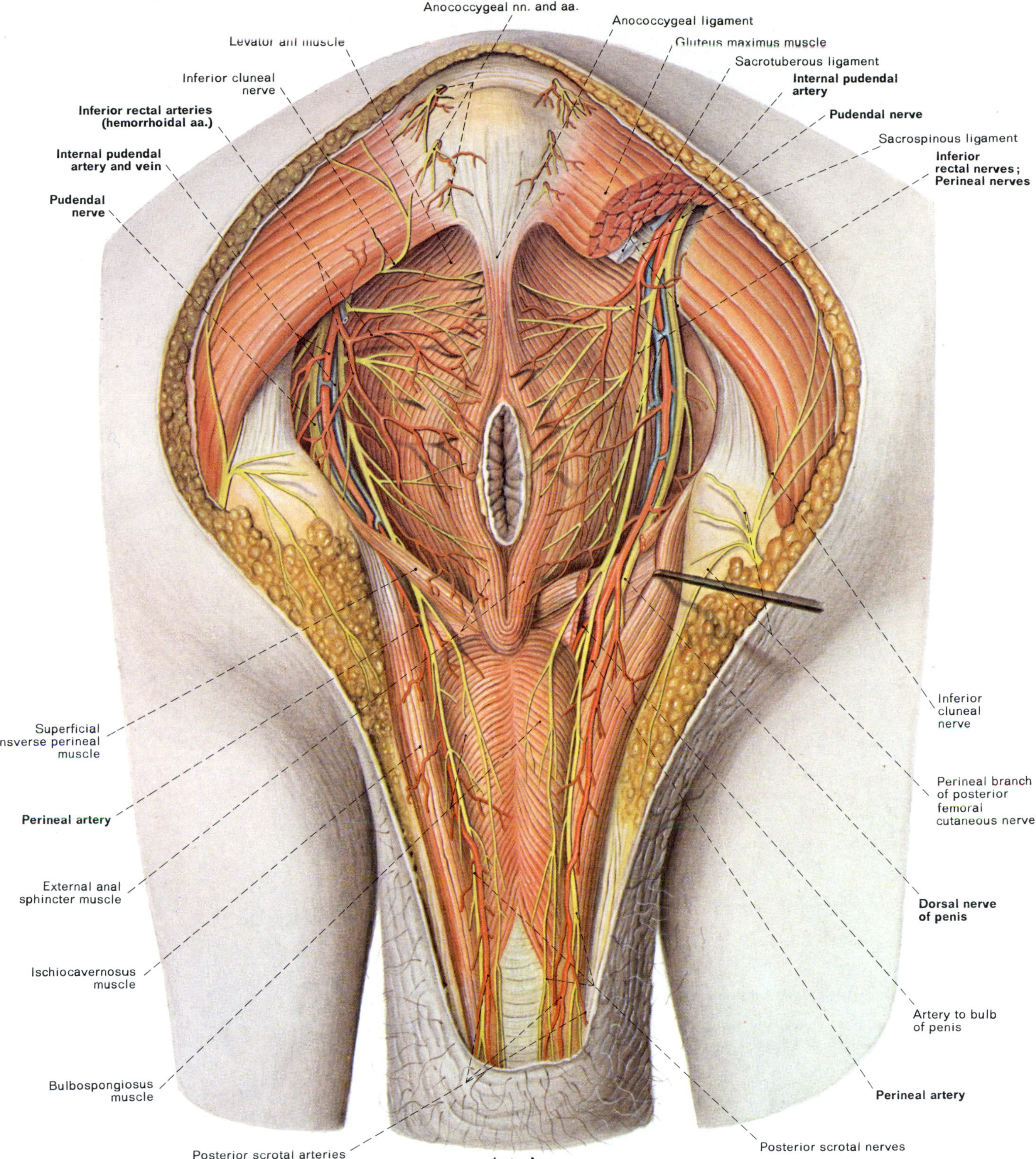
Posterior
Anococcygeal nn. and aa.
Anococcygeal ligament
Levator ani muscle
Gluteus maximus muscle
Sacrotuberous ligament
Inferior cluneal nerve
Internal pudendal artery
Inferior rectal arteries (hemorrhoidal aa.)
Pudendal nerve
Sacrospinous ligament
Internal pudendal artery and vein
Inferior rectal nerves; Perineal nerves
Pudendal nerve
Inferior cluneal nerve
Superficial transverse perineal muscle
Perineal branch of posterior femoral cutaneous nerve
Perineal artery
External anal sphincter muscle
Dorsal nerve of penis
Ischiocavernosus muscle
Artery to bulb of penis
Bulbospongiosus muscle
Perineal artery
Posterior scrotal arteries
Posterior scrotal nerves
Anterior

Fig. 346: Nerves and Blood Vessels of the Male Perineum

NOTE: 1) the skin of the male perineum as well as the fat of the ischiorectal fossae have been removed in order to expose the muscles, vessels and nerves of both the anal and urogenital regions of the perineum.

2) emerging from the pelvis through the lesser sciatic foramen by way of the pudendal canal (Alcock) are the internal pudendal vessels and the pudendal nerve. These structures enter the perineum at the lateral border of the ischiorectal fossa. There is an immediate branching of the inferior rectal (hemorrhoidal) vessels and nerves which course transversely across the ischiorectal fossa to supply the levator ani and external anal sphincter muscles.

3) the internal pudendal structures then continue anteriorly, pierce the urogenital diaphragm, and become the perineal vessels and nerve and the dorsal vessels and nerve of the penis. The muscles of the urogenital triangle are innervated by the perineal nerve, while the dorsal nerve of the penis is the principal sensory nerve of that organ.

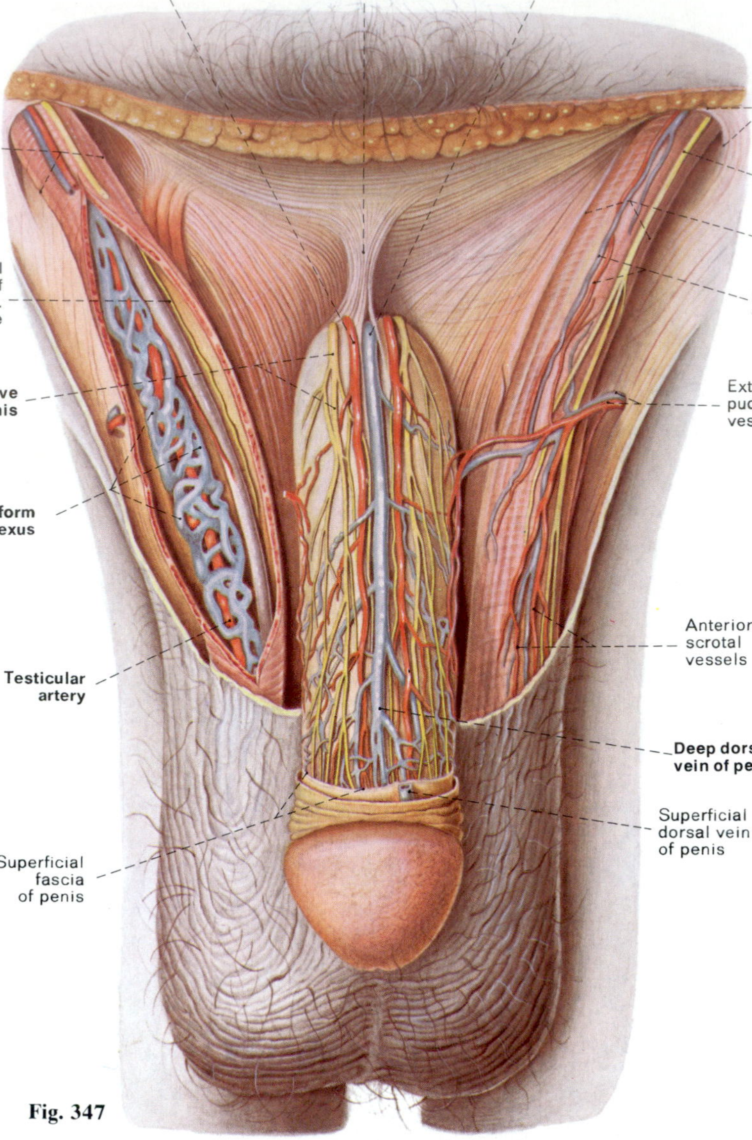

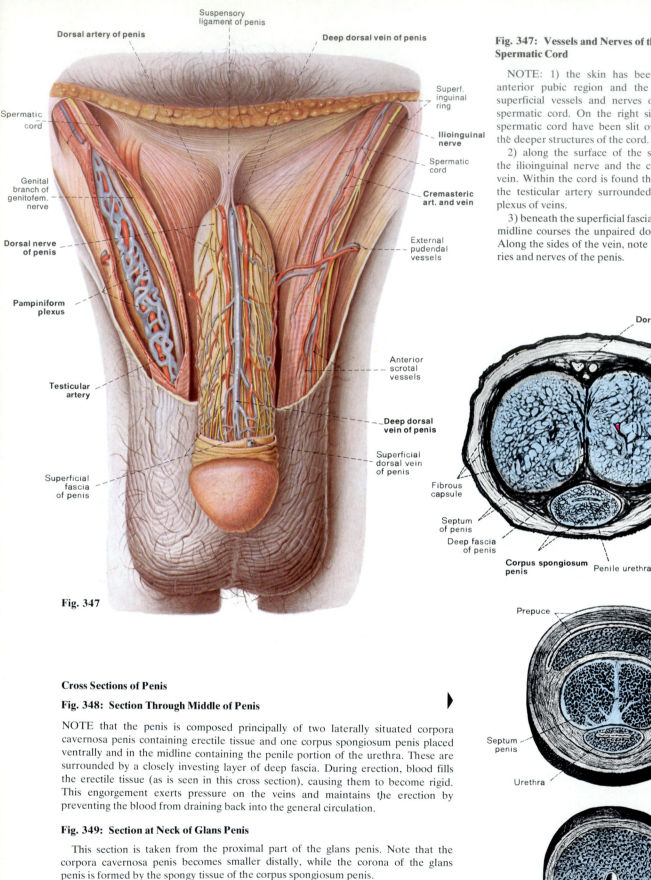

Fig. 347

Fig. 347: Vessels and Nerves of the Penis and Spermatic Cord

NOTE: 1) the skin has been removed from the anterior pubic region and the penis, revealing the superficial vessels and nerves of the penis and left spermatic cord. On the right side, the layers of the spermatic cord have been slit open in order to show the deeper structures of the cord.

2) along the surface of the spermatic cord course the ilioinguinal nerve and the cremasteric artery and vein. Within the cord is found the ductus deferens and the testicular artery surrounded by the pampiniform plexus of veins.

3) beneath the superficial fascia of the penis and in the midline courses the unpaired dorsal vein of the penis. Along the sides of the vein, note the paired dorsal arteries and nerves of the penis.

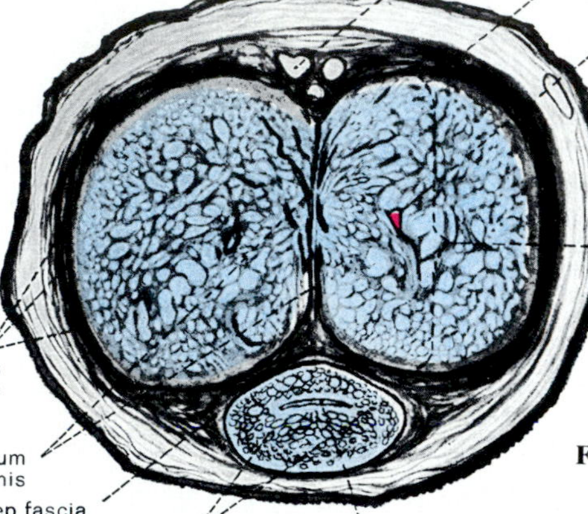

Fig. 348

Cross Sections of Penis

Fig. 348: Section Through Middle of Penis

NOTE that the penis is composed principally of two laterally situated corpora cavernosa penis containing erectile tissue and one corpus spongiosum penis placed ventrally and in the midline containing the penile portion of the urethra. These are surrounded by a closely investing layer of deep fascia. During erection, blood fills the erectile tissue (as is seen in this cross section), causing them to become rigid. This engorgement exerts pressure on the veins and maintains the erection by preventing the blood from draining back into the general circulation.

Fig. 349: Section at Neck of Glans Penis

This section is taken from the proximal part of the glans penis. Note that the corpora cavernosa penis becomes smaller distally, while the corona of the glans penis is formed by the spongy tissue of the corpus spongiosum penis.

Fig. 350: Section Through Glans Penis

The expanded distal extremity of the corpus spongiosum penis is called the glans penis. At its distal end is the opening of the urethra. In the uncircumcised male, the glans penis is covered by a duplication of thin skin, the prepuce, which is attached to the glans penis ventrally by the frenulum.

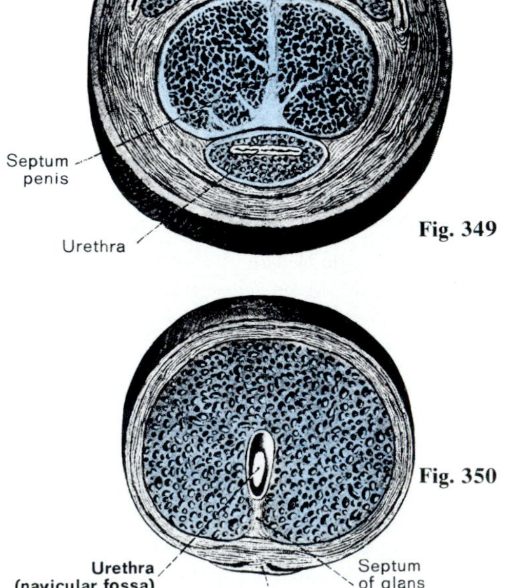
Prepuce
Glans penis (corona)
Septum penis
Urethra
Urethra (navicular fossa)
Frenulum
Septum of glans penis

Fig. 349

Fig. 350

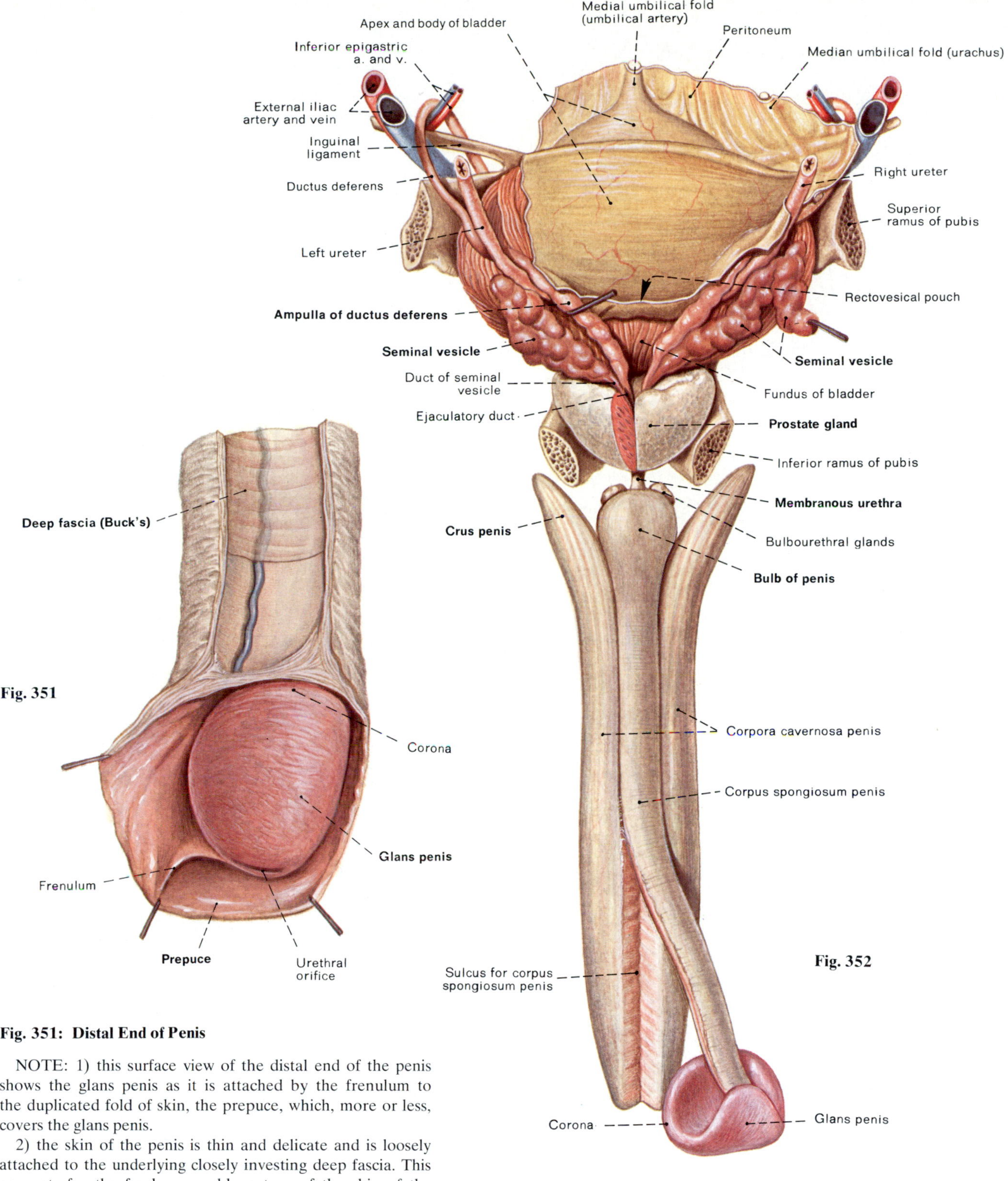
Apex and body of bladder
Medial umbilical fold
(umbilical artery)
Peritoneum
Median umbilical fold (urachus)
Inferior epigastric
a. and v.
External iliac
artery and vein
Inguinal
ligament
Ductus deferens
Right ureter
Superior
ramus of pubis
Left ureter
Rectovesical pouch
Ampulla of ductus deferens
Seminal vesicle
Seminal vesicle
Duct of seminal
vesicle
Fundus of bladder
Ejaculatory duct
Prostate gland
Inferior ramus of pubis
Membranous urethra
Deep fascia (Buck's)
Crus penis
Bulbourethral glands
Bulb of penis
Fig. 351
Corona
Corpora cavernosa penis
Corpus spongiosum penis
Glans penis
Frenulum
Prepuce
Urethral
orifice
Sulcus for corpus
spongiosum penis
Fig. 352
Corona
Glans penis

Fig. 351: Distal End of Penis

NOTE: 1) this surface view of the distal end of the penis shows the glans penis as it is attached by the frenulum to the duplicated fold of skin, the prepuce, which, more or less, covers the glans penis.

2) the skin of the penis is thin and delicate and is loosely attached to the underlying closely investing deep fascia. This accounts for the freely moveable nature of the skin of the penis over the more deeply located bodies of cavernous tissue.

Fig. 352: The Erectile Bodies of the Penis Attached to the Bladder and Other Organs by the Membranous Urethra

NOTE: 1) the firmly investing deep fascia which surrounds the erectile bodies of the penis has been removed, and the distal portion of the corpus spongiosum penis (which contains the penile urethra) has been displaced from its position between the two corpora cavernosa penis.

2) the posterior surface of the bladder and prostate and the associated seminal vesicles, vasa deferens, and bulbourethral glands are also demonstrated. These structures all communicate with the urethra, the membranous portion of which is in continuity with the penile urethra.

3) the tapered crura of the corpora cavernosa penis diverge laterally at their base to become adherent to the ischial and pubic rami. They are surrounded by the fibers of the ischiocavernosus muscles. The base of the corpus spongiosum penis is also expanded and is called the bulb. It is enclosed by the bulbocavernosus muscle.

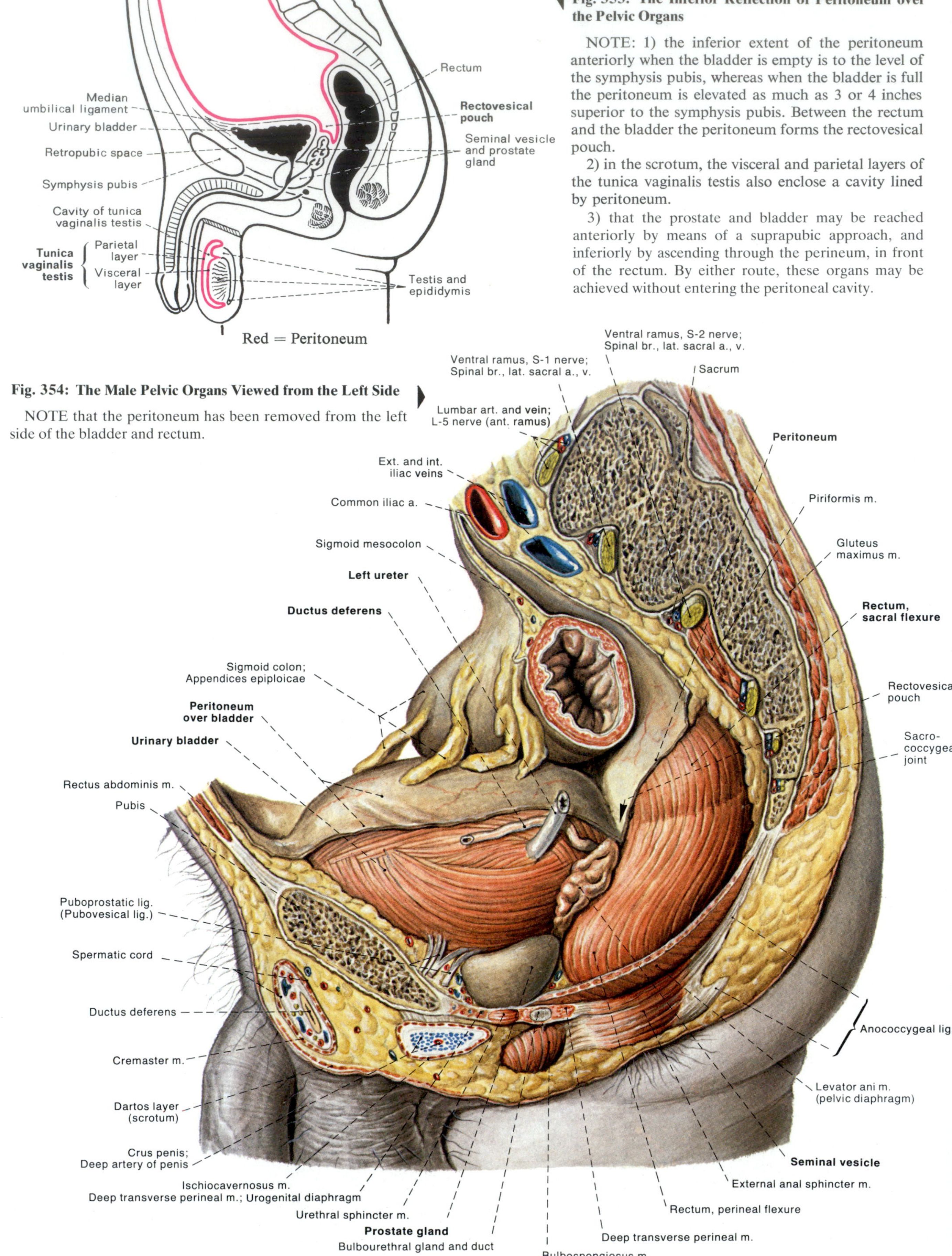

Median
umbilical ligament
Urinary bladder
Retropubic space
Symphysis pubis
Cavity of tunica
vaginalis testis
Tunica
vaginalis
testis
Parietal
layer
Visceral
layer
Rectum
Rectovesical
pouch
Seminal vesicle
and prostate
gland
Testis and
epididymis
Red = Peritoneum
Ventral ramus, S-2 nerve;
Spinal br., lat. sacral a., v.
Ventral ramus, S-1 nerve;
Spinal br., lat. sacral a., v.
Sacrum
Lumbar art. and vein;
L-5 nerve (ant. ramus)
Peritoneum
Ext. and int.
iliac veins
Common iliac a.
Piriformis m.
Sigmoid mesocolon
Gluteus
maximus m.
Left ureter
Ductus deferens
Rectum,
sacral flexure
Sigmoid colon;
Appendices epiploicae
Rectovesical
pouch
Peritoneum
over bladder
Urinary bladder
Sacro-
coccygeal
joint
Rectus abdominis m.
Pubis
Puboprostatic lig.
(Pubovesical lig.)
Spermatic cord
Ductus deferens
Anococcygeal lig.
Cremaster m.
Levator ani m.
(pelvic diaphragm)
Dartos layer
(scrotum)
Crus penis;
Deep artery of penis
Seminal vesicle
Ischiocavernosus m.
Deep transverse perineal m.; Urogenital diaphragm
External anal sphincter m.
Urethral sphincter m.
Rectum, perineal flexure
Prostate gland
Bulbourethral gland and duct
Deep transverse perineal m.
Bulbospongiosus m.

Fig. 353: The Inferior Reflection of Peritoneum over the Pelvic Organs

NOTE: 1) the inferior extent of the peritoneum anteriorly when the bladder is empty is to the level of the symphysis pubis, whereas when the bladder is full the peritoneum is elevated as much as 3 or 4 inches superior to the symphysis pubis. Between the rectum and the bladder the peritoneum forms the rectovesical pouch.

2) in the scrotum, the visceral and parietal layers of the tunica vaginalis testis also enclose a cavity lined by peritoneum.

3) that the prostate and bladder may be reached anteriorly by means of a suprapubic approach, and inferiorly by ascending through the perineum, in front of the rectum. By either route, these organs may be achieved without entering the peritoneal cavity.

Fig. 354: The Male Pelvic Organs Viewed from the Left Side

NOTE that the peritoneum has been removed from the left side of the bladder and rectum.

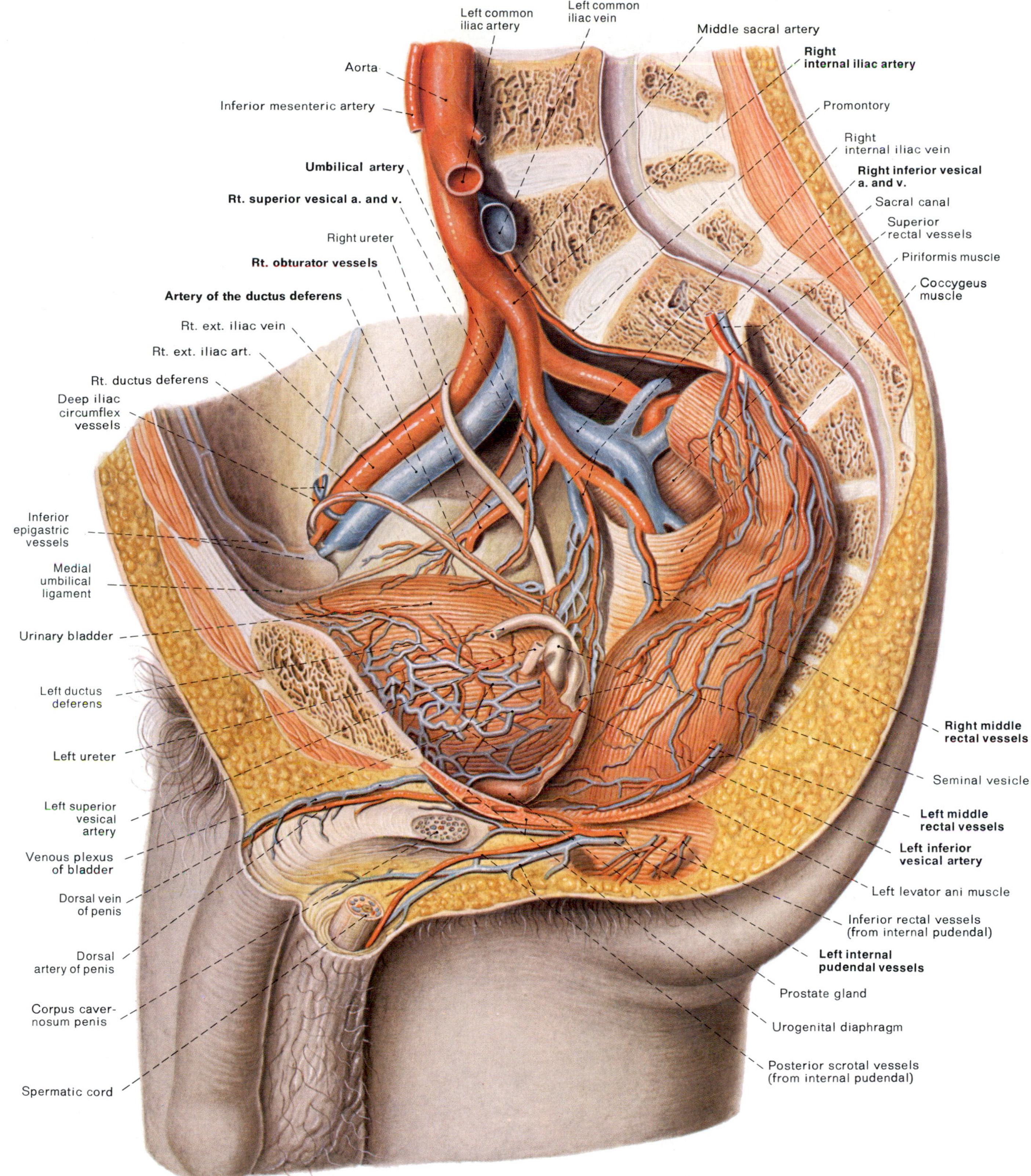

Left common iliac artery
Left common iliac vein
Middle sacral artery
Right internal iliac artery
Aorta
Inferior mesenteric artery
Promontory
Right internal iliac vein
Umbilical artery
Right inferior vesical a. and v.
Rt. superior vesical a. and v.
Sacral canal
Superior rectal vessels
Right ureter
Rt. obturator vessels
Piriformis muscle
Coccygeus muscle
Artery of the ductus deferens
Rt. ext. iliac vein
Rt. ext. iliac art.
Rt. ductus deferens
Deep iliac circumflex vessels
Inferior epigastric vessels
Medial umbilical ligament
Urinary bladder
Left ductus deferens
Right middle rectal vessels
Left ureter
Seminal vesicle
Left superior vesical artery
Left middle rectal vessels
Left inferior vesical artery
Venous plexus of bladder
Left levator ani muscle
Dorsal vein of penis
Inferior rectal vessels (from internal pudendal)
Dorsal artery of penis
Left internal pudendal vessels
Prostate gland
Corpus cavernosum penis
Urogenital diaphragm
Posterior scrotal vessels (from internal pudendal)
Spermatic cord

Fig. 355: Blood Vessels of the Male Pelvis, Perineum and External Genitalia

NOTE: 1) the aorta bifurcates into the common iliac arteries which then divide into the external and internal iliac arteries. The external iliac becomes the chief arterial trunk of the lower extremity, while the internal iliac artery supplies the organs of the pelvis and perineum.

2) the branches of the internal iliac artery include visceral and parietal vessels. The *visceral* branches are (a) the umbilical (from which is derived the superior vesical artery), (b) the inferior vesical, (c) the artery of the vas deferens (uterine in the female) and, (d) the middle rectal. The *parietal* branches include (a) the iliolumbar, (b) lateral sacral, (c) superior gluteal, (d) inferior gluteal, (e) obturator, and (f) internal pudendal.

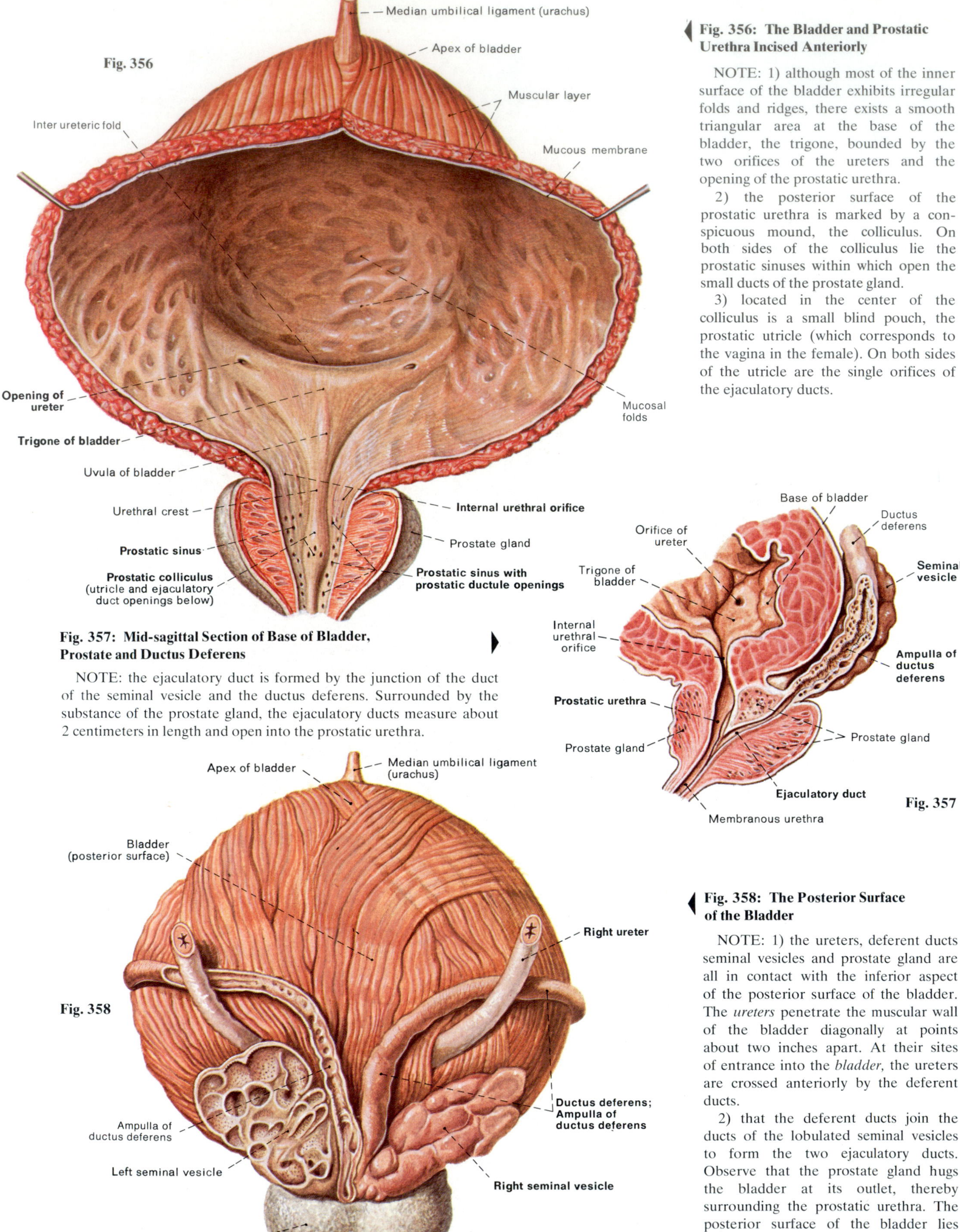

Fig. 356
Median umbilical ligament (urachus)
Apex of bladder
Muscular layer
Inter ureteric fold
Mucous membrane
Opening of ureter
Mucosal folds
Trigone of bladder
Uvula of bladder
Urethral crest
Internal urethral orifice
Prostate gland
Prostatic sinus
Prostatic colliculus (utricle and ejaculatory duct openings below)
Prostatic sinus with prostatic ductule openings
Base of bladder
Ductus deferens
Orifice of ureter
Seminal vesicle
Trigone of bladder
Internal urethral orifice
Ampulla of ductus deferens
Prostatic urethra
Prostate gland
Prostate gland
Ejaculatory duct
Membranous urethra
Fig. 357
Apex of bladder
Median umbilical ligament (urachus)
Bladder (posterior surface)
Right ureter
Fig. 358
Ductus deferens; Ampulla of ductus deferens
Ampulla of ductus deferens
Left seminal vesicle
Right seminal vesicle
Prostate gland

Fig. 356: The Bladder and Prostatic Urethra Incised Anteriorly

NOTE: 1) although most of the inner surface of the bladder exhibits irregular folds and ridges, there exists a smooth triangular area at the base of the bladder, the trigone, bounded by the two orifices of the ureters and the opening of the prostatic urethra.

2) the posterior surface of the prostatic urethra is marked by a conspicuous mound, the colliculus. On both sides of the colliculus lie the prostatic sinuses within which open the small ducts of the prostate gland.

3) located in the center of the colliculus is a small blind pouch, the prostatic utricle (which corresponds to the vagina in the female). On both sides of the utricle are the single orifices of the ejaculatory ducts.

Fig. 357: Mid-sagittal Section of Base of Bladder, Prostate and Ductus Deferens

NOTE: the ejaculatory duct is formed by the junction of the duct of the seminal vesicle and the ductus deferens. Surrounded by the substance of the prostate gland, the ejaculatory ducts measure about 2 centimeters in length and open into the prostatic urethra.

Fig. 358: The Posterior Surface of the Bladder

NOTE: 1) the ureters, deferent ducts seminal vesicles and prostate gland are all in contact with the inferior aspect of the posterior surface of the bladder. The *ureters* penetrate the muscular wall of the bladder diagonally at points about two inches apart. At their sites of entrance into the *bladder,* the ureters are crossed anteriorly by the deferent ducts.

2) that the deferent ducts join the ducts of the lobulated seminal vesicles to form the two ejaculatory ducts. Observe that the prostate gland hugs the bladder at its outlet, thereby surrounding the prostatic urethra. The posterior surface of the bladder lies directly anterior to the rectum in the lower pelvis.

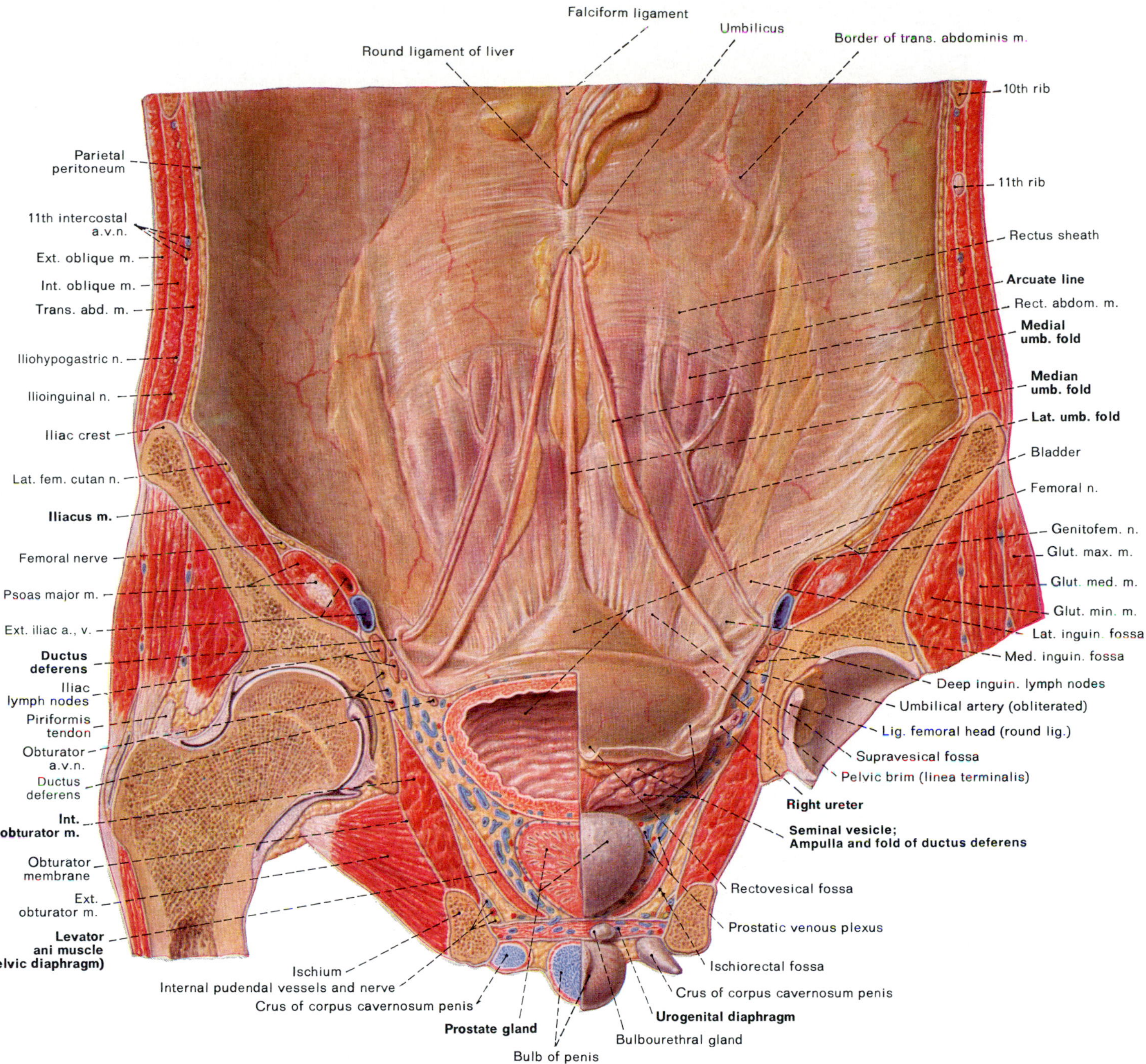
Falciform ligament
Umbilicus
Round ligament of liver
Border of trans. abdominis m.
10th rib
Parietal peritoneum
11th rib
11th intercostal a.v.n.
Rectus sheath
Ext. oblique m.
Int. oblique m.
Arcuate line
Trans. abd. m.
Rect. abdom. m.
Medial umb. fold
Iliohypogastric n.
Median umb. fold
Ilioinguinal n.
Lat. umb. fold
Iliac crest
Bladder
Lat. fem. cutan n.
Femoral n.
Iliacus m.
Genitofem. n.
Femoral nerve
Glut. max. m.
Glut. med. m.
Psoas major m.
Glut. min. m.
Ext. iliac a., v.
Lat. inguin. fossa
Ductus deferens
Med. inguin. fossa
Iliac lymph nodes
Deep inguin. lymph nodes
Piriformis tendon
Umbilical artery (obliterated)
Lig. femoral head (round lig.)
Obturator a.v.n.
Supravesical fossa
Ductus deferens
Pelvic brim (linea terminalis)
Right ureter
Int. obturator m.
Seminal vesicle;
Ampulla and fold of ductus deferens
Obturator membrane
Rectovesical fossa
Ext. obturator m.
Levator ani muscle
(pelvic diaphragm)
Prostatic venous plexus
Ischium
Ischiorectal fossa
Internal pudendal vessels and nerve
Crus of corpus cavernosum penis
Crus of corpus cavernosum penis
Urogenital diaphragm
Prostate gland
Bulbourethral gland
Bulb of penis

Fig. 359: The Posterior Aspect of the Ventral Abdominal Wall and the Male Pelvic Organs

NOTE: 1) the inner aspect of the anterior abdominal wall shows the courses of the umbilical ligaments and the round ligament of the liver, as well as the smooth contours of the rectus abdominis muscle and the arcuate line, which marks the inferior limit of the posterior layer of the rectus sheath.

2) the frontal section through the bones of the pelvis and the femur. Observe the expanded wings (alae) of each ilium which extend laterally to reach the skin as the iliac crest. The wings of the ilia bound the greater (or false) pelvis. On their internal or pelvic surface lie the iliacus and psoas muscles along with the iliac vessels and the various nerves of the lumbar plexus. Their external surface is directed toward the gluteal region.

3) that the lesser (or true) pelvis lies below the pelvic brim. It is restricted by bone and bounds more snugly the pelvic organs (bladder, prostate, seminal vesicle, etc.). Inferiorly the pelvic organs are separated from the perineum by the pelvic diaphragm (levator ani muscles) and, to a lesser extent, the urogenital diaphragm (deep transverse perineal muscles).

4) the reflection of peritoneum as it invests the inner surface of the anterior abdominal wall, curves over the superior surface of the bladder and dips somewhat posterior to the bladder (rectovesical pouch) to come into contact with the seminal vesicles and deferent ducts.

5) the obturator internus muscle which covers much of the inner surface of the lateral wall of the true pelvis. Extending into the perineum, this muscle also forms the lateral boundary of the ischiorectal fossa.

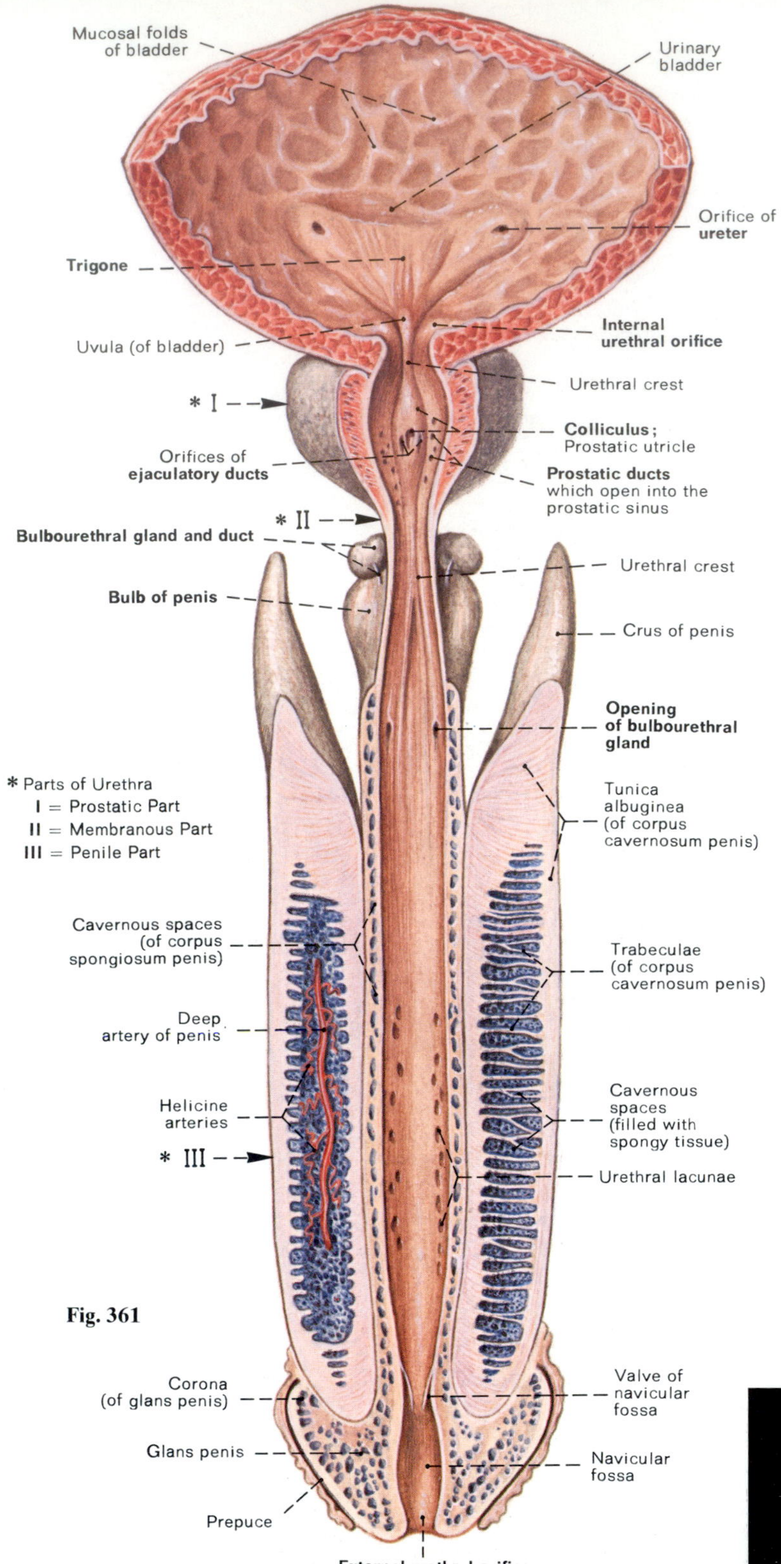

Mucosal folds of bladder
Urinary bladder
Orifice of **ureter**
Trigone
Uvula (of bladder)
Internal urethral orifice
* I
Urethral crest
Colliculus; Prostatic utricle
Orifices of **ejaculatory ducts**
Prostatic ducts which open into the prostatic sinus
* II
Bulbourethral gland and duct
Urethral crest
Bulb of penis
Crus of penis
Opening of bulbourethral gland
* Parts of Urethra
I = Prostatic Part
II = Membranous Part
III = Penile Part
Tunica albuginea (of corpus cavernosum penis)
Cavernous spaces (of corpus spongiosum penis)
Trabeculae (of corpus cavernosum penis)
Deep artery of penis
Helicine arteries
Cavernous spaces (filled with spongy tissue)
* III
Urethral lacunae
Fig. 361
Corona (of glans penis)
Valve of navicular fossa
Glans penis
Navicular fossa
Prepuce
External urethral orifice

Fig. 361: The Male Urethra and its Associated Orifices

NOTE: 1) the male urethra is a canal that extends from its internal urethral orifice at the bladder to its external urethral orifice at the end of the glans penis. Since the male urethra transverses the prostate gland, the urogenital diaphragm (membrane) and the penis, it is divided into three parts: prostatic, membranous and penile.

2) urine passes through the urethra from the bladder. The urethra also transports the seminal fluid which is composed of a mixture of sperm from the testis and the secretions of the seminal vesicles and prostate. Just prior to ejaculation, the urethra is lubricated by a viscous fluid secreted by the bulbourethral glands (of Cowper). These glands are located in the urogenital diaphragm, but their ducts open about one inch or more distally into the penile urethra.

3) the total urethra measures between 7 and 8 inches in length. The prostatic urethra is about $1\frac{1}{2}$ inches long, the membranous urethra about $\frac{1}{2}$ inch and the penile urethra 5 to 6 inches. On its posterior wall, the *prostatic urethra* is marked by a ridge, the urethral crest and a mound, the colliculus. It receives the secretions of the ejaculatory ducts along with those of the prostate. Enlargement of the prostate, often occurring in older men, tends to constrict the urethra at this site and frequently results in difficulty in urination.

4) the *membranous urethra* is short and narrow. In its course through the urogenital diaphragm, it is completely surrounded by the circular fibers of the urethral sphincter muscle (see Figure 366). Since this important sphincter is under voluntary control, its relaxation initiates urination while its tonic contraction constricts the urethra and maintains urinary continence.

5) the penile (or spongy) portion of the urethra at once is surrounded by the bulb of the penis and, thus, the bulbospongiosus muscle. It traverses the penile shaft embedded within the corpus spongiosum penis. In its course, it receives the ducts of the bulbourethral glands. The internal surface of its distal half is marked by a number of small recesses, the urethral lacunae. The distal portion of the penile urethra is somewhat narrowed and its external urethral orifice simply a vertical slit.

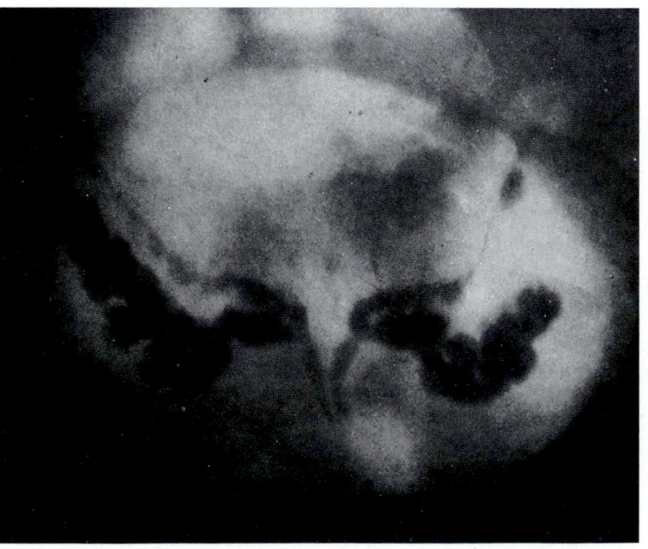

Fig. 360: Radiograph of Bladder, Seminal Vesicles, Deferent Ducts and Ejaculatory Ducts

NOTE: 1) this figure is a positive print of an X-ray. The bladder has been filled with air and appears light, whereas the images of the seminal vesicles, deferent ducts and ejaculatory ducts stand out as dark.

2) that the convoluted seminal vesicles, which secrete the seminal fluid, consist of coiled tubes 4–5 mm in diameter and 2 to 3 inches in length.

3) the shadows of the deferent ducts. These cross the bladder surface toward the midline to join the seminal vesicles forming the ejaculatory ducts.

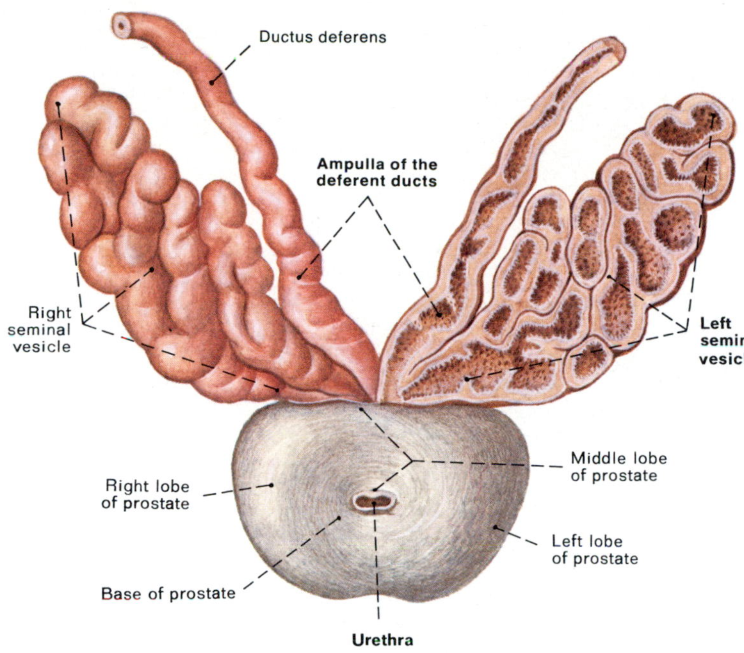
Ductus deferens
Ampulla of the deferent ducts
Right seminal vesicle
Left seminal vesicle
Middle lobe of prostate
Right lobe of prostate
Left lobe of prostate
Base of prostate
Urethra

Fig. 362

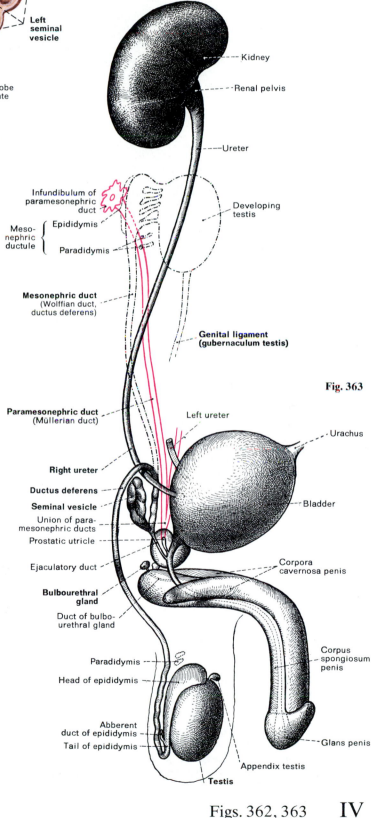

Fig. 362: The Prostate Gland, Seminal Vesicles and Ampullae of the Deferent Ducts (Superior View)

NOTE: the left seminal vesicle and ductus deferens were cut longitudinally, while the urethra was cut transversely just distal to the bladder. The prostate gland is conical in shape and normally measures just over 1½ inches across, 1 inch in thickness and slightly longer than 1 inch vertically. In the young adult male it weighs about 25 grams and is formed by two lateral lobes surrounding a middle lobe.

Fig. 363: Diagram of the Male Genitourinary System

NOTE: 1) this figure shows:

a) the organs of the adult male genitourinary system;

b) the structures of the genital system prior to the descent of the testis (interrupted black lines);

c) those structures which partially or entirely became atrophic and disappeared during development (red lines).

2) the urinary system includes the *kidneys* which produce urine by filtration of the blood, the *ureters* which convey the urine to the *bladder,* where it is stored, and the *urethra* through which urine is discharged.

3) the adult male genital system includes the *testis* where sperm are generated, the *epididymis* and *ductus deferens* which transport the sperm to the *ejaculatory duct* where the *seminal vesicle* joins the genital system. The *prostate* and *bulbourethral glands* along with the ejaculatory ducts join the *urethra,* which then courses through the male copulatory organ, the *penis.*

4) embryologically, structures capable of developing into both male and female genital systems exist in all individuals. In the male, the Wolffian or mesonephric duct (which becomes the epididymis, vas deferens ejaculatory duct and seminal vesicle) and the penis, develop, while the paramesonephric or Müllerian duct becomes vestigial.

5) the testes are developed on the posterior abdominal wall and are attached by a fibrous genital ligament called the gubernaculum testis to the abdominal wall. As development continues, each testis gradually "migrates" from its site of formation, so that by the 5th month it comes to lie adjacent to the abdominal inguinal ring. The gubernaculum testis is still attached to anterior abdominal wall tissue, which by this time has evaginated as the developing scrotum. The testes then commence their descent through the inguinal canal, so that by the 8th month they usually lie in the scrotum. During this "migration", the testes pass behind the peritoneum but are attached to it by its peritoneal reflection, the processus vaginalis testis. This becomes the *tunica vaginalis testis.*

Kidney
Renal pelvis
Ureter
Infundibulum of paramesonephric duct
Meso-nephric ductule
Epididymis
Paradidymis
Developing testis
Mesonephric duct (Wolffian duct, ductus deferens)
Genital ligament (gubernaculum testis)
Paramesonephric duct (Müllerian duct)
Left ureter
Urachus
Right ureter
Ductus deferens
Seminal vesicle
Bladder
Union of paramesonephric ducts
Prostatic utricle
Ejaculatory duct
Corpora cavernosa penis
Bulbourethral gland
Duct of bulbourethral gland
Paradidymis
Head of epididymis
Corpus spongiosum penis
Abberent duct of epididymis
Tail of epididymis
Glans penis
Appendix testis
Testis

Fig. 363

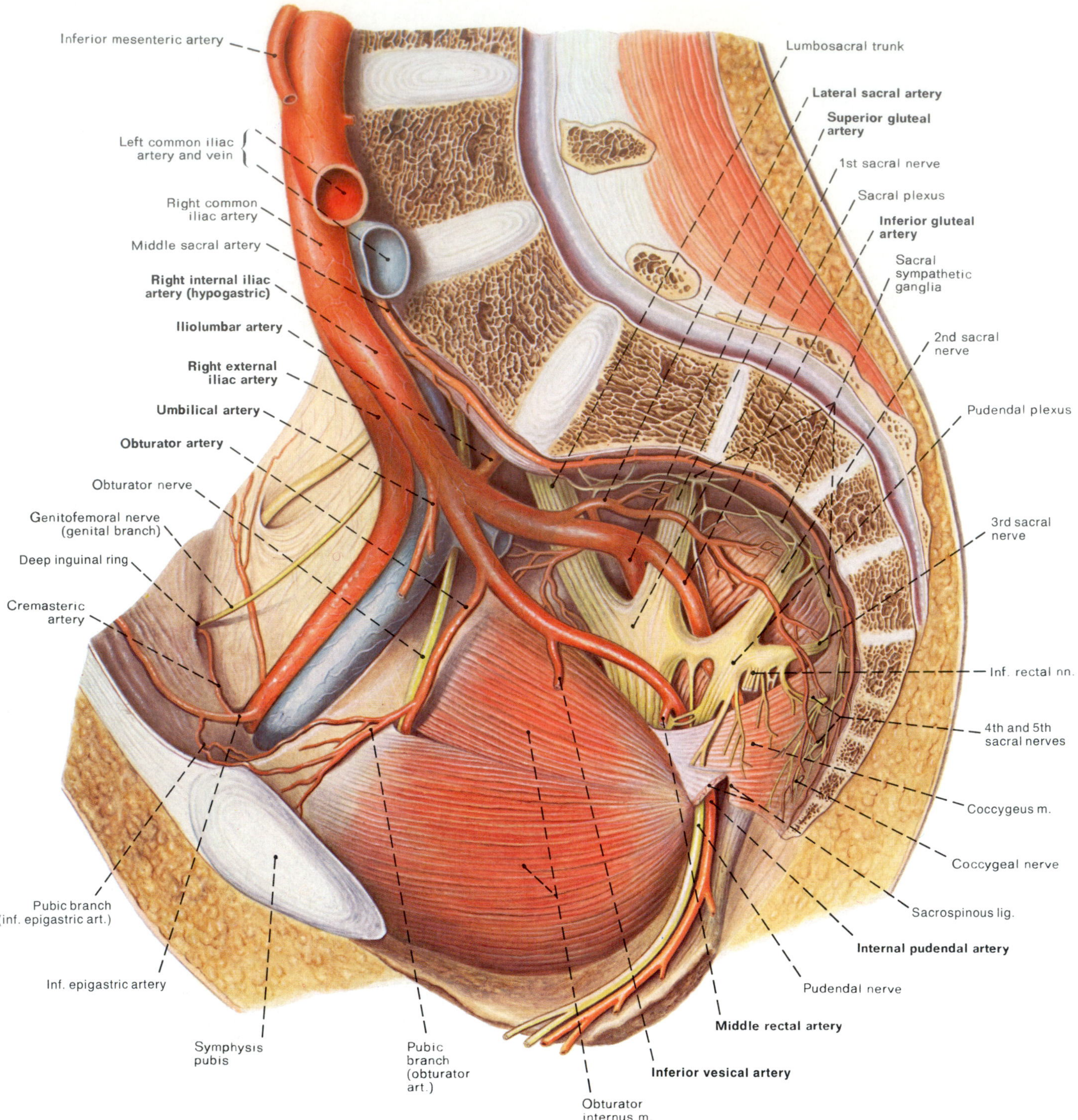
Inferior mesenteric artery
Left common iliac artery and vein
Right common iliac artery
Middle sacral artery
Right internal iliac artery (hypogastric)
Iliolumbar artery
Right external iliac artery
Umbilical artery
Obturator artery
Obturator nerve
Genitofemoral nerve (genital branch)
Deep inguinal ring
Cremasteric artery
Pubic branch (inf. epigastric art.)
Inf. epigastric artery
Symphysis pubis
Pubic branch (obturator art.)
Obturator internus m.
Inferior vesical artery
Middle rectal artery
Pudendal nerve
Internal pudendal artery
Sacrospinous lig.
Coccygeal nerve
Coccygeus m.
4th and 5th sacral nerves
Inf. rectal nn.
3rd sacral nerve
Pudendal plexus
2nd sacral nerve
Sacral sympathetic ganglia
Inferior gluteal artery
Sacral plexus
1st sacral nerve
Superior gluteal artery
Lateral sacral artery
Lumbosacral trunk

Fig. 364: Blood Vessels and Nerves of the Pelvic Wall

NOTE: 1) this is a mid-sagittal view of the right pelvic wall viewed from the left side. The pelvic viscera have been removed and the parietal blood vessels and nerves demonstrated.

2) the principal arteries of the pelvic wall are derived from the internal iliac artery. Although the branches of this vessel are quite variable, in at least 50% of cadavers it courses about 1½ inches toward the greater sciatic foramen before branching into a posterior division of 4 vessels and an anterior division of 6 vessels.

3) the four posterior division vessels include a) the *iliolumbar* which courses superiorly toward the iliac fossa, b) the *lateral sacral* which courses inferiorly, anastomoses with the middle sacral artery and supplies branches to the upper sacral foramina, c) the *superior gluteal* which is the largest of the branches of the internal iliac and which leaves the pelvis above the border of the piriformis muscle, and d) the *inferior gluteal* which leaves the pelvis below the piriformis muscle.

4) the anterior division of the internal iliac artery gives rise to four visceral arteries (umbilical, inferior vesical, middle rectal and uterine or deferential; see in Fig. 355). These are simply indicated by cut stumps in the figure. The two parietal vessels of this anterior division are the *obturator* artery which courses through the obturator canal to the medial thigh, and the long *internal pudendal* artery which leaves the pelvis through the greater sciatic foramen, crosses the ischial spine to reenter the pelvis by way of the lesser sciatic foramen. It then courses toward the perineum by way of the pudendal canal to supply the anal and urogenital triangles.

Fig. 365: The Muscular Floor of the Pelvis: Pelvic Diaphragm

NOTE: 1) with the pelvic organs removed, this superior view of the floor of the pelvis emphasizes the muscular nature of the pelvic outlet. The *pelvic diaphragm* consists of the levator ani and coccygeus muscles along with two fascial layers which cover the pelvic (supra-anal fascia) and perineal (infraanal fascia) surfaces of these muscles.

2) the muscles and fascial layers composing the pelvic diaphragm stretch across the pelvic floor in a concave sling-like manner, thereby forming a separation between the structures of the pelvis and those of the perineum below.

3) the more caudal part of this "sling" is formed by the two coccygeus muscles. Since these muscles stretch from the ischial spines to the sacrum and coccyx, they are sometimes referred to as the ischiococcygeus muscles. The mid and anterior portions of the "sling" are formed by the iliococcygeus and pubococcygeus muscles. These latter combine to form the levator ani muscle. The levator ani arises from the tendinous arc along the pelvic surface of the obturator internus and inserts into the anococcygeal raphé, the central point of the perineum, the external spincter and the lower segments of the coccyx.

4) in the male, the pelvic diaphragm is perforated by the anal canal and urethra.

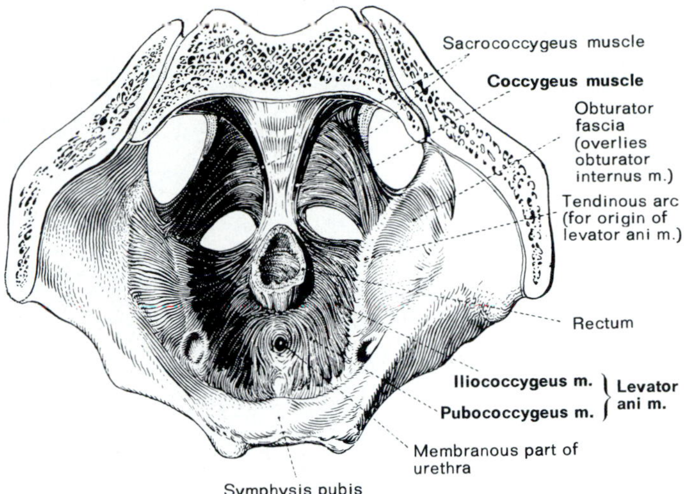

Sacrococcygeus muscle
Coccygeus muscle
Obturator fascia (overlies obturator internus m.)
Tendinous arc (for origin of levator ani m.)
Rectum
Iliococcygeus m.
Pubococcygeus m.
Levator ani m.
Membranous part of urethra
Symphysis pubis

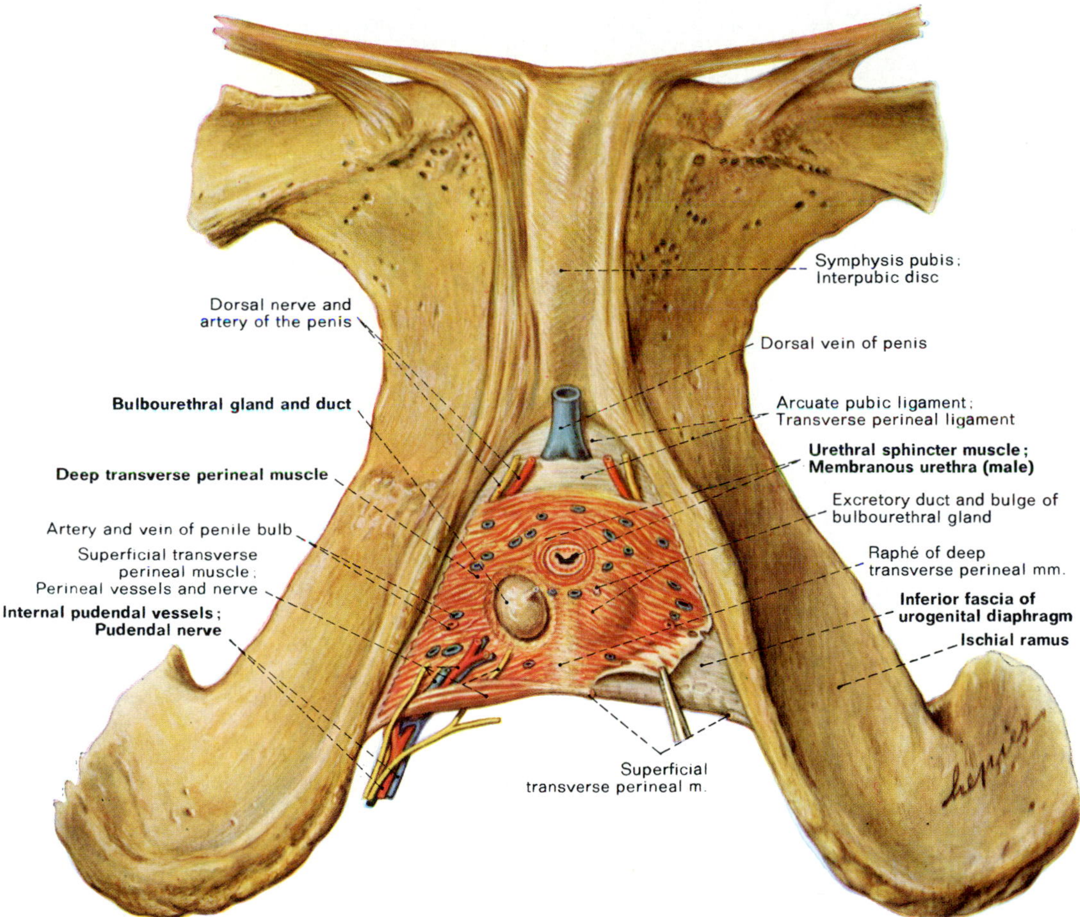

Symphysis pubis; Interpubic disc
Dorsal nerve and artery of the penis
Dorsal vein of penis
Bulbourethral gland and duct
Arcuate pubic ligament; Transverse perineal ligament
Deep transverse perineal muscle
Urethral sphincter muscle; Membranous urethra (male)
Excretory duct and bulge of bulbourethral gland
Artery and vein of penile bulb
Superficial transverse perineal muscle; Perineal vessels and nerve
Raphé of deep transverse perineal mm.
Internal pudendal vessels; Pudendal nerve
Inferior fascia of urogenital diaphragm
Ischial ramus
Superficial transverse perineal m.

Fig. 366: The Urogenital Diaphragm; Deep Transverse Perineal Muscle (Male)

NOTE: 1) the deep transverse perineal muscle, which stretches between the rami of the ischium, is covered by fascia on both its internal (pelvic or superior) surface and its external (perineal or inferior) surface. As such, this muscle and these two fascial sheaths constitute the urogenital diaphragm.

2) the region between the superior and inferior fascial planes is frequently referred to as the deep perineal compartment (pouch, cleft or space). In the male it contains a) the deep transverse perineal muscle, b) the membranous sphincter of the urethra, c) the bulbourethral glands and ducts, d) the membranous urethra and e) the various branches of the internal pudendal vessels and nerves.

3) the urogenital diaphragm assists somewhat in strengthening the anterior portion of the levator ani muscle where, in the midline at the so-called genital hiatus, it is penetrated by the urethra in the male and the urethra and vagina in the female.

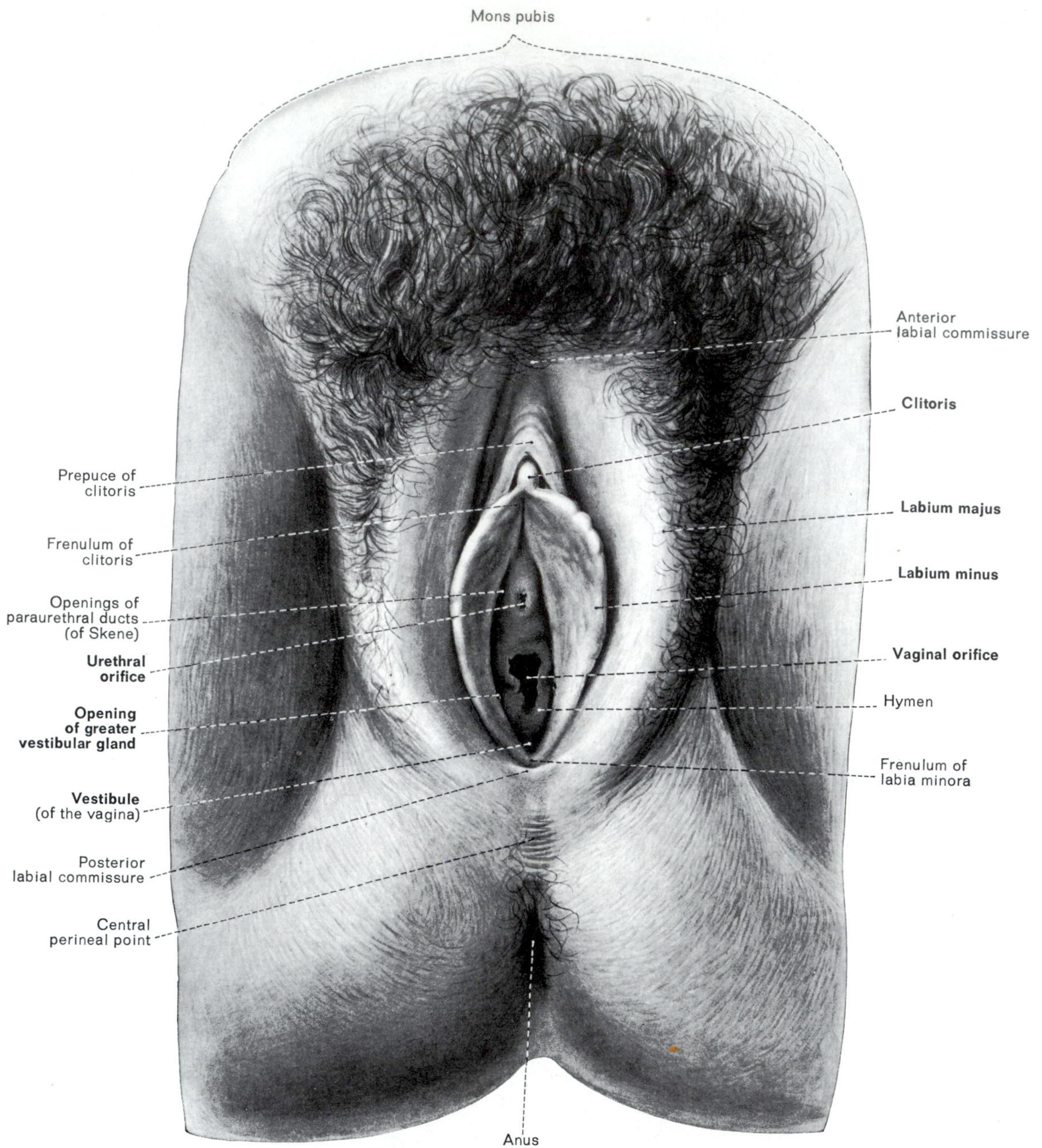

Mons pubis
Anterior labial commissure
Clitoris
Prepuce of clitoris
Labium majus
Frenulum of clitoris
Labium minus
Openings of paraurethral ducts (of Skene)
Urethral orifice
Vaginal orifice
Hymen
Opening of greater vestibular gland
Frenulum of labia minora
Vestibule (of the vagina)
Posterior labial commissure
Central perineal point
Anus

Fig. 367: The External Genitalia of an 18 Years Old Virgin

NOTE: 1) the female external genitalia include a) the labia majora, b) the labia minora, c) the clitoris and d) the vestibule of the vagina. The mons pubis, a rounded mound of adipose tissue anterior to the symphysis pubis and covered with hair in the adult, might also be considered an external genital structure. The orifices of the female perineum include the urethral orifice, the vaginal orifice, the two openings of the ducts of the greater vestibular glands, the orifices of the small paraurethral ducts (of Skene) and the anus.

2) the labia majora are two elongated folds of skin which form natural extensions from the mons pubis toward the anus. Although there is some variation in size and thickness dependent on age and obesity, the labia majora are in contact laterally with the thighs, unite both anteriorly and posteriorly to form integumentary commissures and represent the female homologous structures to the male scrotum. The anterior end of each labium majus receives the fibrous round ligament of the uterus as it leaves the superficial inguinal ring.

3) the labia minora are two thin folds of skin situated between the labia majora. Anteriorly, the labia minora commence at the glans clitoris, although small extensions of the labia pass over the dorsum of the clitoris to unite and form the prepuce. Posteriorly, the labia minora meet at the midline to form the frenulum of the labia.

4) the clitoris, homologous to the male penis, is an erectile organ and measures one inch or less in length. It is composed of two corpora cavernosa which are bound by dense connective tissue and which are attached by crura to the pubic rami; the crura are covered by small ischiocavernosus muscles. The clitoris is maintained by a suspensory ligament and capped by the glans.

5) the vestibule (of the vagina) is the region between the two labia minora. Into it open the urethra ventrally, the ducts of the greater vestibular glands bilaterally and the vagina. The vaginal orifice is partially closed in the virgin by a thin mucous membrane, the hymen, which separates the vestibule from the vagina proper. Generally, the hymen is ruptured at first copulation, however since its form and extent are quite variable, the establishment of virginity in this manner cannot be absolutely ascertained.

Fig. 368: The Dissected Female External Genitalia

NOTE: 1) in this dissection the skin and fascia of the mons pubis and labia majora have been removed in order to demonstrate more clearly the erectile bodies and other female external genital structures. Observe especially the crura, body and glans clitoris, the vestibular bulbs on either side of the vagina and the position of the greater vestibular glands.

2) each of the two crura of the clitoris is covered by an ischiocavernosus muscle and the vestibular bulbs are surrounded by bulbospongiosus muscles in a manner quite similar to that seen in male. The greater vestibular glands (of Bartholin) located on both sides just posterior to the vestibular bulbs are comparable to the bulbourethral glands (of Cowper) in the male. During sexual stimulation, the greater vestibular glands secrete a viscous fluid which lubricates the vaginal orifice.

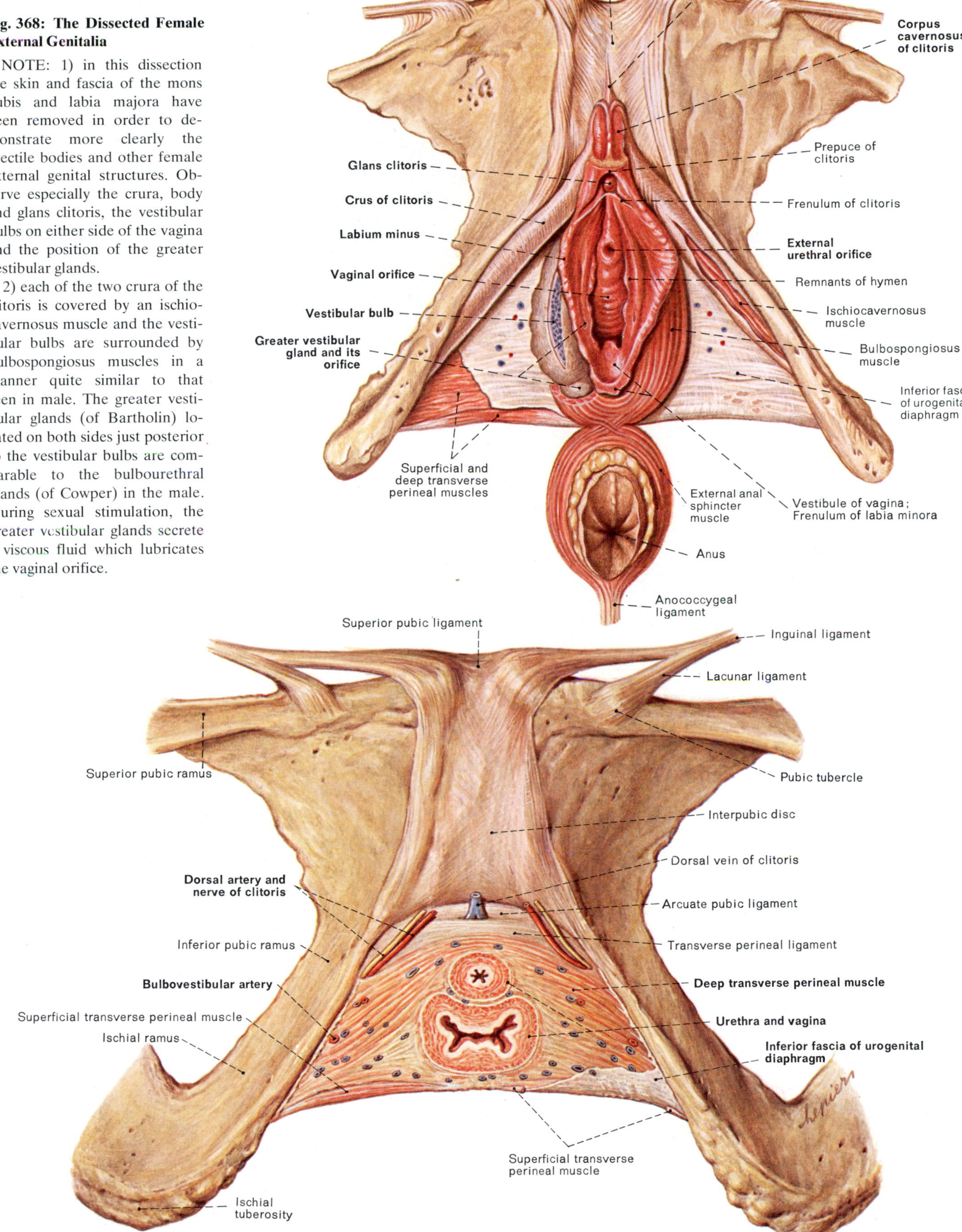
Symphysis pubis
Suspensory ligament of clitoris
Corpus cavernosus of clitoris
Prepuce of clitoris
Glans clitoris
Crus of clitoris
Frenulum of clitoris
Labium minus
External urethral orifice
Vaginal orifice
Remnants of hymen
Vestibular bulb
Ischiocavernosus muscle
Greater vestibular gland and its orifice
Bulbospongiosus muscle
Inferior fascia of urogenital diaphragm
Superficial and deep transverse perineal muscles
External anal sphincter muscle
Vestibule of vagina; Frenulum of labia minora
Anus
Anococcygeal ligament
Superior pubic ligament
Inguinal ligament
Lacunar ligament
Superior pubic ramus
Pubic tubercle
Interpubic disc
Dorsal vein of clitoris
Dorsal artery and nerve of clitoris
Arcuate pubic ligament
Inferior pubic ramus
Transverse perineal ligament
Bulbovestibular artery
Deep transverse perineal muscle
Superficial transverse perineal muscle
Urethra and vagina
Ischial ramus
Inferior fascia of urogenital diaphragm
Superficial transverse perineal muscle
Ischial tuberosity

Fig. 369: The Urogenital Diaphragm in the Female

NOTE that in the female both the urethra and the vagina pass through the urogenital diaphragm. As in the male, the female urogenital diaphragm consists of the deep transverse perineal muscles and a layer of deep fascia on each of their superior and inferior surfaces. Observe the circular sphincter surrounding the membranous urethra.

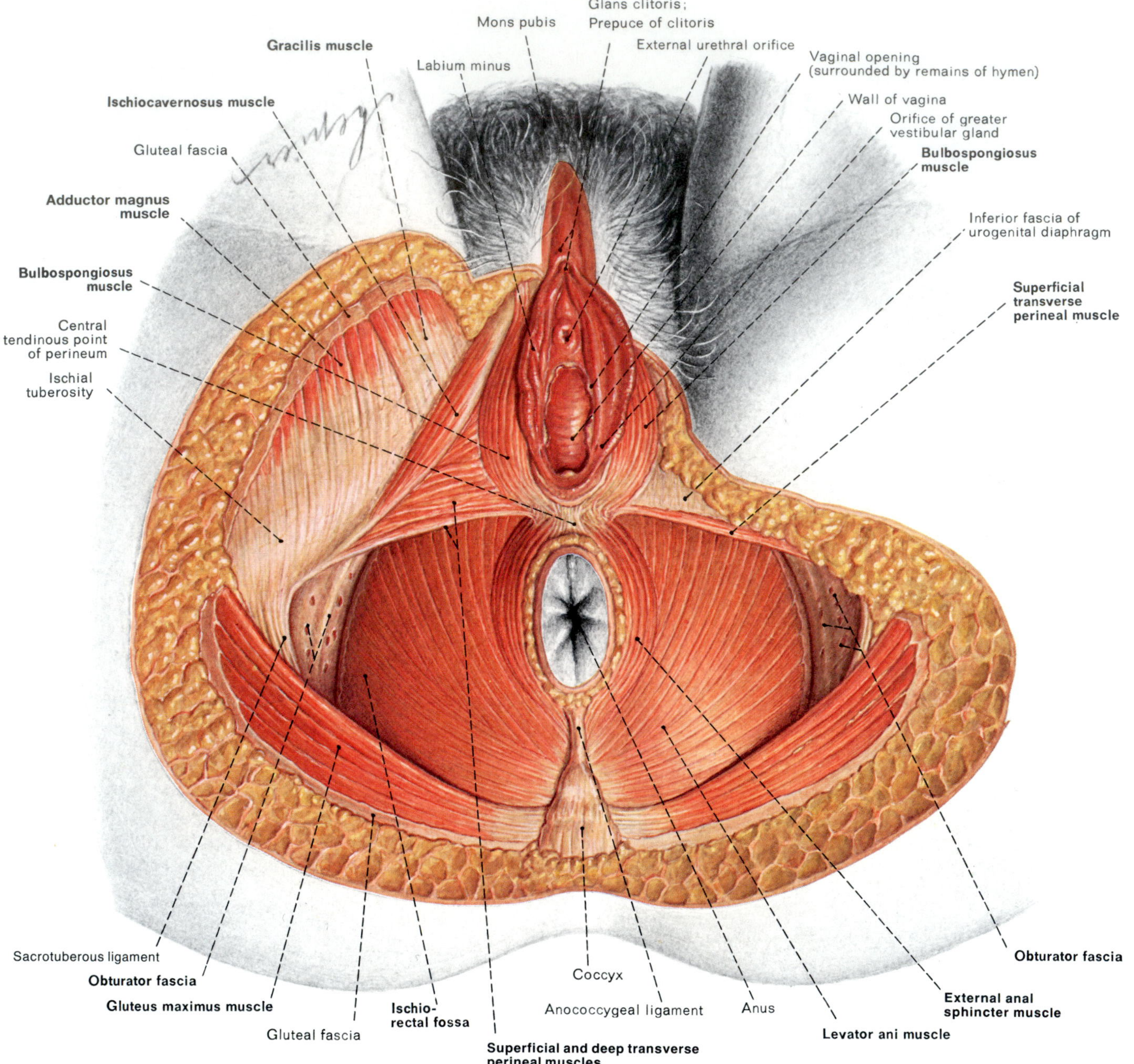

Glans clitoris;
Prepuce of clitoris
Mons pubis
Gracilis muscle
External urethral orifice
Labium minus
Vaginal opening
(surrounded by remains of hymen)
Ischiocavernosus muscle
Wall of vagina
Orifice of greater
vestibular gland
Gluteal fascia
Bulbospongiosus
muscle
Adductor magnus
muscle
Inferior fascia of
urogenital diaphragm
Bulbospongiosus
muscle
Superficial
transverse
perineal muscle
Central
tendinous point
of perineum
Ischial
tuberosity
Sacrotuberous ligament
Obturator fascia
Obturator fascia
Coccyx
External anal
sphincter muscle
Gluteus maximus muscle
Ischio-
rectal fossa
Anococcygeal ligament
Anus
Gluteal fascia
Levator ani muscle
Superficial and deep transverse
perineal muscles

Fig. 370: Superficial Muscles of the Female Perineum

NOTE: 1) the perineum is a diamond-shaped region which is located inferior to the pelvis and separated from it by the muscular pelvic diaphragm. Four points limit the perineum: the symphysis pubis anteriorly; the tip of the coccyx posteriorly; and the two ischial tuberosities laterally. A line drawn transversely across the perineum between the two ischial tuberosities, passing anterior to the anus through the central point of the perineum, divides the diamond-shaped region into an anterior urogenital region and a posterior anal region.

2) the urogenital region contains the external genital organs and the associated muscles and glands. A description of this region frequently refers to a superficial and deep perineal compartment (space, pouch). Simply stated, the superficial perineal compartment lies superficial to the inferior layer of fascia of the urogenital diaphragm and contains the ischiocavernosus, bulbocavernosus and superficial transverse perineal muscle, plus a number of other structures related to the external genitalia. It is traversed by the perineal vessels and nerves and is limited superficially by a layer of deep fascia, the external perineal fascia which stretches just deep to Colles' fascia. The deep perineal compartment is that space enclosed between the superior and inferior layers of fascia of the urogenital diaphragm. Thus, it contains the deep transverse perineal und urethral sphincter muscles (plus the bulbourethral glands in the male) and is traversed by the urethra and vagina in the female and the urethra in the male.

3) the anal region is situated posterior to the urogenital region. The anus, surrounded by the external anal sphincter muscle is located about 1½ inches anterior to the tip of the coccyx. A large portion of the anal region of the perineum is occupied on each side by the fat-filled ischiorectal fossae. Each fossa is wedge-shaped with the base of the wedge directed inferiorly toward the skin. The medial wall of the ischiorectal fossa is formed by the levator ani muscle, the lateral wall being the fascia over the obturator internus muscle. Recesses of each ischiorectal fossa extend anteriorly, adjacent (superior) to the urogenital diaphragm.

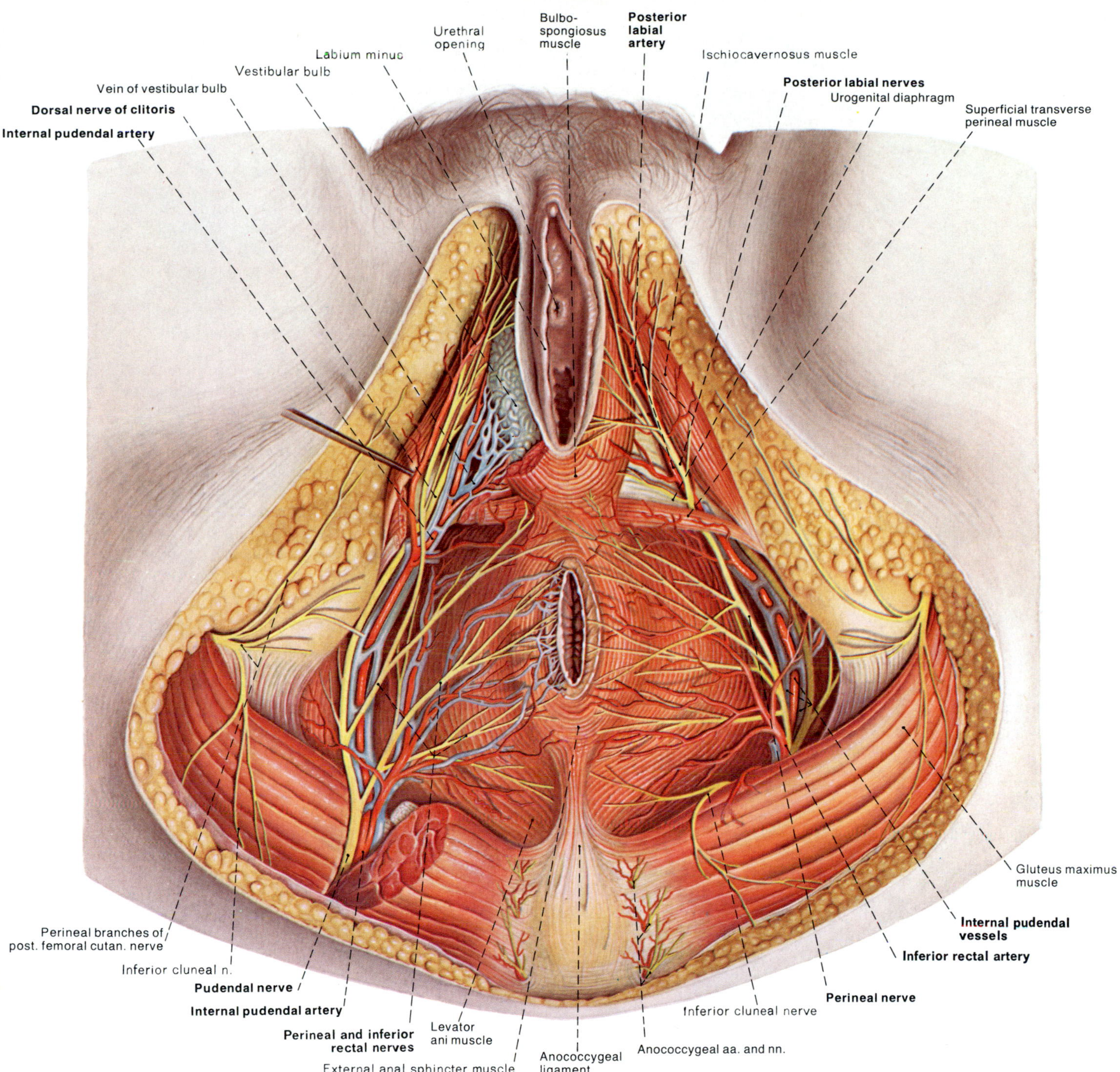
ANTERIOR
Urethral opening
Bulbo-spongiosus muscle
Posterior labial artery
Labium minuc
Vestibular bulb
Ischiocavernosus muscle
Vein of vestibular bulb
Posterior labial nerves
Urogenital diaphragm
Dorsal nerve of clitoris
Superficial transverse perineal muscle
Internal pudendal artery
Gluteus maximus muscle
Internal pudendal vessels
Perineal branches of post. femoral cutan. nerve
Inferior rectal artery
Inferior cluneal n.
Pudendal nerve
Perineal nerve
Internal pudendal artery
Inferior cluneal nerve
Levator ani muscle
Perineal and inferior rectal nerves
Anococcygeal aa. and nn.
Anococcygeal ligament
External anal sphincter muscle
POSTERIOR

Fig. 371: Nerves and Blood Vessels of the Female Perineum

NOTE: 1) in this dissection the skin and superficial fascia have been removed from both the urogenital and anal regions of the female perineum. The ischiorectal fossae have been cleared of fat as well.

2) the principal nerve from which most of the branches which innervate the perineum are derived is the *pudendal nerve*. It originates in the sacral cord and contains fibers of the S_2, S_3 and S_4 segments. In its course within the pelvis, it is joined by the *internal pudendal artery* and *vein* at the lower border of the piriformis muscle at the greater sciatic foramen. Together the vessels and nerve leave the pelvis through the greater sciatic foramen, cross the ischial spine (sacrospinous ligament) to enter the pelvis once again through the lesser sciatic foramen.

3) the pudendal structures achieve the perineum from the pelvis by way of the pudendal canal (of Alcock) which courses beneath the fascia of the obturator internus muscle. Within the perineum, the pudendal structures are first seen at the lateral wall of the ischiorectal fossa. At this point the *inferior rectal vessels* and nerves branch and cross the ischiorectal fossa toward the midline, thereby supplying the levator ani and external anal sphincter muscles along with the other structures in the anal region.

4) continuing anteriorly the pudendal nerve and internal pudendal vessels approach the urogenital diaphragm to become the perineal vessels and nerve. Upon entering the urogenital region, superficial and deep branches supply the structures of the superficial and deep perineal compartments.

5) the superficial perineal branches in the female supply the labia majora and the external genital structures, while the deep branches supply the muscles of the urogenital region, the vestibular bulb and the clitoris.

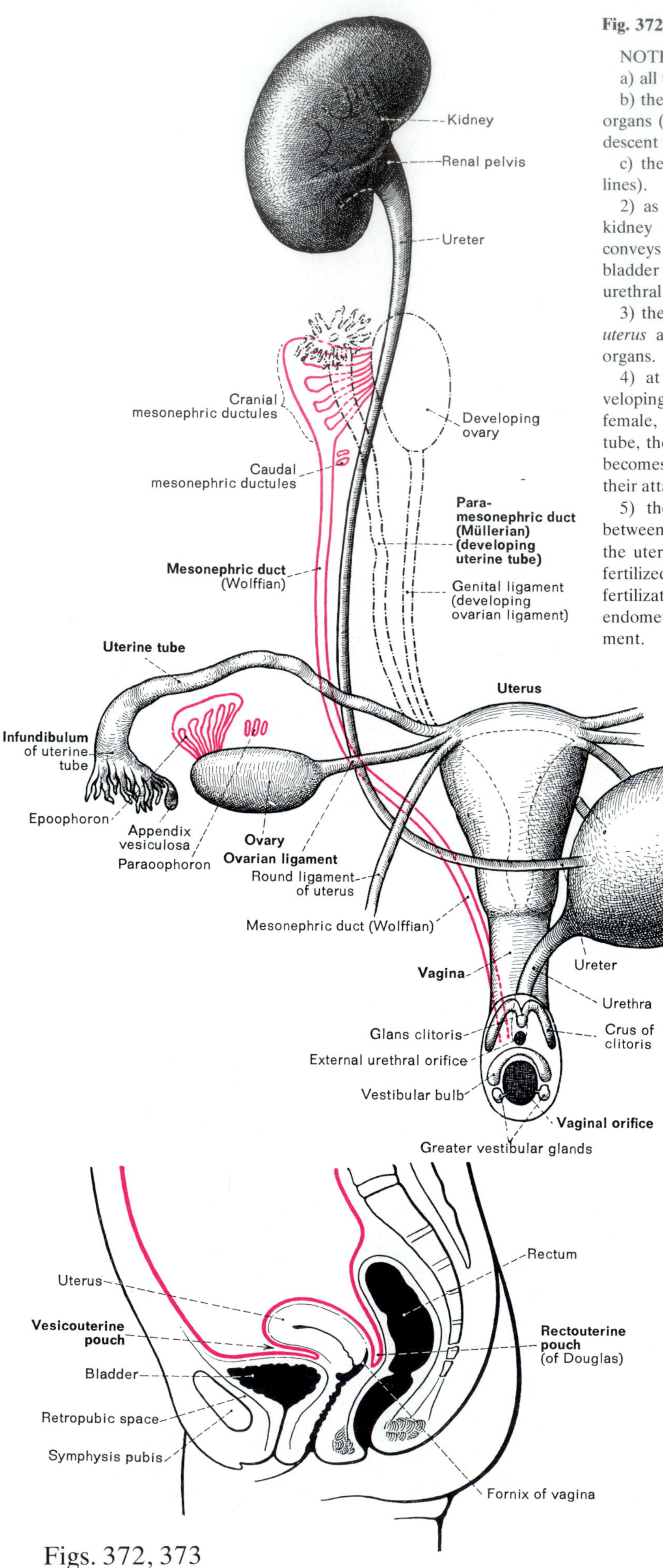

Kidney
Renal pelvis
Ureter
Cranial mesonephric ductules
Developing ovary
Caudal mesonephric ductules
Para-mesonephric duct (Müllerian) (developing uterine tube)
Mesonephric duct (Wolffian)
Genital ligament (developing ovarian ligament)
Uterine tube
Uterus
Infundibulum of uterine tube
Bladder
Urachus
Epoophoron
Appendix vesiculosa
Paraoophoron
Ovary
Ovarian ligament
Round ligament of uterus
Mesonephric duct (Wolffian)
Vagina
Ureter
Urethra
Glans clitoris
Crus of clitoris
External urethral orifice
Vestibular bulb
Vaginal orifice
Greater vestibular glands
Rectum
Uterus
Vesicouterine pouch
Rectouterine pouch (of Douglas)
Bladder
Retropubic space
Symphysis pubis
Fornix of vagina

Fig. 372: Diagram of the Female Genitourinary System

NOTE: 1) this figure shows:

a) all the organs of the adult female genitourinary system;

b) the structures and relevant positions of the female genital organs (gonad and ovarian ligament and uterine tube) prior to their descent into the pelvis (interrupted black lines); and

c) the structures which became atrophic during development (red lines).

2) as in the male, the urinary system of the female includes the kidney which produces urine from the blood, the ureter which conveys the urine to the bladder where it is stored. Leading from the bladder is the urethra through which urine passes to the external urethral orifice during micturition.

3) the adult female genital system includes the *ovary, uterine tube, uterus* and *vagina,* plus the associated glands and external genital organs.

4) at one time during development, structures capable of developing into both male and female genital systems existed. In the female, the Müllerian or paramesonephric duct forms the uterine tube, the uterus and vagina, while the Wolffian or mesonephric duct becomes vestigial. Also the developing gonads become ovaries, while their attachments become the ovarian ligaments.

5) the ovaries produce ova which are discharged periodically between adolescence and the menopause. The ova are captured by the uterine tube where fertilization may occur. If this happens, the fertilized ovum is transported to the uterus and about a week after fertilization, implantation occurs in the wall of the uterus. The endometrium nourishes the embryo in this early period of development.

Fig. 373: Diagram of Peritoneal Reflections Over Female Pelvic Organs (Mid-Sagittal Section)

Peritoneum in red

NOTE: 1) that the parietal peritoneum is reflected over the free abdominal surfaces of the pelvic organs. Observe that as the uterus and vagina are interposed between the bladder and rectum, that peritoneal pouches are formed between the bladder and the uterus (vesicouterine) and between the rectum and the uterus (rectouterine pouch of Douglas).

2) the vesicouterine pouch is relatively shallow. The forward tilt or inclination of the uterus (anteversion) toward the superior surface of the bladder reduces the potential size of the vesicouterine pouch. This fossa does not extend as far inferiorly as the vagina, whereas the deeper rectouterine pouch dips to the posterior surface of the fornix of the vagina. This important anatomical relationship stresses the fact that the fornix of the vagina is separated from the peritoneal cavity only by the thin vaginal wall and the peritoneum.

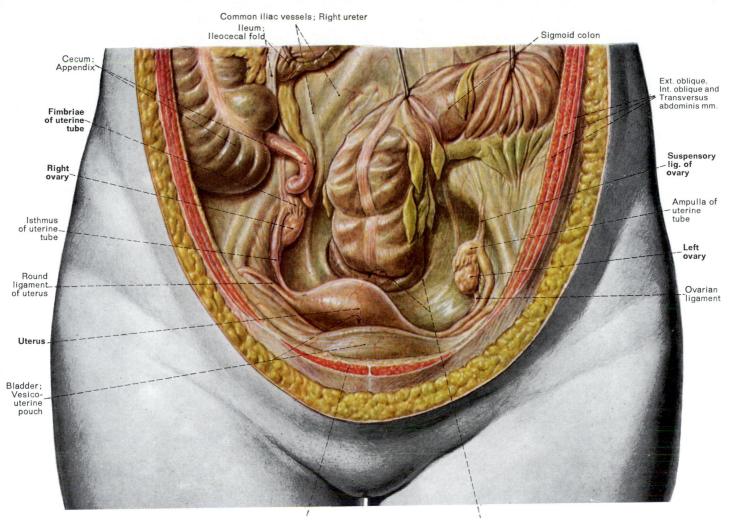

Common iliac vessels; Right ureter
Ileum;
Ileocecal fold
Sigmoid colon
Cecum;
Appendix
Ext. oblique,
Int. oblique and
Transversus
abdominis mm.
Fimbriae of uterine tube
Suspensory lig. of ovary
Right ovary
Ampulla of uterine tube
Isthmus of uterine tube
Left ovary
Round ligament of uterus
Ovarian ligament
Uterus
Bladder;
Vesico-uterine pouch
Rectus abdominis muscle
Rectouterine fold and pouch

Fig. 374: Pelvic Viscera of an Adult Female: Anterior View

NOTE: 1) that the ovaries are situated on the posterolateral aspect of the true pelvis on each side. Having descended from the posterior abdominal wall to their location just below the pelvic brim, the ovaries are held in position by ligamentous attachments. The suspensory ligament transmits the ovarian vessels and nerves.

2) the position of the uterus interposed between the bladder and rectum. Observe that the uterus is frequently located somewhat to one or the other side of the midline.

3) the fimbriae of the uterine tubes as they extend from the ampullae of the tubes to encircle the upper medial surfaces of the ovaries. These tubes vary from 3 to 6 inches in length and, as extensions of the uterus, they convey the ova to the uterus. It is within the uterine tube that fertilization of the ovum usually occurs.

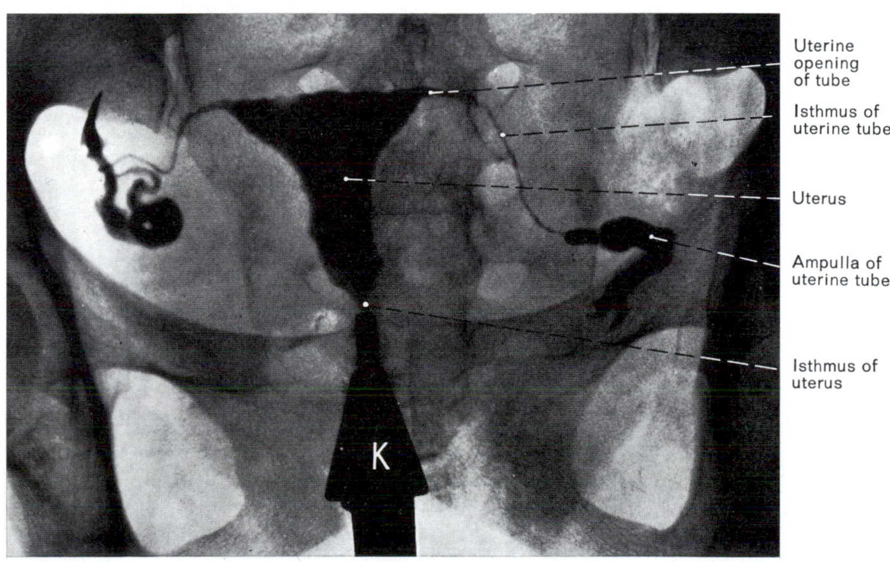

Uterine opening of tube
Isthmus of uterine tube
Uterus
Ampulla of uterine tube
Isthmus of uterus
K

Fig. 375: Utero-Salpingogram

By means of a cannula (K) placed in the vagina, radiopaque material has been injected into the uterus and uterine tubes (salpinx). Observe the narrow lumen of the isthmus of the uterine tubes and note how the tubes enlarge at the ampullae. On the specimen's left side (reader's right side), even the fimbriated end of the tube is discernible, while on the specimen's right side (reader's left side) a small portion of the radiopaque material has been forced into the pelvis through the opening of the uterine tube.

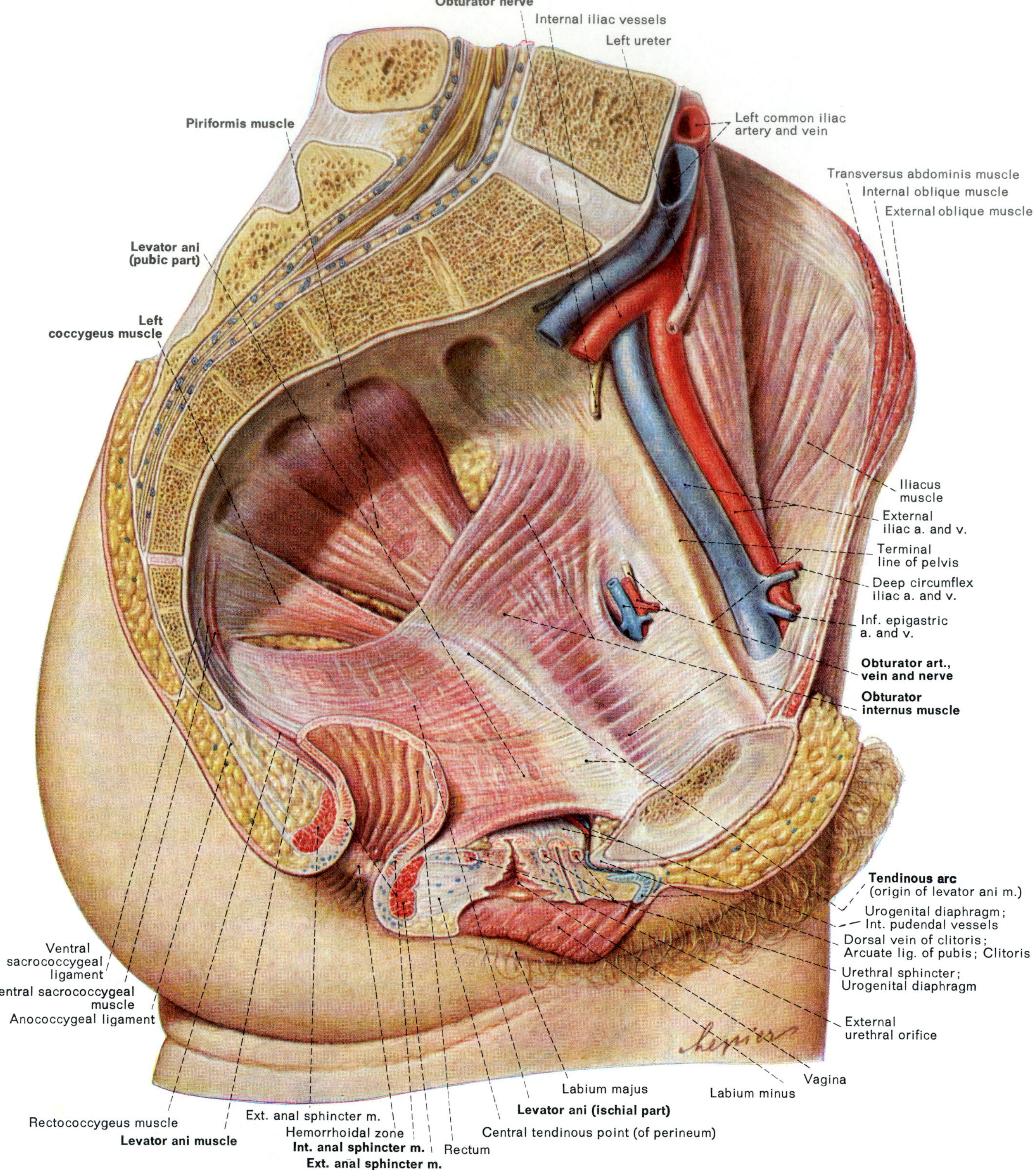

Obturator nerve
Internal iliac vessels
Left ureter
Left common iliac artery and vein
Piriformis muscle
Transversus abdominis muscle
Internal oblique muscle
External oblique muscle
Levator ani (pubic part)
Left coccygeus muscle
Iliacus muscle
External iliac a. and v.
Terminal line of pelvis
Deep circumflex iliac a. and v.
Inf. epigastric a. and v.
Obturator art., vein and nerve
Obturator internus muscle
Tendinous arc (origin of levator ani m.)
Urogenital diaphragm; Int. pudendal vessels
Dorsal vein of clitoris; Arcuate lig. of pubis; Clitoris
Urethral sphincter; Urogenital diaphragm
External urethral orifice
Ventral sacrococcygeal ligament
Ventral sacrococcygeal muscle
Anococcygeal ligament
Vagina
Labium majus
Labium minus
Levator ani (ischial part)
Rectococcygeus muscle
Ext. anal sphincter m.
Levator ani muscle
Hemorrhoidal zone
Central tendinous point (of perineum)
Int. anal sphincter m.
Rectum
Ext. anal sphincter m.

Fig. 376: Muscles of the Lateral Wall and Floor of the Female Pelvis (Left Side)

NOTE: 1) the lateral wall of the true pelvis is covered principally by the piriformis and obturator internus muscles while the floor of the pelvis is formed by both the pubic and ischial portions of the levator ani muscle and, more posteriorly, by the coccygeus muscle.

2) although the *piriformis muscle* arises from the ventral surface of the 2nd, 3rd and 4th sacral vertebrae, it is frequently studied with the gluteal muscles because its fibers converge and leave the pelvis through the greater sciatic foramen. The *obturator internus muscle,* covered by its fascia, has an extensive origin on the inner surface of the lateral and anterior wall of the true bony pelvis. It surrounds the obturator foramen (note the obturator vessels and nerve) and its fibers converge to form a tendon which passes out of the pelvis to enter the gluteal region through the lesser sciatic foramen.

3) both the pubic and ischial portions of the *levator ani* muscle arise principally from the tendinous arc of the obturator internus fascia. Observe the *internal* and *external* anal sphincters surrounding the anal orifice.

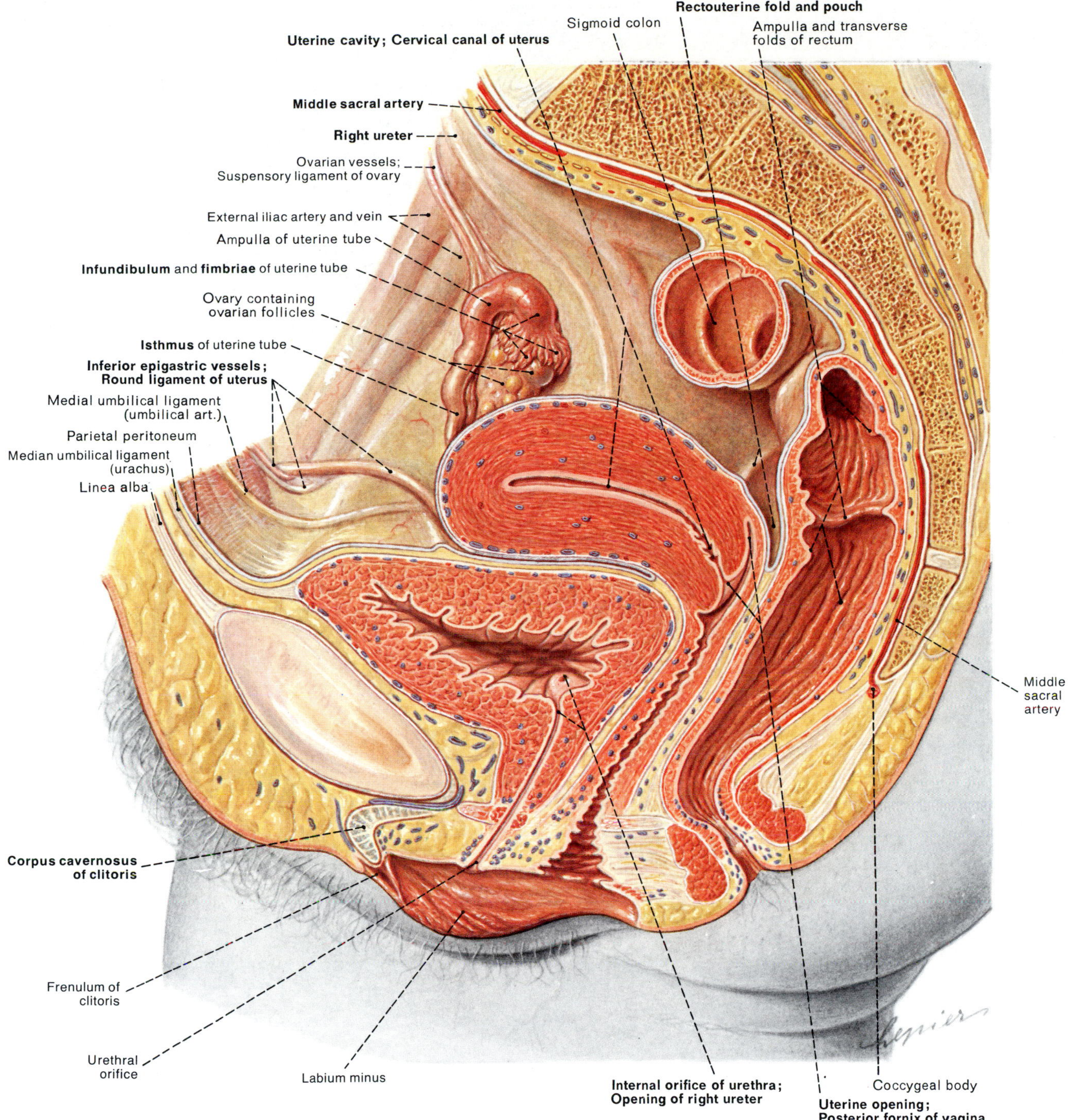
Rectouterine fold and pouch
Sigmoid colon
Ampulla and transverse folds of rectum
Uterine cavity; Cervical canal of uterus
Middle sacral artery
Right ureter
Ovarian vessels; Suspensory ligament of ovary
External iliac artery and vein
Ampulla of uterine tube
Infundibulum and fimbriae of uterine tube
Ovary containing ovarian follicles
Isthmus of uterine tube
Inferior epigastric vessels; Round ligament of uterus
Medial umbilical ligament (umbilical art.)
Parietal peritoneum
Median umbilical ligament (urachus)
Linea alba
Middle sacral artery
Corpus cavernosus of clitoris
Frenulum of clitoris
Urethral orifice
Labium minus
Internal orifice of urethra; Opening of right ureter
Coccygeal body
Uterine opening; Posterior fornix of vagina

Fig. 377: The Adult Female Pelvis. Median Sagittal Section

NOTE: 1) this medial view of the right half of the female pelvis illustrates the relationships among the bladder, uterus and vagina, rectum, ovary and uterine tube. Observe the immediate retropubic position of the empty *bladder* and the relatively short course of the female *urethra,* which leads from the bladder through the urogenital diaphragm to open in the midline, anterior to the vagina and between the labia minora.

2) the opening between the vagina and the uterus. An extension of the vagina, the posterior fornix, reaches enough superiorly to lie just in front of the rectouterine pouch (of Douglas) and separated from it only by the vaginal wall. Note the interposition of the vagina and uterus between the bladder and rectum. The pear-shaped uterus is so positioned over the superior surface of the empty bladder that when the woman is standing erect, the uterus is horizontal.

3) that the round ligament is directed laterally and anteriorly to enter the abdominal inguinal ring and the course of the inferior epigastric vessels in relation to this ligament. Likewise observe the course of the ovarian vessels within the suspensory ligament of the ovary and their important relationship to the descending ureter on the posterolateral wall of the pelvis.

4) the sigmoid flexure of the large bowel and the relatively direct course of the rectum toward the anal canal. The peritoneum is reflected over the anterior surface of the rectum, thereby lining the rectouterine pouch.

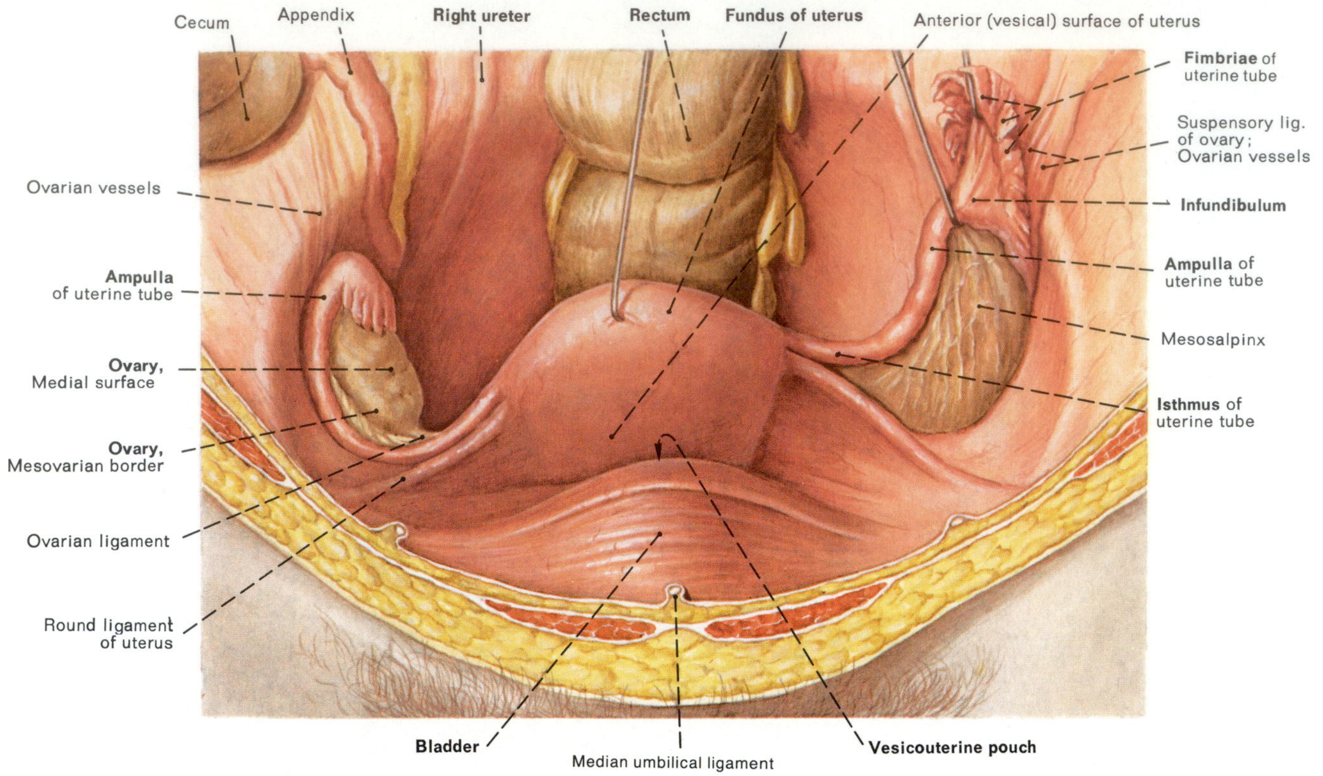
Cecum
Appendix
Right ureter
Rectum
Fundus of uterus
Anterior (vesical) surface of uterus
Fimbriae of uterine tube
Suspensory lig. of ovary; Ovarian vessels
Infundibulum
Ampulla of uterine tube
Mesosalpinx
Isthmus of uterine tube
Ovarian vessels
Ampulla of uterine tube
Ovary, Medial surface
Ovary, Mesovarian border
Ovarian ligament
Round ligament of uterus
Bladder
Median umbilical ligament
Vesicouterine pouch

Fig. 378: The Female Pelvic Organs: Anterosuperior View

Observe that the body of the uterus has been elevated, thereby exposing the vesicouterine pouch and demonstrating the broad ligaments.

Fig. 379: The Internal Genital Organs of an Immature Girl

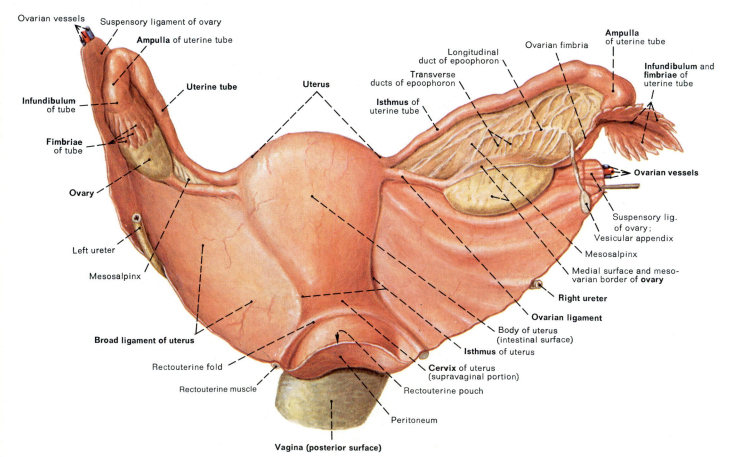

Ovarian vessels
Suspensory ligament of ovary
Ampulla of uterine tube
Uterine tube
Infundibulum of tube
Fimbriae of tube
Ovary
Left ureter
Mesosalpinx
Broad ligament of uterus
Rectouterine fold
Rectouterine muscle
Uterus
Longitudinal duct of epoophoron
Transverse ducts of epoophoron
Isthmus of uterine tube
Ovarian fimbria
Ampulla of uterine tube
Infundibulum and fimbriae of uterine tube
Ovarian vessels
Suspensory lig. of ovary; Vesicular appendix
Mesosalpinx
Medial surface and mesovarian border of ovary
Right ureter
Ovarian ligament
Body of uterus (intestinal surface)
Isthmus of uterus
Cervix of uterus (supravaginal portion)
Rectouterine pouch
Peritoneum
Vagina (posterior surface)

Stroma of ovary
Ovarian ligament
Mesosalpinx
Fundus of uterus
Isthmus of uterine tube
Longitudinal duct of epoophoron
Ampulla of uterine tube
Uterine tube
Transverse ducts of epoophoron
Round ligament of uterus
Ovarian ligament
Mucosal folds of tube
Perimetrium
Uterine cavity; Endometrium
Fimbriae of uterine tube
Myometrium
Accessory hydatid
Ovarian vessels
Ovarian fimbria
Uterosacral ligament
Ovarian follicles
Vesicular appendix
Corpus luteum
Corpus albicans
Cervical canal
Broad ligament
Intramural portion of uterine tube
Cervix of uterus (vaginal part)
Body of uterus
Vagina
Isthmus of uterus
Uterine opening
Vaginal rugae
Cervix of uterus (supravaginal part)

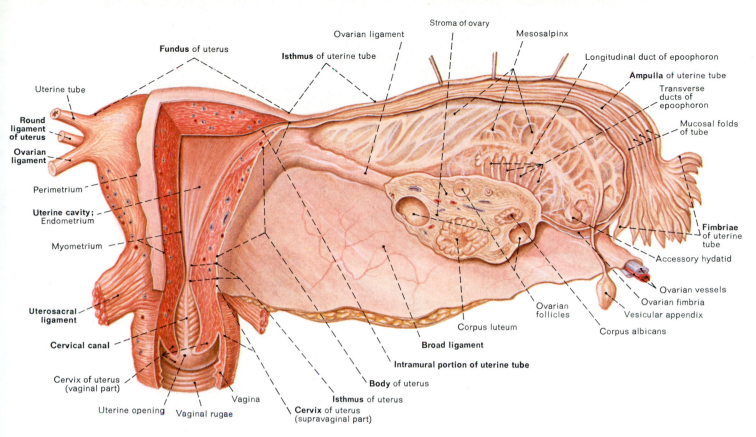

Fig. 380: Frontal Section of Uterus, Uterine Tube and Ovary

NOTE: 1) the vagina communicates with the pelvic cavity through the uterus and the uterine tube. The lumen of this pathway varies in diameter, and its most narrow sites are the isthmus of the uterus and the intrauterine (intramural) portion of the uterine tube.

2) the uterus consists of the cervix (vaginal and supravaginal portions) and the body of the uterus. These are interconnected by the isthmus. The attachments of the uterus include a) the broad ligaments which are mesentery-like attachments to the lateral margins of the uterus, b) the fibrous round ligament of the uterus and the ligament of the ovary attached just below the uterine tube, c) the uterosacral ligaments and d) the lateral cervical or cardinal ligaments (not shown in this figure). The cardinal ligaments are generally regarded as the principal ligamentous support of the uterus and upper vagina.

Fig. 381: Arterial Supply to Female Pelvic Genital Organs

NOTE: 1) the principal vessels supplying the female pelvic genital organs are the uterine arteries from the internal iliac vessels and the ovarian arteries which stem directly from the aorta. The uterine and ovarian arteries anastomose freely along both lateral borders of the uterus.

2) inferiorly, the uterine artery also anastomoses with the arterial supply to the vagina. Frequently, the vaginal arteries branch directly from the uterine, however, they may branch from the inferior vesical artery or even directly from the internal iliac artery.

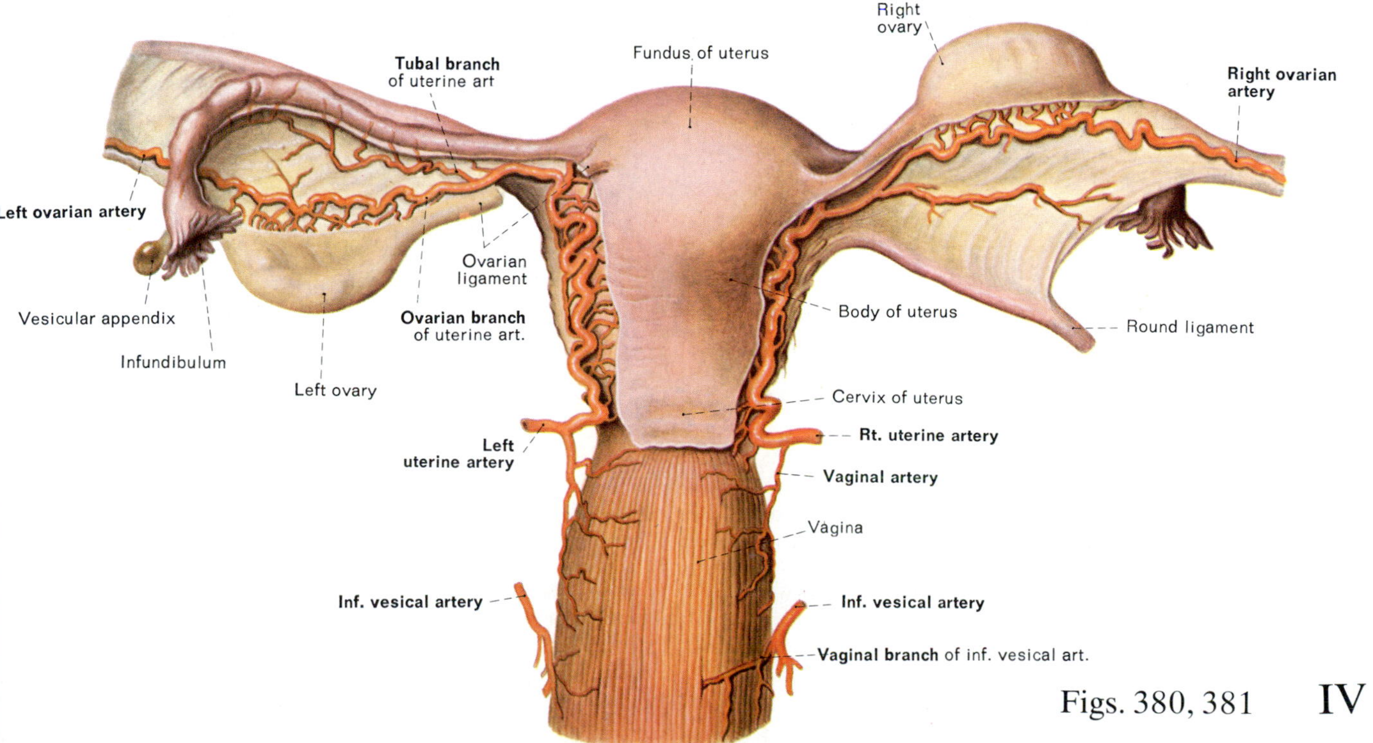
Right ovary
Fundus of uterus
Tubal branch of uterine art
Right ovarian artery
Left ovarian artery
Ovarian ligament
Body of uterus
Vesicular appendix
Ovarian branch of uterine art.
Round ligament
Infundibulum
Left ovary
Cervix of uterus
Left uterine artery
Rt. uterine artery
Vaginal artery
Vagina
Inf. vesical artery
Inf. vesical artery
Vaginal branch of inf. vesical art.

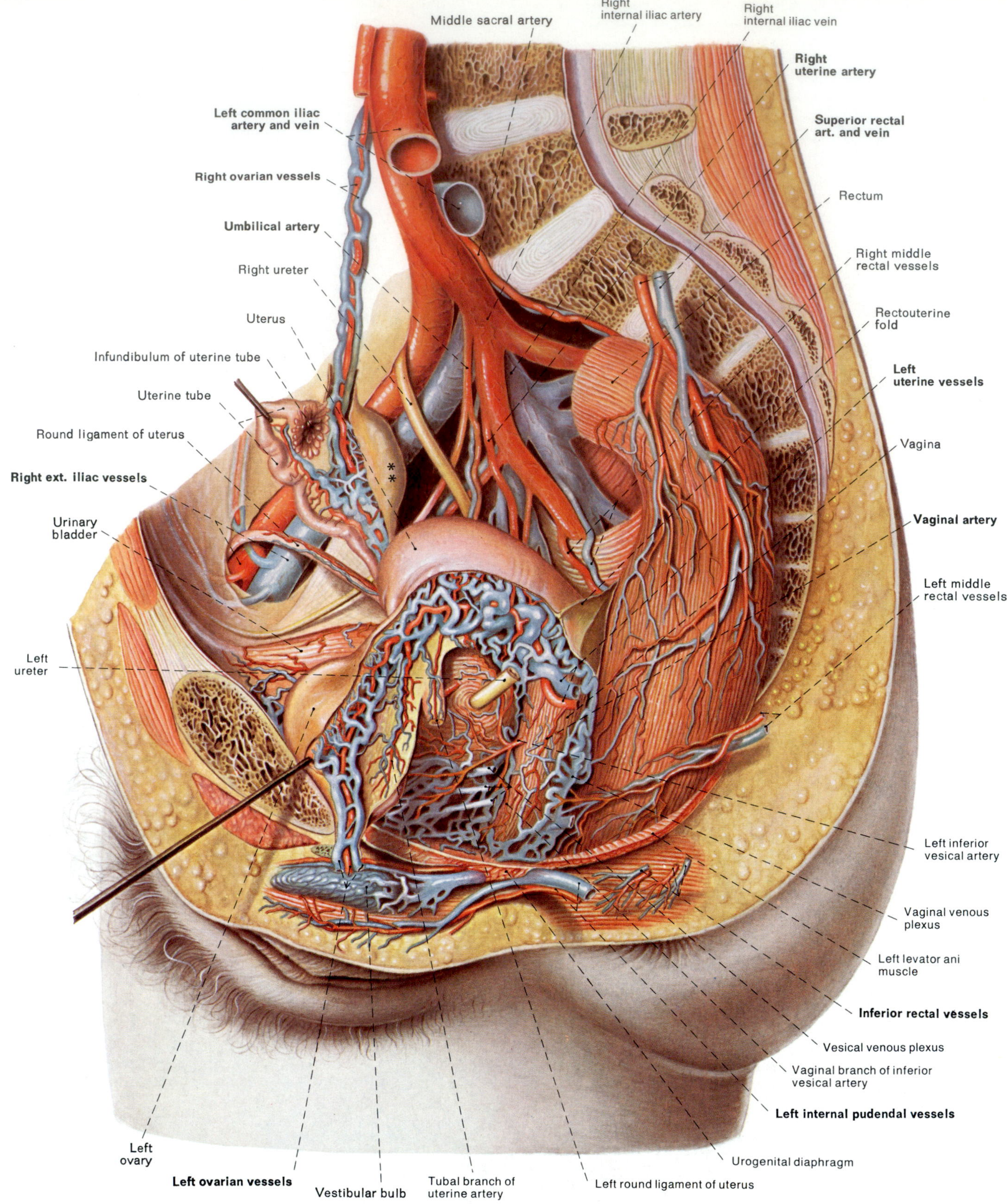

Middle sacral artery
Right internal iliac artery
Right internal iliac vein
Right uterine artery
Superior rectal art. and vein
Rectum
Right middle rectal vessels
Rectouterine fold
Left uterine vessels
Vagina
Vaginal artery
Left middle rectal vessels
Left common iliac artery and vein
Right ovarian vessels
Umbilical artery
Right ureter
Uterus
Infundibulum of uterine tube
Uterine tube
Round ligament of uterus
Right ext. iliac vessels
Urinary bladder
Left ureter
**
Left inferior vesical artery
Vaginal venous plexus
Left levator ani muscle
Inferior rectal vessels
Vesical venous plexus
Vaginal branch of inferior vesical artery
Left internal pudendal vessels
Urogenital diaphragm
Left ovary
Left ovarian vessels
Vestibular bulb
Tubal branch of uterine artery
Left round ligament of uterus

Fig. 382: Blood Vessels of the Female Pelvis and Genital System

NOTE: 1) the left half of the pelvis has been removed while most of the female pelvic organs are still in place. Observe the dense plexuses of veins. These include the ovarian, uterine, vaginal and vesical plexuses which accompany their respective arteries and which drain the pelvic organs.

2) with the exception of the *ovarian artery* which is derived from the aorta and the *superior rectal artery* (hemorrhoidal) which branches from the inferior mesenteric, all the other arteries supplying blood to the pelvic organs, perineum and genital tract are derived from the *internal iliac artery* or its branches. Observe the anastomosis among the superior, middle and inferior rectal vessels.

3) the descending course of the ureter over the pelvic brim from the posterior abdominal wall. In its path it crosses the external iliac artery and vein, as does the round ligament of the uterus more inferiorly.

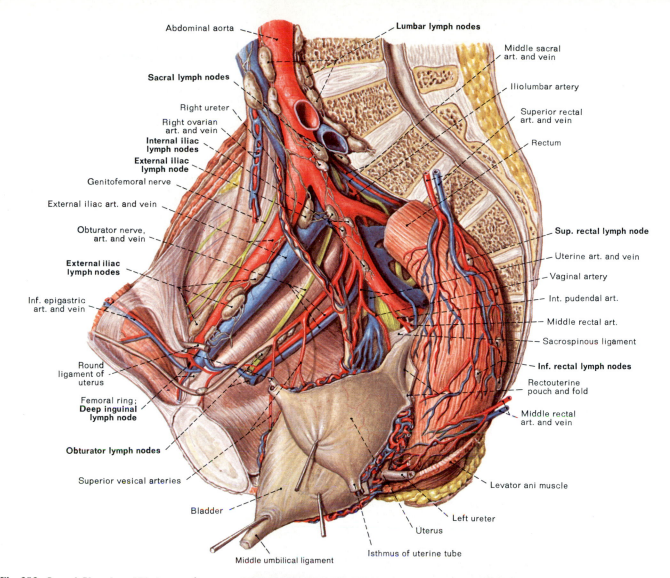

Abdominal aorta
Lumbar lymph nodes
Middle sacral art. and vein
Sacral lymph nodes
Iliolumbar artery
Right ureter
Superior rectal art. and vein
Right ovarian art. and vein
Internal iliac lymph nodes
Rectum
External iliac lymph node
Genitofemoral nerve
External iliac art. and vein
Obturator nerve, art. and vein
Sup. rectal lymph node
Uterine art. and vein
External iliac lymph nodes
Vaginal artery
Inf. epigastric art. and vein
Int. pudendal art.
Middle rectal art.
Sacrospinous ligament
Round ligament of uterus
Inf. rectal lymph nodes
Rectouterine pouch and fold
Femoral ring; **Deep inguinal lymph node**
Middle rectal art. and vein
Obturator lymph nodes
Superior vesical arteries
Levator ani muscle
Bladder
Left ureter
Uterus
Isthmus of uterine tube
Middle umbilical ligament

Fig. 383: Lymph Vessels and Nodes of the Female Pelvis ▲

NOTE: 1) as a rule, the lymph nodes of the pelvis lie along the course of the major vessels. Generally, the lymphatics drain superiorly and posteriorly to achieve the right und left lumbar lymphatic chains of nodes which lie upon the psoas major muscles on both sides of the aorta.

2) the lymphatics of the bladder drain laterally to the external iliac nodes and posteriorly to the internal iliac nodes. These latter lymphatic channels and nodes also receive lymph from the fundus and body of the uterus in the female and the prostate and seminal vesicles in the male. Lymphatics of the cervix and vagina drain into both the external and internal iliac nodes.

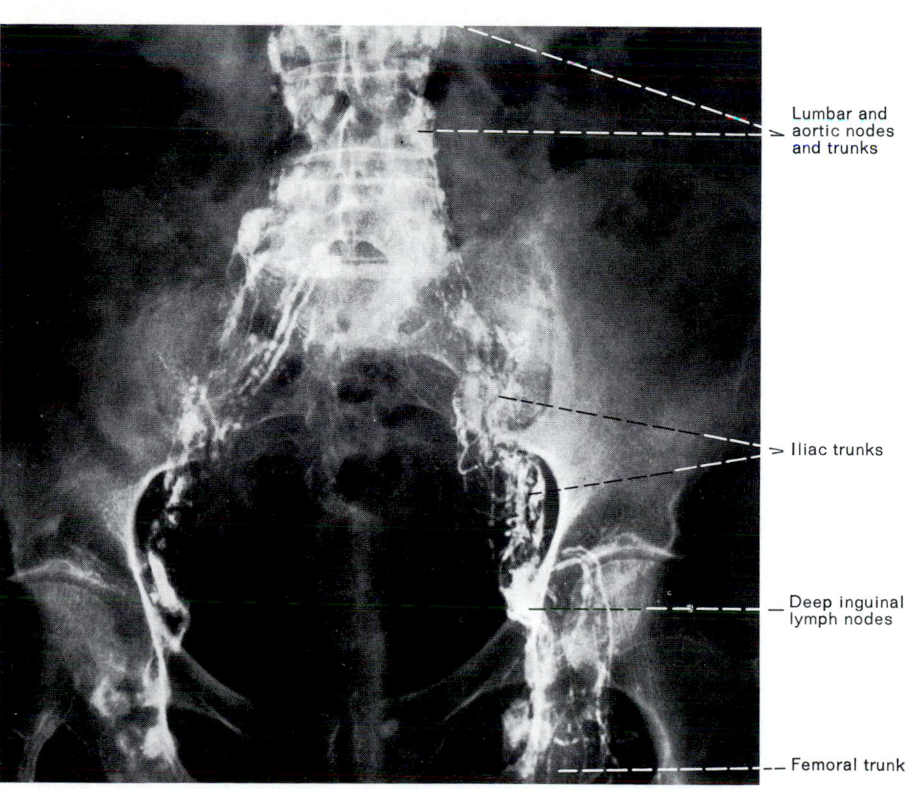
Lumbar and aortic nodes and trunks
Iliac trunks
Deep inguinal lymph nodes
Femoral trunk

Fig. 384: Lymphograph of Pelvis and Lumbar Region ▶

This lymphograph displays the lymphatic channels from the deeper femoral vessels and nodes, which course superiorly through to the lumbar and aortic nodes. Observe the profuse network along the iliac vessels and the concentration of nodes in the deep inguinal region.

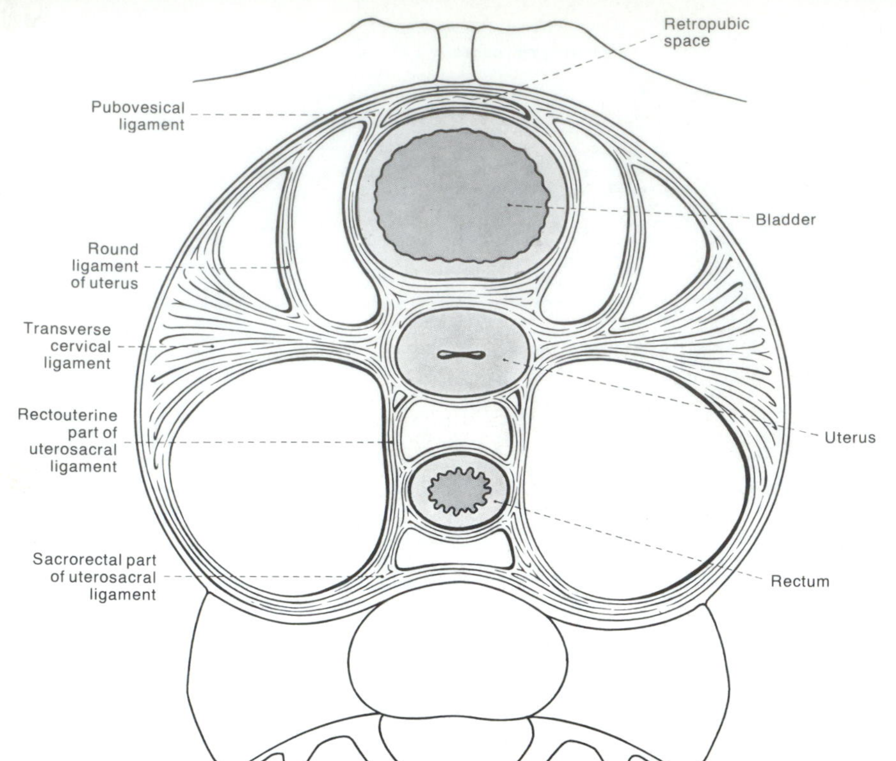
Retropubic space
Pubovesical ligament
Bladder
Round ligament of uterus
Transverse cervical ligament
Uterus
Rectouterine part of uterosacral ligament
Sacrorectal part of uterosacral ligament
Rectum

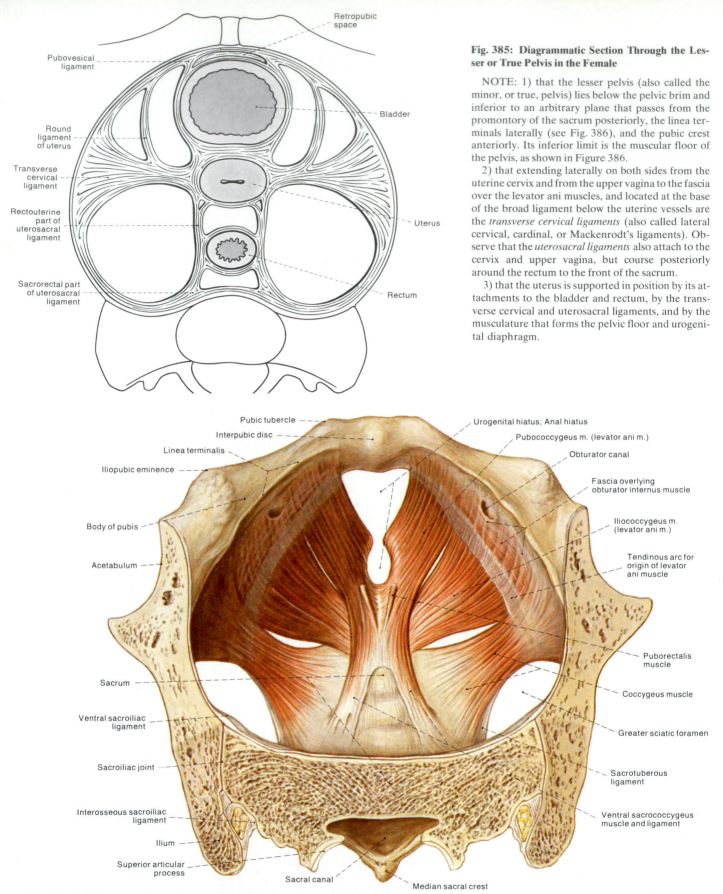

Fig. 385: Diagrammatic Section Through the Lesser or True Pelvis in the Female

NOTE: 1) that the lesser pelvis (also called the minor, or true, pelvis) lies below the pelvic brim and inferior to an arbitrary plane that passes from the promontory of the sacrum posteriorly, the linea terminals laterally (see Fig. 386), and the pubic crest anteriorly. Its inferior limit is the muscular floor of the pelvis, as shown in Figure 386.

2) that extending laterally on both sides from the uterine cervix and from the upper vagina to the fascia over the levator ani muscles, and located at the base of the broad ligament below the uterine vessels are the *transverse cervical ligaments* (also called lateral cervical, cardinal, or Mackenrodt's ligaments). Observe that the *uterosacral ligaments* also attach to the cervix and upper vagina, but course posteriorly around the rectum to the front of the sacrum.

3) that the uterus is supported in position by its attachments to the bladder and rectum, by the transverse cervical and uterosacral ligaments, and by the musculature that forms the pelvic floor and urogenital diaphragm.

Pubic tubercle
Interpubic disc
Linea terminalis
Iliopubic eminence
Body of pubis
Acetabulum
Sacrum
Ventral sacroiliac ligament
Sacroiliac joint
Interosseous sacroiliac ligament
Ilium
Superior articular process
Sacral canal
Median sacral crest
Urogenital hiatus; Anal hiatus
Pubococcygeus m. (levator ani m.)
Obturator canal
Fascia overlying obturator internus muscle
Iliococcygeus m. (levator ani m.)
Tendinous arc for origin of levator ani muscle
Puborectalis muscle
Coccygeus muscle
Greater sciatic foramen
Sacrotuberous ligament
Ventral sacrococcygeus muscle and ligament

Fig. 386: The Muscular Floor of the Female Pelvis, Viewed from Above

NOTE that the muscular floor of the female pelvis closely resembles that of the male pelvis (see Fig. 365 and Notes) *exept* anteriorly at the urogenital hiatus. Realize that at this site the structure of the urogenital diaphragm (deep transverse perineal muscle and the membranous sphincter of the urethra which lie between two fascial layers) and the organs that must penetrate it (the urethra in the male and the urethra and vagina in the female) account for this morphologic difference in the two genders (see Fig. 343, 365, and 366 for the male and Figs. 368-371, 377, and 387 for the female).

Pubic tubercle
Pubic symphysis
Arcuate pubic ligament
Urethral and vaginal sphincter mm.
Transverse perineal ligament
Inferior fascia of the urogenital diaphragm
Deep transverse perineal muscle
Superf. transverse perineal muscle
Ant. inf. iliac spine
Ant. sup. iliac spine
Iliac crest
Acetabulum
Ilium
Lunate surface of acetabulum
Transverse acetabular lig.
Superf. transverse perineal muscle
Tendon and fibers of obt. int. m. (leaving pelvis)
Inferior fascia of the urogenital diaphragm
Ischial tuberosity
Obturator fascia in the perineum
Obturator fascia in the perineum
Sacrotuberous ligament
Inferior fascia of the urogenital diaphragm
Pubococcygeus part of levator ani m.
Spine of the ischium
Probe in pudendal canal, extending anteriorly to the urogenital triangle
Iliococcygeus part of levator ani m.
Coccygeus muscle
Sacrotuberous ligament
Anococcygeal ligament
Sacrotuberous ligament
Central tendinous point of perineum
Post. sup. iliac spine
External anal sphincter muscle
Coccyx
Anus
Sacrum

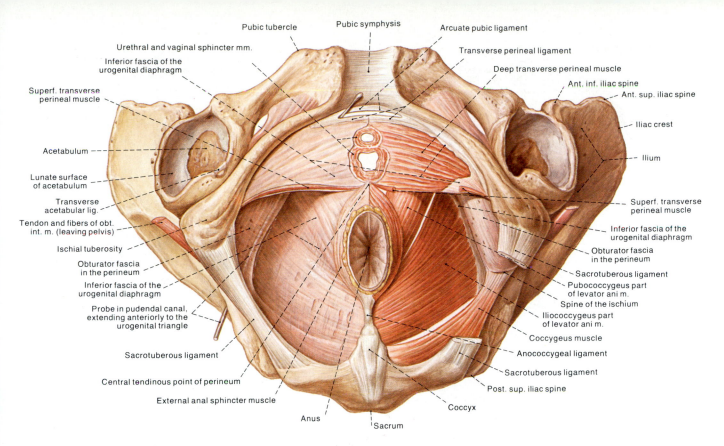

Fig. 387: The Musculature of the Floor of the Female Pelvis, Viewed from the Inferior or Perineal Aspect

NOTE: 1) that on the specimen's left side (reader's right) the inferior fasciae of the pelvic and urogenital diaphragms have been removed.

2) the manner in which the musculature of the urogenital diaphragm completes the anterior part of the female pelvic floor, still allowing the urethra and vagina to pass through the urogenital hiatus (see Fig. 386). Observe also the central tendinous point of the perineum interposed in the midline between the urogenital diaphragm and the anterior limit of the raphé formed by the two external anal sphincters.

3) that the anal hiatus is surrounded in the midline by fibers of the pubococcygeus portions of the two levator ani muscles (see also Fig. 386). These are reinforced superiorly by the puborectalis muscle and inferiorly by the external anal sphincters. Observe also that the iliococcygeus parts of the levator ani muscles sweep medially to the coccyx, but some fibers also insert into a short midline anococcygeal raphé and ligament.

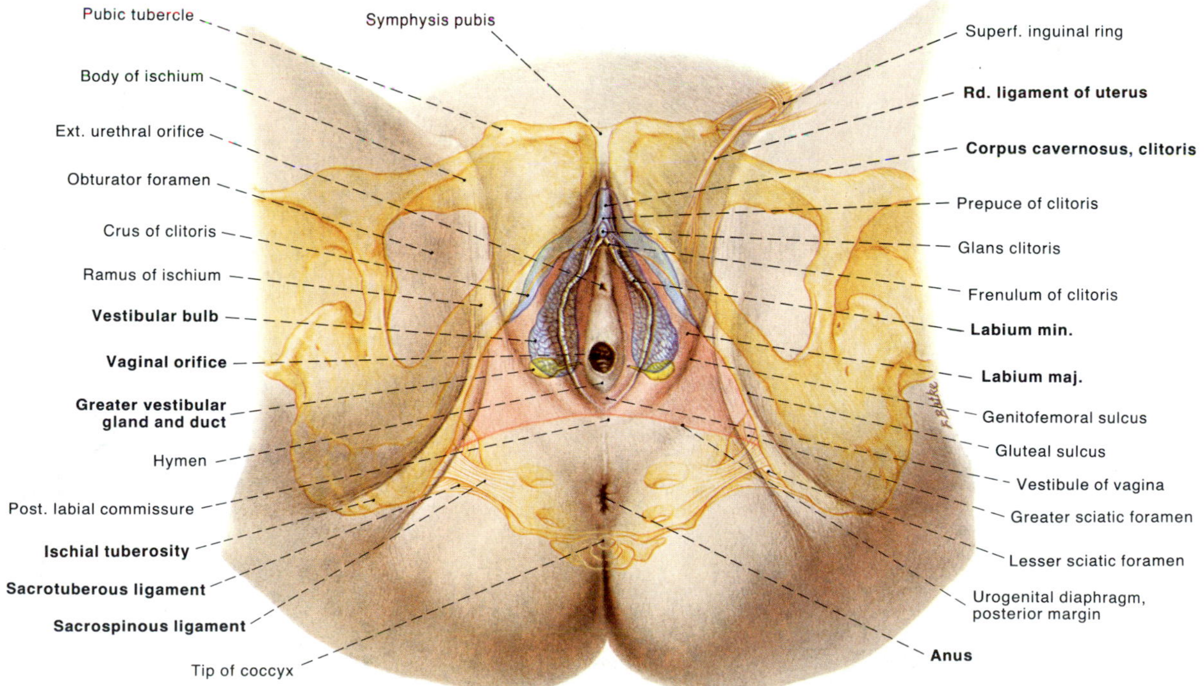
Pubic tubercle
Symphysis pubis
Superf. inguinal ring
Body of ischium
Rd. ligament of uterus
Ext. urethral orifice
Corpus cavernosus, clitoris
Obturator foramen
Prepuce of clitoris
Crus of clitoris
Glans clitoris
Ramus of ischium
Frenulum of clitoris
Vestibular bulb
Labium min.
Vaginal orifice
Labium maj.
Greater vestibular gland and duct
Genitofemoral sulcus
Gluteal sulcus
Hymen
Vestibule of vagina
Post. labial commissure
Greater sciatic foramen
Ischial tuberosity
Lesser sciatic foramen
Sacrotuberous ligament
Urogenital diaphragm, posterior margin
Sacrospinous ligament
Anus
Tip of coccyx

Fig. 388: Projection of External Female Genitalia on Bony Structures of the Pelvis

Compare the locations of the external genitalia (located in the superficial perineal pouch) with the urogenital diaphragm as seen in Figure 387. Observe the external urethral, vaginal, and anal orifices within their locations in the urogenital and anal hiatuses (Fig. 387).

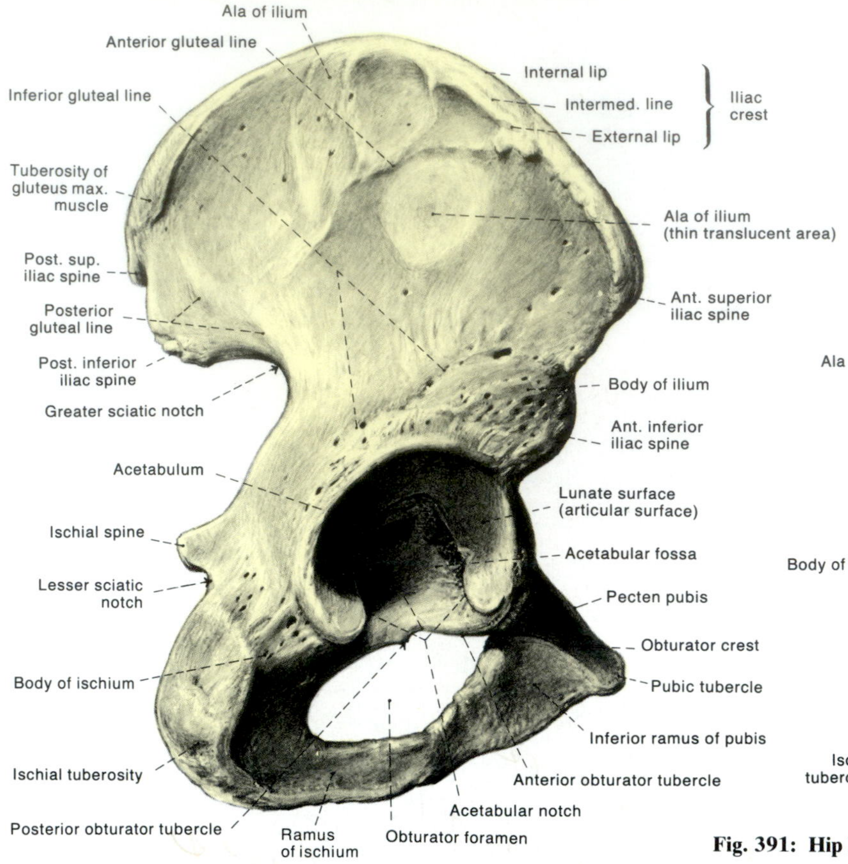

Ala of ilium
Anterior gluteal line
Inferior gluteal line
Internal lip
Intermed. line
External lip
Iliac crest
Tuberosity of gluteus max. muscle
Ala of ilium (thin translucent area)
Post. sup. iliac spine
Posterior gluteal line
Ant. superior iliac spine
Post. inferior iliac spine
Body of ilium
Greater sciatic notch
Ant. inferior iliac spine
Acetabulum
Lunate surface (articular surface)
Ischial spine
Acetabular fossa
Lesser sciatic notch
Pecten pubis
Obturator crest
Body of ischium
Pubic tubercle
Inferior ramus of pubis
Ischial tuberosity
Anterior obturator tubercle
Acetabular notch
Posterior obturator tubercle
Ramus of ischium
Obturator foramen

Fig. 389: Lateral View of the Adult Right Hip Bone

Iliac crest (cartilaginous)
Ala of ilium
Body of ilium
Iliopubic eminence
Body of pubis
Body of ischium
Superior pubic ramus
Ischial tuberosity
Inferior pubic ramus
Ischial ramus

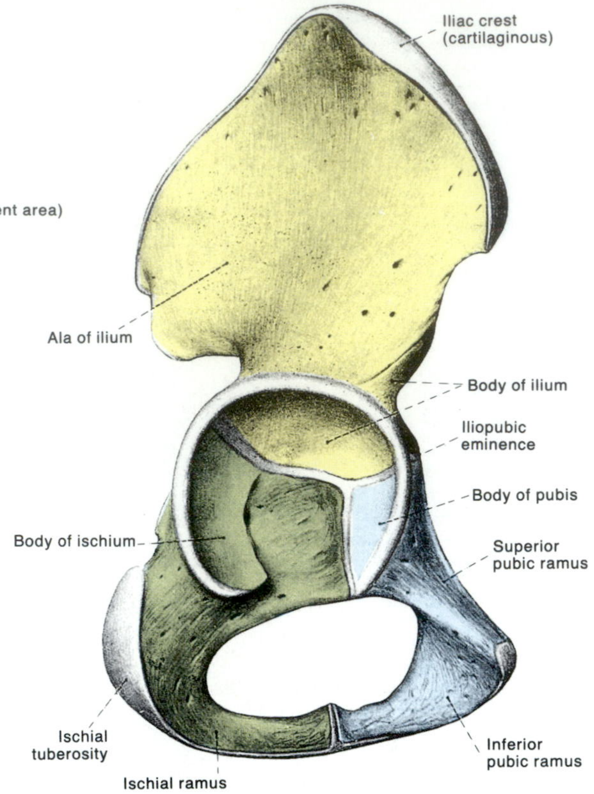

Fig. 391: Hip Bone of 5 Year Old Child: Lateral View

NOTE that the hip bone is formed by a fusion of the ilium (yellow), ischium (green) and pubis (blue). Although ossification of the inferior pubic ramus occurs during the 7th or 8th year, complete fusion of the three bones at the acetabulum occurs sometime between the 15th and 20th year.

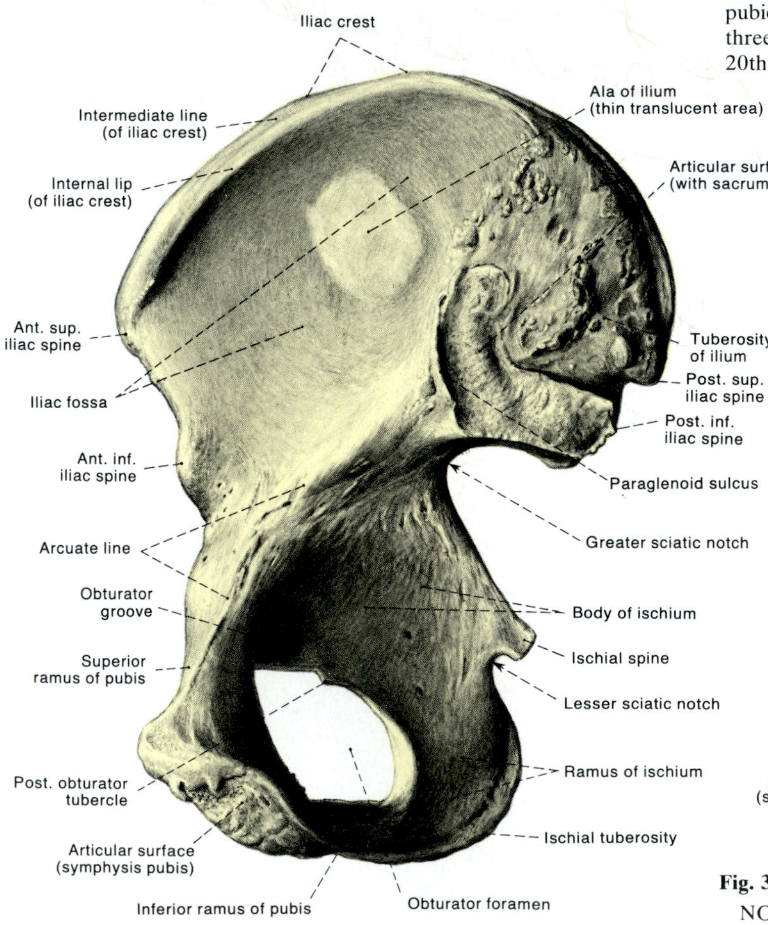

Iliac crest
Ala of ilium (thin translucent area)
Intermediate line (of iliac crest)
Internal lip (of iliac crest)
Articular surface (with sacrum)
Ant. sup. iliac spine
Tuberosity of ilium
Post. sup. iliac spine
Iliac fossa
Post. inf. iliac spine
Ant. inf. iliac spine
Paraglenoid sulcus
Greater sciatic notch
Arcuate line
Obturator groove
Body of ischium
Ischial spine
Superior ramus of pubis
Lesser sciatic notch
Ramus of ischium
Post. obturator tubercle
Ischial tuberosity
Articular surface (symphysis pubis)
Inferior ramus of pubis
Obturator foramen

Fig. 390: Medial View of the Adult Right Hip Bone

Iliac crest (cartilaginous)
Ant. sup. iliac spine
Articular surface (with sacrum)
Body of ilium
Iliopubic surface
Body of ischium
Body of pubis
Spine of ischium
Sup. ramus of pubis
Body of ischium
Articular surface (symphysis pubis)
Inf. ramus of pubis
Ramus of ischium

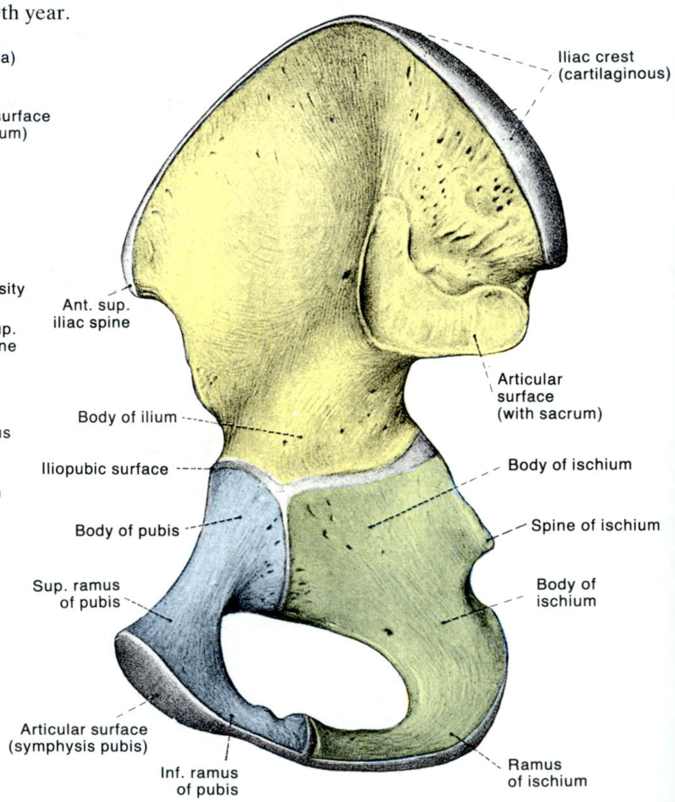

Fig. 392: Hip Bone of 5 Year Old Child: Medial View

NOTE the lines of fusion of the three bones above the obturator foramen and the fusion of the inferior pubic ramus and the ischial ramus below that foramen.

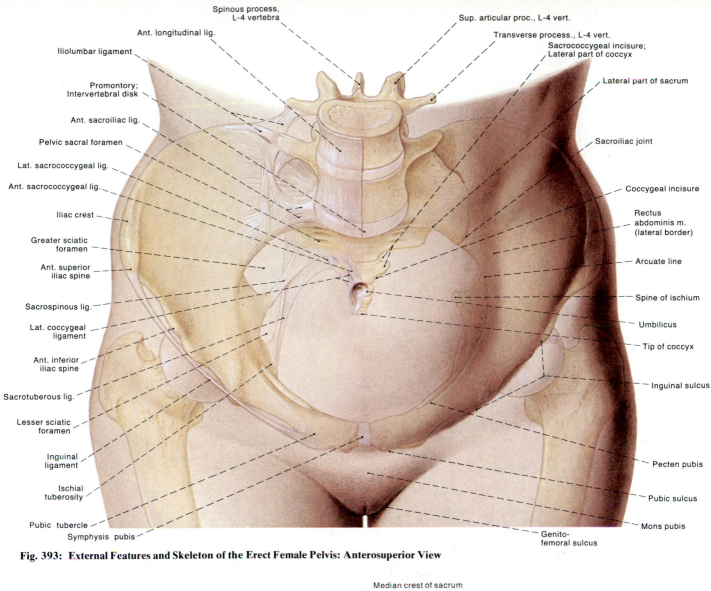

Spinous process, L-4 vertebra
Sup. articular proc., L-4 vert.
Ant. longitudinal lig.
Transverse process., L-4 vert.
Sacrococcygeal incisure; Lateral part of coccyx
Iliolumbar ligament
Lateral part of sacrum
Promontory; Intervertebral disk
Ant. sacroiliac lig.
Pelvic sacral foramen
Sacroiliac joint
Lat. sacrococcygeal lig.
Ant. sacrococcygeal lig.
Coccygeal incisure
Iliac crest
Rectus abdominis m. (lateral border)
Greater sciatic foramen
Arcuate line
Ant. superior iliac spine
Spine of ischium
Sacrospinous lig.
Umbilicus
Lat. coccygeal ligament
Tip of coccyx
Ant. inferior iliac spine
Inguinal sulcus
Sacrotuberous lig.
Lesser sciatic foramen
Pecten pubis
Inguinal ligament
Ischial tuberosity
Pubic sulcus
Pubic tubercle
Mons pubis
Symphysis pubis
Genito-femoral sulcus

Fig. 393: External Features and Skeleton of the Erect Female Pelvis: Anterosuperior View

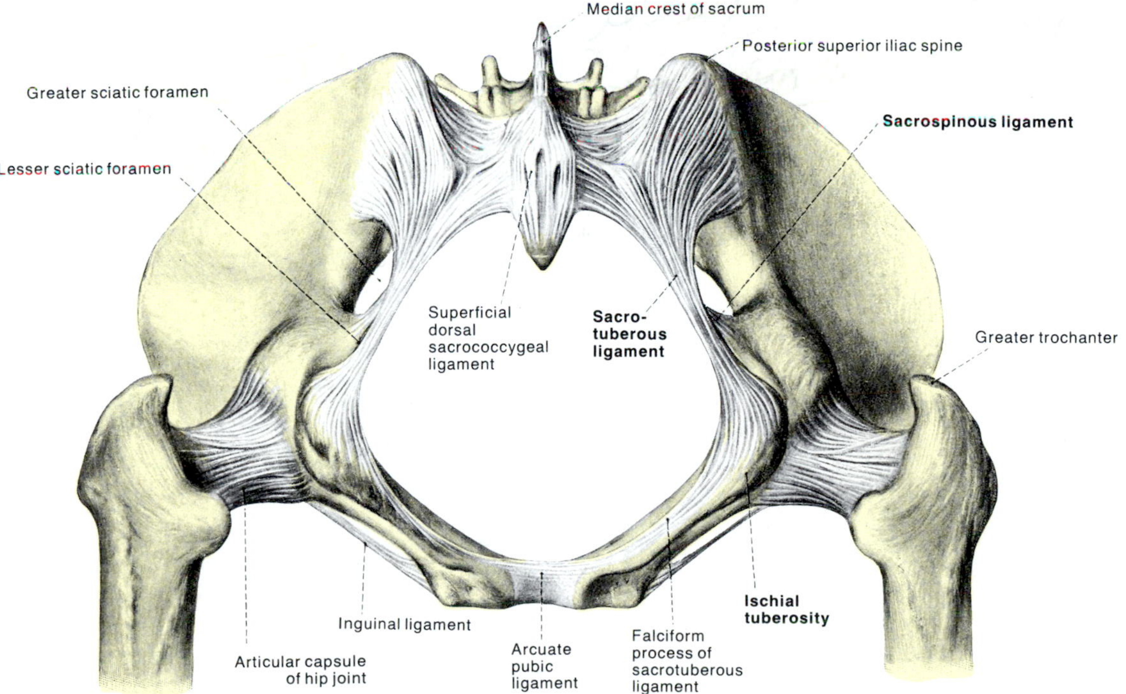
Median crest of sacrum
Posterior superior iliac spine
Greater sciatic foramen
Sacrospinous ligament
Lesser sciatic foramen
Superficial dorsal sacrococcygeal ligament
Sacro-tuberous ligament
Greater trochanter
Ischial tuberosity
Inguinal ligament
Articular capsule of hip joint
Arcuate pubic ligament
Falciform process of sacrotuberous ligament

Fig. 394: The Female Pelvic Outlet Showing the Pelvic Ligaments and Viewed from Below

NOTE the sacrotuberous, sacrospinous, and arcuate pubic ligaments in this inferior view of the pelvic outlet.

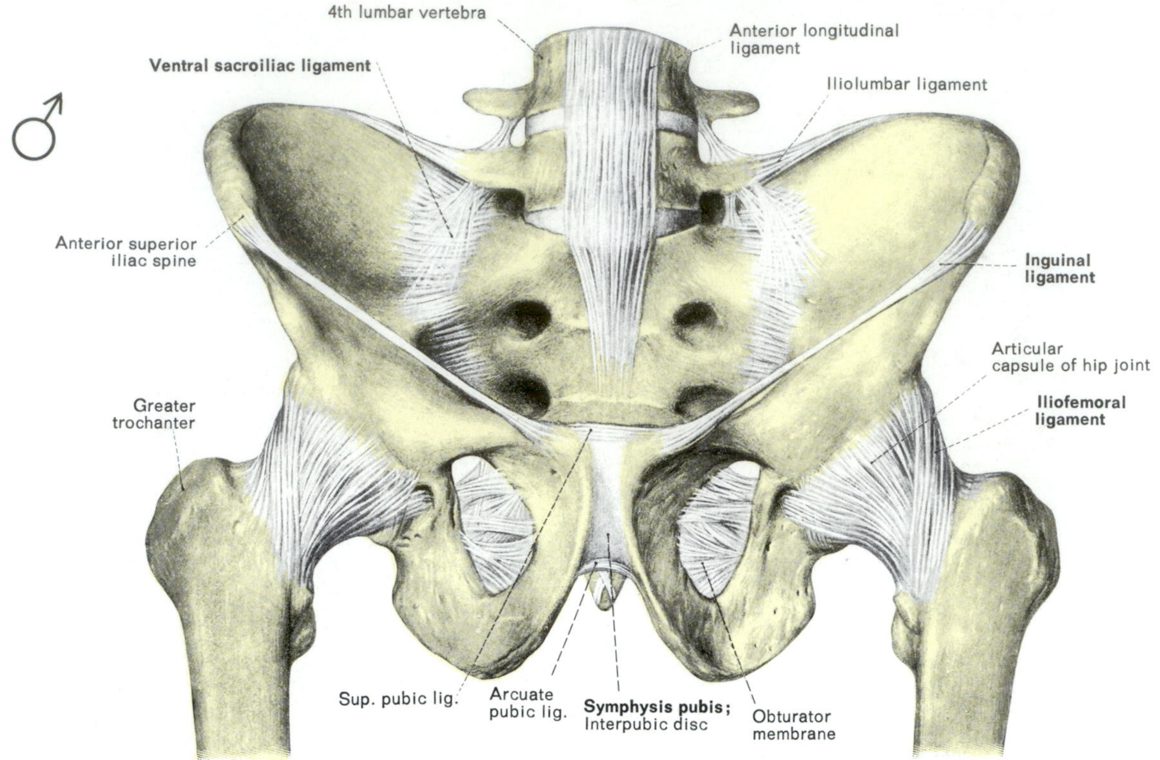
4th lumbar vertebra
Ventral sacroiliac ligament
Anterior longitudinal ligament
Iliolumbar ligament
Anterior superior iliac spine
Inguinal ligament
Articular capsule of hip joint
Greater trochanter
Iliofemoral ligament
Sup. pubic lig.
Arcuate pubic lig.
Symphysis pubis; Interpubic disc
Obturator membrane

Fig. 395: The Male Pelvis and Associated Ligaments: Anterior Aspect

NOTE that the pelvis is formed by the articulation of the left and right hip bones anteriorly at the symphysis pubis and posteriorly with the sacrum and coccyx of the vertebral column. The articulations inferiorly of the pelvis with the two femora allow the weight of the head, trunk and upper extremities to be transmitted to the lower limbs, thereby maintaining the upright posture characteristic of the human being.

Fig. 396: The Female Pelvis with Joints and Ligaments: Posterior Aspect

NOTE that broad ligamentous bands articulate the two hip bones posteriorly with the sacrum and coccyx. This sacroiliac joint is bound by the extremely strong dorsal sacroiliac ligament. Attaching the sacrum to the ischial tuberosity is the broad sacrotuberous ligament. Additionally, the sacrospinous ligament stretches between the sacrum and the ischium (ischial spine).

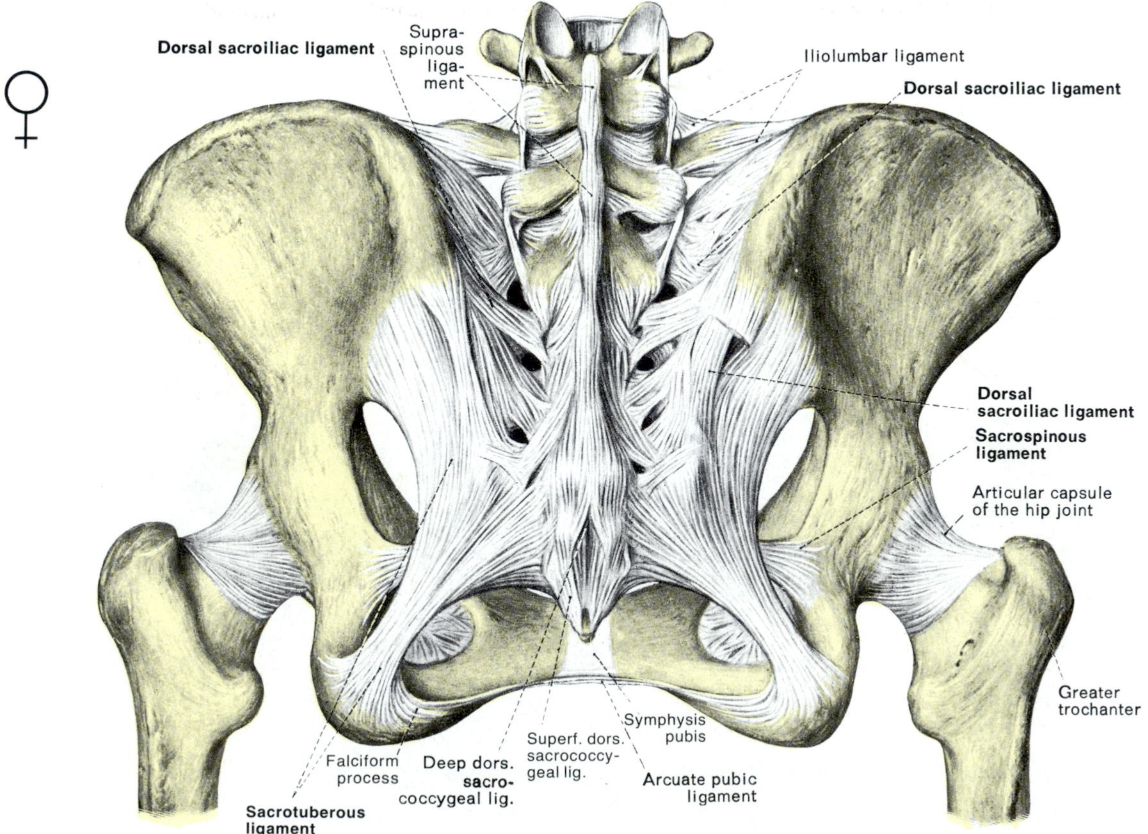
Dorsal sacroiliac ligament
Supra-spinous liga-ment
Iliolumbar ligament
Dorsal sacroiliac ligament
Dorsal sacroiliac ligament
Sacrospinous ligament
Articular capsule of the hip joint
Greater trochanter
Symphysis pubis
Superf. dors. sacrococcy-geal lig.
Arcuate pubic ligament
Falciform process
Deep dors. sacro-coccygeal lig.
Sacrotuberous ligament

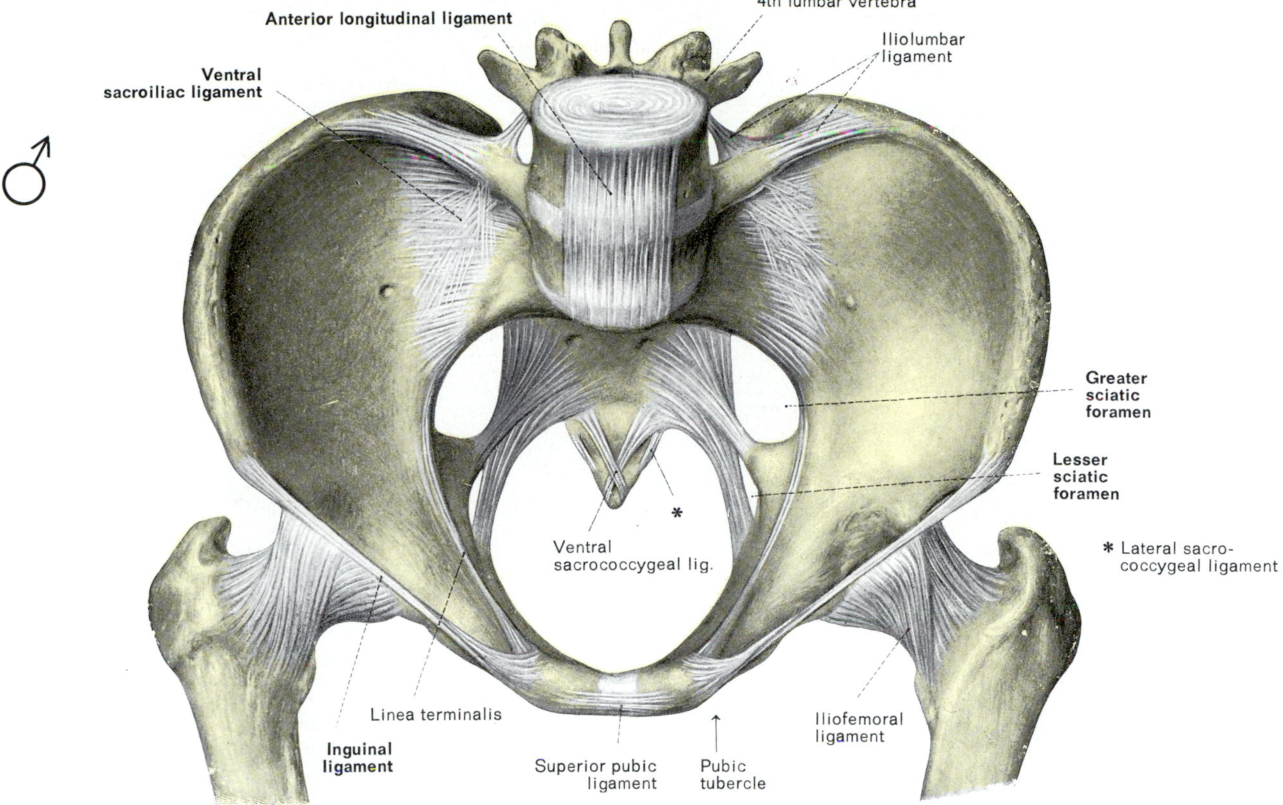

♂
Anterior longitudinal ligament
4th lumbar vertebra
Ventral sacroiliac ligament
Iliolumbar ligament
Greater sciatic foramen
Lesser sciatic foramen
*
Ventral sacrococcygeal lig.
* Lateral sacro-coccygeal ligament
Linea terminalis
Inguinal ligament
Superior pubic ligament
Pubic tubercle
Iliofemoral ligament

Fig. 397: The Male Pelvis and Ligaments Viewed from Above

NOTE that the size of both the pelvic inlet (superior aperture of the minor pelvis) and inferior outlet of the male pelvis is smaller than that in the female (see Figure 398 below). Thus, the minor pelvis is deeper and narrower in the male and its cavity has a smaller capacity than a female. In the male, however, the pelvic bones are thicker and heavier and generally the major pelvis (above the pelvic brim) is larger than in the female.

Fig. 398: The Female Pelvis and Ligaments Viewed from Above

NOTE that in addition to having wider diameters, both the pelvic inlet and outlet of the female pelvis minor are more circular in shape than in the male. The female pelvic bones are more delicate and the sacrum less curved. The larger capacity of the true pelvis in the female, and the fact that the female hormones of pregnancy tend to relax the pelvic ligaments serve to facilitate the function of child bearing.

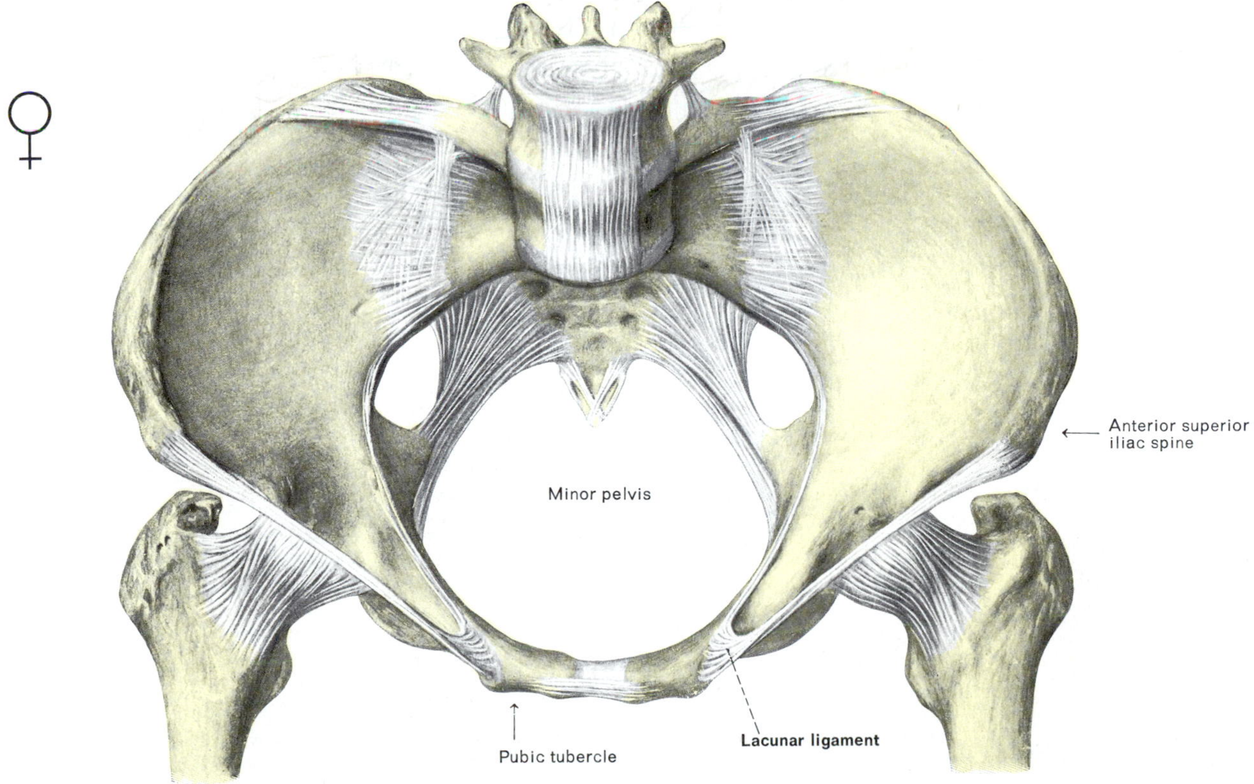

♀
Anterior superior iliac spine
Minor pelvis
Pubic tubercle
Lacunar ligament

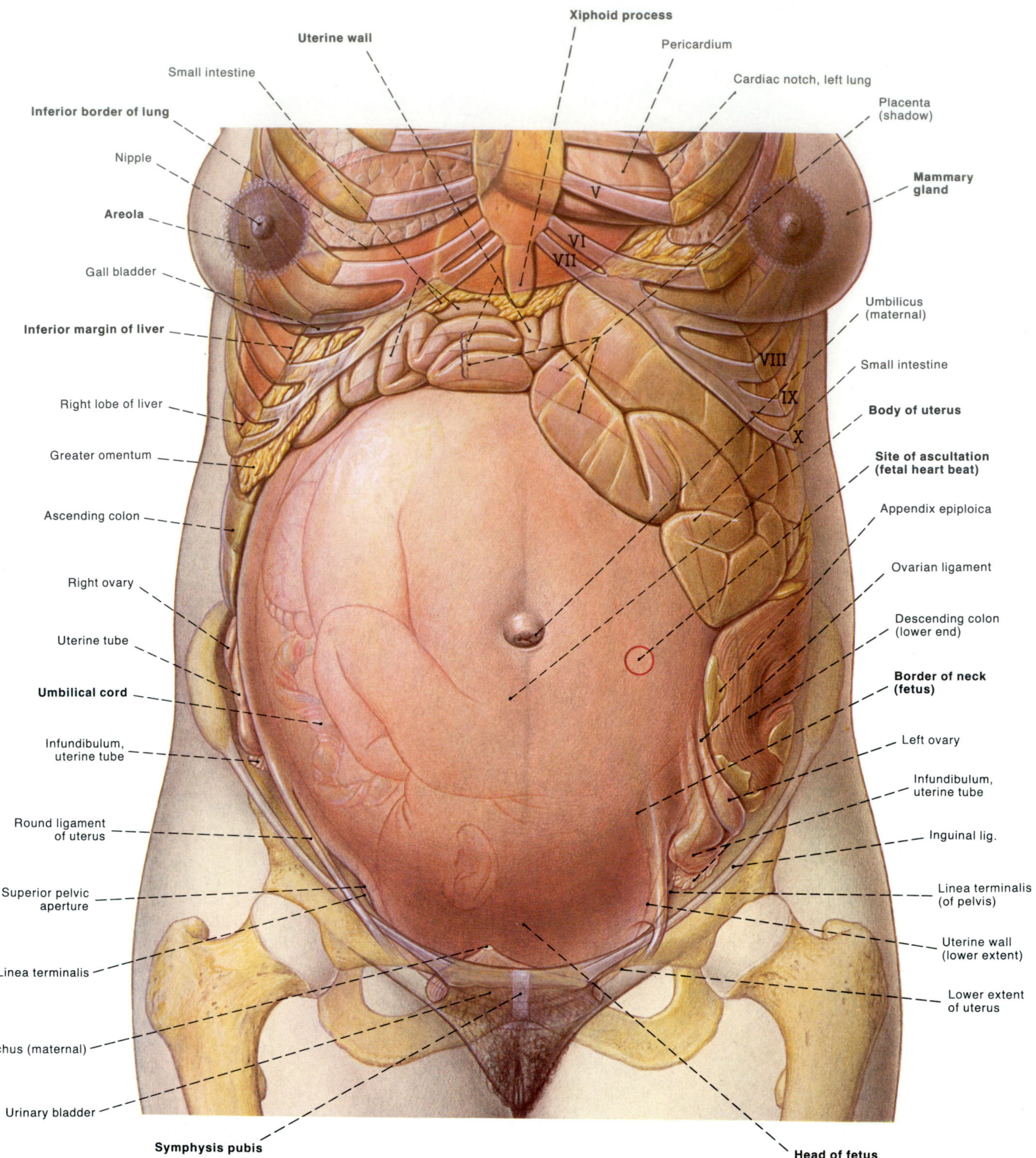
Xiphoid process
Uterine wall
Pericardium
Small intestine
Cardiac notch, left lung
Inferior border of lung
Placenta (shadow)
Nipple
Mammary gland
Areola
V
VI
VII
Gall bladder
Umbilicus (maternal)
Inferior margin of liver
VIII
Small intestine
IX
Right lobe of liver
Body of uterus
X
Greater omentum
Site of ascultation (fetal heart beat)
Ascending colon
Appendix epiploica
Ovarian ligament
Right ovary
Descending colon (lower end)
Uterine tube
Umbilical cord
Border of neck (fetus)
Left ovary
Infundibulum, uterine tube
Infundibulum, uterine tube
Round ligament of uterus
Inguinal lig.
Superior pelvic aperture
Linea terminalis (of pelvis)
Uterine wall (lower extent)
Linea terminalis
Lower extent of uterus
Urachus (maternal)
Urinary bladder
Symphysis pubis
Head of fetus

Fig. 399: Diagrammatic Projection of Abdominal and Pelvic Organs in a Pregnant Woman Shortly Before Giving Birth: Anterior View

NOTE that: 1) in the maternal abdomen, the full-term fetus is in a characteristically longitudinal posture, with the dorsum of the fetal head and back oriented toward the mother's anterior abdominal wall. This type of cephalic longitudinal presentation occurs in 95% of births, while longitudinal pelvic presentation (breech) occurs in about 3% of births. In about 1% of births, a transverse presentation of the fetus occurs, with one of the shoulders as the presenting part *(Reid, 1962, A Textbook of Obstetrics, W. B. Saunders Co. Publ.);*

2) as the fetus grows, the maternal uterus enlarges. By the end of the 3rd month, the uterus occupies most of the pelvis, extending as gestation continues, higher and higher within the abdomen. Near the end of pregnancy, it occupies most of the abdomen, reaching above the costal margin, nearly to the xiphoid process;

3) the maternal liver, stomach and intestines are displaced upward, while the diaphragm is elevated and the dimensions of the thoracic cavity broadened. The breasts enlarge considerably through a proliferation of its glandular tissue in advance of lactation, and the areolar region around the nipple becomes more darkly pigmented.

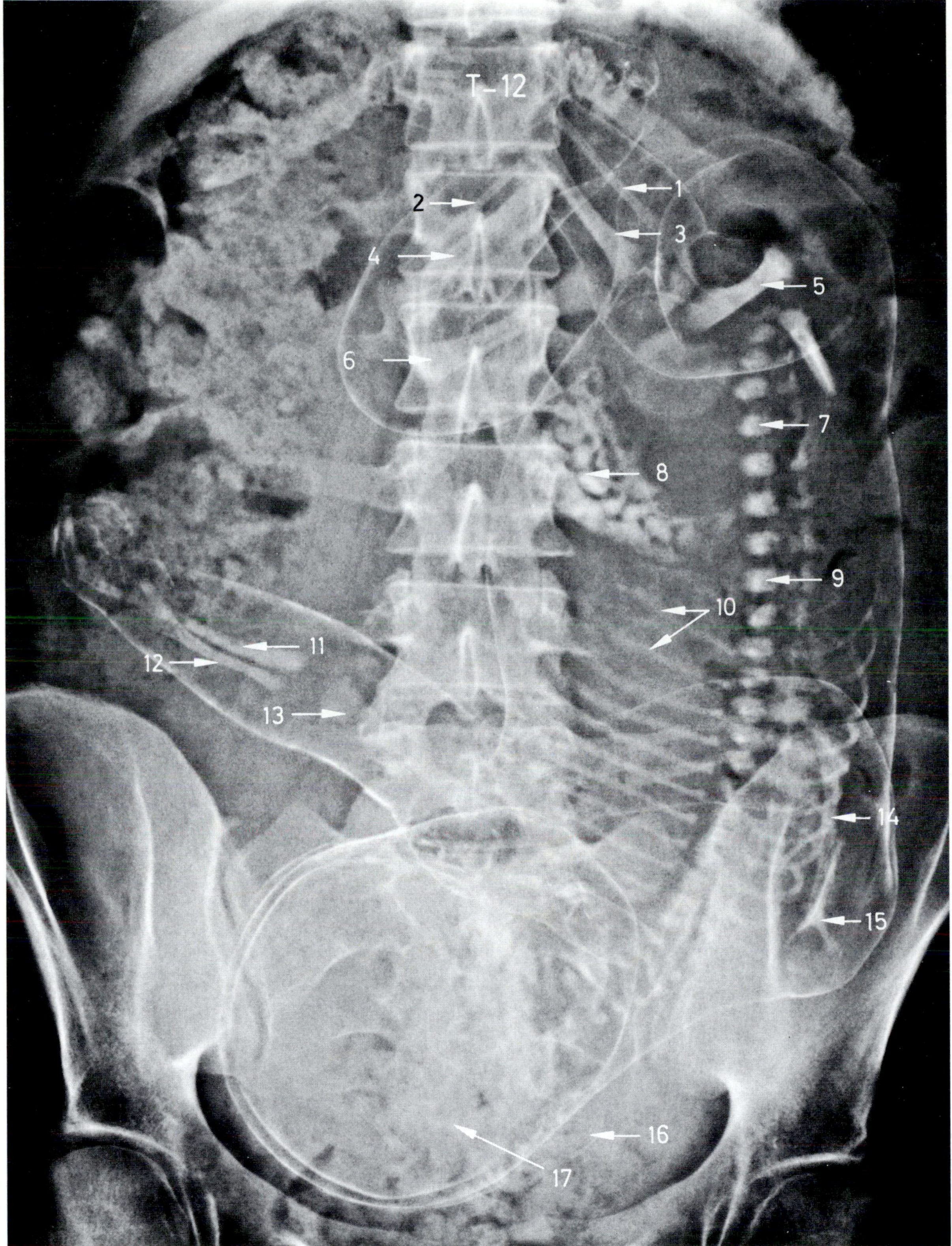
T-12
1
2
3
4
5
6
7
8
9
10
11
12
13
14
15
16
17

Fig. 400: Fetal Roentgenogram

NOTE the body contours of this near-term fetus *in utero* and a number of the ossifying fetal bones. Observe that the uterus extends to the maternal T-12 vertebral body level. This figure is from Wicke, L., *Atlas of Radiologic Anatomy,* 3rd Edition, Urban and Schwarzenberg, Baltimore, 1982.

1. Right fibula
2. Left fibula
3. Right tibia
4. Left tibia
5. Right femur
6. Left femur
7. L-5 vertebra
8. Small intestine (fetal)
9. L-1 vertebra
10. Ribs
11. Left ulna
12. Left radius
13. Left humerus
14. Right humerus
15. Right scapula
16. External ear
17. Fetal head

Fig. 401

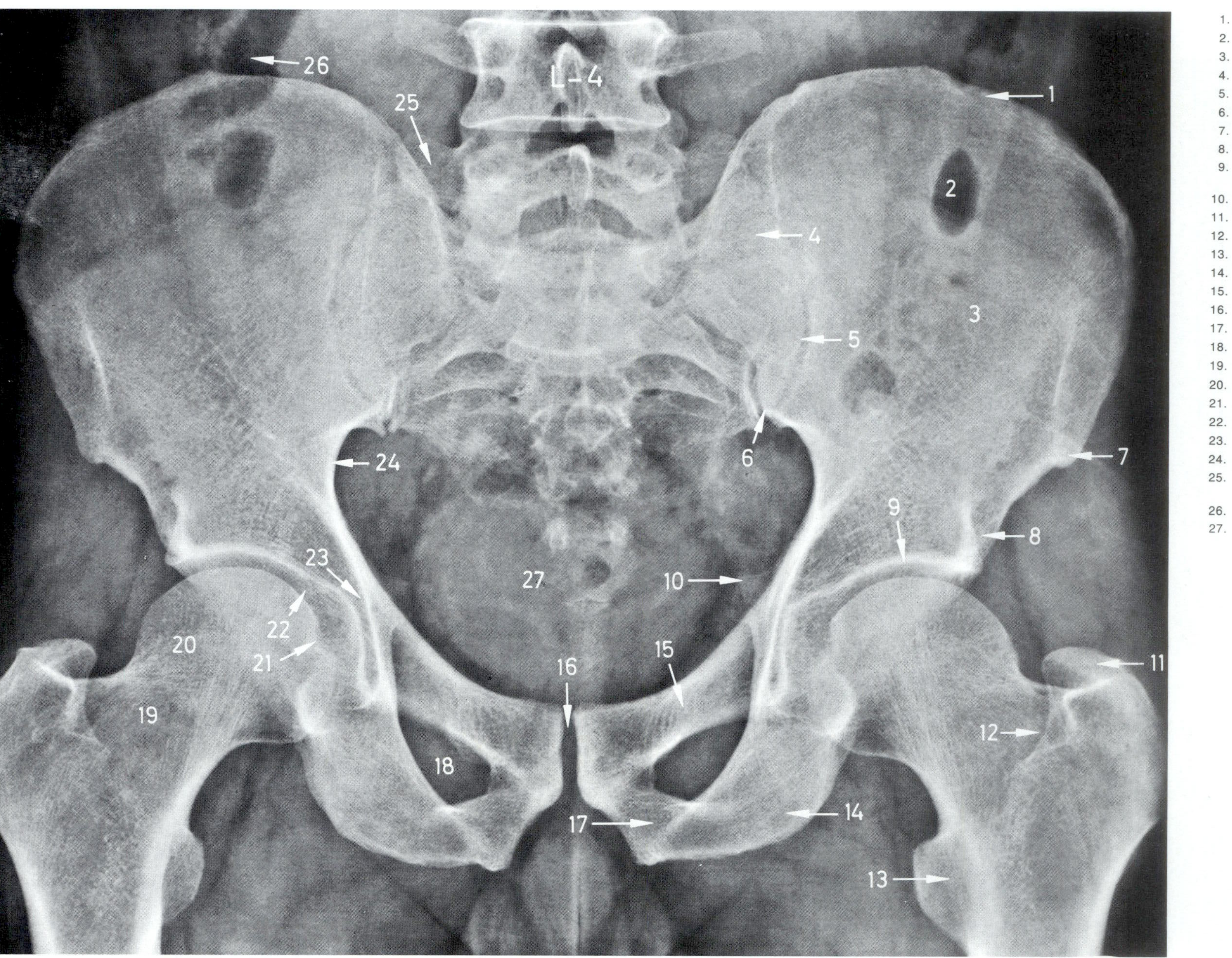

L-4
1
2
3
4
5
6
7
8
9
10
11
12
13
14
15
16
17
18
19
20
21
22
23
24
25
26
27

1. Iliac crest
2. Gas bubble in colon
3. Ala of ilium
4. Lateral part of sacrum
5. Sacroiliac joint
6. Post. inf. iliac spine
7. Ant. sup. iliac spine
8. Ant. inf. iliac spine
9. Lunate surface of acetabulum
10. Spine of ischium
11. Greater trochanter
12. Intertrochanteric crest
13. Lesser trochanter
14. Ischial tuberosity
15. Superior ramus of pubis
16. Symphysis pubis
17. Inferior ramus of pubis
18. Obturator foramen
19. Neck of femur
20. Head of femur
21. Fovea on head of femur
22. Acetabular fossa
23. Iliopubic eminence
24. Greater sciatic notch
25. Transverse process L-5 vertebra
26. Gas bubble in colon
27. Urinary bladder

Fig. 401: Radiograph of the Pelvis and the Sacroiliac and Hip Joints

This radiograph is taken from Wicke, L., *Atlas of Radiologic Anatomy*, 3rd Edition, Urban and Schwarzenberg, Baltimore, 1982.

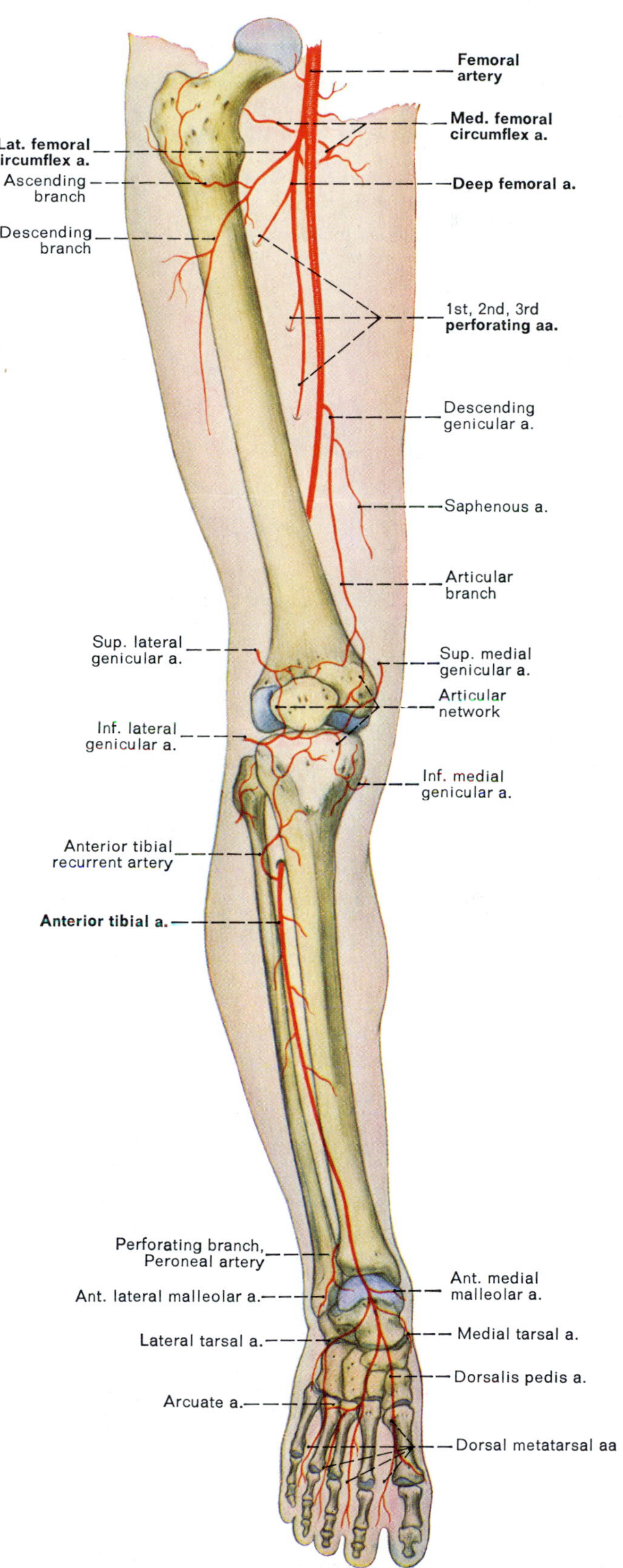
Femoral artery
Med. femoral circumflex a.
Lat. femoral circumflex a.
Ascending branch
Deep femoral a.
Descending branch
1st, 2nd, 3rd perforating aa.
Descending genicular a.
Saphenous a.
Articular branch
Sup. lateral genicular a.
Sup. medial genicular a.
Articular network
Inf. lateral genicular a.
Inf. medial genicular a.
Anterior tibial recurrent artery
Anterior tibial a.
Perforating branch, Peroneal artery
Ant. lateral malleolar a.
Ant. medial malleolar a.
Lateral tarsal a.
Medial tarsal a.
Dorsalis pedis a.
Arcuate a.
Dorsal metatarsal aa

Fig. 402: Arteries and Bones of the Lower Limb (Anterior View)

NOTE the anastomoses in the hip and knee regions, and the perforating branches of the deep femoral artery. In the anterior leg, the anterior tibial artery descends between the tibia and fibula to achieve the malleolar region and dorsum of the foot.

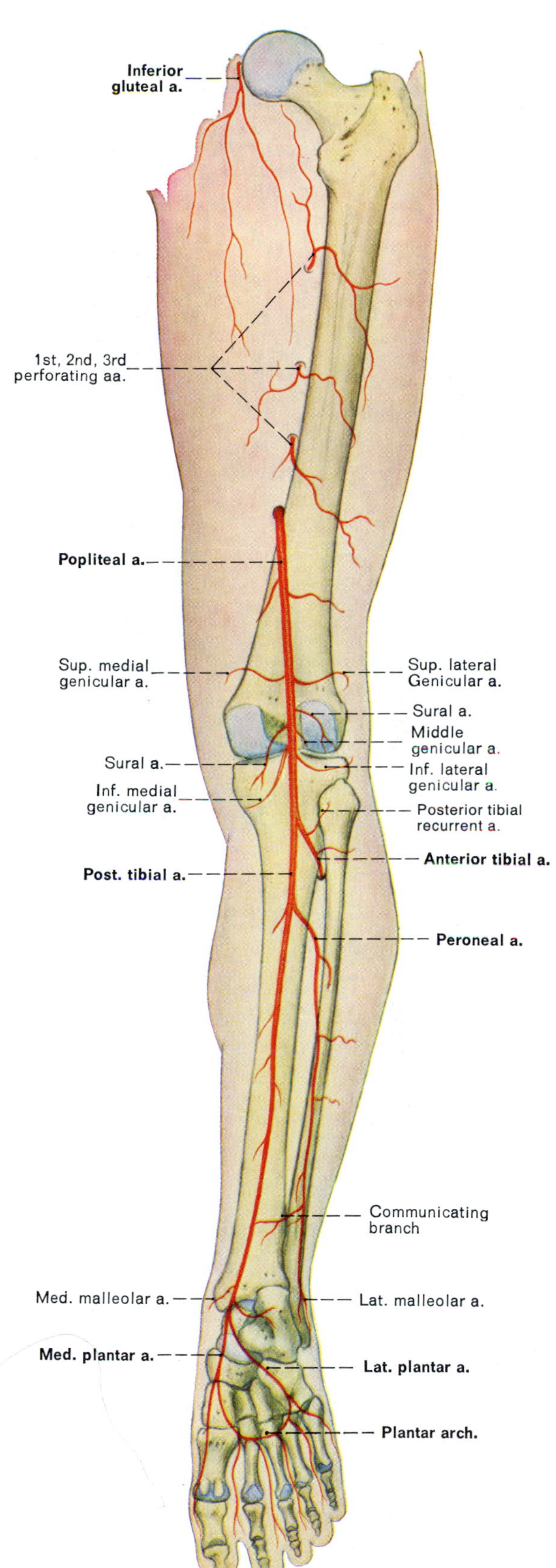
Inferior gluteal a.
1st, 2nd, 3rd perforating aa.
Popliteal a.
Sup. medial genicular a.
Sup. lateral Genicular a.
Sural a.
Middle genicular a.
Sural a.
Inf. lateral genicular a.
Inf. medial genicular a.
Posterior tibial recurrent a.
Anterior tibial a.
Post. tibial a.
Peroneal a.
Communicating branch
Med. malleolar a.
Lat. malleolar a.
Med. plantar a.
Lat. plantar a.
Plantar arch.

Fig. 403: Arteries and Bones of the Lower Limb (Posterior View)

NOTE the branches of the popliteal artery at the knee, and the artery's continuation as the posterior tibial. In the foot this vessel becomes the medial and lateral plantar arteries, which join the plantar arch.

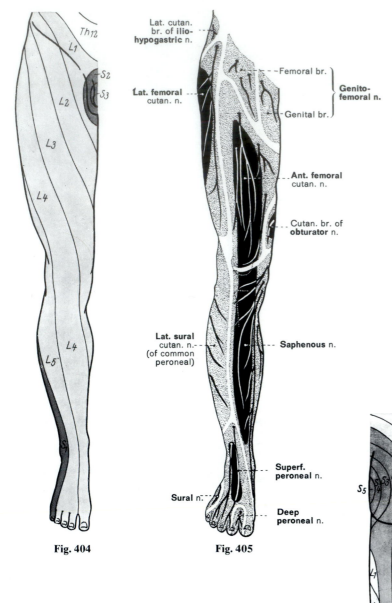

Fig. 404: Dermatomes of the Anterior Aspect of the Lower Extremity

NOTE that as a rule the lumbar segments of the spinal cord supply the cutaneous innervation to the anterior aspect of the lower extremity, and that the dermatomes are segmentally arranged in order from L-1 to S-1. Observe that the genital region is supplied by the sacral segments.

Fig. 405: The Distribution of Cutaneous Nerves: Anterior Aspect of the Lower Extremity

The segmental distribution of the cutaneous nerves supplying the anterior aspect of the lower extremity is as follows:

iliohypogastric nerve:	(T_{12}), L_1
genitofemoral nerve:	L_1, L_2
lateral femoral cutaneous nerve:	L_2, L_3
femoral nerve:	L_2, L_3, L_4
obturator nerve:	L_2, L_3, L_4
saphenous nerve (femoral):	L_2, L_3, L_4
deep peroneal (cutaneous br.):	L_4, L_5
superficial peroneal:	L_4, L_5, S_1
lateral sural cutaneous n.: (common peroneal n.)	L_5, S_1, S_2
sural nerve (tibial):	S_1, S_2

Fig. 404

Fig. 405

Fig. 406: Dermatomes of the Posterior Aspect of the Lower Extremity

NOTE that the skin on the posterior aspect of the lower extremity receives its sensory innervation principally from L5, S1 and S2. Observe, however, how the posterior medial border of the limb consecutively has the L1, L2, L3 and L4 segments represented. Segments S3, S4 and S5 are more limited to the perineal and anal regions.

Fig. 407: The Distribution of Cutaneous Nerves: Posterior Aspect of the Lower Extremity

NOTE: 1) the principal nerve supplying cutaneous innervation to the posterior aspect of the thigh is the posterior femoral cutaneous nerve (S1, S2, S3). The skin of the medial calf is supplied by the saphenous (femoral) nerve (L2, L3, L4), while the lateral calf receives the sural nerve (S1, S2).

2) the heel of the foot is innervated by the tibial nerve through S1 and S2 segments, and the plantar surface of the foot receives L4 and L5 fibers medially (medial plantar nerve) and S1 and S2 laterally (lateral plantar nerve).

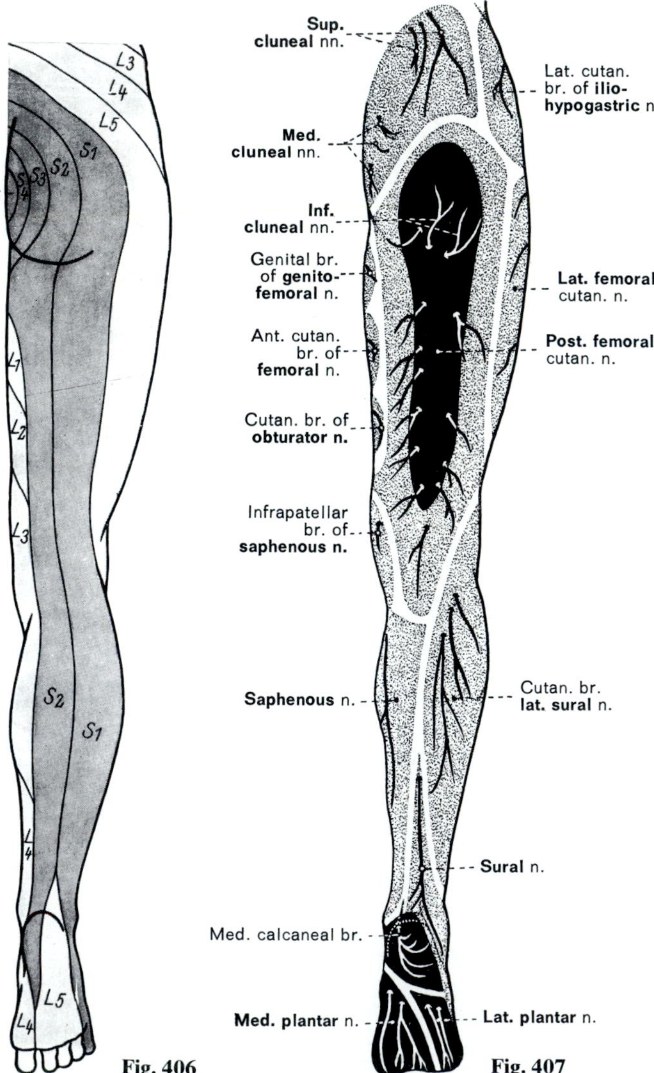

Sup. cluneal nn.
Lat. cutan. br. of ilio-hypogastric n.
Med. cluneal nn.
Inf. cluneal nn.
Genital br. of genito-femoral n.
Lat. femoral cutan. n.
Ant. cutan. br. of femoral n.
Post. femoral cutan. n.
Cutan. br. of obturator n.
Infrapatellar br. of saphenous n.
Saphenous n.
Cutan. br. lat. sural n.
Sural n.
Med. calcaneal br.
Med. plantar n.
Lat. plantar n.

Fig. 406

Fig. 407

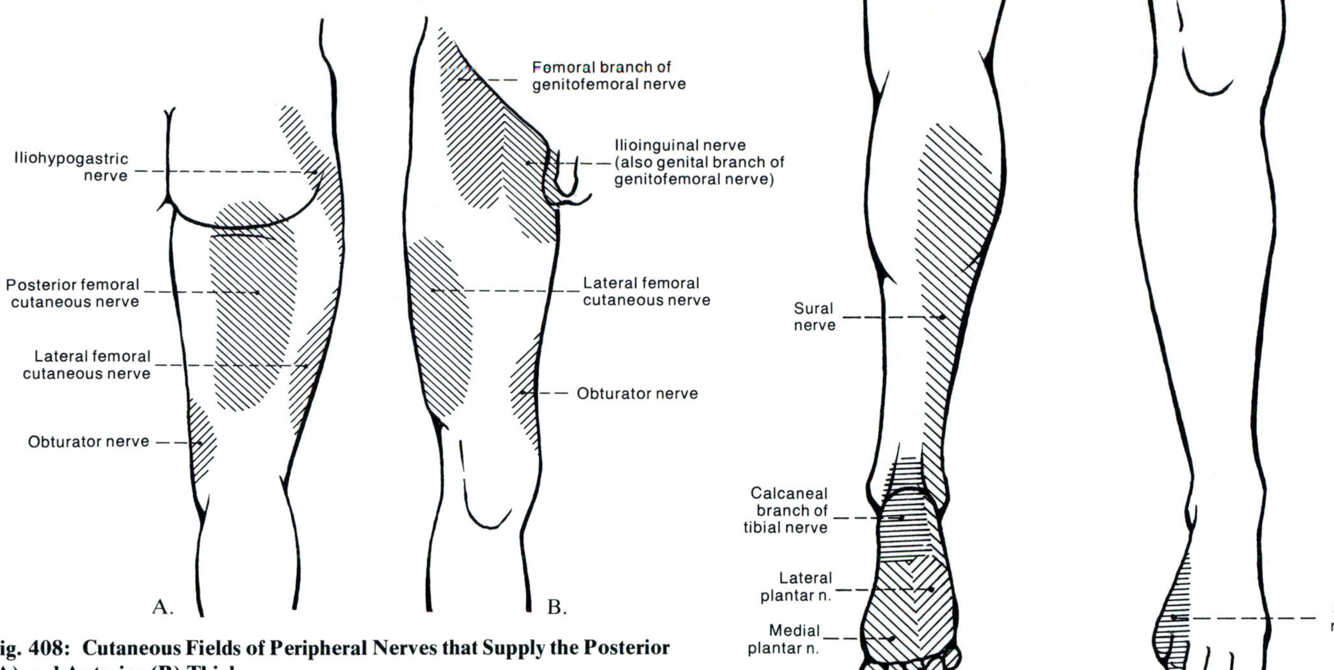
Femoral branch of genitofemoral nerve
Iliohypogastric nerve
Ilioinguinal nerve (also genital branch of genitofemoral nerve)
Posterior femoral cutaneous nerve
Lateral femoral cutaneous nerve
Lateral femoral cutaneous nerve
Obturator nerve
Obturator nerve
A.
B.
Sural nerve
Calcaneal branch of tibial nerve
Lateral plantar n.
Medial plantar n.
Sural nerve
A.
B.

Fig. 408: Cutaneous Fields of Peripheral Nerves that Supply the Posterior (A) and Anterior (B) Thigh

NOTE that *except* for the posterior femoral cutaneous nerve (S1, 2, 3), all of the cutaneous nerves of the thigh are derived from lumbar segments L1 to L4. Observe that the obturator nerve supplies a field on the lower medial aspect of the thigh just above the knee.

Fig. 410: The Cutaneous Fields of the Sural Nerve in the Calf and Foot and the Fields of the Calcaneal and Plantar Branches of the Tibial Nerve

NOTE that the sural nerve supplies much of the posterolateral aspect of the calf (A) and that its field extends beyond the lateral malleolus into the lateral heel and lateral part of the dorsum of the foot (B). Observe that the tibial nerve through its medial calcaneal branch supplies the skin over most of the heel, while the lateral and medial plantar branches of the tibial nerve supply the plantar surface of the foot.

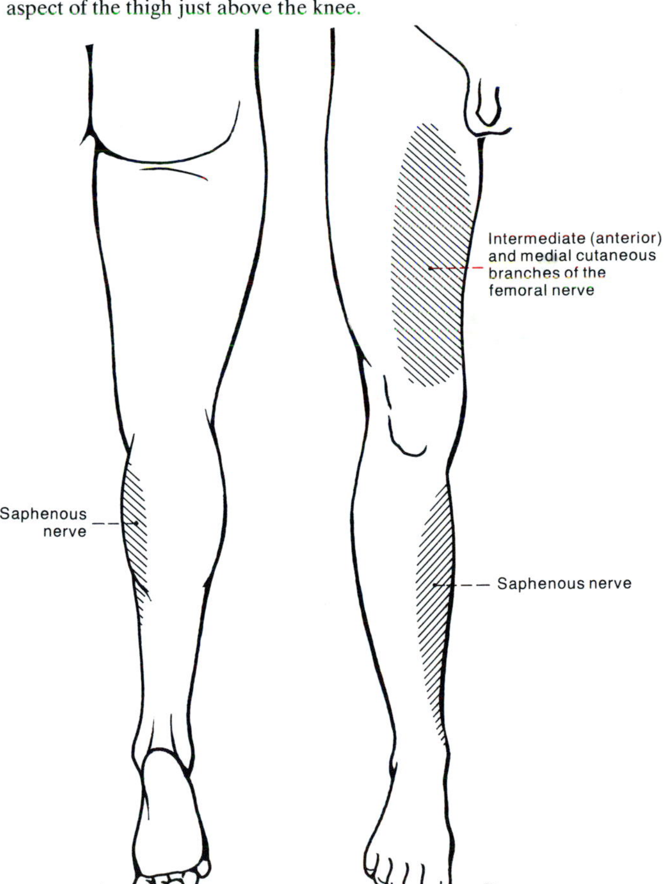
Intermediate (anterior) and medial cutaneous branches of the femoral nerve
Saphenous nerve
Saphenous nerve
A.
B.

Fig. 409: Cutaneous Fields Supplied by the Femoral Nerve and Its Saphenous Branch

NOTE that the femoral nerve, by way of its intermediate (anterior) and medial cutaneous branches, supplies the anteromedial aspect of the thigh (B) below the fields of the ilioinguinal and genitofemoral nerves. The saphenous nerve is the largest cutaneous branch of the femoral, and it supplies the anteromedial (B) and posteromedial aspects of the leg (A).

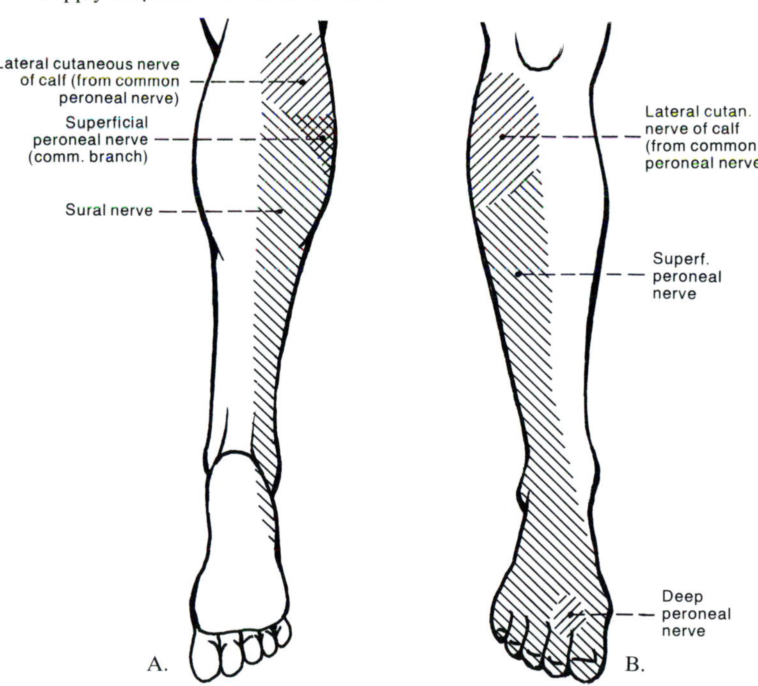
Lateral cutaneous nerve of calf (from common peroneal nerve)
Superficial peroneal nerve (comm. branch)
Sural nerve
Lateral cutan. nerve of calf (from common peroneal nerve)
Superf. peroneal nerve
Deep peroneal nerve
A.
B.

Fig. 411: Cutaneous Fields of the Peroneal and Sural Nerves on the Posterolateral and Anterolateral Aspects of the Leg and the Dorsum of the Foot

NOTE that on the posterolateral aspect of the leg (A), above the cutaneous field of the sural nerve, are areas of skin supplied by the lateral cutaneous nerve of the calf (from common peroneal nerve; also called lateral sural nerve) and the sural communicating branch (often from the superficial peroneal nerve), which joins the sural nerve, a branch of the tibial nerve. Observe the large field supplied by the superficial peroneal nerve on the anterior leg and foot dorsum, and the small field supplied by the deep peroneal nerve (B).

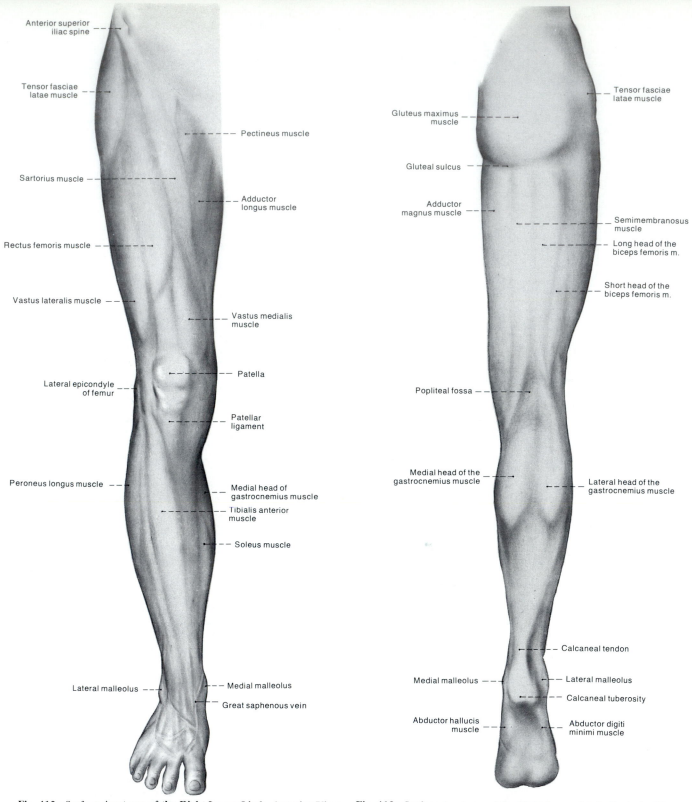

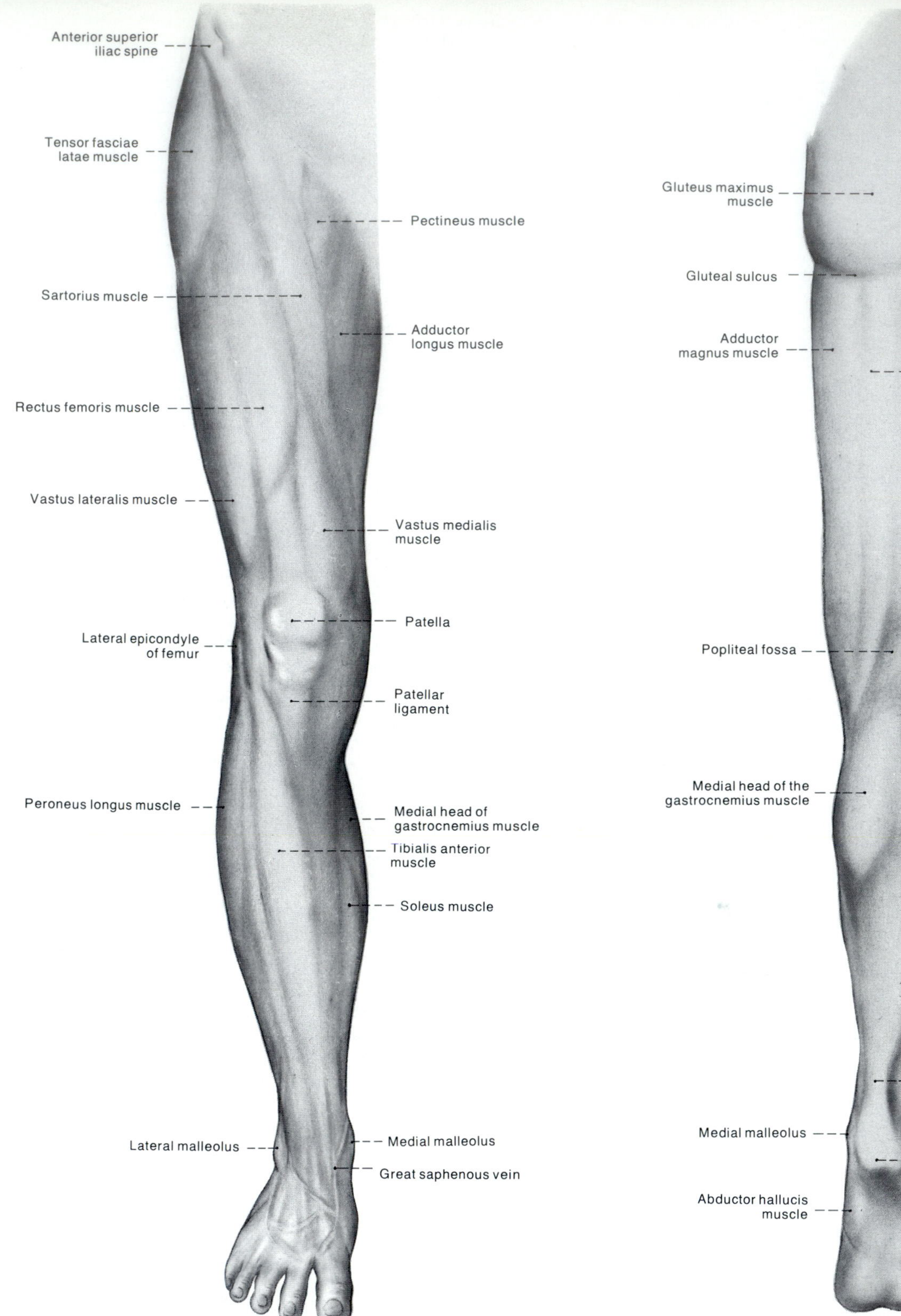
Anterior superior iliac spine
Tensor fasciae latae muscle
Pectineus muscle
Sartorius muscle
Adductor longus muscle
Rectus femoris muscle
Vastus lateralis muscle
Vastus medialis muscle
Patella
Lateral epicondyle of femur
Patellar ligament
Peroneus longus muscle
Medial head of gastrocnemius muscle
Tibialis anterior muscle
Soleus muscle
Lateral malleolus
Medial malleolus
Great saphenous vein
Tensor fasciae latae muscle
Gluteus maximus muscle
Gluteal sulcus
Adductor magnus muscle
Semimembranosus muscle
Long head of the biceps femoris m.
Short head of the biceps femoris m.
Popliteal fossa
Medial head of the gastrocnemius muscle
Lateral head of the gastrocnemius muscle
Calcaneal tendon
Medial malleolus
Lateral malleolus
Calcaneal tuberosity
Abductor hallucis muscle
Abductor digiti minimi muscle

Fig. 412: Surface Anatomy of the Right Lower Limb, Anterior View

NOTE: 1) the muscular features of the anterior thigh. Observe the pectineus and adductor longus muscles which form the floor of the femoral triangle, the sartorius muscle which courses obliquely inferomedially, three of the four components of the quadriceps muscle (vastus lateralis and medialis and the rectus femoris), and the tensor fasciae latae which forms the rounded upper lateral contour of the thigh.

2) that laterally the leg is shaped by the peroneus longus (and brevis) and anteriorly by the tibialis anterior. The medial contour of the leg consists of the gastrocnemius and soleus muscles.

3) the prominent osseous features: the anterior superior iliac spine, the patella and femoral condyles, the tibia (just medial to the tibialis anterior muscle), and the medial and lateral malleoli.

Fig. 413: Surface Anatomy of the Right Lower Limb, Posterior View

NOTE: 1) that the rounded contour of the buttock is formed by the gluteus maximus muscle and an overlying layer of fatty tissue.

2) that the outline of the hamstring muscles (semitendinosus, semimembranosus, and biceps femoris) is visible in the mid-portion of the posterior thigh between the adductor region medially and the quadriceps (not labelled) laterally.

3) the popliteal fossa behind the knee joint, below which are the prominent heads of the gastrocnemius muscle. This muscle helps to form the calcaneal tendon which inserts between the two malleoli into the calcaneal bone. Observe also the muscular contours formed medially by the abductor hallucis muscle and laterally by the abductor digiti minimi muscle on the plantar surface of the foot.

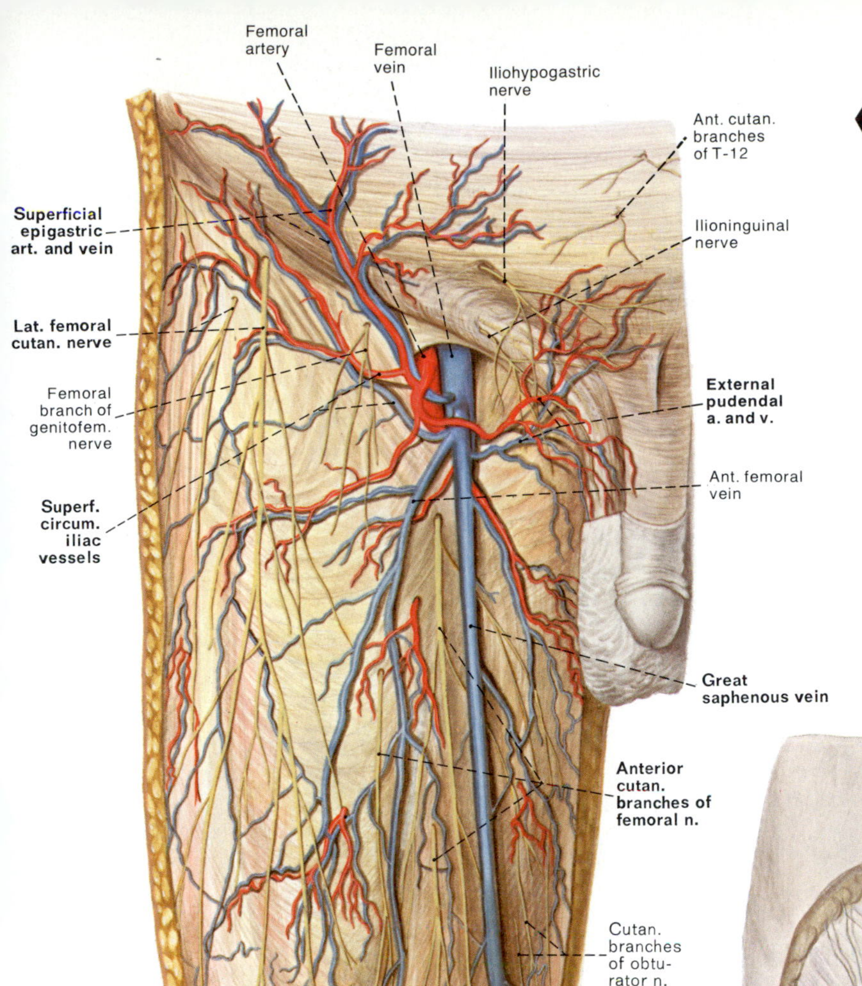

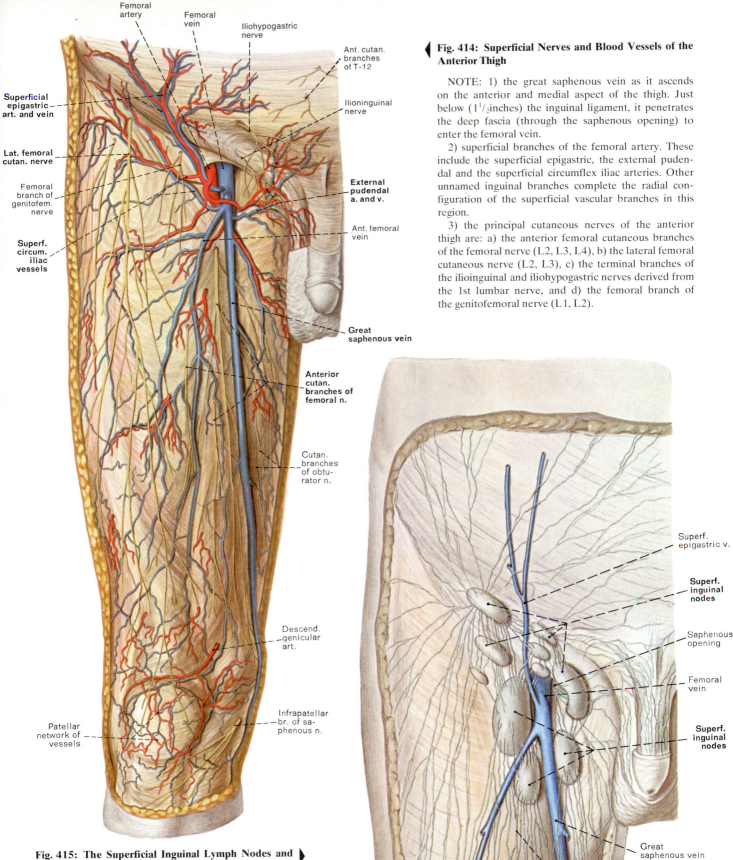

Femoral artery
Femoral vein
Iliohypogastric nerve
Ant. cutan. branches of T-12
Superficial epigastric art. and vein
Ilioninguinal nerve
Lat. femoral cutan. nerve
Femoral branch of genitofem. nerve
External pudendal a. and v.
Ant. femoral vein
Superf. circum. iliac vessels
Great saphenous vein
Anterior cutan. branches of femoral n.
Cutan. branches of obtu- rator n.
Descend. genicular art.
Infrapatellar br. of sa- phenous n.
Patellar network of vessels

Fig. 414: Superficial Nerves and Blood Vessels of the Anterior Thigh

NOTE: 1) the great saphenous vein as it ascends on the anterior and medial aspect of the thigh. Just below ($1^1/_2$inches) the inguinal ligament, it penetrates the deep fascia (through the saphenous opening) to enter the femoral vein.

2) superficial branches of the femoral artery. These include the superficial epigastric, the external pudendal and the superficial circumflex iliac arteries. Other unnamed inguinal branches complete the radial configuration of the superficial vascular branches in this region.

3) the principal cutaneous nerves of the anterior thigh are: a) the anterior femoral cutaneous branches of the femoral nerve (L2, L3, L4), b) the lateral femoral cutaneous nerve (L2, L3), c) the terminal branches of the ilioinguinal and iliohypogastric nerves derived from the 1st lumbar nerve, and d) the femoral branch of the genitofemoral nerve (L1, L2).

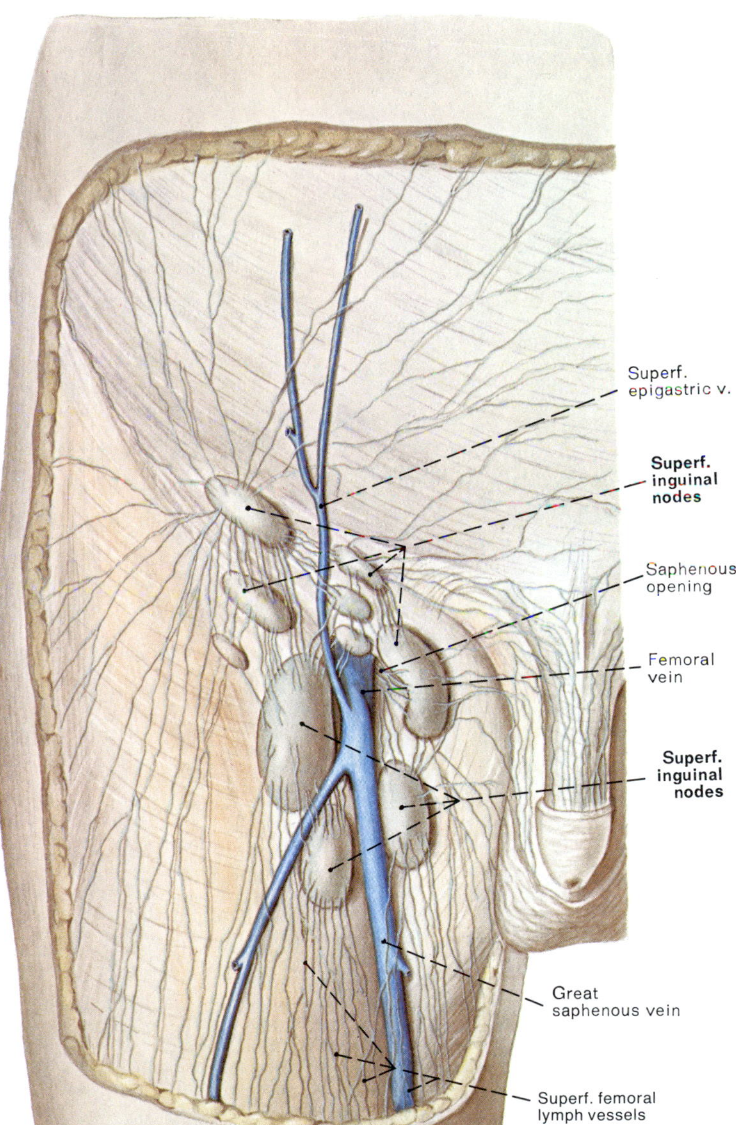

Superf. epigastric v.
Superf. inguinal nodes
Saphenous opening
Femoral vein
Superf. inguinal nodes
Great saphenous vein
Superf. femoral lymph vessels

Fig. 415: The Superficial Inguinal Lymph Nodes and Channels

NOTE that the superficial tissues of the genitalia, lower anterior abdominal wall, inguinal region and anterior thigh drain into the superficial inguinal lymph nodes. These are located around the femoral vessels just below the inguinal ligament and usually number between 10 and 15. In turn these nodes drain into the external iliac nodes within the pelvis.

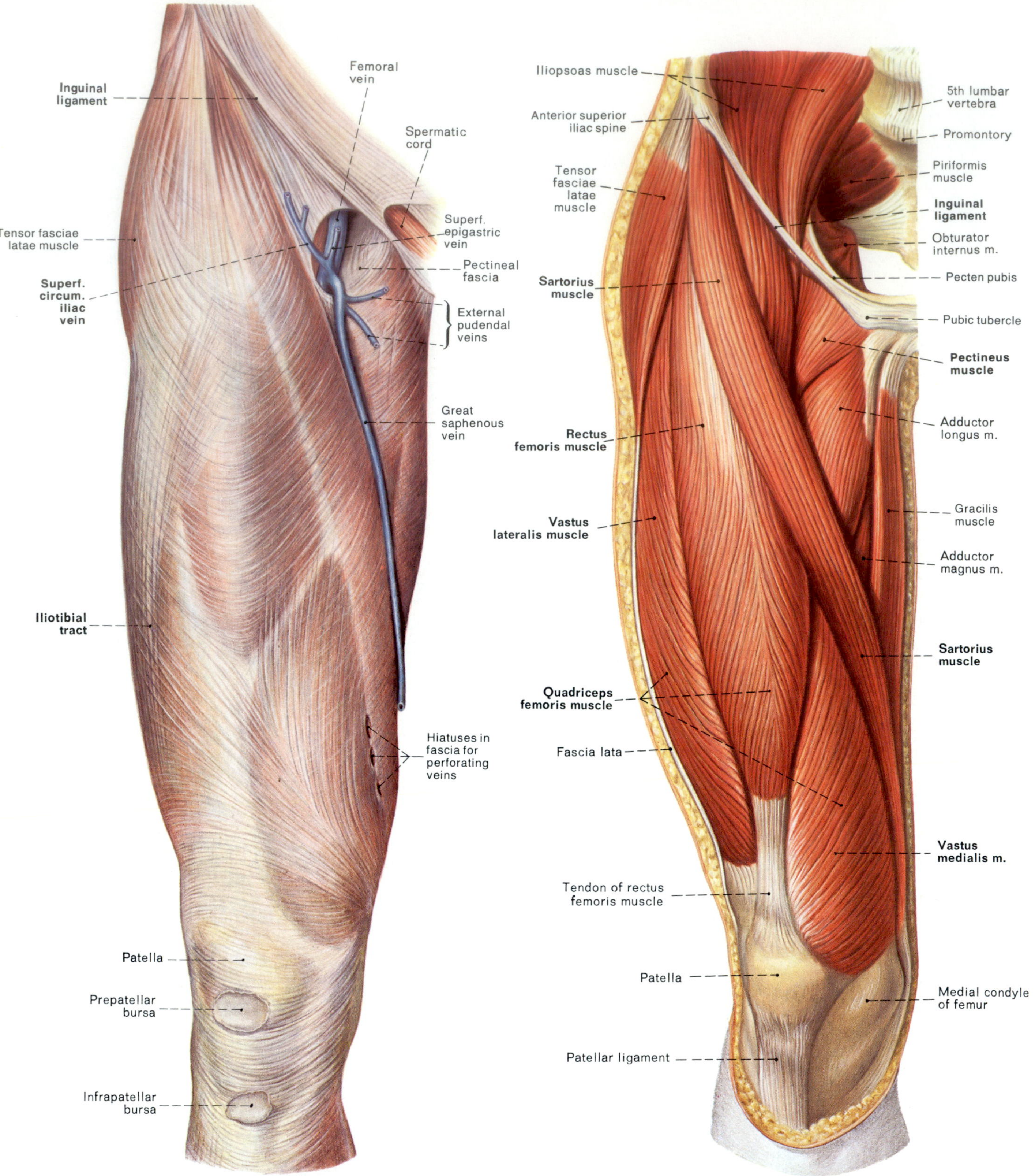

Femoral vein
Inguinal ligament
Spermatic cord
Tensor fasciae latae muscle
Superf. epigastric vein
Pectineal fascia
Superf. circum. iliac vein
External pudendal veins
Great saphenous vein
Iliotibial tract
Hiatuses in fascia for perforating veins
Patella
Prepatellar bursa
Infrapatellar bursa
Iliopsoas muscle
5th lumbar vertebra
Anterior superior iliac spine
Promontory
Tensor fasciae latae muscle
Piriformis muscle
Inguinal ligament
Obturator internus m.
Sartorius muscle
Pecten pubis
Pubic tubercle
Pectineus muscle
Adductor longus m.
Rectus femoris muscle
Gracilis muscle
Vastus lateralis muscle
Adductor magnus m.
Sartorius muscle
Quadriceps femoris muscle
Fascia lata
Vastus medialis m.
Tendon of rectus femoris muscle
Patella
Medial condyle of femur
Patellar ligament

Fig. 416: The Fascia of the Anterior Thigh, the Fascia Lata (right)

NOTE: 1) the dense fascia which invests the muscles of the hip and thigh is called the fascia lata. It is attached above to the ischial and pubic rami and inguinal ligament anteriorly, the crest of the ilium laterally and the ischial tuberosity, sacrotuberous ligament, sacrum and coccyx posteriorly.

2) in the inguinal region the fascia lata is pierced by the greater saphenous vein, and inferiorly it extends to the investing fascia below the knee.

Fig. 417: The Anterior Muscles of the Thigh: Superficial View (right)

NOTE: 1) the long narrow sartorius muscle which arises on the anterior superior iliac spine and passes obliquely across the anterior femoral muscles to insert on the medial aspect of the body of the tibia. The sartorius flexes the thigh and rotates it laterally. It also flexes the knee and rotates it medially.

2) the quadriceps muscle forms the bulk of the anterior femoral muscles, and both the sartorius and quadriceps are innervated by the femoral nerve.

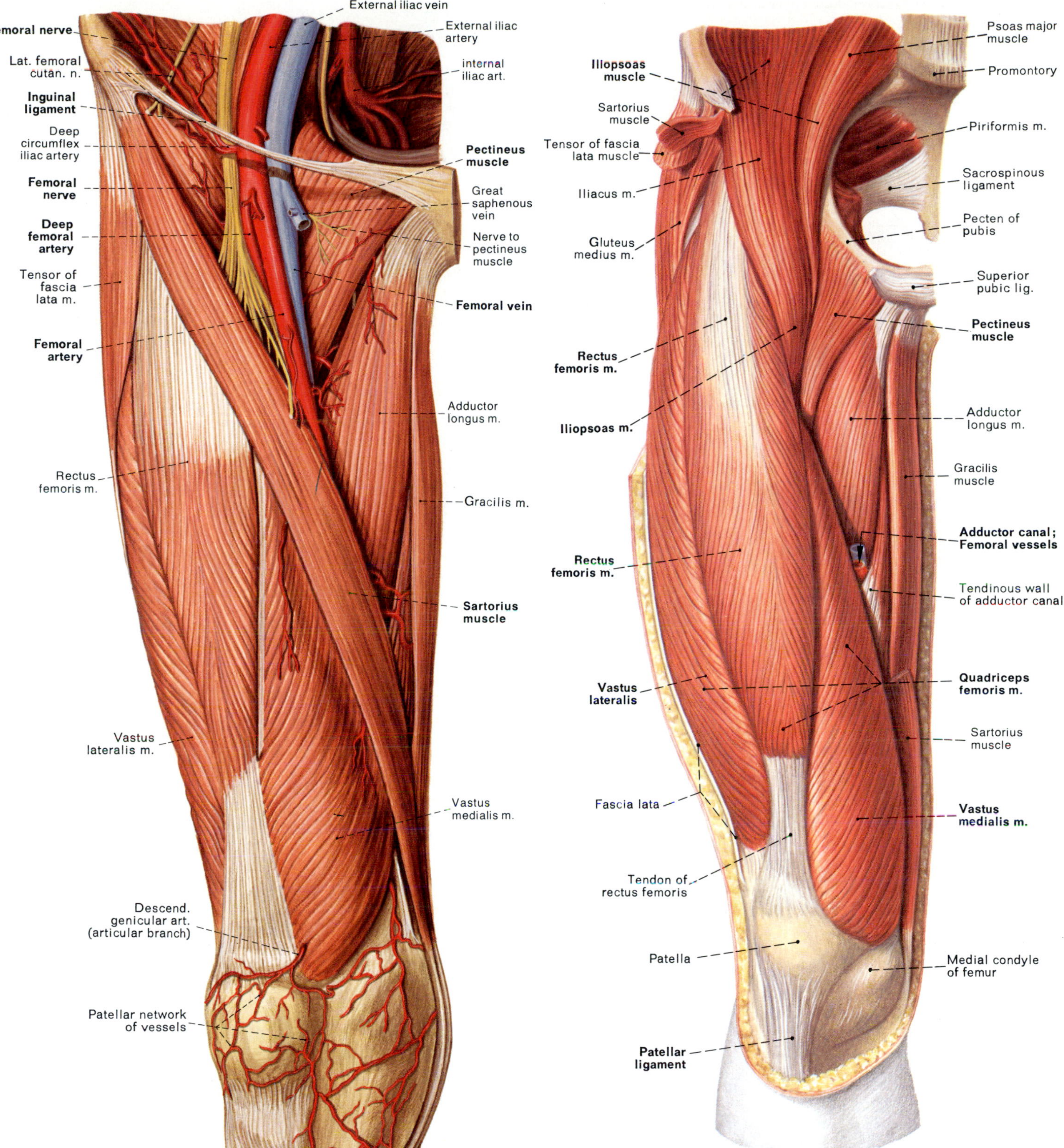

External iliac vein
Femoral nerve
External iliac artery
Lat. femoral cutan. n.
internal iliac art.
Inguinal ligament
Deep circumflex iliac artery
Pectineus muscle
Femoral nerve
Great saphenous vein
Deep femoral artery
Nerve to pectineus muscle
Tensor of fascia lata m.
Femoral vein
Femoral artery
Adductor longus m.
Rectus femoris m.
Gracilis m.
Sartorius muscle
Vastus lateralis m.
Vastus medialis m.
Descend. genicular art. (articular branch)
Patellar network of vessels
Patellar ligament
Psoas major muscle
Iliopsoas muscle
Promontory
Sartorius muscle
Piriformis m.
Tensor of fascia lata muscle
Sacrospinous ligament
Iliacus m.
Pecten of pubis
Gluteus medius m.
Superior pubic lig.
Pectineus muscle
Rectus femoris m.
Adductor longus m.
Iliopsoas m.
Gracilis muscle
Adductor canal; Femoral vessels
Rectus femoris m.
Tendinous wall of adductor canal
Quadriceps femoris m.
Vastus lateralis
Sartorius muscle
Fascia lata
Vastus medialis m.
Tendon of rectus femoris
Patella
Medial condyle of femur
Patellar ligament

Fig. 418: The Femoral Triangle

NOTE: the boundaries of the femoral triangle are the inguinal ligament, the sartorius muscle and the medial border of the adductor longus. The floor is formed by the iliopsoas and pectineus muscles. The femoral nerve, artery and vein descend beneath the inguinal ligament and traverse the femoral triangle.

Fig. 419: The Quadriceps Femoris, Iliopsoas, and Pectineus Muscles

NOTE: 1) the quadriceps femoris (rectus femoris and vastus lateralis, intermedius and medialis) as it converges inferiorly to form a powerful tendon which encases the patella and which is inserted onto the tuberosity of the tibia. The entire quadriceps extends the leg, while the rectus femoris also flexes the thigh.

2) the iliopsoas muscle is a powerful flexor of the thigh and inserts on the lesser trochanter.

Fig. 420: The Intermediate Layer of Anterior and Medial Thigh Muscles

NOTE: 1) that the rectus femoris and iliopsoas muscles are cut to reveal the underlying vastus intermedius situated between the vastus medialis and lateralis.

2) the adductor longus has been reflected. This displays the pectineus, adductors brevis and magnus and the long gracilis muscles.

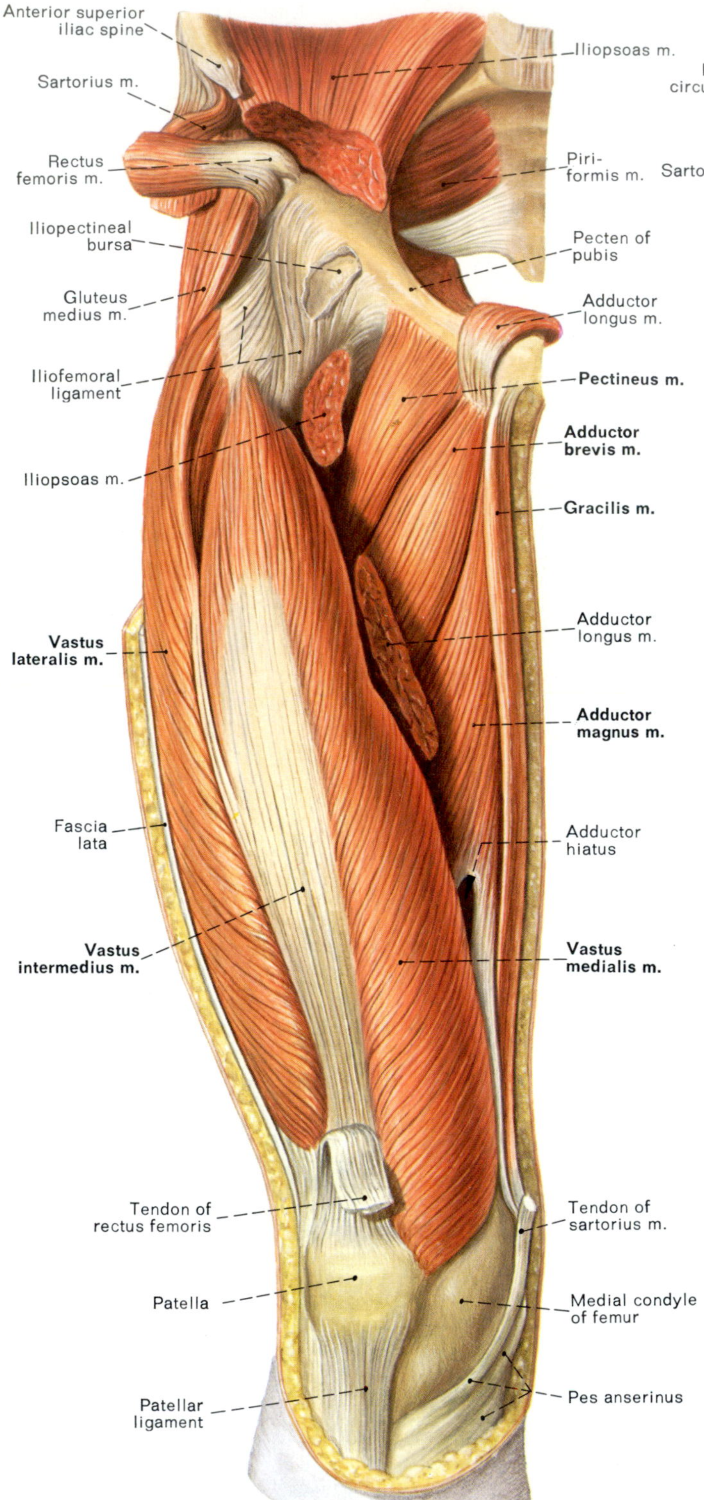

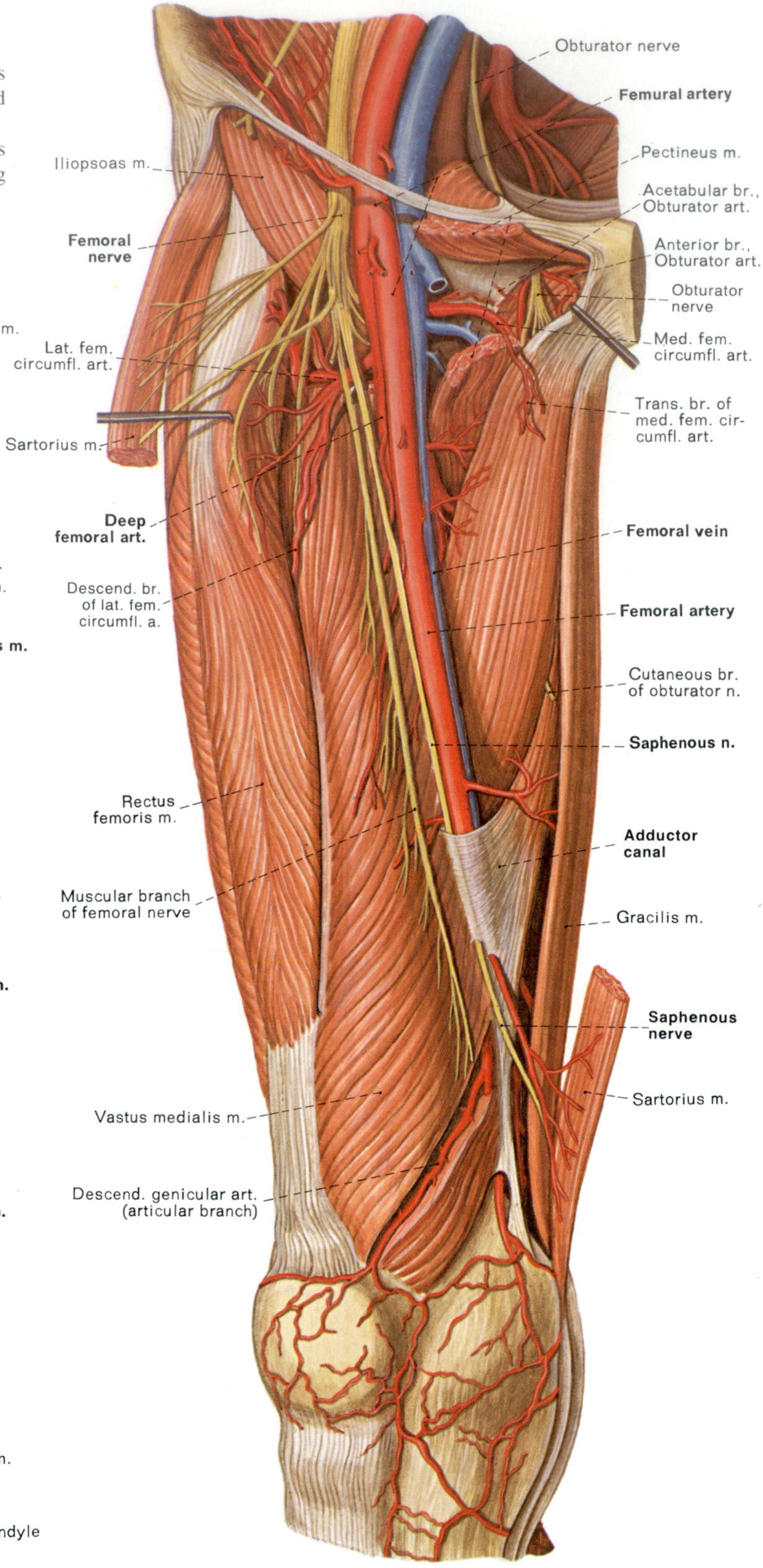

Anterior superior iliac spine
Sartorius m.
Rectus femoris m.
Iliopectineal bursa
Gluteus medius m.
Iliofemoral ligament
Iliopsoas m.
Vastus lateralis m.
Fascia lata
Vastus intermedius m.
Tendon of rectus femoris
Patella
Patellar ligament
Iliopsoas m.
Piri-formis m.
Pecten of pubis
Adductor longus m.
Pectineus m.
Adductor brevis m.
Gracilis m.
Adductor longus m.
Adductor magnus m.
Adductor hiatus
Vastus medialis m.
Tendon of sartorius m.
Medial condyle of femur
Pes anserinus

Iliopsoas m.
Femoral nerve
Lat. fem. circumfl. art.
Sartorius m.
Deep femoral art.
Descend. br. of lat. fem. circumfl. a.
Rectus femoris m.
Muscular branch of femoral nerve
Vastus medialis m.
Descend. genicular art. (articular branch)
Obturator nerve
Femural artery
Pectineus m.
Acetabular br., Obturator art.
Anterior br., Obturator art.
Obturator nerve
Med. fem. circumfl. art.
Trans. br. of med. fem. cir-cumfl. art.
Femoral vein
Femoral artery
Cutaneous br. of obturator n.
Saphenous n.
Adductor canal
Gracilis m.
Saphenous nerve
Sartorius m.

Fig. 421: The Femoral Vessels and Nerves

NOTE that the femoral vessels and the saphenous branch of the femoral nerve descend in the thigh and enter the adductor canal (of Hunter). Whereas the saphenous nerve then penetrates the overlying fascia to reach the superficial leg region, the vessels continue through the adductor magnus to reach the popliteal fossa on the posterior aspect.

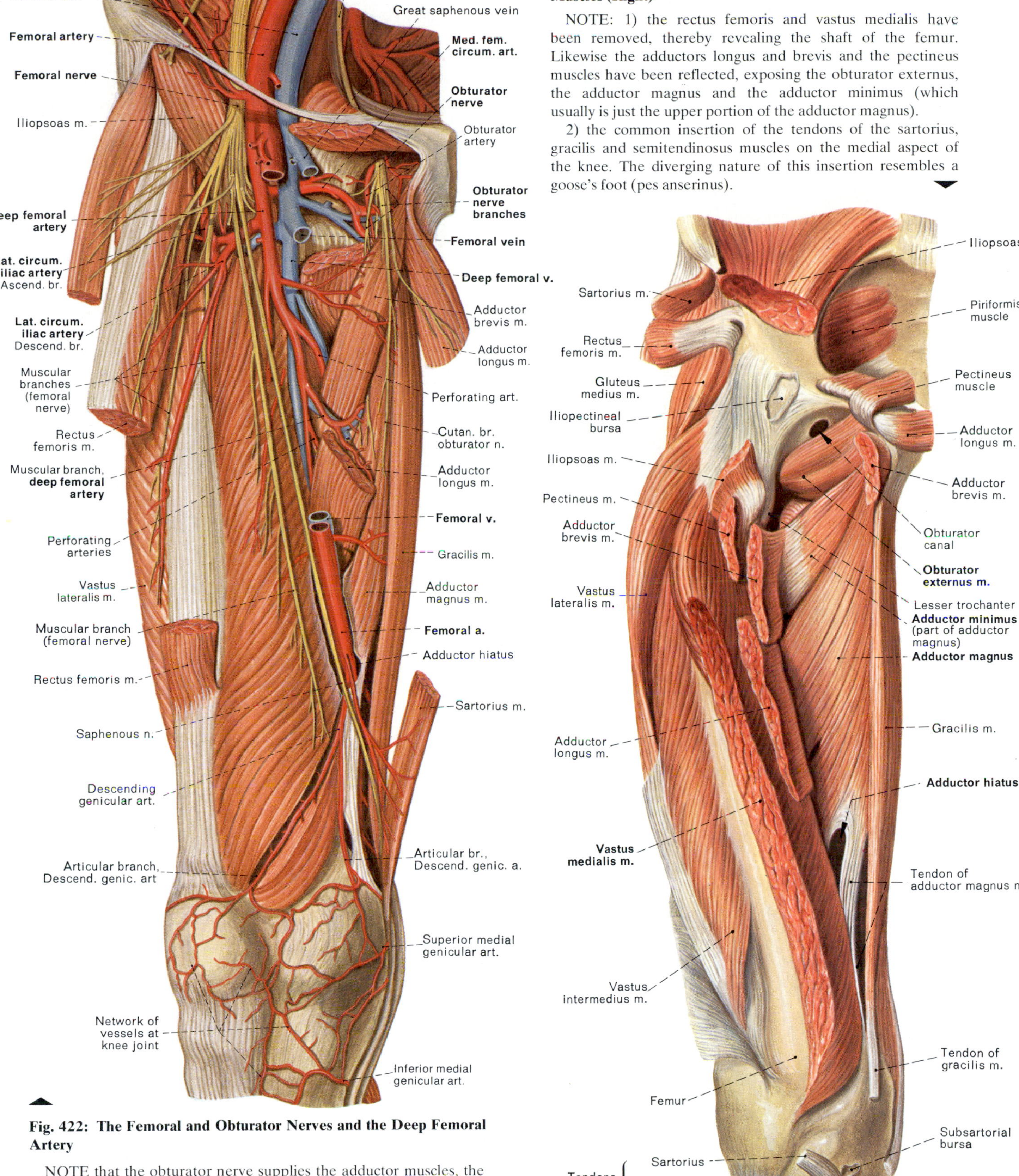

Femoral vein
Femoral artery
Femoral nerve
Iliopsoas m.
Deep femoral artery
Lat. circum. iliac artery Ascend. br.
Lat. circum. iliac artery Descend. br.
Muscular branches (femoral nerve)
Rectus femoris m.
Muscular branch, deep femoral artery
Perforating arteries
Vastus lateralis m.
Muscular branch (femoral nerve)
Rectus femoris m.
Saphenous n.
Descending genicular art.
Articular branch, Descend. genic. art
Network of vessels at knee joint
Obturator nerve
Great saphenous vein
Med. fem. circum. art.
Obturator nerve
Obturator artery
Obturator nerve branches
Femoral vein
Deep femoral v.
Adductor brevis m.
Adductor longus m.
Perforating art.
Cutan. br. obturator n.
Adductor longus m.
Femoral v.
Gracilis m.
Adductor magnus m.
Femoral a.
Adductor hiatus
Sartorius m.
Articular br., Descend. genic. a.
Superior medial genicular art.
Inferior medial genicular art.
Sartorius m.
Rectus femoris m.
Gluteus medius m.
Iliopectineal bursa
Iliopsoas m.
Pectineus m.
Adductor brevis m.
Vastus lateralis m.
Adductor longus m.
Vastus medialis m.
Vastus intermedius m.
Femur
Tendons form pes anserinus
Sartorius
Gracilis
Semitend.
Iliopsoas m.
Piriformis muscle
Pectineus muscle
Adductor longus m.
Adductor brevis m.
Obturator canal
Obturator externus m.
Lesser trochanter
Adductor minimus (part of adductor magnus)
Adductor magnus
Gracilis m.
Adductor hiatus
Tendon of adductor magnus m.
Tendon of gracilis m.
Subsartorial bursa
Subanserina bursa

Fig. 423: The Deep Layer of Anterior and Medial Thigh Muscles (Right)

NOTE: 1) the rectus femoris and vastus medialis have been removed, thereby revealing the shaft of the femur. Likewise the adductors longus and brevis and the pectineus muscles have been reflected, exposing the obturator externus, the adductor magnus and the adductor minimus (which usually is just the upper portion of the adductor magnus).

2) the common insertion of the tendons of the sartorius, gracilis and semitendinosus muscles on the medial aspect of the knee. The diverging nature of this insertion resembles a goose's foot (pes anserinus).

Fig. 422: The Femoral and Obturator Nerves and the Deep Femoral Artery

NOTE that the obturator nerve supplies the adductor muscles, the gracilis and the obturator externus (not shown), while the femoral nerve innervates all the other anterior thigh muscles. The largest branch of the femoral artery is the deep femoral artery from which generally both the medial and lateral femoral circumflex arteries arise. Observe the femoral vessels disappearing in the femoral canal.

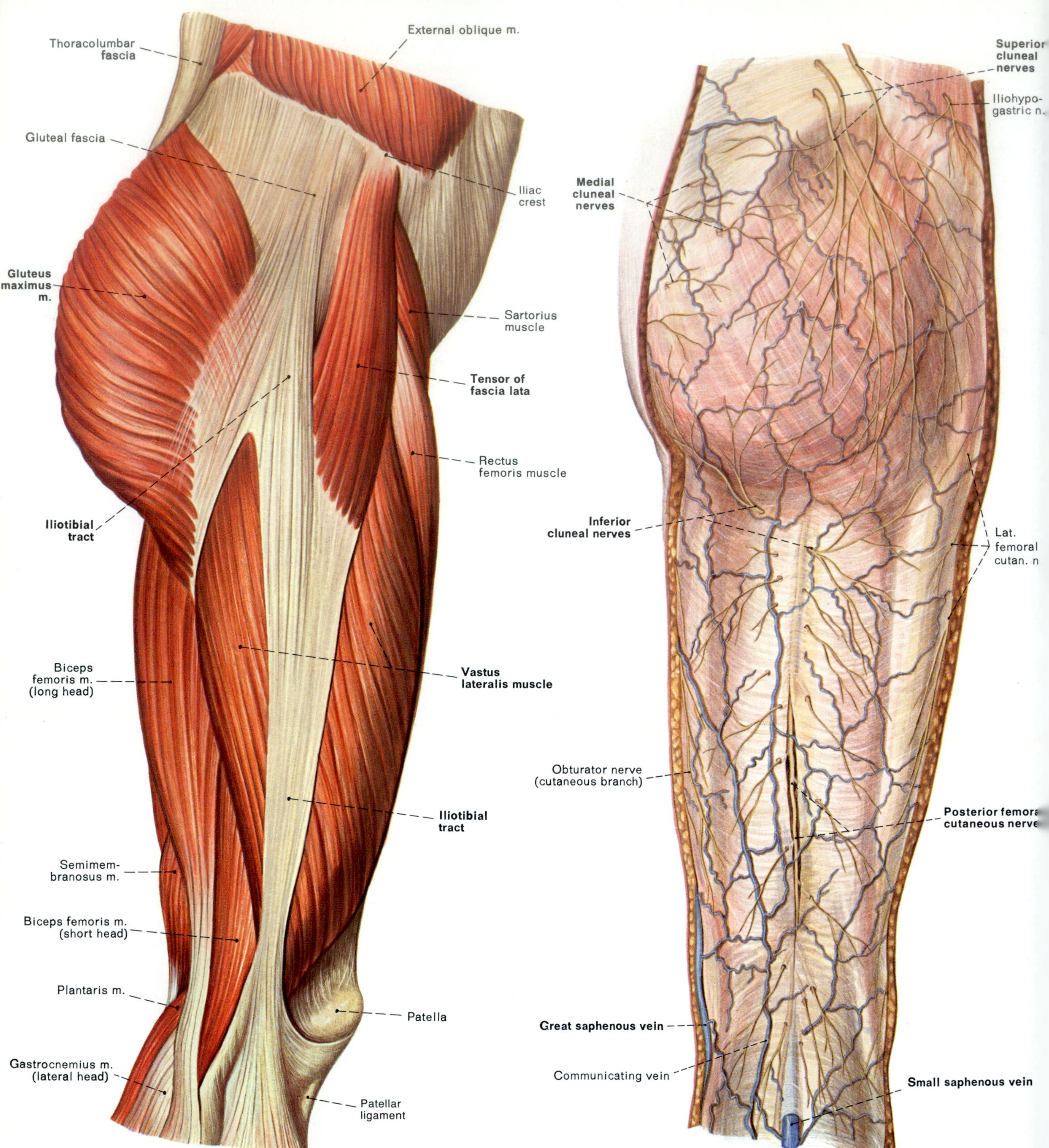

Thoracolumbar fascia
External oblique m.
Gluteal fascia
Iliac crest
Gluteus maximus m.
Sartorius muscle
Tensor of fascia lata
Rectus femoris muscle
Iliotibial tract
Biceps femoris m. (long head)
Vastus lateralis muscle
Iliotibial tract
Semimembranosus m.
Biceps femoris m. (short head)
Plantaris m.
Patella
Gastrocnemius m. (lateral head)
Patellar ligament
Superior cluneal nerves
Iliohypogastric n.
Medial cluneal nerves
Inferior cluneal nerves
Lat. femoral cutan. n
Obturator nerve (cutaneous branch)
Great saphenous vein
Communicating vein
Small saphenous vein

Fig. 424: Superficial Thigh and Gluteal Muscles (Lateral View)

NOTE the massive nature of the vastus lateralis and gluteus maximus muscles. Superficially, the iliotibial tract (band) stretches the length of the thigh, and its muscle, the tensor of the fascia lata, helps to keep the dense fascial sheet taut, thereby assisting in the maintenance of an erect posture.

Fig. 425: Superficial Veins and Nerves of the Gluteal Region and Posterior Thigh

NOTE that the principal cutaneous nerves of the gluteal region are the a) superior cluneal (L1, L2, L3, posterior rami), b) medial cluneal (S1, S2, S3, posterior rami), and c) inferior cluneal (S1, S2, S3 from posterior femoral cutaneous nerve). The skin of the posterior aspect of the thigh is innervated primarily by the posterior femoral cutaneous nerve (S1, S2, S3).

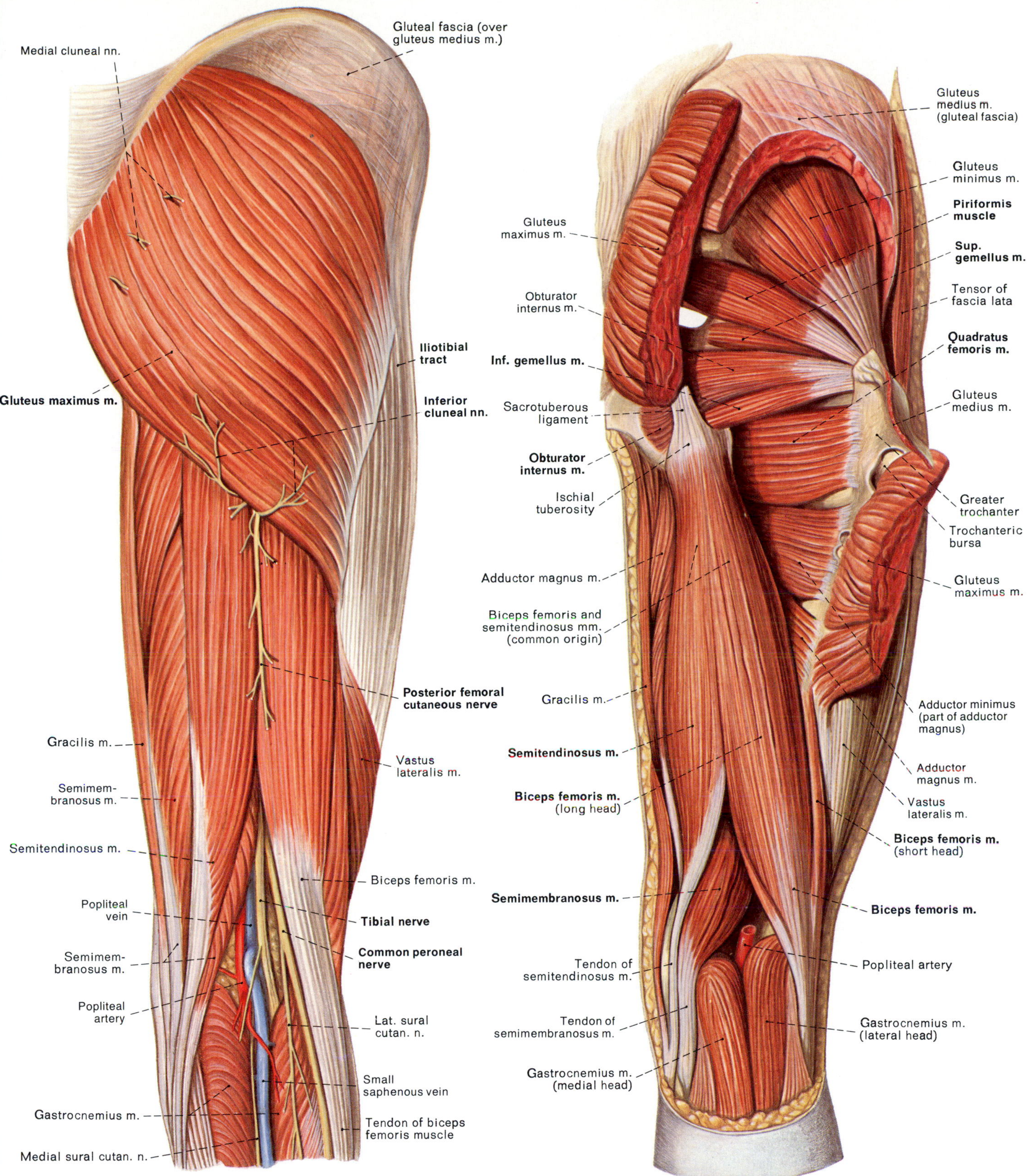

Medial cluneal nn.
Gluteal fascia (over gluteus medius m.)
Iliotibial tract
Gluteus maximus m.
Inferior cluneal nn.
Posterior femoral cutaneous nerve
Gracilis m.
Vastus lateralis m.
Semimembranosus m.
Semitendinosus m.
Biceps femoris m.
Popliteal vein
Tibial nerve
Semimembranosus m.
Common peroneal nerve
Popliteal artery
Lat. sural cutan. n.
Small saphenous vein
Gastrocnemius m.
Tendon of biceps femoris muscle
Medial sural cutan. n.
Gluteus medius m. (gluteal fascia)
Gluteus minimus m.
Piriformis muscle
Gluteus maximus m.
Sup. gemellus m.
Obturator internus m.
Tensor of fascia lata
Quadratus femoris m.
Inf. gemellus m.
Gluteus medius m.
Sacrotuberous ligament
Obturator internus m.
Greater trochanter
Ischial tuberosity
Trochanteric bursa
Adductor magnus m.
Gluteus maximus m.
Biceps femoris and semitendinosus mm. (common origin)
Gracilis m.
Adductor minimus (part of adductor magnus)
Semitendinosus m.
Adductor magnus m.
Biceps femoris m. (long head)
Vastus lateralis m.
Biceps femoris m. (short head)
Semimembranosus m.
Biceps femoris m.
Popliteal artery
Tendon of semitendinosus m.
Tendon of semimembranosus m.
Gastrocnemius m. (lateral head)
Gastrocnemius m. (medial head)

Fig. 426: Nerves and Vessels of the Posterior Thigh

NOTE: 1) the emergence of the posterior femoral cutaneous nerve below the inferior border of the gluteus maximus muscle and the inferior cluneal nerves reflecting superiorly at that point.

2) the appearance of the major vessels (popliteal artery and vein) and the sciatic nerve (tibial and common peroneal nerves) in the popliteal fossa.

Fig. 427: Gluteal and Posterior Thigh Muscles

NOTE: 1) with the gluteus maximus and medius reflected, observe the underlying gluteus minimus, piriformis, superior gemellus, obturator internus, inferior gemellus and quadratus femoris muscles.

2) the three hamstring muscles: the biceps femoris (long head) and semitendinosus overlying the semimembranosus muscle. These muscles arise from the ischial tuberosity.

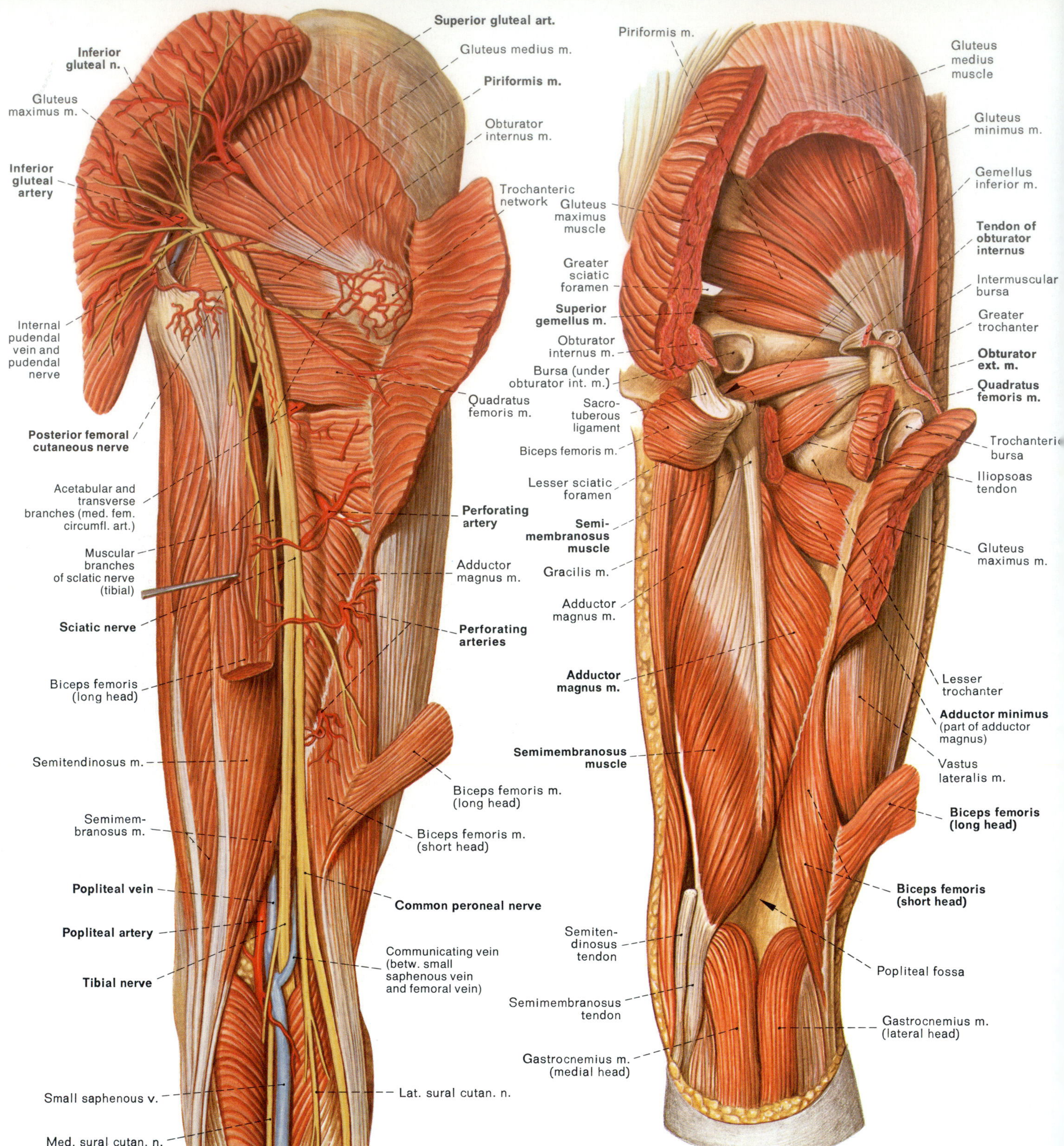
Superior gluteal art.
Inferior gluteal n.
Gluteus medius m.
Gluteus maximus m.
Piriformis m.
Obturator internus m.
Inferior gluteal artery
Trochanteric network
Internal pudendal vein and pudendal nerve
Quadratus femoris m.
Posterior femoral cutaneous nerve
Acetabular and transverse branches (med. fem. circumfl. art.)
Perforating artery
Muscular branches of sciatic nerve (tibial)
Adductor magnus m.
Sciatic nerve
Perforating arteries
Biceps femoris (long head)
Semitendinosus m.
Biceps femoris m. (long head)
Semimembranosus m.
Biceps femoris m. (short head)
Popliteal vein
Common peroneal nerve
Popliteal artery
Communicating vein (betw. small saphenous vein and femoral vein)
Tibial nerve
Small saphenous v.
Lat. sural cutan. n.
Med. sural cutan. n.
Piriformis m.
Gluteus medius muscle
Gluteus minimus m.
Gemellus inferior m.
Gluteus maximus muscle
Tendon of obturator internus
Greater sciatic foramen
Intermuscular bursa
Superior gemellus m.
Greater trochanter
Obturator internus m.
Obturator ext. m.
Bursa (under obturator int. m.)
Quadratus femoris m.
Sacro-tuberous ligament
Trochanteric bursa
Biceps femoris m.
Iliopsoas tendon
Lesser sciatic foramen
Semi-membranosus muscle
Gluteus maximus m.
Gracilis m.
Adductor magnus m.
Adductor magnus m.
Lesser trochanter
Adductor minimus (part of adductor magnus)
Semimembranosus muscle
Vastus lateralis m.
Biceps femoris (long head)
Biceps femoris (short head)
Semiten-dinosus tendon
Popliteal fossa
Semimembranosus tendon
Gastrocnemius m. (lateral head)
Gastrocnemius m. (medial head)

Fig. 428: Deep Nerves and Vessels of the Gluteal Region and Posterior Thigh

NOTE: 1) the course of the sciatic nerve as it passes through the greater sciatic foramen in the gluteal region, inferior to the piriformis muscle, lateral to the ischial tuberosity and under cover of the gluteus maximus muscle.

2) the superior and inferior gluteal arteries and the posterior femoral cutaneous nerve. In the thigh, observe the perforating arteries and the fact that the sciatic nerve splits to become the tibial and common peroneal nerves.

Fig. 429: Deep Muscles of the Gluteal Region and Posterior Thigh

NOTE: 1) in the gluteal region, the quadratus femoris has been reflected, revealing the obturator externus muscle. Also the tendon of the obturator internus muscle (between the gemelli) has been severed.

2) in the thigh, the common tendon of the long head of the biceps femoris and semitendinosus has been cut, thereby exposing the origin of the semimembranosus, the breadth of the adductor magnus, and the short head of the biceps femoris.

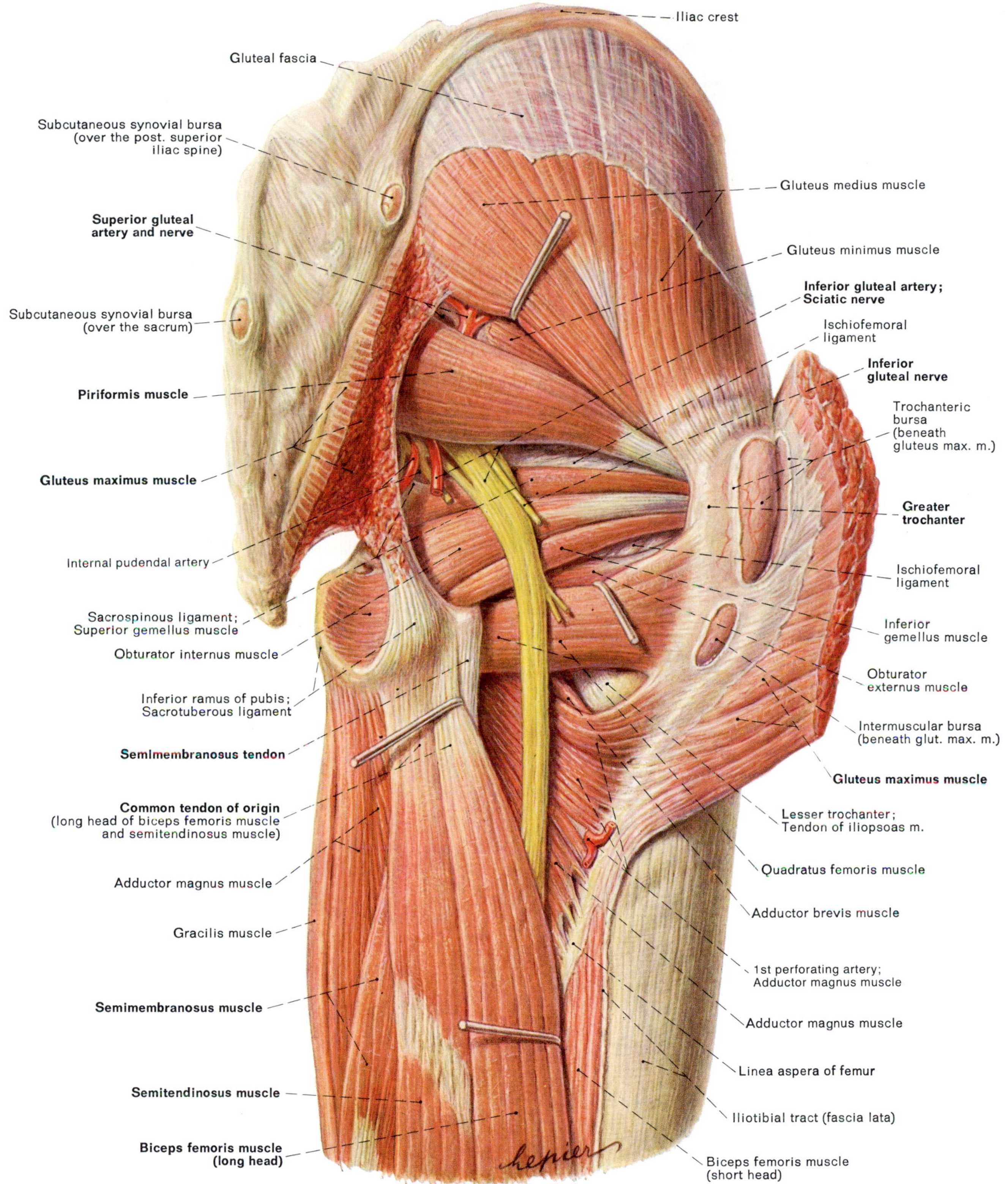
Iliac crest
Gluteal fascia
Subcutaneous synovial bursa
(over the post. superior
iliac spine)
Gluteus medius muscle
Superior gluteal
artery and nerve
Gluteus minimus muscle
Inferior gluteal artery;
Sciatic nerve
Subcutaneous synovial bursa
(over the sacrum)
Ischiofemoral
ligament
Inferior
gluteal nerve
Piriformis muscle
Trochanteric
bursa
(beneath
gluteus max. m.)
Gluteus maximus muscle
Greater
trochanter
Internal pudendal artery
Ischiofemoral
ligament
Sacrospinous ligament;
Superior gemellus muscle
Inferior
gemellus muscle
Obturator internus muscle
Obturator
externus muscle
Inferior ramus of pubis;
Sacrotuberous ligament
Intermuscular bursa
(beneath glut. max. m.)
Semimembranosus tendon
Gluteus maximus muscle
Common tendon of origin
(long head of biceps femoris muscle
and semitendinosus muscle)
Lesser trochanter;
Tendon of iliopsoas m.
Quadratus femoris muscle
Adductor magnus muscle
Adductor brevis muscle
Gracilis muscle
1st perforating artery;
Adductor magnus muscle
Semimembranosus muscle
Adductor magnus muscle
Linea aspera of femur
Semitendinosus muscle
Iliotibial tract (fascia lata)
Biceps femoris muscle
(long head)
Biceps femoris muscle
(short head)

Fig. 430: The Middle and Deep Gluteal Muscles and the Sciatic Nerve

NOTE: 1) with the gluteal muscles exposed by the reflection of the gluteus maximus muscle, one can appreciate that the centrally located piriformis muscle is the key structure in understanding the anatomy of this region. This muscle, as do most of the other structures which leave the pelvis to enter the gluteal region, passes through the greater sciatic foramen. The nerves and vessels which enter the gluteal region from the pelvis are situated either above the piriformis or below it. The important sciatic nerve enters the gluteal region below the piriformis muscle.

2) through their insertions on or around the greater trochanter, many of the gluteal muscles laterally rotate the femur. The most powerful lateral rotator is the gluteus maximus which is also the most powerful extensor of the thigh at the hip joint.

3) the origin of the hamstring muscles from the ischial tuberosity. The semitendinosus and long head of the biceps femoris have a common tendon of origin deep to which arises the semimembranosus muscle.

Quadratus lumborum m.
Internal oblique m.
Transversus abdominis m.
Iliacus m.
Psoas major m.
Piriformis m.
Sartorius m.
Obturator internus m.
Sacrospinous ligament; Coccygeus m.
Iliopectineal bursa
Rectus femoris m.
Pectineus m.
Obturator externus m.
Gluteus minimus m.
Vastus lateralis m.
Vastus intermedius m.
Iliopsoas m.
Vastus medialis m.
Quadratus femoris m.
Semimembranosus m.
Biceps femoris m.
Adductor longus m.
Adductor brevis m.
Gracilis m.
Adductor magnus m.
Articularis genu m.
Vastus medialis m.
Adductor magnus m.
Quadriceps femoris m.
Fibular collateral ligament
Biceps femoris m.
Lateral patellar retinaculum
Sartorius m.
Gracilis m.
Semitendinosus m.
Patellar ligament
Extensor digitorum longus m.
Tibialis anterior m.
Peroneus longus m.

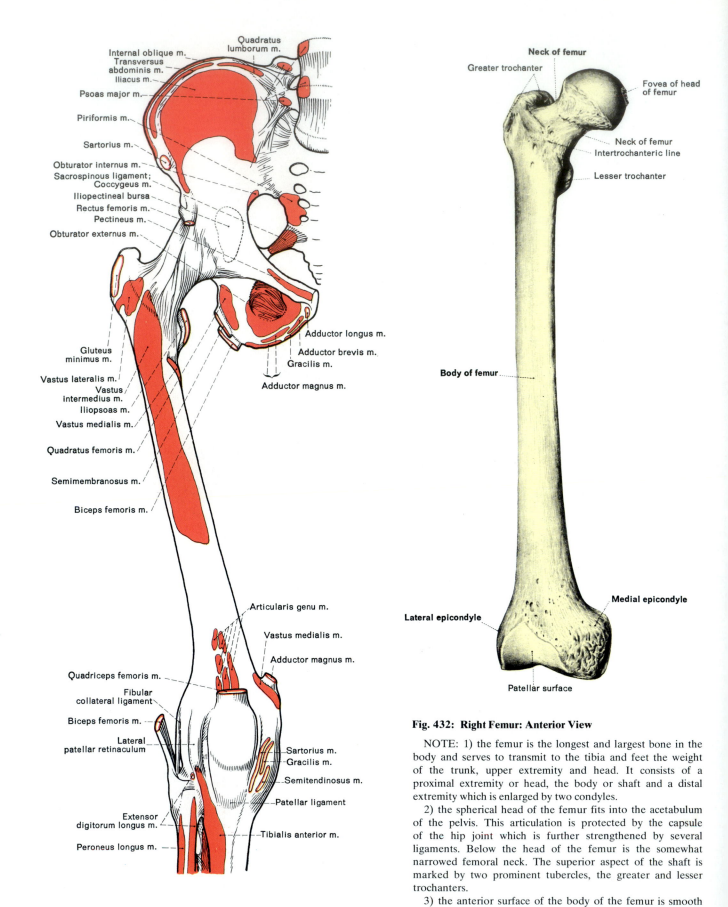

Fig. 431: Anterior View of Right Pelvis and Femur Showing Muscle Attachments

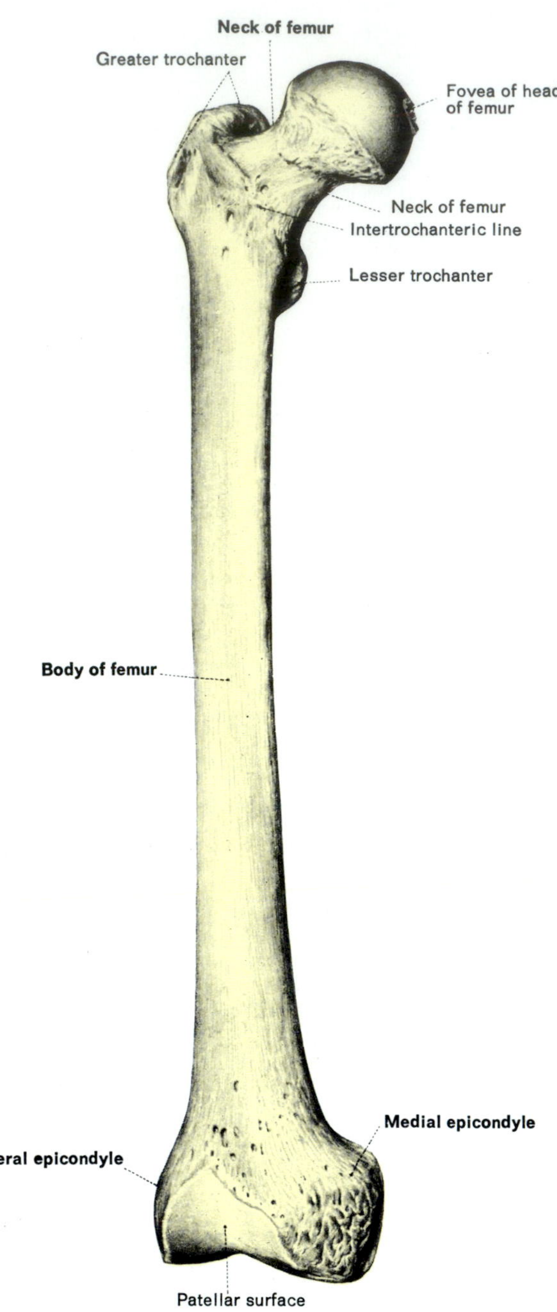

Neck of femur
Greater trochanter
Fovea of head of femur
Neck of femur
Intertrochanteric line
Lesser trochanter
Body of femur
Medial epicondyle
Lateral epicondyle
Patellar surface

Fig. 432: Right Femur: Anterior View

NOTE: 1) the femur is the longest and largest bone in the body and serves to transmit to the tibia and feet the weight of the trunk, upper extremity and head. It consists of a proximal extremity or head, the body or shaft and a distal extremity which is enlarged by two condyles.

2) the spherical head of the femur fits into the acetabulum of the pelvis. This articulation is protected by the capsule of the hip joint which is further strengthened by several ligaments. Below the head of the femur is the somewhat narrowed femoral neck. The superior aspect of the shaft is marked by two prominent tubercles, the greater and lesser trochanters.

3) the anterior surface of the body of the femur is smooth and its proximal 2/3rds gives origin to the vastus intermedius muscle.

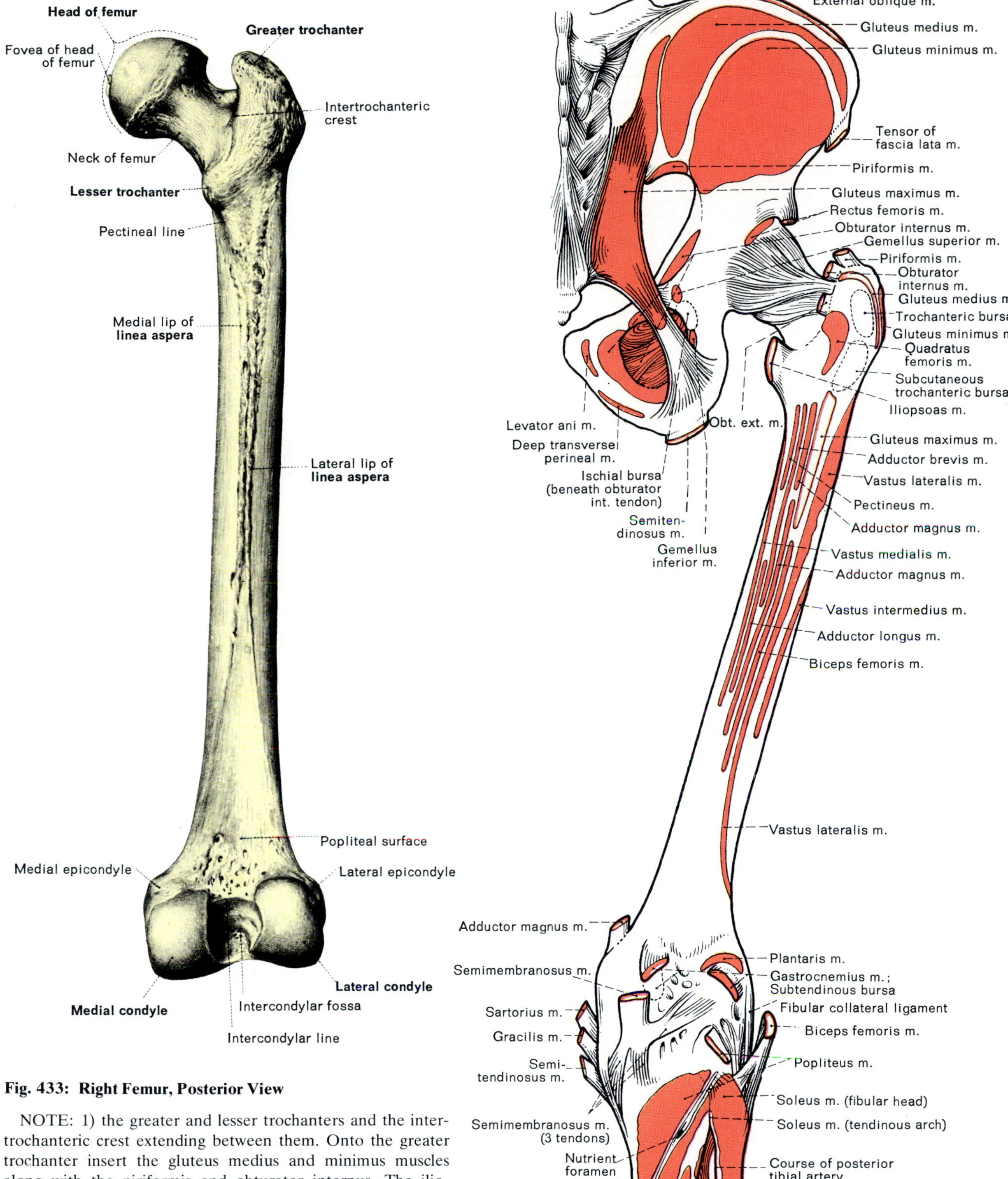
Head of femur
Fovea of head of femur
Greater trochanter
Intertrochanteric crest
Neck of femur
Lesser trochanter
Pectineal line
Medial lip of linea aspera
Lateral lip of linea aspera
Popliteal surface
Medial epicondyle
Lateral epicondyle
Lateral condyle
Medial condyle
Intercondylar fossa
Intercondylar line
External oblique m.
Gluteus medius m.
Gluteus minimus m.
Tensor of fascia lata m.
Piriformis m.
Gluteus maximus m.
Rectus femoris m.
Obturator internus m.
Gemellus superior m.
Piriformis m.
Obturator internus m.
Gluteus medius m.
Trochanteric bursa
Gluteus minimus m.
Quadratus femoris m.
Subcutaneous trochanteric bursa
Iliopsoas m.
Levator ani m.
Obt. ext. m.
Deep transverse perineal m.
Ischial bursa (beneath obturator int. tendon)
Semitendinosus m.
Gemellus inferior m.
Gluteus maximus m.
Adductor brevis m.
Vastus lateralis m.
Pectineus m.
Adductor magnus m.
Vastus medialis m.
Adductor magnus m.
Vastus intermedius m.
Adductor longus m.
Biceps femoris m.
Vastus lateralis m.
Adductor magnus m.
Plantaris m.
Semimembranosus m.
Gastrocnemius m.; Subtendinous bursa
Fibular collateral ligament
Sartorius m.
Biceps femoris m.
Gracilis m.
Semi-tendinosus m.
Popliteus m.
Soleus m. (fibular head)
Semimembranosus m. (3 tendons)
Soleus m. (tendinous arch)
Nutrient foramen
Course of posterior tibial artery
Soleus m. (tibial head)
Tibialis posterior m.
Soleus m. (fibular head)
Flexor digitorum longus m.

Fig. 433: Right Femur, Posterior View

NOTE: 1) the greater and lesser trochanters and the intertrochanteric crest extending between them. Onto the greater trochanter insert the gluteus medius and minimus muscles along with the piriformis and obturator internus. The iliopsoas muscle inserts on the lesser trochanter while the quadratus femoris attaches along the intertrochanteric crest.

2) the prominent, longitudinally oriented ridge, called the linea aspera along the posterior surface of the body of the femur. This likewise serves for muscle attachments.

3) the medial and lateral condyles and epicondyles inferiorly. The condyles articulate with the tibia and the intercondyloid fossa affords attachment to the cruciate ligaments.

Fig. 434: Posterior View of Right Pelvis and Femur Showing Muscle Attachments

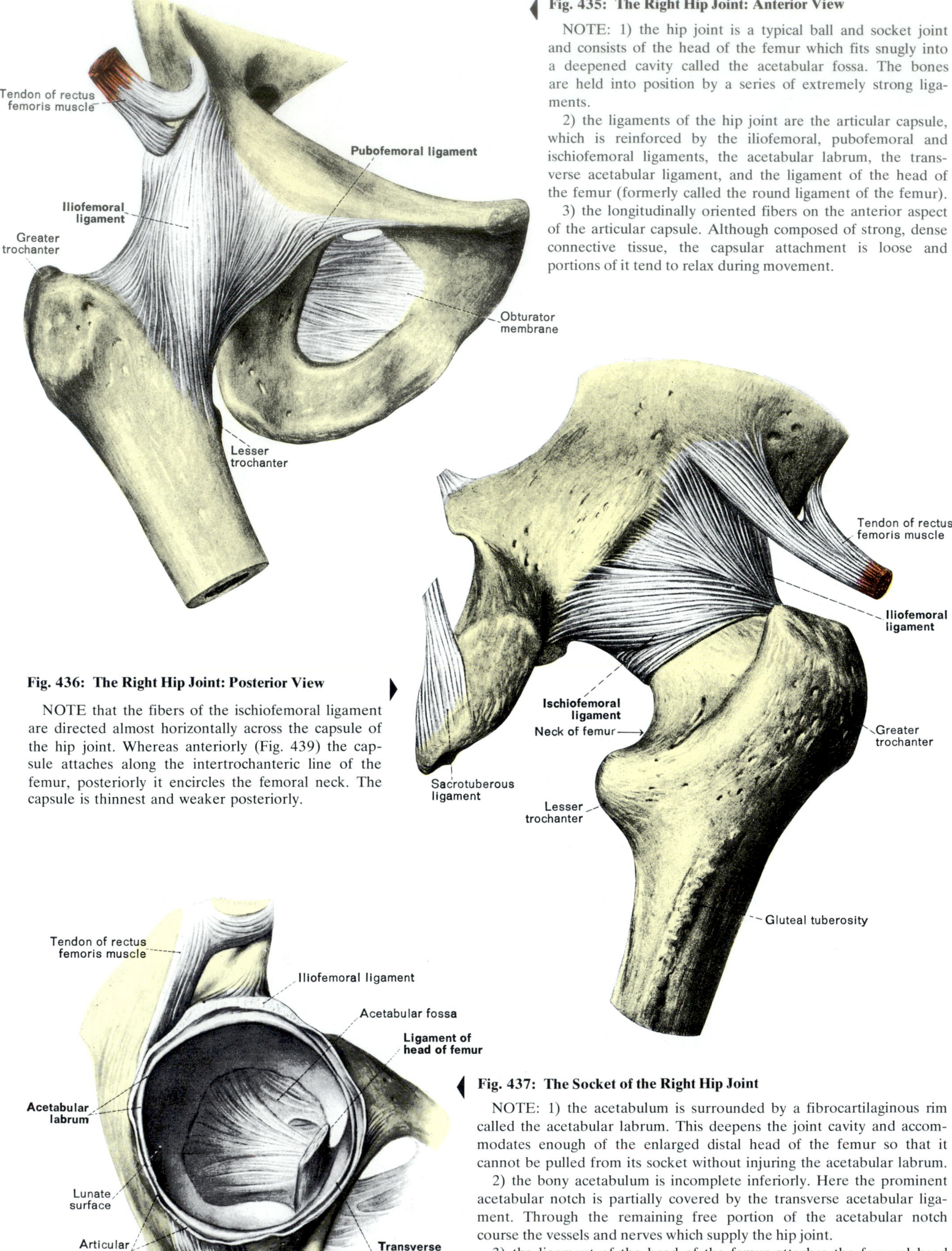
Tendon of rectus femoris muscle
Pubofemoral ligament
Iliofemoral ligament
Greater trochanter
Obturator membrane
Lesser trochanter
Tendon of rectus femoris muscle
Iliofemoral ligament
Ischiofemoral ligament
Neck of femur
Greater trochanter
Sacrotuberous ligament
Lesser trochanter
Gluteal tuberosity
Tendon of rectus femoris muscle
Iliofemoral ligament
Acetabular fossa
Ligament of head of femur
Acetabular labrum
Lunate surface
Articular capsule
Transverse acetabular ligament

Fig. 435: The Right Hip Joint: Anterior View

NOTE: 1) the hip joint is a typical ball and socket joint and consists of the head of the femur which fits snugly into a deepened cavity called the acetabular fossa. The bones are held into position by a series of extremely strong ligaments.

2) the ligaments of the hip joint are the articular capsule, which is reinforced by the iliofemoral, pubofemoral and ischiofemoral ligaments, the acetabular labrum, the transverse acetabular ligament, and the ligament of the head of the femur (formerly called the round ligament of the femur).

3) the longitudinally oriented fibers on the anterior aspect of the articular capsule. Although composed of strong, dense connective tissue, the capsular attachment is loose and portions of it tend to relax during movement.

Fig. 436: The Right Hip Joint: Posterior View

NOTE that the fibers of the ischiofemoral ligament are directed almost horizontally across the capsule of the hip joint. Whereas anteriorly (Fig. 439) the capsule attaches along the intertrochanteric line of the femur, posteriorly it encircles the femoral neck. The capsule is thinnest and weaker posteriorly.

Fig. 437: The Socket of the Right Hip Joint

NOTE: 1) the acetabulum is surrounded by a fibrocartilaginous rim called the acetabular labrum. This deepens the joint cavity and accommodates enough of the enlarged distal head of the femur so that it cannot be pulled from its socket without injuring the acetabular labrum.

2) the bony acetabulum is incomplete inferiorly. Here the prominent acetabular notch is partially covered by the transverse acetabular ligament. Through the remaining free portion of the acetabular notch course the vessels and nerves which supply the hip joint.

3) the ligament of the head of the femur attaches the femoral head by two bands to either side of the acetabular notch.

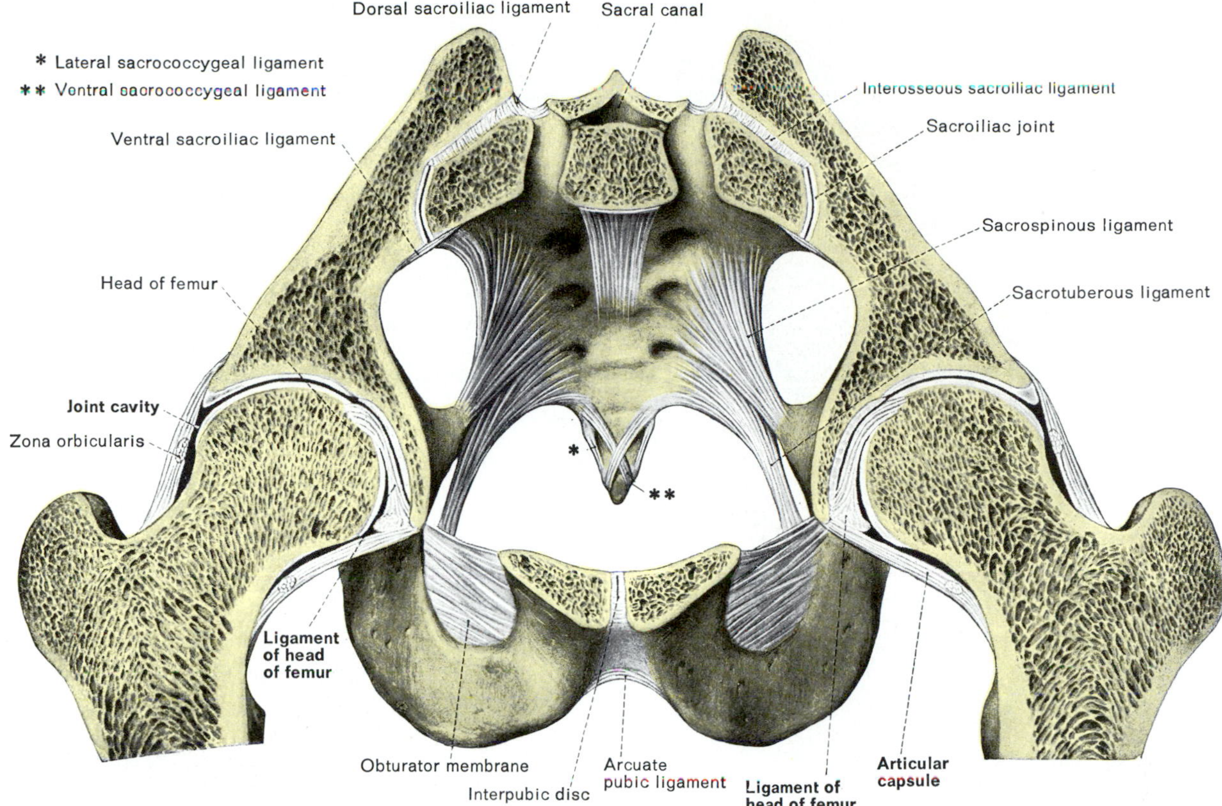
Dorsal sacroiliac ligament
Sacral canal
* Lateral sacrococcygeal ligament
** Ventral sacrococcygeal ligament
Interosseous sacroiliac ligament
Sacroiliac joint
Ventral sacroiliac ligament
Sacrospinous ligament
Head of femur
Sacrotuberous ligament
Joint cavity
Zona orbicularis
*
**
Ligament of head of femur
Obturator membrane
Interpubic disc
Arcuate pubic ligament
Ligament of head of femur
Articular capsule

Fig. 438: Frontal Section of the Pelvis Showing Both Hip Joints

Fig. 439: Anterior Exposure of the Right Hip Joint

NOTE that the anterior aspect of the articular capsule of the hip joint has been opened close to the rounded acetabular labrum. This exposes the cartilage covered head of the femur within the joint cavity. Observe the ligament of the femoral head which is attached to the femur at a site where the cartilage is lacking. This depression is called the fovea of the femoral head.

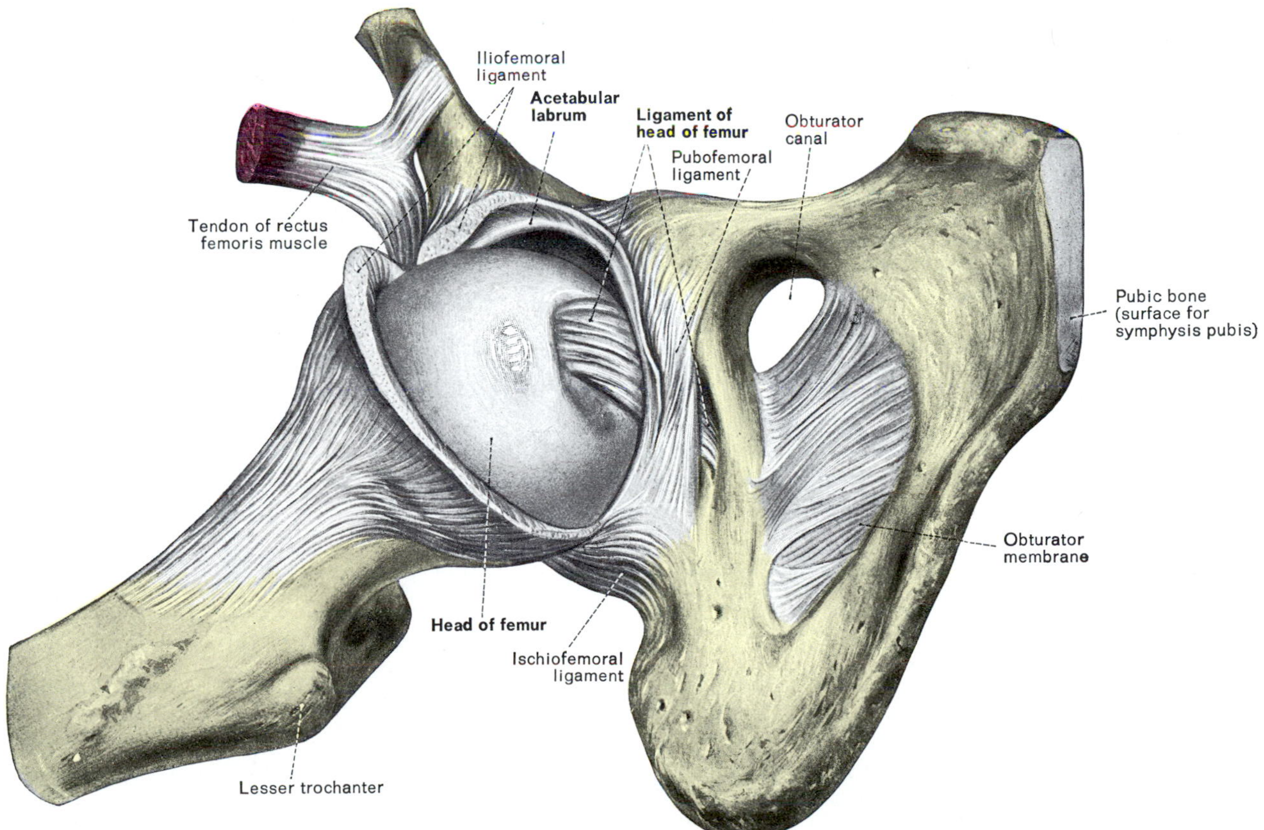
Iliofemoral ligament
Acetabular labrum
Ligament of head of femur
Obturator canal
Pubofemoral ligament
Tendon of rectus femoris muscle
Pubic bone (surface for symphysis pubis)
Obturator membrane
Head of femur
Ischiofemoral ligament
Lesser trochanter

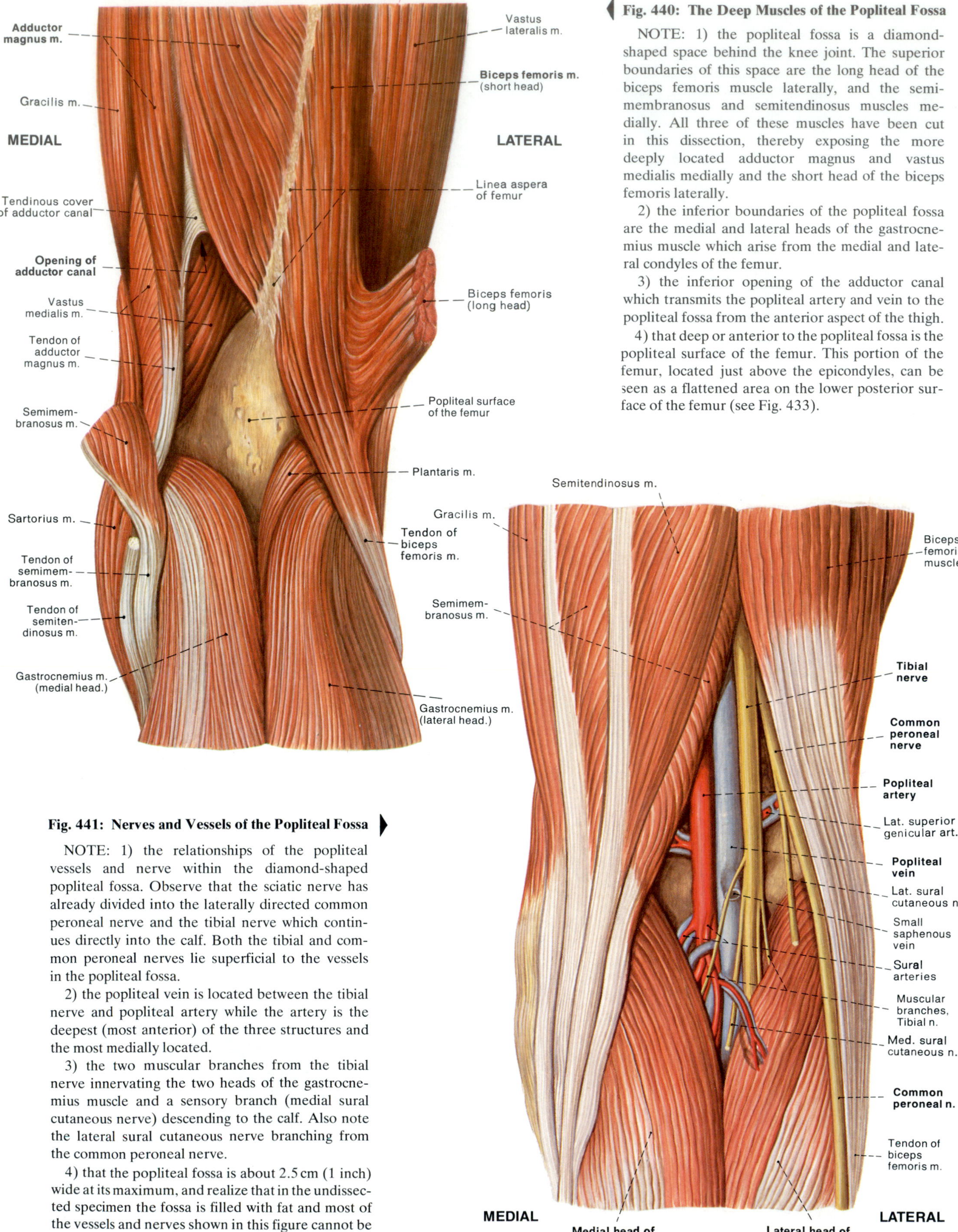

Adductor magnus m.
Gracilis m.
MEDIAL
Tendinous cover of adductor canal
Opening of adductor canal
Vastus medialis m.
Tendon of adductor magnus m.
Semimembranosus m.
Sartorius m.
Tendon of semimembranosus m.
Tendon of semitendinosus m.
Gastrocnemius m. (medial head.)
Vastus lateralis m.
Biceps femoris m. (short head)
LATERAL
Linea aspera of femur
Biceps femoris (long head)
Popliteal surface of the femur
Plantaris m.
Tendon of biceps femoris m.
Gastrocnemius m. (lateral head.)
Semitendinosus m.
Gracilis m.
Semimembranosus m.
Biceps femoris muscle
Tibial nerve
Common peroneal nerve
Popliteal artery
Lat. superior genicular art.
Popliteal vein
Lat. sural cutaneous n.
Small saphenous vein
Sural arteries
Muscular branches, Tibial n.
Med. sural cutaneous n.
Common peroneal n.
Tendon of biceps femoris m.
MEDIAL
LATERAL
Medial head of gastrocnemius m.
Lateral head of gastrocnemius m.

Fig. 440: The Deep Muscles of the Popliteal Fossa

NOTE: 1) the popliteal fossa is a diamond-shaped space behind the knee joint. The superior boundaries of this space are the long head of the biceps femoris muscle laterally, and the semimembranosus and semitendinosus muscles medially. All three of these muscles have been cut in this dissection, thereby exposing the more deeply located adductor magnus and vastus medialis medially and the short head of the biceps femoris laterally.

2) the inferior boundaries of the popliteal fossa are the medial and lateral heads of the gastrocnemius muscle which arise from the medial and lateral condyles of the femur.

3) the inferior opening of the adductor canal which transmits the popliteal artery and vein to the popliteal fossa from the anterior aspect of the thigh.

4) that deep or anterior to the popliteal fossa is the popliteal surface of the femur. This portion of the femur, located just above the epicondyles, can be seen as a flattened area on the lower posterior surface of the femur (see Fig. 433).

Fig. 441: Nerves and Vessels of the Popliteal Fossa

NOTE: 1) the relationships of the popliteal vessels and nerve within the diamond-shaped popliteal fossa. Observe that the sciatic nerve has already divided into the laterally directed common peroneal nerve and the tibial nerve which continues directly into the calf. Both the tibial and common peroneal nerves lie superficial to the vessels in the popliteal fossa.

2) the popliteal vein is located between the tibial nerve and popliteal artery while the artery is the deepest (most anterior) of the three structures and the most medially located.

3) the two muscular branches from the tibial nerve innervating the two heads of the gastrocnemius muscle and a sensory branch (medial sural cutaneous nerve) descending to the calf. Also note the lateral sural cutaneous nerve branching from the common peroneal nerve.

4) that the popliteal fossa is about 2.5 cm (1 inch) wide at its maximum, and realize that in the undissected specimen the fossa is filled with fat and most of the vessels and nerves shown in this figure cannot be seen.

Fig. 442: Arteriogram of Lower Femoral and Popliteal Arteries

NOTE: 1) this arteriogram shows the arterial tree of the lower third of the thigh and the upper third of the calf. Note the course of the femoral artery as it becomes the popliteal artery just above the intercondylar fossa in the popliteal space. The genicular arteries supplying the knee joint, and the sural arteries descending to supply both superficial and muscular tissues in the calf can be seen arising from the popliteal artery.

2) below the popliteal fossa the popliteal artery becomes the posterior tibial artery. Soon, the anterior tibial artery branches from the posterior tibial to course toward the anterior compartment in the leg, while the posterior tibial continues to descend in the posterior compartment. About 3 inches below the knee the peroneal artery arises from the posterior tibial artery and descends in the lateral portion of the deep calf.

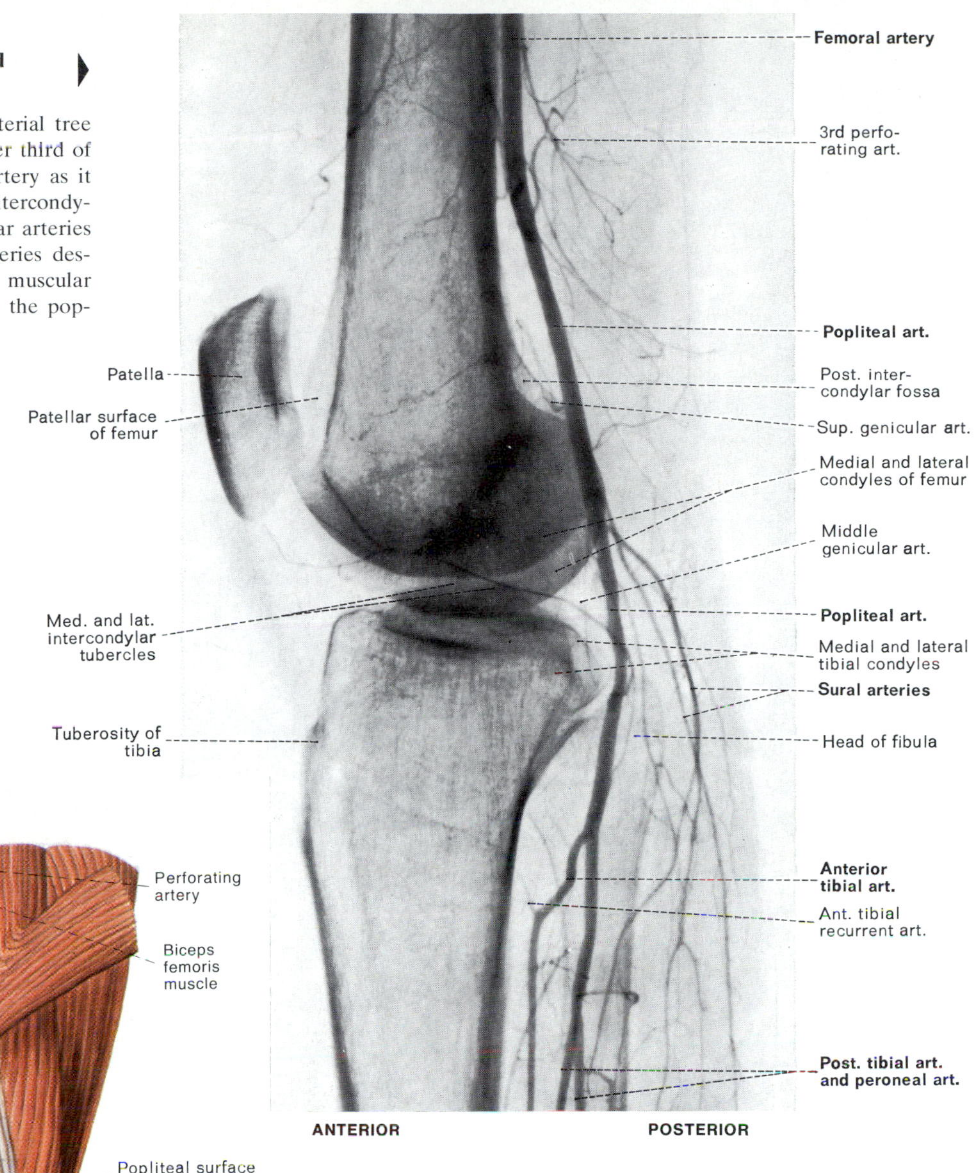
Femoral artery
3rd perforating art.
Popliteal art.
Patella
Patellar surface of femur
Post. intercondylar fossa
Sup. genicular art.
Medial and lateral condyles of femur
Middle genicular art.
Med. and lat. intercondylar tubercles
Popliteal art.
Medial and lateral tibial condyles
Sural arteries
Tuberosity of tibia
Head of fibula
Anterior tibial art.
Ant. tibial recurrent art.
Post. tibial art. and peroneal art.
ANTERIOR
POSTERIOR

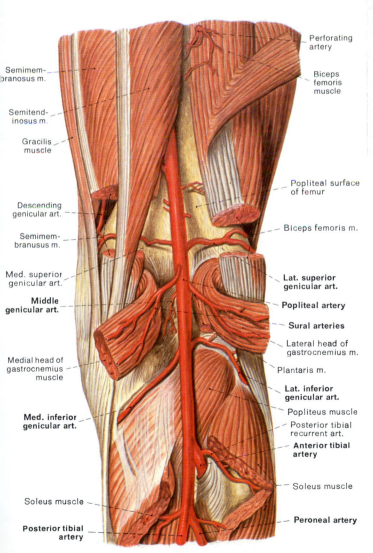

Perforating artery
Semimembranosus m.
Biceps femoris muscle
Semitendinosus m.
Gracilis muscle
Popliteal surface of femur
Descending genicular art.
Biceps femoris m.
Semimembranusus m.
Med. superior genicular art.
Lat. superior genicular art.
Middle genicular art.
Popliteal artery
Sural arteries
Lateral head of gastrocnemius m.
Medial head of gastrocnemius muscle
Plantaris m.
Lat. inferior genicular art.
Med. inferior genicular art.
Popliteus muscle
Posterior tibial recurrent art.
Anterior tibial artery
Soleus muscle
Soleus muscle
Peroneal artery
Posterior tibial artery

Fig. 443: The Branches of the Popliteal Artery

NOTE: 1) within the popliteal fossa, the popliteal artery most frequently gives rise to five genicular arteries. There are usually two superior (lateral and medial), one middle and two inferior (lateral and medial) genicular arteries.

2) the genicular vessels course around the bones comprising the knee joint and supply the joint itself. Further, they form an anastomosis with each other and with descending branches from the femoral artery as well as recurrent vessels from the posterior and anterior tibial arteries.

3) the sural arteries branching from the popliteal to supply the two heads of the gastrocnemius muscle.

4) the anterior tibial artery branching from the posterior tibial and, penetrating an aperture above the interosseous membrane, achieves the anterior compartment. Somewhat lower, the peroneal artery also branches from the posterior tibial.

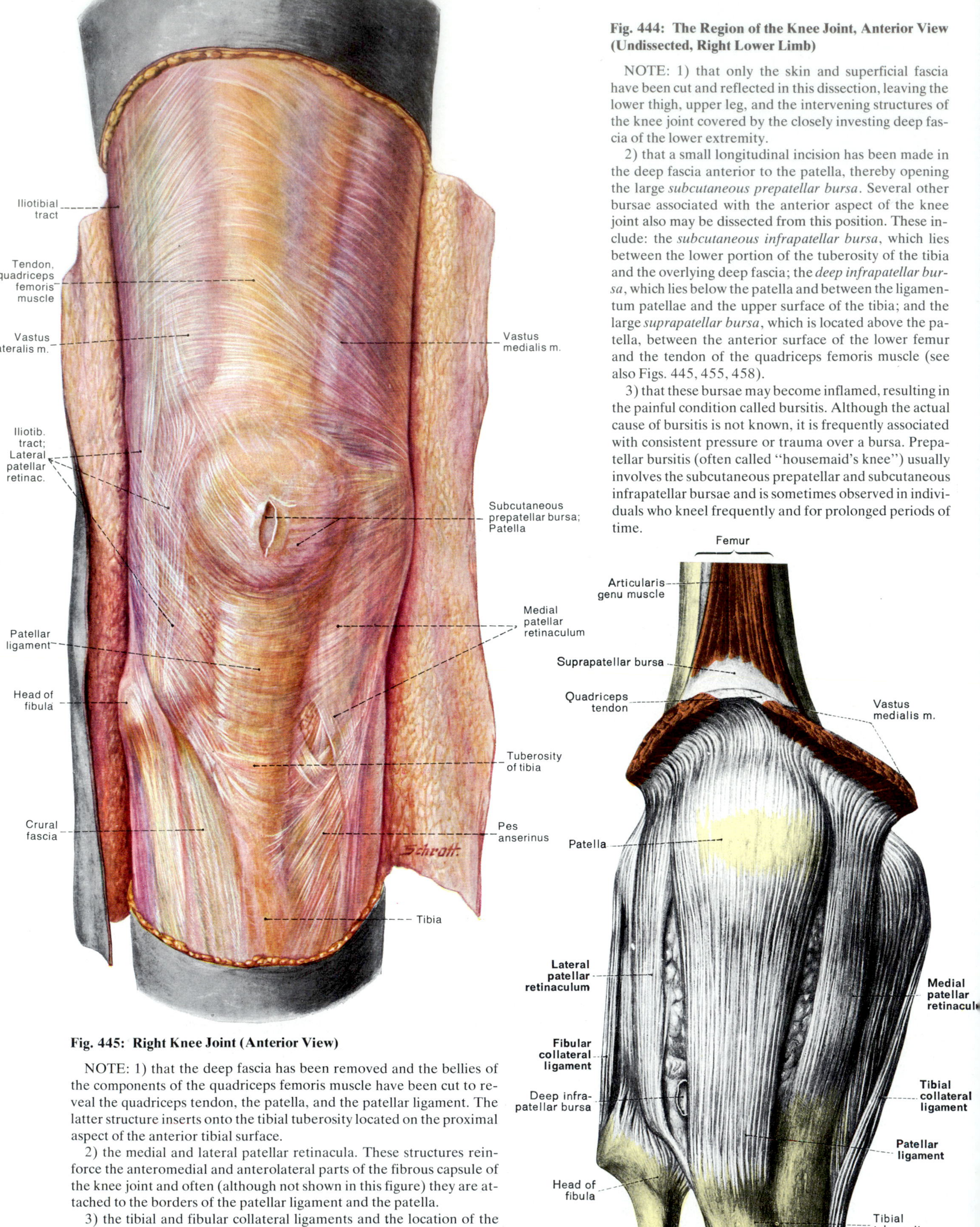

Iliotibial tract
Tendon, quadriceps femoris muscle
Vastus lateralis m.
Vastus medialis m.
Iliotib. tract; Lateral patellar retinac.
Subcutaneous prepatellar bursa; Patella
Medial patellar retinaculum
Patellar ligament
Head of fibula
Tuberosity of tibia
Crural fascia
Pes anserinus
Schrott
Tibia
Femur
Articularis genu muscle
Suprapatellar bursa
Quadriceps tendon
Vastus medialis m.
Patella
Lateral patellar retinaculum
Medial patellar retinacul
Fibular collateral ligament
Deep infrapatellar bursa
Tibial collateral ligament
Patellar ligament
Head of fibula
Tibial tuberosity

Fig. 444: The Region of the Knee Joint, Anterior View (Undissected, Right Lower Limb)

NOTE: 1) that only the skin and superficial fascia have been cut and reflected in this dissection, leaving the lower thigh, upper leg, and the intervening structures of the knee joint covered by the closely investing deep fascia of the lower extremity.

2) that a small longitudinal incision has been made in the deep fascia anterior to the patella, thereby opening the large *subcutaneous prepatellar bursa*. Several other bursae associated with the anterior aspect of the knee joint also may be dissected from this position. These include: the *subcutaneous infrapatellar bursa*, which lies between the lower portion of the tuberosity of the tibia and the overlying deep fascia; the *deep infrapatellar bursa*, which lies below the patella and between the ligamentum patellae and the upper surface of the tibia; and the large *suprapatellar bursa*, which is located above the patella, between the anterior surface of the lower femur and the tendon of the quadriceps femoris muscle (see also Figs. 445, 455, 458).

3) that these bursae may become inflamed, resulting in the painful condition called bursitis. Although the actual cause of bursitis is not known, it is frequently associated with consistent pressure or trauma over a bursa. Prepatellar bursitis (often called "housemaid's knee") usually involves the subcutaneous prepatellar and subcutaneous infrapatellar bursae and is sometimes observed in individuals who kneel frequently and for prolonged periods of time.

Fig. 445: Right Knee Joint (Anterior View)

NOTE: 1) that the deep fascia has been removed and the bellies of the components of the quadriceps femoris muscle have been cut to reveal the quadriceps tendon, the patella, and the patellar ligament. The latter structure inserts onto the tibial tuberosity located on the proximal aspect of the anterior tibial surface.

2) the medial and lateral patellar retinacula. These structures reinforce the anteromedial and anterolateral parts of the fibrous capsule of the knee joint and often (although not shown in this figure) they are attached to the borders of the patellar ligament and the patella.

3) the tibial and fibular collateral ligaments and the location of the deep infrapatellar bursa.

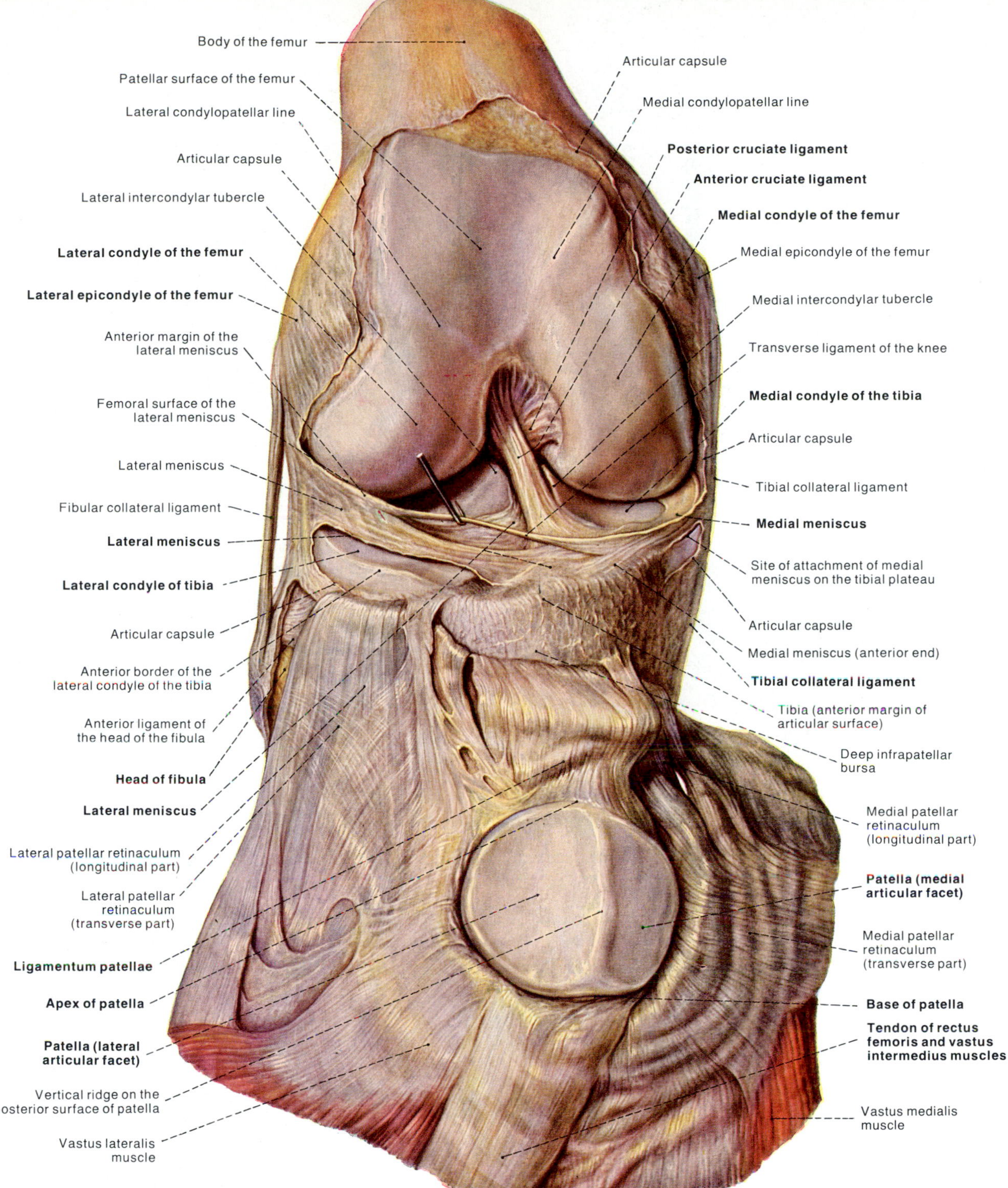
Body of the femur
Patellar surface of the femur
Lateral condylopatellar line
Articular capsule
Lateral intercondylar tubercle
Lateral condyle of the femur
Lateral epicondyle of the femur
Anterior margin of the lateral meniscus
Femoral surface of the lateral meniscus
Lateral meniscus
Fibular collateral ligament
Lateral meniscus
Lateral condyle of tibia
Articular capsule
Anterior border of the lateral condyle of the tibia
Anterior ligament of the head of the fibula
Head of fibula
Lateral meniscus
Lateral patellar retinaculum (longitudinal part)
Lateral patellar retinaculum (transverse part)
Ligamentum patellae
Apex of patella
Patella (lateral articular facet)
Vertical ridge on the posterior surface of patella
Vastus lateralis muscle
Articular capsule
Medial condylopatellar line
Posterior cruciate ligament
Anterior cruciate ligament
Medial condyle of the femur
Medial epicondyle of the femur
Medial intercondylar tubercle
Transverse ligament of the knee
Medial condyle of the tibia
Articular capsule
Tibial collateral ligament
Medial meniscus
Site of attachment of medial meniscus on the tibial plateau
Articular capsule
Medial meniscus (anterior end)
Tibial collateral ligament
Tibia (anterior margin of articular surface)
Deep infrapatellar bursa
Medial patellar retinaculum (longitudinal part)
Patella (medial articular facet)
Medial patellar retinaculum (transverse part)
Base of patella
Tendon of rectus femoris and vastus intermedius muscles
Vastus medialis muscle

Fig. 446: Anterior View of the Right Knee Joint, Opened from the Front to Expose the Ligaments and Menisci

NOTE: 1) that the bellies and tendons of the quadriceps muscle have been severed and reflected downward to expose the inner (posterior) surface of the patella which is attached (a) to the joined tendons of the rectus femoris and vastus medialis superiorly and (b) to the patellar ligament inferiorly.

2) that the knee joint is a bicondylar synovial joint with a single joint cavity between the condyles of the femur and the condyles of the tibia.

3) within the notch between the femoral condyles are attached the posterior and anterior cruciate ligaments. The *posterior cruciate ligament* can be seen to be fixed to the medial condyle of the femur, while the *anterior cruciate ligament* is attached to the lateral condyle of the femur deep in the intercondylar notch. The attachment of the anterior cruciate ligament onto the anterior part of the tibial plateau is well shown in this figure, and the tibial attachment of the posterior cruciate ligament can be seen in Figures 448 and 453. These ligaments are essential for anterior-posterior stability of the knee joint.

4) that the anterior cruciate ligament is best dissected from this frontal approach. Although this ligament helps to prevent posterior displacement of the femur on the upper tibial surface, more importantly, as the ligament becomes tense, it serves to limit the extension (produced by the quadriceps muscle) of the lateral femoral condyle. Further extension beyond this limit places the knee joint in a locked position. This additional extension can be achieved at the knee joint because (a) the medial femoral condyle has a longer articular surface and a greater curvature than the lateral condyle and (b) after the anterior cruciate ligament becomes taut, the lateral condyle can rotate around „the radius of the ligament. This forces the medial condyle to glide backwards into its own full extension" (Last, R. J., *Anatomy, Regional and Applied*, Churchill Livingstone, Edinburgh, 1978, p. 164).

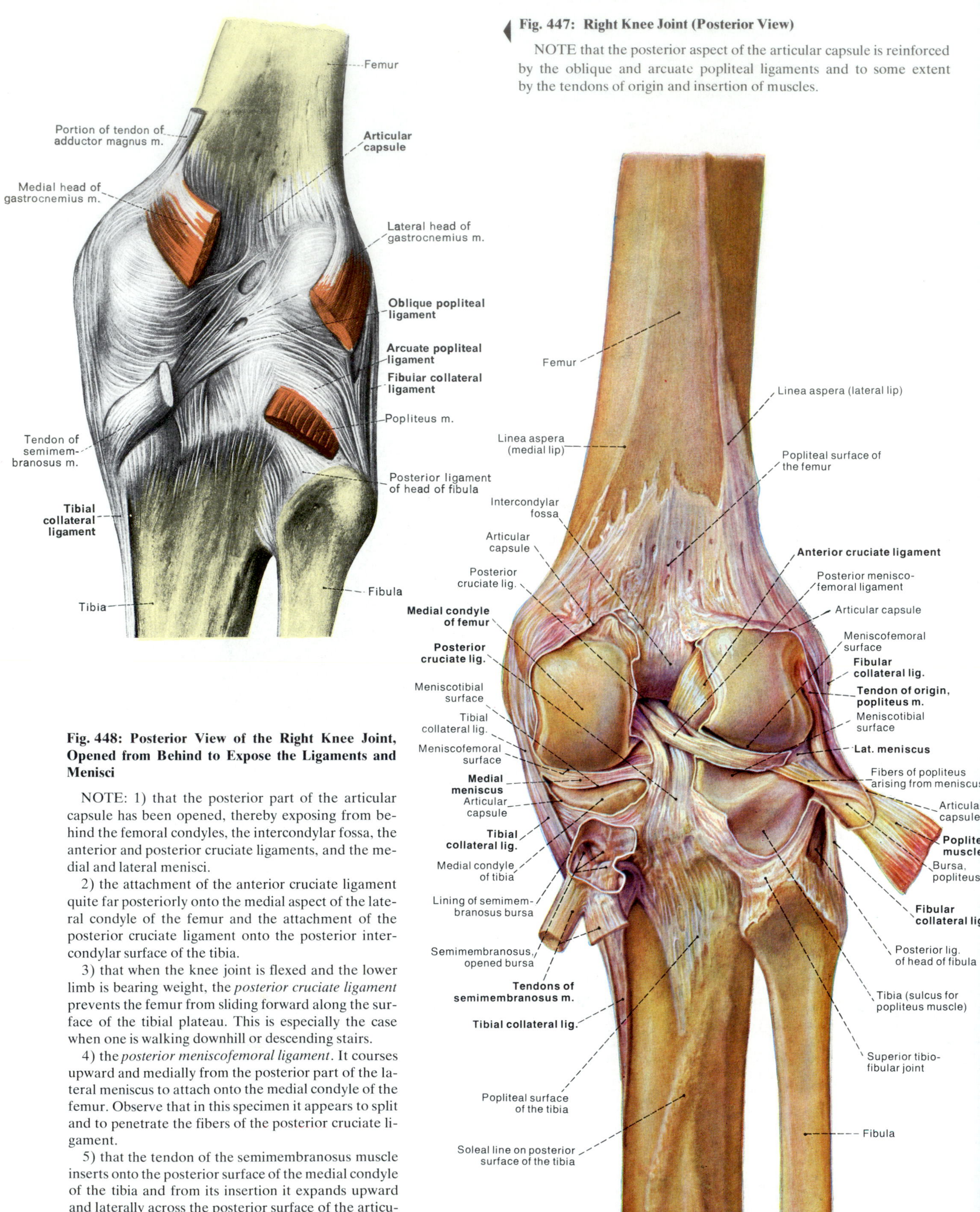

Femur
Portion of tendon of adductor magnus m.
Articular capsule
Medial head of gastrocnemius m.
Lateral head of gastrocnemius m.
Oblique popliteal ligament
Arcuate popliteal ligament
Fibular collateral ligament
Popliteus m.
Tendon of semimembranosus m.
Posterior ligament of head of fibula
Tibial collateral ligament
Fibula
Tibia
Femur
Linea aspera (lateral lip)
Linea aspera (medial lip)
Popliteal surface of the femur
Intercondylar fossa
Articular capsule
Anterior cruciate ligament
Posterior cruciate lig.
Posterior menisco-femoral ligament
Medial condyle of femur
Articular capsule
Posterior cruciate lig.
Meniscofemoral surface
Fibular collateral lig.
Meniscotibial surface
Tendon of origin, popliteus m.
Tibial collateral lig.
Meniscotibial surface
Meniscofemoral surface
Lat. meniscus
Medial meniscus
Fibers of popliteus arising from meniscus
Articular capsule
Articular capsule
Tibial collateral lig.
Poplite muscle
Medial condyle of tibia
Bursa, popliteus
Lining of semimem-branosus bursa
Fibular collateral lig
Semimembranosus, opened bursa
Posterior lig. of head of fibula
Tendons of semimembranosus m.
Tibia (sulcus for popliteus muscle)
Tibial collateral lig.
Superior tibio-fibular joint
Popliteal surface of the tibia
Fibula
Soleal line on posterior surface of the tibia

Fig. 447: Right Knee Joint (Posterior View)

NOTE that the posterior aspect of the articular capsule is reinforced by the oblique and arcuate popliteal ligaments and to some extent by the tendons of origin and insertion of muscles.

Fig. 448: Posterior View of the Right Knee Joint, Opened from Behind to Expose the Ligaments and Menisci

NOTE: 1) that the posterior part of the articular capsule has been opened, thereby exposing from behind the femoral condyles, the intercondylar fossa, the anterior and posterior cruciate ligaments, and the medial and lateral menisci.

2) the attachment of the anterior cruciate ligament quite far posteriorly onto the medial aspect of the lateral condyle of the femur and the attachment of the posterior cruciate ligament onto the posterior intercondylar surface of the tibia.

3) that when the knee joint is flexed and the lower limb is bearing weight, the *posterior cruciate ligament* prevents the femur from sliding forward along the surface of the tibial plateau. This is especially the case when one is walking downhill or descending stairs.

4) the *posterior meniscofemoral ligament*. It courses upward and medially from the posterior part of the lateral meniscus to attach onto the medial condyle of the femur. Observe that in this specimen it appears to split and to penetrate the fibers of the posterior cruciate ligament.

5) that the tendon of the semimembranosus muscle inserts onto the posterior surface of the medial condyle of the tibia and from its insertion it expands upward and laterally across the posterior surface of the articular capsule of the knee joint as the oblique popliteal ligament (see Fig. 447).

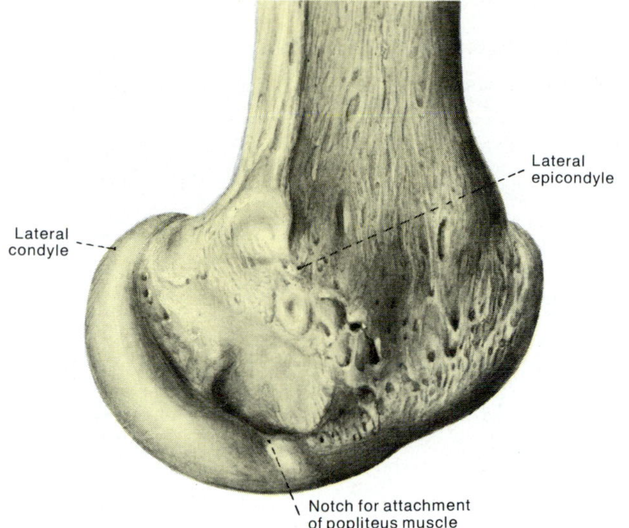
Lateral epicondyle
Lateral condyle
Notch for attachment of popliteus muscle

Fig. 449: The Distal End of the Right Femur, Lateral View

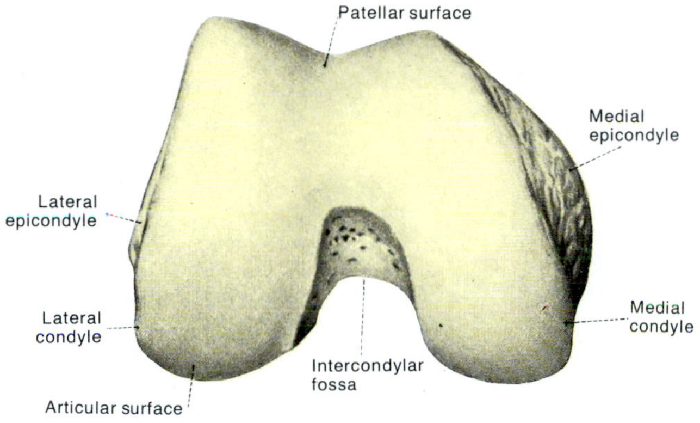
Patellar surface
Medial epicondyle
Lateral epicondyle
Lateral condyle
Medial condyle
Intercondylar fossa
Articular surface

Fig. 450: The Distal End of the Right Femur, Viewed from Below

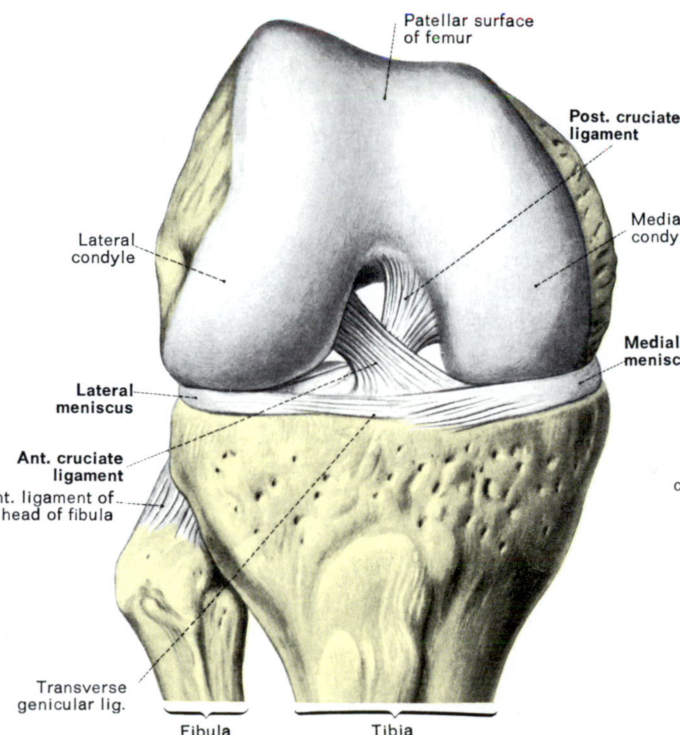
Patellar surface of femur
Post. cruciate ligament
Medial condyle
Lateral condyle
Medial meniscus
Lateral meniscus
Ant. cruciate ligament
nt. ligament of head of fibula
Transverse genicular lig.
Fibula
Tibia

Fig. 451 Flexed Right Knee Joint (Anterior View) Showing Cruciate Ligaments

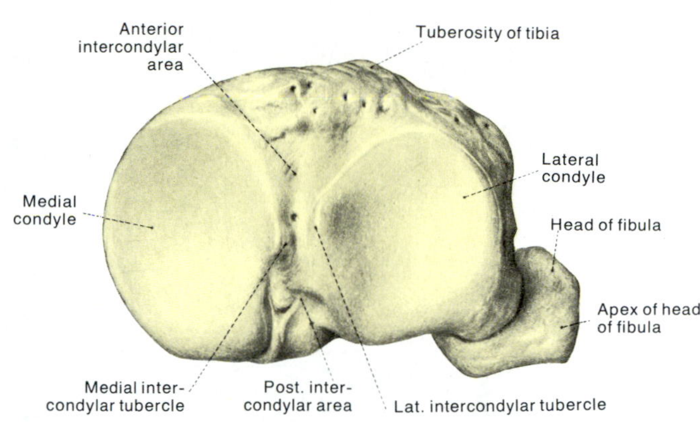
Anterior intercondylar area
Tuberosity of tibia
Lateral condyle
Medial condyle
Head of fibula
Apex of head of fibula
Medial inter-condylar tubercle
Post. inter-condylar area
Lat. intercondylar tubercle

Fig. 452: The Proximal Ends of the Right Tibia and Fibula, Viewed from Above

NOTE that the tibial plateau consists of two concave surfaces, the lateral and medial tibial condyles, on which rest the menisci and the femoral condyles. Between the tibial condyles is the intercondylar area onto which attach the cruciate ligaments and the menisci.

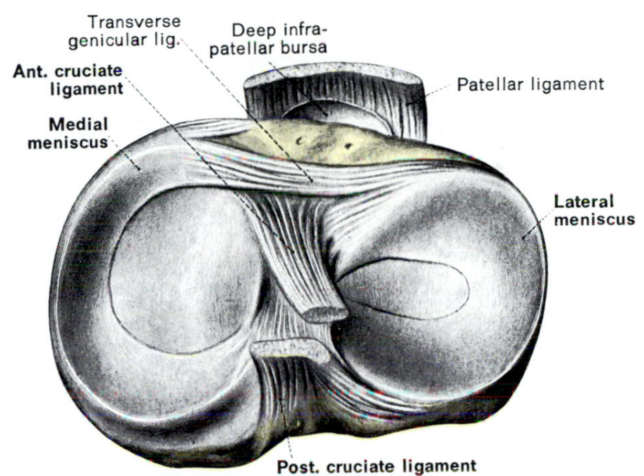
Transverse genicular lig.
Deep infra-patellar bursa
Ant. cruciate ligament
Patellar ligament
Medial meniscus
Lateral meniscus
Post. cruciate ligament

Fig. 453: The Condyles of the Right Tibia (from Above), Showing the Menisci and the Origins of the Cruciate Ligaments

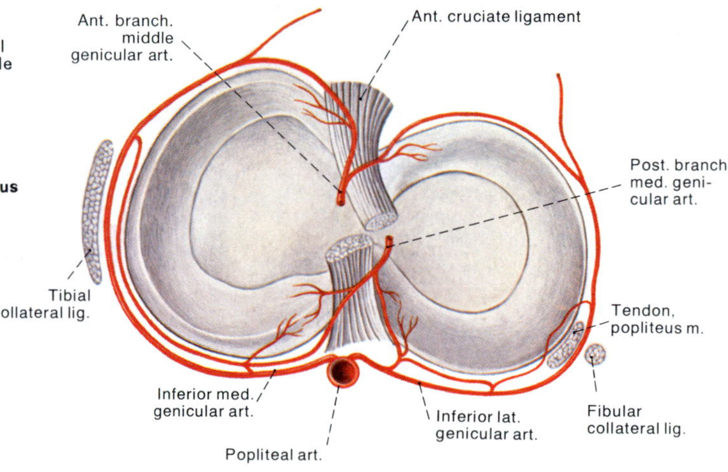
Ant. branch. middle genicular art.
Ant. cruciate ligament
Post. branch, med. genicular art.
Tibial collateral lig.
Tendon, popliteus m.
Inferior med. genicular art.
Inferior lat. genicular art.
Fibular collateral lig.
Popliteal art.

Fig. 454: The Arterial Supply of the Menisci, Right Knee

NOTE that the medial and lateral *inferior* genicular arteries encircle the tibia and supply the menisci. The *middle* genicular (not shown in this figure) supplies the cruciate ligaments. (From Sick, H. and Kortiké, J. G., Z. Anat. Entwickl.-Gesch., *129*: 359–379, 1969.)

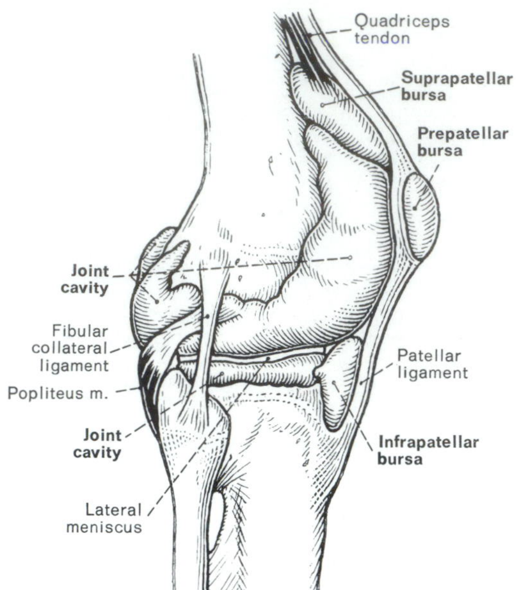
Quadriceps tendon
Suprapatellar bursa
Prepatellar bursa
Joint cavity
Fibular collateral ligament
Popliteus m.
Joint cavity
Lateral meniscus
Patellar ligament
Infrapatellar bursa

Fig. 455: Cast of Knee Joint (Distended) Showing Bursae and Joint Cavity

NOTE: 1) that this lateral view of the distended synovial cavity of the right knee joint demonstrates well that the synovial membrane of this joint is the most extensive of any in the body.

2) that the synovial membrane extends superiorly several centimeters above the patella to form a large pouch called the suprapatellar bursa. Laterally it courses deep to the popliteus tendon and fibular collateral ligament, while posteriorly it extends above the menisci as high as the origins of the heads of the gastrocnemius muscle. Inferiorly, the joint cavity is seen to descend below the lateral meniscus (and medial meniscus, see Fig. 456).

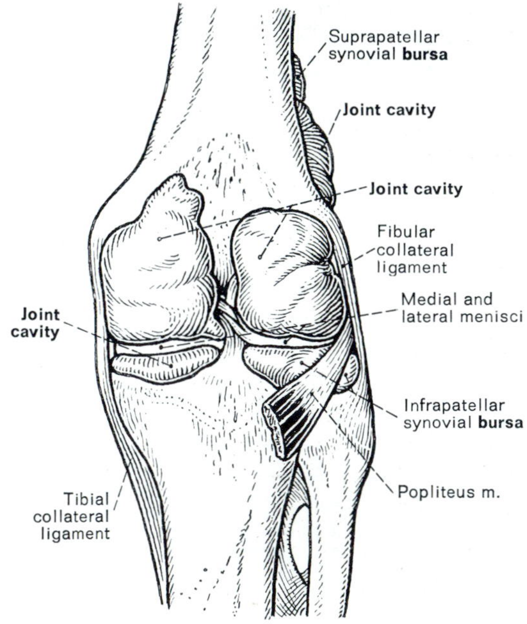
Suprapatellar synovial bursa
Joint cavity
Joint cavity
Fibular collateral ligament
Medial and lateral menisci
Joint cavity
Infrapatellar synovial bursa
Popliteus m.
Tibial collateral ligament

Fig. 456: Cast of Knee Joint Showing Bursae and Joint Cavity (Posterior View)

NOTE that in this posterior dissection of the right knee joint, the fibrous articular capsule has been removed to expose the joint cavity. The synovial membrane extends above the menisci deep to the heads of the gastrocnemius muscle, and below the menisci deep to the popliteus muscle laterally and the semimembranosus medially.

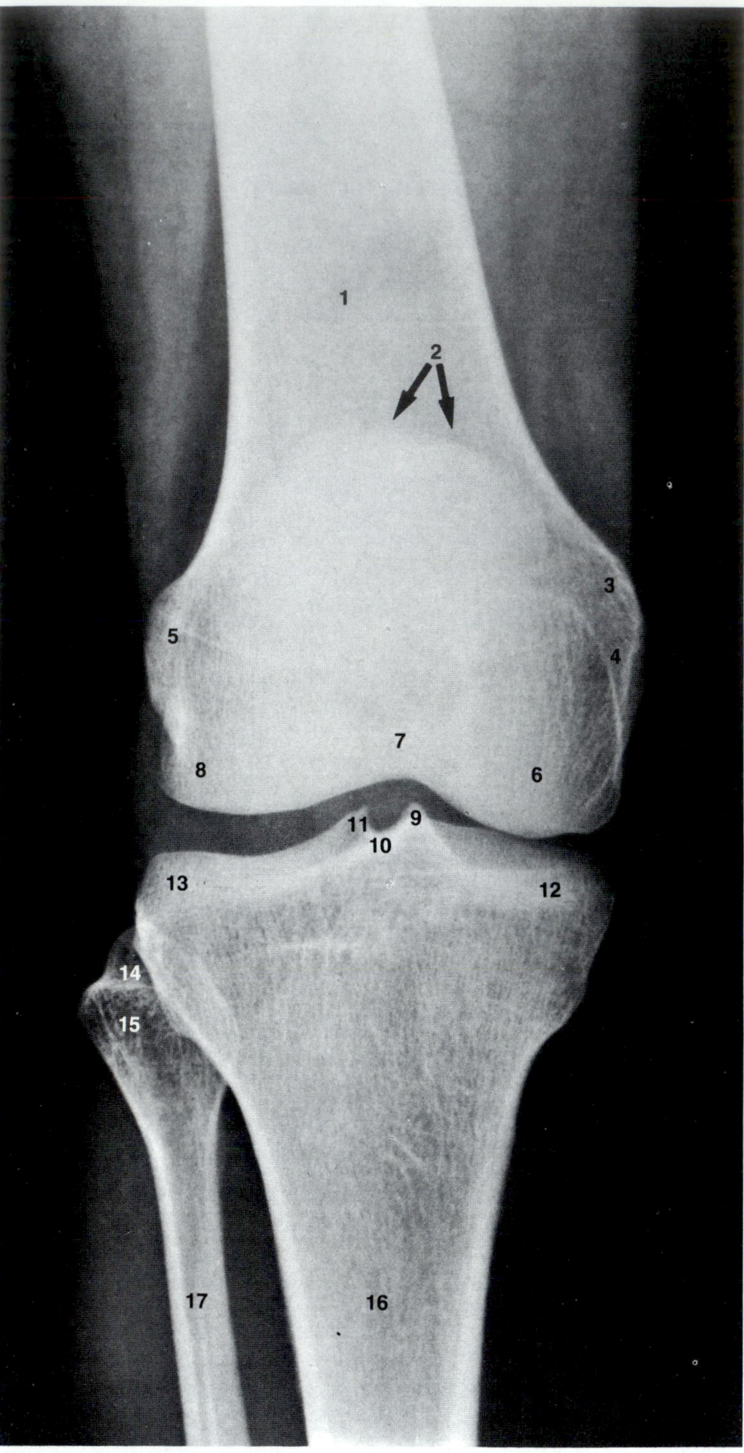
1
2
3
4
5
6
7
8
9
10
11
12
13
14
15
16
17

Fig. 457: Radiograph of the Right Knee (Ant.-Post. Projection)

NOTE the following bony structures on the femur, tibia, and fibula in the region of the knee.

1. Body of femur
2. Margin of patella
3. Adductor tubercle
4. Medial epicondyle
5. Lateral epicondyle
6. Medial condyle of femur
7. Intercondylar fossa
8. Lateral condyle of femur
9. Med. intercondylar tubercle
10. Ant. intercondylar area
11. Lat. intercondylar tubercle
12. Med. condyle of tibia
13. Lat. condyle of tibia
14. Apex of head of fibula
15. Head of fibula
16. Body of tibia
17. Body of fibula

This radiograph comes from Wicke, L., *Atlas of Radiologic Anatomy*, 3rd Edition, Urban and Schwarzenberg, Baltimore, 1982.

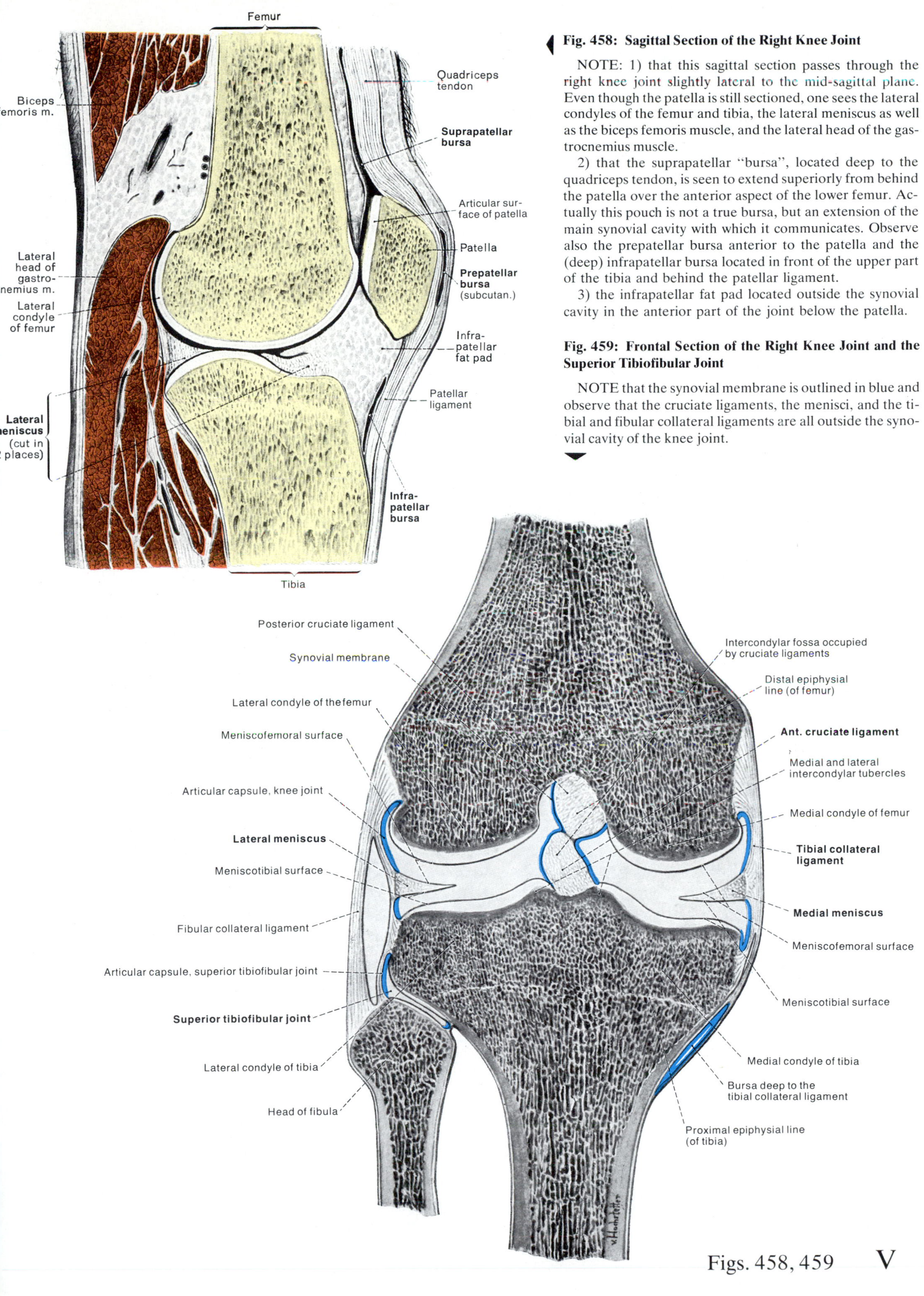

Femur
Biceps femoris m.
Quadriceps tendon
Suprapatellar bursa
Articular surface of patella
Patella
Lateral head of gastrocnemius m.
Prepatellar bursa (subcutan.)
Lateral condyle of femur
Infrapatellar fat pad
Patellar ligament
Lateral meniscus (cut in 2 places)
Infrapatellar bursa
Tibia
Posterior cruciate ligament
Synovial membrane
Lateral condyle of the femur
Meniscofemoral surface
Articular capsule, knee joint
Lateral meniscus
Meniscotibial surface
Fibular collateral ligament
Articular capsule, superior tibiofibular joint
Superior tibiofibular joint
Lateral condyle of tibia
Head of fibula
Intercondylar fossa occupied by cruciate ligaments
Distal epiphysial line (of femur)
Ant. cruciate ligament
Medial and lateral intercondylar tubercles
Medial condyle of femur
Tibial collateral ligament
Medial meniscus
Meniscofemoral surface
Meniscotibial surface
Medial condyle of tibia
Bursa deep to the tibial collateral ligament
Proximal epiphysial line (of tibia)

Fig. 458: Sagittal Section of the Right Knee Joint

NOTE: 1) that this sagittal section passes through the right knee joint slightly lateral to the mid-sagittal plane. Even though the patella is still sectioned, one sees the lateral condyles of the femur and tibia, the lateral meniscus as well as the biceps femoris muscle, and the lateral head of the gastrocnemius muscle.

2) that the suprapatellar "bursa", located deep to the quadriceps tendon, is seen to extend superiorly from behind the patella over the anterior aspect of the lower femur. Actually this pouch is not a true bursa, but an extension of the main synovial cavity with which it communicates. Observe also the prepatellar bursa anterior to the patella and the (deep) infrapatellar bursa located in front of the upper part of the tibia and behind the patellar ligament.

3) the infrapatellar fat pad located outside the synovial cavity in the anterior part of the joint below the patella.

Fig. 459: Frontal Section of the Right Knee Joint and the Superior Tibiofibular Joint

NOTE that the synovial membrane is outlined in blue and observe that the cruciate ligaments, the menisci, and the tibial and fibular collateral ligaments are all outside the synovial cavity of the knee joint.

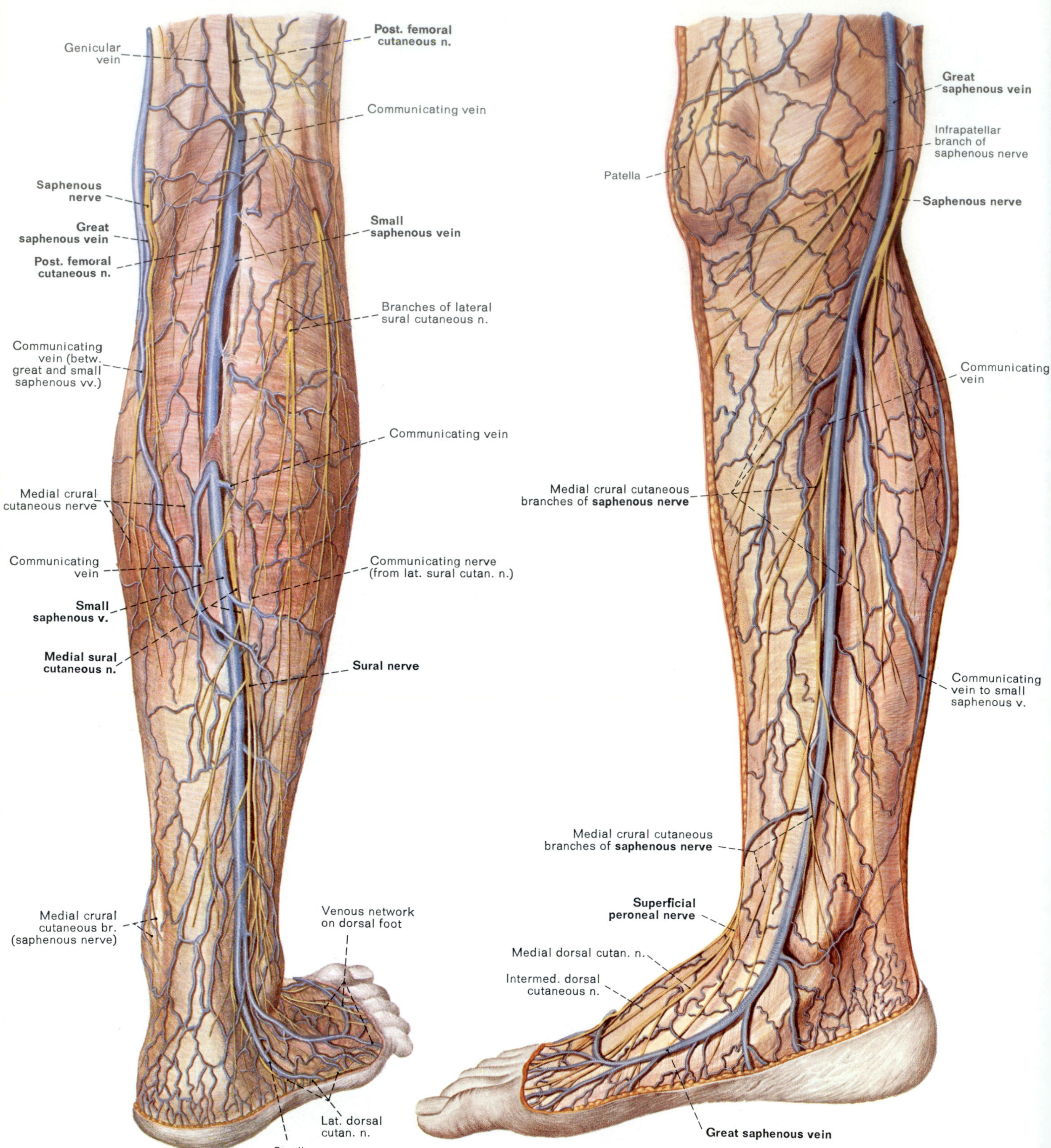
Genicular vein
Post. femoral cutaneous n.
Communicating vein
Saphenous nerve
Great saphenous vein
Post. femoral cutaneous n.
Small saphenous vein
Branches of lateral sural cutaneous n.
Communicating vein (betw. great and small saphenous vv.)
Communicating vein
Medial crural cutaneous nerve
Communicating vein
Communicating nerve (from lat. sural cutan. n.)
Small saphenous v.
Medial sural cutaneous n.
Sural nerve
Medial crural cutaneous br. (saphenous nerve)
Venous network on dorsal foot
Lat. dorsal cutan. n.
Small saphenous vein
Great saphenous vein
Infrapatellar branch of saphenous nerve
Patella
Saphenous nerve
Communicating vein
Medial crural cutaneous branches of saphenous nerve
Communicating vein to small saphenous v.
Medial crural cutaneous branches of saphenous nerve
Superficial peroneal nerve
Medial dorsal cutan. n.
Intermed. dorsal cutaneous n.
Great saphenous vein

Fig. 460: Superficial Veins and Nerves of the Posterior Leg and Foot

NOTE the small saphenous vein which forms on the dorsolateral aspect of the foot and ascends to the popliteal fossa, and the sural nerve which is formed by the junction of the medial sural cutaneous nerve and a communicating branch from the lateral sural cutaneous nerve.

Fig. 461: Superficial Veins and Nerves on the Medial Aspect of the Leg and Foot

NOTE the formation of the great saphenous vein on the medial aspect of the foot and its course anterior to the medial malleolus and up the medial aspect of the leg. Branches of the saphenous nerve accompany the great saphenous vein below the knee. The saphenous nerve is the largest branch of the femoral nerve.

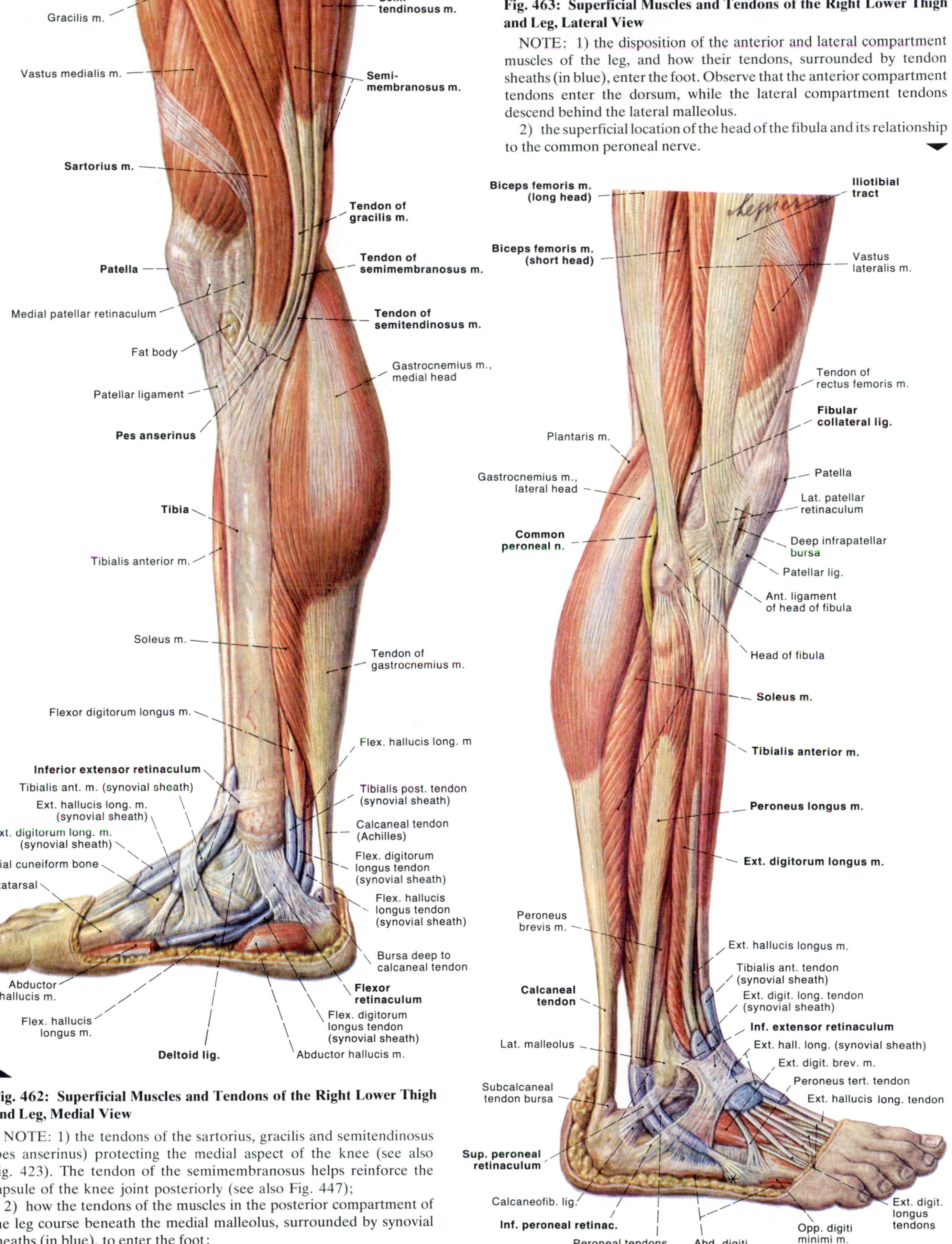
Gracilis m.
Semi-tendinosus m.
Vastus medialis m.
Semi-membranosus m.
Sartorius m.
Tendon of gracilis m.
Patella
Tendon of semimembranosus m.
Medial patellar retinaculum
Tendon of semitendinosus m.
Fat body
Gastrocnemius m., medial head
Patellar ligament
Pes anserinus
Tibia
Tibialis anterior m.
Soleus m.
Tendon of gastrocnemius m.
Flexor digitorum longus m.
Flex. hallucis long. m
Inferior extensor retinaculum
Tibialis ant. m. (synovial sheath)
Tibialis post. tendon (synovial sheath)
Ext. hallucis long. m. (synovial sheath)
Calcaneal tendon (Achilles)
Ext. digitorum long. m. (synovial sheath)
Flex. digitorum longus tendon (synovial sheath)
Medial cuneiform bone
1st metatarsal
Flex. hallucis longus tendon (synovial sheath)
Bursa deep to calcaneal tendon
Abductor hallucis m.
Flexor retinaculum
Flex. hallucis longus m.
Flex. digitorum longus tendon (synovial sheath)
Deltoid lig.
Abductor hallucis m.
Biceps femoris m. (long head)
Iliotibial tract
Biceps femoris m. (short head)
Vastus lateralis m.
Tendon of rectus femoris m.
Fibular collateral lig.
Plantaris m.
Patella
Gastrocnemius m., lateral head
Lat. patellar retinaculum
Common peroneal n.
Deep infrapatellar bursa
Patellar lig.
Ant. ligament of head of fibula
Head of fibula
Soleus m.
Tibialis anterior m.
Peroneus longus m.
Ext. digitorum longus m.
Peroneus brevis m.
Ext. hallucis longus m.
Tibialis ant. tendon (synovial sheath)
Calcaneal tendon
Ext. digit. long. tendon (synovial sheath)
Inf. extensor retinaculum
Lat. malleolus
Ext. hall. long. (synovial sheath)
Ext. digit. brev. m.
Peroneus tert. tendon
Subcalcaneal tendon bursa
Ext. hallucis long. tendon
Sup. peroneal retinaculum
Calcaneofib. lig.
Ext. digit. longus tendons
Inf. peroneal retinac.
Opp. digiti minimi m.
Peroneal tendons (synovial sheath)
Abd. digiti minimi m.
* Tuberosity of 5th metatarsal

Fig. 462: Superficial Muscles and Tendons of the Right Lower Thigh and Leg, Medial View

NOTE: 1) the tendons of the sartorius, gracilis and semitendinosus (pes anserinus) protecting the medial aspect of the knee (see also Fig. 423). The tendon of the semimembranosus helps reinforce the capsule of the knee joint posteriorly (see also Fig. 447);

2) how the tendons of the muscles in the posterior compartment of the leg course beneath the medial malleolus, surrounded by synovial sheaths (in blue), to enter the foot;

3) the superficial location of most of the tibia in the leg.

Fig. 463: Superficial Muscles and Tendons of the Right Lower Thigh and Leg, Lateral View

NOTE: 1) the disposition of the anterior and lateral compartment muscles of the leg, and how their tendons, surrounded by tendon sheaths (in blue), enter the foot. Observe that the anterior compartment tendons enter the dorsum, while the lateral compartment tendons descend behind the lateral malleolus.

2) the superficial location of the head of the fibula and its relationship to the common peroneal nerve.

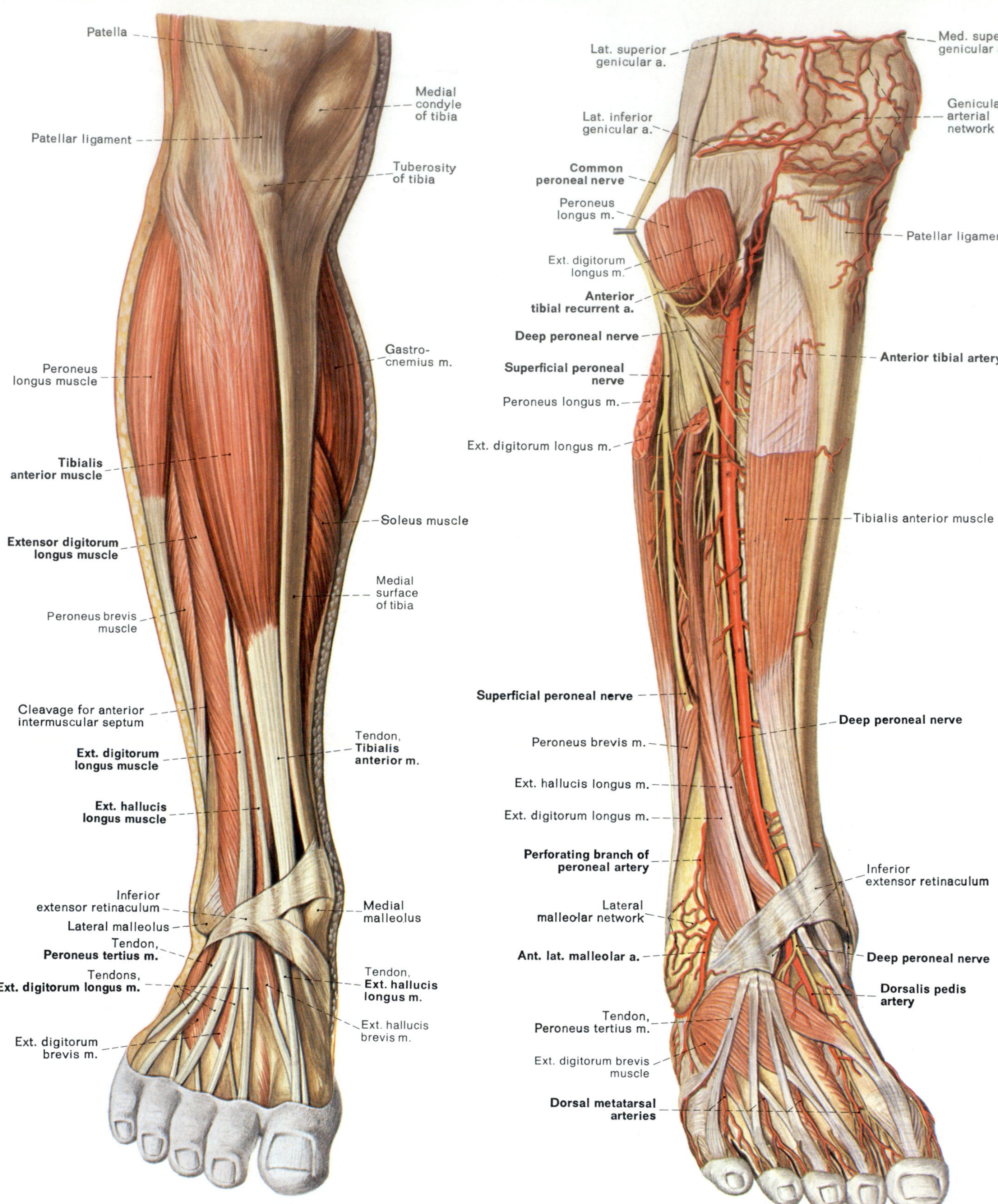

Patella
Medial condyle of tibia
Patellar ligament
Tuberosity of tibia
Peroneus longus muscle
Gastro-cnemius m.
Tibialis anterior muscle
Soleus muscle
Extensor digitorum longus muscle
Medial surface of tibia
Peroneus brevis muscle
Cleavage for anterior intermuscular septum
Tendon, Tibialis anterior m.
Ext. digitorum longus muscle
Ext. hallucis longus muscle
Inferior extensor retinaculum
Medial malleolus
Lateral malleolus
Tendon, Peroneus tertius m.
Tendons, Ext. digitorum longus m.
Tendon, Ext. hallucis longus m.
Ext. hallucis brevis m.
Ext. digitorum brevis m.
Lat. superior genicular a.
Med. superior genicular a.
Lat. inferior genicular a.
Genicular arterial network
Common peroneal nerve
Peroneus longus m.
Patellar ligament
Ext. digitorum longus m.
Anterior tibial recurrent a.
Deep peroneal nerve
Anterior tibial artery
Superficial peroneal nerve
Peroneus longus m.
Ext. digitorum longus m.
Tibialis anterior muscle
Superficial peroneal nerve
Deep peroneal nerve
Peroneus brevis m.
Ext. hallucis longus m.
Ext. digitorum longus m.
Perforating branch of peroneal artery
Inferior extensor retinaculum
Lateral malleolar network
Ant. lat. malleolar a.
Deep peroneal nerve
Dorsalis pedis artery
Tendon, Peroneus tertius m.
Ext. digitorum brevis muscle
Dorsal metatarsal arteries

Fig. 464: Muscles of the Anterior Compartment of the Leg

NOTE: 1) the four anterior compartment muscles are the tibialis anterior, extensor hallucis longus, extensor digitorum longus and peroneus tertius.

2) the tibialis anterior dorsally flexes and supinates the foot. The other muscles extend the toes as well as dorsiflex the foot. Additionally, the extensor hallucis longus assists in supination, while the extensor digitorum longus and peroneus tertius are pronators.

Fig. 465: Nerves and Arteries of the Anterior and Lateral Compartments of the Leg

NOTE: 1) as the common peroneal nerve courses laterally around the head of the fibula, it divides into the superficial and deep peroneal nerves which innervate the muscles of the lateral and anterior compartments.

2) the deep peroneal nerve is joined by the anterior tibial artery which descends toward the foot.

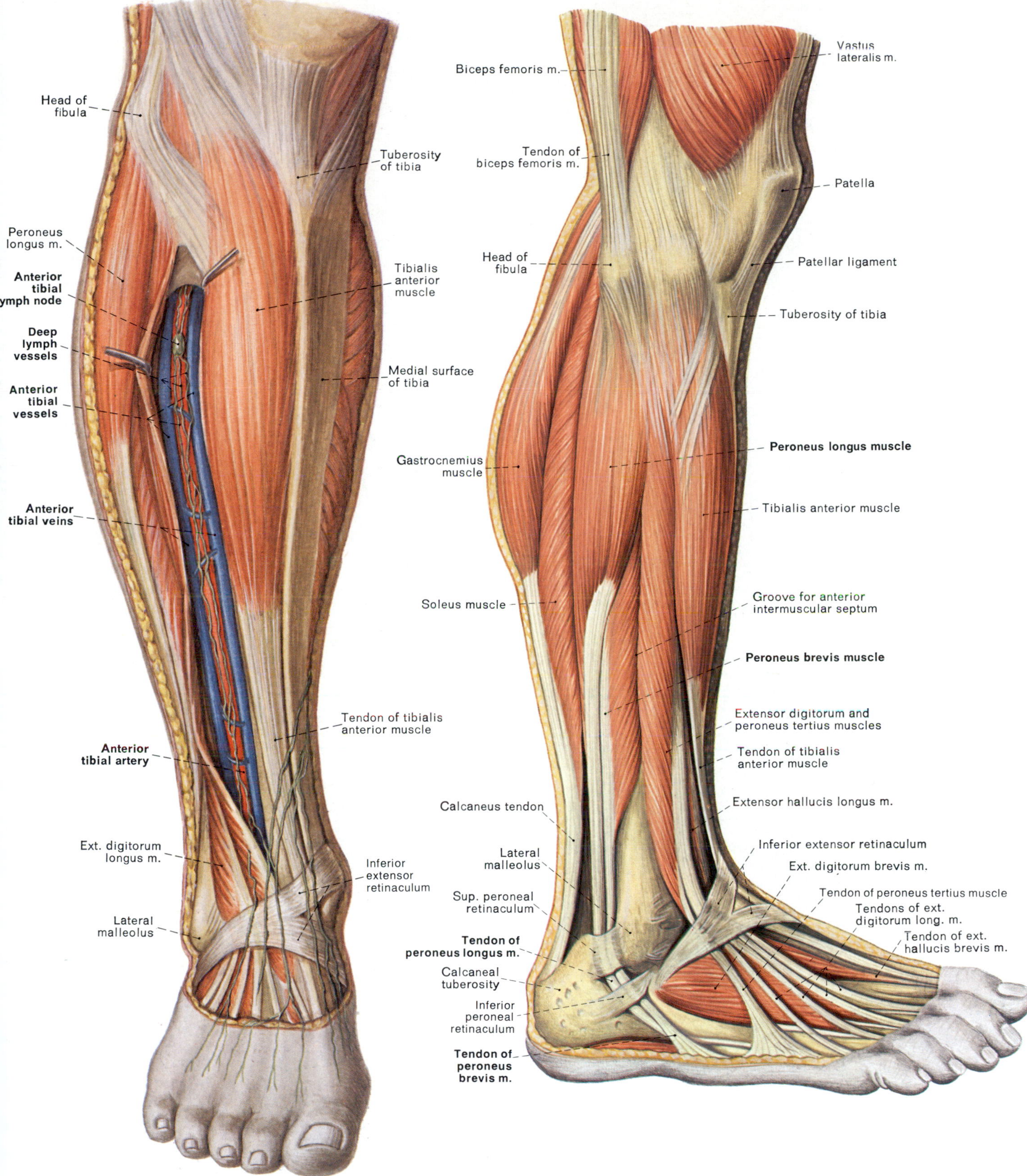
Head of fibula
Tuberosity of tibia
Peroneus longus m.
Anterior tibial lymph node
Tibialis anterior muscle
Deep lymph vessels
Anterior tibial vessels
Medial surface of tibia
Anterior tibial veins
Tendon of tibialis anterior muscle
Anterior tibial artery
Ext. digitorum longus m.
Inferior extensor retinaculum
Lateral malleolus
Vastus lateralis m.
Biceps femoris m.
Tendon of biceps femoris m.
Patella
Head of fibula
Patellar ligament
Tuberosity of tibia
Peroneus longus muscle
Gastrocnemius muscle
Tibialis anterior muscle
Soleus muscle
Groove for anterior intermuscular septum
Peroneus brevis muscle
Extensor digitorum and peroneus tertius muscles
Tendon of tibialis anterior muscle
Calcaneus tendon
Extensor hallucis longus m.
Inferior extensor retinaculum
Lateral malleolus
Ext. digitorum brevis m.
Sup. peroneal retinaculum
Tendon of peroneus tertius muscle
Tendons of ext. digitorum long. m.
Tendon of peroneus longus m.
Tendon of ext. hallucis brevis m.
Calcaneal tuberosity
Inferior peroneal retinaculum
Tendon of peroneus brevis m.

Fig. 466: Deep Lymphatic Channels and Nodes of the Anterior Leg

NOTE that lymphatic channels from the dorsum of the foot course superiorly and collect along the path of the more deeply situated anterior tibial vessels and nerve. At times a lymph node can be found just ventral to the anterior tibial artery below the knee.

Fig. 467: Muscles of the Lateral Compartment of the Leg

NOTE that the peroneus longus and brevis occupy the lateral compartment of the leg. Their tendons descend into the foot behind the lateral malleolus. The peroneus longus tendon crosses the sole of the foot to insert on the base of the 1st metatarsal bone, while the peroneus brevis inserts directly onto the 5th metatarsal bone.

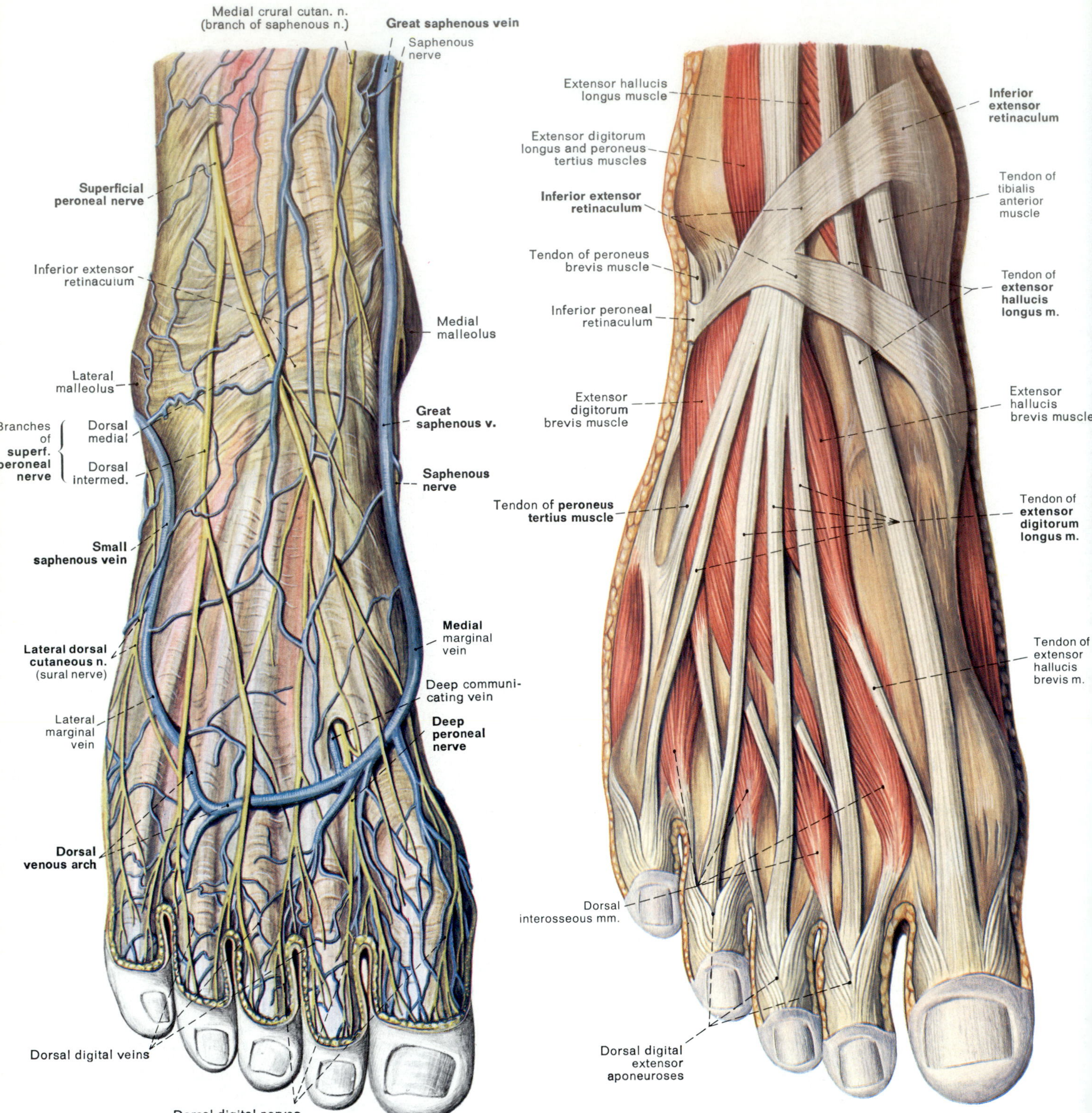

Medial crural cutan. n.
(branch of saphenous n.)
Great saphenous vein
Saphenous nerve
Superficial peroneal nerve
Inferior extensor retinaculum
Medial malleolus
Lateral malleolus
Branches of superf. peroneal nerve
Dorsal medial
Dorsal intermed.
Great saphenous v.
Saphenous nerve
Small saphenous vein
Medial marginal vein
Lateral dorsal cutaneous n.
(sural nerve)
Deep communicating vein
Lateral marginal vein
Deep peroneal nerve
Dorsal venous arch
Dorsal digital veins
Dorsal digital nerves
Extensor hallucis longus muscle
Extensor digitorum longus and peroneus tertius muscles
Inferior extensor retinaculum
Tendon of peroneus brevis muscle
Inferior peroneal retinaculum
Extensor digitorum brevis muscle
Tendon of peroneus tertius muscle
Dorsal interosseous mm.
Dorsal digital extensor aponeuroses
Inferior extensor retinaculum
Tendon of tibialis anterior muscle
Tendon of extensor hallucis longus m.
Extensor hallucis brevis muscle
Tendon of extensor digitorum longus m.
Tendon of extensor hallucis brevis m.

Fig. 468: Superficial Nerves and Veins of the Dorsal Right Foot

NOTE: 1) cutaneous innervation of the dorsal foot is supplied principally by the superficial peroneal nerve (L 4, L 5, S 1). Additionally, the deep peroneal nerve (L 4, L 5) supplies the adjacent sides of the 1st and 2nd toes, while the lateral dorsal cutaneous nerve (S 1, S 2; terminal branch of the sural nerve in the foot) supplies the lateral and dorsal aspect of the 5th digit.

2) the digital and metatarsal veins drain back from the toes to form the dorsal venous arch of the foot. From this arch, the great saphenous vein ascends medially and the small saphenous vein laterally on the foot dorsum.

Fig. 469: Muscles and Tendons of the Dorsal Right Foot (Superficial View)

NOTE: 1) the tendons of the tibialis anterior, extensor hallucis longus and extensor digitorum longus are bound by the Y-shaped inferior extensor retinaculum as they enter the dorsum of the foot at the level of the ankle joint.

2) the long extensor tendons insert onto the dorsal aspect of the distal phalanx of each toe. In addition, the tendons of the extensor digitorum longus also insert onto the dorsum of the middle phalanx of the lateral four toes.

3) the tendon of the peroneus tertius insert on the base of the 5th metatarsal bone (and at times the 4th also).

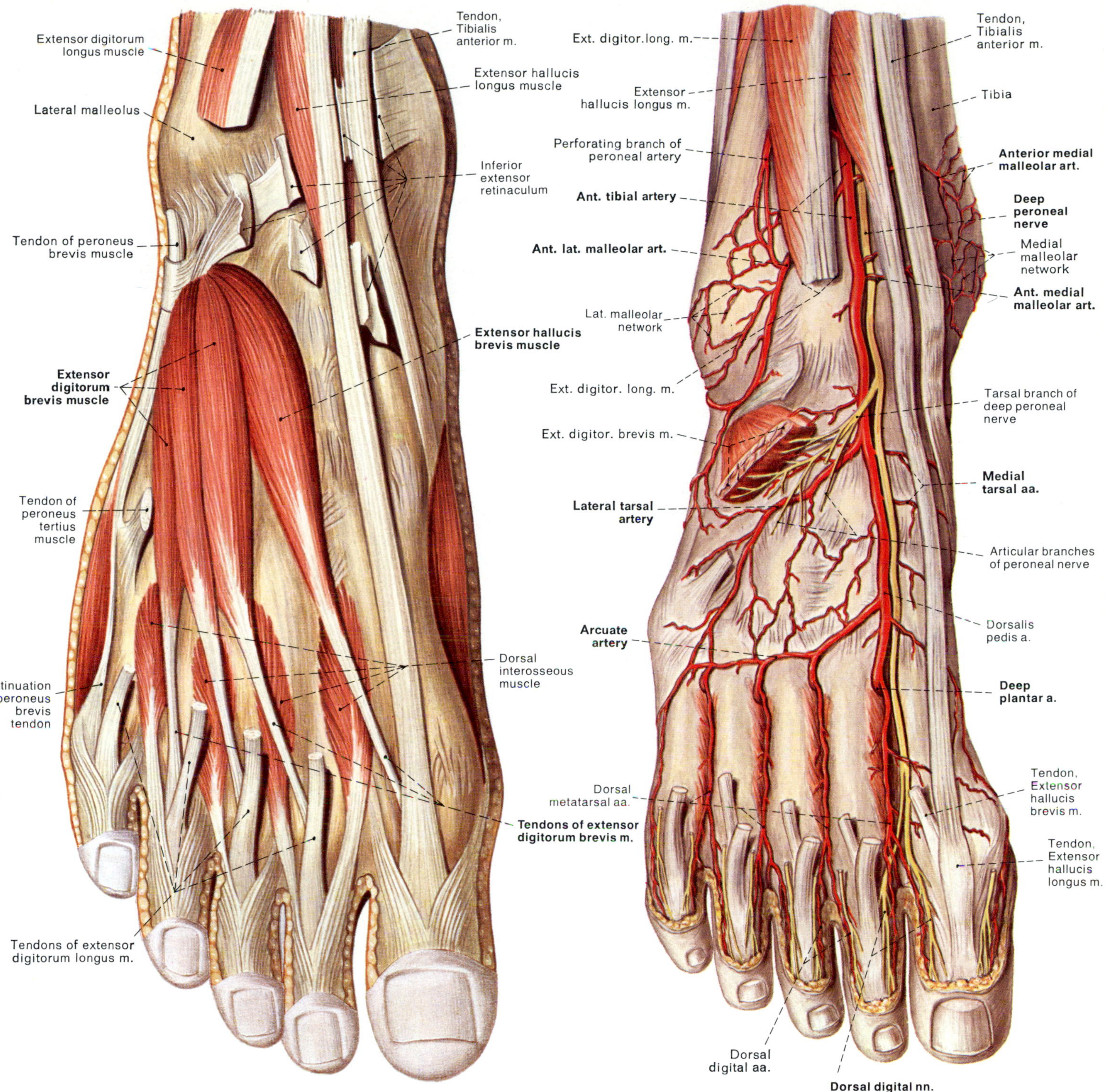

Extensor digitorum longus muscle
Tendon, Tibialis anterior m.
Extensor hallucis longus muscle
Lateral malleolus
Inferior extensor retinaculum
Tendon of peroneus brevis muscle
Extensor hallucis brevis muscle
Extensor digitorum brevis muscle
Tendon of peroneus tertius muscle
Dorsal interosseous muscle
ntinuation f peroneus brevis tendon
Tendons of extensor digitorum brevis m.
Tendons of extensor digitorum longus m.
Ext. digitor. long. m.
Tendon, Tibialis anterior m.
Extensor hallucis longus m.
Tibia
Perforating branch of peroneal artery
Anterior medial malleolar art.
Ant. tibial artery
Deep peroneal nerve
Ant. lat. malleolar art.
Medial malleolar network
Ant. medial malleolar art.
Lat. malleolar network
Ext. digitor. long. m.
Tarsal branch of deep peroneal nerve
Ext. digitor. brevis m.
Medial tarsal aa.
Lateral tarsal artery
Articular branches of peroneal nerve
Dorsalis pedis a.
Arcuate artery
Deep plantar a.
Dorsal metatarsal aa.
Tendon, Extensor hallucis brevis m.
Tendon, Extensor hallucis longus m.
Dorsal digital aa.
Dorsal digital nn. (deep peroneal n.)

Fig. 470: The Intrinsic Muscles of the Dorsal Foot (right)

NOTE: 1) the inferior extensor retinaculum has been opened and the tendons of the extensor digitorum longus and peroneus tertius have been severed.

2) the extensor hallucis brevis and the three small bellies of the extensor digitorum brevis. The delicate tendons of these muscles insert on the proximal phalanx of the medial four toes.

3) the four dorsal interosseous muscles. These muscles abduct the toes from the longitudinal axis of the foot (down the middle of the 2nd toe). The first dorsal interosseous muscle inserts on the medial side of the 2nd toe, while the remaining three insert on the lateral side of the 2nd, 3rd, and 4th toes.

Fig. 471: Deep Nerves and Arteries of the Dorsal Foot

NOTE: 1) the deep coursing anterior tibial artery and deep peroneal nerve and their branches have been exposed. They enter the foot between the tendons of the extensor hallucis longus and extensor digitorum.

2) the anterior tibial artery becomes the dorsalis pedis artery below the ankle joint. Observe the malleolar, tarsal, arcuate, dorsal metatarsal and digital arteries.

3) the deep peroneal nerve supplies the extensor brevis muscle in the foot and continues distally to terminate as two dorsal digital nerves supplying the adjacent sides of the 1st and 2nd toes.

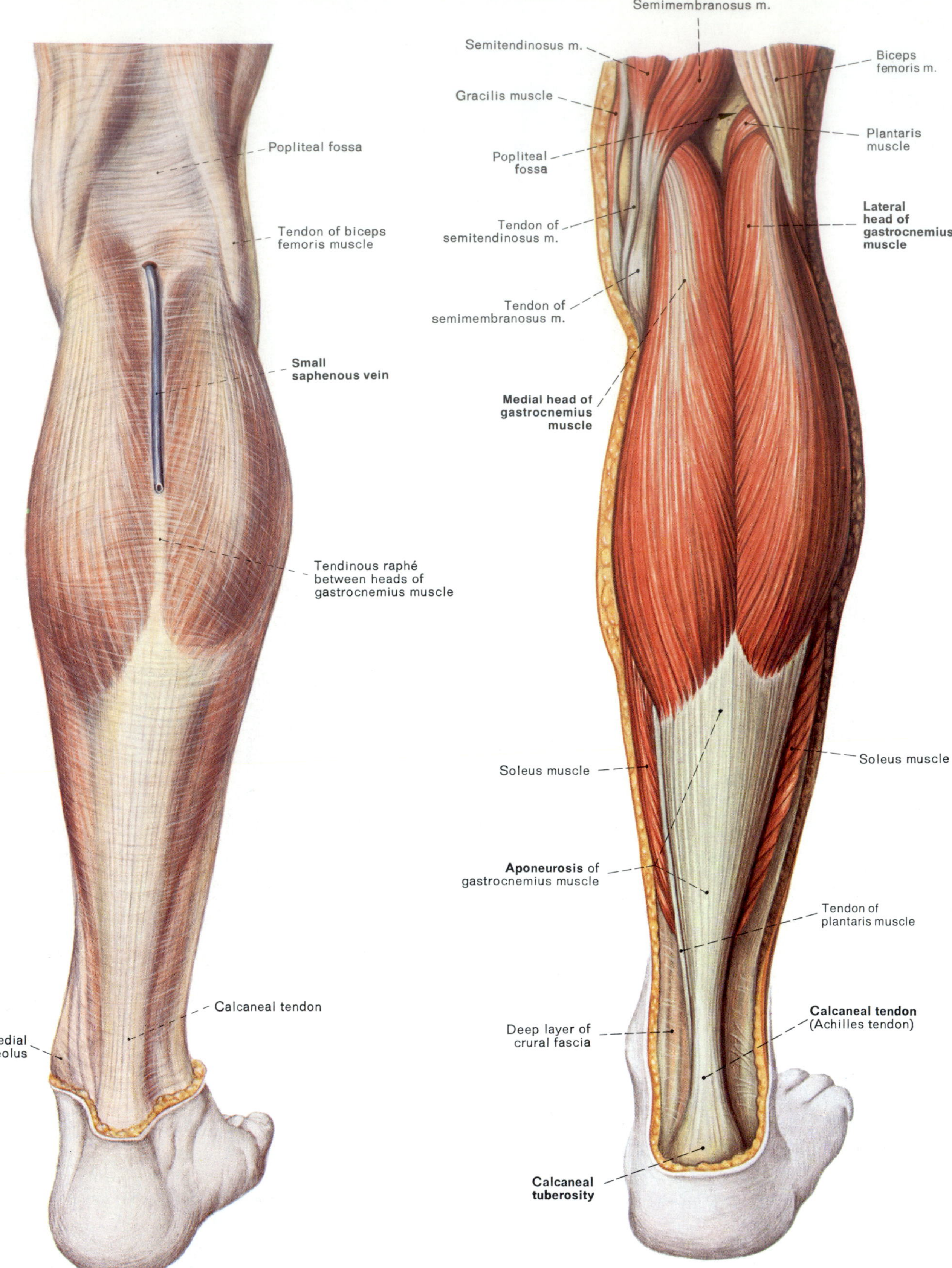
Popliteal fossa
Tendon of biceps femoris muscle
Small saphenous vein
Tendinous raphé between heads of gastrocnemius muscle
Calcaneal tendon
Medial malleolus
Semimembranosus m.
Semitendinosus m.
Biceps femoris m.
Gracilis muscle
Plantaris muscle
Popliteal fossa
Lateral head of gastrocnemius muscle
Tendon of semitendinosus m.
Tendon of semimembranosus m.
Medial head of gastrocnemius muscle
Soleus muscle
Soleus muscle
Aponeurosis of gastrocnemius muscle
Tendon of plantaris muscle
Calcaneal tendon (Achilles tendon)
Deep layer of crural fascia
Calcaneal tuberosity

Fig. 472: The Deep Fascia of the Leg (the Crural Fascia), Posterior View

NOTE that the deep fascia of the leg closely invests all of the muscles between the knee and ankle and forms the fascial covering over the popliteal fossa. It is continuous above with the fascia lata of the thigh and inferiorly with the retinacula which bind the tendons close to the bones.

Fig. 473: Muscles of the Posterior Leg: Superficial Calf Muscles

NOTE: 1) the gastrocnemius muscle arises by two heads from the condyles and posterior popliteal surface of the femur and its fibers are oriented inferomedially toward a central tendinous raphe. It inserts by means of the strong calcaneal tendon onto the calcaneal tuberosity.

2) the gastrocnemius is a powerful plantar flexor of the foot and its continued action also tends to flex the leg at the knee joint.

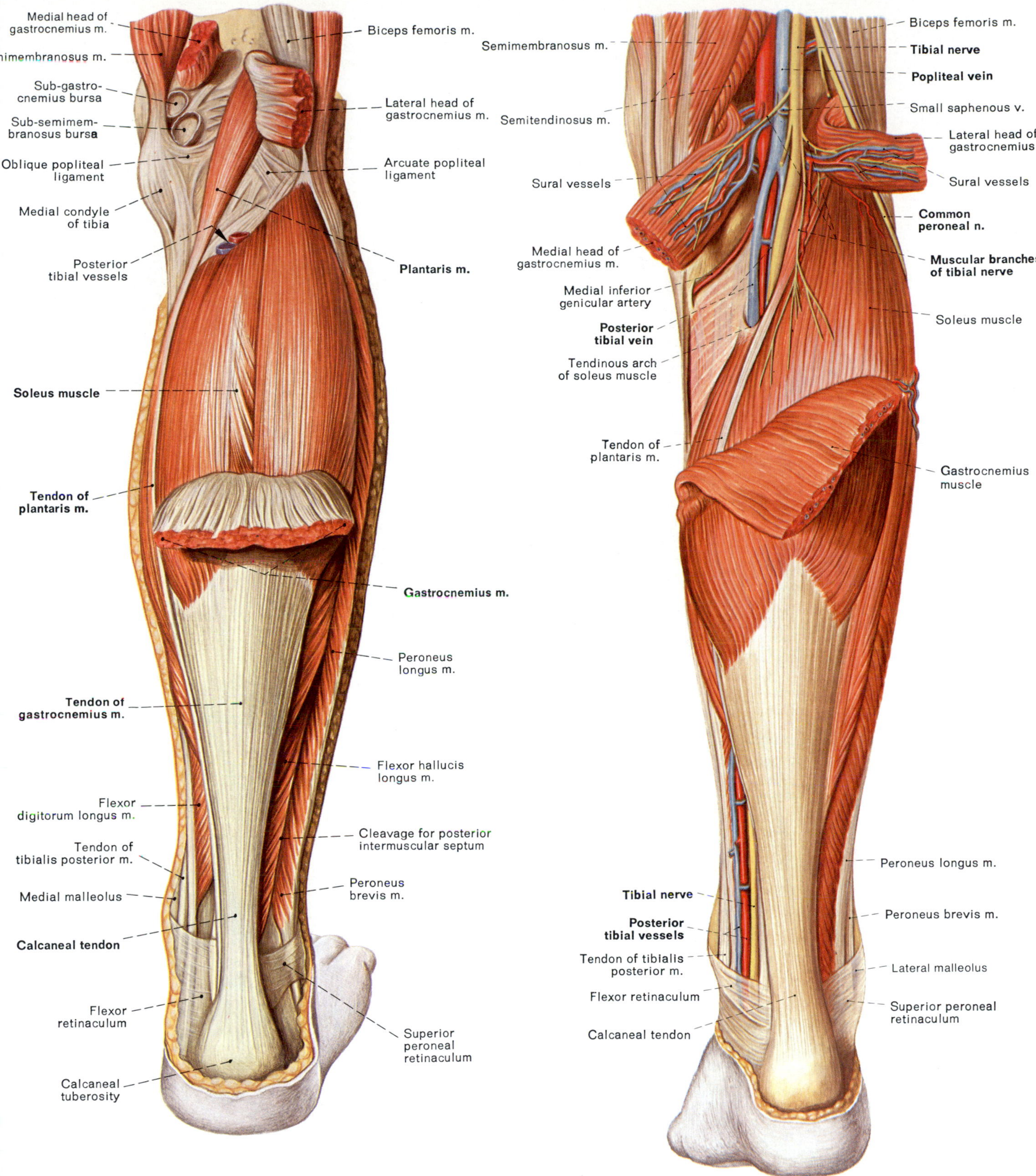
Medial head of gastrocnemius m.
Semimembranosus m.
Sub-gastrocnemius bursa
Sub-semimembranosus bursa
Oblique popliteal ligament
Medial condyle of tibia
Posterior tibial vessels
Biceps femoris m.
Lateral head of gastrocnemius m.
Arcuate popliteal ligament
Plantaris m.
Soleus muscle
Tendon of plantaris m.
Gastrocnemius m.
Peroneus longus m.
Tendon of gastrocnemius m.
Flexor hallucis longus m.
Flexor digitorum longus m.
Tendon of tibialis posterior m.
Cleavage for posterior intermuscular septum
Medial malleolus
Peroneus brevis m.
Calcaneal tendon
Flexor retinaculum
Superior peroneal retinaculum
Calcaneal tuberosity
Semimembranosus m.
Semitendinosus m.
Biceps femoris m.
Tibial nerve
Popliteal vein
Small saphenous v.
Lateral head of gastrocnemius m.
Sural vessels
Sural vessels
Common peroneal n.
Medial head of gastrocnemius m.
Muscular branches of tibial nerve
Medial inferior genicular artery
Soleus muscle
Posterior tibial vein
Tendinous arch of soleus muscle
Tendon of plantaris m.
Gastrocnemius muscle
Peroneus longus m.
Tibial nerve
Peroneus brevis m.
Posterior tibial vessels
Tendon of tibialis posterior m.
Lateral malleolus
Flexor retinaculum
Superior peroneal retinaculum
Calcaneal tendon

Fig. 474: Muscles of the Posterior Leg, Second Layer of Calf Muscles

NOTE: 1) both heads of the gastrocnemius have been severed to uncover the underlying soleus and plantaris muscles. The soleus is broad and thick, arising from the posterior surface of the fibula, the intermuscular septum and the dorsal aspect of the tibia. Its fibers join the calcaneal tendon.

2) the plantaris courses between the gastrocnemius and soleus. Both muscles are plantar flexors of the foot.

Fig. 475: Nerves and Vessels of the Posterior Leg: Superficial Layer

NOTE that the popliteal vessels and tibial nerve, descending from the popliteal fossa into the posterior compartment of the leg, commence in the middle of the leg and course medially in a gradual fashion so that at the ankle they lie behind the medial malleolus.

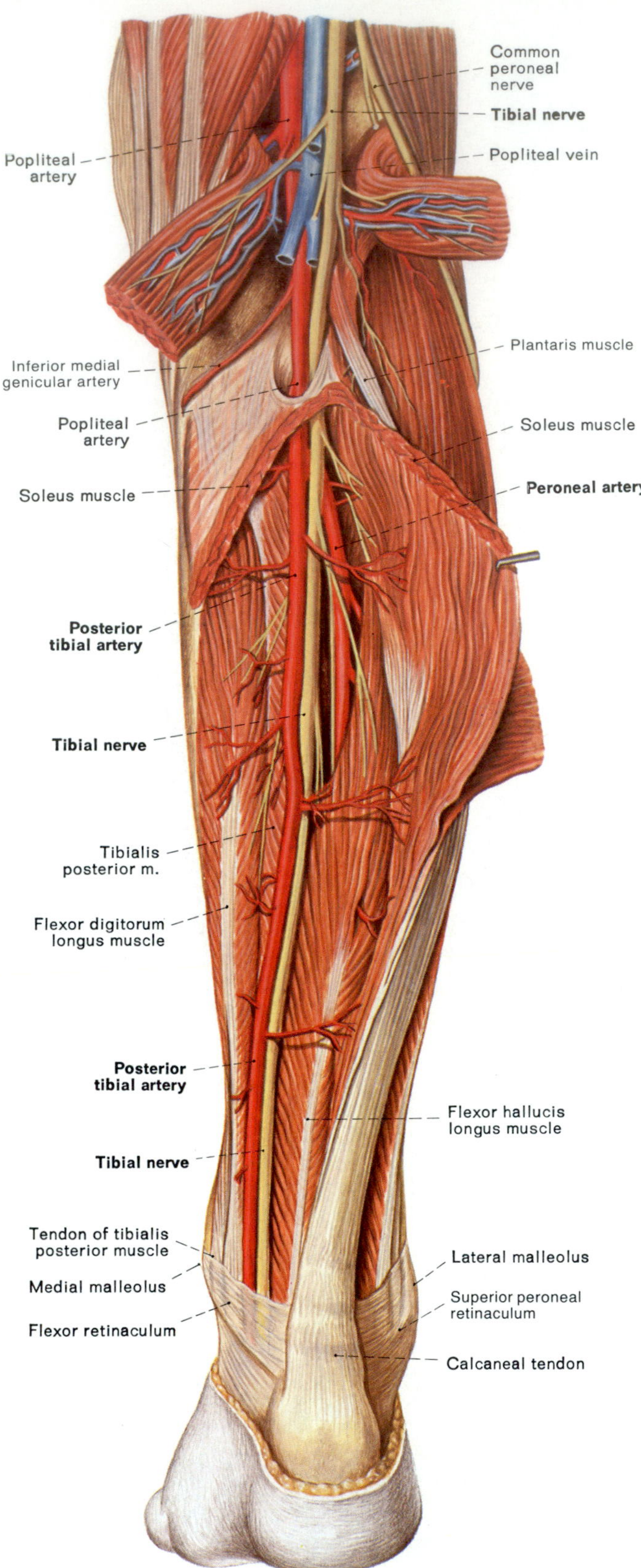

Common peroneal nerve
Tibial nerve
Popliteal artery
Popliteal vein
Inferior medial genicular artery
Plantaris muscle
Popliteal artery
Soleus muscle
Soleus muscle
Peroneal artery
Posterior tibial artery
Tibial nerve
Tibialis posterior m.
Flexor digitorum longus muscle
Posterior tibial artery
Flexor hallucis longus muscle
Tibial nerve
Tendon of tibialis posterior muscle
Lateral malleolus
Medial malleolus
Superior peroneal retinaculum
Flexor retinaculum
Calcaneal tendon

Fig. 476: Nerves and Vessels of the Right Posterior Leg, Intermediate Layer

NOTE: 1) the soleus muscle has been severed and reflected laterally, exposing the course of the tibial artery and posterior tibial nerve as far as the medial malleolus.

2) that this vessel and nerve descend in the leg between the flexor hallucis longus and the flexor digitorum longus and dorsal to the tibialis posterior muscle.

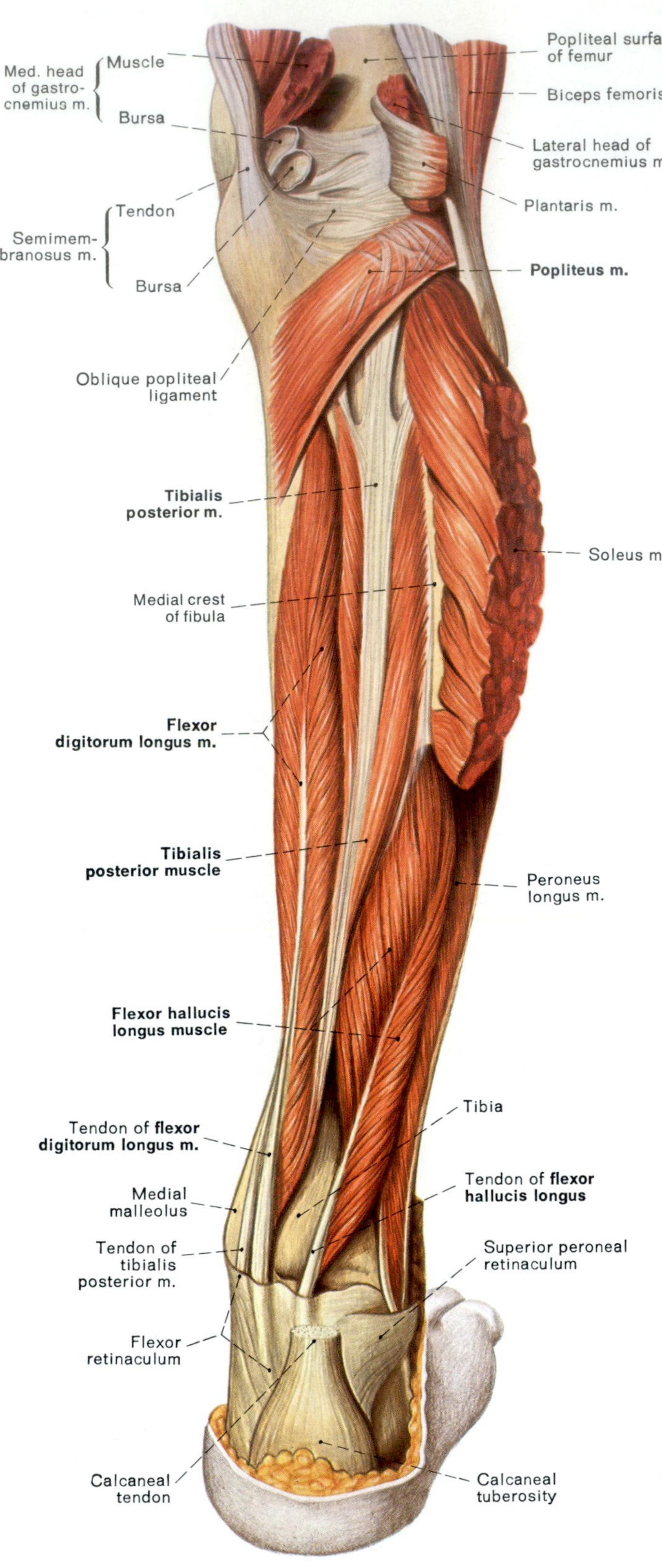

Med. head of gastrocnemius m.
Muscle
Bursa
Popliteal surface of femur
Biceps femoris m.
Lateral head of gastrocnemius m.
Semimembranosus m.
Tendon
Bursa
Plantaris m.
Popliteus m.
Oblique popliteal ligament
Tibialis posterior m.
Soleus m.
Medial crest of fibula
Flexor digitorum longus m.
Tibialis posterior muscle
Peroneus longus m.
Flexor hallucis longus muscle
Tibia
Tendon of flexor digitorum longus m.
Medial malleolus
Tendon of flexor hallucis longus
Tendon of tibialis posterior m.
Superior peroneal retinaculum
Flexor retinaculum
Calcaneal tendon
Calcaneal tuberosity

Fig. 477: Muscles of the Posterior Compartment of the Leg: Deep Group: Four Muscles

NOTE: 1) the four deep posterior compartment muscles are: a) the popliteus, b) the flexor digitorum longus, c) the tibialis posterior and d) the flexor hallucis longus.

2) the popliteus is a femorotibial muscle and tends to rotate the leg medially and flex the leg at the knee joint. The other three muscles are cruropedal muscles and, as a group, they invert the foot, flex the toes and assist in plantarflexion at the ankle joint.

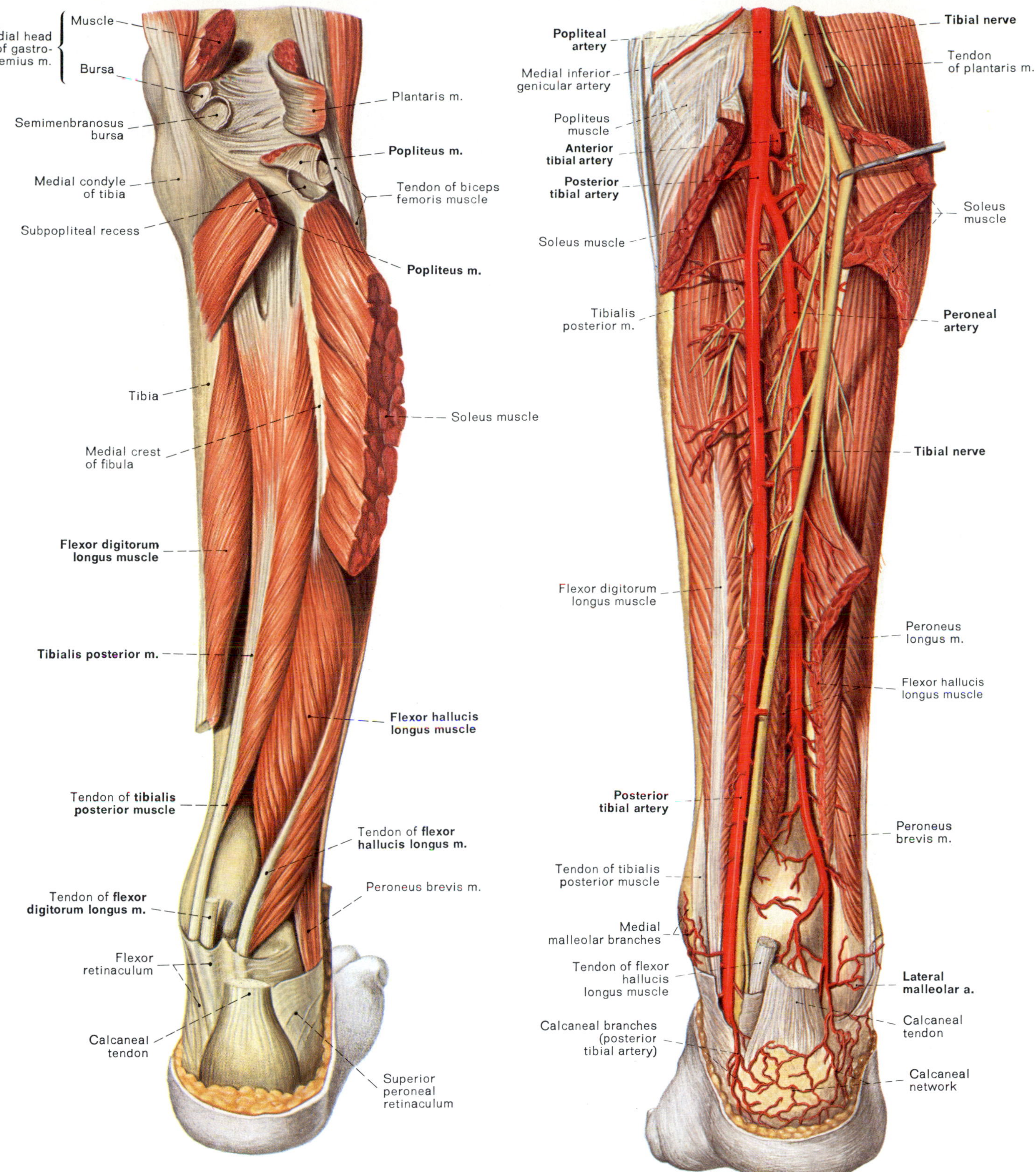
Medial head of gastrocnemius m.
Muscle
Bursa
Plantaris m.
Semimenbranosus bursa
Popliteus m.
Medial condyle of tibia
Tendon of biceps femoris muscle
Subpopliteal recess
Popliteus m.
Tibia
Soleus muscle
Medial crest of fibula
Flexor digitorum longus muscle
Tibialis posterior m.
Flexor hallucis longus muscle
Tendon of tibialis posterior muscle
Tendon of flexor hallucis longus m.
Peroneus brevis m.
Tendon of flexor digitorum longus m.
Flexor retinaculum
Calcaneal tendon
Superior peroneal retinaculum
Popliteal artery
Tibial nerve
Tendon of plantaris m.
Medial inferior genicular artery
Popliteus muscle
Anterior tibial artery
Posterior tibial artery
Soleus muscle
Soleus muscle
Tibialis posterior m.
Peroneal artery
Tibial nerve
Flexor digitorum longus muscle
Peroneus longus m.
Flexor hallucis longus muscle
Posterior tibial artery
Peroneus brevis m.
Tendon of tibialis posterior muscle
Medial malleolar branches
Tendon of flexor hallucis longus muscle
Lateral malleolar a.
Calcaneal tendon
Calcaneal branches (posterior tibial artery)
Calcaneal network

Fig. 478: The Tibialis Posterior and Flexor Hallucis Longus Muscles

NOTE: 1) the tendon of the flexor digitorum longus and the popliteus muscle have been severed. The tendon of the tibialis posterior crosses beneath that of the flexor digitorum longus just before entering the foot.

2) the flexor hallucis longus muscle arises from the distal 2/3rds of the fibula and the intermuscular septa. Its tendon lies in a groove on the posterior surface of the talus.

Fig. 479: Nerves and Muscles of the Deep Posterior Leg

NOTE: 1) the soleus muscle was resected and the tibial nerve pulled aside. Observe the branching of the peroneal artery from the posterior tibial, and its descending course toward the lateral malleolus.

2) in the popliteal fossa, the tibial nerve courses superficial to the popliteal artery, whereas at the ankle, the posterior tibial artery is superficial to the tibial nerve.

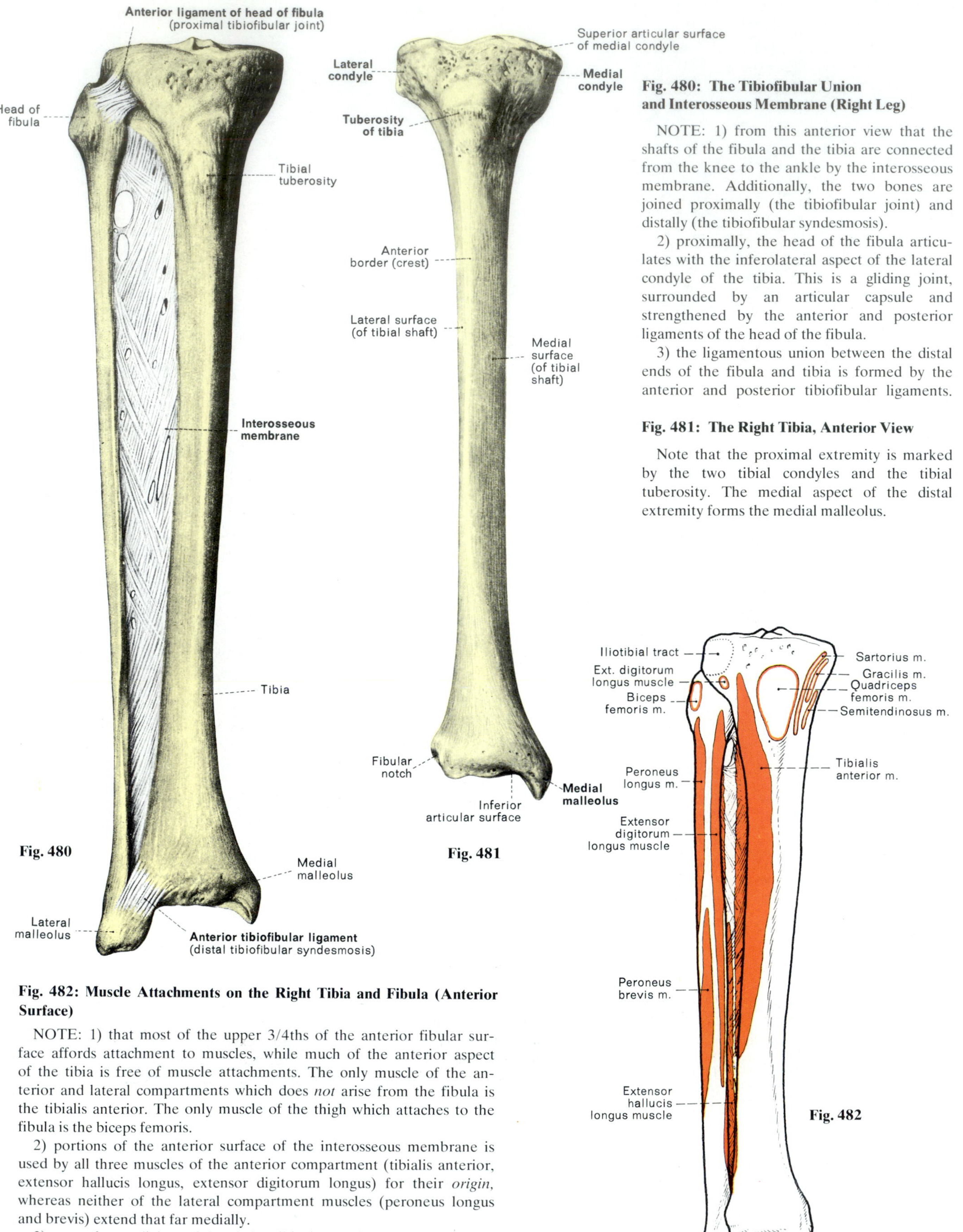

Anterior ligament of head of fibula
(proximal tibiofibular joint)
Head of fibula
Tibial tuberosity
Interosseous membrane
Tibia
Medial malleolus
Lateral malleolus
Anterior tibiofibular ligament
(distal tibiofibular syndesmosis)
Fig. 480
Superior articular surface of medial condyle
Lateral condyle
Medial condyle
Tuberosity of tibia
Anterior border (crest)
Lateral surface (of tibial shaft)
Medial surface (of tibial shaft)
Fibular notch
Inferior articular surface
Medial malleolus
Fig. 481
Iliotibial tract
Ext. digitorum longus muscle
Biceps femoris m.
Sartorius m.
Gracilis m.
Quadriceps femoris m.
Semitendinosus m.
Peroneus longus m.
Tibialis anterior m.
Extensor digitorum longus muscle
Peroneus brevis m.
Extensor hallucis longus muscle
Fig. 482

Fig. 480: The Tibiofibular Union and Interosseous Membrane (Right Leg)

NOTE: 1) from this anterior view that the shafts of the fibula and the tibia are connected from the knee to the ankle by the interosseous membrane. Additionally, the two bones are joined proximally (the tibiofibular joint) and distally (the tibiofibular syndesmosis).

2) proximally, the head of the fibula articulates with the inferolateral aspect of the lateral condyle of the tibia. This is a gliding joint, surrounded by an articular capsule and strengthened by the anterior and posterior ligaments of the head of the fibula.

3) the ligamentous union between the distal ends of the fibula and tibia is formed by the anterior and posterior tibiofibular ligaments.

Fig. 481: The Right Tibia, Anterior View

Note that the proximal extremity is marked by the two tibial condyles and the tibial tuberosity. The medial aspect of the distal extremity forms the medial malleolus.

Fig. 482: Muscle Attachments on the Right Tibia and Fibula (Anterior Surface)

NOTE: 1) that most of the upper 3/4ths of the anterior fibular surface affords attachment to muscles, while much of the anterior aspect of the tibia is free of muscle attachments. The only muscle of the anterior and lateral compartments which does *not* arise from the fibula is the tibialis anterior. The only muscle of the thigh which attaches to the fibula is the biceps femoris.

2) portions of the anterior surface of the interosseous membrane is used by all three muscles of the anterior compartment (tibialis anterior, extensor hallucis longus, extensor digitorum longus) for their *origin*, whereas neither of the lateral compartment muscles (peroneus longus and brevis) extend that far medially.

3) onto the medial condyle of the tibia insert the sartorius, gracilis and semitendinosus muscles forming a tendinous expansion sometimes called the pes anserinus (goose's foot), while onto the tibial tuberosity inserts the massive quadriceps femoris muscle.

Fig. 483: Muscle Attachments on the Right Tibia and Fibula (Posterior Surface)

NOTE: 1) that of the posterior compartment muscles, only the gastrocnemius and plantaris muscles do not attach to the posterior surface of the tibia, fibula or interosseous membrane.

2) that virtually the entire posterior surface of the fibula serves for the origin of muscles. The soleus arises from the upper one-third of the posterior fibular surface and along the soleal line of the tibia. Inferior to the origin of the soleus which spans across both bones, the flexor hallucis longus arises principally from the fibula, while the flexor digitorum longus arises primarily from the tibia.

3) the tibialis posterior is interposed between the flexors hallucis longus and digitorum longus, thereby arising from the posterior surface of the interosseous membrane.

4) the arrows indicating the course of the tendons into the foot from the posterior surface. Observe that the tendon of the tibialis posterior crosses from lateral to medial beneath the tendon of the flexor digitorum longus and enters the foot immediately behind the medial malleolus.

Fig. 483

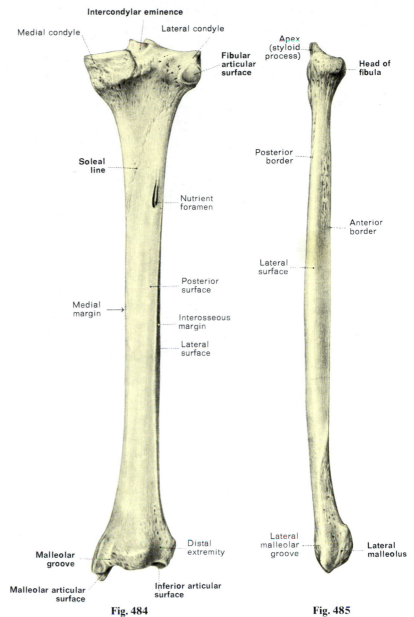

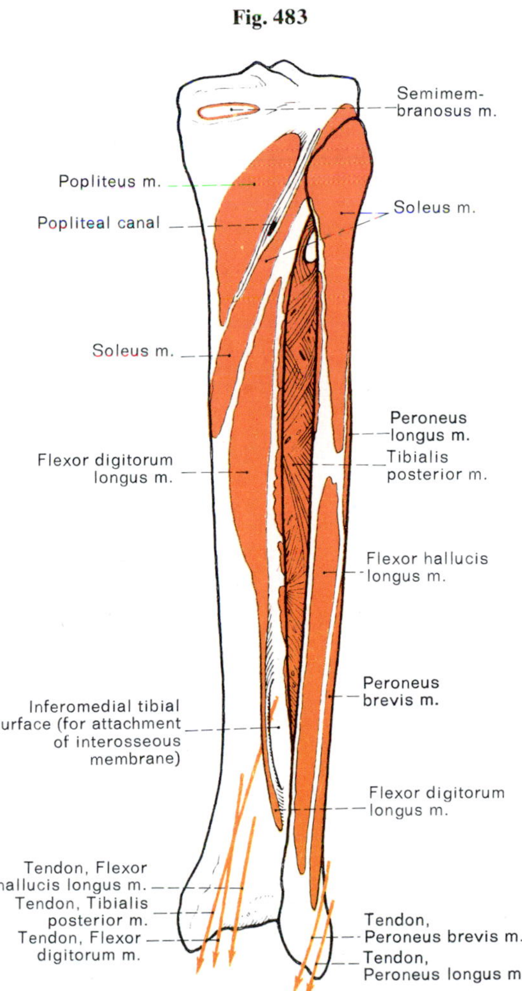
Semimembranosus m.
Popliteus m.
Popliteal canal
Soleus m.
Soleus m.
Peroneus longus m.
Flexor digitorum longus m.
Tibialis posterior m.
Flexor hallucis longus m.
Peroneus brevis m.
Inferomedial tibial surface (for attachment of interosseous membrane)
Flexor digitorum longus m.
Tendon, Flexor hallucis longus m.
Tendon, Tibialis posterior m.
Tendon, Flexor digitorum m.
Tendon, Peroneus brevis m.
Tendon, Peroneus longus m.

Intercondylar eminence
Medial condyle
Lateral condyle
Fibular articular surface
Soleal line
Nutrient foramen
Posterior surface
Medial margin
Interosseous margin
Lateral surface
Distal extremity
Malleolar groove
Malleolar articular surface
Inferior articular surface

Fig. 484

Apex (styloid process)
Head of fibula
Posterior border
Anterior border
Lateral surface
Lateral malleolar groove
Lateral malleolus

Fig. 485

Fig. 484: The Right Tibia, Posterior View

NOTE: 1) the smooth posterior surface of the shaft of the tibia is marked by a prominent ridge (the soleal line) and a large oblong foramen (the nutrient foramen). The tibial shaft tapers toward a larger proximal extremity and somewhat less pronounced distal extremity.

2) at the proximal extremity the rounded medial and lateral condyles are separated by the intercondylar eminence, anterior and posterior to which attach the cruciate ligaments. At its distal extremity, the tibia articulates with the talus and, on this posterior surface, presents grooves for the passage of the tendons of the tibialis posterior, flexor digitorum longus and flexor hallucis longus.

Fig. 485: The Right Fibula, Lateral View

NOTE: 1) the fibula is a long slender bone situated lateral to the tibia to which it articulates proximally (see Figure 480). Distally, the fibula expands to form the lateral malleolus. The medial aspect of its inferior articular surface participates with the tibia in forming the talocrural or ankle joint.

2) although the fibula does not bear any weight of the trunk (since it does not participate in the knee joint articulation), it is important because of the numerous muscles wich attach to its surface (see Figures 482 and 483) and because it assists in the formation of the ankle joint.

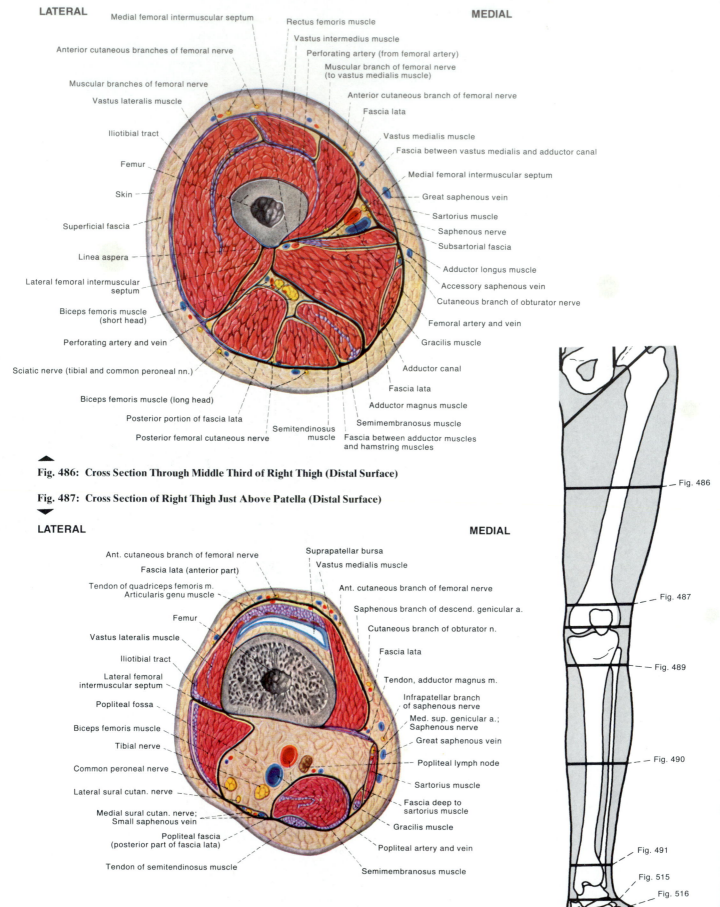

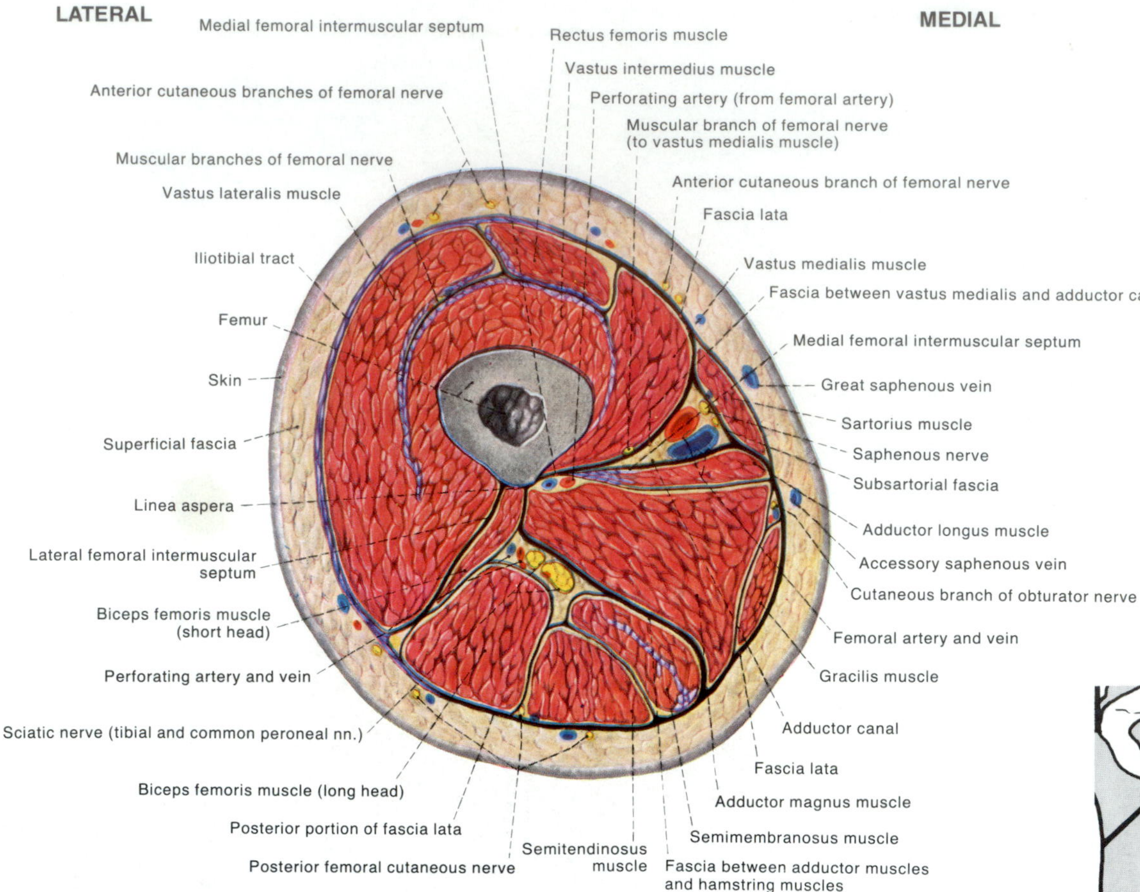

LATERAL
MEDIAL
Medial femoral intermuscular septum
Rectus femoris muscle
Vastus intermedius muscle
Anterior cutaneous branches of femoral nerve
Perforating artery (from femoral artery)
Muscular branch of femoral nerve
(to vastus medialis muscle)
Muscular branches of femoral nerve
Vastus lateralis muscle
Anterior cutaneous branch of femoral nerve
Fascia lata
Iliotibial tract
Vastus medialis muscle
Fascia between vastus medialis and adductor canal
Femur
Medial femoral intermuscular septum
Skin
Great saphenous vein
Sartorius muscle
Superficial fascia
Saphenous nerve
Subsartorial fascia
Linea aspera
Adductor longus muscle
Lateral femoral intermuscular
septum
Accessory saphenous vein
Cutaneous branch of obturator nerve
Biceps femoris muscle
(short head)
Femoral artery and vein
Perforating artery and vein
Gracilis muscle
Sciatic nerve (tibial and common peroneal nn.)
Adductor canal
Fascia lata
Biceps femoris muscle (long head)
Adductor magnus muscle
Posterior portion of fascia lata
Semimembranosus muscle
Semitendinosus
muscle
Posterior femoral cutaneous nerve
Fascia between adductor muscles
and hamstring muscles

Fig. 486: Cross Section Through Middle Third of Right Thigh (Distal Surface)

Fig. 487: Cross Section of Right Thigh Just Above Patella (Distal Surface)

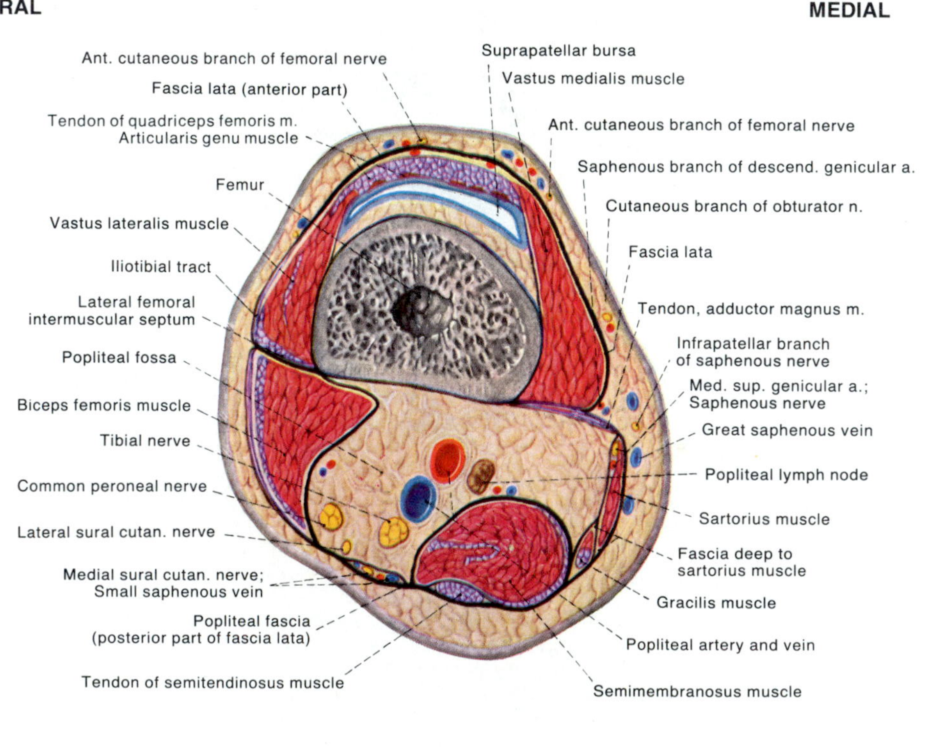

LATERAL
MEDIAL
Ant. cutaneous branch of femoral nerve
Suprapatellar bursa
Vastus medialis muscle
Fascia lata (anterior part)
Tendon of quadriceps femoris m.
Articularis genu muscle
Ant. cutaneous branch of femoral nerve
Saphenous branch of descend. genicular a.
Femur
Cutaneous branch of obturator n.
Vastus lateralis muscle
Fascia lata
Iliotibial tract
Lateral femoral
intermuscular septum
Tendon, adductor magnus m.
Infrapatellar branch
of saphenous nerve
Popliteal fossa
Med. sup. genicular a.;
Saphenous nerve
Biceps femoris muscle
Great saphenous vein
Tibial nerve
Popliteal lymph node
Common peroneal nerve
Sartorius muscle
Lateral sural cutan. nerve
Fascia deep to
sartorius muscle
Medial sural cutan. nerve;
Small saphenous vein
Gracilis muscle
Popliteal fascia
(posterior part of fascia lata)
Popliteal artery and vein
Tendon of semitendinosus muscle
Semimembranosus muscle

Fig. 486
Fig. 487
Fig. 489
Fig. 490
Fig. 491
Fig. 515
Fig. 516

Fig. 488: Cross Sectional Planes of Lower Extremity Shown in Seven Figures of This Atlas

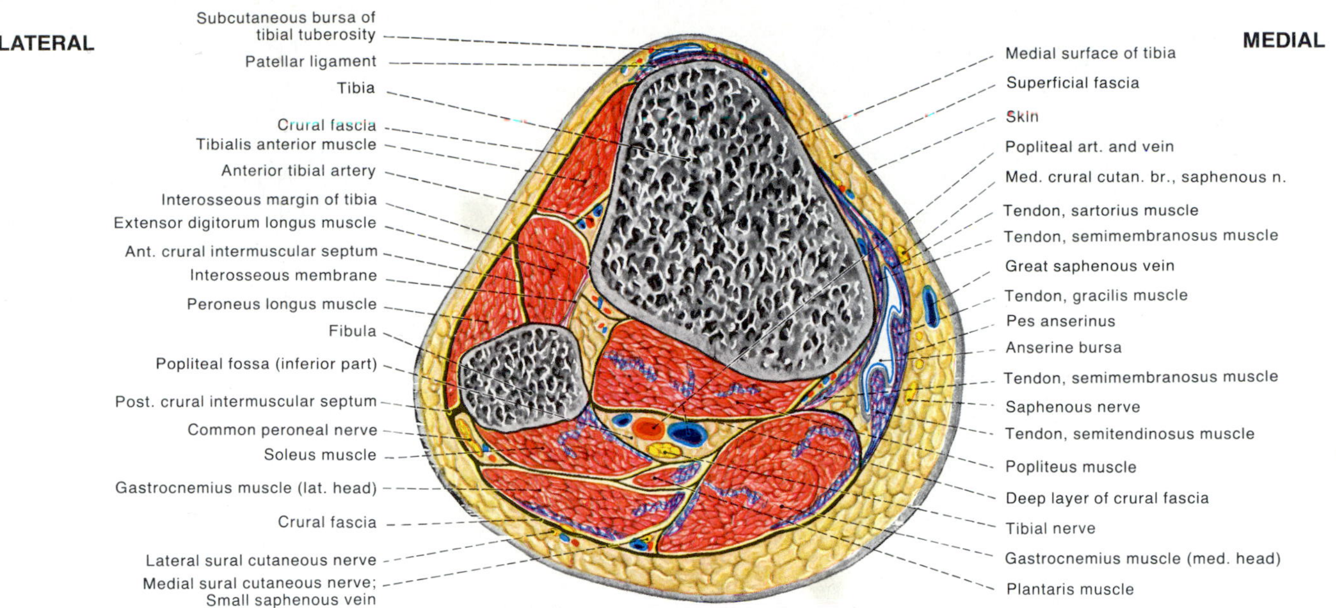

LATERAL
MEDIAL
Subcutaneous bursa of tibial tuberosity
Patellar ligament
Tibia
Crural fascia
Tibialis anterior muscle
Anterior tibial artery
Interosseous margin of tibia
Extensor digitorum longus muscle
Ant. crural intermuscular septum
Interosseous membrane
Peroneus longus muscle
Fibula
Popliteal fossa (inferior part)
Post. crural intermuscular septum
Common peroneal nerve
Soleus muscle
Gastrocnemius muscle (lat. head)
Crural fascia
Lateral sural cutaneous nerve
Medial sural cutaneous nerve; Small saphenous vein
Medial surface of tibia
Superficial fascia
Skin
Popliteal art. and vein
Med. crural cutan. br., saphenous n.
Tendon, sartorius muscle
Tendon, semimembranosus muscle
Great saphenous vein
Tendon, gracilis muscle
Pes anserinus
Anserine bursa
Tendon, semimembranosus muscle
Saphenous nerve
Tendon, semitendinosus muscle
Popliteus muscle
Deep layer of crural fascia
Tibial nerve
Gastrocnemius muscle (med. head)
Plantaris muscle

Fig. 489: Cross Section Through Upper Third of Right Leg (Distal Surface)

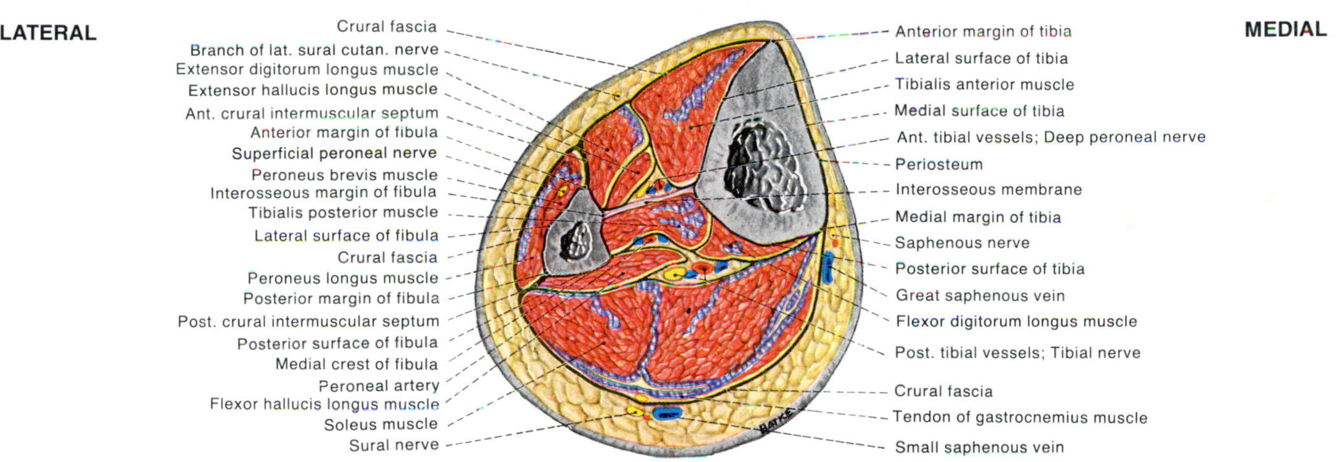

LATERAL
MEDIAL
Crural fascia
Branch of lat. sural cutan. nerve
Extensor digitorum longus muscle
Extensor hallucis longus muscle
Ant. crural intermuscular septum
Anterior margin of fibula
Superficial peroneal nerve
Peroneus brevis muscle
Interosseous margin of fibula
Tibialis posterior muscle
Lateral surface of fibula
Crural fascia
Peroneus longus muscle
Posterior margin of fibula
Post. crural intermuscular septum
Posterior surface of fibula
Medial crest of fibula
Peroneal artery
Flexor hallucis longus muscle
Soleus muscle
Sural nerve
Anterior margin of tibia
Lateral surface of tibia
Tibialis anterior muscle
Medial surface of tibia
Ant. tibial vessels; Deep peroneal nerve
Periosteum
Interosseous membrane
Medial margin of tibia
Saphenous nerve
Posterior surface of tibia
Great saphenous vein
Flexor digitorum longus muscle
Post. tibial vessels; Tibial nerve
Crural fascia
Tendon of gastrocnemius muscle
Small saphenous vein

Fig. 490: Cross Section Through the Middle of Right Leg (Distal Surface)

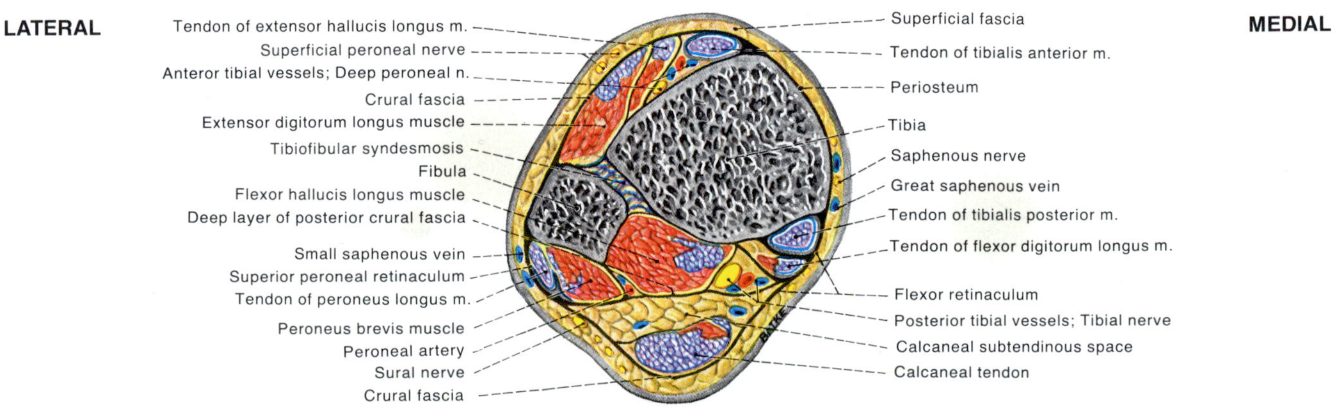

LATERAL
MEDIAL
Tendon of extensor hallucis longus m.
Superficial peroneal nerve
Anteror tibial vessels; Deep peroneal n.
Crural fascia
Extensor digitorum longus muscle
Tibiofibular syndesmosis
Fibula
Flexor hallucis longus muscle
Deep layer of posterior crural fascia
Small saphenous vein
Superior peroneal retinaculum
Tendon of peroneus longus m.
Peroneus brevis muscle
Peroneal artery
Sural nerve
Crural fascia
Superficial fascia
Tendon of tibialis anterior m.
Periosteum
Tibia
Saphenous nerve
Great saphenous vein
Tendon of tibialis posterior m.
Tendon of flexor digitorum longus m.
Flexor retinaculum
Posterior tibial vessels; Tibial nerve
Calcaneal subtendinous space
Calcaneal tendon

Fig. 491: Cross Section Through the Right Leg Just Above the Malleoli (Distal Surface)

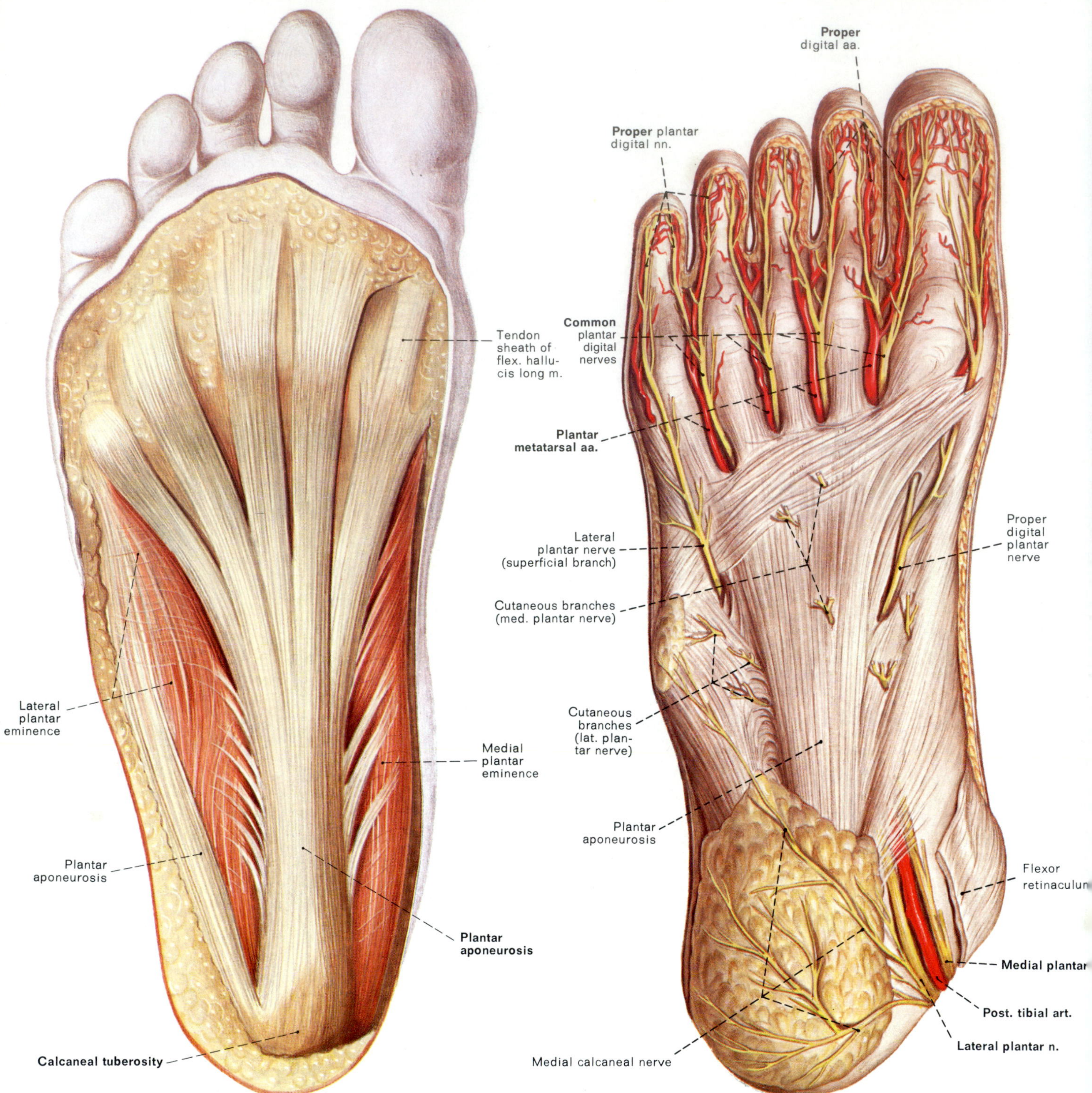

Tendon sheath of flex. hallucis long m.
Lateral plantar eminence
Medial plantar eminence
Plantar aponeurosis
Plantar aponeurosis
Calcaneal tuberosity
Proper digital aa.
Proper plantar digital nn.
Common plantar digital nerves
Plantar metatarsal aa.
Lateral plantar nerve (superficial branch)
Cutaneous branches (med. plantar nerve)
Proper digital plantar nerve
Cutaneous branches (lat. plantar nerve)
Plantar aponeurosis
Flexor retinaculum
Medial plantar
Post. tibial art.
Lateral plantar n.
Medial calcaneal nerve

Fig. 492: The Sole of the Right Foot: Plantar Aponeurosis

NOTE: 1) the plantar aponeurosis which stretches across the sole of the foot. Similar to the palmar aponeurosis in the hand, the plantar aponeurosis is a thickened layer of deep fascia which serves both a protective and supportive function to the underlying muscles, vessels and nerves.

2) the longitudinal orientation of the fibers of the plantar aponeurosis and their attachment to the calcaneal tuberosity. Distally, the aponeurosis divides into five digital slips, one of which courses to each toe. Fibers extend from the margins of the aponeurosis to cover partially both the medial and lateral plantar eminences.

Fig. 493: The Sole of the Right Foot: Superficial Nerves and Arteries

NOTE: 1) the medial and lateral plantar nerves and posterior tibial artery as they enter the foot behind the medial malleolus and immediately course beneath the plantar aponeurosis toward the digits. Sensory branches of the nerves penetrate the aponeurosis to innervate the overlying skin and superficial fascia.

2) between the digital slips of the plantar aponeurosis, the vessels and the nerves course superficially toward the toes. Metatarsal arteries and common plantar digital nerves divide to supply adjacent portions of the toes as proper plantar digital arteries and nerves.

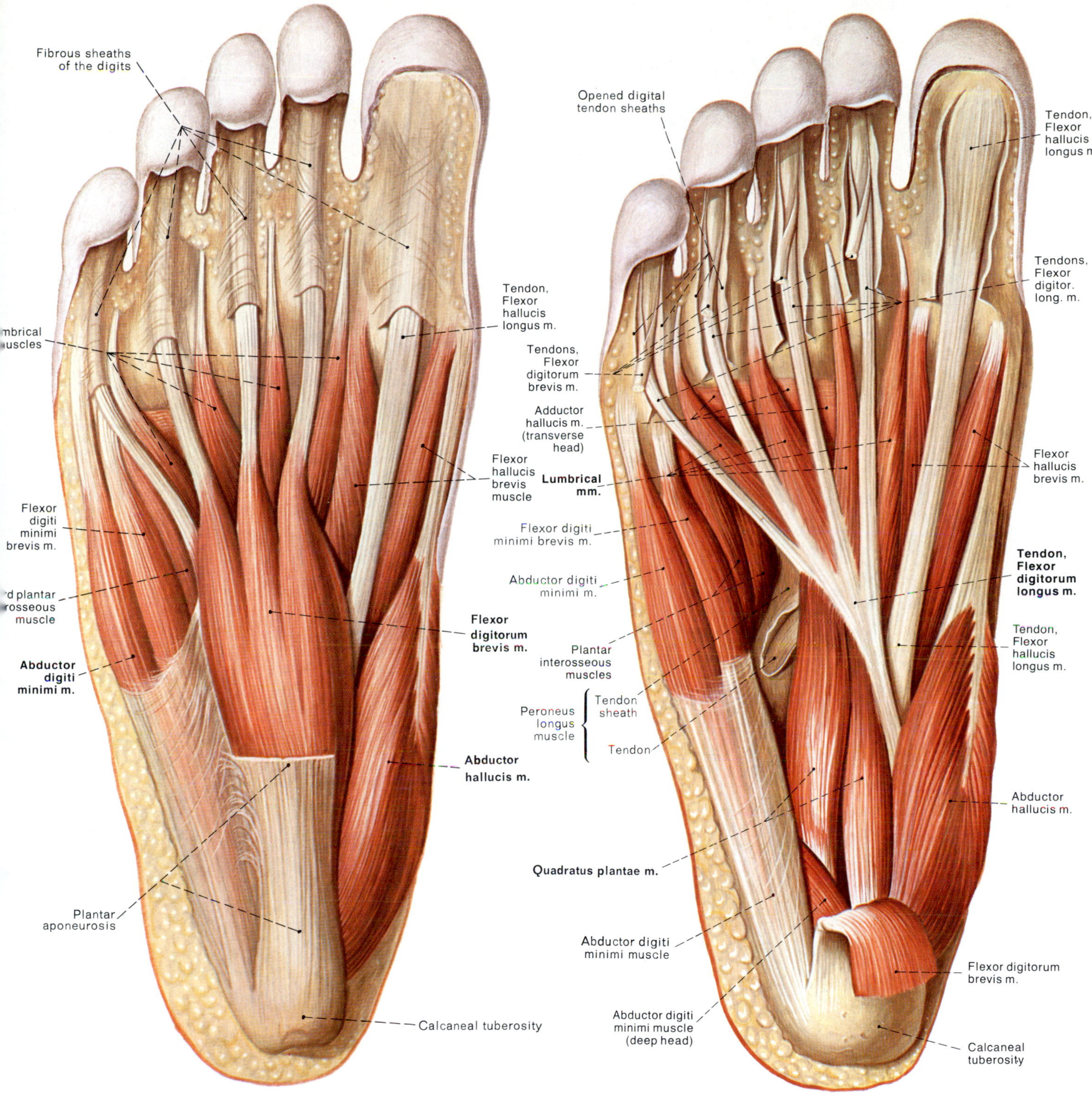
Fibrous sheaths of the digits
Tendon, Flexor hallucis longus m.
mbrical muscles
Flexor hallucis brevis muscle
Flexor digiti minimi brevis m.
d plantar rosseous muscle
Flexor digitorum brevis m.
Abductor digiti minimi m.
Abductor hallucis m.
Plantar aponeurosis
Calcaneal tuberosity
Opened digital tendon sheaths
Tendon, Flexor hallucis longus m.
Tendons, Flexor digitor. long. m.
Tendons, Flexor digitorum brevis m.
Adductor hallucis m. (transverse head)
Flexor hallucis brevis m.
Lumbrical mm.
Flexor digiti minimi brevis m.
Tendon, Flexor digitorum longus m.
Abductor digiti minimi m.
Tendon, Flexor hallucis longus m.
Plantar interosseous muscles
Peroneus longus muscle
Tendon sheath
Tendon
Abductor hallucis m.
Quadratus plantae m.
Abductor digiti minimi muscle
Flexor digitorum brevis m.
Abductor digiti minimi muscle (deep head)
Calcaneal tuberosity

Fig. 494: The Sole of the Right Foot: First Layer of Plantar Muscles

NOTE: 1) with most of the plantar aponeurosis removed, three muscles comprising the first layer of the sole are exposed. These are the abductor hallucis, the flexor digitorum brevis, and the abductor digiti minimi.

2) all three muscles arise from the tuberosity of the calcaneus. The abductor hallucis inserts on the proximal phalanx of the large toe. The flexor digitorum brevis separates into four tendons which insert onto the middle phalanges of the lateral four toes. The abductor digiti minimi inserts onto the proximal phalanx of the little toe.

Fig. 495: The Sole of the Right Foot: Second Layer of Plantar Muscles

NOTE: 1) the tendons of the flexor digitorum brevis muscle were severed and removed, thereby exposing the underlying tendons of the flexor digitorum longus muscle.

2) the muscles of the second layer in the plantar foot include the quadratus plantae muscle and the four lumbrical muscles. The quadratus plantae arises by two heads from the calcaneus and inserts into the tendon of the flexor digitorum longus.

3) the four lumbrical muscles arising from the tendons of the flexor digitorum longus muscle. They insert on the medial aspect of the first phalanx of the lateral four toes as well as on the dorsal extensor hoods.

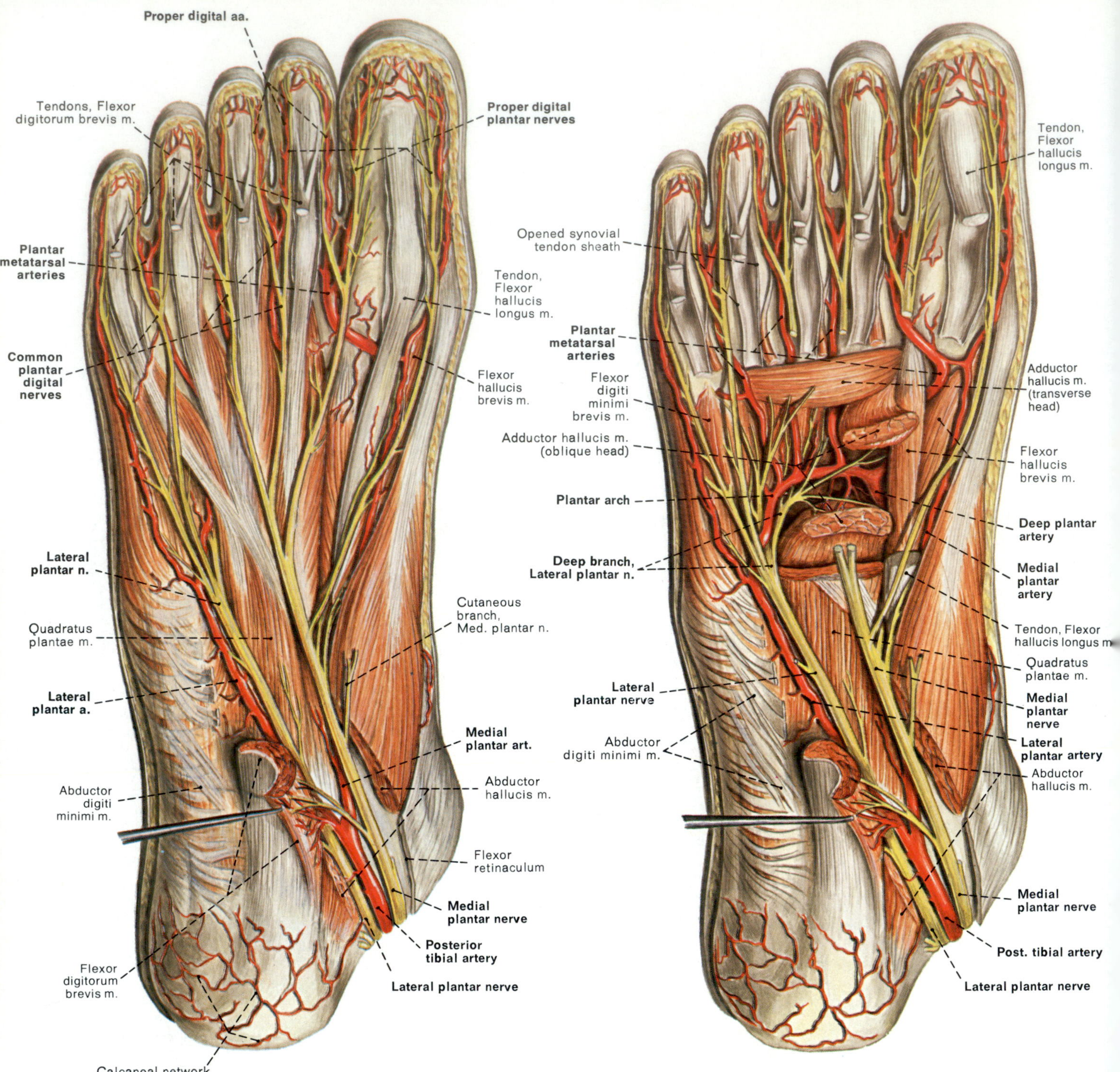

Proper digital aa.
Tendons, Flexor digitorum brevis m.
Proper digital plantar nerves
Plantar metatarsal arteries
Tendon, Flexor hallucis longus m.
Common plantar digital nerves
Flexor hallucis brevis m.
Lateral plantar n.
Quadratus plantae m.
Cutaneous branch, Med. plantar n.
Lateral plantar a.
Medial plantar art.
Abductor digiti minimi m.
Abductor hallucis m.
Flexor retinaculum
Medial plantar nerve
Posterior tibial artery
Flexor digitorum brevis m.
Lateral plantar nerve
Calcaneal network
Tendon, Flexor hallucis longus m.
Opened synovial tendon sheath
Plantar metatarsal arteries
Adductor hallucis m. (transverse head)
Flexor digiti minimi brevis m.
Adductor hallucis m. (oblique head)
Flexor hallucis brevis m.
Plantar arch
Deep plantar artery
Deep branch, Lateral plantar n.
Medial plantar artery
Tendon, Flexor hallucis longus m.
Quadratus plantae m.
Lateral plantar nerve
Medial plantar nerve
Abductor digiti minimi m.
Lateral plantar artery
Abductor hallucis m.
Medial plantar nerve
Post. tibial artery
Lateral plantar nerve

Fig. 496: The Sole of the Right Foot: the Plantar Nerves and Arteries

NOTE: 1) whereas the tibial nerve divides into medial and lateral plantar nerves just inferior to the medial malleolus, the posterior tibial artery enters the plantar surface of the foot as a single vessel and divides into medial and lateral plantar arteries beneath or at the medial border of the abductor hallucis muscle.

2) the lateral plantar nerve supplies the lateral 1−1/2 digits while the medial plantar nerve supplies the medial 3−1/2 digits. Observe the formation of the common digital plantar nerves which then divide into the proper digital plantar nerves.

3) that the main trunks of the plantar vessels and nerves cross the sole of the foot from medial to lateral deep to the flexor digitorum brevis and abductor hallucis muscles (first layer), but superficial to the quadratus plantae and lumbrical muscles (second layer).

Fig. 497: The Sole of the Right Foot: the Plantar Arch and Deep Vessels and Nerves

NOTE: 1) the formation of the deep plantar arch principally from the lateral plantar artery, and the junction of the deep plantar arch with the deep plantar artery from the foot dorsum (see Fig. 471). From the plantar arch branch plantar metatarsal arteries which then divide into proper digital arteries.

2) the muscles of the foot are innervated in the following manner:

	medial plantar nerve	*lateral plantar nerve*
1st layer	abductor hallucis flexor digitorum brevis	abductor digiti minimi
2nd layer	1st lumbrical	quadratus plantae 2nd, 3rd and 4th lumbrical
3rd layer	flexor hallucis brevis	adductor hallucis flexor digiti minimi brevis
4th layer	———	plantar interossei dorsal interossei

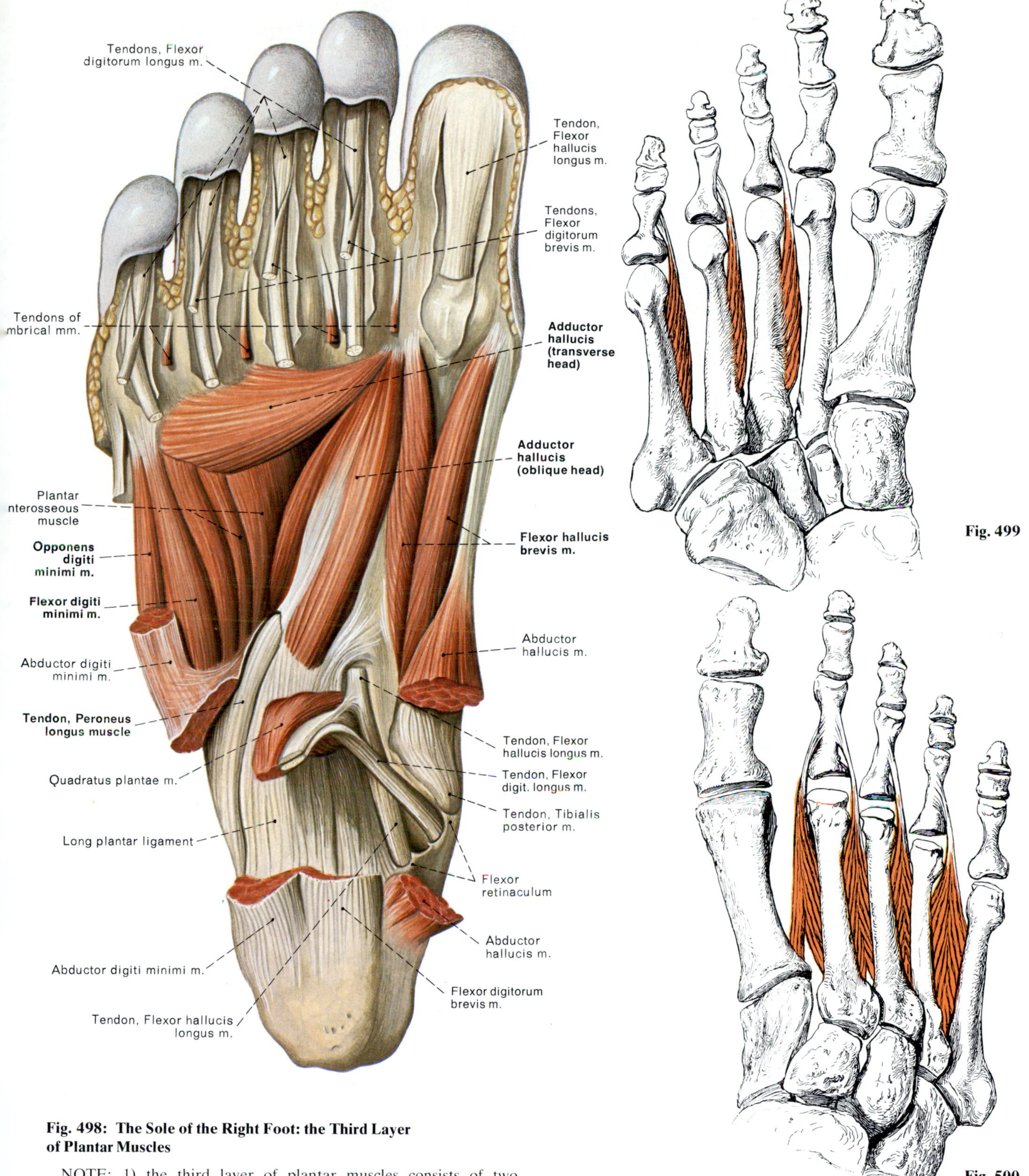

Tendons, Flexor digitorum longus m.
Tendon, Flexor hallucis longus m.
Tendons, Flexor digitorum brevis m.
Tendons of mbrical mm.
Adductor hallucis (transverse head)
Adductor hallucis (oblique head)
Plantar nterosseous muscle
Opponens digiti minimi m.
Flexor hallucis brevis m.
Flexor digiti minimi m.
Abductor hallucis m.
Abductor digiti minimi m.
Tendon, Peroneus longus muscle
Tendon, Flexor hallucis longus m.
Quadratus plantae m.
Tendon, Flexor digit. longus m.
Tendon, Tibialis posterior m.
Long plantar ligament
Flexor retinaculum
Abductor hallucis m.
Abductor digiti minimi m.
Flexor digitorum brevis m.
Tendon, Flexor hallucis longus m.
Fig. 499
Fig. 500

Fig. 498: The Sole of the Right Foot: the Third Layer of Plantar Muscles

NOTE: 1) the third layer of plantar muscles consists of two flexors and an adductor, in contrast to the first layer which contains one flexor and two abductors. Thus, the flexor hallucis brevis, flexor digiti minimi brevis and the two heads (oblique and transverse) of the adductor hallucis form the third layer of plantar muscles.

2) at times those fibers of the flexor digiti minimi brevis muscle which insert on the lateral side of the first phalanx of the 5th toe are referred to as a separate muscle, the opponens digiti minimi.

3) the tendon of the peroneus longus muscle which crosses the plantar aspect of the foot obliquely to insert on the lateral side of the base of the first metatarsal and the first (medial) cuneiform bone.

Fig. 499: The Plantar Interossei

NOTE that there are three plantar interossei. These muscles adduct the 3rd, 4th and 5th toes toward the 2nd toe which acts as the longitudinal axis of the foot.

Fig. 500: The Dorsal Interossei

NOTE that there are four dorsal interossei. These muscles abduct the toes from the reference axis. Both plantar and dorsal interossei flex the metatarsophalangeal joints and extend the interphalangeal joints.

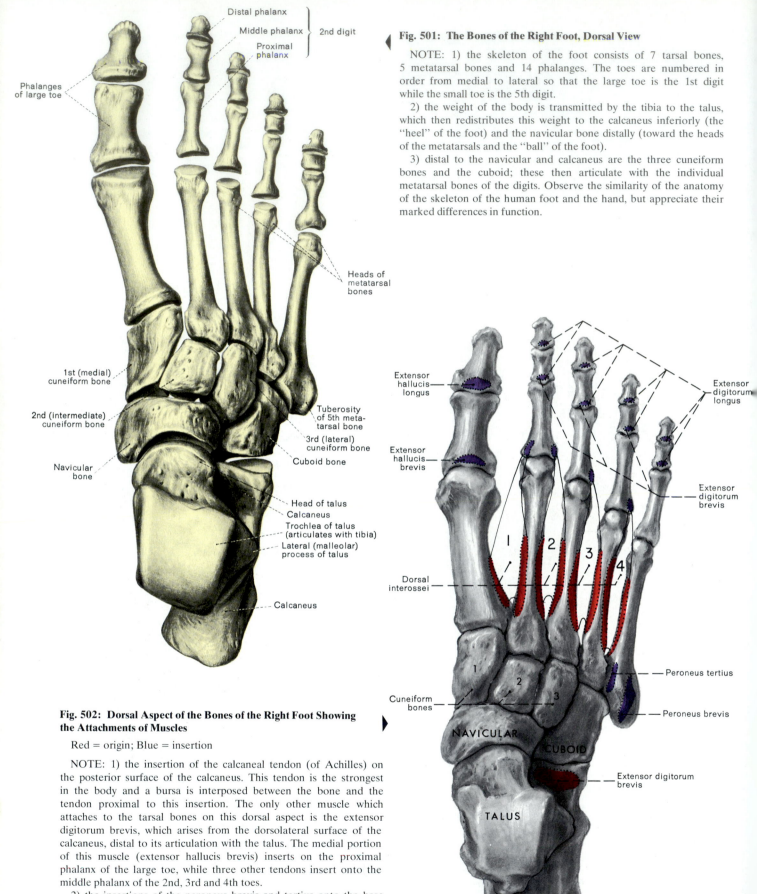

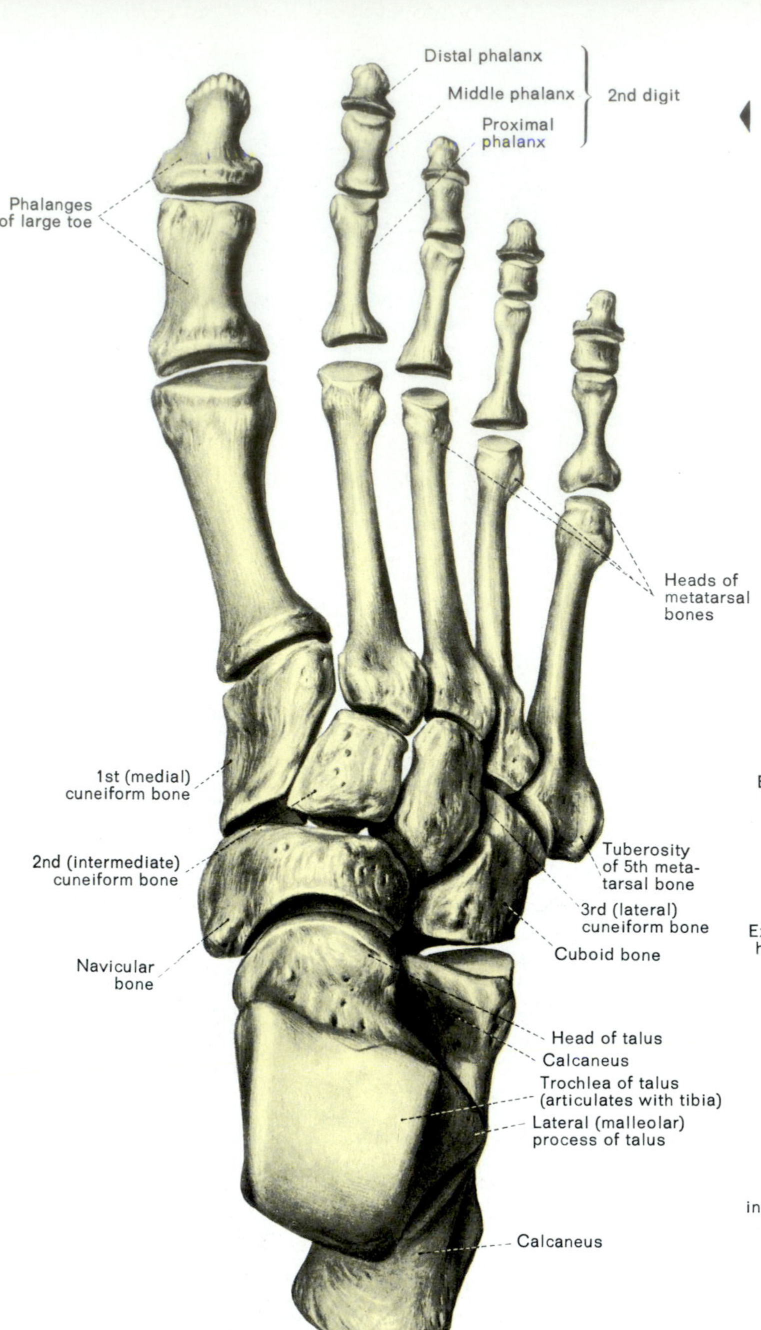

Distal phalanx
Middle phalanx
Proximal phalanx
2nd digit
Phalanges of large toe
Heads of metatarsal bones
1st (medial) cuneiform bone
2nd (intermediate) cuneiform bone
Navicular bone
Tuberosity of 5th metatarsal bone
3rd (lateral) cuneiform bone
Cuboid bone
Head of talus
Calcaneus
Trochlea of talus (articulates with tibia)
Lateral (malleolar) process of talus
Calcaneus

Fig. 501: The Bones of the Right Foot, Dorsal View

NOTE: 1) the skeleton of the foot consists of 7 tarsal bones, 5 metatarsal bones and 14 phalanges. The toes are numbered in order from medial to lateral so that the large toe is the 1st digit while the small toe is the 5th digit.

2) the weight of the body is transmitted by the tibia to the talus, which then redistributes this weight to the calcaneus inferiorly (the "heel" of the foot) and the navicular bone distally (toward the heads of the metatarsals and the "ball" of the foot).

3) distal to the navicular and calcaneus are the three cuneiform bones and the cuboid; these then articulate with the individual metatarsal bones of the digits. Observe the similarity of the anatomy of the skeleton of the human foot and the hand, but appreciate their marked differences in function.

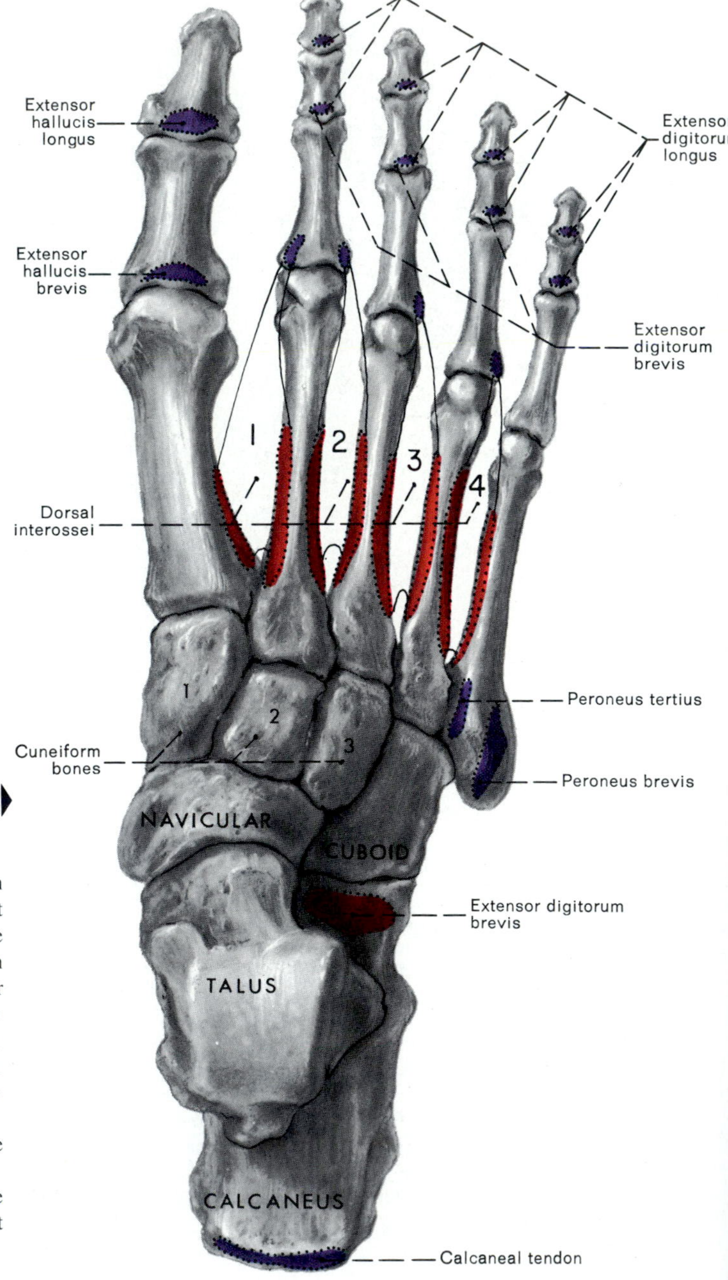

Extensor hallucis longus
Extensor digitorum longus
Extensor hallucis brevis
Extensor digitorum brevis
1
2
3
4
Dorsal interossei
Peroneus tertius
Cuneiform bones
1
2
3
Peroneus brevis
NAVICULAR
CUBOID
Extensor digitorum brevis
TALUS
CALCANEUS
Calcaneal tendon

Fig. 502: Dorsal Aspect of the Bones of the Right Foot Showing the Attachments of Muscles

Red = origin; Blue = insertion

NOTE: 1) the insertion of the calcaneal tendon (of Achilles) on the posterior surface of the calcaneus. This tendon is the strongest in the body and a bursa is interposed between the bone and the tendon proximal to this insertion. The only other muscle which attaches to the tarsal bones on this dorsal aspect is the extensor digitorum brevis, which arises from the dorsolateral surface of the calcaneus, distal to its articulation with the talus. The medial portion of this muscle (extensor hallucis brevis) inserts on the proximal phalanx of the large toe, while three other tendons insert onto the middle phalanx of the 2nd, 3rd and 4th toes.

2) the insertions of the peroneus brevis and tertius onto the base of the 5th metatarsal.

3) the 1st and 2nd dorsal interosseus muscles inserting onto the 2nd toe, while the 3rd and 4th insert onto the dorsolateral aspect of the 3rd and 4th digits. These muscles serve as abductors.

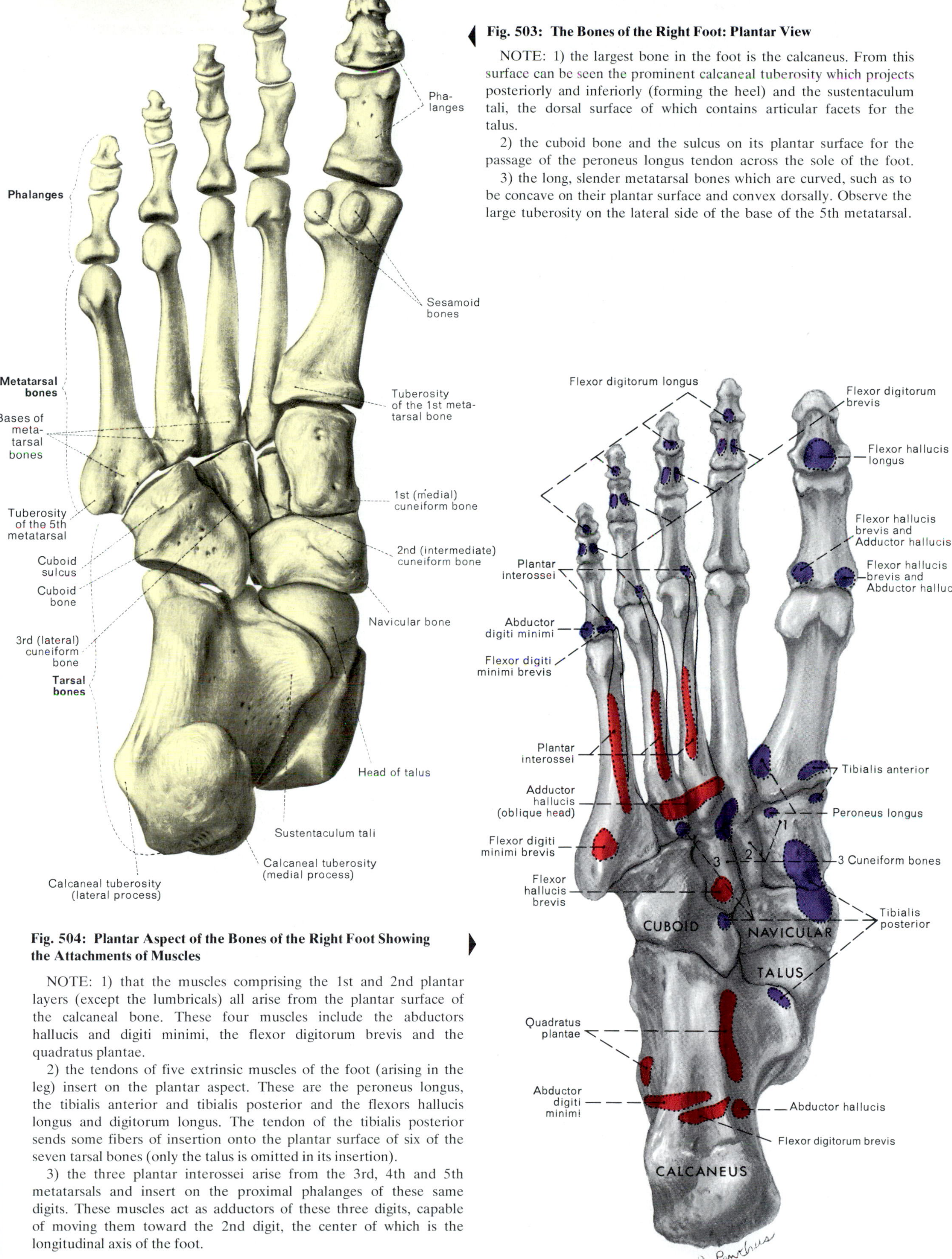

Pha-
langes
Phalanges
Sesamoid
bones
Metatarsal
bones
Bases of
meta-
tarsal
bones
Tuberosity
of the 1st meta-
tarsal bone
1st (medial)
cuneiform bone
Tuberosity
of the 5th
metatarsal
Cuboid
sulcus
2nd (intermediate)
cuneiform bone
Cuboid
bone
Navicular bone
3rd (lateral)
cuneiform
bone
Tarsal
bones
Head of talus
Sustentaculum tali
Calcaneal tuberosity
(medial process)
Calcaneal tuberosity
(lateral process)
Flexor digitorum longus
Flexor digitorum
brevis
Flexor hallucis
longus
Flexor hallucis
brevis and
Adductor hallucis
Plantar
interossei
Flexor hallucis
brevis and
Abductor hallucis
Abductor
digiti minimi
Flexor digiti
minimi brevis
Plantar
interossei
Tibialis anterior
Adductor
hallucis
(oblique head)
Peroneus longus
1
2
3
Flexor digiti
minimi brevis
3 Cuneiform bones
Flexor
hallucis
brevis
Tibialis
posterior
CUBOID
NAVICULAR
TALUS
Quadratus
plantae
Abductor
digiti
minimi
Abductor hallucis
Flexor digitorum brevis
CALCANEUS

Fig. 503: The Bones of the Right Foot: Plantar View

NOTE: 1) the largest bone in the foot is the calcaneus. From this surface can be seen the prominent calcaneal tuberosity which projects posteriorly and inferiorly (forming the heel) and the sustentaculum tali, the dorsal surface of which contains articular facets for the talus.

2) the cuboid bone and the sulcus on its plantar surface for the passage of the peroneus longus tendon across the sole of the foot.

3) the long, slender metatarsal bones which are curved, such as to be concave on their plantar surface and convex dorsally. Observe the large tuberosity on the lateral side of the base of the 5th metatarsal.

Fig. 504: Plantar Aspect of the Bones of the Right Foot Showing the Attachments of Muscles

NOTE: 1) that the muscles comprising the 1st and 2nd plantar layers (except the lumbricals) all arise from the plantar surface of the calcaneal bone. These four muscles include the abductors hallucis and digiti minimi, the flexor digitorum brevis and the quadratus plantae.

2) the tendons of five extrinsic muscles of the foot (arising in the leg) insert on the plantar aspect. These are the peroneus longus, the tibialis anterior and tibialis posterior and the flexors hallucis longus and digitorum longus. The tendon of the tibialis posterior sends some fibers of insertion onto the plantar surface of six of the seven tarsal bones (only the talus is omitted in its insertion).

3) the three plantar interossei arise from the 3rd, 4th and 5th metatarsals and insert on the proximal phalanges of these same digits. These muscles act as adductors of these three digits, capable of moving them toward the 2nd digit, the center of which is the longitudinal axis of the foot.

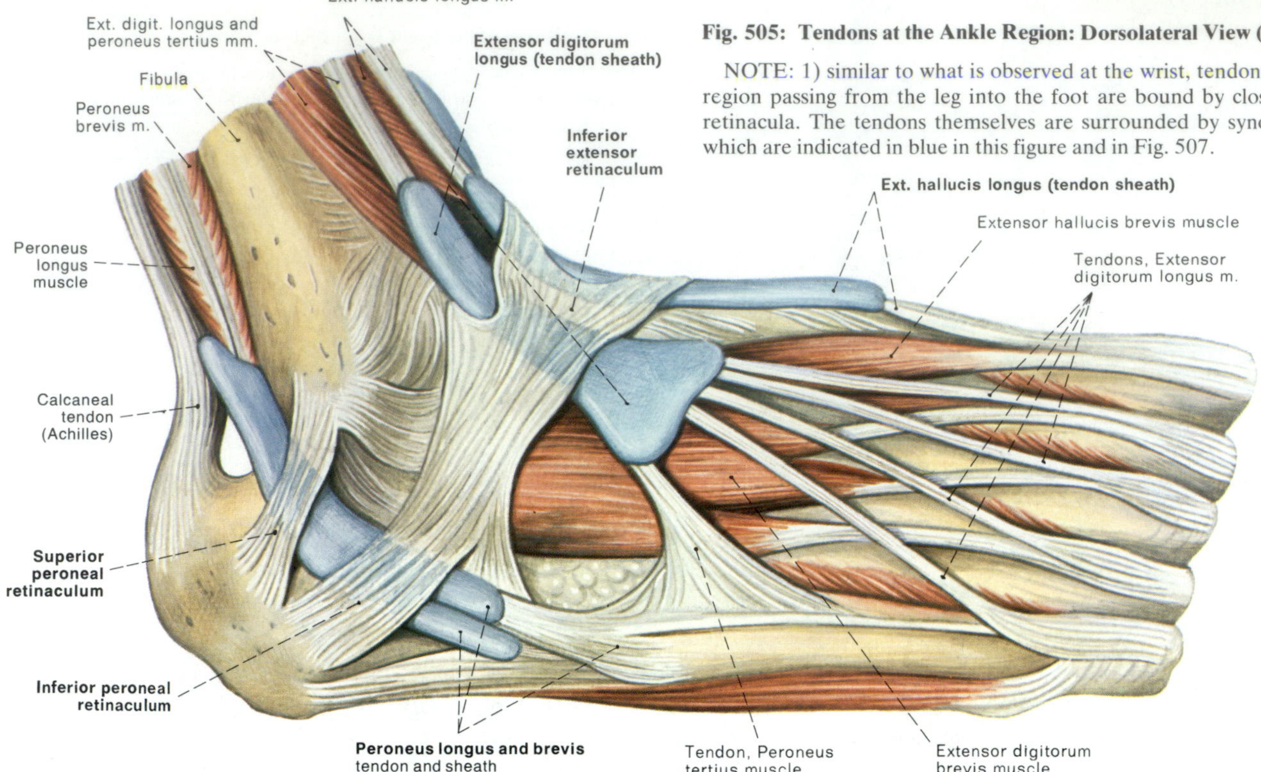
Ext. hallucis longus m.
Ext. digit. longus and peroneus tertius mm.
Fibula
Peroneus brevis m.
Extensor digitorum longus (tendon sheath)
Inferior extensor retinaculum
Ext. hallucis longus (tendon sheath)
Extensor hallucis brevis muscle
Tendons, Extensor digitorum longus m.
Peroneus longus muscle
Calcaneal tendon (Achilles)
Superior peroneal retinaculum
Inferior peroneal retinaculum
Peroneus longus and brevis tendon and sheath
Tendon, Peroneus tertius muscle
Extensor digitorum brevis muscle

Fig. 505: Tendons at the Ankle Region: Dorsolateral View (Right Foot)

NOTE: 1) similar to what is observed at the wrist, tendons at the ankle region passing from the leg into the foot are bound by closely investing retinacula. The tendons themselves are surrounded by synovial sheaths, which are indicated in blue in this figure and in Fig. 507.

2) anterior to the ankle joint and on the dorsum of the foot are three separate synovial sheaths. One is for the extensor digitorum longus and peroneus tertius, a second is for the extensor hallucis longus and the third surrounds the tibialis anterior (see Fig. 507). Behind the lateral malleolus is a single tendon sheath for the peroneus longus and brevis which then splits distally to continue along each individual tendon for some distance.

3) the inferior extensor retinaculum and the superior and inferior peroneal retinacula which bind the tendons and their sheaths close to bone.

Fig. 506: The Tendons of the Peroneus Longus and Tibialis Anterior Muscle

NOTE that the tendons of the tibialis anterior and peroneus longus muscles insert on the medial aspect of the plantar surface of the foot. The peroneus longus muscle achieves this insertion by traversing the sole of the foot from lateral to medial. In this manner, the two muscles form a tendinous sling under the foot which serves to support the transverse arch. Also assisting in this support is the tendon of the tibialis posterior muscle.

Fig. 507: Tendons at the Ankle Region: Medial View (Right Foot)

NOTE: 1) from this medial view can be seen the synovial sheaths and tendons of the tibialis anterior and extensor hallucis longus on the dorsum of the foot, as well as the three tendons which course beneath the medial malleolus from the posterior compartment of the leg into the plantar foot: tibialis posterior, flexor digitorum longus and flexor hallucis longus.

2) the bifurcating nature of the inferior extensor retinaculum, and the manner in which the flexor retinaculum secures the structures beneath the medial malleolus.

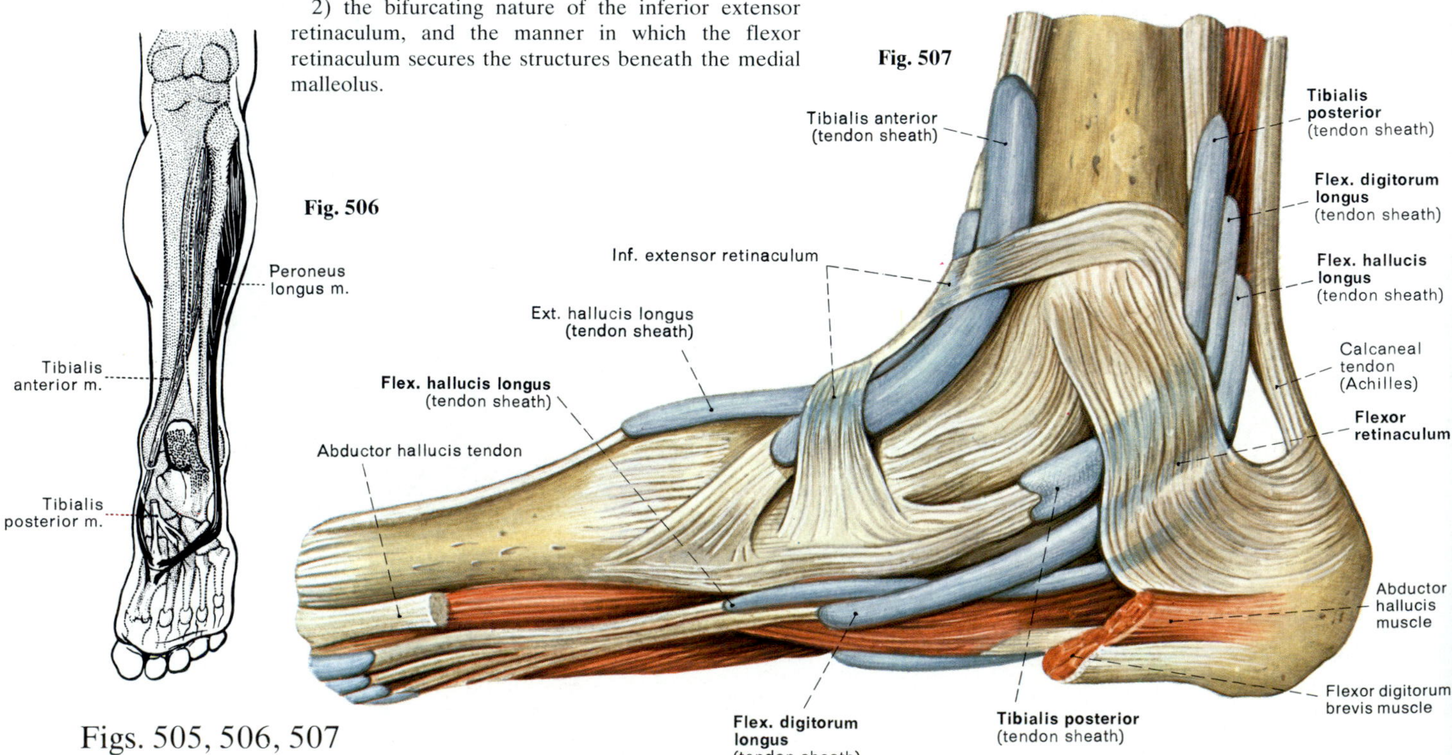
Fig. 506
Peroneus longus m.
Tibialis anterior m.
Tibialis posterior m.
Fig. 507
Tibialis anterior (tendon sheath)
Tibialis posterior (tendon sheath)
Flex. digitorum longus (tendon sheath)
Inf. extensor retinaculum
Flex. hallucis longus (tendon sheath)
Ext. hallucis longus (tendon sheath)
Calcaneal tendon (Achilles)
Flex. hallucis longus (tendon sheath)
Flexor retinaculum
Abductor hallucis tendon
Abductor hallucis muscle
Flexor digitorum brevis muscle
Flex. digitorum longus (tendon sheath)
Tibialis posterior (tendon sheath)

Figs. 505, 506, 507

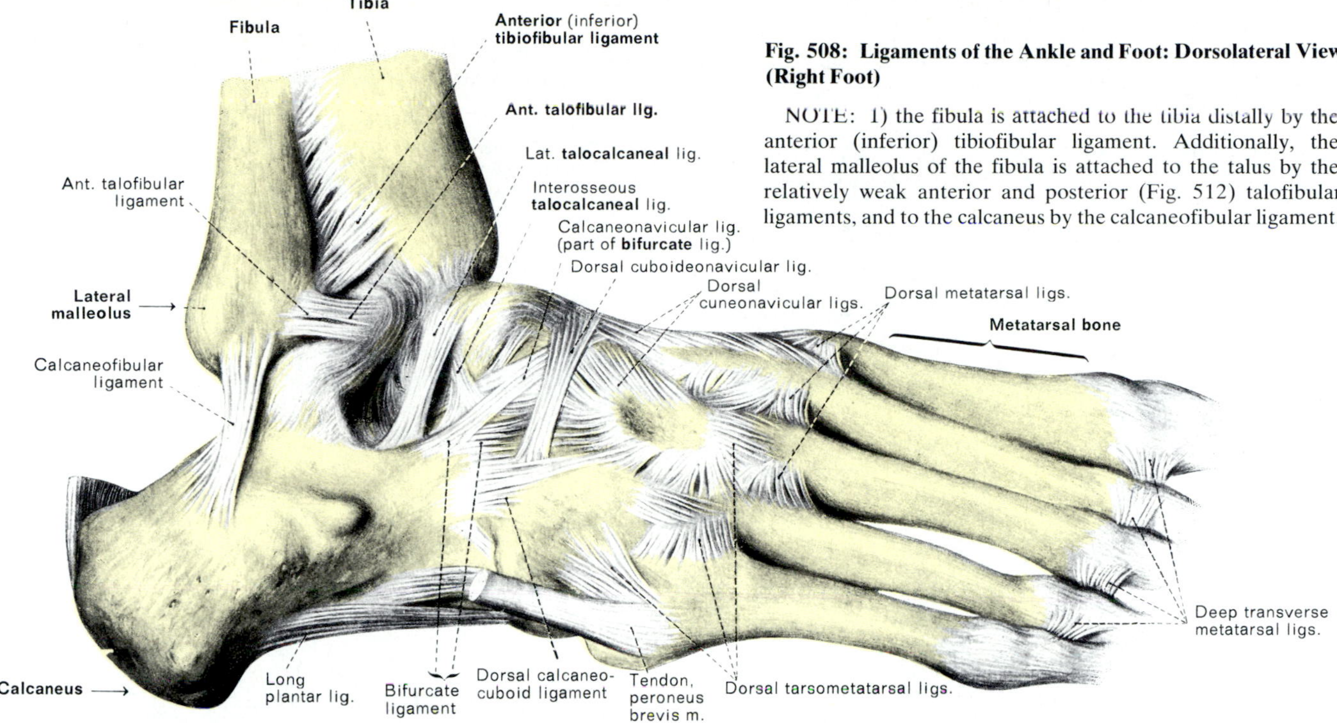
Tibia
Fibula
Anterior (inferior) tibiofibular ligament
Ant. talofibular lig.
Lat. talocalcaneal lig.
Ant. talofibular ligament
Interosseous talocalcaneal lig.
Calcaneonavicular lig. (part of bifurcate lig.)
Dorsal cuboideonavicular lig.
Lateral malleolus
Dorsal cuneonavicular ligs.
Dorsal metatarsal ligs.
Metatarsal bone
Calcaneofibular ligament
Deep transverse metatarsal ligs.
Calcaneus
Long plantar lig.
Bifurcate ligament
Dorsal calcaneo-cuboid ligament
Tendon, peroneus brevis m.
Dorsal tarsometatarsal ligs.

Fig. 508: Ligaments of the Ankle and Foot: Dorsolateral View (Right Foot)

NOTE: 1) the fibula is attached to the tibia distally by the anterior (inferior) tibiofibular ligament. Additionally, the lateral malleolus of the fibula is attached to the talus by the relatively weak anterior and posterior (Fig. 512) talofibular ligaments, and to the calcaneus by the calcaneofibular ligament.

2) the joint between the talus and calcaneus (subtalar joint) is principally strenghtened by the interosseous talocalcaneal ligament. The talocalcaneonavicular joint more anteriorly is of important clinical significance since the weight of the body tends to push the head of the talus down between the navicular and calcaneus. The stability of this joint is assisted dorsolaterally by the calcaneonavicular ligament (a part of the bifurcate ligament); however, the thick plantar calcaneonavicular or spring ligament (Figs. 509, 511, 513, 514) is the principal support for this joint in the maintenance of the longitudinal arch of the foot.

3) the bifurcate ligament consists of the calcaneonavicular ligament and the calcaneocuboid ligament.

Fig. 509: Ligaments of the Ankle and Foot: Medial View (Right Foot)

NOTE: 1) the medial aspect of the ankle joint is protected by the deltoid ligament, which is triangular in shape and which connects the tibia (medial malleolus) to the navicular, calcaneus and talus. The deltoid ligament consists of 4 parts: a) an anterior part which attaches the medial malleolus to the navicular (tibionavicular part), b) a superficial part attaching the malleolus to the sustentaculum tali of the calcaneus (tibiocalcaneal part), and c) and d) the anterior and posterior tibiotalar parts which lie more deeply and attach the malleolus to the adjacent talus.

2) the insertions of the tendons of the tibialis anterior and tibialis posterior muscles which attach on this medial aspect of the foot. Observe also the long plantar and plantar calcaneonavicular ligaments on the plantar surface. These are shown more clearly in Figures 511, 513 and 514.

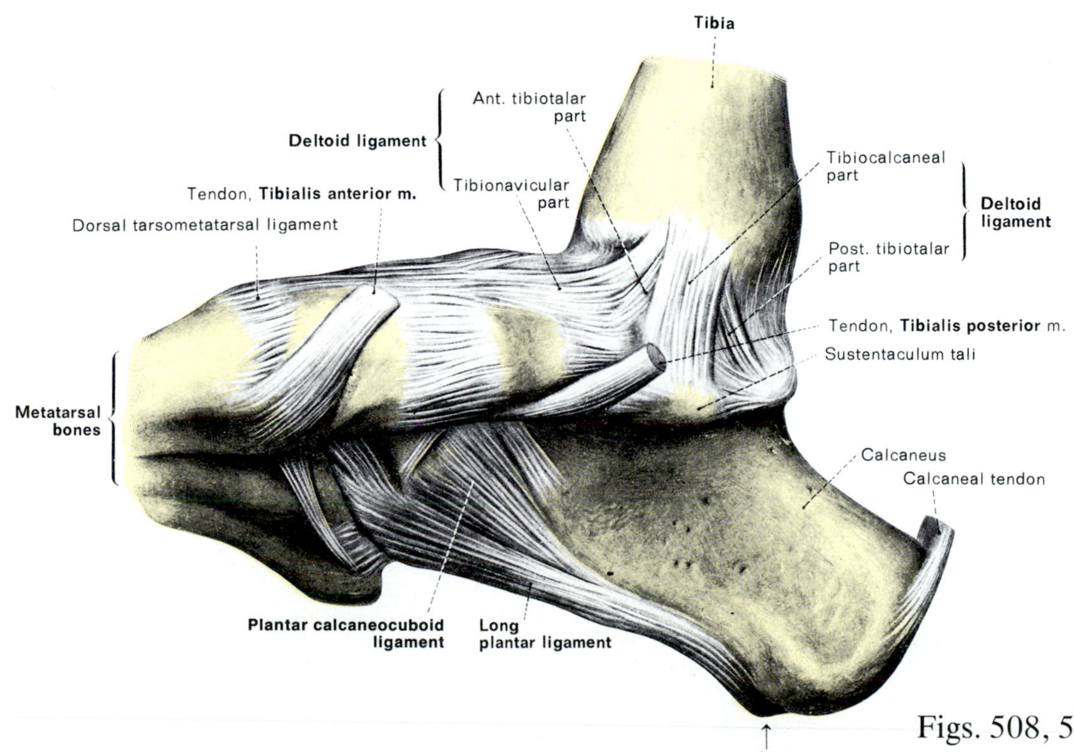
Tibia
Ant. tibiotalar part
Deltoid ligament
Tibionavicular part
Tendon, Tibialis anterior m.
Dorsal tarsometatarsal ligament
Tibiocalcaneal part
Deltoid ligament
Post. tibiotalar part
Tendon, Tibialis posterior m.
Sustentaculum tali
Metatarsal bones
Calcaneus
Calcaneal tendon
Plantar calcaneocuboid ligament
Long plantar ligament
Tuberosity of calcaneus

1st metatarsal (of large toe)
2nd metatarsal
3rd metatarsal
4th metatarsal
5th metatarsal
Tarsometatarsal joint of large toe
Interosseous metatarsal lig.
Tarsometatarsal joint
1st (med.) cuneiform
2nd (intermed.) cuneiform
Cuneonavicular joint
Navicular bone
Tarsometatarsal joint
Tuberosity of 5th metatarsal
Interosseous tarsal lig.
3rd (lat.) cuneiform
Cuboid bone
Talonavicular joint
Calcaneocuboid joint
Talus
Interosseous talocalcaneal lig.
Subtalar joint
Calcaneus

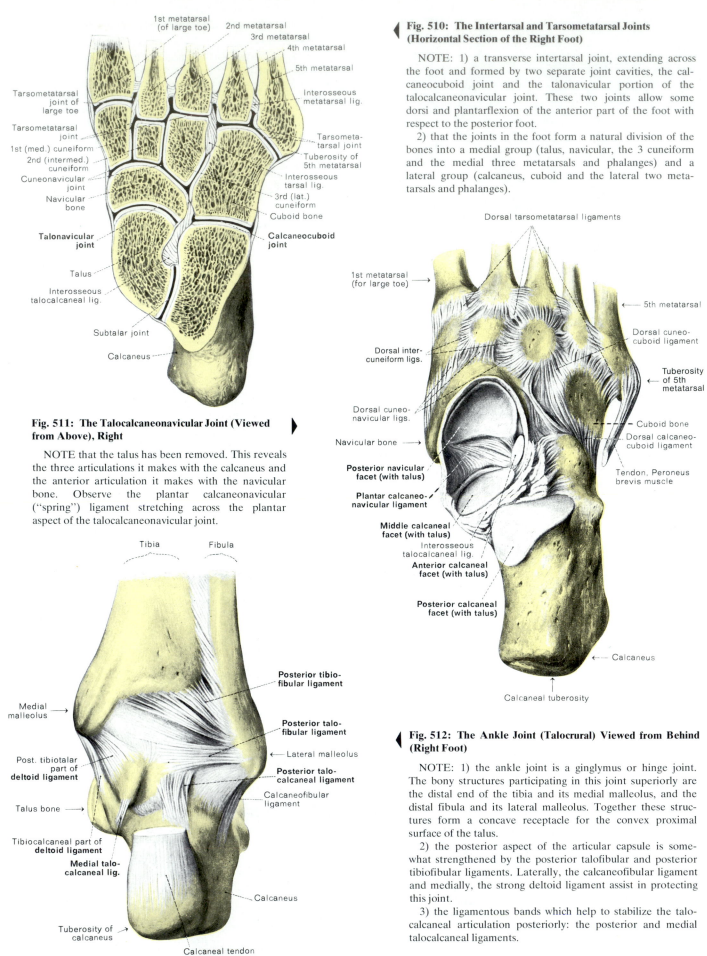

Fig. 510: The Intertarsal and Tarsometatarsal Joints (Horizontal Section of the Right Foot)

NOTE: 1) a transverse intertarsal joint, extending across the foot and formed by two separate joint cavities, the calcaneocuboid joint and the talonavicular portion of the talocalcaneonavicular joint. These two joints allow some dorsi and plantarflexion of the anterior part of the foot with respect to the posterior foot.

2) that the joints in the foot form a natural division of the bones into a medial group (talus, navicular, the 3 cuneiform and the medial three metatarsals and phalanges) and a lateral group (calcaneus, cuboid and the lateral two metatarsals and phalanges).

Dorsal tarsometatarsal ligaments
1st metatarsal (for large toe)
5th metatarsal
Dorsal cuneocuboid ligament
Dorsal intercuneiform ligs.
Tuberosity of 5th metatarsal
Dorsal cuneonavicular ligs.
Cuboid bone
Navicular bone
Dorsal calcaneocuboid ligament
Posterior navicular facet (with talus)
Tendon, Peroneus brevis muscle
Plantar calcaneonavicular ligament
Middle calcaneal facet (with talus)
Interosseous talocalcaneal lig.
Anterior calcaneal facet (with talus)
Posterior calcaneal facet (with talus)
Calcaneus
Calcaneal tuberosity

Fig. 511: The Talocalcaneonavicular Joint (Viewed from Above), Right

NOTE that the talus has been removed. This reveals the three articulations it makes with the calcaneus and the anterior articulation it makes with the navicular bone. Observe the plantar calcaneonavicular ("spring") ligament stretching across the plantar aspect of the talocalcaneonavicular joint.

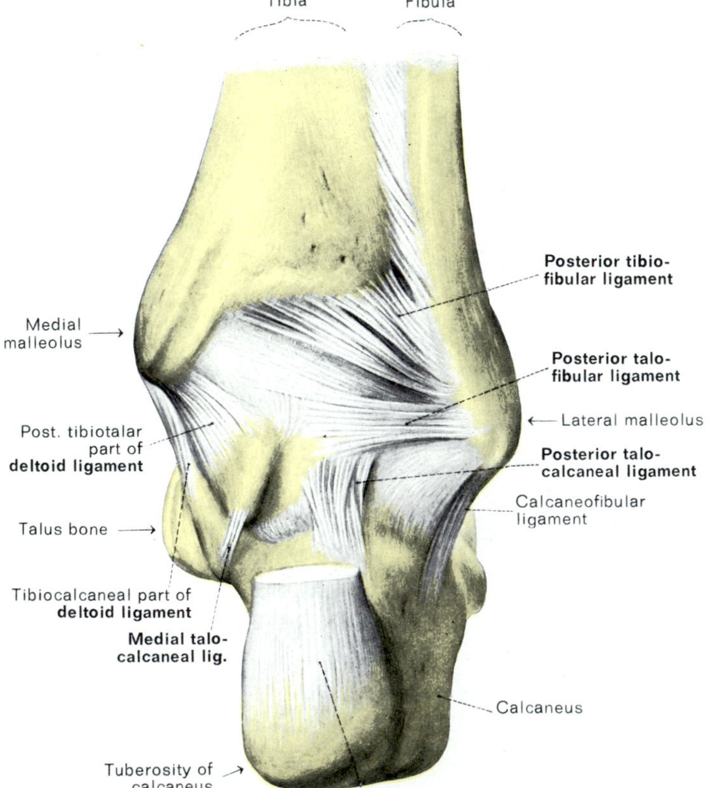
Tibia
Fibula
Posterior tibiofibular ligament
Medial malleolus
Posterior talofibular ligament
Lateral malleolus
Post. tibiotalar part of deltoid ligament
Posterior talocalcaneal ligament
Calcaneofibular ligament
Talus bone
Tibiocalcaneal part of deltoid ligament
Medial talocalcaneal lig.
Calcaneus
Tuberosity of calcaneus
Calcaneal tendon

Fig. 512: The Ankle Joint (Talocrural) Viewed from Behind (Right Foot)

NOTE: 1) the ankle joint is a ginglymus or hinge joint. The bony structures participating in this joint superiorly are the distal end of the tibia and its medial malleolus, and the distal fibula and its lateral malleolus. Together these structures form a concave receptacle for the convex proximal surface of the talus.

2) the posterior aspect of the articular capsule is somewhat strengthened by the posterior talofibular and posterior tibiofibular ligaments. Laterally, the calcaneofibular ligament and medially, the strong deltoid ligament assist in protecting this joint.

3) the ligamentous bands which help to stabilize the talocalcaneal articulation posteriorly: the posterior and medial talocalcaneal ligaments.

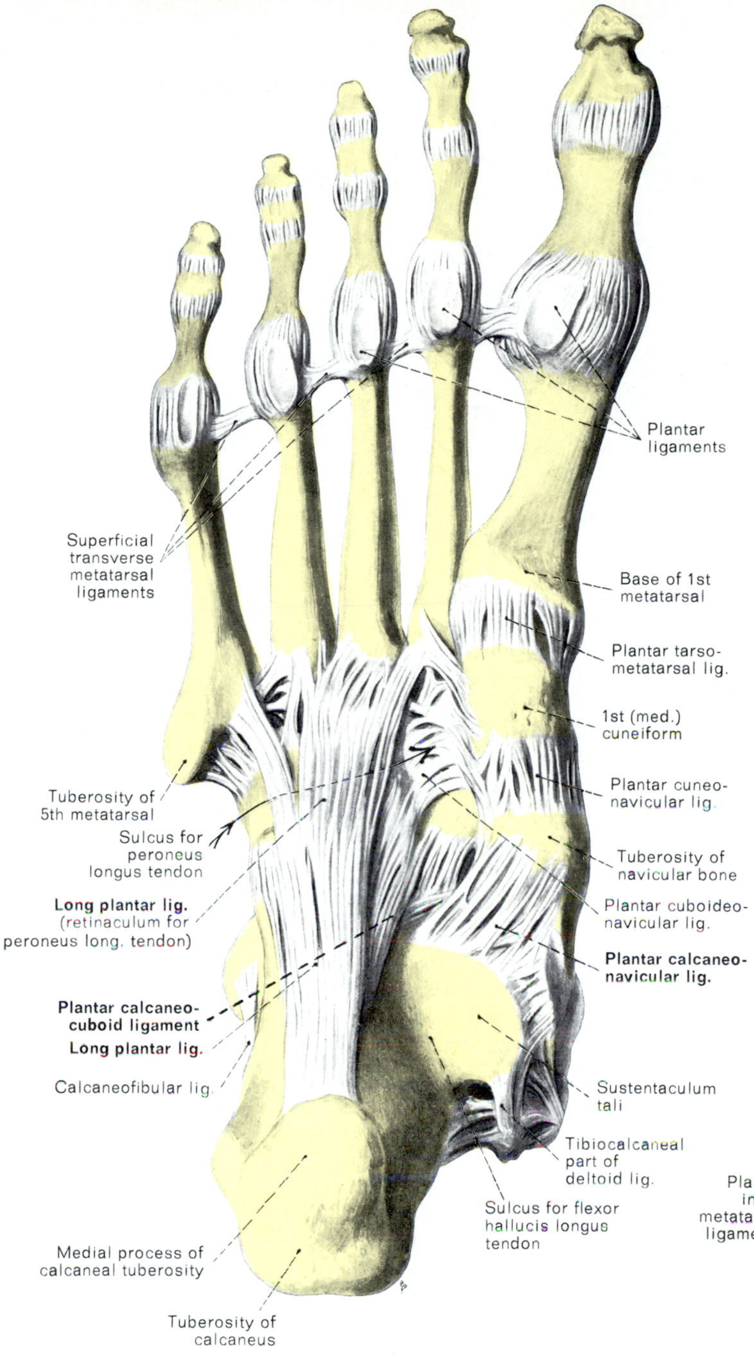

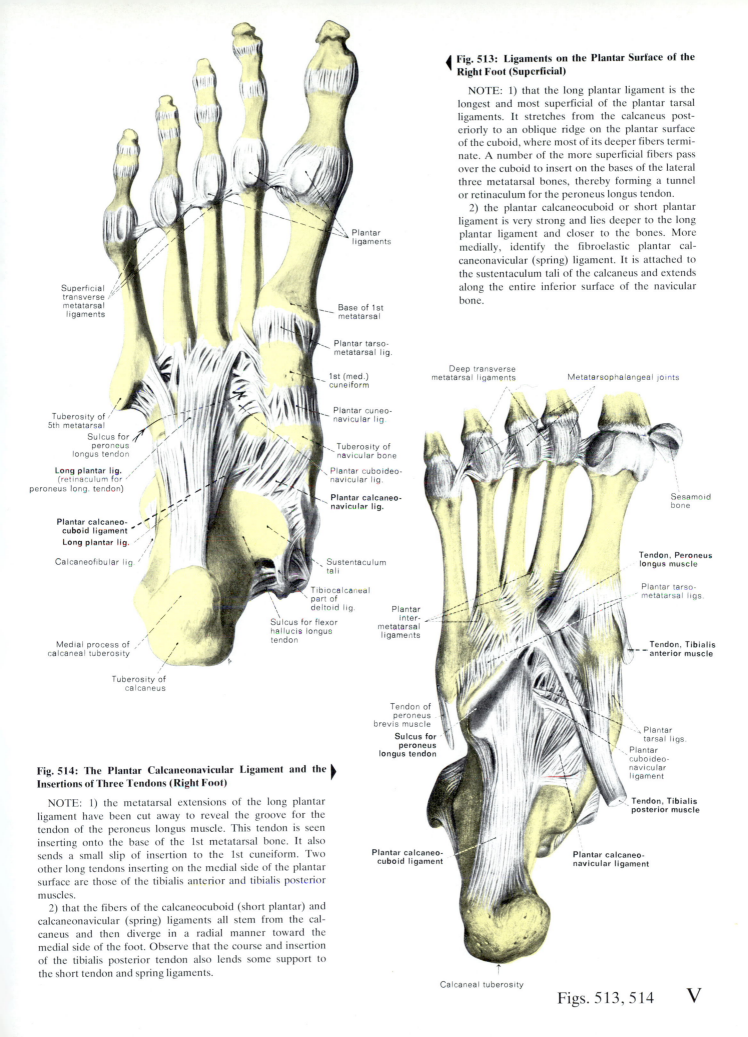

Plantar ligaments
Superficial transverse metatarsal ligaments
Base of 1st metatarsal
Plantar tarso-metatarsal lig.
1st (med.) cuneiform
Tuberosity of 5th metatarsal
Plantar cuneo-navicular lig.
Sulcus for peroneus longus tendon
Tuberosity of navicular bone
Long plantar lig. (retinaculum for peroneus long. tendon)
Plantar cuboideo-navicular lig.
Plantar calcaneo-navicular lig.
Plantar calcaneo-cuboid ligament
Long plantar lig.
Calcaneofibular lig.
Sustentaculum tali
Tibiocalcaneal part of deltoid lig.
Sulcus for flexor hallucis longus tendon
Medial process of calcaneal tuberosity
Tuberosity of calcaneus

Fig. 513: Ligaments on the Plantar Surface of the Right Foot (Superficial)

NOTE: 1) that the long plantar ligament is the longest and most superficial of the plantar tarsal ligaments. It stretches from the calcaneus posteriorly to an oblique ridge on the plantar surface of the cuboid, where most of its deeper fibers terminate. A number of the more superficial fibers pass over the cuboid to insert on the bases of the lateral three metatarsal bones, thereby forming a tunnel or retinaculum for the peroneus longus tendon.

2) the plantar calcaneocuboid or short plantar ligament is very strong and lies deeper to the long plantar ligament and closer to the bones. More medially, identify the fibroelastic plantar calcaneonavicular (spring) ligament. It is attached to the sustentaculum tali of the calcaneus and extends along the entire inferior surface of the navicular bone.

Fig. 514: The Plantar Calcaneonavicular Ligament and the Insertions of Three Tendons (Right Foot)

NOTE: 1) the metatarsal extensions of the long plantar ligament have been cut away to reveal the groove for the tendon of the peroneus longus muscle. This tendon is seen inserting onto the base of the 1st metatarsal bone. It also sends a small slip of insertion to the 1st cuneiform. Two other long tendons inserting on the medial side of the plantar surface are those of the tibialis anterior and tibialis posterior muscles.

2) that the fibers of the calcaneocuboid (short plantar) and calcaneonavicular (spring) ligaments all stem from the calcaneus and then diverge in a radial manner toward the medial side of the foot. Observe that the course and insertion of the tibialis posterior tendon also lends some support to the short tendon and spring ligaments.

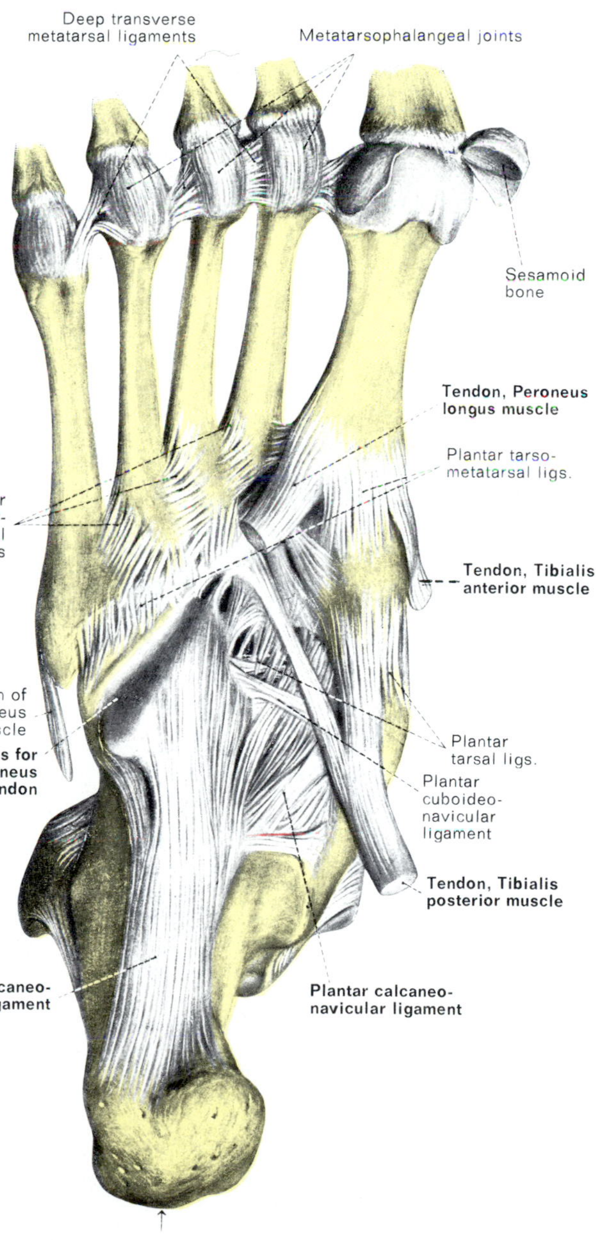
Deep transverse metatarsal ligaments
Metatarsophalangeal joints
Sesamoid bone
Tendon, Peroneus longus muscle
Plantar tarso-metatarsal ligs.
Plantar inter-metatarsal ligaments
Tendon, Tibialis anterior muscle
Tendon of peroneus brevis muscle
Sulcus for peroneus longus tendon
Plantar tarsal ligs.
Plantar cuboideo-navicular ligament
Tendon, Tibialis posterior muscle
Plantar calcaneo-cuboid ligament
Plantar calcaneo-navicular ligament
Calcaneal tuberosity

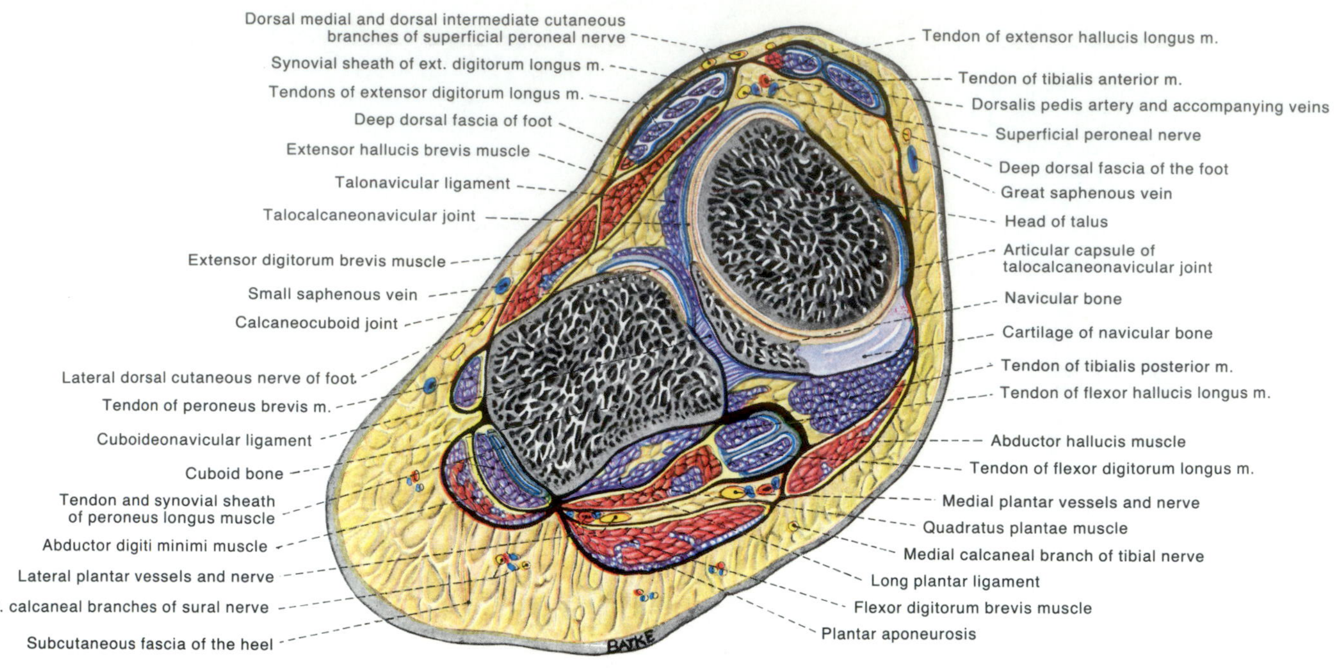
Dorsal medial and dorsal intermediate cutaneous branches of superficial peroneal nerve
Synovial sheath of ext. digitorum longus m.
Tendons of extensor digitorum longus m.
Deep dorsal fascia of foot
Extensor hallucis brevis muscle
Talonavicular ligament
Talocalcaneonavicular joint
Extensor digitorum brevis muscle
Small saphenous vein
Calcaneocuboid joint
Lateral dorsal cutaneous nerve of foot
Tendon of peroneus brevis m.
Cuboideonavicular ligament
Cuboid bone
Tendon and synovial sheath of peroneus longus muscle
Abductor digiti minimi muscle
Lateral plantar vessels and nerve
Lat. calcaneal branches of sural nerve
Subcutaneous fascia of the heel
Tendon of extensor hallucis longus m.
Tendon of tibialis anterior m.
Dorsalis pedis artery and accompanying veins
Superficial peroneal nerve
Deep dorsal fascia of the foot
Great saphenous vein
Head of talus
Articular capsule of talocalcaneonavicular joint
Navicular bone
Cartilage of navicular bone
Tendon of tibialis posterior m.
Tendon of flexor hallucis longus m.
Abductor hallucis muscle
Tendon of flexor digitorum longus m.
Medial plantar vessels and nerve
Quadratus plantae muscle
Medial calcaneal branch of tibial nerve
Long plantar ligament
Flexor digitorum brevis muscle
Plantar aponeurosis
BATKE

Fig. 515: Cross Section Through the Right Foot at the Level of the Head of the Talus (Distal Surface)

NOTE that the tendon sheaths and synovial membranes are shown in light blue. The level of this cross section is shown in the diagram Figure 488.

Tendons of extensor digitorum longus m.
Tendons of extensor digitorum brevis m.
Dorsal digital nerve
Fourth tendon of extensor digitorum longus m.
Dorsal metatarsal vein
Dorsal interfascial space
Second plantar interosseous muscle
Deep layer of dorsal fascia of foot
Fourth dorsal interosseous muscle
Tendon of peroneus tertius m.
Lateral dorsal cutaneous nerve (sural); Small saphenous vein
Third plantar metatarsal a. and v.
Abductor digiti minimi muscle
Opponens digiti minimi muscle
Third plantar interosseous muscle
Deep fascia over lateral plantar eminence
Flexor digiti minimi brevis m.
Superficial fascia
Proper plantar digital n. and a.
Deep branch, lateral plantar nerve
Plantar metatarsal artery
Common plantar digital nerve
Tendon of flexor digitorum longus m.; Lumbrical muscle
Common plantar digital nerve
Plantar aponeurosis
Cutaneous branch of lateral plantar nerve
Tendon of flexor digitorum brevis m.
First plantar interosseous muscle
Dorsal digital nerve
Dorsal metatarsal vein
Dorsal metatarsal artery
Tendon of extensor digitorum longus m.
Deep dorsal fascia of foot
Dorsal digital nerves; Dorsal metatarsal vein
Dorsal metatarsal artery and vein
Tendon of extensor hallucis brevis m.
Tendon of extensor hallucis longus m.
Second dorsal interosseous muscle
First dorsal interosseous muscle; Perforating br., plantar metatarsal a.
Plantar metatarsal artery
Abductor hallucis muscle
Adductor hallucis muscle (oblique head)
Flexor hallucis brevis muscle
Plantar proper digital a., v., n.
Tendon of flexor hallucis longus m.
Tendon of flexor digitorum longus m.; Lumbrical muscle
Cutaneous branch of medial plantar nerve
Tendon of flexor digitorum longus m.; Lumbrical muscle
Tendon of flexor digitorum brevis m.
Third dorsal interosseous muscle
Common plantar digital nerve
Tendon of flexor digitorum brevis m.
Tendon of flexor digitorum longus m.; Lumbrical muscle
Second common plantar digital nerve
V
IV
III
II
I
BATKE

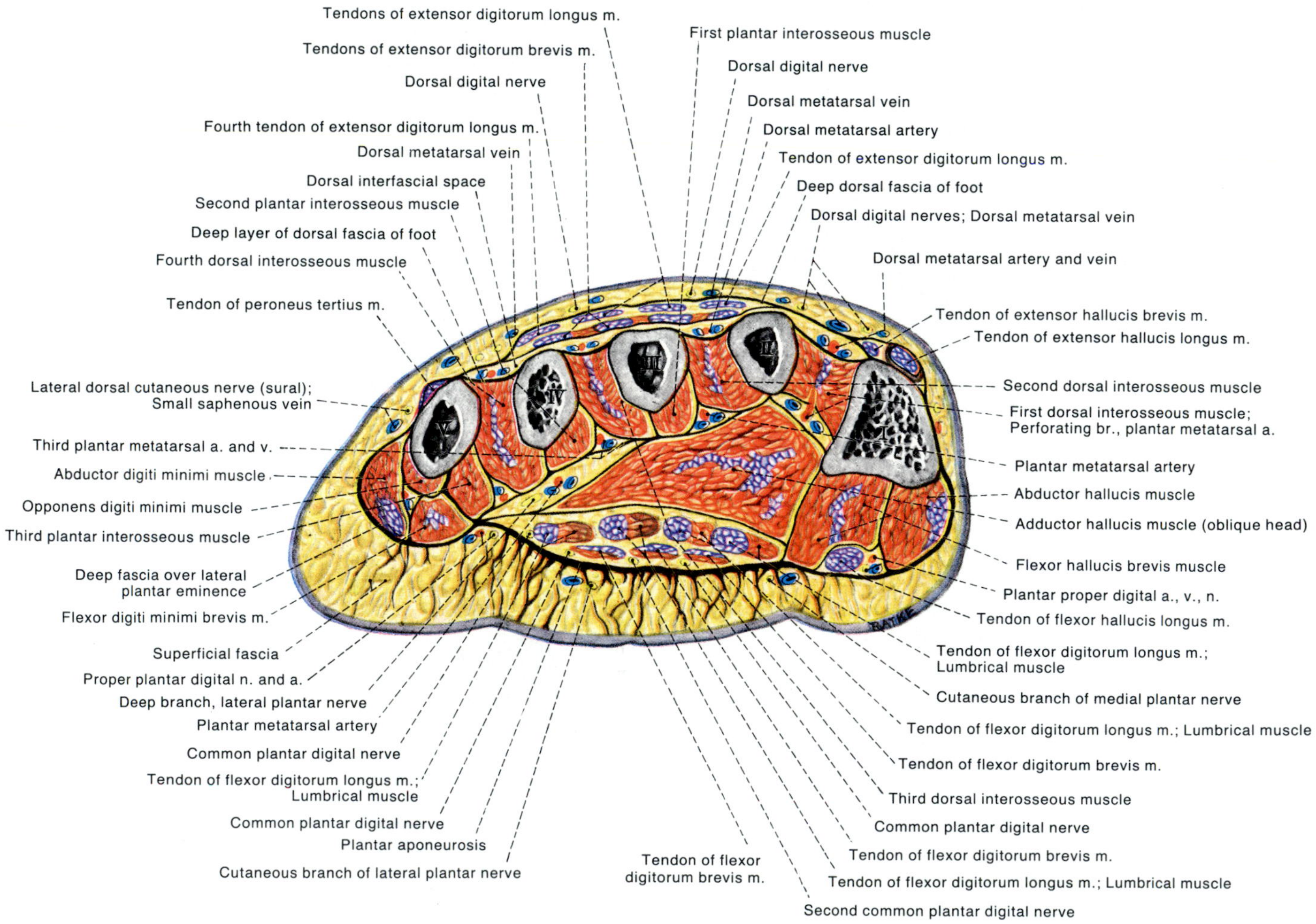

Fig. 516: Cross Section of the Right Foot Through the Metatarsal Bones (Distal Surface)

NOTE that the level of this cross section is shown in the diagram Figure 488.

PART VI: THE BACK, VERTEBRAL COLUMN AND SPINAL CORD

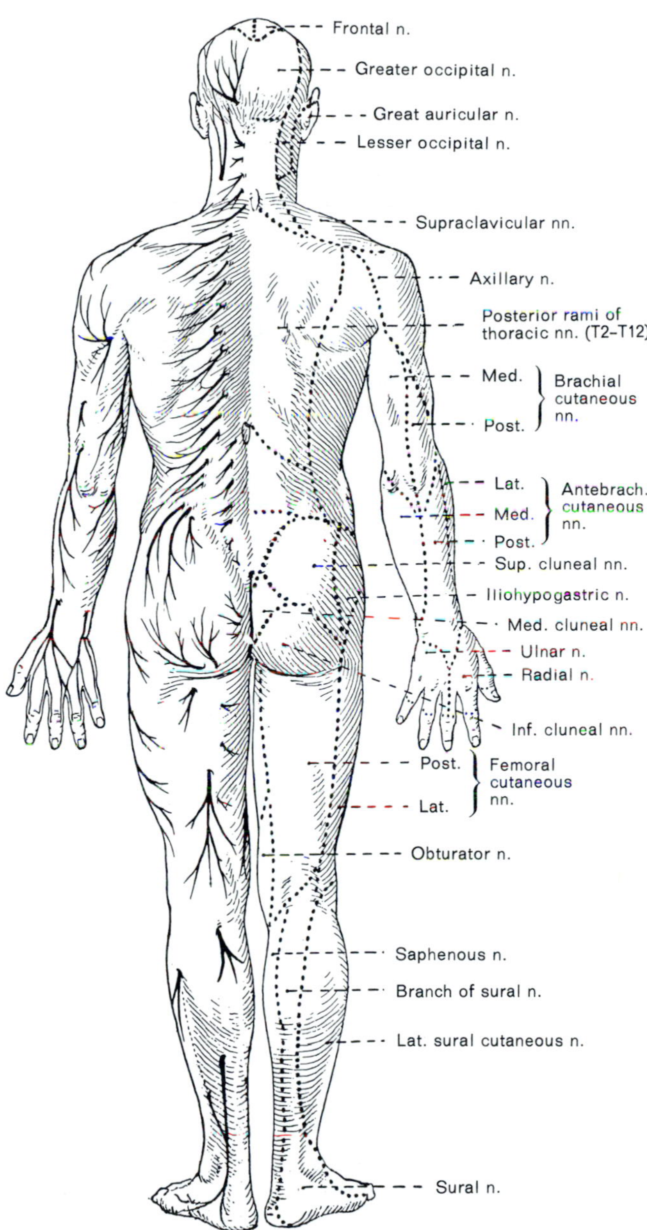

Frontal n.
Greater occipital n.
Great auricular n.
Lesser occipital n.
Supraclavicular nn.
Axillary n.
Posterior rami of thoracic nn. (T2–T12)
Med.
Post.
Brachial cutaneous nn.
Lat.
Med.
Post.
Antebrach. cutaneous nn.
Sup. cluneal nn.
Iliohypogastric n.
Med. cluneal nn.
Ulnar n.
Radial n.
Inf. cluneal nn.
Post.
Lat.
Femoral cutaneous nn.
Obturator n.
Saphenous n.
Branch of sural n.
Lat. sural cutaneous n.
Sural n.

Fig. 517: The Cutaneous Nerve Surface Areas, Posterior Aspect of the Body

NOTE that on the left side the position and course of the spinal cutaneous nerves are shown, while on the right side the surface areas of distribution are indicated. Demonstrated are the posterior surface zones for the cervical, thoracic, lumbar and sacral segmental nerves.

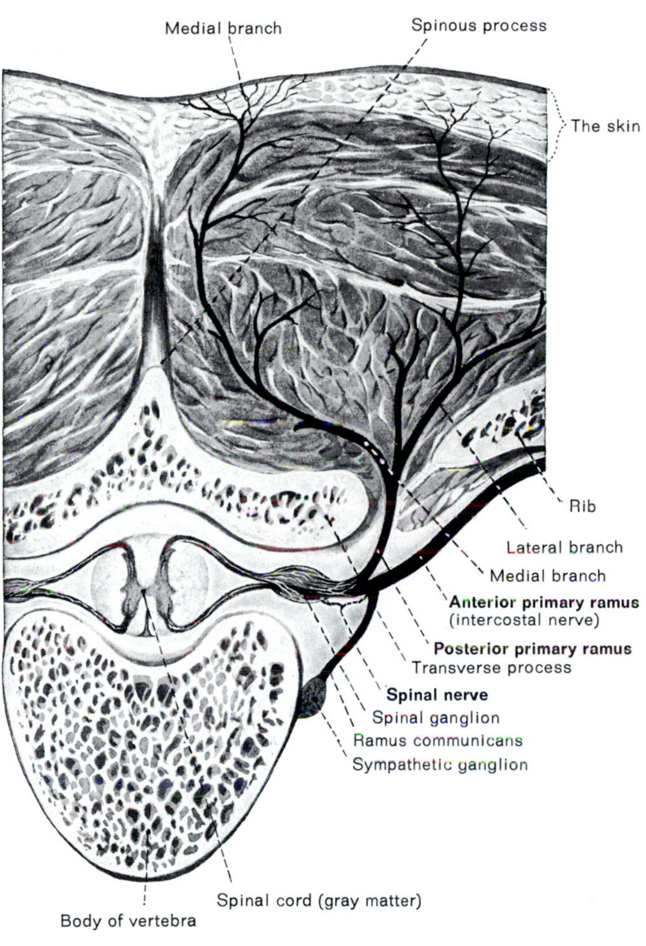

Medial branch
Spinous process
The skin
Rib
Lateral branch
Medial branch
Anterior primary ramus (intercostal nerve)
Posterior primary ramus
Transverse process
Spinal nerve
Spinal ganglion
Ramus communicans
Sympathetic ganglion
Spinal cord (gray matter)
Body of vertebra

Fig. 518: The Branching of a Typical Spinal Nerve

NOTE: 1) fibers from both dorsal and motor roots join to form a spinal nerve which soon divides into a posterior and an anterior primary ramus. The posterior primary ramus courses dorsally to innervate the muscles and skin of the back. The anterior primary ramus courses laterally and anteriorly around the body, to innervate the remainder of the segment.

2) the posterior primary divisions of typical spinal nerves are smaller in diameter than the anterior divisions, and each usually divides into medial and lateral branches which contain both motor and sensory fibers innervating structures in the back.

3) unlike the anterior primary divisions of the spinal nerves which join to form the cervical, brachial and lumbosacral plexuses, the peripheral nerves derived from posterior primary divisions, as a rule, do not intercommunicate and form plexuses. There is, however, some segmental overlap of the peripheral sensory fields, as occurs with anterior primary division nerves.

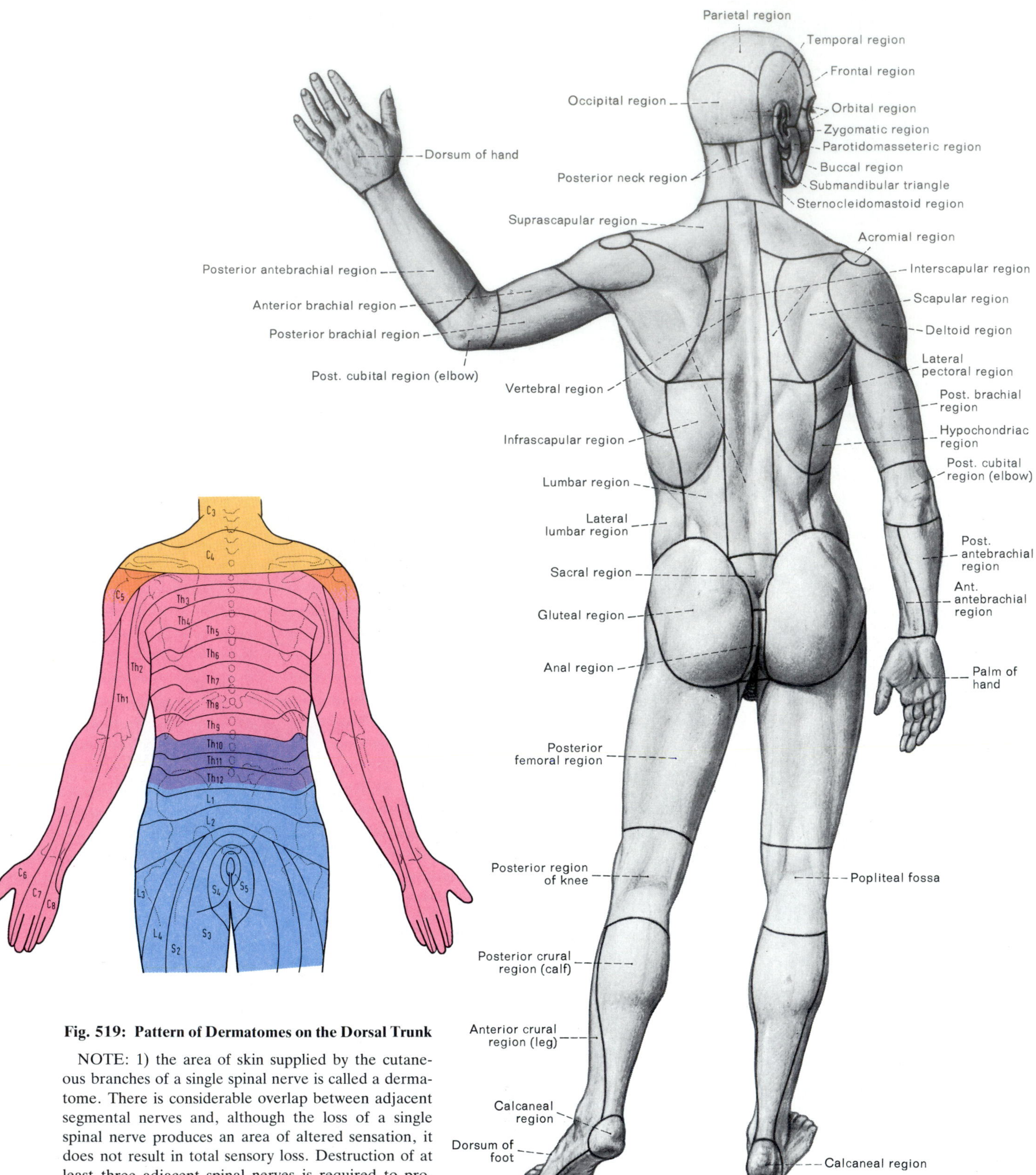

C3
C4
C5
Th3
Th4
Th5
Th6
Th7
Th8
Th9
Th10
Th11
Th12
Th2
Th1
L1
L2
L3
L4
S2
S3
S4
S5
C6
C7
C8
Parietal region
Temporal region
Frontal region
Occipital region
Orbital region
Zygomatic region
Parotidomasseteric region
Dorsum of hand
Buccal region
Posterior neck region
Submandibular triangle
Sternocleidomastoid region
Suprascapular region
Acromial region
Posterior antebrachial region
Interscapular region
Anterior brachial region
Scapular region
Posterior brachial region
Deltoid region
Post. cubital region (elbow)
Lateral pectoral region
Vertebral region
Post. brachial region
Infrascapular region
Hypochondriac region
Post. cubital region (elbow)
Lumbar region
Lateral lumbar region
Post. antebrachial region
Sacral region
Ant. antebrachial region
Gluteal region
Anal region
Palm of hand
Posterior femoral region
Posterior region of knee
Popliteal fossa
Posterior crural region (calf)
Anterior crural region (leg)
Calcaneal region
Dorsum of foot
Calcaneal region
Sole of foot

Fig. 519: Pattern of Dermatomes on the Dorsal Trunk

NOTE: 1) the area of skin supplied by the cutaneous branches of a single spinal nerve is called a dermatome. There is considerable overlap between adjacent segmental nerves and, although the loss of a single spinal nerve produces an area of altered sensation, it does not result in total sensory loss. Destruction of at least three adjacent spinal nerves is required to produce a total sensory loss of the dermatome supplied by the middle nerve of the three.

2) mapping of skin areas affected by herpes zoster (shingles) has also allowed the mapping of dermatomes. Another method is the method of "remaining sensibility". In the latter, dermatome areas are established after the destruction of several roots above and below the intact root whose dermatome is being studied.

Fig. 520: The Regions of the Body: Posterior View

NOTE: the posterior aspect of the head, trunk and limbs is subdivided into many topographic regions to allow more exact anatomical localization and communication. Although the boundaries between the regions are somewhat arbitrary, it can be observed that the regions assume the names of bony structures, muscles, organs, joints and orifices comparable to those observed on the anterior aspect of the body (Figure 1).

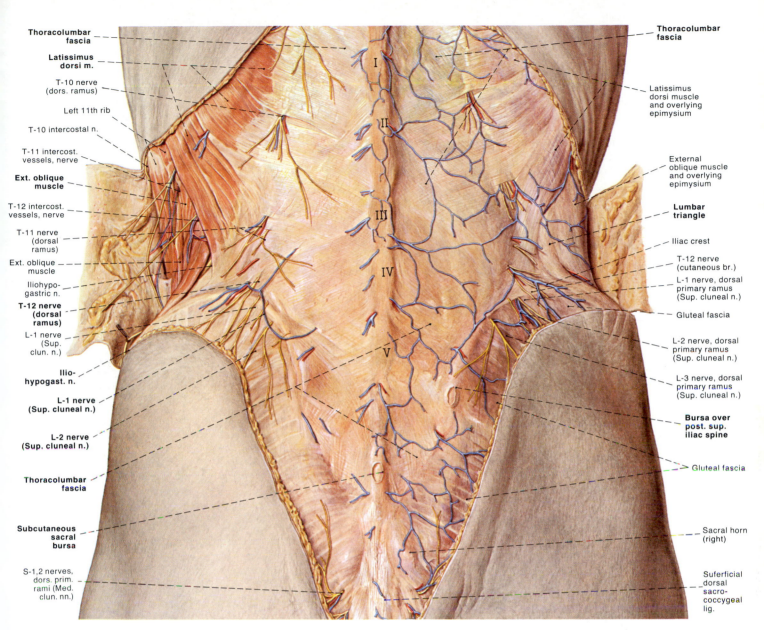

Thoracolumbar fascia
Latissimus dorsi m.
T-10 nerve (dors. ramus)
Left 11th rib
T-10 intercostal n.
T-11 intercost. vessels, nerve
Ext. oblique muscle
T-12 intercost. vessels, nerve
T-11 nerve (dorsal ramus)
Ext. oblique muscle
Iliohypo-gastric n.
T-12 nerve (dorsal ramus)
L-1 nerve (Sup. clun. n.)
Ilio-hypogast. n.
L-1 nerve (Sup. cluneal n.)
L-2 nerve (Sup. cluneal n.)
Thoracolumbar fascia
Subcutaneous sacral bursa
S-1,2 nerves, dors. prim. rami (Med. clun. nn.)
I
II
III
IV
V
Thoracolumbar fascia
Latissimus dorsi muscle and overlying epimysium
External oblique muscle and overlying epimysium
Lumbar triangle
Iliac crest
T-12 nerve (cutaneous br.)
L-1 nerve, dorsal primary ramus (Sup. cluneal n.)
Gluteal fascia
L-2 nerve, dorsal primary ramus (Sup. cluneal n.)
L-3 nerve, dorsal primary ramus (Sup. cluneal n.)
Bursa over post. sup. iliac spine
Gluteal fascia
Sacral horn (right)
Suferficial dorsal sacro-coccygeal lig.

▲

Fig. 521: The Lumbosacral Region of the Back, Superficial View

NOTE that the spinous processes of the five lumbar vertebrae are numbered and that the superficial vessels and nerves penetrate the thoracolumbar fascia. Observe the lumbar triangle (labeled on the right) which is bounded by the crest of the ilium and the external oblique and latissimus dorsi muscles (see also Figure 523).

Fig. 522: Cross Section of Back, Lumbar Region

NOTE: the sacrospinalis muscle is ensheathed by the thick lumbar part of the thoracolumbar fascia which attaches medially to the spinous and transverse processes of the lumbar vertebrae, and which becomes continuous laterally with the aponeuroses and fasciae of the latissimus dorsi and anterior abdominal muscles.

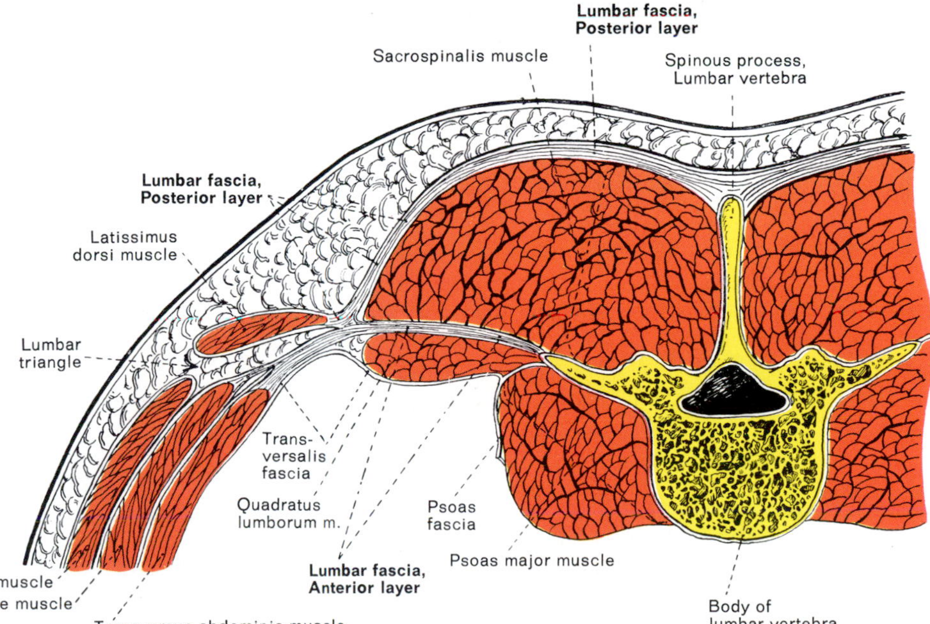

Lumbar fascia, Posterior layer
Sacrospinalis muscle
Spinous process, Lumbar vertebra
Lumbar fascia, Posterior layer
Latissimus dorsi muscle
Lumbar triangle
Trans-versalis fascia
Quadratus lumborum m.
Psoas fascia
Psoas major muscle
Lumbar fascia, Anterior layer
External oblique muscle
Internal oblique muscle
Transversus abdominis muscle
Body of lumbar vertebra

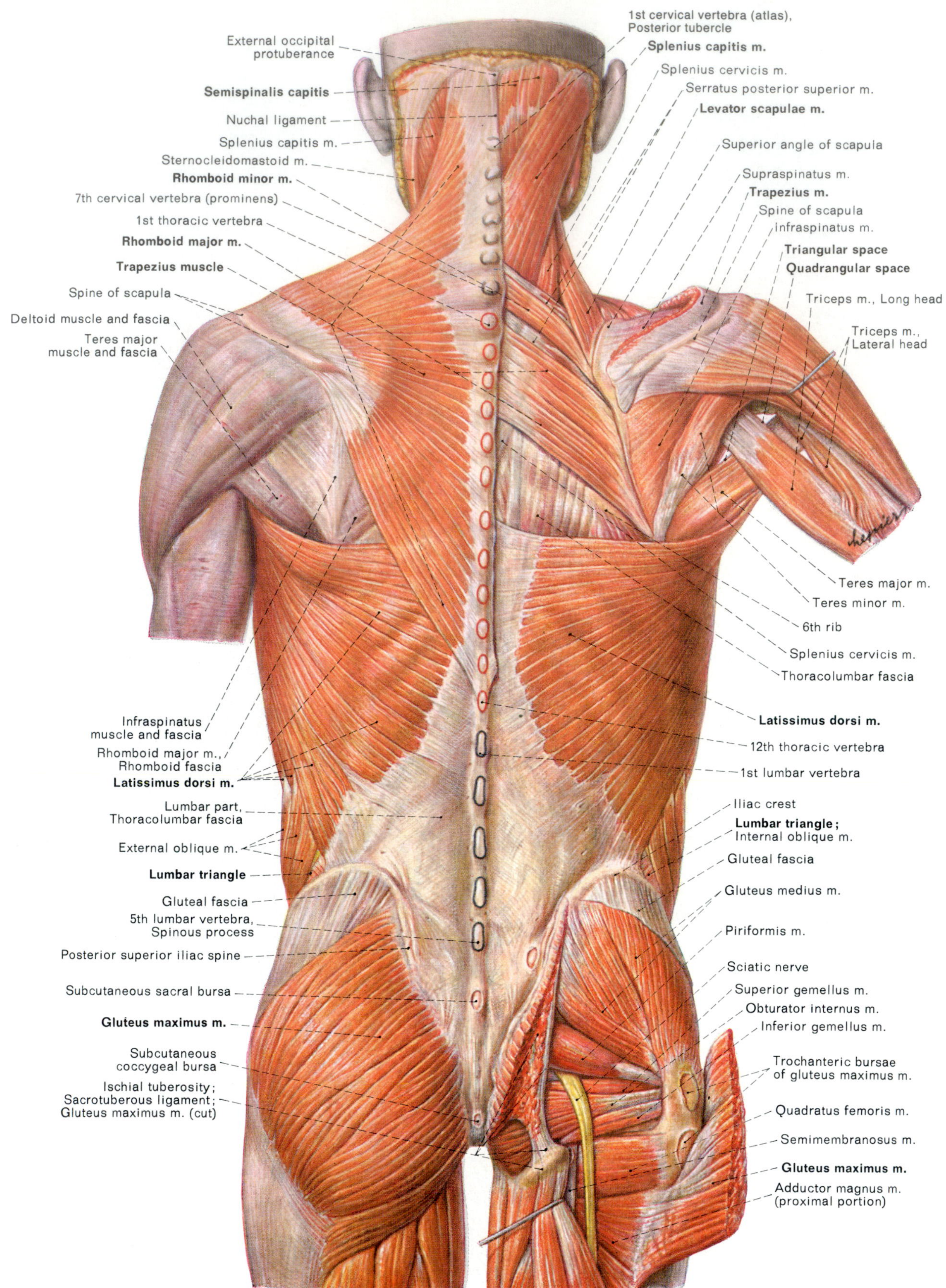

External occipital protuberance
Semispinalis capitis
Nuchal ligament
Splenius capitis m.
Sternocleidomastoid m.
Rhomboid minor m.
7th cervical vertebra (prominens)
1st thoracic vertebra
Rhomboid major m.
Trapezius muscle
Spine of scapula
Deltoid muscle and fascia
Teres major muscle and fascia
Infraspinatus muscle and fascia
Rhomboid major m., Rhomboid fascia
Latissimus dorsi m.
Lumbar part, Thoracolumbar fascia
External oblique m.
Lumbar triangle
Gluteal fascia
5th lumbar vertebra, Spinous process
Posterior superior iliac spine
Subcutaneous sacral bursa
Gluteus maximus m.
Subcutaneous coccygeal bursa
Ischial tuberosity; Sacrotuberous ligament; Gluteus maximus m. (cut)
1st cervical vertebra (atlas), Posterior tubercle
Splenius capitis m.
Splenius cervicis m.
Serratus posterior superior m.
Levator scapulae m.
Superior angle of scapula
Supraspinatus m.
Trapezius m.
Spine of scapula
Infraspinatus m.
Triangular space
Quadrangular space
Triceps m., Long head
Triceps m., Lateral head
Teres major m.
Teres minor m.
6th rib
Splenius cervicis m.
Thoracolumbar fascia
Latissimus dorsi m.
12th thoracic vertebra
1st lumbar vertebra
Iliac crest
Lumbar triangle; Internal oblique m.
Gluteal fascia
Gluteus medius m.
Piriformis m.
Sciatic nerve
Superior gemellus m.
Obturator internus m.
Inferior gemellus m.
Trochanteric bursae of gluteus maximus m.
Quadratus femoris m.
Semimembranosus m.
Gluteus maximus m.
Adductor magnus m. (proximal portion)

Fig. 523: Muscles of the Posterior Neck, Shoulder, Back and Gluteal Region

NOTE that the most superficial layer of back muscles includes the latissimus dorsi and trapezius. Beneath the trapezius, observe the levator scapulae and rhomboid major and minor muscles attaching along the vertebral border of the scapula. In the neck, the splenius capitis and semispinalis capitis muscles lie directly under the trapezius.

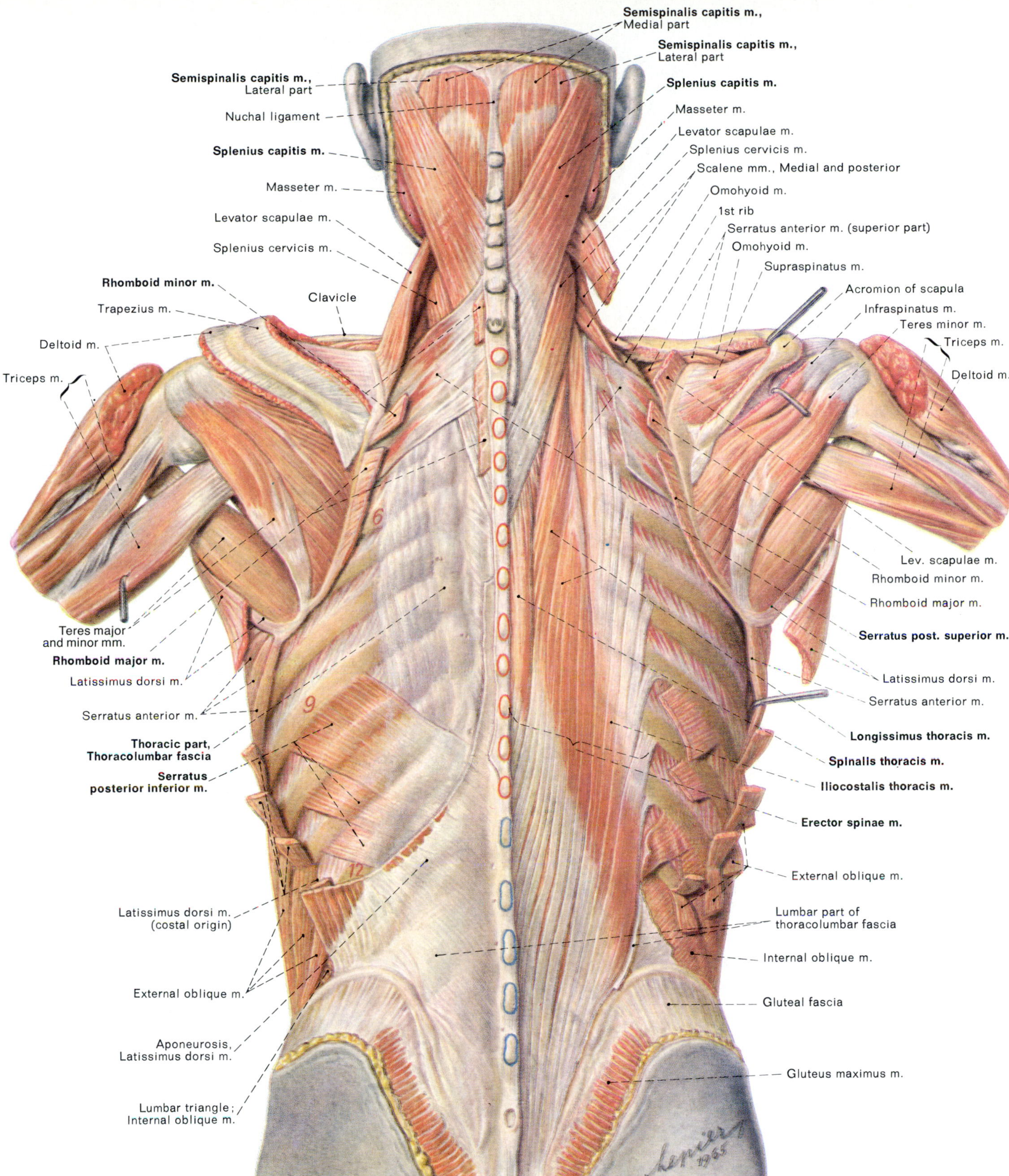
Semispinalis capitis m.,
Medial part
Semispinalis capitis m.,
Lateral part
Semispinalis capitis m.,
Lateral part
Splenius capitis m.
Nuchal ligament
Masseter m.
Levator scapulae m.
Splenius capitis m.
Splenius cervicis m.
Scalene mm., Medial and posterior
Masseter m.
Omohyoid m.
Levator scapulae m.
1st rib
Serratus anterior m. (superior part)
Splenius cervicis m.
Omohyoid m.
Supraspinatus m.
Rhomboid minor m.
Clavicle
Acromion of scapula
Trapezius m.
Infraspinatus m.
Teres minor m.
Deltoid m.
Triceps m.
Triceps m.
Deltoid m.
Lev. scapulae m.
Rhomboid minor m.
Rhomboid major m.
Teres major
and minor mm.
Serratus post. superior m.
Rhomboid major m.
Latissimus dorsi m.
Latissimus dorsi m.
Serratus anterior m.
Serratus anterior m.
Longissimus thoracis m.
Thoracic part,
Thoracolumbar fascia
Spinalis thoracis m.
Serratus
posterior inferior m.
Iliocostalis thoracis m.
Erector spinae m.
External oblique m.
Latissimus dorsi m.
(costal origin)
Lumbar part of
thoracolumbar fascia
Internal oblique m.
External oblique m.
Gluteal fascia
Aponeurosis,
Latissimus dorsi m.
Gluteus maximus m.
Lumbar triangle;
Internal oblique m.
6
9
12
1965

Fig. 524: Muscles of the Back: Intermediate Layer (left), Deep Layer (right)

NOTE: 1) *on the left side* the superficial back muscles (trapezius and latissimus dorsi) have been cut, as have the rhomboid major and minor which attach the vertebral border of the scapula to the vertebral column. Observe the underlying serratus posterior superior and serratus posterior inferior muscles.

2) *on the right side* the serratus posterior muscles and the thoracolumbar fascia have been removed, revealing the erector spinae muscle (formerly called the sacrospinalis muscle).

3) *in the neck* the splenius cervicis, splenius capitus and semispinalis capitis underlie the trapezius.

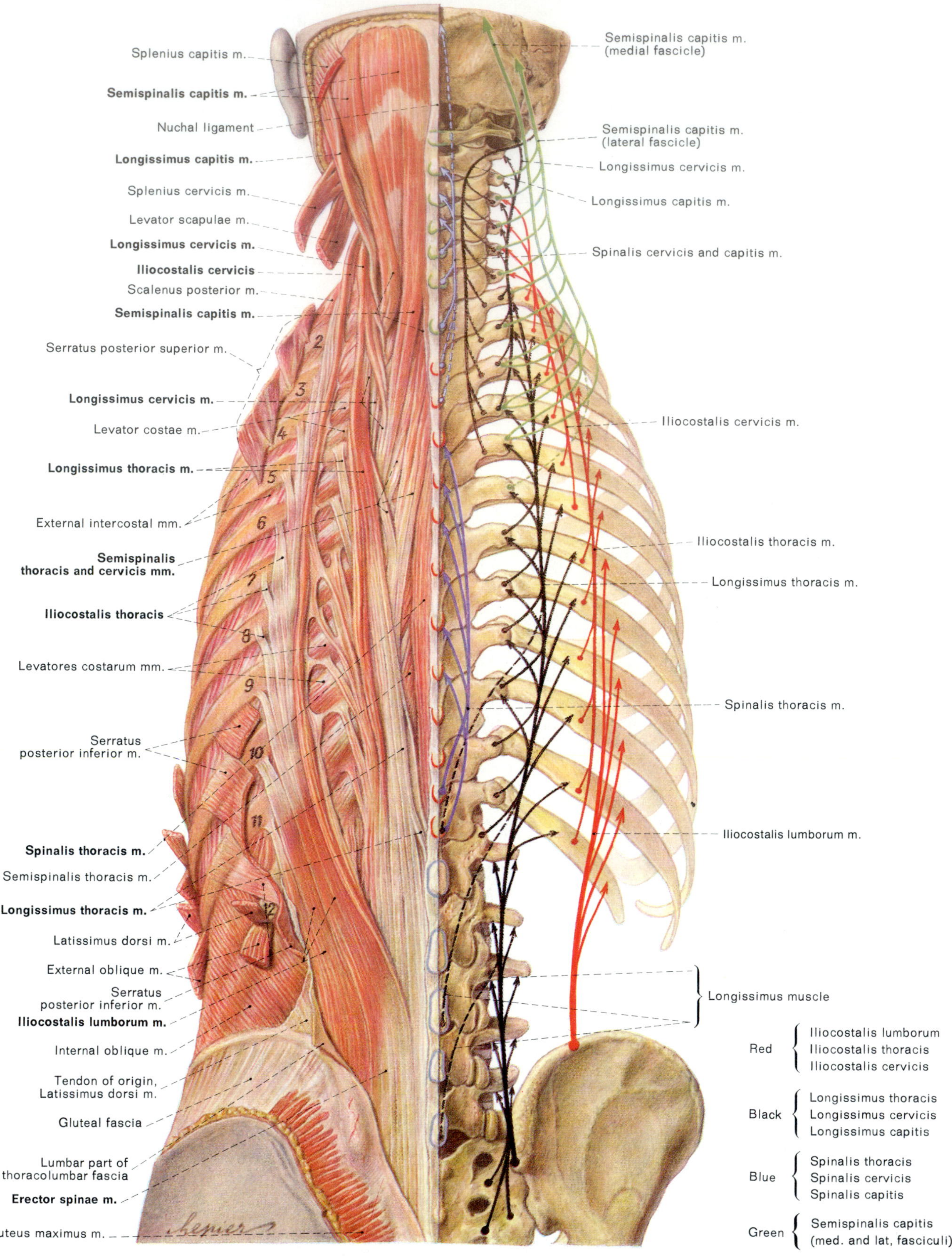

Splenius capitis m.
Semispinalis capitis m.
Nuchal ligament
Longissimus capitis m.
Splenius cervicis m.
Levator scapulae m.
Longissimus cervicis m.
Iliocostalis cervicis
Scalenus posterior m.
Semispinalis capitis m.
Serratus posterior superior m.
Longissimus cervicis m.
Levator costae m.
Longissimus thoracis m.
External intercostal mm.
Semispinalis thoracis and cervicis mm.
Iliocostalis thoracis
Levatores costarum mm.
Serratus posterior inferior m.
Spinalis thoracis m.
Semispinalis thoracis m.
Longissimus thoracis m.
Latissimus dorsi m.
External oblique m.
Serratus posterior inferior m.
Iliocostalis lumborum m.
Internal oblique m.
Tendon of origin, Latissimus dorsi m.
Gluteal fascia
Lumbar part of thoracolumbar fascia
Erector spinae m.
Gluteus maximus m.
Semispinalis capitis m. (medial fascicle)
Semispinalis capitis m. (lateral fascicle)
Longissimus cervicis m.
Longissimus capitis m.
Spinalis cervicis and capitis m.
Iliocostalis cervicis m.
Iliocostalis thoracis m.
Longissimus thoracis m.
Spinalis thoracis m.
Iliocostalis lumborum m.
Longissimus muscle
Red: Iliocostalis lumborum, Iliocostalis thoracis, Iliocostalis cervicis
Black: Longissimus thoracis, Longissimus cervicis, Longissimus capitis
Blue: Spinalis thoracis, Spinalis cervicis, Spinalis capitis
Green: Semispinalis capitis (med. and lat. fasciculi)

Fig. 525: Deep Muscles of the Back and Neck: the Erector Spinae Muscle

NOTE: 1) *on the left,* the erector spinae (sacrospinalis) muscle is separated into its iliocostalis, longissimus and spinalis portions. In the neck observe the semispinalis capitis which has both medial and lateral fascicles. The semispinalis cervicis and thoracis extend inferiorly from above. and lie deep to the sacrospinalis layer of musculature.

2) *on the right,* all of the muscles have been removed and their attachments have been diagrammed by means of colored lines.

Fig. 525

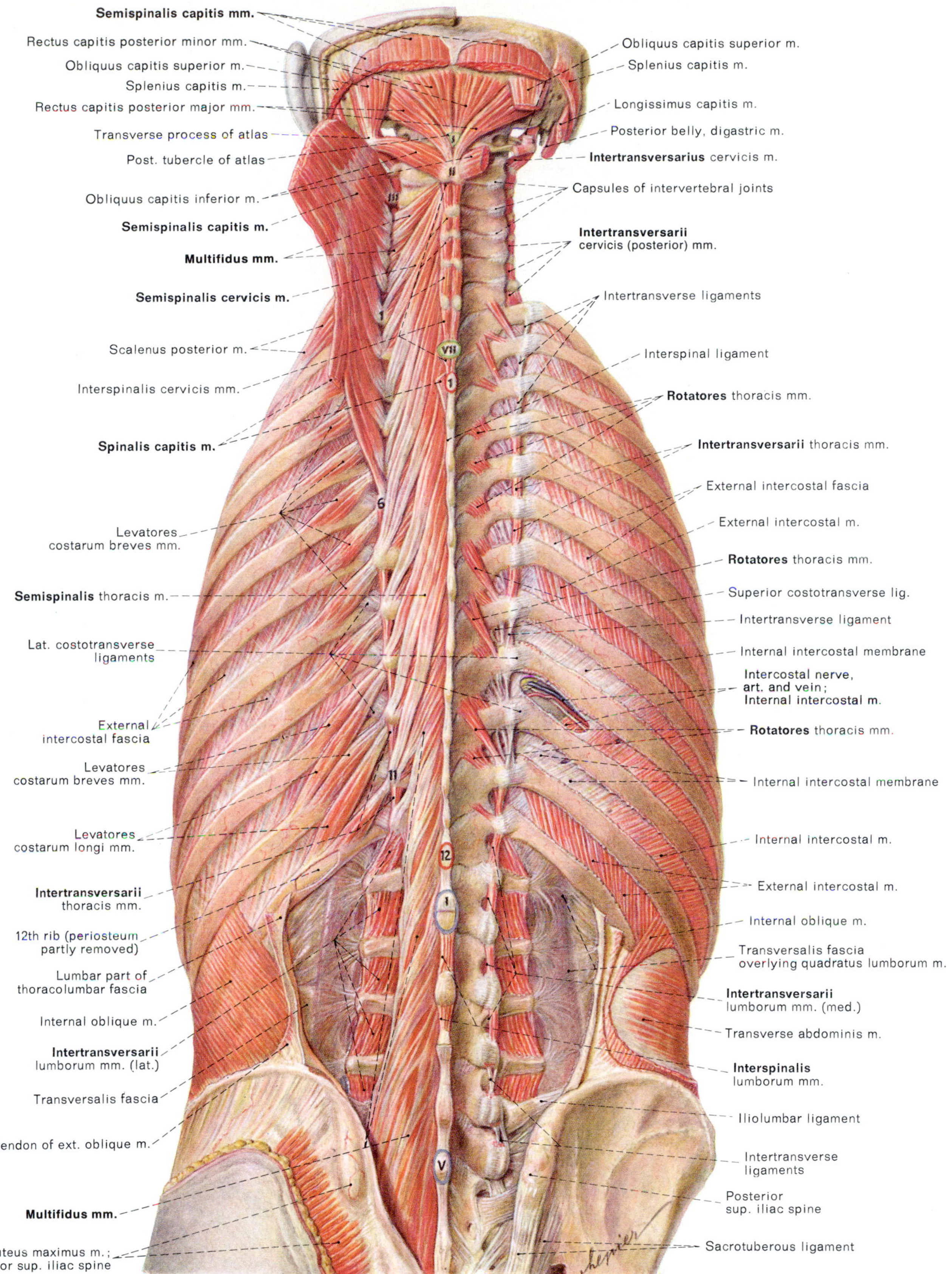
Semispinalis capitis mm.
Rectus capitis posterior minor mm.
Obliquus capitis superior m.
Splenius capitis m.
Rectus capitis posterior major mm.
Transverse process of atlas
Post. tubercle of atlas
Obliquus capitis inferior m.
Semispinalis capitis m.
Multifidus mm.
Semispinalis cervicis m.
Scalenus posterior m.
Interspinalis cervicis mm.
Spinalis capitis m.
Levatores costarum breves mm.
Semispinalis thoracis m.
Lat. costotransverse ligaments
External intercostal fascia
Levatores costarum breves mm.
Levatores costarum longi mm.
Intertransversarii thoracis mm.
12th rib (periosteum partly removed)
Lumbar part of thoracolumbar fascia
Internal oblique m.
Intertransversarii lumborum mm. (lat.)
Transversalis fascia
Tendon of ext. oblique m.
Multifidus mm.
Gluteus maximus m.; Posterior sup. iliac spine
Obliquus capitis superior m.
Splenius capitis m.
Longissimus capitis m.
Posterior belly, digastric m.
Intertransversarius cervicis m.
Capsules of intervertebral joints
Intertransversarii cervicis (posterior) mm.
Intertransverse ligaments
Interspinal ligament
Rotatores thoracis mm.
Intertransversarii thoracis mm.
External intercostal fascia
External intercostal m.
Rotatores thoracis mm.
Superior costotransverse lig.
Intertransverse ligament
Internal intercostal membrane
Intercostal nerve, art. and vein; Internal intercostal m.
Rotatores thoracis mm.
Internal intercostal membrane
Internal intercostal m.
External intercostal m.
Internal oblique m.
Transversalis fascia overlying quadratus lumborum m.
Intertransversarii lumborum mm. (med.)
Transverse abdominis m.
Interspinalis lumborum mm.
Iliolumbar ligament
Intertransverse ligaments
Posterior sup. iliac spine
Sacrotuberous ligament

Fig. 526: Deep Muscles of the Back and Neck: Transversospinal Group

NOTE: 1) the transversospinal groups of muscles lie deep to the erector spinae muscles and generally extend between the transverse processes of the vertebrae to the spinous processes of more superior vertebrae. Their actions extend the vertebral column, or upon acting individually and on one side, they bend and rotate the vertebrae.

2) within this group of muscles are the semispinalis (thoracis, cervicis, and capitis), the multifidus, the rotatores (lumborum, thoracis, cervicis), the interspinales (lumborum, thoracis, cervicis) and the intertransversarii.

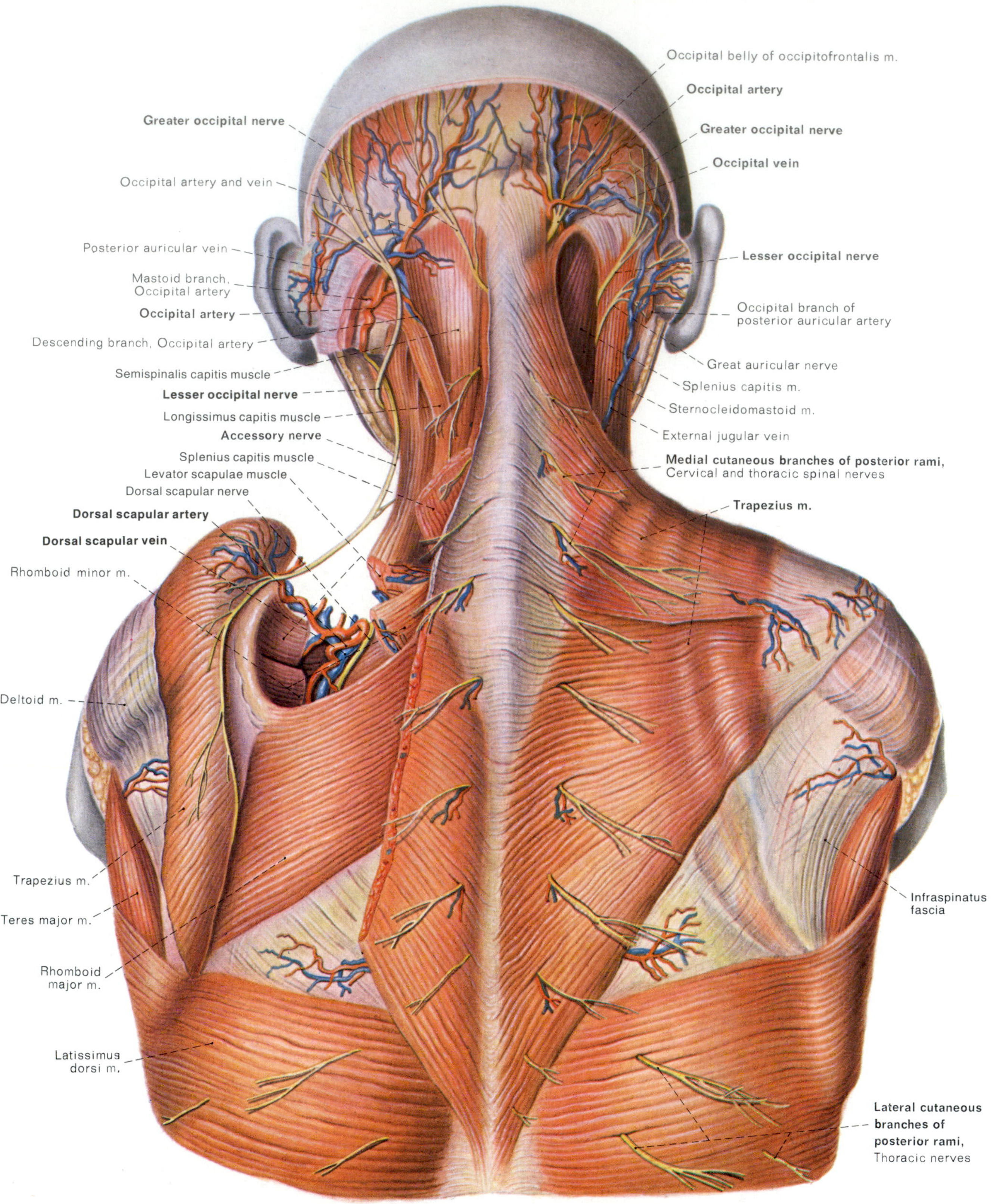

Occipital belly of occipitofrontalis m.
Occipital artery
Greater occipital nerve
Occipital vein
Lesser occipital nerve
Occipital branch of
posterior auricular artery
Great auricular nerve
Splenius capitis m.
Sternocleidomastoid m.
External jugular vein
Medial cutaneous branches of posterior rami,
Cervical and thoracic spinal nerves
Trapezius m.
Greater occipital nerve
Occipital artery and vein
Posterior auricular vein
Mastoid branch,
Occipital artery
Occipital artery
Descending branch, Occipital artery
Semispinalis capitis muscle
Lesser occipital nerve
Longissimus capitis muscle
Accessory nerve
Splenius capitis muscle
Levator scapulae muscle
Dorsal scapular nerve
Dorsal scapular artery
Dorsal scapular vein
Rhomboid minor m.
Deltoid m.
Trapezius m.
Teres major m.
Rhomboid
major m.
Latissimus
dorsi m.
Infraspinatus
fascia
Lateral cutaneous
branches of
posterior rami,
Thoracic nerves

Fig. 527: Nerves and Vessels of the Superficial and Intermediate Muscle Layers of the Upper Back and Posterior Neck

NOTE: 1) the segmental distribution of the cutaneous branches of the posterior primary rami of the cervical and thoracic nerves over the posterior neck and back. Observe the accessory nerve (XI) as it descends to innervate the trapezius and sternocleidomastoid muscles.

2) the greater occipital nerve, which is a sensory nerve, ascending to the posterior scalp. It derives from the posterior primary ramus of the C-2 spinal nerve and is accompanied in its course by the occipital artery and vein. Observe also the lesser occipital nerve which courses to the lateral posterior scalp and which arises from the anterior primary ramus of C-2.

3) the dorsal scapular nerve, artery and vein which course beneath the levator scapulae and the rhomboid muscles.

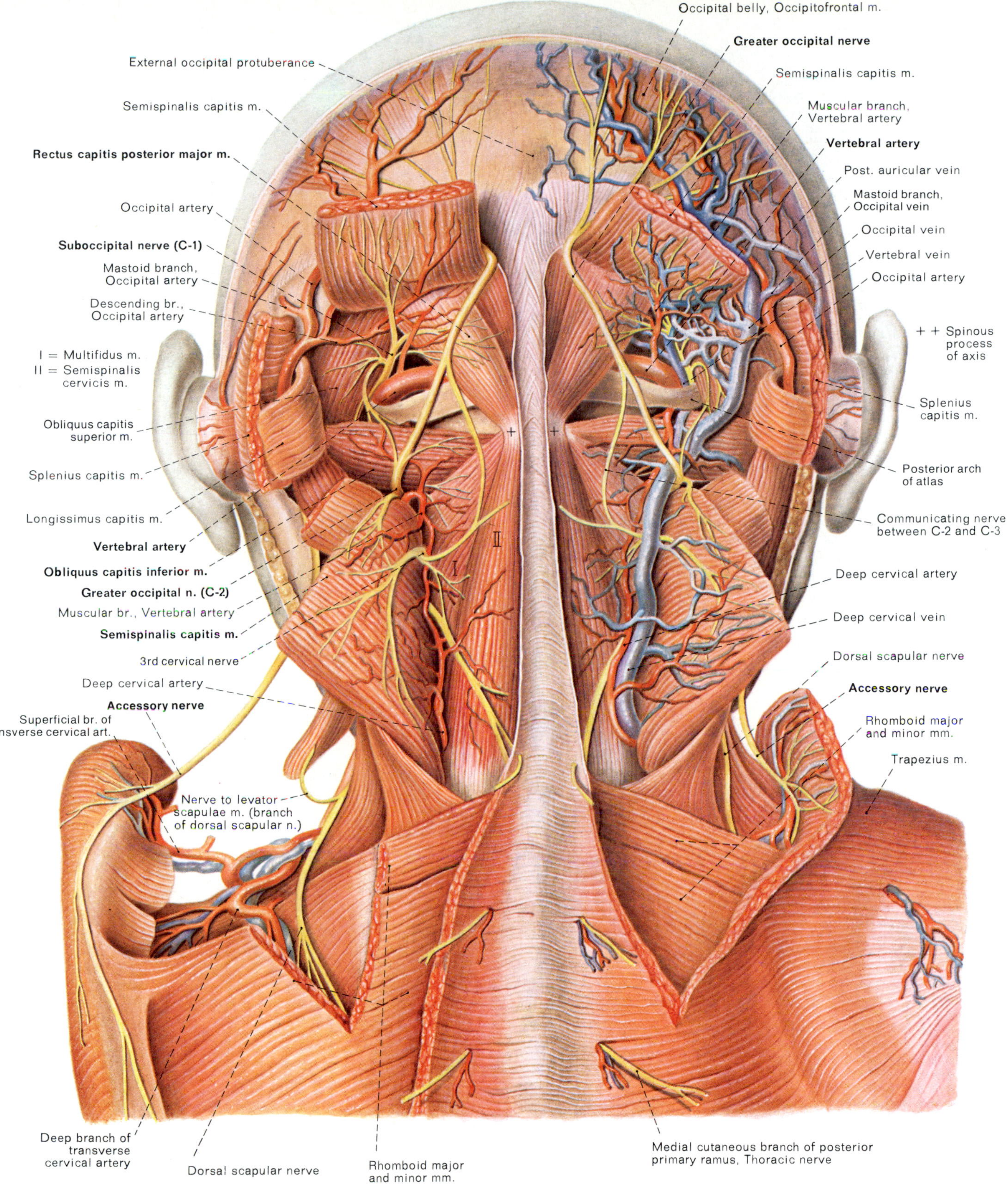
Occipital belly, Occipitofrontal m.
Greater occipital nerve
External occipital protuberance
Semispinalis capitis m.
Semispinalis capitis m.
Muscular branch, Vertebral artery
Rectus capitis posterior major m.
Vertebral artery
Post. auricular vein
Occipital artery
Mastoid branch, Occipital vein
Suboccipital nerve (C-1)
Occipital vein
Mastoid branch, Occipital artery
Vertebral vein
Occipital artery
Descending br., Occipital artery
+ + Spinous process of axis
I = Multifidus m.
II = Semispinalis cervicis m.
Splenius capitis m.
Obliquus capitis superior m.
Splenius capitis m.
Posterior arch of atlas
Longissimus capitis m.
Communicating nerve between C-2 and C-3
Vertebral artery
Obliquus capitis inferior m.
Deep cervical artery
Greater occipital n. (C-2)
Muscular br., Vertebral artery
Deep cervical vein
Semispinalis capitis m.
3rd cervical nerve
Dorsal scapular nerve
Deep cervical artery
Accessory nerve
Accessory nerve
Superficial br. of transverse cervical art.
Rhomboid major and minor mm.
Trapezius m.
Nerve to levator scapulae m. (branch of dorsal scapular n.)
Deep branch of transverse cervical artery
Dorsal scapular nerve
Rhomboid major and minor mm.
Medial cutaneous branch of posterior primary ramus, Thoracic nerve

Fig. 528: The Deep Vessels and Nerves of the Suboccipital Region and Upper Back; the Suboccipital Triangle

NOTE: 1) the suboccipital triangle lies directly beneath the semispinalis capitis muscle and is bounded by the rectus capitis posterior major, obliquus capitis superior and the obliquus capitis inferior.

2) the vertebral artery crosses (from lateral to medial) the suboccipital triangle, while the suboccipital nerve (posterior primary ramus of C-1) emerges through the triangle to distribute motor innervation to the three muscles which bound the triangle, as well as the rectus capitis posterior minor and the overlying semispinalis capitis muscle.

3) the greater occipital nerve (posterior primary ramus of C-2) is a sensory nerve which makes its appearance caudal to the obliquus capitis inferior, and then courses medially and superiorly to become subcutaneous just lateral and below the external occipital protuberance.

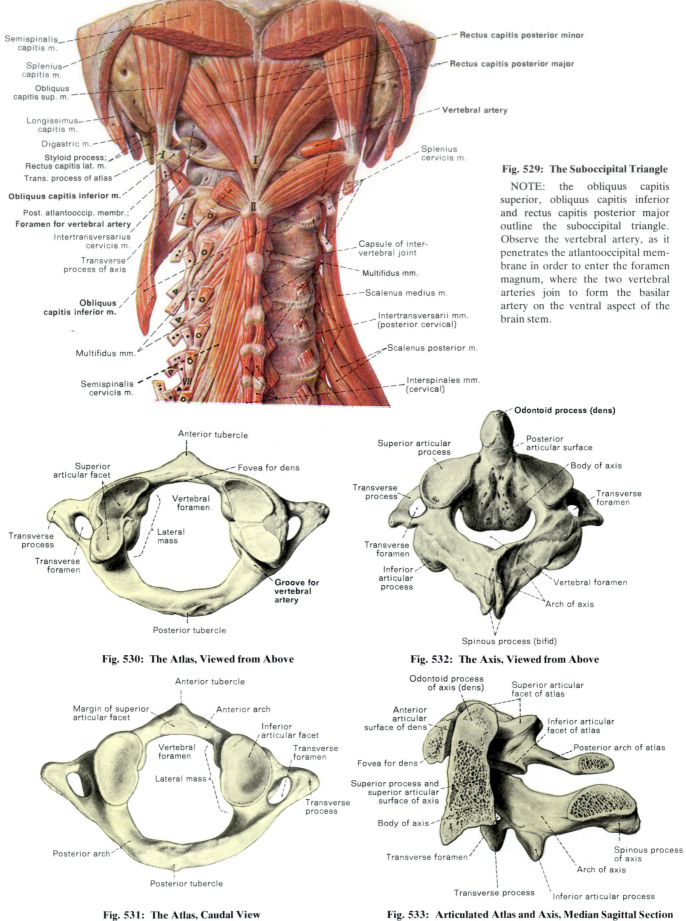

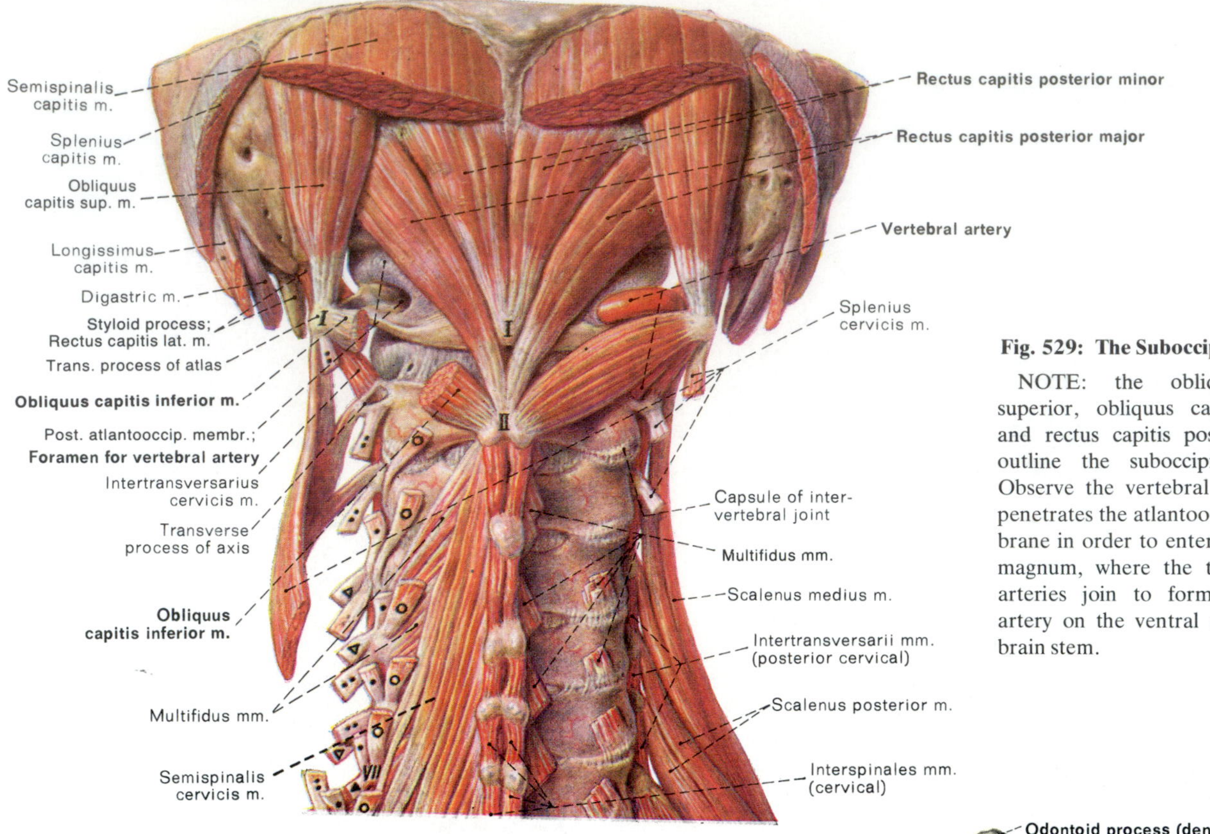

Semispinalis capitis m.
Splenius capitis m.
Obliquus capitis sup. m.
Longissimus capitis m.
Digastric m.
Styloid process; Rectus capitis lat. m.
Trans. process of atlas
Obliquus capitis inferior m.
Post. atlantooccip. membr.;
Foramen for vertebral artery
Intertransversarius cervicis m.
Transverse process of axis
Obliquus capitis inferior m.
Multifidus mm.
Semispinalis cervicis m.
Rectus capitis posterior minor
Rectus capitis posterior major
Vertebral artery
Splenius cervicis m.
Capsule of inter-vertebral joint
Multifidus mm.
Scalenus medius m.
Intertransversarii mm. (posterior cervical)
Scalenus posterior m.
Interspinales mm. (cervical)
I
II
VII

Fig. 529: The Suboccipital Triangle

NOTE: the obliquus capitis superior, obliquus capitis inferior and rectus capitis posterior major outline the suboccipital triangle. Observe the vertebral artery, as it penetrates the atlantooccipital membrane in order to enter the foramen magnum, where the two vertebral arteries join to form the basilar artery on the ventral aspect of the brain stem.

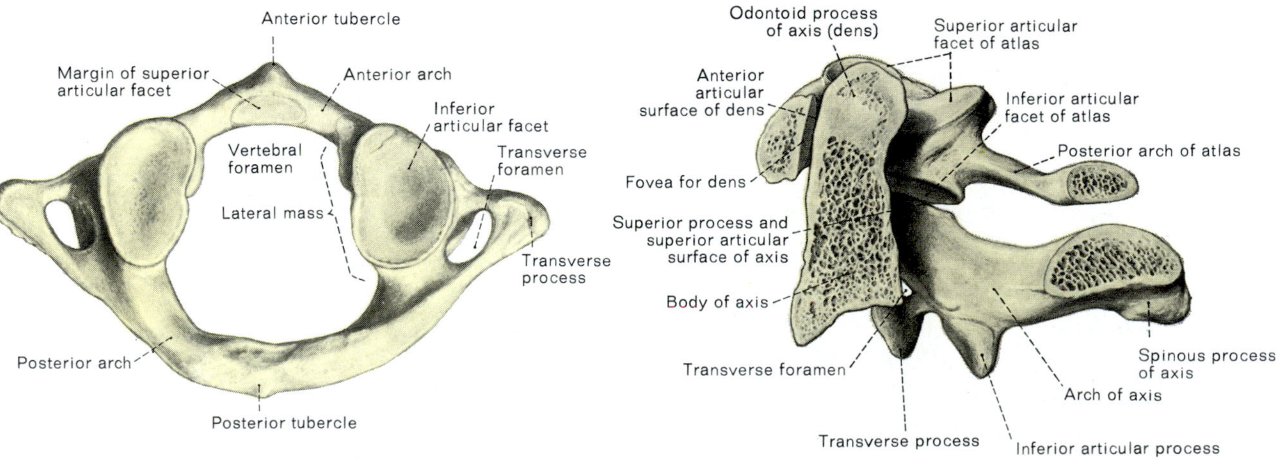

Anterior tubercle
Superior articular facet
Fovea for dens
Vertebral foramen
Lateral mass
Transverse process
Transverse foramen
Groove for vertebral artery
Posterior tubercle
Odontoid process (dens)
Superior articular process
Posterior articular surface
Body of axis
Transverse process
Transverse foramen
Transverse foramen
Inferior articular process
Vertebral foramen
Arch of axis
Spinous process (bifid)
Anterior tubercle
Margin of superior articular facet
Anterior arch
Inferior articular facet
Vertebral foramen
Transverse foramen
Lateral mass
Transverse process
Posterior arch
Posterior tubercle
Odontoid process of axis (dens)
Superior articular facet of atlas
Anterior articular surface of dens
Inferior articular facet of atlas
Posterior arch of atlas
Fovea for dens
Superior process and superior articular surface of axis
Body of axis
Transverse foramen
Spinous process of axis
Arch of axis
Transverse process
Inferior articular process

Fig. 530: The Atlas, Viewed from Above

Fig. 532: The Axis, Viewed from Above

Fig. 531: The Atlas, Caudal View

Fig. 533: Articulated Atlas and Axis, Median Sagittal Section

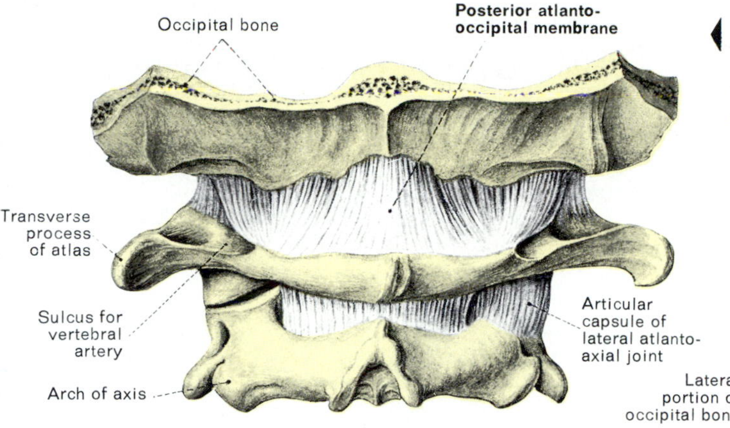
Occipital bone
Posterior atlanto-occipital membrane
Transverse process of atlas
Sulcus for vertebral artery
Arch of axis
Articular capsule of lateral atlanto-axial joint

Fig. 534: Atlantooccipital and Atlantoaxial Joints (Posterior View)

NOTE: the posterior atlantoccipital membrane extends from the posterior margin of the foramen magnum to the upper border of the posterior arch of the atlas.

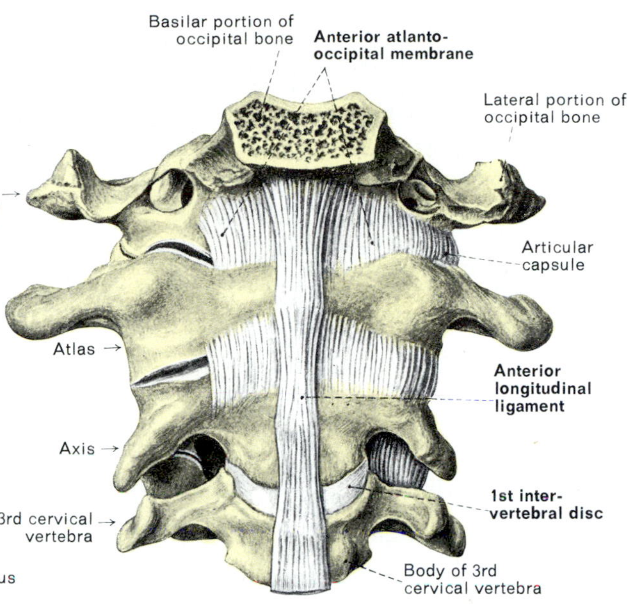
Basilar portion of occipital bone
Anterior atlanto-occipital membrane
Lateral portion of occipital bone
Lateral portion of occipital bone
Articular capsule
Atlas
Anterior longitudinal ligament
Axis
3rd cervical vertebra
1st inter-vertebral disc
Body of 3rd cervical vertebra

Fig. 535: Articulations of Occipital Bone and First Three Cervical Vertebrae (Anterior View)

NOTE: the anterior atlantooccipital membrane extending between the occipital bone and the anterior arch of the atlas and continuing laterally to join the articular capsules. Also observe the anterior longitudinal ligament.

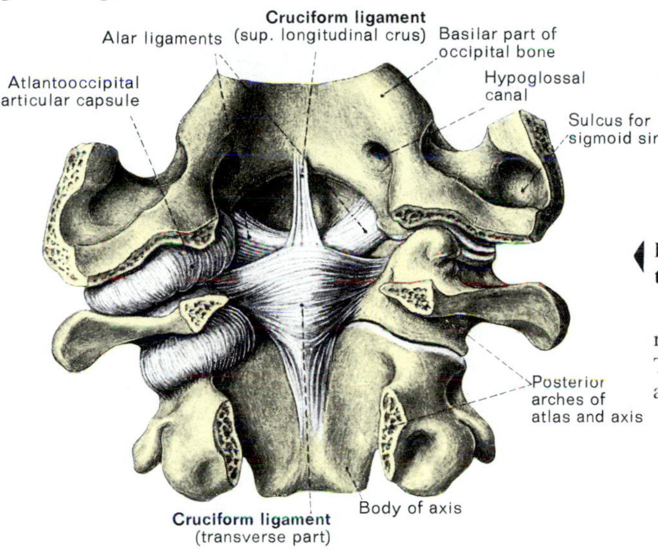
Cruciform ligament (sup. longitudinal crus)
Alar ligaments
Basilar part of occipital bone
Atlantooccipital articular capsule
Hypoglossal canal
Sulcus for sigmoid sinus
Posterior arches of atlas and axis
Body of axis
Cruciform ligament (transverse part)

Fig. 536: The Atlantooccipital and Atlantoaxial Joints Showing the Cruciform Ligament (Posterior View)

NOTE: the posterior arches of the atlas and axis have been removed and the cruciform ligament is seen from this posterior view. This ligament consists of the transverse ligament (see Figure 537) and the longitudinal fascicles extending superiorly and inferiorly.

Fig. 537: Median Atlantoaxial Joint (from Above)

NOTE: that the odontoid process (dens) of the axis articulates with the anterior arch of the atlas, thereby forming the median atlantoaxial joint, and that the thick strong transverse ligament (part of the cruciform ligament) of the atlas retains the dens on its posterior surface.

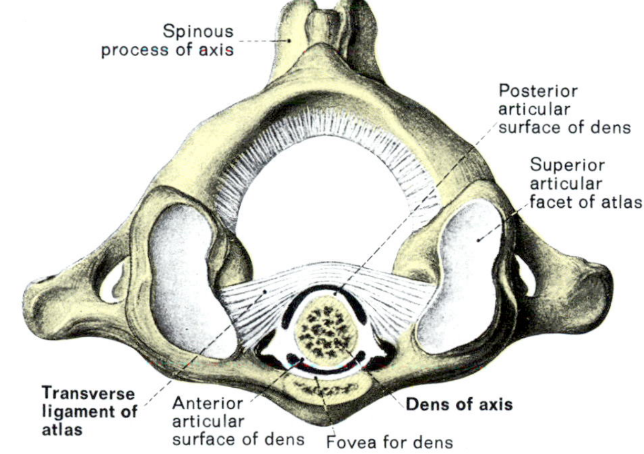
Spinous process of axis
Posterior articular surface of dens
Superior articular facet of atlas
Transverse ligament of atlas
Anterior articular surface of dens
Fovea for dens
Dens of axis

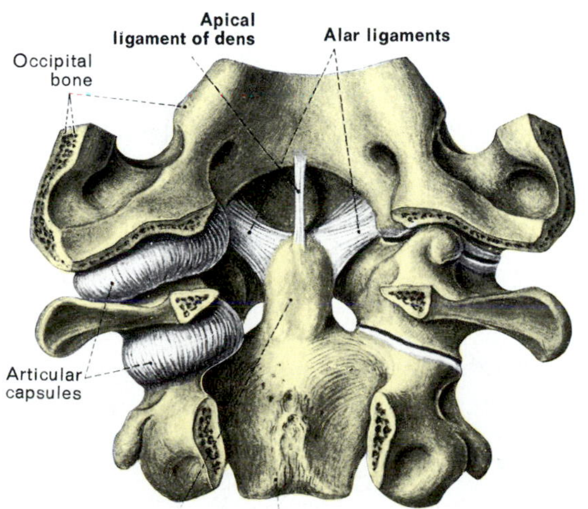
Apical ligament of dens
Alar ligaments
Occipital bone
Articular capsules
Dens of axis
Body of axis

Fig. 538: The Alar and Apical Ligaments (Posterior View)

NOTE: this figure's orientation is similar to Figure 536. The posterior arches of the atlas and axis have been removed as well as the cruciform ligament (both transverse and longitudinal parts). This reveals the odontoid process of the axis, which is attached superiorly to the occipital bone by the two alar ligaments and the apical ligament of the dens. These ligaments tend to limit lateral rotation of the skull.

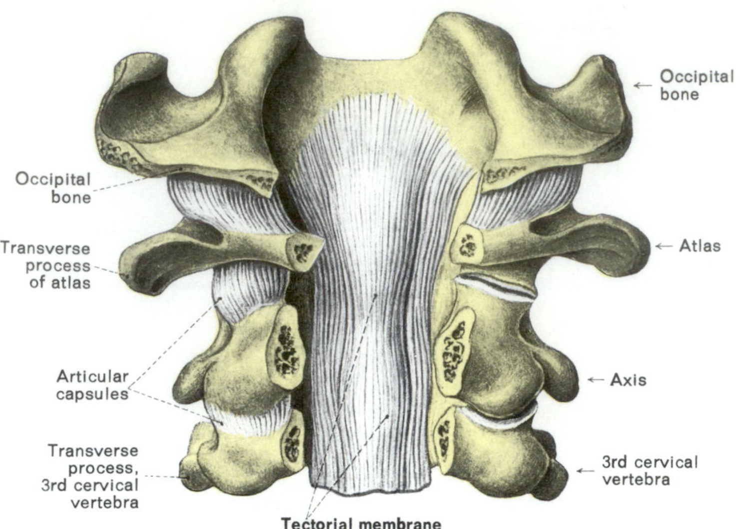
Occipital bone
Occipital bone
Transverse process of atlas
Atlas
Articular capsules
Axis
Transverse process, 3rd cervical vertebra
3rd cervical vertebra
Tectorial membrane

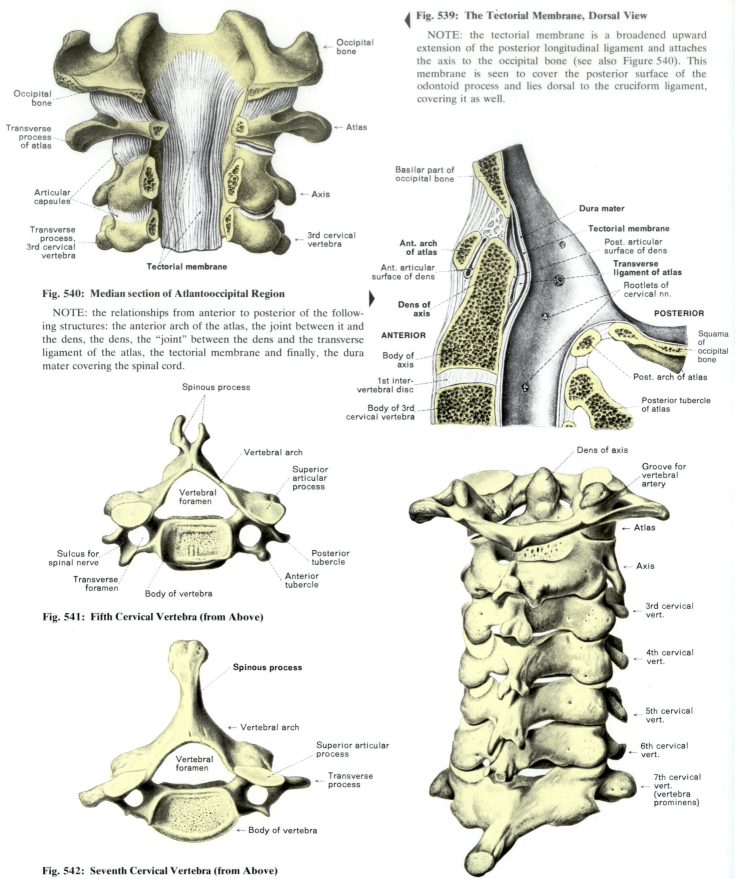

Fig. 539: The Tectorial Membrane, Dorsal View

NOTE: the tectorial membrane is a broadened upward extension of the posterior longitudinal ligament and attaches the axis to the occipital bone (see also Figure 540). This membrane is seen to cover the posterior surface of the odontoid process and lies dorsal to the cruciform ligament, covering it as well.

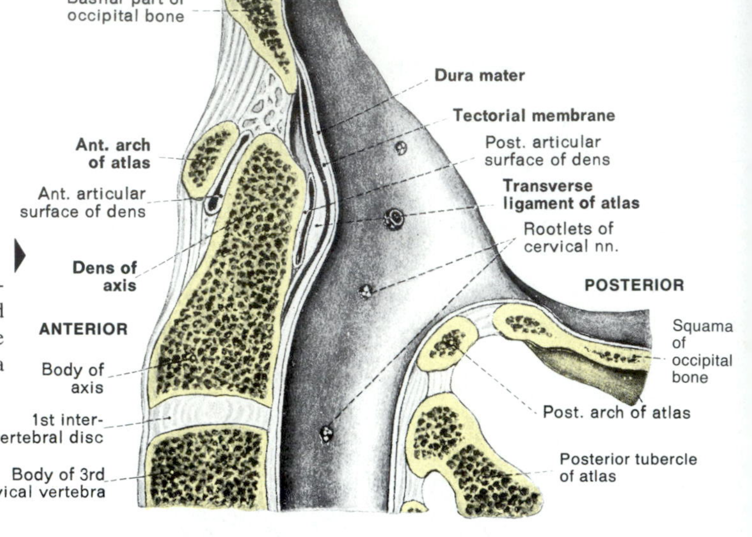
Basilar part of occipital bone
Dura mater
Tectorial membrane
Ant. arch of atlas
Post. articular surface of dens
Ant. articular surface of dens
Transverse ligament of atlas
Rootlets of cervical nn.
Dens of axis
POSTERIOR
ANTERIOR
Squama of occipital bone
Body of axis
Post. arch of atlas
1st inter-vertebral disc
Posterior tubercle of atlas
Body of 3rd cervical vertebra

Fig. 540: Median section of Atlantooccipital Region

NOTE: the relationships from anterior to posterior of the following structures: the anterior arch of the atlas, the joint between it and the dens, the dens, the "joint" between the dens and the transverse ligament of the atlas, the tectorial membrane and finally, the dura mater covering the spinal cord.

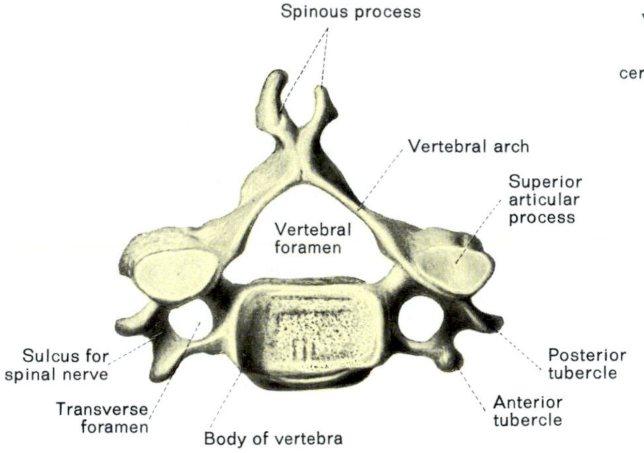
Spinous process
Vertebral arch
Superior articular process
Vertebral foramen
Sulcus for spinal nerve
Posterior tubercle
Transverse foramen
Anterior tubercle
Body of vertebra

Fig. 541: Fifth Cervical Vertebra (from Above)

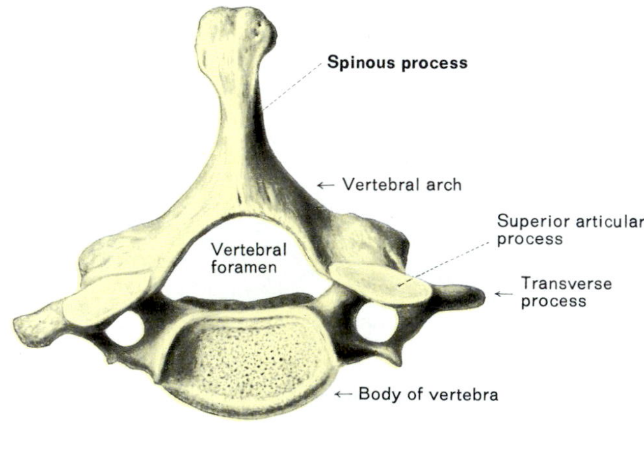
Spinous process
Vertebral arch
Superior articular process
Vertebral foramen
Transverse process
Body of vertebra

Fig. 542: Seventh Cervical Vertebra (from Above)

NOTE: the 7th cervical vertebra, being transitional between the cervical and thoracic vertebrae, has a transverse foramen similar to the cervical and a large spinous process similar to the thoracic. The latter gives it the name vertebra prominens.

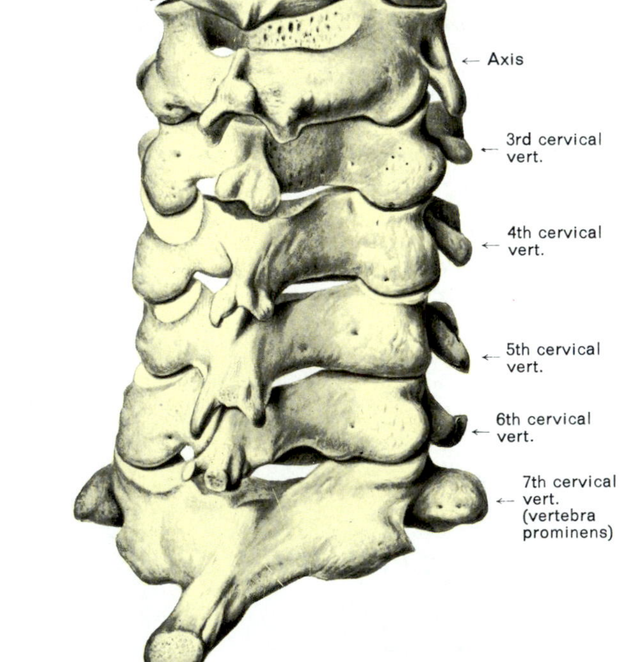
Dens of axis
Groove for vertebral artery
Atlas
Axis
3rd cervical vert.
4th cervical vert.
5th cervical vert.
6th cervical vert.
7th cervical vert. (vertebra prominens)

Fig. 543: The Cervical Spinal Column (Dorsal)

NOTE: while flexion and extension of the head are performed at the atlantooccipital joint, rotation of the head is the result of rotation of the atlas on the axis.

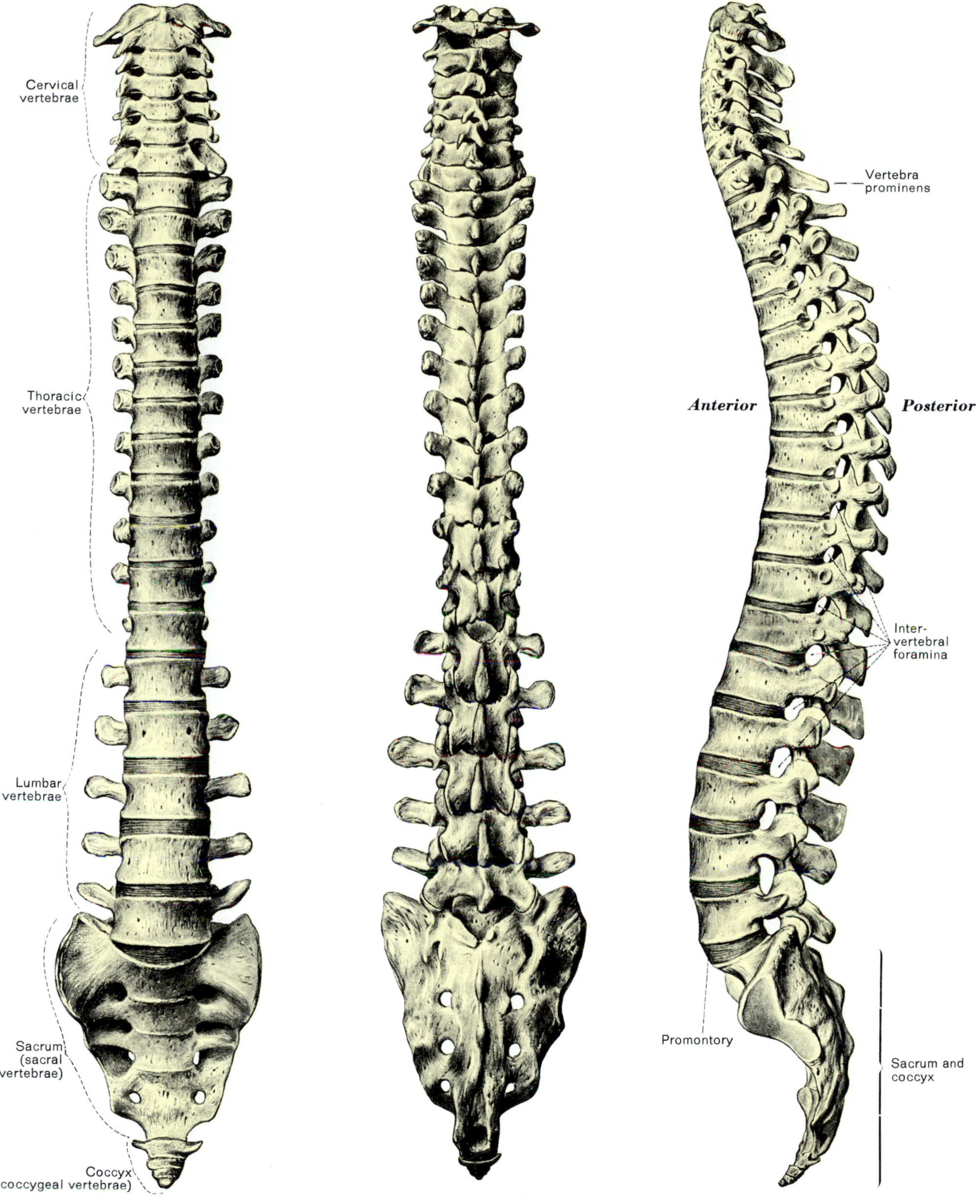
Cervical vertebrae
Thoracic vertebrae
Lumbar vertebrae
Sacrum (sacral vertebrae)
Coccyx (coccygeal vertebrae)
Vertebra prominens
Anterior
Posterior
Inter-vertebral foramina
Promontory
Sacrum and coccyx

Fig. 544: Anterior View **Fig. 545: Posterior View** **Fig. 546: Left Lateral View**

Figs. 544, 545 and 546: The Vertebral Column, Including the Sacrum and Coccyx

NOTE: 1) the vertebral column generally consists of 7 cervical, 12 thoracic and 5 lumbar vertebrae and the sacrum and coccyx, thus, 26 bones in all. Its principal functions are to assist in the maintenance of the erect posture in man, to encase and protect the spinal cord and to allow attachments of the musculature important for movements of the head and trunk.

2) from a dorsal or ventral view, the normal spinal column is straight. When viewed from the side, the spinal column presents two ventrally convex curvatures (cervical and lumbar) and two dorsally convex curvatures (thoracic and sacral).

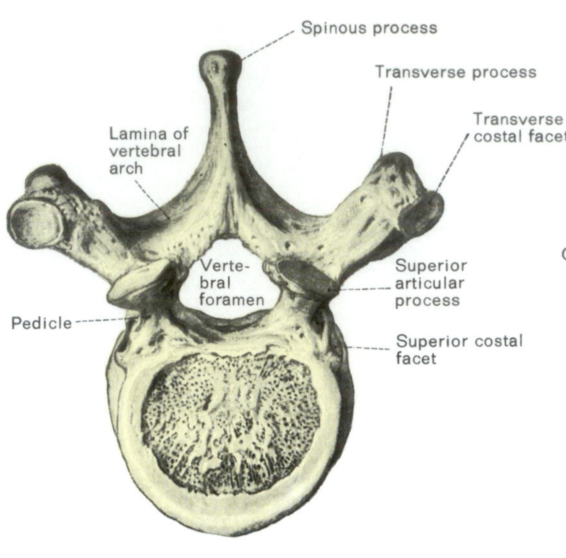
Spinous process
Transverse process
Transverse costal facet
Lamina of vertebral arch
Verte-bral foramen
Superior articular process
Pedicle
Superior costal facet

Fig. 547: Sixth Thoracic Vertebra (from Above)

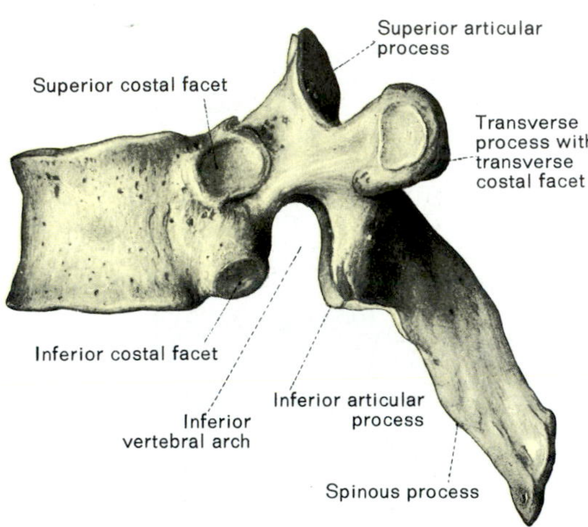
Superior articular process
Superior costal facet
Transverse process with transverse costal facet
Inferior costal facet
Inferior articular process
Inferior vertebral arch
Spinous process

Fig. 548: Sixth Thoracic Vertebra (from Left Lateral Side)

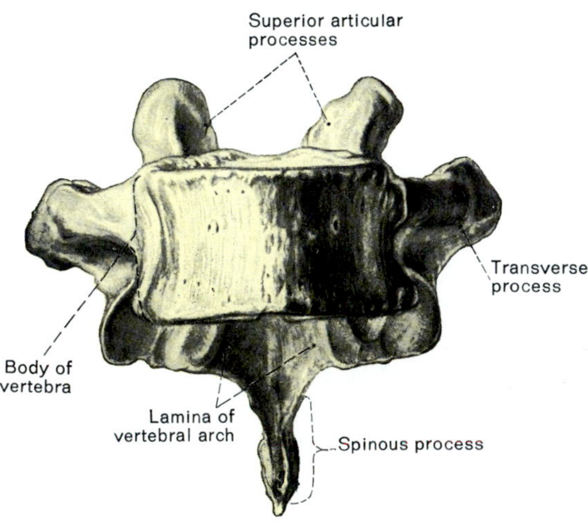
Superior articular processes
Transverse process
Body of vertebra
Lamina of vertebral arch
Spinous process

Fig. 549: Tenth Thoracic Vertebra (Ventral View)

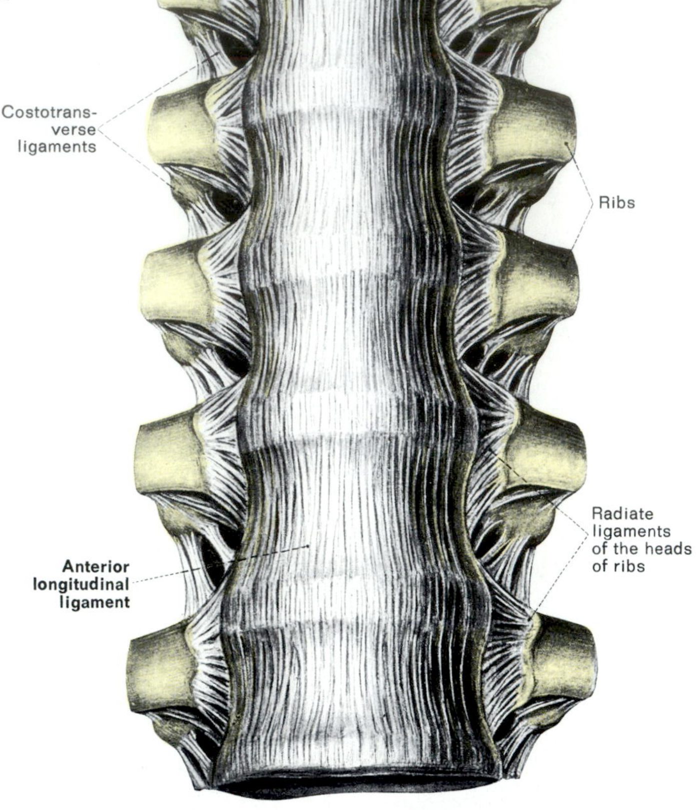
Intervertebral disc
Costotrans-verse ligaments
Ribs
Radiate ligaments of the heads of ribs
Anterior longitudinal ligament

Fig. 550: Anterior Longitudinal Ligament (Ventral View)

NOTE: the anterior longitudinal ligament extends from the axis to the sacrum along the anterior aspect of the bodies of the vertebrae and the intervertebral discs to which it is firmly attached. Its longitudinal fibers are white and glistening.

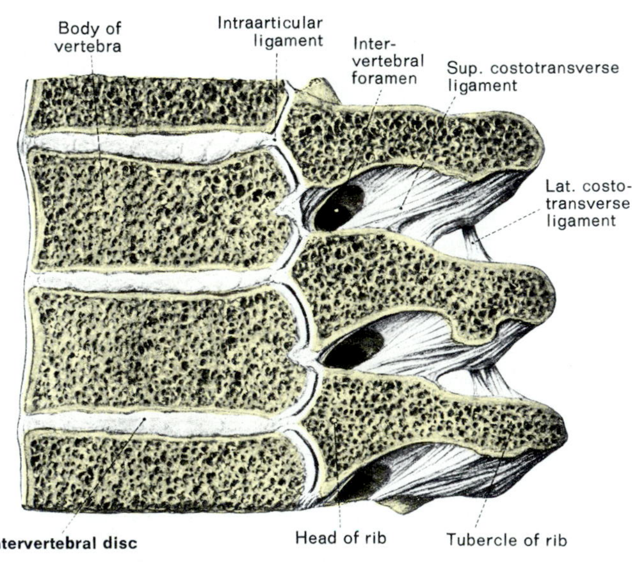
Body of vertebra
Intraarticular ligament
Inter-vertebral foramen
Sup. costotransverse ligament
Lat. costo-transverse ligament
Intervertebral disc
Head of rib
Tubercle of rib

Fig. 551: Sagittal Section (at an Angle) through Spinal Column Showing Costovertebral Articulations

NOTE: the intervertebral discs, the intraarticular and costotransverse ligaments, and the intervertebral foramina which transmit the spinal nerves.

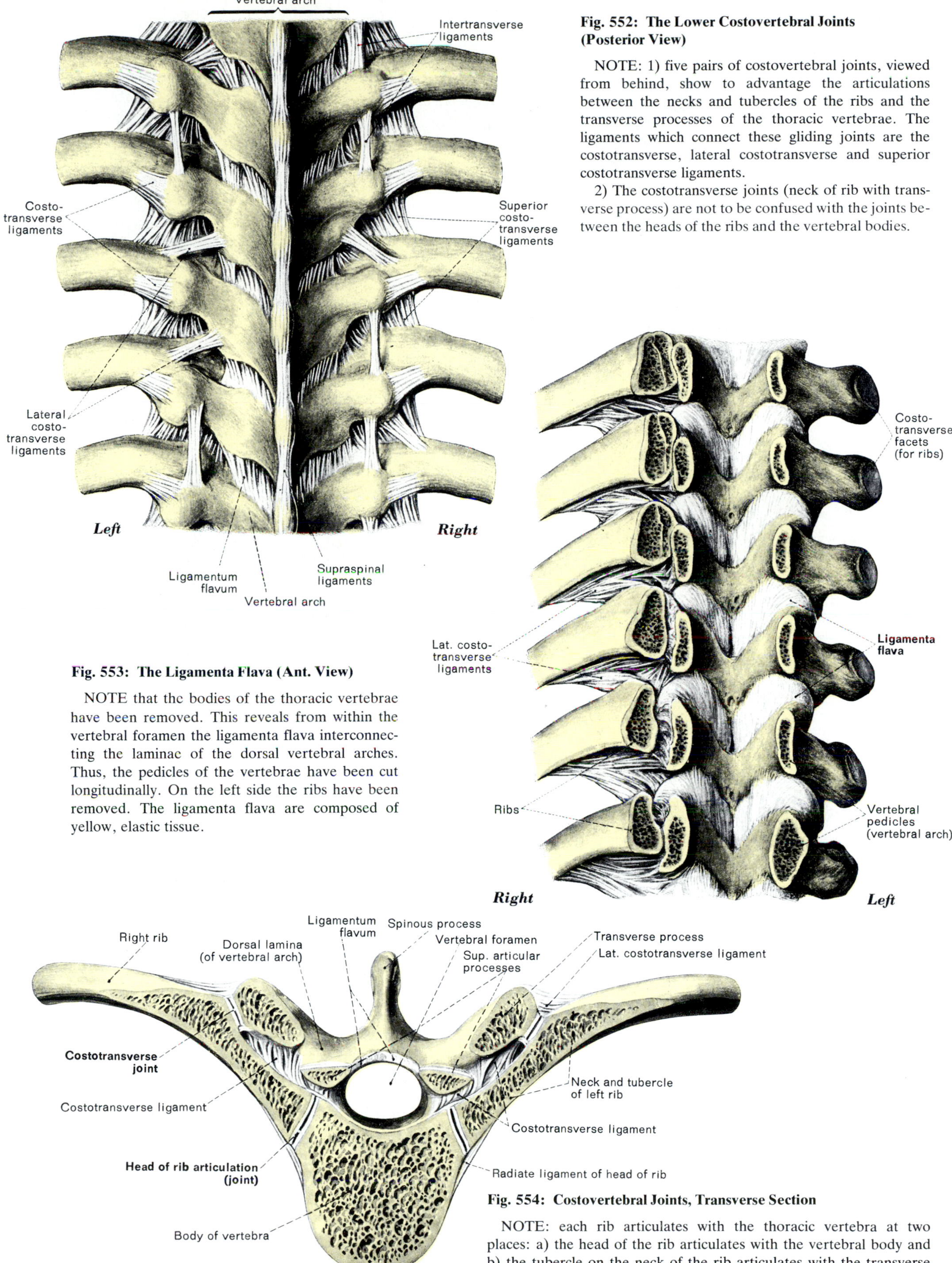
Vertebral arch
Intertransverse ligaments
Costo-transverse ligaments
Superior costo-transverse ligaments
Lateral costo-transverse ligaments
Left
Right
Ligamentum flavum
Supraspinal ligaments
Vertebral arch
Costo-transverse facets (for ribs)
Lat. costo-transverse ligaments
Ligamenta flava
Ribs
Vertebral pedicles (vertebral arch)
Right
Left
Right rib
Dorsal lamina (of vertebral arch)
Ligamentum flavum
Spinous process
Vertebral foramen
Sup. articular processes
Transverse process
Lat. costotransverse ligament
Costotransverse joint
Costotransverse ligament
Neck and tubercle of left rib
Costotransverse ligament
Head of rib articulation (joint)
Radiate ligament of head of rib
Body of vertebra

Fig. 552: The Lower Costovertebral Joints (Posterior View)

NOTE: 1) five pairs of costovertebral joints, viewed from behind, show to advantage the articulations between the necks and tubercles of the ribs and the transverse processes of the thoracic vertebrae. The ligaments which connect these gliding joints are the costotransverse, lateral costotransverse and superior costotransverse ligaments.

2) The costotransverse joints (neck of rib with transverse process) are not to be confused with the joints between the heads of the ribs and the vertebral bodies.

Fig. 553: The Ligamenta Flava (Ant. View)

NOTE that the bodies of the thoracic vertebrae have been removed. This reveals from within the vertebral foramen the ligamenta flava interconnecting the laminae of the dorsal vertebral arches. Thus, the pedicles of the vertebrae have been cut longitudinally. On the left side the ribs have been removed. The ligamenta flava are composed of yellow, elastic tissue.

Fig. 554: Costovertebral Joints, Transverse Section

NOTE: each rib articulates with the thoracic vertebra at two places: a) the head of the rib articulates with the vertebral body and b) the tubercle on the neck of the rib articulates with the transverse process of the vertebra.

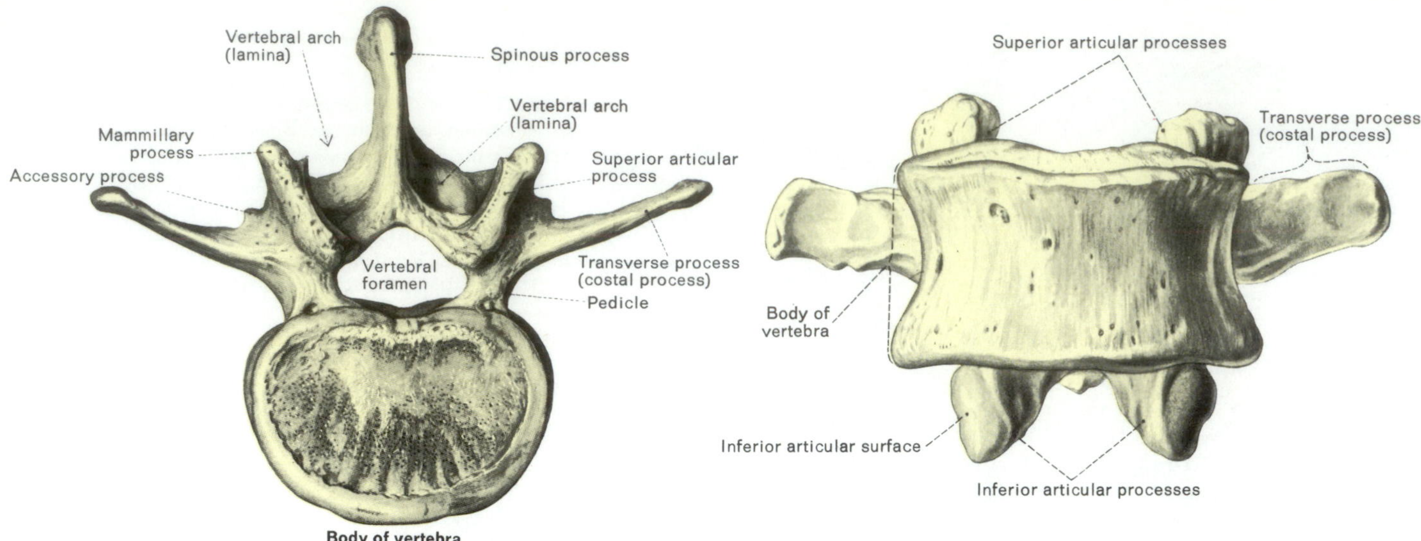
Vertebral arch (lamina)
Spinous process
Vertebral arch (lamina)
Mammillary process
Accessory process
Superior articular process
Vertebral foramen
Transverse process (costal process)
Pedicle
Body of vertebra
Superior articular processes
Transverse process (costal process)
Body of vertebra
Inferior articular surface
Inferior articular processes

Fig. 555: Lumbar Vertebra (Cranial View)

NOTE: the large bodies characteristic of the lumbar vertebrae.

Fig. 556: Lumbar Vertebra (Anterior View)

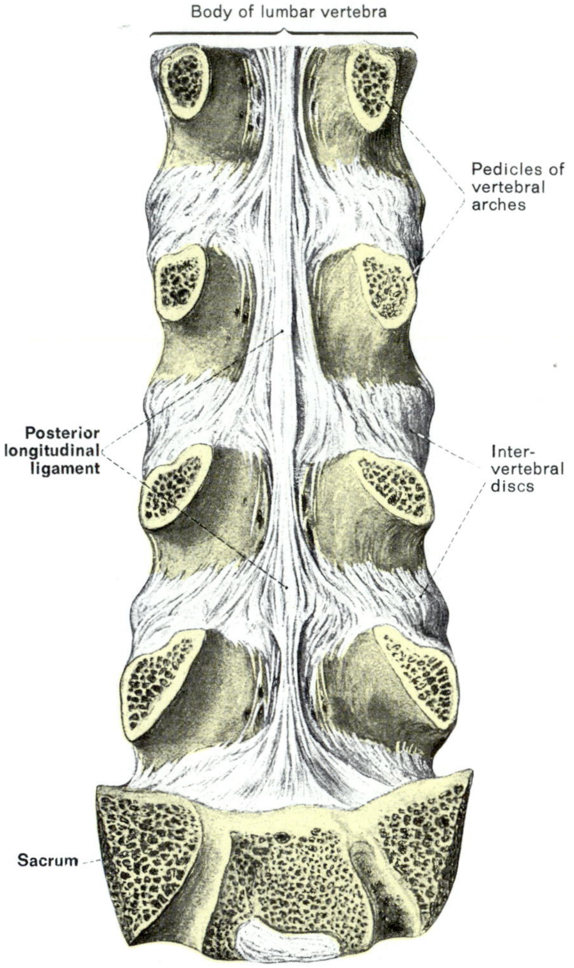

Body of lumbar vertebra
Pedicles of vertebral arches
Posterior longitudinal ligament
Inter-vertebral discs
Sacrum

Fig. 557: Posterior Longitudinal Ligament and Intervertebral Discs, Lumbosacral Region

NOTE: this dorsal view of the spinal column shows the pedicles of the vertebral arches of the lumbar vertebrae severed to expose the posterior longitudinal ligament coursing along the posterior aspect of the bodies of the lumbar vertebrae within the vertebral canal. This ligament extends from the axis (where it joins the tectorial membrane) to the sacrum, which is shown in this figure.

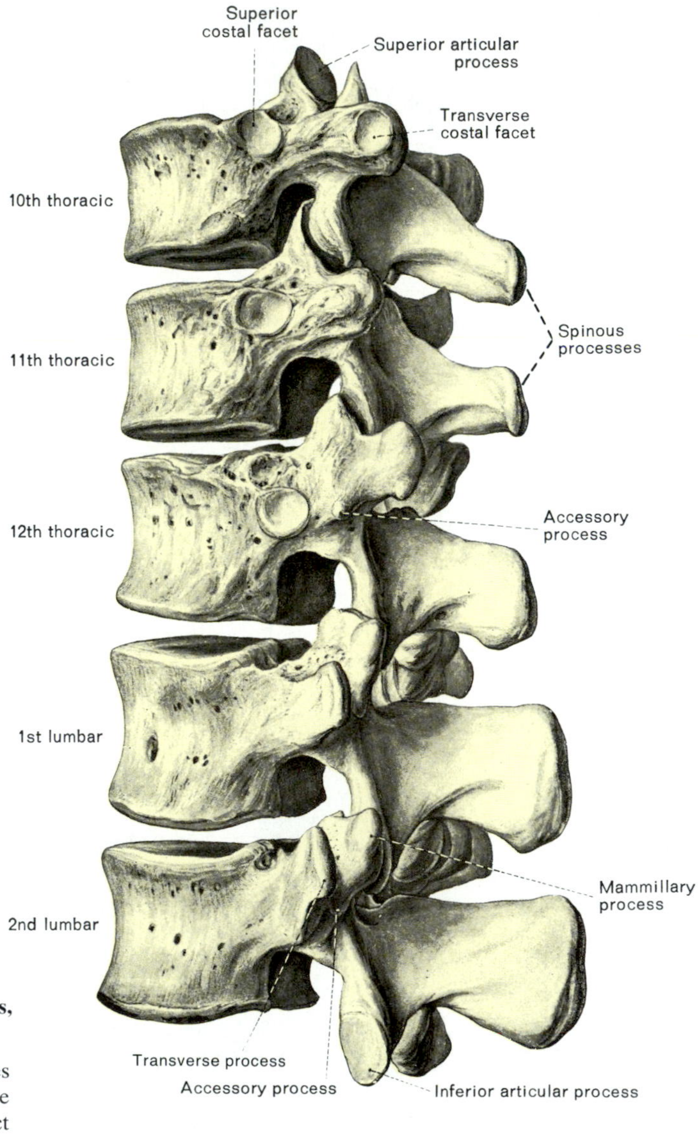

Superior costal facet
Superior articular process
Transverse costal facet
10th thoracic
11th thoracic
12th thoracic
1st lumbar
2nd lumbar
Spinous processes
Accessory process
Mammillary process
Transverse process
Accessory process
Inferior articular process

Fig. 558: Last Three Thoracic and First Two Lumbar Vertebrae (Lateral View)

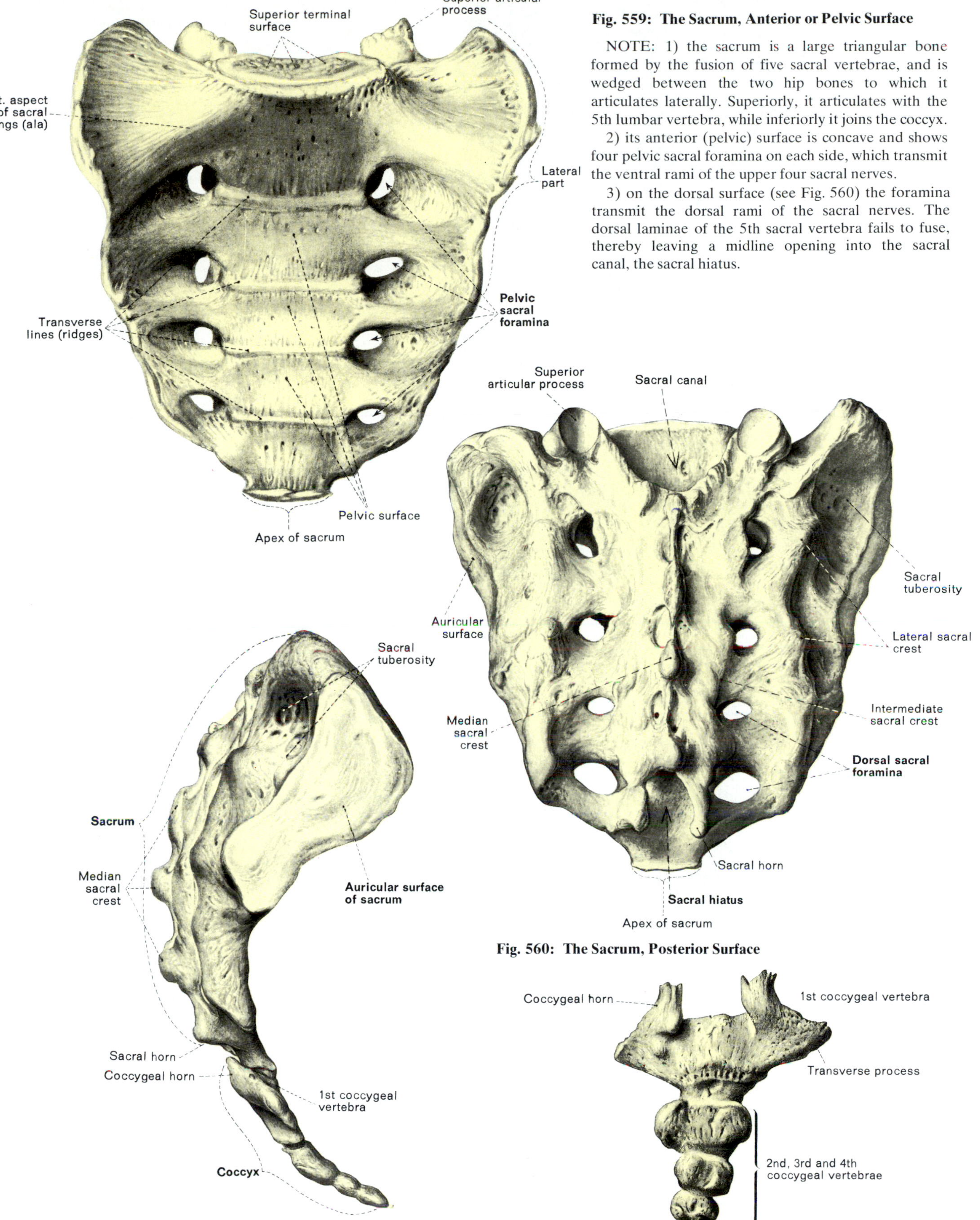

Superior terminal surface
Superior articular process
Ant. aspect of sacral wings (ala)
Lateral part
Transverse lines (ridges)
Pelvic sacral foramina
Pelvic surface
Apex of sacrum
Superior articular process
Sacral canal
Sacral tuberosity
Auricular surface
Lateral sacral crest
Intermediate sacral crest
Median sacral crest
Dorsal sacral foramina
Sacral horn
Sacral hiatus
Apex of sacrum
Sacral tuberosity
Sacrum
Median sacral crest
Auricular surface of sacrum
Sacral horn
Coccygeal horn
1st coccygeal vertebra
Coccyx
Coccygeal horn
1st coccygeal vertebra
Transverse process
2nd, 3rd and 4th coccygeal vertebrae

Fig. 559: The Sacrum, Anterior or Pelvic Surface

NOTE: 1) the sacrum is a large triangular bone formed by the fusion of five sacral vertebrae, and is wedged between the two hip bones to which it articulates laterally. Superiorly, it articulates with the 5th lumbar vertebra, while inferiorly it joins the coccyx.

2) its anterior (pelvic) surface is concave and shows four pelvic sacral foramina on each side, which transmit the ventral rami of the upper four sacral nerves.

3) on the dorsal surface (see Fig. 560) the foramina transmit the dorsal rami of the sacral nerves. The dorsal laminae of the 5th sacral vertebra fails to fuse, thereby leaving a midline opening into the sacral canal, the sacral hiatus.

Fig. 560: The Sacrum, Posterior Surface

Fig. 561: The Sacrum and Coccyx, Lateral View

NOTE: the auricular (ear-like) surface of the sacrum articulates with the iliac portion of the pelvis. Inferiorly, the sacral apex joins the coccyx.

Fig. 562: The Coccyx, Dorsal View

NOTE: this coccyx has 4 segments, but in many instances there are 3 or 5.

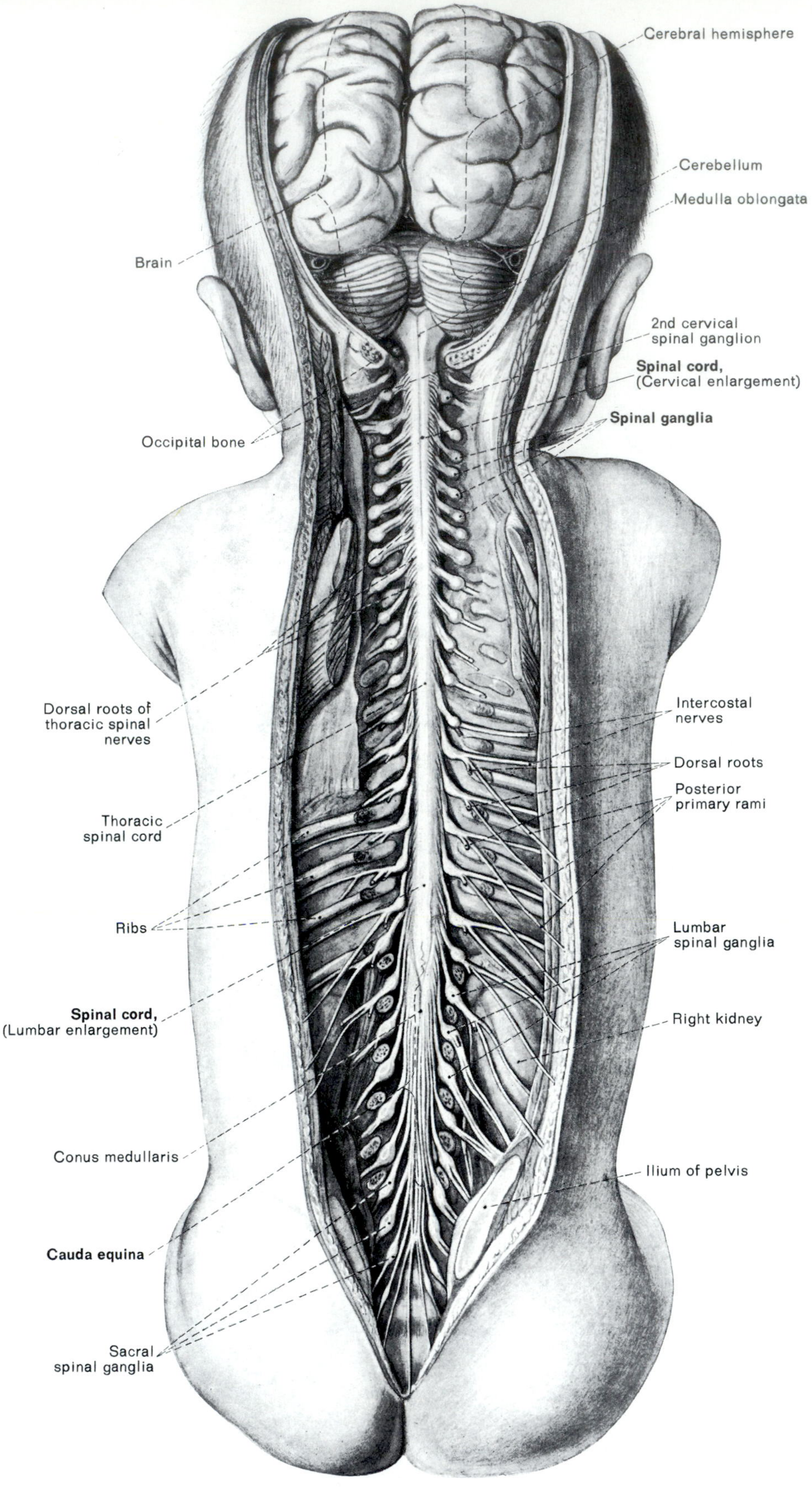

Cerebral hemisphere
Cerebellum
Medulla oblongata
Brain
2nd cervical spinal ganglion
Spinal cord, (Cervical enlargement)
Spinal ganglia
Occipital bone
Dorsal roots of thoracic spinal nerves
Intercostal nerves
Dorsal roots
Posterior primary rami
Thoracic spinal cord
Ribs
Lumbar spinal ganglia
Spinal cord, (Lumbar enlargement)
Right kidney
Conus medullaris
Ilium of pelvis
Cauda equina
Sacral spinal ganglia

Fig. 563: The Spinal Cord and Brain of a Newborn Child (Posterior View)

NOTE: 1) the central nervous system has been exposed by removal of the spinal column and dorsal cranium. The spinal ganglia have been dissected as well as their corresponding spinal nerves.

2) although in this dissection it appears as though the substance of the spinal cord terminates at about L-1, it is more usual in the newborn for the cord to end at about L-3 or L-4, thereby filling the spinal canal more completely than in the adult.

Yellow: Cervical segments (C1–C8)
Red: Thoracic segments (T1–T12)
Blue: Lumbar segments (L1–L5)
Black: Sacral segments (S1–S5)
White: Coccygeal segments (C1–C2)

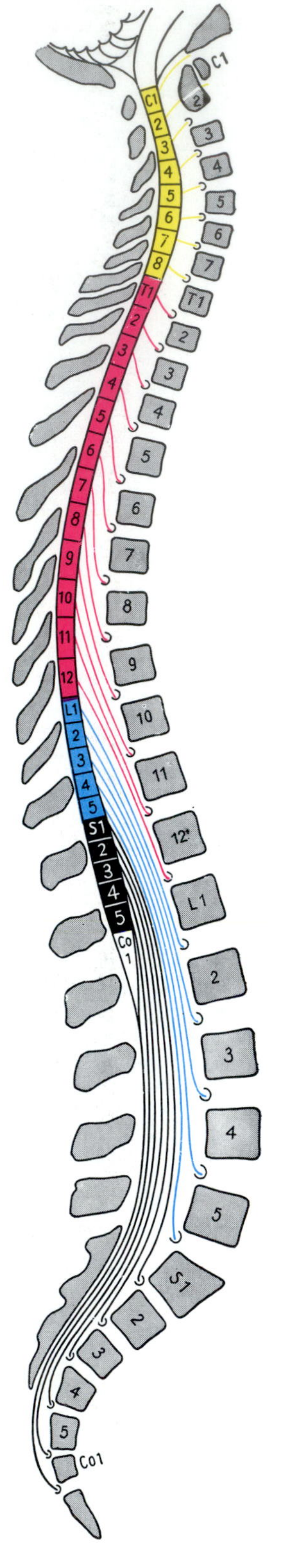

Fig. 564: The Emerging Spinal Nerves and their Segments in the Adult

NOTE: many spinal nerves travel considerable distances before leaving the vertebral canal in the adult.

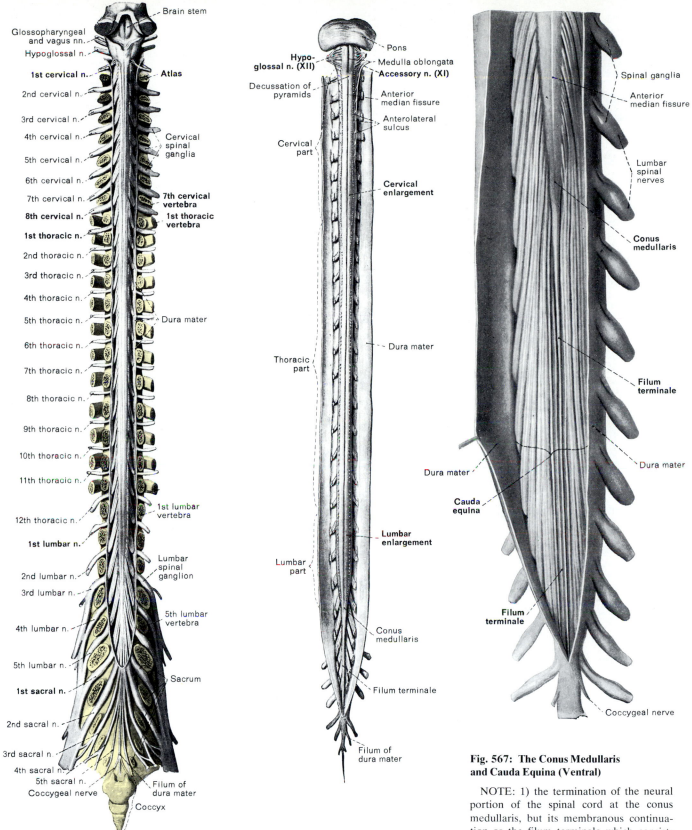

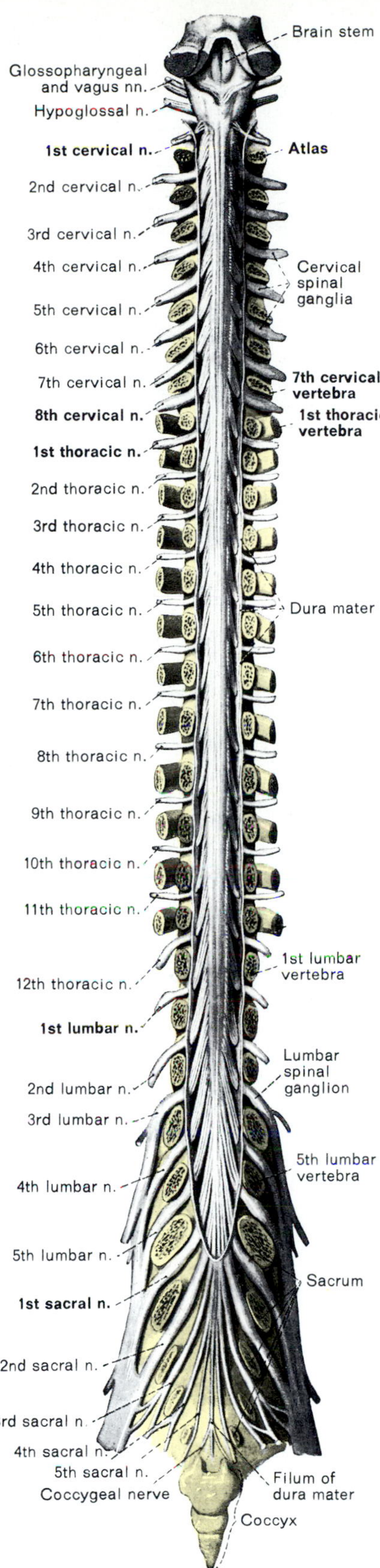
Brain stem
Glossopharyngeal and vagus nn.
Hypoglossal n.
1st cervical n.
Atlas
2nd cervical n.
3rd cervical n.
4th cervical n.
Cervical spinal ganglia
5th cervical n.
6th cervical n.
7th cervical n.
7th cervical vertebra
8th cervical n.
1st thoracic vertebra
1st thoracic n.
2nd thoracic n.
3rd thoracic n.
4th thoracic n.
5th thoracic n.
Dura mater
6th thoracic n.
7th thoracic n.
8th thoracic n.
9th thoracic n.
10th thoracic n.
11th thoracic n.
1st lumbar vertebra
12th thoracic n.
1st lumbar n.
Lumbar spinal ganglion
2nd lumbar n.
3rd lumbar n.
5th lumbar vertebra
4th lumbar n.
5th lumbar n.
Sacrum
1st sacral n.
2nd sacral n.
3rd sacral n.
4th sacral n.
5th sacral n.
Coccygeal nerve
Filum of dura mater
Coccyx

Fig. 565: The Spinal Cord Within the Vertebral Canal (Dorsal View)

NOTE: 1) the 1st cervical nerve emerges above the first vertebra and the 8th cervical nerve emerges below the 7th vertebra.

2) the cervical spinal cord is continuous above with the medulla oblongata of the brain stem.

Pons
Hypoglossal n. (XII)
Medulla oblongata
Accessory n. (XI)
Decussation of pyramids
Anterior median fissure
Anterolateral sulcus
Cervical part
Cervical enlargement
Dura mater
Thoracic part
Lumbar enlargement
Lumbar part
Conus medullaris
Filum terminale
Filum of dura mater

Fig. 566: The Spinal Cord (Ventral View)

NOTE: 1) the origin of the spinal portion of the accessory nerve (Cranial XI) arising from the cervical spinal cord and ascending to join the bulbar portion of that nerve.

2) the alignment of the rootlets of the hypoglossal nerve (Cranial XII) with the ventral roots of the spinal cord.

Spinal ganglia
Anterior median fissure
Lumbar spinal nerves
Conus medullaris
Filum terminale
Dura mater
Dura mater
Cauda equina
Filum terminale
Coccygeal nerve

Fig. 567: The Conus Medullaris and Cauda Equina (Ventral)

NOTE: 1) the termination of the neural portion of the spinal cord at the conus medullaris, but its membranous continuation as the filum terminale which consists principally of pia mater.

2) the long (approx. 8 inches) cauda equina (horse's tail) consisting of the intravertebral portions of the lower spinal nerves.

3) prolongations of the dura mater continue to cover the spinal nerves for some distance as they enter the intervertebral foramina.

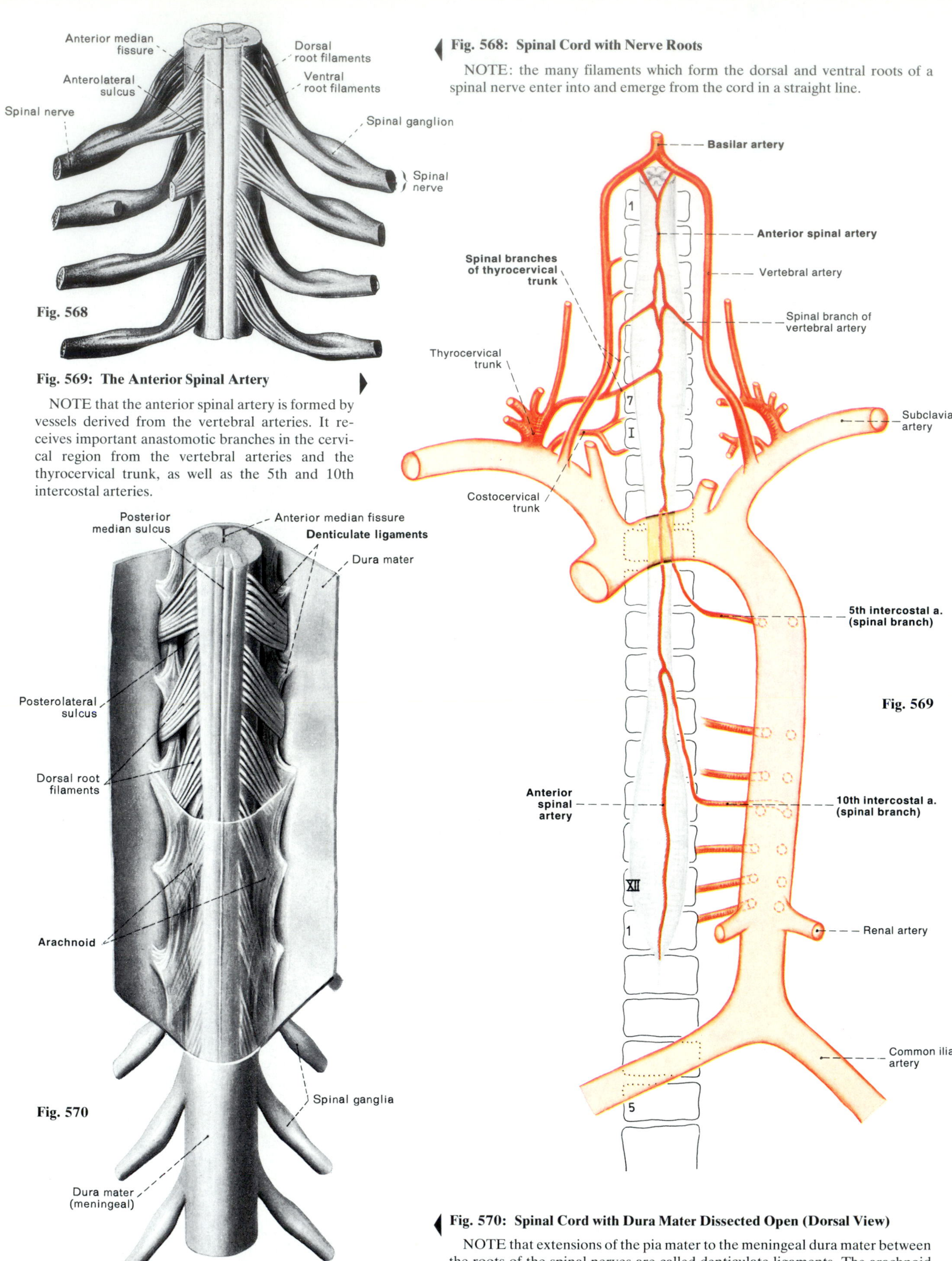
Anterior median fissure
Dorsal root filaments
Anterolateral sulcus
Ventral root filaments
Spinal nerve
Spinal ganglion
Spinal nerve
Fig. 568
Basilar artery
Anterior spinal artery
Spinal branches of thyrocervical trunk
Vertebral artery
Spinal branch of vertebral artery
Thyrocervical trunk
Subclavian artery
Costocervical trunk
5th intercostal a. (spinal branch)
Fig. 569
Anterior spinal artery
10th intercostal a. (spinal branch)
Renal artery
Common iliac artery
Posterior median sulcus
Anterior median fissure
Denticulate ligaments
Dura mater
Posterolateral sulcus
Dorsal root filaments
Arachnoid
Spinal ganglia
Fig. 570
Dura mater (meningeal)

Fig. 568: Spinal Cord with Nerve Roots

NOTE: the many filaments which form the dorsal and ventral roots of a spinal nerve enter into and emerge from the cord in a straight line.

Fig. 569: The Anterior Spinal Artery

NOTE that the anterior spinal artery is formed by vessels derived from the vertebral arteries. It receives important anastomotic branches in the cervical region from the vertebral arteries and the thyrocervical trunk, as well as the 5th and 10th intercostal arteries.

Fig. 570: Spinal Cord with Dura Mater Dissected Open (Dorsal View)

NOTE that extensions of the pia mater to the meningeal dura mater between the roots of the spinal nerves are called denticulate ligaments. The arachnoid sends fine attachments to both the pia and dura.

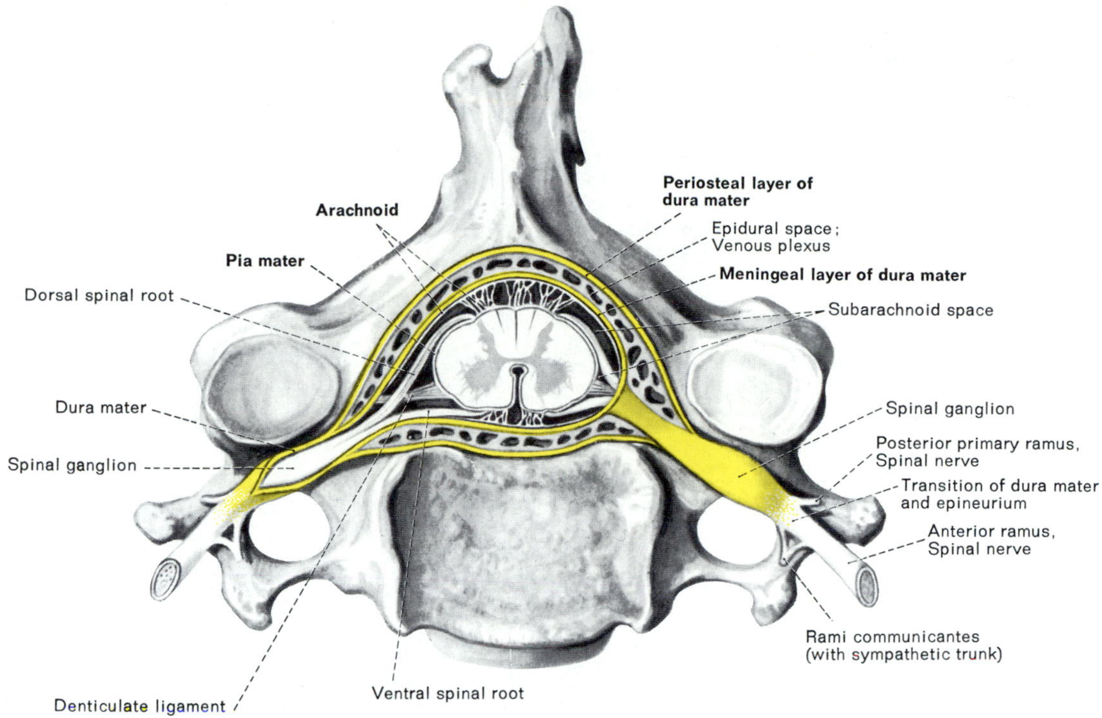
Periosteal layer of
dura mater
Arachnoid
Epidural space;
Venous plexus
Pia mater
Meningeal layer of dura mater
Dorsal spinal root
Subarachnoid space
Dura mater
Spinal ganglion
Spinal ganglion
Posterior primary ramus,
Spinal nerve
Transition of dura mater
and epineurium
Anterior ramus,
Spinal nerve
Rami communicantes
(with sympathetic trunk)
Ventral spinal root
Denticulate ligament

Fig. 571: Meninges of the Spinal Cord Shown at Cervical Level (Transverse Section)

NOTE: 1) the meningeal dura mater (inner layer of yellow) surrounds the spinal cord and continues along the spinal nerve through the intervertebral foramen. Its outer periosteal layer (outer layer of yellow) is formed of connective tissue which closely adheres to the bone of the vertebrae forming the vertebral canal.

2) the delicate, film-like arachnoid which lies between the meningeal layer of dura mater, and the vascularized pia mater which is closely applied to the cord.

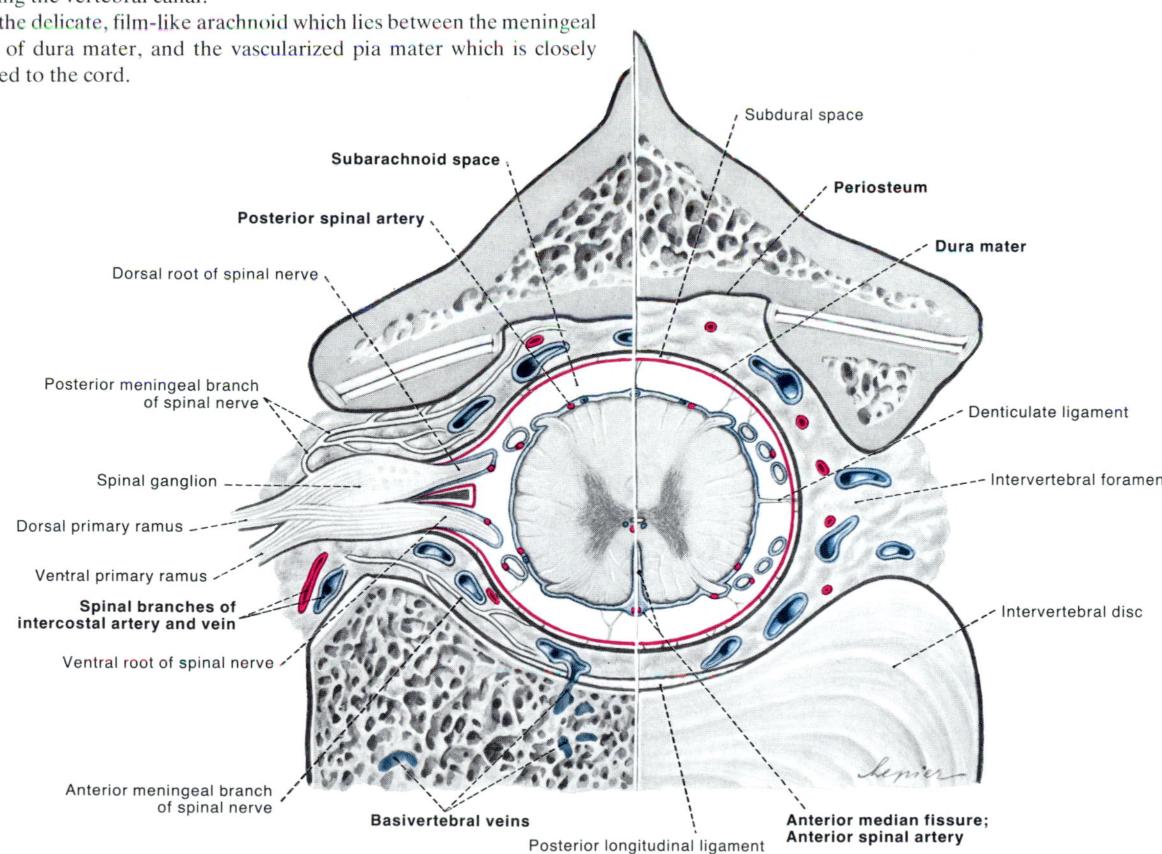
Subdural space
Subarachnoid space
Periosteum
Posterior spinal artery
Dura mater
Dorsal root of spinal nerve
Posterior meningeal branch
of spinal nerve
Denticulate ligament
Spinal ganglion
Intervertebral foramen
Dorsal primary ramus
Ventral primary ramus
Spinal branches of
intercostal artery and vein
Intervertebral disc
Ventral root of spinal nerve
Anterior meningeal branch
of spinal nerve
Basivertebral veins
Posterior longitudinal ligament
Anterior median fissure;
Anterior spinal artery

Fig. 572: The Spinal Cord Within the Vertebral Canal of the Spinal Column (Transverse Section)

NOTE: 1) that this transverse section is taken through the thoracic region of the spinal cord and shows one level *on the left* through the body of a vertebra and another level *on the right* through an intervertebral disc;

2) that the arachnoid membrane and the arteries are in red; the pia mater and the veins are in blue; the dura mater and the periosteum are in black;

3) the anterior and posterior spinal arteries as well as the intradurally located spinal veins that drain the cord. Observe also the extensive system of extradurally located vertebral veins within the spinal canal which anastomose with the basivertebral veins draining the bodies of the vertebrae.

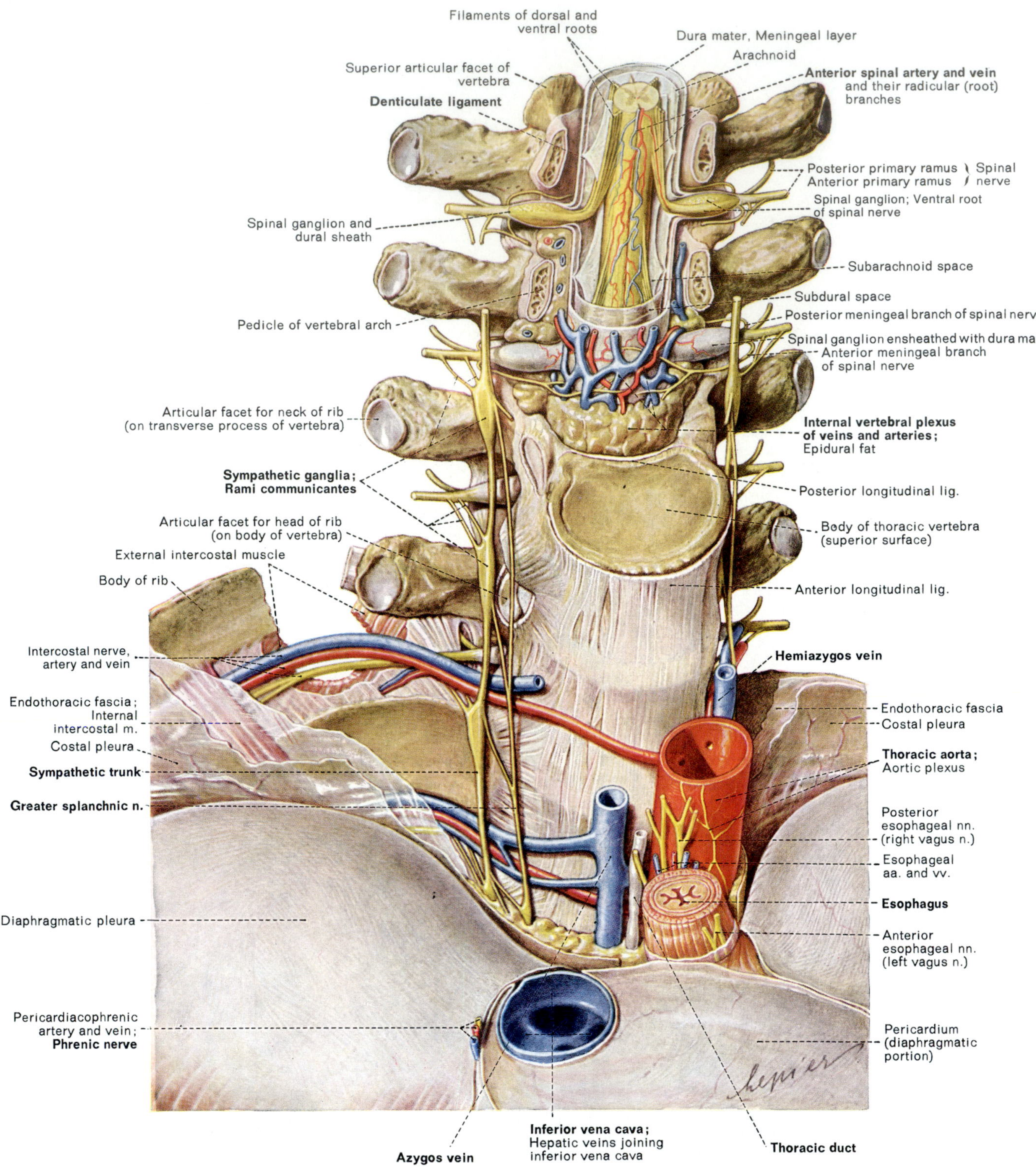

Filaments of dorsal and ventral roots
Dura mater, Meningeal layer
Arachnoid
Superior articular facet of vertebra
Anterior spinal artery and vein and their radicular (root) branches
Denticulate ligament
Posterior primary ramus
Anterior primary ramus
Spinal nerve
Spinal ganglion; Ventral root of spinal nerve
Spinal ganglion and dural sheath
Subarachnoid space
Subdural space
Pedicle of vertebral arch
Posterior meningeal branch of spinal nerve
Spinal ganglion ensheathed with dura mater
Anterior meningeal branch of spinal nerve
Articular facet for neck of rib (on transverse process of vertebra)
Internal vertebral plexus of veins and arteries; Epidural fat
Sympathetic ganglia; Rami communicantes
Posterior longitudinal lig.
Articular facet for head of rib (on body of vertebra)
Body of thoracic vertebra (superior surface)
External intercostal muscle
Body of rib
Anterior longitudinal lig.
Intercostal nerve, artery and vein
Hemiazygos vein
Endothoracic fascia; Internal intercostal m.
Endothoracic fascia
Costal pleura
Costal pleura
Thoracic aorta; Aortic plexus
Sympathetic trunk
Greater splanchnic n.
Posterior esophageal nn. (right vagus n.)
Esophageal aa. and vv.
Esophagus
Diaphragmatic pleura
Anterior esophageal nn. (left vagus n.)
Pericardiacophrenic artery and vein; Phrenic nerve
Pericardium (diaphragmatic portion)
Azygos vein
Inferior vena cava; Hepatic veins joining inferior vena cava
Thoracic duct

Fig. 573: Anterior Dissection of Vertebral Column, Spinal Cord and Prevertebral Structures at a Lower Thoracic Level

NOTE: 1) the internal vertebral plexus of veins and arteries which lie in the epidural space, where also is found the epidural fat. These should not be confused with the spinal vessels which are situated in the pia mater and which are seen to be intimately applied to the spinal cord tissue.

2) the ganglionated sympathetic chain observable here in the thoracic region receiving and giving communicating rami with the spinal nerves. Note also the formation of the greater splanchnic nerve and its descent prevertebrally into the abdomen.

3) the aorta, inferior vena cava, azygos and hemiazygos veins, esophagus and thoracic duct all lying anterior or somewhat to the left of the vertebral column and passing through the diaphragm.

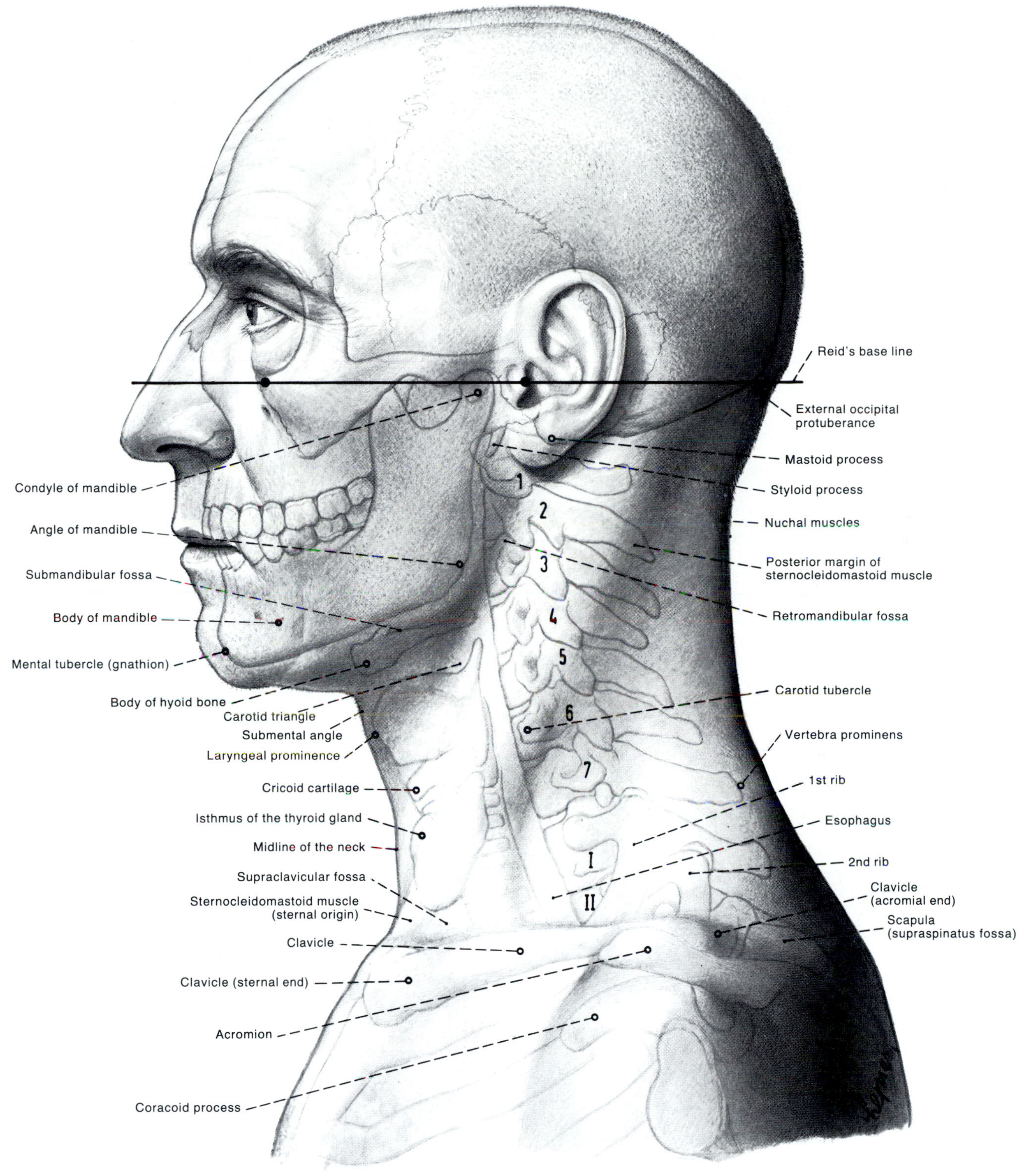
Reid's base line
External occipital protuberance
Mastoid process
Styloid process
Nuchal muscles
Posterior margin of sternocleidomastoid muscle
Retromandibular fossa
Carotid tubercle
Vertebra prominens
1st rib
Esophagus
2nd rib
Clavicle (acromial end)
Scapula (supraspinatus fossa)
Condyle of mandible
Angle of mandible
Submandibular fossa
Body of mandible
Mental tubercle (gnathion)
Body of hyoid bone
Carotid triangle
Submental angle
Laryngeal prominence
Cricoid cartilage
Isthmus of the thyroid gland
Midline of the neck
Supraclavicular fossa
Sternocleidomastoid muscle (sternal origin)
Clavicle
Clavicle (sternal end)
Acromion
Coracoid process
1
2
3
4
5
6
7
I
II

Fig. 574: External Features of the Neck Shown in Relation to Underlying Skeletal and Visceral Structures

NOTE: 1) that the palpable bony points are indicated by open circles at the end of certain leader lines;

2) the seven cervical and two upper thoracic vertebrae as well as the prominent skeletal features of the pectoral girdle. Observe the laryngeal cartilages and hyoid bone in the anterior neck and their respective cervical vertebral planes;

3) that a horizontal line (Reid's base line), drawn from the inferior margin of the orbit through the center of the external auditory canal and continuing backward to the center of the occipital bone, is frequently used for cranial topography and cephalometric studies.

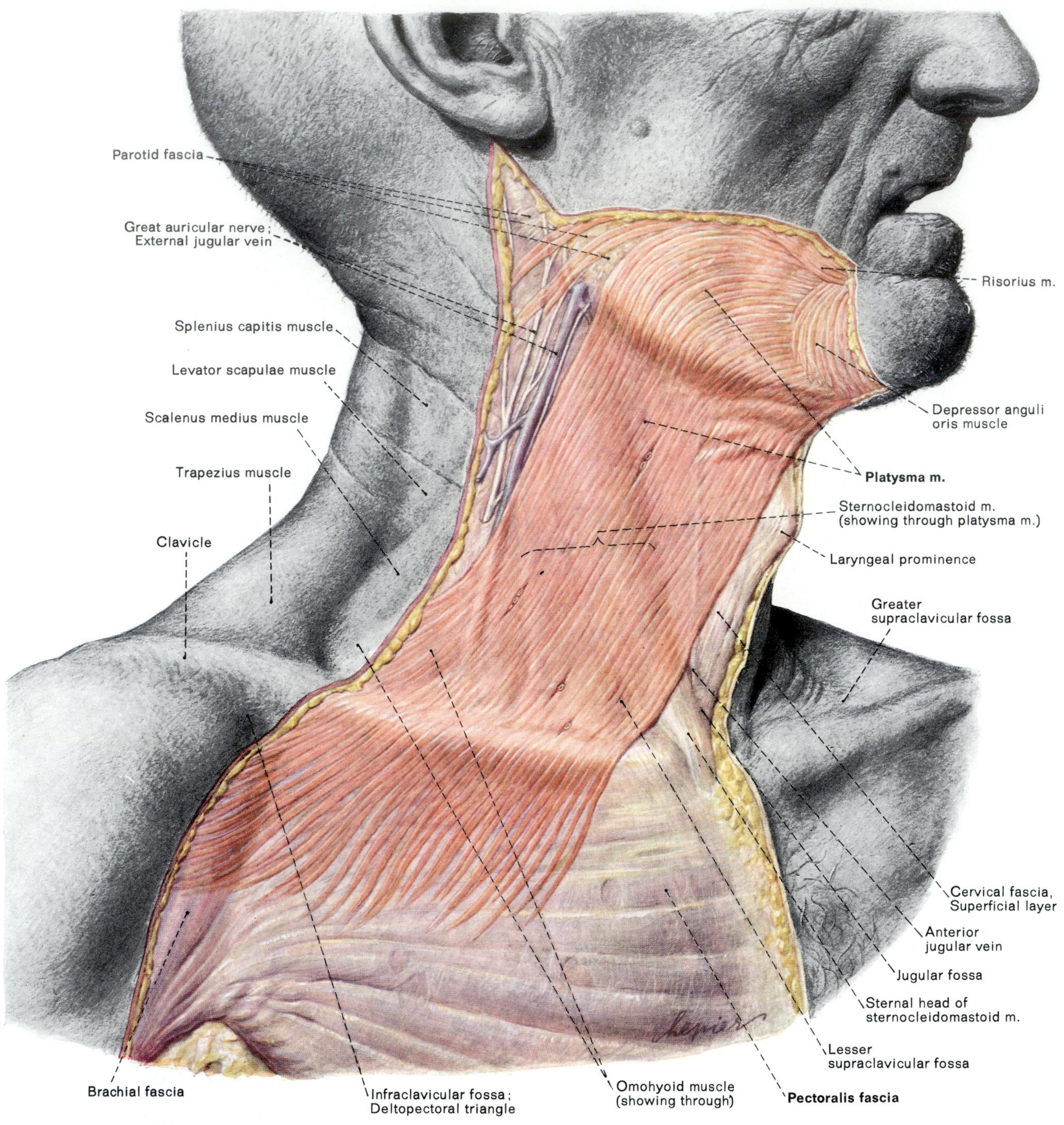
Parotid fascia
Great auricular nerve;
External jugular vein
Splenius capitis muscle
Levator scapulae muscle
Scalenus medius muscle
Trapezius muscle
Clavicle
Risorius m.
Depressor anguli
oris muscle
Platysma m.
Sternocleidomastoid m.
(showing through platysma m.)
Laryngeal prominence
Greater
supraclavicular fossa
Cervical fascia,
Superficial layer
Anterior
jugular vein
Jugular fossa
Sternal head of
sternocleidomastoid m.
Lesser
supraclavicular fossa
Brachial fascia
Infraclavicular fossa;
Deltopectoral triangle
Omohyoid muscle
(showing through)
Pectoralis fascia

Fig. 575: The Right Platysma Muscle and Pectoral Fascia

NOTE: 1) the platysma muscle is a broad, thin quadrangular muscle located in the superficial fascia on the anterolateral aspect of the neck, extending from the angle of the mouth and chin across the clavicle to the upper part of the thorax and anterior shoulder.

2) the platysma muscle can be considered as one of the muscles of facial expression, which characteristically do not arise and insert on bony structures, but within superficial fascia instead. Upon contraction, the platysma tends to depress the angle of the mouth and wrinkle the skin of the neck, thereby participating in the formation of facial expressions of anxiety, sadness, dissatisfaction and suffering.

3) similar to the other muscles of facial expression, the platysma is innervated by the cervical branch of the facial or VIIth cranial nerve.

4) overlying the pectoralis major is the well developed pectoralis fascia which extends from the midline in the thorax laterally to the axilla. Observe the external jugular vein and great auricular nerve exposed in the upper lateral aspect of the neck.

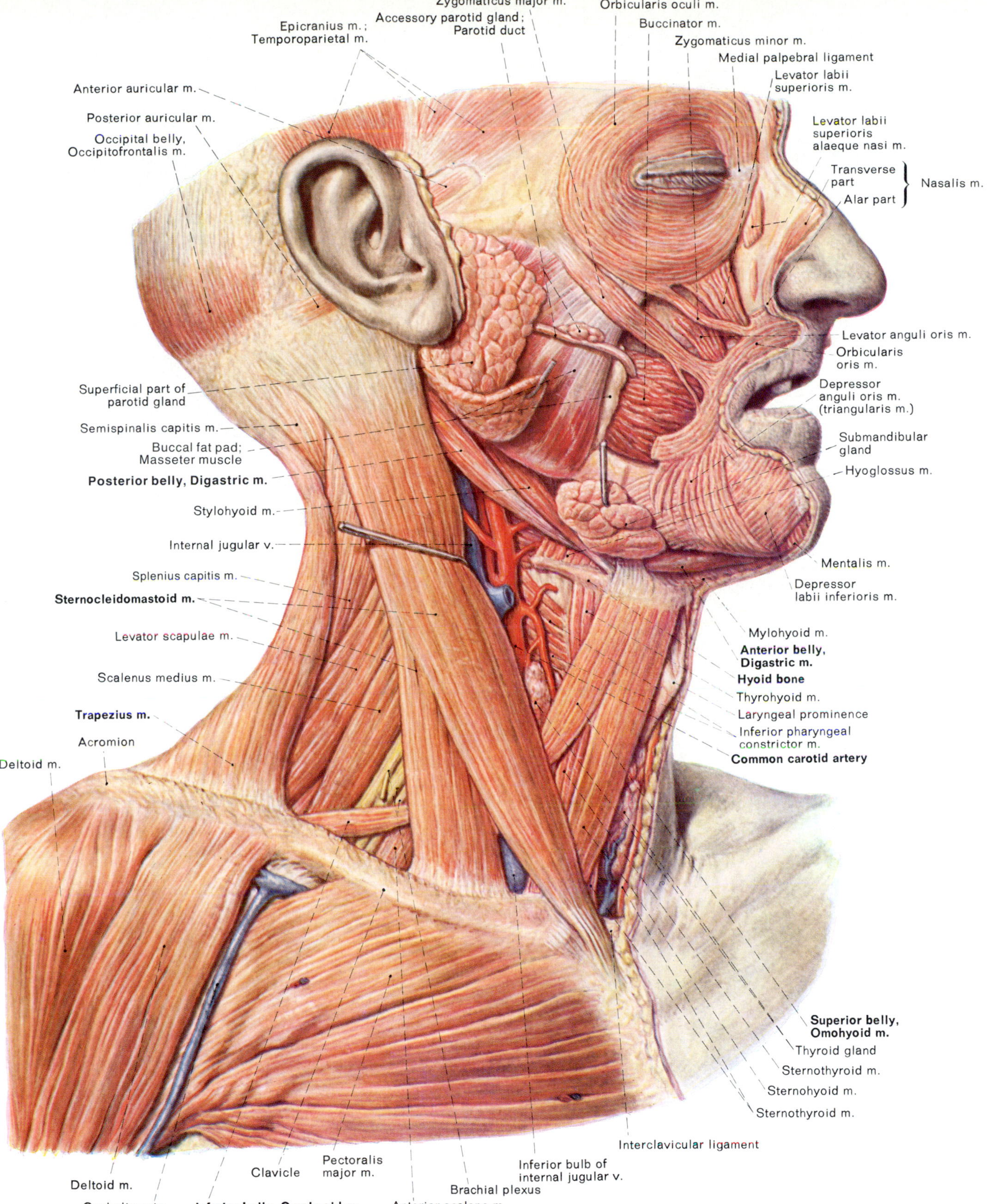

Zygomaticus major m.
Orbicularis oculi m.
Accessory parotid gland; Parotid duct
Epicranius m.; Temporoparietal m.
Buccinator m.
Zygomaticus minor m.
Medial palpebral ligament
Levator labii superioris m.
Anterior auricular m.
Posterior auricular m.
Occipital belly, Occipitofrontalis m.
Levator labii superioris alaeque nasi m.
Transverse part
Alar part
Nasalis m.
Levator anguli oris m.
Orbicularis oris m.
Superficial part of parotid gland
Depressor anguli oris m. (triangularis m.)
Semispinalis capitis m.
Buccal fat pad; Masseter muscle
Submandibular gland
Hyoglossus m.
Posterior belly, Digastric m.
Stylohyoid m.
Internal jugular v.
Mentalis m.
Splenius capitis m.
Depressor labii inferioris m.
Sternocleidomastoid m.
Levator scapulae m.
Mylohyoid m.
Anterior belly, Digastric m.
Hyoid bone
Scalenus medius m.
Thyrohyoid m.
Laryngeal prominence
Trapezius m.
Inferior pharyngeal constrictor m.
Acromion
Common carotid artery
Deltoid m.
Superior belly, Omohyoid m.
Thyroid gland
Sternothyroid m.
Sternohyoid m.
Sternothyroid m.
Interclavicular ligament
Pectoralis major m.
Inferior bulb of internal jugular v.
Clavicle
Deltoid m.
Brachial plexus
Cephalic vein
Inferior belly, Omohyoid m.
Anterior scalene m.

Fig. 576: The Anterior and Posterior Triangles of the Neck

NOTE: 1) the *anterior triangle* of the neck is bounded by the midline of the neck, the anterior border of the sternocleidomastoid and the mandible. This area is further subdivided by the superior belly of the omohyoid and the two bellies of the digastric into the following: a) *muscular triangle* (midline, superior belly of omohyoid, sternocleidomastoid), b) *carotid triangle* (superior belly of omohyoid, sternocleidomastoid and posterior belly of digastric), c) *submandibular triangle* (anterior and posterior bellies of digastric and the inferior margin of the mandible), and d) *suprahyoid triangle* (midline, anterior belly of digastric and hyoid bone).

2) the *posteror triangle* of the neck is bounded by the posterior border of the sternocleidomastoid, the trapezius and the clavicle. This area is subdivided into the *occipital triangle* above, and *subclavian triangle* below, by the inferior belly of the omohyoid muscle.

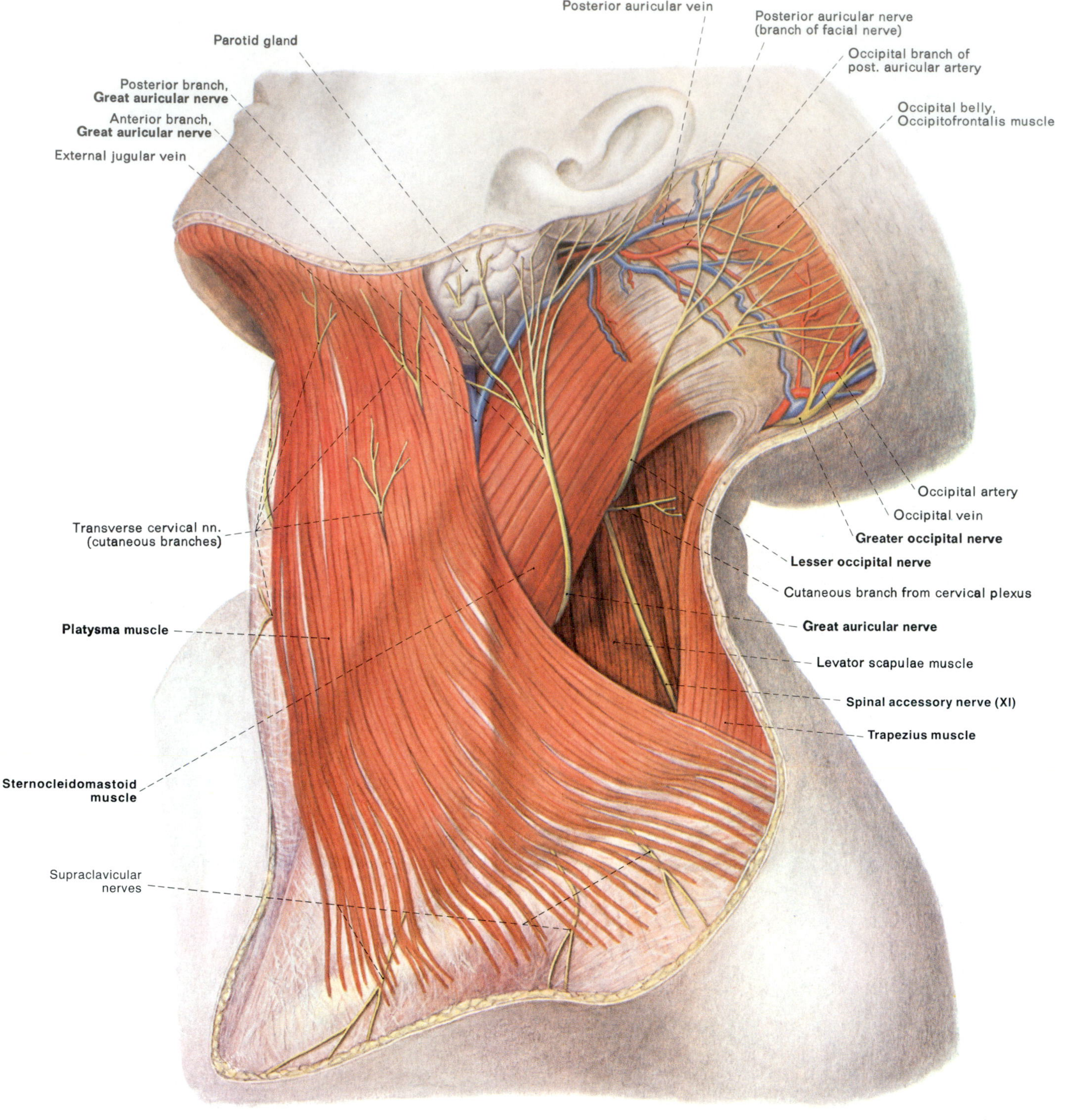
Posterior auricular vein
Posterior auricular nerve
(branch of facial nerve)
Parotid gland
Occipital branch of
post. auricular artery
Posterior branch,
Great auricular nerve
Occipital belly,
Occipitofrontalis muscle
Anterior branch,
Great auricular nerve
External jugular vein
Occipital artery
Occipital vein
Transverse cervical nn.
(cutaneous branches)
Greater occipital nerve
Lesser occipital nerve
Cutaneous branch from cervical plexus
Platysma muscle
Great auricular nerve
Levator scapulae muscle
Spinal accessory nerve (XI)
Trapezius muscle
Sternocleidomastoid
muscle
Supraclavicular
nerves

Fig. 577: Nerves and Blood Vessels of the Neck, Stage 1: Platysma Layer

NOTE: 1) the skin has been removed from both anterior and posterior triangles of the neck to reveal the platysma muscle. Observe the cutaneous branches of the transverse cervical nerves emanating from the cervical plexus and penetrating through the platysma (and superficial fascia) to reach the skin of the anterolateral aspect of the neck.

2) four other nerves: a) the great auricular (C-2, C-3), b) the lesser occipital (C-2), c) the greater occipital (C-2) and d) the spinal accessory (XI).

3) after it has supplied the sternocleidomastoid muscle, the spinal accessory nerve (XI) descends in the posterior triangle to reach the trapezius muscle which it also supplies with motor innervation.

4) the supraclavicular nerves, which descend in the lateral part of the neck, under cover of the deep fascia and the superficial fascia and platysma muscle. These nerves become superficial just above the clavicle and they then cross that bone as anterior (medial), middle (intermediate), and posterior (lateral) supraclavicular nerves (see also Fig. 578). They are derived from the 3rd and 4th cervical nerves, and they supply the skin over the clavicle and over the upper trunk (down to the 2nd rib) and shoulder from the acromion laterally to the midline anteriorly.

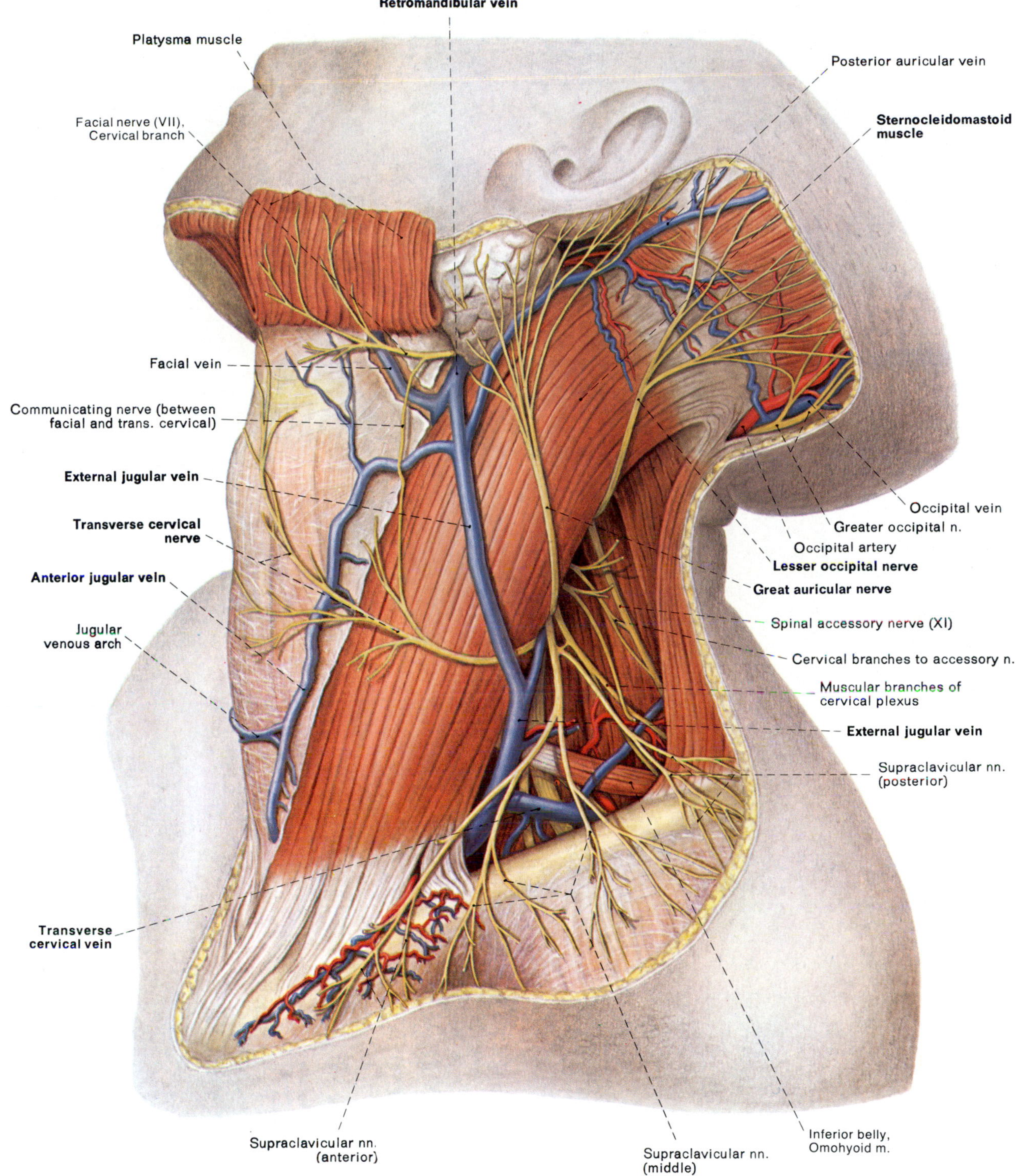
Retromandibular vein
Platysma muscle
Posterior auricular vein
Facial nerve (VII), Cervical branch
Sternocleidomastoid muscle
Facial vein
Communicating nerve (between facial and trans. cervical)
External jugular vein
Transverse cervical nerve
Anterior jugular vein
Jugular venous arch
Occipital vein
Greater occipital n.
Occipital artery
Lesser occipital nerve
Great auricular nerve
Spinal accessory nerve (XI)
Cervical branches to accessory n.
Muscular branches of cervical plexus
External jugular vein
Supraclavicular nn. (posterior)
Transverse cervical vein
Supraclavicular nn. (anterior)
Supraclavicular nn. (middle)
Inferior belly, Omohyoid m.

Fig. 578: Nerves and Blood Vessels of the Neck, Stage 2: Sternocleidomastoid Layer

NOTE: 1) with the platysma reflected upward, the full extent of the sternocleidomastoid muscle is exposed. Observe that the nerves of the cervical plexus diverge at the posterior border of the sternocleidomastoid muscle: the great auricular and lesser occipital ascend to the head, the transverse colli course across the neck toward the midline, while the supraclavicular descend over the clavicle.

2) the external jugular vein formed by the junction of the retromandibular and posterior auricular veins. The external jugular crosses the sternocleidomastoid muscle obliquely and receives tributaries from the anterior jugular, posterior external jugular, transverse cervical and suprascapular veins before terminating into the subclavian vein.

3) the cervical branch of the facial (VII) nerve supplying the inner surface of the platysma muscle with motor innervation.

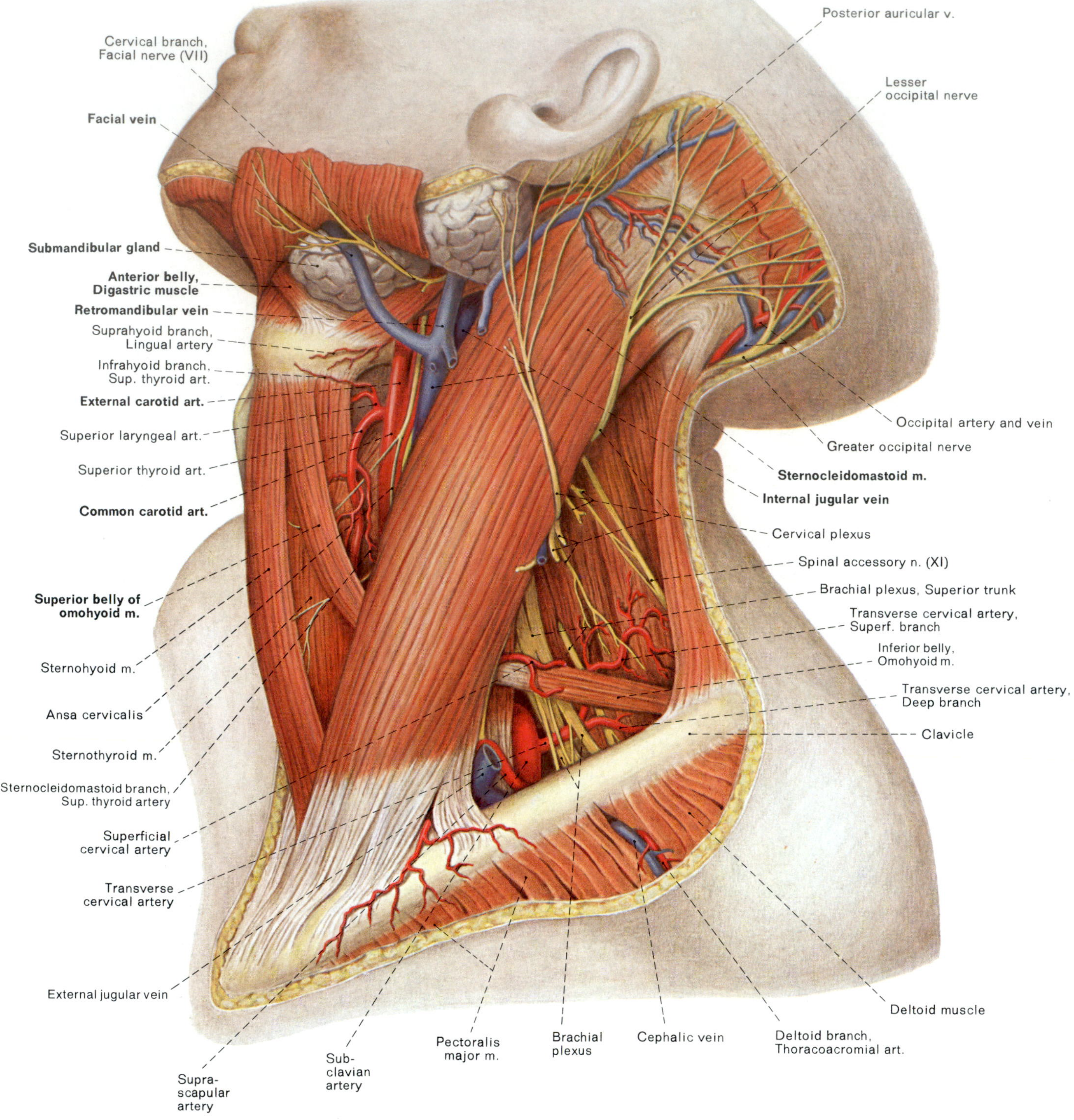

Cervical branch, Facial nerve (VII)
Facial vein
Submandibular gland
Anterior belly, Digastric muscle
Retromandibular vein
Suprahyoid branch, Lingual artery
Infrahyoid branch, Sup. thyroid art.
External carotid art.
Superior laryngeal art.
Superior thyroid art.
Common carotid art.
Superior belly of omohyoid m.
Sternohyoid m.
Ansa cervicalis
Sternothyroid m.
Sternocleidomastoid branch, Sup. thyroid artery
Superficial cervical artery
Transverse cervical artery
External jugular vein
Supra-scapular artery
Sub-clavian artery
Pectoralis major m.
Brachial plexus
Cephalic vein
Deltoid branch, Thoracoacromial art.
Deltoid muscle
Clavicle
Transverse cervical artery, Deep branch
Inferior belly, Omohyoid m.
Transverse cervical artery, Superf. branch
Brachial plexus, Superior trunk
Spinal accessory n. (XI)
Cervical plexus
Internal jugular vein
Sternocleidomastoid m.
Greater occipital nerve
Occipital artery and vein
Lesser occipital nerve
Posterior auricular v.

Fig. 579: Nerves and Blood Vessels of the Neck, Stage 3: the Anterior Triangle

NOTE: 1) with the investing fascia removed, the outlines of the muscular, carotid and submandibular triangles, which subdivide the anterior neck region, are revealed.

2) the infrahyoid (strap) muscles which cover the thyroid gland in the *muscular triangle* (bounded by sternocleidomastoid, midline and superior belly of omohyoid).

3) the carotid vessels and jugular vein in the *carotid triangle* (bounded by superior belly of omohyoid, posterior belly of digastric and sternocleidomastoid).

4) the submandibular gland in the *submandibular triangle* (bounded by anterior and posterior bellies of digastric and the inferior border of the mandible).

Fig. 579

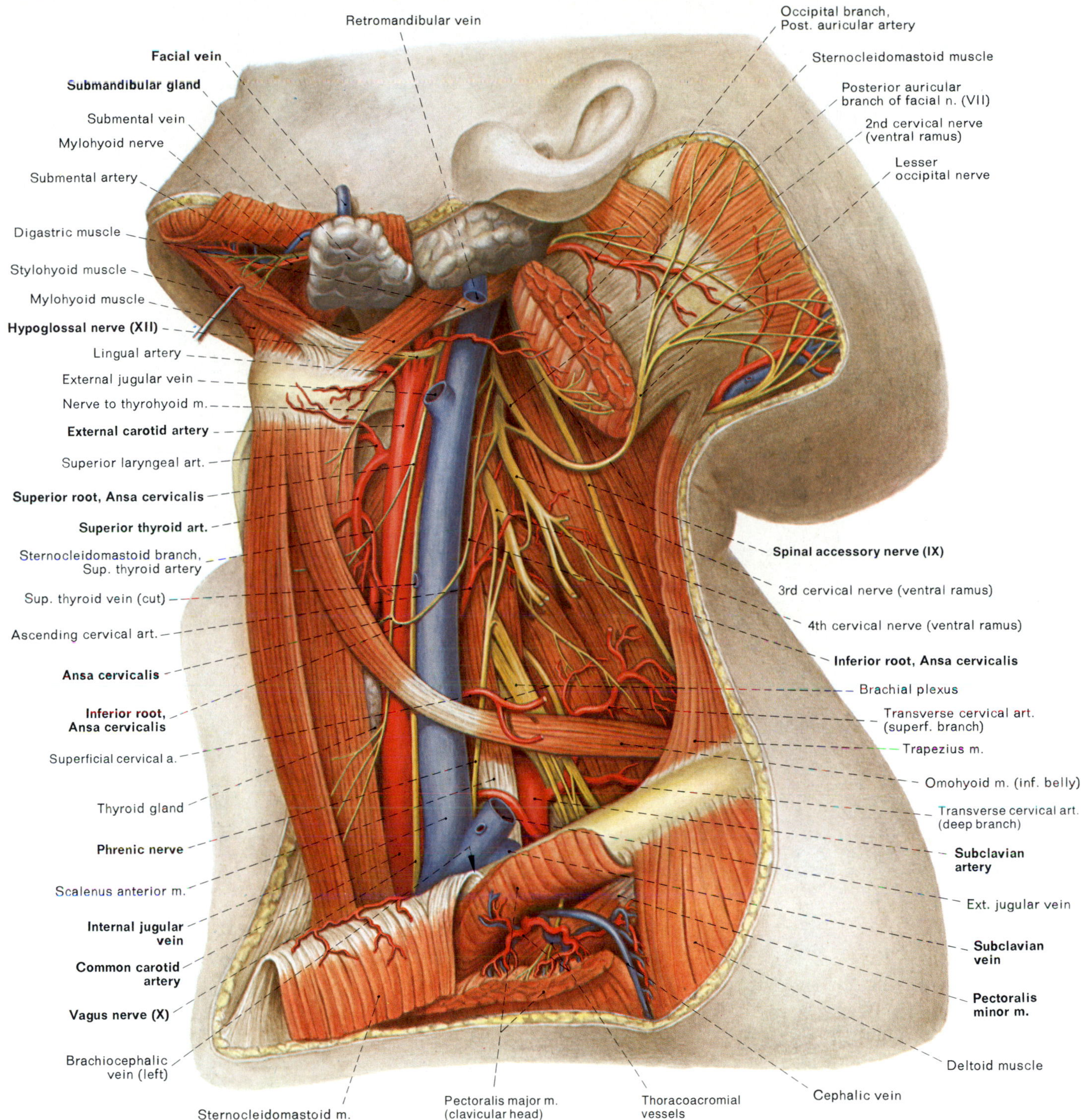

Retromandibular vein
Occipital branch,
Post. auricular artery
Facial vein
Sternocleidomastoid muscle
Submandibular gland
Posterior auricular
branch of facial n. (VII)
Submental vein
2nd cervical nerve
(ventral ramus)
Mylohyoid nerve
Lesser
occipital nerve
Submental artery
Digastric muscle
Stylohyoid muscle
Mylohyoid muscle
Hypoglossal nerve (XII)
Lingual artery
External jugular vein
Nerve to thyrohyoid m.
External carotid artery
Superior laryngeal art.
Superior root, Ansa cervicalis
Superior thyroid art.
Sternocleidomastoid branch,
Sup. thyroid artery
Spinal accessory nerve (IX)
3rd cervical nerve (ventral ramus)
Sup. thyroid vein (cut)
Ascending cervical art.
4th cervical nerve (ventral ramus)
Inferior root, Ansa cervicalis
Ansa cervicalis
Brachial plexus
Inferior root,
Ansa cervicalis
Transverse cervical art.
(superf. branch)
Superficial cervical a.
Trapezius m.
Omohyoid m. (inf. belly)
Thyroid gland
Transverse cervical art.
(deep branch)
Phrenic nerve
Subclavian
artery
Scalenus anterior m.
Ext. jugular vein
Internal jugular
vein
Subclavian
vein
Common carotid
artery
Pectoralis
minor m.
Vagus nerve (X)
Brachiocephalic
vein (left)
Deltoid muscle
Cephalic vein
Sternocleidomastoid m.
Pectoralis major m.
(clavicular head)
Thoracoacromial
vessels

Fig. 580: Nerves and Blood Vessels of the Neck, Stage 4: the Large Vessels

NOTE: 1) the sternocleidomastoid muscle and the superficial veins and nerves have been removed to expose the internal and external carotid arteries, the internal jugular vein, both bellies of the omohyoid muscle, the vagus nerve and the ansa cervicalis.

2) superiorly, the facial vein has been cut and the submandibular gland elevated, thereby exposing the hypoglossal nerve (XII). Observe that nerve fibers (originating from C-1 and travelling for a short distance) leave the hypoglossal nerve to descend in the neck. These form the superior root of the ansa cervicalis and are joined by other descending fibers from C-2 and C-3 which are called the inferior root of the ansa cervicalis. These two roots form the ansa cervicalis, from which innervation for a number of the strap muscles is derived.

3) the left phrenic nerve descending toward the thorax, where it and the right phrenic nerve are the only motor nerves that supply the musculature of the diaphragm. This nerve arises principally from the fourth cervical nerve, but it also receives fibers from C-3 and C-5. Observe that it courses deep to the omohyoid muscle and superficial to the anterior scalene muscle.

4) the common carotid artery, internal jugular vein, and vagus nerve, forming a vertically oriented neurovascular bundle in the neck that is normally surrounded by the carotid sheath of deep fascia. Observe that the common carotid artery bifurcates at about the level of the hyoid bone (or a bit lower) to form the external and internal carotid arteries.

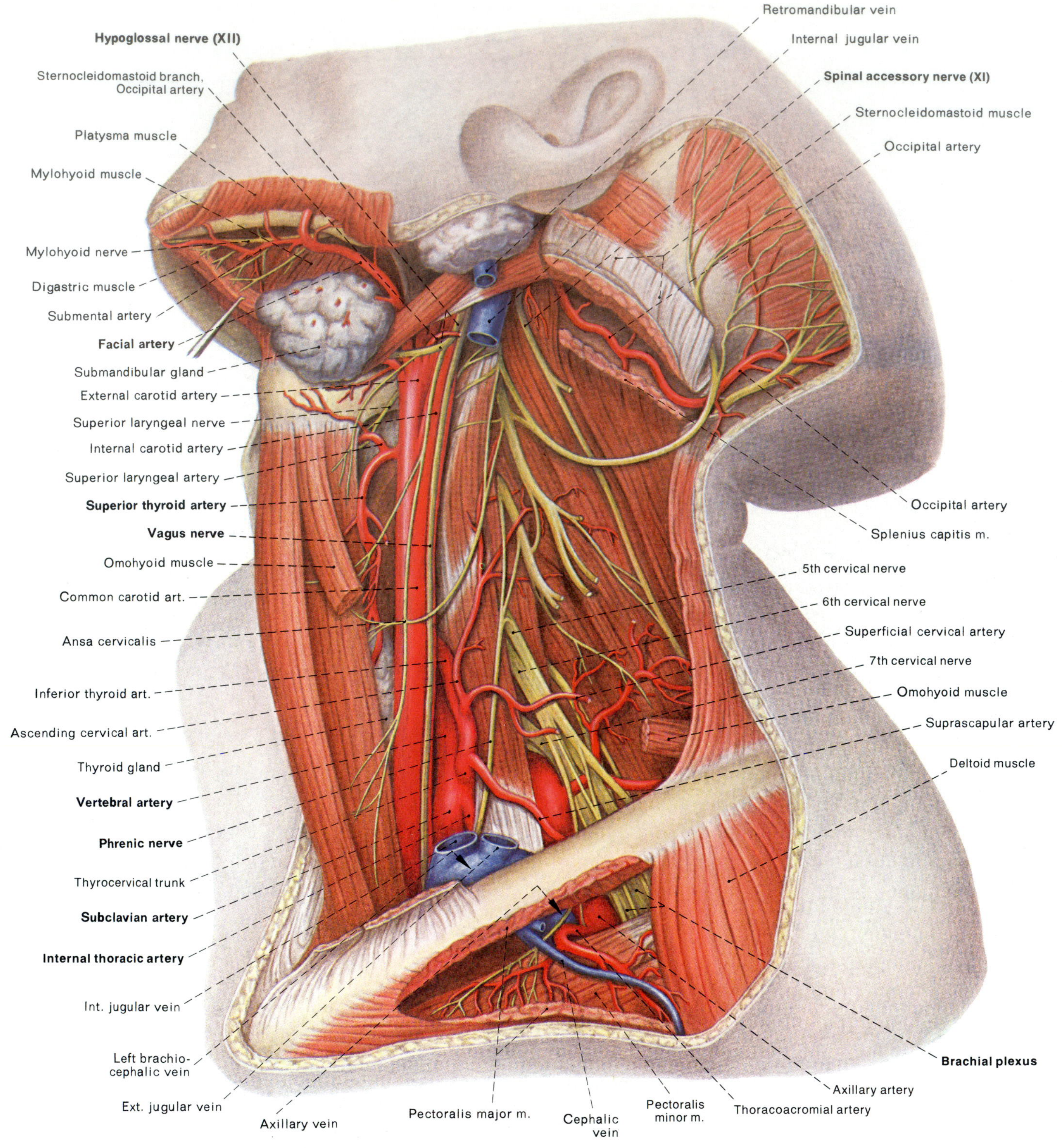

Retromandibular vein
Hypoglossal nerve (XII)
Internal jugular vein
Sternocleidomastoid branch, Occipital artery
Spinal accessory nerve (XI)
Sternocleidomastoid muscle
Platysma muscle
Occipital artery
Mylohyoid muscle
Mylohyoid nerve
Digastric muscle
Submental artery
Facial artery
Submandibular gland
External carotid artery
Superior laryngeal nerve
Internal carotid artery
Superior laryngeal artery
Superior thyroid artery
Occipital artery
Vagus nerve
Splenius capitis m.
Omohyoid muscle
5th cervical nerve
Common carotid art.
6th cervical nerve
Ansa cervicalis
Superficial cervical artery
7th cervical nerve
Inferior thyroid art.
Omohyoid muscle
Ascending cervical art.
Suprascapular artery
Thyroid gland
Deltoid muscle
Vertebral artery
Phrenic nerve
Thyrocervical trunk
Subclavian artery
Internal thoracic artery
Int. jugular vein
Left brachio-cephalic vein
Brachial plexus
Axillary artery
Ext. jugular vein
Pectoralis major m.
Pectoralis minor m.
Thoracoacromial artery
Axillary vein
Cephalic vein

Fig. 581: Nerves and Blood Vessels of the Neck, Stage 5: the Subclavian Artery

NOTE: 1) with the internal and external jugular veins removed, the subclavian artery becomes exposed as it ascends from the thorax and loops in the subclavian triangle of the neck to descend beneath the clavicle into the axilla. Observe the vertebral, thyrocervical and internal thoracic branches and the transverse cervical artery, which comes off separately in this dissection and which divides into an ascending superficial branch and a descending deep branch (not labelled).

2) the vagus nerve coursing with the internal and common carotid arteries and the phrenic nerve descending in the neck along the surface of the anterior scalene muscle toward the thorax.

3) some of the vessels derived from the external carotid artery; well illustrated are the superior thyroid, facial, and occipital arteries. The lingual artery is also shown but is not labelled. The occipital artery courses posteriorly, deep to the sternocleidomastoid and splenius capitis muscle, and it becomes superficial on the posterior aspect of the scalp, just lateral to the superior nuchal line.

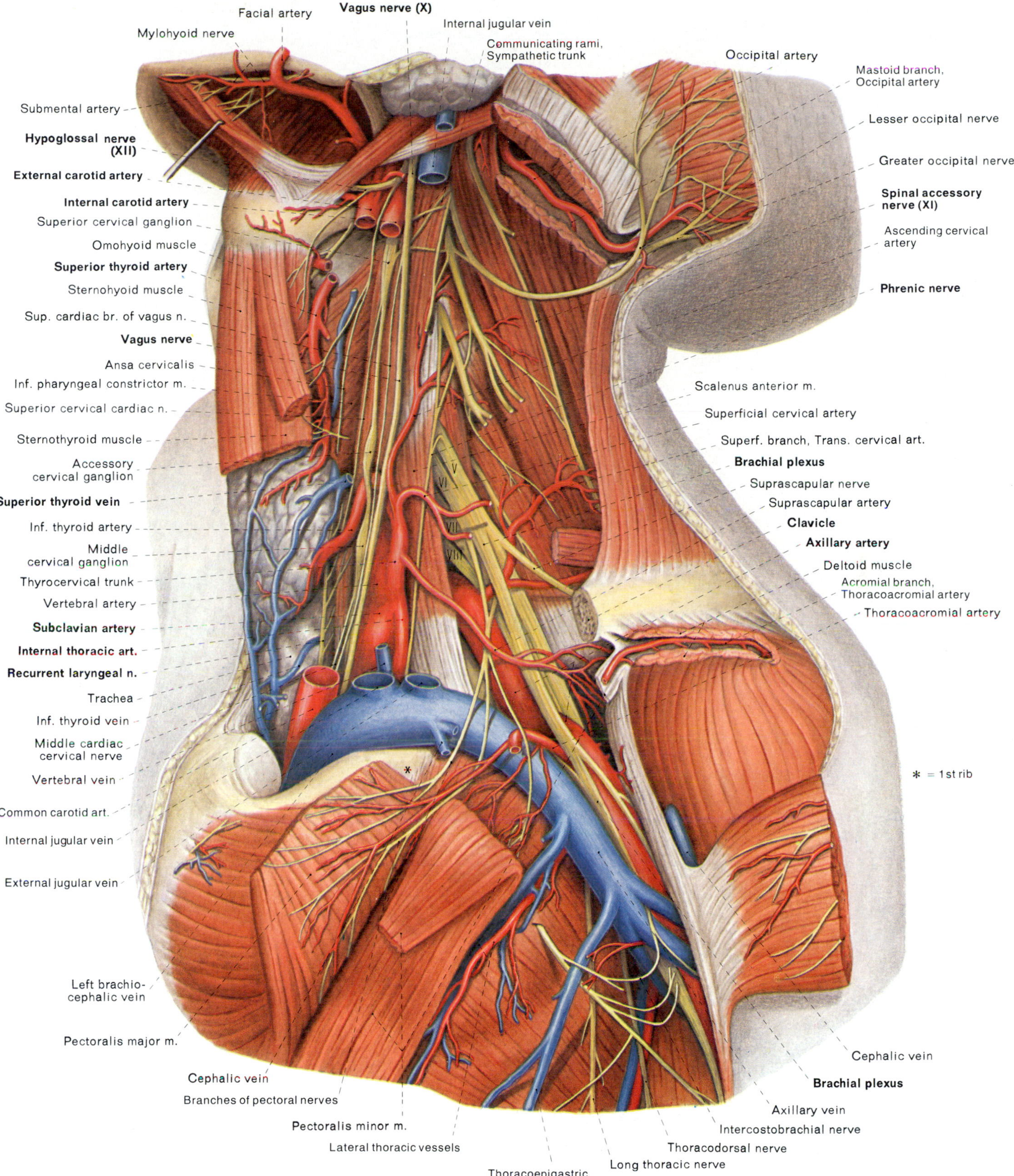

Facial artery
Vagus nerve (X)
Mylohyoid nerve
Internal jugular vein
Communicating rami, Sympathetic trunk
Occipital artery
Mastoid branch, Occipital artery
Submental artery
Lesser occipital nerve
Hypoglossal nerve (XII)
Greater occipital nerve
External carotid artery
Spinal accessory nerve (XI)
Internal carotid artery
Superior cervical ganglion
Ascending cervical artery
Omohyoid muscle
Superior thyroid artery
Phrenic nerve
Sternohyoid muscle
Sup. cardiac br. of vagus n.
Vagus nerve
Ansa cervicalis
Inf. pharyngeal constrictor m.
Scalenus anterior m.
Superior cervical cardiac n.
Superficial cervical artery
Sternothyroid muscle
Superf. branch, Trans. cervical art.
Accessory cervical ganglion
Brachial plexus
Suprascapular nerve
Superior thyroid vein
Suprascapular artery
Inf. thyroid artery
Clavicle
Middle cervical ganglion
Axillary artery
Deltoid muscle
Thyrocervical trunk
Acromial branch, Thoracoacromial artery
Vertebral artery
Thoracoacromial artery
Subclavian artery
Internal thoracic art.
Recurrent laryngeal n.
Trachea
Inf. thyroid vein
Middle cardiac cervical nerve
* = 1st rib
Vertebral vein
Common carotid art.
Internal jugular vein
External jugular vein
V
VI
VII
VIII
Left brachio-cephalic vein
Pectoralis major m.
Cephalic vein
Cephalic vein
Brachial plexus
Branches of pectoral nerves
Axillary vein
Pectoralis minor m.
Intercostobrachial nerve
Lateral thoracic vessels
Thoracodorsal nerve
Long thoracic nerve
Thoracoepigastric vein

Fig. 582: Nerves and Blood Vessels of the Neck, Stage 6: the Brachial Plexus

NOTE: 1) with the carotid arteries, jugular vein and clavicle removed, the roots forming the trunks of the brachial plexus are exposed as they divide and descend into the axilla to surround the axillary artery.

2) the sympathetic trunk lying deep to the carotid arteries and coursing with the vagus nerve and the superior cardiac branch of the vagus nerve.

3) the thyroid gland receiving the superior and inferior thyroid arteries and being drained by the thyroid veins. Observe also the proximity of the recurrent laryngeal nerve to the thyroid gland.

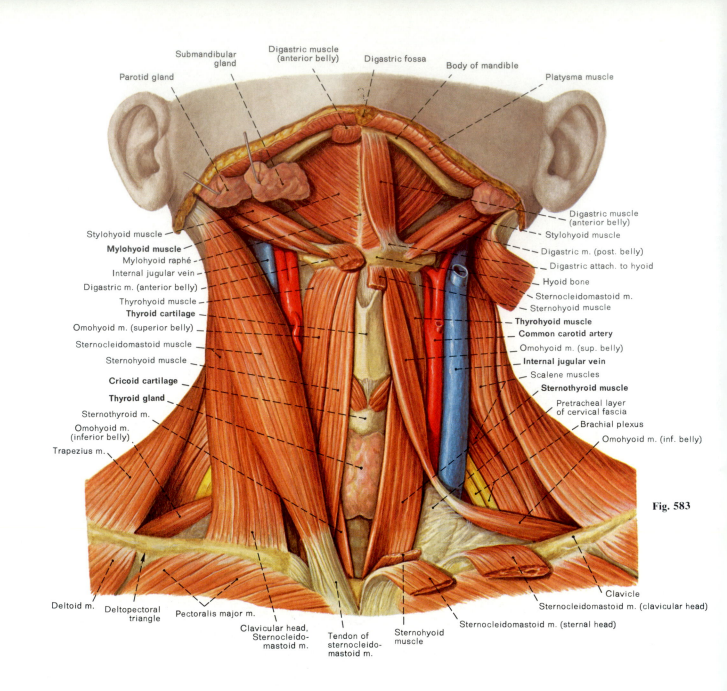

Parotid gland
Submandibular gland
Digastric muscle (anterior belly)
Digastric fossa
Body of mandible
Platysma muscle
Stylohyoid muscle
Mylohyoid muscle
Mylohyoid raphé
Internal jugular vein
Digastric m. (anterior belly)
Thyrohyoid muscle
Thyroid cartilage
Omohyoid m. (superior belly)
Sternocleidomastoid muscle
Sternohyoid muscle
Cricoid cartilage
Thyroid gland
Sternothyroid m.
Omohyoid m. (inferior belly)
Trapezius m.
Digastric muscle (anterior belly)
Stylohyoid muscle
Digastric m. (post. belly)
Digastric attach. to hyoid
Hyoid bone
Sternocleidomastoid m.
Sternohyoid muscle
Thyrohyoid muscle
Common carotid artery
Omohyoid m. (sup. belly)
Internal jugular vein
Scalene muscles
Sternothyroid muscle
Pretracheal layer of cervical fascia
Brachial plexus
Omohyoid m. (inf. belly)
Clavicle
Sternocleidomastoid m. (clavicular head)
Sternocleidomastoid m. (sternal head)
Sternohyoid muscle
Tendon of sternocleidomastoid m.
Clavicular head, Sternocleidomastoid m.
Pectoralis major m.
Deltopectoral triangle
Deltoid m.

Fig. 583

Fig. 584

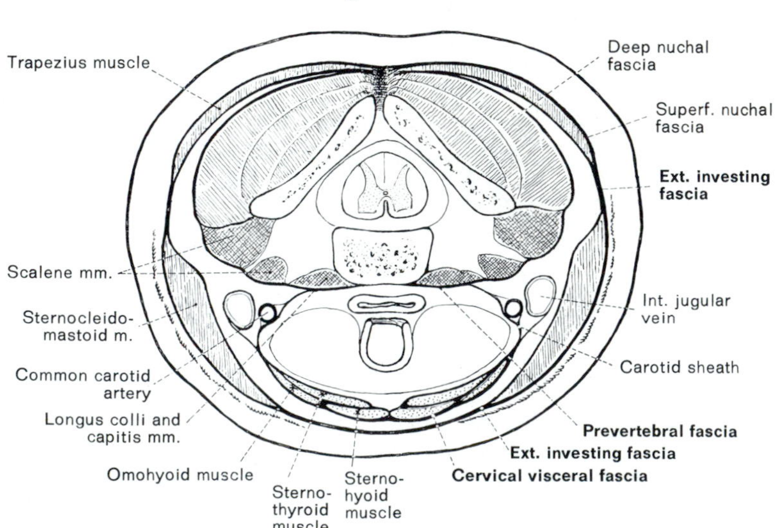

Trapezius muscle
Deep nuchal fascia
Superf. nuchal fascia
Ext. investing fascia
Scalene mm.
Int. jugular vein
Sternocleido-mastoid m.
Carotid sheath
Common carotid artery
Longus colli and capitis mm.
Prevertebral fascia
Ext. investing fascia
Cervical visceral fascia
Omohyoid muscle
Sterno-thyroid muscle
Sterno-hyoid muscle

Fig. 583: Anterior View of the Musculature of the Neck

NOTE: 1) the right superior belly of the digastric muscle was removed and submandibular gland elevated in order to show the mylohyoid muscle. On the left side, the sternocleidomastoid and sternohyoid muscles have been transected and the submandibular gland removed.

2) the relationship of the strap muscles to the thyroid gland and realize that inferior to the thyroid gland and above the suprasternal notch, the trachea lies immediately under the skin.

Fig. 584: Fascial Planes of the Neck in a Newborn Child (Cross Section)

NOTE: the external investing fascia splits to encase the sternocleidomastoid and trapezius muscles. The prevertebral fascia encloses the vertebral column and its muscles, while the cervical visceral fascia encloses the esophagus, trachea, thyroid gland and strap muscles.

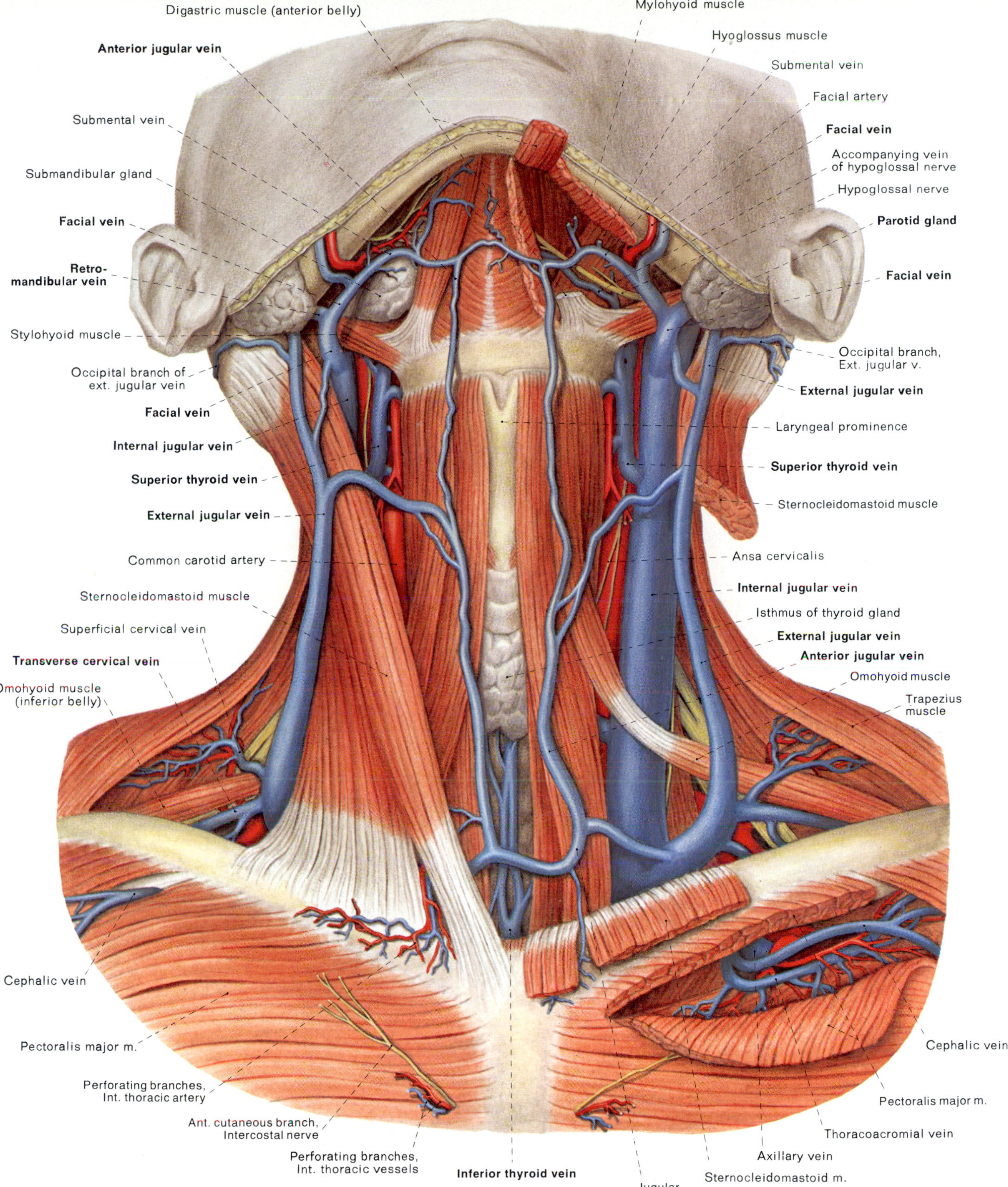
Digastric muscle (anterior belly)
Mylohyoid muscle
Hyoglossus muscle
Anterior jugular vein
Submental vein
Facial artery
Submental vein
Facial vein
Accompanying vein of hypoglossal nerve
Submandibular gland
Hypoglossal nerve
Parotid gland
Facial vein
Retro-mandibular vein
Facial vein
Stylohyoid muscle
Occipital branch, Ext. jugular v.
Occipital branch of ext. jugular vein
External jugular vein
Facial vein
Laryngeal prominence
Internal jugular vein
Superior thyroid vein
Superior thyroid vein
Sternocleidomastoid muscle
External jugular vein
Common carotid artery
Ansa cervicalis
Internal jugular vein
Sternocleidomastoid muscle
Isthmus of thyroid gland
Superficial cervical vein
External jugular vein
Transverse cervical vein
Anterior jugular vein
Omohyoid muscle
Omohyoid muscle (inferior belly)
Trapezius muscle
Cephalic vein
Pectoralis major m.
Cephalic vein
Perforating branches, Int. thoracic artery
Pectoralis major m.
Ant. cutaneous branch, Intercostal nerve
Thoracoacromial vein
Perforating branches, Int. thoracic vessels
Axillary vein
Inferior thyroid vein
Sternocleidomastoid m.
Jugular venous arch

Fig. 585: Veins of the Neck and Infraclavicular Region

NOTE: 1) the jugular system of veins, although somewhat variable, generally consists of an anterior jugular, external jugular and internal jugular vein, all of which are shown on the left side where the sternocleidomastoid muscle has been removed.

2) the anterior jugular descends close to the midline, is frequently small and drains laterally through several tributaries into the external jugular vein. The external jugular courses along the surface of the sternocleidomastoid muscle. It commences usually within the parotid gland and enlarges through the junction of occipital, retromandibular and facial branches. It flows into the subclavian vein with the internal jugular after it receives branches from the shoulder and clavicular regions.

3) the internal jugular is large and collects blood from the brain, face and neck. At its junction with the subclavian, the brachiocephalic vein is formed.

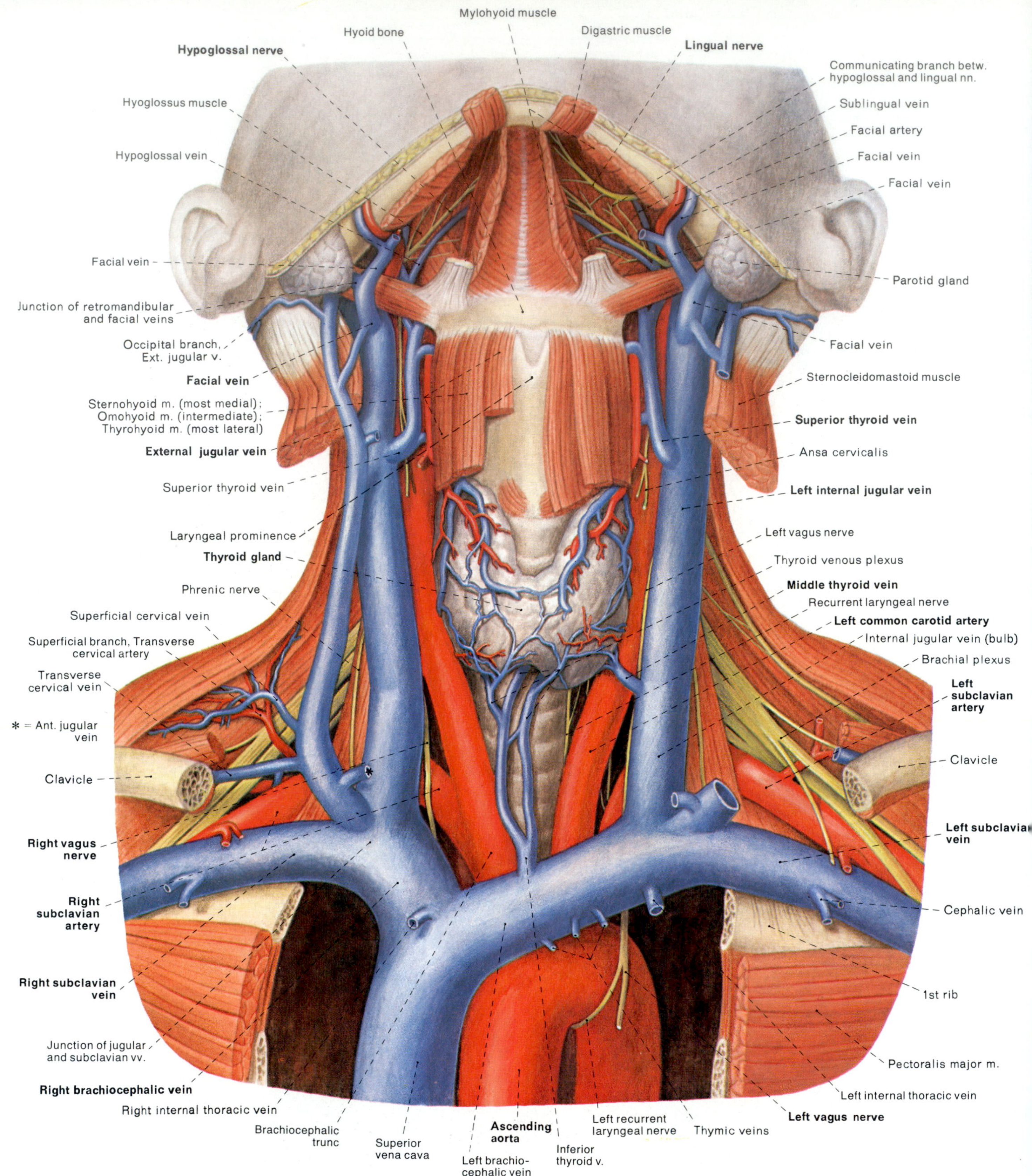

Mylohyoid muscle
Hyoid bone
Digastric muscle
Hypoglossal nerve
Lingual nerve
Communicating branch betw. hypoglossal and lingual nn.
Hyoglossus muscle
Sublingual vein
Facial artery
Hypoglossal vein
Facial vein
Facial vein
Facial vein
Parotid gland
Junction of retromandibular and facial veins
Facial vein
Occipital branch, Ext. jugular v.
Facial vein
Sternocleidomastoid muscle
Sternohyoid m. (most medial); Omohyoid m. (intermediate); Thyrohyoid m. (most lateral)
Superior thyroid vein
External jugular vein
Ansa cervicalis
Superior thyroid vein
Left internal jugular vein
Left vagus nerve
Laryngeal prominence
Thyroid gland
Thyroid venous plexus
Middle thyroid vein
Phrenic nerve
Recurrent laryngeal nerve
Superficial cervical vein
Left common carotid artery
Superficial branch, Transverse cervical artery
Internal jugular vein (bulb)
Brachial plexus
Transverse cervical vein
Left subclavian artery
* = Ant. jugular vein
Clavicle
Clavicle
Right vagus nerve
Left subclavian vein
Right subclavian artery
Cephalic vein
Right subclavian vein
1st rib
Junction of jugular and subclavian vv.
Pectoralis major m.
Right brachiocephalic vein
Left internal thoracic vein
Right internal thoracic vein
Left vagus nerve
Brachiocephalic trunc
Superior vena cava
Ascending aorta
Left brachio-cephalic vein
Inferior thyroid v.
Left recurrent laryngeal nerve
Thymic veins

Fig. 586: Deep Arteries and Veins of the Neck and Great Vessels of the Thorax

NOTE: 1) in the neck, the sternocleidomastoid and strap muscles have been removed, thereby exposing the carotid arteries, internal jugular veins and thyroid gland.

2) the middle portion of the anterior thoracic wall has been resected in order to show the aortic arch and its branches, the brachiocephalic veins, the superior vena cava and the vagus nerves.

3) in the submandibular region, the mylohyoid and anterior digastric muscles have been cut, revealing the lingual and hypoglossal nerves.

Fig. 586

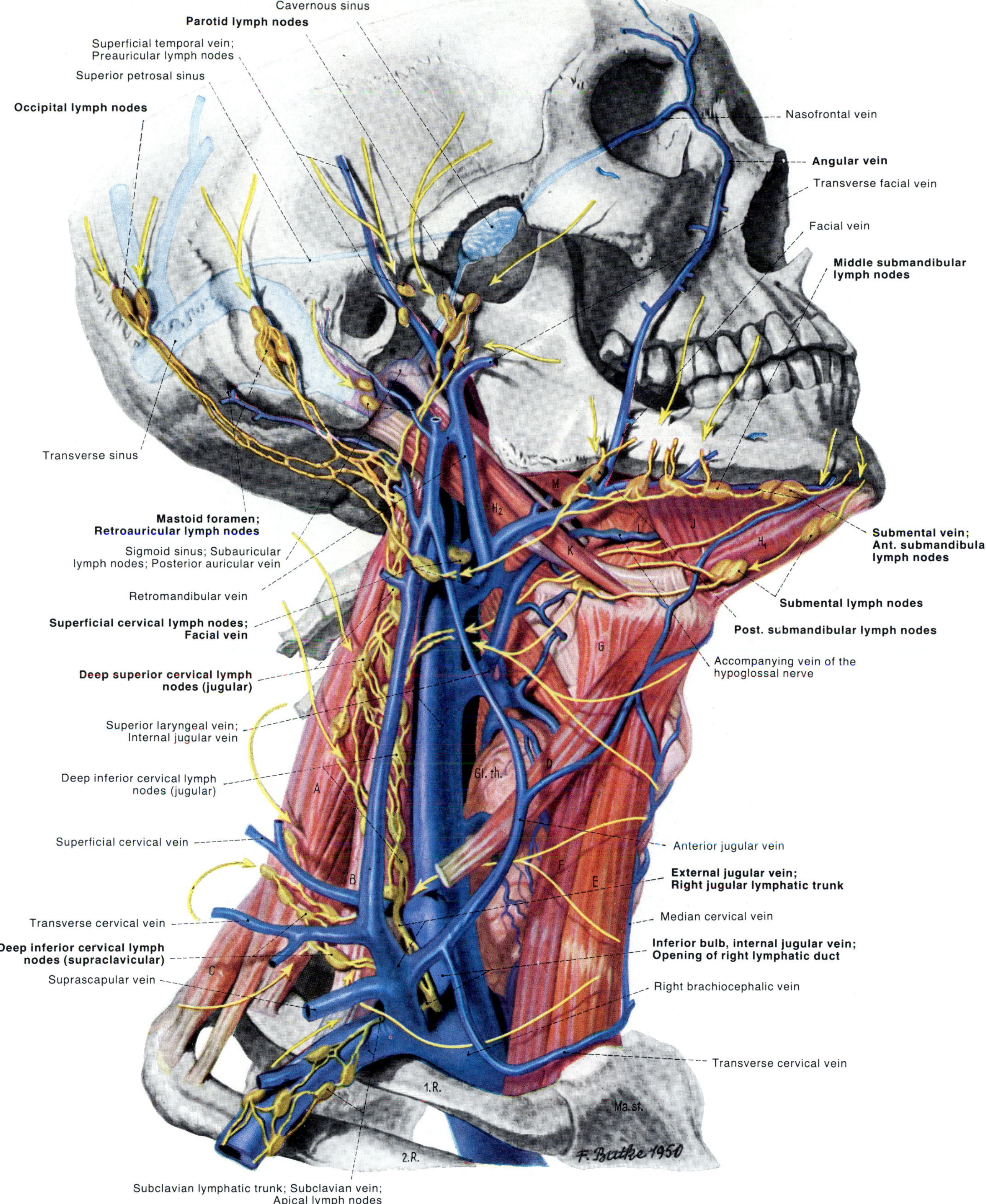

Cavernous sinus
Parotid lymph nodes
Superficial temporal vein;
Preauricular lymph nodes
Superior petrosal sinus
Occipital lymph nodes
Nasofrontal vein
Angular vein
Transverse facial vein
Facial vein
Middle submandibular
lymph nodes
Transverse sinus
Mastoid foramen;
Retroauricular lymph nodes
Sigmoid sinus; Subauricular
lymph nodes; Posterior auricular vein
Retromandibular vein
Superficial cervical lymph nodes;
Facial vein
Deep superior cervical lymph
nodes (jugular)
Superior laryngeal vein;
Internal jugular vein
Deep inferior cervical lymph
nodes (jugular)
Superficial cervical vein
Transverse cervical vein
Deep inferior cervical lymph
nodes (supraclavicular)
Suprascapular vein
Subclavian lymphatic trunk; Subclavian vein;
Apical lymph nodes
Submental vein;
Ant. submandibular
lymph nodes
Submental lymph nodes
Post. submandibular lymph nodes
Accompanying vein of the
hypoglossal nerve
Anterior jugular vein
External jugular vein;
Right jugular lymphatic trunk
Median cervical vein
Inferior bulb, internal jugular vein;
Opening of right lymphatic duct
Right brachiocephalic vein
Transverse cervical vein
M
H_2
L
J
K
H_1
G
D
Gl. th.
A
B
C
F
E
1.R.
2.R.
Ma.st.
F. Batke 1950

Fig. 587: The Superficial Lymph Nodes, Lymphatic Channels and Veins of the Face, Lower Head and Neck (Right Lateral View)

NOTE that the superficial lymphatic channels of the face and the temporal and occipital regions of the head drain inferiorly and posteriorly to the superficial and deeper cervical nodes accompanying the internal jugular vein. These in turn drain into the right jugular lymphatic trunk.

A = Middle scalene m.
B = Ant. scalene m.
C = Post. scalene m.
D = Omohyoid m.

E = Sternohyoid m.
F = Sternothyroid m.
G = Thyrohyoid m.
Gl. th. = Thyroid gland

H_1 = Digastric m. (ant. belly)
H_2 = Digastric m. (post belly)
J = Mylohyoid m.
K = Stylohyoid m.

L = Hyoglossus m.
M = Styloglossus m.
Ma. st. = Manubrium sterni
1., 2. R = 1st and 2nd ribs

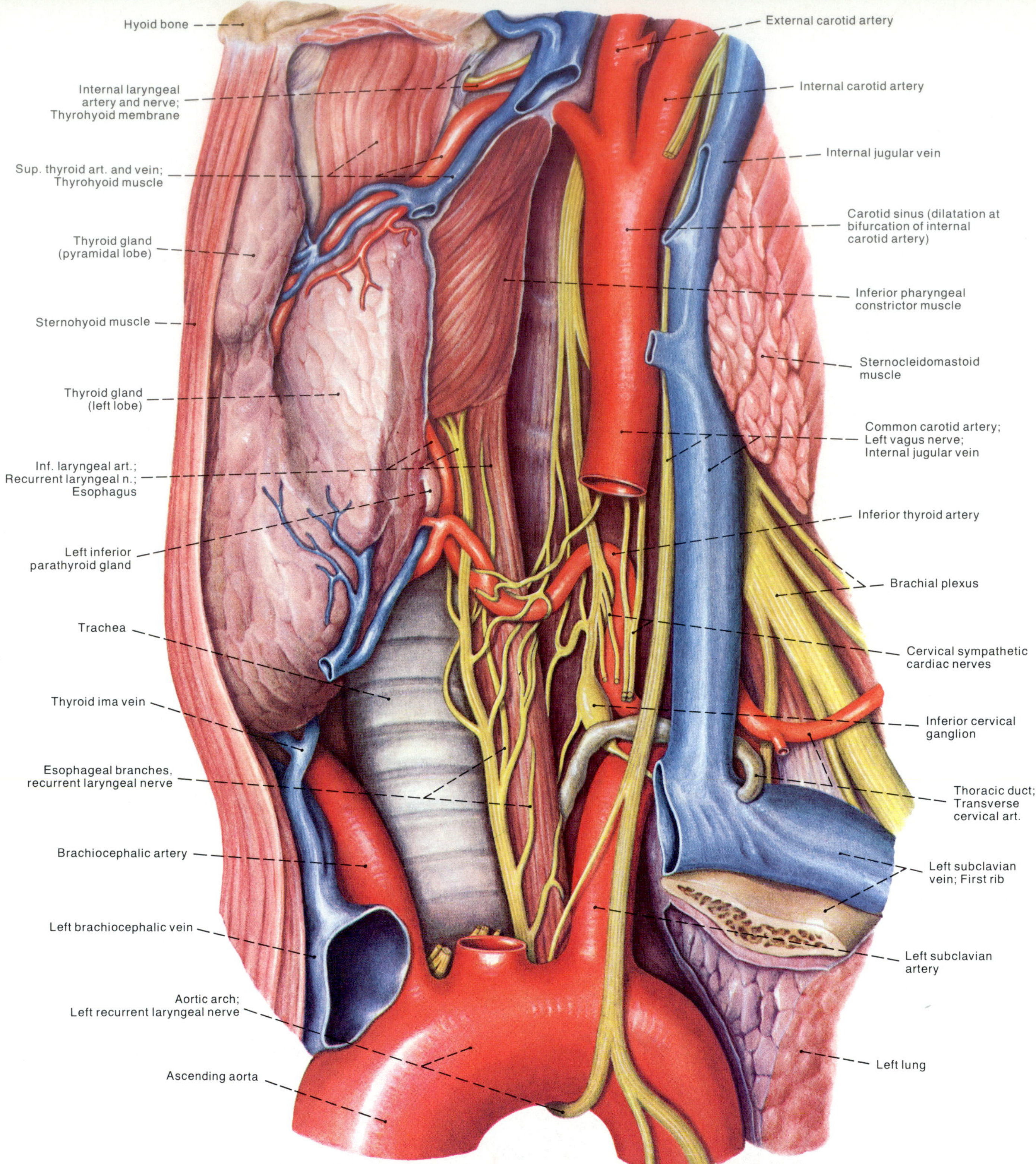

Hyoid bone
Internal laryngeal artery and nerve; Thyrohyoid membrane
Sup. thyroid art. and vein; Thyrohyoid muscle
Thyroid gland (pyramidal lobe)
Sternohyoid muscle
Thyroid gland (left lobe)
Inf. laryngeal art.; Recurrent laryngeal n.; Esophagus
Left inferior parathyroid gland
Trachea
Thyroid ima vein
Esophageal branches, recurrent laryngeal nerve
Brachiocephalic artery
Left brachiocephalic vein
Aortic arch; Left recurrent laryngeal nerve
Ascending aorta
External carotid artery
Internal carotid artery
Internal jugular vein
Carotid sinus (dilatation at bifurcation of internal carotid artery)
Inferior pharyngeal constrictor muscle
Sternocleidomastoid muscle
Common carotid artery; Left vagus nerve; Internal jugular vein
Inferior thyroid artery
Brachial plexus
Cervical sympathetic cardiac nerves
Inferior cervical ganglion
Thoracic duct; Transverse cervical art.
Left subclavian vein; First rib
Left subclavian artery
Left lung

Fig. 588: The Nerves and Vessels in the Deep Part of the Lower Neck, Anterolateral View

NOTE: 1) that this is an anterolateral view of the *left side* of the lower neck, so that the lateral aspect of the left lobe of the thyroid gland and the vessels entering and leaving the gland, the left superior and inferior thyroid arteries and veins and the thyroid ima vein, are clearly seen. Observe a parathyroid gland located behind the thyroid adjacent to the inferior thyroid vessels.

2) that in this dissection the lower part of the left common carotid artery, near its origin from the aortic arch, has been cut away, while the left subclavian artery has been left intact. Additionally, the left brachiocephalic vein has been removed from its point of formation at the junction of the left subclavian and internal jugular veins to the site at which the thyroid ima vein joins it.

3) that with these large vessels removed, the left vagus nerve with its recurrent laryngeal branch is exposed. Observe the surgically important relationship between the two principal terminal branches of the inferior thyroid artery and the recurrent laryngeal nerve near the point of entrance of these vessels into the thyroid gland. On the left side the recurrent nerve courses between these two branches (as shown in this figure) about 33% of the time. The nerve ascends deep to these vessels 50 to 55% of the time and superficial, or in front of them, 11 to 12% of the time (Hollinshead, W. H., Surg. Clin. North Amer., *32*: 1115, 1952). Care must be taken that the recurrent laryngeal nerves are not injured during thyroidectomy, because they are the source of motor innervation to nearly all of the muscles of the larynx.

Lesser horn, Hyoid bone
Body of hyoid bone
Greater horn, Hyoid bone
Superior horn, Thyroid cartilage
Thyrohyoid membrane
Laryngeal prominence
Lamina of thyroid cartilage
Cricothyroid ligament
Cricothyroid muscle, Straight and oblique parts
Thyroid gland, Left lobe
Thyroid gland, Right lobe
Thyroid gland, Isthmus
Trachea

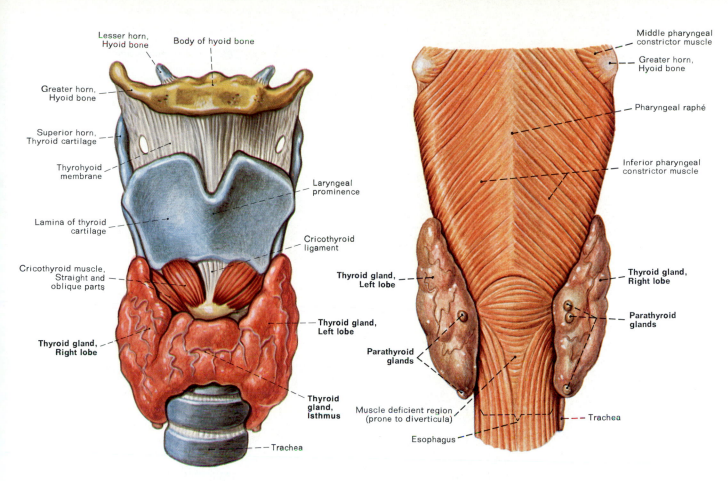

Fig. 589: Ventral View of Thyroid Gland Showing Relation to Larynx and Trachea

Middle pharyngeal constrictor muscle
Greater horn, Hyoid bone
Pharyngeal raphé
Inferior pharyngeal constrictor muscle
Thyroid gland, Left lobe
Thyroid gland, Right lobe
Parathyroid glands
Parathyroid glands
Muscle deficient region (prone to diverticula)
Trachea
Esophagus

Fig. 590: Dorsal View of Thyroid Gland Showing Relation to Pharynx and Parathyroids

Fig. 591: Cross Section of Cervical Viscera Showing Relation of Thyroid Gland to Trachea, Esophagus, Parathyroid Glands and Cervical Vessels and Nerves

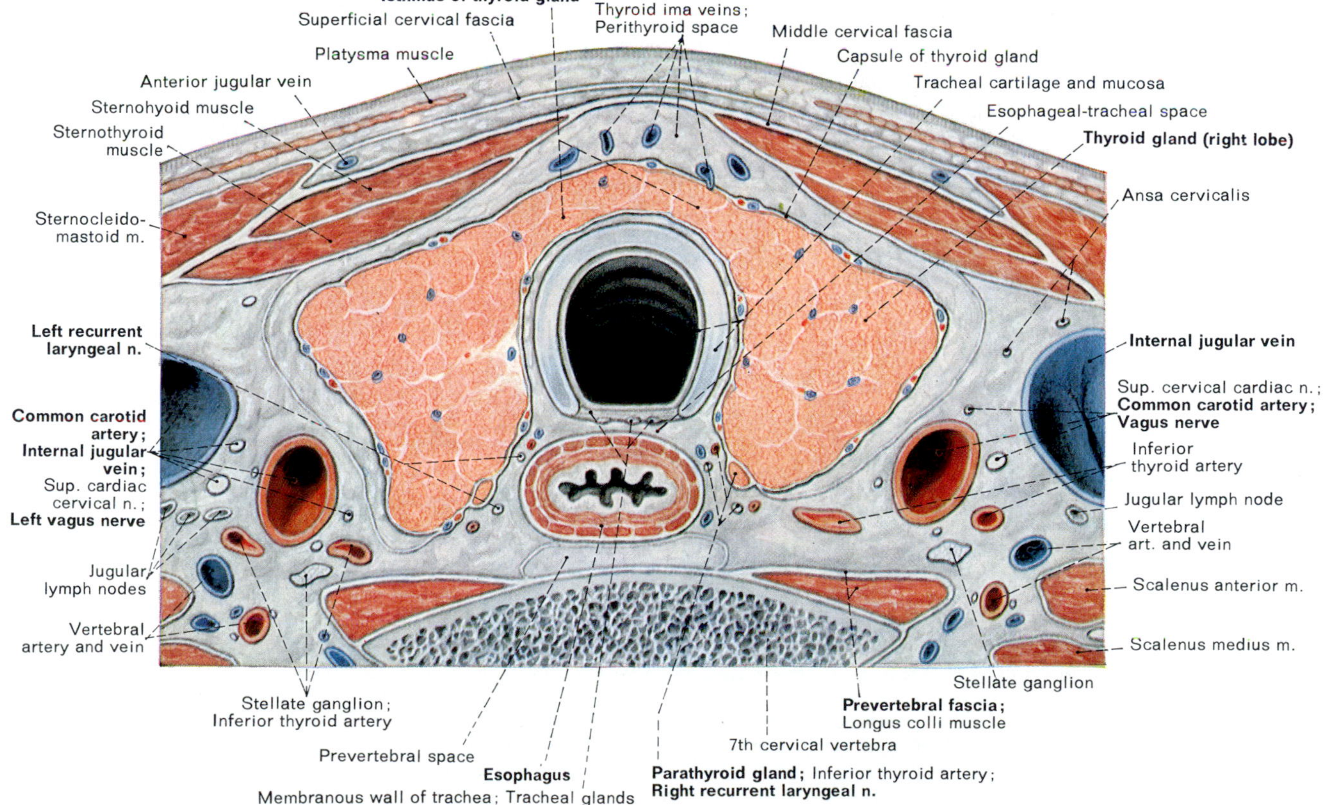
Isthmus of thyroid gland
Superficial cervical fascia
Thyroid ima veins; Perithyroid space
Middle cervical fascia
Platysma muscle
Capsule of thyroid gland
Anterior jugular vein
Tracheal cartilage and mucosa
Sternohyoid muscle
Esophageal-tracheal space
Sternothyroid muscle
Thyroid gland (right lobe)
Ansa cervicalis
Sternocleido-mastoid m.
Left recurrent laryngeal n.
Internal jugular vein
Sup. cervical cardiac n.; **Common carotid artery; Vagus nerve**
Common carotid artery; Internal jugular vein; Sup. cardiac cervical n.; **Left vagus nerve**
Inferior thyroid artery
Jugular lymph node
Vertebral art. and vein
Jugular lymph nodes
Scalenus anterior m.
Vertebral artery and vein
Scalenus medius m.
Stellate ganglion
Stellate ganglion; Inferior thyroid artery
Prevertebral fascia; Longus colli muscle
7th cervical vertebra
Prevertebral space
Esophagus
Parathyroid gland; Inferior thyroid artery; **Right recurrent laryngeal n.**
Membranous wall of trachea; Tracheal glands

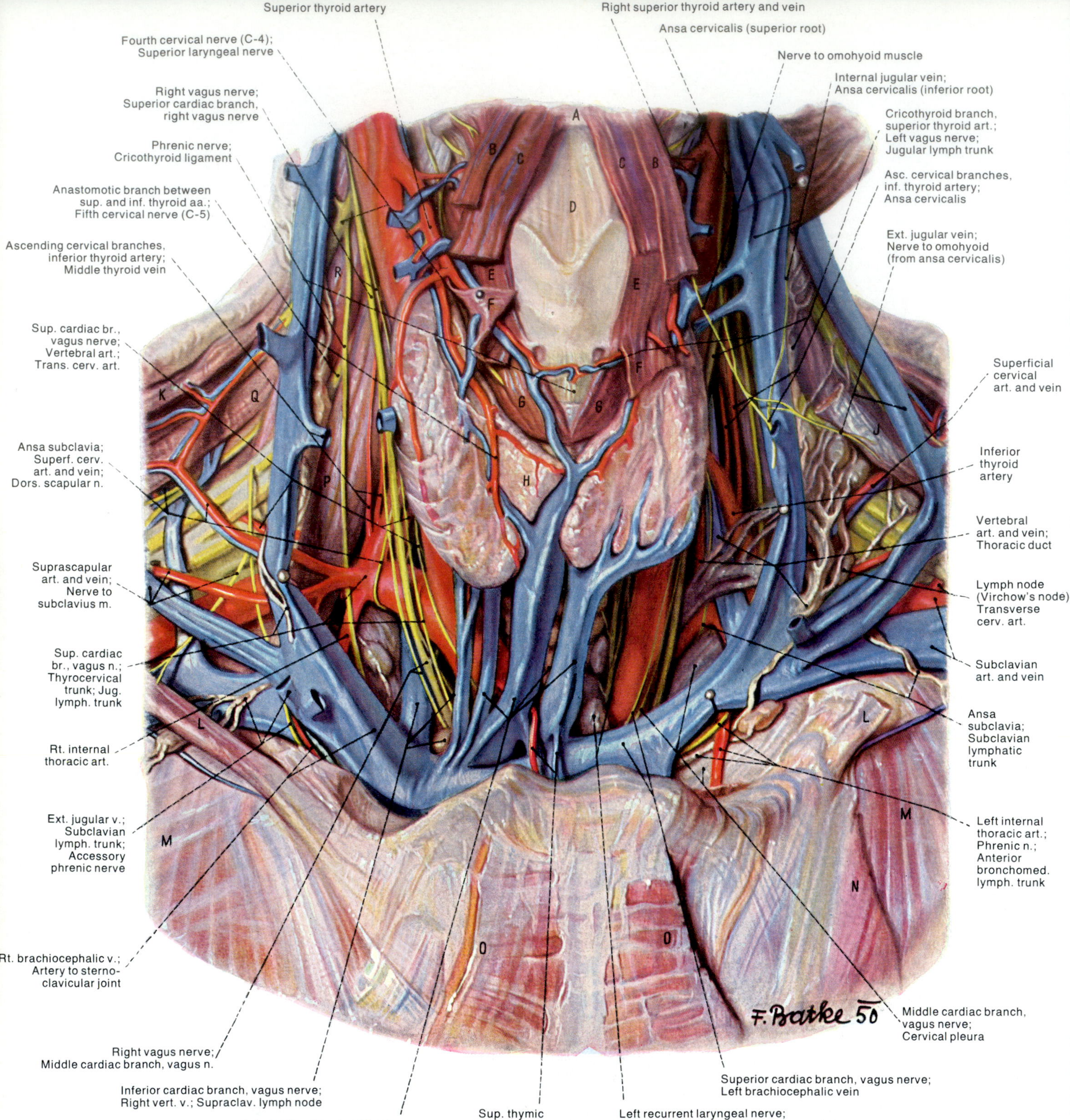

Superior thyroid artery
Right superior thyroid artery and vein
Ansa cervicalis (superior root)
Fourth cervical nerve (C-4); Superior laryngeal nerve
Nerve to omohyoid muscle
Internal jugular vein; Ansa cervicalis (inferior root)
Right vagus nerve; Superior cardiac branch, right vagus nerve
Cricothyroid branch, superior thyroid art.; Left vagus nerve; Jugular lymph trunk
Phrenic nerve; Cricothyroid ligament
Asc. cervical branches, inf. thyroid artery; Ansa cervicalis
Anastomotic branch between sup. and inf. thyroid aa.; Fifth cervical nerve (C-5)
Ascending cervical branches, inferior thyroid artery; Middle thyroid vein
Ext. jugular vein; Nerve to omohyoid (from ansa cervicalis)
Sup. cardiac br., vagus nerve; Vertebral art.; Trans. cerv. art.
Superficial cervical art. and vein
Ansa subclavia; Superf. cerv. art. and vein; Dors. scapular n.
Inferior thyroid artery
Vertebral art. and vein; Thoracic duct
Suprascapular art. and vein; Nerve to subclavius m.
Lymph node (Virchow's node); Transverse cerv. art.
Sup. cardiac br., vagus n.; Thyrocervical trunk; Jug. lymph. trunk
Subclavian art. and vein
Ansa subclavia; Subclavian lymphatic trunk
Rt. internal thoracic art.
Ext. jugular v.; Subclavian lymph. trunk; Accessory phrenic nerve
Left internal thoracic art.; Phrenic n.; Anterior bronchomed. lymph. trunk
Rt. brachiocephalic v.; Artery to sterno-clavicular joint
Middle cardiac branch, vagus nerve; Cervical pleura
Right vagus nerve; Middle cardiac branch, vagus n.
Superior cardiac branch, vagus nerve; Left brachiocephalic vein
Inferior cardiac branch, vagus nerve; Right vert. v.; Supraclav. lymph node
Inferior thyroid veins; Trachea
Sup. thymic art. and vein
Left recurrent laryngeal nerve; Paratracheal lymph node
A
B
C
D
E
F
G
H
J
K
L
M
N
O
P
Q
R
F. Batke 50

Fig. 592: The Superficial Structures at the Root of the Neck and Thoracic Inlet, Anterior View

NOTE: 1) that in this dissection the two clavicles have been removed, thereby exposing both brachiocephalic veins and their principal tributaries: the internal and external jugular veins, the subclavian veins, and the thyroid veins. Compare the venous pattern in this realistic figure with the more graphic pattern of the veins in the anterior neck shown in Figure 586.

2) that the right subclavian artery and its thyrocervical trunk have been exposed by severing the superior and middle thyroid veins and pulling the right internal jugular vein laterally. Observe the right vagus nerve and its cardiac branches and, posterolaterally, the right vagus nerve. Also note the right ansa subclavia, which loops around the subclavian artery. The ansa subclavia consists principally of communicating sympathetic fibers interconnecting the middle cervical ganglion with either the inferior cervical ganglion or the cervicothoracic (stellate) ganglion.

A = Hyoid bone
B = Omohyoid muscle
C = Sternohyoid muscle
D = Thyrohyoid membrane
E = Thyrohyoid muscle
F = Sternothyroid muscle
G = Cricothyroid muscle
H = Isthmus, thyroid gland
J = Omohyoid muscle
K = Trapezius muscle
L = Subclavius muscle
M = Int. intercostal muscle
N = Ext. intercostal muscle
O = Pectoralis major muscle
P = Anterior scalene muscle
Q = Levator scapulae muscle
R = Middle scalene muscle

Fig. 592

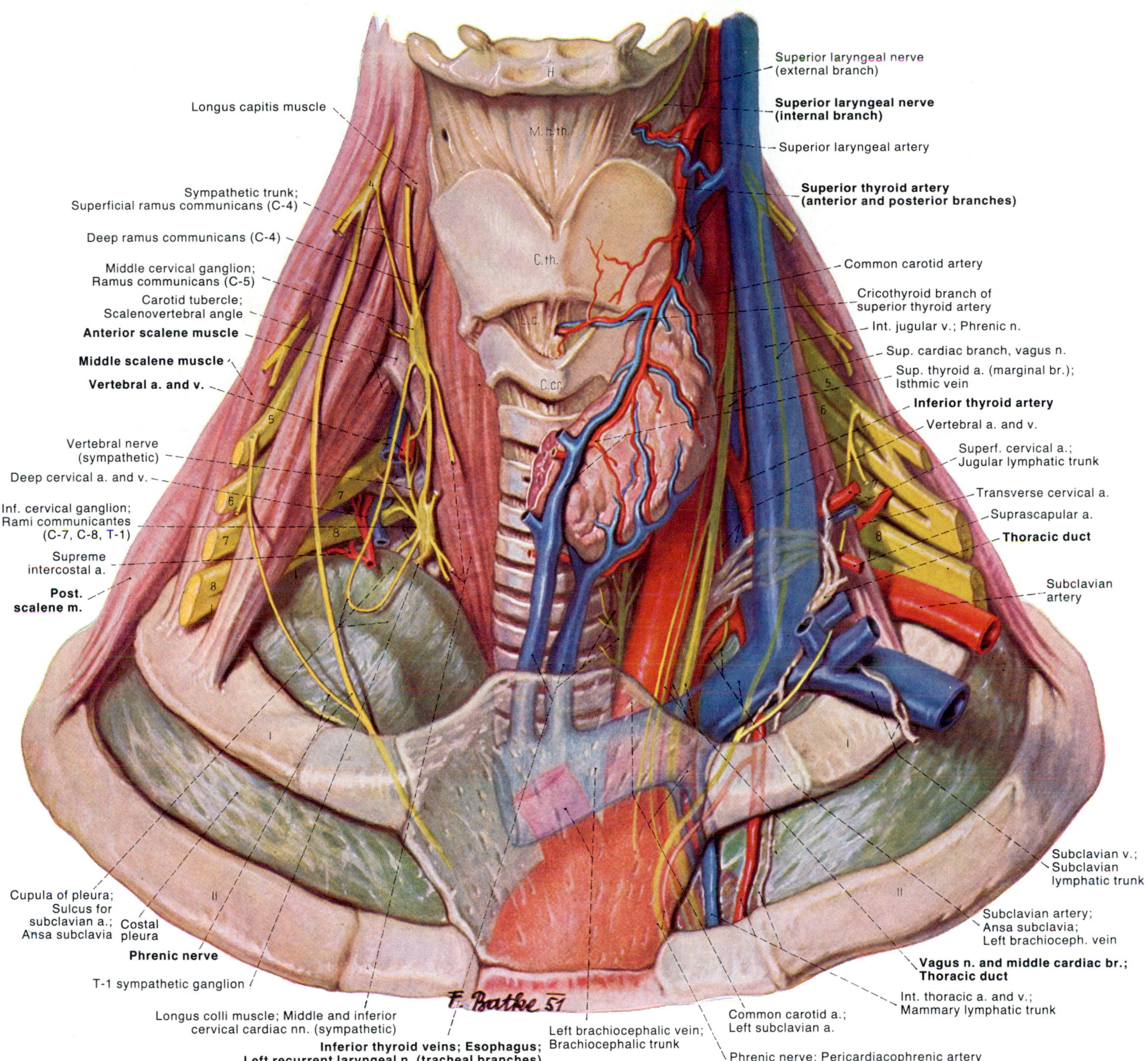
Superior laryngeal nerve (external branch)
Superior laryngeal nerve (internal branch)
Superior laryngeal artery
Superior thyroid artery (anterior and posterior branches)
Common carotid artery
Cricothyroid branch of superior thyroid artery
Int. jugular v.; Phrenic n.
Sup. cardiac branch, vagus n.
Sup. thyroid a. (marginal br.); Isthmic vein
Inferior thyroid artery
Vertebral a. and v.
Superf. cervical a.; Jugular lymphatic trunk
Transverse cervical a.
Suprascapular a.
Thoracic duct
Subclavian artery
Subclavian v.; Subclavian lymphatic trunk
Subclavian artery; Ansa subclavia; Left brachioceph. vein
Vagus n. and middle cardiac br.; Thoracic duct
Int. thoracic a. and v.; Mammary lymphatic trunk
Common carotid a.; Left subclavian a.
Phrenic nerve; Pericardiacophrenic artery
Left brachiocephalic vein; Brachiocephalic trunk
Inferior thyroid veins; Esophagus; Left recurrent laryngeal n. (tracheal branches)
Longus colli muscle; Middle and inferior cervical cardiac nn. (sympathetic)
T-1 sympathetic ganglion
Phrenic nerve
Cupula of pleura; Sulcus for subclavian a.; Ansa subclavia
Costal pleura
Post. scalene m.
Supreme intercostal a.
Inf. cervical ganglion; Rami communicantes (C-7, C-8, T-1)
Deep cervical a. and v.
Vertebral nerve (sympathetic)
Vertebral a. and v.
Middle scalene muscle
Anterior scalene muscle
Carotid tubercle; Scalenovertebral angle
Middle cervical ganglion; Ramus communicans (C-5)
Deep ramus communicans (C-4)
Sympathetic trunk; Superficial ramus communicans (C-4)
Longus capitis muscle
H
M. h. th.
C. th.
L. c.
C. cr.
F. Batke 51

Fig. 593: The Root of the Neck and the Thoracic Inlet, Anterior View

NOTE: 1) that the sternocleidomastoid and strap muscles have been removed. Additionally, *on the right side* (reader's left), the large vessels, vagus nerve, and right lobe of the thyroid gland have been resected to reveal the scalene muscles, the roots of the brachial plexus, the inferior and middle cervical ganglia and cupula of the right pleura.

2) *on the left side,* the course of the vagus nerve and its recurrent laryngeal branch. Observe also the thoracic duct as it ascends into the root of the neck to open into the venous system at the angle of junction between the left internal jugular and left subclavian veins.

3) the descending course of the phrenic nerves. These nerves are derived from the 3rd, 4th, and 5th cervical segments, lie along the ventral surface of the anterior scalene muscle, and enter the thorax deep to the sternocostal joints of the first rib.

4) the cupula of the pleura overlying the apex of the lung. On each side these project 3 to 5 cm above the sternal end of the first rib and, therefore, are unprotected by bone.

5) the superior thyroid vessels and their superior laryngeal and cricothyroid branches. Observe the rich vascularity of the thyroid gland and the veins (of considerable size) which course anterior to the trachea at the site where tracheostomy is frequently performed.

C. cr. = Cricoid cartilage
C. th. = Thyroid cartilage
H = Hyoid bone

L. c. = Cricothyroid ligament
M. h. th. = Thyrohyoid membrane
I, II = 1st and 2nd ribs

4–8 = Ventral rami of C-4 to C-8 nerves
I = Ventral ramus of T-1 nerve

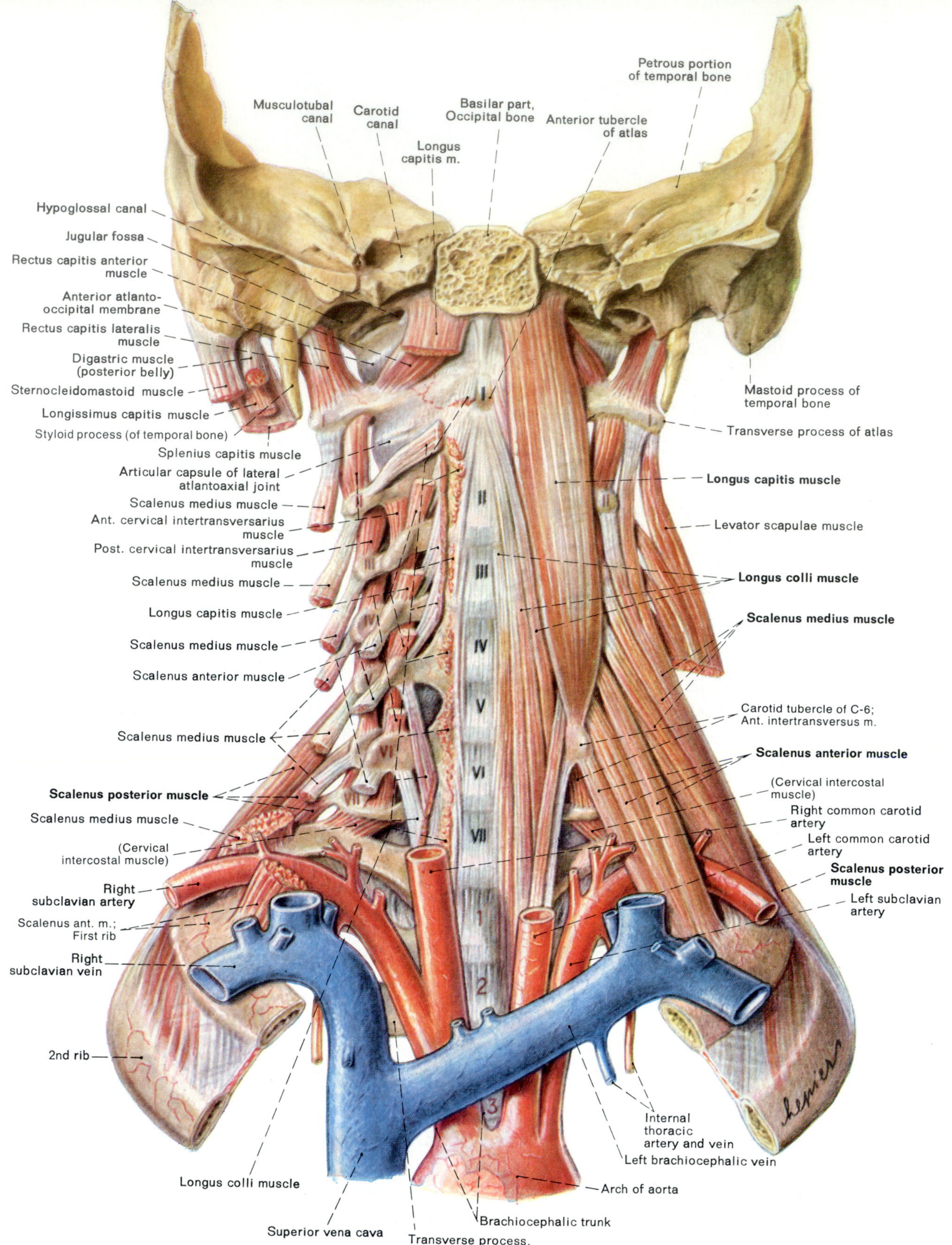

Petrous portion
of temporal bone
Musculotubal
canal
Carotid
canal
Basilar part,
Occipital bone
Anterior tubercle
of atlas
Longus
capitis m.
Hypoglossal canal
Jugular fossa
Rectus capitis anterior
muscle
Anterior atlanto-
occipital membrane
Rectus capitis lateralis
muscle
Digastric muscle
(posterior belly)
Sternocleidomastoid muscle
Longissimus capitis muscle
Styloid process (of temporal bone)
Mastoid process of
temporal bone
Transverse process of atlas
Splenius capitis muscle
Articular capsule of lateral
atlantoaxial joint
Longus capitis muscle
Scalenus medius muscle
Ant. cervical intertransversarius
muscle
Levator scapulae muscle
Post. cervical intertransversarius
muscle
Longus colli muscle
Scalenus medius muscle
Longus capitis muscle
Scalenus medius muscle
Scalenus medius muscle
Scalenus anterior muscle
Carotid tubercle of C-6;
Ant. intertransversus m.
Scalenus medius muscle
Scalenus anterior muscle
Scalenus posterior muscle
(Cervical intercostal
muscle)
Scalenus medius muscle
Right common carotid
artery
(Cervical
intercostal muscle)
Left common carotid
artery
Scalenus posterior
muscle
Right
subclavian artery
Left subclavian
artery
Scalenus ant. m.;
First rib
Right
subclavian vein
2nd rib
Internal
thoracic
artery and vein
Left brachiocephalic vein
Longus colli muscle
Arch of aorta
Brachiocephalic trunk
Superior vena cava
Transverse process,
3rd thoracic vertrebra

Fig. 594: The Prevertebral Region and Root of the Neck, Anterior View

NOTE: 1) on the specimen's right side, the longus capitis, longus colli and scalene muscles have been removed, exposing the transverse processes of the cervical vertebrae onto which these muscles are seen to attach.

2) the scalenus posterior muscle inserts on the 2nd rib, whereas both the scalenus anterior and medius insert onto the 1st rib. Thus, when these muscles are fixed superiorly, they would act as inspiratory muscles by elevating the first two ribs. Fixed inferiorly, they assist in bending the cervical vertebral column to one or the other side. Between the scalenus anterior and medius emerge the roots which form the trunks of the brachial plexus (see Figs. 593 and 596).

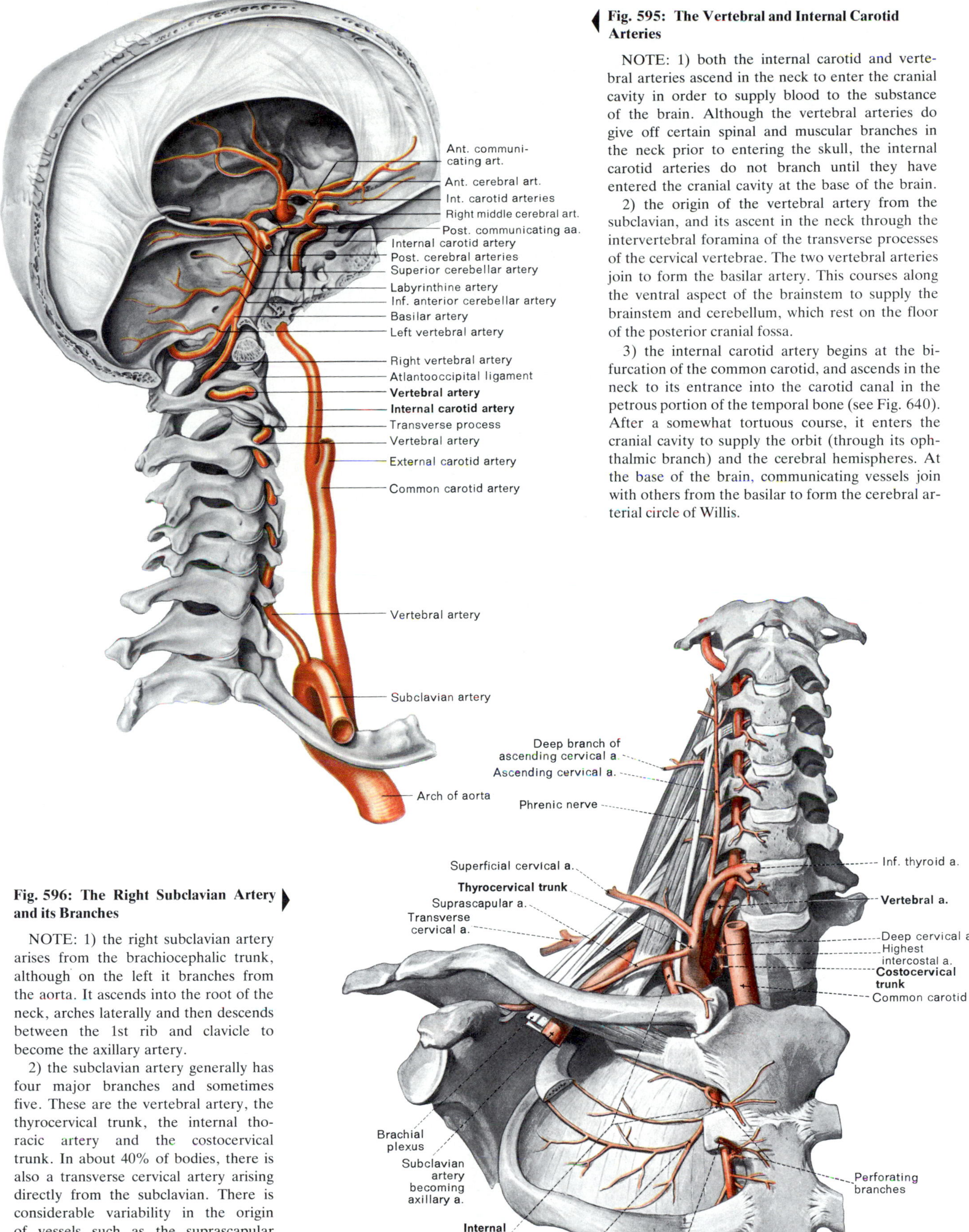

Ant. communi-
cating art.
Ant. cerebral art.
Int. carotid arteries
Right middle cerebral art.
Post. communicating aa.
Internal carotid artery
Post. cerebral arteries
Superior cerebellar artery
Labyrinthine artery
Inf. anterior cerebellar artery
Basilar artery
Left vertebral artery
Right vertebral artery
Atlantooccipital ligament
Vertebral artery
Internal carotid artery
Transverse process
Vertebral artery
External carotid artery
Common carotid artery
Vertebral artery
Subclavian artery
Arch of aorta
Deep branch of
ascending cervical a.
Ascending cervical a.
Phrenic nerve
Superficial cervical a.
Thyrocervical trunk
Suprascapular a.
Transverse
cervical a.
Inf. thyroid a.
Vertebral a.
Deep cervical a.
Highest
intercostal a.
Costocervical
trunk
Common carotid a.
Brachial
plexus
Subclavian
artery
becoming
axillary a.
Internal
thoracic a.
Anterior
intercostal a.
Anterior
intercostal a.
Perforating
branches
Vierling

Fig. 595: The Vertebral and Internal Carotid Arteries

NOTE: 1) both the internal carotid and vertebral arteries ascend in the neck to enter the cranial cavity in order to supply blood to the substance of the brain. Although the vertebral arteries do give off certain spinal and muscular branches in the neck prior to entering the skull, the internal carotid arteries do not branch until they have entered the cranial cavity at the base of the brain.

2) the origin of the vertebral artery from the subclavian, and its ascent in the neck through the intervertebral foramina of the transverse processes of the cervical vertebrae. The two vertebral arteries join to form the basilar artery. This courses along the ventral aspect of the brainstem to supply the brainstem and cerebellum, which rest on the floor of the posterior cranial fossa.

3) the internal carotid artery begins at the bifurcation of the common carotid, and ascends in the neck to its entrance into the carotid canal in the petrous portion of the temporal bone (see Fig. 640). After a somewhat tortuous course, it enters the cranial cavity to supply the orbit (through its ophthalmic branch) and the cerebral hemispheres. At the base of the brain, communicating vessels join with others from the basilar to form the cerebral arterial circle of Willis.

Fig. 596: The Right Subclavian Artery and its Branches

NOTE: 1) the right subclavian artery arises from the brachiocephalic trunk, although on the left it branches from the aorta. It ascends into the root of the neck, arches laterally and then descends between the 1st rib and clavicle to become the axillary artery.

2) the subclavian artery generally has four major branches and sometimes five. These are the vertebral artery, the thyrocervical trunk, the internal thoracic artery and the costocervical trunk. In about 40% of bodies, there is also a transverse cervical artery arising directly from the subclavian. There is considerable variability in the origin of vessels such as the suprascapular artery, the transverse cervical artery and the superficial and deep cervical branches.

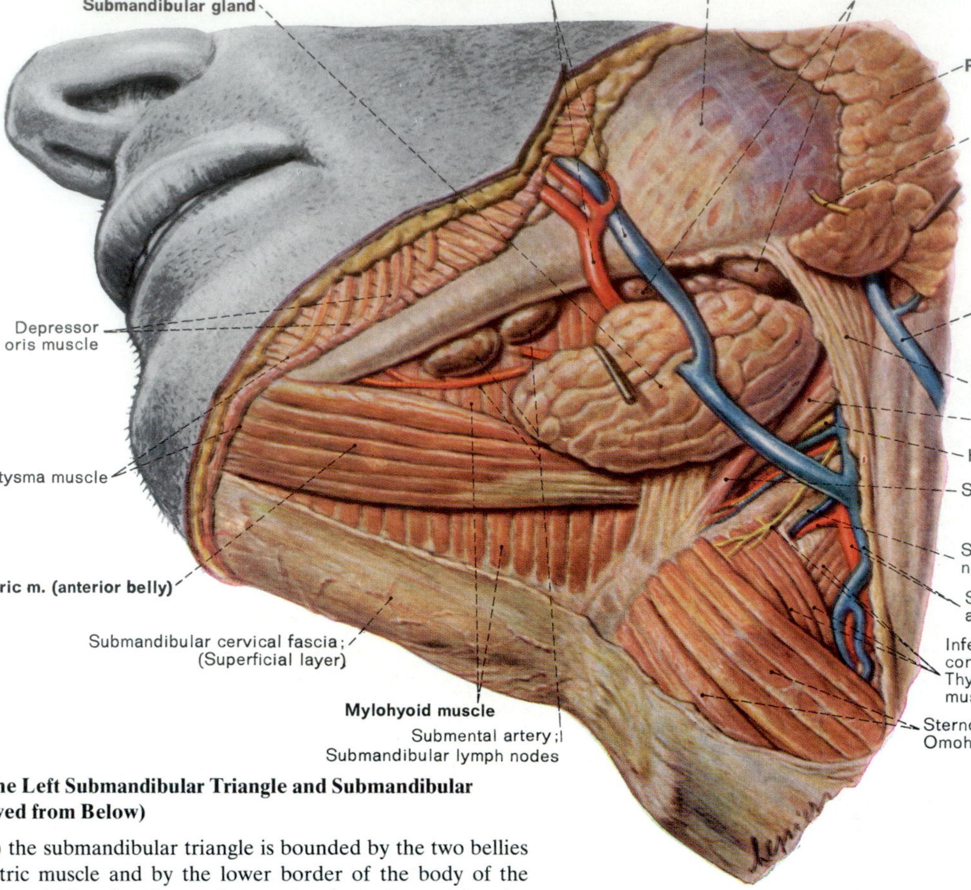

Masseter muscle and fascia
Facial artery and vein
Submandibular gland
Submandibular lymph nodes
Parotid gland
Cervical branch of facial nerve
Depressor anguli oris muscle
External jugular vein
Cervical fascia (angular band)
Digastric m. (post. belly)
Hypoglossal nerve
Platysma muscle
Stylohyoid muscle
Superior laryngeal nerve and vessels
Digastric m. (anterior belly)
Superior thyroid art. and vein
Submandibular cervical fascia; (Superficial layer)
Inferior pharyngeal constrictor m.; Thyrohyoid muscle and nerve
Mylohyoid muscle
Sternohyoid muscle (medial); Omohyoid muscle (lateral)
Submental artery;
Submandibular lymph nodes

Fig. 597: The Left Submandibular Triangle and Submandibular Gland (Viewed from Below)

NOTE: 1) the submandibular triangle is bounded by the two bellies of the digastric muscle and by the lower border of the body of the mandible. Anteriorly, the floor of the triangle is formed by the mylohyoid muscle, whereas posteriorly it is the hyoglossus muscle.

2) the submandibular gland is situated in the anterior part of the triangle. Crossing obliquely are the anterior facial vein and facial artery. Overlying the posterior part of the triangle is the inferior extension of the parotid gland. Likewise, arranged along the lower border of the body of the mandible are a number of submandibular lymph nodes.

Fig. 598: The Suprahyoid Muscles and Hyoid Bone (Viewed from Below)

NOTE that indicated on the mandible are the inner attachments of the mylohyoid muscle (broken line) and the anterior belly of the digastric muscle (circle). Observe the attachments of the mylohyoid, digastric and stylohyoid muscles as well as the stylohyoid ligament onto the hyoid bone.

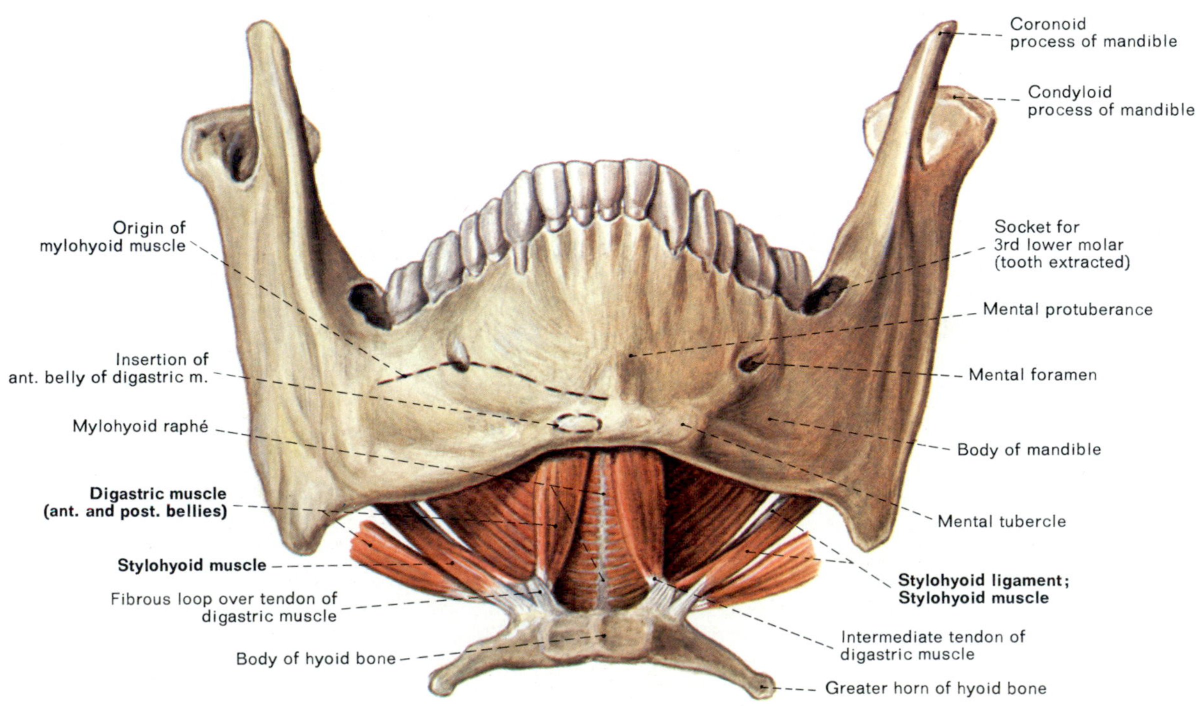

Coronoid process of mandible
Condyloid process of mandible
Origin of mylohyoid muscle
Socket for 3rd lower molar (tooth extracted)
Mental protuberance
Insertion of ant. belly of digastric m.
Mental foramen
Mylohyoid raphé
Body of mandible
Digastric muscle (ant. and post. bellies)
Mental tubercle
Stylohyoid muscle
Stylohyoid ligament; Stylohyoid muscle
Fibrous loop over tendon of digastric muscle
Intermediate tendon of digastric muscle
Body of hyoid bone
Greater horn of hyoid bone

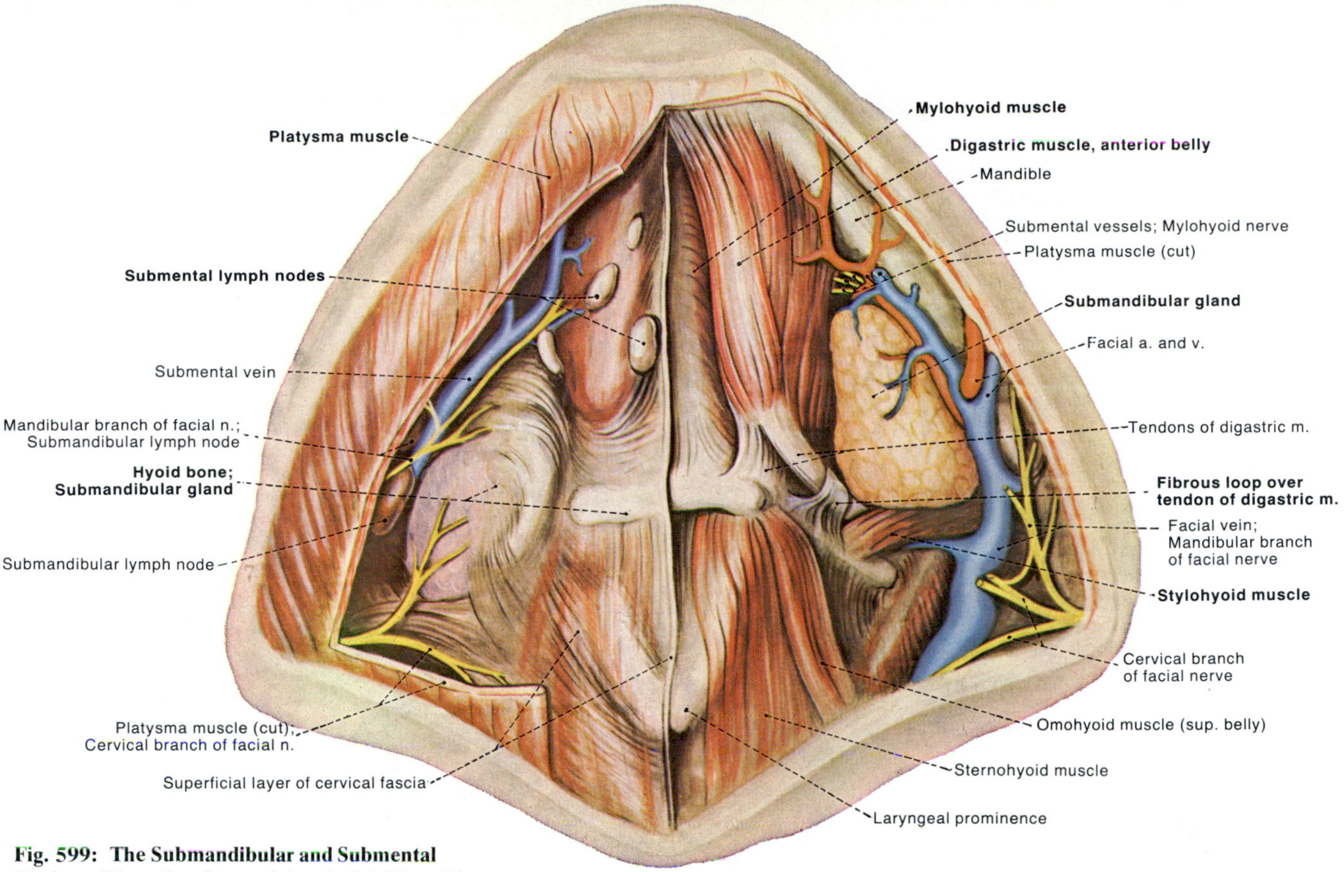

Platysma muscle
Mylohyoid muscle
Digastric muscle, anterior belly
Mandible
Submental vessels; Mylohyoid nerve
Platysma muscle (cut)
Submental lymph nodes
Submandibular gland
Facial a. and v.
Submental vein
Mandibular branch of facial n.;
Submandibular lymph node
Tendons of digastric m.
Hyoid bone;
Submandibular gland
Fibrous loop over
tendon of digastric m.
Facial vein;
Mandibular branch
of facial nerve
Submandibular lymph node
Stylohyoid muscle
Cervical branch
of facial nerve
Platysma muscle (cut);
Cervical branch of facial n.
Omohyoid muscle (sup. belly)
Superficial layer of cervical fascia
Sternohyoid muscle
Laryngeal prominence

Fig. 599: The Submandibular and Submental Regions, Dissection Stages 1 (reader's left) and 2

NOTE that in Dissection Stage 1 the superficial fascia containing the platysma has been opened, revealing the submandibular gland and the superficial lymph nodes. In Stage 2 (reader's right), the superficial layer of cervical fascia has been removed, thereby exposing the substance of the submandibular gland and the digastric (anterior belly), mylohyoid and stylohyoid muscles.

Fig. 600: The Submandibular and Submental Regions, Dissection Stages 3 (reader's left) and 4

NOTE that in Dissection Stage 3 much of the submandibular gland has been removed, revealing the posterior border of the mylohyoid muscle and the submental vessels and mylohyoid nerve. In Stage 4, the anterior belly of the digastric and the mylohyoid muscles have been partially removed, exposing the hypoglossal nerve and its accompanying vein.

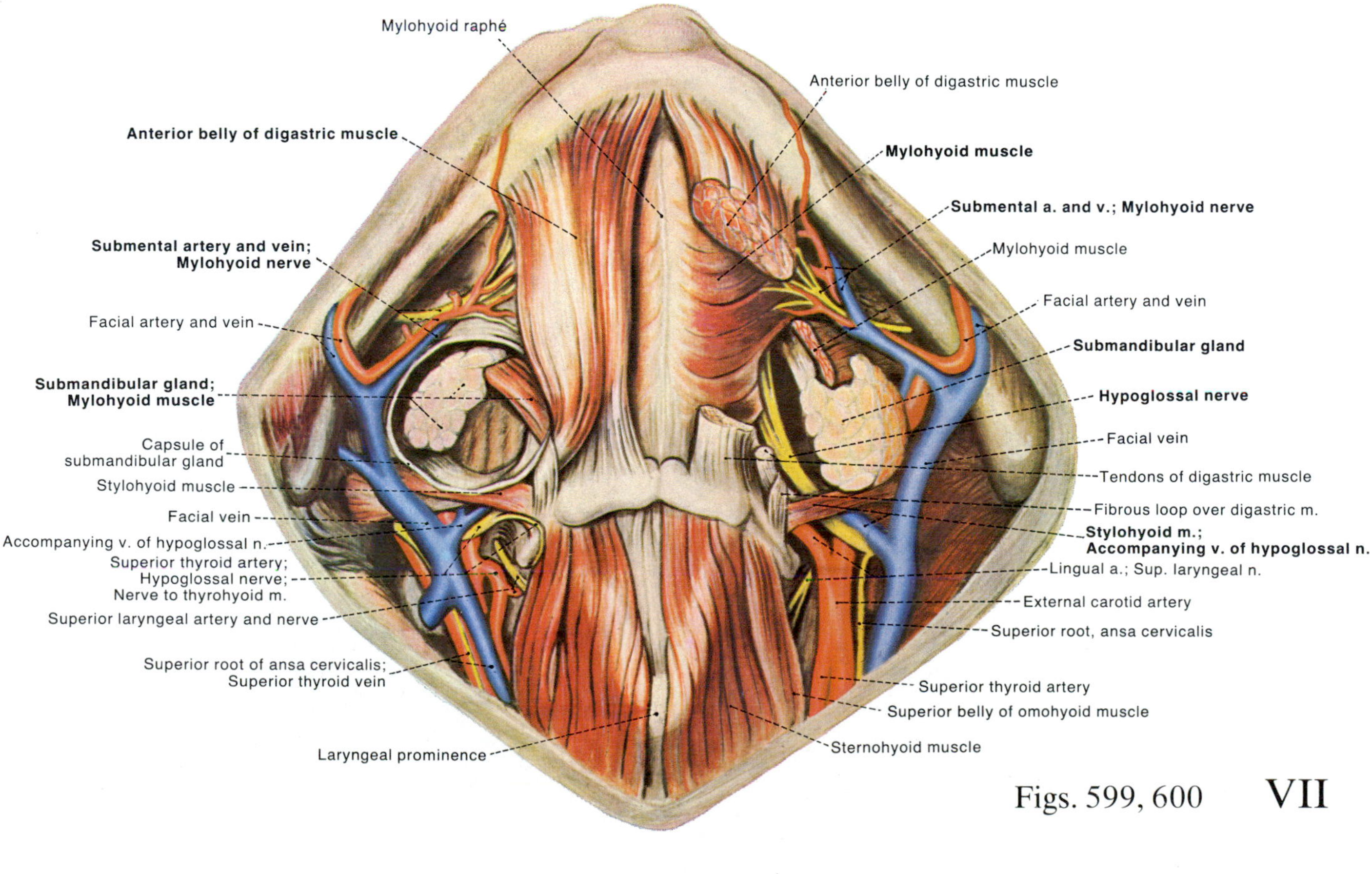

Mylohyoid raphé
Anterior belly of digastric muscle
Anterior belly of digastric muscle
Mylohyoid muscle
Submental a. and v.; Mylohyoid nerve
Submental artery and vein;
Mylohyoid nerve
Mylohyoid muscle
Facial artery and vein
Facial artery and vein
Submandibular gland
Submandibular gland;
Mylohyoid muscle
Hypoglossal nerve
Capsule of
submandibular gland
Facial vein
Tendons of digastric muscle
Stylohyoid muscle
Fibrous loop over digastric m.
Facial vein
Stylohyoid m.;
Accompanying v. of hypoglossal n.
Accompanying v. of hypoglossal n.
Superior thyroid artery;
Hypoglossal nerve;
Nerve to thyrohyoid m.
Lingual a.; Sup. laryngeal n.
External carotid artery
Superior laryngeal artery and nerve
Superior root, ansa cervicalis
Superior root of ansa cervicalis;
Superior thyroid vein
Superior thyroid artery
Superior belly of omohyoid muscle
Laryngeal prominence
Sternohyoid muscle

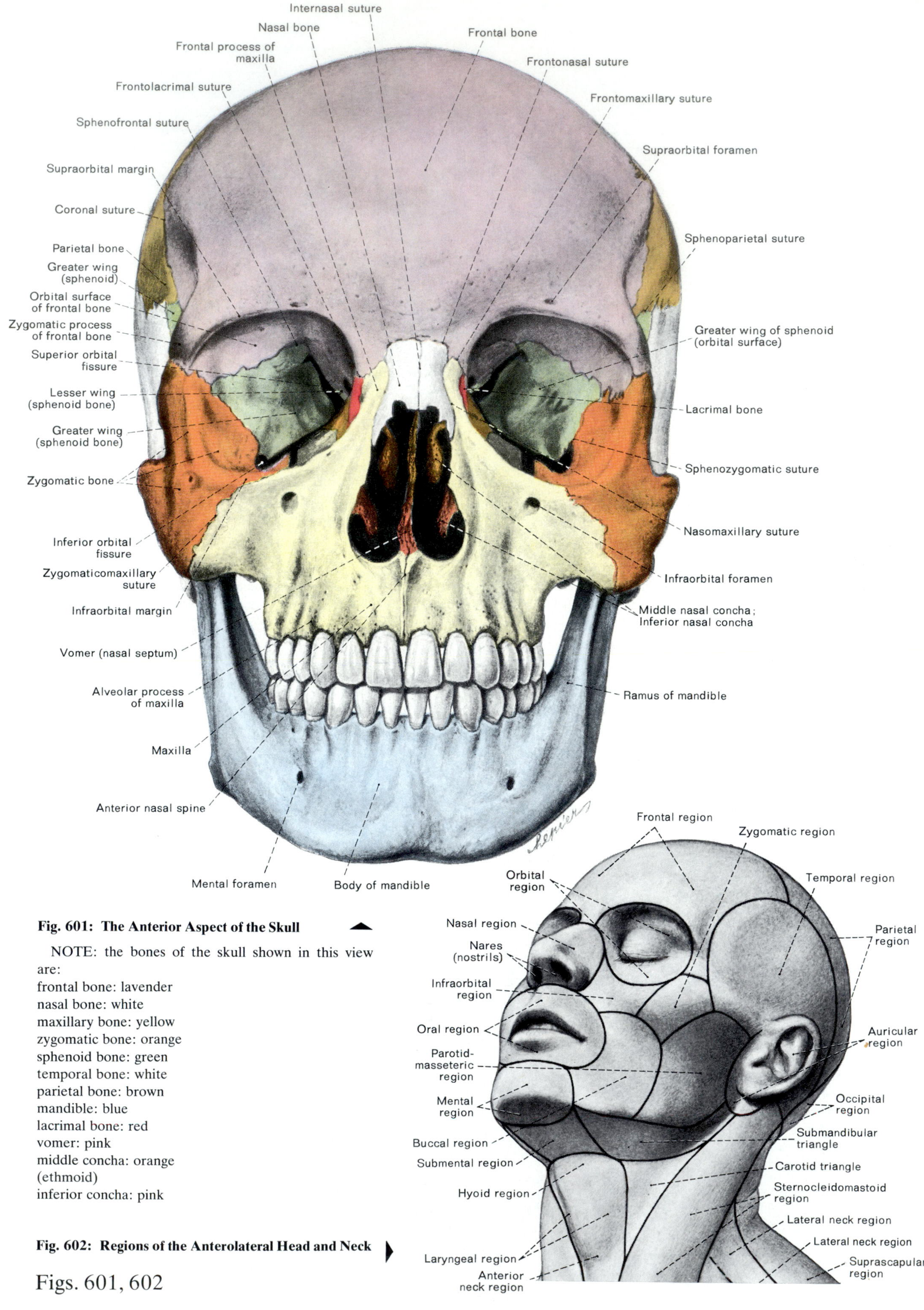
Internasal suture
Nasal bone
Frontal process of maxilla
Frontal bone
Frontonasal suture
Frontolacrimal suture
Frontomaxillary suture
Sphenofrontal suture
Supraorbital foramen
Supraorbital margin
Coronal suture
Sphenoparietal suture
Parietal bone
Greater wing (sphenoid)
Orbital surface of frontal bone
Zygomatic process of frontal bone
Greater wing of sphenoid (orbital surface)
Superior orbital fissure
Lesser wing (sphenoid bone)
Lacrimal bone
Greater wing (sphenoid bone)
Sphenozygomatic suture
Zygomatic bone
Nasomaxillary suture
Inferior orbital fissure
Infraorbital foramen
Zygomaticomaxillary suture
Infraorbital margin
Middle nasal concha; Inferior nasal concha
Vomer (nasal septum)
Alveolar process of maxilla
Ramus of mandible
Maxilla
Anterior nasal spine
Mental foramen
Body of mandible
Frontal region
Zygomatic region
Orbital region
Temporal region
Nasal region
Parietal region
Nares (nostrils)
Infraorbital region
Oral region
Auricular region
Parotid-masseteric region
Mental region
Occipital region
Submandibular triangle
Buccal region
Submental region
Carotid triangle
Sternocleidomastoid region
Hyoid region
Lateral neck region
Lateral neck region
Laryngeal region
Suprascapular region
Anterior neck region

Fig. 601: The Anterior Aspect of the Skull

NOTE: the bones of the skull shown in this view are:
frontal bone: lavender
nasal bone: white
maxillary bone: yellow
zygomatic bone: orange
sphenoid bone: green
temporal bone: white
parietal bone: brown
mandible: blue
lacrimal bone: red
vomer: pink
middle concha: orange (ethmoid)
inferior concha: pink

Fig. 602: Regions of the Anterolateral Head and Neck

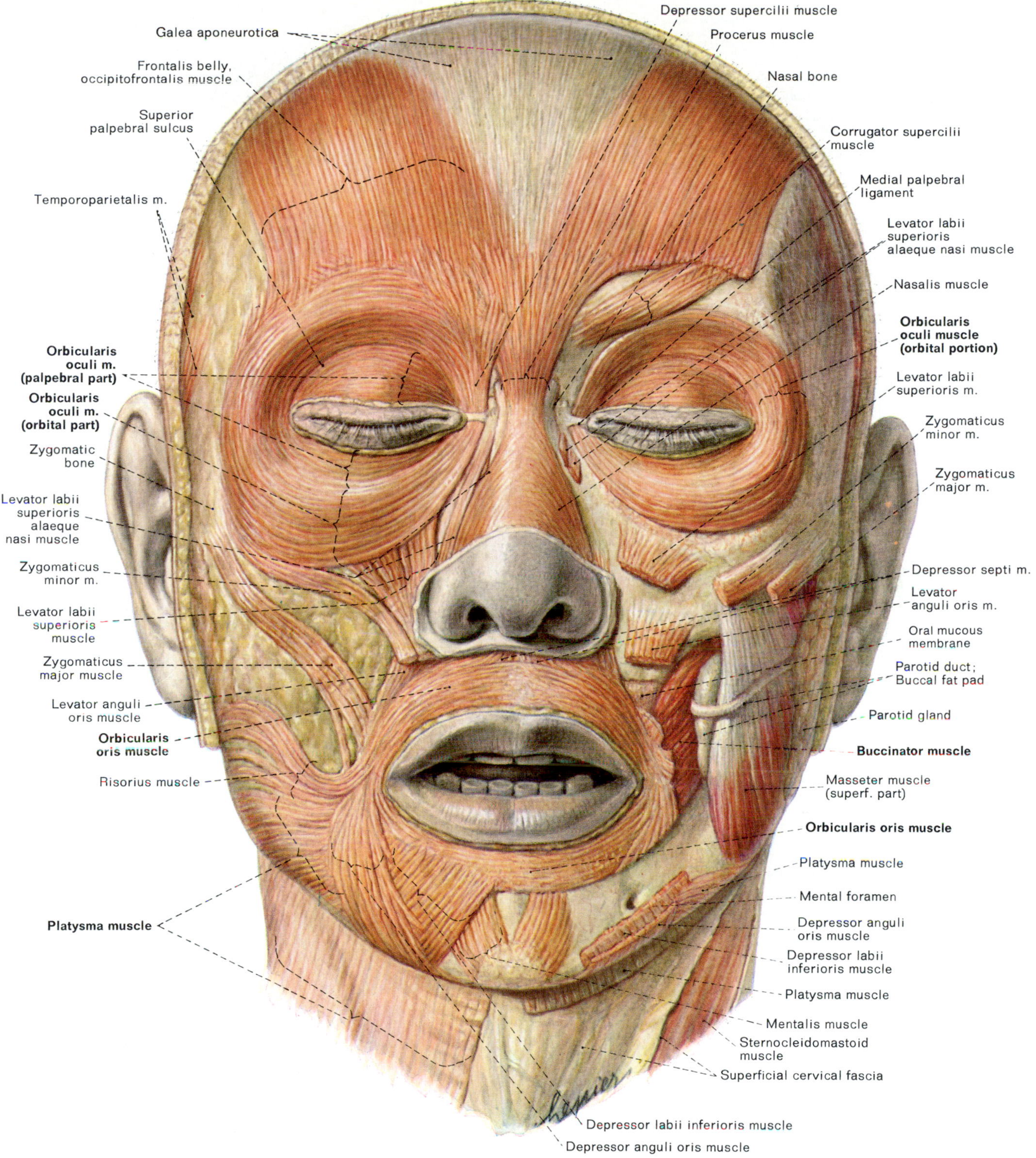

Galea aponeurotica
Frontalis belly, occipitofrontalis muscle
Superior palpebral sulcus
Temporoparietalis m.
Orbicularis oculi m. (palpebral part)
Orbicularis oculi m. (orbital part)
Zygomatic bone
Levator labii superioris alaeque nasi muscle
Zygomaticus minor m.
Levator labii superioris muscle
Zygomaticus major muscle
Levator anguli oris muscle
Orbicularis oris muscle
Risorius muscle
Platysma muscle
Depressor supercilii muscle
Procerus muscle
Nasal bone
Corrugator supercilii muscle
Medial palpebral ligament
Levator labii superioris alaeque nasi muscle
Nasalis muscle
Orbicularis oculi muscle (orbital portion)
Levator labii superioris m.
Zygomaticus minor m.
Zygomaticus major m.
Depressor septi m.
Levator anguli oris m.
Oral mucous membrane
Parotid duct; Buccal fat pad
Parotid gland
Buccinator muscle
Masseter muscle (superf. part)
Orbicularis oris muscle
Platysma muscle
Mental foramen
Depressor anguli oris muscle
Depressor labii inferioris muscle
Platysma muscle
Mentalis muscle
Sternocleidomastoid muscle
Superficial cervical fascia
Depressor labii inferioris muscle
Depressor anguli oris muscle

Fig. 603: The Muscles of Facial Expression (Anterior View)

NOTE: 1) the muscles of facial expression are superficial muscles located within the layers of subcutaneous fascia. Having developed from the mesoderm of the 2nd branchial arch, they are all innervated by the nerve of that arch, the seventh cranial or facial nerve.

2) the facial muscles may be grouped into: a) the muscles of the scalp, b) the muscles of external ear, c) the muscles of the eyelid, d) the nasal muscles and e) the oral muscles. Frequently, the limits of the facial muscles are not easily defined and there is a tendency for them to merge. The platysma muscle also belongs in the facial group, even though it extends over the neck.

3) the circular muscles surrounding the eyes (orbicularis oculi) and the mouth (orbicularis oris) assist in closure of the orbital and oral apertures, and thus contribute to functions such as blinking of the eyelids and the oral ingestion of liquids and food.

4) since facial muscles can respond to thoughts and emotions, they are of assistance in communicative functions. The buccinator muscles are flat and are situated on the lateral aspects of the oral cavity. They assist in mastication by pressing the checks against the teeth, and thus, prevent food from accumulating in the oral vestibule.

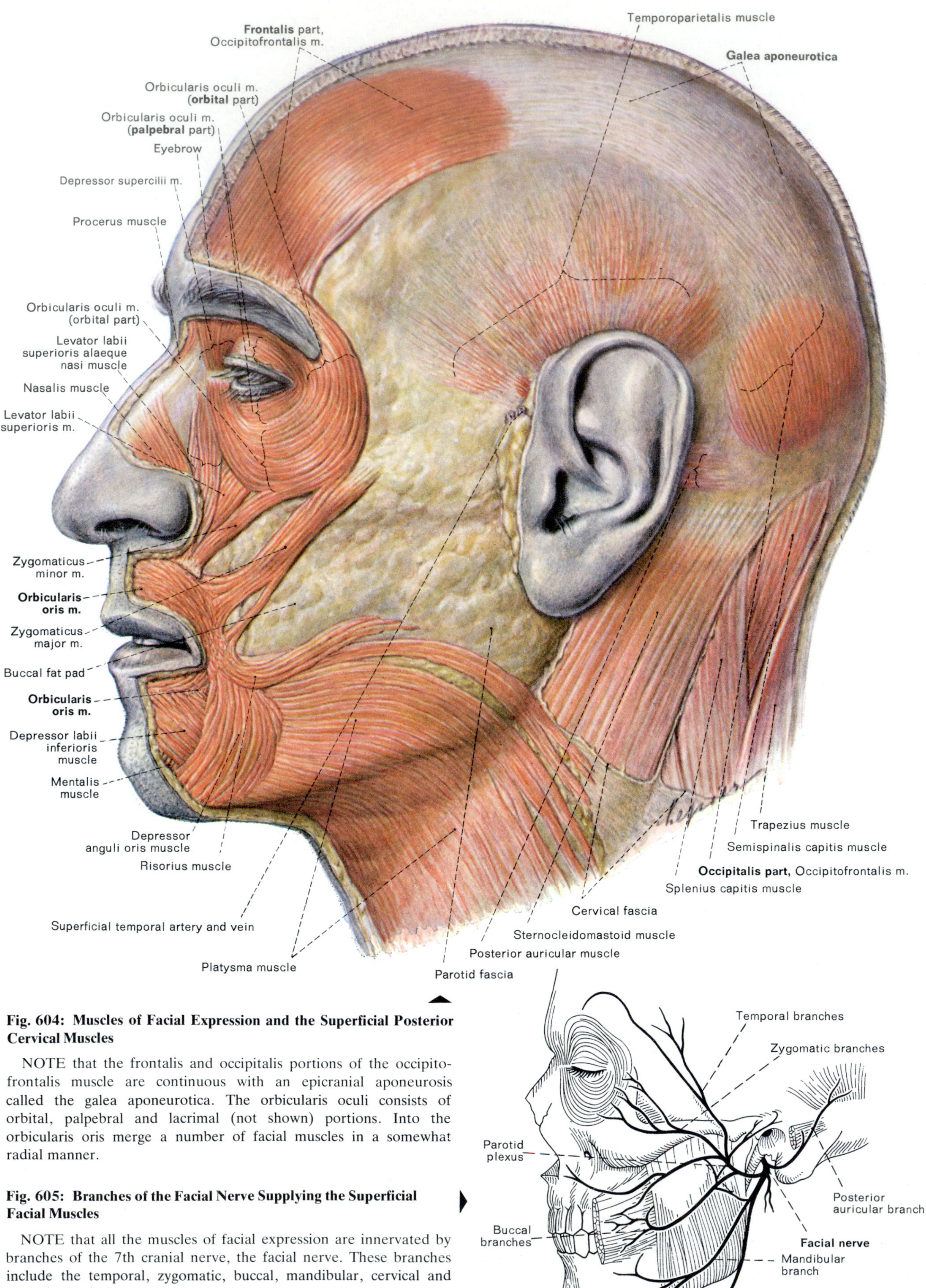
Frontalis part,
Occipitofrontalis m.
Temporoparietalis muscle
Galea aponeurotica
Orbicularis oculi m.
(orbital part)
Orbicularis oculi m.
(palpebral part)
Eyebrow
Depressor supercilii m.
Procerus muscle
Orbicularis oculi m.
(orbital part)
Levator labii
superioris alaeque
nasi muscle
Nasalis muscle
Levator labii
superioris m.
Zygomaticus
minor m.
Orbicularis
oris m.
Zygomaticus
major m.
Buccal fat pad
Orbicularis
oris m.
Depressor labii
inferioris
muscle
Mentalis
muscle
Depressor
anguli oris muscle
Risorius muscle
Superficial temporal artery and vein
Platysma muscle
Parotid fascia
Posterior auricular muscle
Sternocleidomastoid muscle
Cervical fascia
Splenius capitis muscle
Occipitalis part, Occipitofrontalis m.
Semispinalis capitis muscle
Trapezius muscle
Temporal branches
Zygomatic branches
Parotid
plexus
Posterior
auricular branch
Facial nerve
Buccal
branches
Mandibular
branch
Cervical branch

Fig. 604: Muscles of Facial Expression and the Superficial Posterior Cervical Muscles

NOTE that the frontalis and occipitalis portions of the occipitofrontalis muscle are continuous with an epicranial aponeurosis called the galea aponeurotica. The orbicularis oculi consists of orbital, palpebral and lacrimal (not shown) portions. Into the orbicularis oris merge a number of facial muscles in a somewhat radial manner.

Fig. 605: Branches of the Facial Nerve Supplying the Superficial Facial Muscles

NOTE that all the muscles of facial expression are innervated by branches of the 7th cranial nerve, the facial nerve. These branches include the temporal, zygomatic, buccal, mandibular, cervical and posterior auricular nerves.

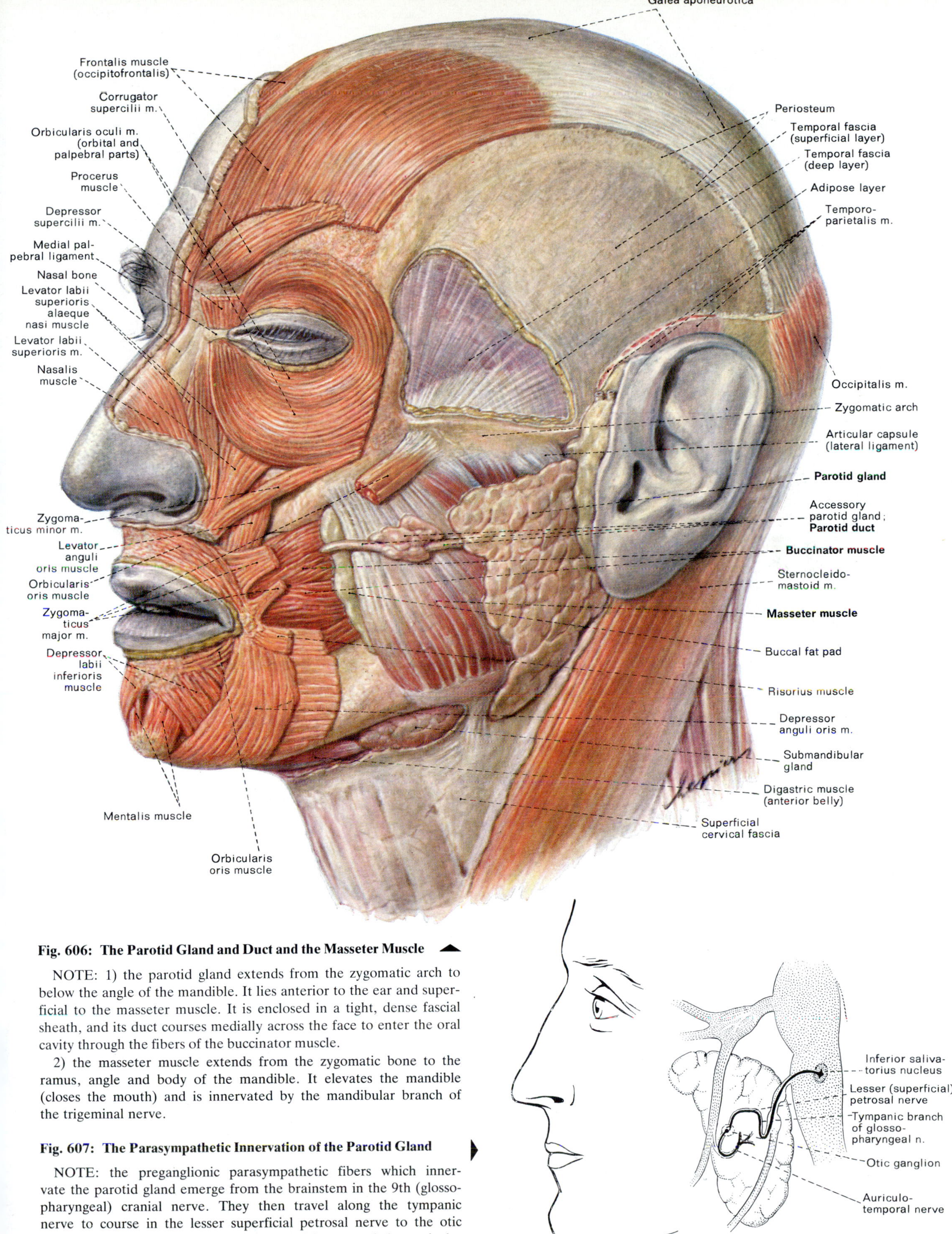
Galea aponeurotica
Frontalis muscle (occipitofrontalis)
Corrugator supercilii m.
Orbicularis oculi m. (orbital and palpebral parts)
Procerus muscle
Depressor supercilii m.
Medial palpebral ligament
Nasal bone
Levator labii superioris alaeque nasi muscle
Levator labii superioris m.
Nasalis muscle
Zygomaticus minor m.
Levator anguli oris muscle
Orbicularis oris muscle
Zygomaticus major m.
Depressor labii inferioris muscle
Mentalis muscle
Orbicularis oris muscle
Periosteum
Temporal fascia (superficial layer)
Temporal fascia (deep layer)
Adipose layer
Temporoparietalis m.
Occipitalis m.
Zygomatic arch
Articular capsule (lateral ligament)
Parotid gland
Accessory parotid gland; Parotid duct
Buccinator muscle
Sternocleidomastoid m.
Masseter muscle
Buccal fat pad
Risorius muscle
Depressor anguli oris m.
Submandibular gland
Digastric muscle (anterior belly)
Superficial cervical fascia
Inferior salivatorius nucleus
Lesser (superficial) petrosal nerve
Tympanic branch of glossopharyngeal n.
Otic ganglion
Auriculotemporal nerve

Fig. 606: The Parotid Gland and Duct and the Masseter Muscle ▲

NOTE: 1) the parotid gland extends from the zygomatic arch to below the angle of the mandible. It lies anterior to the ear and superficial to the masseter muscle. It is enclosed in a tight, dense fascial sheath, and its duct courses medially across the face to enter the oral cavity through the fibers of the buccinator muscle.

2) the masseter muscle extends from the zygomatic bone to the ramus, angle and body of the mandible. It elevates the mandible (closes the mouth) and is innervated by the mandibular branch of the trigeminal nerve.

Fig. 607: The Parasympathetic Innervation of the Parotid Gland ▶

NOTE: the preganglionic parasympathetic fibers which innervate the parotid gland emerge from the brainstem in the 9th (glossopharyngeal) cranial nerve. They then travel along the tympanic nerve to course in the lesser superficial petrosal nerve to the otic ganglion. Postganglionic fibers then travel by way of the auriculotemporal nerve to reach the parotid gland.

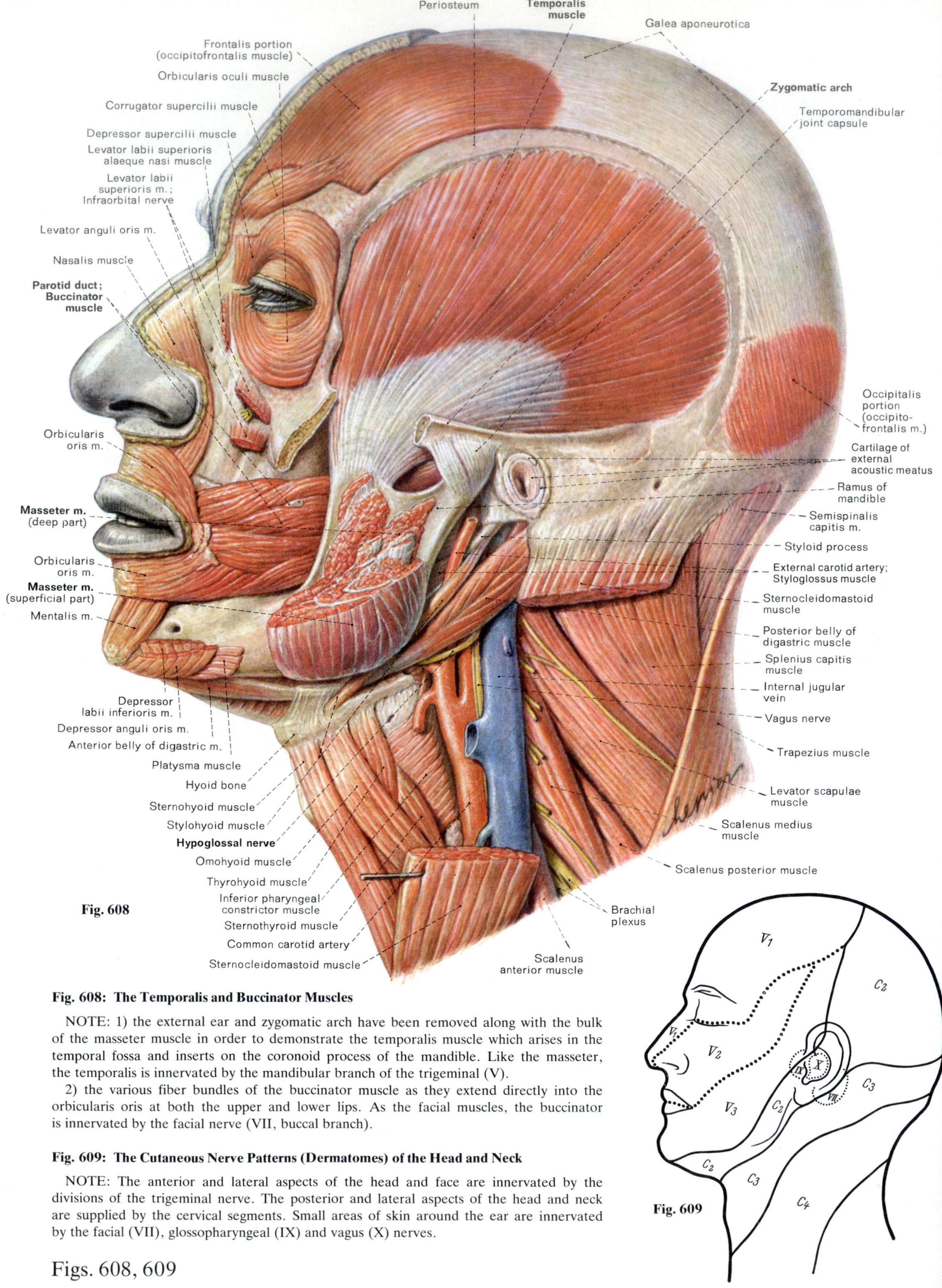
Periosteum
Temporalis muscle
Galea aponeurotica
Frontalis portion (occipitofrontalis muscle)
Orbicularis oculi muscle
Zygomatic arch
Temporomandibular joint capsule
Corrugator supercilii muscle
Depressor supercilii muscle
Levator labii superioris alaeque nasi muscle
Levator labii superioris m.; Infraorbital nerve
Levator anguli oris m.
Nasalis muscle
Parotid duct; Buccinator muscle
Occipitalis portion (occipito-frontalis m.)
Orbicularis oris m.
Cartilage of external acoustic meatus
Ramus of mandible
Masseter m. (deep part)
Semispinalis capitis m.
Styloid process
Orbicularis oris m.
External carotid artery; Styloglossus muscle
Masseter m. (superficial part)
Sternocleidomastoid muscle
Mentalis m.
Posterior belly of digastric muscle
Splenius capitis muscle
Internal jugular vein
Depressor labii inferioris m.
Vagus nerve
Depressor anguli oris m.
Anterior belly of digastric m.
Trapezius muscle
Platysma muscle
Hyoid bone
Levator scapulae muscle
Sternohyoid muscle
Stylohyoid muscle
Scalenus medius muscle
Hypoglossal nerve
Omohyoid muscle
Scalenus posterior muscle
Thyrohyoid muscle
Inferior pharyngeal constrictor muscle
Brachial plexus
Sternothyroid muscle
Common carotid artery
Scalenus anterior muscle
Sternocleidomastoid muscle
Fig. 608
V1
V2
V3
C2
C3
C4
IX
X
VII
Fig. 609

Fig. 608: The Temporalis and Buccinator Muscles

NOTE: 1) the external ear and zygomatic arch have been removed along with the bulk of the masseter muscle in order to demonstrate the temporalis muscle which arises in the temporal fossa and inserts on the coronoid process of the mandible. Like the masseter, the temporalis is innervated by the mandibular branch of the trigeminal (V).

2) the various fiber bundles of the buccinator muscle as they extend directly into the orbicularis oris at both the upper and lower lips. As the facial muscles, the buccinator is innervated by the facial nerve (VII, buccal branch).

Fig. 609: The Cutaneous Nerve Patterns (Dermatomes) of the Head and Neck

NOTE: The anterior and lateral aspects of the head and face are innervated by the divisions of the trigeminal nerve. The posterior and lateral aspects of the head and neck are supplied by the cervical segments. Small areas of skin around the ear are innervated by the facial (VII), glossopharyngeal (IX) and vagus (X) nerves.

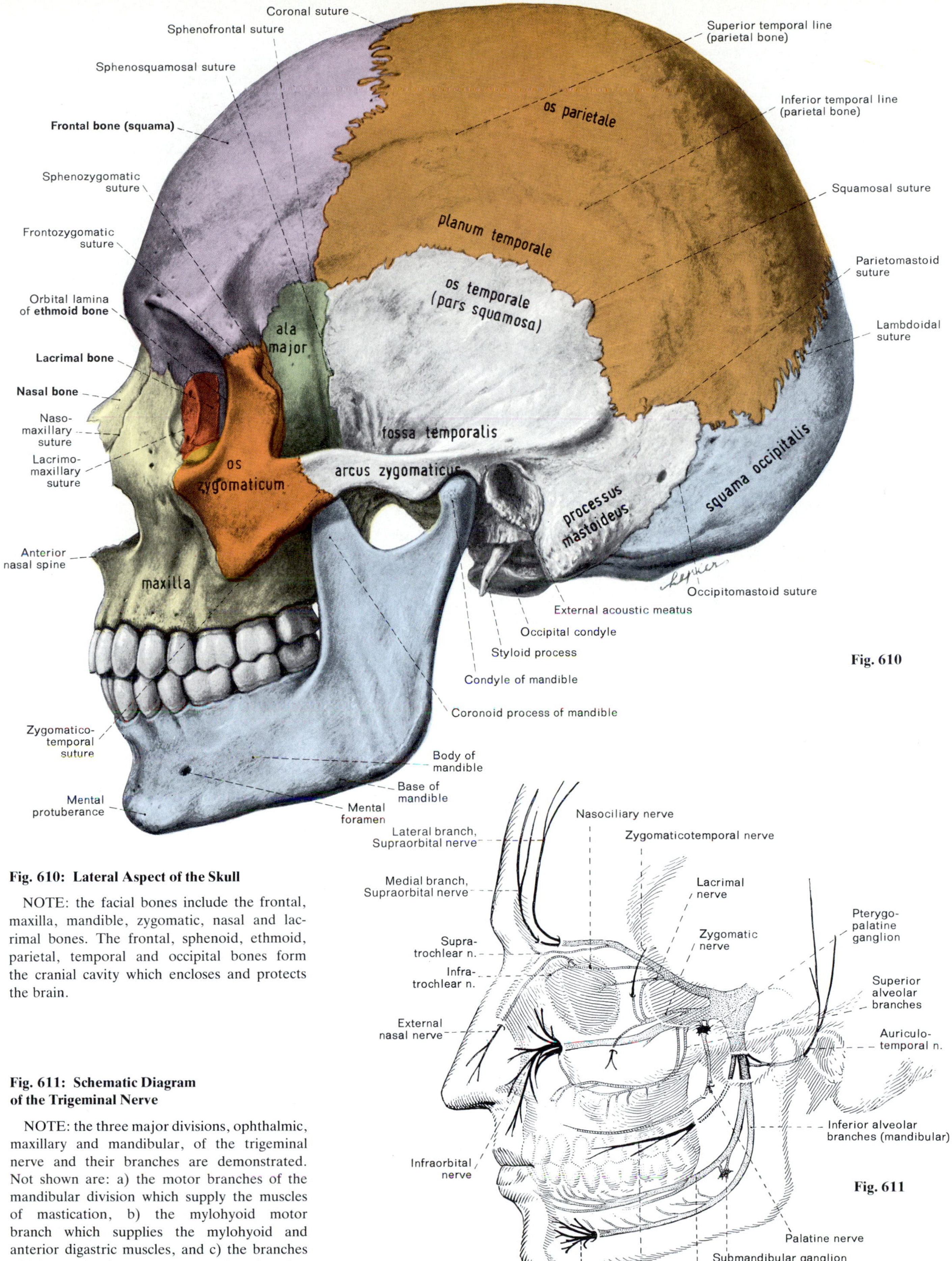

Coronal suture
Sphenofrontal suture
Sphenosquamosal suture
Frontal bone (squama)
Sphenozygomatic suture
Frontozygomatic suture
Orbital lamina of ethmoid bone
Lacrimal bone
Nasal bone
Naso-maxillary suture
Lacrimo-maxillary suture
Anterior nasal spine
Zygomatico-temporal suture
Mental protuberance
Superior temporal line (parietal bone)
Inferior temporal line (parietal bone)
Squamosal suture
Parietomastoid suture
Lambdoidal suture
os parietale
planum temporale
os temporale (pars squamosa)
ala major
os zygomaticum
fossa temporalis
arcus zygomaticus
processus mastoideus
squama occipitalis
maxilla
Occipitomastoid suture
External acoustic meatus
Occipital condyle
Styloid process
Condyle of mandible
Coronoid process of mandible
Body of mandible
Base of mandible
Mental foramen
Fig. 610
Nasociliary nerve
Zygomaticotemporal nerve
Lateral branch, Supraorbital nerve
Medial branch, Supraorbital nerve
Lacrimal nerve
Zygomatic nerve
Pterygo-palatine ganglion
Supra-trochlear n.
Infra-trochlear n.
Superior alveolar branches
External nasal nerve
Auriculo-temporal n.
Inferior alveolar branches (mandibular)
Infraorbital nerve
Fig. 611
Palatine nerve
Submandibular ganglion
Lingual nerve
Buccal nerve
Mental nerve

Fig. 610: Lateral Aspect of the Skull

NOTE: the facial bones include the frontal, maxilla, mandible, zygomatic, nasal and lacrimal bones. The frontal, sphenoid, ethmoid, parietal, temporal and occipital bones form the cranial cavity which encloses and protects the brain.

Fig. 611: Schematic Diagram of the Trigeminal Nerve

NOTE: the three major divisions, ophthalmic, maxillary and mandibular, of the trigeminal nerve and their branches are demonstrated. Not shown are: a) the motor branches of the mandibular division which supply the muscles of mastication, b) the mylohyoid motor branch which supplies the mylohyoid and anterior digastric muscles, and c) the branches which supply the tensor tympani and tensor palati muscles.

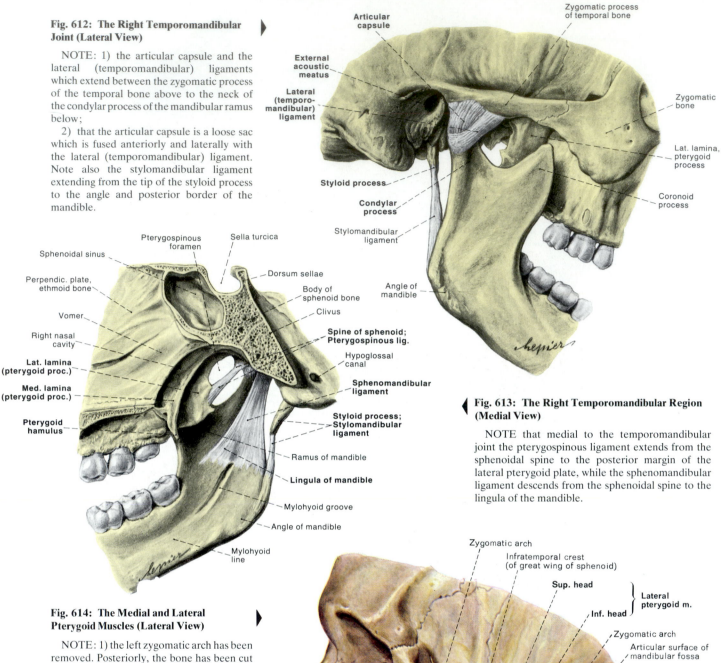

Fig. 612: The Right Temporomandibular Joint (Lateral View)

NOTE: 1) the articular capsule and the lateral (temporomandibular) ligaments which extend between the zygomatic process of the temporal bone above to the neck of the condylar process of the mandibular ramus below;

2) that the articular capsule is a loose sac which is fused anteriorly and laterally with the lateral (temporomandibular) ligament. Note also the stylomandibular ligament extending from the tip of the styloid process to the angle and posterior border of the mandible.

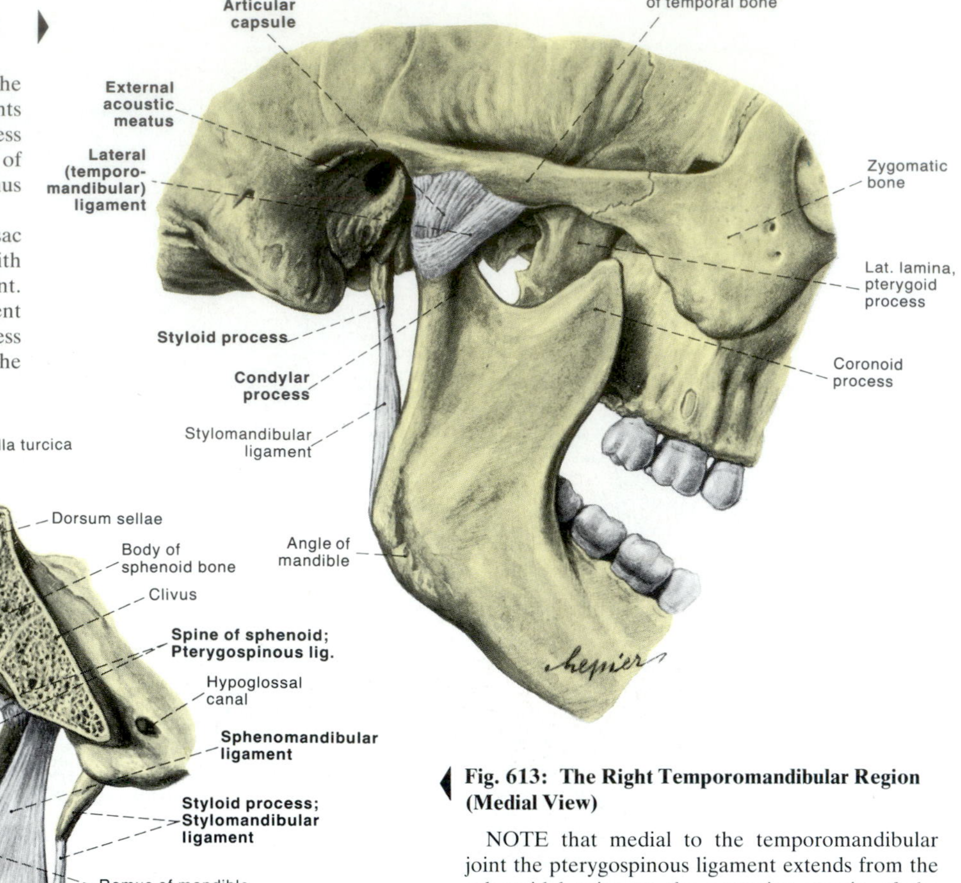

Articular capsule
Zygomatic process of temporal bone
External acoustic meatus
Lateral (temporo-mandibular) ligament
Zygomatic bone
Lat. lamina, pterygoid process
Styloid process
Coronoid process
Condylar process
Stylomandibular ligament
Angle of mandible

Sphenoidal sinus
Pterygospinous foramen
Sella turcica
Dorsum sellae
Perpendic. plate, ethmoid bone
Body of sphenoid bone
Clivus
Vomer
Right nasal cavity
Spine of sphenoid; Pterygospinous lig.
Hypoglossal canal
Lat. lamina (pterygoid proc.)
Med. lamina (pterygoid proc.)
Sphenomandibular ligament
Pterygoid hamulus
Styloid process; Stylomandibular ligament
Ramus of mandible
Lingula of mandible
Mylohyoid groove
Angle of mandible
Mylohyoid line

Fig. 613: The Right Temporomandibular Region (Medial View)

NOTE that medial to the temporomandibular joint the pterygospinous ligament extends from the sphenoidal spine to the posterior margin of the lateral pterygoid plate, while the sphenomandibular ligament descends from the sphenoidal spine to the lingula of the mandible.

Fig. 614: The Medial and Lateral Pterygoid Muscles (Lateral View)

NOTE: 1) the left zygomatic arch has been removed. Posteriorly, the bone has been cut through the temporomandibular joint, revealing the articular disc. The location of the medial pterygoid muscle and part of the lateral pterygoid muscle on the inner aspect of the ramus of the mandible is represented as though the bone were transparent.

2) the lateral pterygoid muscle arises by two heads, a superior from the great wing of the sphenoid bone and an inferior from the lateral surface of the lateral pterygoid plate of the sphenoid. The two heads insert posteriorly on the neck of the condyle of the mandible. The lateral pterygoid muscle opens and protracts the mandible and also moves it from side to side.

3) the medial pterygoid muscle arises from the medial surface of the lateral pterygoid plate of the sphenoid as well as from the palatine bone, and inserts on the medial surface of the ramus and angle of the mandible. It assists the masseter and temporalis in closing the jaw.

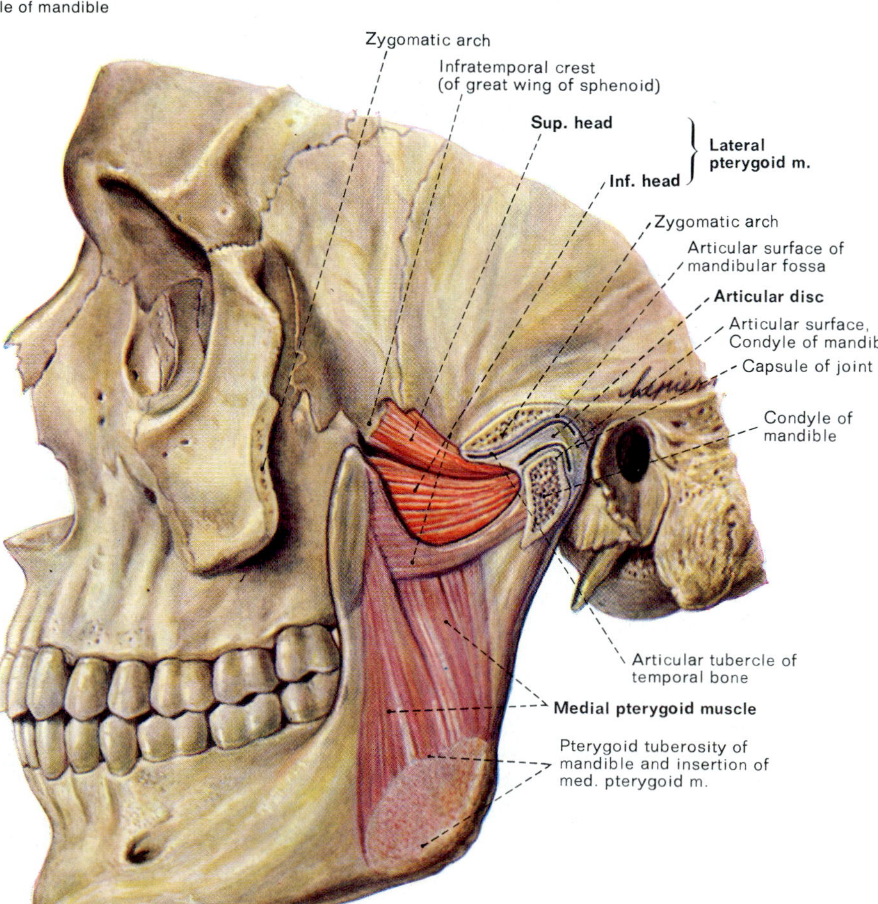

Zygomatic arch
Infratemporal crest (of great wing of sphenoid)
Sup. head
Inf. head
Lateral pterygoid m.
Zygomatic arch
Articular surface of mandibular fossa
Articular disc
Articular surface, Condyle of mandible
Capsule of joint
Condyle of mandible
Articular tubercle of temporal bone
Medial pterygoid muscle
Pterygoid tuberosity of mandible and insertion of med. pterygoid m.

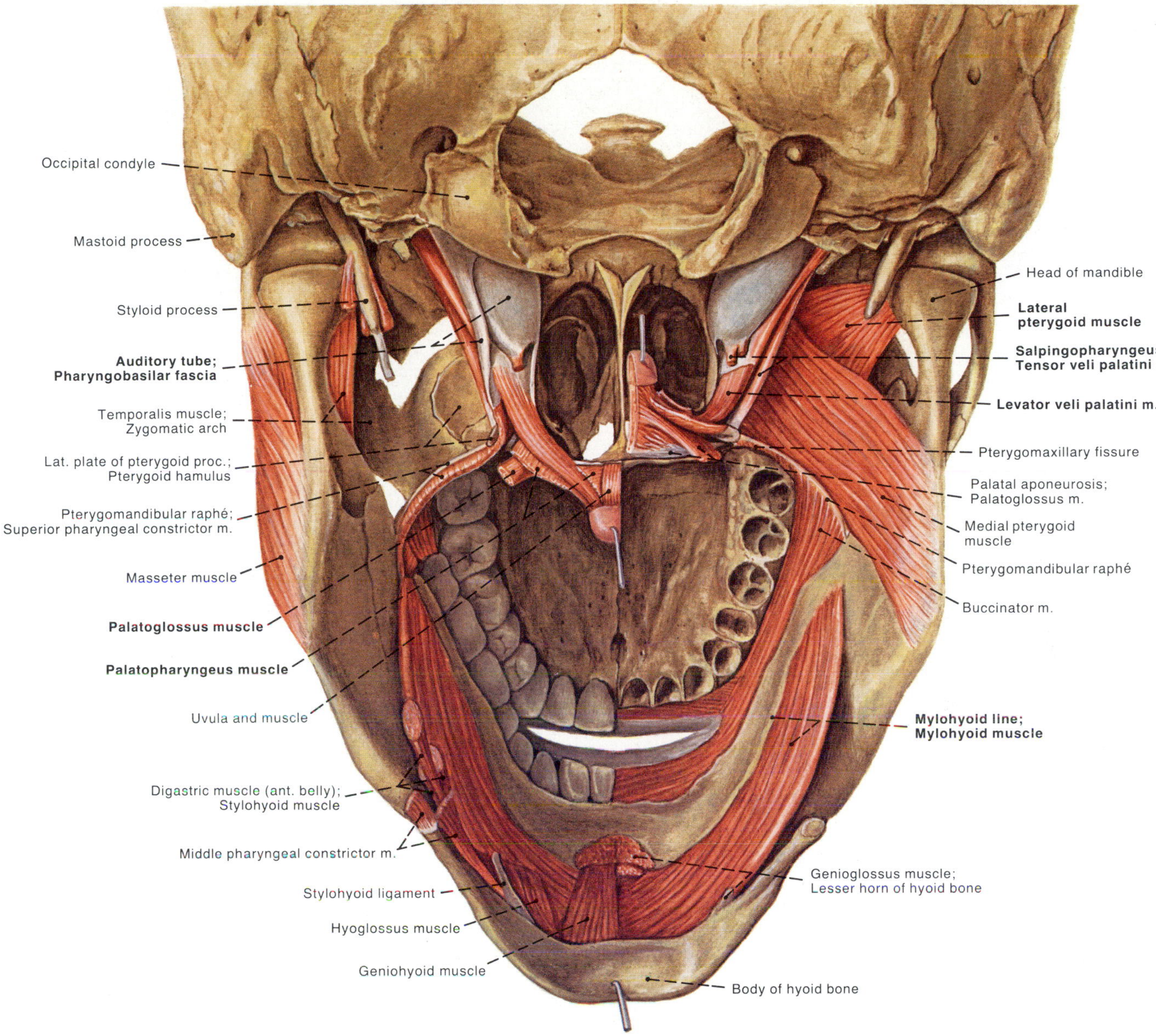

Occipital condyle
Mastoid process
Styloid process
Auditory tube; Pharyngobasilar fascia
Temporalis muscle; Zygomatic arch
Lat. plate of pterygoid proc.; Pterygoid hamulus
Pterygomandibular raphé; Superior pharyngeal constrictor m.
Masseter muscle
Palatoglossus muscle
Palatopharyngeus muscle
Uvula and muscle
Digastric muscle (ant. belly); Stylohyoid muscle
Middle pharyngeal constrictor m.
Stylohyoid ligament
Hyoglossus muscle
Geniohyoid muscle
Head of mandible
Lateral pterygoid muscle
Salpingopharyngeus m.; Tensor veli palatini m.
Levator veli palatini m.
Pterygomaxillary fissure
Palatal aponeurosis; Palatoglossus m.
Medial pterygoid muscle
Pterygomandibular raphé
Buccinator m.
Mylohyoid line; Mylohyoid muscle
Genioglossus muscle; Lesser horn of hyoid bone
Body of hyoid bone

Fig. 615: The Pterygoid, Buccinator and Mylohyoid Muscles as Seen from Below and Behind

NOTE: 1) that a muscular sling is formed around the posterior border of the ramus of the mandible as far as its angle by the insertions of the medial pterygoid muscle (seen on the right) and the masseter muscle (seen on the left). The medial pterygoid muscle descends to attach along the medial aspect of the mandible, while the fibers of the masseter course downward to insert on the outer aspect of the jaw.

2) that the fibers of the lateral pterygoid muscle (right side) course principally in the horizontal plane. Observe that the mylohyoid and geniohyoid muscles attach the mandible to the hyoid bone. Other muscles shown to advantage in this figure are the tensor and levator veli palatini muscles.

3) the right pterygomandibular raphé extending between the pterygoid hamulus and the mylohyoid line. This raphé gives attachment to the buccinator muscle which forms the soft lateral wall of the oral cavity on each side.

Figs. 616 and 617: Sagittal Sections of Temporomandibular Joint

NOTE: 1) the articular disc (stippled structure) is an oval plate interposed between the mandibular fossa of the temporal bone and the condyle of the mandible. Thus, there exists a joint cavity between the disc and the mandibular fossa and another between the disc and the condyle.

2) with the jaw closed (616), the head of the condyle of the mandible and the articular disc lie totally within the mandibular fossa. When the jaw is opened (617), the condyle turns in a hinge-fashion on the disc, and both bone and disc glide forward within the joint capsule to lie opposite the articular tubercle of the temporal bone.

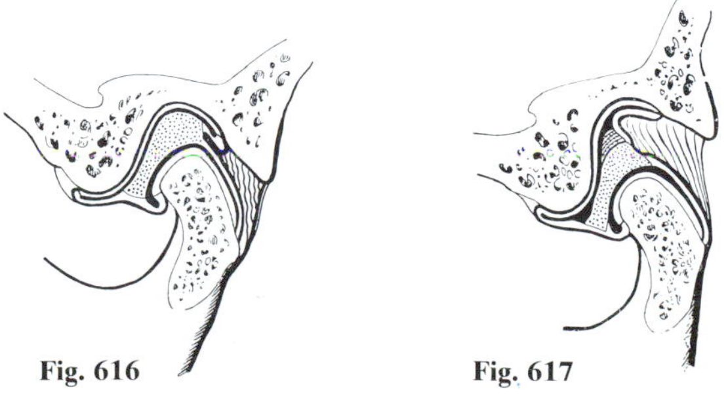

Fig. 616 **Fig. 617**

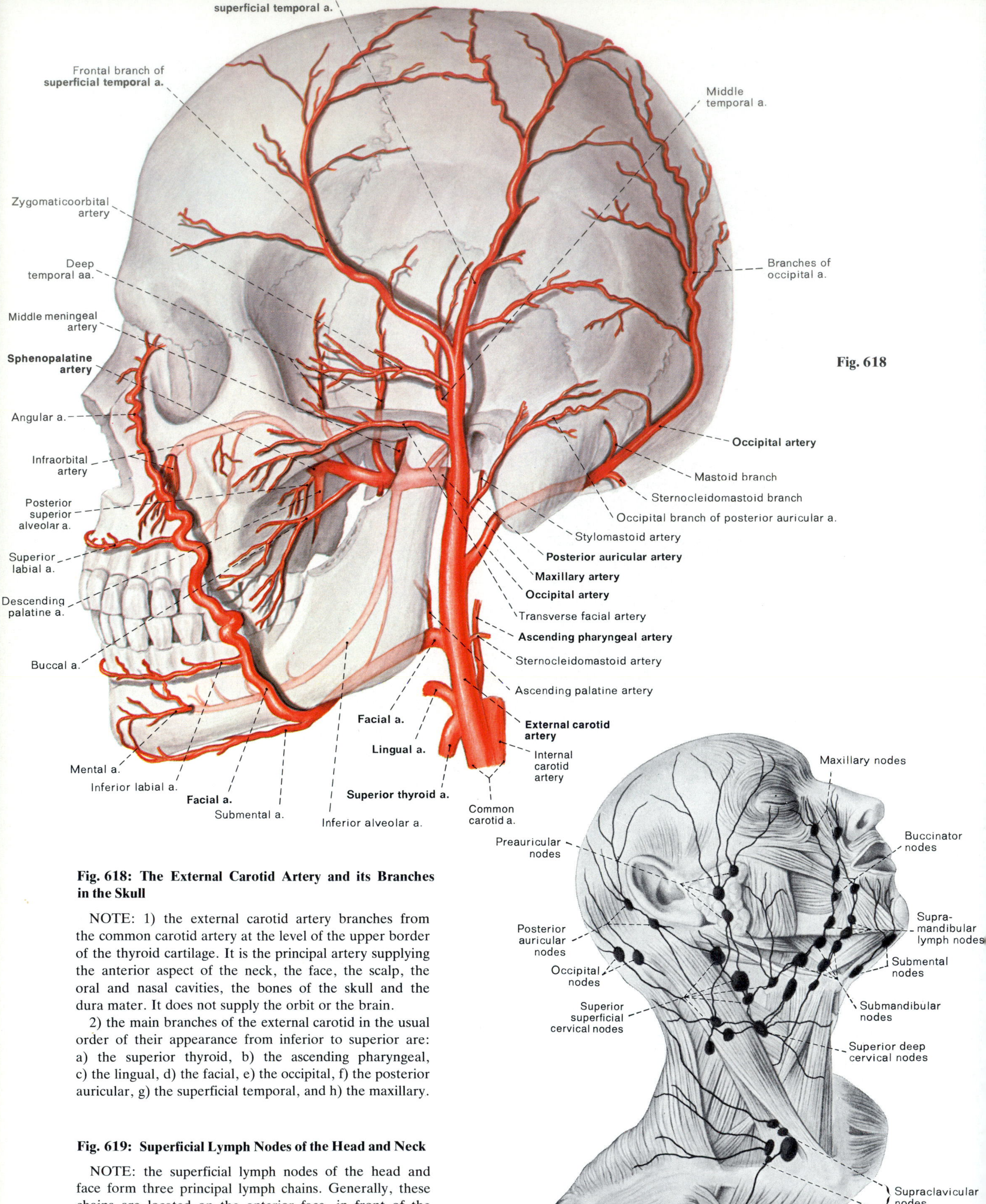

Parietal branch of
superficial temporal a.
Frontal branch of
superficial temporal a.
Middle
temporal a.
Zygomaticoorbital
artery
Deep
temporal aa.
Branches of
occipital a.
Middle meningeal
artery
Sphenopalatine
artery
Fig. 618
Angular a.
Occipital artery
Infraorbital
artery
Mastoid branch
Sternocleidomastoid branch
Posterior
superior
alveolar a.
Occipital branch of posterior auricular a.
Stylomastoid artery
Superior
labial a.
Posterior auricular artery
Maxillary artery
Descending
palatine a.
Occipital artery
Transverse facial artery
Ascending pharyngeal artery
Buccal a.
Sternocleidomastoid artery
Ascending palatine artery
Facial a.
External carotid
artery
Lingual a.
Internal
carotid
artery
Mental a.
Inferior labial a.
Facial a.
Submental a.
Superior thyroid a.
Inferior alveolar a.
Common
carotid a.
Maxillary nodes
Preauricular
nodes
Buccinator
nodes
Posterior
auricular
nodes
Supra-
mandibular
lymph nodes
Occipital
nodes
Submental
nodes
Superior
superficial
cervical nodes
Submandibular
nodes
Superior deep
cervical nodes
Supraclavicular
nodes
Fig. 619

Fig. 618: The External Carotid Artery and its Branches in the Skull

NOTE: 1) the external carotid artery branches from the common carotid artery at the level of the upper border of the thyroid cartilage. It is the principal artery supplying the anterior aspect of the neck, the face, the scalp, the oral and nasal cavities, the bones of the skull and the dura mater. It does not supply the orbit or the brain.

2) the main branches of the external carotid in the usual order of their appearance from inferior to superior are: a) the superior thyroid, b) the ascending pharyngeal, c) the lingual, d) the facial, e) the occipital, f) the posterior auricular, g) the superficial temporal, and h) the maxillary.

Fig. 619: Superficial Lymph Nodes of the Head and Neck

NOTE: the superficial lymph nodes of the head and face form three principal lymph chains. Generally, these chains are located on the anterior face, in front of the ear and in the occipital region. Observe that they all drain inferiorly and posteriorly to the superficial and deeper nodes of the neck.

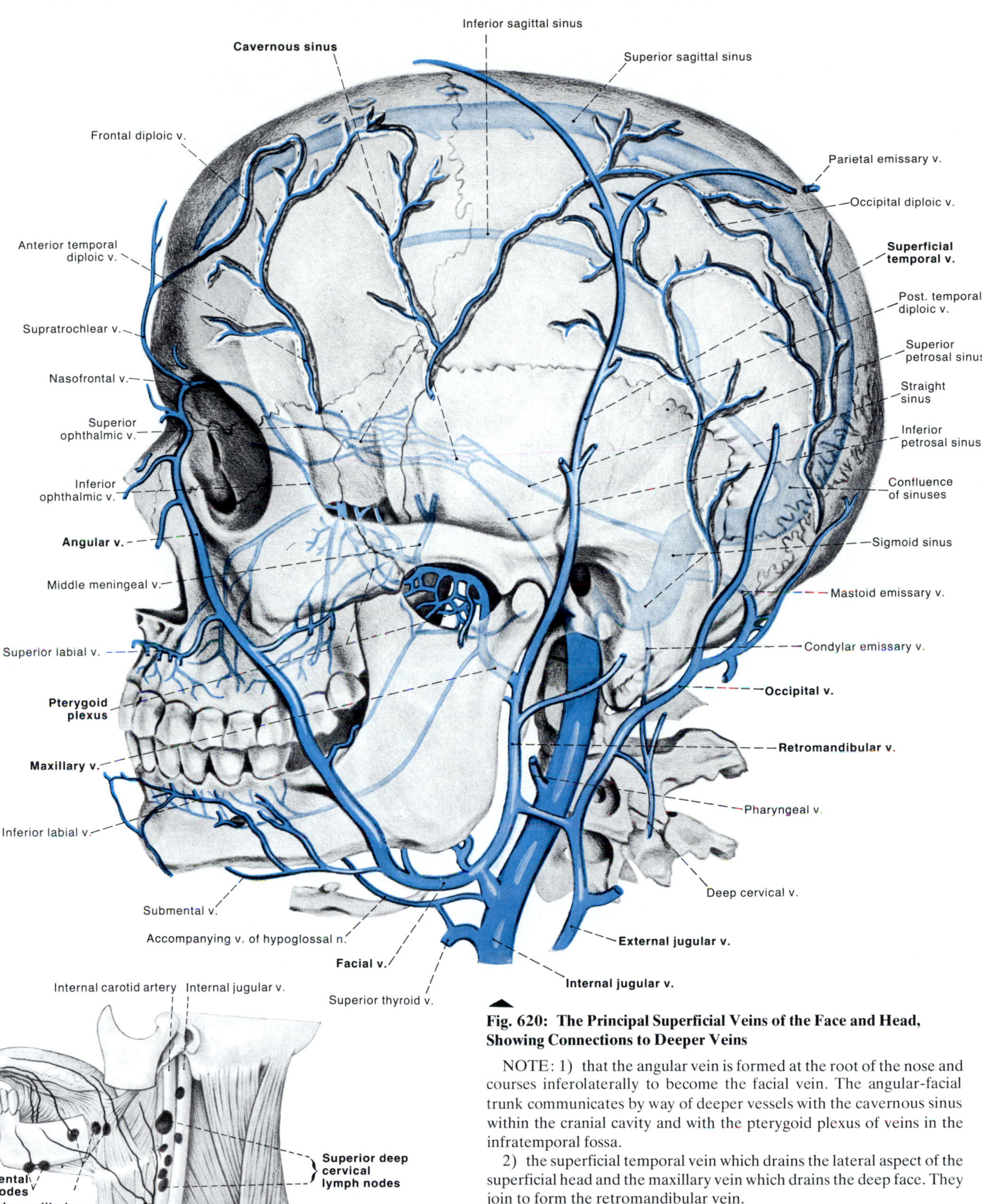
Inferior sagittal sinus
Cavernous sinus
Superior sagittal sinus
Frontal diploic v.
Parietal emissary v.
Occipital diploic v.
Anterior temporal diploic v.
Superficial temporal v.
Post. temporal diploic v.
Supratrochlear v.
Superior petrosal sinus
Nasofrontal v.
Straight sinus
Superior ophthalmic v.
Inferior petrosal sinus
Inferior ophthalmic v.
Confluence of sinuses
Angular v.
Sigmoid sinus
Middle meningeal v.
Mastoid emissary v.
Superior labial v.
Condylar emissary v.
Pterygoid plexus
Occipital v.
Maxillary v.
Retromandibular v.
Pharyngeal v.
Inferior labial v.
Deep cervical v.
Submental v.
Accompanying v. of hypoglossal n.
External jugular v.
Facial v.
Internal jugular v.
Superior thyroid v.
Internal carotid artery
Internal jugular v.
Tongue
Superior deep cervical lymph nodes
Submental lymph nodes
Submandibular lymph nodes
Inferior deep cervical lymph nodes
Common carotid artery
Supraclavicular deep cervical lymph nodes

Fig. 620: The Principal Superficial Veins of the Face and Head, Showing Connections to Deeper Veins

NOTE: 1) that the angular vein is formed at the root of the nose and courses inferolaterally to become the facial vein. The angular-facial trunk communicates by way of deeper vessels with the cavernous sinus within the cranial cavity and with the pterygoid plexus of veins in the infratemporal fossa.

2) the superficial temporal vein which drains the lateral aspect of the superficial head and the maxillary vein which drains the deep face. They join to form the retromandibular vein.

3) the occipital vein which forms on the posterolateral aspect of the scalp and which courses downward into the external jugular vein. Observe that the diploic veins and the various emissary veins (condylar, mastoid and parietal) which interconnect the superficial veins with the dural sinuses.

Fig. 621: Lymphatic Drainage from the Tongue and Lower Oral Cavity

NOTE that lymph from the tongue and lower oral cavity drains downward and posteriorly into submandibular and superior deep cervical nodes along the internal jugular vein.

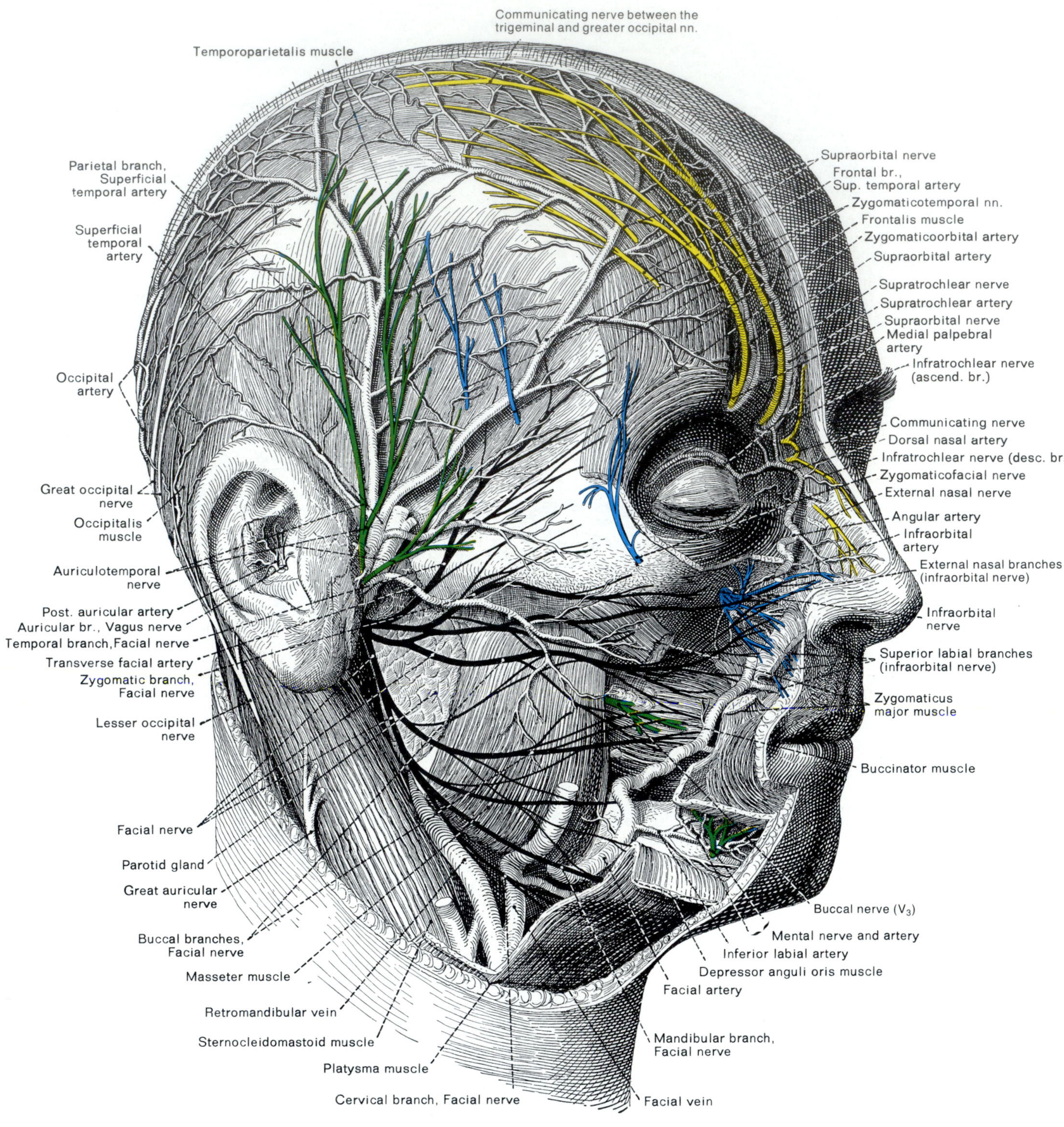

Communicating nerve between the trigeminal and greater occipital nn.
Temporoparietalis muscle
Parietal branch, Superficial temporal artery
Superficial temporal artery
Occipital artery
Great occipital nerve
Occipitalis muscle
Auriculotemporal nerve
Post. auricular artery
Auricular br., Vagus nerve
Temporal branch, Facial nerve
Transverse facial artery
Zygomatic branch, Facial nerve
Lesser occipital nerve
Facial nerve
Parotid gland
Great auricular nerve
Buccal branches, Facial nerve
Masseter muscle
Retromandibular vein
Sternocleidomastoid muscle
Platysma muscle
Cervical branch, Facial nerve
Supraorbital nerve
Frontal br., Sup. temporal artery
Zygomaticotemporal nn.
Frontalis muscle
Zygomaticoorbital artery
Supraorbital artery
Supratrochlear nerve
Supratrochlear artery
Supraorbital nerve
Medial palpebral artery
Infratrochlear nerve (ascend. br.)
Communicating nerve
Dorsal nasal artery
Infratrochlear nerve (desc. br.)
Zygomaticofacial nerve
External nasal nerve
Angular artery
Infraorbital artery
External nasal branches (infraorbital nerve)
Infraorbital nerve
Superior labial branches (infraorbital nerve)
Zygomaticus major muscle
Buccinator muscle
Buccal nerve (V_3)
Mental nerve and artery
Inferior labial artery
Depressor anguli oris muscle
Facial artery
Mandibular branch, Facial nerve
Facial vein

Fig. 622: Superficial Nerves and Vessels of the Face and Head

NOTE: 1) a portion of the parotid gland has been removed in order to reveal the branches of the facial nerve (black) which emerge from within the substance of the gland to supply the superficial muscles of facial expression. Identify the temporal, zygomatic, buccal, mandibular, and cervical branches. The posterior auricular branch is not shown.

2) the superficial branches of the trigeminal nerve are indicated in various colors; ophthlamic (yellow), maxillary (blue) and mandibular (green). The trigeminal nerve serves as the cutaneous nerve of the anterior and lateral face, but is also the motor nerve to the muscles of mastication. The cervical sensory nerves (white, not colored) supply the occipital region and much of the ear and neck.

3) the general distribution of the superficial temporal artery and its branches, the zygomaticoorbital and transverse facial arteries. Follow the course of the facial artery across the face to become the angular artery. Among other structures, the facial supplies the chin and the upper and lower lips and anastomoses with vessels emerging from the orbit.

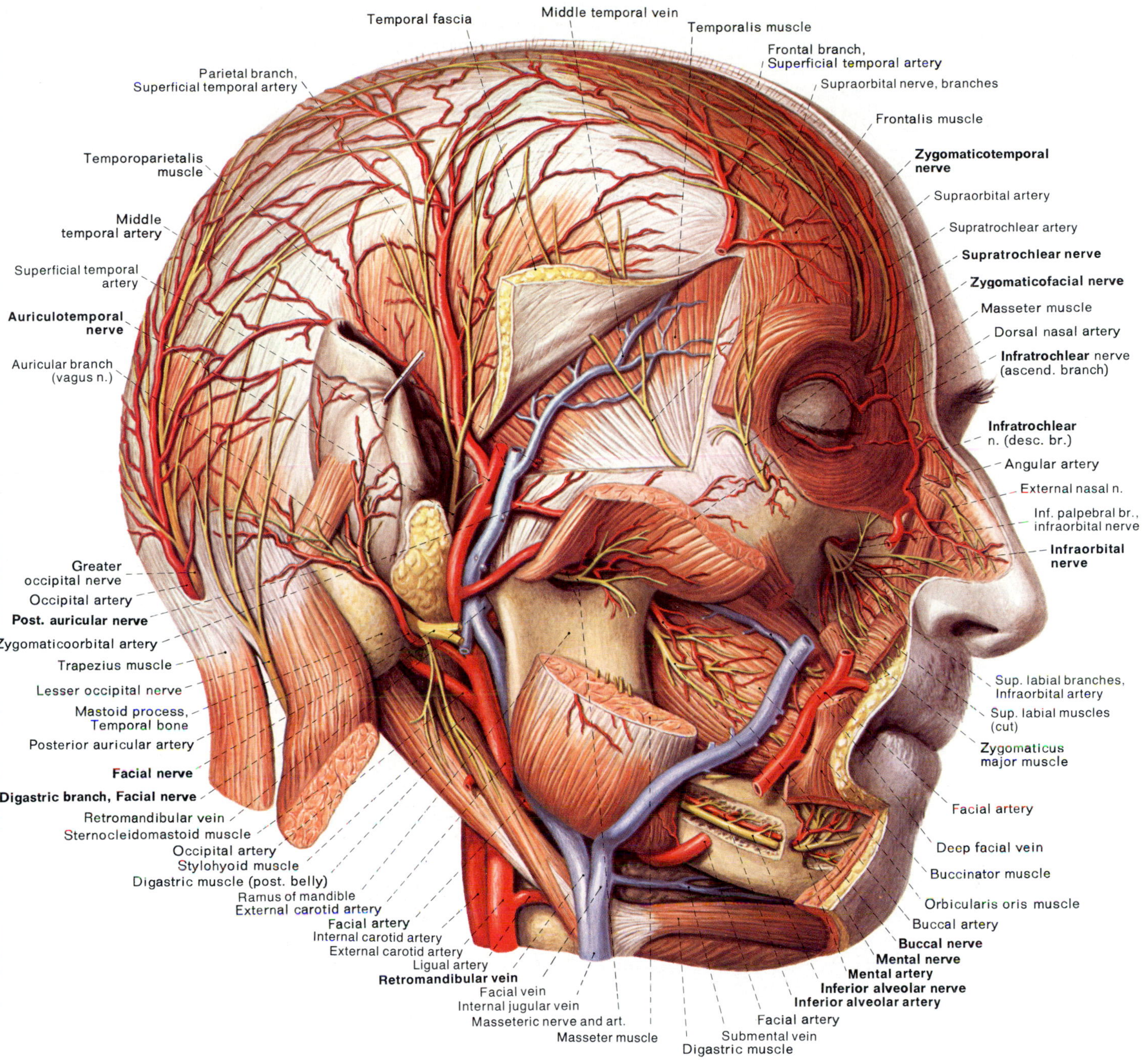

Temporal fascia
Middle temporal vein
Temporalis muscle
Frontal branch,
Superficial temporal artery
Supraorbital nerve, branches
Frontalis muscle
Zygomaticotemporal nerve
Supraorbital artery
Supratrochlear artery
Supratrochlear nerve
Zygomaticofacial nerve
Masseter muscle
Dorsal nasal artery
Infratrochlear nerve (ascend. branch)
Infratrochlear n. (desc. br.)
Angular artery
External nasal n.
Inf. palpebral br., infraorbital nerve
Infraorbital nerve
Sup. labial branches, Infraorbital artery
Sup. labial muscles (cut)
Zygomaticus major muscle
Facial artery
Deep facial vein
Buccinator muscle
Orbicularis oris muscle
Buccal artery
Buccal nerve
Mental nerve
Mental artery
Inferior alveolar nerve
Inferior alveolar artery
Facial artery
Submental vein
Digastric muscle
Masseter muscle
Masseteric nerve and art.
Internal jugular vein
Facial vein
Retromandibular vein
Ligual artery
External carotid artery
Internal carotid artery
Facial artery
External carotid artery
Ramus of mandible
Digastric muscle (post. belly)
Stylohyoid muscle
Occipital artery
Sternocleidomastoid muscle
Retromandibular vein
Digastric branch, Facial nerve
Facial nerve
Posterior auricular artery
Mastoid process, Temporal bone
Lesser occipital nerve
Trapezius muscle
Zygomaticoorbital artery
Post. auricular nerve
Occipital artery
Greater occipital nerve
Auricular branch (vagus n.)
Auriculotemporal nerve
Superficial temporal artery
Middle temporal artery
Temporoparietalis muscle
Parietal branch, Superficial temporal artery

Fig. 623: Nerves and Vessels of the Superficial Head and Deeper Face

NOTE: 1) in this dissection, the temporal fascia has been cut and partially reflected. The superficial muscles on the side of the face have been removed along with the parotid gland. The main trunk of the facial nerve has been cut and its branches across the face removed. The masseter muscle has been severed and its upper half reflected upward.

2a) the supraorbital, supratrochlear, infratrochlear and external nasal branches of the ophthalmic division of the trigeminal nerve.

b) zygomaticotemporal, zygomaticofacial and infraorbital branches of the maxillary division of the trigeminal nerve.

c) auriculotemporal, masseteric, buccal, inferior alveolar and mental branches of the mandibular division of the trigeminal nerve.

3) the posterior auricular, digastric and stylohyoid branches of the facial nerve which arise from the main nerve trunk prior to its division within the parotid gland.

4) the anastomosis of arteries above and at the medial aspect of the orbit. The vessels involved include the frontal branch of the superficial temporal, the supraorbital, supratrochlear, dorsal nasal and angular arteries. Observe the palpebral branches (not labelled) supplying the upper and lower eyelids.

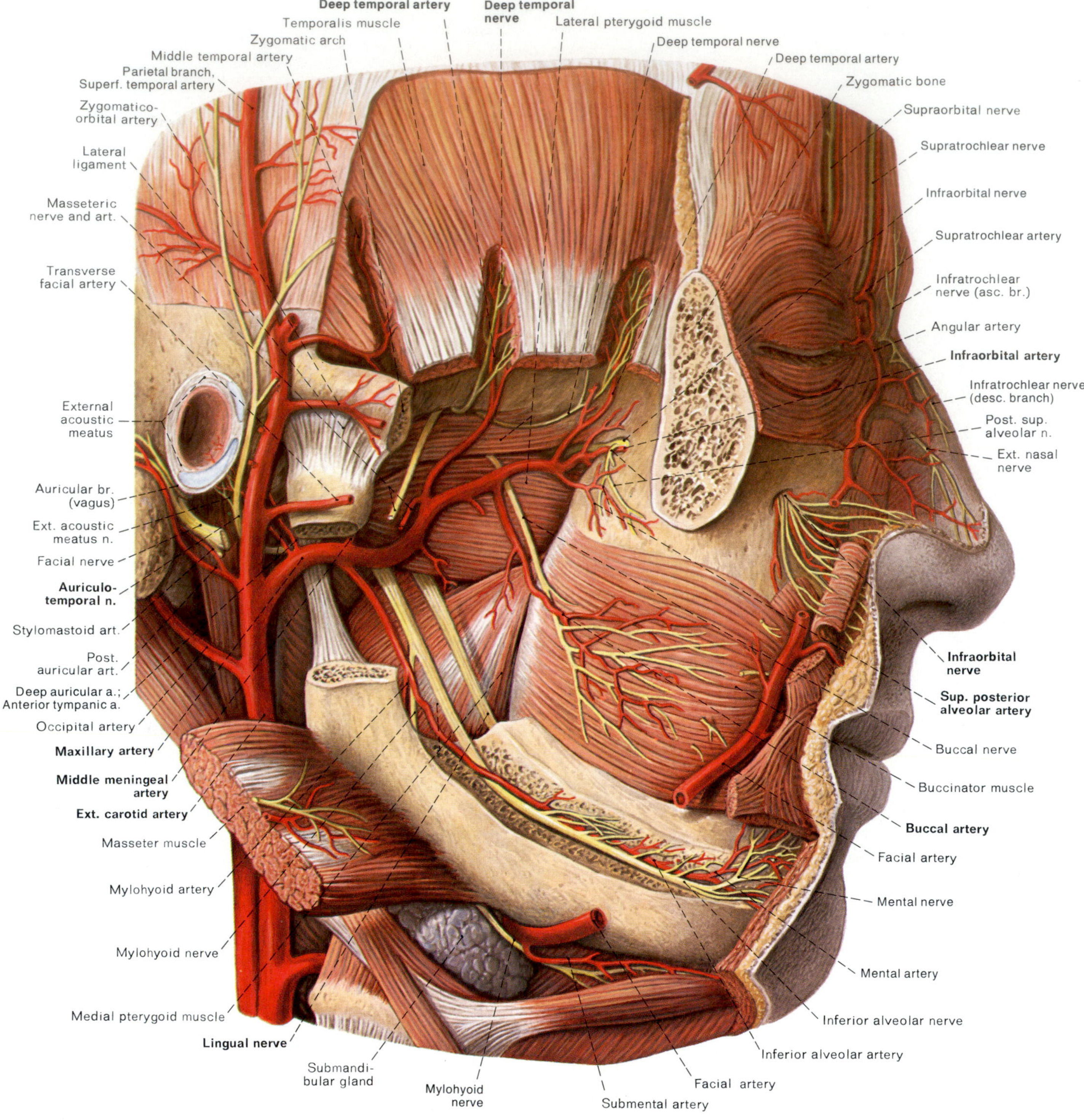

Deep temporal artery
Deep temporal nerve
Temporalis muscle
Lateral pterygoid muscle
Zygomatic arch
Deep temporal nerve
Middle temporal artery
Deep temporal artery
Parietal branch,
Superf. temporal artery
Zygomatic bone
Zygomatico-orbital artery
Supraorbital nerve
Supratrochlear nerve
Lateral ligament
Infraorbital nerve
Masseteric nerve and art.
Supratrochlear artery
Transverse facial artery
Infratrochlear nerve (asc. br.)
Angular artery
Infraorbital artery
External acoustic meatus
Infratrochlear nerve (desc. branch)
Post. sup. alveolar n.
Ext. nasal nerve
Auricular br. (vagus)
Ext. acoustic meatus n.
Facial nerve
Auriculo-temporal n.
Stylomastoid art.
Post. auricular art.
Infraorbital nerve
Deep auricular a.; Anterior tympanic a.
Sup. posterior alveolar artery
Occipital artery
Buccal nerve
Maxillary artery
Middle meningeal artery
Buccinator muscle
Ext. carotid artery
Buccal artery
Masseter muscle
Facial artery
Mylohyoid artery
Mental nerve
Mylohyoid nerve
Mental artery
Medial pterygoid muscle
Inferior alveolar nerve
Lingual nerve
Inferior alveolar artery
Submandibular gland
Mylohyoid nerve
Facial artery
Submental artery

Fig. 624: The Infratemporal Region and the Maxillary Artery

NOTE: 1) the zygomatic arch and a portion of the ramus of the mandible have been removed and the temporalis and masseter muscles have been cut and partially reflected. The pterygoid venous plexus has been removed revealing the maxillary artery and its branches.

2) the infratemporal fossa lies deep to the zygomatic arch and posterior to the maxilla. It contains the medial and lateral pterygoid muscles, the inferior part of the temporalis muscle, the maxillary artery, the pterygoid plexus of veins (not shown) and the mandibular division of the trigeminal nerve.

3) the maxillary artery and a number of its branches. Seen in this dissection are the following branches of the maxillary artery: a) deep auricular, b) anterior tympanic, c) inferior alveolar, d) middle meningeal, e) masseteric (cut), f) deep temporal, g) pterygoid (not labelled), h) buccal, i) posterior superior alveolar, j) infraorbital. Branches of the maxillary artery not seen from this view are the greater palatine, the artery of the pterygoid canal, pharyngeal and the sphenopalatine.

4) the auriculotemporal, lingual, inferior alveolar, mylohyoid, masseteric and deep temporal branches of the mandibular division of the trigeminal nerve. Observe the course of the inferior alveolar nerve, accompanied by the inferior alveolar artery within the mandible.

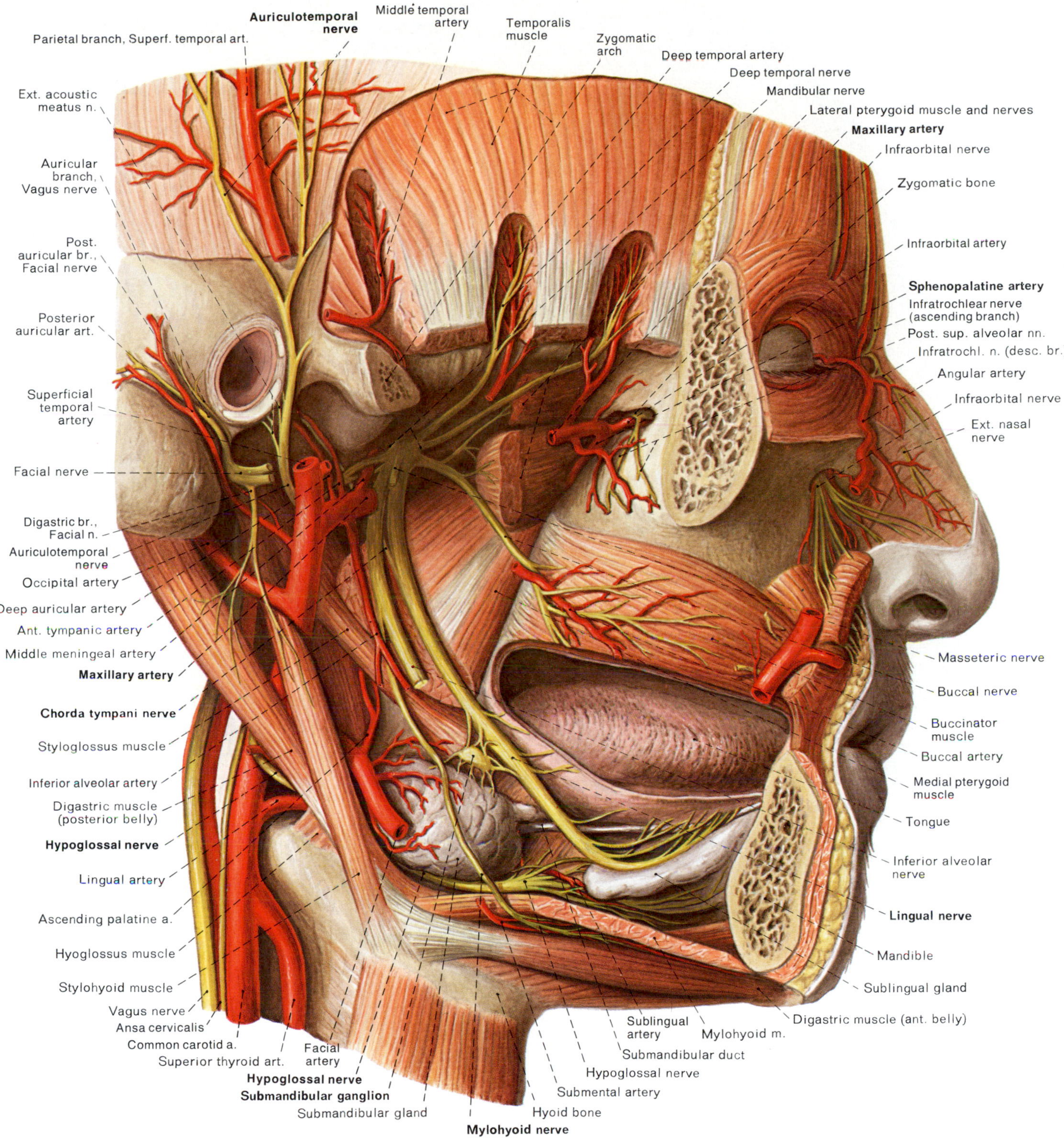
Auriculotemporal nerve
Middle temporal artery
Temporalis muscle
Zygomatic arch
Parietal branch, Superf. temporal art.
Deep temporal artery
Deep temporal nerve
Mandibular nerve
Lateral pterygoid muscle and nerves
Maxillary artery
Infraorbital nerve
Zygomatic bone
Infraorbital artery
Sphenopalatine artery
Infratrochlear nerve (ascending branch)
Post. sup. alveolar nn.
Infratrochl. n. (desc. br.)
Angular artery
Infraorbital nerve
Ext. nasal nerve
Ext. acoustic meatus n.
Auricular branch, Vagus nerve
Post. auricular br., Facial nerve
Posterior auricular art.
Superficial temporal artery
Facial nerve
Digastric br., Facial n.
Auriculotemporal nerve
Occipital artery
Deep auricular artery
Ant. tympanic artery
Middle meningeal artery
Maxillary artery
Chorda tympani nerve
Styloglossus muscle
Inferior alveolar artery
Digastric muscle (posterior belly)
Hypoglossal nerve
Lingual artery
Ascending palatine a.
Hyoglossus muscle
Stylohyoid muscle
Vagus nerve
Ansa cervicalis
Common carotid a.
Superior thyroid art.
Facial artery
Hypoglossal nerve
Submandibular ganglion
Submandibular gland
Mylohyoid nerve
Hyoid bone
Submental artery
Hypoglossal nerve
Submandibular duct
Sublingual artery
Mylohyoid m.
Digastric muscle (ant. belly)
Sublingual gland
Mandible
Lingual nerve
Inferior alveolar nerve
Tongue
Medial pterygoid muscle
Buccal artery
Buccinator muscle
Buccal nerve
Masseteric nerve

Fig. 625: The Infratemporal Region and the Branches of the Mandibular Nerve

NOTE: 1) this dissection of the deep face has removed the zygomatic arch, much of the right mandible and the lateral pterygoid muscle. A portion of the maxillary artery has been cut away along with the distal part of the inferior alveolar nerve beyond the point where the mylohyoid nerve branches.

2) the lingual nerve as it courses to the tongue. High in the infratemporal fossa, the chorda tympani nerve (which is a branch of the facial) joins the lingual. The chorda tympani not only carries special sensory fibers for taste for the anterior two-thirds of the tongue, but additionally carries the preganglionic parasympathetic fibers from the facial nerve to the submandibular ganglion.

3) the distal portion of the maxillary artery as it courses toward the sphenopalatine foramen. After giving off the infraorbital artery, the vessel passes through the foramen and enters the nasal cavity as the sphenopalatine artery, becoming the principal vessel supplying the mucosa overlying the conchae.

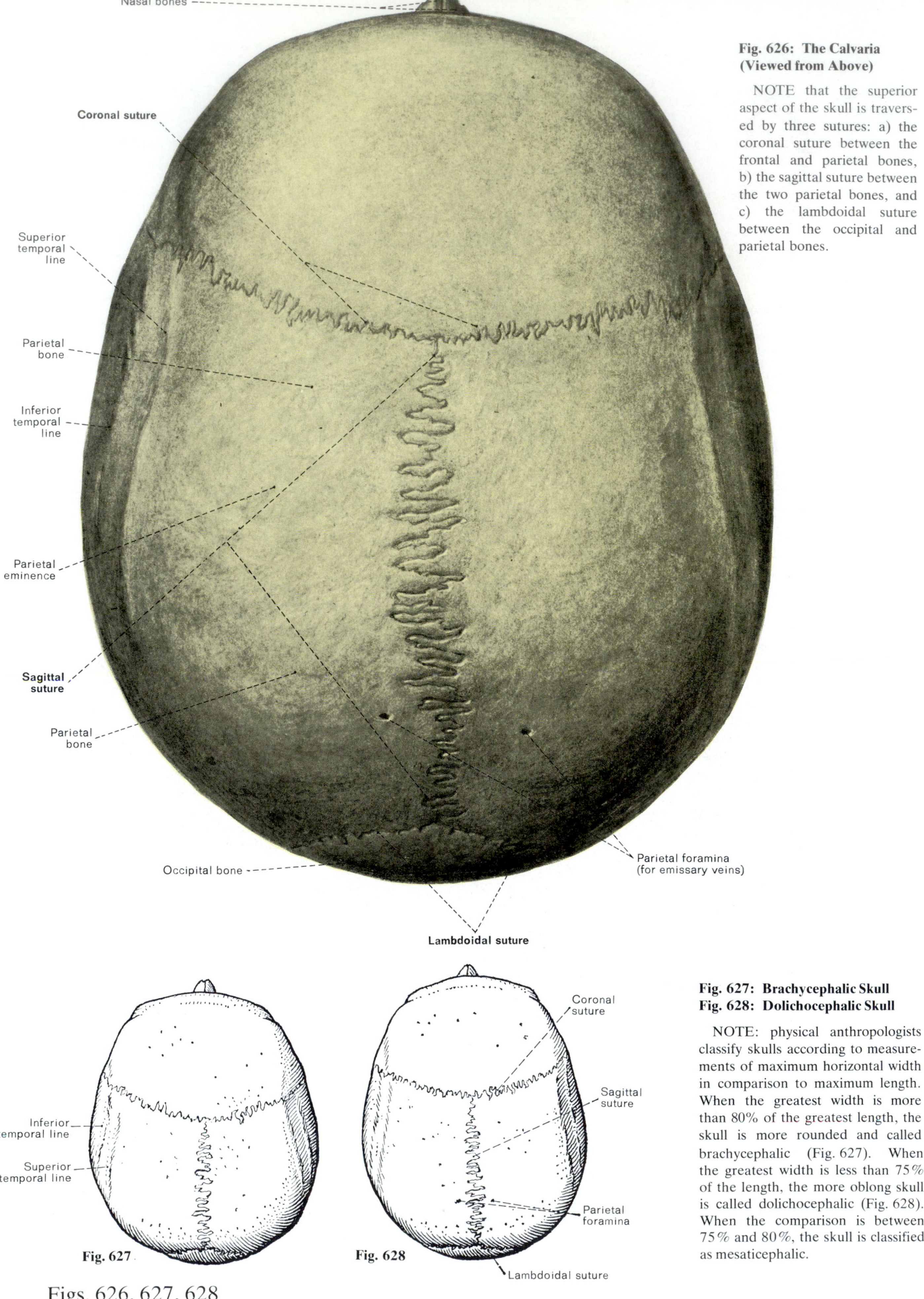

Nasal bones
Coronal suture
Superior temporal line
Parietal bone
Inferior temporal line
Parietal eminence
Sagittal suture
Parietal bone
Occipital bone
Parietal foramina (for emissary veins)
Lambdoidal suture
Coronal suture
Sagittal suture
Inferior temporal line
Superior temporal line
Parietal foramina
Lambdoidal suture
Fig. 627
Fig. 628

Fig. 626: The Calvaria (Viewed from Above)

NOTE that the superior aspect of the skull is traversed by three sutures: a) the coronal suture between the frontal and parietal bones, b) the sagittal suture between the two parietal bones, and c) the lambdoidal suture between the occipital and parietal bones.

Fig. 627: Brachycephalic Skull
Fig. 628: Dolichocephalic Skull

NOTE: physical anthropologists classify skulls according to measurements of maximum horizontal width in comparison to maximum length. When the greatest width is more than 80% of the greatest length, the skull is more rounded and called brachycephalic (Fig. 627). When the greatest width is less than 75% of the length, the more oblong skull is called dolichocephalic (Fig. 628). When the comparison is between 75% and 80%, the skull is classified as mesaticephalic.

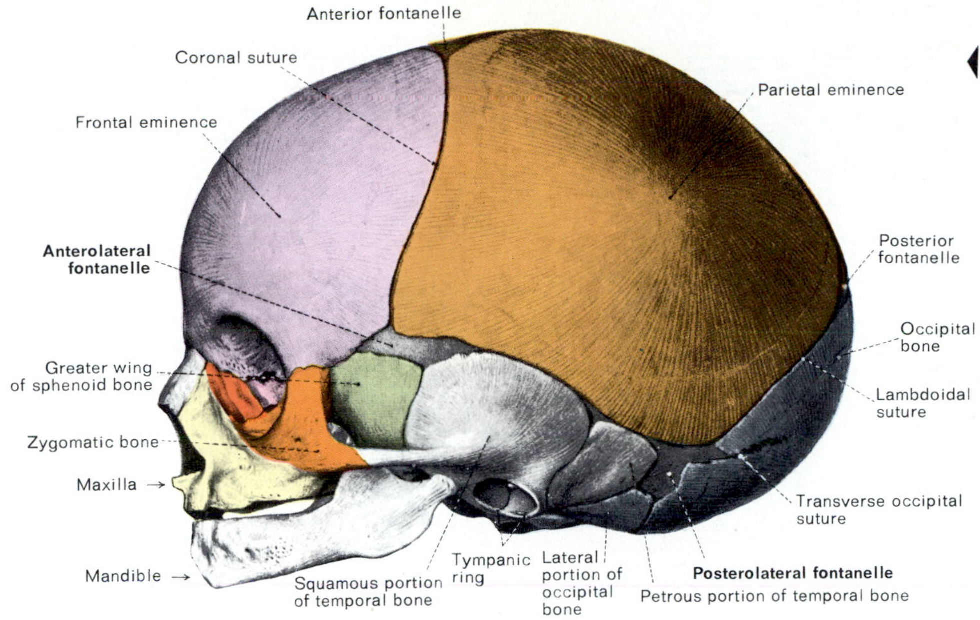
Anterior fontanelle
Coronal suture
Parietal eminence
Frontal eminence
Posterior fontanelle
Anterolateral fontanelle
Occipital bone
Greater wing of sphenoid bone
Lambdoidal suture
Zygomatic bone
Maxilla →
Transverse occipital suture
Mandible →
Tympanic ring
Squamous portion of temporal bone
Lateral portion of occipital bone
Posterolateral fontanelle
Petrous portion of temporal bone

Fig. 629: The Skull at Birth (Lateral View)

NOTE: 1) ossification of the maturing flat bones of the skull is accomplished by the intramembranous process of bone formation. At birth this process is incomplete, thereby leaving softened membranous sites between the growing bones.

2) the nature of the skull just prior to birth is of some benefit, however, since the mobility of the bones permits some changes in its shape, as might be required during the birth process.

3) the soft sites on the skull of the newborn infant are called fontanelles. From this lateral view can be seen at least two such fontanelles, the anterolateral (or sphenoid) which is at the pterion, and the posterolateral (or mastoid) at the asterion.

Fig. 630: The Skull at Birth (Seen from Above)

NOTE: 1) the largest of the fontanelles at birth is the anterior fontanelle located at the bregma and interconnecting the frontal and parietal bones. It is approximately diamond-shaped and is situated at the junction of the coronal and sagittal sutures.

2) following the sagittal suture to its junction with the occipital bone will locate the posterior fontanelle (at the lambda). This is generally triangular in shape and is relatively small at birth.

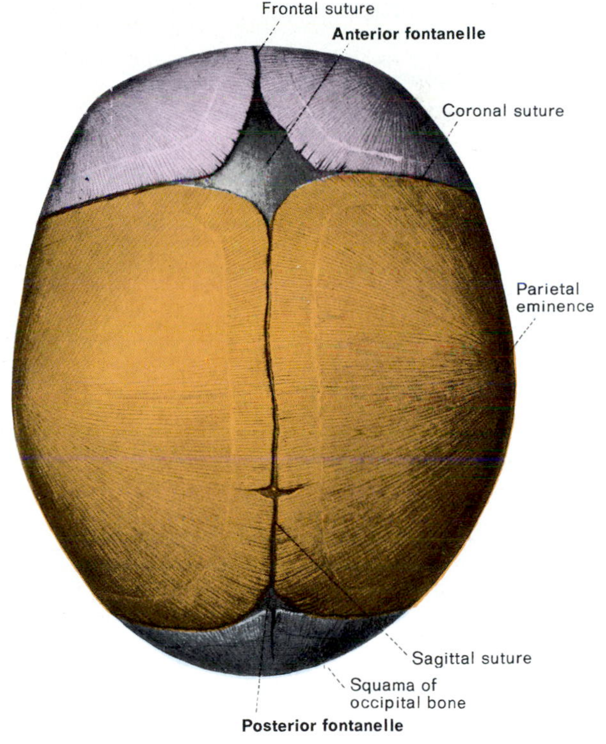
Frontal suture
Anterior fontanelle
Coronal suture
Parietal eminence
Sagittal suture
Squama of occipital bone
Posterior fontanelle

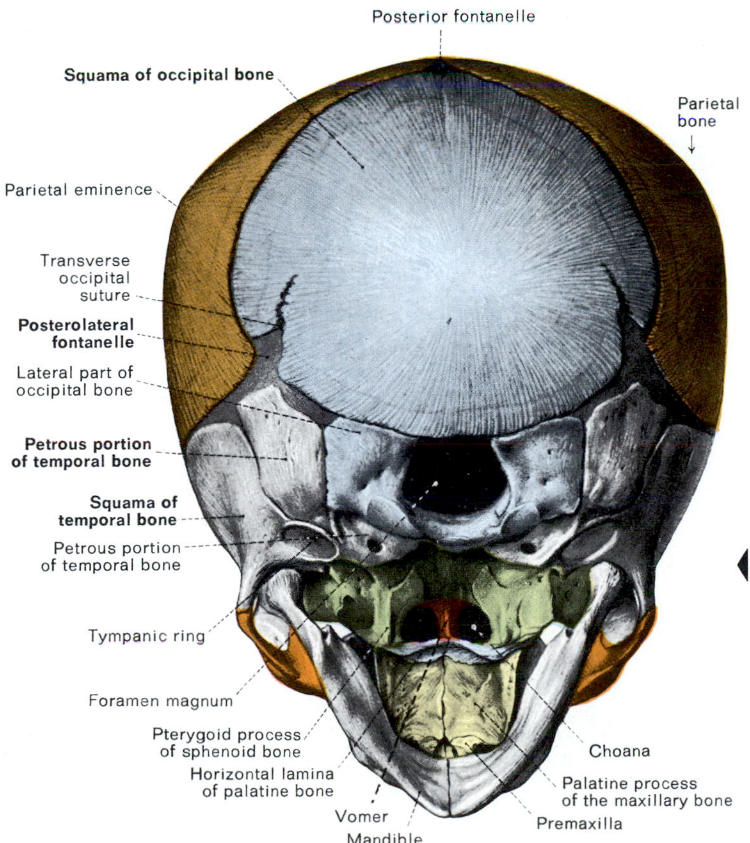
Posterior fontanelle
Squama of occipital bone
Parietal bone
Parietal eminence
Transverse occipital suture
Posterolateral fontanelle
Lateral part of occipital bone
Petrous portion of temporal bone
Squama of temporal bone
Petrous portion of temporal bone
Tympanic ring
Foramen magnum
Pterygoid process of sphenoid bone
Choana
Horizontal lamina of palatine bone
Palatine process of the maxillary bone
Vomer
Premaxilla
Mandible

Fig. 631: The Skull at Birth (Posterior-Inferior View)

NOTE: 1) the separate ossification of the petrous and squamous portions of the temporal bone as well as the basilar and squamous portions of the occipital bone. The posterolateral fontanelles are found at the articulation of the occipital, temporal and parietal bones.

2) the skull is large at birth in comparison to the size of the rest of the body, however, the facial bones (see Fig. 629) are still rudimentary and not well developed. The teeth have yet to erupt, and the sinuses and nasal cavity are small.

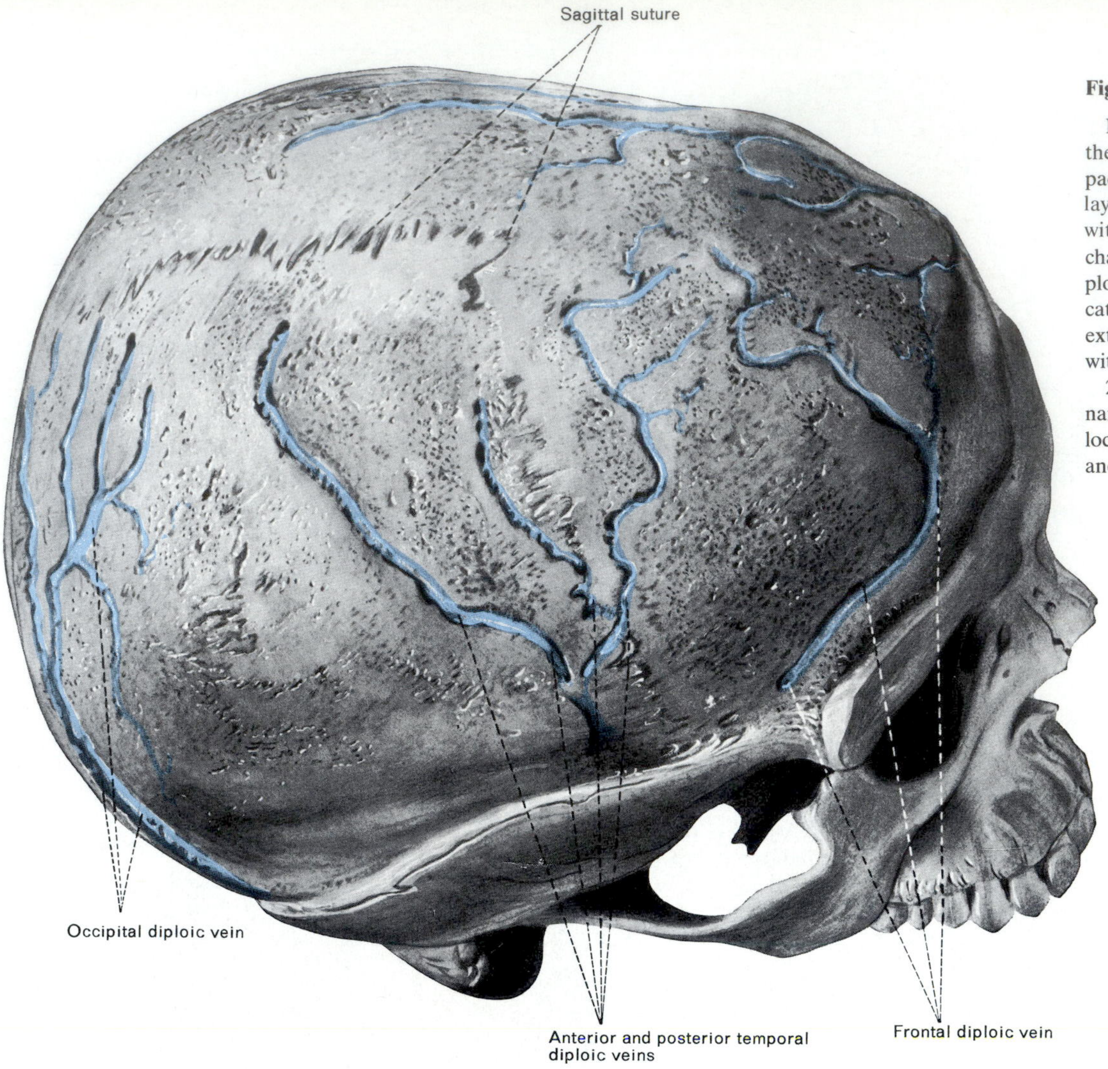

Sagittal suture
Occipital diploic vein
Anterior and posterior temporal diploic veins
Frontal diploic vein

Fig. 632: The Diploic Veins

NOTE: 1) by removal of the outermost table of compact bone, a more spongy layer of bone is encountered, within which course venous channels known as the diploic veins. These communicate with the scalp on the exterior and the dural sinuses within the skull.

2) the diploic veins are named according to their location: frontal, temporal and occipital.

Fig. 633: The Scalp, Skull, Meninges and Brain

NOTE: 1) this frontal section through the cranium and upper cerebrum depicts the bony and soft coverings of the brain. The veins and dural sinuses are shown in blue. The layers of the scalp and the bony tissue of the skull lie superficial to the dura mater, arachnoid and pia mater coverings of the neural tissue of the brain.

2) the arachnoid granulations which project into the dural sinuses. These tufts of arachnoid lie next to the endothelium of the sinuses and allow the passage of cerebrospinal fluid from the subarachnoid space into the venous system.

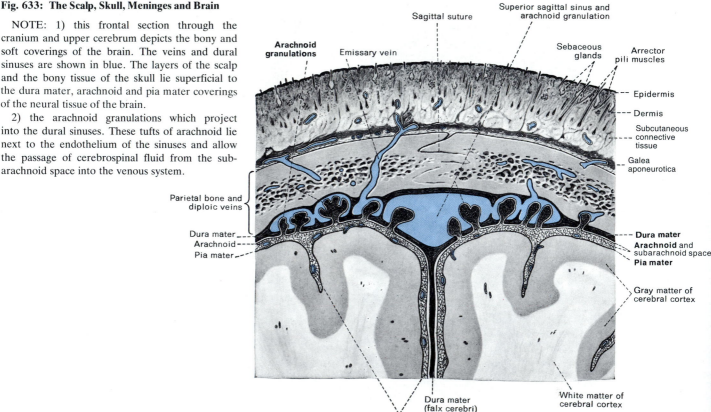

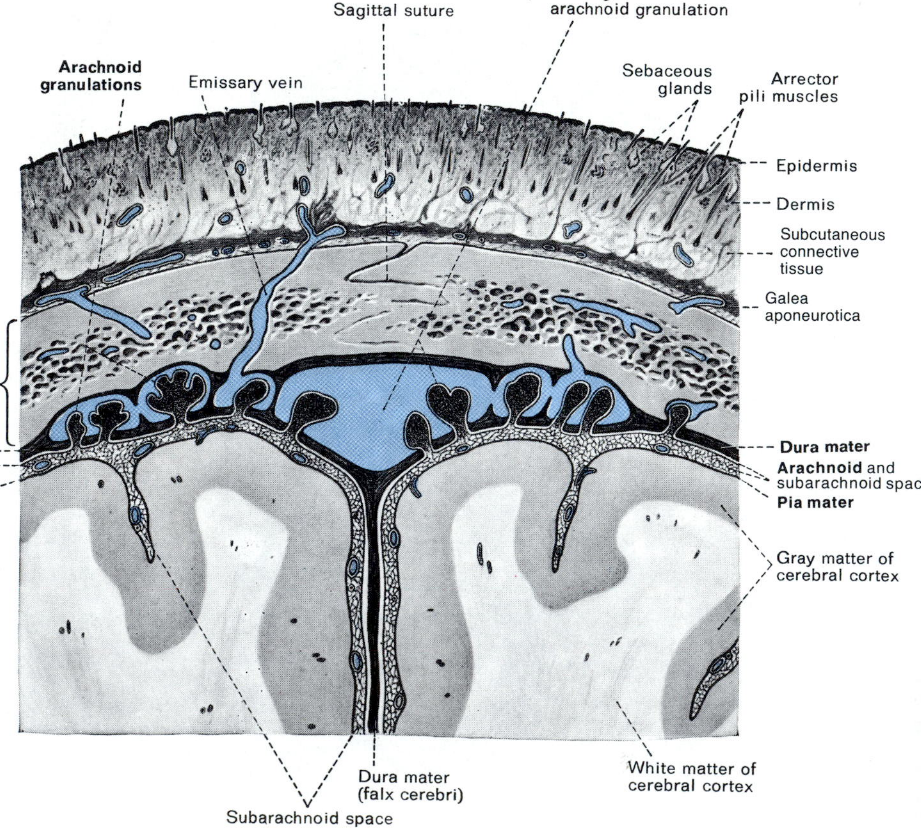
Sagittal suture
Superior sagittal sinus and arachnoid granulation
Arachnoid granulations
Emissary vein
Sebaceous glands
Arrector pili muscles
Epidermis
Dermis
Subcutaneous connective tissue
Galea aponeurotica
Parietal bone and diploic veins
Dura mater
Arachnoid
Pia mater
Dura mater
Arachnoid and subarachnoid space
Pia mater
Gray matter of cerebral cortex
White matter of cerebral cortex
Dura mater (falx cerebri)
Subarachnoid space

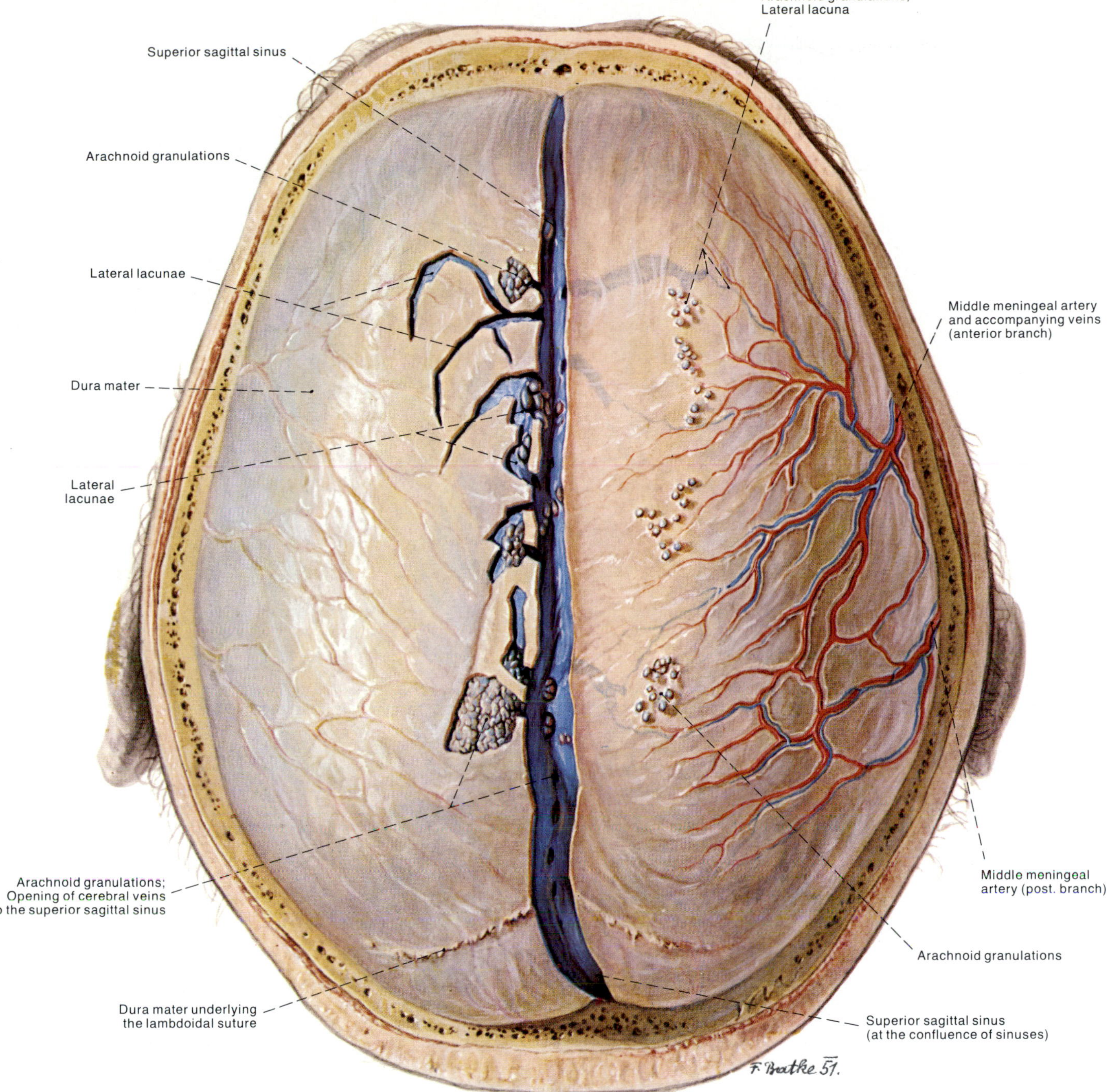
Arachnoid granulations;
Lateral lacuna
Superior sagittal sinus
Arachnoid granulations
Lateral lacunae
Middle meningeal artery
and accompanying veins
(anterior branch)
Dura mater
Lateral
lacunae
Arachnoid granulations;
Opening of cerebral veins
into the superior sagittal sinus
Middle meningeal
artery (post. branch)
Arachnoid granulations
Dura mater underlying
the lambdoidal suture
Superior sagittal sinus
(at the confluence of sinuses)
F. Batke 51.

Fig. 634: The Surface of the Dura Mater with the Superior Sagittal Sinus Opened; Viewed from Above

NOTE: 1) that the skull cap (also called calvaria) has been removed. This was done by making a circular saw cut around the skull, but through the bone only and by carefully separating the dura mater from the internal surface of the skull cap.

2) that although the dura mater is often described as a two layered structure (an internal or *meningeal layer* and an external or *periosteal layer*), the two layers are so closely fused that they are inseparable throughout much of their expanse. After removal of the skull cap, the "two layers" are stripped from the inner surface of the skull as a single membrane.

3) that in some regions the meningeal and periosteal layers are separate, and in this way they form the cavities of the venous sinsus in the dura mater. Observe that in this dissection the longitudinally oriented superior sagittal sinus has been opened, as have a number of lateral venous lacunae that communicate with this venous dural sinus.

4) that the outer or periosteal layer of the dura mater is continuous through foramina in the skull and through the sutures between the bones of the skull with the periosteum that lines the outer surface of the skull and which is called the pericranium (see the dura mater at the lambdoidal suture). Anteriorly, the periosteal layer of dura is continuous through the superior orbital fissure with the periosteum that lines the bones forming the orbital cavity.

5) the *arachnoid granulations*. These are elevated masses that grow in size from infancy through childhood, so that they eventually cause some bone resorption and thereby form visible pits on the inner surface of the skull cap. The arachnoid granulations are collections of small projections of the arachnoid membrane which extend through openings in the dura mater (see Fig. 633). These projections are called the *arachnoid villi*, and they appear as diverticula of the subarachnoid space into the venous sinus. Cerebrospinal fluid appears to pass by osmosis from the subarachnoid space through the arachnoid villi into the venous blood of the dural sinus.

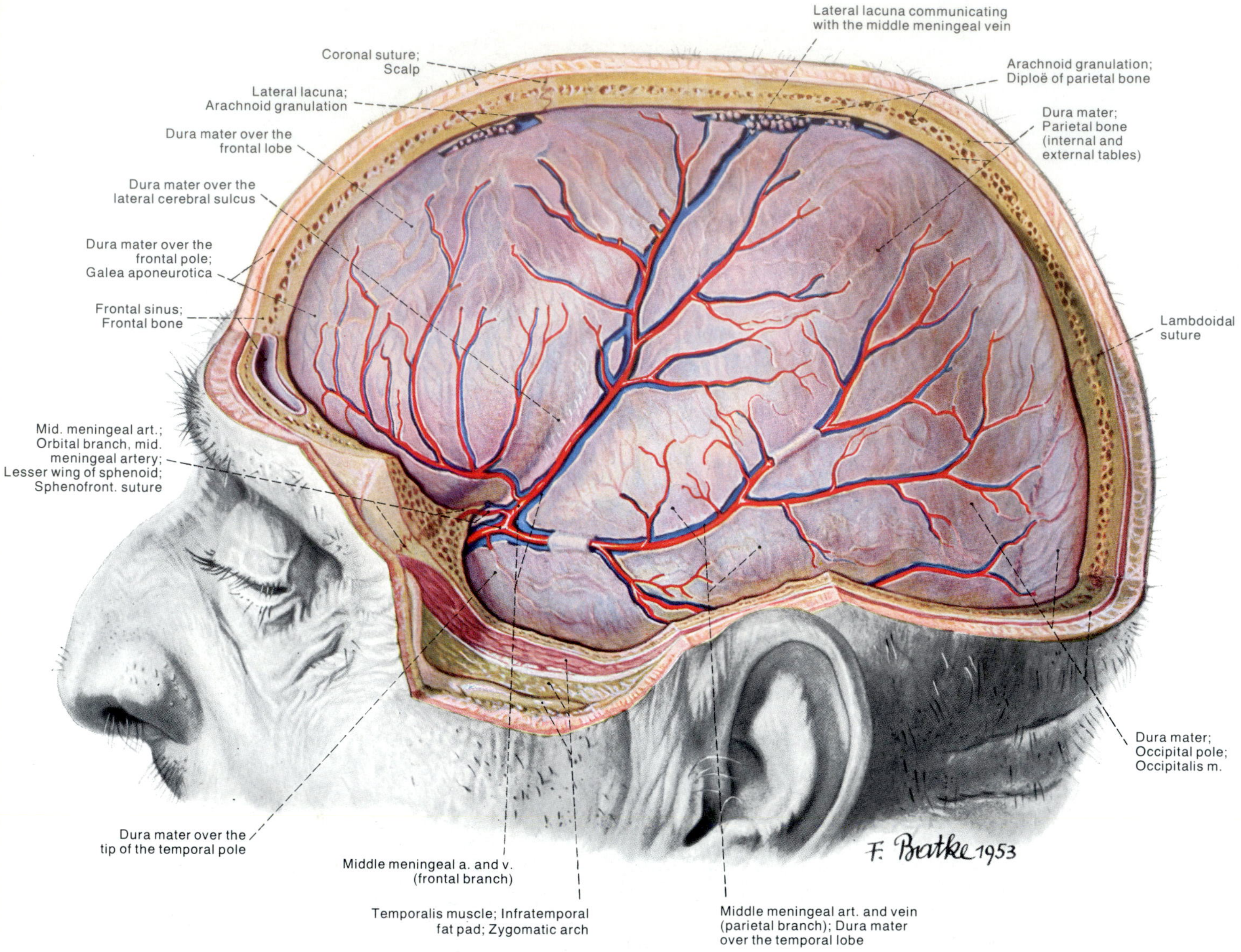

Lateral lacuna communicating with the middle meningeal vein
Coronal suture; Scalp
Arachnoid granulation; Diploë of parietal bone
Lateral lacuna; Arachnoid granulation
Dura mater over the frontal lobe
Dura mater; Parietal bone (internal and external tables)
Dura mater over the lateral cerebral sulcus
Dura mater over the frontal pole; Galea aponeurotica
Frontal sinus; Frontal bone
Lambdoidal suture
Mid. meningeal art.; Orbital branch, mid. meningeal artery; Lesser wing of sphenoid; Sphenofront. suture
Dura mater; Occipital pole; Occipitalis m.
Dura mater over the tip of the temporal pole
F. Batke 1953
Middle meningeal a. and v. (frontal branch)
Temporalis muscle; Infratemporal fat pad; Zygomatic arch
Middle meningeal art. and vein (parietal branch); Dura mater over the temporal lobe

Fig. 635: The Dura Mater and the Middle Meningeal Vessels, Viewed Following a Lateral Exposure of the Cranial Cavity

NOTE: 1) that most of the left half of the skull has been removed in order to expose the dura mater covering the left cerebral hemisphere and the middle meningeal artery and veins. The middle meningeal artery arises in the infratemporal fossa from the first part of the maxillary artery (see Figs. 624 and 625), and it ascends between the lateral pterygoid muscle and the sphenomandibular ligament directly to the foramen spinosum at the base of the skull.

2) that the middle meningeal vessels may be located deep to the squamous part of the temporal bone just above the mid-point of the zygomatic arch. A short distance (1 to 2 cm) above this site, the vessels divide into frontal (or anterior) and parietal (or posterior) branches. The dura mater requires only a small amount of blood. Most of the blood in these vessels supplies the bones of the skull, as do meningeal branches from several other arteries.

3) that the cerebral dura is supplied by sensory nerve fibers from all three divisions of the trigeminal nerve and also from the upper three cervical spinal nerves.

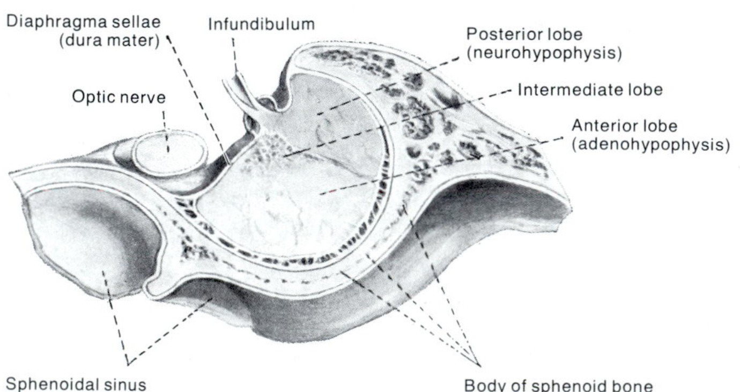
Diaphragma sellae (dura mater)
Infundibulum
Posterior lobe (neurohypophysis)
Optic nerve
Intermediate lobe
Anterior lobe (adenohypophysis)
Sphenoidal sinus
Body of sphenoid bone

Fig. 636: A Median Sagittal Section Through the Pituitary Gland and the Sella Turcica of the Sphenoid Bone

NOTE: 1) that the pituitary gland (or hypophysis cerebri) lies within the cup-shaped sella turcica or pituitary (hypophyseal) fossa of the sphenoid bone and that this fossa is lined by dura mater. Observe that the pituitary gland is covered superiorly by a reflection of dura mater called the diaphragma sellae and that this horizontal sheet of dura separates the sella turcica and the pituitary gland (except for the infundibulum) from the hypothalamus at the base of the brain (see Figs. 637 to 639 and 642).

2) that the three lobes of the pituitary gland (anterior, intermediate, and posterior) form a single organ that lies below and often slightly behind the optic chiasma.

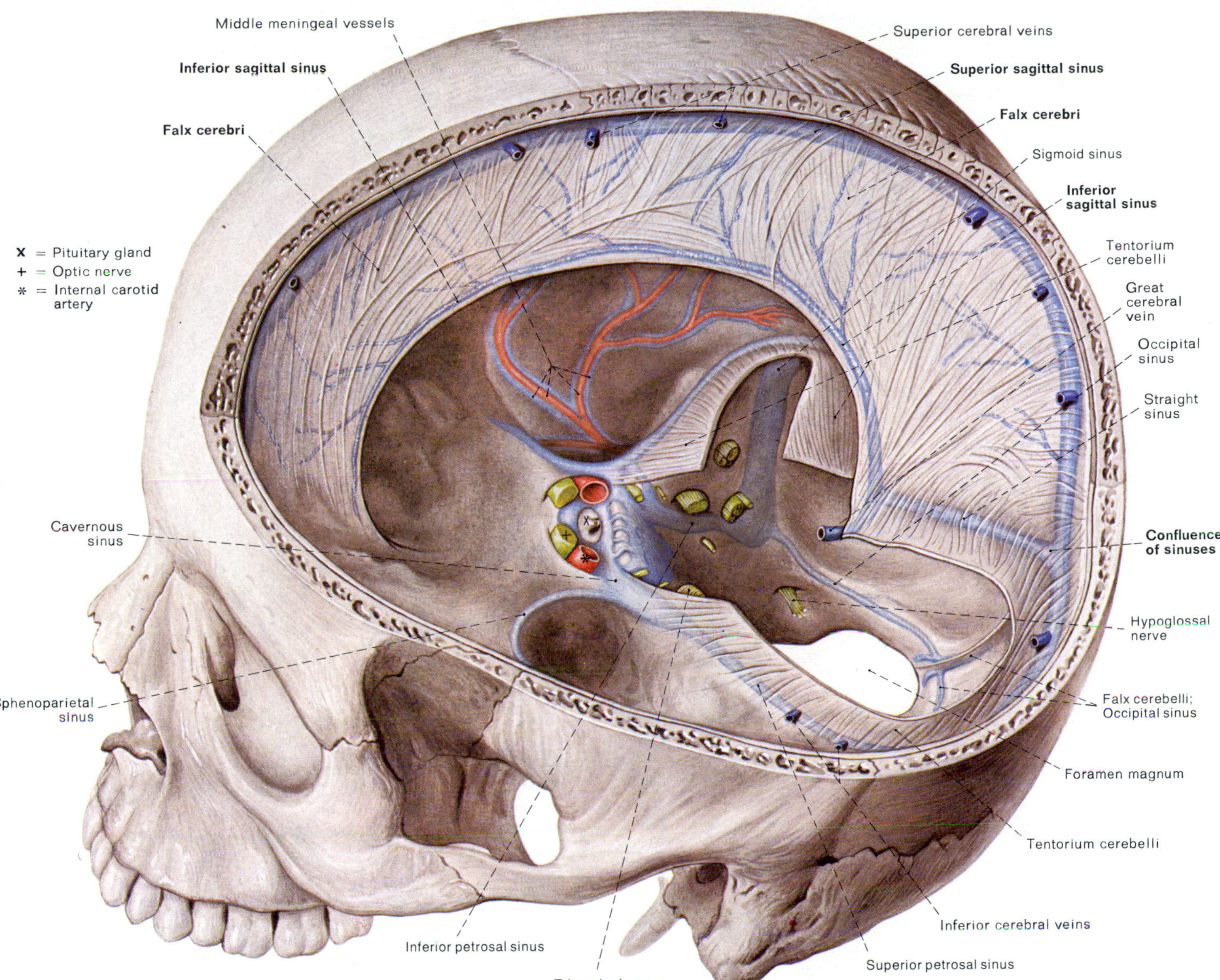
Middle meningeal vessels
Inferior sagittal sinus
Falx cerebri
x = Pituitary gland
+ = Optic nerve
* = Internal carotid artery
Cavernous sinus
Sphenoparietal sinus
Inferior petrosal sinus
Trigeminal nerve
Superior petrosal sinus
Inferior cerebral veins
Tentorium cerebelli
Foramen magnum
Falx cerebelli; Occipital sinus
Hypoglossal nerve
Confluence of sinuses
Straight sinus
Occipital sinus
Great cerebral vein
Tentorium cerebelli
Inferior sagittal sinus
Sigmoid sinus
Falx cerebri
Superior sagittal sinus
Superior cerebral veins

Fig. 637: The Intracranial Dura Mater and the Dural Sinuses

NOTE: 1) with the left skull opened and the brain removed, the reflections of the dura mater and its venous sinuses are exposed. The sinuses are colored blue while the arteries are red. Most of the left tentorium cerebelli and a portion of the right were cut away to open the posterior cranial fossa.

2) the five *unpaired* dural sinuses: the superior sagittal sinus, the inferior sagittal sinus, the straight sinus. Two other unpaired sinuses (not labelled) at the base of the skull include the intercavernous sinus and the basilar sinus. These can be seen in Fig. 638.

3) that the sphenoparietal sinuses course near the posterior margin of the lesser wings of the sphenoid bone and help to form the boundary between the anterior and middle cranial fossae. Similarly, the superior petrosal sinuses course along the superior margins of the petrous portions of the temporal bones at the boundary between the middle and posterior cranial fossae.

4) the seven *paired* dural sinuses: transverse sinus, sigmoid sinus, occipital sinus, superior petrosal sinus, inferior petrosal sinus, cavernous sinus and sphenoparietal sinus. The dural sinuses consist of spaces between the two layers of dura mater which drain the cerebral blood, returning it to the internal jugular veins.

5) the sickle-shaped falx cerebri. This double layer, midline reflection of dura mater extends from the crista galli anteriorly to the tentorium cerebelli posteriorly. It also extends vertically between the two cerebral hemispheres. Within the layers of the falx, observe the superior and inferior sagittal sinuses and the straight sinus which meet at the confluence of sinuses.

6) the tentorium cerebelli is a tent-like reflection of dura mater which forms a partition separating the occipital lobes of the cerebral cortex and the surface of the cerebellum. The falx cerebelli extends vertically between the two cerebellar hemispheres.

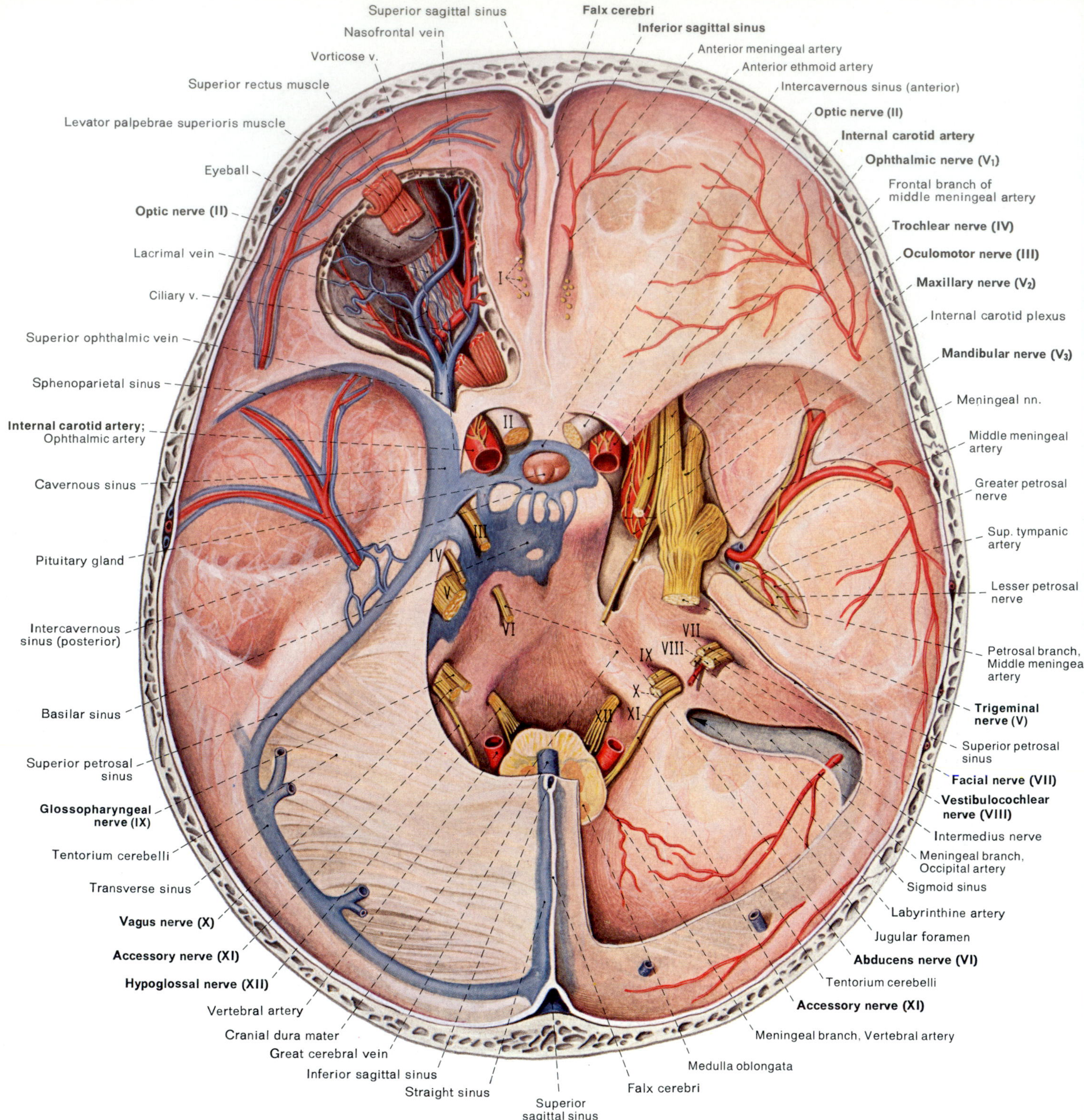

Superior sagittal sinus
Falx cerebri
Inferior sagittal sinus
Nasofrontal vein
Anterior meningeal artery
Vorticose v.
Anterior ethmoid artery
Intercavernous sinus (anterior)
Superior rectus muscle
Optic nerve (II)
Levator palpebrae superioris muscle
Internal carotid artery
Ophthalmic nerve (V_1)
Eyeball
Frontal branch of middle meningeal artery
Optic nerve (II)
Trochlear nerve (IV)
Lacrimal vein
Oculomotor nerve (III)
Maxillary nerve (V_2)
Ciliary v.
Internal carotid plexus
Superior ophthalmic vein
Mandibular nerve (V_3)
Sphenoparietal sinus
Meningeal nn.
Internal carotid artery; Ophthalmic artery
Middle meningeal artery
Cavernous sinus
Greater petrosal nerve
Sup. tympanic artery
Pituitary gland
Lesser petrosal nerve
Intercavernous sinus (posterior)
Petrosal branch, Middle meningeal artery
Basilar sinus
Trigeminal nerve (V)
Superior petrosal sinus
Superior petrosal sinus
Facial nerve (VII)
Glossopharyngeal nerve (IX)
Vestibulocochlear nerve (VIII)
Intermedius nerve
Tentorium cerebelli
Meningeal branch, Occipital artery
Transverse sinus
Sigmoid sinus
Labyrinthine artery
Vagus nerve (X)
Jugular foramen
Accessory nerve (XI)
Abducens nerve (VI)
Hypoglossal nerve (XII)
Tentorium cerebelli
Vertebral artery
Accessory nerve (XI)
Cranial dura mater
Meningeal branch, Vertebral artery
Great cerebral vein
Medulla oblongata
Inferior sagittal sinus
Falx cerebri
Straight sinus
Superior sagittal sinus
I
II
III
IV
V
VI
VII
VIII
IX
X
XI
XII

Fig. 638: The Base of the Cranial Cavity: Vessels, Nerves and Dura Mater

NOTE: 1) the base of the cranial cavity displays anterior, middle and posterior cranial fossae. The anterior fossa sustains the frontal lobes while the temporal lobes of the brain rest in the middle fossa. Posteriorly, the brain stem and the overlying cerebellum rest in the posterior fossa.

2) the dura mater and the orbital plate of the left frontal bone have been chipped away to expose the structures in the left orbit. The superior ophthalmic vein drains posteriorly and the optic nerve is seen to course from the orbit through the optic canal. Caudal to the optic foramina, observe the pituitary gland within the sella turcica.

3) the medial aspect of the middle cranial fossa shows the internal carotid artery, the 3rd, 4th, 5th and 6th cranial nerves coursing anteriorly or inferiorly toward the orbit or face and the middle meningeal artery traversing the foramen spinosum.

4) the posterior cranial fossa is marked by foramina for the last six pairs of cranial nerves. The 7th and 8th nerves pass through the internal acoustic meatus while the 9th, 10th and 11th nerves emerge through the jugular foramen. The 12th nerve courses through the hypoglossal canal in the occipital bone.

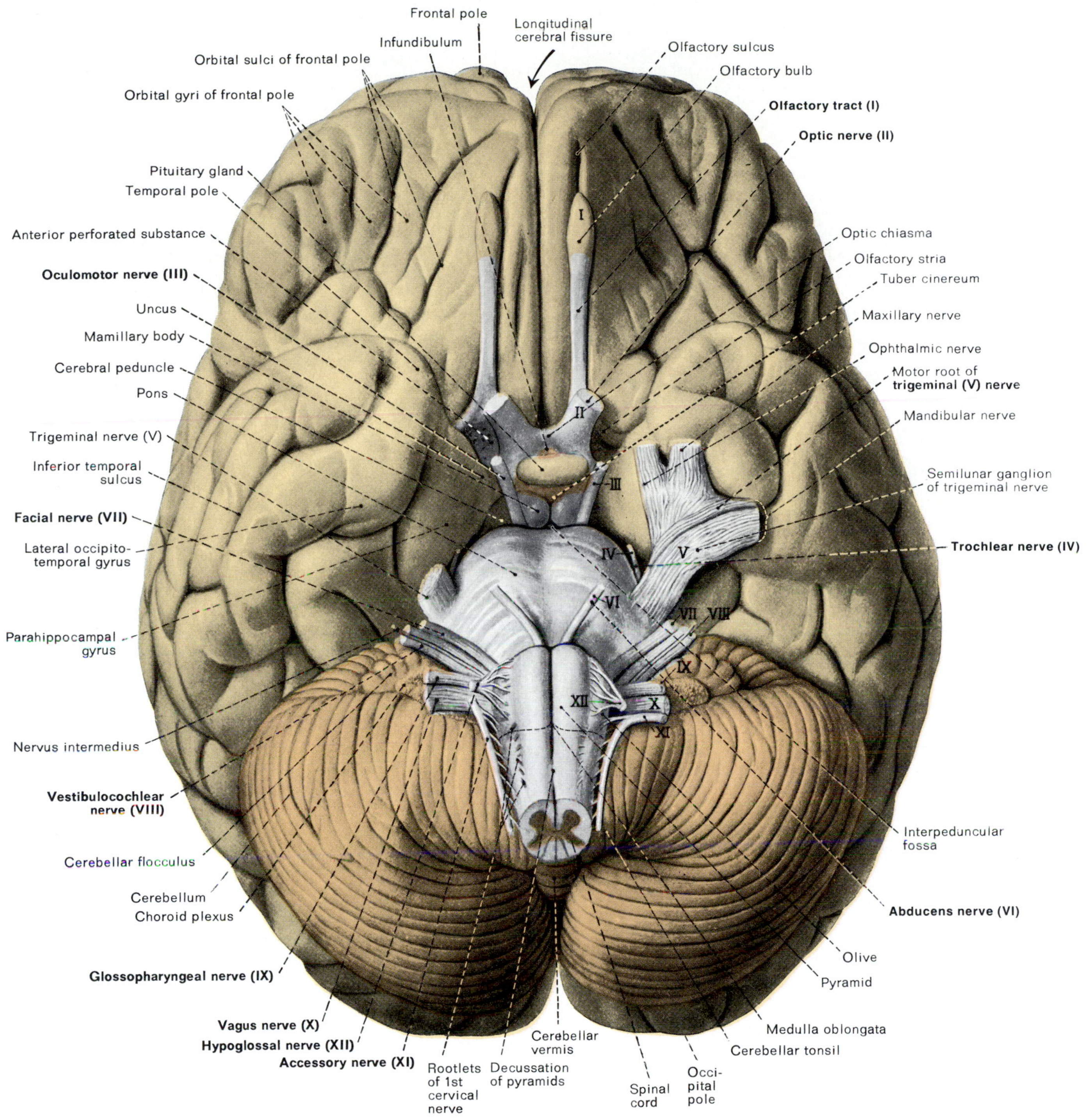
Frontal pole
Longitudinal cerebral fissure
Infundibulum
Orbital sulci of frontal pole
Orbital gyri of frontal pole
Olfactory sulcus
Olfactory bulb
Olfactory tract (I)
Optic nerve (II)
Pituitary gland
Temporal pole
Anterior perforated substance
Oculomotor nerve (III)
Uncus
Mamillary body
Cerebral peduncle
Pons
Trigeminal nerve (V)
Inferior temporal sulcus
Facial nerve (VII)
Lateral occipito-temporal gyrus
Parahippocampal gyrus
Nervus intermedius
Vestibulocochlear nerve (VIII)
Cerebellar flocculus
Cerebellum
Choroid plexus
Glossopharyngeal nerve (IX)
Vagus nerve (X)
Hypoglossal nerve (XII)
Accessory nerve (XI)
Rootlets of 1st cervical nerve
Decussation of pyramids
Cerebellar vermis
Spinal cord
Occi-pital pole
Cerebellar tonsil
Medulla oblongata
Pyramid
Olive
Abducens nerve (VI)
Interpeduncular fossa
Trochlear nerve (IV)
Semilunar ganglion of trigeminal nerve
Mandibular nerve
Motor root of trigeminal (V) nerve
Ophthalmic nerve
Maxillary nerve
Tuber cinereum
Olfactory stria
Optic chiasma
I
II
III
IV
V
VI
VII VIII
IX
X
XI
XII

Fig. 639: Ventral View of the Brain Showing the Origins of the Cranial Nerves

NOTE: 1) the cranial nerves are numbered sequentially with Roman numerals. The olfactory tracts and optic nerves (I and II) subserve receptors of special sense in the nose and eye and, as cranial nerve trunks, they attach to the base of the forebrain in contrast to all the other cranial nerves which attach at midbrain, pontine or medullary levels.

2) the oculomotor (III), trochlear (IV) and abducens (VI) nerves are motor nerves to the extraocular muscles. While the trigeminal nerve (V) is the largest of all the cranial nerves, the trochlear is the smallest. The abducens nerve emerges from the brainstem at the pontomedullary junction at about the same level, but somewhat more laterally, as the attachments of the facial (VII) and vestibulocochlear (VIII) nerves.

3) the glossopharyngeal (IX) and vagus (X) nerves emerge from the medulla laterally, in a line roughly comparable to the spinal and medullary portions of the accessory nerve (XI). In contrast, the hypoglossal (XII) nerve rootlets emerge from the ventral medulla in a line consistent with the motor ventral rootlets of the cervical segments of the spinal cord.

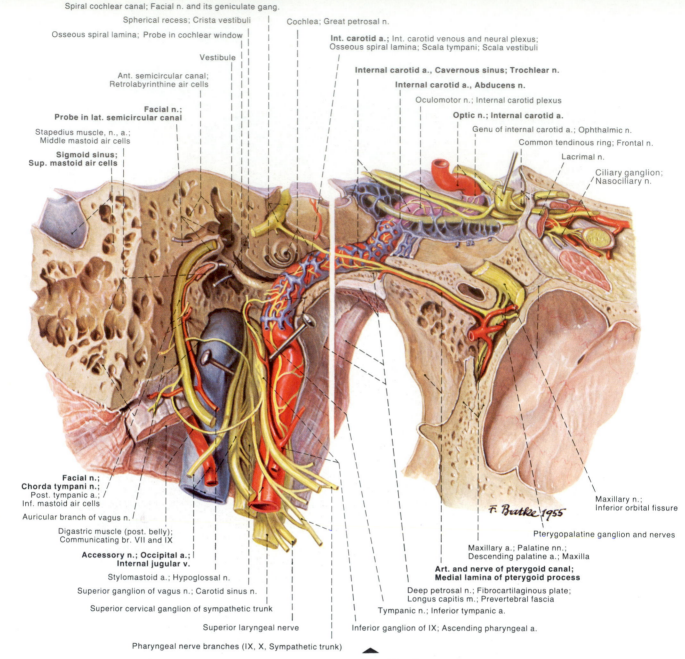

Spiral cochlear canal; Facial n. and its geniculate gang.
Spherical recess; Crista vestibuli
Osseous spiral lamina; Probe in cochlear window
Cochlea; Great petrosal n.
Int. carotid a.; Int. carotid venous and neural plexus; Osseous spiral lamina; Scala tympani; Scala vestibuli
Vestibule
Ant. semicircular canal; Retrolabyrinthine air cells
Internal carotid a., Cavernous sinus; Trochlear n.
Internal carotid a., Abducens n.
Oculomotor n.; Internal carotid plexus
Facial n.; Probe in lat. semicircular canal
Optic n.; Internal carotid a.
Genu of internal carotid a.; Ophthalmic n.
Stapedius muscle, n., a.; Middle mastoid air cells
Common tendinous ring; Frontal n.
Sigmoid sinus; Sup. mastoid air cells
Lacrimal n.
Ciliary ganglion; Nasociliary n.
F. Batke 1955
Facial n.; Chorda tympani n.; Post. tympanic a.; Inf. mastoid air cells
Maxillary n.; Inferior orbital fissure
Auricular branch of vagus n.
Digastric muscle (post. belly); Communicating br. VII and IX
Pterygopalatine ganglion and nerves
Accessory n.; Occipital a.; Internal jugular v.
Maxillary a.; Palatine nn.; Descending palatine a.; Maxilla
Art. and nerve of pterygoid canal; Medial lamina of pterygoid process
Stylomastoid a.; Hypoglossal n.
Superior ganglion of vagus n.; Carotid sinus n.
Deep petrosal n.; Fibrocartilaginous plate; Longus capitis m.; Prevertebral fascia
Superior cervical ganglion of sympathetic trunk
Tympanic n.; Inferior tympanic a.
Superior laryngeal nerve
Inferior ganglion of IX; Ascending pharyngeal a.
Pharyngeal nerve branches (IX, X, Sympathetic trunk)

Fig. 640: The Petrous, Cavernous and Cerebral Portions of the Internal Carotid Artery

NOTE: 1) a vertical section oriented along the superior margin of the petrous portion of the right temporal bone has been made, thereby opening the carotid and facial canals and the inner ear. More anteriorly, the section goes through the sphenoid bone, opening its pterygoid canal and meeting the petrous section in a V-shaped manner.

2) that the *petrous portion* of the internal carotid ascends vertically in the carotid canal and then curves medially and anteriorly. Surrounded by a plexus of small veins and sympathetic nerve fibers, the *cavernous portion* of the artery ascends into the cavernous sinus and curves forward in the carotid sulcus along the side of the body of the sphenoid bone. It then, once again, curves upward to perforate the dura mater just lateral to the optic nerve. This *cerebral portion* immediately passes backward between the optic and oculomotor nerves, where it divides into its terminal branches.

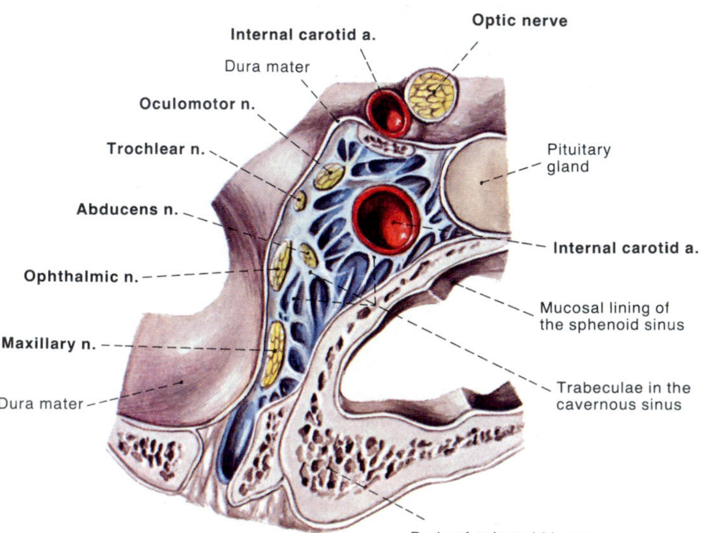

Internal carotid a.
Optic nerve
Dura mater
Oculomotor n.
Trochlear n.
Pituitary gland
Abducens n.
Internal carotid a.
Ophthalmic n.
Mucosal lining of the sphenoid sinus
Maxillary n.
Trabeculae in the cavernous sinus
Dura mater
Body of sphenoid bone

Fig. 641: Frontal Section Through the Left Cavernous Sinus

NOTE that the internal carotid artery (which is seen to have turned back on itself above) and the oculomotor, trochlear, V_1, V_2, and abducens nerves pass through the cavernous sinus.

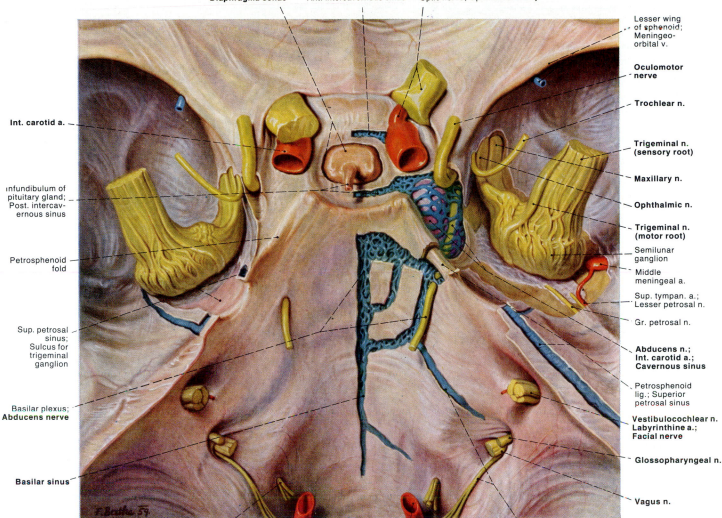

Pituitary gland;
Diaphragma sellae
Ant. intercavernous sinus
Optic nerve; Ophthalmic artery
Lesser wing of sphenoid; Meningeo-orbital v.
Oculomotor nerve
Trochlear n.
Int. carotid a.
Trigeminal n. (sensory root)
Maxillary n.
Infundibulum of pituitary gland; Post. intercavernous sinus
Ophthalmic n.
Trigeminal n. (motor root)
Petrosphenoid fold
Semilunar ganglion
Middle meningeal a.
Sup. tympan. a.; Lesser petrosal n.
Gr. petrosal n.
Sup. petrosal sinus; Sulcus for trigeminal ganglion
Abducens n.; Int. carotid a.; Cavernous sinus
Petrosphenoid lig.; Superior petrosal sinus
Basilar plexus; Abducens nerve
Vestibulocochlear n. Labyrinthine a.; Facial nerve
Glossopharyngeal n.
Basilar sinus
Vagus n.
F. Batke 59
Hypoglossal nerve
Vertebral artery
Accessory n.; Inferior petrosal sinus

Fig. 642: Blood Vessels and Nerves in the Region of the Sella Turcica Viewed from Above

NOTE: 1) that portions of the dura mater have been opened to expose the right cavernous sinus, the anterior and posterior intercavernous sinuses, the basilar plexus of veins and basilar sinus, as well as the right superior and inferior petrosal sinuses.

2) that the stumps of the oculomotor, trochlear and trigeminal nerves have been pulled forward or laterally, and the right optic nerve has been elevated to reveal the origin of the ophthalmic artery from the cerebral portion of the internal carotid artery.

3) the infundibulum, or stalk, of the pituitary gland by which the gland was attached to the hypothalamus of the brain. Also observe the stumps of all the cranial nerves, (except the olfactory tracts), traversing their respective foramina.

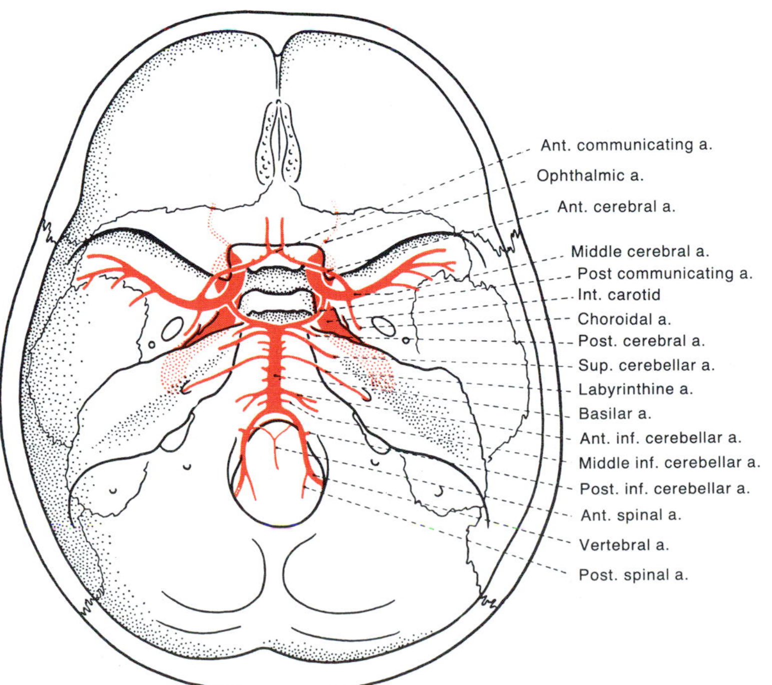

Ant. communicating a.
Ophthalmic a.
Ant. cerebral a.
Middle cerebral a.
Post communicating a.
Int. carotid
Choroidal a.
Post. cerebral a.
Sup. cerebellar a.
Labyrinthine a.
Basilar a.
Ant. inf. cerebellar a.
Middle inf. cerebellar a.
Post. inf. cerebellar a.
Ant. spinal a.
Vertebral a.
Post. spinal a.

Fig. 643: The Relationship of the Arteries Supplying the Brain to the Bones at the Base of the Cranial Cavity

From the two vertebral arteries which form the basilar artery and from the two internal carotid arteries branch all of the named vessels shown in this diagram. For more details, see Figure 644.

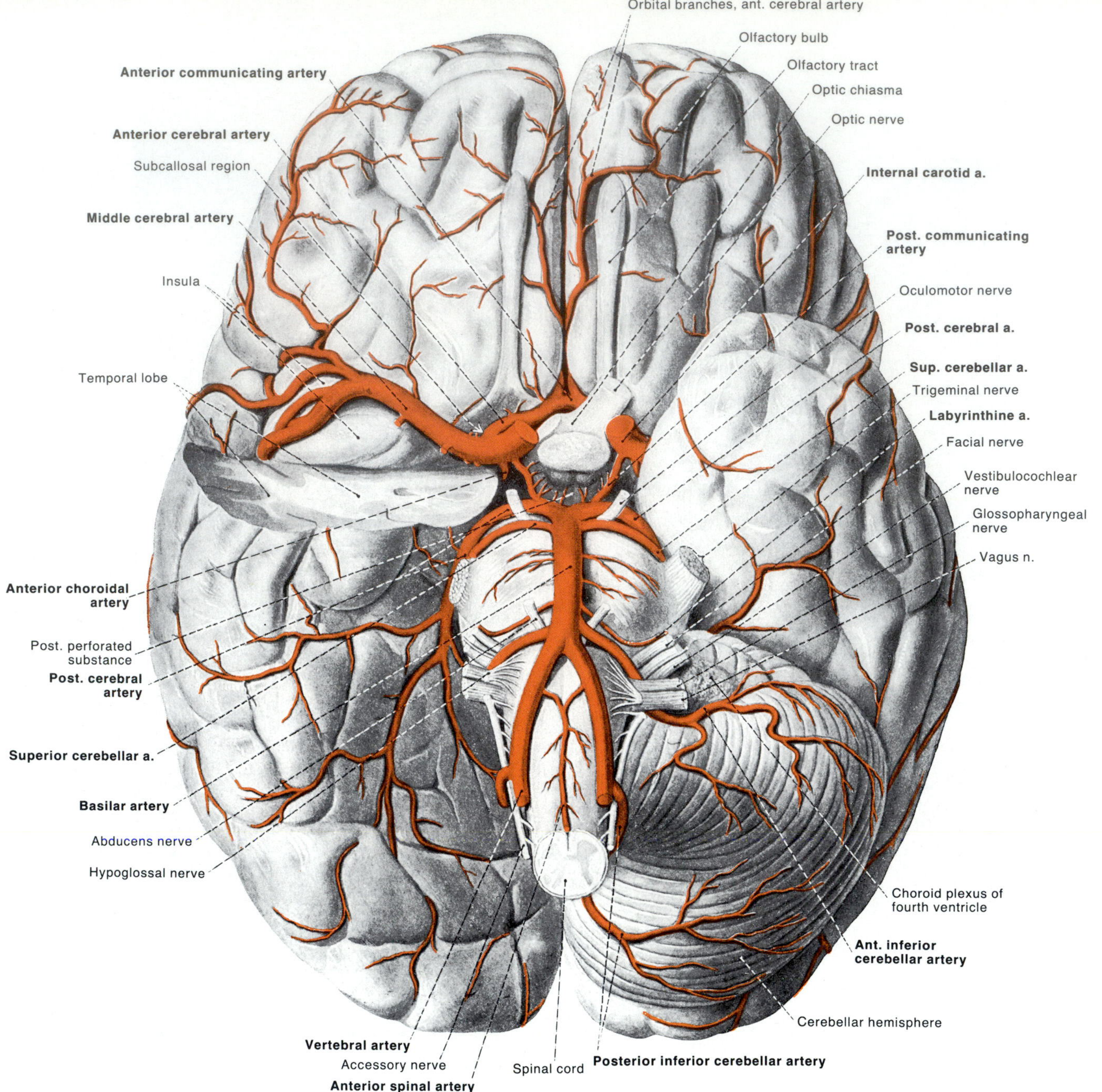

Orbital branches, ant. cerebral artery
Olfactory bulb
Olfactory tract
Optic chiasma
Optic nerve
Internal carotid a.
Post. communicating artery
Oculomotor nerve
Post. cerebral a.
Sup. cerebellar a.
Trigeminal nerve
Labyrinthine a.
Facial nerve
Vestibulocochlear nerve
Glossopharyngeal nerve
Vagus n.
Anterior communicating artery
Anterior cerebral artery
Subcallosal region
Middle cerebral artery
Insula
Temporal lobe
Anterior choroidal artery
Post. perforated substance
Post. cerebral artery
Superior cerebellar a.
Basilar artery
Abducens nerve
Hypoglossal nerve
Choroid plexus of fourth ventricle
Ant. inferior cerebellar artery
Cerebellar hemisphere
Vertebral artery
Accessory nerve
Anterior spinal artery
Spinal cord
Posterior inferior cerebellar artery

Fig. 644: The Arteries at the Base of the Brain

NOTE: 1) that branches of the vertebral arteries form the anterior spinal artery medially and the posterior inferior cerebellar arteries laterally.

2) the basilar artery is formed near the pontomedullary junction and successively gives off the anterior inferior cerebellar, labyrinthine, pontine (not labelled), superior cerebellar and posterior cerebral arteries.

3) the internal carotid arteries interconnect with the posterior cerebral by way of the posterior communicating arteries, and then give off the middle and anterior cerebral arteries. The anterior cerebral arteries are joined by the anterior communicating artery.

4) the circle of Willis surrounds the optic chiasma and is formed by the posterior cerebral, posterior communicating, internal carotid, anterior cerebral and anterior communicating arteries.

Fig. 645 con't.

Middle Cranial Fossa: a) *optic foramen:* optic nerve and ophthalmic artery. – b) *superior orbital fissure:* oculomotor nerve, trochlear nerve, ophthalmic nerve, abducens nerve, sympathetic fibers, superior ophthalmic vein, orbital branches of the middle meningeal artery and a dural recurrent branch from the lacrimal artery. – c) *foramen rotundum:* maxillary nerve. – d) *foramen ovale:* mandibular nerve, accessory meningeal artery. – e) *foramen spinosum:* middle meningeal artery, recurrent branch from mandibular nerve. – f) *foramen lacerum:* internal carotid artery passes across the superior part of the foramen but does not traverse it; nerve of pterygoid canal and meningeal branch of ascending pharyngeal artery traverse foramen lacerum.

Posterior Cranial Fossa: a) *internal acoustic meatus:* facial nerve, vestibulocochlear nerve, labyrinthine artery. – b) *jugular foramen:* inferior petrosal sinus and transverse sinus which together form the jugular vein; meningeal branches of occipital and ascending pharyngeal arteries; glossopharyngeal, vagus and accessory nerves. – c) *hypoglossal canal:* hypoglossal nerve. – d) *foramen magnum:* spinal cord; accessory nerve; anterior and posterior spinal arteries; vertebral arteries; tectorial membrane.

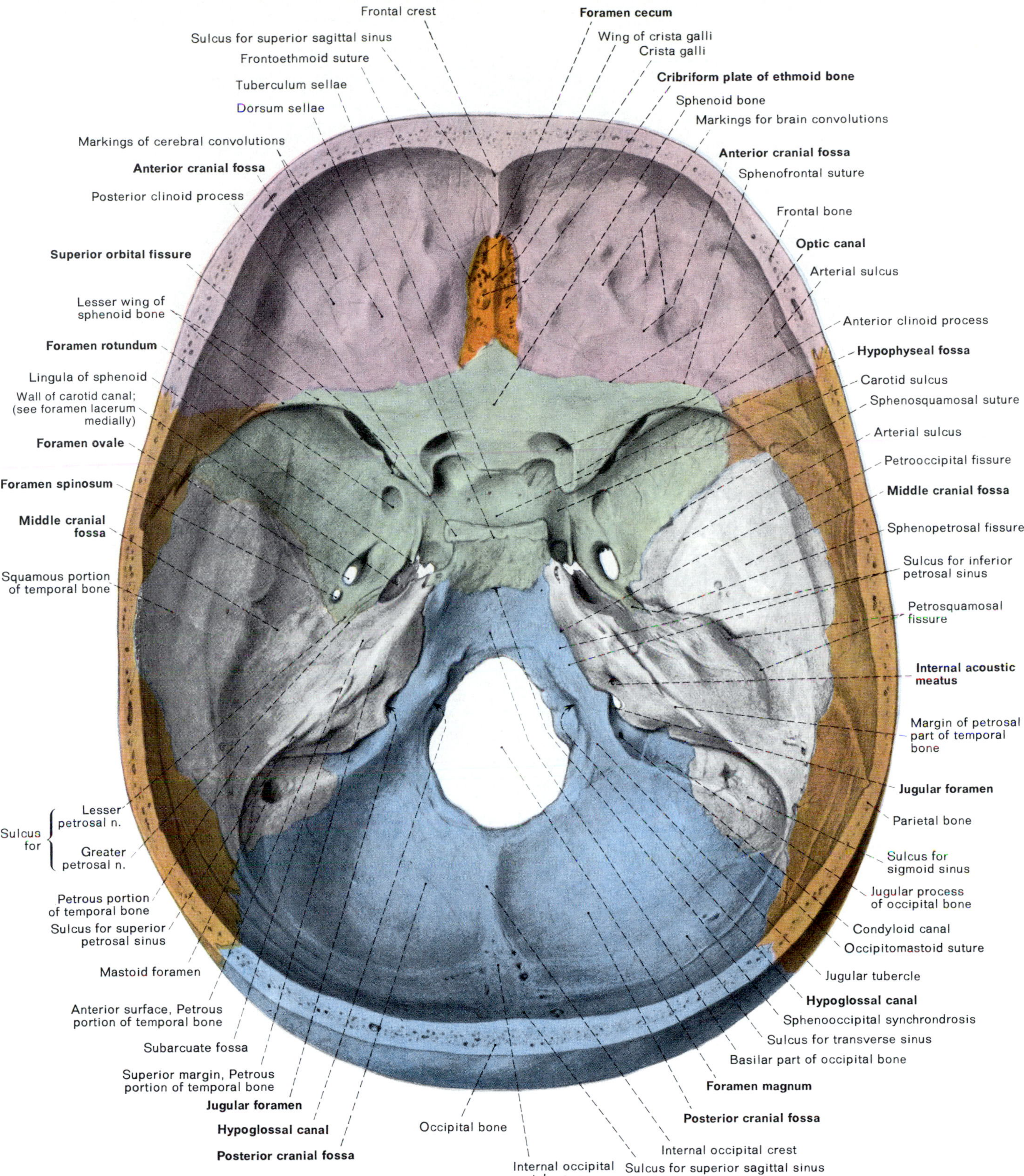
Frontal crest
Foramen cecum
Sulcus for superior sagittal sinus
Wing of crista galli
Crista galli
Frontoethmoid suture
Cribriform plate of ethmoid bone
Tuberculum sellae
Sphenoid bone
Dorsum sellae
Markings for brain convolutions
Markings of cerebral convolutions
Anterior cranial fossa
Anterior cranial fossa
Sphenofrontal suture
Posterior clinoid process
Frontal bone
Optic canal
Superior orbital fissure
Arterial sulcus
Lesser wing of sphenoid bone
Anterior clinoid process
Foramen rotundum
Hypophyseal fossa
Lingula of sphenoid
Carotid sulcus
Wall of carotid canal; (see foramen lacerum medially)
Sphenosquamosal suture
Foramen ovale
Arterial sulcus
Petrooccipital fissure
Foramen spinosum
Middle cranial fossa
Middle cranial fossa
Sphenopetrosal fissure
Squamous portion of temporal bone
Sulcus for inferior petrosal sinus
Petrosquamosal fissure
Internal acoustic meatus
Margin of petrosal part of temporal bone
Jugular foramen
Sulcus for Lesser petrosal n.
Sulcus for Greater petrosal n.
Parietal bone
Sulcus for sigmoid sinus
Petrous portion of temporal bone
Jugular process of occipital bone
Sulcus for superior petrosal sinus
Condyloid canal
Occipitomastoid suture
Mastoid foramen
Jugular tubercle
Anterior surface, Petrous portion of temporal bone
Hypoglossal canal
Sphenooccipital synchrondrosis
Sulcus for transverse sinus
Subarcuate fossa
Basilar part of occipital bone
Superior margin, Petrous portion of temporal bone
Foramen magnum
Jugular foramen
Posterior cranial fossa
Hypoglossal canal
Occipital bone
Posterior cranial fossa
Internal occipital crest
Internal occipital protuberance
Sulcus for superior sagittal sinus

Fig. 645: The Base of the Skull: Internal Aspect (Superior View)

NOTE: 1) the base of the cranial cavity is composed principally of the frontal (lavender) bones, the ethmoid (orange) bone, the sphenoid (green) bone, the temporal (gray) bones and the occipital (blue) bone. Additionally, the sphenoidal angle of each parietal (tan) bone is interposed between the frontal and temporal bones.

2) Important structures traverse the various foramina through the base of the skull:

Anterior Cranial Fossa: a) *foramen cecum:* a small vein. b) *foramina of cribriform plate:* filaments of olfactory receptor neurons to olfactory bulb. c) *anterior ethmoid foramen:* anterior ethmoidal vessels and nasociliary nerve. d) *posterior ethmoid foramen:* posterior ethmoidal vessels and nerve.

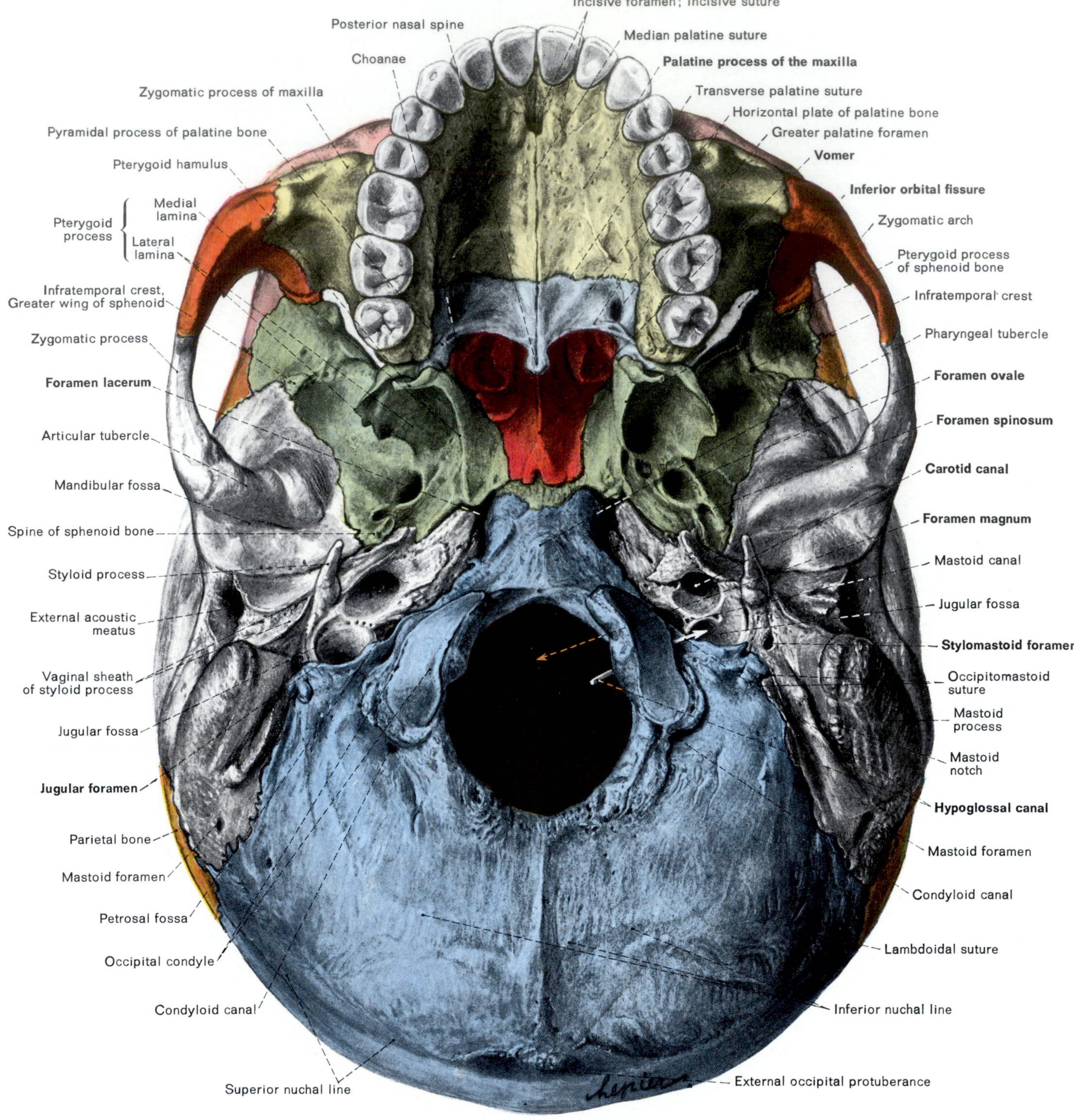

Incisive foramen; Incisive suture
Posterior nasal spine
Median palatine suture
Choanae
Palatine process of the maxilla
Zygomatic process of maxilla
Transverse palatine suture
Horizontal plate of palatine bone
Pyramidal process of palatine bone
Greater palatine foramen
Vomer
Pterygoid hamulus
Inferior orbital fissure
Pterygoid process
Medial lamina
Lateral lamina
Zygomatic arch
Pterygoid process of sphenoid bone
Infratemporal crest, Greater wing of sphenoid
Infratemporal crest
Zygomatic process
Pharyngeal tubercle
Foramen lacerum
Foramen ovale
Foramen spinosum
Articular tubercle
Carotid canal
Mandibular fossa
Foramen magnum
Spine of sphenoid bone
Mastoid canal
Styloid process
Jugular fossa
External acoustic meatus
Stylomastoid foramen
Vaginal sheath of styloid process
Occipitomastoid suture
Mastoid process
Jugular fossa
Mastoid notch
Jugular foramen
Hypoglossal canal
Parietal bone
Mastoid foramen
Mastoid foramen
Condyloid canal
Petrosal fossa
Lambdoidal suture
Occipital condyle
Inferior nuchal line
Condyloid canal
External occipital protuberance
Superior nuchal line

Fig. 646: The Base of the Skull: External Aspect (Inferior View)

NOTE: 1) the base of the skull reveals a posterior region comprised of the occipital and temporal bones which are attached by muscles to the thoracic and vertebral skeleton. More anteriorly are the facial bones consisting of the maxilla, palatine, zygomatic and vomer. Interposed between these two groups of bones is the sphenoid bone.

2) the bony palate and the cruciform suture where the transverse processes of the maxillary bones meet the horizontal plates of the palatine bones. Observe the incisive and greater palatine foramina. The lesser palatine foramina are shown but not labelled.

3) The posterior aspect of the vomer (in red) bounding the posterior nasal apertures, or choanae. Observe the medial and lateral laminae of the pterygoid process of the sphenoid bone (in green), behind which are the foramen ovale and foramen spinosum in the greater wings of that bone.

4) The petrous and mastoid parts of the temporal bone. Identify the foramen lacerum interposed between the sphenoid and temporal bones, as well as the carotid canal and jugular foramen located medial to the styloid process of the temporal bone.

5) That the foramen magnum lies completely within the occipital bone. An arrow has been placed through the hypoglossal canal located just above the occipital condyles.

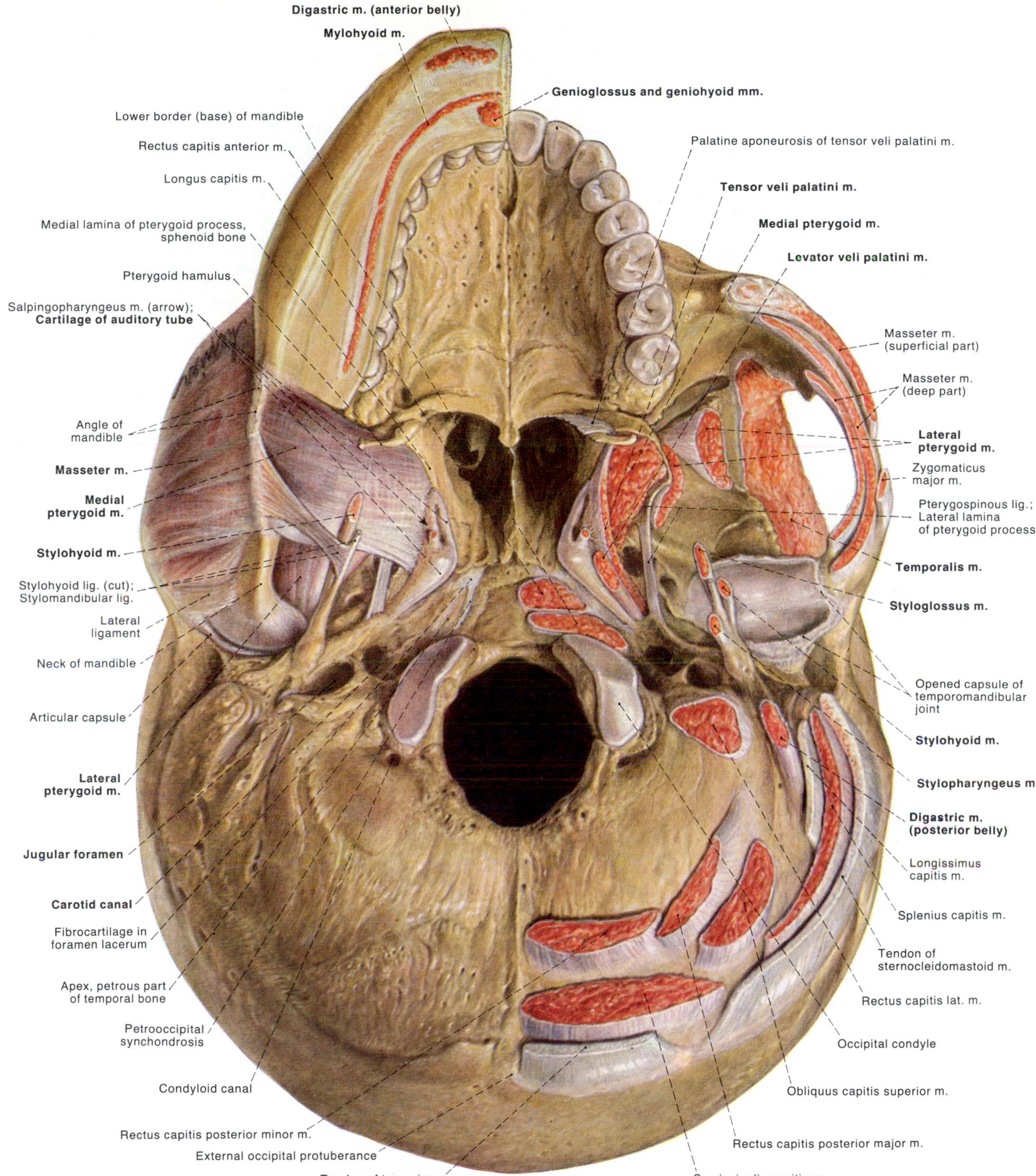
Digastric m. (anterior belly)
Mylohyoid m.
Genioglossus and geniohyoid mm.
Lower border (base) of mandible
Rectus capitis anterior m.
Palatine aponeurosis of tensor veli palatini m.
Longus capitis m.
Tensor veli palatini m.
Medial lamina of pterygoid process, sphenoid bone
Medial pterygoid m.
Levator veli palatini m.
Pterygoid hamulus
Salpingopharyngeus m. (arrow); Cartilage of auditory tube
Masseter m. (superficial part)
Masseter m. (deep part)
Angle of mandible
Lateral pterygoid m.
Masseter m.
Zygomaticus major m.
Medial pterygoid m.
Pterygospinous lig.; Lateral lamina of pterygoid process
Stylohyoid m.
Temporalis m.
Stylohyoid lig. (cut); Stylomandibular lig.
Styloglossus m.
Lateral ligament
Neck of mandible
Opened capsule of temporomandibular joint
Articular capsule
Stylohyoid m.
Lateral pterygoid m.
Stylopharyngeus m.
Digastric m. (posterior belly)
Jugular foramen
Longissimus capitis m.
Carotid canal
Splenius capitis m.
Fibrocartilage in foramen lacerum
Tendon of sternocleidomastoid m.
Apex, petrous part of temporal bone
Rectus capitis lat. m.
Petrooccipital synchondrosis
Occipital condyle
Condyloid canal
Obliquus capitis superior m.
Rectus capitis posterior minor m.
Rectus capitis posterior major m.
External occipital protuberance
Tendon of trapezius m.
Semispinalis capitis m.

Fig. 647: Muscle Origins and Other Structures on the Base of the Skull

NOTE: 1) that the left mandible (reader's right) has been removed while the right muscles of mastication (reader's left), covered by their fascial sheaths, have been retained. Observe the attachments of the superficial and deep parts of the *masseter* muscle, the *temporalis,* the two heads of the *lateral pterygoid* and the *medial pterygoid* muscles.

2) the cartilage of the auditory tube as it opens inferomedially in the lateral wall of the nasopharynx, and the origins of the *tensor* and *levator veli palatini* muscles. The levator arises from the inferior surface of the apex of the petrous temporal bone, anterior to the opening of the carotid canal. The tensor veli palatini lies lateral and anterior to the levator, and it arises from the scaphoid fossa at the base of the medial pterygoid plate, the sphenoidal spine and the lateral aspect of the cartilaginous auditory tube.

3) that the muscles which pass from the skull to cervical and back structures arise from the occipital bone and the mastoid and styloid processes of the temporal bone.

Fig. 648: The Right Eye and Eyelids

NOTE: 1) the eyeball, protected anteriorly by two thin, movable eyelids or palpebrae, is covered by a transparent mucous membrane, the conjunctiva, which is continuous along the inner surface of both eyelids as the palpebral conjunctiva.

2) at the medial angle of the eye is located a small, reddish island of skin called the caruncula lacrimalis. It contains sebaceous and sweat glands which secrete a whitish substance.

3) the pubil is the opening in the iris. Constriction and dilatation of the pupil is controlled autonomically. Parasympathetic fibers in the oculomotor nerve innervate the constrictor muscle of the pupil, while sympathetic fibers from the superior cervical ganglion supply the pupillary dilator muscle.

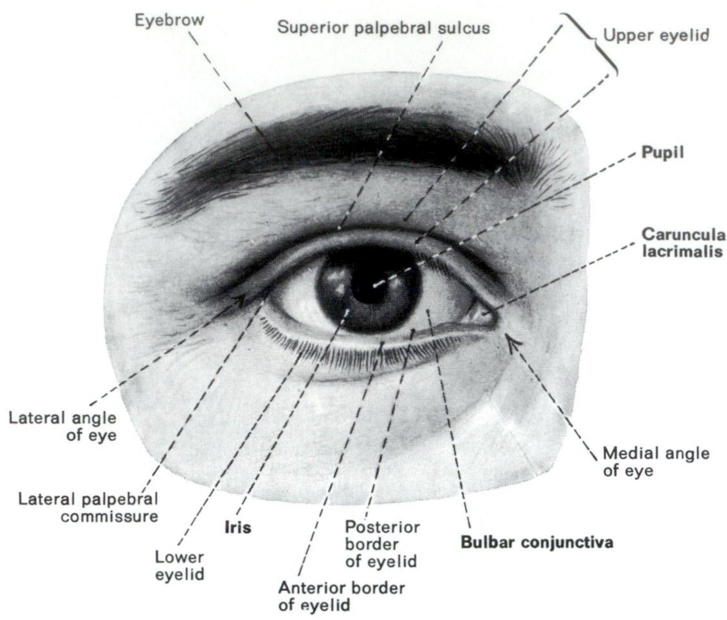

Eyebrow
Superior palpebral sulcus
Upper eyelid
Pupil
Caruncula lacrimalis
Lateral angle of eye
Medial angle of eye
Lateral palpebral commissure
Iris
Posterior border of eyelid
Bulbar conjunctiva
Lower eyelid
Anterior border of eyelid

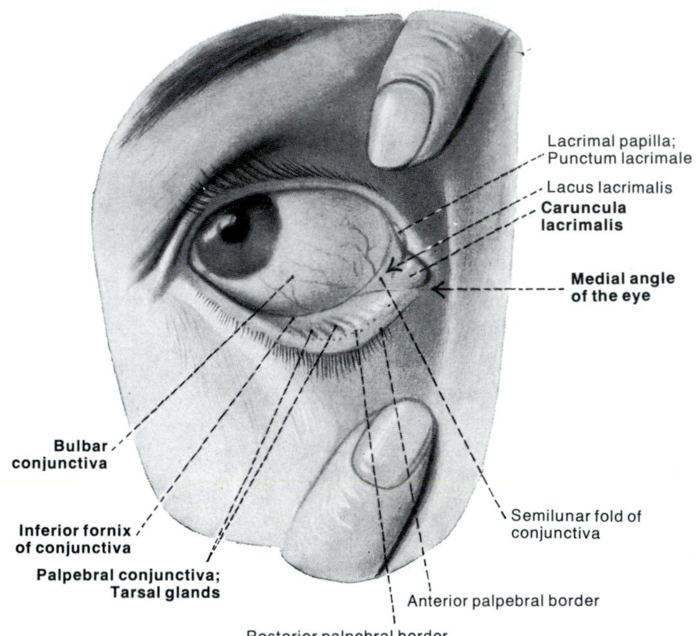

Lacrimal papilla; Punctum lacrimale
Lacus lacrimalis
Caruncula lacrimalis
Medial angle of the eye
Bulbar conjunctiva
Inferior fornix of conjunctiva
Palpebral conjunctiva; Tarsal glands
Semilunar fold of conjunctiva
Anterior palpebral border
Posterior palpebral border

Fig. 649: The Right Lower Eyelid and Medial Angle

NOTE: 1) that the right lower eyelid has been pulled downward in order to display the inner surface of the lower lid (i.e., the palpebral conjunctiva), and to enlarge the exposure of the medial angle (also called medial canthus). The gaze is oriented superiorly and laterally.

2) that the conjunctiva is highly vascular and its bulbar part (over the eyeball) and inferior palpebral part (on the inner surface of the lower eyelid) are continuous along a line of reflection called the inferior conjunctival fornix. A similar reflection line, the superior conjunctival fornix, lies between the eyeball and the upper eyelid.

3) that when the medial angle is more completely exposed, a pair of small openings, the punctae lacrimali, can be found located above and below the caruncula lacrimalis. These openings lead into the small lacrimal canals, and through them the tear fluid secreted by the lacrimal gland enters the nasolacrimal duct (see Figure 660).

Fig. 650: The Innervation of the Eyelids; Anterior View, Right Eye

NOTE: 1) the rich cutaneous innervation that is found around the anterior aspect of the orbit. These cutaneous nerves are derived from either the ophthalmic or maxillary divisions of the trigeminal nerve and, for the most part, achieve the anterior orbital region through foramina in the frontal, zygomatic, and maxillary bones.

2) that superomedially are found the large medial and lateral rami of the supraorbital branch of the frontal nerve (ophthalmic division of V), which appears above the orbit through the supraorbital foramen or notch. Also observe the supratrochlear branch of the frontal nerve, which becomes superficial through a small foramen above the trochlea of the superior oblique muscle.

3) the infratrochlear nerve, which is the terminal branch of the nasociliary nerve (ophthalmic division of V). It becomes superficial below the trochlea of the superior oblique muscle and, along with the palpebral branches of the infraorbital nerve (maxillary division of V), sends fibers to the lower eyelid.

4) the palpebral branches of the lacrimal nerve (ophthalmic division of V) superolaterally, which helps to supply the upper eyelid. Observe also a cutaneous ramus of the zygomaticofacial branch of the maxillary nerve that helps to supply the lateral aspect of the lower eyelid and the skin over the cheekbone. The temporal branch of the facial nerve is a motor nerve that supplies the orbicularis oculi muscle.

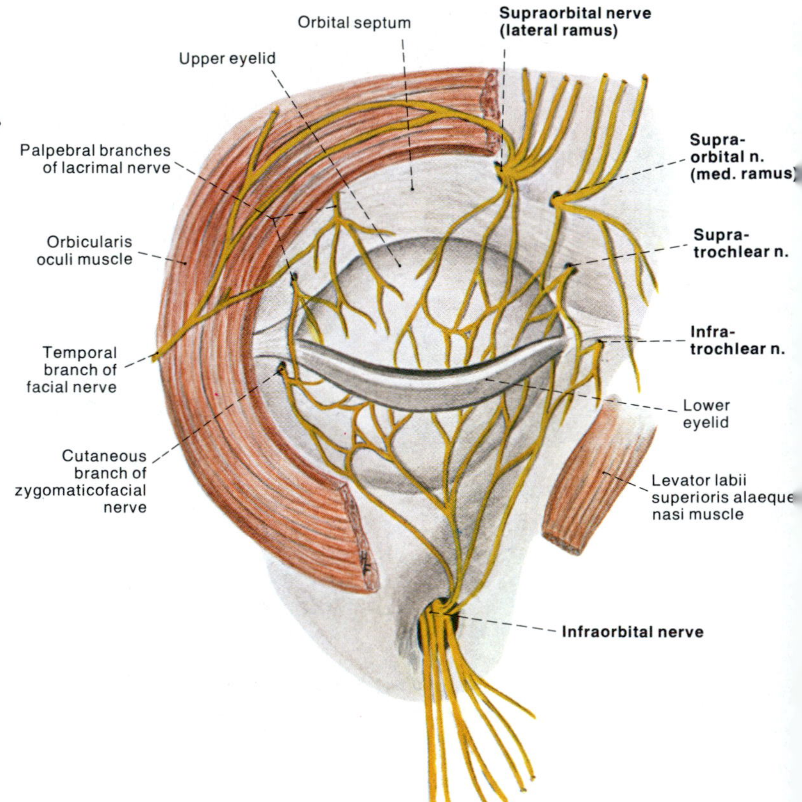

Orbital septum
Supraorbital nerve (lateral ramus)
Upper eyelid
Palpebral branches of lacrimal nerve
Supra-orbital n. (med. ramus)
Orbicularis oculi muscle
Supra-trochlear n.
Temporal branch of facial nerve
Infra-trochlear n.
Lower eyelid
Cutaneous branch of zygomaticofacial nerve
Levator labii superioris alaeque nasi muscle
Infraorbital nerve

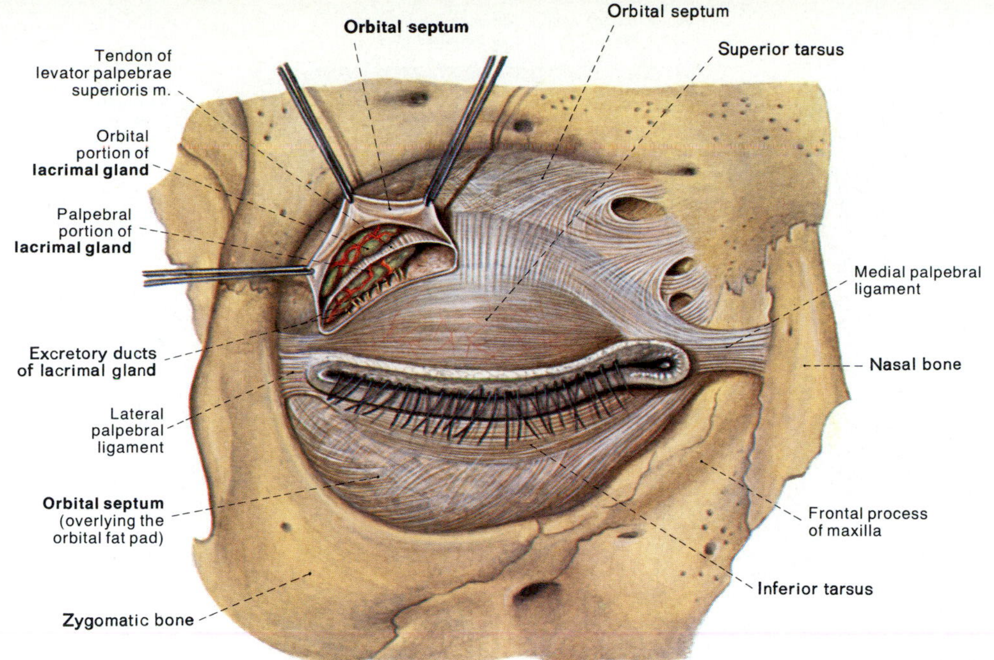

Orbital septum
Orbital septum
Superior tarsus
Tendon of levator palpebrae superioris m.
Orbital portion of lacrimal gland
Palpebral portion of lacrimal gland
Medial palpebral ligament
Excretory ducts of lacrimal gland
Nasal bone
Lateral palpebral ligament
Orbital septum (overlying the orbital fat pad)
Frontal process of maxilla
Inferior tarsus
Zygomatic bone

Fig. 651: The Orbital Septum, Lacrimal Gland and Tarsi of the Right Eye

NOTE: With the skin, superficial fascia and orbicularis oculi muscle removed, the orbital septum is exposed anteriorly. It attaches to the periosteum of the bone peripherally around the orbit and to the tarsi of the eyelids centrally. Observe the lacrimal gland in the upper lateral, aspect of the anterior orbit, lying just beneath the orbital septum.

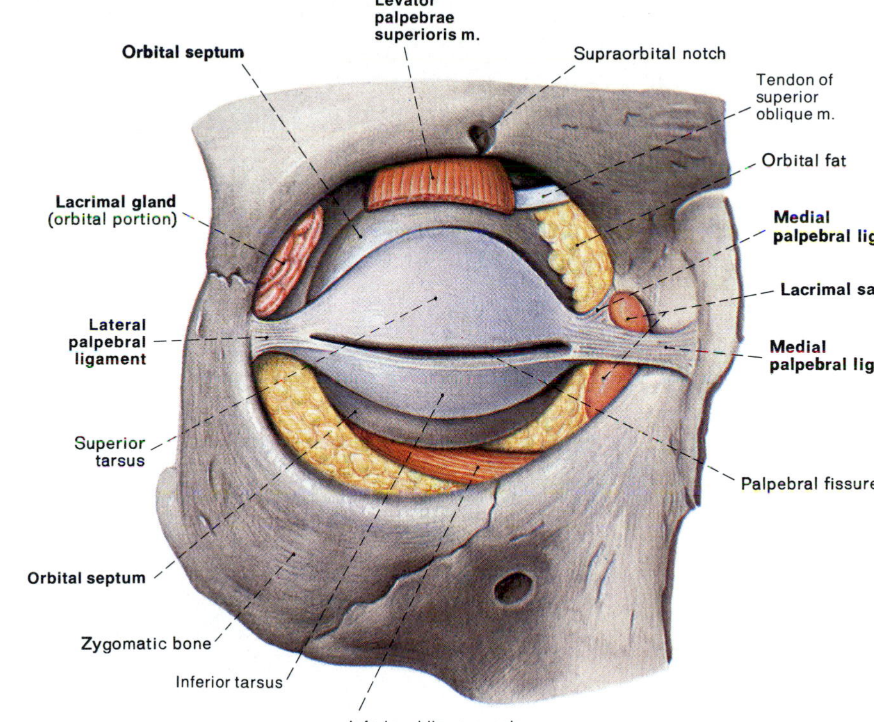

Levator palpebrae superioris m.
Orbital septum
Supraorbital notch
Tendon of superior oblique m.
Orbital fat
Lacrimal gland (orbital portion)
Medial palpebral lig.
Lacrimal sac
Lateral palpebral ligament
Medial palpebral lig.
Superior tarsus
Palpebral fissure
Orbital septum
Zygomatic bone
Inferior tarsus
Inferior oblique muscle

Fig. 652: The Palpebral Ligaments and Tarsal Plates

NOTE: 1) the superficial structures of the anterior orbit have been removed along with the orbital septum and the tendon of the levator palpebrae superioris muscle.

2) the lateral and medial margins of the tarsal plates are attached to the lateral and medial palpebral ligaments which in turn are attached to bone. Observe that the medial ligament is located just anterior to the lacrimal sac.

3) from this anterior view, both the tendon of the superior oblique muscle and the inferior oblique muscle can be visualized. Note also the location of the orbital portion of the lacrimal gland. Its secretions course from lateral to medial across the surface of the eye.

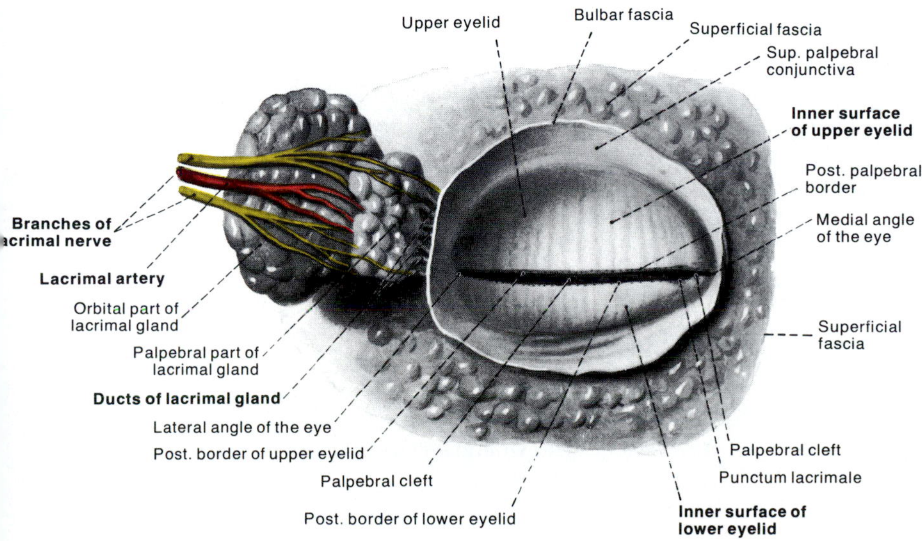

Upper eyelid
Bulbar fascia
Superficial fascia
Sup. palpebral conjunctiva
Inner surface of upper eyelid
Post. palpebral border
Branches of lacrimal nerve
Medial angle of the eye
Lacrimal artery
Orbital part of lacrimal gland
Superficial fascia
Palpebral part of lacrimal gland
Ducts of lacrimal gland
Lateral angle of the eye
Palpebral cleft
Post. border of upper eyelid
Punctum lacrimale
Palpebral cleft
Post. border of lower eyelid
Inner surface of lower eyelid

Fig. 653: The Eyelids and Lacrimal Gland, Viewed from Within the Orbit

NOTE: 1) the inner surface of the eyelids and the vessels and nerves that supply the lacrimal gland. Observe that the lacrimal gland within this left orbit is located superior and lateral to the upper eyelid (see also Figs. 651 and 652).

2) the lacrimal artery and nerve. The lacrimal artery is a thin, tortuous branch of the ophthalmic artery. Close to the gland, the lacrimal nerve contains postganglionic fibers from the pterygopalatine ganglion (see Fig. 659 and Note 2, Fig. 659), some sensory fibers that course in the ophthalmic nerve, and a few sympathetic fibers.

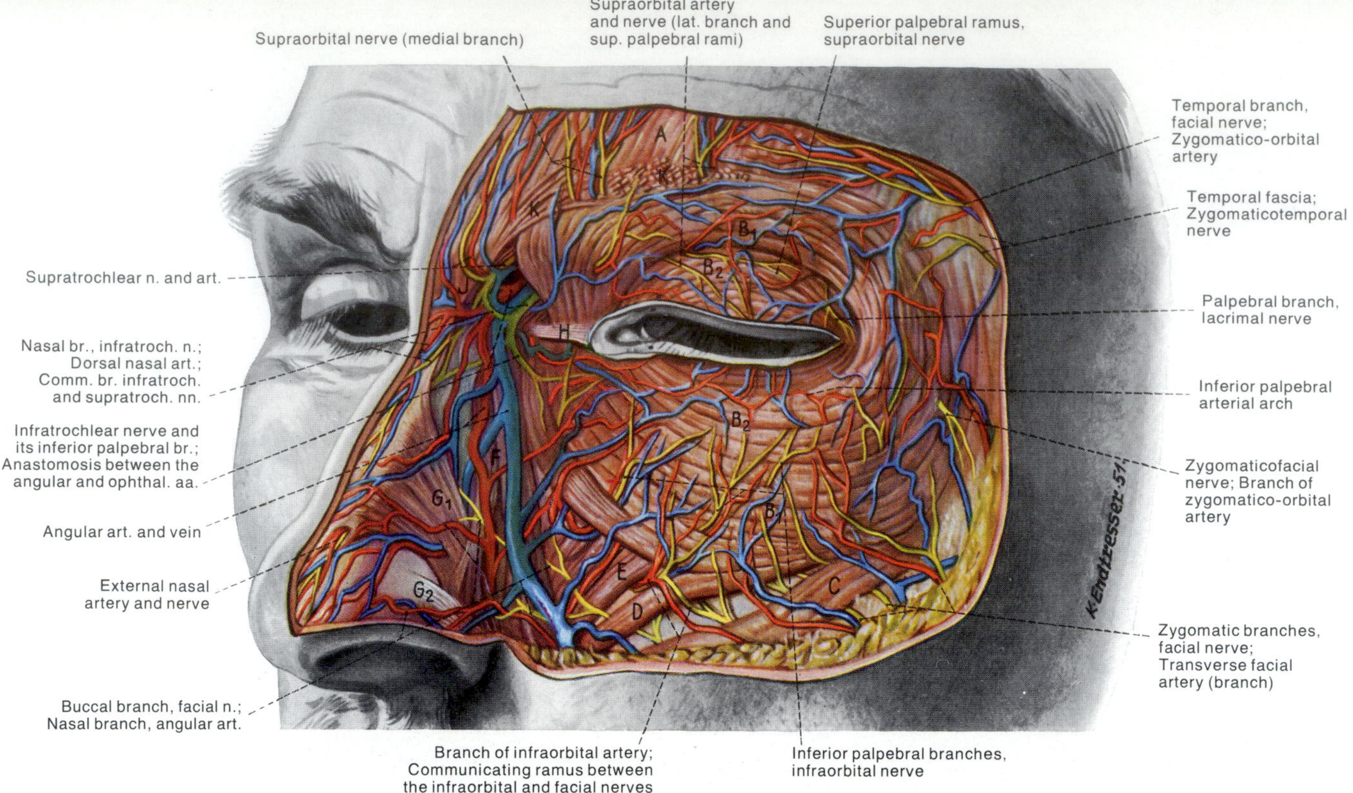

Supraorbital nerve (medial branch)
Supraorbital artery and nerve (lat. branch and sup. palpebral rami)
Superior palpebral ramus, supraorbital nerve
Temporal branch, facial nerve; Zygomatico-orbital artery
Temporal fascia; Zygomaticotemporal nerve
Supratrochlear n. and art.
Palpebral branch, lacrimal nerve
Nasal br., infratroch. n.; Dorsal nasal art.; Comm. br. infratroch. and supratroch. nn.
Inferior palpebral arterial arch
Infratrochlear nerve and its inferior palpebral br.; Anastomosis between the angular and ophthal. aa.
Zygomaticofacial nerve; Branch of zygomatico-orbital artery
Angular art. and vein
External nasal artery and nerve
Zygomatic branches, facial nerve; Transverse facial artery (branch)
Buccal branch, facial n.; Nasal branch, angular art.
Branch of infraorbital artery; Communicating ramus between the infraorbital and facial nerves
Inferior palpebral branches, infraorbital nerve
A
K
B1
B2
H
F
G1
G2
E
D
C
K. Endtresser 51

Fig. 654: Anterior Dissection of the Orbit, Stage 1: Superficial Vessels, Nerves and Muscles After Removal of the Skin and the Thin Layer of Superficial Fascia

A = Frontalis part, Occipitofrontalis muscle
B_1 = Orbital part, orbicularis oculi m.
B_2 = Palpebral part, Orbicularis oculi m.
C = Zygomaticus major m.
D = Zygomaticus minor m.
E = Levator labii sup. m.
F = Levator labii sup. alaeque nasi m.
G_1 = Transverse part, Nasalis muscle
G_2 = Alar part, Nasalis m.
H = Medial palpebral lig.
J = Procerus muscle
K = Corrugator supercilii m.

Supratrochlear nerve
Supratrochlear art.; Supraorbital nerve (medial branch) and its sup. palpebr. ramus
Supraorbital artery and nerve; Levator palpebrae sup. m.
Orbital septum; Superior tarsus; Sup. palpebral arterial arch
Infratrochlear n.; Corrugator supercilii m.
Palpebral branch, lacrimal nerve; Zygomatico-orbital art.
Angular art. and vein; Communicating br. betw. infra. and supratrochlear nn.; Procerus muscle
Lat. palpebral lig.; Inferior palpebral arterial arch
Caruncula lacrimalis; Orbicularis oculi m.
Zygomaticofacial branch of zygomatic nerve (V_2)
Med. palpebral lig.; Palpebral branch, infratrochlear nerve
Ext. nasal art. and n.
Zygomatic branches of the facial nerve
Lateral nasal cartilage; Angular art. and vein; Inf. palpebral branch, infraorbital nerve
Greater alar cartilage of the nose
Zygomaticus minor muscle
Zygomaticus major muscle
Infraorbital art. and nerve
Levator labii superioris alaeque nasi muscle; Buccal branch, facial n.
Inferior palpebral br., infraorbital n.; Levator anguli oris muscle

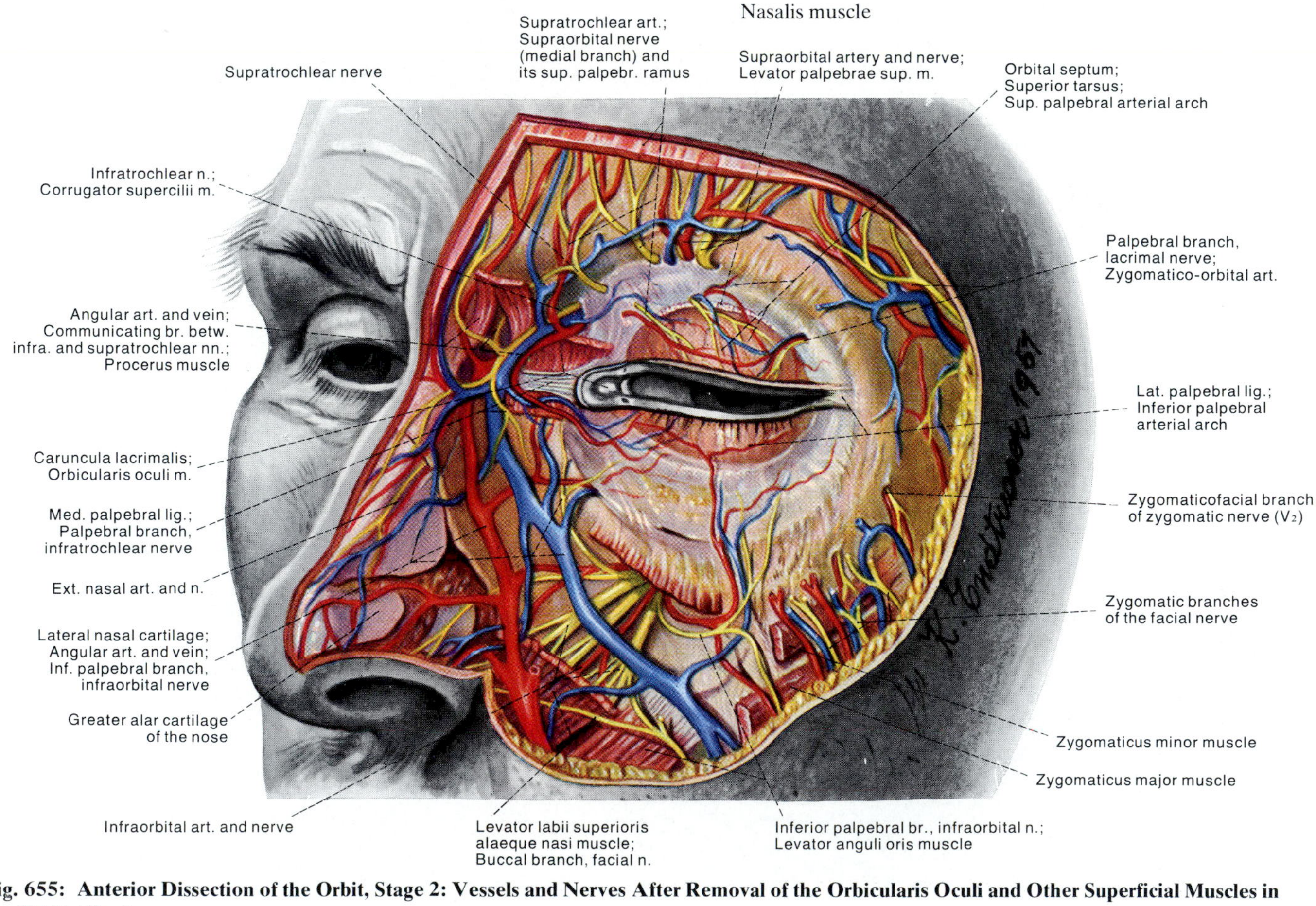

Fig. 655: Anterior Dissection of the Orbit, Stage 2: Vessels and Nerves After Removal of the Orbicularis Oculi and Other Superficial Muscles in the Orbital Region

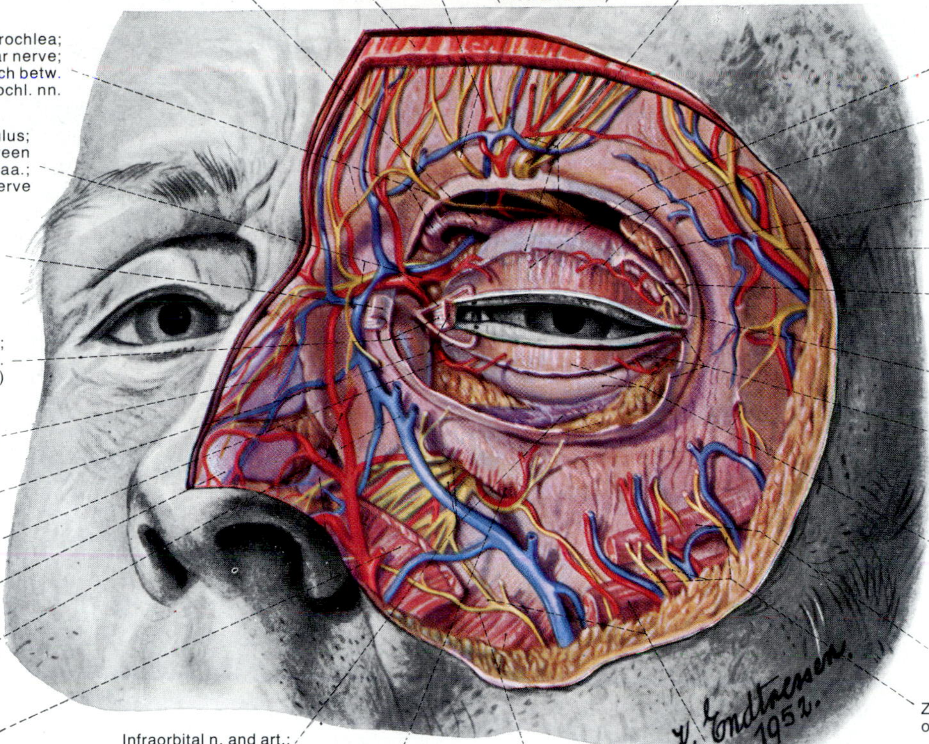

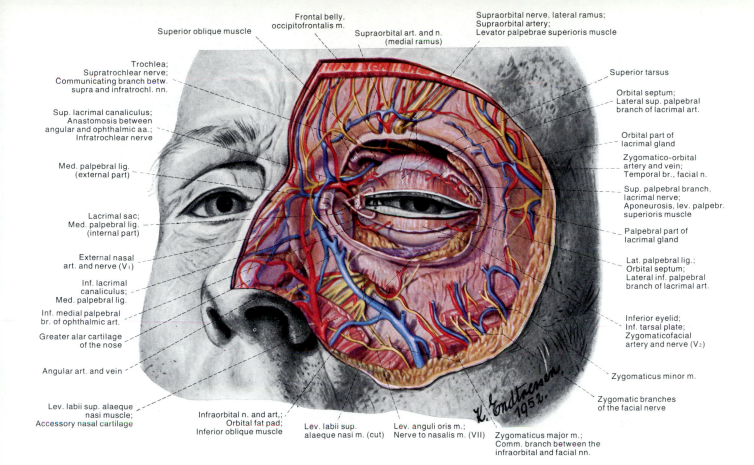

Superior oblique muscle
Frontal belly, occipitofrontalis m.
Supraorbital art. and n. (medial ramus)
Supraorbital nerve, lateral ramus; Supraorbital artery; Levator palpebrae superioris muscle
Trochlea; Supratrochlear nerve; Communicating branch betw. supra and infratrochl. nn.
Superior tarsus
Orbital septum; Lateral sup. palpebral branch of lacrimal art.
Sup. lacrimal canaliculus; Anastomosis between angular and ophthalmic aa.; Infratrochlear nerve
Orbital part of lacrimal gland
Zygomatico-orbital artery and vein; Temporal br., facial n.
Med. palpebral lig. (external part)
Sup. palpebral branch, lacrimal nerve; Aponeurosis, lev. palpebr. superioris muscle
Lacrimal sac; Med. palpebral lig. (internal part)
Palpebral part of lacrimal gland
External nasal art. and nerve (V_1)
Lat. palpebral lig.; Orbital septum; Lateral inf. palpebral branch of lacrimal art.
Inf. lacrimal canaliculus; Med. palpebral lig.
Inf. medial palpebral br. of ophthalmic art.
Inferior eyelid; Inf. tarsal plate; Zygomaticofacial artery and nerve (V_2)
Greater alar cartilage of the nose
Angular art. and vein
Zygomaticus minor m.
Zygomatic branches of the facial nerve
Lev. labii sup. alaeque nasi muscle; Accessory nasal cartilage
Infraorbital n. and art.; Orbital fat pad; Inferior oblique muscle
Lev. labii sup. alaeque nasi m. (cut)
Lev. anguli oris m.; Nerve to nasalis m. (VII)
Zygomaticus major m.; Comm. branch between the infraorbital and facial nn.

Fig. 656: Anterior Dissection of the Orbit, Stage 3: Visualization of More Deeply Located Structures After Cutting the Orbital Septum Circumferentially and Severing the Medial Palpebral Ligament

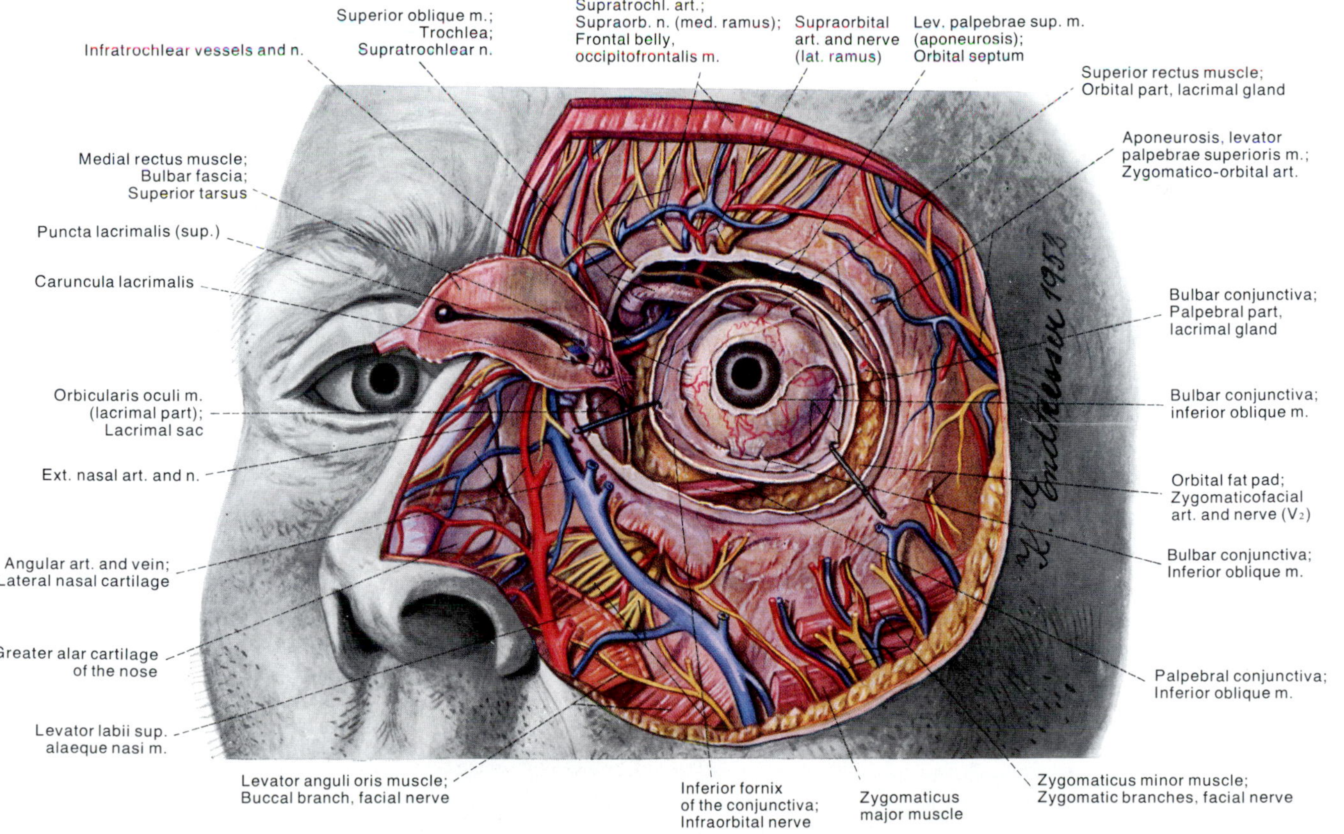

Infratrochlear vessels and n.
Superior oblique m.; Trochlea; Supratrochlear n.
Supratrochl. art.; Supraorb. n. (med. ramus); Frontal belly, occipitofrontalis m.
Supraorbital art. and nerve (lat. ramus)
Lev. palpebrae sup. m. (aponeurosis); Orbital septum
Superior rectus muscle; Orbital part, lacrimal gland
Medial rectus muscle; Bulbar fascia; Superior tarsus
Aponeurosis, levator palpebrae superioris m.; Zygomatico-orbital art.
Puncta lacrimalis (sup.)
Caruncula lacrimalis
Bulbar conjunctiva; Palpebral part, lacrimal gland
Orbicularis oculi m. (lacrimal part); Lacrimal sac
Bulbar conjunctiva; inferior oblique m.
Ext. nasal art. and n.
Orbital fat pad; Zygomaticofacial art. and nerve (V_2)
Angular art. and vein; Lateral nasal cartilage
Bulbar conjunctiva; Inferior oblique m.
Greater alar cartilage of the nose
Palpebral conjunctiva; Inferior oblique m.
Levator labii sup. alaeque nasi m.
Levator anguli oris muscle; Buccal branch, facial nerve
Inferior fornix of the conjunctiva; Infraorbital nerve
Zygomaticus major muscle
Zygomaticus minor muscle; Zygomatic branches, facial nerve

Fig. 657: Anterior Dissection of the Orbit, Stage 4: Exposure of the Eyeball After Cutting the Lateral Palpebral Ligament and Reflecting the Eyelids Medially

Fig. 658: The Lacrimal Canaliculi and Lacrimal Sac: Left Side, Superficial Dissection

NOTE: 1) that the skin and superficial fascia have been removed from the upper lateral nasal region and over the medial angle of the orbit. Observe that the orbital and lacrimal parts of the orbicularis oculi muscle have been cut and reflected inferolaterally. Observe also that the tarsal attachments of the medial palpebral ligament have been cut, but this ligament remains attached to the frontal process of the maxilla.

2) the inferior oblique muscle, which is located deep to the infraorbital margin of the orbital cavity. In order to visualize better the orientation of the left inferior oblique muscle in this dissection, compare this figure with Figure 671.

3) that severance of the medial palpebral ligament exposes the underlying sac; the relationship of these two structures is more clearly seen in Figure 652. The lacrimal sac is located in a small fossa formed by the maxilla and the lacrimal bone. This bony sulcus is limited by the anterior lacrimal crest of the maxilla and the posterior lacrimal crest of the lacrimal bone (see Figs. 661 and 662). The lacrimal sac receives a lacrimal canaliculus from each eyelid and these two ducts are each about 1 cm in length (see the deeper dissection in Fig. 660).

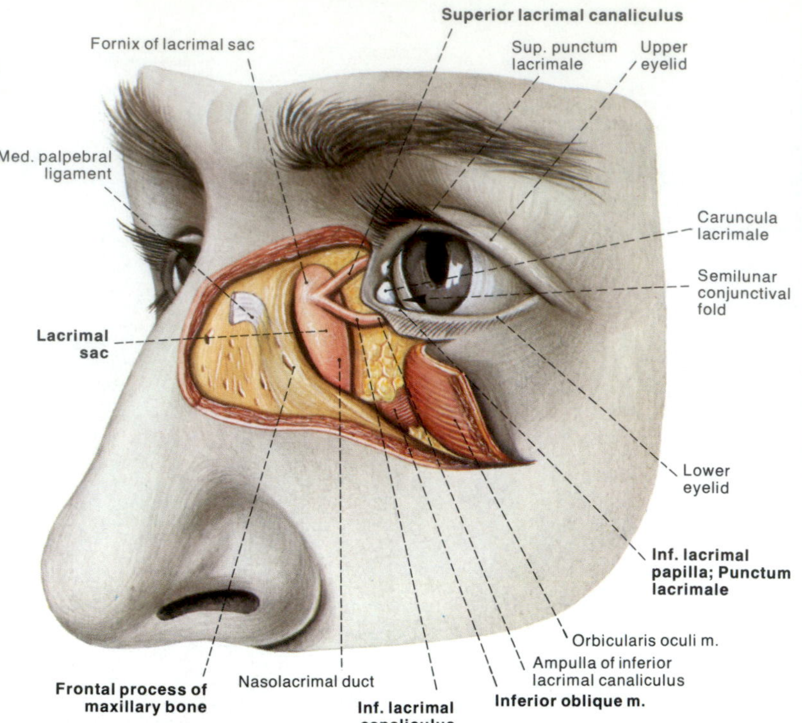
Superior lacrimal canaliculus
Fornix of lacrimal sac
Sup. punctum lacrimale
Upper eyelid
Med. palpebral ligament
Caruncula lacrimale
Semilunar conjunctival fold
Lacrimal sac
Lower eyelid
Inf. lacrimal papilla; Punctum lacrimale
Orbicularis oculi m.
Ampulla of inferior lacrimal canaliculus
Inferior oblique m.
Frontal process of maxillary bone
Nasolacrimal duct
Inf. lacrimal canaliculus

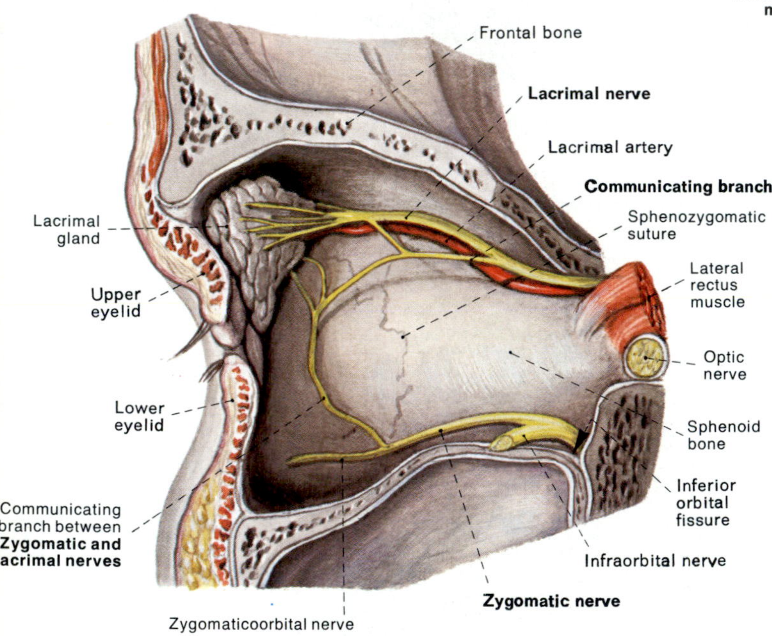
Frontal bone
Lacrimal nerve
Lacrimal artery
Communicating branch
Sphenozygomatic suture
Lacrimal gland
Lateral rectus muscle
Upper eyelid
Optic nerve
Lower eyelid
Sphenoid bone
Inferior orbital fissure
Communicating branch between Zygomatic and lacrimal nerves
Infraorbital nerve
Zygomatic nerve
Zygomaticoorbital nerve

Fig. 659: The Innervation of the Lacrimal Gland

NOTE: 1) the lacrimal gland receives postganglionic parasympathetic nerve fibers which are secretomotor in nature. The preganglionic fibers are generally said to emerge from the brain with the facial nerve (VII). The synapse between pre and postganglionic fibers occurs in the pterygopalatine ganglion.

2) the preganglionic parasympathetic fibers reach the pterygopalatine ganglion by way of the greater petrosal nerve which becomes the nerve of the pterygoid canal. The postganglionic fibers leave the ganglion and travel for a short distance with the zygomatic branch of the infraorbital nerve. From this point in the inferior aspect of the orbit, the parasympathetic fibers travel by way of a communicating branch to the lacrimal nerve by which they achieve the lacrimal gland.

Fig. 660: The Lacrimal Canaliculi, Lacrimal Sac and Nasolacrimal Duct: Left Side, Deep Dissection

NOTE: 1) at the medial edge of both the upper and lower eyelids are found single minute orifices (puncta lacrimalia) of small ducts, the lacrimal canaliculi, which lead from the eyelids to the lacrimal sac. The sac forms the upper dilated end of the nasolacrimal duct, which then extends a distance of about $3^1/_2$ inches into the inferior meatus of the nasal cavity.

2) secretions from the lacrimal gland pass medially across the surface of the eye toward the canaliculi and then are transported to the nasal cavity by way of the nasolacrimal duct. Excessive secretions (such as during crying) cannot be handled in this manner and, thus, roll over the edge of the lower eyelid as tears.

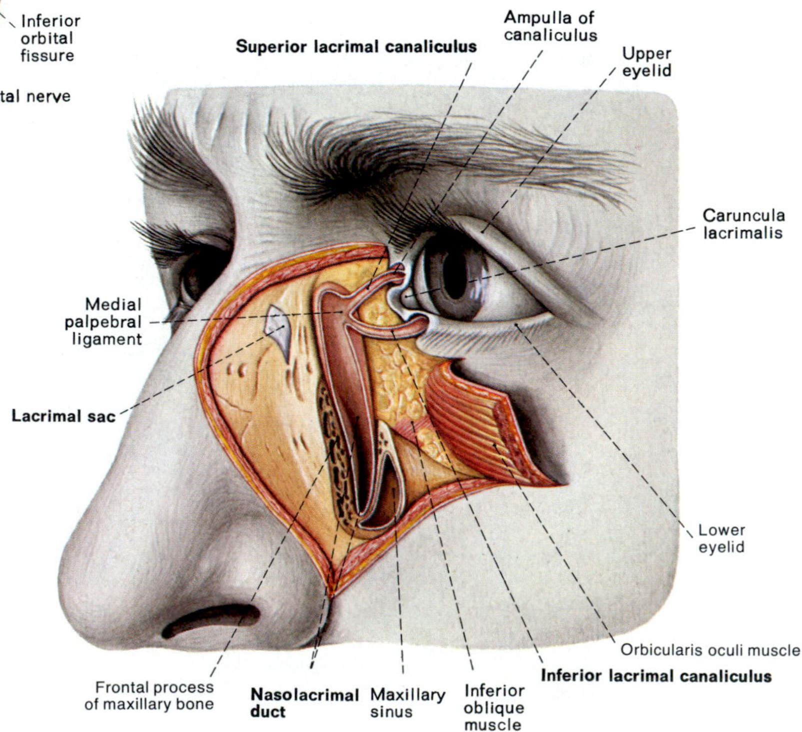
Superior lacrimal canaliculus
Ampulla of canaliculus
Upper eyelid
Caruncula lacrimalis
Medial palpebral ligament
Lacrimal sac
Lower eyelid
Orbicularis oculi muscle
Inferior lacrimal canaliculus
Frontal process of maxillary bone
Nasolacrimal duct
Maxillary sinus
Inferior oblique muscle

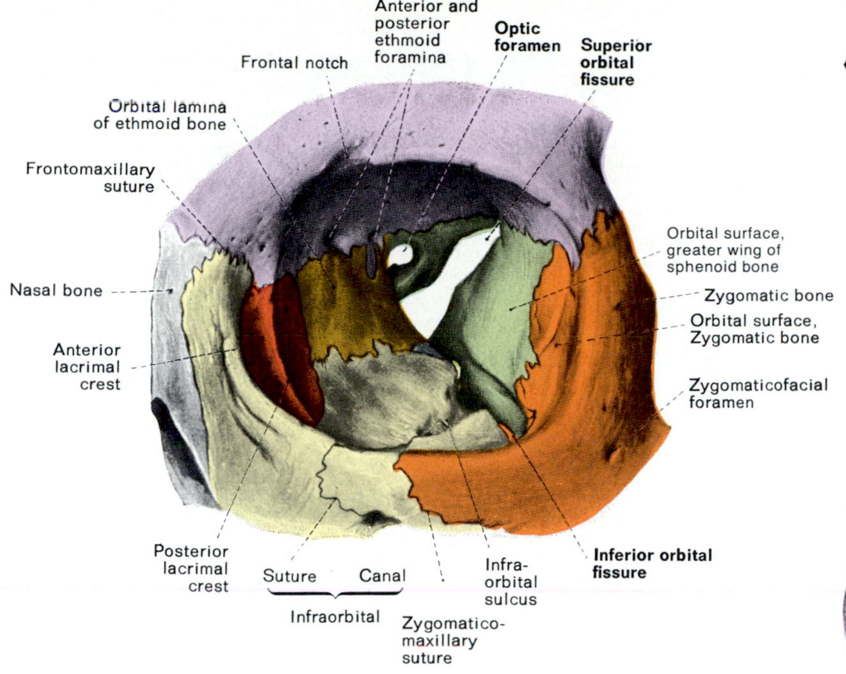
Anterior and posterior ethmoid foramina
Optic foramen
Superior orbital fissure
Frontal notch
Orbital lamina of ethmoid bone
Frontomaxillary suture
Nasal bone
Anterior lacrimal crest
Orbital surface, greater wing of sphenoid bone
Zygomatic bone
Orbital surface, Zygomatic bone
Zygomaticofacial foramen
Posterior lacrimal crest
Suture
Canal
Infraorbital
Infra-orbital sulcus
Inferior orbital fissure
Zygomatico-maxillary suture

Fig. 661: The Left Bony Orbital Cavity (Anterior View)

NOTE: 1) the bony structure of the orbit is composed of parts of seven bones: the maxilla, zygomatic, frontal, lacrimal, palatine, ethmoid and sphenoid.

2) the *roof* of the orbit is formed by the orbital plate of the frontal bone; the *floor* consists of the orbital plate of the maxilla, the palatine (blue, not labelled) and zygomatic bones; the *medial wall* is thin and delicate and is formed by the frontal process of the maxilla, the orbital lamina of the ethmoid and the lacrimal bone; the strong *lateral wall* consists of the orbital processes of the sphenoid and zygomatic bones.

3) the optic foramen and the superior and inferior orbital fissures.

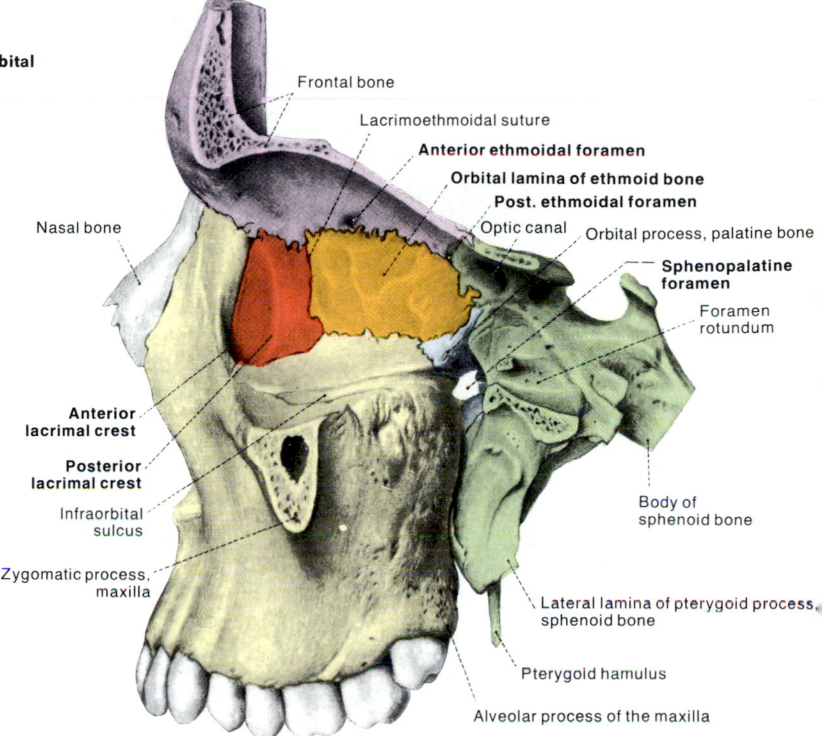
Frontal bone
Lacrimoethmoidal suture
Anterior ethmoidal foramen
Orbital lamina of ethmoid bone
Post. ethmoidal foramen
Nasal bone
Optic canal
Orbital process, palatine bone
Sphenopalatine foramen
Foramen rotundum
Anterior lacrimal crest
Posterior lacrimal crest
Infraorbital sulcus
Zygomatic process, maxilla
Body of sphenoid bone
Lateral lamina of pterygoid process, sphenoid bone
Pterygoid hamulus
Alveolar process of the maxilla

Fig. 662: The Medial Wall of the Left Orbital Cavity and a Lateral View of the Pterygopalatine Fossa

NOTE: 1) that anteriorly on the thin medial wall of the orbital cavity is found the lacrimal fossa for the lacrimal sac. This fossa is limited in front by the anterior lacrimal crest of the maxilla and behind by the posterior lacrimal crest of the lacrimal bone.

2) that the largest contribution to the medial wall of the orbit is made by the orbital lamina of the ethmoid bone. It is nearly rectangular in shape and articulates with the frontal bone above, the lacrimal bone anteriorly, the maxilla below, and the sphenoid and palatine bones posteriorly. Observe the anterior and posterior ethmoidal foramina through which course the anterior and posterior ethmoidal vessels and nerves.

3) that below the apex of the orbital cavity (posteriorly) is located the pterygopalatine fossa and the sphenopalatine foramen.

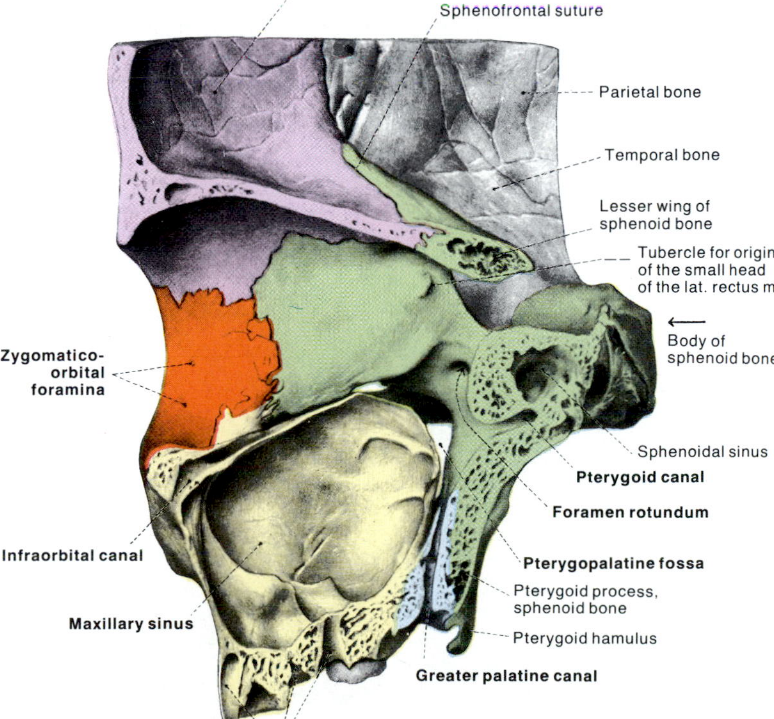
Internal surface of frontal bone
Sphenofrontal suture
Parietal bone
Temporal bone
Lesser wing of sphenoid bone
Tubercle for origin of the small head of the lat. rectus m.
Body of sphenoid bone
Zygomatico-orbital foramina
Sphenoidal sinus
Pterygoid canal
Foramen rotundum
Infraorbital canal
Pterygopalatine fossa
Pterygoid process, sphenoid bone
Maxillary sinus
Pterygoid hamulus
Greater palatine canal
Sockets for teeth (dental alveoli)

Fig. 663: The Lateral Wall of the Right Orbital Cavity and a Medial View of the Pterygopalatine Fossa

NOTE: 1) that the lateral wall of the orbit is formed posteriorly by the orbital surface of the greater wing of the sphenoid bone and anteriorly by the frontal process (i. e., the orbital surface) of the zygomatic bone. In this part of the zygomatic bone are found the small zygomatico-orbital foramina for the zygomaticofacial and zygomaticotemporal nerves. These two sensory nerves are derived from the zygomatic branch of the maxillary nerve. The zygomaticofacial nerve supplies the skin over the prominence of the cheek, while the terminal branch of the zygomaticotemporal nerve supplies the skin over the temple.

2) the foramen rotundum, through which the maxillary division of the trigeminal nerve leaves the cranial cavity. Also observe the infraorbital canal (on the floor of the orbit), along which the infraorbital branch of the maxillary nerve courses to the infraorbital foramen on the anterior aspect of the face.

3) the maxillary sinus, below most of the orbital floor, and the pterygopalatine fossa, inferior to the posterior part of the orbit. Observe the openings of the pterygoid canal and the greater palatine canal.

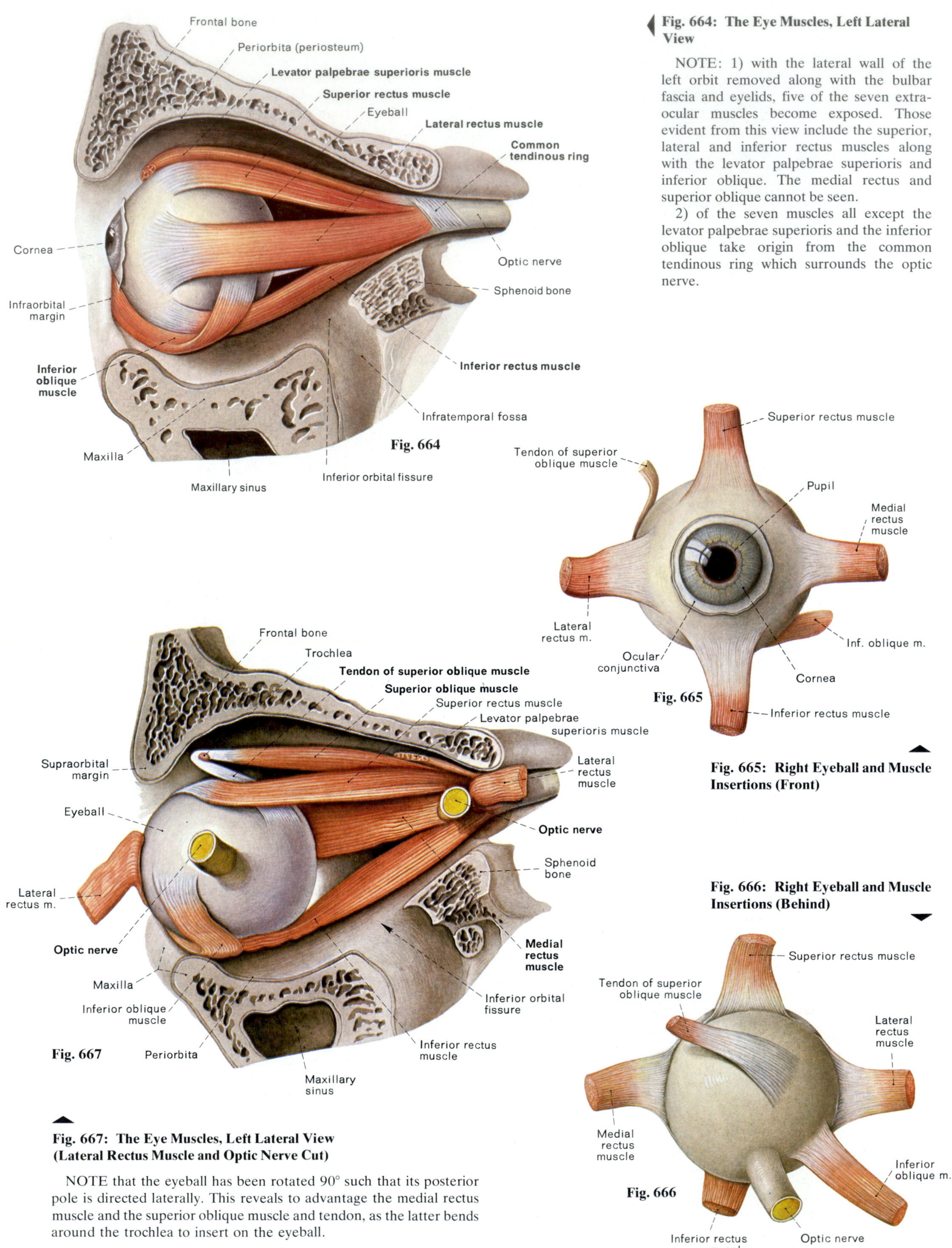
Frontal bone
Periorbita (periosteum)
Levator palpebrae superioris muscle
Superior rectus muscle
Eyeball
Lateral rectus muscle
Common tendinous ring
Cornea
Optic nerve
Sphenoid bone
Infraorbital margin
Inferior oblique muscle
Inferior rectus muscle
Infratemporal fossa
Maxilla
Maxillary sinus
Inferior orbital fissure
Superior rectus muscle
Tendon of superior oblique muscle
Pupil
Medial rectus muscle
Lateral rectus m.
Ocular conjunctiva
Inf. oblique m.
Cornea
Inferior rectus muscle
Frontal bone
Trochlea
Tendon of superior oblique muscle
Superior oblique muscle
Superior rectus muscle
Levator palpebrae superioris muscle
Supraorbital margin
Lateral rectus muscle
Eyeball
Optic nerve
Sphenoid bone
Lateral rectus m.
Optic nerve
Medial rectus muscle
Maxilla
Inferior orbital fissure
Inferior oblique muscle
Periorbita
Inferior rectus muscle
Maxillary sinus
Superior rectus muscle
Tendon of superior oblique muscle
Lateral rectus muscle
Medial rectus muscle
Inferior oblique m.
Inferior rectus muscle
Optic nerve

Fig. 664

Fig. 664: The Eye Muscles, Left Lateral View

NOTE: 1) with the lateral wall of the left orbit removed along with the bulbar fascia and eyelids, five of the seven extraocular muscles become exposed. Those evident from this view include the superior, lateral and inferior rectus muscles along with the levator palpebrae superioris and inferior oblique. The medial rectus and superior oblique cannot be seen.

2) of the seven muscles all except the levator palpebrae superioris and the inferior oblique take origin from the common tendinous ring which surrounds the optic nerve.

Fig. 665

Fig. 665: Right Eyeball and Muscle Insertions (Front)

Fig. 666: Right Eyeball and Muscle Insertions (Behind)

Fig. 666

Fig. 667

Fig. 667: The Eye Muscles, Left Lateral View (Lateral Rectus Muscle and Optic Nerve Cut)

NOTE that the eyeball has been rotated 90° such that its posterior pole is directed laterally. This reveals to advantage the medial rectus muscle and the superior oblique muscle and tendon, as the latter bends around the trochlea to insert on the eyeball.

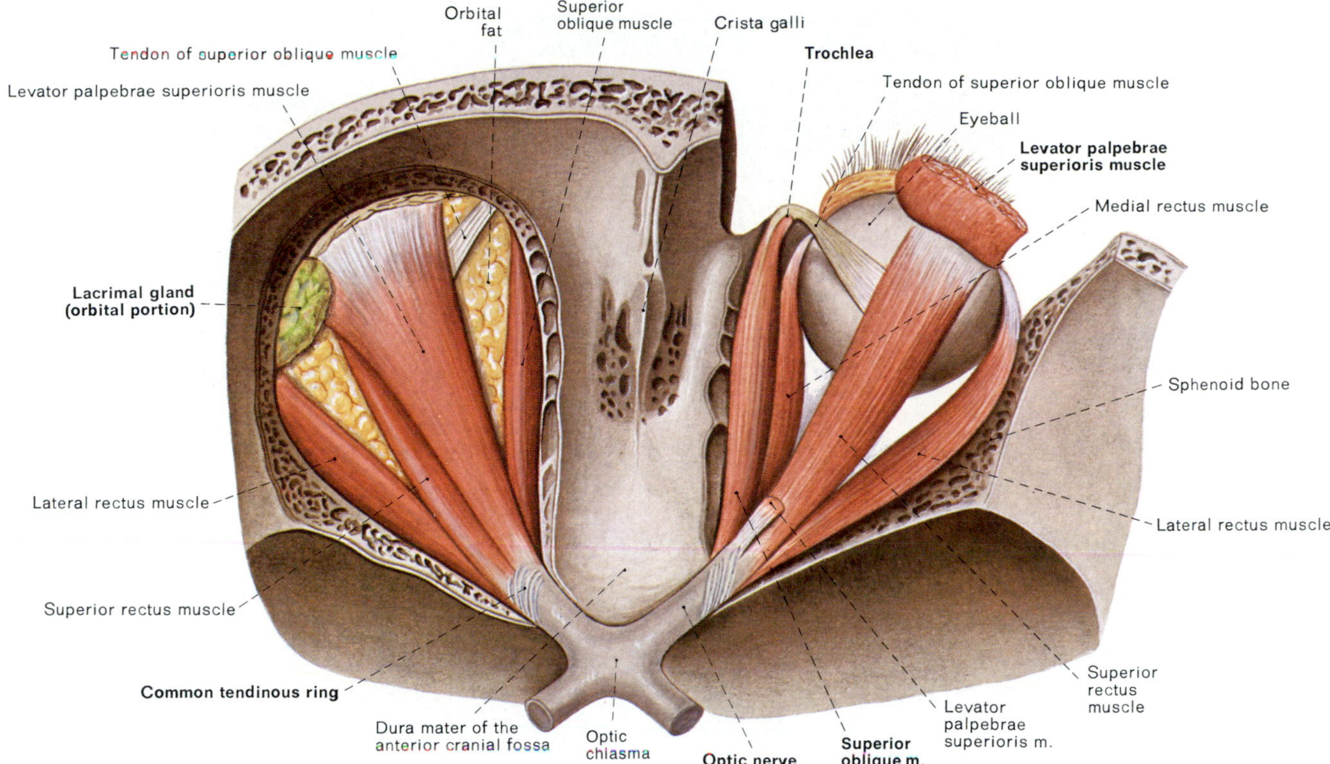
Orbital fat
Superior oblique muscle
Crista galli
Tendon of superior oblique muscle
Trochlea
Levator palpebrae superioris muscle
Tendon of superior oblique muscle
Eyeball
Levator palpebrae superioris muscle
Medial rectus muscle
Lacrimal gland (orbital portion)
Sphenoid bone
Lateral rectus muscle
Lateral rectus muscle
Superior rectus muscle
Superior rectus muscle
Common tendinous ring
Levator palpebrae superioris m.
Dura mater of the anterior cranial fossa
Optic chiasma
Optic nerve
Superior oblique m.

Fig. 668: The Muscles of the Orbital Cavity as seen from above

NOTE: 1) the orbital plates of the frontal bone have been removed from within the cranium. On the left side, only the bony roof of the orbit has been opened and the muscles, orbital fat and lacrimal gland have been left intact.

2) on the right side, the levator palpebrae superioris muscle has been resected and the orbital fat removed in order to expose the ocular muscles.

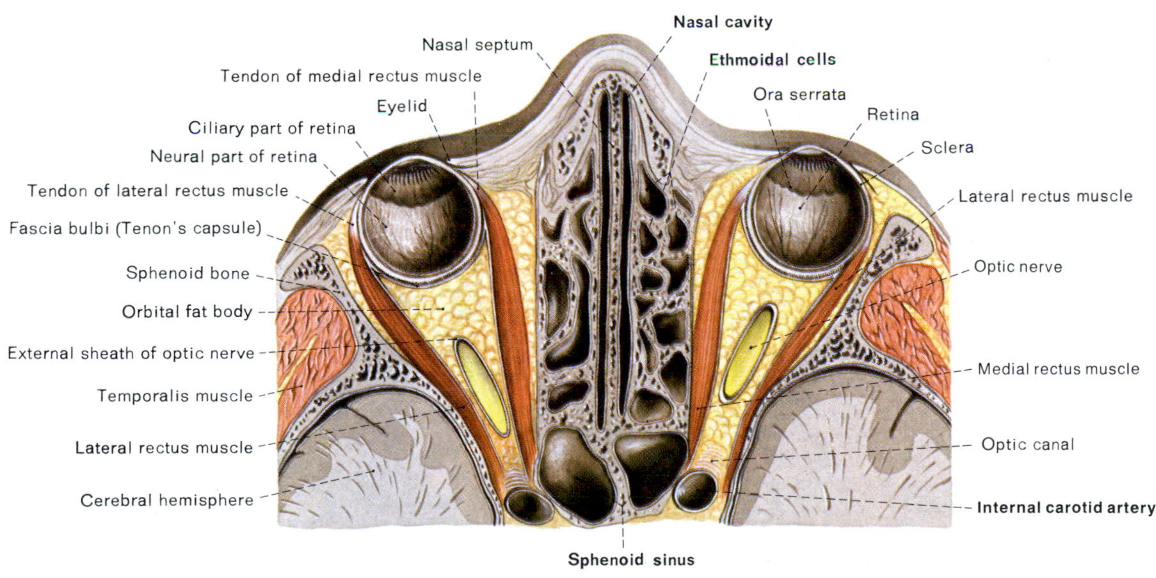
Nasal cavity
Nasal septum
Ethmoidal cells
Tendon of medial rectus muscle
Ora serrata
Eyelid
Retina
Ciliary part of retina
Sclera
Neural part of retina
Tendon of lateral rectus muscle
Lateral rectus muscle
Fascia bulbi (Tenon's capsule)
Optic nerve
Sphenoid bone
Orbital fat body
External sheath of optic nerve
Medial rectus muscle
Temporalis muscle
Lateral rectus muscle
Optic canal
Cerebral hemisphere
Internal carotid artery
Sphenoid sinus

Fig. 669: A Horizontal Section Through Both Orbits at the Level of the Sphenoid Sinus

NOTE: 1) between the orbital cavities is situated the ethmoid bone containing the ethmoidal air sinuses (air cells). The vertically oriented perpendicular plate of the ethmoid serves as the nasal septum, which subdivides the nasal cavity into two symmetrical chambers.

2) the posterior portion of the orbits are separated by the sphenoidal sinuses, located within the body of the sphenoid bone. These sinuses usually are not symmetrical.

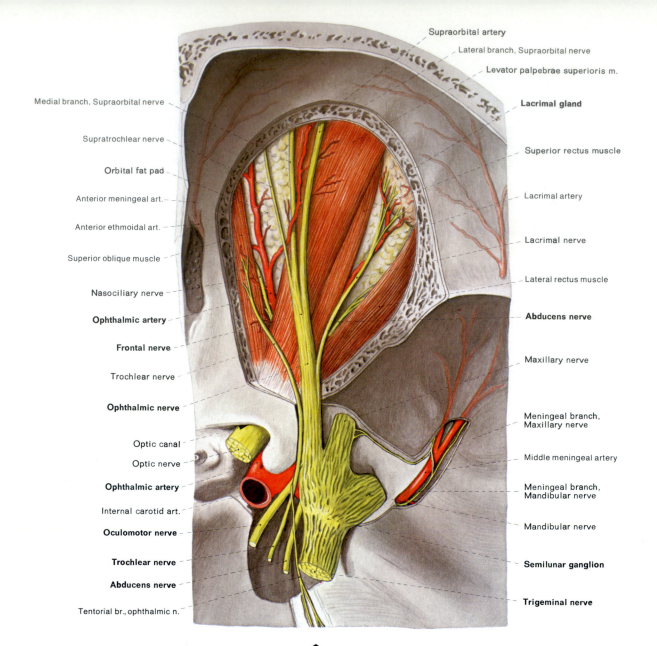

Supraorbital artery
Lateral branch, Supraorbital nerve
Levator palpebrae superioris m.
Medial branch, Supraorbital nerve
Lacrimal gland
Supratrochlear nerve
Superior rectus muscle
Orbital fat pad
Anterior meningeal art.
Lacrimal artery
Anterior ethmoidal art.
Lacrimal nerve
Superior oblique muscle
Lateral rectus muscle
Nasociliary nerve
Ophthalmic artery
Abducens nerve
Frontal nerve
Maxillary nerve
Trochlear nerve
Ophthalmic nerve
Meningeal branch, Maxillary nerve
Optic canal
Middle meningeal artery
Optic nerve
Ophthalmic artery
Meningeal branch, Mandibular nerve
Internal carotid art.
Mandibular nerve
Oculomotor nerve
Trochlear nerve
Semilunar ganglion
Abducens nerve
Trigeminal nerve
Tentorial br., ophthalmic n.

Fig. 670: Nerves and Arteries of the Orbit (Stage 1) Superior View: Ophthalmic Nerve and Artery

NOTE: 1) that the orbital plate of the frontal bone has been removed to expose the structures in the right orbit from above. Observe that the ophthalmic division of the trigeminal nerve divides into lacrimal, frontal and nasociliary branches. The lacrimal nerve courses anteriorly and laterally in the orbit and accompanies the lacrimal branch of the ophthalmic artery to supply the lacrimal gland.

2) that the frontal nerve overlies the levator palpebrae superioris and soon divides into a delicate supratrochlear branch and larger medial and lateral branches. These course toward the front of the orbit where they emerge on the forehead. Observe that the nasociliary nerve crosses the orbit anteromedially, deep to the superior rectus muscle, and accompanies the ophthalmic artery for a short distance.

3) the trochlear nerve entering the orbit medial to the ophthalmic nerve to supply the superior oblique muscle.

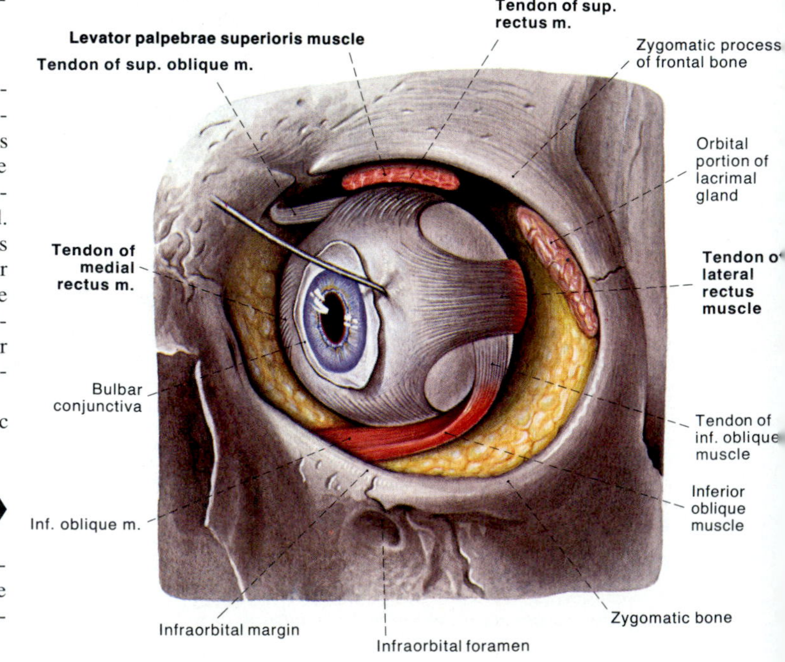
Tendon of sup. rectus m.
Levator palpebrae superioris muscle
Tendon of sup. oblique m.
Zygomatic process of frontal bone
Orbital portion of lacrimal gland
Tendon of medial rectus m.
Tendon of lateral rectus muscle
Bulbar conjunctiva
Tendon of inf. oblique muscle
Inferior oblique muscle
Inf. oblique m.
Zygomatic bone
Infraorbital margin
Infraorbital foramen

Fig. 671: The Extrinsic Muscle of the Left Eyeball, Anterolateral View

NOTE that the skin, eyelids and fascia have been removed. Observe that the tendon of the inferior oblique muscle inserts onto the lateral part of the eyeball, behind its equator and between the insertions of the superior and lateral recti muscles.

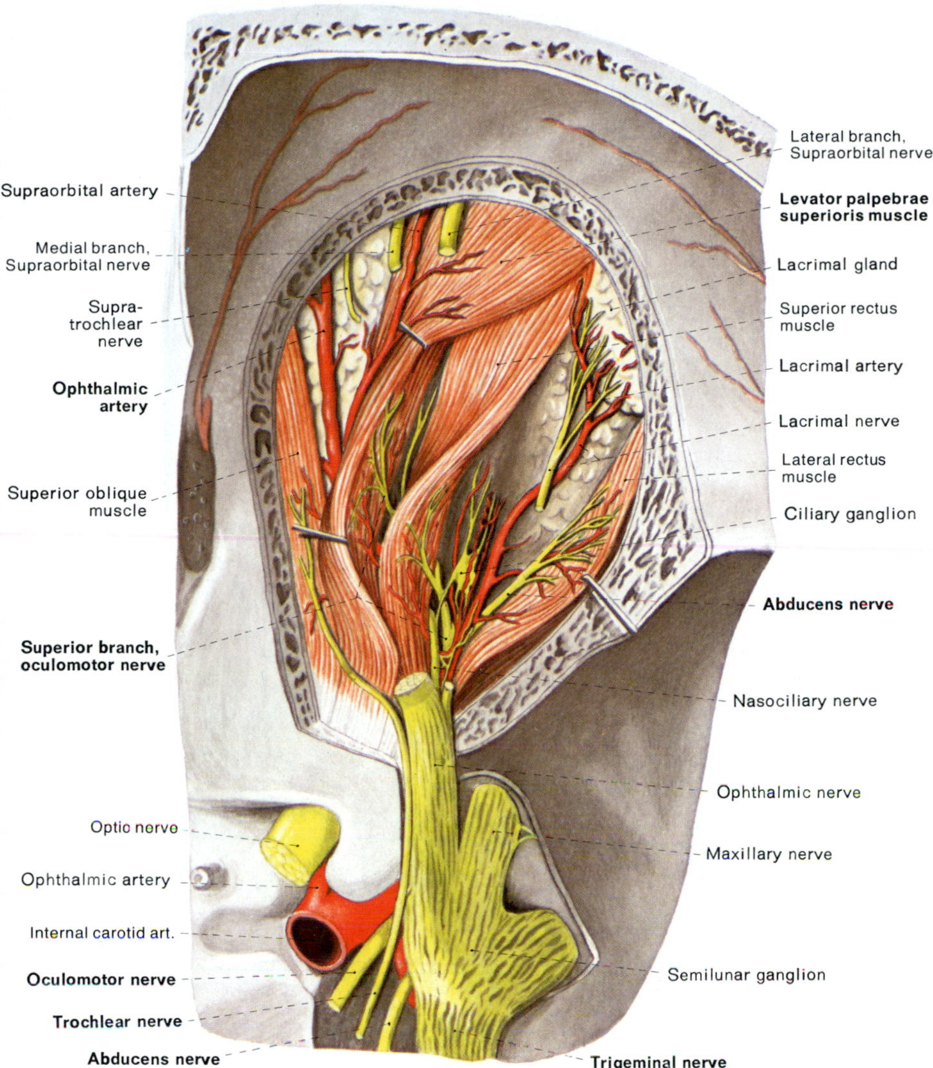

Lateral branch, Supraorbital nerve
Supraorbital artery
Levator palpebrae superioris muscle
Medial branch, Supraorbital nerve
Lacrimal gland
Supra-trochlear nerve
Superior rectus muscle
Lacrimal artery
Ophthalmic artery
Lacrimal nerve
Lateral rectus muscle
Superior oblique muscle
Ciliary ganglion
Abducens nerve
Superior branch, oculomotor nerve
Nasociliary nerve
Ophthalmic nerve
Optic nerve
Maxillary nerve
Ophthalmic artery
Internal carotid art.
Oculomotor nerve
Semilunar ganglion
Trochlear nerve
Abducens nerve
Trigeminal nerve

Fig. 672: Nerves and Arteries of the Orbit (Stage 2) Superior View: The Trochlear and Abducens Nerves

NOTE: 1) that with the right orbit opened from above, the ophthalmic division of the trigeminal nerve and its lacrimal, supratrochlear and frontal branches have been cut. The levator palpebrae superioris and superior rectus muscles have been pulled medially to reveal their inferior surface where filaments from the *superior branch of the oculomotor nerve* innervate the two muscles.

2) that the *nasociliary branch* of the ophthalmic nerve is still intact, and it is seen turning medially deep to the superior rectus muscle. Observe that a fine communicating filament containing sensory fibers interconnects the ciliary ganglion with the nasociliary nerve.

3) the *trochlear nerve* as it supplies the superior oblique muscle along its upper surface. If this nerve is injured, the patient has difficulty turning the eyeball laterally and downward. When asked to look inferolaterally, the affected eye rotates medially, resulting in double vision or diplopia.

4) the *abducens nerve* as it supplies the lateral rectus muscle along its inner or medial surface. After emerging from the brainstem at the pontomedullary junction, the abducens nerve follows a long course in the floor of the cranial cavity to achieve the orbit by way of the superior orbital fissure. Injury to this nerve produces a diminished ability to move the eyeball laterally. From the resulting medial or convergent gaze of the affected eyeball, the patient complains of diplopia (double vision).

Fig. 673: Sagittal View of the Orbital Cavity and Eyeball

NOTE: 1) that bulbar fascia (also seen in Figure 677) is a thin membrane which encloses the posterior 3/4ths of the eyeball, separating the eyeball from the orbital fat and other contents of the orbital cavity. This fascia is pierced by the tendons of the ocular muscles, over which the fascia is prolonged like a tubular sheath.

2) the insertion of the levator palpebrae superioris is trilaminar. The superficial layer inserts into the upper eyelid, the middle into the superior tarsus and the deep layer inserts into the superior fornix of the conjunctiva.

3) the palpebral conjunctiva is a thin transparent mucous membrane on the innermost aspect of the eyelid. At the conjunctival angle (fornix), it reflects over the eyeball as far as the sclerocorneal junction.

4) that the optic nerve is located on the posterior aspect of the eyeball about 3 mm to the nasal side of the macula lutea and fovea centralis (see Fig. 680). The point at which the optic nerve pierces the eyeball is called the optic disc (optic papilla) and, because this site is devoid of neural retina, it is often referred to as the blind spot. The blood vessels that supply and drain the retina pass through the optic disc along with the axons of the ganglion cells of the retina that form the optic nerve.

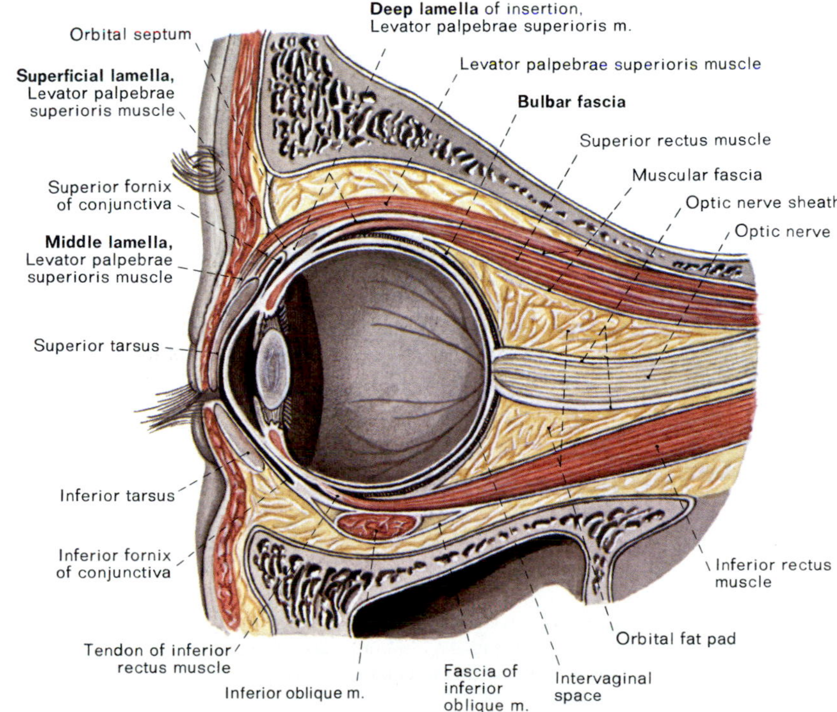

Deep lamella of insertion, Levator palpebrae superioris m.
Orbital septum
Levator palpebrae superioris muscle
Superficial lamella, Levator palpebrae superioris muscle
Bulbar fascia
Superior rectus muscle
Muscular fascia
Superior fornix of conjunctiva
Optic nerve sheath
Optic nerve
Middle lamella, Levator palpebrae superioris muscle
Superior tarsus
Inferior tarsus
Inferior fornix of conjunctiva
Inferior rectus muscle
Orbital fat pad
Tendon of inferior rectus muscle
Inferior oblique m.
Fascia of inferior oblique m.
Intervaginal space

Fig. 674: Nerves and Arteries of the Orbit (Stage 3) Superior View: The Optic Nerve and Ciliary Ganglion

NOTE: 1) that with the levator palpebrae superioris, superior rectus and superior oblique muscles cut and reflected, the nasociliary nerve and ophthalmic artery are seen crossing over the optic nerve from lateral to medial.

2) the optic nerve and the longitudinally oriented long ciliary arteries (from the ophthalmic) and long ciliary nerves (usually 2 or 3 branches of the nasociliary nerve).

3) the ciliary ganglion lateral to the optic nerve. Its *parasymphathetic root* comes from the oculomotor nerve and its *sensory root* from the nasociliary nerve. Postganglionic parasymphathetic fibers reach the eyeball by way of the short ciliary nerves.

4) that postganglionic parasympathetic nerve fibers supply the sphincter of the pupil and the muscle responsible for accommodation of the lens – the ciliary muscle.

5) that some sympathetic fibers (that arrive in the orbit along the ophthalmic artery from the cavernous plexus) also course through the ciliary ganglion. These are principally vasoconstrictor fibers to arteries that supply the eyeball. Sympathetic fibers that supply the dilator muscle of the pupil course to the posterior pole of the eyeball by way of the long ciliary nerves.

6) that although the supratrochlear nerve is derived from the frontal branch of the ophthalmic nerve, the infratrochlear nerve (as well as the anterior and posterior ethmoidal nerves) are derived from the nasociliary branch of the ophthalmic nerve.

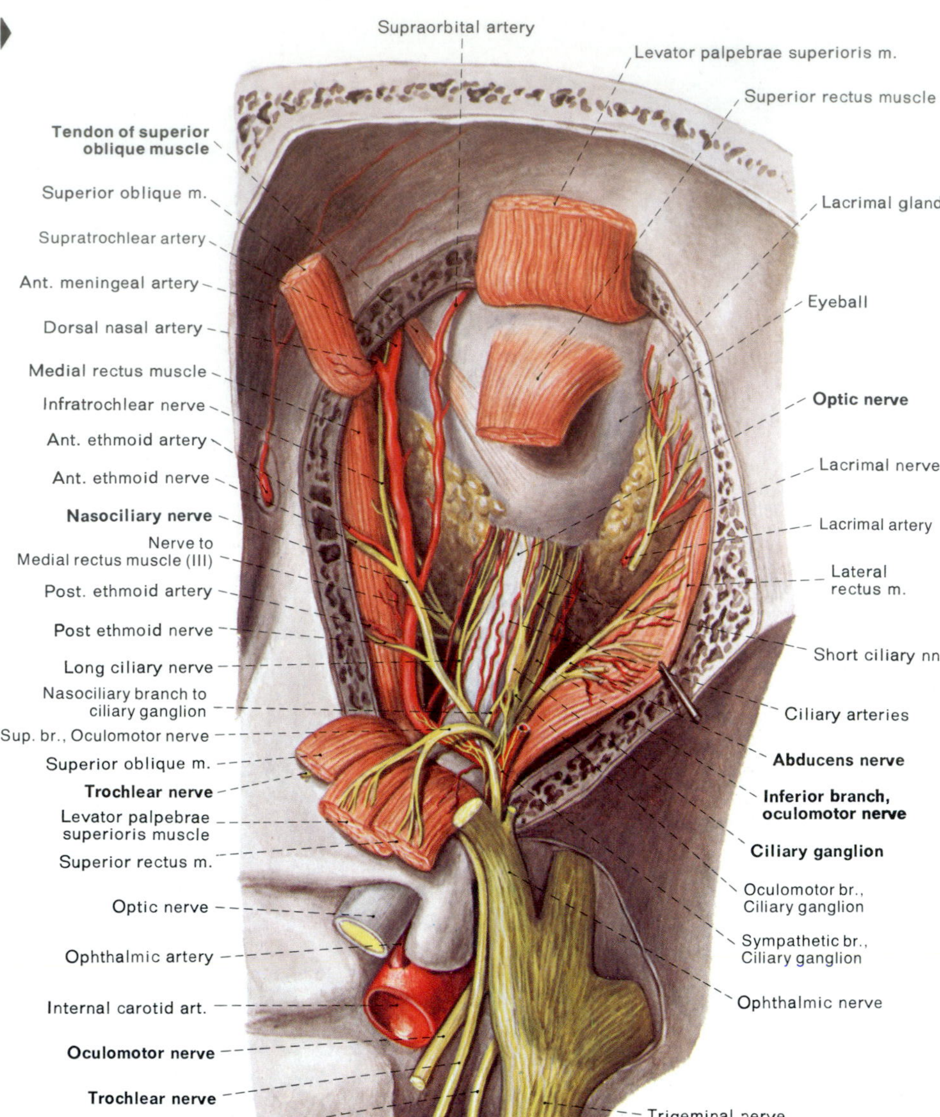
Supraorbital artery
Levator palpebrae superioris m.
Superior rectus muscle
Lacrimal gland
Eyeball
Optic nerve
Lacrimal nerve
Lacrimal artery
Lateral rectus m.
Short ciliary nn.
Ciliary arteries
Abducens nerve
Inferior branch, oculomotor nerve
Ciliary ganglion
Oculomotor br., Ciliary ganglion
Sympathetic br., Ciliary ganglion
Ophthalmic nerve
Trigeminal nerve
Tendon of superior oblique muscle
Superior oblique m.
Supratrochlear artery
Ant. meningeal artery
Dorsal nasal artery
Medial rectus muscle
Infratrochlear nerve
Ant. ethmoid artery
Ant. ethmoid nerve
Nasociliary nerve
Nerve to Medial rectus muscle (III)
Post. ethmoid artery
Post ethmoid nerve
Long ciliary nerve
Nasociliary branch to ciliary ganglion
Sup. br., Oculomotor nerve
Superior oblique m.
Trochlear nerve
Levator palpebrae superioris muscle
Superior rectus m.
Optic nerve
Ophthalmic artery
Internal carotid art.
Oculomotor nerve
Trochlear nerve
Abducens nerve

Fig. 675: Origins of the Ocular Muscles, Apex of Left Orbit

NOTE: 1) this anterior view of the apex of the left orbit shows the stumps of the ocular muscles which have been cut close to their origins.

2) the four rectus muscles arise from a common tendinous ring surrounding the optic canal. The levator palpebrae superioris and superior oblique muscles arise from the sphenoid bone close to the tendinous ring, while the inferior oblique (not shown) arises from the orbital surface of the maxilla.

3) a. The **lateral rectus** *abducts* the eyeball;
   b. the **superior oblique** *abducts, depresses and medially rotates* the eyeball;
   c. the **inferior oblique** *abducts, elevates* and *laterally rotates* the eyeball;
   d. the **medial rectus** *adducts* the eyeball;
   e. the **inferior rectus** *adducts, depresses* and *laterally rotates* the eyeball;
   f. the **superior rectus** *adducts, elevates* and *medially rotates* the eyeball.

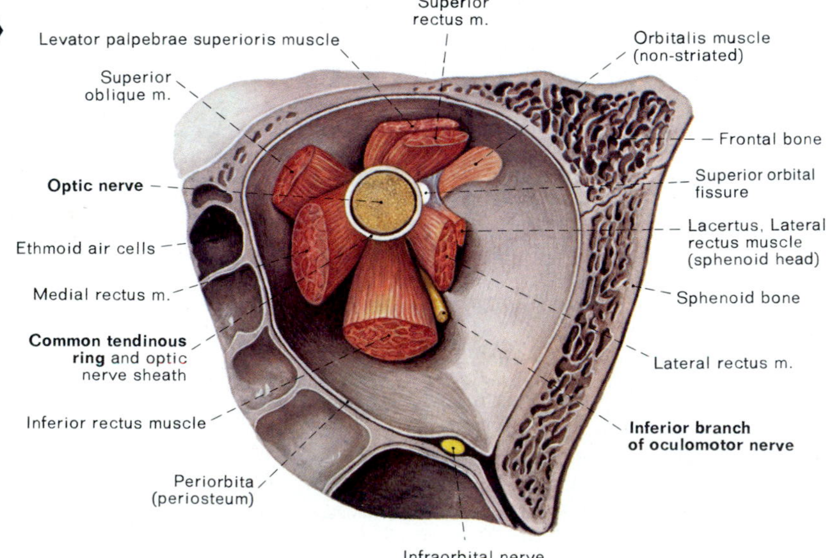
Superior rectus m.
Levator palpebrae superioris muscle
Orbitalis muscle (non-striated)
Superior oblique m.
Frontal bone
Superior orbital fissure
Optic nerve
Lacertus, Lateral rectus muscle (sphenoid head)
Ethmoid air cells
Sphenoid bone
Medial rectus m.
Common tendinous ring and optic nerve sheath
Lateral rectus m.
Inferior rectus muscle
Inferior branch of oculomotor nerve
Periorbita (periosteum)
Infraorbital nerve

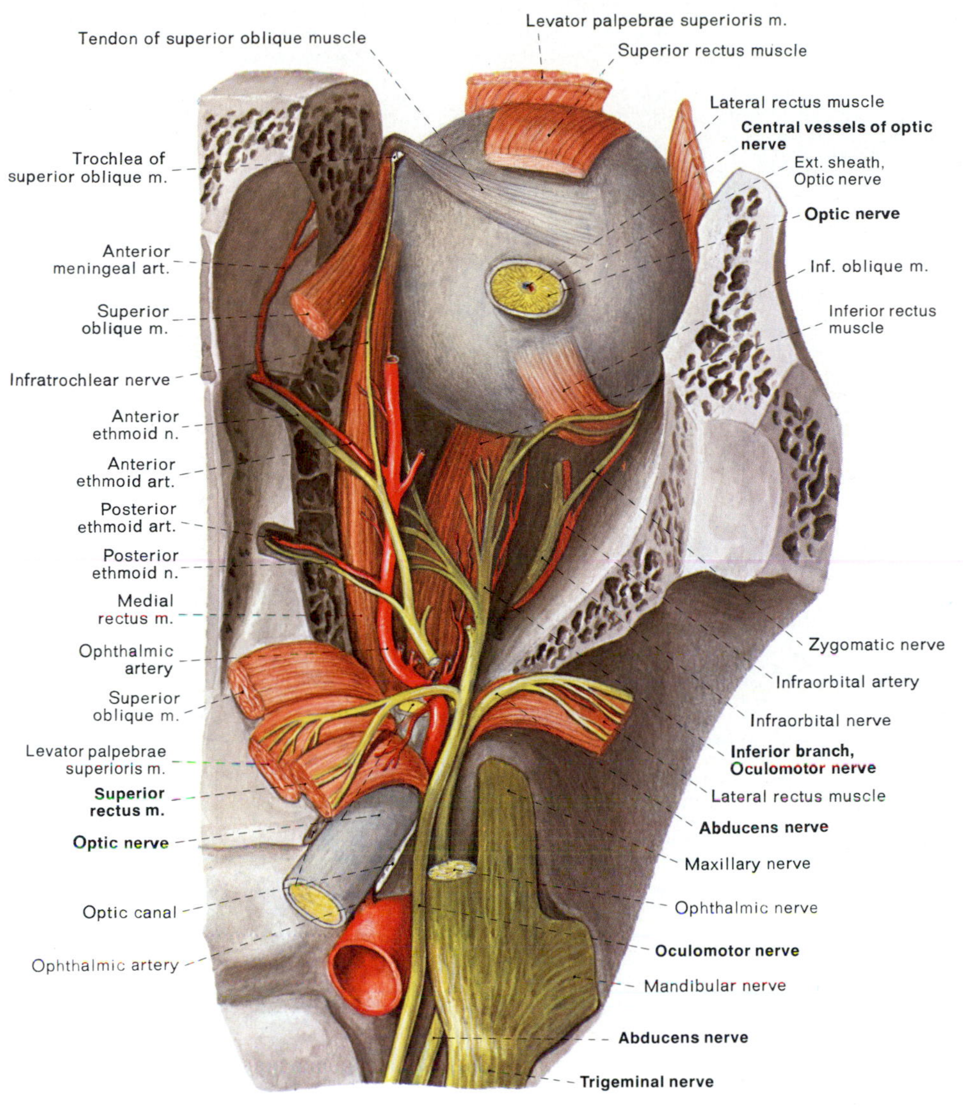

Tendon of superior oblique muscle
Levator palpebrae superioris m.
Superior rectus muscle
Lateral rectus muscle
Central vessels of optic nerve
Ext. sheath, Optic nerve
Optic nerve
Trochlea of superior oblique m.
Anterior meningeal art.
Inf. oblique m.
Superior oblique m.
Inferior rectus muscle
Infratrochlear nerve
Anterior ethmoid n.
Anterior ethmoid art.
Posterior ethmoid art.
Posterior ethmoid n.
Medial rectus m.
Zygomatic nerve
Ophthalmic artery
Infraorbital artery
Superior oblique m.
Infraorbital nerve
Levator palpebrae superioris m.
Inferior branch, Oculomotor nerve
Superior rectus m.
Lateral rectus muscle
Abducens nerve
Optic nerve
Maxillary nerve
Optic canal
Ophthalmic nerve
Ophthalmic artery
Oculomotor nerve
Mandibular nerve
Abducens nerve
Trigeminal nerve

Fig. 676: Nerves and Arteries of the Orbit (Stage 4) Superior View: The Oculomotor Nerve (Inferior Branch)

NOTE: 1) that the levator palpebrae superioris, superior rectus, superior oblique and lateral rectus muscles have been cut and reflected; the optic nerve has also been severed. The anterior half of the eyeball has been depressed and its posterior pole directed upward. Observe the central vessels of the optic nerve, as well as the insertions of the superior oblique and inferior oblique muscles.

2) that the *oculomotor nerve* enters the orbit through the superior orbital fissure and within the common tendinous ring. It quickly gives off its superior branch which courses upward in the orbit to the superior oblique and levator palpebrae superioris muscles. The inferior branch of the oculomotor nerve courses anteriorly in the deep part of the orbit to supply the inferior rectus, medial rectus and inferior oblique muscles.

3) the anterior and posterior ethmoidal arteries and nerves and the infratrochlear nerve medially in the orbit, and the infraorbit to supply the inferior rectus, medial rectus and inferior oblique muscles.

4) that the ophthalmic artery is the first branch of the internal carotid artery within the cranial cavity and it immediately enters the orbit through the optic canal alongside the optic nerve. Although nearly all the blood vessels in the orbit are branches from the ophthalmic artery, perhaps the most important is the central artery of the optic nerve (or retina), seen with its accompanying vein in the center of the optic nerve. This artery is the only source of blood to the neural retina and increased pressure on the posterior part of the orbital cavity or edema of the optic nerve caused by an inflammatory process can seriously compromise vision either by blockage of the artery or by diminishing the flow in the central retinal vein.

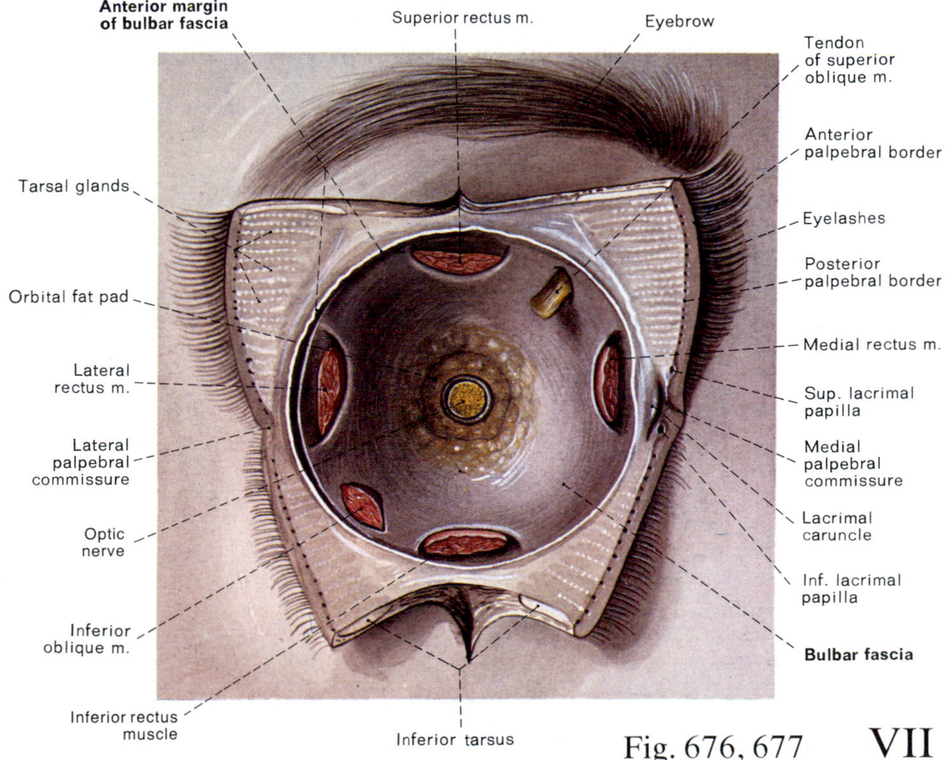

Anterior margin of bulbar fascia
Superior rectus m.
Eyebrow
Tendon of superior oblique m.
Anterior palpebral border
Tarsal glands
Eyelashes
Posterior palpebral border
Orbital fat pad
Medial rectus m.
Lateral rectus m.
Sup. lacrimal papilla
Lateral palpebral commissure
Medial palpebral commissure
Lacrimal caruncle
Optic nerve
Inf. lacrimal papilla
Inferior oblique m.
Bulbar fascia
Inferior rectus muscle
Inferior tarsus

Fig. 677: The Bulbar Fascia (Capsule of Tenon), Right Eye

NOTE: 1) longitudinal incisions have been made down the middle of each eyelid and the flaps have been reflected to expose the orbital cavity anteriorly.

2) the optic nerve has been severed at the optic disc and the eyeball, along with the insertions of the ocular muscles, has been removed from the orbital cavity.

3) the bulbar fascia which envelopes the posterior aspect of the eyeball (from the sclerocorneal junction to the optic nerve) has been left within the orbit. Observe how the bulbar fascia is perforated by the tendons of the ocular muscles. It is also pierced from behind by the ciliary vessels and nerves.

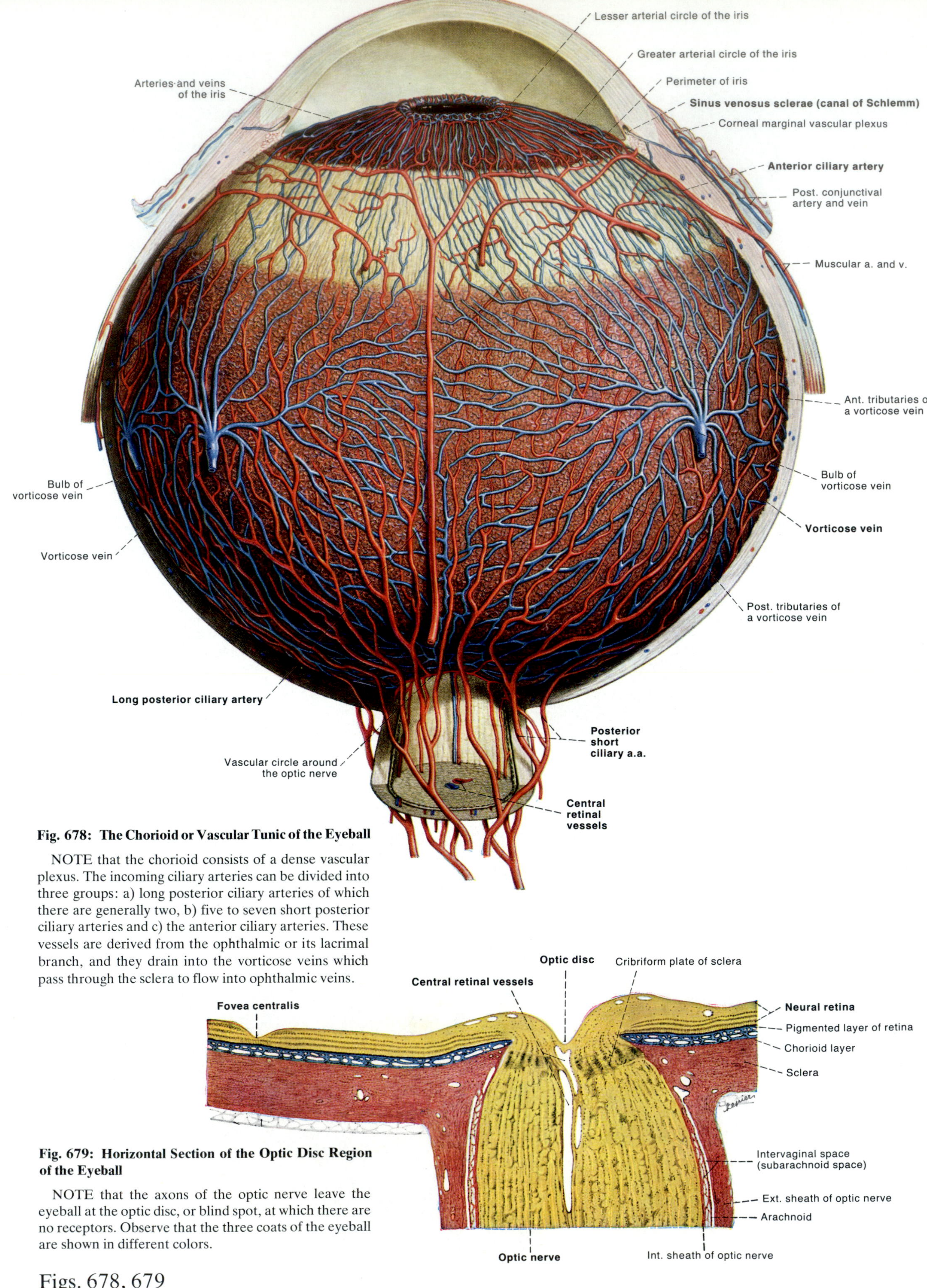
Lesser arterial circle of the iris
Greater arterial circle of the iris
Perimeter of iris
Sinus venosus sclerae (canal of Schlemm)
Corneal marginal vascular plexus
Arteries and veins of the iris
Anterior ciliary artery
Post. conjunctival artery and vein
Muscular a. and v.
Ant. tributaries of a vorticose vein
Bulb of vorticose vein
Bulb of vorticose vein
Vorticose vein
Vorticose vein
Post. tributaries of a vorticose vein
Long posterior ciliary artery
Posterior short ciliary a.a.
Vascular circle around the optic nerve
Central retinal vessels
Optic disc
Cribriform plate of sclera
Central retinal vessels
Fovea centralis
Neural retina
Pigmented layer of retina
Chorioid layer
Sclera
Intervaginal space (subarachnoid space)
Ext. sheath of optic nerve
Arachnoid
Optic nerve
Int. sheath of optic nerve

Fig. 678: The Chorioid or Vascular Tunic of the Eyeball

NOTE that the chorioid consists of a dense vascular plexus. The incoming ciliary arteries can be divided into three groups: a) long posterior ciliary arteries of which there are generally two, b) five to seven short posterior ciliary arteries and c) the anterior ciliary arteries. These vessels are derived from the ophthalmic or its lacrimal branch, and they drain into the vorticose veins which pass through the sclera to flow into ophthalmic veins.

Fig. 679: Horizontal Section of the Optic Disc Region of the Eyeball

NOTE that the axons of the optic nerve leave the eyeball at the optic disc, or blind spot, at which there are no receptors. Observe that the three coats of the eyeball are shown in different colors.

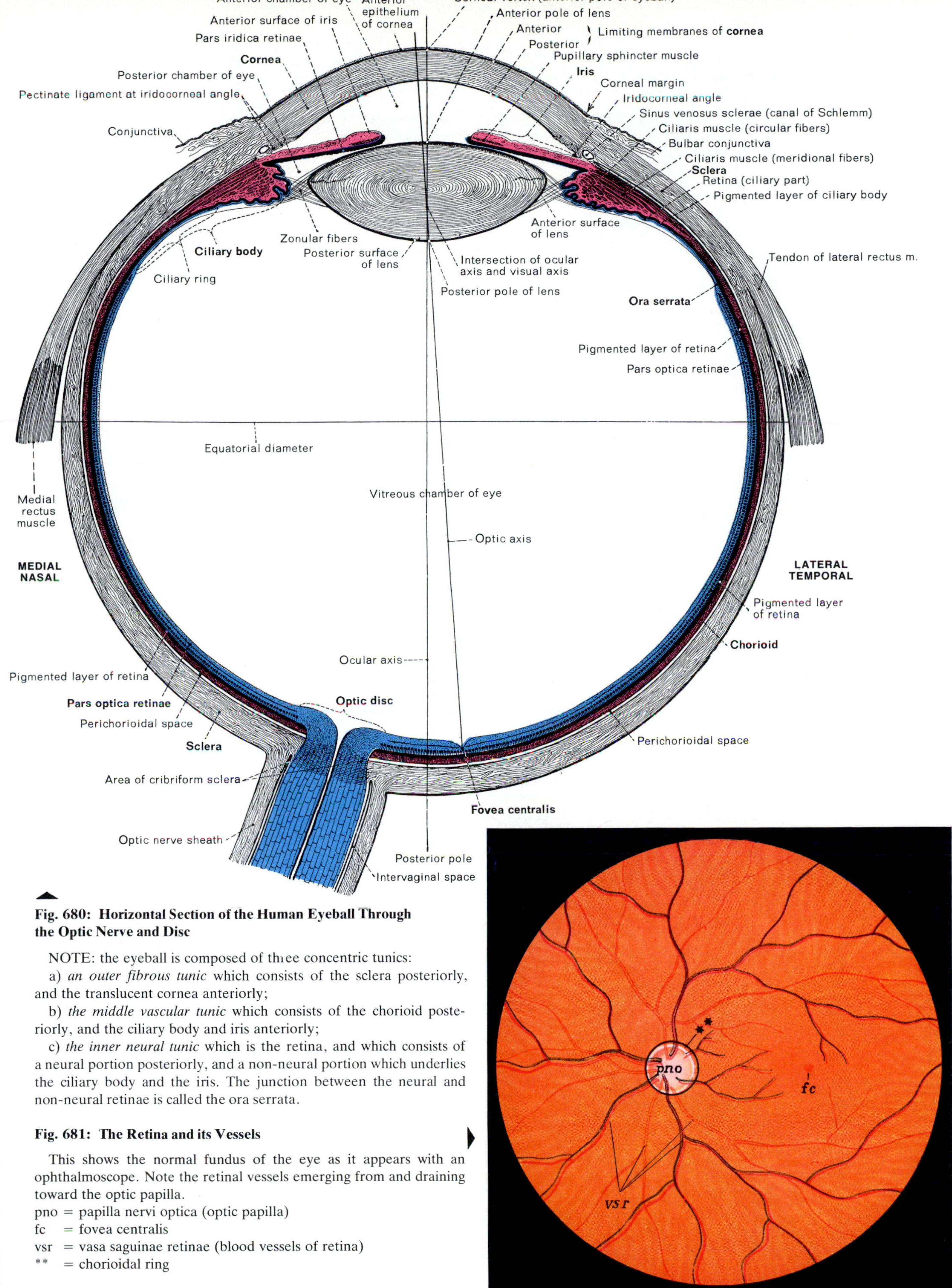

Anterior chamber of eye
Anterior epithelium of cornea
Corneal vertex (anterior pole of eyeball)
Anterior surface of iris
Anterior pole of lens
Pars iridica retinae
Anterior
Posterior
Limiting membranes of cornea
Cornea
Pupillary sphincter muscle
Posterior chamber of eye
Iris
Pectinate ligament at iridocorneal angle
Corneal margin
Iridocorneal angle
Sinus venosus sclerae (canal of Schlemm)
Conjunctiva
Ciliaris muscle (circular fibers)
Bulbar conjunctiva
Ciliaris muscle (meridional fibers)
Sclera
Retina (ciliary part)
Pigmented layer of ciliary body
Anterior surface of lens
Zonular fibers
Ciliary body
Posterior surface of lens
Intersection of ocular axis and visual axis
Tendon of lateral rectus m.
Ciliary ring
Posterior pole of lens
Ora serrata
Pigmented layer of retina
Pars optica retinae
Equatorial diameter
Medial rectus muscle
Vitreous chamber of eye
Optic axis
MEDIAL NASAL
LATERAL TEMPORAL
Pigmented layer of retina
Chorioid
Ocular axis
Pigmented layer of retina
Optic disc
Pars optica retinae
Perichorioidal space
Perichorioidal space
Sclera
Area of cribriform sclera
Fovea centralis
Optic nerve sheath
Posterior pole
Intervaginal space
pno
fc
vsr
**

Fig. 680: Horizontal Section of the Human Eyeball Through the Optic Nerve and Disc

NOTE: the eyeball is composed of three concentric tunics:

a) *an outer fibrous tunic* which consists of the sclera posteriorly, and the translucent cornea anteriorly;

b) *the middle vascular tunic* which consists of the chorioid posteriorly, and the ciliary body and iris anteriorly;

c) *the inner neural tunic* which is the retina, and which consists of a neural portion posteriorly, and a non-neural portion which underlies the ciliary body and the iris. The junction between the neural and non-neural retinae is called the ora serrata.

Fig. 681: The Retina and its Vessels

This shows the normal fundus of the eye as it appears with an ophthalmoscope. Note the retinal vessels emerging from and draining toward the optic papilla.

pno = papilla nervi optica (optic papilla)
fc = fovea centralis
vsr = vasa saguinae retinae (blood vessels of retina)
** = chorioidal ring

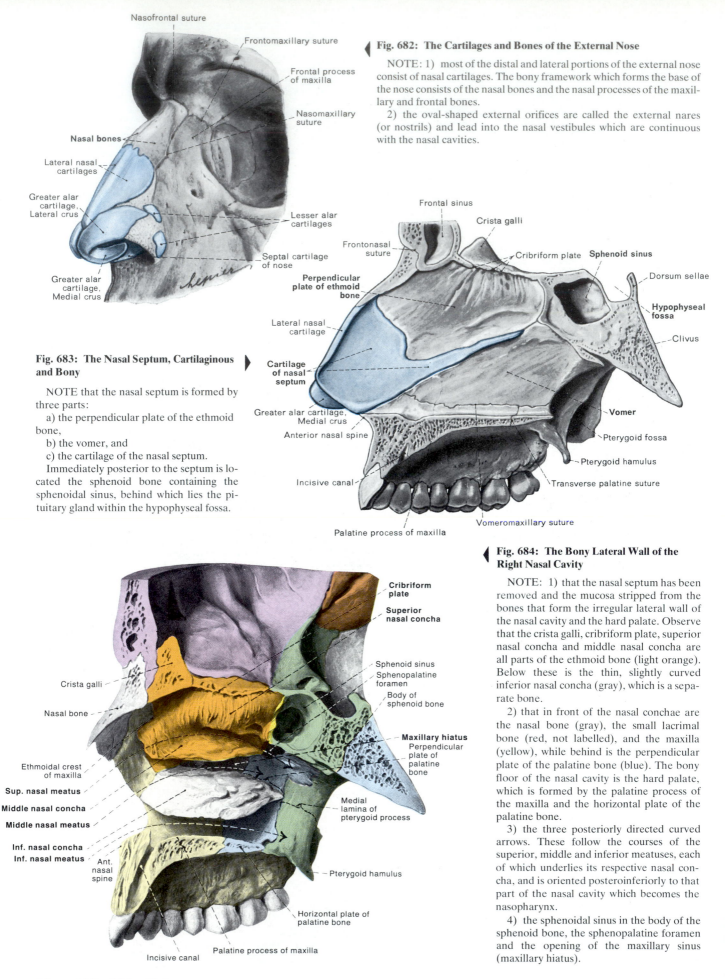

Fig. 682: The Cartilages and Bones of the External Nose

NOTE: 1) most of the distal and lateral portions of the external nose consist of nasal cartilages. The bony framework which forms the base of the nose consists of the nasal bones and the nasal processes of the maxillary and frontal bones.

2) the oval-shaped external orifices are called the external nares (or nostrils) and lead into the nasal vestibules which are continuous with the nasal cavities.

Fig. 683: The Nasal Septum, Cartilaginous and Bony

NOTE that the nasal septum is formed by three parts:

a) the perpendicular plate of the ethmoid bone,

b) the vomer, and

c) the cartilage of the nasal septum.

Immediately posterior to the septum is located the sphenoid bone containing the sphenoidal sinus, behind which lies the pituitary gland within the hypophyseal fossa.

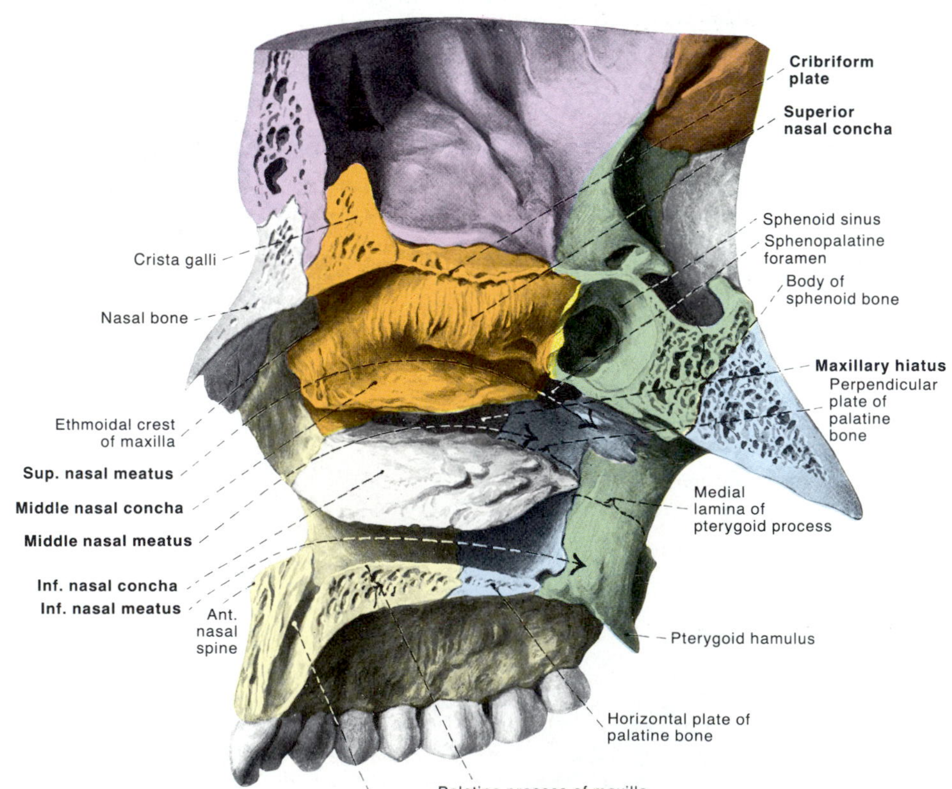
Cribriform plate
Superior nasal concha
Sphenoid sinus
Sphenopalatine foramen
Body of sphenoid bone
Crista galli
Nasal bone
Maxillary hiatus
Perpendicular plate of palatine bone
Ethmoidal crest of maxilla
Sup. nasal meatus
Middle nasal concha
Middle nasal meatus
Medial lamina of pterygoid process
Inf. nasal concha
Inf. nasal meatus
Ant. nasal spine
Pterygoid hamulus
Horizontal plate of palatine bone
Palatine process of maxilla
Incisive canal

Fig. 684: The Bony Lateral Wall of the Right Nasal Cavity

NOTE: 1) that the nasal septum has been removed and the mucosa stripped from the bones that form the irregular lateral wall of the nasal cavity and the hard palate. Observe that the crista galli, cribriform plate, superior nasal concha and middle nasal concha are all parts of the ethmoid bone (light orange). Below these is the thin, slightly curved inferior nasal concha (gray), which is a separate bone.

2) that in front of the nasal conchae are the nasal bone (gray), the small lacrimal bone (red, not labelled), and the maxilla (yellow), while behind is the perpendicular plate of the palatine bone (blue). The bony floor of the nasal cavity is the hard palate, which is formed by the palatine process of the maxilla and the horizontal plate of the palatine bone.

3) the three posteriorly directed curved arrows. These follow the courses of the superior, middle and inferior meatuses, each of which underlies its respective nasal concha, and is oriented posteroinferiorly to that part of the nasal cavity which becomes the nasopharynx.

4) the sphenoidal sinus in the body of the sphenoid bone, the sphenopalatine foramen and the opening of the maxillary sinus (maxillary hiatus).

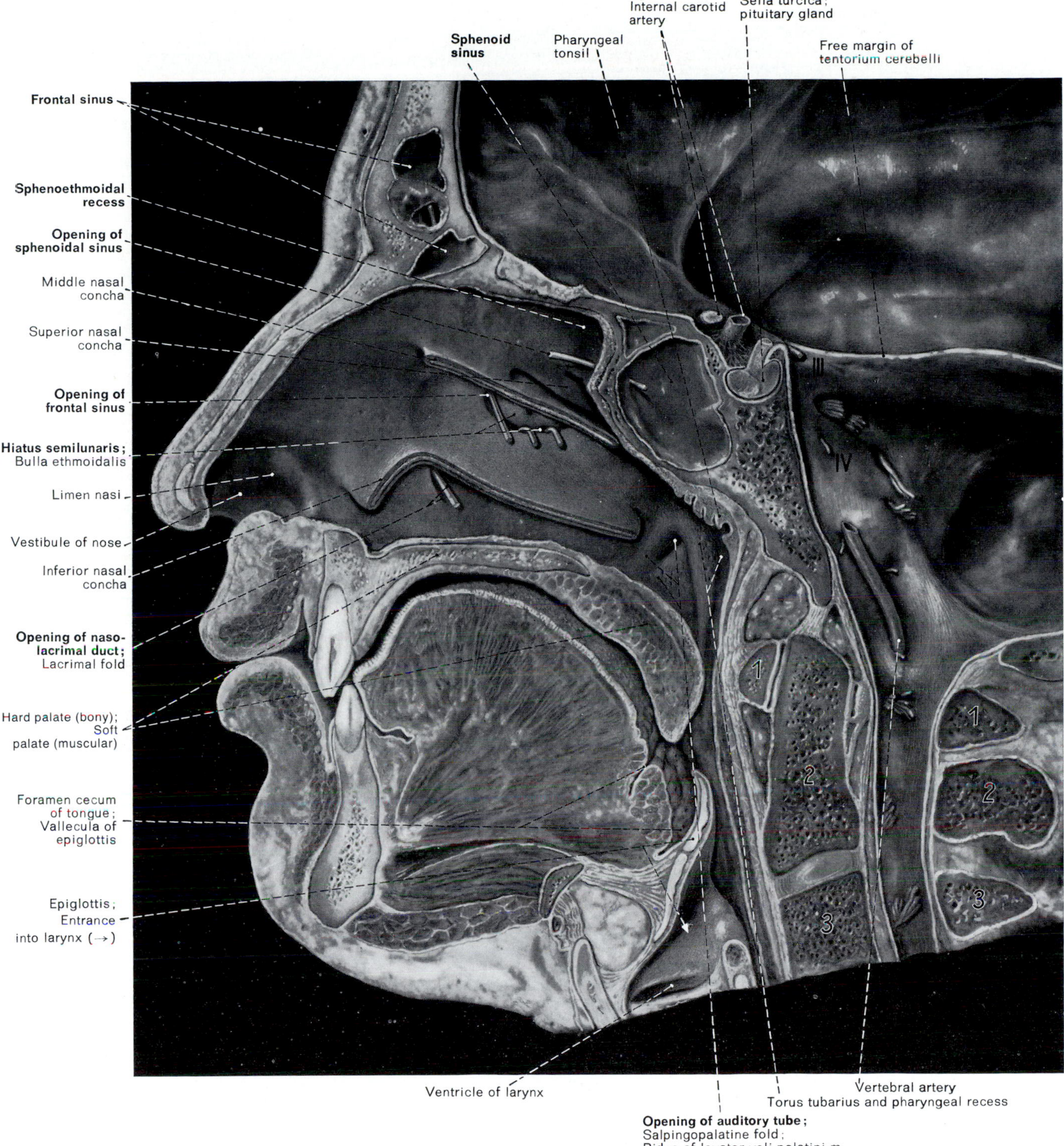

Optic nerve;
Internal carotid
artery
Sella turcica;
pituitary gland
Sphenoid
sinus
Pharyngeal
tonsil
Free margin of
tentorium cerebelli
Frontal sinus
Sphenoethmoidal
recess
Opening of
sphenoidal sinus
Middle nasal
concha
Superior nasal
concha
Opening of
frontal sinus
Hiatus semilunaris;
Bulla ethmoidalis
Limen nasi
Vestibule of nose
Inferior nasal
concha
Opening of naso-
lacrimal duct;
Lacrimal fold
Hard palate (bony);
Soft
palate (muscular)
Foramen cecum
of tongue;
Vallecula of
epiglottis
Epiglottis;
Entrance
into larynx (→)
III
IV
1
2
3
1
2
3
Ventricle of larynx
Vertebral artery
Torus tubarius and pharyngeal recess
Opening of auditory tube;
Salpingopalatine fold;
Ridge of levator veli palatini m.

Fig. 685: Lateral Wall of the Right Nasal Cavity Showing Openings of the Paranasal Air Sinuses and the Nasopharynx

NOTE: 1) this median sagittal section of the adult head displays the right nasal cavity with the middle and inferior nasal conchae removed. The nasal cavity communicates anteriorly with the environment through the vestibule and nostril, and posteriorly with the pharynx (nasopharynx).

2) opening of the various paranasal sinuses and other structures. These include:

a) the *sphenoid sinus* which opens into the sphenoethmoidal recess above the superior concha.

b) the *frontal sinus* and *maxillary sinus* both of which open into a groove called the hiatus semilunaris in the middle meatus below the middle concha.

c) the *nasolacrimal duct* which opens into the inferior meatus below the inferior concha.

d) the *auditory tube* which opens into the nasopharynx just behind the inferior concha. This tube stretches between the cavity of the middle ear (tympanic cavity) and the nasopharynx, thereby allowing the cavity of the middle ear to alter its air pressure consistent with the environment. This mechanism equalizes the air pressure on both sides of the tympanic membrane.

3) the nasal cavity, oral cavity and laryngeal cavity all communicate with the pharynx, forming, in turn, the nasopharynx, oral pharynx and laryngeal pharynx.

Crista galli;
Cribriform plate
Foramen cecum; Right frontal sinus
Left frontal sinus
Cribriform plate
Opening of right frontal sinus
Left anterior ethmoidal air cells (sinus)
Ant. ethmoidal air cells (sinus)
Post. ethmoidal air cells (sinus)
Optic foramen;
Ant. clinoid process;
Sup. orbital fissure
Right sphenoid sinus
Left posterior ethmoidal air cells (sinus)
Posterior clinoid process;
Foramen rotundum
Left sphenoid sinus
Dorsum sellae; Foramen ovale
Hypophyseal fossa

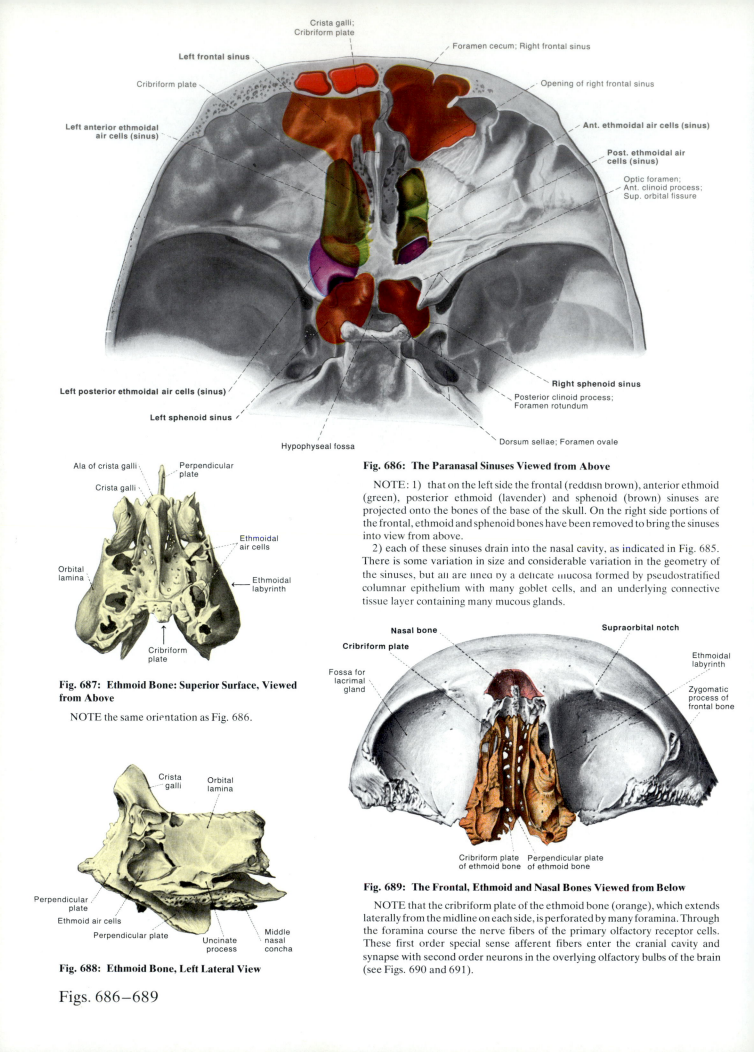

Fig. 686: The Paranasal Sinuses Viewed from Above

NOTE: 1) that on the left side the frontal (reddish brown), anterior ethmoid (green), posterior ethmoid (lavender) and sphenoid (brown) sinuses are projected onto the bones of the base of the skull. On the right side portions of the frontal, ethmoid and sphenoid bones have been removed to bring the sinuses into view from above.

2) each of these sinuses drain into the nasal cavity, as indicated in Fig. 685. There is some variation in size and considerable variation in the geometry of the sinuses, but all are lined by a delicate mucosa formed by pseudostratified columnar epithelium with many goblet cells, and an underlying connective tissue layer containing many mucous glands.

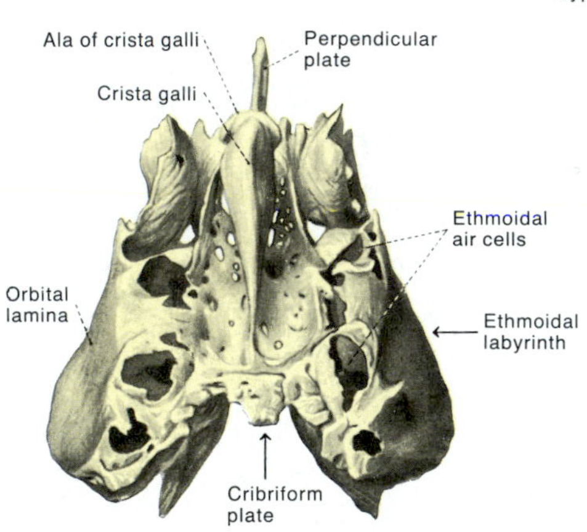
Ala of crista galli
Perpendicular plate
Crista galli
Ethmoidal air cells
Orbital lamina
Ethmoidal labyrinth
Cribriform plate

Fig. 687: Ethmoid Bone: Superior Surface, Viewed from Above

NOTE the same orientation as Fig. 686.

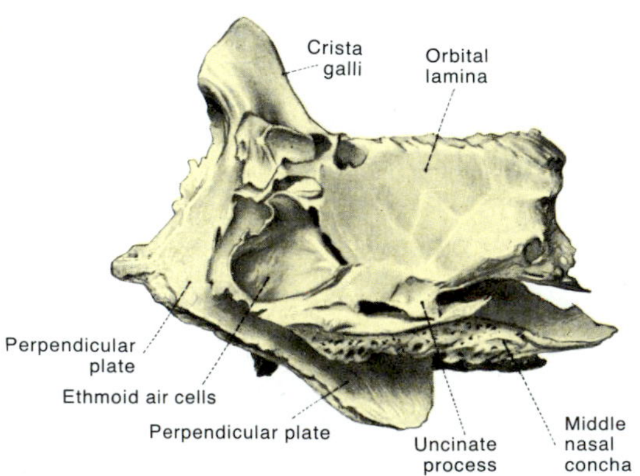
Crista galli
Orbital lamina
Perpendicular plate
Ethmoid air cells
Perpendicular plate
Uncinate process
Middle nasal concha

Fig. 688: Ethmoid Bone, Left Lateral View

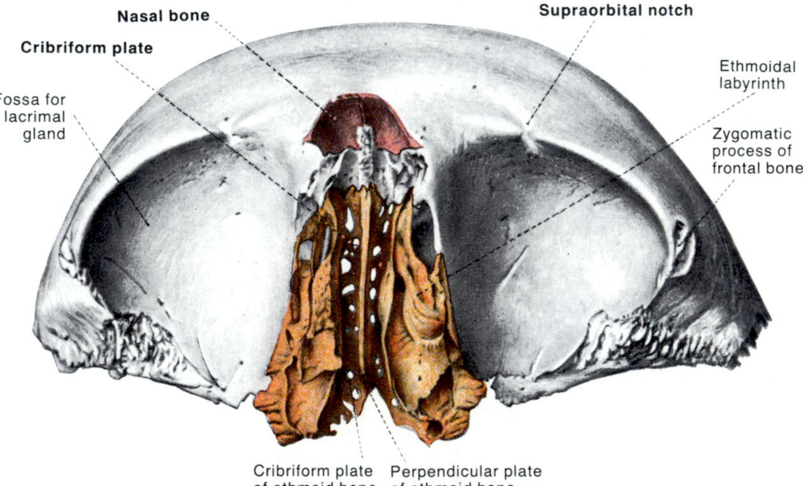
Nasal bone
Supraorbital notch
Cribriform plate
Ethmoidal labyrinth
Fossa for lacrimal gland
Zygomatic process of frontal bone
Cribriform plate of ethmoid bone
Perpendicular plate of ethmoid bone

Fig. 689: The Frontal, Ethmoid and Nasal Bones Viewed from Below

NOTE that the cribriform plate of the ethmoid bone (orange), which extends laterally from the midline on each side, is perforated by many foramina. Through the foramina course the nerve fibers of the primary olfactory receptor cells. These first order special sense afferent fibers enter the cranial cavity and synapse with second order neurons in the overlying olfactory bulbs of the brain (see Figs. 690 and 691).

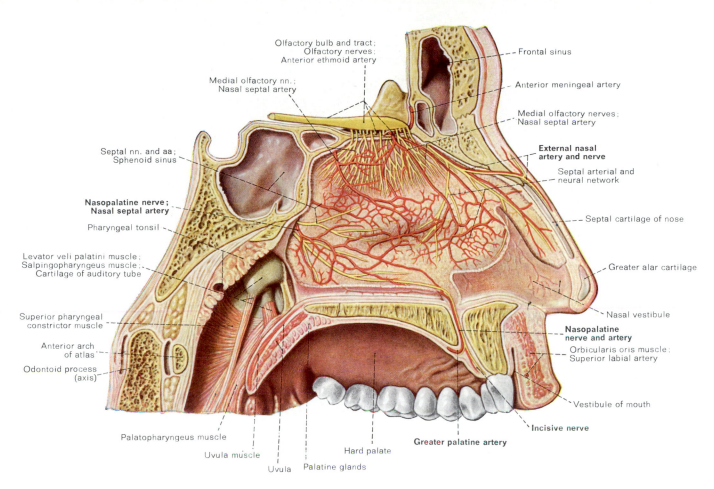

Olfactory bulb and tract;
Olfactory nerves;
Anterior ethmoid artery
Frontal sinus
Medial olfactory nn.;
Nasal septal artery
Anterior meningeal artery
Medial olfactory nerves;
Nasal septal artery
Septal nn. and aa;
Sphenoid sinus
External nasal artery and nerve
Septal arterial and neural network
Nasopalatine nerve;
Nasal septal artery
Pharyngeal tonsil
Septal cartilage of nose
Levator veli palatini muscle;
Salpingopharyngeus muscle;
Cartilage of auditory tube
Greater alar cartilage
Superior pharyngeal constrictor muscle
Nasal vestibule
Nasopalatine nerve and artery
Anterior arch of atlas
Orbicularis oris muscle;
Superior labial artery
Odontoid process (axis)
Vestibule of mouth
Incisive nerve
Palatopharyngeus muscle
Greater palatine artery
Uvula muscle
Hard palate
Uvula
Palatine glands

Fig. 690: Arteries and Nerves of the Nasal Septum

NOTE that the mucous membrane has been removed from the nasal septum and nasopharynx revealing the spetal vessels and nerves and the nasopharyngeal muscles.

Fig. 691: The Lateral Wall of the Left Nasal Cavity

NOTE: the mucous membrane overlying the lateral olfactory nerves has been removed. The lateral wall of the nasal cavity is marked by the superior, middle and inferior nasal conchae. Beneath each concha courses its corresponding nasal passage or meatus.

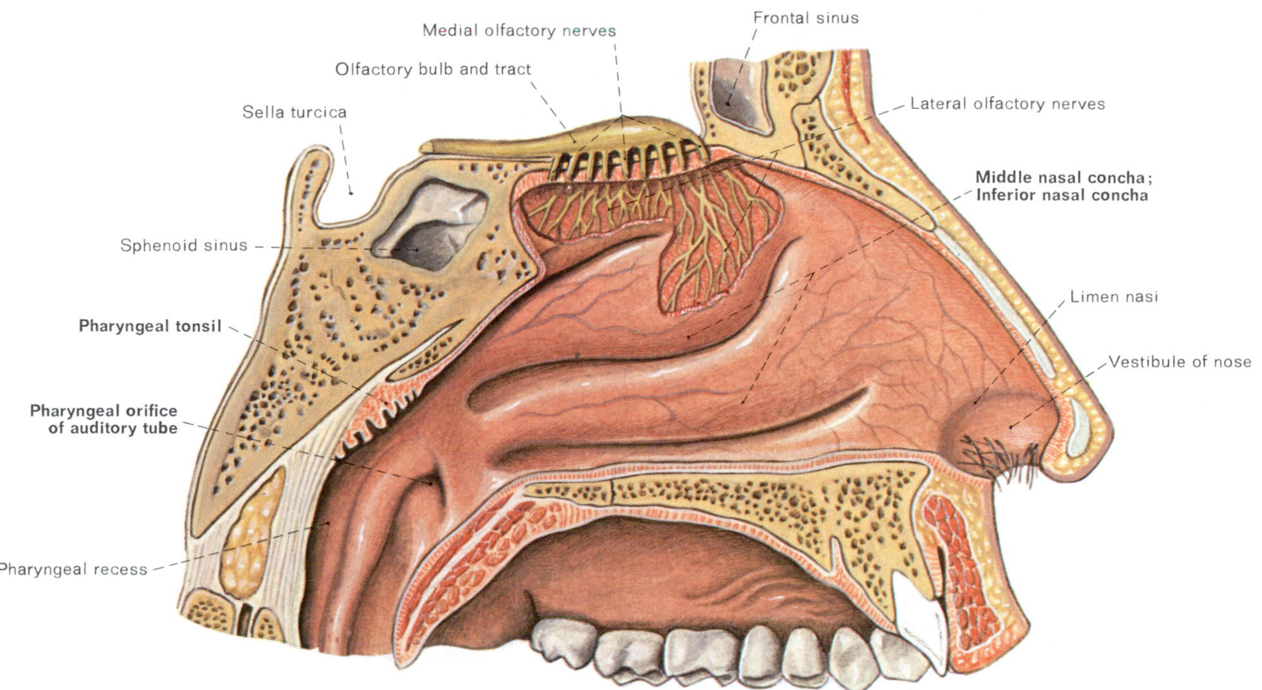
Frontal sinus
Medial olfactory nerves
Olfactory bulb and tract
Lateral olfactory nerves
Sella turcica
Middle nasal concha;
Inferior nasal concha
Sphenoid sinus
Limen nasi
Pharyngeal tonsil
Vestibule of nose
Pharyngeal orifice of auditory tube
Pharyngeal recess

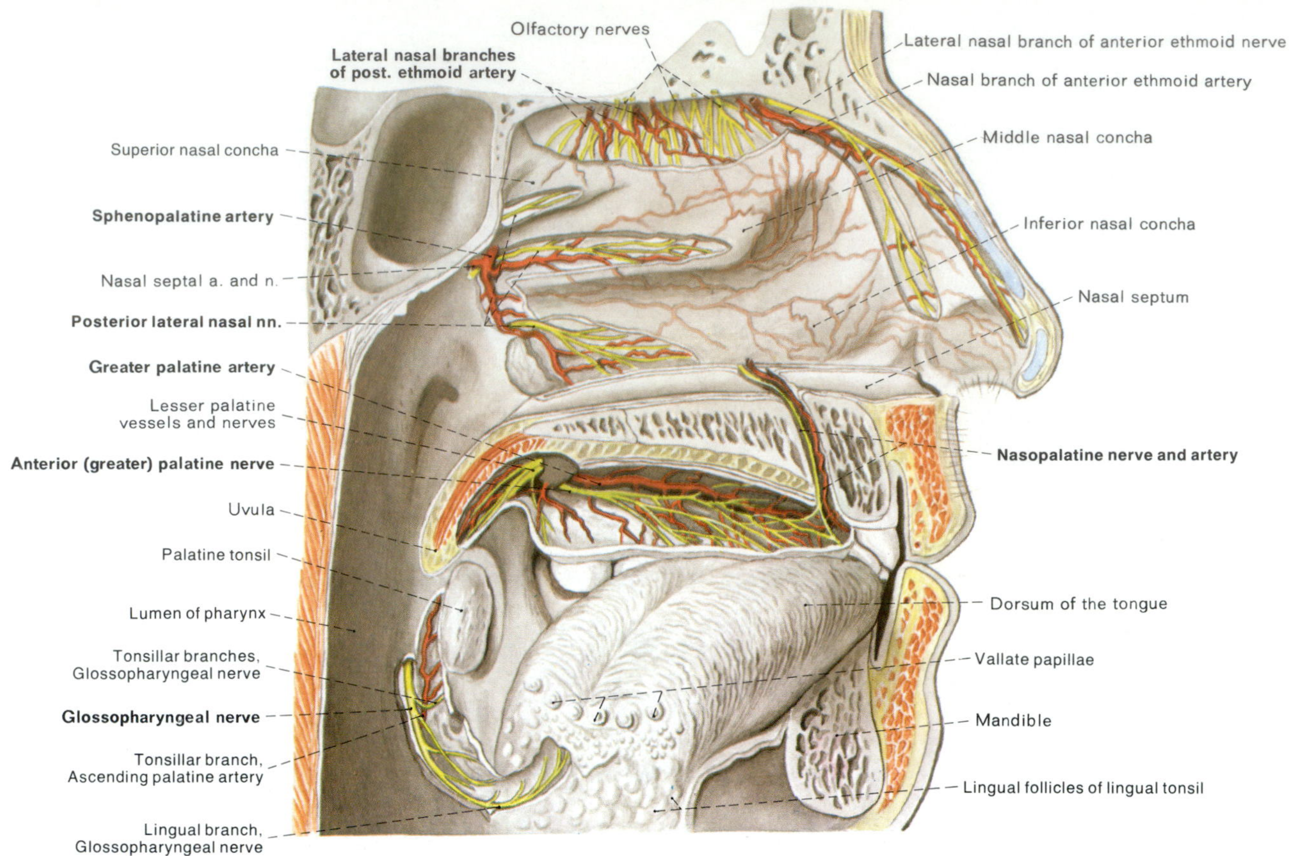
Olfactory nerves
Lateral nasal branches of post. ethmoid artery
Lateral nasal branch of anterior ethmoid nerve
Nasal branch of anterior ethmoid artery
Middle nasal concha
Superior nasal concha
Sphenopalatine artery
Inferior nasal concha
Nasal septal a. and n.
Nasal septum
Posterior lateral nasal nn.
Greater palatine artery
Lesser palatine vessels and nerves
Nasopalatine nerve and artery
Anterior (greater) palatine nerve
Uvula
Palatine tonsil
Dorsum of the tongue
Lumen of pharynx
Tonsillar branches, Glossopharyngeal nerve
Vallate papillae
Glossopharyngeal nerve
Mandible
Tonsillar branch, Ascending palatine artery
Lingual follicles of lingual tonsil
Lingual branch, Glossopharyngeal nerve

Fig. 692: Nerves and Arteries of the Palate and Lateral Wall of the Nasal Cavity

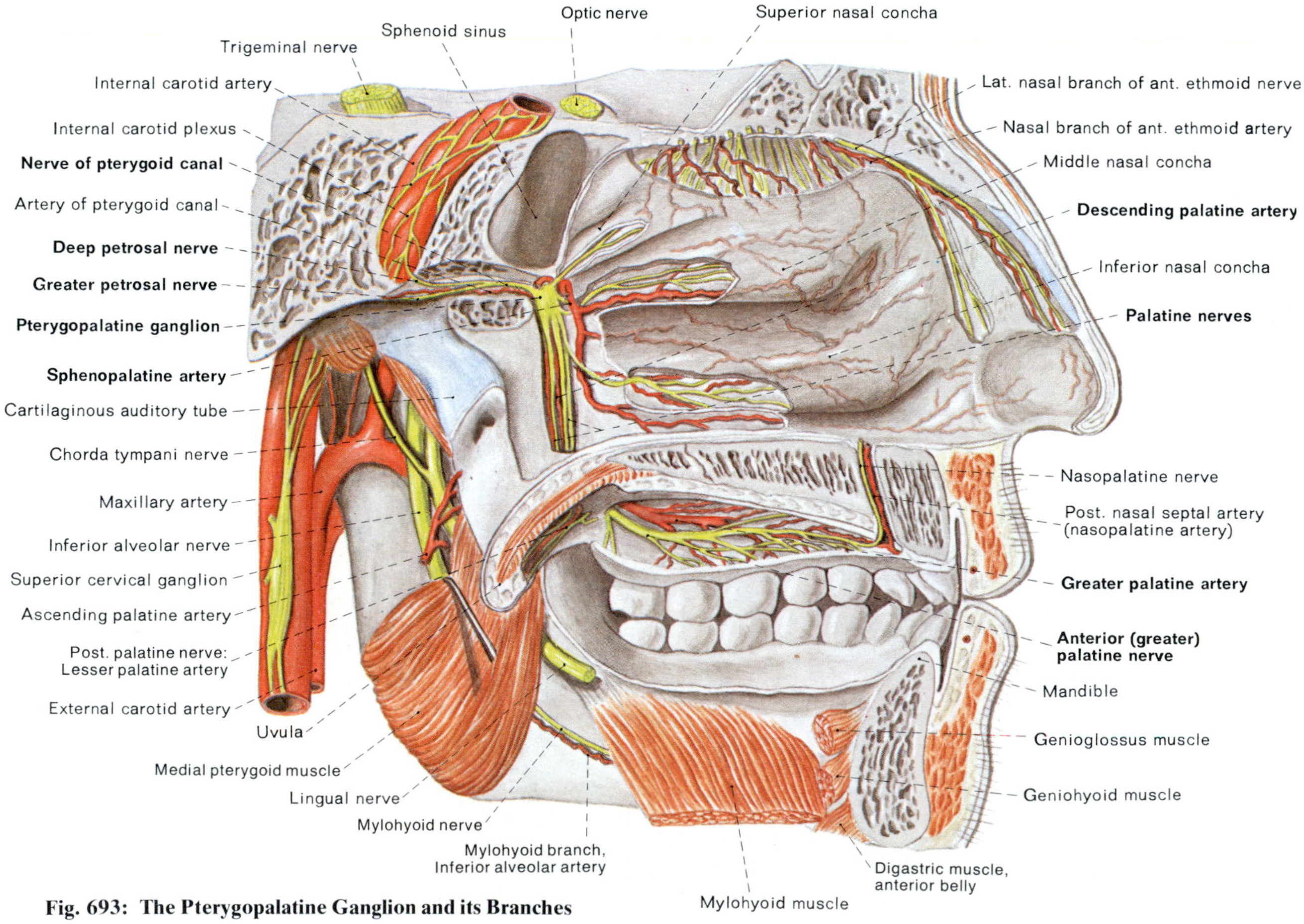

Optic nerve
Superior nasal concha
Sphenoid sinus
Trigeminal nerve
Internal carotid artery
Lat. nasal branch of ant. ethmoid nerve
Internal carotid plexus
Nasal branch of ant. ethmoid artery
Nerve of pterygoid canal
Middle nasal concha
Artery of pterygoid canal
Descending palatine artery
Deep petrosal nerve
Inferior nasal concha
Greater petrosal nerve
Palatine nerves
Pterygopalatine ganglion
Sphenopalatine artery
Cartilaginous auditory tube
Chorda tympani nerve
Nasopalatine nerve
Maxillary artery
Post. nasal septal artery (nasopalatine artery)
Inferior alveolar nerve
Superior cervical ganglion
Greater palatine artery
Ascending palatine artery
Anterior (greater) palatine nerve
Post. palatine nerve: Lesser palatine artery
Mandible
External carotid artery
Uvula
Genioglossus muscle
Medial pterygoid muscle
Geniohyoid muscle
Lingual nerve
Mylohyoid nerve
Mylohyoid branch, Inferior alveolar artery
Digastric muscle, anterior belly
Mylohyoid muscle

Fig. 693: The Pterygopalatine Ganglion and its Branches

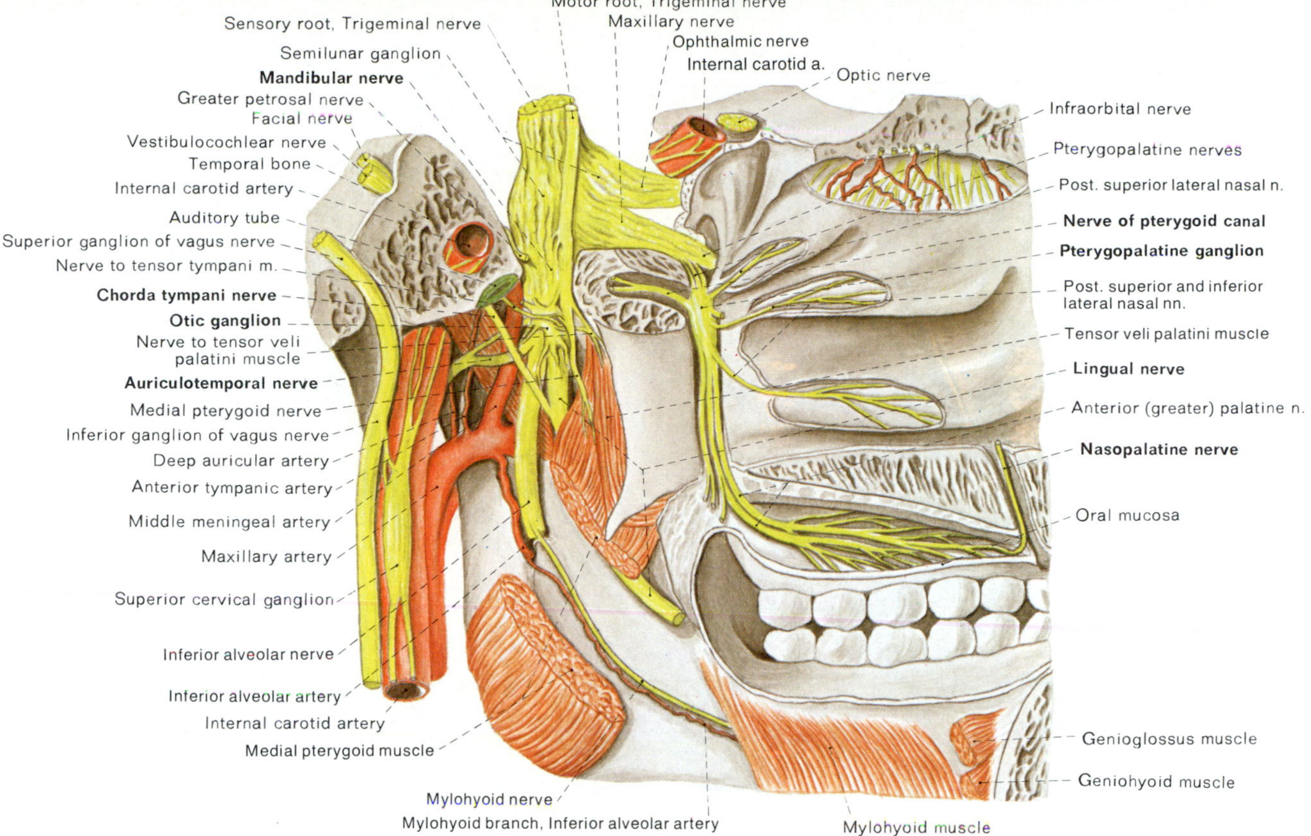

Sensory root, Trigeminal nerve
Motor root, Trigeminal nerve
Maxillary nerve
Ophthalmic nerve
Internal carotid a.
Optic nerve
Semilunar ganglion
Mandibular nerve
Greater petrosal nerve
Facial nerve
Vestibulocochlear nerve
Temporal bone
Internal carotid artery
Auditory tube
Superior ganglion of vagus nerve
Nerve to tensor tympani m.
Chorda tympani nerve
Otic ganglion
Nerve to tensor veli palatini muscle
Auriculotemporal nerve
Medial pterygoid nerve
Inferior ganglion of vagus nerve
Deep auricular artery
Anterior tympanic artery
Middle meningeal artery
Maxillary artery
Superior cervical ganglion
Inferior alveolar nerve
Inferior alveolar artery
Internal carotid artery
Medial pterygoid muscle
Mylohyoid nerve
Mylohyoid branch, Inferior alveolar artery
Mylohyoid muscle
Infraorbital nerve
Pterygopalatine nerves
Post. superior lateral nasal n.
Nerve of pterygoid canal
Pterygopalatine ganglion
Post. superior and inferior lateral nasal nn.
Tensor veli palatini muscle
Lingual nerve
Anterior (greater) palatine n.
Nasopalatine nerve
Oral mucosa
Genioglossus muscle
Geniohyoid muscle

Fig. 694: Nerves of the Nasal and Oral Cavities and the Otic Ganglion

NOTE the junction of the chorda tympani nerve with the lingual nerve and the position of the otic ganglion in relation to the mandibular branch of the trigeminal. This ganglion receives preganglionic parasympathetic fibers by way of the tympanic branch of the glossopharyngeal nerve. Its postganglionic fibers supply the parotid gland.

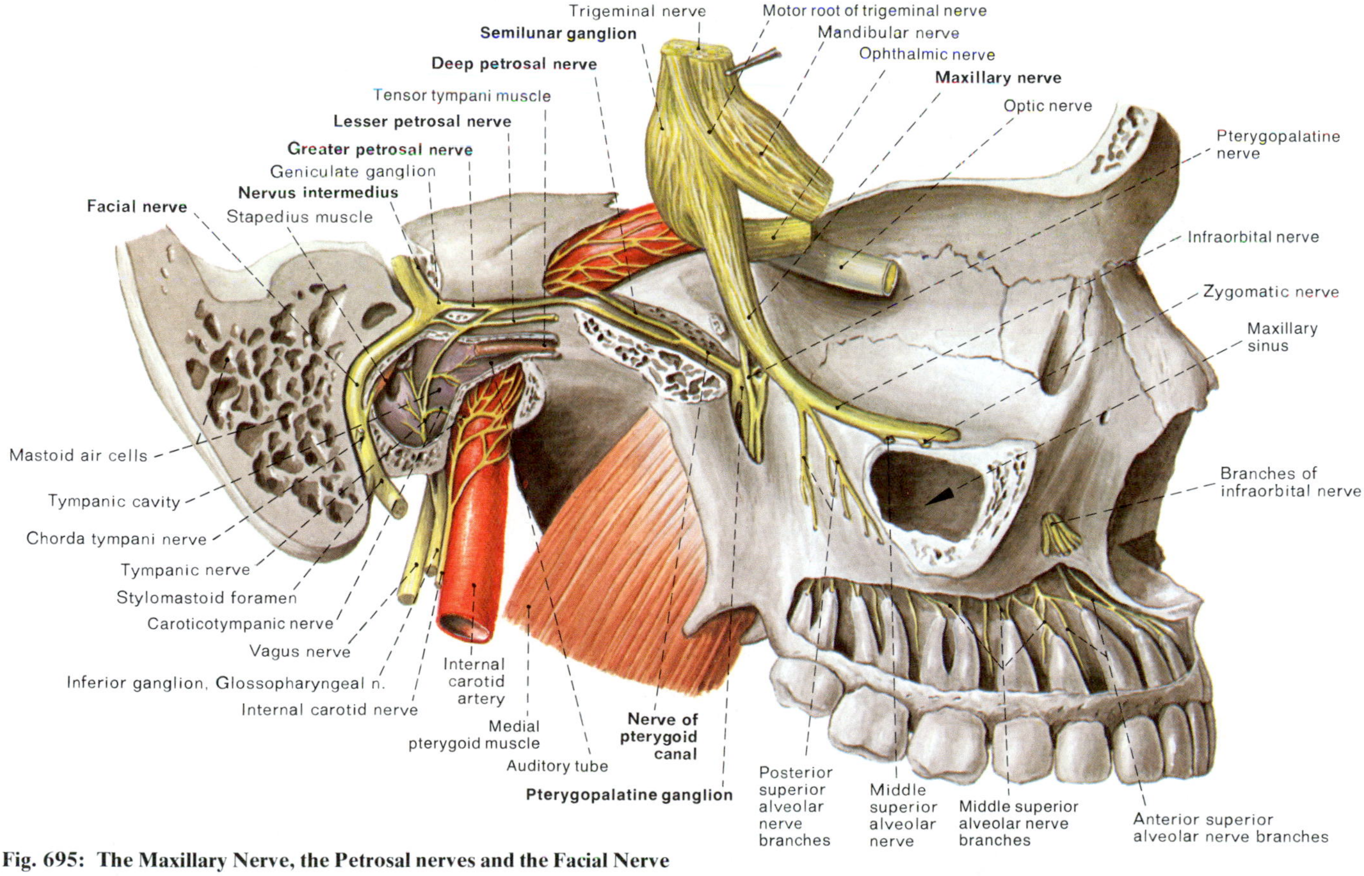

Trigeminal nerve
Motor root of trigeminal nerve
Semilunar ganglion
Mandibular nerve
Ophthalmic nerve
Deep petrosal nerve
Maxillary nerve
Tensor tympani muscle
Optic nerve
Lesser petrosal nerve
Greater petrosal nerve
Geniculate ganglion
Nervus intermedius
Facial nerve
Stapedius muscle
Pterygopalatine nerve
Infraorbital nerve
Zygomatic nerve
Maxillary sinus
Mastoid air cells
Tympanic cavity
Chorda tympani nerve
Tympanic nerve
Stylomastoid foramen
Caroticotympanic nerve
Vagus nerve
Inferior ganglion, Glossopharyngeal n.
Internal carotid nerve
Internal carotid artery
Medial pterygoid muscle
Auditory tube
Nerve of pterygoid canal
Pterygopalatine ganglion
Posterior superior alveolar nerve branches
Middle superior alveolar nerve
Middle superior alveolar nerve branches
Anterior superior alveolar nerve branches
Branches of infraorbital nerve

Fig. 695: The Maxillary Nerve, the Petrosal nerves and the Facial Nerve

NOTE that the nerve of the pterygoid canal is formed by the union of the deep petrosal and greater petrosal nerves. The deep petrosal nerve transmits postganglionic sympathetic fibers, while the greater petrosal nerve contains sensory fibers from the geniculate ganglion of the facial nerve and preganglionic parasympathetic fibers from the nervus intermedius portion of the facial nerve. The lesser petrosal nerve carries preganglionic fibers of the glossopharyngeal nerve to the otic ganglion.

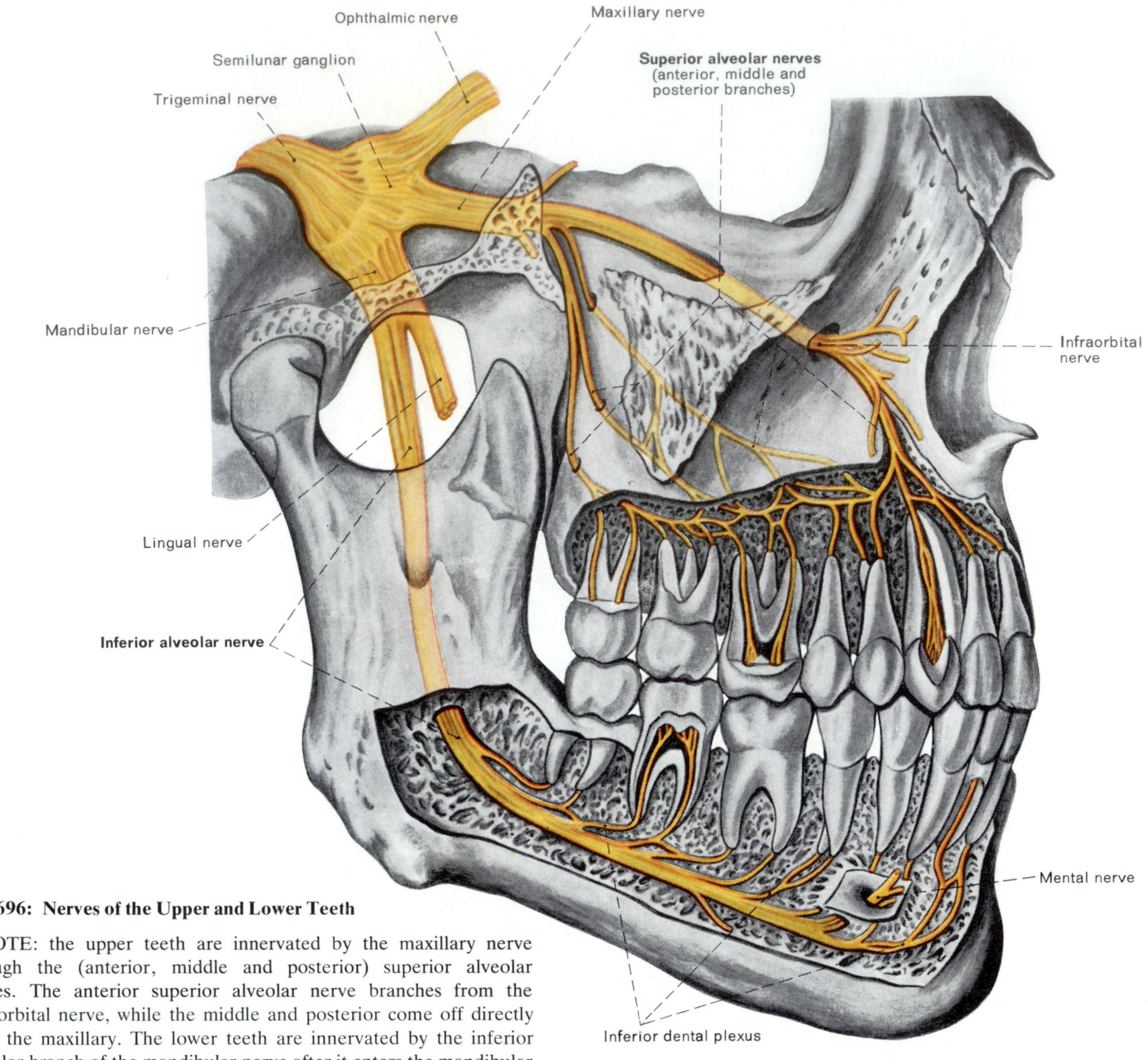
Ophthalmic nerve
Maxillary nerve
Semilunar ganglion
Superior alveolar nerves
(anterior, middle and posterior branches)
Trigeminal nerve
Mandibular nerve
Infraorbital nerve
Lingual nerve
Inferior alveolar nerve
Mental nerve
Inferior dental plexus

Fig. 696: Nerves of the Upper and Lower Teeth

NOTE: the upper teeth are innervated by the maxillary nerve through the (anterior, middle and posterior) superior alveolar nerves. The anterior superior alveolar nerve branches from the infraorbital nerve, while the middle and posterior come off directly from the maxillary. The lower teeth are innervated by the inferior alveolar branch of the mandibular nerve after it enters the mandibular foramen.

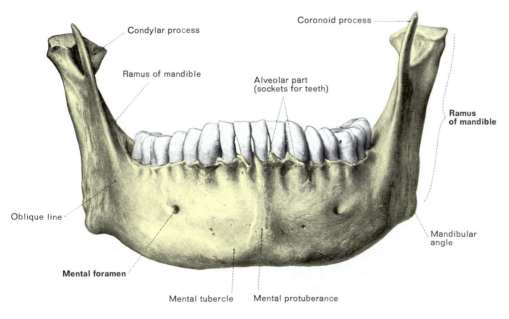

Coronoid process
Condylar process
Ramus of mandible
Alveolar part (sockets for teeth)
Ramus of mandible
Oblique line
Mandibular angle
Mental foramen
Mental tubercle
Mental protuberance

Fig. 697: The Mandible as seen from the Front

NOTE: 1) the anterior aspect of the mandible forms the bony substructure of the mentum or chin. The two vertical rami of the mandible are continuous with the body of the mandible at the mandibular angle.

2) the mental foramen transmits the mental branch of the inferior alveolar nerve to the skin of the chin and lower lip on each side. The mental branch of the inferior alveolar artery accompanies the mental nerve and participates in the vascular supply of the lower lip.

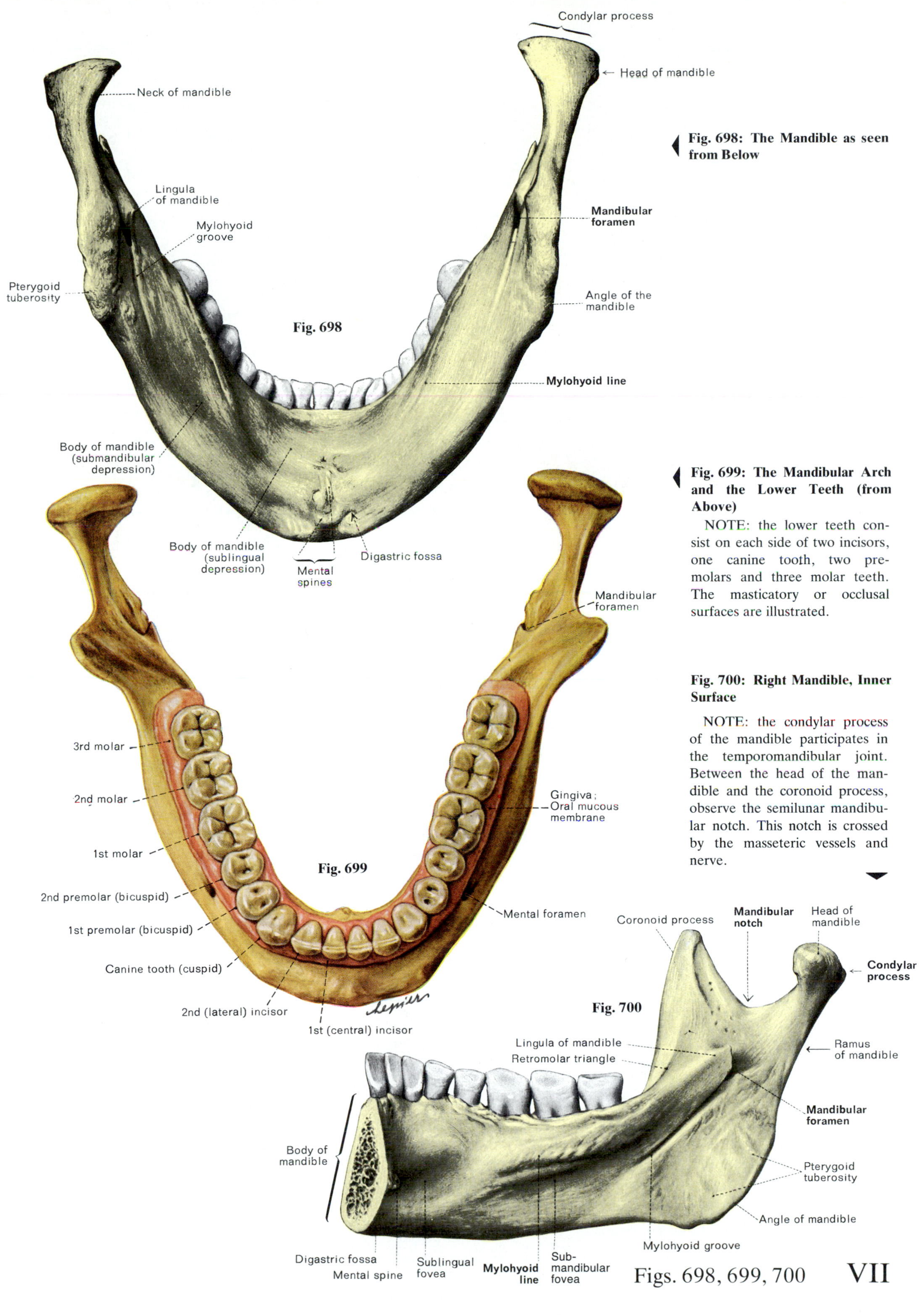
Condylar process
Head of mandible
Neck of mandible
Lingula of mandible
Mylohyoid groove
Mandibular foramen
Pterygoid tuberosity
Angle of the mandible
Fig. 698
Mylohyoid line
Body of mandible (submandibular depression)
Body of mandible (sublingual depression)
Mental spines
Digastric fossa
Mandibular foramen
3rd molar
2nd molar
Gingiva; Oral mucous membrane
1st molar
Fig. 699
2nd premolar (bicuspid)
1st premolar (bicuspid)
Mental foramen
Canine tooth (cuspid)
2nd (lateral) incisor
1st (central) incisor
Coronoid process
Mandibular notch
Head of mandible
Condylar process
Fig. 700
Lingula of mandible
Retromolar triangle
Ramus of mandible
Mandibular foramen
Body of mandible
Pterygoid tuberosity
Angle of mandible
Mylohyoid groove
Digastric fossa
Mental spine
Sublingual fovea
Mylohyoid line
Sub-mandibular fovea

Fig. 698: The Mandible as seen from Below

Fig. 699: The Mandibular Arch and the Lower Teeth (from Above)

NOTE: the lower teeth consist on each side of two incisors, one canine tooth, two premolars and three molar teeth. The masticatory or occlusal surfaces are illustrated.

Fig. 700: Right Mandible, Inner Surface

NOTE: the condylar process of the mandible participates in the temporomandibular joint. Between the head of the mandible and the coronoid process, observe the semilunar mandibular notch. This notch is crossed by the masseteric vessels and nerve.

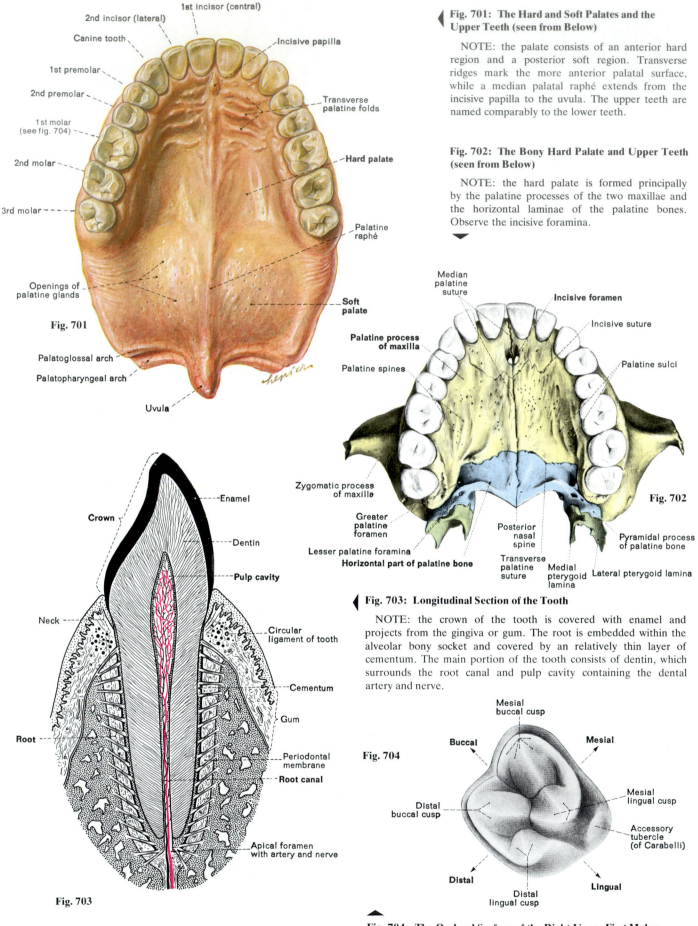

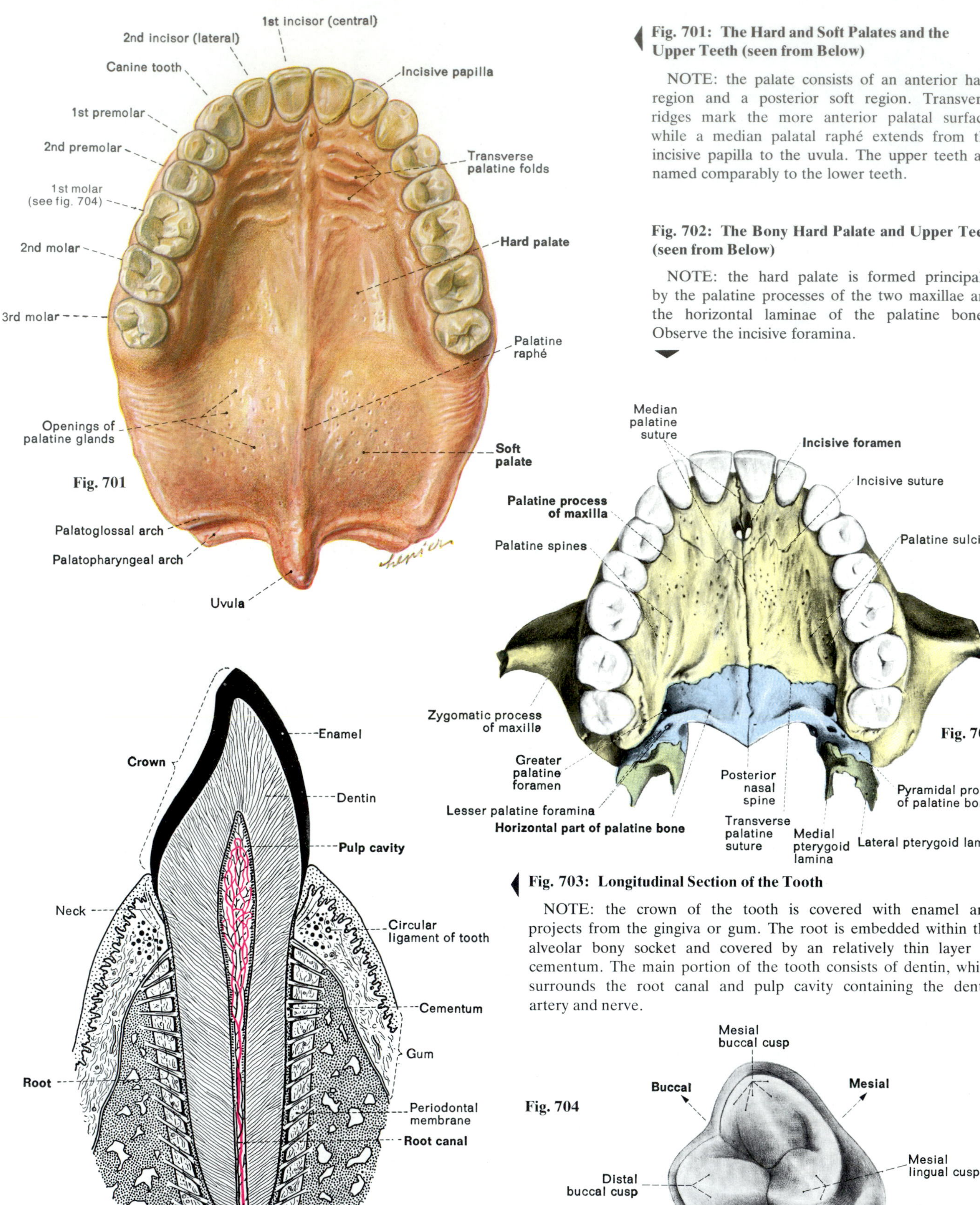
1st incisor (central)
2nd incisor (lateral)
Canine tooth
Incisive papilla
1st premolar
2nd premolar
Transverse palatine folds
1st molar (see fig. 704)
2nd molar
Hard palate
3rd molar
Palatine raphé
Openings of palatine glands
Soft palate
Palatoglossal arch
Palatopharyngeal arch
Uvula
Median palatine suture
Incisive foramen
Incisive suture
Palatine process of maxilla
Palatine spines
Palatine sulci
Zygomatic process of maxilla
Greater palatine foramen
Lesser palatine foramina
Horizontal part of palatine bone
Posterior nasal spine
Transverse palatine suture
Medial pterygoid lamina
Lateral pterygoid lamina
Pyramidal process of palatine bone
Crown
Enamel
Dentin
Pulp cavity
Neck
Circular ligament of tooth
Cementum
Gum
Root
Periodontal membrane
Root canal
Apical foramen with artery and nerve
Mesial buccal cusp
Buccal
Mesial
Mesial lingual cusp
Distal buccal cusp
Accessory tubercle (of Carabelli)
Distal
Lingual
Distal lingual cusp

Fig. 701

Fig. 702

Fig. 703

Fig. 704

Fig. 701: The Hard and Soft Palates and the Upper Teeth (seen from Below)

NOTE: the palate consists of an anterior hard region and a posterior soft region. Transverse ridges mark the more anterior palatal surface, while a median palatal raphé extends from the incisive papilla to the uvula. The upper teeth are named comparably to the lower teeth.

Fig. 702: The Bony Hard Palate and Upper Teeth (seen from Below)

NOTE: the hard palate is formed principally by the palatine processes of the two maxillae and the horizontal laminae of the palatine bones. Observe the incisive foramina.

Fig. 703: Longitudinal Section of the Tooth

NOTE: the crown of the tooth is covered with enamel and projects from the gingiva or gum. The root is embedded within the alveolar bony socket and covered by an relatively thin layer of cementum. The main portion of the tooth consists of dentin, which surrounds the root canal and pulp cavity containing the dental artery and nerve.

Fig. 704: The Occlusal Surface of the Right Upper First Molar

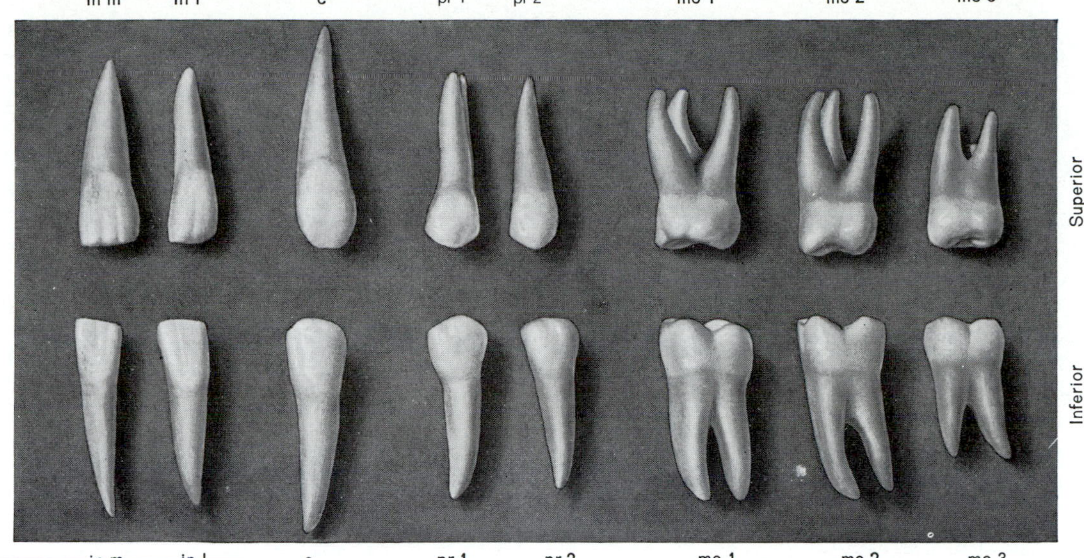
in m
in l
c
pr 1
pr 2
mo 1
mo 2
mo 3
Superior
Inferior
in m
in l
c
pr 1
pr 2
mo 1
mo 2
mo 3

in m = Medial incisor
in l = Lateral incisor
c = Canine
pr 1 = 1st premolar
pr 2 = 2nd premolar
mo 1 = 1st molar
mo 2 = 2nd molar
mo 3 = 3rd molar

Fig. 705

Fig. 705: The Maxillary and Mandibular Permanent Teeth (Buccal Surface)

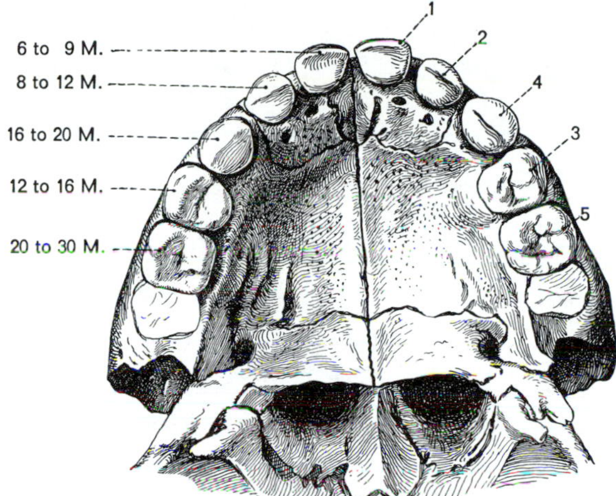
6 to 9 M.
8 to 12 M.
16 to 20 M.
12 to 16 M.
20 to 30 M.
1
2
4
3
5

Fig. 707

Fig. 707: Diagram Showing Eruption Times of Deciduous Upper Teeth

NOTE: on the left side of the illustration the times of eruption are shown in *months* for each tooth, while the sequential order of appearance of the erupted *deciduous* teeth is indicated on the right side by numbers.

Fig. 706: Diagram Showing Eruption Times of Permanent Upper Teeth

NOTE: on the left side of the illustration the times of eruption are shown in *years* for each tooth while the sequential order of appearance of the erupted *permanent* teeth is indicated on the right side by numbers.

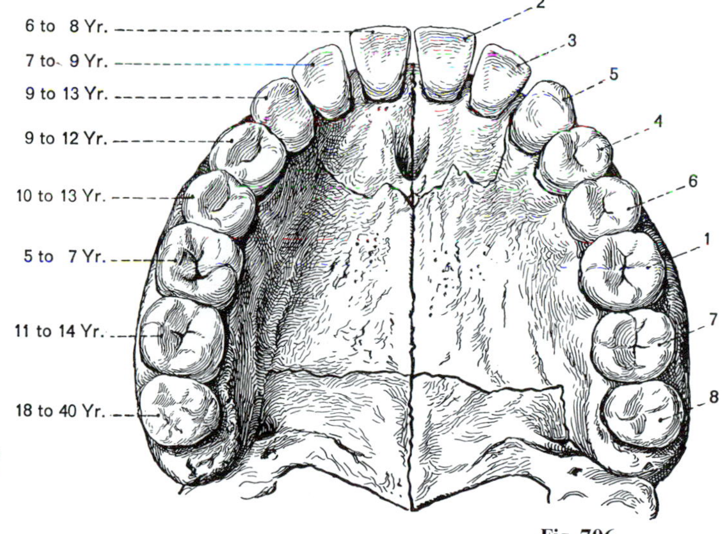
6 to 8 Yr.
7 to 9 Yr.
9 to 13 Yr.
9 to 12 Yr.
10 to 13 Yr.
5 to 7 Yr.
11 to 14 Yr.
18 to 40 Yr.
2
3
5
4
6
1
7
8

Fig. 706

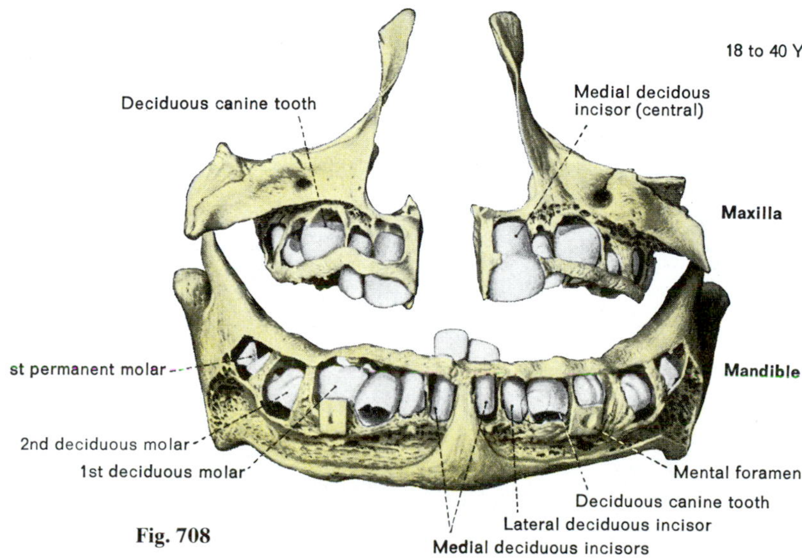
Deciduous canine tooth
Medial decidous incisor (central)
Maxilla
Mandible
st permanent molar
2nd deciduous molar
1st deciduous molar
Mental foramen
Deciduous canine tooth
Lateral deciduous incisor
Medial deciduous incisors

Fig. 708

Fig. 708: Dentition of a Child of Nearly One Year of Age

NOTE: 1) the deciduous or "milk teeth" number 20 in all, including two incisors, one canine and two molars in each jaw quadrant. The earliest of these teeth to erupt are the incisors, which generally penetrate through the gum line before the end of the first year.

2) generally speaking:
a 1 year old child has 6 teeth,
a 1½ year old child has 12 teeth,
a 2 year old child has 16 teeth, and
a 2½ year old child has 20 teeth.

Fig. 709: The Facial Skeleton of a Five Year Old Child Showing Full Deciduous Dentition (20 Teeth)

NOTE: 1) the deciduous teeth are shown as white, while the rudiments of the permanent teeth, shown in blue, have been exposed by removing the outer walls of the alveolar processes of both maxillae and the mandible.

2) all 20 deciduous teeth have erupted: eight incisors, four canines and eight molars. Normally, all the deciduous teeth are replaced by the 12th year.

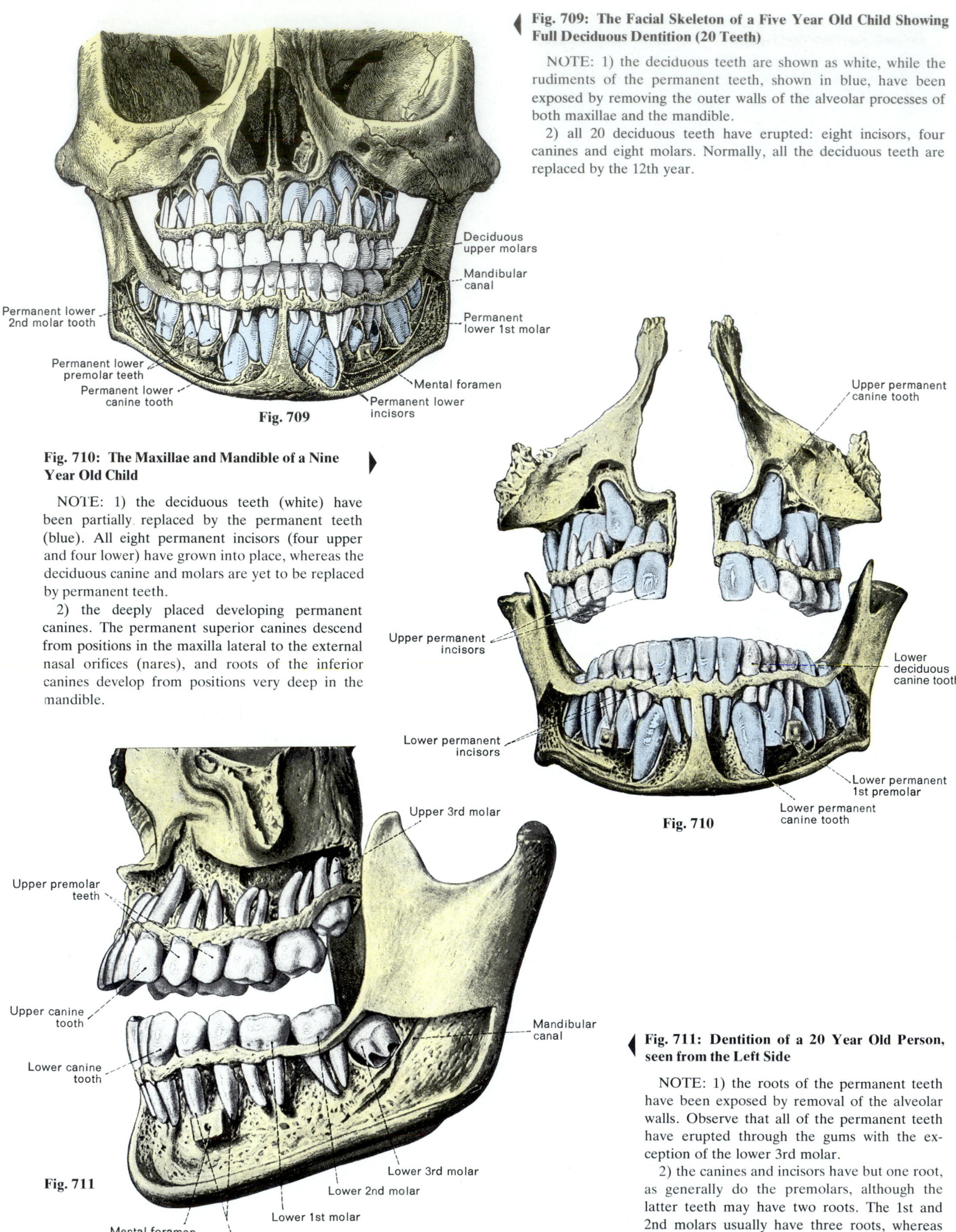

Deciduous upper molars
Mandibular canal
Permanent lower 1st molar
Mental foramen
Permanent lower incisors
Permanent lower canine tooth
Permanent lower premolar teeth
Permanent lower 2nd molar tooth
Upper permanent canine tooth
Upper permanent incisors
Lower deciduous canine tooth
Lower permanent incisors
Lower permanent 1st premolar
Lower permanent canine tooth
Upper 3rd molar
Upper premolar teeth
Upper canine tooth
Lower canine tooth
Mandibular canal
Lower 3rd molar
Lower 2nd molar
Lower 1st molar
Mental foramen
Lower premolars

Fig. 709

Fig. 710: The Maxillae and Mandible of a Nine Year Old Child

NOTE: 1) the deciduous teeth (white) have been partially replaced by the permanent teeth (blue). All eight permanent incisors (four upper and four lower) have grown into place, whereas the deciduous canine and molars are yet to be replaced by permanent teeth.

2) the deeply placed developing permanent canines. The permanent superior canines descend from positions in the maxilla lateral to the external nasal orifices (nares), and roots of the inferior canines develop from positions very deep in the mandible.

Fig. 710

Fig. 711

Fig. 711: Dentition of a 20 Year Old Person, seen from the Left Side

NOTE: 1) the roots of the permanent teeth have been exposed by removal of the alveolar walls. Observe that all of the permanent teeth have erupted through the gums with the exception of the lower 3rd molar.

2) the canines and incisors have but one root, as generally do the premolars, although the latter teeth may have two roots. The 1st and 2nd molars usually have three roots, whereas the smaller 3rd molar may have less than three and may even be single rooted.

UPPER TEETH

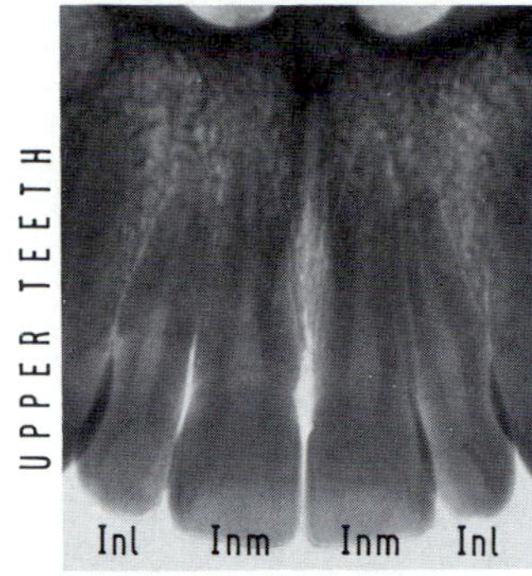
Inl
Inm
Inm
Inl

Fig. 712: The Four Incisors

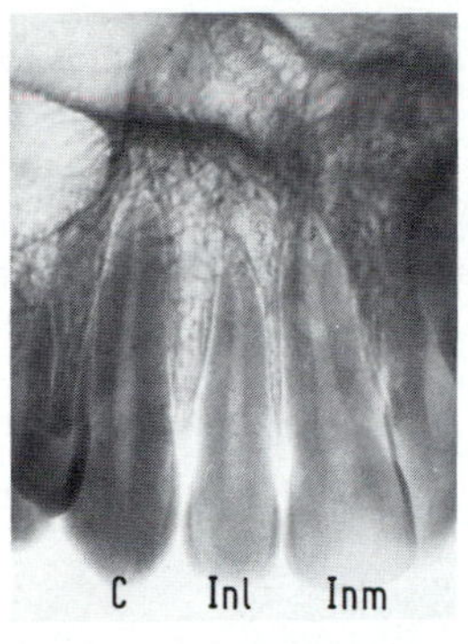
C
Inl
Inm

Fig. 713: The Two Right Incisors and Canine Teeth

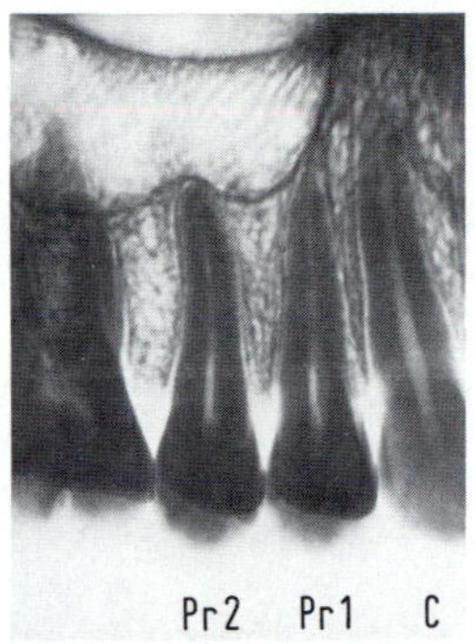
Pr2
Pr1
C

Fig. 714: The Right Canine Tooth and Two Premolars

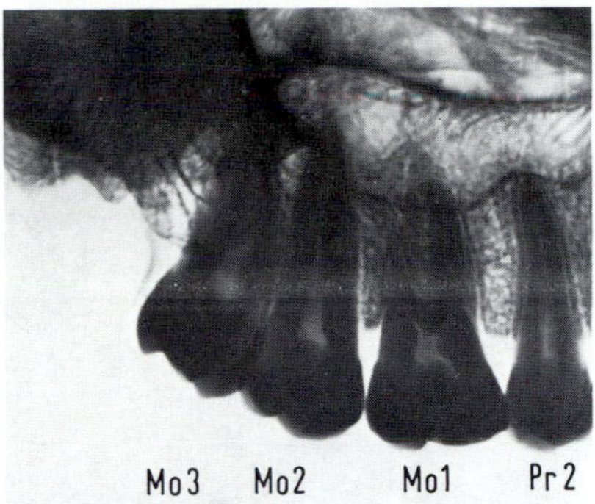
Mo3
Mo2
Mo1
Pr2

Fig. 715: The Right Second Premolar and Three Molars

LOWER TEETH

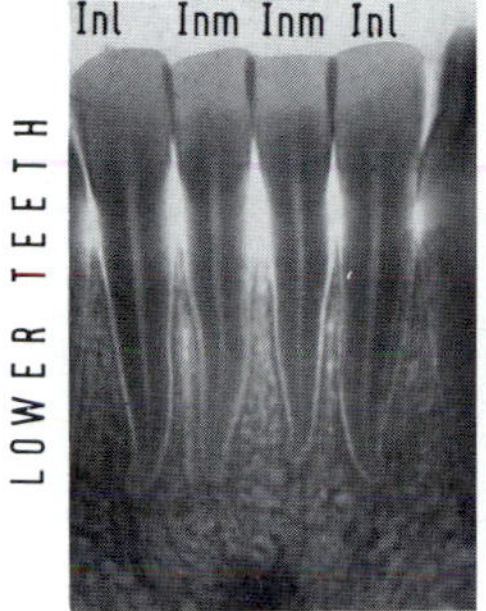
Inl
Inm
Inm
Inl

Fig. 716: The Four Incisors

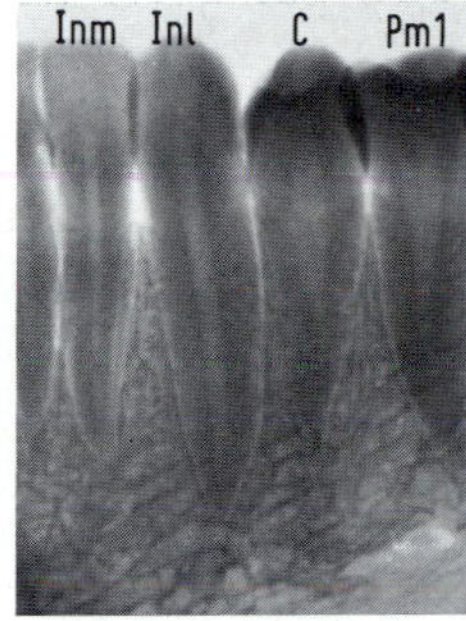
Inm
Inl
C
Pm1

Fig. 717: The Two Incisors, Canine and First Premolar

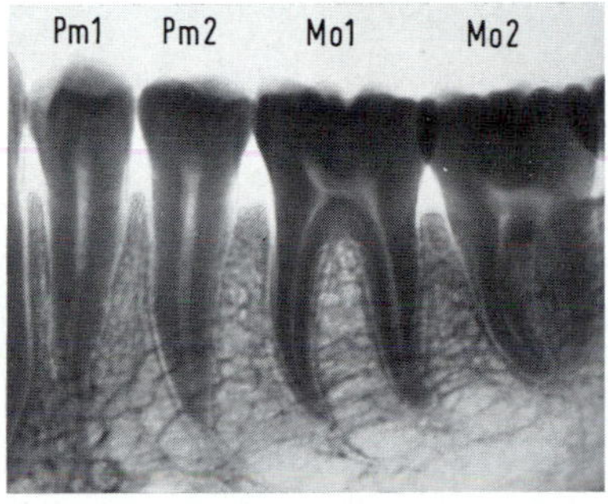
Pm1
Pm2
Mo1
Mo2

Fig. 718: The Two Premolars and First Two Molar Teeth

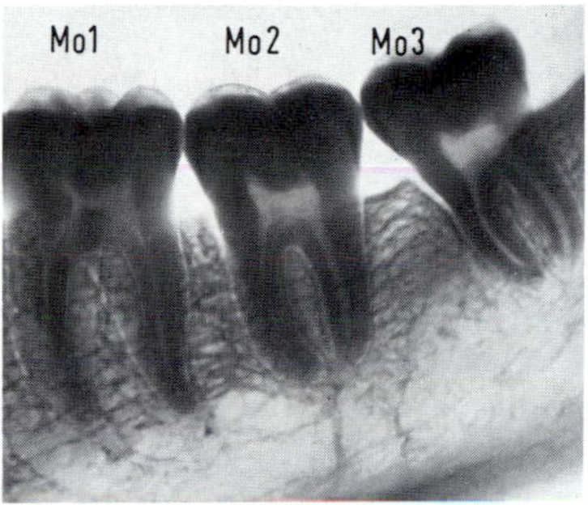
Mo1
Mo2
Mo3

Fig. 719: The Three Molar Teeth

Fig. 720: The Right Upper and Lower Teeth, Lateral View

NOTE: 1) that the right upper eight and right lower eight teeth have been completely exposed by removal of the outer bony layers of the alveolar processes from both the maxilla and mandible. Each tooth has been cut along its longitudinal axis to reveal the pulp cavity and root canal.

2) the fine canaliculi leading to the root canals along which course the delicate vessels and nerves supplying the individual teeth.

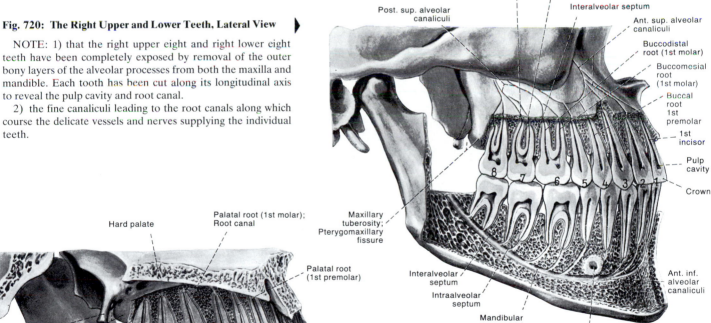

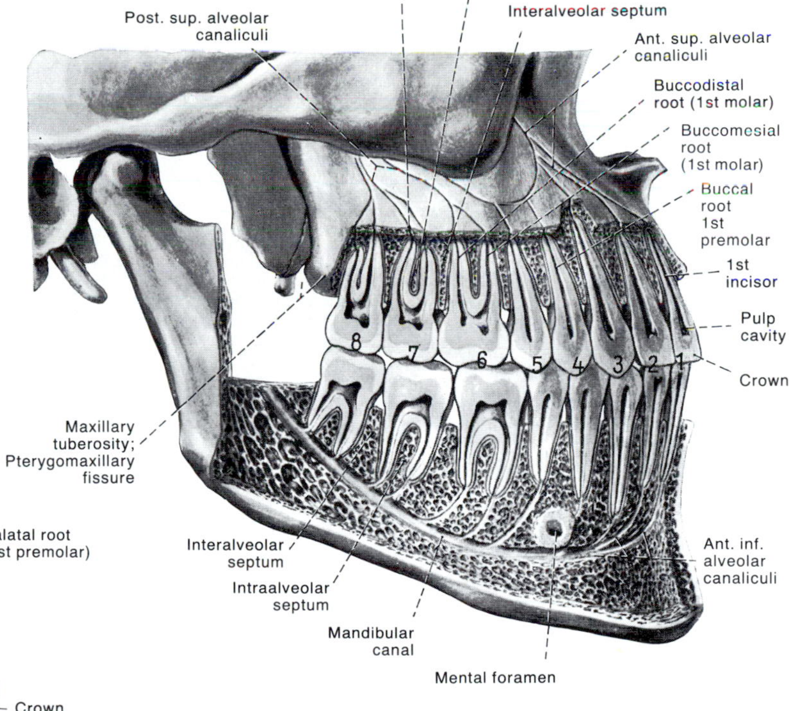
Alveolar foramen
Intraalveolar septum
Interalveolar septum
Post. sup. alveolar canaliculi
Ant. sup. alveolar canaliculi
Buccodistal root (1st molar)
Buccomesial root (1st molar)
Buccal root 1st premolar
1st incisor
Pulp cavity
Crown
8
7
6
5
4
3
2
1
Maxillary tuberosity; Pterygomaxillary fissure
Interalveolar septum
Intraalveolar septum
Mandibular canal
Mental foramen
Ant. inf. alveolar canaliculi

Hard palate
Palatal root (1st molar); Root canal
Palatal root (1st premolar)
Pterygoid hamulus
Maxillary tuberosity; Pterygomax. fissure
Mardibular canal
8 7 6 5 4 3 2 1
Crown
Pulp cavity
Root
Apex of root
Interalveolar septum
Intraalveolar septum
Apex of root
Ant. inf. alveolar canaliculi
L.SCHROTT
Apex of root
Inf. alveolar canaliculus
Mental canal

Fig. 721: The Left Upper and Lower Teeth, Medial View

NOTE that the inner aspect of the roots of the teeth on the left side have been exposed. Observe that the sockets for the canine teeth are the deepest, while those for the molar teeth are the widest. Interalveolar bony septa separate the sockets of individual teeth, and in the case of the multiple rooted molar teeth, intraalveolar septa separate each root.

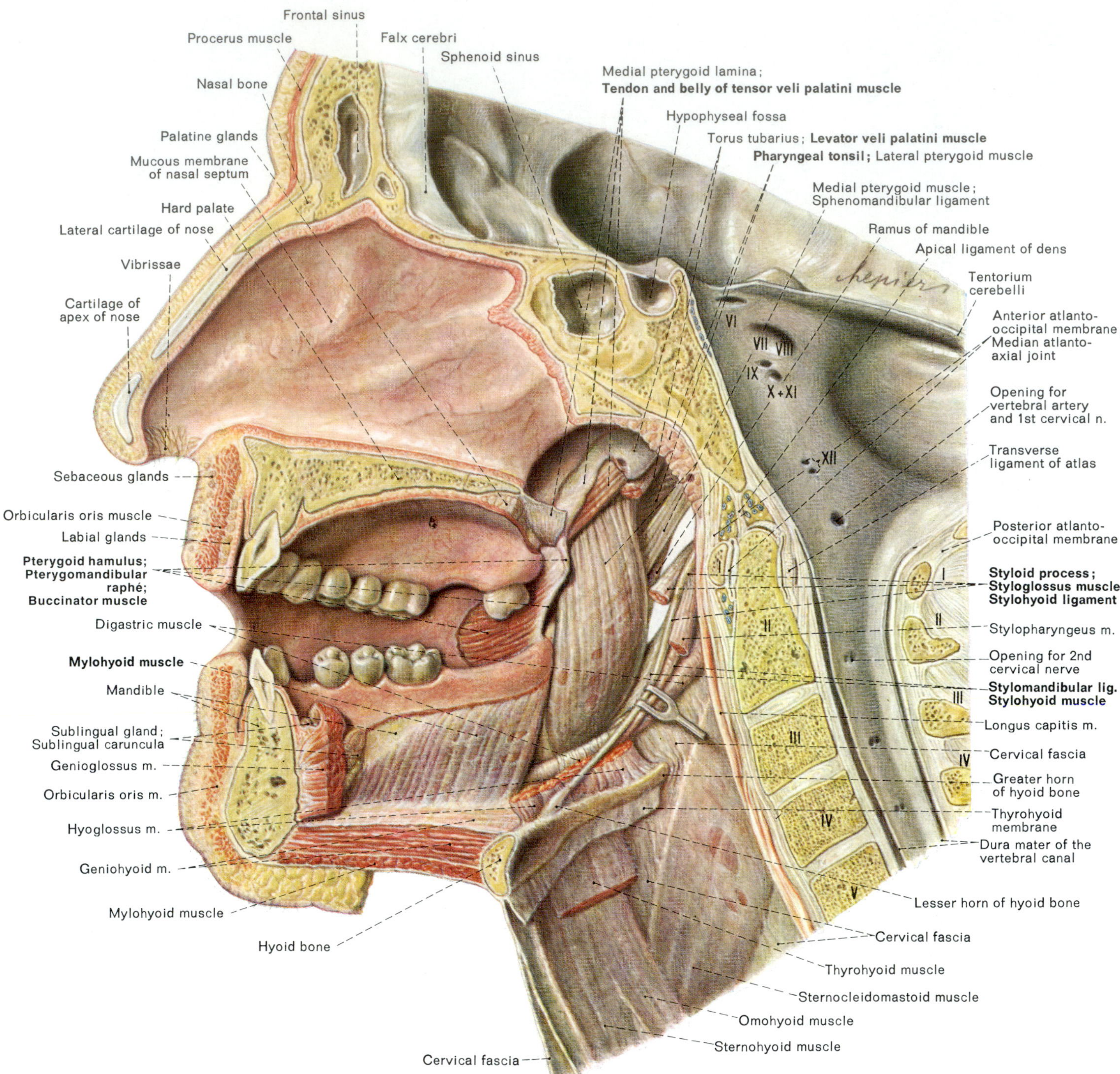

Frontal sinus
Procerus muscle
Falx cerebri
Sphenoid sinus
Nasal bone
Medial pterygoid lamina;
Tendon and belly of tensor veli palatini muscle
Hypophyseal fossa
Palatine glands
Torus tubarius; **Levator veli palatini muscle**
Mucous membrane of nasal septum
Pharyngeal tonsil; Lateral pterygoid muscle
Medial pterygoid muscle; Sphenomandibular ligament
Hard palate
Lateral cartilage of nose
Ramus of mandible
Apical ligament of dens
Vibrissae
Tentorium cerebelli
Cartilage of apex of nose
Anterior atlanto-occipital membrane; Median atlanto-axial joint
Opening for vertebral artery and 1st cervical n.
Transverse ligament of atlas
Sebaceous glands
Orbicularis oris muscle
Labial glands
Posterior atlanto-occipital membrane
Pterygoid hamulus; Pterygomandibular raphé; Buccinator muscle
Styloid process; Styloglossus muscle; Stylohyoid ligament
Digastric muscle
Stylopharyngeus m.
Mylohyoid muscle
Opening for 2nd cervical nerve
Mandible
Stylomandibular lig.; Stylohyoid muscle
Longus capitis m.
Sublingual gland; Sublingual caruncula
Cervical fascia
Genioglossus m.
Greater horn of hyoid bone
Orbicularis oris m.
Thyrohyoid membrane
Hyoglossus m.
Dura mater of the vertebral canal
Geniohyoid m.
Lesser horn of hyoid bone
Mylohyoid muscle
Hyoid bone
Cervical fascia
Thyrohyoid muscle
Sternocleidomastoid muscle
Omohyoid muscle
Sternohyoid muscle
Cervical fascia
VI
VII VIII
IX
X+XI
XII
I
II
III
IV
V

Fig. 722: Paramedian Sagittal View of the Face and Neck

NOTE: 1) this dissection has exposed the right half of the oral cavity and the nasal septum. The mucous membrane has been removed from the floor of the mouth exposing the mylohyoid muscle; the pharyngeal constrictors have been removed. Observe the pterygomandibular raphé and the buccinator muscle.

2) the tendon of the tensor veli palatini muscle as it turns medially around the pterygoid hamulus. The tendon has been severed at its insertion into the palatine aponeurosis. This muscle tightens the soft palate and, being derived from the first or mandibular branchial arch, it is innervated by the trigeminal nerve.

3) the stylohyoid ligament and the styloglossus and stylopharyngeus muscles which have been cut. These three structures all attach to the styloid process, as do the stylomandibular ligament and the stylohyoid muscles which have not been cut.

4) that the hyoid bone lies at the level of the third cervical vertebra and that the atlas (1st vertebra) and axis (2nd vertebra), which surround the uppermost part of the cervical spinal cord, lie directly behind the oral cavity and oropharynx. The spinomedullary junction and the lower medulla oblongata lie posterior to the palate and nasopharynx.

5) that the genioglossus and hyoglossus muscles have been cut. These attach the base and sides of the tongue to the mandible and hyoid bone.

2) the *trigeminal field* (nerve of 1st branchial arch). The *dark blue* zones are supplied by the ophthalmic division of the trigeminal nerve through its anterior and posterior ethmoidal branches. This includes the mucosa of the anterior nasal cavity and the frontal sinus. At times the posterior ethmoidal branch reaches the sphenoid sinus.

The *blue-gray* zones, which include the posterior part of the nasal cavity, the upper nasopharynx, most of the palate and the upper gums of the oral cavity, are supplied by the maxillary division of the trigeminal nerve. This is achieved by branches from the greater and lesser palatine nerves and mucosal branches from the superior alveolar and superior labial nerves. Observe the *white dots* outlining the superior part of the nasal cavity. The mucosa at this site contains olfactory receptor cells for the special sense of smell.

The *blue-green* zones are supplied by the mandibular division of the trigeminal nerve. These include the anterior two-thirds of the tongue, the lower gums and lateral walls of the oral cavity by way of the lingual, buccal and inferior alveolar nerves. Note the *red dots* on the anterior tongue. These represent the field of the chorda tympani nerve which supplies the region with special sensory taste fibers.

3) the *glossopharyngeal field* (nerve of 3rd branchial arch). The *yellow* zones indicate the general sensory supply of the lingual and pharyngeal branches of the glossopharyngeal nerve. These include the posterior one-third of the tongue, part of the soft palate, the oropharynx and the lower part of the nasopharynx. The *white dots* indicate that this nerve also supplies the posterior one-third of the tongue with special sensory fibers of taste.

4) the *vagal field* (nerve of 4th and 6th branchial arches). The green zones represent fields supplied by fibers of general sensation from pharyngeal and laryngeal branches of the vagus nerve. These zones include the laryngopharynx, the larynx and trachea, as well as the esophagus.

Fig. 723: The Sensory Fields of Cranial Nerves Supplying the Nasal and Oral Cavities, the Pharynx, Esophagus and Larynx

NOTE: 1) various shades of *blue* = trigeminal nerve; *yellow* = glossopharyngeal nerve; *green* = vagus nerve.

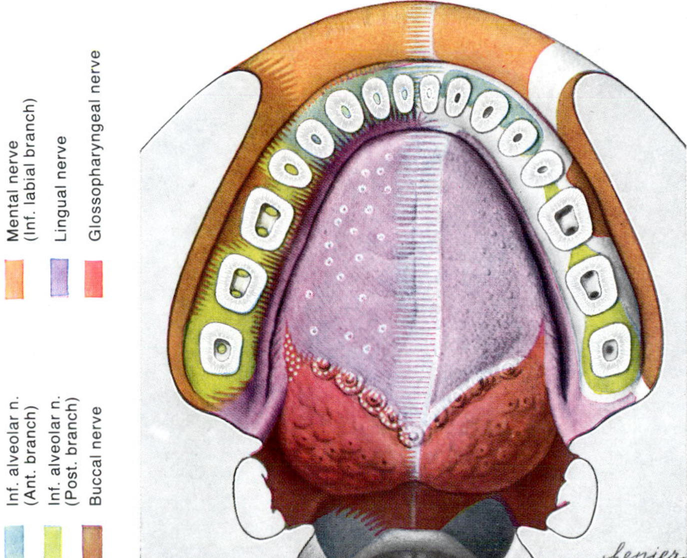
Mental nerve (Inf. labial branch)
Lingual nerve
Glossopharyngeal nerve
Inf. alveolar n. (Ant. branch)
Inf. alveolar n. (Post. branch)
Buccal nerve
Lepier

Fig. 724: Sensory Innervation of the Tongue. Cheeks and Floor of the Oral Cavity

NOTE that on the right side are specific areas of distribution; on the left side, the overlapping fields. *White circles:* taste zone of chorda tympani nerve; *white dots:* taste zone of glossopharyngeal nerve.

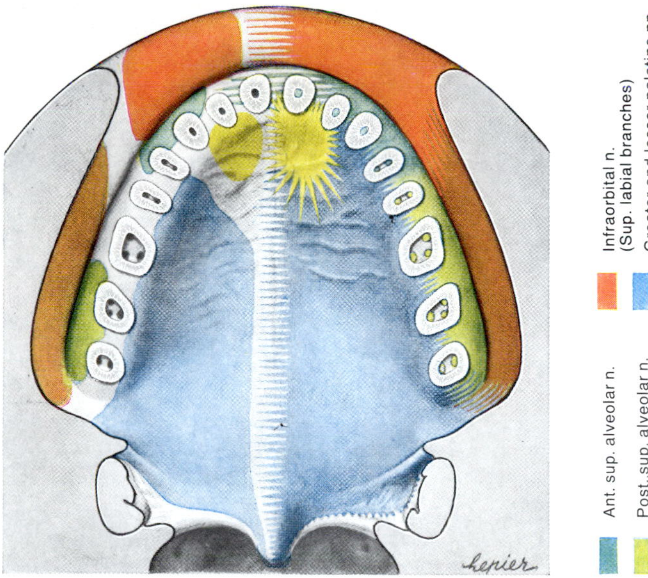
Infraorbital n. (Sup. labial branches)
Greater and lesser palatine nn.
Nasopalatine nerve
Ant. sup. alveolar n.
Post. sup. alveolar n.
Buccal nerve
Lepier

Fig. 725: Sensory Innervation of the Palate, Cheeks and Upper Gums of the Oral Cavity

NOTE that on the (reader's) left side are shown specific fields of distribution of the individual sensory nerves, while on the right side the overlapping fields of innervation are shown.

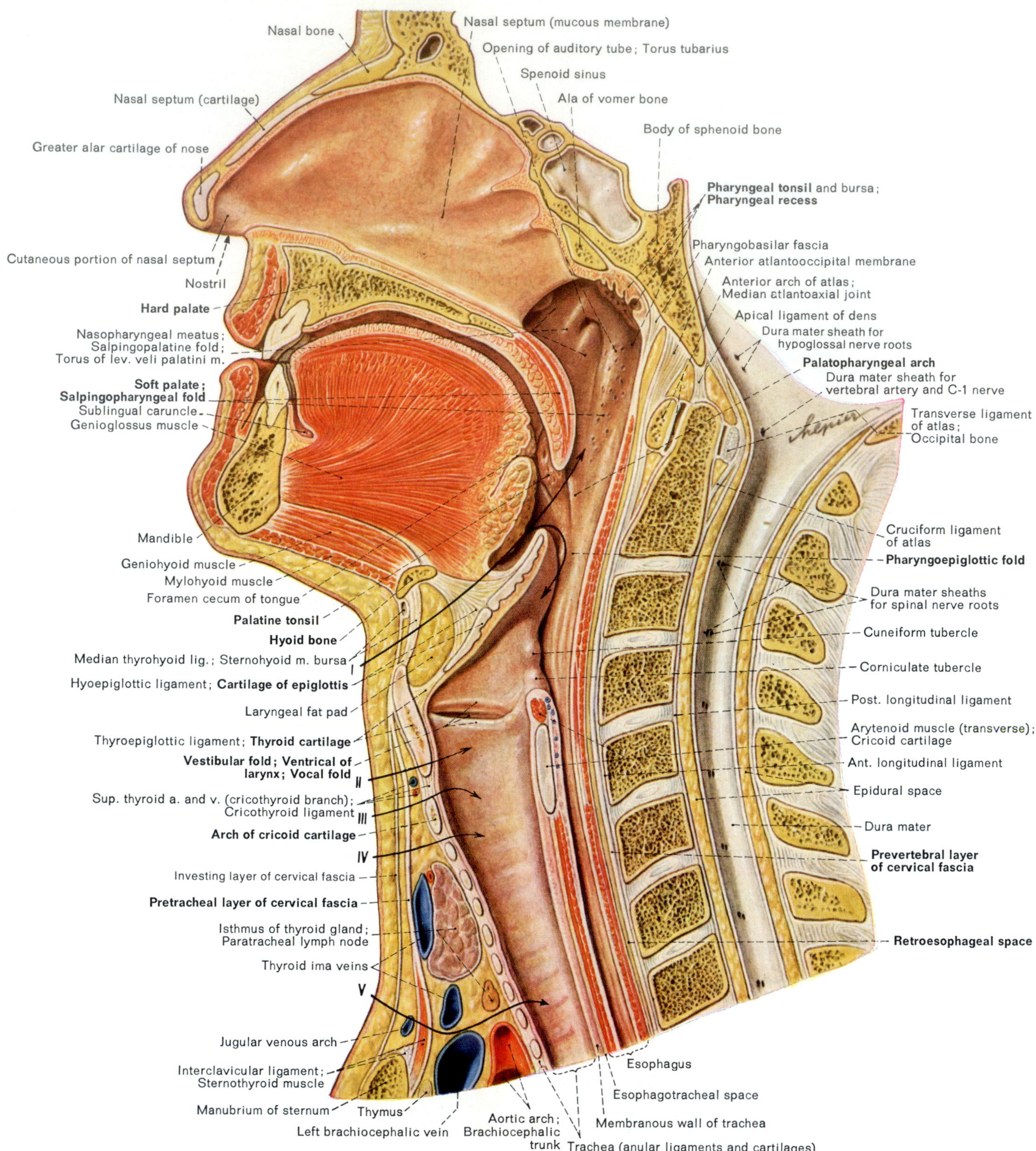

Nasal bone
Nasal septum (mucous membrane)
Opening of auditory tube; Torus tubarius
Spenoid sinus
Ala of vomer bone
Nasal septum (cartilage)
Body of sphenoid bone
Greater alar cartilage of nose
Pharyngeal tonsil and bursa;
Pharyngeal recess
Pharyngobasilar fascia
Anterior atlantooccipital membrane
Cutaneous portion of nasal septum
Nostril
Anterior arch of atlas;
Median atlantoaxial joint
Hard palate
Apical ligament of dens
Nasopharyngeal meatus;
Salpingopalatine fold;
Torus of lev. veli palatini m.
Dura mater sheath for
hypoglossal nerve roots
Palatopharyngeal arch
Dura mater sheath for
vertebral artery and C-1 nerve
Soft palate;
Salpingopharyngeal fold
Sublingual caruncle
Genioglossus muscle
Transverse ligament
of atlas;
Occipital bone
Mandible
Cruciform ligament
of atlas
Geniohyoid muscle
Pharyngoepiglottic fold
Mylohyoid muscle
Foramen cecum of tongue
Dura mater sheaths
for spinal nerve roots
Palatine tonsil
Hyoid bone
Cuneiform tubercle
Median thyrohyoid lig.; Sternohyoid m. bursa
I
Corniculate tubercle
Hyoepiglottic ligament; Cartilage of epiglottis
Laryngeal fat pad
Post. longitudinal ligament
Thyroepiglottic ligament; Thyroid cartilage
Arytenoid muscle (transverse);
Cricoid cartilage
Vestibular fold; Ventrical of
larynx; Vocal fold
II
Ant. longitudinal ligament
Sup. thyroid a. and v. (cricothyroid branch);
Cricothyroid ligament
III
Epidural space
Arch of cricoid cartilage
Dura mater
IV
Prevertebral layer
of cervical fascia
Investing layer of cervical fascia
Pretracheal layer of cervical fascia
Isthmus of thyroid gland;
Paratracheal lymph node
Retroesophageal space
Thyroid ima veins
V
Jugular venous arch
Esophagus
Interclavicular ligament;
Sternothyroid muscle
Esophagotracheal space
Manubrium of sternum
Thymus
Aortic arch;
Brachiocephalic
trunk
Membranous wall of trachea
Left brachiocephalic vein
Trachea (anular ligaments and cartilages)

Fig. 726: The Viscera of the Head and Neck: Mid-sagittal Section

NOTE: 1) the closed oral cavity is occupied principally by the tongue. Observe that the posterior aspect of the oral cavity opens into the oropharanyx. Superiorly, the posterior nasal cavities are continuous with the nasopharynx, whereas inferiorly the laryngeal portion of the pharynx (between the levels of the epiglottis and cricoid cartilages) opens into the larynx.

2) the pharynx continues inferiorly as the esophagus, while the larynx becomes the trachea below the level of the cricoid cartilage.

3) during the act of swallowing (deglutition), as food is directed toward the posterior part of the oral cavity, the soft palate is both elevated and tensed (by the levator and tensor veli palatini muscles) and directed toward the posterior wall of the pharynx, thereby closing off the nasopharynx. Simultaneously the larynx is drawn superiorly toward the epiglottis to the level of the hyoid bone, and the pharynx also ascends. This action closes the laryngeal orifice, thereby preventing food from entering the larynx.

4) arrows I to V: surgical approaches to pharynx, larynx and trachea.

Fig. 726

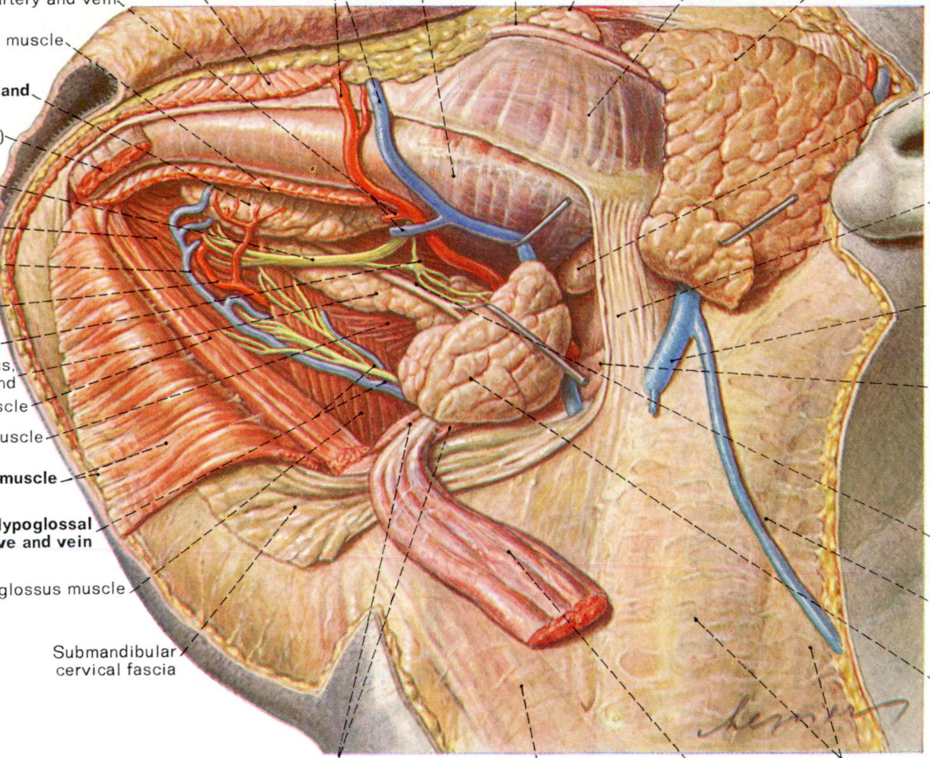

Facial artery and vein
Med. pterygoid m.
Parotid duct
Accessory parotid gland
Masseter muscle
Parotid gland
Platysma muscle
Submental artery and vein
Mylohyoid muscle
Sublingual gland
Digastric m. (ant. belly)
Lingual nerve
Genioglossus muscle
Sublingual vein
Lingual artery
Submandibular ganglion
Deep process, Submandibular gland
Geniohyoid muscle
Styloglossus muscle
Mylohyoid muscle
Hypoglossal nerve and vein
Hyoglossus muscle
Submandibular cervical fascia
Dorsal submandibular lymph nodes
Cervical fascia (angular band)
Retromandibular vein
Digastric m. (post. belly); Stylohyoid muscle
Submandibular duct
External jugular vein
Submandibular gland
Stylohyoid muscle
Sternohyoid muscle
Digastric muscle (anterior belly)
Superficial investing layer of cervical fascia; Sternocleidomastoid muscle

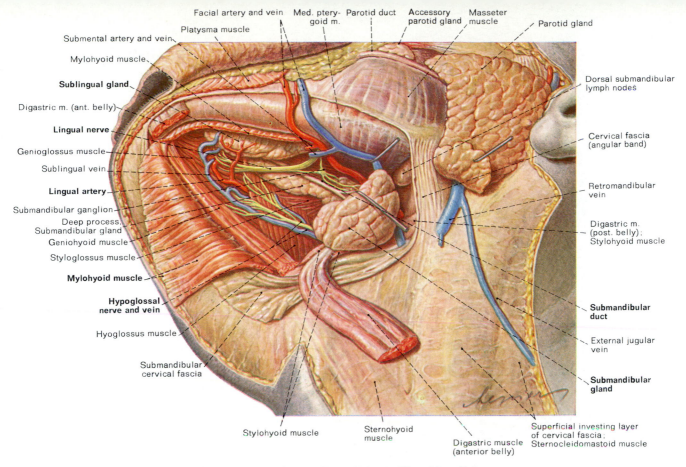

Fig. 727: The Floor of the Oral Cavity Opened from the Submandibular Triangle, Viewed from Below

NOTE: 1) the anterior belly of the digastric muscle and the mylohyoid muscle have been reflected in order to reveal the following structures: the sublingual gland, the lingual nerve, the submandibular ganglion and duct, the hypoglossal nerve and vein and the lingual artery.

2) presynaptic parasympathetic fibers (VII) accompany the lingual nerve (V) to reach the submandibular ganglion, where they synapse with postganglionic neurons whose fibers innervate both the submandibular and sublingual glands.

3) the hypoglossal (XII) nerve is the motor nerve to all tongue muscles, except the palatoglossus, while the lingual nerve (V) supplies the anterior 2/3rds of the tongue with general sensation.

Fig. 728: The Mylohyoid and Geniohyoid Muscles (Viewed from Above)

NOTE that the mylohyoid muscles along with the geniohyoid muscles form the muscular floor of the oral cavity. The mylohyoids arise along the mylohyoid lines of the mandible and insert into the median fibrous raphé which extends from the hyoid bone to the symphysis menti. Observe that the genioglossal muscles have been severed near their origin.

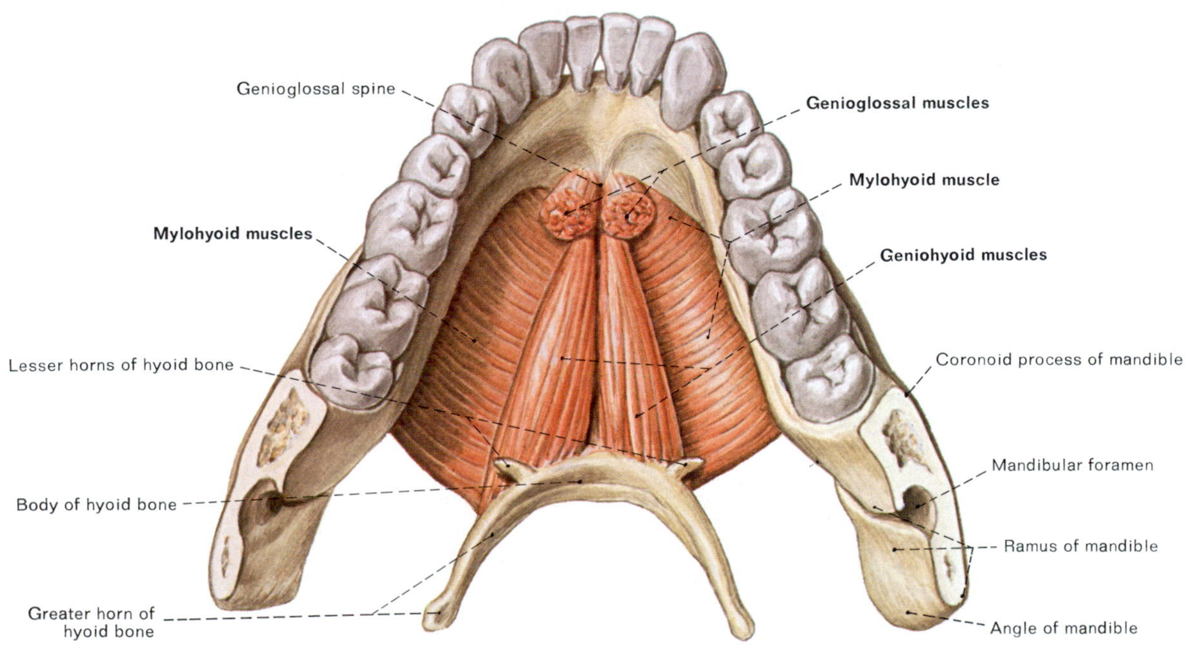

Genioglossal spine
Genioglossal muscles
Mylohyoid muscle
Mylohyoid muscles
Geniohyoid muscles
Lesser horns of hyoid bone
Coronoid process of mandible
Mandibular foramen
Body of hyoid bone
Ramus of mandible
Greater horn of hyoid bone
Angle of mandible

Frenulum of upper lip
Uvula and soft palate
Hard palate (palatine raphé)
Supratonsillar fossa
Palatopharyngeal arch
Posterior wall of oral pharynx
Buccinator m.
Palatoglossal arch
Isthmus of the fauces
Buccal fat pad
Cheek (cut)
Dorsum of tongue
Palatine tonsil
Gum (gingiva)
Vestibule of mouth
Frenulum of lower lip

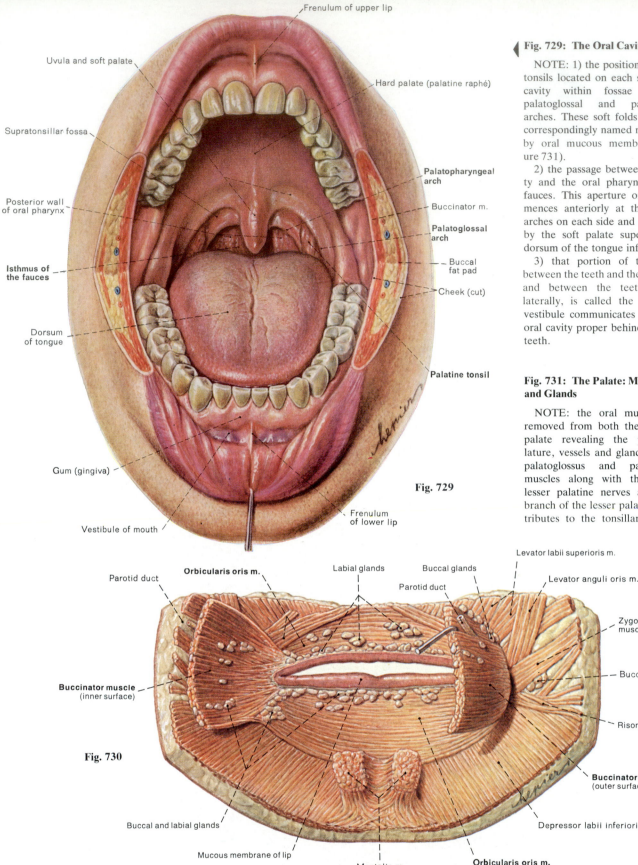

Fig. 729

Fig. 729: The Oral Cavity

NOTE: 1) the position of the palatine tonsils located on each side of the oral cavity within fossae between the palatoglossal and palatopharyngeal arches. These soft folds are formed by correspondingly named muscles covered by oral mucous membrane (see Figure 731).

2) the passage between the oral cavity and the oral pharynx is called the fauces. This aperture or isthmus commences anteriorly at the palatoglossal arches on each side and is also bounded by the soft palate superiorly and the dorsum of the tongue inferiorly.

3) that portion of the oral cavity between the teeth and the lips anteriorly, and between the teeth and cheeks laterally, is called the vestibule. The vestibule communicates with the larger oral cavity proper behind the 3rd molar teeth.

Fig. 731: The Palate: Muscular Folds and Glands

NOTE: the oral mucosa has been removed from both the hard and soft palate revealing the palatal musculature, vessels and glands. Observe the palatoglossus and palatopharyngeus muscles along with the greater and lesser palatine nerves and vessels. A branch of the lesser palatine artery contributes to the tonsillar blood supply.

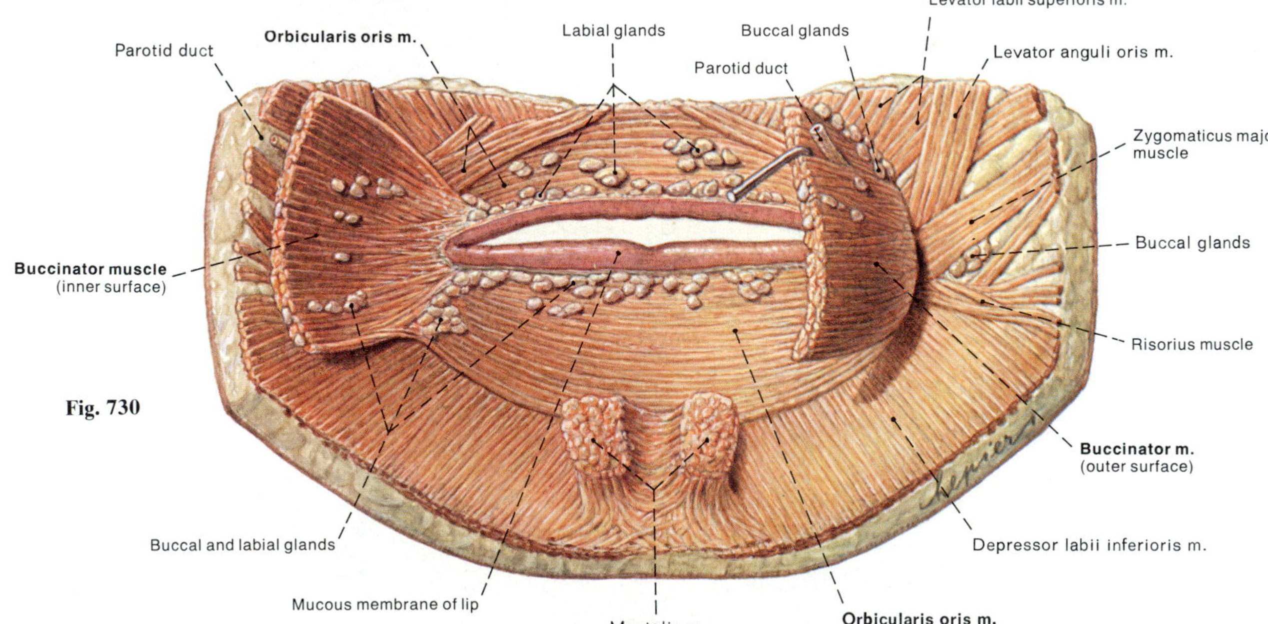

Parotid duct
Orbicularis oris m.
Labial glands
Buccal glands
Levator labii superioris m.
Parotid duct
Levator anguli oris m.
Zygomaticus major muscle
Buccinator muscle (inner surface)
Buccal glands
Risorius muscle
Buccinator m. (outer surface)
Buccal and labial glands
Depressor labii inferioris m.
Mucous membrane of lip
Mentalis m.
Orbicularis oris m.

Fig. 730

Fig. 730: The Lips Viewed from Within the Oral Cavity

NOTE: 1) this dissection demonstrates the muscles of the mouth from within the oral cavity after the oral mucous membrane has been removed. Observe the numerous small labial glands.

2) the contour of the lips as they surround the oral orifice depends on the arrangement of the muscular bundles which interlace at the labial margins. These include the elevators and depressors of the lips and their angles, along with the orbicularis oris and buccinator muscles.

3) the fibers of the buccinator muscle converging at the angle of the mouth. Many of these fibers decussate, becoming continuous with the fibers of the orbicularis oris muscle of both the upper and lower lips.

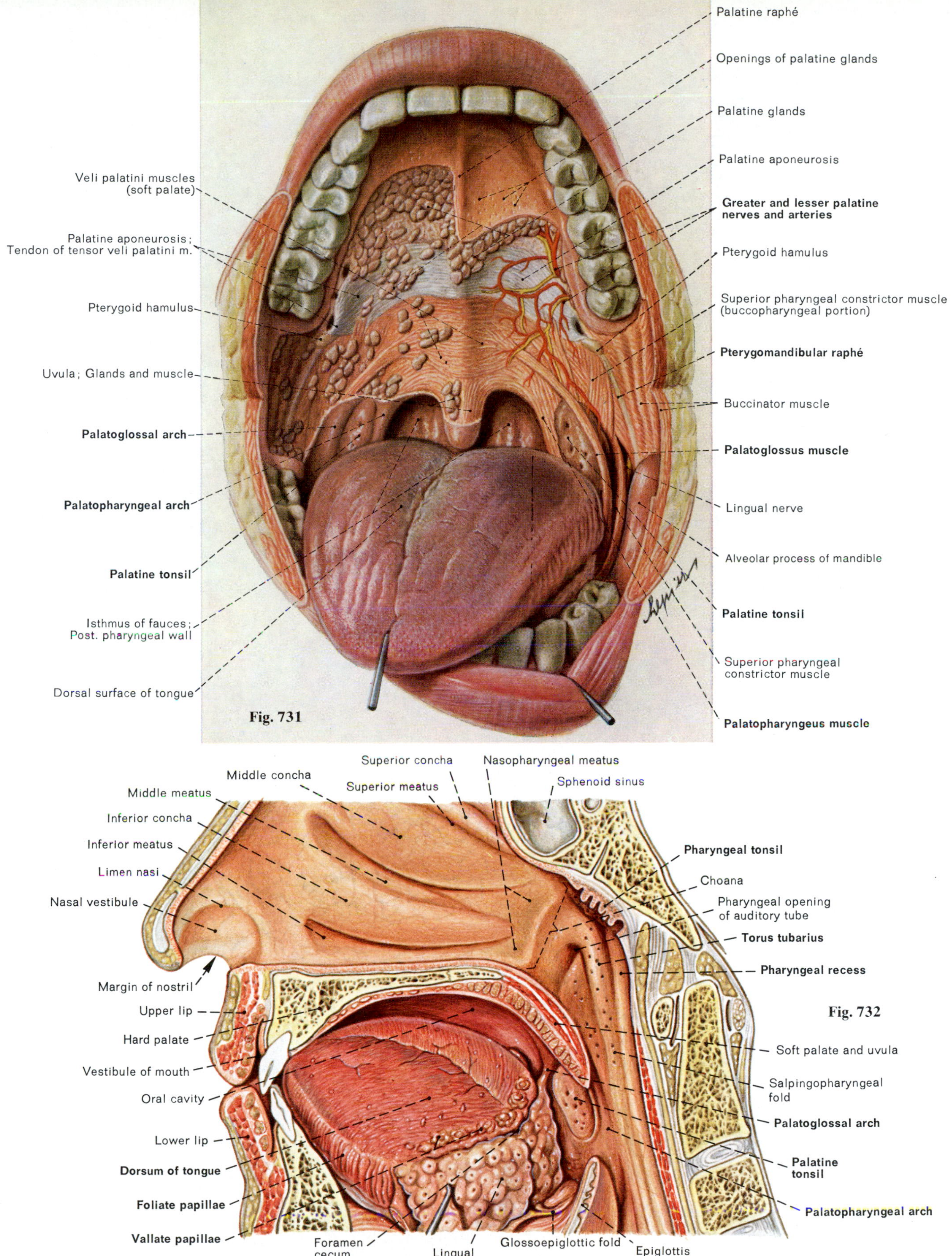
Palatine raphé
Openings of palatine glands
Palatine glands
Palatine aponeurosis
Greater and lesser palatine nerves and arteries
Pterygoid hamulus
Superior pharyngeal constrictor muscle (buccopharyngeal portion)
Pterygomandibular raphé
Buccinator muscle
Palatoglossus muscle
Lingual nerve
Alveolar process of mandible
Palatine tonsil
Superior pharyngeal constrictor muscle
Palatopharyngeus muscle
Veli palatini muscles (soft palate)
Palatine aponeurosis; Tendon of tensor veli palatini m.
Pterygoid hamulus
Uvula; Glands and muscle
Palatoglossal arch
Palatopharyngeal arch
Palatine tonsil
Isthmus of fauces; Post. pharyngeal wall
Dorsal surface of tongue
Fig. 731
Superior concha
Nasopharyngeal meatus
Middle concha
Superior meatus
Sphenoid sinus
Middle meatus
Inferior concha
Inferior meatus
Limen nasi
Nasal vestibule
Pharyngeal tonsil
Choana
Pharyngeal opening of auditory tube
Torus tubarius
Pharyngeal recess
Margin of nostril
Upper lip
Hard palate
Vestibule of mouth
Oral cavity
Lower lip
Dorsum of tongue
Foliate papillae
Vallate papillae
Fig. 732
Soft palate and uvula
Salpingopharyngeal fold
Palatoglossal arch
Palatine tonsil
Palatopharyngeal arch
Foramen cecum
Lingual tonsil
Glossoepiglottic fold
Epiglottis

Fig. 732: The Tongue, Palatine Tonsil and the Oropharynx

NOTE: 1) in this sagittal view, the tongue has been deviated to demonstrate the right palatoglossal arch and right palatine tonsil. Observe the large vallate papillae.

2) the opening of the auditory tube in the nasopharynx behind which is a cartilaginous elevation of the tube called the torus tubarius. Note also the pharyngeal tonsil (adenoids).

Fig. 733: A Lateral View of the Salivary Glands and Their Ducts

NOTE that the body of the left half of the mandible has been removed, thereby exposing the salivary glands in the oral cavity. In addition to the parotid, submandibular, and sublingual glands, note also the more variable anterior lingual and accessory parotid glands.

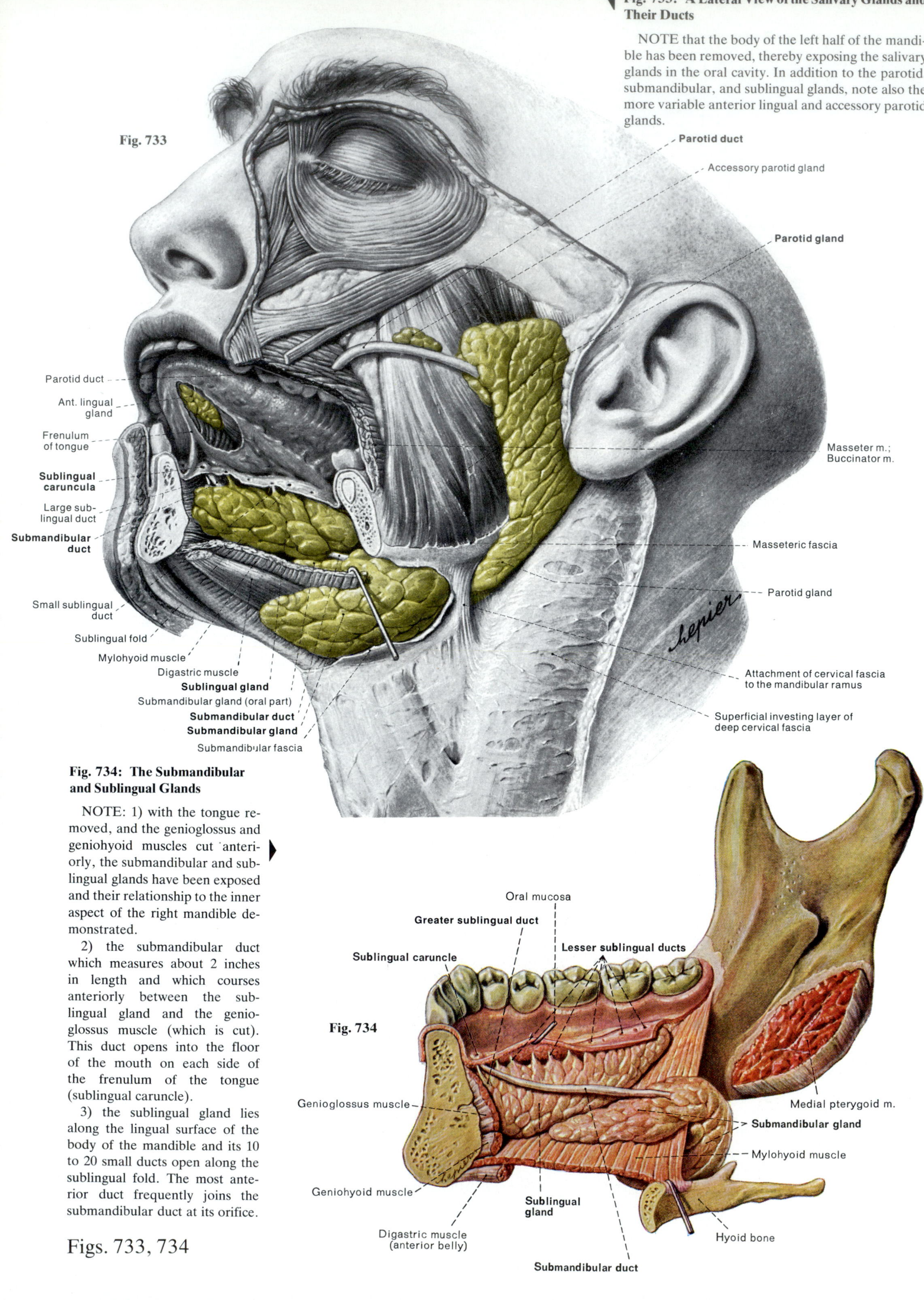

Fig. 733
Parotid duct
Accessory parotid gland
Parotid gland
Parotid duct
Ant. lingual gland
Frenulum of tongue
Sublingual caruncula
Large sublingual duct
Submandibular duct
Masseter m.; Buccinator m.
Masseteric fascia
Parotid gland
Small sublingual duct
Sublingual fold
Mylohyoid muscle
Digastric muscle
Sublingual gland
Submandibular gland (oral part)
Submandibular duct
Submandibular gland
Submandibular fascia
Attachment of cervical fascia to the mandibular ramus
Superficial investing layer of deep cervical fascia
Lepier
Fig. 734
Oral mucosa
Greater sublingual duct
Sublingual caruncle
Lesser sublingual ducts
Genioglossus muscle
Medial pterygoid m.
Submandibular gland
Mylohyoid muscle
Geniohyoid muscle
Sublingual gland
Digastric muscle (anterior belly)
Hyoid bone
Submandibular duct

Fig. 734: The Submandibular and Sublingual Glands

NOTE: 1) with the tongue removed, and the genioglossus and geniohyoid muscles cut anteriorly, the submandibular and sublingual glands have been exposed and their relationship to the inner aspect of the right mandible demonstrated.

2) the submandibular duct which measures about 2 inches in length and which courses anteriorly between the sublingual gland and the genioglossus muscle (which is cut). This duct opens into the floor of the mouth on each side of the frenulum of the tongue (sublingual caruncle).

3) the sublingual gland lies along the lingual surface of the body of the mandible and its 10 to 20 small ducts open along the sublingual fold. The most anterior duct frequently joins the submandibular duct at its orifice.

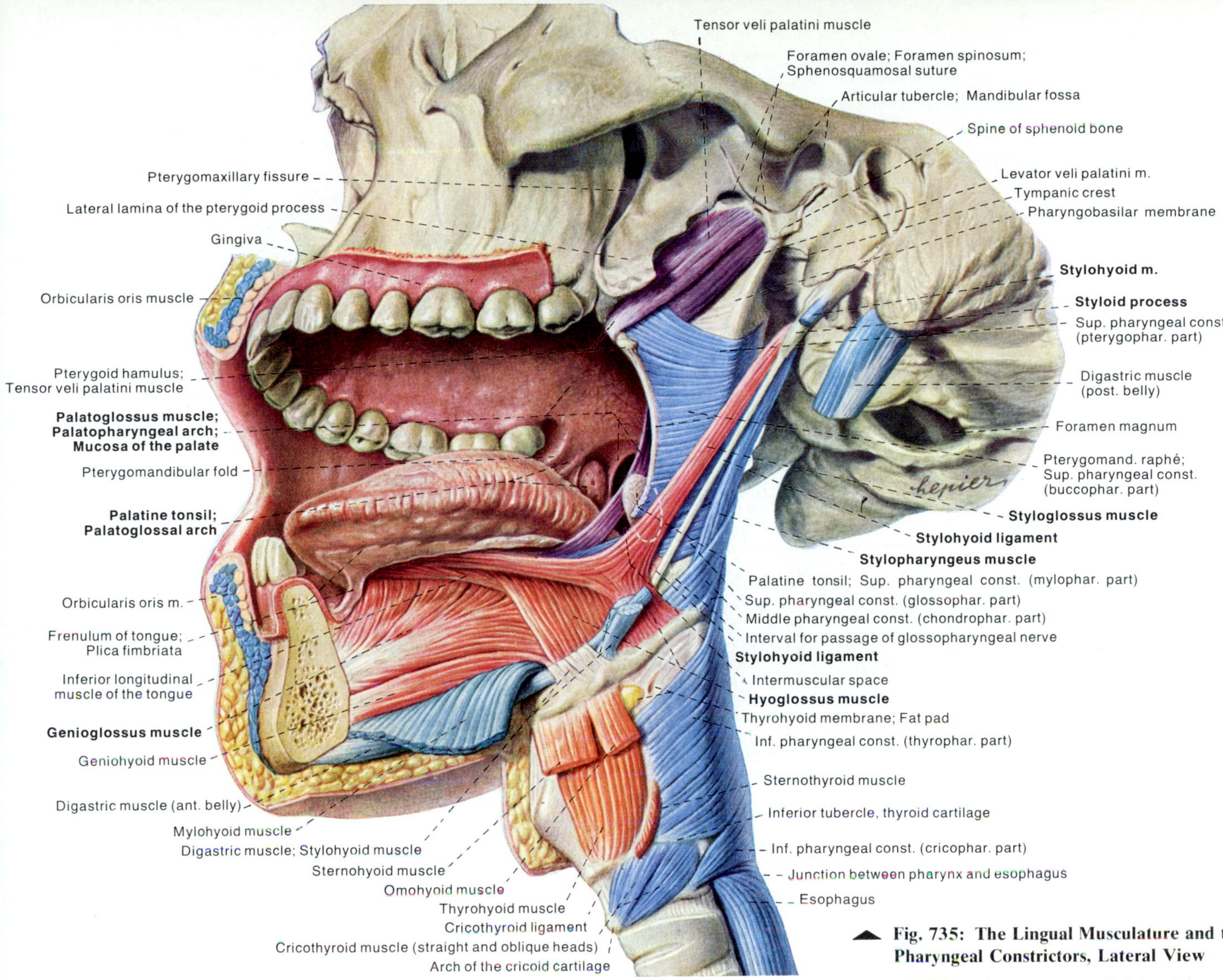

Tensor veli palatini muscle
Foramen ovale; Foramen spinosum; Sphenosquamosal suture
Articular tubercle; Mandibular fossa
Spine of sphenoid bone
Pterygomaxillary fissure
Lateral lamina of the pterygoid process
Levator veli palatini m.
Tympanic crest
Pharyngobasilar membrane
Gingiva
Stylohyoid m.
Orbicularis oris muscle
Styloid process
Sup. pharyngeal const. (pterygophar. part)
Pterygoid hamulus; Tensor veli palatini muscle
Digastric muscle (post. belly)
Palatoglossus muscle; Palatopharyngeal arch; Mucosa of the palate
Foramen magnum
Pterygomandibular fold
Pterygomand. raphé; Sup. pharyngeal const. (buccophar. part)
Palatine tonsil; Palatoglossal arch
Styloglossus muscle
Stylohyoid ligament
Stylopharyngeus muscle
Palatine tonsil; Sup. pharyngeal const. (mylophar. part)
Orbicularis oris m.
Sup. pharyngeal const. (glossophar. part)
Middle pharyngeal const. (chondrophar. part)
Frenulum of tongue; Plica fimbriata
Interval for passage of glossopharyngeal nerve
Stylohyoid ligament
Inferior longitudinal muscle of the tongue
Intermuscular space
Hyoglossus muscle
Genioglossus muscle
Thyrohyoid membrane; Fat pad
Geniohyoid muscle
Inf. pharyngeal const. (thyrophar. part)
Sternothyroid muscle
Digastric muscle (ant. belly)
Inferior tubercle, thyroid cartilage
Mylohyoid muscle
Digastric muscle; Stylohyoid muscle
Inf. pharyngeal const. (cricophar. part)
Sternohyoid muscle
Junction between pharynx and esophagus
Omohyoid muscle
Esophagus
Thyrohyoid muscle
Cricothyroid ligament
Cricothyroid muscle (straight and oblique heads)
Arch of the cricoid cartilage

Fig. 735: The Lingual Musculature and the Pharyngeal Constrictors, Lateral View

NOTE: 1) that the superficial and deep facial structures on the left side have been removed, as has the left half of the mandible in order to expose the tongue, the lingual muscles, and the pharyngeal constrictors. Observe that the buccinator has been removed but that the superior pharyngeal constrictor is still attached to the pterygomandibular raphé.

2) that the stylohyoid muscle has been cut, but the stylohyoid ligament and the stylopharyngeus and styloglossus muscles are still intact. Observe the hyoglossus, geniohyoid, and genioglossus muscles and that the palatoglossus descends into the tongue from the soft palate.

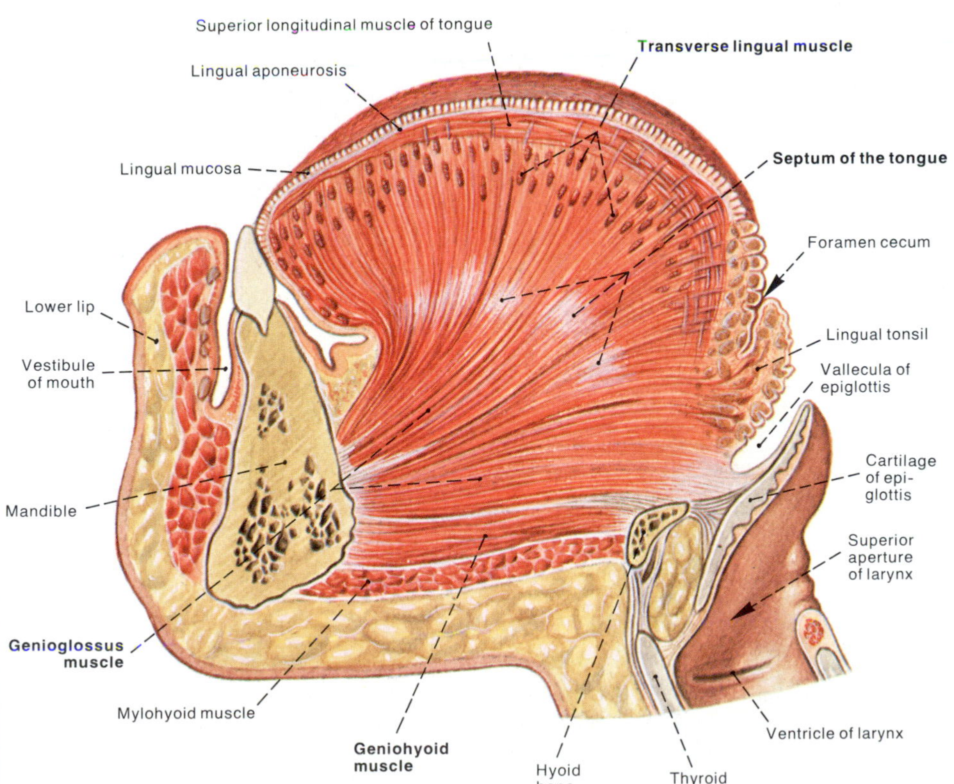

Superior longitudinal muscle of tongue
Transverse lingual muscle
Lingual aponeurosis
Lingual mucosa
Septum of the tongue
Foramen cecum
Lower lip
Lingual tonsil
Vestibule of mouth
Vallecula of epiglottis
Cartilage of epiglottis
Mandible
Superior aperture of larynx
Genioglossus muscle
Mylohyoid muscle
Ventricle of larynx
Geniohyoid muscle
Hyoid bone
Thyroid cartilage

Fig. 736: The Genioglossus and the Intrinsic Muscles of the Tongue

NOTE: 1) that in this mid-sagittal section can be seen the median fibrous septum of the tongue and that the intrinsic tongue musculature includes the longitudinal, transverse, and vertical muscles of the tongue.

2) that the genioglossus constitutes most of the tongue musculature and that its fibers radiate backward and upward in a fan-like manner from the uppermost of the mental spines (genial tubercles) on the inner surface of the mandible, just above the origin of the geniohyoid muscles.

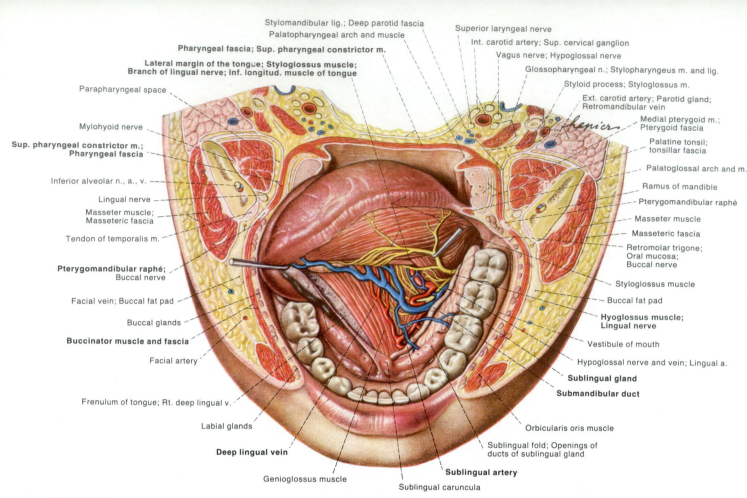

Stylomandibular lig.; Deep parotid fascia
Palatopharyngeal arch and muscle
Pharyngeal fascia; Sup. pharyngeal constrictor m.
Lateral margin of the tongue; Styloglossus muscle;
Branch of lingual nerve; Inf. longitud. muscle of tongue
Parapharyngeal space
Mylohyoid nerve
Sup. pharyngeal constrictor m.;
Pharyngeal fascia
Inferior alveolar n., a., v.
Lingual nerve
Masseter muscle;
Masseteric fascia
Tendon of temporalis m.
Pterygomandibular raphé;
Buccal nerve
Facial vein; Buccal fat pad
Buccal glands
Buccinator muscle and fascia
Facial artery
Frenulum of tongue; Rt. deep lingual v.
Labial glands
Deep lingual vein
Genioglossus muscle
Sublingual caruncula
Sublingual artery
Sublingual fold; Openings of
ducts of sublingual gland
Orbicularis oris muscle
Submandibular duct
Sublingual gland
Hypoglossal nerve and vein; Lingual a.
Vestibule of mouth
Hyoglossus muscle;
Lingual nerve
Buccal fat pad
Styloglossus muscle
Retromolar trigone;
Oral mucosa;
Buccal nerve
Masseteric fascia
Masseter muscle
Pterygomandibular raphé
Ramus of mandible
Palatoglossal arch and m.
Palatine tonsil;
tonsillar fascia
Medial pterygoid m.;
Pterygoid fascia
Ext. carotid artery; Parotid gland;
Retromandibular vein
Styloid process; Styloglossus m.
Glossopharyngeal n.; Stylopharyngeus m. and lig.
Vagus nerve; Hypoglossal nerve
Int. carotid artery; Sup. cervical ganglion
Superior laryngeal nerve
Lepier

Fig. 737: Vessels and Nerves in the Floor of the Oral Cavity, Viewed from Above and Anteriorly, Stage 1: With Salivary Glands in Place

NOTE that the tongue has been deflected to the (specimen's) right, and the left lingual nerve and vessels and the hypoglossal nerve, sublingual gland and submandibular duct exposed in the floor of the oral cavity. Observe also that the lateral walls of the oral cavity and the oropharynx have been cut in cross section, showing the muscles of mastication laterally, the buccinator muscles medially and the superior pharyngeal constrictor posteriorly.

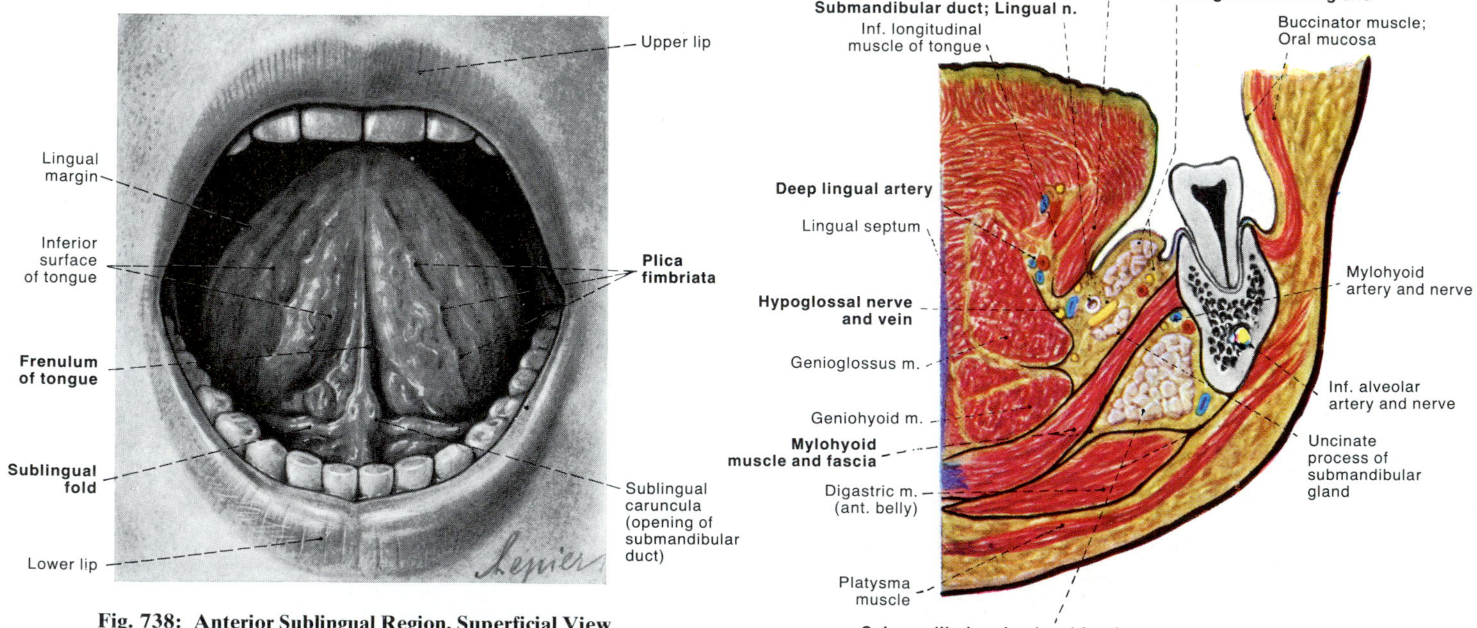
Upper lip
Lingual margin
Inferior surface of tongue
Plica fimbriata
Frenulum of tongue
Sublingual fold
Sublingual caruncula (opening of submandibular duct)
Lower lip
Lepier
Hyoglossus muscle
Submandibular duct; Lingual n.
Sublingual fold and gland
Inf. longitudinal muscle of tongue
Buccinator muscle; Oral mucosa
Deep lingual artery
Lingual septum
Mylohyoid artery and nerve
Hypoglossal nerve and vein
Genioglossus m.
Inf. alveolar artery and nerve
Geniohyoid m.
Mylohyoid muscle and fascia
Uncinate process of submandibular gland
Digastric m. (ant. belly)
Platysma muscle
Submandibular gland and fascia

Fig. 738: Anterior Sublingual Region, Superficial View

NOTE that the mucous membrane covering the floor of the oral cavity continues over the inferior surface of the tongue, meeting at the midline as an elevated fold called the frenulum of the tongue. Observe the sublingual folds. Along these open the ducts of the sublingual glands, and at their anterior end on each side is an orifice for the submandibular duct called the sublingual caruncula.

Fig. 739: Frontal Section Through the Tongue, Sublingual and Submandibular Regions

NOTE the locations of the lingual nerve, submandibular duct and hypoglossal nerve and its accompanying vein. Observe that the mylohyoid muscle separates the sublingual and submandibular regions.

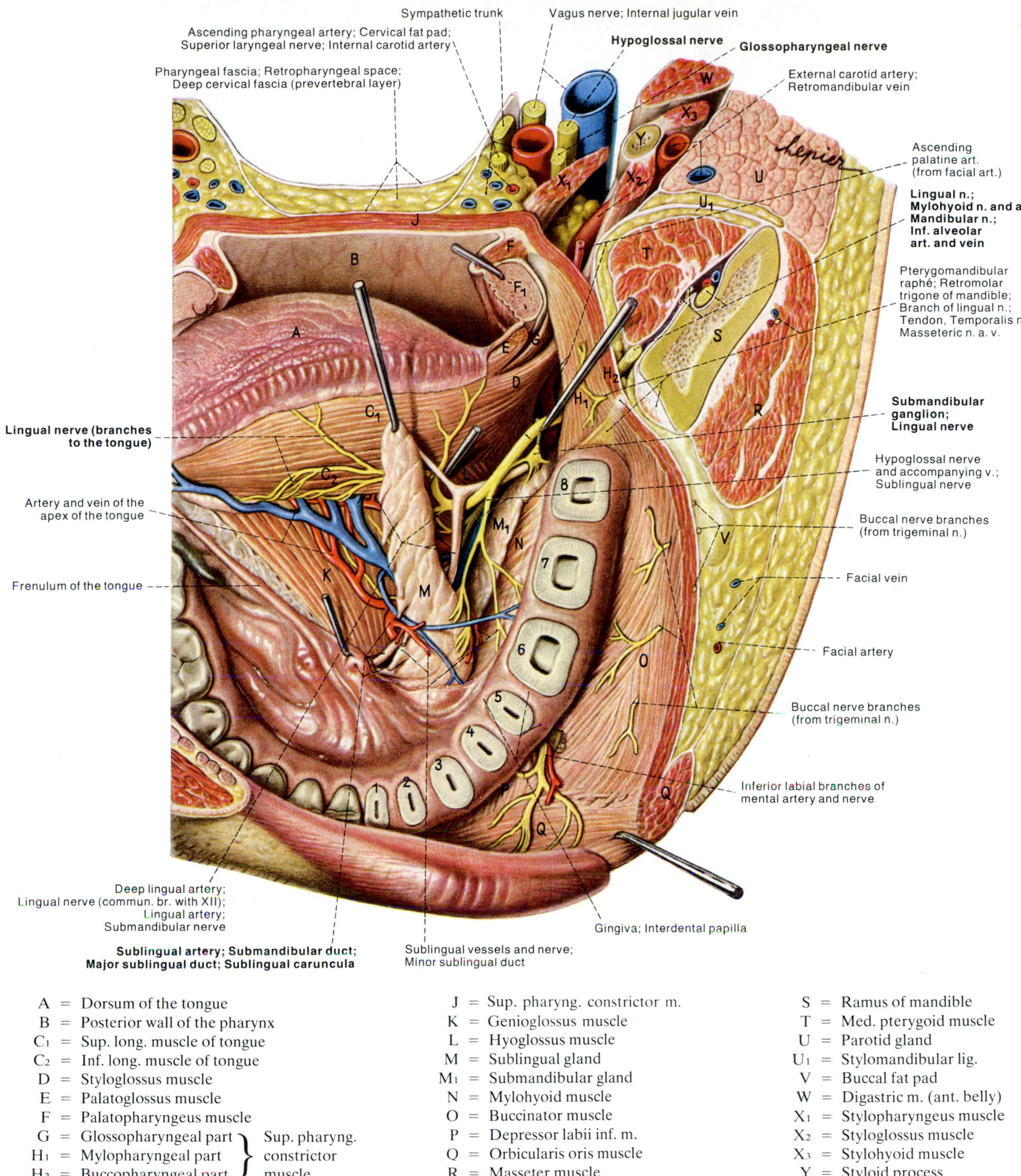

Sympathetic trunk
Vagus nerve; Internal jugular vein
Ascending pharyngeal artery; Cervical fat pad;
Superior laryngeal nerve; Internal carotid artery
Hypoglossal nerve
Glossopharyngeal nerve
Pharyngeal fascia; Retropharyngeal space;
Deep cervical fascia (prevertebral layer)
External carotid artery;
Retromandibular vein
Ascending
palatine art.
(from facial art.)
Lingual n.;
Mylohyoid n. and art.;
Mandibular n.;
Inf. alveolar
art. and vein
Pterygomandibular
raphé; Retromolar
trigone of mandible;
Branch of lingual n.;
Tendon, Temporalis m.;
Masseteric n. a. v.
Submandibular
ganglion;
Lingual nerve
Lingual nerve (branches
to the tongue)
Hypoglossal nerve
and accompanying v.;
Sublingual nerve
Artery and vein of the
apex of the tongue
Buccal nerve branches
(from trigeminal n.)
Frenulum of the tongue
Facial vein
Facial artery
Buccal nerve branches
(from trigeminal n.)
Inferior labial branches of
mental artery and nerve
Deep lingual artery;
Lingual nerve (commun. br. with XII);
Lingual artery;
Submandibular nerve
Gingiva; Interdental papilla
Sublingual artery; Submandibular duct;
Major sublingual duct; Sublingual caruncula
Sublingual vessels and nerve;
Minor sublingual duct
Lepier

A = Dorsum of the tongue
B = Posterior wall of the pharynx
C_1 = Sup. long. muscle of tongue
C_2 = Inf. long. muscle of tongue
D = Styloglossus muscle
E = Palatoglossus muscle
F = Palatopharyngeus muscle
G = Glossopharyngeal part } Sup. pharyng. constrictor muscle
H_1 = Mylopharyngeal part } Sup. pharyng. constrictor muscle
H_2 = Buccopharyngeal part } Sup. pharyng. constrictor muscle
J = Sup. pharyng. constrictor m.
K = Genioglossus muscle
L = Hyoglossus muscle
M = Sublingual gland
M_1 = Submandibular gland
N = Mylohyoid muscle
O = Buccinator muscle
P = Depressor labii inf. m.
Q = Orbicularis oris muscle
R = Masseter muscle
S = Ramus of mandible
T = Med. pterygoid muscle
U = Parotid gland
U_1 = Stylomandibular lig.
V = Buccal fat pad
W = Digastric m. (ant. belly)
X_1 = Stylopharyngeus muscle
X_2 = Styloglossus muscle
X_3 = Stylohyoid muscle
Y = Styloid process

Fig. 740: Vessels and Nerves in the Floor of the Oral Cavity, Viewed from Above and Anteriorly, Stage 2: With the Sublingual Gland Mobilized and the Vestibule of the Mouth Exposed

NOTE: 1) that the tongue has been directed to the right, and the sublingual gland (M) has been elevated from its bed; this exposes the intraoral portion of the submandibular gland (M_1) and the mylohyoid muscle (N). Observe the submandibular ganglion attached to the lingual nerve. Preganglionic parasympathetic fibers course to this ganglion successively by way of the nervus intermedius (facial nerve), chorda tympani, and lingual nerves. These fibers synapse in the ganglion and then the postganglionic fibers course forward again through the lingual nerve to the sublingual and submandibular glands.

2) the branches of the buccal nerve that supply the mucous membrane lining the vestibule of the mouth. These are sensory nerves and are derived from the mandibular division of the trigeminal nerve. Observe that the vestibule is limited by the cheeks and lips externally and by the teeth and gums internally.

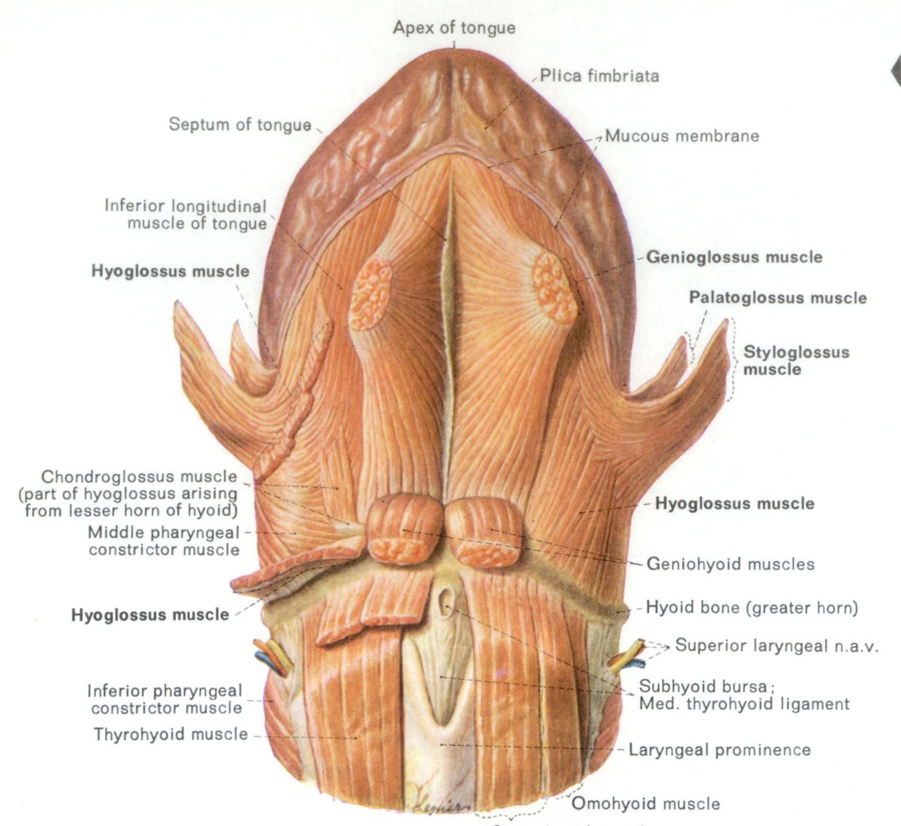

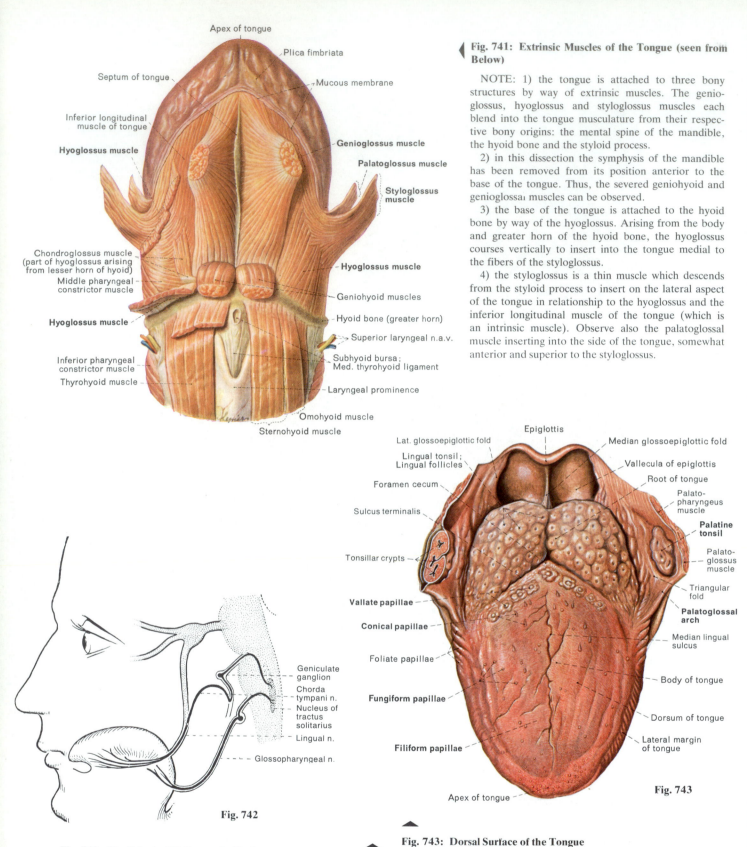

Apex of tongue
Plica fimbriata
Septum of tongue
Mucous membrane
Inferior longitudinal muscle of tongue
Hyoglossus muscle
Genioglossus muscle
Palatoglossus muscle
Styloglossus muscle
Chondroglossus muscle (part of hyoglossus arising from lesser horn of hyoid)
Middle pharyngeal constrictor muscle
Hyoglossus muscle
Geniohyoid muscles
Hyoglossus muscle
Hyoid bone (greater horn)
Superior laryngeal n.a.v.
Inferior pharyngeal constrictor muscle
Subhyoid bursa; Med. thyrohyoid ligament
Thyrohyoid muscle
Laryngeal prominence
Omohyoid muscle
Sternohyoid muscle

Fig. 741: Extrinsic Muscles of the Tongue (seen from Below)

NOTE: 1) the tongue is attached to three bony structures by way of extrinsic muscles. The genioglossus, hyoglossus and styloglossus muscles each blend into the tongue musculature from their respective bony origins: the mental spine of the mandible, the hyoid bone and the styloid process.

2) in this dissection the symphysis of the mandible has been removed from its position anterior to the base of the tongue. Thus, the severed geniohyoid and genioglossaı muscles can be observed.

3) the base of the tongue is attached to the hyoid bone by way of the hyoglossus. Arising from the body and greater horn of the hyoid bone, the hyoglossus courses vertically to insert into the tongue medial to the fibers of the styloglossus.

4) the styloglossus is a thin muscle which descends from the styloid process to insert on the lateral aspect of the tongue in relationship to the hyoglossus and the inferior longitudinal muscle of the tongue (which is an intrinsic muscle). Observe also the palatoglossal muscle inserting into the side of the tongue, somewhat anterior and superior to the styloglossus.

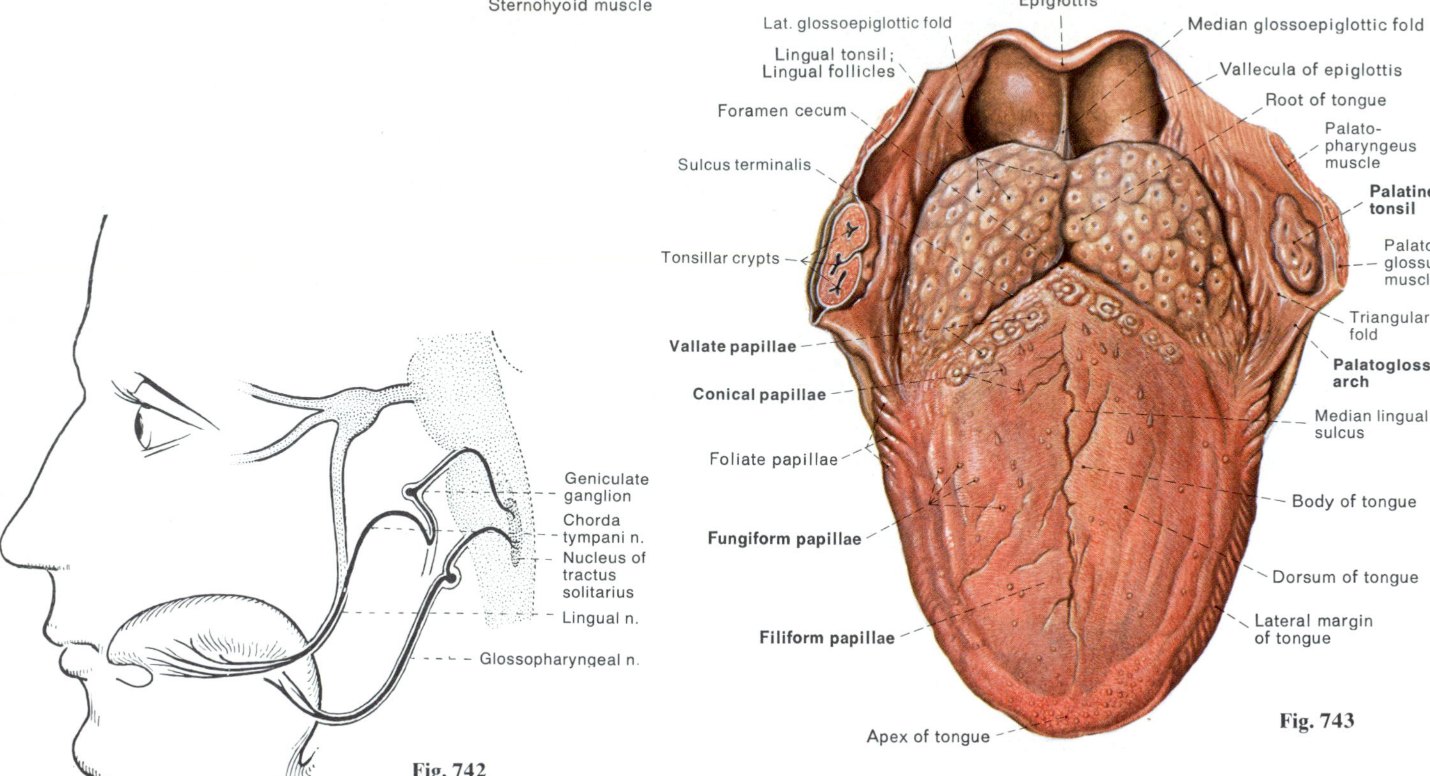

Geniculate ganglion
Chorda tympani n.
Nucleus of tractus solitarius
Lingual n.
Glossopharyngeal n.
Fig. 742
Epiglottis
Lat. glossoepiglottic fold
Median glossoepiglottic fold
Lingual tonsil; Lingual follicles
Vallecula of epiglottis
Foramen cecum
Root of tongue
Palato-pharyngeus muscle
Sulcus terminalis
Palatine tonsil
Tonsillar crypts
Palato-glossus muscle
Triangular fold
Vallate papillae
Palatoglossal arch
Conical papillae
Median lingual sulcus
Foliate papillae
Body of tongue
Fungiform papillae
Dorsum of tongue
Filiform papillae
Lateral margin of tongue
Apex of tongue
Fig. 743

Fig. 742: The Principal Pathways for Taste

NOTE: 1) that the two principal pathways for taste are by means of the chorda tympani nerve for the anterior two-thirds of the tongue and the glossopharyngeal nerve for the posterior one-third of the tongue;

2) that two lesser pathways also exist (not shown in this figure). From the epiglottis, taste fibers enter the brain by way of the internal laryngeal branch of the vagus, and taste fibers from the palate course through the palatine nerves and nerve of the pterygoid canal to the greater petrosal nerve and nervus intermedius of the facial nerve.

Fig. 743: Dorsal Surface of the Tongue

NOTE:1) the dorsum of the tongue is marked by numerous elevations called papillae. These serve as receptor sites for the special sense of taste. Observe the inverted V-shaped group of large vallate papillae.

2) the fungiform papillae which are found principally at the sides and apex of the tongue. These are relatively large and round and have a deep red color.

3) the filiform (conical) papillae. These are very small and generally arranged in rows which course parallel to the vallate papillae.

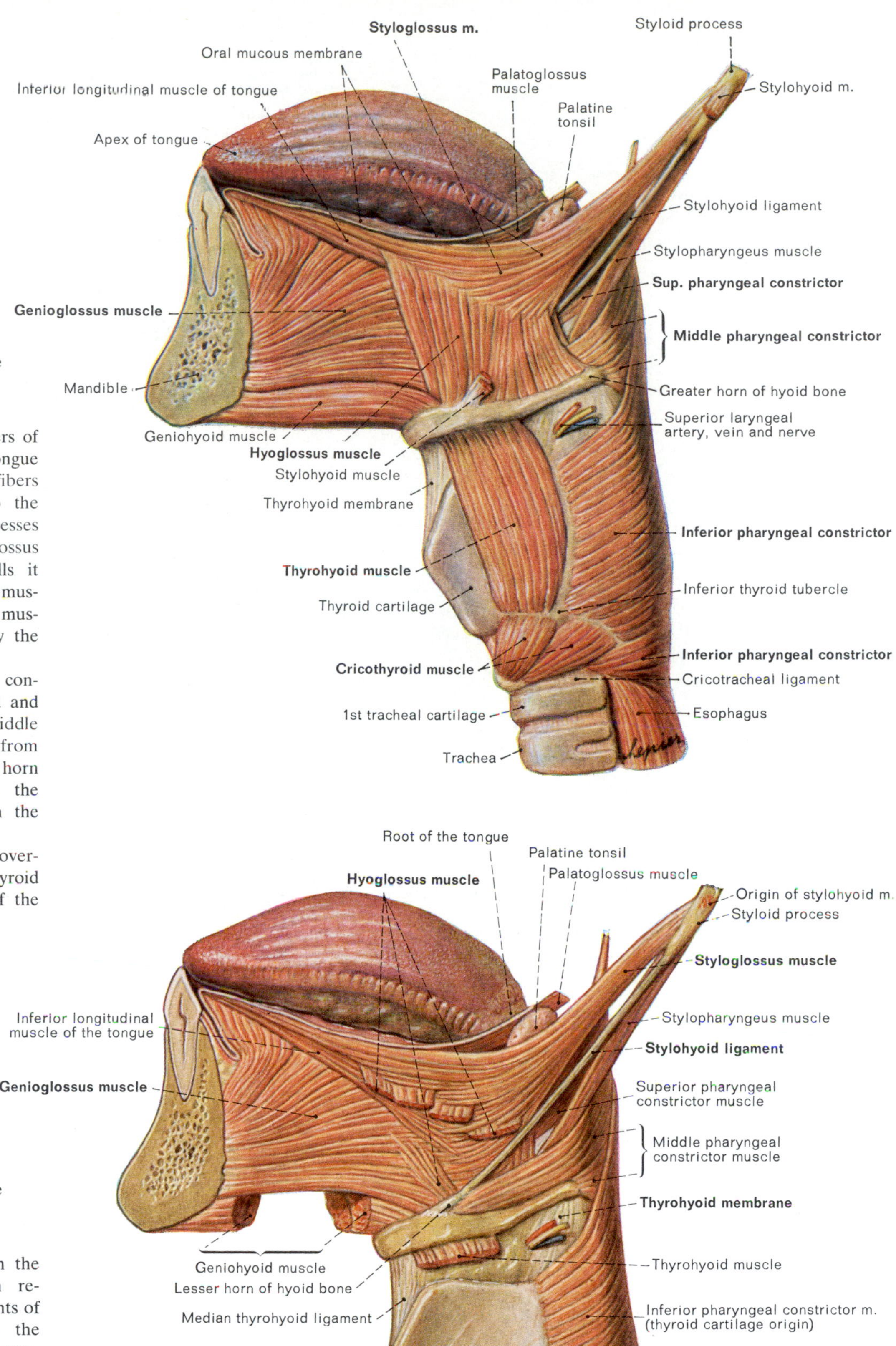
Styloglossus m.
Styloid process
Oral mucous membrane
Palatoglossus muscle
Inferior longitudinal muscle of tongue
Stylohyoid m.
Palatine tonsil
Apex of tongue
Stylohyoid ligament
Stylopharyngeus muscle
Sup. pharyngeal constrictor
Genioglossus muscle
Middle pharyngeal constrictor
Mandible
Greater horn of hyoid bone
Superior laryngeal artery, vein and nerve
Geniohyoid muscle
Hyoglossus muscle
Stylohyoid muscle
Thyrohyoid membrane
Inferior pharyngeal constrictor
Thyrohyoid muscle
Inferior thyroid tubercle
Thyroid cartilage
Inferior pharyngeal constrictor
Cricothyroid muscle
Cricotracheal ligament
1st tracheal cartilage
Esophagus
Trachea
Root of the tongue
Palatine tonsil
Hyoglossus muscle
Palatoglossus muscle
Origin of stylohyoid m.
Styloid process
Styloglossus muscle
Stylopharyngeus muscle
Inferior longitudinal muscle of the tongue
Stylohyoid ligament
Genioglossus muscle
Superior pharyngeal constrictor muscle
Middle pharyngeal constrictor muscle
Thyrohyoid membrane
Geniohyoid muscle
Thyrohyoid muscle
Lesser horn of hyoid bone
Median thyrohyoid ligament
Inferior pharyngeal constrictor m. (thyroid cartilage origin)
Thyroid cartilage
Inf. thyroid tubercle (origin of thyrohyoid m.)
Inferior pharyngeal constrictor m. (cricoid cartilage origin)
Cricothyroid muscle
Esophagus
1st and 2nd tracheal cartilages

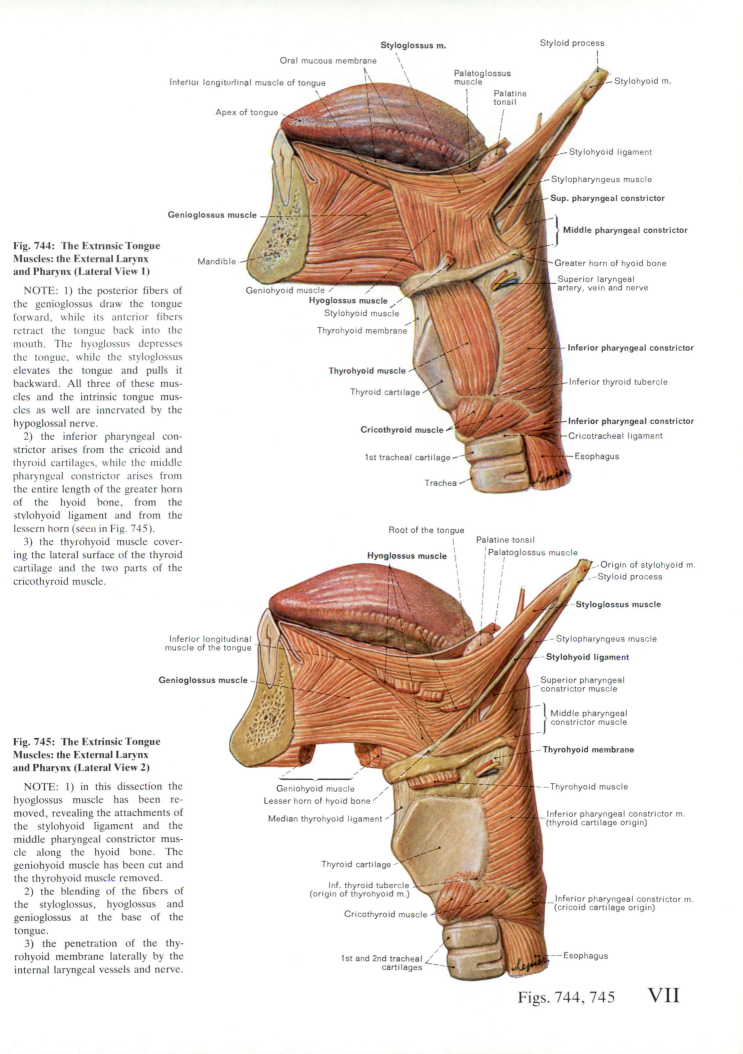

Fig. 744: The Extrinsic Tongue Muscles: the External Larynx and Pharynx (Lateral View 1)

NOTE: 1) the posterior fibers of the genioglossus draw the tongue forward, while its anterior fibers retract the tongue back into the mouth. The hyoglossus depresses the tongue, while the styloglossus elevates the tongue and pulls it backward. All three of these muscles and the intrinsic tongue muscles as well are innervated by the hypoglossal nerve.

2) the inferior pharyngeal constrictor arises from the cricoid and thyroid cartilages, while the middle pharyngeal constrictor arises from the entire length of the greater horn of the hyoid bone, from the stylohyoid ligament and from the lessern horn (seen in Fig. 745).

3) the thyrohyoid muscle covering the lateral surface of the thyroid cartilage and the two parts of the cricothyroid muscle.

Fig. 745: The Extrinsic Tongue Muscles: the External Larynx and Pharynx (Lateral View 2)

NOTE: 1) in this dissection the hyoglossus muscle has been removed, revealing the attachments of the stylohyoid ligament and the middle pharyngeal constrictor muscle along the hyoid bone. The geniohyoid muscle has been cut and the thyrohyoid muscle removed.

2) the blending of the fibers of the styloglossus, hyoglossus and genioglossus at the base of the tongue.

3) the penetration of the thyrohyoid membrane laterally by the internal laryngeal vessels and nerve.

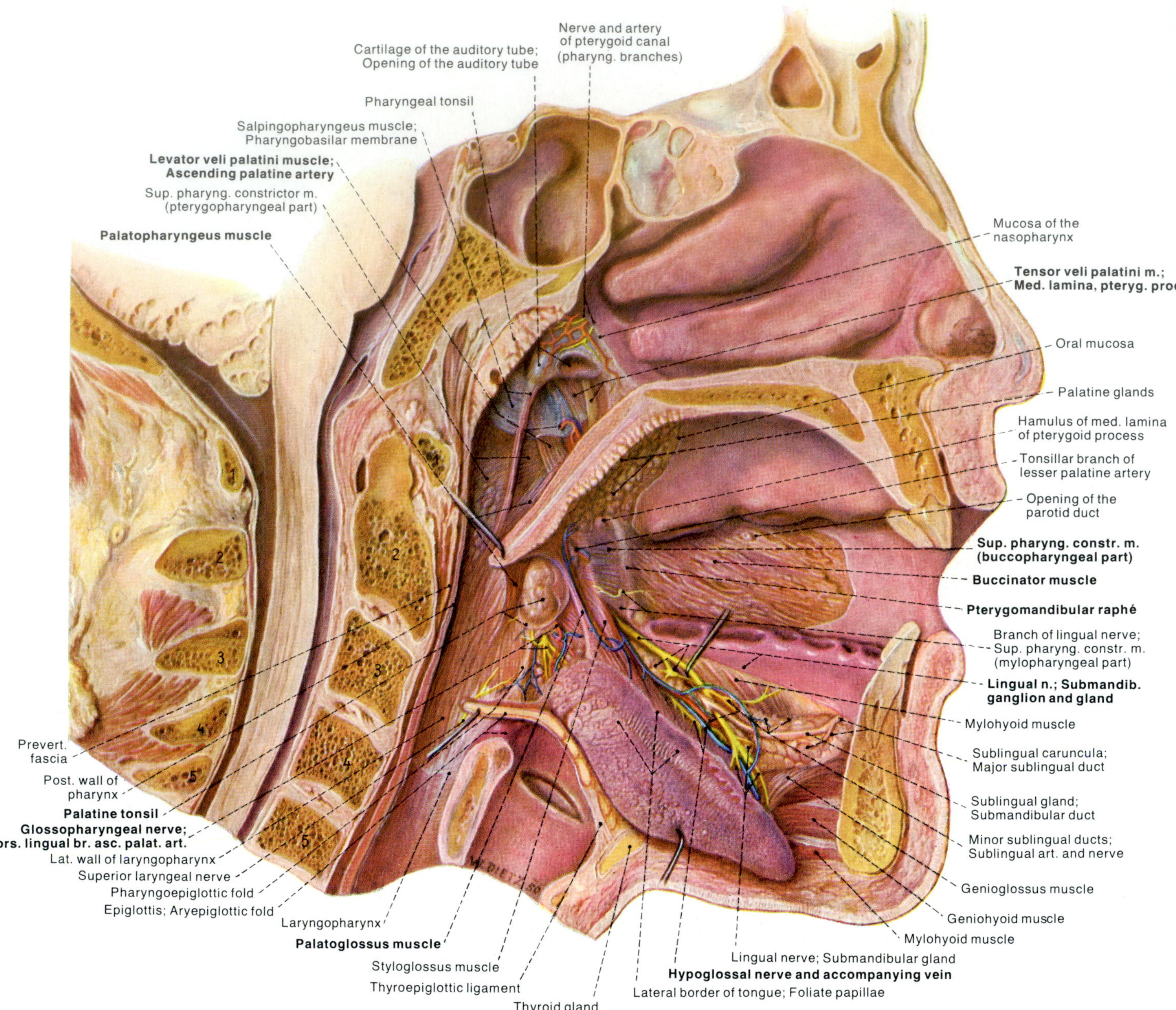

Nerve and artery of pterygoid canal (pharyng. branches)
Cartilage of the auditory tube; Opening of the auditory tube
Pharyngeal tonsil
Salpingopharyngeus muscle; Pharyngobasilar membrane
Levator veli palatini muscle; Ascending palatine artery
Sup. pharyng. constrictor m. (pterygopharyngeal part)
Palatopharyngeus muscle
Mucosa of the nasopharynx
Tensor veli palatini m.; Med. lamina, pteryg. proc.
Oral mucosa
Palatine glands
Hamulus of med. lamina of pterygoid process
Tonsillar branch of lesser palatine artery
Opening of the parotid duct
Sup. pharyng. constr. m. (buccopharyngeal part)
Buccinator muscle
Pterygomandibular raphé
Branch of lingual nerve; Sup. pharyng. constr. m. (mylopharyngeal part)
Lingual n.; Submandib. ganglion and gland
Mylohyoid muscle
Sublingual caruncula; Major sublingual duct
Sublingual gland; Submandibular duct
Minor sublingual ducts; Sublingual art. and nerve
Genioglossus muscle
Geniohyoid muscle
Mylohyoid muscle
Lingual nerve; Submandibular gland
Hypoglossal nerve and accompanying vein
Lateral border of tongue; Foliate papillae
Thyroid gland
Thyroepiglottic ligament
Styloglossus muscle
Palatoglossus muscle
Laryngopharynx
Epiglottis; Aryepiglottic fold
Pharyngoepiglottic fold
Superior laryngeal nerve
Lat. wall of laryngopharynx
Dors. lingual br. asc. palat. art.
Glossopharyngeal nerve;
Palatine tonsil
Post. wall of pharynx
Prevert. fascia
1
2
3
4
5

Fig. 746: The Left Side of the Oral Cavity and Pharynx, Viewed From the Right, Stage 1: Muscles in the Nasopharynx and Oropharynx and the Nerves and Vessels at the Base of the Tongue and Within the Oral Cavity

NOTE: 1) that the mucosa has been stripped from the surface of the palatoglossus and styloglossus muscles anterior to the palatine tonsil. Observe the glossopharyngeal nerve which has been dissected in the oropharynx below the palatine tonsil, near the base of the tongue. At this point, the glossopharyngeal nerve is entirely sensory, supplying both general sensation and the special sense of taste to the posterior one-third of the tongue.

2) the levator and tensor veli palatini muscles coursing toward the soft palate on the lateral side of the posterior opening of the nasal cavity. Observe the cartilaginous auditory tube and the delicate salpingopharyngeus muscle. This muscle arises from the inferior part of the cartilage of the auditory tube and its fibers descend to blend with the palatopharyngeus muscle prior to inserting into the pharyngeal wall. The salpingopharyngeus muscle elevates the lateral wall of the pharynx and with other muscles helps to displace the pharynx upward during swallowing.

3) the buccinator and the superior pharyngeal constrictor muscle (buccopharyngeal part), which becomes visible after removal of the mucosa on the inner surface of the cheek. See the pterygomandibular raphé at which these two muscles are juxtaposed. Turn to Figure 722 and observe that this raphé is formed by decussating tendon-like fibers and that it extends from the hook-like hamulus of the medial pterygoid plate of the sphenoid bone to the posterior end of the mylohyoid line on the inner surface of the mandible.

4) the hypoglossal nerve coursing toward the tongue with the accompanying vein to the hypoglossal nerve. This nerve supplies all of the intrinsic muscles of the tongue (i.e., the longitudinal, vertical, and transverse) and all of the extrinsic muscles of the tongue except the palatoglossus muscle. Thus, the styloglossus, hyoglossus (including the chondroglossus), and the genioglossus muscles are all supplied by the hypoglossal nerve, while the palatoglossus is supplied by the pharyngeal branch of the vagus nerve.

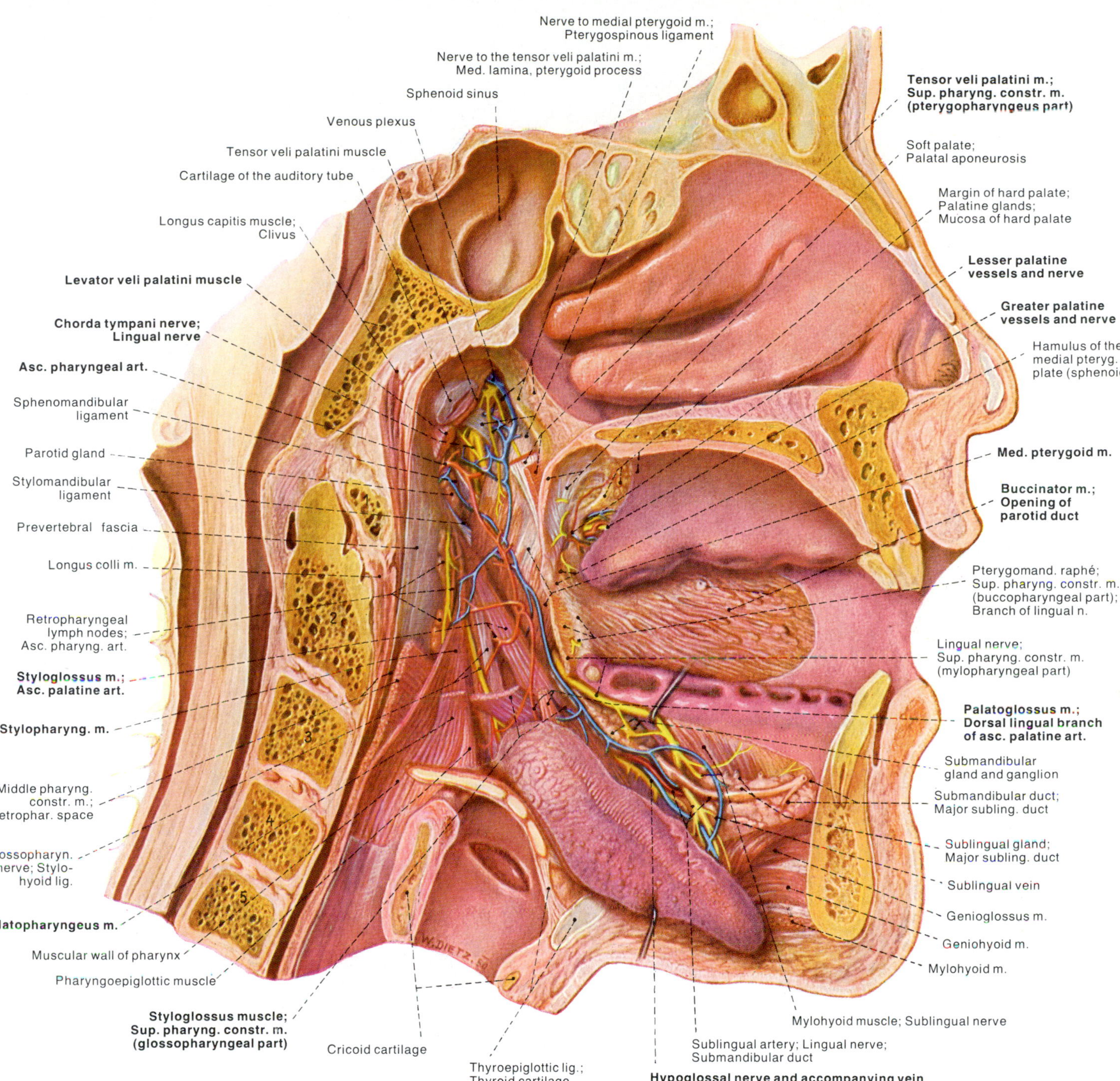

Nerve to medial pterygoid m.;
Pterygospinous ligament
Nerve to the tensor veli palatini m.;
Med. lamina, pterygoid process
Sphenoid sinus
Venous plexus
Tensor veli palatini muscle
Cartilage of the auditory tube
Longus capitis muscle;
Clivus
Levator veli palatini muscle
Chorda tympani nerve;
Lingual nerve
Asc. pharyngeal art.
Sphenomandibular
ligament
Parotid gland
Stylomandibular
ligament
Prevertebral fascia
Longus colli m.
Retropharyngeal
lymph nodes;
Asc. pharyng. art.
Styloglossus m.;
Asc. palatine art.
Stylopharyng. m.
Middle pharyng.
constr. m.;
Retrophar. space
Glossopharyn.
nerve; Stylo-
hyoid lig.
Palatopharyngeus m.
Muscular wall of pharynx
Pharyngoepiglottic muscle
Styloglossus muscle;
Sup. pharyng. constr. m.
(glossopharyngeal part)
Cricoid cartilage
Thyroepiglottic lig.;
Thyroid cartilage
Hypoglossal nerve and accompanying vein
Sublingual artery; Lingual nerve;
Submandibular duct
Mylohyoid muscle; Sublingual nerve
Tensor veli palatini m.;
Sup. pharyng. constr. m.
(pterygopharyngeus part)
Soft palate;
Palatal aponeurosis
Margin of hard palate;
Palatine glands;
Mucosa of hard palate
Lesser palatine
vessels and nerve
Greater palatine
vessels and nerve
Hamulus of the
medial pteryg.
plate (sphenoid)
Med. pterygoid m.
Buccinator m.;
Opening of
parotid duct
Pterygomand. raphé;
Sup. pharyng. constr. m.
(buccopharyngeal part);
Branch of lingual n.
Lingual nerve;
Sup. pharyng. constr. m.
(mylopharyngeal part)
Palatoglossus m.;
Dorsal lingual branch
of asc. palatine art.
Submandibular
gland and ganglion
Submandibular duct;
Major subling. duct
Sublingual gland;
Major subling. duct
Sublingual vein
Genioglossus m.
Geniohyoid m.
Mylohyoid m.
1
2
3
4
5
W. DIETZ

Fig. 747: The Left Side of the Oral Cavity and Pharynx, Viewed From the Right, Stage 2: Exposure of the Pharyngeal, Palatine, and Oral Vessels and Nerves After Dissection of the Pharyngeal Muscles and Mucosa

NOTE: 1) that in the lateral wall of the nasopharynx, the cartilaginous auditory tube and the salpingopharyngeus muscle have been cut and removed, as have the levator and tensor veli palatini muscles. Additionally, the superior constrictor muscle (all of its parts: pterygo, bucco, mylo, and glosso; see Figs. 749 and 750) has been cut in such a way that the medial pterygoid muscle becomes visible from this medial view (see also Fig. 722).

2) that above and behind the medial pterygoid muscle, the lingual nerve has been exposed at the site at which it is joined at an acute angle by the chorda tympani nerve. The lingual nerve descends to the oral cavity, coursing lateral to the medial pterygoid muscle. It then enters the oral cavity just below the pterygomandibular raphé, a short distance behind the last molar tooth, where it is covered only by mucous membrane. The lingual nerve can again be seen lateral to the styloglossus, where it passes forward toward the tip of the tongue. The lingual nerve is the general sensory nerve to the anterior two-thirds of the tongue, but also it supplies the floor of the oral cavity and the inner surface of the mandibular gums.

3) that in the oropharynx, both the palatoglossus and palatopharyngeus muscles have been cut and the palatine tonsil has been removed. This allows a more complete exposure of the glossopharyneal nerve deep in the tonsillar bed. Observe also the ascending pharyngeal artery (a branch of the external carotid artery) and the ascending palatine artery (a branch of the facial artery) in the oropharynx.

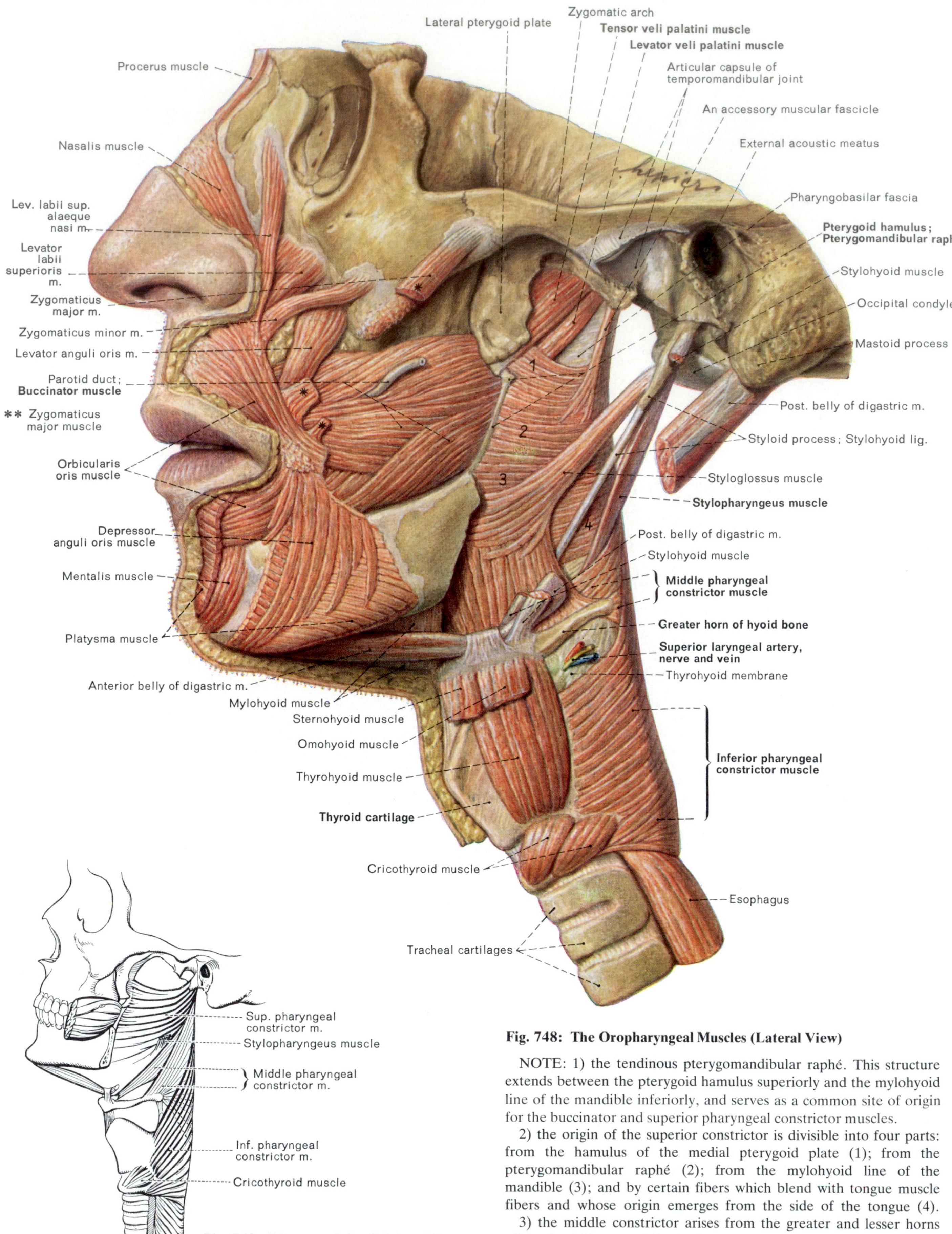

Lateral pterygoid plate
Zygomatic arch
Tensor veli palatini muscle
Levator veli palatini muscle
Articular capsule of temporomandibular joint
An accessory muscular fascicle
External acoustic meatus
Procerus muscle
Nasalis muscle
Lev. labii sup. alaeque nasi m.
Levator labii superioris m.
Zygomaticus major m.
Zygomaticus minor m.
Levator anguli oris m.
Parotid duct; Buccinator muscle
** Zygomaticus major muscle
Orbicularis oris muscle
Depressor anguli oris muscle
Mentalis muscle
Platysma muscle
Anterior belly of digastric m.
Mylohyoid muscle
Sternohyoid muscle
Omohyoid muscle
Thyrohyoid muscle
Thyroid cartilage
Cricothyroid muscle
Tracheal cartilages
Pharyngobasilar fascia
Pterygoid hamulus; Pterygomandibular raphé
Stylohyoid muscle
Occipital condyle
Mastoid process
Post. belly of digastric m.
Styloid process; Stylohyoid lig.
Styloglossus muscle
Stylopharyngeus muscle
Post. belly of digastric m.
Stylohyoid muscle
Middle pharyngeal constrictor muscle
Greater horn of hyoid bone
Superior laryngeal artery, nerve and vein
Thyrohyoid membrane
Inferior pharyngeal constrictor muscle
Esophagus
1
2
3
4
Sup. pharyngeal constrictor m.
Stylopharyngeus muscle
Middle pharyngeal constrictor m.
Inf. pharyngeal constrictor m.
Cricothyroid muscle

Fig. 748: The Oropharyngeal Muscles (Lateral View)

NOTE: 1) the tendinous pterygomandibular raphé. This structure extends between the pterygoid hamulus superiorly and the mylohyoid line of the mandible inferiorly, and serves as a common site of origin for the buccinator and superior pharyngeal constrictor muscles.

2) the origin of the superior constrictor is divisible into four parts: from the hamulus of the medial pterygoid plate (1); from the pterygomandibular raphé (2); from the mylohyoid line of the mandible (3); and by certain fibers which blend with tongue muscle fibers and whose origin emerges from the side of the tongue (4).

3) the middle constrictor arises from the greater and lesser horns of the hyoid bone, while the larger and thicker inferior constrictor arises from the thyroid and cricoid cartilages.

Fig. 749: Diagram of the Origins of the Pharyngeal Constrictor Muscles

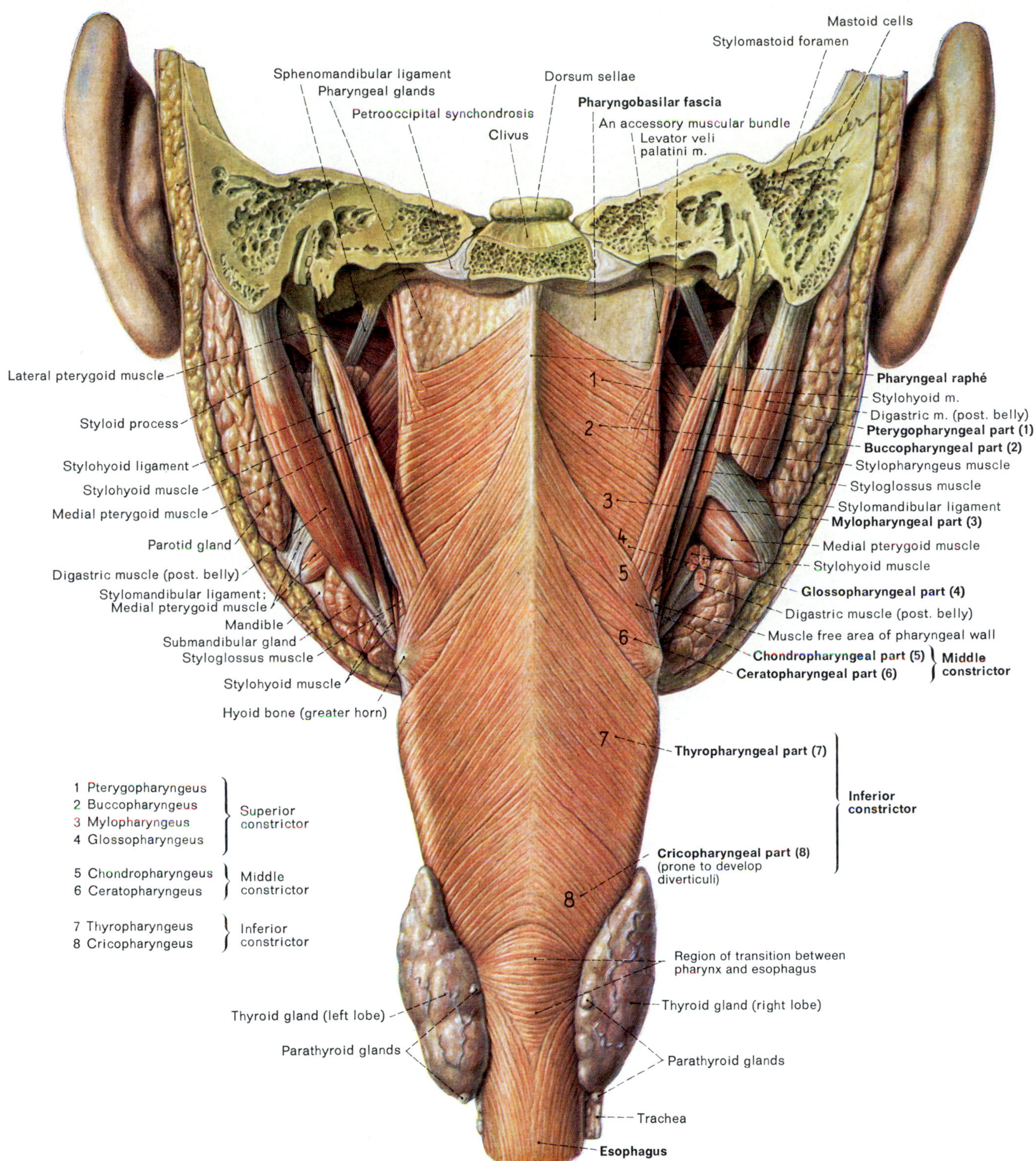

Mastoid cells
Stylomastoid foramen
Sphenomandibular ligament
Pharyngeal glands
Dorsum sellae
Petrooccipital synchondrosis
Clivus
Pharyngobasilar fascia
An accessory muscular bundle
Levator veli palatini m.
Lateral pterygoid muscle
Styloid process
Stylohyoid ligament
Stylohyoid muscle
Medial pterygoid muscle
Parotid gland
Digastric muscle (post. belly)
Stylomandibular ligament; Medial pterygoid muscle
Mandible
Submandibular gland
Styloglossus muscle
Stylohyoid muscle
Hyoid bone (greater horn)
Pharyngeal raphé
Stylohyoid m.
Digastric m. (post. belly)
Pterygopharyngeal part (1)
Buccopharyngeal part (2)
Stylopharyngeus muscle
Styloglossus muscle
Stylomandibular ligament
Mylopharyngeal part (3)
Medial pterygoid muscle
Stylohyoid muscle
Glossopharyngeal part (4)
Digastric muscle (post. belly)
Muscle free area of pharyngeal wall
Chondropharyngeal part (5)
Ceratopharyngeal part (6)
Middle constrictor
1
2
3
4
5
6
7
8
Thyropharyngeal part (7)
Cricopharyngeal part (8) (prone to develop diverticuli)
Inferior constrictor
1 Pterygopharyngeus
2 Buccopharyngeus
3 Mylopharyngeus
4 Glossopharyngeus
Superior constrictor
5 Chondropharyngeus
6 Ceratopharyngeus
Middle constrictor
7 Thyropharyngeus
8 Cricopharyngeus
Inferior constrictor
Region of transition between pharynx and esophagus
Thyroid gland (left lobe)
Thyroid gland (right lobe)
Parathyroid glands
Parathyroid glands
Trachea
Esophagus

Fig. 750: Dorsal View of the Pharyngeal Muscles

NOTE: 1) this posterior view of the pharynx was achieved by making a frontal transection through the petrous and mastoid portions of the temporal bones and through the body of the occipital bone. The styloid processes and their muscular attachments have been left intact.

2) the arrangement of the divisions of the pharyngeal constrictors. Their muscle fibers arise laterally to insert in a posterior median pharyngeal raphé. The superior constrictor is divisible into four parts while the middle and inferior constrictor are each divisible into two. Above the superior constrictor, observe the fibrous pharyngobasilar fascia which attaches to the basal portion of the occipital bone and to the temporal bones. Below the inferior constrictor, the pharynx is continuous with the muscular esophagus.

3) the *superior* and *middle constrictor* muscles and the *thyropharyngeal part of the inferior constrictor* are innervated by the pharyngeal branch of the vagus nerve. These fibers have their cell bodies in the nucleus ambiguus in the brain stem, emerge from the brain with the rootlets of the bulbar part of the accessory nerve and then, by a communicating branch, join the vagus. The cricopharyngeal part of the inferior constrictor is supplied by the recurrent laryngeal branch of the vagus nerve.

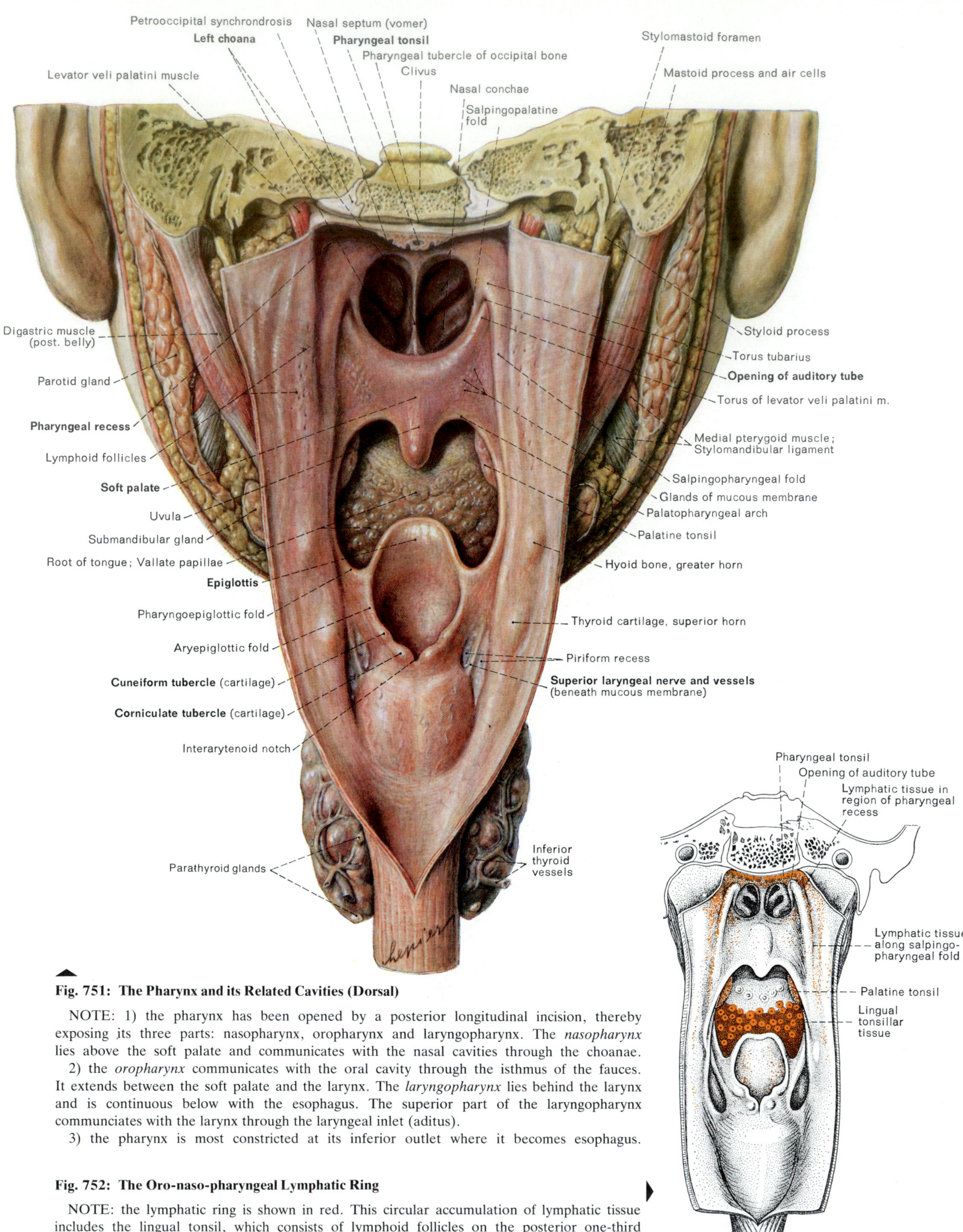
Petrooccipital synchrondrosis
Nasal septum (vomer)
Left choana
Pharyngeal tonsil
Pharyngeal tubercle of occipital bone
Clivus
Levator veli palatini muscle
Nasal conchae
Salpingopalatine fold
Stylomastoid foramen
Mastoid process and air cells
Digastric muscle (post. belly)
Parotid gland
Pharyngeal recess
Lymphoid follicles
Soft palate
Uvula
Submandibular gland
Root of tongue; Vallate papillae
Epiglottis
Pharyngoepiglottic fold
Aryepiglottic fold
Cuneiform tubercle (cartilage)
Corniculate tubercle (cartilage)
Interarytenoid notch
Parathyroid glands
Styloid process
Torus tubarius
Opening of auditory tube
Torus of levator veli palatini m.
Medial pterygoid muscle; Stylomandibular ligament
Salpingopharyngeal fold
Glands of mucous membrane
Palatopharyngeal arch
Palatine tonsil
Hyoid bone, greater horn
Thyroid cartilage, superior horn
Piriform recess
Superior laryngeal nerve and vessels (beneath mucous membrane)
Inferior thyroid vessels
Pharyngeal tonsil
Opening of auditory tube
Lymphatic tissue in region of pharyngeal recess
Lymphatic tissue along salpingo-pharyngeal fold
Palatine tonsil
Lingual tonsillar tissue

Fig. 751: The Pharynx and its Related Cavities (Dorsal)

NOTE: 1) the pharynx has been opened by a posterior longitudinal incision, thereby exposing its three parts: nasopharynx, oropharynx and laryngopharynx. The *nasopharynx* lies above the soft palate and communicates with the nasal cavities through the choanae.

2) the *oropharynx* communicates with the oral cavity through the isthmus of the fauces. It extends between the soft palate and the larynx. The *laryngopharynx* lies behind the larynx and is continuous below with the esophagus. The superior part of the laryngopharynx communciates with the larynx through the laryngeal inlet (aditus).

3) the pharynx is most constricted at its inferior outlet where it becomes esophagus.

Fig. 752: The Oro-naso-pharyngeal Lymphatic Ring

NOTE: the lymphatic ring is shown in red. This circular accumulation of lymphatic tissue includes the lingual tonsil, which consists of lymphoid follicles on the posterior one-third of the tongue, the palatine tonsil, the pharyngeal tonsil and more diffuse lymphoid tissue in the wall of the nasopharynx along the salpingopharyngeal fold.

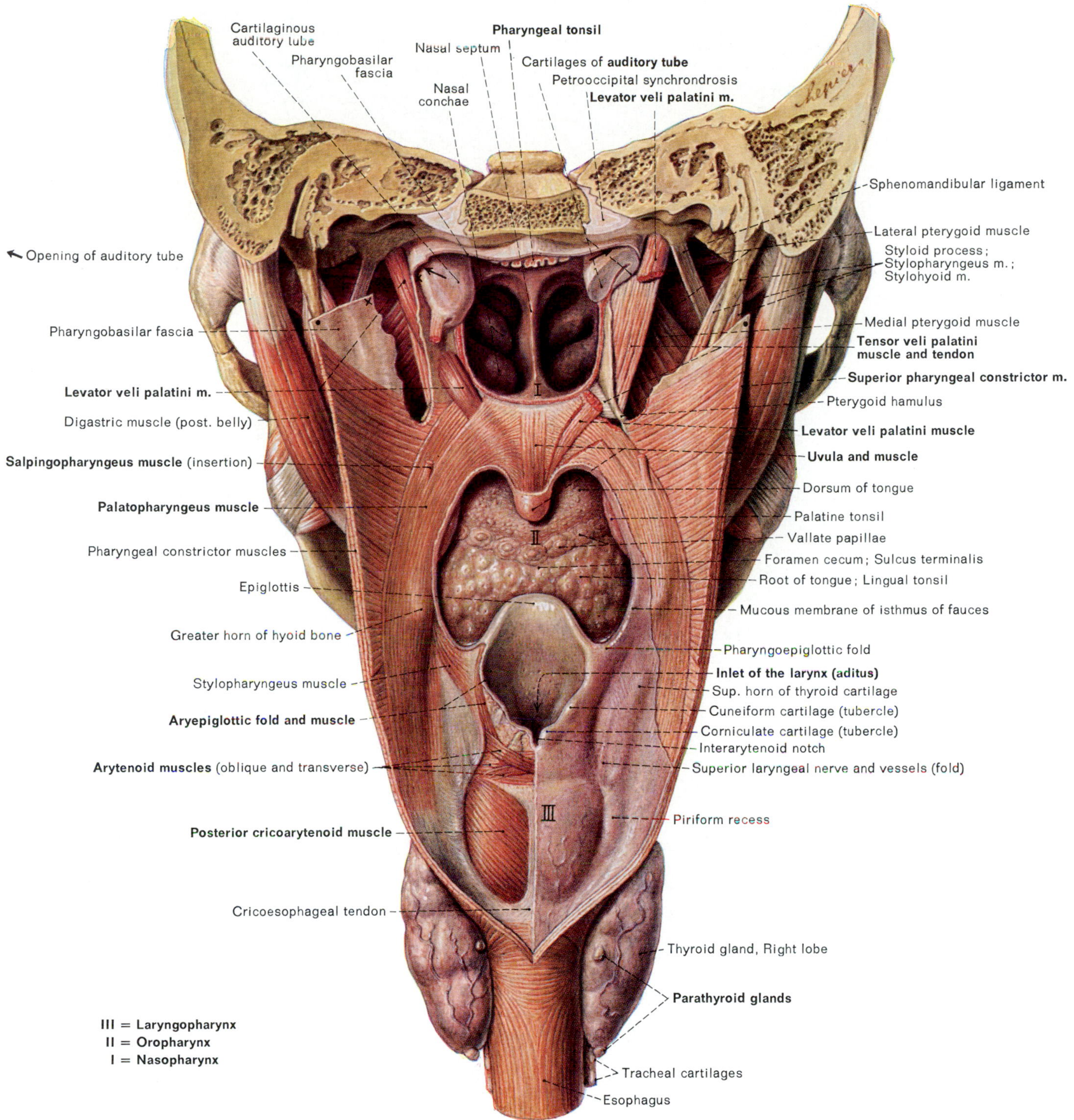
Cartilaginous auditory tube
Pharyngobasilar fascia
Nasal septum
Pharyngeal tonsil
Cartilages of auditory tube
Petrooccipital synchrondrosis
Nasal conchae
Levator veli palatini m.
Sphenomandibular ligament
Lateral pterygoid muscle
Opening of auditory tube
Styloid process; Stylopharyngeus m.; Stylohyoid m.
Pharyngobasilar fascia
Medial pterygoid muscle
Tensor veli palatini muscle and tendon
Superior pharyngeal constrictor m.
Levator veli palatini m.
Pterygoid hamulus
Digastric muscle (post. belly)
Levator veli palatini muscle
Salpingopharyngeus muscle (insertion)
Uvula and muscle
Dorsum of tongue
Palatopharyngeus muscle
Palatine tonsil
Vallate papillae
Pharyngeal constrictor muscles
Foramen cecum; Sulcus terminalis
Epiglottis
Root of tongue; Lingual tonsil
Mucous membrane of isthmus of fauces
Greater horn of hyoid bone
Pharyngoepiglottic fold
Inlet of the larynx (aditus)
Stylopharyngeus muscle
Sup. horn of thyroid cartilage
Cuneiform cartilage (tubercle)
Aryepiglottic fold and muscle
Corniculate cartilage (tubercle)
Interarytenoid notch
Arytenoid muscles (oblique and transverse)
Superior laryngeal nerve and vessels (fold)
I
II
III
Piriform recess
Posterior cricoarytenoid muscle
Cricoesophageal tendon
Thyroid gland, Right lobe
Parathyroid glands
III = Laryngopharynx
II = Oropharynx
I = Nasopharynx
Tracheal cartilages
Esophagus

Fig. 753: Muscles of the Soft Palate, Pharynx, and Posterior Larynx

NOTE: 1) the dissection in this figure is similar to that in Figure 751. The pharynx has been opened dorsally by a midline incision and the mucous membrane has been removed from the soft palate, pharynx and the left posterior larynx. On the right side, a portion of the levator veli palatini muscle has been removed in order to expose the adjacent belly and tendon of the tensor veli palatini muscle.

2) the muscles of the soft palate. Both the muscle of the uvula and the levator veli palatini muscle are innervated by the vagus and accessory nerve contributions to the pharyngeal plexus, whereas the tensor veli palatini muscle is innervated by the mandibular division of the trigeminal nerve.

3) the palatopharyngeus muscle arising by two fascicles from the soft palate. The muscle fibers of these two fascicles arise posterior and anterior to the insertion of the levator veli palatini muscle. The fascicles descend and merge and then insert into the posterior border of the thyroid cartilage and onto the adjacent pharyngeal wall.

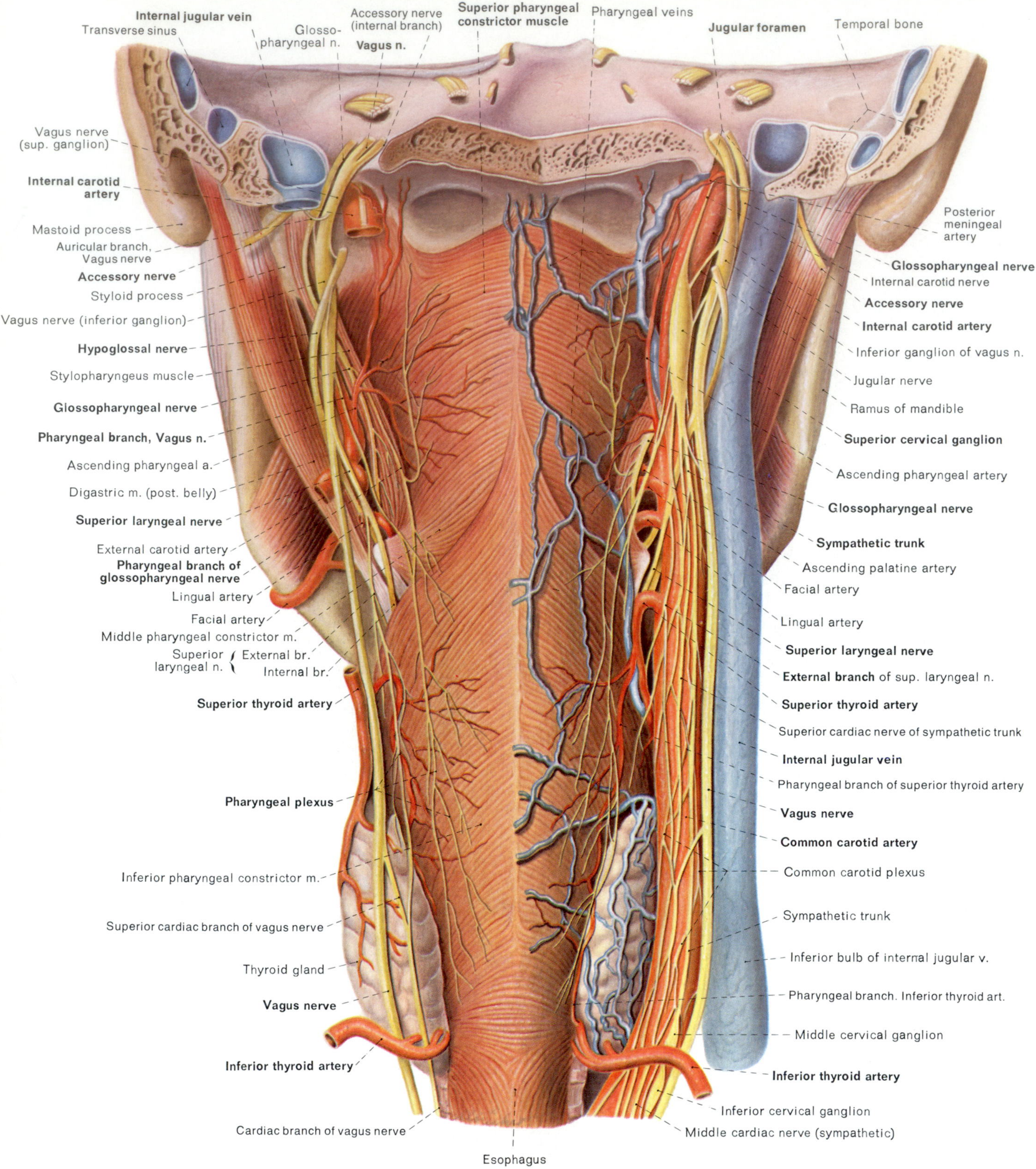
Internal jugular vein
Transverse sinus
Glosso-pharyngeal n.
Accessory nerve (internal branch)
Vagus n.
Superior pharyngeal constrictor muscle
Pharyngeal veins
Jugular foramen
Temporal bone
Vagus nerve (sup. ganglion)
Internal carotid artery
Mastoid process
Auricular branch, Vagus nerve
Accessory nerve
Styloid process
Vagus nerve (inferior ganglion)
Hypoglossal nerve
Stylopharyngeus muscle
Glossopharyngeal nerve
Pharyngeal branch, Vagus n.
Ascending pharyngeal a.
Digastric m. (post. belly)
Superior laryngeal nerve
External carotid artery
Pharyngeal branch of glossopharyngeal nerve
Lingual artery
Facial artery
Middle pharyngeal constrictor m.
Superior laryngeal n.
External br.
Internal br.
Superior thyroid artery
Pharyngeal plexus
Inferior pharyngeal constrictor m.
Superior cardiac branch of vagus nerve
Thyroid gland
Vagus nerve
Inferior thyroid artery
Cardiac branch of vagus nerve
Esophagus
Posterior meningeal artery
Glossopharyngeal nerve
Internal carotid nerve
Accessory nerve
Internal carotid artery
Inferior ganglion of vagus n.
Jugular nerve
Ramus of mandible
Superior cervical ganglion
Ascending pharyngeal artery
Glossopharyngeal nerve
Sympathetic trunk
Ascending palatine artery
Facial artery
Lingual artery
Superior laryngeal nerve
External branch of sup. laryngeal n.
Superior thyroid artery
Superior cardiac nerve of sympathetic trunk
Internal jugular vein
Pharyngeal branch of superior thyroid artery
Vagus nerve
Common carotid artery
Common carotid plexus
Sympathetic trunk
Inferior bulb of internal jugular v.
Pharyngeal branch. Inferior thyroid art.
Middle cervical ganglion
Inferior thyroid artery
Inferior cervical ganglion
Middle cardiac nerve (sympathetic)

Fig. 754: Nerves and Vessels on the Dorsal and Lateral Walls of the Pharynx

NOTE: 1) the head has been split longitudinally. The pharynx, larynx and facial structures were separated from the vertebral column and its associated muscles. This posterior view of the pharynx also shows the large nerves and blood vessels which course through the neck. On the right side, observe the carotid artery, internal jugular vein, vagus nerve and sympathetic trunk.

2) on the left side, the glossopharyngeal and hypoglossal nerves were exposed by the removal of the carotid arteries and internal jugular vein. Along with the jugular vein, the jugular foramen transmits the 9th, 10th and 11th cranial nerves.

3) the thyroid gland and its superior and inferior thyroid arteries. The superior and middle thyroid veins drain into the internal jugular vein, while the inferior thyroid veins (not shown) usually drain into the left brachiocephalic vein.

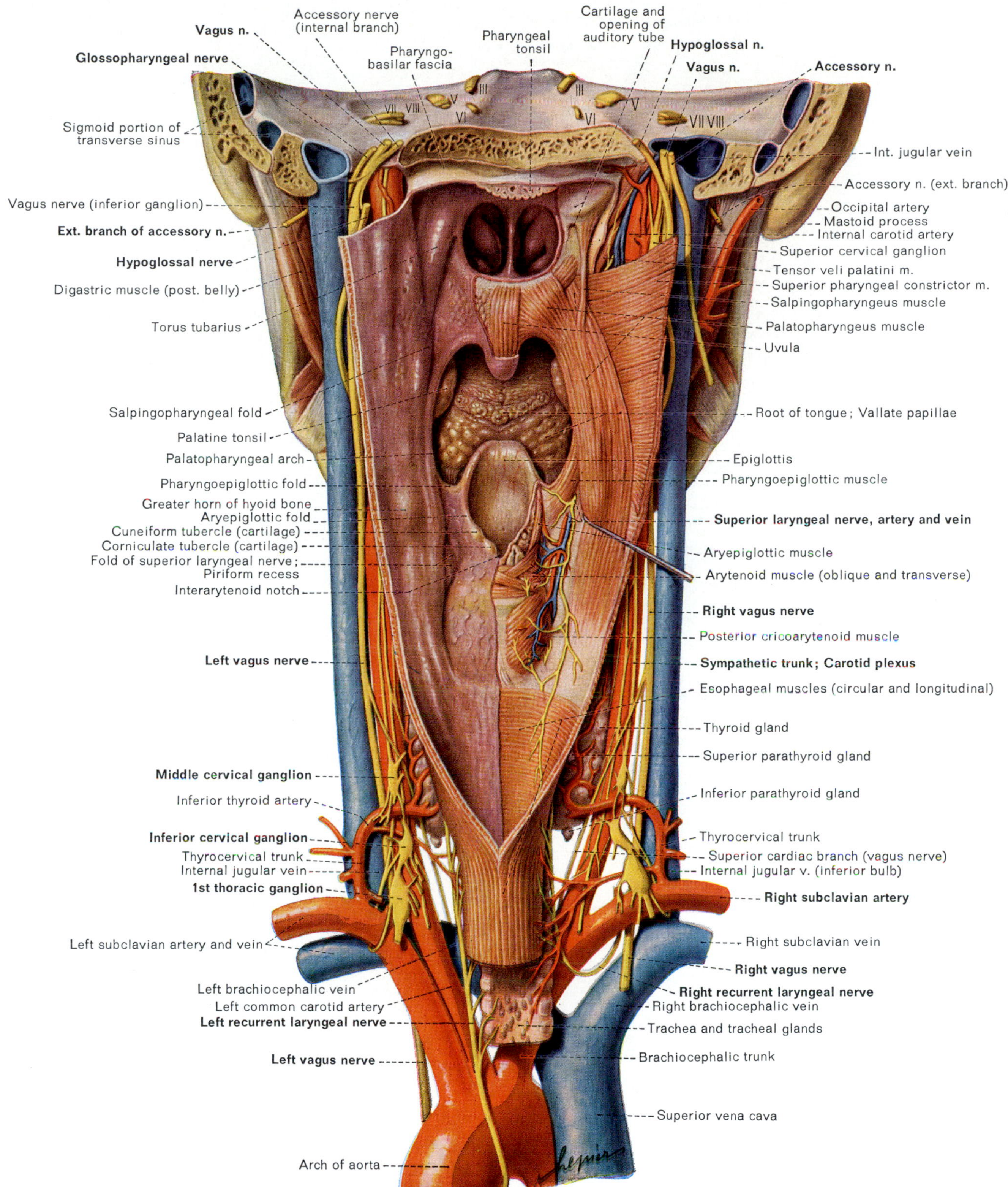

Vagus n.
Accessory nerve (internal branch)
Glossopharyngeal nerve
Pharyngo-basilar fascia
Pharyngeal tonsil
Cartilage and opening of auditory tube
Hypoglossal n.
Vagus n.
Accessory n.
Sigmoid portion of transverse sinus
Int. jugular vein
Accessory n. (ext. branch)
Vagus nerve (inferior ganglion)
Occipital artery
Mastoid process
Ext. branch of accessory n.
Internal carotid artery
Hypoglossal nerve
Superior cervical ganglion
Tensor veli palatini m.
Digastric muscle (post. belly)
Superior pharyngeal constrictor m.
Salpingopharyngeus muscle
Torus tubarius
Palatopharyngeus muscle
Uvula
Salpingopharyngeal fold
Root of tongue; Vallate papillae
Palatine tonsil
Palatopharyngeal arch
Epiglottis
Pharyngoepiglottic fold
Pharyngoepiglottic muscle
Greater horn of hyoid bone
Aryepiglottic fold
Superior laryngeal nerve, artery and vein
Cuneiform tubercle (cartilage)
Corniculate tubercle (cartilage)
Aryepiglottic muscle
Fold of superior laryngeal nerve;
Piriform recess
Arytenoid muscle (oblique and transverse)
Interarytenoid notch
Right vagus nerve
Posterior cricoarytenoid muscle
Left vagus nerve
Sympathetic trunk; Carotid plexus
Esophageal muscles (circular and longitudinal)
Thyroid gland
Superior parathyroid gland
Middle cervical ganglion
Inferior thyroid artery
Inferior parathyroid gland
Inferior cervical ganglion
Thyrocervical trunk
Thyrocervical trunk
Superior cardiac branch (vagus nerve)
Internal jugular vein
Internal jugular v. (inferior bulb)
1st thoracic ganglion
Right subclavian artery
Left subclavian artery and vein
Right subclavian vein
Right vagus nerve
Left brachiocephalic vein
Right recurrent laryngeal nerve
Left common carotid artery
Right brachiocephalic vein
Left recurrent laryngeal nerve
Trachea and tracheal glands
Left vagus nerve
Brachiocephalic trunk
Superior vena cava
Arch of aorta

Fig. 755: Posterior View of Cervical Viscera: Muscles, Vessels and Nerves

NOTE: 1) in this posterior view of the vessels and nerves of the neck, the dorsal wall of the pharynx has been opened longitudinally, revealing the nasal, oral and laryngeal orifices which communicate with the pharynx. Observe the superior laryngeal artery, vein and nerve entering the larynx from above.

2) the recurrent laryngeal nerves ascending from the thorax to the larynx in the tracheoesophageal groove. On the left side, the recurrent laryngeal nerve courses around the arch of the aorta to reach this groove, while on the right side the recurrent laryngeal nerve curves around the subclavian artery.

3) the inferior cervical ganglion (at the level of 7th cervical vertebra). This ganglion is fused with the 1st thoracic sympathetic ganglion in about 80% of cases. The fused 1st thoracic and inferior cervical ganglion is called the stellate ganglion.

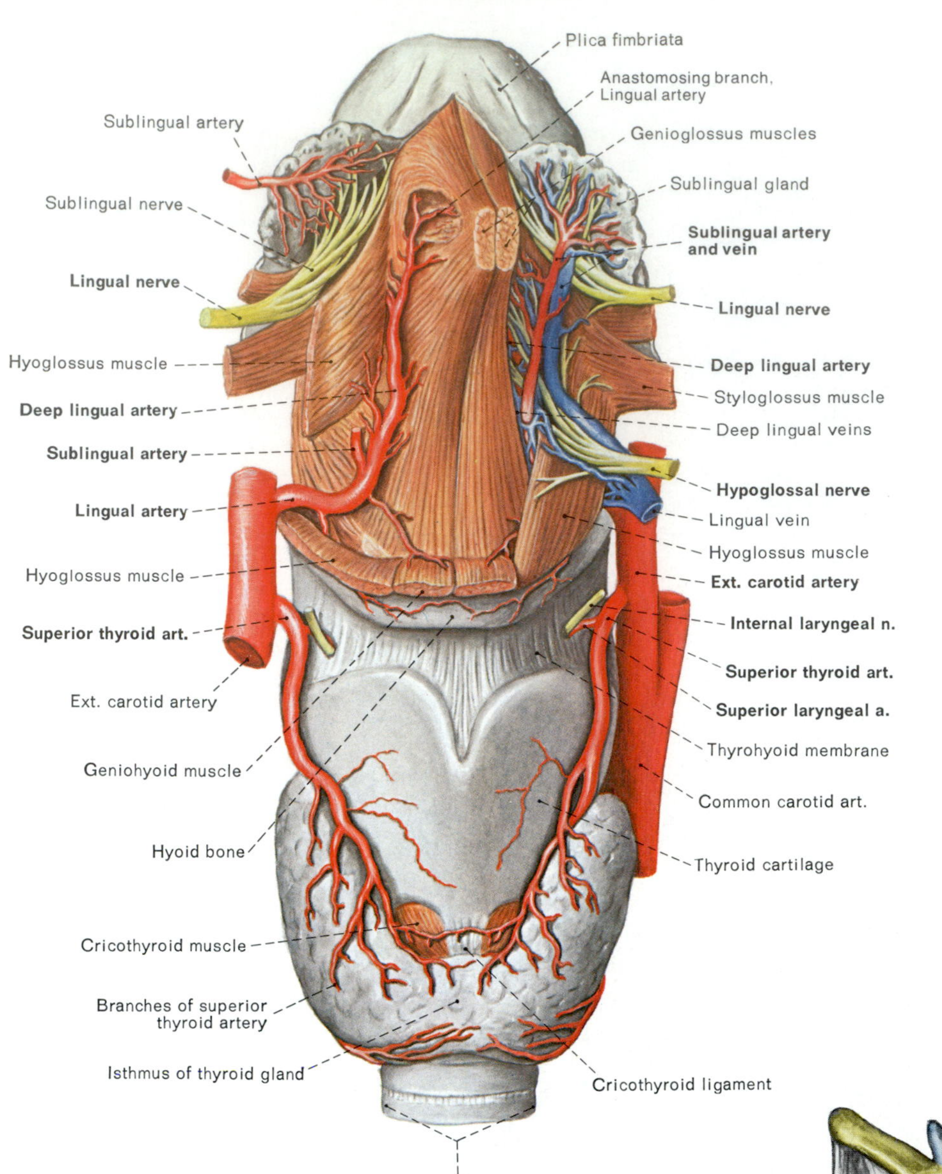

Plica fimbriata
Anastomosing branch, Lingual artery
Sublingual artery
Genioglossus muscles
Sublingual gland
Sublingual nerve
Sublingual artery and vein
Lingual nerve
Lingual nerve
Hyoglossus muscle
Deep lingual artery
Deep lingual artery
Styloglossus muscle
Deep lingual veins
Sublingual artery
Hypoglossal nerve
Lingual artery
Lingual vein
Hyoglossus muscle
Hyoglossus muscle
Ext. carotid artery
Superior thyroid art.
Internal laryngeal n.
Superior thyroid art.
Ext. carotid artery
Superior laryngeal a.
Thyrohyoid membrane
Geniohyoid muscle
Common carotid art.
Hyoid bone
Thyroid cartilage
Cricothyroid muscle
Branches of superior thyroid artery
Isthmus of thyroid gland
Cricothyroid ligament
Trachea

Fig. 756: Ventral View of Larynx, Tongue and Thyroid Gland: Vessels and Nerves

NOTE: 1) the superior thyroid arteries descending to the thyroid gland. In their course they give off the superior laryngeal arteries, which penetrate the thyrohyoid membrane to gain entrance to the interior of the larynx, accompanied by the internal laryngeal branch of the superior laryngeal nerve. Observe also the cranial and medial course of the lingual artery deep to the hyoglossus muscle and its suprahyoid, sublingual and deep lingual branches.

2) the lingual nerves as they enter the tongue to supply its anterior two-thirds with general sensation. The motor nerve to the tongue is the hypoglossal, and it is seen coursing along with accompanying veins (venae comitantes). The hypoglossal nerve enters the base of the tongue just above the hyoid bone, passing anteriorly across the external carotid and lingual arteries.

3) that the common carotid artery bifurcates at about the level of the upper border of the thyroid cartilage. Observe also that the lingual artery branches from the external carotid above the hyoid bone, while the superior laryngeal artery arises at the level of the thyrohyoid membrane.

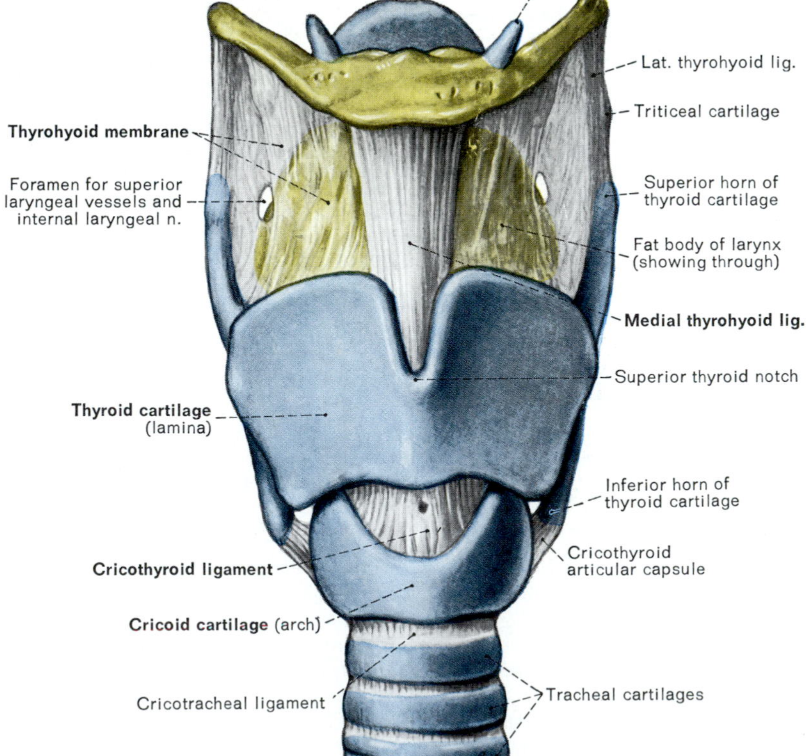

Epiglottis
Lesser horn of hyoid bone
Greater horn of hyoid bone
Lat. thyrohyoid lig.
Triticeal cartilage
Thyrohyoid membrane
Foramen for superior laryngeal vessels and internal laryngeal n.
Superior horn of thyroid cartilage
Fat body of larynx (showing through)
Medial thyrohyoid lig.
Superior thyroid notch
Thyroid cartilage (lamina)
Inferior horn of thyroid cartilage
Cricothyroid ligament
Cricothyroid articular capsule
Cricoid cartilage (arch)
Cricotracheal ligament
Tracheal cartilages

Fig. 757: The Cartilages and Ligaments of the Larynx (Ventral View)

NOTE: 1) the laryngeal cartilages comprise the skeletal framework of the larynx and they are interconnected by ligaments and membranes. There are *three larger unpaired* cartilages, the cricoid, thyroid and epiglottis and *three sets of paired* cartilages, the arytenoid, cuneiform and corniculate. In this anterior view the unpaired cricoid, thyroid and epiglottis cartilages are all visible.

2) the thyrohyoid membrane and the centrally located thyrohyoid ligament. Attached at the cranial border of the thyroid cartilage, this membrane stretches across the posterior surfaces of the greater horns of the hyoid bone. The medial thyrohyoid ligament extends from the thyroid notch to the body of the hyoid bone. The membrane is pierced by the superior laryngeal vessels and the internal laryngeal nerve.

3) the cricothyroid ligament attaching the contiguous margins of the cricoid and thyroid cartilages. This is a strong ligament and its lateral portions underlie the cricothyroid muscles. Below, the cricotracheal ligament connects the inferior margin of the cricoid cartilage with the first tracheal ring.

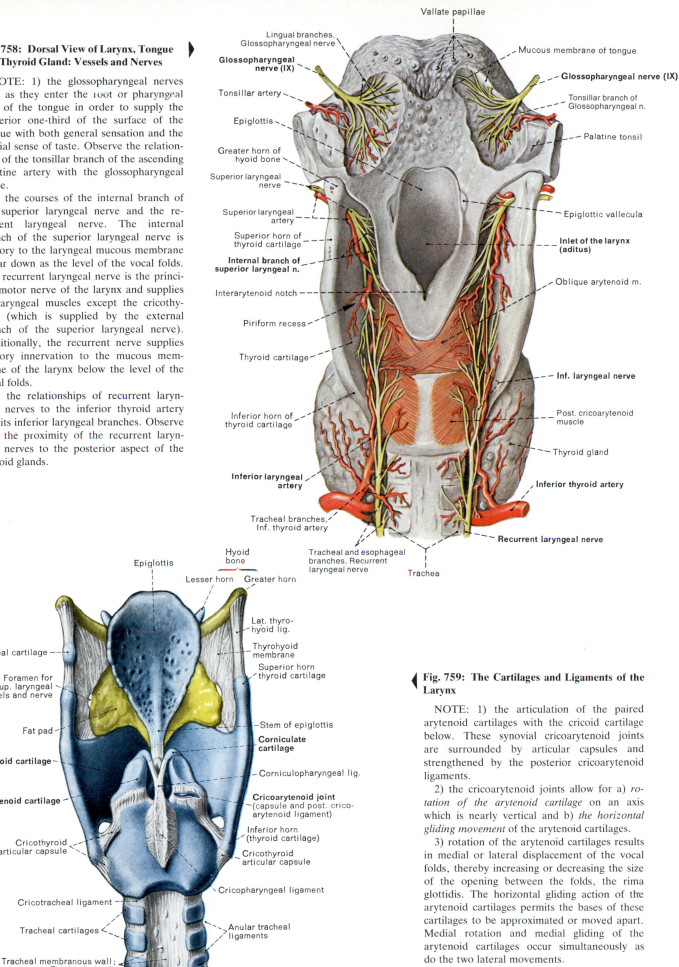

Fig. 758: Dorsal View of Larynx, Tongue and Thyroid Gland: Vessels and Nerves

NOTE: 1) the glossopharyngeal nerves (IX) as they enter the root or pharyngeal part of the tongue in order to supply the posterior one-third of the surface of the tongue with both general sensation and the special sense of taste. Observe the relationship of the tonsillar branch of the ascending palatine artery with the glossopharyngeal nerve.

2) the courses of the internal branch of the superior laryngeal nerve and the recurrent laryngeal nerve. The internal branch of the superior laryngeal nerve is sensory to the laryngeal mucous membrane as far down as the level of the vocal folds. The recurrent laryngeal nerve is the principal motor nerve of the larynx and supplies all laryngeal muscles except the cricothyroid (which is supplied by the external branch of the superior laryngeal nerve). Additionally, the recurrent nerve supplies sensory innervation to the mucous membrane of the larynx below the level of the vocal folds.

3) the relationships of recurrent laryngeal nerves to the inferior thyroid artery and its inferior laryngeal branches. Observe also the proximity of the recurrent laryngeal nerves to the posterior aspect of the thyroid glands.

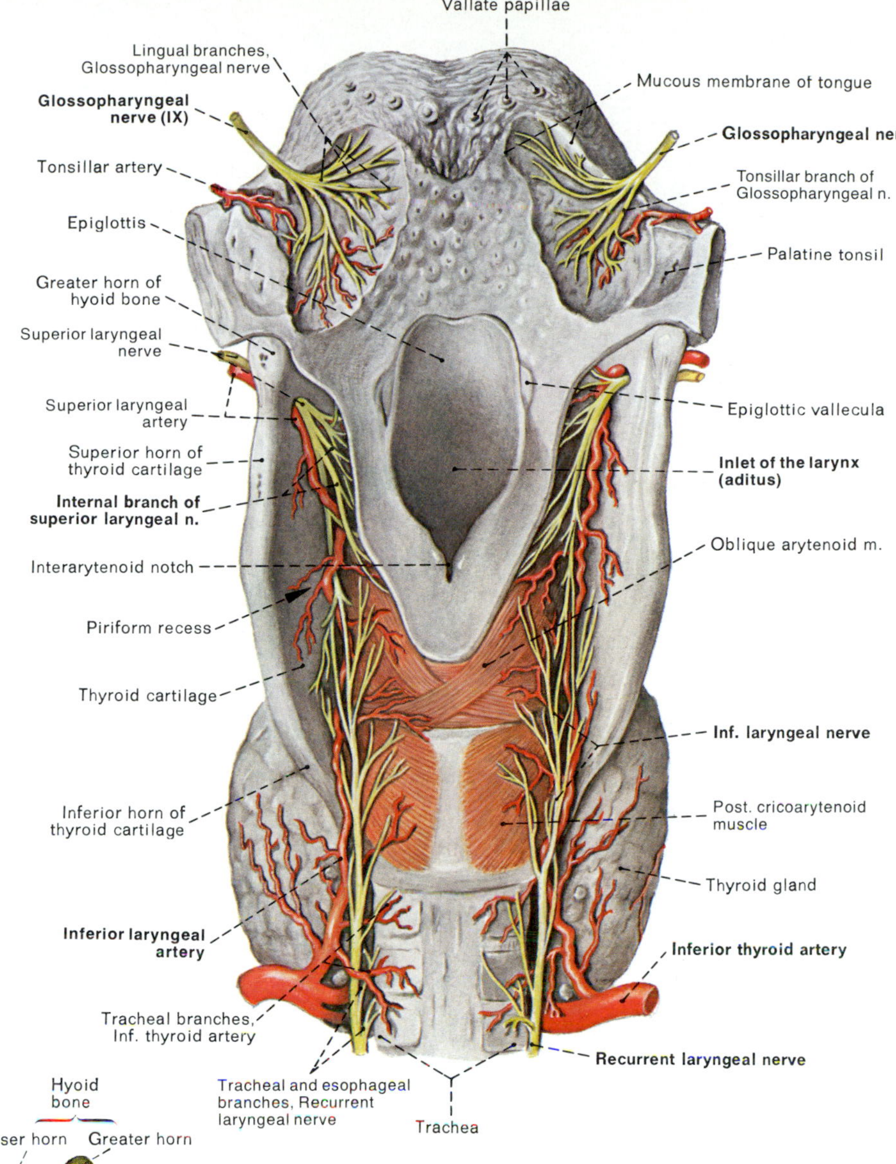

Vallate papillae
Lingual branches, Glossopharyngeal nerve
Mucous membrane of tongue
Glossopharyngeal nerve (IX)
Glossopharyngeal nerve (IX)
Tonsillar artery
Tonsillar branch of Glossopharyngeal n.
Epiglottis
Palatine tonsil
Greater horn of hyoid bone
Superior laryngeal nerve
Epiglottic vallecula
Superior laryngeal artery
Superior horn of thyroid cartilage
Inlet of the larynx (aditus)
Internal branch of superior laryngeal n.
Oblique arytenoid m.
Interarytenoid notch
Piriform recess
Thyroid cartilage
Inf. laryngeal nerve
Inferior horn of thyroid cartilage
Post. cricoarytenoid muscle
Thyroid gland
Inferior laryngeal artery
Inferior thyroid artery
Tracheal branches, Inf. thyroid artery
Recurrent laryngeal nerve
Tracheal and esophageal branches, Recurrent laryngeal nerve
Trachea

Epiglottis
Hyoid bone
Lesser horn
Greater horn
Lat. thyrohyoid lig.
Triticeal cartilage
Thyrohyoid membrane
Superior horn thyroid cartilage
Foramen for sup. laryngeal vessels and nerve
Fat pad
Stem of epiglottis
Corniculate cartilage
Thyroid cartilage
Corniculopharyngeal lig.
Arytenoid cartilage
Cricoarytenoid joint (capsule and post. cricoarytenoid ligament)
Inferior horn (thyroid cartilage)
Cricothyroid articular capsule
Cricothyroid articular capsule
Cricopharyngeal ligament
Cricotracheal ligament
Tracheal cartilages
Anular tracheal ligaments
Tracheal membranous wall; Tracheal glands

Fig. 759: The Cartilages and Ligaments of the Larynx

NOTE: 1) the articulation of the paired arytenoid cartilages with the cricoid cartilage below. These synovial cricoarytenoid joints are surrounded by articular capsules and strengthened by the posterior cricoarytenoid ligaments.

2) the cricoarytenoid joints allow for a) *rotation of the arytenoid cartilage* on an axis which is nearly vertical and b) *the horizontal gliding movement* of the arytenoid cartilages.

3) rotation of the arytenoid cartilages results in medial or lateral displacement of the vocal folds, thereby increasing or decreasing the size of the opening between the folds, the rima glottidis. The horizontal gliding action of the arytenoid cartilages permits the bases of these cartilages to be approximated or moved apart. Medial rotation and medial gliding of the arytenoid cartilages occur simultaneously as do the two lateral movements.

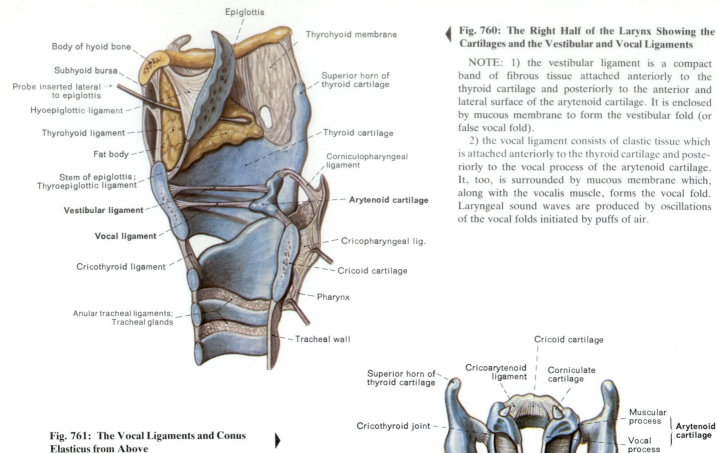

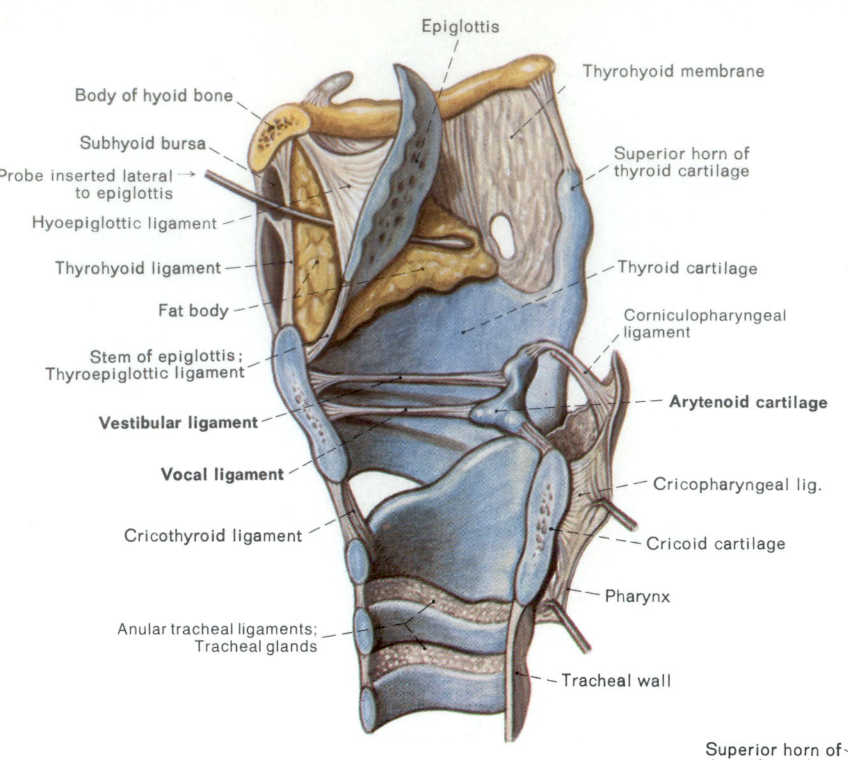
Epiglottis
Body of hyoid bone
Thyrohyoid membrane
Subhyoid bursa
Superior horn of thyroid cartilage
Probe inserted lateral to epiglottis
Hyoepiglottic ligament
Thyrohyoid ligament
Thyroid cartilage
Fat body
Corniculopharyngeal ligament
Stem of epiglottis; Thyroepiglottic ligament
Arytenoid cartilage
Vestibular ligament
Vocal ligament
Cricopharyngeal lig.
Cricothyroid ligament
Cricoid cartilage
Pharynx
Anular tracheal ligaments; Tracheal glands
Tracheal wall

Fig. 760: The Right Half of the Larynx Showing the Cartilages and the Vestibular and Vocal Ligaments

NOTE: 1) the vestibular ligament is a compact band of fibrous tissue attached anteriorly to the thyroid cartilage and posteriorly to the anterior and lateral surface of the arytenoid cartilage. It is enclosed by mucous membrane to form the vestibular fold (or false vocal fold).

2) the vocal ligament consists of elastic tissue which is attached anteriorly to the thyroid cartilage and posteriorly to the vocal process of the arytenoid cartilage. It, too, is surrounded by mucous membrane which, along with the vocalis muscle, forms the vocal fold. Laryngeal sound waves are produced by oscillations of the vocal folds initiated by puffs of air.

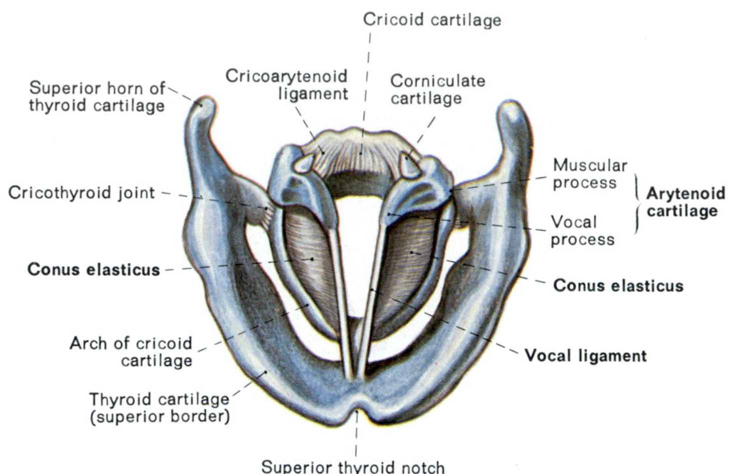
Cricoid cartilage
Superior horn of thyroid cartilage
Cricoarytenoid ligament
Corniculate cartilage
Muscular process
Arytenoid cartilage
Cricothyroid joint
Vocal process
Conus elasticus
Conus elasticus
Arch of cricoid cartilage
Vocal ligament
Thyroid cartilage (superior border)
Superior thyroid notch

Fig. 761: The Vocal Ligaments and Conus Elasticus from Above

NOTE that the conus elasticus is a membrane consisting principally of yellow elastic fibers which interconnects the thyroid, cricoid and arytenoid cartilages. It underlies the mucous membrane below the vocal folds and is overlain to some extent by the circothyroid muscle on the exterior of the larynx. Observe the symmetry of the arytenoid cartilages and their related vocal ligaments.

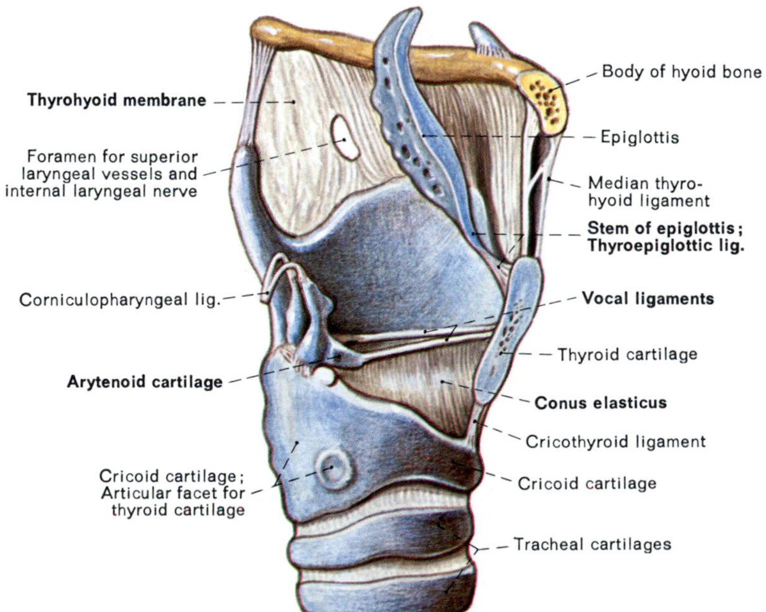
Body of hyoid bone
Thyrohyoid membrane
Epiglottis
Foramen for superior laryngeal vessels and internal laryngeal nerve
Median thyrohyoid ligament
Stem of epiglottis; Thyroepiglottic lig.
Corniculopharyngeal lig.
Vocal ligaments
Thyroid cartilage
Arytenoid cartilage
Conus elasticus
Cricothyroid ligament
Cricoid cartilage; Articular facet for thyroid cartilage
Cricoid cartilage
Tracheal cartilages

Fig. 762: The Upper Left Part of the Larynx

NOTE: 1) the right halves of the hyoid bone, epiglottis and thyroid cartilage have been removed to reveal the interior of the upper left portion of the larynx. The two vocal ligaments, the arytenoid cartilages and the conus elasticus are also displayed.

2) the broad thyrohyoid membrane and its foramen. Observe the attachment of the stem of the epiglottis to the thyroid cartilage by means of the thyroepiglottic ligament.

3) the conus elasticus as it attaches to the vocal fold and the arytenoid, thyroid and cricoid cartilages.

4) although laryngeal sounds are initiated at the vocal folds, the pitch, quality, volume, range, tone and overtone characteristics of the human voice incorporate not only the vocal folds but also the structures in the mouth (tongue, teeth, and palate), the nasal cavities (sinuses), the pharynx, the rest of the larynx, the lungs, diaphragm and abdominal musculature.

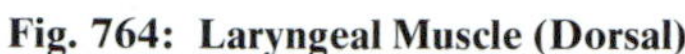

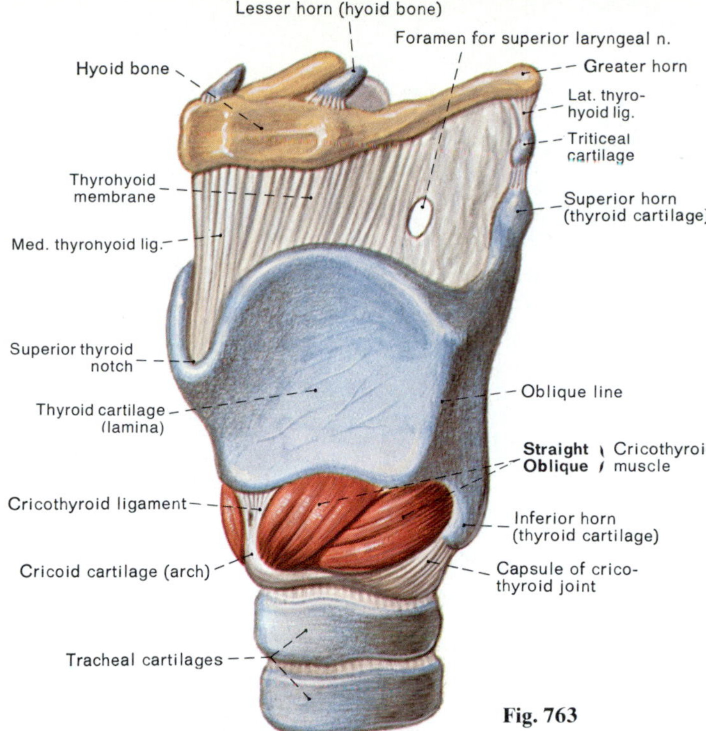
Lesser horn (hyoid bone)
Foramen for superior laryngeal n.
Hyoid bone
Greater horn
Lat. thyro-hyoid lig.
Triticeal cartilage
Thyrohyoid membrane
Superior horn (thyroid cartilage)
Med. thyrohyoid lig.
Superior thyroid notch
Thyroid cartilage (lamina)
Oblique line
Straight
Oblique
Cricothyroid muscle
Cricothyroid ligament
Inferior horn (thyroid cartilage)
Cricoid cartilage (arch)
Capsule of crico-thyroid joint
Tracheal cartilages
Fig. 763

Fig. 763: Ventrolateral View of the Exterior Larynx and the Cricothyroid Muscle

NOTE: 1) the cricothyroid muscle consists of straight and oblique heads. The *straight head* is more vertical and inserts into the lower border of the lamina of the thyroid cartilage, while the *oblique head is* more horizontal and inserts onto the inferior horn of the thyroid cartilage.

2) the cricothyroid muscle tilts the anterior part of the cricoid cartilage superiorly. In so doing, the arytenoid cartilages which are attached to the cricoid are pulled dorsally. In addition, the thyroid cartilage is pulled forward and downward. These actions increase the distance between the arytenoid and thyroid cartilages, thereby increasing the tension of and elongating the vocal folds.

Fig. 764: Laryngeal Muscle (Dorsal)

NOTE: 1) the *arytenoid muscle* consisting of a *transverse portion* which spans the zone between the two arytenoid cartilages horizontally, and an *oblique portion* which consists of two muscular fascicles that cross one another. Thus, each of the two fascicles of the oblique portion extends from the base of one arytenoid cartilage to the apex of the other cartilage. Some oblique arytenoid fibers continue to the epiglottis along the aryepiglottic fold. These constitute the *aryepiglottis muscle.*

2) the *transverse arytenoid* approximates the arytenoid cartilages, and, therefore, closes the posterior part of the rima glottis. The *oblique arytenoid and aryepiglottic muscles* tend to close the inlet into the larynx by pulling the aryepiglottic folds together and approximating the arytenoid cartilages and the epiglottis.

3) the *posterior cricoarytenoid* muscle extends from the lamina of the cricoid cartilage to the muscular process of the arytenoid cartilage while the *lateral cricoarytenoid* muscle arises laterally from the arch of the cricoid cartilage to insert with the posterior cricoarytenoid muscle onto the arytenoid cartilage.

4) the *posterior cricoarytenoids* are the only abductors of the vocal folds, while the *lateral cricoarytenoids* act as antagonists, and adduct the vocal folds. The posterior muscle abducts by pulling the base of the arytenoid cartilages medially and posteriorly, while the lateral muscle adducts by pulling these same cartilages anteriorly and laterally.

5) the *thyroarytenoid muscle* (Fig. 765) is a thin sheet of muscle radiating from the thyroid cartilage principally backward toward the arytenoid cartilage. Its upper fibers continue to the epiglottis and, joining the aryepiglottic fibers, become the *thyroepiglottic muscle.* Its deepest (and most medial) fibers form the *vocalis muscle* which is attached to the lateral aspect of the vocal fold.

6) the thyroarytenoid muscles draw the arytenoid cartilages toward the thyroid cartilage and, thus, shorten (relax) the vocal folds.

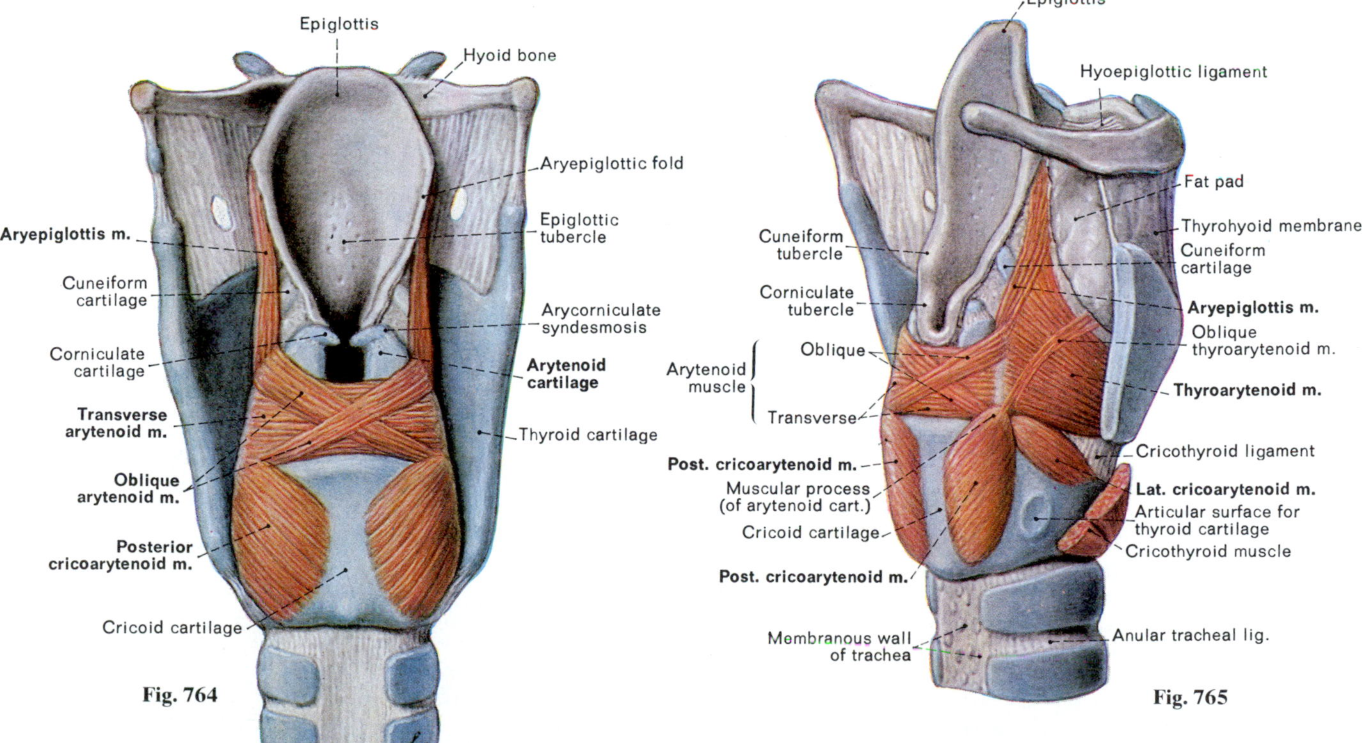
Epiglottis
Hyoid bone
Aryepiglottic fold
Aryepiglottis m.
Epiglottic tubercle
Cuneiform cartilage
Arycorniculate syndesmosis
Corniculate cartilage
Arytenoid cartilage
Transverse arytenoid m.
Thyroid cartilage
Oblique arytenoid m.
Posterior cricoarytenoid m.
Cricoid cartilage
Fig. 764
Epiglottis
Hyoepiglottic ligament
Fat pad
Thyrohyoid membrane
Cuneiform tubercle
Cuneiform cartilage
Corniculate tubercle
Aryepiglottis m.
Oblique
Oblique thyroarytenoid m.
Arytenoid muscle
Thyroarytenoid m.
Transverse
Cricothyroid ligament
Post. cricoarytenoid m.
Lat. cricoarytenoid m.
Muscular process (of arytenoid cart.)
Articular surface for thyroid cartilage
Cricoid cartilage
Cricothyroid muscle
Post. cricoarytenoid m.
Membranous wall of trachea
Anular tracheal lig.
Fig. 765

Fig. 765: Laryngeal Muscles (Lateral)

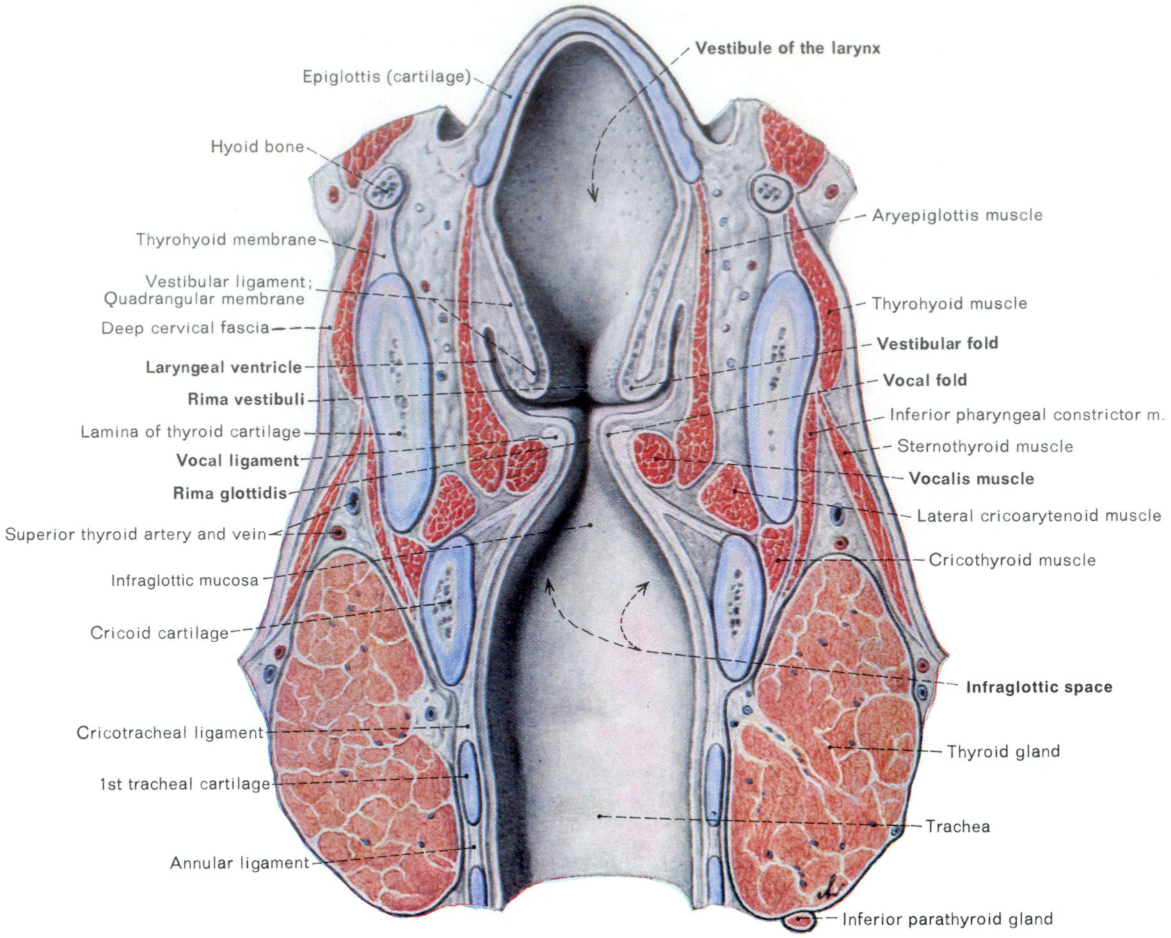

Vestibule of the larynx
Epiglottis (cartilage)
Hyoid bone
Aryepiglottis muscle
Thyrohyoid membrane
Vestibular ligament;
Quadrangular membrane
Thyrohyoid muscle
Deep cervical fascia
Vestibular fold
Laryngeal ventricle
Vocal fold
Rima vestibuli
Inferior pharyngeal constrictor m.
Lamina of thyroid cartilage
Sternothyroid muscle
Vocal ligament
Vocalis muscle
Rima glottidis
Lateral cricoarytenoid muscle
Superior thyroid artery and vein
Cricothyroid muscle
Infraglottic mucosa
Cricoid cartilage
Infraglottic space
Cricotracheal ligament
Thyroid gland
1st tracheal cartilage
Trachea
Annular ligament
Inferior parathyroid gland

Fig. 766: Frontal Section Through the Larynx Showing the Laryngeal Folds and Cavities in its Anterior One-Half

NOTE: 1) the paired vocal folds consisting of mucous membrane overlying the vocal ligaments and vocalis muscles. Just superior to the vocal folds observe the vestibular folds. On each side, the vestibular fold is separated from the vocal folds by a recess called the laryngeal ventricle (or sinus).

2) that above the vestibular folds is the vestibule of the larynx which lies just below the laryngeal inlet. Below the vocal folds observe the infraglottic space. This space communicates with the trachea below and is limited by the rima glottidis between the vocal folds above.

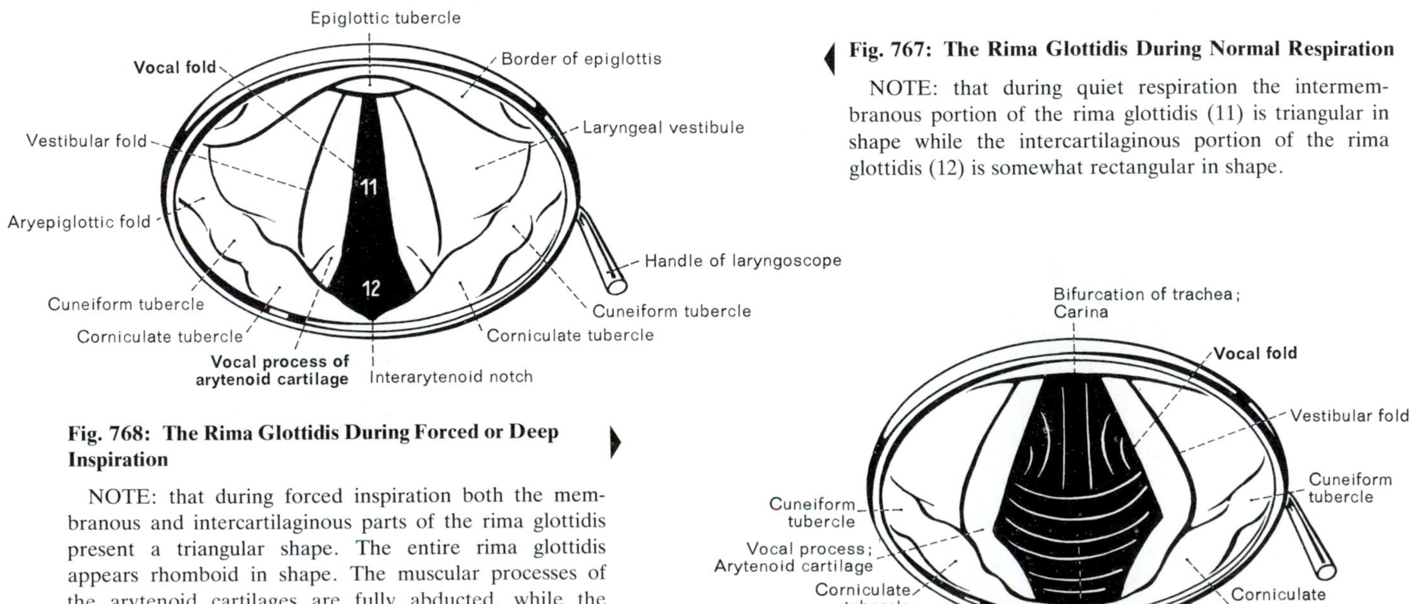

Epiglottic tubercle
Vocal fold
Border of epiglottis
Laryngeal vestibule
Vestibular fold
11
Aryepiglottic fold
Handle of laryngoscope
12
Cuneiform tubercle
Cuneiform tubercle
Corniculate tubercle
Corniculate tubercle
Vocal process of arytenoid cartilage
Interarytenoid notch
Bifurcation of trachea; Carina
Vocal fold
Vestibular fold
Cuneiform tubercle
Cuneiform tubercle
Vocal process; Arytenoid cartilage
Corniculate tubercle
Corniculate tubercle
Interarytenoid fold
Posterior wall of trachea

Fig. 767: The Rima Glottidis During Normal Respiration

NOTE: that during quiet respiration the intermembranous portion of the rima glottidis (11) is triangular in shape while the intercartilaginous portion of the rima glottidis (12) is somewhat rectangular in shape.

Fig. 768: The Rima Glottidis During Forced or Deep Inspiration

NOTE: that during forced inspiration both the membranous and intercartilaginous parts of the rima glottidis present a triangular shape. The entire rima glottidis appears rhomboid in shape. The muscular processes of the arytenoid cartilages are fully abducted, while the vocal processes of the cartilages are laterally rotated and retracted (principally by the posterior cricoarytenoids).

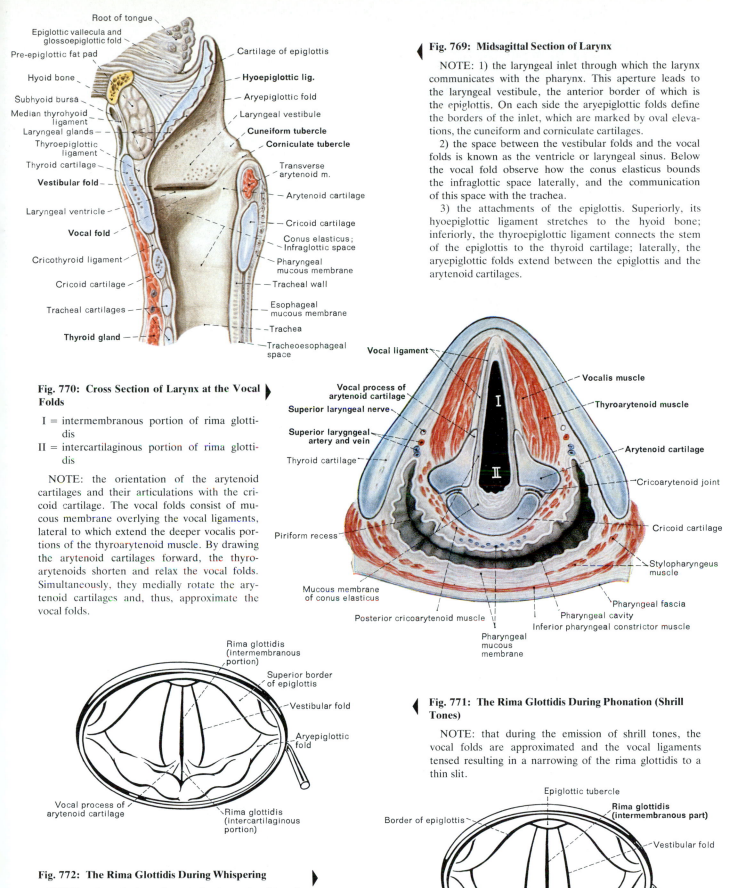

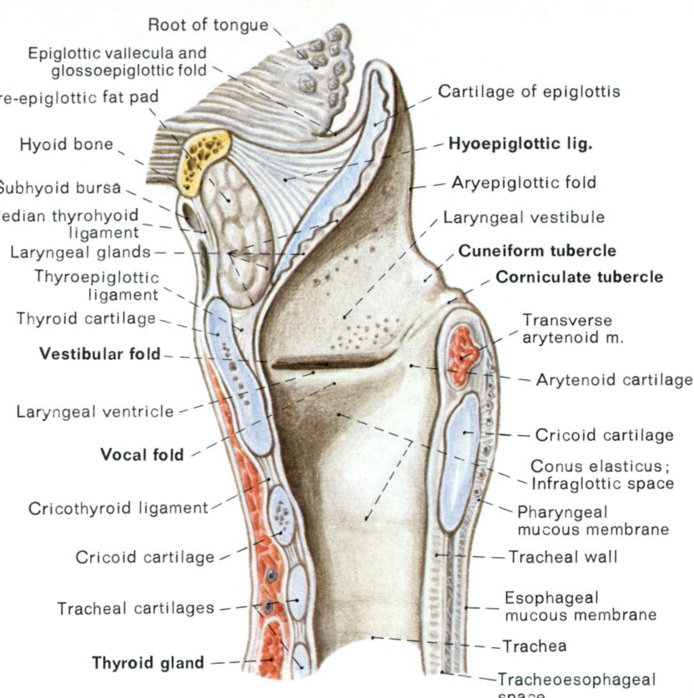
Root of tongue
Epiglottic vallecula and glossoepiglottic fold
Pre-epiglottic fat pad
Hyoid bone
Subhyoid bursa
Median thyrohyoid ligament
Laryngeal glands
Thyroepiglottic ligament
Thyroid cartilage
Vestibular fold
Laryngeal ventricle
Vocal fold
Cricothyroid ligament
Cricoid cartilage
Tracheal cartilages
Thyroid gland
Cartilage of epiglottis
Hyoepiglottic lig.
Aryepiglottic fold
Laryngeal vestibule
Cuneiform tubercle
Corniculate tubercle
Transverse arytenoid m.
Arytenoid cartilage
Cricoid cartilage
Conus elasticus; Infraglottic space
Pharyngeal mucous membrane
Tracheal wall
Esophageal mucous membrane
Trachea
Tracheoesophageal space

Fig. 769: Midsagittal Section of Larynx

NOTE: 1) the laryngeal inlet through which the larynx communicates with the pharynx. This aperture leads to the laryngeal vestibule, the anterior border of which is the epiglottis. On each side the aryepiglottic folds define the borders of the inlet, which are marked by oval elevations, the cuneiform and corniculate cartilages.

2) the space between the vestibular folds and the vocal folds is known as the ventricle or laryngeal sinus. Below the vocal fold observe how the conus elasticus bounds the infraglottic space laterally, and the communication of this space with the trachea.

3) the attachments of the epiglottis. Superiorly, its hyoepiglottic ligament stretches to the hyoid bone; inferiorly, the thyroepiglottic ligament connects the stem of the epiglottis to the thyroid cartilage; laterally, the aryepiglottic folds extend between the epiglottis and the arytenoid cartilages.

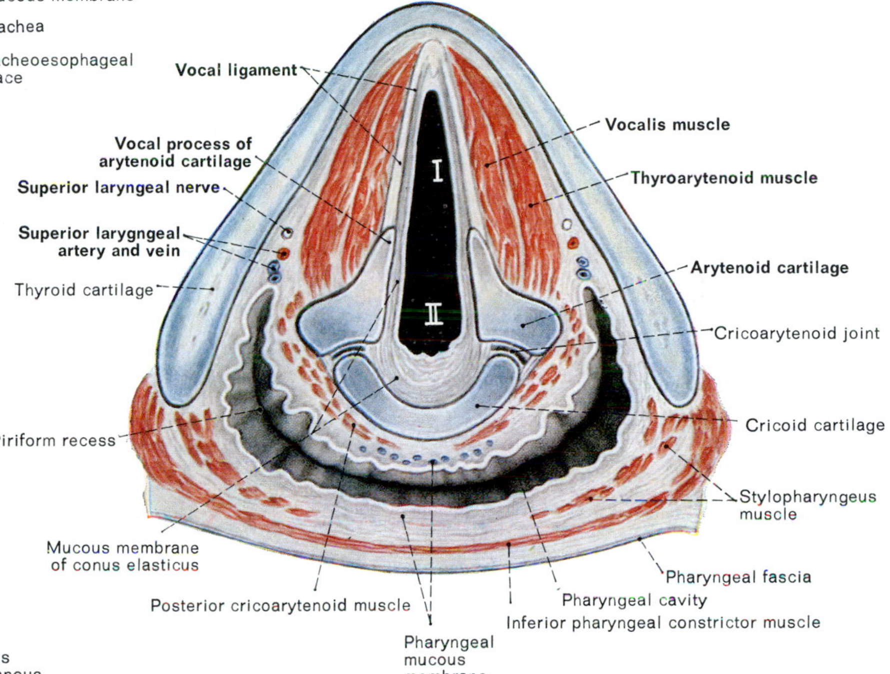
Vocal ligament
Vocal process of arytenoid cartilage
Superior laryngeal nerve
Superior larygngeal artery and vein
Thyroid cartilage
I
II
Vocalis muscle
Thyroarytenoid muscle
Arytenoid cartilage
Cricoarytenoid joint
Cricoid cartilage
Piriform recess
Stylopharyngeus muscle
Mucous membrane of conus elasticus
Pharyngeal fascia
Pharyngeal cavity
Posterior cricoarytenoid muscle
Inferior pharyngeal constrictor muscle
Pharyngeal mucous membrane

Fig. 770: Cross Section of Larynx at the Vocal Folds

I = intermembranous portion of rima glottidis

II = intercartilaginous portion of rima glottidis

NOTE: the orientation of the arytenoid cartilages and their articulations with the cricoid cartilage. The vocal folds consist of mucous membrane overlying the vocal ligaments, lateral to which extend the deeper vocalis portions of the thyroarytenoid muscle. By drawing the arytenoid cartilages forward, the thyroarytenoids shorten and relax the vocal folds. Simultaneously, they medially rotate the arytenoid cartilages and, thus, approximate the vocal folds.

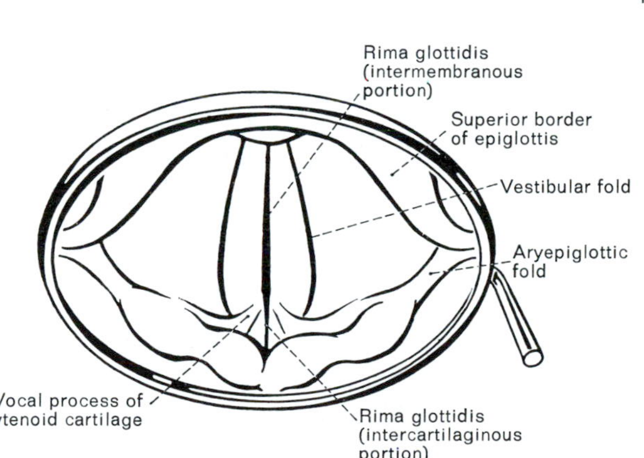
Rima glottidis (intermembranous portion)
Superior border of epiglottis
Vestibular fold
Aryepiglottic fold
Vocal process of arytenoid cartilage
Rima glottidis (intercartilaginous portion)

Fig. 772: The Rima Glottidis During Whispering

NOTE: that during the production of whispering sounds the intermembranous portion of the vocal folds are approximated, while the intercartilaginous portion of the folds are kept separated. The anterior part of the rima glottidis is narrow and slit-like and the posterior, intercartilaginous portion remains open.

Fig. 771: The Rima Glottidis During Phonation (Shrill Tones)

NOTE: that during the emission of shrill tones, the vocal folds are approximated and the vocal ligaments tensed resulting in a narrowing of the rima glottidis to a thin slit.

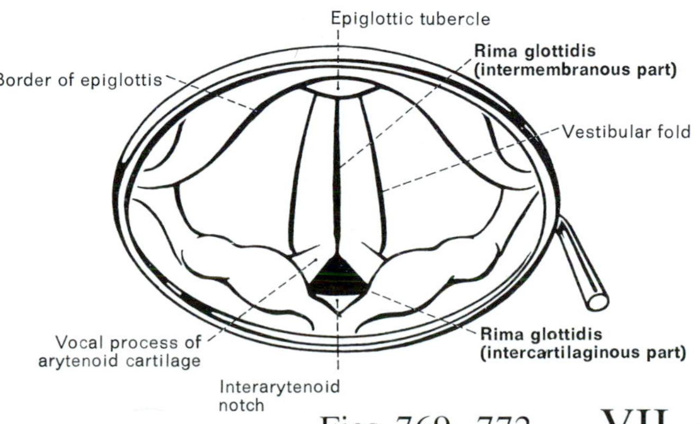
Epiglottic tubercle
Rima glottidis (intermembranous part)
Border of epiglottis
Vestibular fold
Vocal process of arytenoid cartilage
Rima glottidis (intercartilaginous part)
Interarytenoid notch

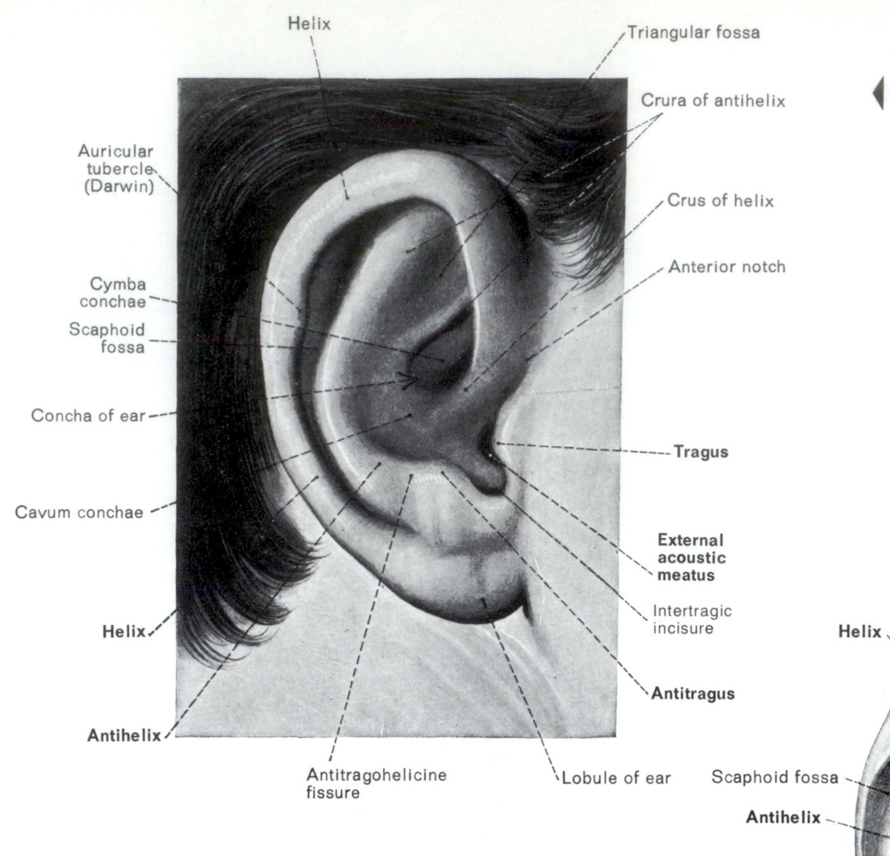
Helix
Triangular fossa
Crura of antihelix
Auricular tubercle (Darwin)
Crus of helix
Anterior notch
Cymba conchae
Scaphoid fossa
Concha of ear
Tragus
Cavum conchae
External acoustic meatus
Intertragic incisure
Helix
Antitragus
Antihelix
Antitragohelicine fissure
Lobule of ear

Fig. 773: The Right External Ear (Lateral View)

NOTE: 1) the external ear (or auricle) consists of skin overlying elastic fibrocartilage. It is irregularly shaped and the external acoustic meatus courses through the external ear to the tympanic membrane.

2) the external rim of the auricle is called the helix. Another curved prominence anterior to the helix is the antihelix. Inferiorly, a notch in the cartilage (the intertragic incisure) separates the tragus anteriorly from the antitragus posteriorly.

3) the ear lobe or lobule does not contain cartilage. Instead, it is soft and contains connective tissue and fat.

4) the external acoustic meatus is an oval canal which extends for about 2.5 cm in a S-shaped curve from the concha to the tympanic membrane. It consists of an outer cartilaginous part (1 cm) and an narrower more medial osseous part (1.5 cm).

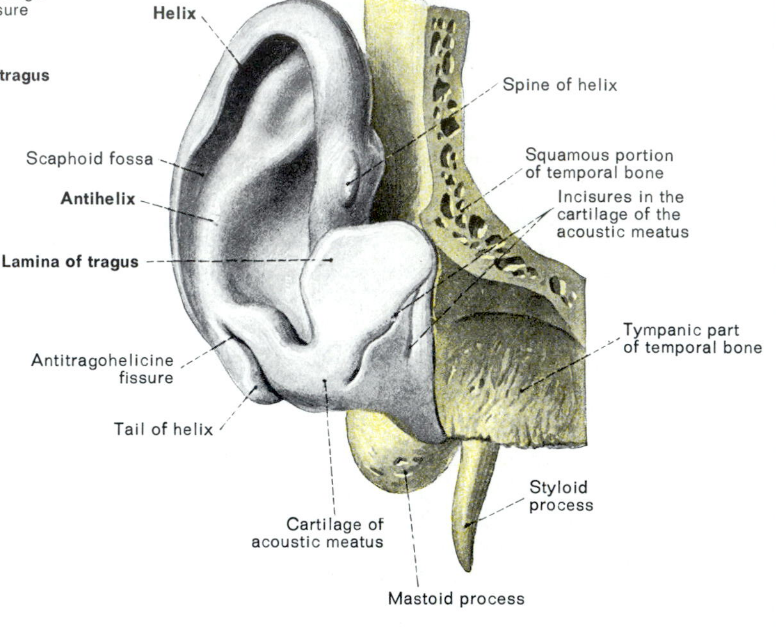
Helix
Spine of helix
Squamous portion of temporal bone
Scaphoid fossa
Incisures in the cartilage of the acoustic meatus
Antihelix
Lamina of tragus
Tympanic part of temporal bone
Antitragohelicine fissure
Tail of helix
Styloid process
Cartilage of acoustic meatus
Mastoid process

Fig. 774: The Cartilage of the Right External Ear (seen from Front)

NOTE: 1) with the skin of the external ear removed, the contours of the single cartilage conform generally with those of the intact auricle. The cartilage is seen to be absent inferiorly at the site of the ear lobe.

2) the attachment of the auricular cartilage to the temporal bone. This attachment is strengthened by the anterior and posterior ligaments (not shown in this figure) which articulate the tragus and spine of the helix to the temporal bone anteriorly, and the concha to the mastoid process posteriorly. The continuous overlying skin completes the attachment.

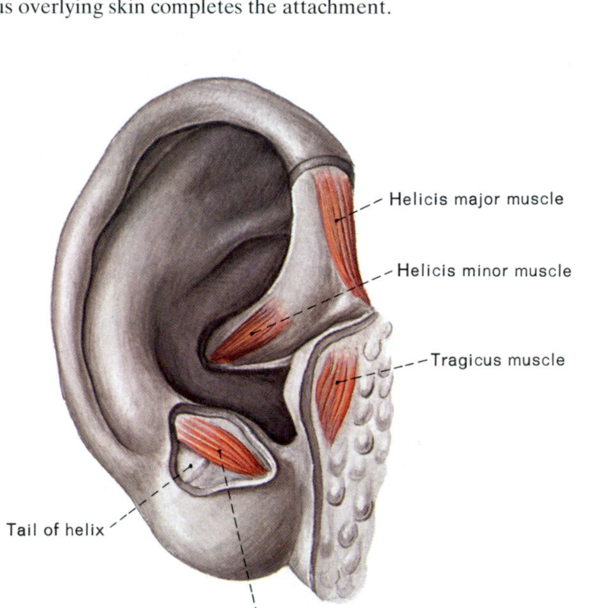
Helicis major muscle
Helicis minor muscle
Tragicus muscle
Tail of helix
Antitragicus muscle

Fig. 775: Intrinsic Muscles of External Ear (Lateral Surface)

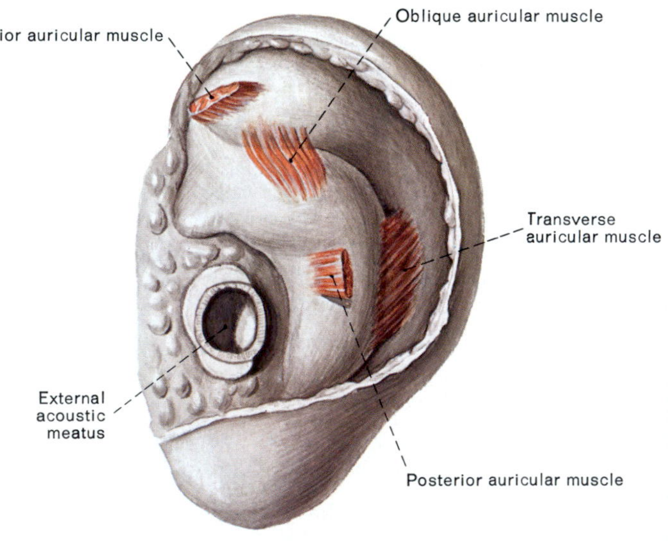
Superior auricular muscle
Oblique auricular muscle
Transverse auricular muscle
External acoustic meatus
Posterior auricular muscle

Fig. 776: Muscles Attaching to the Medial Surface of External Ear

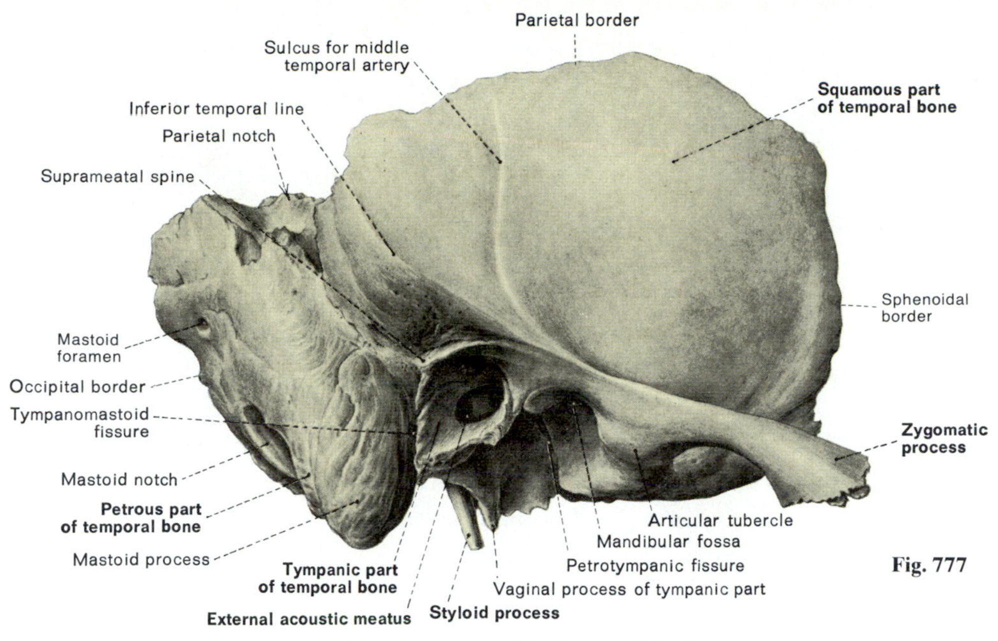

Parietal border
Sulcus for middle temporal artery
Inferior temporal line
Parietal notch
Suprameatal spine
Squamous part of temporal bone
Sphenoidal border
Mastoid foramen
Occipital border
Tympanomastoid fissure
Mastoid notch
Petrous part of temporal bone
Mastoid process
Tympanic part of temporal bone
External acoustic meatus
Styloid process
Vaginal process of tympanic part
Petrotympanic fissure
Mandibular fossa
Articular tubercle
Zygomatic process

Fig. 777

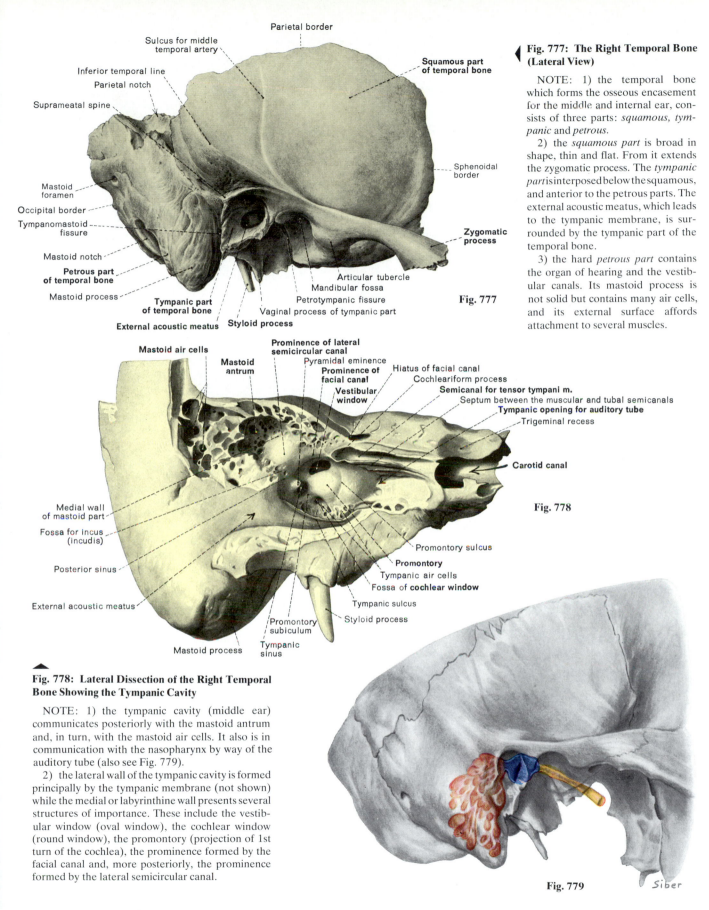

Fig. 777: The Right Temporal Bone (Lateral View)

NOTE: 1) the temporal bone which forms the osseous encasement for the middle and internal ear, consists of three parts: *squamous, tympanic* and *petrous.*

2) the *squamous part* is broad in shape, thin and flat. From it extends the zygomatic process. The *tympanic part* is interposed below the squamous, and anterior to the petrous parts. The external acoustic meatus, which leads to the tympanic membrane, is surrounded by the tympanic part of the temporal bone.

3) the hard *petrous part* contains the organ of hearing and the vestibular canals. Its mastoid process is not solid but contains many air cells, and its external surface affords attachment to several muscles.

Mastoid air cells
Mastoid antrum
Prominence of lateral semicircular canal
Pyramidal eminence
Prominence of facial canal
Vestibular window
Hiatus of facial canal
Cochleariform process
Semicanal for tensor tympani m.
Septum between the muscular and tubal semicanals
Tympanic opening for auditory tube
Trigeminal recess
Carotid canal
Medial wall of mastoid part
Fossa for incus (incudis)
Posterior sinus
External acoustic meatus
Mastoid process
Promontory subiculum
Tympanic sinus
Styloid process
Tympanic sulcus
Fossa of cochlear window
Tympanic air cells
Promontory
Promontory sulcus

Fig. 778

Fig. 778: Lateral Dissection of the Right Temporal Bone Showing the Tympanic Cavity

NOTE: 1) the tympanic cavity (middle ear) communicates posteriorly with the mastoid antrum and, in turn, with the mastoid air cells. It also is in communication with the nasopharynx by way of the auditory tube (also see Fig. 779).

2) the lateral wall of the tympanic cavity is formed principally by the tympanic membrane (not shown) while the medial or labyrinthine wall presents several structures of importance. These include the vestibular window (oval window), the cochlear window (round window), the promontory (projection of 1st turn of the cochlea), the prominence formed by the facial canal and, more posteriorly, the prominence formed by the lateral semicircular canal.

Fig. 779

Siber

Fig. 779: A Projection of the Middle Ear Air Spaces on the Lateral Surface of the Skull

NOTE that the mastoid air cells within the mastoid process of the temporal bone (shown in red) communicate with the tympanic cavity (blue). Coursing between the middle ear and the nasopharynx is the auditory tube (yellow). This communication allows the air pressure in the tympanic cavity to be the same as external atmospheric pressure and, thus, equivalent on both sides of the tympanic membrane.

Fig. 780: Frontal Section Through Left External, Middle and Internal Ear

NOTE: 1) the external acoustic meatus commences at the auricle and leads to the external surface of the tympanic membrane. Through the meatus course the sound waves which cause vibration of the tympanum.

2) the middle ear (or tympanic cavity) contains three ossicles (malleus, incus and stapes) and two muscles (tensor tympani and stapedius [not shown]). The cavity of the middle ear communicates with the mastoid antrum and mastoid air cells posteriorly, and the nasopharynx by way of the auditory tube. From the middle ear this tube courses downward, forward and medially. The ossicles interconnect the tympanic membrane with the inner ear.

3) the inner ear contains the coiled cochlea (or organ of hearing) and the three semicircular canals (the vestibular organ) and their associated structures and nerves.

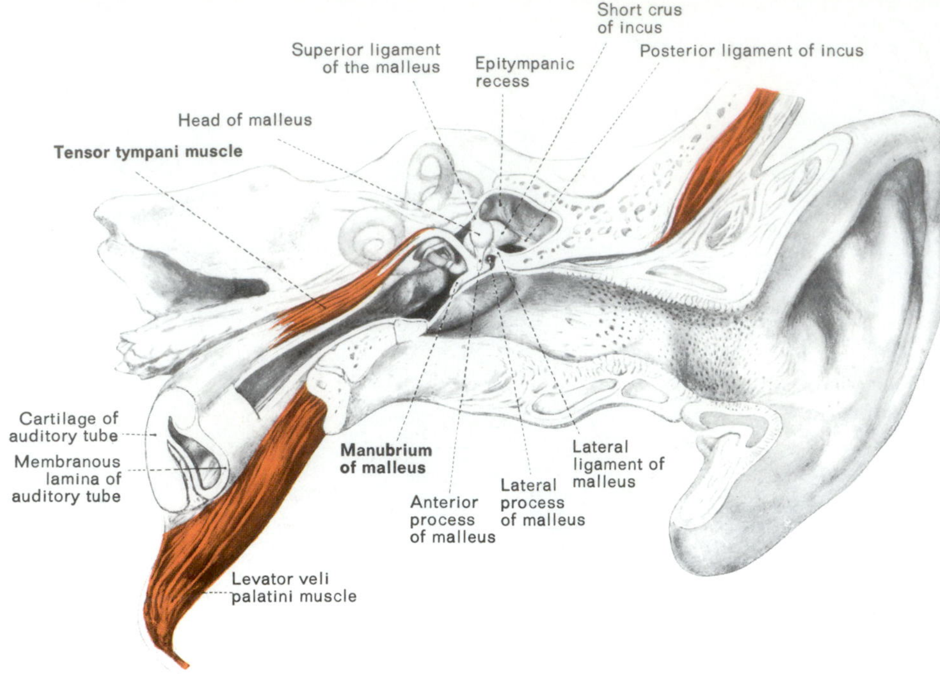

Short crus of incus
Superior ligament of the malleus
Epitympanic recess
Posterior ligament of incus
Head of malleus
Tensor tympani muscle
Cartilage of auditory tube
Membranous lamina of auditory tube
Manubrium of malleus
Lateral ligament of malleus
Lateral process of malleus
Anterior process of malleus
Levator veli palatini muscle

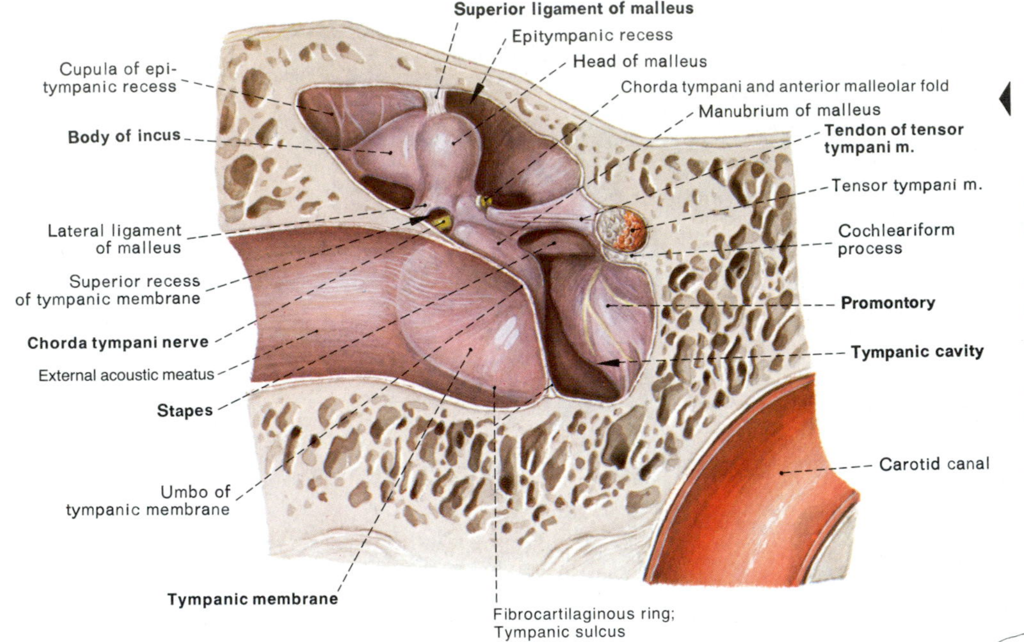

Superior ligament of malleus
Epitympanic recess
Head of malleus
Chorda tympani and anterior malleolar fold
Manubrium of malleus
Tendon of tensor tympani m.
Tensor tympani m.
Cochleariform process
Promontory
Tympanic cavity
Carotid canal
Cupula of epitympanic recess
Body of incus
Lateral ligament of malleus
Superior recess of tympanic membrane
Chorda tympani nerve
External acoustic meatus
Stapes
Umbo of tympanic membrane
Tympanic membrane
Fibrocartilaginous ring; Tympanic sulcus

Fig. 781: Frontal Section Through Right External and Middle Ear

NOTE: 1) the slender tendon of the tensor tympani muscle, as it turns sharply upon reaching the tympanic cavity to terminate on the manubrium of the malleus.

2) the tympanic cavity is extended superiorly by the epitympanic recess and, inferiorly, by the hypotympanic recess. On the medial wall of the middle ear observe the promontory which protrudes into the tympanic cavity and which contains the spiral cochlea.

3) the lateral and superior ligaments attaching to the head of the malleus.

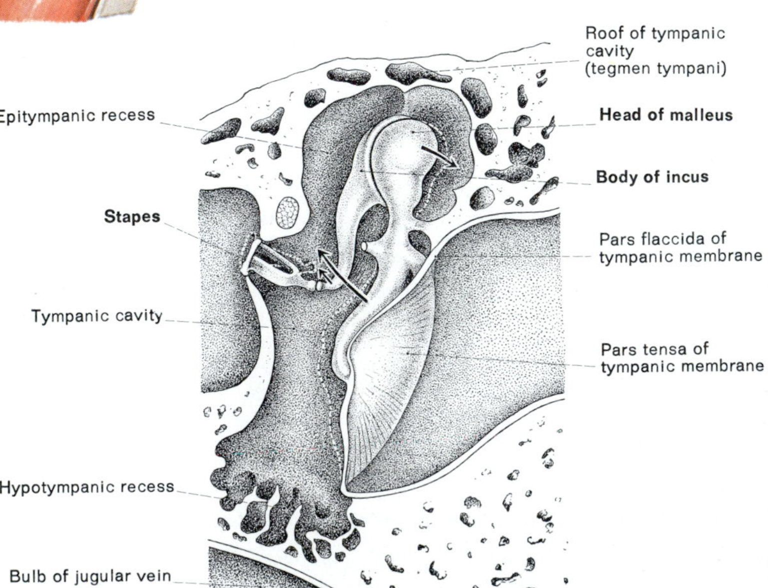

Roof of tympanic cavity (tegmen tympani)
Epitympanic recess
Head of malleus
Body of incus
Stapes
Pars flaccida of tympanic membrane
Tympanic cavity
Pars tensa of tympanic membrane
Hypotympanic recess
Bulb of jugular vein

Fig. 782: Frontal Section of Left Ear Through the Middle Ear Bones

NOTE: 1) when sound waves are received at the tympanic membrane, they cause medial displacements of the manubrium of the malleus. The head of the malleus is tilted laterally, pulling with it the body of the incus. At the same time the long process of the incus is displaced medially, as is the articulation between the incus and the stapes.

2) the base of the stapes rocks as though it were on a fulcrum at the vestibular window, thereby establishing waves in the perilymph. These waves stimulate the auditory receptors and become dissipated at the secondary tympanic membrane covering the cochlear window.

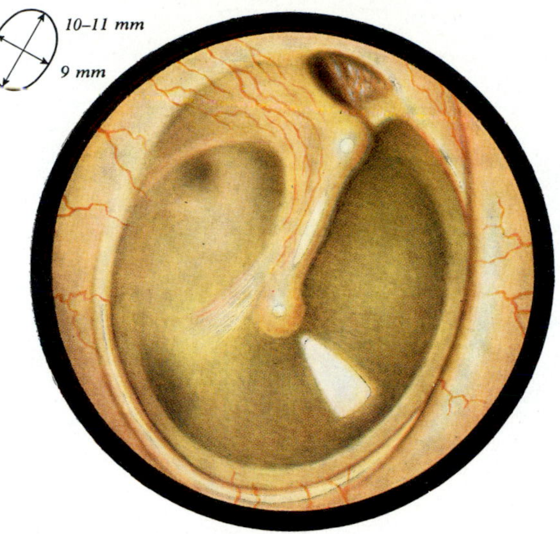
10–11 mm
9 mm

Fig. 783: Right Tympanic Membrane as seen with an Otoscope in a Living Person

NOTE: that the tympanic membrane is oval in shape and measures about 9 mm across and from 10 to 11 mm vertically (upper inset, actual size). Its blood supply is derived from the deep auricular and anterior tympanic branches of the maxillary artery, and the stylomastoid branch of the posterior auricular artery.

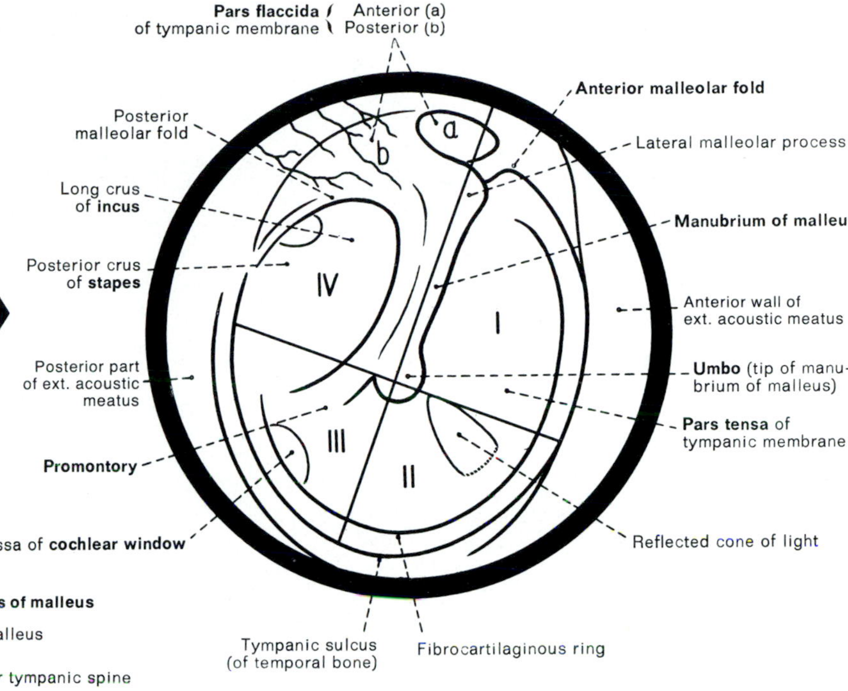
Pars flaccida of tympanic membrane
Anterior (a)
Posterior (b)
Anterior malleolar fold
Posterior malleolar fold
Lateral malleolar process
Long crus of incus
Manubrium of malleus
Posterior crus of stapes
IV
I
Anterior wall of ext. acoustic meatus
Posterior part of ext. acoustic meatus
Umbo (tip of manubrium of malleus)
III
II
Pars tensa of tympanic membrane
Promontory
Fossa of cochlear window
Reflected cone of light
Tympanic sulcus (of temporal bone)
Fibrocartilaginous ring

Fig. 784: Labeled Diagram of Fig. 783

NOTE: 1) the tympanic membrane has been divided into quadrants. The most depressed point is the umbo and is found at the rounded extremity of the handle of the malleus.

2) the anterior and posterior malleolar folds. The more lax part (pars flaccida) of the tympanic membrane lies above and between these folds, while the rest is more tightly stretched (pars tensa).

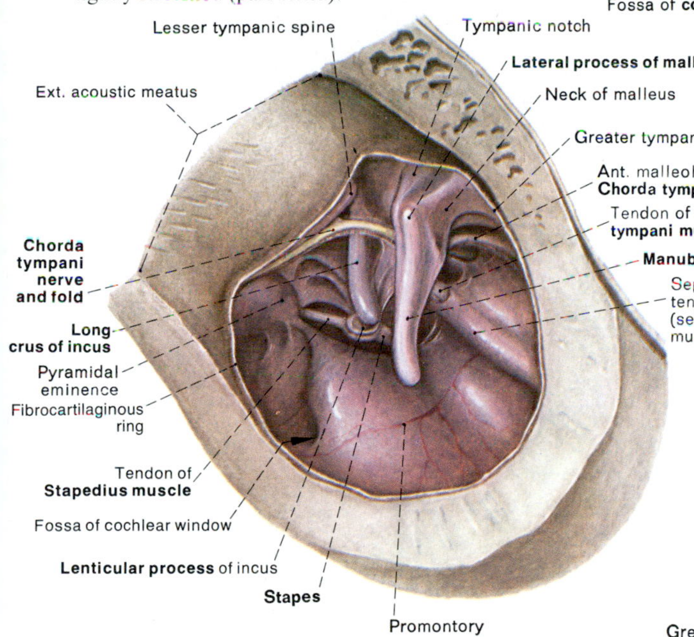
Lesser tympanic spine
Tympanic notch
Lateral process of malleus
Ext. acoustic meatus
Neck of malleus
Greater tympanic spine
Ant. malleolar fold; Chorda tympani nerve
Tendon of tensor tympani muscle
Chorda tympani nerve and fold
Manubrium of malleus
Septum covering tensor tympani m (septum canalis musculotubarii)
Long crus of incus
Pyramidal eminence
Fibrocartilaginous ring
Tendon of Stapedius muscle
Fossa of cochlear window
Lenticular process of incus
Stapes
Promontory

Fig. 785: Right Tympanic Cavity seen After Removal of Tympanic Membrane

NOTE: that from this lateral view, the lateral process and manubrium of the malleus, the lenticular process and long crus of the incus, as well as the stapes can be observed upon the removal of the tympanic membrane. Note also the chorda tympani nerve and the tendons of the stapedius and tensor tympani muscles.

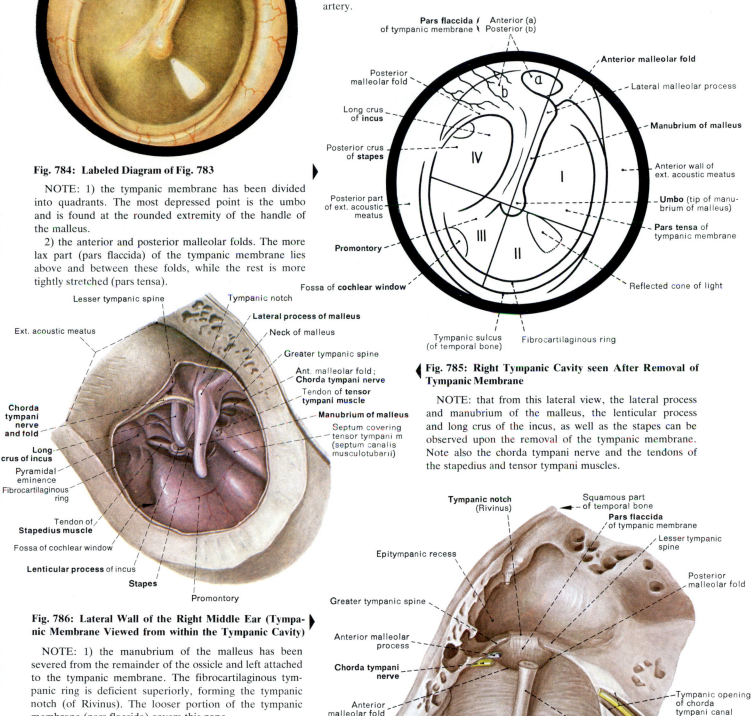

Fig. 786: Lateral Wall of the Right Middle Ear (Tympanic Membrane Viewed from within the Tympanic Cavity)

NOTE: 1) the manubrium of the malleus has been severed from the remainder of the ossicle and left attached to the tympanic membrane. The fibrocartilaginous tympanic ring is deficient superiorly, forming the tympanic notch (of Rivinus). The looser portion of the tympanic membrane (pars flaccida) covers this zone.

2) the tympanic membrane below the malleolar folds is the pars tensa. This portion is made taut by the tensor tympani muscle which attaches to the manubrium of the malleus.

3) the innervation of external surface of the tympanic membrane comes from the auriculotemporal branch of the mandibular nerve (V) and the auricular branch of the vagus (X). The inner surface of the membrane receives innervation from the tympanic branch of the glossopharyngeal nerve (IX).

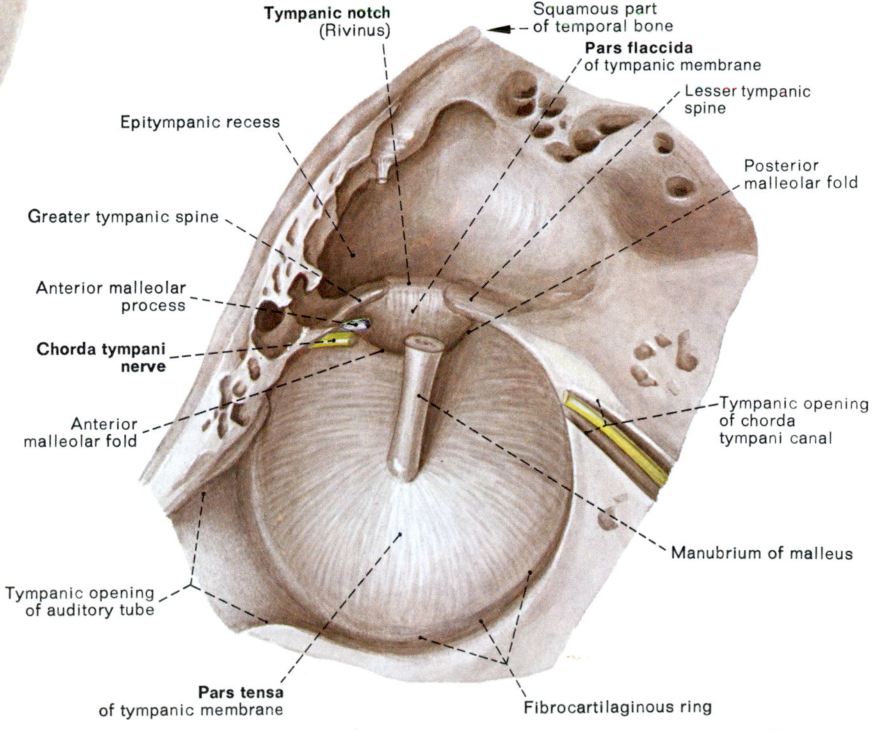
Tympanic notch (Rivinus)
Squamous part of temporal bone
Pars flaccida of tympanic membrane
Lesser tympanic spine
Epitympanic recess
Posterior malleolar fold
Greater tympanic spine
Anterior malleolar process
Chorda tympani nerve
Tympanic opening of chorda tympani canal
Anterior malleolar fold
Manubrium of malleus
Tympanic opening of auditory tube
Pars tensa of tympanic membrane
Fibrocartilaginous ring

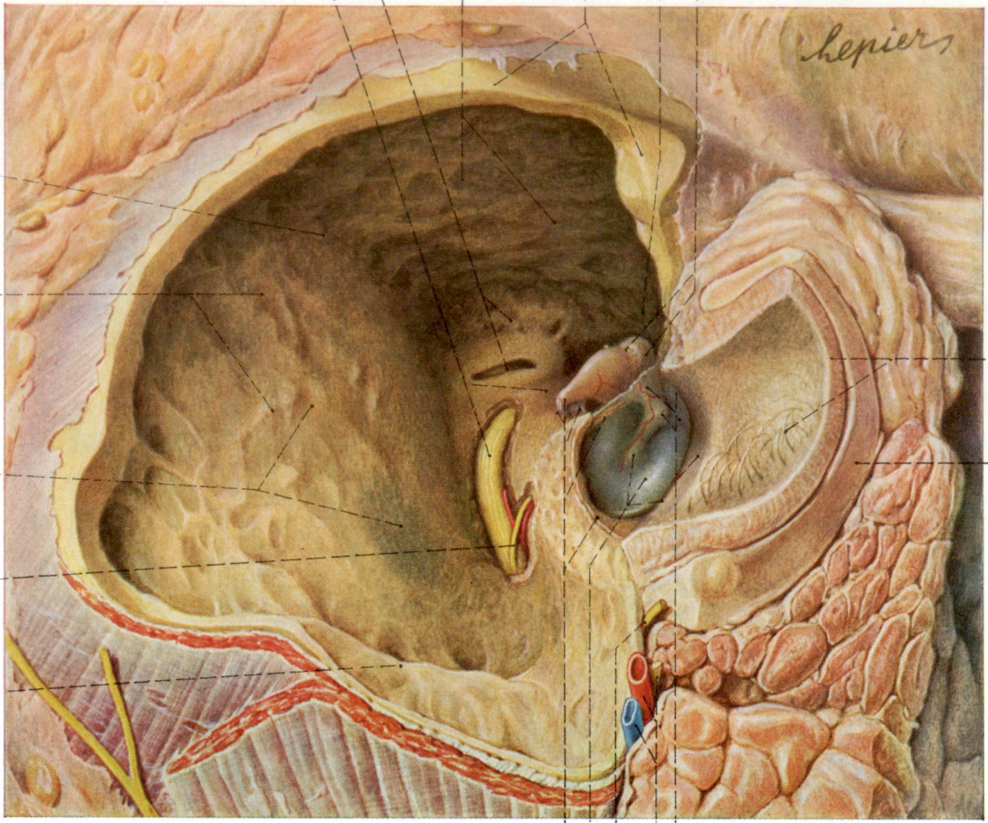

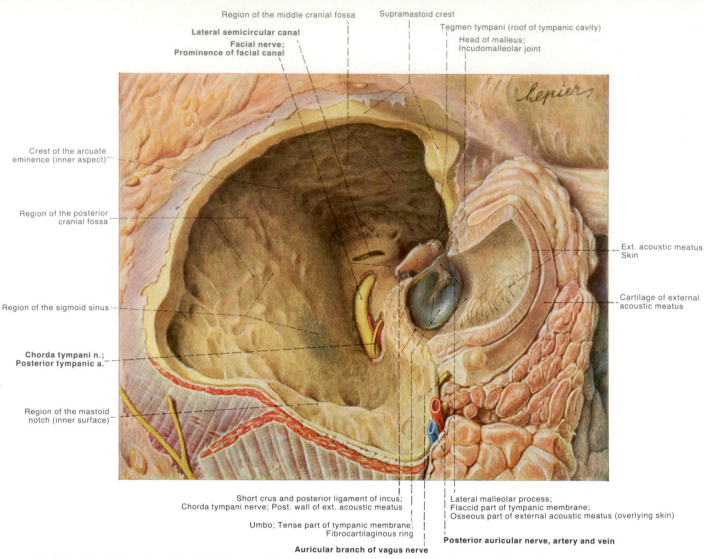

Region of the middle cranial fossa
Supramastoid crest
Lateral semicircular canal
Tegmen tympani (roof of tympanic cavity)
Facial nerve;
Prominence of facial canal
Head of malleus;
Incudomalleolar joint
Crest of the arcuate eminence (inner aspect)
Region of the posterior cranial fossa
Ext. acoustic meatus; Skin
Cartilage of external acoustic meatus
Region of the sigmoid sinus
Chorda tympani n.; Posterior tympanic a.
Region of the mastoid notch (inner surface)
Short crus and posterior ligament of incus; Chorda tympani nerve; Post. wall of ext. acoustic meatus
Lateral malleolar process; Flaccid part of tympanic membrane; Osseous part of external acoustic meatus (overlying skin)
Umbo; Tense part of tympanic membrane; Fibrocartilaginous ring
Posterior auricular nerve, artery and vein
Auricular branch of vagus nerve

Fig. 787: Lateral Dissection of the Right Temporal Bone: The Tympanic Membrane, Middle Ear, and Facial Canal

NOTE: 1) that this dissection has proceeded through the mastoid region of the temporal bone, much of which has been removed, as has the posterior wall of the external acoustic meatus. The descending part of the facial canal and the lateral semicircular canal have been opened.

2) the lateral or outer surface of the tympanic membrane. Observe the manubrium of the malleus on the inner surface of the membrane, and the head of the malleus, the incus and the incudomalleolar joint within the tympanic cavity.

3) the junction of the chorda tympani nerve with the facial nerve.

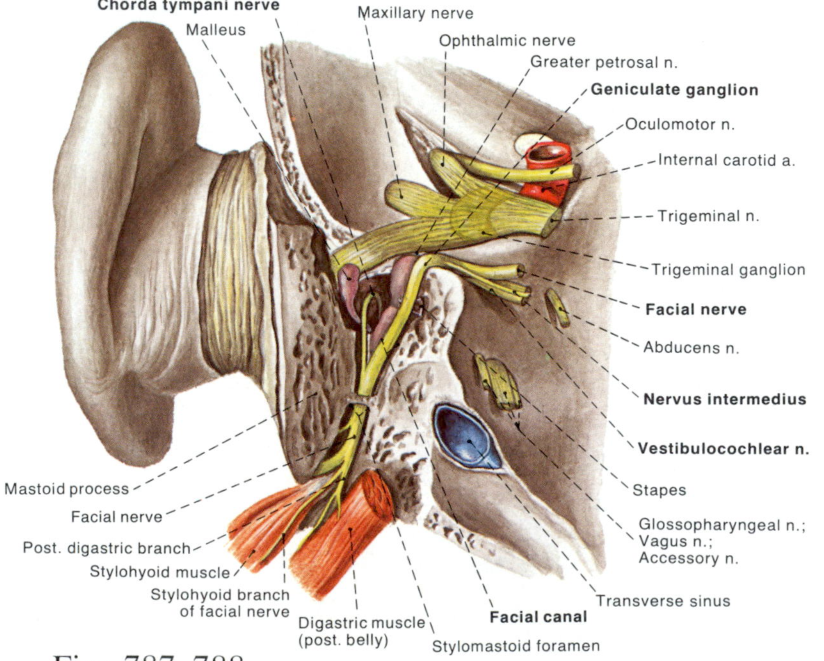
Chorda tympani nerve
Malleus
Maxillary nerve
Ophthalmic nerve
Greater petrosal n.
Geniculate ganglion
Oculomotor n.
Internal carotid a.
Trigeminal n.
Trigeminal ganglion
Facial nerve
Abducens n.
Nervus intermedius
Vestibulocochlear n.
Stapes
Glossopharyngeal n.; Vagus n.; Accessory n.
Transverse sinus
Mastoid process
Facial nerve
Post. digastric branch
Stylohyoid muscle
Stylohyoid branch of facial nerve
Digastric muscle (post. belly)
Facial canal
Stylomastoid foramen

Fig. 788: The Intracranial Course of the Facial Nerve, View from Behind

NOTE: 1) that a frontal section has been made through the temporal bone at an angle which opens the facial canal and tympanic cavity from behind. Observe the chorda tympani nerve coursing anteroposteriorly across the middle ear cavity to its junction with the facial nerve.

2) that the internal acoustic meatus in the floor of the cranial cavity can be seen transmitting the facial nerve with its nervous intermedius, as well as the vestibulocochlear nerve. Peripheral to the geniculate ganglion (which serves as the sensory ganglion of VII), the facial nerve enters the facial canal. Within the canal, the facial nerve initially courses laterally and then turns sharply backward and inferiorly (see Figs. 787, 793 and 794).

3) that beyond its junction with the chorda tympani nerve, the main trunk of the facial nerve continues its descent in the temporal bone to emerge on the lateral aspect of the face through the stylomastoid foramen.

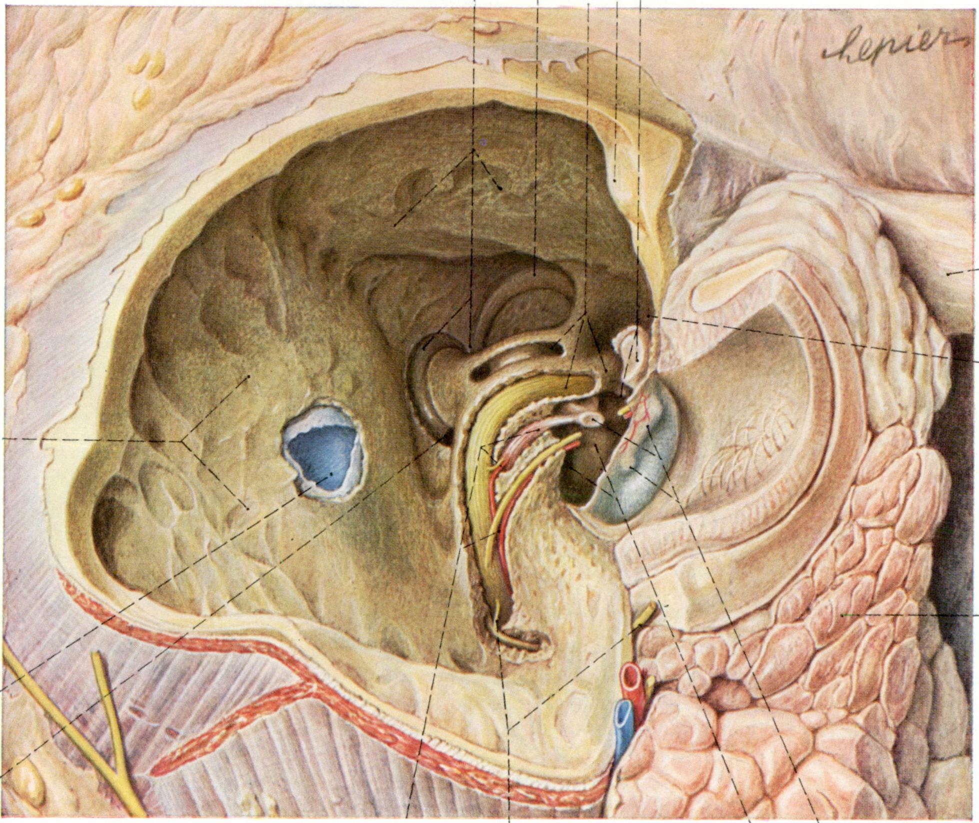

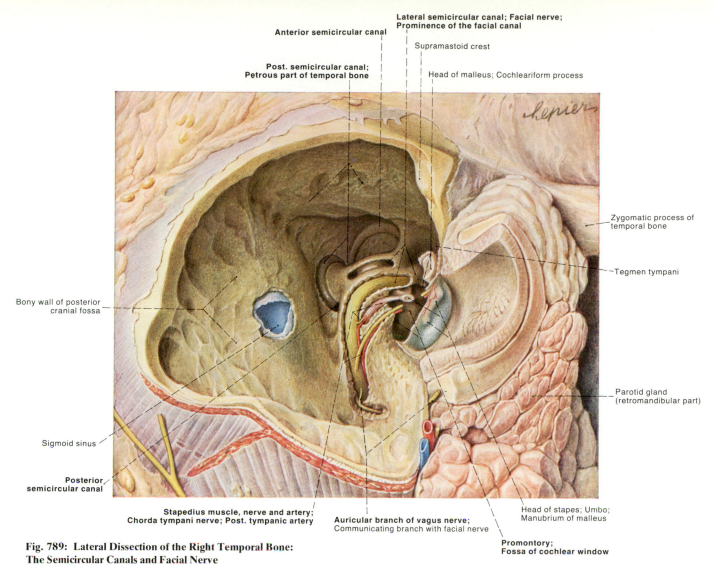

Lateral semicircular canal; Facial nerve; Prominence of the facial canal
Anterior semicircular canal
Supramastoid crest
Post. semicircular canal; Petrous part of temporal bone
Head of malleus; Cochleariform process
Zygomatic process of temporal bone
Tegmen tympani
Bony wall of posterior cranial fossa
Parotid gland (retromandibular part)
Sigmoid sinus
Posterior semicircular canal
Stapedius muscle, nerve and artery; Chorda tympani nerve; Post. tympanic artery
Auricular branch of vagus nerve; Communicating branch with facial nerve
Head of stapes; Umbo; Manubrium of malleus
Promontory; Fossa of cochlear window

Fig. 789: Lateral Dissection of the Right Temporal Bone: The Semicircular Canals and Facial Nerve

NOTE: 1) that the facial canal has been opened more extensively than in Figure 787, exposing further the chorda tympani nerve and the posterior tympanic artery. This vessel branches from the stylomastoid artery and helps to supply the internal surface of the tympanic membrane. Observe the nerve to the stapedius muscle which arises from the facial nerve near the pyramidal eminence on the posterior wall of the tympanic cavity. Coursing with the nerve is the stapedial artery, and more inferiorly is found the auricular branch of the vagus nerve.

2) that the incus has been removed, thereby revealing the head of the stapes, which is directed laterally. The stapedius muscle inserts onto the posterior surface of the head of the stapes.

3) the anterior, posterior and lateral semicircular canals. These become exposed after removal of the bony prominence over the lateral canal, and additional bone above and behind that prominence.

Fig. 790: The Right Temporal Bone Opened Laterally to Demonstrate the Medial Wall of the Tympanic Cavity

NOTE: 1) that the irregularly shaped mastoid portion of the temporal bone has been dissected laterally and that its interior is marked by a number of cavities called the mastoid cells. These are organized into three groups: anterosuperior, middle, and apical. The anterosuperior and middle mastoid cells are lined with mucous membrane, filled with air, and communicate with the tympanic cavity of the middle ear. Because of this, middle ear infections can spread directly to the mastoid air cells. The apical mastoid cells are quite different for they are small and contain bone marrow instead of air.

2) that the medial wall of the tympanic cavity contains a convex elevation of bone called the promontory, which is formed by the projection laterally of the first turn of the cochlea. Behind the middle ear cavity the curved facial canal has been opened, as have the lateral semicircular canal and the sigmoid sinus.

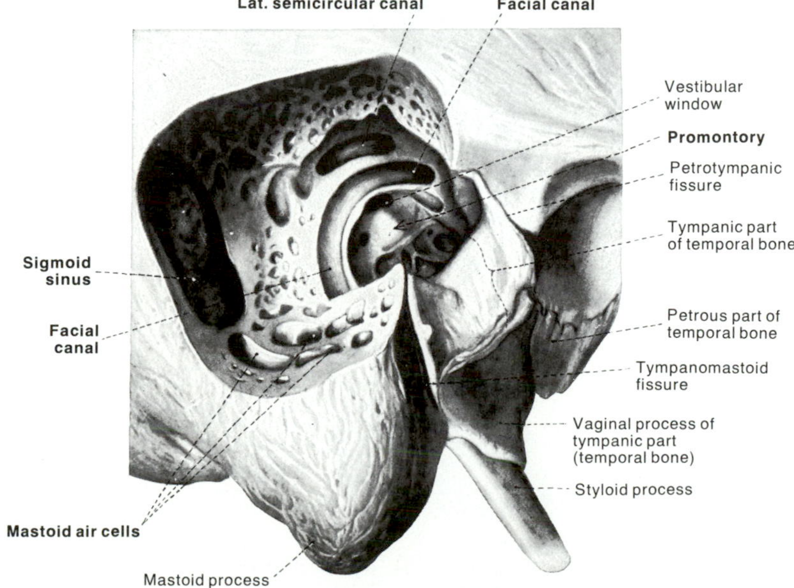
Lat. semicircular canal
Facial canal
Vestibular window
Promontory
Petrotympanic fissure
Tympanic part of temporal bone
Sigmoid sinus
Facial canal
Petrous part of temporal bone
Tympanomastoid fissure
Vaginal process of tympanic part (temporal bone)
Styloid process
Mastoid air cells
Mastoid process

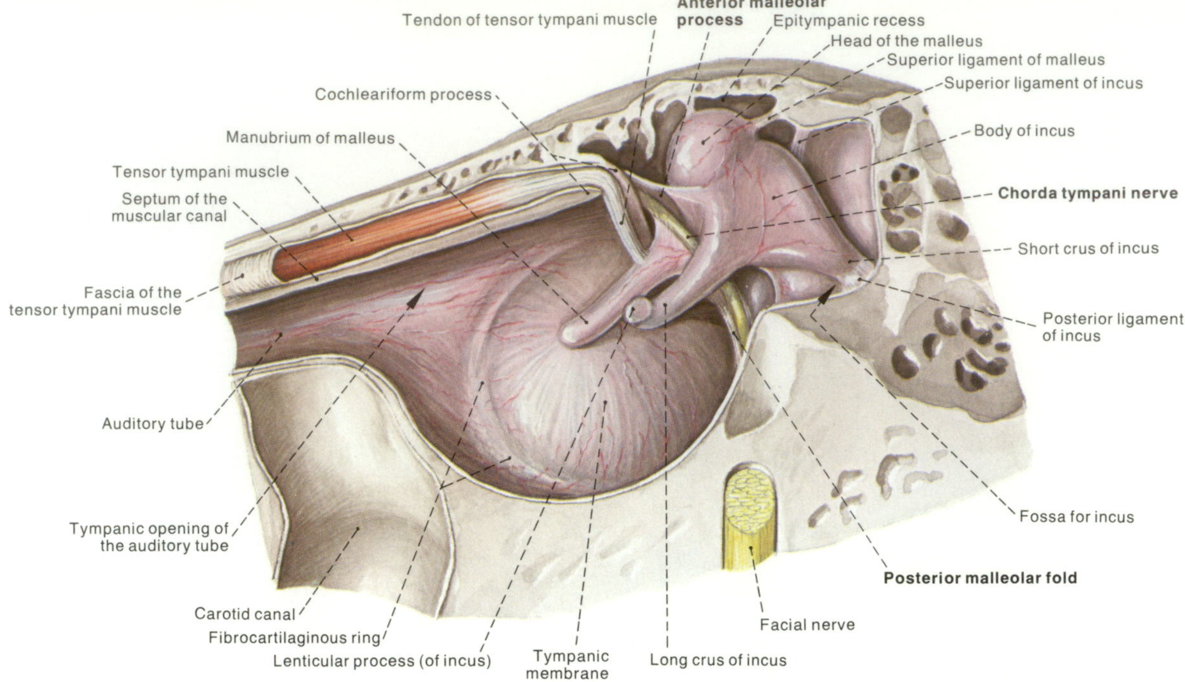
Anterior malleolar process
Tendon of tensor tympani muscle
Epitympanic recess
Head of the malleus
Superior ligament of malleus
Superior ligament of incus
Cochleariform process
Body of incus
Manubrium of malleus
Tensor tympani muscle
Chorda tympani nerve
Septum of the muscular canal
Short crus of incus
Fascia of the tensor tympani muscle
Posterior ligament of incus
Auditory tube
Fossa for incus
Tympanic opening of the auditory tube
Posterior malleolar fold
Carotid canal
Facial nerve
Fibrocartilaginous ring
Lenticular process (of incus)
Tympanic membrane
Long crus of incus

Fig. 791: Lateral Wall of the Right Tympanic Cavity (Viewed From the Medial Aspect)

NOTE: 1) that the tympanic cavity is lined completely with a mucous membrane that attaches onto the surface of all of the structures of the middle ear. This tympanic mucosa is continuous with that lining the mastoid air cells posteriorly and the auditory tube anteriorly.

2) that reflections of the tympanic mucous membrane form the anterior and posterior malleolar folds. These are also reflected around the chorda tympani nerve as it curves along the medial side of the manubrium of the malleus.

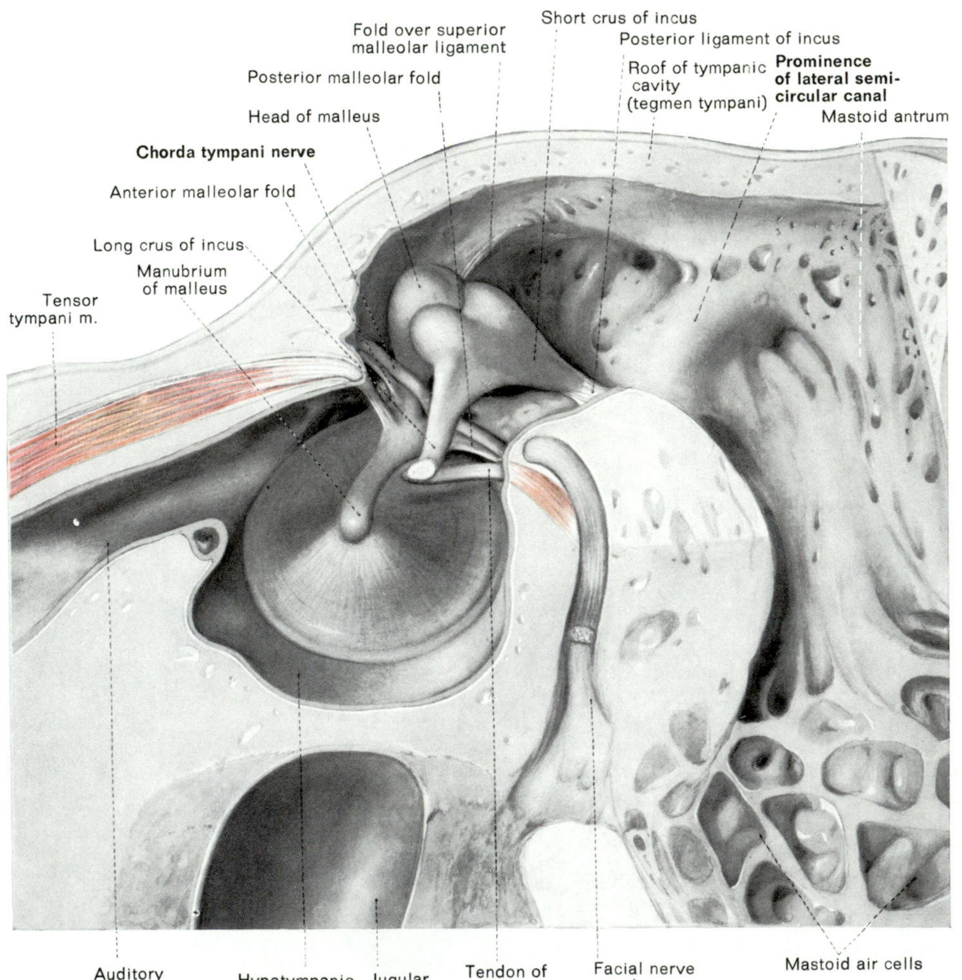
Short crus of incus
Fold over superior malleolar ligament
Posterior ligament of incus
Posterior malleolar fold
Roof of tympanic cavity (tegmen tympani)
Prominence of lateral semicircular canal
Head of malleus
Mastoid antrum
Chorda tympani nerve
Anterior malleolar fold
Long crus of incus
Manubrium of malleus
Tensor tympani m.
Auditory tube
Hypotympanic recess
Jugular fossa
Tendon of stapedius m.
Facial nerve canal
Mastoid air cells

Fig. 792: The Muscles, Chorda Tympani Nerve, and Ossicles of the Right Middle Ear

NOTE: 1) the tendon of the tensor tympani muscle inserting on the manubrium of the malleus and the short tendon of the stapedius as it inserts onto the neck of the stapes close to its articulation with the incus. The tensor tympani draws the manubrium medially, thereby making the tympanic membrane taut. It is innervated by the mandibular division of the trigeminal nerve.

2) the stapedius is the smallest of the skeletal muscles in the body and it is supplied by the facial nerve. It pulls the head of the stapes posteriorly, thereby tilting the base of the stapes in the vestibular window.

3) the direct communication between the nasopharynx, tympanic cavity and mastoid air cells has important clinical meaning, since oro-respiratory tract infections reach the middle ear readily by way of the auditory tube.

Auditory tube; Nerve to tens. tymp. m.; Sup. tymp. art.
Greater petrosal nerve
Lesser petrosal nerve
Geniculate ganglion (facial nerve)
Caroticotympanic n. and art.; Tensor tymp. m.
Tendon, tensor tympani muscle
Stapes; Stapedial membrane
Anterior semicircular canal
Lateral semicircular canal
Mastoid air cells
Facial nerve
Stapedius nerve and muscle
Mastoid air cells
Lesser petrosal nerve
Dura mater; Trigeminal ganglion; Oculomotor nerve
Motor root, trigeminal nerve
Sphenoid sinus; Pterygoid plexus
Optic nerve; Internal carotid artery; Middle cerebral veins
Pterygopalatine ganglion and nerves; Infraorbital nerve
Frontal and lacrimal branches, ophthalmic nerve
Ciliary gangl.; Oculomotor n.; Central retinal artery
Nasociliary n. and branch; Ophthalmic artery
Optic nerve and dural sheath
VII
V
V_1
V_2
V_3
VII
XI
XII
IX
X
F. Batke 1954
Branches of tympanic plexus to cochlear and vest. windows
Facial n.; chorda tymp. n.; Posterior tymp. art.
Auricular br., vagus n.
Nerve to digastric m. (VII); Stylomastoid artery
Communicating nerve (VII and IX)
Tympanic n.; Inf. tympanic art.
Int. jugular v.; Occipital art.
Int. carotid art. and plexus; Inf. ganglion, IX
Inferior gangl., vagus n.; Comm. br. with XII
Superior cervical sympathetic ganglion
Superior laryngeal nerve
Ascending pharyngeal artery
Pharyngeal branches, glossopharyngeal and vagus nerves
Tympanic plexus; Mucosa of auditory tube
Levator veli palatini m.; Ascend. pharyngeal art.; Otic gangl. rami to chorda tymp. n. and auriculotemp. n
Tensor veli palatini muscle and nerve; Nerve to medial pterygoid muscle
Medial pterygoid muscle
Otic ganglion and sympathetic root; Access. mening. art.
Maxillary sinus

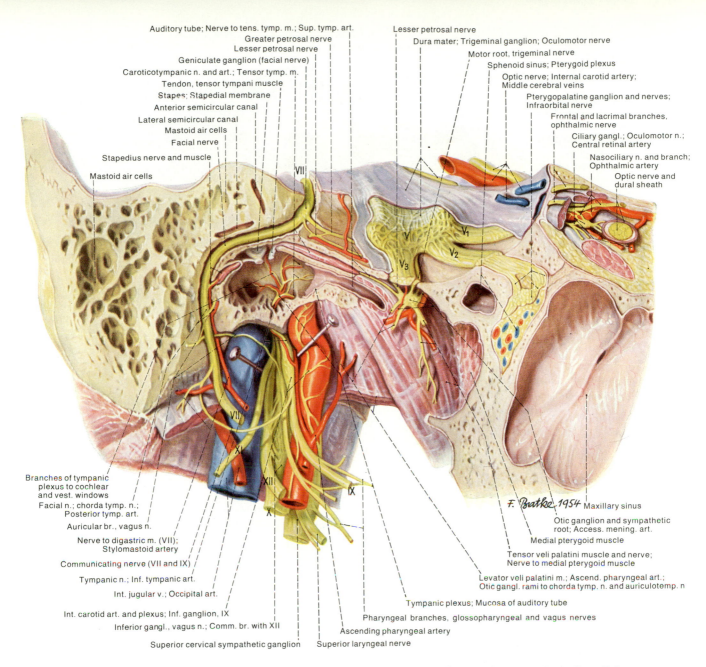

Fig. 793: Lateral View of the Right Facial Canal, Medial Wall of the Tympanic Cavity, the Semicircular Canals and the Auditory Tube

Fig. 794: The Facial, Glossopharyngeal and Vagus Nerves Projected Onto the Temporal Bone, Lateral View

NOTE: 1) the tympanic branch of the glossopharyngeal nerve which, along with sympathetic fibers from the caroticotympanic nerve, contributes to the formation of the tympanic plexus (see Figure 799). From this plexus emerges the lesser petrosal nerve which courses to the otic ganglion.

2) the greater petrosal nerve joining with sympathetic branches of the internal carotid plexus (actually the deep petrosal nerve) to form the nerve of the pterygoid canal.

3) the auricular branch of the vagus nerve which is distributed to the upper surface of the external auricle. to the posterior wall and floor of the external acoustic meatus and part of the lateral (outer) surface of the tympanic membrane.

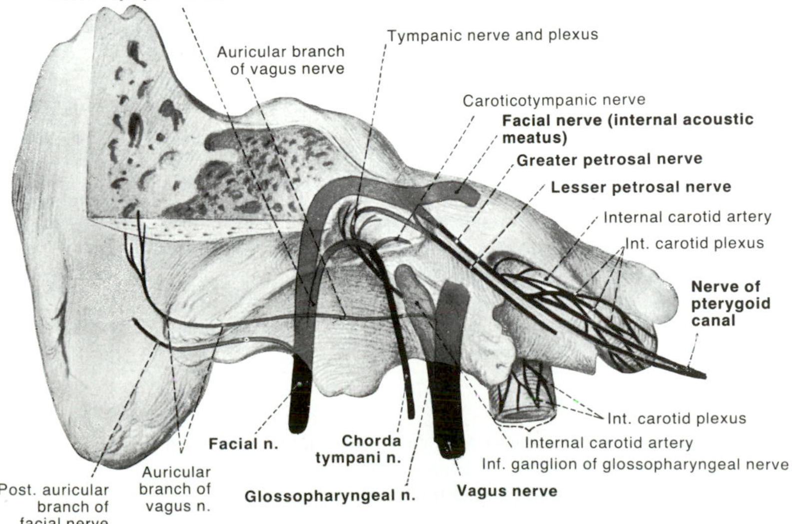

Chorda tympani nerve
Auricular branch of vagus nerve
Tympanic nerve and plexus
Caroticotympanic nerve
Facial nerve (internal acoustic meatus)
Greater petrosal nerve
Lesser petrosal nerve
Internal carotid artery
Int. carotid plexus
Nerve of pterygoid canal
Int. carotid plexus
Internal carotid artery
Inf. ganglion of glossopharyngeal nerve
Vagus nerve
Glossopharyngeal n.
Chorda tympani n.
Facial n.
Auricular branch of vagus n.
Post. auricular branch of facial nerve

Fig. 795: Medial Wall of the Right Tympanic Cavity (Viewed from Lateral Aspect)

NOTE: 1) the tympanic membrane has been removed along with the bony roof of the tympanic cavity. The malleus and incus have also been removed and the tendon of the tensor tympani severed. Observe the stapes with its base directed toward the vestibular window and the stapedius muscle still attached to its neck.

2) several bony markings: a) the prominence containing the lateral semicircular canal, b) the curved prominence of the facial canal with its facial nerve, c) the rounded promontory, which is the thin bony covering over the cochlea and d) the hollow pyramidal eminence from which arises the stapedius muscle.

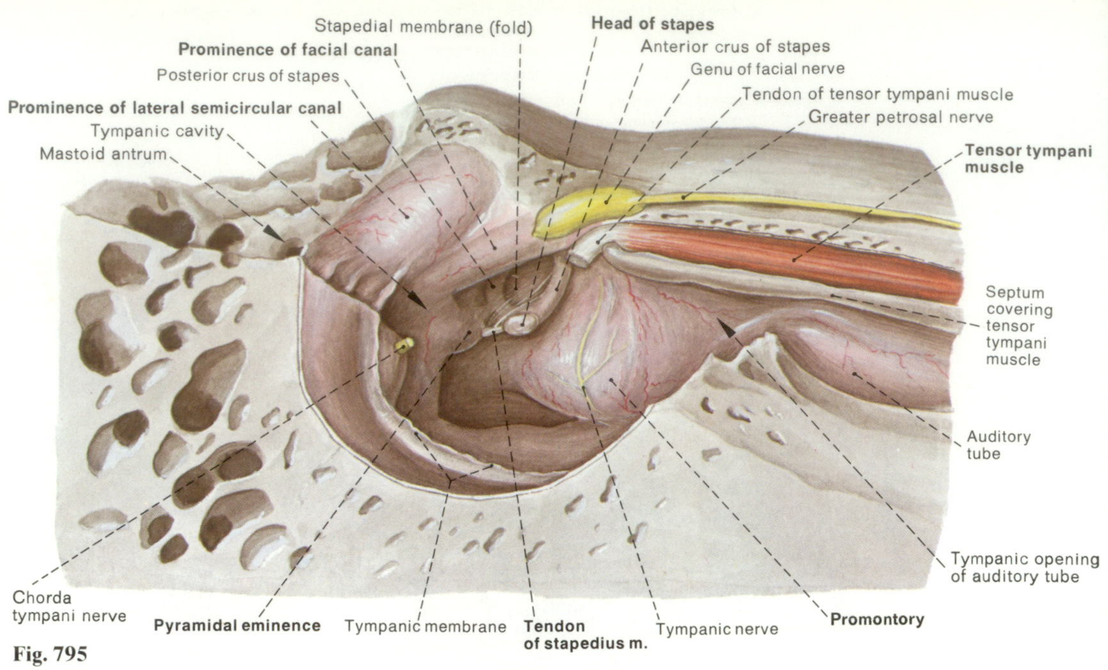

Stapedial membrane (fold)
Head of stapes
Prominence of facial canal
Anterior crus of stapes
Posterior crus of stapes
Genu of facial nerve
Tendon of tensor tympani muscle
Prominence of lateral semicircular canal
Greater petrosal nerve
Tympanic cavity
Mastoid antrum
Tensor tympani muscle
Septum covering tensor tympani muscle
Auditory tube
Tympanic opening of auditory tube
Chorda tympani nerve
Pyramidal eminence
Tympanic membrane
Tendon of stapedius m.
Tympanic nerve
Promontory

Fig. 795

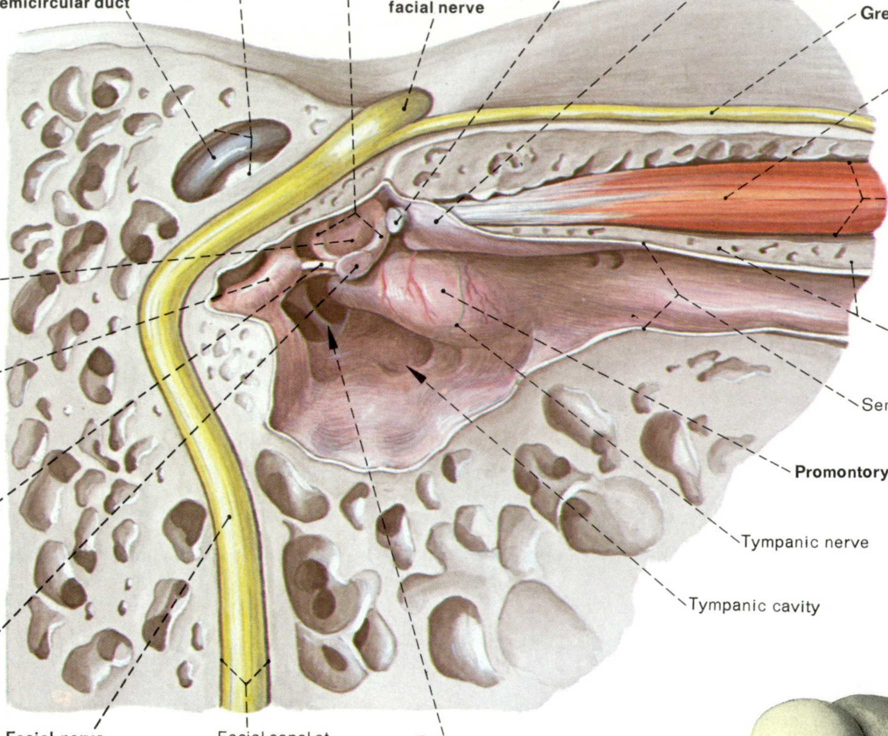

Lateral semicircular canal
Lateral semicircular duct
The crura of the stapes
Genu of facial nerve
Tendon of tensor tympani muscle
Cochleariform process
Greater petrosal nerve
Tensor tympani muscle
Semicanal for tensor tympani muscle
Stapedial membrane
Septum covering tensor tympani m. (septum canalis musculotubarii)
Pyramidal eminence
Semicanal of auditory tube
Promontory
Tendon of stapedius muscle
Tympanic nerve
Tympanic cavity
Head of stapes
Facial nerve
Facial canal at stylomastoid foramen
Tympanic sinus

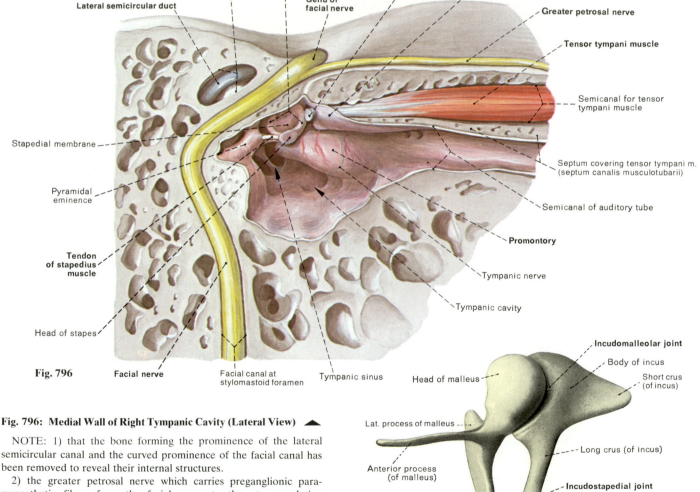

Fig. 796

Fig. 796: Medial Wall of Right Tympanic Cavity (Lateral View) ▲

NOTE: 1) that the bone forming the prominence of the lateral semicircular canal and the curved prominence of the facial canal has been removed to reveal their internal structures.

2) the greater petrosal nerve which carries preganglionic parasympathetic fibers from the facial nerve to the pterygopalatine ganglion and many taste fibers from the soft palate.

Fig. 797: Right Auditory Ossicles ▶

NOTE: that the malleus, incus and stapes articulate to form an osseous bridge for transmission of sound waves from the tympanic membrane to the vestibular window of the inner ear.

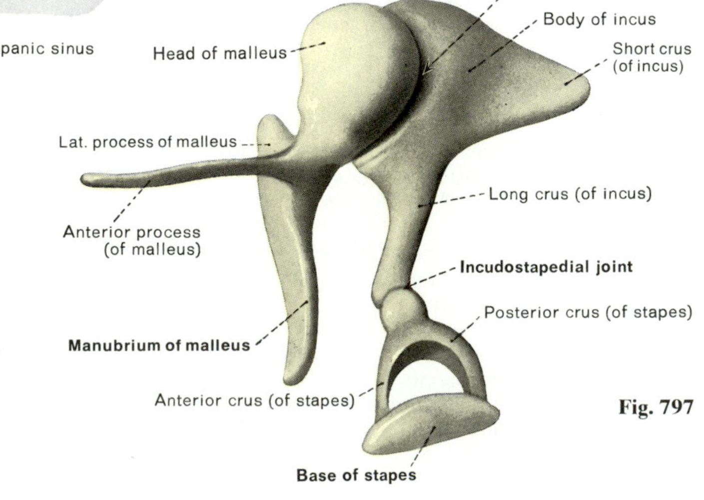

Incudomalleolar joint
Body of incus
Head of malleus
Short crus (of incus)
Lat. process of malleus
Long crus (of incus)
Anterior process (of malleus)
Incudostapedial joint
Posterior crus (of stapes)
Manubrium of malleus
Anterior crus (of stapes)
Base of stapes

Fig. 797

Fig. 798: Medial Wall of the Right Tympanic Cavity Showing the Stapedius Muscle (Lateral View)

NOTE: 1) the lateral aspect of the pyramidal eminence was removed in order to demonstrate the stapedius muscle. Its tendon emerges through the apex of the eminence and inserts on the neck of the stapes. This muscle measures about 4 mm in length and its action on the base of the stapes (pulling it laterally) protects the inner ear from damage caused by loud sounds.

2) the tympanic branch of the glossopharyngeal nerve (IX) coursing along the promontory. This nerve is sensory to the mucous membrane of the middle ear and is also known as the nerve of Jacobson. Its fibers are joined by sympathetic fibers and by branches from the facial nerve to form the tympanic plexus.

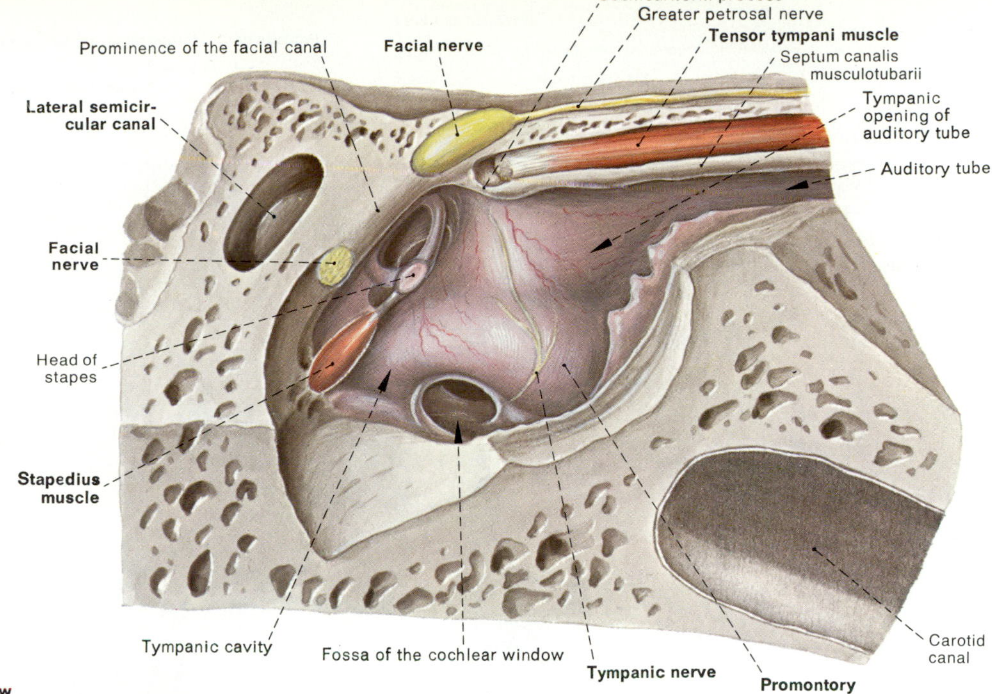

Cochleariform process
Greater petrosal nerve
Tensor tympani muscle
Septum canalis musculotubarii
Tympanic opening of auditory tube
Auditory tube
Prominence of the facial canal
Facial nerve
Lateral semicircular canal
Facial nerve
Head of stapes
Stapedius muscle
Tympanic cavity
Fossa of the cochlear window
Tympanic nerve
Promontory
Carotid canal

Vestibular window
Prominence of facial canal
Fossa of vestibular window
Subiculum of promontory
Facial nerve
Pyramidal eminence
Tympanic sinus
Fossa of cochlear window
Secondary tympanic membrane
Promontory
Tympanic nerve
Branches of tympanic vessels
Bony separation between tympanic cavity and carotid canal

Fig. 799: The Vestibular and Cochlear Windows of the Right Tympanic Cavity

NOTE: the three ossicles and the stapedius and tensor tympani muscles have been removed. The pyramidal eminence has been opened. On the medial wall of the tympanic cavity observe the oval vestibular window and the round cochlear window which communicate with the internal ear. Coursing along the surface of the promontory can be seen the tympanic vessels and nerve.

Fig. 800: The Right Membranous Labyrinth Partially Exposed within the Temporal Bone

NOTE: 1) the membranous labyrinth is in blue, while the roots of the vestibulocochlear nerve are in yellow. The membranous labyrinth lies within the bony labyrinth and generally conforms to it in shape. Between them is found the perilymphatic fluid.

2) within the bony cochlea lie the coils of the cochlear duct. The bony labyrinth consists of the cochlea and the semicircular canals interconnected by the vestibule. In the lateral wall of the vestibule is situated the vestibular window where the stapes has access to the perilymph.

3) the basal end of the cochlear duct communicates with the saccule through the ductus reuniens. The saccule communicates with the utricle and the semicircular canals through the utriculosaccular duct.

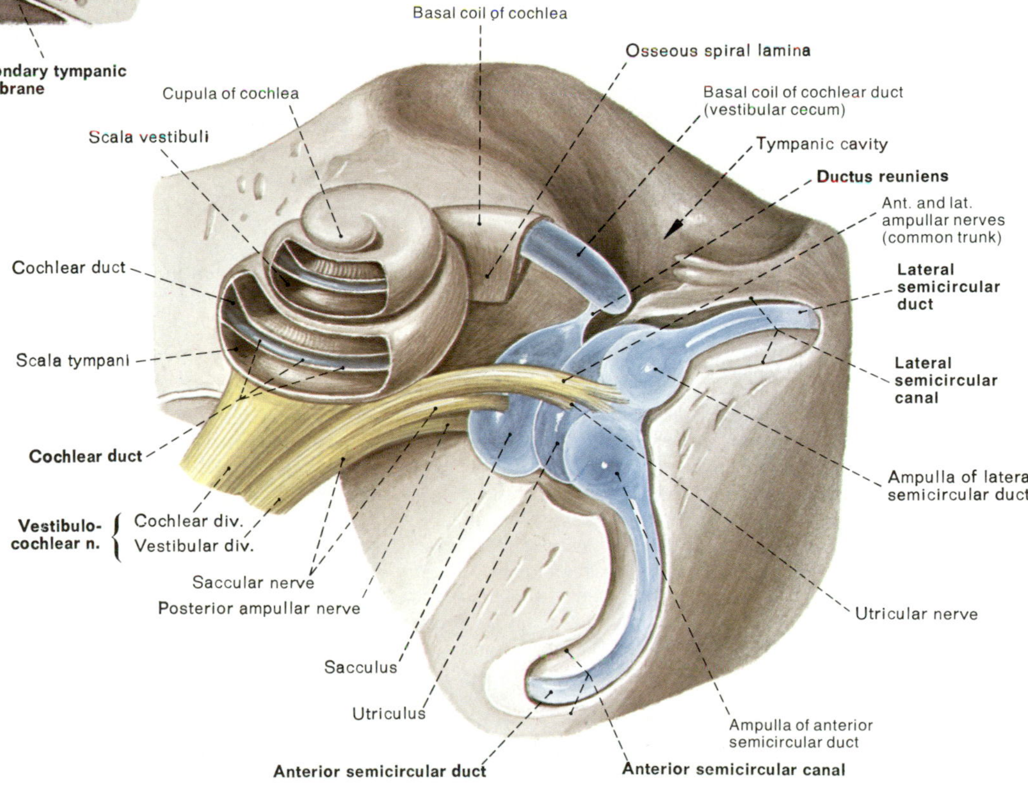

Basal coil of cochlea
Osseous spiral lamina
Cupula of cochlea
Basal coil of cochlear duct (vestibular cecum)
Scala vestibuli
Tympanic cavity
Ductus reuniens
Ant. and lat. ampullar nerves (common trunk)
Cochlear duct
Lateral semicircular duct
Scala tympani
Lateral semicircular canal
Cochlear duct
Ampulla of lateral semicircular duct
Vestibulocochlear n.
Cochlear div.
Vestibular div.
Saccular nerve
Posterior ampullar nerve
Utricular nerve
Sacculus
Utriculus
Ampulla of anterior semicircular duct
Anterior semicircular duct
Anterior semicircular canal

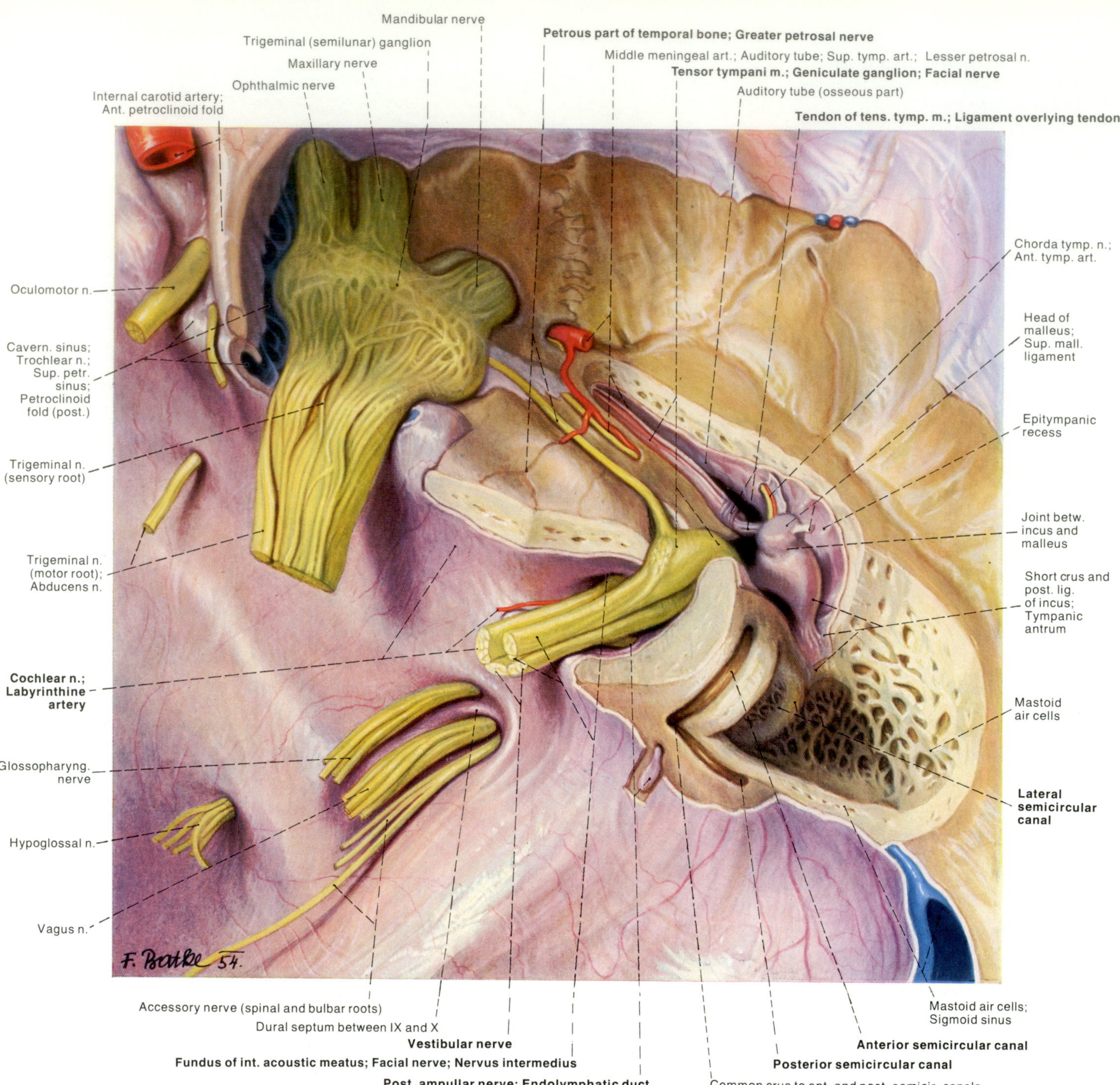
Mandibular nerve
Trigeminal (semilunar) ganglion
Maxillary nerve
Ophthalmic nerve
Internal carotid artery;
Ant. petroclinoid fold
Petrous part of temporal bone; Greater petrosal nerve
Middle meningeal art.; Auditory tube; Sup. tymp. art.; Lesser petrosal n.
Tensor tympani m.; Geniculate ganglion; Facial nerve
Auditory tube (osseous part)
Tendon of tens. tymp. m.; Ligament overlying tendon
Chorda tymp. n.;
Ant. tymp. art.
Head of
malleus;
Sup. mall.
ligament
Epitympanic
recess
Joint betw.
incus and
malleus
Short crus and
post. lig.
of incus;
Tympanic
antrum
Mastoid
air cells
Lateral
semicircular
canal
Oculomotor n.
Cavern. sinus;
Trochlear n.;
Sup. petr.
sinus;
Petroclinoid
fold (post.)
Trigeminal n.
(sensory root)
Trigeminal n.
(motor root);
Abducens n.
Cochlear n.;
Labyrinthine
artery
Glossopharyng.
nerve
Hypoglossal n.
Vagus n.
F. Batke 54.
Accessory nerve (spinal and bulbar roots)
Dural septum between IX and X
Vestibular nerve
Fundus of int. acoustic meatus; Facial nerve; Nervus intermedius
Post. ampullar nerve; Endolymphatic duct
Mastoid air cells;
Sigmoid sinus
Anterior semicircular canal
Posterior semicircular canal
Common crus to ant. and post. semicir. canals

Fig. 801: The Right Tympanic Cavity, Auditory Tube, and Semicircular Canals Opened from the Floor of the Cranial Cavity and View from Above

NOTE: 1) that the orientation of this dissection is the same as that seen in Figure 802. The petrous portion of the right temporal bone (including the arcuate eminence), the tegmen tympani (which is the thin roof of the tympanic cavity), and the mastoid part of the temporal bone have all been dissected from the surface of the floor of the cranial cavity. This dissection has also opened the internal acoustic meatus, exposing within it the facial nerve and its accompanying nervus intermedius, the cochlear and vestibular divisions of the vestibulocochlear nerve, and the labyrinthine artery.

2) the geniculate ganglion at the fundus of the internal acoustic meatus. This sensory ganglion of the facial nerve contains the cell bodies of the primary afferent neurons (both general sensory and the special sense of taste) that course in the facial nerve. Attached to the ganglion are: a) the main trunk of the facial nerve, which descends into the facial canal behind the tympanic cavity (it cannot be seen in this figure, but is shown well in Figures 640, 793, and 796), and b) the greater petrosal nerve, which courses along a shallow groove (called the hiatus of the facial canal) that extends between the geniculate ganglion and the foramen lacerum (through which the greater petrosal nerve reaches the pterygoid canal, see Figure 640).

3) the auditory tube and its tympanic opening. Observe also that the tensor tympani muscle lies in a thin, bony canal just above the auditory tube. Its tendon bends around the cochleariform process to attach to the handle of the malleus. This muscle is supplied by a small nerve, the fibers of which are derived from the mandibular division of the trigeminal nerve but course through the otic ganglion (no synapse) before reaching the muscle (see Figure 793).

4) that the anterior and posterior semicircular canals have been opened, but the bone that forms the lateral semicircular canal is still intact. Observe that behind these canals is found an opening for the small endolymphatic sac called the aqueduct of the vestibule. Also note the direct communication that exists between the tympanic cavity, its antrum, and the mastoid air cells.

Fig. 802: The Structures of the Right Inner Ear and the Facial Nerve, Dissected from Above

NOTE that the upper part of the petrous portion of the right temporal bone has been removed in order to expose the membranous labyrinth, the cochlear duct, the facial and vestibulocochlear nerves and the labyrinthine artery. The orientation of the semicircular ducts and the cochlear duct is nearly the same as that in Figure 803. Observe the geniculate ganglion from which extends the greater petrosal nerve inferomedially. The facial nerve can be seen entering the facial canal lateral and posterior to the ganglion and in relationship to the tympanic cavity.

Fig. 803: The Right Membranous Labyrinth (Medial View)

NOTE: that the right membranous labyrinth and the branches of the vestibulocochlear nerve have been isolated from the bony labyrinth and shown somewhat diagrammatically. Observe the ampullae of the three semicircular ducts, the sacculus, the utriculus and the cochlear duct. The sites of connection of the endolymphatic duct to the utriculus and sacculus is also indicated.

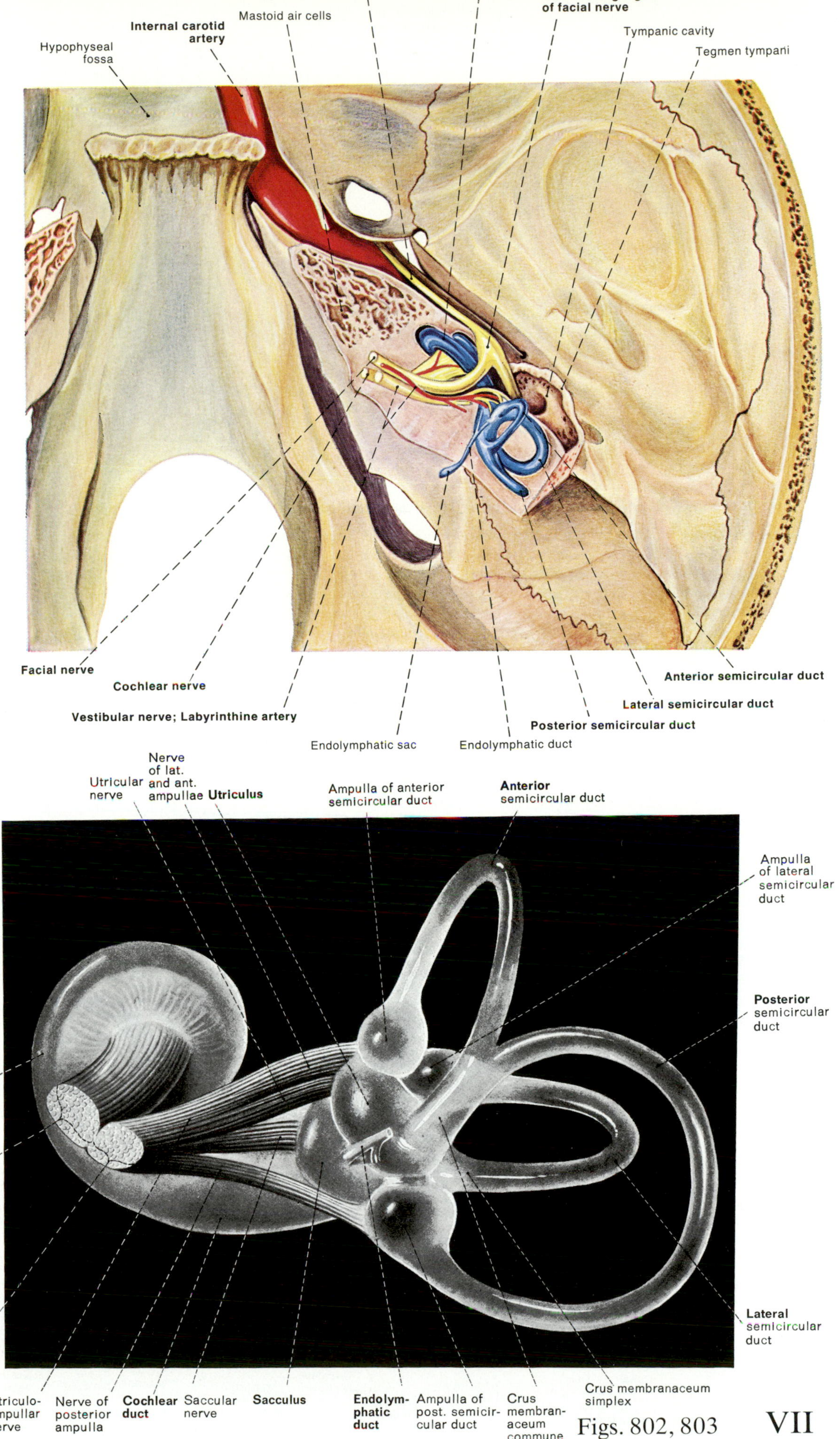

Greater petrosal nerve
Cochlear duct
Geniculate ganglion of facial nerve
Mastoid air cells
Internal carotid artery
Tympanic cavity
Tegmen tympani
Hypophyseal fossa
Facial nerve
Cochlear nerve
Vestibular nerve; Labyrinthine artery
Anterior semicircular duct
Lateral semicircular duct
Posterior semicircular duct
Endolymphatic sac
Endolymphatic duct
Utricular nerve
Nerve of lat. and ant. ampullae
Utriculus
Ampulla of anterior semicircular duct
Anterior semicircular duct
Ampulla of lateral semicircular duct
Posterior semicircular duct
Cochlear duct
Cochlear nerve
Vestibular nerve
Lateral semicircular duct
Utriculo-ampullar nerve
Nerve of posterior ampulla
Cochlear duct
Saccular nerve
Sacculus
Endolymphatic duct
Ampulla of post. semicircular duct
Crus membranaceum commune
Crus membranaceum simplex

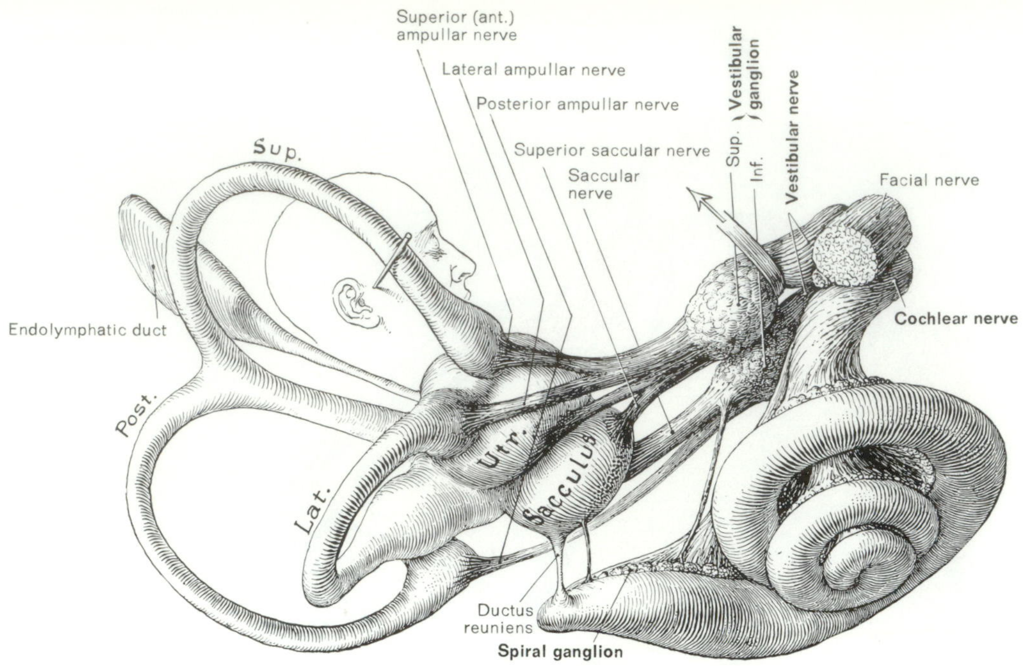

Superior (ant.) ampullar nerve
Lateral ampullar nerve
Posterior ampullar nerve
Superior saccular nerve
Saccular nerve
Sup.
Inf.
Vestibular ganglion
Vestibular nerve
Facial nerve
Cochlear nerve
Endolymphatic duct
Sup.
Post.
Lat.
Utr.
Sacculus
Ductus reuniens
Spiral ganglion

Fig. 804: The Right Membranous Labyrinth Viewed from the Lateral Aspect (Drawing by Max Brödel, 1934)

NOTE: 1) nerve branches from each of the ampullae of the semicircular ducts and from the utriculus and sacculus join to form the vestibular division of the vestibulocochlear nerve. The ganglion containing the sensory nerve cell bodies for this division is the vestibular (Scarpa's) ganglion and it has two parts, superior and inferior.

2) the cochlear division of the vestibulocochlear nerve serves the cochlear duct. Its ganglion, the spiral ganglion, is long and coiled. Observe the close relationship between the facial nerve and the vestibulocochlear nerve within the internal acoustic meatus. (Note that the superior semicircular duct is synonymous with the anterior semicircular duct.)

Fig. 805: Diagram of the Membranous Labyrinth Showing the Stapes

NOTE: 1) the bony structures are shown as cross hatched, while the perilymphatic spaces are diagrammed as white. The membranous labyrinthine duct system (endolymphatic spaces) which contains the receptors is shown as black.

2) the stapes at the vestibular window. The cochlear window, shown close by, in life is covered by the second tympanic membrane.

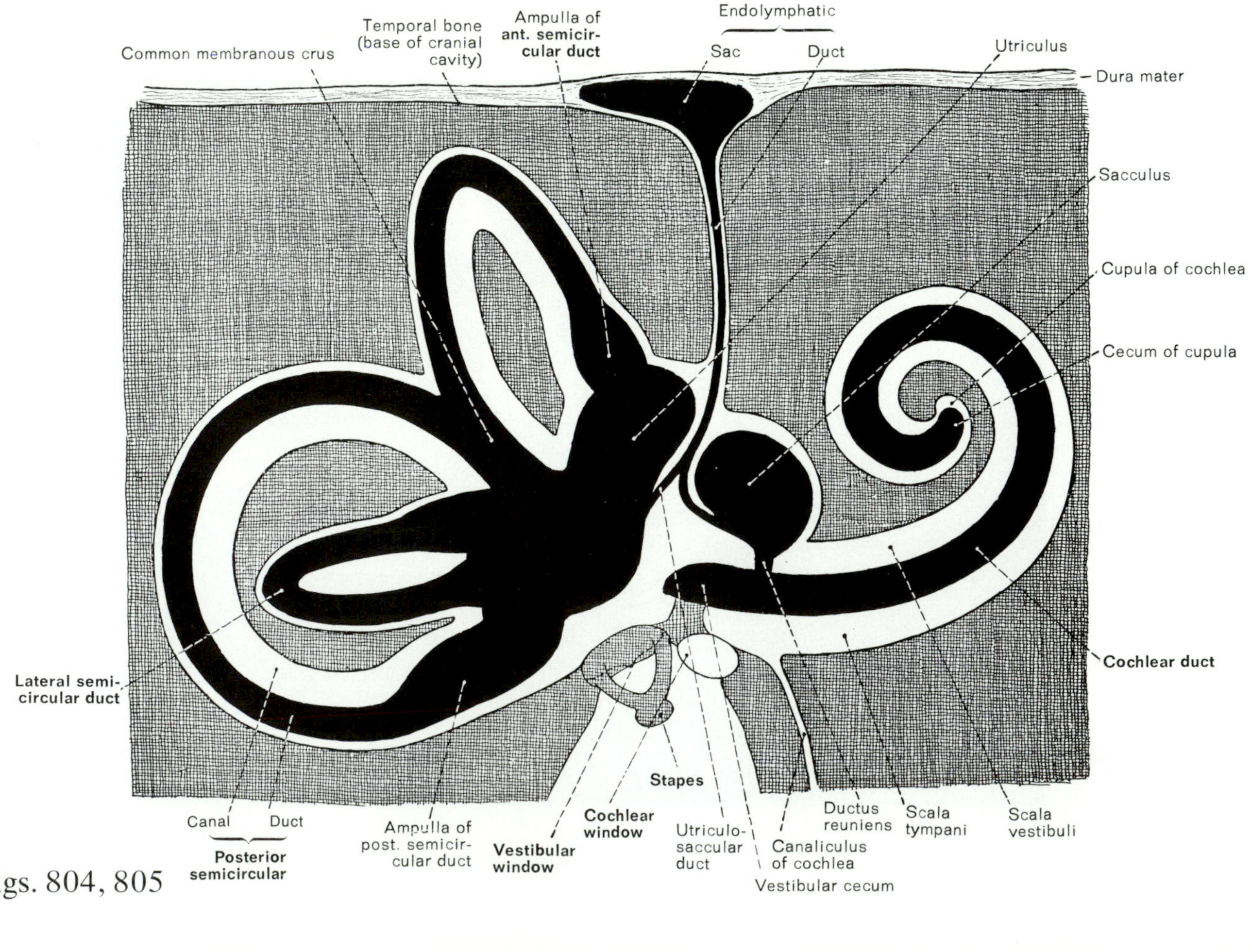

Common membranous crus
Temporal bone (base of cranial cavity)
Ampulla of ant. semicircular duct
Endolymphatic
Sac
Duct
Utriculus
Dura mater
Sacculus
Cupula of cochlea
Cecum of cupula
Cochlear duct
Lateral semicircular duct
Canal
Duct
Posterior semicircular
Ampulla of post. semicircular duct
Vestibular window
Cochlear window
Stapes
Utriculosaccular duct
Vestibular cecum
Canaliculus of cochlea
Ductus reuniens
Scala tympani
Scala vestibuli

Figs. 804, 805

Index

(Numbers refer to Figures in this *Atlas*.)

Abdomen, 215, 231–342
inferior, nerves of, 340
vessels of, 340
muscles of, deep layer, 238
superficial layer, 12
section of, frontal, 215
transverse, at L-2, 331
at L-3, 332
at T-10, 299
between T-12 and L-1, 300
upper, sympathetic trunk of, 224
vagus nerve in, 224
Abdominal cavity, 260, 265, 266, 297, 301–303, 305, 316–318, 320, 321, 399
during pregnancy, 399
organs of, male retroperitoneal, 317
structures of, posterior abdominal retroperitoneal, in lumbar region, 320
viscera of, intact, 257–259
wall of, posterior abdominal, 321
Abdominal region, lateral view, 1
Abdominal wall, above umbilicus, 241
anterior, inner aspect, 242, 243
anastomosis of, epigastric, 240
region of, inguinal, 244–246
section of, transverse, 241
vessels and nerves of, cutaneous, 239
below arcuate line, 241
musculature of, superficial, 231
nerves and vessels of, 232, 239
superficial, 232
posterior, 321, 337
ventral, posterior aspect, 243, 359
Acetabulum, 235, 253, 339, 386, 387, 389, 401
Acromion, 15, 21, 23, 26–31, 33–37, 51, 61, 137, 138, 151, 233, 257, 258, 524, 574, 576
articular surface of, 27, 138
Adminiculum, of linea alba, 240, 242, 243
Air cells, ethmoid, 595, 669, 675, 686–688, 750
mastoid, 595, 640, 695, 750, 751, 778, 779, 790, 792, 793, 801, 802
tympanic, 778
Amnion, 181
cavity of, 181
Ampulla, of canaliculus, 658, 660
of duodenum, 331
of rectum, 312, 314, 343, 377
of semicircular duct, anterior, 800, 803, 805
lateral, 800, 803
posterior, 803, 805
of uterine tube, 310, 316, 318, 374, 375, 377–380
of vas deferens, 340, 352, 357–359, 362
Anastomosis, at elbow joint, 62
at wrist, 76
cubital, 58, 60, 76
epigastric, 240
Angle, costal, 143
costodiaphragmatic, 136, 227, 257
iridocorneal, 680
of eye, lateral, 648, 653
medial, 648, 649, 653
of mandible, 574, 612, 613, 647, 697, 698, 700, 728
of scapula, inferior, 26, 27, 34, 36, 258
lateral, 26, 27
superior, 21, 26, 27, 34, 36, 51, 258, 523
scalenovertebral, 593
sternal, 141, 142, 144
submental, 574
Ankle, 512
Ansa, thyroid, 226
Ansa cervicalis, 579–582, 585, 591, 592, 625
branch of, inferior, 20
root of, inferior, 580, 592
superior, 580, 592, 600
Ansa subclavia, 219, 220, 224, 226, 592, 593, 755
strand of, deep, 228
superficial, 228
Antebrachium, 1, 98, 99
Antihelix, 773, 774
Antitragus, 773
Antrum, cardiac, 216
mastoid, 778, 792, 795
pyloric, 268–270, 331
tympanic, 801
Anus, 258, 313, 314, 344, 367, 368, 370, 387, 388
Aorta, 158, 178, 202, 204, 206, 229, 264, 285, 300, 319, 320, 331, 332, 355
abdominal, 182, 215, 216, 218, 224, 262, 265, 272, 303, 314, 317, 333, 335, 338, 340–342, 383
arch of, 136, 150, 157, 172, 176, 178, 182, 186, 188–190, 192, 194, 195, 208, 209, 212, 214–219, 257, 258, 588, 594, 595, 726, 755
ascending, 136, 172, 176, 181, 188, 190, 195, 207–209, 212, 214, 218, 224, 586, 588
bifurcation of, 303, 364
descending, 159, 214, 219, 230, 299
groove for, 158
hiatus of, 170, 214, 216, 285, 317, 320, 333–335
isthmus of, 176
thoracic, 182, 185, 216–218, 224, 573
descending, 195
Aperture, vena caval, in diaphragm, 299
Aponeurosis, bicipital, 47, 49, 53, 54, 65, 67, 71, 98
extensor, dorsal digital, 123, 124, 469
lingual, 736
lumbodorsal, 320
of gastrocnemius muscle, 473
of latissimus dorsi muscle, 524
of levator labii superioris muscle, 657
of levator palpebrae superioris muscle, 656, 657
of oblique muscle, external, 12, 151, 232, 235, 236, 239–241, 244–246, 248, 251, 253, 255
internal, 233, 238, 239, 241
of rectus abdominis muscle, 241
palatal, 615, 647, 731, 747
palatine, 615, 647, 731, 747
palmar, 106, 120, 121
plantar, 492–494, 515, 516
Appendices epiploicae, 260, 307, 317, 318, 354, 399
Appendix, vermiform, 257, 258, 297, 301, 302, 304–306, 309–311, 374, 378
blood supply to, 309
location of, variations in, 310
orifice of, 306
vesicular, 372, 379–381
Appendix epididymis, 247, 249, 253
Appendix fibrosa hepatis, 273, 282
Appendix testis, 247–249, 253, 363
Arachnoid, 343, 570, 571, 573, 633, 679
Arachnoid granulations, 633–635
Arc, tendinous, 365, 376, 386
Arch(es), aortic. See *Aorta, arch of.*
costal, 233, 242
iliopectineal, 318
jugular venous, 585, 726
lumbocostal, lateral, 335
medial, 219, 335
mandibular, 699
of atlas, anterior, 531, 540, 690, 726
posterior, 528, 531, 536, 540
of axis, 532–534
posterior, 536
of cricoid cartilage, 726, 735, 761
palatoglossal, 729, 731, 732, 735, 737, 743
palatopharyngeal, 726, 729, 731, 732, 735, 737, 751, 755
palmar, deep, 61, 111, 113, 182
superficial, 62, 110, 111, 113, 182, 183
palpebral arterial, inferior, 654, 655
superior, 655
plantar, 403, 497
tendinous, of psoas major muscle, 333
of soleus muscle, 434, 475
venous, dorsal, of foot, 468
vertebral, 543, 552, 553
Arch(es), vertebral, *continued*
inferior, 548
lamina of, 547, 549, 554
lumbar, 565
of 7th cervical vertebra, 542
pedicles of, 557, 573
zygomatic, 606, 608, 614, 615, 624, 625, 635, 646, 748
Area, cardiac, 153
of superficial cardiac dullness, 170
interpleural, inferior, 153
superior, 153
thymic, 153
Area cribrosa, 325
Areola, 4, 257
Arm, 1, 37–40, 42, 47, 48, 53–64, 67, 68
anatomy of, surface, 37
arteries of, 54, 56, 58, 60, 62, 67, 68, 73
arteriogram of, 73
muscles of, 37, 47–49, 51, 53, 55, 57, 59
nerves of, 38–40, 42, 47, 48, 54, 56, 58, 60, 67, 68
cutaneous, distribution of, 39
upper, deep fascia, 47, 48
veins of, 40, 42, 47, 54
Arterial supply, of genitalia, female, 367, 368, 378, 381, 388
of menisci, 454
Arterial system, coronary, 196, 197
Arteriogram, brachial, 73
of the hand, 113
Artery (arteries), alveolar, inferior, 618, 623–625, 694, 737, 739, 740
posterior superior, 618, 624
angular, 618, 622–625, 654–657
nasal branch of, 654
anococcygeal, 346, 371
appendicular, 296, 298, 309
arcuate, of foot, 182, 402, 471
of kidney, 325
auricular, deep, 624, 625, 694
posterior, 618, 622–625, 787
occipital branch of, 527, 577, 580, 618
axillary, 17, 20, 22, 54, 56, 62, 63, 182, 185, 187, 188, 227, 581, 582, 596
basilar, 569, 595, 643, 644
brachial, 17, 18, 21, 22, 51, 54, 56, 58, 60, 62, 64, 67, 68, 70–73, 82, 98, 182
profunda, 54, 62, 64, 73, 182
brachiocephalic, 172, 177, 182, 186, 189, 194, 195, 216–218, 228, 586, 588
bronchial, 184, 188, 218, 220, 229
buccal, 618, 623–625
bulbovestibular, 369
cardiac, anterior, 190, 191
caroticotympanic, 793
carotid, 674
common, 17, 20, 182, 188, 216, 224, 228, 576, 579–585, 588, 591, 593, 595, 596, 608, 618, 621, 625, 754, 756
left, 151, 152, 172, 177, 178, 185, 186, 189, 194, 195, 212, 214, 216, 217, 219, 586
right, 172, 217, 594
external, 182, 579–582, 588, 595, 600, 608, 618, 623, 624, 693, 737, 740, 754, 756
branches in skull, 618
internal, 182, 581, 582, 588, 595, 618, 621, 623, 637, 638, 640–644, 669, 670, 672, 685, 693–695, 737, 740, 754, 755, 788, 793, 794, 801, 802
cerebellar, inferior, anterior, 595, 643, 644
middle, 643
posterior, 643, 644
superior, 595, 643, 644
cerebral, anterior, 595, 643, 644
middle, left, 595, 643, 644
posterior, 595, 643, 644
cervical, ascending, 580–582, 596
deep branch, 596

Artery (arteries), cervical, *continued*
deep, 218, 528, 593, 596
superficial, 11, 579–582, 592, 593, 596
transverse, 579, 588, 592, 593, 596
deep branch, 528, 579, 580
superficial branch, 528, 579, 580, 582, 586
choroidal, 643, 644
ciliary, 674
anterior, 678
posterior, 678
long, 678
circumflex, femoral, lateral, 182, 402, 421
ascending branch, 402
descending branch, 402, 421
medial, 182, 402, 421, 422
acetabular branch, 428
transverse branch, 421, 428
humeral, anterior, 17, 22, 54, 62, 182
posterior, 17, 19, 22, 24, 60, 62, 63, 182
superficial branches, 42
iliac, deep, 239, 245, 317, 338, 342, 376, 418
lateral, 422
ascending branch, 422
descending branch, 422
superficial, 11, 232, 239, 240, 245, 246
scapular, 19, 22, 24
colic, 320
left, 296, 297, 303, 304, 317, 320, 321, 332
middle, 277, 285, 296, 298, 302
right, 285, 296–298
communicating, anterior, 595, 643, 644
posterior, 595, 643, 644
conjunctival, 678
coronary, complete arterial system, 196
distribution of, balanced, 197
left dominant, to heart wall, 198
right dominant, to heart wall, 198
left, 190, 191, 196–201, 215
atrial branch, 196–199
atrioventricular branch, 197, 198
branch to atrioventricular node, 198
branch to pulmonary trunk, 196
circumflex branch, 190–192, 194–200
diagonal branch, 196–199
distribution of, dominant, to heart wall, 198
interventricular branch, anterior, 190, 191, 194–199
posterior, 198
lateral branch, 198, 199
marginal branch, 192, 196–199
opening of, 212
posterolateral branch, 197, 198
septal branch, anterior, 196–199
posterior, 198
right, 190–192, 194–199, 201, 207, 214
atrial branch, 196–199
atrioventricular branch, 197, 199
branch to atrioventricular node, 196, 197, 199
branch to pulmonary trunk, 196–199
branch to sinoatrial node, 196–199
cardiac branch, anterior, 191
distribution of, dominant, to heart wall, 198
interventricular branch, posterior, 191, 192, 195–197, 199
marginal branch, 196–199
opening of, 212
posterolateral branch, 196, 197, 199
septal branch, posterior, 197, 199
ventricular branch, 196–199
posterior, 200
cremasteric, 239, 246, 254, 347, 365
cystic, 265
deep, of penis, 248, 348, 361
digital, dorsal, of hand, 112, 122
of foot, 182, 402, 465, 471, 515
palmar, common, 106, 110, 113
proper, 106, 110, 111, 113, 122–124
proper, of foot, 493, 496
dorsal, of clitoris, 245, 369
of penis, 347, 355, 366
dorsalis pedis, 182, 402, 465, 471, 515
epigastric, inferior, 11, 149, 182, 239, 243, 245, 253, 256, 297, 317, 318, 321, 338, 340, 342, 352, 364, 376, 383
pubic branch, 243, 364
superficial, 11, 232, 239, 240, 245, 246, 414
superior, 11, 149, 230, 243, 255
esophageal, 218, 220, 573
ethmoid, anterior, 638, 670, 674, 676, 690, 693
posterior, 674, 676

Artery (arteries), *continued*
facial, 182, 581, 582, 585, 586, 597, 599, 600, 618, 622–625, 727, 737, 754
transverse, 618, 622, 624, 654
femoral, 182, 335, 341, 342, 402, 414, 418, 421, 422, 442, 486
deep, 182, 342, 402, 418, 421, 422
gastric, left, 182, 265, 266, 270–273, 277, 285, 300, 317, 318, 341
right, 265, 270, 271, 300
short, 266, 270
gastroduodenal, 265, 266, 271, 331
gastroepiploic, 273
left, 265, 271
right, 266, 270–272, 316
genicular, descending, 182, 402, 414, 421, 422, 443, 486
articular branch, 418, 422
lateral, inferior, 402, 403, 443, 454, 465
superior, 402, 441–443, 465
medial, inferior, 402, 403, 422, 443, 454, 479
superior, 402, 403, 422, 443, 465, 476, 487
middle, 403, 442, 443
anterior branch, 455
posterior branch, 454
gluteal, inferior, 182, 364, 403, 428, 430
superior, 182, 314, 342, 364, 428, 430
helicine, 361
hemorrhoidal, 346
hepatic, common, 182, 265, 266, 270, 271, 285, 300, 317
left, 271
middle, 271
peritoneal fold over, 300
proper, 270, 272, 277, 285
right, 271, 300
hypogastric, 364. See also *Artery, iliac, internal*
ileal, 296–298
ileocolic, 296–298, 309
iliac, common, 178, 182, 304, 314, 316, 317, 320, 321, 338, 340–342, 354, 355, 364, 376, 382, 569
external, 149, 178, 182, 243, 314, 317, 338, 340, 342, 352, 355, 359, 364, 376, 377, 383, 418
internal, 178, 182, 314, 321, 342, 355, 364, 382, 418
iliolumbar, 321, 338, 342, 364
infraorbital, 618, 622–625, 655, 656, 676
branch of, 654
intercostal, 182, 184, 185, 195, 214, 217, 218, 227, 229, 230, 239, 240, 243, 255, 299, 526, 573
anterior, 149, 596
spinal branch of, 226, 569, 572
supreme, 218, 219, 226, 593, 596
interlobar, of kidney, 330, 332
interosseous, anterior, 62, 70, 73, 76, 77, 100, 105, 113
common, 62, 68, 70, 73, 99
posterior, 62, 70, 73, 76, 78, 100
recurrent, 62, 72, 73, 78
interventricular, anterior, 204
jejunal, 296–298, 332
labial, anterior, 244, 245
inferior, 618, 622
posterior, 244, 245, 371
superior, 618, 690
labyrinthine, 595, 638, 642–644, 801, 802
lacrimal, 653, 659, 670, 672, 674
palpebral branch, lateral inferior, 656
lateral superior, 656
laryngeal, inferior, 588, 758
internal, 588
superior, 581, 593, 600, 741, 744, 748, 755, 756, 758, 770
lingual, 182, 580, 600, 618, 623, 625, 727, 737, 739, 740, 754, 756
deep, 740
suprahyoid branch, 579
lumbar, 182, 317, 321, 332, 338, 340, 342, 354
malleolar, lateral, 402, 403, 465, 479
medial, 402, 403, 471
mammary, external, 11
marginal, 296, 298
masseteric, 623, 624, 740
maxillary, 182, 618, 624, 640, 693, 694
median, 62, 70, 100
meningeal, accessory, 793
anterior, 638, 670, 674, 676, 690
middle, 182, 618, 624, 625, 634, 635, 638, 642, 670, 694, 801
anterior branch, 634
frontal branch, 635, 638

Artery (arteries), meningeal, *continued*
orbital branch, 635
parietal branch, 635
petrosal branch, 638
posterior branch, 634
posterior, 754
mental, 618, 622–624
labial branches, 740
mesenteric, inferior, 182, 285, 303, 304, 314, 316, 317, 320, 321, 338, 339, 355, 364
superior, 182, 218, 226, 266, 272, 285, 286, 296–298, 303, 316, 317, 321, 331, 332, 338, 339, 341
metacarpal, dorsal, 112
palmar, 111
metatarsal, dorsal, 182, 402, 465, 471, 516
plantar, 493, 496, 497, 516
perforating branch, 516
musculophrenic, 149, 230
mylohyoid, 624, 693, 694, 739, 740
nasal, dorsal, 622, 623, 654, 674
external, 654–657, 690
nasopalatine, 690, 692, 693
obturator, 182, 243, 314, 340, 342, 359, 364, 376, 383, 421, 422
acetabular branch, 421
anterior branch, 421
pubic branch, 243, 364
occipital, 527, 528, 577, 579, 581, 582, 618, 622–625, 638, 640, 755, 793
descending branch, 527, 528
mastoid branch, 527, 528, 582, 618
sternocleidomastoid branch, 581, 618
of ductus deferens, 254, 340, 355
of foot, 471, 493, 496, 497
of forearm, 67, 68, 70–72
of hand, 112, 113
of nasal septum, 690
of palm, 62, 106, 110, 111
of penile bulb, 346, 366
of penis, 248, 348, 354, 361
of pterygoid canal, 640, 693, 746
of tongue, apex of, 740
of upper extremity, 62
ophthalmic, 638, 642, 643, 670, 672, 674, 676, 793
palpebral branch, inferior medial, 656
ovarian, left, 285, 318, 381, 383
right, 318
palatine, ascending, 618, 625, 693, 740, 746, 747, 754
dorsal lingual branch, 746, 747
descending, 618, 640, 693
greater, 690, 692, 693, 731
lesser, 693, 731
tonsillar branch, 746
palmar digital, common, 106, 110, 113
proper, 106, 110, 111, 113, 122–124
to thumb, 106, 111
palpebral, medial, 622
pancreaticoduodenal, inferior, 266, 296, 303
superior, 266, 272
perforating, 182, 402, 403, 422, 428, 430, 442, 443, 486
anterior, 232
pericardiacophrenic, 149, 175, 184, 186–188, 230, 573, 593
perineal, 346
peroneal, 182, 403, 442, 443, 476, 479, 490, 491
perforating branch, 402, 465, 471
pharyngeal, ascending, 618, 640, 740, 747, 754, 793
phrenic, inferior, 216, 243, 265, 266, 300, 321, 338, 341
superior, 230
plantar, 471, 496
lateral, 403, 496, 497, 515
medial, 182, 403, 496, 497, 515
popliteal, 182, 403, 426–428, 441–443, 454, 476, 479, 487, 489
princeps pollicis, 105, 111, 113, 121
profunda brachii, 56, 58, 60, 73
deltoid branch, 58, 60
pudendal, external, 11, 245, 246, 414
superficial, 244
internal, 314, 342, 346, 364, 371, 383, 428, 430
pulmonary, 136, 164, 165, 176, 184, 185, 187–190, 192, 194, 195, 204, 208, 209, 212, 214, 215, 227
radial, 41, 62, 67, 68, 70–73, 99–101, 105, 108, 110–113, 120, 182
carpal branch, 62, 105, 111
dorsal, 112

Artery (arteries), radial, *continued*
carpal network of, dorsal, 112
palmar branch, superficial, 67, 68, 70, 110–112, 120
perforating branch, 105, 112
radial collateral, 58, 60, 62, 67, 70, 72, 73, 76, 78, 98
radial recurrent, 62, 67, 68, 70, 72, 73
radialis indicis, 112, 113
rectal, inferior, 314, 315, 346
middle, 314, 315, 364, 383
superior, 303, 304, 314, 316, 317, 321, 382, 383
renal, 178, 182, 226, 272, 285, 321–323, 325, 326, 330–332, 339, 341, 569
accessory, 321
retinal, central, 678, 793
sacral, lateral, 342, 354, 364
middle, 182, 303, 314, 317, 320, 321, 338, 342, 355, 364, 377, 382, 383
scaphenous, 402
scapular, descending, 20
dorsal, 20, 527
scrotal, anterior, 246
posterior, 246, 346
sigmoid, 303, 304, 316, 317, 321
spermatic, 182, 248, 285
sphenopalatine, 618, 625, 692, 693
spinal, anterior, 226, 569, 572, 573, 643, 644
posterior, 572, 643
splenic, 182, 265, 266, 271, 272, 275, 278, 285, 300, 317
stapedius, 640, 789
sternocleidomastoid, 618
stylomastoid, 618, 624, 640, 793
subclavian, 17, 20, 177, 182, 220, 569, 579–582, 592, 593, 595, 596
left, 151, 172, 176, 177, 185, 188, 189, 194, 195, 215, 216, 218–220, 224, 586, 588, 593, 594, 755
right, 18, 172, 216–219, 224, 586, 594, 596, 755
subcostal, 182, 218, 220, 300
sublingual, 625, 737, 740, 746, 747, 756
submental, 580–582, 597, 599, 600, 618, 624, 625, 727
subscapular, 17, 22, 182
acromial branch, 24
supraorbital, 622, 623, 654–657, 670, 672, 674
suprarenal, 320, 321
inferior, 318, 320, 322, 323
middle, 320, 322, 323
superior, 322, 323
suprascapular, 22, 24, 579–582, 592, 593, 596
supratrochlear, 622–625, 654, 655, 657, 674
sural, 403, 441–443
tarsal, 402, 471
temporal, deep, 618, 624, 625
middle, 618, 623–625
superficial, 604, 618, 622–625
testicular, 243, 254, 255, 317, 320, 321, 332, 338, 340, 347
thoracic, internal, 11, 17, 20, 149, 182, 184, 186–188, 218, 220, 243, 581, 582, 585, 592–594, 596
perforating branches, 13
lateral, 11, 13, 14, 17, 19, 20, 182
supreme, 17, 182
thoracoacromial, 13, 15, 17, 20, 24, 42, 182, 579, 581, 582
acromial branch, 582
thoracodorsal, 7, 13, 14, 17, 19, 20, 22
thymic, 175
superior, 592
thyroid, inferior, 226, 581, 582, 588, 591–593, 596, 754–756, 758
cervical branches, ascending, 592
esophageal branch, 219
superior, 152, 579–582, 588, 592, 593, 597, 600, 618, 625, 726, 754, 756, 766
cricothyroid branch, 592, 593
tibial, anterior, 182, 402, 403, 442, 443, 465, 466, 471, 479, 489–491
posterior, 182, 403, 434, 442, 443, 476, 479, 490, 491, 493, 496, 497
tibial recurrent, anterior, 402, 442, 465
posterior, 403, 443
to sternoclavicular joint, 592
tonsillar, 758
tympanic, anterior, 624, 625, 694, 801
inferior, 640, 754, 793
posterior, 640, 787, 789, 793
superior, 638, 642, 793, 801
ulnar, 62, 67, 68, 70, 71, 73, 99–101, 105, 106, 108, 110, 111, 113, 120, 182

Artery (arteries), *continued*
carpal branch, dorsal, 62, 67
palmar, 62
palmar branch, deep, 113
ulnar collateral, inferior, 54, 56, 58, 60, 62, 67, 70, 71, 73, 78, 98
superior, 54, 56, 62, 64, 67, 70, 71, 73, 98
ulnar recurrent, 60, 62, 68, 70, 71, 73, 76, 99
umbilical, 178, 181, 253, 255, 256, 259, 261, 319, 352, 355, 364, 377, 382
obliterated, 243, 340, 359
uterine, 342, 381–383
ovarian branch, 381
tubal branch, 381, 382
uteroplacental, 181
vaginal, 381–383
vertebral, 17, 20, 151, 182, 528, 529, 569, 581, 582, 591–593, 595, 596, 638, 642–644, 685
groove for, 530, 534, 543
meningeal branch, 638
spinal branch, 569
vesicle, inferior, 243, 355, 364, 381, 382
vaginal branch, 381, 382
superior, 240, 243, 255, 340, 355, 383
zygomaticofacial, 656, 657
zygomatico-orbital, 618, 623, 624, 645–657
zygomaticotemporal, 622
Articular capsule, cricothyoid, 757, 759
of atlas, 539
of axis, 539
of elbow joint, 82, 91, 98
of hip joint, 394–396, 437, 438
of intervertebral joint, 526, 529
of knee joint, 446, 448, 459
of metacarpophalangeal joint, 114, 129
of shoulder joint, 24, 30, 32–35, 49–51
of talocalcaneonavicular joint, 515
of tibiofibular joint, 459
temporomandibular, 606, 608, 612, 614, 647, 748
Articulations, costosternal, 147
costovertebral, 551
of rib, head of, 554
Aryepiglottic fold, 746, 751, 753, 755, 764, 767, 769, 771
Atlas, 151, 530, 531, 533–540, 543, 565
arch of, anterior, 531, 540, 690, 726
posterior, 528, 531, 533, 536, 540
body of, 530, 531, 533–536, 540
facet of, articular, inferior, 531, 533
superior, 530, 531, 533, 537
foramen of, transverse, 530, 531
vertebral, 530, 531
fovea for dens, 530, 533, 537
ligament of, transverse, 537, 540, 722, 726
mass of, lateral, 530, 531
process of, transverse, 258, 526, 529–531, 534, 539, 594
tubercle of, anterior, 530, 531, 594
posterior, 523, 526, 530, 531, 540
Atrioventricular bundle, 191, 210, 213
left branch, 210–213
right branch, 213
Atrioventricular opening, 213, 215
left, 206, 212
right, 202–204, 214
Atrium, of heart, left, 136, 158, 159, 181, 190, 192, 195, 201, 206, 207, 211, 214, 215, 257
right, 136, 150, 158, 159, 181, 190, 192, 194, 195, 201–203, 257
Auditory tube. See *Tube, auditory*
Auricle, of heart, left, 150, 176, 188, 190, 192, 194, 195, 204, 206, 208, 210, 212, 215
right, 136, 176, 188, 190, 194, 202, 208, 209
Axilla, 14, 19, 20, 63
nerves and vessels of, 37
Axis, 532–540, 543
arch of, anterior, 531–534, 540, 690, 726
posterior, 528, 531, 536, 540
body of, 532, 533, 535–540
capsule of, articular, 539
dens of, 532, 533, 538, 540, 543, 690
foramen of, transverse, 532, 533
process of, articular, 532, 533
inferior, 532, 533
superior, 532
odontoid, 532, 533, 538, 690
spinous, 528, 533, 537
bifid, 532
transverse, 529, 532, 533
surface of, articular, posterior, 532
superior, 533
Azygos system of veins, 221

Back, cross section of, 522
muscles of, 523–526
deep, 524–526
intermediate layer, 524
region of, lumbosacral, 521
vessels and nerves of, 527, 528
Bicuspid tooth, 699
Bifurcation, aortic, 303, 364
tracheal, 157, 168, 214–216, 768
Bile duct, common, 265, 268, 272, 283, 284, 286–293, 300, 317, 331
Bladder, urinary, 240, 243, 253, 255–257, 305, 310, 316, 317, 319, 321, 335, 340, 353–355, 359–361, 363, 372–374, 378, 382, 383, 385, 399
apex of, 318, 343, 352, 356, 358
base of, 357
body of, 352
fundus of, 352
surface of, posterior, 358
uvula of, 356
Blood vessels. See *Artery (Arteries); Vein(s); Vessel(s)*
Bone(s), calcaneus, 501, 502, 504, 508, 511
tuberosity of, 413, 467, 473, 474, 477, 492, 494, 495, 503, 509, 511, 513, 514
capitate, 118, 120, 129, 131–135
carpal, 118
clavicle, 6, 16, 20–22, 24, 25, 29, 31, 34, 36, 49, 51–53, 55, 57, 61, 136–138, 140, 144, 148, 150–152, 176, 184–186, 188, 227, 231, 233, 256–258, 524, 574–576, 579, 582, 583, 586
border of, anterior, 138
extremity of, acromial, 33, 34, 139
sternal, 138, 139
foramen of, nutrient, 138
coccyx, 25, 344, 370, 387, 393, 544, 546, 561, 562
body of, 377
horn of, 561, 562
vertebral column, part of, 544–546
cuboid, 501–504, 510, 511, 515
cuneiform of foot, 462, 501–504, 510, 513
ethmoid, 601, 687–689
lamina of, orbital, 610, 661, 662, 687, 688
plate of, cribriform, 645, 683, 684, 686, 687, 689
perpendicular, 613, 683, 687–689
process of, uncinate, 688
femur, 25, 423, 431–434, 445, 447–450, 457, 458, 486, 487
body of, 432, 446, 457
condyle of, lateral, 433, 442, 446, 449–451, 457–459
medial, 417, 419, 420, 433, 442, 446, 448, 450, 451, 457, 459, 464
end of, distal, 449, 450
epicondyle of, 432, 433
lateral, 412, 446, 449, 450, 457
medial, 446, 450, 457
foramen of, nutrient, 434
fossa, intercondylar, 433, 442, 448, 450, 457, 459
head of, 401, 433, 438, 439
fovea of, 401, 432, 433
line of, epiphysial, distal, 459
linea aspera, 430, 433, 440, 448, 486
neck of, 401, 432, 433, 436
of fetus, 400
popliteus muscle, notch for attachment of, 449
surface of, articular, 450
meniscofemoral, 448, 459
patellar, 432, 442, 446, 450, 451
popliteal, 433, 440, 443, 448, 477
fibula, 447, 448, 451, 452, 457, 482, 483, 485, 489, 491, 505, 508, 512
body of, 457
crest of, medial, 477, 478, 490
end of, proximal, 452
head of, 442, 444–446, 452, 457, 459, 463, 466, 467, 480, 485
apex of, 452, 457
ligament of, anterior, 446
posterior, 448
margin of, anterior, 490
interosseous, 490
posterior, 485, 490
of fetus, 400
process of, styloid, 485
surface of, lateral, 485, 490
posterior, 490
frontal, 601, 635, 645, 659, 662, 664, 667, 675, 689
process of, zygomatic, 671, 689

Bone(s), frontal, *continued*
squama of, 610
surface of, internal, 663
orbital, 728
hamate, 118, 120, 129, 132–135
hamulus of, 108, 118, 131, 133, 135
humerus, 17, 30, 31, 34, 49, 52, 63, 64, 83–86
body of, 34, 51, 55, 82, 83, 91, 92, 96, 97
capitulum of, 84, 86, 94–97
diaphysis of, 94, 95
epicondyle of, lateral, 42, 48, 51, 57, 58, 60, 61, 74–77, 81, 83, 84, 86, 91, 92, 97, 98, 303
medial, 37, 49, 50, 53, 55, 56, 65–68, 70, 71, 76, 81, 83, 84, 86, 91, 92, 96–98, 303
epiphysis of, 94
extremity of, distal, 86
foramen of, nutrient, 52, 84
fossa of, coronoid, 82, 84, 98
olecranon, 82, 83, 96, 97
radial, 84, 96
groove, for ulnar nerve, 86
radial, 83
head of, 6, 25, 26, 34, 50, 63, 83, 85
muscle attachment, sites of, 85
neck of, anatomical, 34, 36, 83–85
surgical, 23, 34, 36, 59, 83
of fetus, 400
process of, coronoid, 84
sulcus of, intertubercular, 84, 85
trochlea of, 82–84, 86, 96–98
tubercle of, greater, 30, 32, 36, 83–85
lesser, 84, 85
crest of, 36, 84
tuberosity of, deltoid, 84
hyoid, 151, 217, 233, 257, 576, 583, 586, 588, 592, 593, 598, 599, 608, 625, 722, 726, 734, 736, 756, 763, 764, 766, 769
body of, 574, 589, 598, 615, 728, 760
horn of, greater, 589, 590, 598, 722, 728, 741, 744, 748, 750, 751, 753, 755, 757–759, 763
lesser, 589, 615, 722, 728, 745, 757, 759, 763
ilium, 25, 386, 387
ala of, 389–391, 401
body of, 389–392
crest of, 232, 257, 258, 298, 333, 335, 339, 359, 387, 389–393, 401, 424, 430, 521, 523
of newborn, 563
tuberosity of, 390
incus, body of, 781, 782, 791, 797
crus of, long, 784, 785, 791, 792, 797
short, 780, 787, 791, 792, 797, 801
fossa for, 778, 791
ligaments of, 801
superior 791
process of, lenticular, 785, 791
ischium, 359, 388–393
lacrimal, 601, 728
lunate, 118, 120, 131–135
malleus, head of, 780–782, 787–789, 791, 792, 797, 801
ligament of, 791, 801
manubrium of, 789, 791
neck of, 785
process of, lateral, 780, 787, 797
mandible, 599, 601, 625, 629, 631, 692, 693, 697, 698, 708, 722, 726, 736, 744, 750
angle of, 574, 612, 613, 647, 697, 698, 700, 728
arch of, 699
base of, 610, 647
body of, 574, 583, 598, 601, 700, 728
depression of, sublingual, 698
submandibular, 698
canal of, 709, 711, 720, 721
head of, 615, 698, 700
line of, oblique, 697
lingula of, 613, 698, 700
neck of, 647, 698
notch of, 700
of 9-year-old child, 710
part of, alveolar, 697
process of, condyloid, 574, 598, 610, 612, 614, 697, 698, 700
coronoid, 598, 610, 612, 697, 700, 728
ramus of, 601, 608, 613, 697, 700, 722, 728, 737, 740, 754
trigone of, retromolar, 740
tuberosity of pterygoid, 614, 730
maxilla, 601, 629, 640, 664, 667, 708, 710
crest of, ethmoidal, 684
hiatus of, 684
process of, alveolar, 601, 662, 731
frontal, 601, 651, 658, 660, 682

Bone(s), maxilla, *continued*
palatine, 631, 683, 684, 702
zygomatic, 646, 662, 702
metacarpal, 79, 109, 112, 118, 121, 132–135
base of, 132, 133, 135
body of, 132
head of, 132, 135
metatarsal, 462, 501, 503, 508–511
base of, 510, 511, 513
tuberosity of, 501, 503, 510, 511, 513
nasal, 601, 603, 606, 610, 626, 651, 661, 662, 682, 684, 689, 722, 726
navicular, 501–504, 510, 511, 515
tuberosity of, 513
occipital, 534, 535, 538, 539, 563, 626, 629, 637, 645, 726
condyle of, 615, 647
part of, basilar, 535, 536, 540, 594, 645
lateral, 535, 629, 631
process of, jugular, 645
squama of, 540, 630, 631
tubercle of, pharyngeal, 646, 751
of fetus, 400
of skull. See names of specific bones
palatine, 631, 646, 702
lamina of, horizontal, 631, 646, 684, 702
plate of, perpendicular, 684
process of, orbital, 662
pyramidal, 646, 702
parietal, 601, 626, 631, 633, 635, 645, 663
diploë of, 635
line of, superior temporal, 610
table of, external, 635
internal, 635
patella, 412, 416, 417, 419, 420, 424, 442, 444–446, 457, 458, 461, 468
apex of, 446
base of, 446
surface of, articular, 458
posterior, ridge of, 446
pelvic, 389–392. See also *Pelvis*
phalanx (phalanges), of foot, 501, 503
of hand, 109, 118, 119, 123, 124, 132, 135
distal phalanx, base of, 4th finger, 132
tuberosity of, 118, 132
middle phalanx, 4th finger, 132
middle finger, 124
head of, 4th finger, 132
tuberosity of, little finger, 132
proximal, 4th finger, 132
middle finger, 123
base of, little finger, 132
pisiform, 108, 110, 114, 120, 131, 133, 135
pubic, surface for symphysis pubis, 439
radius, 48, 61, 75, 77, 79, 89–93, 96, 97, 99–101, 114, 118, 129, 131–135
body of, 79, 90, 96, 97
crest of, interosseous, 89, 90
diaphysis of, 95
epiphysis of, 94
extremity of, distal, 69
foramen of, nutrient, 89
head of, 69, 79, 88, 90, 91, 93, 95–97
margin of, interosseous, 132
neck of, 88, 90, 96, 97
notch of, ulnar, 133
of fetus, 400
process of, styloid, 37, 89, 90, 118, 129, 131, 133
surface of, posterior, 132
tuberosity of, 89, 96, 97
rib(s), 146, 157, 158, 518, 550, 553
1st, 137, 140, 143, 145, 146, 150, 152, 184, 185, 219, 227, 524, 574, 582, 586–588, 593, 594
2nd, 25, 36, 137, 140, 143, 146, 148, 149, 175, 574, 587, 593, 594
3rd, 143, 146, 148, 149
4th, 137, 146, 148, 149, 186
5th, 146, 186
6th, 137, 227, 231, 523
7th, 137, 148
8th, 143, 146
9th, 146, 238
10th, 146, 238, 299, 359
11th, 25, 137, 140, 320, 339, 359, 521
12th, 16, 25, 137, 140, 218, 320, 333, 526
angle of, 146
costodiaphragmatic, 170
body of, 143, 573
cartilage of, costal, 144, 146–148, 158, 237, 238, 255
crest of, 143
extremity of, sternal, 143

Bone(s), rib(s), *continued*
head of, 143, 551, 573
ligaments of, 299
neck of, 143, 554, 573
ligaments of, 299
of fetus, 298
osseous part of, 146
spaces of, intercostal, 146
tubercle of, 143, 551, 554
sacrum, 25, 258, 320, 354, 386, 387, 544, 546, 557, 561, 565
ala of, anterior aspect, 559
apex of, 559, 560
coccyx and, lateral view, 561
crest of, intermediate, 560
lateral, 560
median, 386, 394, 560, 561
foramina of, dorsal, 560
horn of, 521, 560, 561
lateral part, 401, 559
lines of, transverse, 559
process of, articular, superior, 386, 560
promontory of, 319, 335
ridges of, 559
surface of, anterior, 559
articular, 559–561
pelvic, 559
posterior, 560
terminal, superior, 559
tuberosity of, 560, 561
vertebral column, part of, 544–546
scaphoid, 118, 120, 129, 132–134
tuberosity of, 135
scapula, 6, 25–28, 32, 50, 52, 140, 157
acromion of, 15, 21, 23, 26–31, 33–37, 51, 61, 137, 140, 233, 257, 258, 524, 574, 576
angle of, inferior, 26, 27, 34, 36, 258
lateral, 26, 27
superior, 21, 26, 27, 34, 36, 51, 258, 523
border of, lateral, 26–28, 34, 36
medial, 26, 27, 34, 36, 205
superior, 26, 27
cavity of, glenoid, 26–30, 34–36, 137
foramen of, nutrient, 34
fossa of, infraspinatus, 26, 34
supraspinatus, 26, 34, 574
labrum of, glenoid, 34
neck of, 26, 27, 34
of fetus, 398
process of, coracoid, 6, 15, 16, 21, 22, 26–29, 31, 32, 35, 36, 49, 53, 136, 137, 151, 257, 574
spine of, 23, 24, 26, 30, 32–34, 36, 51, 61, 258, 523
surface of, dorsal, 26, 28, 34
ventral, 27, 28
tubercle of, infraglenoid, 6, 26, 28
view of, posterior, 52
sesamoid, 118, 131, 132, 135, 503, 504
sphenoid, 601, 636, 645, 659, 661, 662, 667–669, 675
body of, 613, 636, 640, 662, 663, 684, 726
lingula of, 645
process of, pterygoid, 612, 613, 615, 631, 646, 647, 663, 684
lateral lamina of, 662
section through, sagittal, 636
sella turcica, 613, 636, 685, 691
spine of, 735
tuberosity of, pterygoid, 698, 700
wing of, greater, 601, 614, 629, 646, 728
lesser, 601, 635, 642, 645, 663
stapes, 781, 782, 785, 788, 793, 805
base of, 797
crus of, anterior, 795–797
posterior, 784, 795–797
head of, 789, 795, 796, 798
sternum, 141, 145, 157, 158, 177, 188, 208, 209, 227, 243
body of, 141, 142, 145, 214
manubrium of, 141, 142, 144, 145, 147, 149, 175, 188, 212, 587, 726
notch of, clavicular, 141, 142
costal, 141, 142
jugular, 141
ossification, centers of, 145
process of, xiphoid, 141, 142, 145, 176, 186, 188, 230, 255, 299, 399
surface of, articular, 138
talus, 502, 504, 510, 512, 515
tarsal, 503
temporal, 601, 663, 694, 754, 779, 787, 789, 790, 794, 800

Bone(s), temporal, *continued*
border of, occipital, 777
parietal, 777
crest of, supramastoid, 787, 789
notch of, mastoid, 777
part of, petrous, 594, 629, 645, 647, 749, 777, 790, 801
squamous, 629, 631, 645, 669, 777, 786
tympanic, 774, 777, 790
vaginal process of, 790
process of, mastoid, 574, 594, 615, 623, 646, 748, 751, 754, 755, 774, 777–779, 788, 790
styloid, 529, 574, 608, 610, 612, 613, 615, 646, 722, 737, 740, 744, 745, 748, 751, 753, 754, 774, 776, 778, 790
vaginal, 790
zygomatic, 612, 777, 789
tubercle of, articular, 614, 646, 735, 777
thoracic, anterior, 148. See also *Thorax*
tibia, 444, 446–448, 451, 452, 457, 458, 462, 471, 477, 478, 480–483, 491, 508, 509, 512
area of, intercondylar, anterior, 452, 457
posterior, 452
body of, 457
condyle of, lateral, 442, 446, 452, 453, 457, 459, 481, 484
medial, 442, 446, 448, 452, 453, 457, 459, 474, 478, 481, 484
condyles of, showing menisci, and origins of cruciate ligaments, 453
eminence of, intercondylar, 484
end of, proximal, 452
extremity of, distal, 484
groove of, malleolar, 484
line of, epiphysial, proximal, 459
soleal, 448, 484
margin of, anterior, 490
interosseous, 484, 489
medial, 484, 490
of fetus, 400
shaft of, 481
surface of, articular, inferior, 481
lateral, 490
medial, 464, 466, 489, 490
meniscotibial, 448, 459
popliteal, 448
posterior, 484, 490
tubercle of, intercondylar, lateral, 446, 452, 457
medial, 442, 446, 452, 457, 459
tuberosity of, 442, 444, 446, 452, 464, 466, 467, 480, 481
trapezium, 118, 132–135
trapezoid, 118, 129, 132–135
triquetral, 118, 120, 129, 132–135
ulna, 77, 79, 91–93, 97, 99–101, 114, 118, 129–135
circumference of, articular, 87
crest of, interosseous, 87, 88
supinator, 88
diaphysis, 94, 95
disc, articular, 132
epicondyle, medial, 67
extremity of, distal, 74
foramen of, nutrient, 87
head of, 50, 79, 87, 88, 118, 132, 134
margin of, interosseous, 132
posterior, 51, 99, 100
notch of, trochlear, 87, 88, 93, 96, 133
of fetus, 451
olecranon, 96, 97
styloid, 37, 87, 88, 114, 118, 129, 131–133, 135
surface of, anterior, 87, 100
lateral, 88
medial, 100
posterior, 87, 88, 100, 132
tuberosity of, 87, 88
vertebral, See *Atlas; Axis; Vertebra(e)*
vomer, ala of, 726
zygomatic, 601, 603, 612, 624, 625, 629, 651, 652, 661, 671
surface of, orbital, 661
Brain, 633
cerebellum, 639, 644
convolutions of, markings of, 645
cortex, cerebral, 633
gyrus. See *Gyrus*
hemisphere, cerebral, 669
lobe of, frontal, 635
mammillary body, 639
medulla oblongata, 563, 566, 638, 639
Brain, *continued*
of newborn, 563
vessels of, 643, 644
view of, ventral, 639
Brain stem, 565
Breast, female, lymphatic drainage of, 5. See also *Mammary gland*
Bronchial tree, diagram of, 169
divisions of, bronchopulmonary and lobar, 169
Bronchopulmonary segments, of lung, 162, 163, 166, 167, 169
Bronchus (Bronchi), divisions of, segmental, 168
eparterial, to upper lobe of right lung, 165
hyparterial, to middle and lower lobes of right lung, 165
lobar, left, 168
right, 168, 220, 228, 229
primary, 187, 216
left, 164, 168, 185, 218, 220, 224, 258
right, 165, 168, 184, 216–218, 224, 227, 228, 258
view of, with larynx and trachea, anterior aspect, 168
Bulb, inferior, of internal jugular vein, 576, 586, 587, 754, 755, 782
of duodenum (1st part), 269, 294, 331
of penis, 343, 352, 359, 361
olfactory, 639, 644, 690, 691
vestibular, 368, 371, 372, 382, 388
Bulla ethmoidalis, 585
Bundle, atrioventricular, 191, 207, 210–213
Bursa(e), anserine, 489
between pectoralis major and long tendon of biceps, 49
bicipitoradial, 66, 99
coccygeal, subcutaneous, 523
deep to obturator internus muscle, 429
iliopectineal, 420, 423, 431
infrapatellar, 416, 445, 453, 455, 456, 458, 463
deep, 446
infraspinatus, 63
intermuscular, 429
beneath gluteus maximus muscle, 430
ischial, 434
beneath obturator internus tendon, 434
of calcaneal tendon, 462, 463
of gastrocnemius muscle, 477, 478
of knee joint, cast of, 455
of latissimus dorsi muscle, 53
of olecranon, 82, 98
subcutaneous, 48
of popliteus, muscle, 448
of semimembranosus muscle, 448, 477
lining of, 448
of tibial collateral ligament, deep to, 459
of tibial tuberosity, 489
of triceps muscle, 98
omental, 262B, 264, 275, 277, 300, 305, 331, 332
development of, stages in, 263
recess of, inferior, 275
splenic, 275, 300
superior, 264, 275
vestibule of, 275, 277
pharyngeal, 726
prepatellar, 416, 455, 458
subcutaneous 444
sacral, subcutaneous, 521, 523
semimembranosus, 478
subacromial, 13, 30, 31, 35
subanserina, 423
subcoracoid, 21, 34, 35, 49
subdeltoid, 23, 49, 51, 59, 63
subgastrocnemius, 474
subhyoid, 741, 760, 769
subsartorial, 423
subscapular, 34, 35, 63
subsemimembranosus, 448, 474, 477
lining of, 448
suprapatellar, 445, 455, 456, 458, 487
synovial, subcutaneous, over posterior superior iliac spine, 430
over sacrum, 430
trochanteric, 427, 429, 430, 434, 523

Calcaneal region, 1
Calf, of leg, 520
Calvaria, viewed from above, 626
Calyx (Calyces), renal, 325, 327, 331, 332
major, 327
minor, 327
Canal(s), adductor, 386, 419, 421
opening of, 440
tendinous cover of, 419, 440
Canal(s), *continued*
Alcock's, 314, 344, 345
anal, inner surface of, 312
carotid, 594, 646, 647, 778, 781, 791, 798
carpal, 108
cervical, of uterus, 377, 380
cochlear, 640
condyloid, 645–647
facial, 778, 787, 788, 790, 793, 795, 796, 799
hiatus of, 778
prominence of, 778, 787, 789, 795, 799
view of, lateral, 793
gastric, 268
hypoglossal, 536, 594, 613, 645, 646
incisive, 683, 684
infraorbital, 661, 663
inguinal, 237, 245, 248, 253
opened, 245
mandibular, 709, 711, 720, 721
mastoid, 646
mental, 721
musculotubal, 594
septum of, 791
obturator, 243, 386, 423, 439
optic, 645, 662, 669, 670, 676
palatine, greater, 663
pterygoid, 663
pudendal, 314, 344, 345, 387
sacral, 355, 386, 438, 560
Schlemm's, 678, 680
semicircular, 789, 793, 800–805
anterior, 640, 789, 793, 801–805
crus of, common, 801
lateral, 640, 778, 787, 789, 790, 792, 793, 795–798, 800–805
posterior, 789, 801–805
view of, lateral, 793
vertebral, spinal cord within, dorsal view, 230, 565
Canaliculus (Canaliculi), alveolar, 720, 721
cochlear, 805
lacrimal, 656, 658, 660
inferior, 656, 658
ampulla of, 658, 660
superior, 656, 658
Canine tooth, 699, 701, 705, 708–711, 713, 714, 717
Capitulum, of humerus, 84, 86, 94–97
Capsule, articular. See *Articular capsule*
fibrous, of kidney, 322, 323, 325, 331, 332
of penis, 348
of spleen, 278
of perirenal fat, 332
of suprarenal gland, 300, 324
of thyroid, 591
Tenon's, 669, 677
Cardiac antrum, 216
Cardiac area, 153
Cardiac dullness, superficial, area of, 170
Cardiac notch, 153, 170
Carina, 768
Cartilage(s), alar, greater, of nose, 655–657, 682, 683, 690, 726
lesser, of nose, 682
annular, 726
arytenoid, 759, 760, 762, 764, 768–770
process of, muscular, 761, 765
vocal, 761, 767, 770–772
bronchial, 168
corniculate, 759, 761, 764
tubercle of, 726, 751, 753, 755, 765, 767–769
costal, 144, 146–148, 158, 237, 238, 255
cricoid, 136, 168, 257, 574, 583, 593, 726, 747, 757, 760–762, 764–766, 769, 770
arch of, 726, 735, 761, 763
origin of, 745
cuneiform 753, 764, 765
of acoustic meatus, external, 608, 787
of auditory tube, 647, 690, 693, 746, 747, 753, 780
of ear, external, 774
of epiglottis, 726, 736, 769
of larynx, 757, 759, 760
of nose, accessory, 656
at apex of, 722
external, 682
lateral, 655, 657, 682, 683, 722
septal, 682, 683, 690, 726
thyroid, 136, 152, 168, 228, 257, 583, 593, 726, 736, 744, 745, 747, 748, 756–760, 762–764, 769, 770
border of, superior, 761
horn of, inferior, 757–759, 763
superior, 589, 751, 753, 757–761, 763

Cartilage(s), thyroid, *continued*
lamina of, 217, 583, 652
origin of, 745
surface of, articular, 762, 765
tubercle of, inferior, 735, 744, 745
tracheal, 168, 217, 591, 744, 745, 748, 753, 757, 759, 762, 763, 766, 769
triteceal, 757, 759, 763
Caruncle, lacrimal, 648, 649, 655, 657, 658, 660, 677
sublingual, 722, 726, 733, 734, 737, 738, 740, 746
Caruncula lacrimalis, 648, 649, 655, 657, 658, 660, 677
Cauda equina, 340, 563, 567
Cavity, abdominal. See *Abdominal cavity*
cranial, 635
glenoid, 26–30, 34–36, 137
nasal. See *Nasal cavity*
oral, 740, 746, 747
pericardial, 159, 215, 299
peritoneal, 215, 262, 299, 300, 331, 332
pharyngeal, 691, 726, 732, 751, 770
pleural, 152, 157, 158, 299, 300, 331
pulp, of tooth, 703, 720, 721
tympanic. See *Tympanic cavity*
uterine, 377, 380
Cavum conchae, 773
Cecum, 257, 258, 260, 261, 272, 297, 301, 302, 305, 306, 310, 311, 374, 378
of cupula, 805
vestibular, 800, 805
Cells, air, ethmoid, 595, 669, 675, 686–688, 750
mastoid, 595, 640, 695, 750, 751, 778, 779, 790, 792, 793, 801, 802
tympanic, 778
Cementum, 703
Cerebellum 639, 644
of newborn, 563
Cerebral, convolutions, markings of, 645
Cerebral cortex, 633
Cerebral hemisphere, 669
gyrus of, 639
occipitotemporal, 639
orbital, 639
parahippocampal, 639
Cervix, of uterus, 379–381
Chiasma, optic, 639, 644, 668
Choana, 631, 646, 732, 751
Chorda tympani, 781
Chordae tendinae, 204, 206
Chorioid, 678–680
Chorion, 181
Chorion laeve, 181
Chorion villi, 181
Ciliary body, 680
pigmented layer of, 680
Circulation of blood, 178, 181–183
arterial, 182
fetal, 181
of newborn, 178
schema of, 183
Cisterna chyli, 183, 318, 320, 341
Cleft, palpebral, 653
Clinoid process, anterior and posterior, of sphenoid bone, 645, 686
Clitoris, 367, 376
crus of, 368, 372, 388
frenulum of, 367, 368, 377, 388
prepuce of, 244
Clivus, 613, 683, 747, 750, 751
Coat, longitudinal, of rectum, 312, 313
muscular, of gall bladder, 284
of stomach, 268, 270
serous, of gall bladder, 284
of spleen, 278
of stomach, 268
Coccygeal horn, 561, 562
Cochlea, 640
canaliculus of, 805
coil of, basal, 800
cupula of, 800, 805
lamina of, osseous spiral, 640, 800
Cochlear window, 640, 799, 805
Cochleariform process, 778, 781, 789, 791, 796, 798
Colliculus, prostatic, 356, 361
Colon, ascending, 136, 257, 259–261, 272, 277, 295, 298, 301, 304–306, 310, 311, 320, 328, 331, 399
descending, 256–258, 272, 302–304, 310, 311, 328, 332, 399
flexure of, splenic, 260, 311, 331
sigmoid, 257–260, 272, 297, 298, 302–305, 310, 317, 318, 340, 342, 344, 354, 374, 377
Colon, *continued*
transverse, 136, 215, 257, 259–261, 273, 277, 295–297, 301, 302, 305, 307, 310, 311, 332
wall of, free of omentum, 305
Columns anal, 312, 342
renal, 325, 327
Commissure, labial, anterior, 367
posterior, 367, 388
of aortic semilunar valve, 212
palpebral, lateral, 648, 677
medial, 677
Concha(e), nasal, 601, 751
inferior, 601, 684, 685, 691–693, 732
middle, 601, 684, 685, 688, 691–693, 732
superior, 684, 685, 692, 693, 732
of ear, 773
Conjunctiva, 648, 649, 665, 671, 680
bulbar, 649, 657
fold of, semilunar, 649, 658
fornix of, inferior, 649, 657, 673
superior, 673
palpebral, 649, 653, 657
Conus arteriosus, 136, 176, 188, 190, 194, 204, 208
Conus elasticus, 761, 762, 769
Conus medullaris, 563, 566, 567
Convolutions, cerebral, markings of, 645
Cord, of brachial plexus, lateral, 22, 54
medial, 19, 54
posterior, 19, 54
oblique, 50, 80, 91, 93
spermatic, 12, 231, 232, 236–239, 242, 246, 247, 251, 341, 347, 354, 355, 416
nerves of, 347
vessels of, 347
umbilical, 178, 255, 399
Cornea, 664, 665, 680
epithelium of, anterior, 680
membranes of, limiting, 680
plexus of, marginal, 678
Corona, of glans penis, 246, 343, 349, 351, 352, 361
Coronary arterial system, 196, 197
Corpus albicans, 380
Corpus cavernosum penis, 248, 348, 352, 355, 363,
Corpus cavernosus, of clitoris, 368, 377, 388
Corpus luteum 380
Corpum spongiosum penis, 248, 343, 344, 352, 363
Cortex, adrenal, 324
glomerulosa of, 324
reticularis of, 324
cerebral, 633
renal, 325, 327, 331
Costal margin, 7, 61
Cranial cavity, 635
base of, 638, 801
Crease, palmar, longitudinal, medial, 37
radial, 37
transverse, distal, 37
proximal, 37
Crest, infratemporal, 646
intertrochanteric, 400, 401
lacrimal, anterior, 661, 662
posterior, 661, 662
occipital, internal, 645
supraventricular, of right atrium, 204, 209
tympanic, 735
Crista galli, 645, 668, 683, 684, 686–688
ala of, 687
Crista terminalis, 201, 202, 208
Crista vestibuli, 640
Crus (Crura), anterior, of stapes, 795–797
common, anterior and posterior, of semicircular canals, 801
lateral, of greater alar cartilage, 682
of superficial inguinal ring, 231, 235, 236, 244, 245, 248, 253, 333
long, of incus, 784, 785, 791, 792, 797
medial, of greater alar cartilage, 682
of superficial inguinal ring, 12, 231, 235, 236, 239, 244–246, 248, 253
of antihelix, 773
of clitoris, 368, 372, 388
of diaphragm, 215, 216, 221, 223, 230, 299, 329, 333, 335
of helix, 773
of penis, 352, 354, 359, 361
posterior, of stapes, 784, 795–797
short, of incus, 780, 787, 791, 792, 797, 801
Crus membranaceum commune, 803
Crus membranaceum simplex, 803
Crypts tonsillar, 743
Cubital region, anterior, 1
Cupula, cecum of, 805
of cochlea, 800, 805
of diaphragm, 227
of pleura, cervical, 153–155, 170, 215, 219, 227, 258, 593
of recess, epitympanic, 781
Cusp(s), of tooth, buccal, 704
lingual, 704
of valve, mitral, anterior, 206, 207, 212
posterior, 201, 206, 212
tricuspid, dorsal, 201–204, 208, 214
septal, 201–204, 208, 214
ventral, 202–204, 208, 214
Cutaneous nerve fields, in leg and foot, 409–411
in the thigh, 408, 409
of femoral nerve and its saphenous branch, 409
of peripheral nerves in the thigh, 408
of peroneal nerve, 411
of sural nerve, 410, 411
of tibial nerve branches, 410
Cutaneous nerves, distribution of, in lower extremity, 405, 407
in upper extremity, 38, 39
Cuticle, 125–128
Cymba conchae, 773

Dartos tunic, 248, 255, 354
Decidua basalis, 181
Decidua capsularis, 181
Decidua parietalis, 181
Decussation, of pyramids, 566, 639
Dens, fovea for, 530, 533, 537
ligament of, apical, 538
of axis, 532, 533, 538, 540, 543, 690
surface of, articular, anterior, 533, 537
posterior, 537
Dentin, 703
Dentition, at one year of age, 708
deciduous, full, of 5-year old child, 709
of 20-year old person, 711
Dermatomes, of head and neck, 609
of lower extremity, anterior aspect, 404
posterior aspect, 406
of trunk, anterior, 2
dorsal, 519
of upper extremity, 45
Dermis, of scalp, 633
Diaphragm, 136, 148–150, 152, 170, 172, 175–177, 184–186, 188, 200, 212, 214, 215, 218, 219, 224, 229, 230, 243, 256, 257, 264, 265, 273, 281, 285, 299, 300, 320, 331, 333–335, 338
costal part of, 170, 219, 230, 242, 243, 317, 333, 335
crus of, 215, 216, 221, 223, 230, 299, 329, 333, 335
left, 299
right, 335
cupula of, 227
dome of, 136, 170, 227, 257, 258, 269
fascia of, inferior, 299, 300
liver, contact with bare area of, 316
lumbar part, 170, 231, 300, 320, 331, 334
pelvic, 354, 359, 365
sternal part of, 230, 243, 299, 335
tendon of, central, 184, 185, 215, 216, 230, 333–335
urogenital, 343, 354, 355, 359, 366, 371, 376, 382
in female, 369, 388
vena caval aperture in, 299
Diaphragma sellae, 636, 642
Digital sheath, fibrous annular part, 108
cruciform part, 108
Digiti minimi, fibrous sheath of, 108
Disc, articular, 614
of sternoclavicular joint, 144
of wrist, 132, 134
interpubic, 366, 369, 386, 438
intervertebral, 214, 215, 219, 393, 535, 540, 550, 551, 557, 572
optic, 679, 680
Diverticulum, ileal (Meckel's), 308
Dorsum sellae, 613, 645, 683, 686, 750
Duct(s), aberrant, of epididymis, 363
aberrant inferior, 250
bile, common, 265, 268, 272, 283, 284, 286–293, 300, 317, 331
and cystic, union of and variations in, 283
and pancreatic, union of and variations in, 288–293
bulbourethral, 354, 361, 366
cochlear, 800, 802, 803, 805
basal coil of, 800

Duct(s), *continued*
cystic, 270, 272, 283, 285, 286, 300, 317
and common bile, union of and variations in, 283
endolymphatic, 801–805
ejaculatory, 352, 357, 363
excretory of lacrimal gland, 651
hepatic, 270, 272, 285, 286, 317
common, 283, 284
hepatopancreatic, 288–290, 293, 332
lactiferous, 8, 9
longitudinal, of epoophoron, 379, 380
lymphatic, right, 223
mesonephric, 363, 372
Müllerian, 363, 372
nasolacrimal, 658, 660, 685
opening of, 685
of epididymis, aberrant, 363
of epoophoron, 379, 380
of gland, bulbourethral, 363
lacrimal, 653
vestibular, greater, 388
of Santorini, 293
of vesicle, seminal, 352
of Wirsung, 284, 288–293
pancreatic, 268, 284, 287–293
accessory, 268, 287, 293
and common bile, union of and variations in, 288–293
orifice of, 284
paramesonephric, 363, 372
infundibulum of, 363
parotid, 576, 603, 606, 608, 727, 730, 733, 748
opening of, 746, 747
prostatic, 361
salivary, 733
semicircular, anterior, 800, 802, 803, 805
lateral, 796, 800, 802, 803, 805
posterior, 802, 803, 805
sublingual, greater, 733, 734, 737, 740, 746, 747
smaller, 733, 734, 740, 746
submandibular, 625, 727, 733, 734, 739, 740, 746, 747
thoracic, 183, 185, 214, 218–221, 223, 229, 230, 299, 300, 320, 331, 332, 573, 588, 592, 593
origin and course of, 223
transverse, of epoophoron, 379, 380
utriculosaccular, 805
vitelline, 261
Wolffian, 363, 372
Ductule(s), efferent, of epididymis, 250
mesonephric, 363
caudal, 372
cranial, 372
Ductus arteriosus, 176, 178, 181, 208, 209
in neonatal, heart, 179
Ductus deferens, 243, 248, 250, 252, 255, 258, 305, 317, 338, 340, 343, 347, 354
beginning of, 250
Ductus reuniens, 800, 804, 805
Ductus venosus, 178, 181
Duodenal fold, circular, 284, 287, 302, 316
inferior, 316
longitudinal, 268, 284, 287
Duodenojejunal flexure, 258, 287, 294, 296, 303, 305, 331, 332
Duodenojejunal junction, 302, 316
Duodenum, 136, 224, 258, 261, 272, 283, 285, 288–294, 320, 331
1st part (superior), 257, 268, 270, 273, 277, 286, 294, 305, 316, 317, 330, 331
ampulla of, 269, 294, 331
2nd part (descending), 268–270, 285–287, 294, 310, 316, 317, 331, 332
3rd part (horizontal), 258, 268, 270, 285–287, 294, 296, 301, 316, 317
4th part (ascending), 258, 287, 294, 310, 316
covered by peritoneum, 297
pancreas and, 285
Dura mater, 219, 226, 320, 540, 565–567, 570–572, 633–635, 638, 726, 744, 793, 805
cranial, 638
filum of, 565
intracranial, 637
lacuna of, lateral, 634, 635
layer of, meningeal, 570, 571, 573
periosteal, 571
of cranial fossa, anterior, 668
of vertebral canal, 722
spinal ganglion ensheathed with, 573
surface of, 634

Ear, 773–805
concha of, 773
external ear, cartilage, of, 774
muscles of, 775, 776
fossa of, triangular, 773
helix of, 773–775
lobule of, 773
malleolar processes of, 780, 784, 786, 791, 797
middle ear, muscles of, 792
ossicles of, 792, 797
wall of, lateral, 786, 787
notch of, intertragic, 773
of fetus, 400
Elbow, 520
arteries of, lateral view, 72
medial view, 71
joint of, 91, 92
roentgenogram of, 94–97
section of, sagittal, 82
line of, epiphyseal, 95
nerves of, lateral view, 72
medial view, 71
Eminence, arcuate, 787
frontal, 629
hypothenar, 37
iliopubic, 386, 391, 401
parietal, 626, 629–631
pyramidal, 795, 796, 799
thenar, 37
Enamel, 703
Endocardium, 206
Endometrium, 380
Epicardium, 158, 160, 202, 212
Epidermis, of scalp, 633
Epididymis, 242, 250, 319, 353, 363
appendix of, 247, 249, 253
body of, 250, 252
head of, 247–249, 251, 253, 254, 343, 363
ligament of, superior, 249
section of, longitudinal, 251
sinus of, 249, 252, 253
tail of, 249–251, 254, 363
Epidural fat, 573
Epidural space, 343, 571, 726
Epiglottis, 685, 732, 743, 746, 751, 753, 755, 757–760, 762, 764, 765
cartilage of, 736, 766, 767, 771, 722
stem of, 759, 760, 762
vallecula of, 685, 736, 743, 758, 769
Epineurium, 571
Epiphysis, of humerus, 94
of radius, 94
Epitympanic recess, 780–782, 786, 791, 801
Eponychium, 125–128. See also *Finger nail*
Epoophoron, 372
duct of, longitudinal, 379, 380
transverse, 379, 380
Esophageal-tracheal space, 591, 726, 769
Esophagus, 157–159, 184, 185, 214–220, 224, 228–230, 258, 268–270, 294, 299, 335, 573, 574, 588, 590, 591, 593, 723, 726, 735, 744, 745, 748, 750, 753, 754
beneath pericardium, 189
hiatus of, 170, 214, 216, 230, 317, 321, 333, 335, 338
junction with pharynx, 735
lower, at tracheal bifurcation and diaphragm, 216
muscle deficient region, 590
part of, abdominal, 214–216, 224, 299
cervical, 217
thoracic, 216, 217
traversing esophageal hiatus, 335
Ethmoid air cells, 669, 675, 686–688
Expansion, extensor, 123, 124
Eye, 648
angle of, lateral, 648, 653
medial, 648, 649, 653
chamber of, anterior, 680
posterior, 680
vitreous, 680
margin of, infraorbital, 664, 671, 714
muscles of, lateral, view, 664, 667
orbit of, 653–657
septum of, orbital, 650–652, 655–657, 673
space of, intervaginal, 680
tarsi of, 651, 652, 655–657, 673, 677
Eyeball, 638, 664, 665, 667, 668, 674, 678, 679
diameter of, equatorial, 680
muscle insertions of, behind, 666
in front, 665
pole of, anterior, 680
posterior, 680

Eyeball, *continued*
section of, horizontal, 680
tunic of, inner neural, 680
middle vascular, 680
outer fibrous, 680
view of, sagittal, 673
Eyebrow, 604, 648, 677
Eyelashes, 677
Eyelid, 648–650, 653, 658, 660, 664, 669
innervation of, 650
lower, 649, 650, 656, 658
palpebral borders, 648, 649, 653
palpebral cleft, 653
surface of, inner, 653, 656
upper, 650, 658

Face, nerves of, See *Nerve(s), facial*
vessels of, deeper, 623
superficial, 622
Facet(s), articular, of atlas, inferior, 531
superior, 530
of rib tubercle, 143
of vertebra, 573
for head of rib, 568
for neck of rib, 573
superior, of vertebra, 573
calcaneal, anterior, with talus
middle, with talus, 511
posterior, with talus, 511
costal, inferior, 548
superior, 558
of 6th thoracic vertebra, 547, 548
transverse, 558
for ribs, 553
of 6th thoracic vertebra, 547
transverse process with, 548
navicular, posterior, with talus, 511
Facial canal. See *Canal, facial*
Facial expression, muscles of, 603, 604
Falciform process, 394, 396
Falx, inguinal, 234, 239, 240, 243, 245
opened, 245
Falx cerebelli, 637
Falx cerebri, 633, 637, 638, 722
Falx septi, 206, 208
Fascia(e), antebrachial, 51, 53, 55, 57, 59, 65, 74, 77, 98–104
axillary, 7, 12, 13, 47, 63
brachial, 7, 12, 13, 21, 64, 575
buccinator, 737
Buck's, 348, 351
bulbar, 653, 657, 673, 677
cervical, 152, 155, 604, 722
angular band of, 597, 727
attachment to mandibular ramus, 733
deep, 740, 766
superficial investing layer of, 733
middle, 571
pretracheal layer of, 151, 583, 726
prevertebral layer of, 726, 740
submandibular, 597, 617
superficial, 575, 591, 599, 603, 606, 727, 726
clavipectoral, 13, 15
cremasteric, 248, 249, 252, 343
cribriform, 341
crural, 444, 472, 473, 489–491
cubital, 98, 99
deep, of penis, 246, 351
overlying antebrachial muscles, 47, 48
overlying brachial muscles, 47, 48
overlying external oblique muscle, 332
overlying volar carpal ligament, 108, 114
deltoid, 7, 48, 63, 523
diaphragmatic, inferior, 299, 300
dorsal, deep, of foot, 515, 516
of hand, 121
endothoracic, 227, 230, 299, 573
femoral, 344
gluteal, 344, 370, 424, 427, 430, 521, 523–525
overlying gluteus medius muscle, 426
iliac, 243, 317, 345
inferior, of diaphragm, 299
of pelvic diaphragm, 345
of urogenital diaphragm, 344, 366, 368–370, 387
infraspinatus, 57, 61, 527
intercostal, 184
external, 526
lumbar, anterior layer, 332, 522
posterior layer, 522
masseteric, 597, 733, 737
mylohyoid, 739

Fascia(e), *continued*
nuchal, deep, 548, 584
obturator, 344, 345, 370
in perineum, 345, 387
overlying obturator internus muscle, 243, 365
of arm, 47, 48
of eyelid, superficial, 653
of foot, 515, 516
of hand, 121
of inferior oblique muscle, 673
of infraspinatus muscle, 523
of penis, 246, 348, 351
of psoas muscle, 522
of rectum, 313
of spermatic cord, 239
of supinator muscle, 99
of tensor tympani muscle, 791
of teres major muscle, 523
of thigh, anterior, 416
over external oblique muscle, 13, 253, 332
over latissimus dorsi muscle, 7
over obturator internus muscle, 386
over serratus anterior muscle, 13
over triceps muscle, 7
parotid, 575, 604, 737
pectineal, 416
pectoral, 7, 8, 35, 63, 575
platysma muscle and, 575
pelvic, 343
pharyngeal, 737, 740, 770
pharyngobasilar, 615, 625, 748, 750, 753, 755
popliteal, 487
prevertebral, 584, 591, 640, 746, 747
pterygoid, 737
renal, 331, 332
rhomboid, 523
spermatic, external, 239, 246, 248, 253, 255
internal, 246–249, 252, 253, 255
submandibular, 733
subsartorial, 486, 487
subserous, 240, 320
superficial, of eyelid, 653
of penis, 347, 348
of rectum, 313
superior, of pelvic diaphragm, 345
temporal, 654
adipose layer of, 606, 623
deep layer of, 606
superficial layer of, 606
thoracolumbar, 230, 424, 521, 523
anterior layer of, 219, 334
lumbar part of, 523–526
posterior layer of, 334
thoracic part of, 524
tonsillar, 737
transversalis, 149, 240, 241, 243, 245, 253, 332, 335, 522, 526
overlying quadratus lumborum muscle, 526
visceral, cervical, 584
Fascia bulbi, 669
Fascia lata, 244, 246, 416–420, 430, 486, 487
Fascia penis, 234
Fascia propria, subperitoneal, 320
Fascial planes, of neck, in newborn, cross section of, 584
Fasciculata, of adrenal cortex, 324
Fat, cardiac, 189, 194
epidural, 573
orbital, 652, 668, 669. Also see *Fat pad, orbital*
pararenal, 331, 332
perirenal, 300, 320, 321, 331, 332
preperitoneal, 240
subpleural, 185
Fat body, of breast, 8
Fat pad, axillary, 7, 15
buccal, 576, 603, 604, 606, 729, 737, 740
cervical, 740
infrapatellar, 458
infratemporal, 635
laryngeal, 726, 757, 759, 765
mediastinal, 227
orbital, 652, 656, 657, 668–670, 673, 677
pre-epiglottic, 769
Fauces, isthmus of, 729, 731
Fetus, 181, 319, 399, 400
heart of, right atrium, 180
Fibers, intercrural, 12, 233, 235, 236, 244, 246
zonular, 680
Fibrous sheaths, of digits, 108, 494
of flexor hallucis longus tendon, 494
Fibular notch, 481
Filaments, dorsal root, 568, 570, 573
Filaments, *continued*
ventral root, 568, 573
Filum terminale, 343, 566, 567
Fimbria(e), of uterine tube, 374, 377–380
ovarian, 379, 380
Finger(s), arteries of, 112, 113
index finger, 122
articular capsule of, middle finger, 130
bones of. See *Bone(s), phalanx (phalanges)*
expansion of, dorsal digital, of middle finger, 123, 124
extensor, of middle finger, 123, 124
joints of, 109, 114, 129, 131, 132, 134
carpometacarpal, 134
index finger, 132
little finger, 134
middle finger, 129
ligaments of, 108, 129, 131, 132
of middle finger, 129
mesotendon, of middle finger, 123, 124
nerves of. See Nerve(s), digital, of hand
pad of, distal digital, of index finger, 37
periosteum of, middle finger, 123, 124
region of, palmar, 4th digit, 37
section of, cross, middle finger, 123, 124
tendon insertions, index finger, radial side, 109
See also *Hand*
Finger nail, bed of, exposed, 126–128
body of, 125, 126
lunula of, 125, 127, 128
margin of, free, 125, 126
hidden, 128
lateral, 128
matrix of, 126
root of, 126–128
view of, dorsal, 125
Fissure, antitragohelicine, 773, 744
cerebral, longitudinal, 639
horizontal, 165, 170
right lung, 136, 153, 161, 215, 257, 258
median, anterior, of spinal cord, 566–568, 570, 572
oblique, 154, 160, 164, 170
lung, left, 136, 153, 156, 160, 164, 258
right, 136, 153, 161, 165, 215, 257, 258
orbital, inferior, 601, 640, 646, 659, 661, 664, 667
superior, 645, 661, 675, 686, 728
palpebral, 652
petrosquamosal, 645
petrotympanic, 777, 790
pterygomaxillary, 615, 720, 721, 735
sphenopetrosal, 645
tympanomastoid, 777, 790
Flexure, colic, left (splenic), 136, 257, 260, 261, 269, 272, 297, 298, 302–305, 311, 331
right (hepatic), 136, 257, 260, 272, 273, 275, 296–298, 301, 305, 311
duodenal, inferior, 286
superior, 286
duodenojejunal, 258, 287, 294, 296, 303, 305, 331, 332
perineal, 313, 354
sacral, of rectum, 313, 354
Flocculus, cerebellar, 639
Fold(s), aryepiglottic, 746, 751, 753, 755, 764, 767, 769, 771
axillary, anterior, 37
posterior, 37
chorda tympani, 785
conjunctival, 649, 658
duodenal, circular, 284, 287, 302, 316
inferior, 316
longitudinal, 268, 284, 287
gastric, 268, 295
gastropancreatic, 274, 275, 277, 305, 316
for left gastric vessels, 275
glossoepiglottic, lateral, 732, 743, 769
median, 743
ileocecal, 260, 301, 305, 309, 316, 374
inferior, 305, 309
superior, 305, 309
inferior vena caval, 305
interarytenoid, 768
interdigital, 37
lacrimal, 685
laryngeal, 766, 770
malleolar, 781, 784–786, 792
posterior, 791
mucosal, of bladder, 356, 361
of uterine tube, 380
muscular, of palate, 685, 731
of mucous membrane, of gall bladder, 284
Fold(s), *continued*
of ovarian vessels, 316
of ureter, left, 316
of vas deferens, 359
palatine, transverse, 701
peritoneal, 277, 301, 305, 340
over hepatic artery, 300
petroclinoid, anterior, 801
posterior, 801
petrosphenoid, 642
pharyngoepiglottic, 726, 746, 751, 753, 755
pterygomandibular, 735
rectal, transverse, 312, 343, 377
rectouterine, 374, 377, 379, 382, 383
salpingopalatine, 685, 726, 751
salpingopharyngeal, 726, 732, 751, 755
semilunar, 306, 307, 332
between haustrae, 301
spiral, of cystic duct, 284
stapedial, 795
sublingual, 733, 737–739
triangular, 743
umbilical, lateral, 256, 259, 321, 340, 359
medial, 256, 259, 321, 340, 352, 359
median, 243, 256, 320, 321, 340, 352, 359
vestibular, 726, 766–769, 771, 772
vocal, 726, 766–770
Follicle(s), lingual, 692, 743
lymphoid, 268, 751
ovarian, 380
Fontanelle, anterior, 629, 743
anterolateral, 629
posterior, 629–631
posterolateral, 629, 631
Foot, dorsal, 1, 520
arteries of, 471, 493, 496, 497
deep, 471
bones of, 501–503
showing muscle attachments, 502
cross sections of, 515, 516
muscles and tendons of, 468–470
nerves of, deep, 471
superficial, 468
veins of, superficial, 468
cutaneous nerve fields of, 409–411
peroneal neve, 411
sural nerve, 410, 411
tibial nerve, 410
ligaments of, 508–510, 513, 514
dorsolateral view, 508
medial view, 509
plantar aspect, bones of, 503
showing muscle attachments, 504
ligaments of, superficial, 513
section of, cross section, 515, 516
horizontal, 510
sole of, 492–498, 520
arteries of, 493, 496, 497
superficial, 493
muscles of, first layer, 494
second layer, 495
third layer, 498
nerves of, 493, 496, 497
superficial, 493
surface anatomy of, 412, 413
Foramen (Foramina), apical, 703
epiploic, 264, 273, 275, 305, 316
ethmoid, anterior, 661, 662
posterior, 661, 662
for perineal vessels and nerves, 344
for pudendal vessels and nerves, 344
for superior laryngeal vessels and nerve, 757, 759, 762, 763
incisive, 646, 702
inferior vena cava, 216
infraorbital, 601, 671
intervertebral, 546, 551, 572
jugular, 638, 645–647, 754
mandibular, 698–700, 728
mastoid, 587, 645, 646
of temporal bone, 777
mental, 598, 601, 603, 610, 697, 699, 708, 709, 711, 720
nutrient, 484
of clavicle, 138
of femur, 434
of humerus, 52, 84
of radius, 89
of scapula, 34
of ulna, 87
obturator, 388–390, 401
optic, 661, 686

Foramen (Foramina), *continued*
palatine, greater, 646, 702
lesser, 702
parietal, 626, 628
pterygospinous, 613
sacral, dorsal, 560
pelvic, 393, 559
sciatic, greater, 386, 388, 389, 393, 394, 397, 429
lesser, 388, 389, 393, 394, 397, 429
sphenopalatine, 662, 684
stylomastoid, 646, 695, 750, 751, 788
supraorbital, 601
transverse, of atlas, 530, 531
of axis, 532, 533
of 5th cervical vertebra, 541
vena caval, 216, 229, 333–335
vertebral, 529, 532, 541, 554
of atlas, 530, 531
of 7th cervical vertebra, 542
of lumbar vertebra, 555
zygomaticofacial, 661
zygomatico-orbital, 659, 663
Foramen cecum, 645, 686, 732, 736, 743, 753
of tongue, 685, 726, 736
Foramen lacerum, 645–647
Foramen magnum, 631, 637, 645, 646, 735
Foramen ovale, 180, 181, 208, 645, 646, 686, 735
remainder of, 180
Foramen rotundum, 645, 662, 663, 686
Foramen spinosum, 645, 646, 735
Foramina venarum minimarum, 201, 203
Forearm, anterior, arteries of, 62, 67, 68, 70
muscles of, 37, 65, 66, 69, 70, 74
deep, 69
extensor, 37
flexor, 37
nerves of, 67, 68, 70
bones of, showing, muscle attachments, 80, 81
cross section of, 98–101
dermatomes of, 45
fascia of, deep, 47, 48
muscles of, 37, 61, 65, 66, 69, 70, 74–76, 79
lateral view of, 61, 75
nerves of, 38, 39, 41, 43, 67, 68, 70, 76, 78
cutaneous, 38, 39, 41, 43
posterior, arteries of, 43, 62, 76, 78
muscles of, 74–76, 79
nerves of, 43, 76, 78
surface anatomy of, 37
veins of, superficial, 41, 43, 44
Fornix, of conjunctiva, 649, 657, 673
of lacrimal sac, 658
of vagina, 373, 377
Fossa, acetabular, 401, 437
axillary, 1, 37, 231
coronoid, of humerus, 82, 84, 98
cranial, anterior, 645
middle, 645, 787
posterior, 645, 787, 789
cubital, 37
digastric, 583, 698, 700
for incus, 778, 791
for lacrimal gland, 689
glenoid, 34
hypophyseal, 645, 683, 686, 722, 802
iliac, 390
infraclavicular, 37, 575
infraspinatus, of scapula, 26, 34
infratemporal, 664
inguinal, lateral, 340, 359
medial, 340, 359
intercondylar, 433, 442, 448, 450, 457, 459
interpeduncular, 639
intersigmoid, 302
ischiorectal, 344, 346, 359, 370
jugular, 575, 594, 646, 792
mandibular, 614, 646, 735, 777
navicular, 343, 350, 361
of cochlear window, 778, 784, 785, 789, 798, 799
of vestibular window, 799
olecranon, 82, 83, 96, 97
petrosal, 646
popliteal, 413, 429, 440, 441, 472, 473, 487, 489, 520
pterygoid, 683
pterygopalatine, lateral, view, 662
medial view, 663
radial, of humerus, 84, 96
rectovesical, 359
retromandibular, 574
scaphoid, 773, 774
subarcuate, 665
Fossa, *continued*
submandibular, 574
subscapular, 27
supraclavicular, 574, 575
supraspinatus, of scapula, 26, 34, 574
supratonsillar, 729
supravesical, 340, 359
triangular, of ear, 773
uterovesical, 181
Fossa ovalis, 202, 203, 208
Fovea, for dens, 530, 533, 537
of head of femur, 401, 432, 433
sublingual, 700
submandibular, 700
Fovea centralis, 784, 785
Foveola, radial, 108
Frenulum, 306, 350, 351
of clitoris, 367, 368, 377, 388
of labia minora, 367, 368
of lips, 729, 738
of tongue, 733, 735, 737, 740
Frontal, crest, 645
Frontal notch, 661
Frontomaxillary suture, 682
Fundus, of bladder, 352
of gall bladder, 136, 284, 301, 305
of internal acoustic meatus, 801
of stomach, 136, 268–270, 273, 294
of uterus, 318, 378, 380, 381
Furrow, antebrachial, lateral, 37
medial, 37
bicipital, medial, 37, 47

Galea aponeurotica, 603, 604, 606, 608, 633, 635
Gall bladder, 136, 257, 259, 265, 272, 273, 275, 281–283, 296, 310, 331, 332, 399
biliary duct system and, 284
body of, 284
coat of, muscular, 284
serous, 284, 331
fundus of, 136, 284, 301, 305
mucosa of, 331
neck of, 284
Ganglion (Ganglia), abdominal autonomic, 338
aorticorenal, 318
cardiac, 185
celiac, 225, 226, 300, 318, 320, 331, 339
cervical, inferior, 220, 224, 225, 228, 588, 593, 754, 755
middle, 224–226, 228, 582, 593, 754, 755
superior, 225, 582, 640, 693, 694, 737, 754, 755
cervicothoracic, 220, 226
dorsal root, on thoracic nerve, 226
ciliary, 225, 640, 672, 674, 793
nasociliary branch to, 674
oculomotor branch to, 674
sympathetic branch to, 674
geniculate, 640, 695, 742, 788, 793, 801, 802
inferior, of glossopharyngeal nerve, 640, 695, 794
of vagus nerve, 694, 754, 755, 793
mesenteric, inferior, 225
superior, 225, 318, 332, 339
otic, 225, 318, 339, 793
rami to chorda tympani nerve, 793
pelvic, 225
pterygopalatine, 225, 611, 640, 693–695, 793
branches of, 640, 693
renal, 339
semilunar, 642, 670, 672, 694, 696, 788, 801
spinal, 10, 219, 220, 518, 563, 567, 568, 570, 571, 573
cervical, 563, 565
lumbar, 563, 565
sacral, 563
thoracic, 320
spiral, 372, 804
splanchnic, 219, 226
stellate, 220, 224–226, 591
submandibular, 225, 611, 625, 727, 740, 746, 747
superior, of vagus nerve, 640, 694, 754
sympathetic, 10, 218, 219, 224, 320, 518, 573
cervical, 793
lumbar, 320, 339, 670
sacral, 339, 364
thoracic, 184, 185, 224, 226–229, 339, 593, 755
sympathetic trunk, 10, 184, 219, 224, 226, 299, 793
trigeminal, 793, 801
vestibular, inferior, 804
superior, 804
Ganglion impar, 337, 339
Genitalia, female, 367–388
arterial supply of, 381
external, 367, 368, 388
internal, 378
male, 246–254, 347–363
Genitourinary system, female, 372
male, 363
Genu, of facial nerve, 795, 796
of internal carotid. artery, 640
Gingiva, 548, 703, 729, 735, 740
Gland(s), areolar, 4
buccal, 730, 737
bulbourethral, 344, 352, 354, 359, 361, 363, 366
bulge of, 366
labial, 722, 730, 737
lacrimal, 651–653, 656, 657, 659, 670–672, 674
innervation of, 659
orbital part of, 653, 656, 657
palpebral part of, 653, 656, 657
laryngeal, 769
lingual, anterior, 733
mammary. See *Mammary gland*
mucous, in bile ducts, 284
of mucous membrane, 751
of uvula, 731
palatine, 690, 701, 722, 731, 746, 747
parathyroid, 258, 590, 591, 750, 751, 753, 755, 766
left inferior, 588
parotid, 576, 577, 583, 585, 586, 597, 603, 606, 607, 622, 727, 733, 737, 740, 747, 751, 789
accessory, 733
parasympathetic innervation of, 607
pharyngeal, 750
pituitary, 636–639, 641, 642, 685
infundibulum of, 636
lobes of, anterior (adenohypophysis), 636
intermediate, 636
posterior (neurohypophysis), 636
sagittal section through, 636
prostate, 258, 343, 352–359, 362
colliculus of, 356, 361
lobe of, 362
salivary, 733. See also *Gland(s), parotid, sublingual, submandibular*
sebaceous, 633, 722
sublingual, 625, 722, 727, 733, 734, 737, 739, 740, 746, 747, 756
submandibular, 579–581, 583, 585, 597, 599, 600, 606, 624, 625, 727, 733, 734, 737, 739, 740, 746, 747, 750, 751
suprarenal, 258, 277, 285, 300, 316–320, 322–324
capsule of, 300, 324
left, 300
sweat, 7
tarsal, 649, 677
thymus, 152, 153, 175, 177, 184, 185, 256, 726
lobes of, 175
thyroid, 136, 152, 177, 178, 186, 214, 217, 257, 258, 576, 580, 581, 583, 586–592, 746, 750, 753–756, 758, 766, 769
capsule of, 591
isthmus of, 217, 574, 585, 589, 591, 592, 726, 756
lobe of, left, 588
pyramidal, 152, 588
tracheal, 591, 755, 759, 760
vestibular, greater, 368, 371, 373, 388
Glans clitoris, 244, 368, 370, 371, 388
Glans penis, 246, 343, 351, 352, 361, 363
corona of, 246, 343, 349, 351, 352, 361
Glenoid cavity, 26–30, 34–36, 137
Glenoid labrum, 29, 30, 34, 35
Glomerulosa, of adrenal cortex, 324
Gnathion, 574
Granulations, arachnoid, 633–635
Groove(s), costal, 143
deltopectoral, 37
for anterior intermuscular septum, 467
for aorta, 158
for flexor tendons, 131
for vertebral artery, 530, 543
mylohyoid, 613, 678, 700
obturator, 390
radial, 59, 83
Gubernaculum testis, 255, 363
Gum, 699, 703, 725, 729
Gyrus (Gyri), occipitotemporal, 639
orbital, 639
parahippocampal, 639

Hamulus, of hamate bone, 108, 118, 131, 133, 135
pterygoid, 613, 615, 646, 647, 662, 663, 683, 684, 721, 722, 731, 735, 746–748
Hand, 1, 37–39, 45, 61, 62, 102–107, 520
arteries of, 62, 102, 104–107, 110–113
dorsal, 102, 104, 105
palmar, 62, 106, 107
arteriogram of, 113
bones of, 118–119
showing muscle attachments, 119
cross section of, at level of 1st carpal bones, 120
through metacarpal bones, 121
dorsal, arteries of, 102, 104, 105
nerves and veins of, superficial, 102
intertendinous connections of, 103
joints of, dorsal view, 129, 132
palmar view, 131
ligaments of, 129–131
muscles of, 37, 61, 107, 108, 114–117
nerves of, 38, 39, 102, 104, 106, 107, 110–112
cutaneous, 38, 39
palm of, 1, 37–39, 62, 106, 107, 110–112, 114, 115
arteries of, 62, 106, 110, 111
creases of, 37
eminence of, hypothenar, 37
thenar, 37
monticuli of, 37
muscles of, 37, 107, 114, 115
nerves of, 38, 39, 106, 110, 111
tendons of, 107, 112
tendons and synovial, sheaths of, 103, 104, 107, 112
veins of, 102
x-ray of, 133, 135
See also *Finger(s)*
Haustrum (Haustrae), 259, 301, 306, 307, 332
Head, nerves of, superficial, 609, 622, 623
of fetus, 400
vessels of, superficial, 622, 623
viscera of, mid-sagittal, section, 726
Heart, 136, 170–174, 176, 186, 192, 194–199, 257
aorta of. See *Aorta*
apex of, 136, 188, 190, 192, 194, 195, 202, 204, 208, 212, 257, 258
arterial system of, 196–199
arteriogram of, 200
ascultation of, sites on chest wall for, 171
atrium of. See *Atrium, of heart*
auricle of. See *Auricle, of heart*
cardiac dullness, area of superficial, 170
dorsal half of, 201
neonatal, 179
of child, 176, 208, 209
of fetus, right atrium of, 180
position of, during full expiration, 174
during full inspiration, 173
projected onto anterior thoracic wall, 136, 170, 172, 257
section through, frontal, 201, 205
valves of, 171, 180, 190
superior view, 190
Also see *Valve(s), aortic, atrioventricular*
veins of, 180, 193, 215. Also see *Vein(s), cardiac, coronary*
vena cava of, See *Vena cava*
ventricle of. See *Ventricle(s), of heart*
view of, anterior, 136, 170–174, 176, 186, 194, 196–199, 257
posterior, 192, 194, 195
Helix, 773–775
Hemisphere, cerebral, 669
Hemorrhoid, 312
Hiatus, adductor (tendineus), 420, 422, 423
anal, 386
aortic, 170, 214, 216, 285, 317, 320, 333–335
esophageal, 170, 214, 216, 230, 317, 321, 333, 335, 338
for basilic vein, 47
for cutaneous nerve, lateral antebrachial, 47
posterior antebrachial, 47, 48
for posterior interosseous vessels of forearm, 69
maxillary, 684
urogenital, 386
Hiatus semilunaris, 685
Hilum, of lung, 158, 188
of spleen, 274, 277, 278
renal, 323, 329, 332
Hydatid, 360
Hymen, 367, 368, 388
Hypoglossal canal, 594
Hypothalamus, infundibulum of, 639, 642
Hypotympanic recess, 782, 792

Ileocecal fold. See *Fold, ileocecal*
Ileocecal junction, 306
Ileum, 259–261, 272, 295–297, 301, 306, 308, 310, 374
terminal part, 301, 302, 305, 311
Iliotibial tract, 296, 416, 424, 430, 444, 463, 482, 486, 487
Incisors, 699, 701, 705, 708–710, 712, 713, 716, 717, 720
permanent, lower, 709, 710, 716, 717, 720
upper, 710, 712, 713, 720
Incisures, in cartilage of acoustic meatus, 774
sacrococcygeal, 388
Infraglottic space, 766
Infraorbital margin, of eye, 664, 671, 714
Infratemporal crest, 646
Infundibulum of hypothalamus, 639, 642
of paramesonephric duct, 363
of pituitary gland, 636
of uterine tube, 372, 377–379, 381, 382, 399
Inguinal canal, 237, 245, 248, 253
opened, 245
Inguinal, falx, 234, 239, 240, 243, 245
Inguinal ring, deep, 242, 243, 253, 340, 364
superficial 231, 232, 235, 236, 244, 248, 255, 347, 388
crus of, lateral, 231, 235, 236, 244, 245, 248, 253, 333
medial, 12, 231, 235, 236, 239, 244–246, 248, 253
Insula, 644
Interarytenoid notch, 751, 755, 758, 767, 772
Interpleural area, inferior, 153
superior, 153
Intersection, of ocular axis and visual axis, 680
tendinous, 238, 242
Intertragic notch, 773
Intertrochanteric crest, 400, 401
Intestine, large, 257, 305, 311
radiograph of, 311
small, 215, 256–260, 399
fetal, 400
Iridocorneal angle, 680
Iris, 648, 680
arterial circle of, 678
surface of, anterior, 680
Isthmus, of aorta, 176
of fauces, 729, 731
of thyroid gland, 217, 574, 585, 589, 591, 726, 756
of uterine tube, 316, 318, 374, 375, 377–380, 383
of uterus, 375, 379, 380

Jejunum, 260, 261, 269, 285, 295–297, 301, 332
Joint(s), acromioclavicular, 29, 31, 36
ankle, 512
atlantoaxial, 534, 536, 537, 594, 722, 726
atlanto-occipital, 534, 536, 538
calcaneocuboid, 510, 515
carpometacarpal, 134
of thumb, 131, 134
costotransverse, 554
costovertebral, 552, 554
cricoarytenoid, 759, 770
cricothyroid, 761, 763
cuneonavicular, 510
elbow, anterior, 91
articular capsule of, 82, 91, 98
posterolateral, 92
roentgenogram of, 94–97
sagittal section of, 82
finger, 129, 132
hip, anterior view, 282, 287, 435–439
articular capsule of, 394–396, 437, 438
socket of, 401, 437
incudomalleolar, 787, 797, 801
incudostapedial, 797
interphalangeal, distal, 132
proximal, 132
intertarsal, 510
intervertebral, capsule of, 526, 529
knee, 444–459
bursa of, 455, 456
capsule of, articular, 447, 448, 459
cavity of, 455, 456
ligaments of, 448
menisci of, 446, 448, 453
arterial supply of, 454
radiograph of, 457
section of, frontal, 458, 459
sagittal, 458
vessels at, network of, 422
view of, anterior, 444–446, 451, 457
Joint(s), knee, view of, *continued*
posterior, 447, 448, 456
mediocarpal, between two rows of carpal bones, 134
metacarpophalangeal, 109, 114, 129
capsule of, 132
metatarsophalangeal, 514
of hand, 129, 131
radiocarpal, 134
radioulnar, 93, 101, 131, 133, 134
sacrococcygeal, 343, 354
sacroiliac, 386, 393, 401, 438
shoulder, 31–36
capsule of, articular, 24, 30, 32–35, 49–51
section of, frontal, 30
posterior, 34
sternoclavicular, 13, 144, 151
sternocostal, 144
sternomanubrial, 144, 147
subtalar, 510
talocalcaneonavicular, 511, 515
capsule of, 515
viewed from above, 511
talocrural, 512
talonavicular, 510
tarsometatarsal, 510
temporomandibular, 612, 613, 616, 617
capsule of, articular, 608, 612, 614, 647, 748
tibiofibular, proximal, 480, 498
superior, 448, 459
capsule of, articular, 459
wrist, 129, 131–134
Junction, between pharynx and esophagus, 735
duodenojejunal, 302, 316
ileocecal, 306
of inferior mesenteric vein and splenic vein, 272
of jugular and subclavian veins, 586
of retromandibular and facial veins, 586

Kidney, 215, 258, 262, 273, 285, 300, 303, 305, 316, 317, 319–323, 325–327, 329, 331, 332, 340, 363, 372
artery and vein of, 331
border of, lateral, 326
medial, 326
calyx (calyces) of, 325, 327, 331, 332
capsule of, fibrous, 322, 323, 325, 331, 332
cortex of, 325, 327, 331
lobules of, 323
extremities of, 326, 327
hilum of, 323, 329, 332
left, 300, 320
inferior pole of, 302, 316, 328
lymph nodes of, 331
of newborn, 563
pelvis of, 271, 285, 304, 323, 325–328, 331, 332, 363, 372
pyramid (medulla) of, 331, 332
right, 320
superior pole of, 305
Knee joint. See *Joint(s), knee*
Knee region, anterior, 1, 444, 445, 520
Knots, false, 255

Labium majus, 245, 367, 376, 388
skin of, 244
Labium minus, 367, 368, 370, 371, 376, 377, 388
frenulum of, 367, 368
Labrum, acetabular, 437, 439
glenoid, 29, 30, 34, 35
Labyrinth, ethmoidal, 687, 689
membranous, 800, 803–805
Lacertus, of lateral rectus muscle, 675
Lacertus fibrosus, 49
Lacrimal canaliculus, 656, 658, 660
ampulla of, 658
Lacrimal crest, anterior, 661, 662
posterior, 661, 662
Lacrimal gland. See *Gland, lacrimal*
Lacrimal papilla, 649, 658, 677
Lacrimal sac, 652, 656–658, 660
fornix of, 658
Lacuna(e), lateral, 634, 635
urethral, 361
Lacus lacrimale, 649
Large intestine, 305
Laryngeal prominence, 575, 576, 585, 586, 626
Laryngopharynx, 746
Larynx, 151, 168, 257, 723, 756–759, 760, 762, 766
aditus of, 758
aperture of, superior, 736
cartilages of, 757, 759, 760

Larynx, *continued*
cavities of, 766
entrance into, 685, 753
external, lateral view, 744, 745, 763
fat body of, 757
folds of, 766, 770
ligaments of, 757, 759
section of, frontal, 766
ventricle of, 685, 726, 736, 766, 769
Leg, 1, 462, 463, 489–491, 520
compartment of, anterior, muscles of, 464
nerves and arteries of, 465
lateral, muscles of, 463, 467
nerves and arteries of, 465
posterior, muscles of, 473, 474, 477, 479
cross sections of, 489–491
cutaneous nerve fields of, femoral nerve, 409
peroneal nerve, 411
sural nerve, 410, 411
tibial nerve, 410
interosseous membrane of, 480
lymphatic channels and nodes of, 466
posterior, nerves of, 475, 476, 479
vessels of, 475, 476
surface anatomy of, anterior 412
posterior, 413
tibiofibular union of, 480
Lens, pole of, anterior, 680
posterior, 680
Lenticular process, of incus, 785, 791
Ligament(s), acetabular, transverse, 387, 437
acromioclavicular, 29–32, 34
alar, 536, 538
annular, of radius, 79, 91–93
of trachea, 168, 726, 759, 760, 765, 766
anococcygeal, 343, 344, 346, 354, 368, 370, 371, 376, 387
anterior, of head of fibula, 446, 451, 463, 480
apical, of dens, 538, 722, 726
arcuate, of pubis, 245, 366, 369, 376, 387, 394–396, 438
superior, 369, 395, 397, 419
over quadratus lumborum muscle, 333, 334
atlanto-occipital, 595
bifurcate, 508
broad, of uterus, 379, 380
calcaneocuboid, 508, 509, 511, 513, 514
calcaneofibular, 463, 508, 512, 513
calcaneonavicular, 508, 511, 513
plantar, 514
long, 498, 508, 509, 513, 515
superficial, 513
carpal, radiate, 114, 120, 129, 131
carpometacarpal, dorsal, 79, 129
cervical, transverse, 385
circular, of tooth, 703
coccygeal, lateral, 393
collateral, fibular, 431, 434, 445–448, 454–456, 459, 463
of middle finger, 129, 130, 132
radial, 79, 91, 92, 134
tibial, 445–448, 454, 456, 459
ulnar, 91, 134
conoid, 29, 31, 35, 49, 151
coracoacromial, 29, 31–35, 151
coracoclavicular, 31, 34, 49, 53
coracohumeral, 31, 32, 34, 35, 63
corniculopharyngeal, 759, 760, 762
coronary, of liver, 272, 274, 280–282, 299, 316, 318, 320, 331
margin of, 300
costoclavicular, 50, 51, 138, 144
costotransverse, 227, 550, 552, 554
lateral, 526, 551–554
superior, 227, 526, 552
costoxiphoid, 12, 151, 231
cricoarytenoid, 761
cricopharyngeal, 759, 760
cricothyroid, 168, 589, 592, 593, 726, 735, 756, 757, 760, 762, 763, 765, 769
cricotracheal, 744, 757, 759, 766
cruciate, anterior, 336, 446, 448, 453, 454, 459
occupying intercondylar fossa, 459
origins of, 453
posterior, 446, 448, 451, 453, 459
cruciform, of atlas, 536, 726
cuboideonavicular, 508, 513–515
cuneocuboid, dorsal, 511
cuneonavicular, 508, 511, 513
deltoid, 462, 509, 512, 513
denticulate, 226, 570–573
duodenal suspensory, 285, 287
Ligament(s), *continued*
duodenocolic, 277, 316
falciform 243, 256, 259, 261, 262, 272, 273, 280, 281, 299, 300, 305, 316, 331, 332, 359
fibular collateral, 431, 434, 445–448, 454–456, 459, 463
fundiform, of penis, 12, 231–233, 236, 343
gastrocolic, 259, 273, 305, 331, 332
gastrohepatic, 262
gastrolienal, 262
gastrophrenic, 299, 316
gastrosplenic, 273, 274, 277, 278, 300, 305, 331
genicular, transverse, 451, 453
genital, 363, 372
glenohumeral, 31, 34, 63
hamatometacarpal, 131
hepatoduodenal, 270, 273, 275, 277, 283, 305, 316
hepatogastric, 273, 275, 299, 300, 305
hyoepiglottic, 726, 760, 765, 769
iliofemoral, 333, 395, 397, 420, 435–437, 439
iliolumbar, 321, 333, 393, 395–397, 526
inferior, of epididymis, 249
inguinal, 11, 231, 235–237, 239, 242–244, 253, 333, 335, 340, 352, 369, 393–395, 397, 399, 416–418
reflected, 12, 233, 235, 245, 246, 253
intercarpal, dorsal, 120, 129
interclavicular, 144, 151, 576, 726
intercuneiform, dorsal, 511
interfoveolar, 239, 242, 243, 245, 340
intermetatarsal, plantar, 514
interosseous, between ulna and radius, 50
interosseous metacarpal, 134
interosseous metatarsal, 510
interosseous sacroiliac, 386
interosseous tarsal, 510
interspinal, 214, 526
intertransverse, 526, 552
intra-articular, of spine, 551
ischiofemoral, 430, 436, 439
lacunar, 236, 243, 335, 369, 398
lateral, of temporomandibular joint, 612, 647
lienorenal, 277, 300, 331
longitudinal, anterior, 214, 393, 395, 397, 535, 550, 573, 726
posterior, 214, 557, 572, 573, 726
lumbocostal, 331
malleolar, lateral, 780, 781
superior, 780, 781, 791, 801
meniscofemoral, posterior, 448
metacarpal, dorsal, 129, 131
interosseous, 134
palmar, 131
transverse, deep, 108, 131, 132
superficial, 106
metatarsal, dorsal, 508
interosseous, 510
transverse, deep, 508, 514
superficial, 513
nuchal, 523–525
of ankle and foot, 508, 509
of femur, head of, 359, 437, 438
of fibula, head of, 446, 448
of foot, 513
of hand and finger, dorsal, 129, 131
of inferior vena cava, 280, 282
of larynx, 757, 759
of pelvis, 394, 395, 397, 398
of rib, head and neck of, 299
of umbilical artery, 321
of umbilical vein, 243, 332
of wrist, dorsal, 129, 131
ovarian, 372, 374, 378–381, 399
palpebral, lateral, 651, 652, 655, 656
medial, 576, 603, 606, 651, 652, 654–656, 658, 660
patellar, 412, 417–420, 424, 431, 444, 445, 453, 455, 458, 462–465, 467, 487, 489
pectinate, at iridocorneal angle, 680
pectineal, 335
pelvic, female, 394
perineal, transverse, 366, 369, 387
petrospinous, 612, 642
phrenicocolic, 273, 277, 305, 316
pisometacarpal, 131
popliteal, arcuate, 447, 474
oblique, 447, 474, 477
posterior, of head of fibula, 447, 448
of incus, 780, 787, 791, 792, 801
pterygospinous, 647, 747
Ligament(s), *continued*
pubic, arcuate, 245, 366, 369, 376, 387, 394–396, 438
superior, 369, 395, 397, 419
pubofemoral, 435, 439
puboprostatic, 354
pubovesical, 343, 354, 385
pulmonary, 159, 164, 165, 184, 220
radiate, of head of ribs, 227, 550, 554
radiocarpal, dorsal, 129
palmar, 114, 131
radioulnar, anterior, 93
reflected inguinal, 12, 233, 235, 245, 246, 253
round, of liver, 259, 265, 272, 273, 281, 282, 300, 305, 331, 359
of uterus, 244, 245, 316, 318, 372, 374, 377, 378, 380, 381, 383, 385, 388, 399
insertion of, 244
sacrococcygeal, 334, 343, 376, 386, 393, 396, 397, 438, 521
superficial dorsal, 394
sacroiliac, 320, 395–397, 438
interosseous, 386
ventral, 386
sacrospinous, 333, 346, 364, 383, 388, 393, 394, 419, 431, 438
sacrotuberous, 333, 344, 346, 370, 386–388, 393, 394, 396, 427, 429, 430, 436, 438, 523, 526
scapular, transverse, inferior, 21, 22, 24
superior, 24, 31, 32, 34, 49, 53, 151
scrotal, of testis, 253, 255
sphenomandibular, 613, 722, 747, 750, 753
splenorenal, 300
sternoclavicular, anterior, 15, 144
sternocostal, 144, 147
stylohyoid, 598, 615, 647, 722, 735, 744, 745, 747, 748, 750
stylomandibular, 612, 613, 647, 722, 737, 740, 747, 750, 751
superior, of incus, 791
of malleus, 780, 781, 791, 801
supraspinal, 214, 219, 396, 552
suspensory, duodenal, 285, 287
of clitoris, 244, 368
of ovary, 374, 377–379
of penis, 12, 232, 234, 239, 246, 248, 347
talocalcaneal, 508, 510–512
talofibular, 508, 512
talonavicular, 515
tarsal, interosseous, 510
plantar, 514
tarsometatarsal, 508, 509, 511, 513, 514
thyroepiglottic, 726, 746, 747, 760, 762, 769
thyrohyoid, 760
lateral, 757, 759, 763
medial, 741, 757, 763
median, 726, 745, 762
tibial collateral, 445–448, 454, 456, 459
tibiofibular, 480, 508, 512
tracheal, annular, 168, 726, 759, 760, 765, 766
transverse, of atlas, 537, 540, 722, 726
of knee, 446
trapezoid, 29, 31, 35, 49, 151
Treitz's, 285, 287
triangular, of liver, 272, 280, 281, 305
ulnar collateral, 91, 134
ulnocarpal, palmar, 131
umbilical, lateral, 253, 256
medial, 11, 240, 243, 253, 256, 321, 340, 355, 377
median, 243, 256, 343, 353, 356, 358, 377, 378, 383
uterosacral, 380, 385
vestibular, 760, 766
vocal, 760–762, 766, 770
Ligamentum arteriosum, 187, 188, 190, 194, 195, 212, 216, 217
Ligamentum flavum, 219, 552–554
Ligamentum patellae, 446
Ligamentum venosum, 282, 299
Limbus fossa ovalis, 201, 208, 214
Limen nasi, 685, 691, 732
Line, arcuate, 238, 242, 243, 255, 259, 340, 359, 393, 390
condylopatellar, lateral, 446
medial, 446
gluteal, anterior, 389
inferior, 389
posterior, 389
Hilton's, 312
intercondylar, of femur, 433
intermediate, of iliac crest, 390

Line, *continued*
intertrochanteric, 432
mylohyoid, 613, 615, 698, 700
nuchal, inferior, 646
superior, 646
oblique, of larynx, 763
of mandible, 697
pectineal, 433
Reid's base, 574
soleal, of tibia, 448, 484
temporal, inferior, of parietal bone, 626, 627, 777
superior, of parietal bone, 626, 627
tendinous, of levator ani muscle origin, 345
trapezoid, 212
Linea alba, 12, 231, 233, 235–237, 240, 241, 242, 245, 253, 255, 300, 332, 343, 377
adminiculum of, 240, 242, 243
Linea aspera, of femur, 430, 433, 440, 448, 486
Linea semicircularis, 240
Linea semilunaris, of rectus abdominis, 238, 239, 243
Linea terminalis, of pelvis, 333, 359, 376, 386, 397, 399
Lingula, of lung (left), 152, 153, 156, 160, 164, 170
of mandible, 613, 698, 700
of sphenoid, 645
Lip(s), 730, 732, 736, 738
aponeurosis of, 736
mucosa of, 736
upper, frenulum of, 729, 738
Liver, 139, 178, 215, 257, 258, 261, 273, 279, 296, 299, 310, 331, 332
bare area of, 264, 282, 299, 316, 321
colic impression on, 282
calyx of, 331
capsule of, fibrous, 331
impression of, colic, 282
duodenal, 282
esophageal, 282
gastric, 282
renal, 282
suprarenal, 282
ligament of, coronary, 272, 274, 280–282, 299, 300, 316, 318, 320, 331
falciform, 243, 256, 259, 261, 262, 272, 273, 280, 281, 299, 300, 305, 316, 321, 332, 359
round, 259, 265, 272, 273, 281, 282, 300, 305, 331, 359
lobe of, caudate, 265, 272, 273, 275, 277, 280, 282, 299, 300, 305
papillary process of, 305
left, 256, 259, 265, 272, 273, 275, 280–282, 299, 300, 305, 331
quadrate, 272, 273, 275, 282, 300
right, 256, 259, 265, 272, 273, 275, 279, 280–282, 299, 301, 305, 331, 332, 399
margin of, anterior, 299, 300
inferior, 255, 258, 281, 301, 399
surface of, anterior, with diaphragmatic attachment, 281
diaphragmatic, 281, 282, 321
posterior, 282
visceral, 282, 283
tuber omentale, 300
Lobe(s), frontal, 635
mammary, 7, 8
of liver. See *Liver, lobes of*
of lung. See *Lung(s), lobes of*
of prostate, 362
of thymus, 175
of thyroid, left, 588
pyramidal, 152, 158
Lung(s), 136, 151–156, 160, 161, 164, 165, 170, 172, 175–177, 186, 188, 189, 208, 209, 212, 215, 228, 256–258, 299, 320, 388
apex of, 136, 151, 156, 160, 161, 164, 165, 170, 257, 258
bronchopulmonary segments of, 162, 163, 166, 167, 169
cardiac notch of, 152, 153, 156, 160, 164, 170, 257, 270, 399
hilum of, 158, 188
left, 136, 152, 160, 164, 170, 175, 177, 188, 189, 208, 209, 212, 215, 257, 299, 588
cardiac impression on, 164, 170
interlobar fissure of, horizontal, 165, 170
oblique, 136, 153, 154, 156, 160, 161, 164, 165, 170, 215, 257, 258
lobe of, lower, 152, 153, 156, 164, 170, 175, 186, 256
upper, 152, 153, 156, 170, 175, 186, 215, 256
surface of, costal, 160
Lung(s), left, *continued*
diaphragmatic, 164
mediastinal, 164
lingula of, 152, 153, 156, 160, 164, 170
margin of, anterior, 136, 152, 160, 164, 170, 175–177
inferior, left lung, 136, 152, 160, 164, 170, 186, 215, 219, 258, 299
right lung, 136, 154, 161, 170, 219, 257, 258, 299, 399
projected onto thoracic wall, anterior view, 153–156
right, 152, 161, 165, 170, 172, 175, 178, 188, 189, 208, 209, 212, 215, 228, 320
base of, 161
cardiac impression on, 165
interlobar fissure of, horizontal, 136, 153, 161, 215, 257, 258
oblique, 136, 153, 161, 165, 215, 257, 258
lobe of, lower, 152–155, 161, 165, 170, 175, 186, 215, 220, 256
middle, 152, 153, 155, 165, 170, 175, 186, 215, 256
upper, 152–155, 161, 165, 170, 175, 186, 215, 256
surface of, costal, 161
diaphragmatic, 165
mediastinal, 165
Lunula, of finger nail, 125, 127, 128
of semilunar valve, 212
Lymph nodes, aortic, 331, 341, 384
apical, 587
auricular, posterior, 619
axillary, 3, 5–7, 13–15, 63
anterior, 5, 6
central, 5
scapular, 5
bronchopulmonary, 164, 165, 185, 187, 215
buccinator, 619
cervical, 587, 619, 621
deep, 5
cubital, superficial, 3
diaphragmatic, 177, 187
duodenal, 284
gastric, left, 279, 318, 341
right, 279
gastroepiploic, left, 279
right, 279
hepatic, 279, 300
ileal, 296
ileocecal, 296, 306
ileocolic, 296
iliac, 359
external, 243, 340, 341, 383
internal, 340, 341, 383
infracardiac, 230
infraclavicular, 5, 6
inframammary, 5
inguinal, 341, 359
deep, 359, 384
superficial, 244, 246, 341, 415
intercostal, 227
jejunal, 296
jugular, 591
lumbar, 318, 340, 341, 383, 384
maxillary, 619
mediastinal, 175, 177, 187, 188, 219, 220
posterior, 299
mesenteric, 295, 296, 332
mesocolic, 296
obturator, 340, 383
occipital, 587, 619
of face, head, and neck, 587, 621
superficial, 587, 619
of thorax, 3, 5, 6, 187
pancreatic, 300, 331
para-aortic, 332
parasternal, 5
paratracheal, 188, 592, 726
parotid, 587
pectoral, 5, 6
pelvic, 341, 383
popliteal, 487
preaortic, 300
preauricular, 587, 619
pyloric, 279
rectal, 312, 383
renal, 331
retroauricular, 587
retropharyngeal, 747
Rosenmüller's, 341
sacral, 341, 383
Lymph nodes, *continued*
splenic, 300, 331
sternal, 11
subauricular, 587
subinguinal, deep, 341
submandibular, 587, 597, 599, 619, 621, 727
submental, 587, 599, 619, 621
subpleural, 185
subscapular, 6
superficial, of head and neck, 587, 619
supraclavicular, 5, 6, 587, 592, 619, 621
supramandibular, 619
suprarenal, 341
tibial, anterior, 466
tracheal, 187
tracheobronchial, superior, 187, 220
vertebral, 227
Virchow's, 592
Lymph vessels, intercostal, 227
of pancreas, 279
of pelvis, female, 383
of porta hepatis, 279
of stomach, 279
vertebral, 227
Lymphangiogram, of pectoral and axillary nodes, 6
Lymphatic drainage, of female breast, 5
Lymphatic ring, oro-naso-pharyngeal, 752
Lymphatics, of thorax, 3, 5, 6, 187

Malleolar processes, of ear, 780, 784, 786, 791, 797
anterior, 791
Malleolus, lateral, 1, 412, 413, 463, 464, 466–468, 470, 475, 476, 480, 485, 508, 512
medial, 412, 413, 464, 468, 472, 474, 476, 477, 480, 481, 512
Mammillary body, 639
Mammary gland, 3, 4, 7, 8, 257, 399
areola of, 4, 257, 399
fat body of, 8
lobes of, 7, 8
nipple of, 3, 4, 9, 399
Mandibular notch, 700
Manubrium, of malleus, 780, 781, 784–786, 789, 791, 792, 797
of sternum, 141, 142, 144, 145, 147, 149, 175, 188, 212, 587, 726
Mastoid air cells, 640, 695, 751, 778, 779, 790, 792, 793, 801, 802
Mastoid antrum, 778, 792, 795
Mastoid notch, 646, 777, 787
Mastoid process, of temporal bone, 574, 594, 615, 623, 646, 748, 751, 754, 755, 774, 777–779, 788, 790
Matrix, of finger nail, 126
Meatus, acoustic, external, 610, 612, 624, 646, 748, 773, 776–778, 784, 785, 787
cartilage of, 608, 787
osseous part of, 787
internal, 645
fundus of, 801
nasal, inferior, 684, 732
middle, 684, 732
superior, 684, 732
nasopharyngeal, 726, 732
Mediastinum, 157–159, 168, 185, 186, 219
anterior, 158, 159, 227
posterior, 159, 219, 220
superior, 157
Mediastinum testis, 251, 252, 343
Medulla, adrenal, 324
renal, 325, 327
Medulla oblongata, 563, 566, 638, 639
pyramid of, 566, 639
Membrane, arachnoid, 226
atlanto-occipital, anterior, 535, 594, 722, 726
posterior, 529, 534, 609
intercostal, external, 151
internal, 227, 526
interosseous, of forearm, 78, 79, 80, 93, 105, 114, 132
of leg, 480, 489, 490
mucous. See *Mucous membrane*
obturator, 314, 333, 359, 395, 435, 438, 439
periodontal, 703
pharyngobasilar, 735, 746
phrenicoesophageal, 299
quadrangular, 766
stapedial, 793, 795, 796
synovial, 459
tectorial, occipital, 425, 539, 540
thyrohyoid, 217, 588, 592, 593, 722, 735, 744, 745, 756, 757, 759, 760, 762, 763, 765, 766

Membrane, *continued*
tympanic, 783, 784, 786, 787, 791, 795
pars flaccida of, 766, 782, 786
pars tensa of, 782, 784, 786
secondary, 799
Membranous wall, tracheal, 591, 726, 759, 765
Meninges, 570, 571, 633
Meniscus, 453, 454
lateral, 446, 448, 451, 453, 455, 456, 458, 459
arterial supply of, 454
femoral, surface of, 446, 459
medial, 446, 448, 451, 453, 456, 459
site of attachment on tibial plateau, 446
tibial surface of, 459
Mental foramen, 598, 601, 603, 610, 697, 699, 708, 709, 711, 720
Mental protuberance, 598, 697
Mental tubercle, 574, 598, 697
Mesentery (Mesenteries), 264, 295, 302, 305, 308, 344
developing, 261, 262
dorsal, 261
of small intestine, 296, 305, 332
root of, 295, 301, 302, 310, 316
of small intestine, 264
Mesoappendix, 301, 302, 305, 309, 316
Mesocardium, 195
Mesocolon, sigmoid, 264, 302, 305, 317, 318, 321, 341, 354
root of, 316
transverse, 260, 264, 273, 275, 277, 296, 297, 303, 305, 307, 331, 332
root of, 301
Mesosalpinx, 278–280
Mesotendon of digit, 123, 124
Molar teeth, 699, 701, 704, 705, 708, 711, 715, 718, 719
deciduous, 708, 709
permanent, 708, 709, 715, 718, 719
socket for, 598
Mons pubis, 370, 393
Mouth, vestibule of, 690, 729, 732, 736, 737
Mucosa, lingual, 736
nasopharyngeal, 746
of palate, 735, 747
of stomach, 299
oral, 603, 694, 699, 734, 737, 739, 744, 746
tracheal, 591
Mucous membrane, esophageal, 769
of bladder, 356
of conus elasticus, 770
of gall bladder, 284
of hard palate, 747
of nasal septum, 722, 726
of stomach, 267, 268
of tongue, 741, 758
pharyngeal, 769, 770
Muscle(s), abdominal, anterior, 151, 238
posterior, 333
abductor digiti minimi, of foot, 413, 463, 494–498, 504, 515, 516
of hand, 108, 111, 114, 119, 120, 121
abductor hallucis, 413, 462, 494–498, 504, 507, 515, 516
abductor pollicis brevis, 107, 108, 110–112, 114, 119, 120, 121
abductor pollicis longus, 61, 65, 66, 74, 75, 77–79, 81, 100, 119, 120
adductor brevis, 420, 422, 423, 430, 431, 434
adductor hallucis, 497, 504
oblique head of, 497, 498
transverse head of, 495, 497, 498
adductor longus, 412, 417–420, 422, 423, 431, 434, 486
adductor magnus, 344, 413, 417, 420, 422, 423, 427–431, 434, 440, 486, 523
adductor minimus, 423, 427, 429, 430
adductor pollicis, 61, 105, 107, 108, 110–112, 114
oblique head of, 108, 119, 121
transverse head of, 108, 119
anal sphincter, external, 312, 313, 343–346, 354, 368, 370, 371, 376, 387
internal, 312, 343, 376
anconeus, 51, 52, 58–61, 74–79, 81, 98
antitragicus, 775
arm, anterior, 55
arrector pili, 633
articularis cubiti, 50, 80, 81
articularis genu, 431, 445, 487
aryepiglottic, 753, 755, 764–766
arytenoid, oblique, 753, 755, 758, 764, 765
transverse, 726, 753, 755, 764, 765, 769

Muscle(s), *continued*
auricular, anterior, 576
oblique, 776
posterior, 576, 604, 776
superior, 776
transverse, 776
biceps brachii, 13–15, 23, 37, 47, 49, 50, 53–61, 64, 65, 67, 68, 71, 72, 75, 80, 82
long head of, 21–23, 30, 31, 49, 50, 53–55, 63, 233
short head of, 21, 22, 35, 49, 50, 53, 55, 56, 151, 233
biceps femoris, 426–431, 434, 441, 443, 458, 467, 473–475, 477, 482, 487
long head of, 413, 424, 427–430, 440, 463, 486
short head of, 413, 424, 427–430, 440, 463, 486
brachialis, 23, 49–53, 55–61, 64–71, 74, 75, 80, 82, 98
brachioradialis, 47, 50–53, 56, 57, 59, 61, 65–69, 71, 72, 74–78, 80, 81, 98–100
buccinator, 576, 603, 606, 608, 615, 622–625, 722, 729–731, 733, 737, 739, 740, 746–748
bulbocavernosus, 371
bulbospongiosus, 244, 245, 344, 346, 354, 368, 370
calf, 473, 474
chondroglossus, 741
ciliary, 680
coccygeus, 355, 364, 365, 376, 386, 387, 431
coracobrachialis, 13, 14, 16, 21, 22, 35, 37, 49, 50, 53, 55, 56, 63, 64, 151, 233
corrugator supercilii, 603, 606, 608, 654, 655
cremaster, 12, 231–233, 237, 239, 246, 248, 249, 252, 253, 255, 343, 354
cricoarytenoid, lateral, 765, 766
posterior, 753, 755, 758, 764, 765, 770
cricothyroid, 152, 217, 228, 589, 592, 735, 744–749, 756, 763, 765, 766
oblique head of, 735
straight head of, 735
deltoid, 12–16, 20–24, 35, 37, 49–61, 63, 151, 231, 233, 523, 524, 527, 576, 579–583
depressor anguli oris, 575, 576, 597, 603, 604, 606, 608, 622, 748
depressor labii inferioris, 576, 603, 604, 606, 608, 730, 740
depressor septi, 603
depressor supercilii, 603, 604, 606, 608
digastric, 529, 580, 581, 586, 623, 693, 722, 733, 735, 740
anterior belly of, 151, 233, 576, 579, 583, 585, 587, 597–600, 606, 608, 615, 625, 647, 727, 734, 735, 739, 740, 748
posterior belly of, 151, 233, 526, 576, 583, 587, 594, 598, 608, 623, 625, 640, 647, 727, 735, 748, 750, 751, 753–755, 788
epicranius, 576
erector spinae, 331, 332, 524, 525
esophageal, circular, 755
longitudinal, 755
extensor carpi radialis brevis, 37, 47, 50–53, 57, 59, 61, 65, 74–81, 98–100, 120
extensor carpi radialis, longus, 37, 47, 50–53, 57, 61, 65, 66, 69, 74–78, 80, 81, 98, 99, 120
extensor carpi ulnaris, 37, 51, 52, 74–77, 81, 99, 100, 105
extensor digiti minimi, 74, 75, 77, 81, 99, 100
extensor digitorum, of forearm, 51, 52, 74–76, 78, 81, 99–101
extensor digitorum brevis, 463–465, 467, 469–471, 502, 505, 515
extensor digitorum longus, of leg, 431, 463–466, 469, 482, 489–491, 502, 505
extensor hallucis brevis, 463, 464, 469, 470, 502, 505, 515
extensor hallucis longus, 464, 465, 467, 469–471, 482, 490, 502, 505
extensor indicis, 77, 79, 81, 100, 101
extensor pollicis brevis, 61, 74–77, 79, 81, 100, 105, 120
extensor pollicis longus, 75, 77–79, 81, 100, 120
extensor retinaculum, 48, 112
extensors, in forearm, 37
external oblique, 7, 12, 14, 19, 20, 61, 151, 152, 219, 231–233, 235–242, 253, 255, 259, 320, 321, 332, 334, 359, 374, 376, 521–525
aponeurosis of, 239, 240, 244
extrinsic, of tongue, 741, 744, 745
facial, superficial, 603, 604
facial expression, muscles of, 603, 604
flexor carpi radialis, 47, 49, 50, 61, 65–68, 99, 100, 119

Muscle(s), *continued*
flexor carpi ulnaris, 37, 47, 50, 51, 58, 65–67, 69–71, 74, 75, 77, 79–81, 100, 101, 107, 110, 111, 119
flexor digiti minimi brevis, of foot, 494, 495, 497, 498, 504, 516
of hand, 108, 114, 119, 121
flexor digitorum brevis, 494–496, 498, 504, 507, 515
flexor digitorum longus, 434, 462, 474, 476–479, 483, 490, 504
flexor digitorum profundus, 50, 69, 71, 81, 100, 119
flexor digitorum superficialis, 50, 65–69, 80, 99–101, 119, 120
flexor hallucis brevis, 494–498, 504, 516
flexor hallucis longus, 462, 474, 476–479, 483, 490, 491, 504
flexor pollicis brevis, 108, 110, 111, 114, 119, 121
flexor pollicis longus, 61, 65, 66, 68–70, 80, 100, 120
flexor retinaculum, 47
flexors, in forearm, 37
superficial, 47
frontalis, 622, 623, 656, 657
gastrocnemius, 426, 434, 441, 464, 467, 474, 475
lateral head of, 413, 424, 427, 429, 441, 443, 447, 458, 463, 473–475, 489
medial head of, 412, 413, 427–429, 440, 441, 443, 447, 462, 474, 475, 477, 478, 489
gemellus, inferior, 427, 429, 430, 434, 523
superior, 427, 429, 430, 434, 523
genioglossus, 615, 647, 693, 694, 722, 726–728, 734–737, 739–741, 744–747, 756
geniohyoid, 615, 647, 693, 694, 722, 726–728, 734–736, 739, 741, 744–747, 756
gluteal, 424, 427, 523
gluteus maximus, 344, 346, 354, 359, 370, 371, 413, 424, 426–430, 434, 523–526
gluteus medius, 359, 419, 420, 423, 427–430, 434, 523
gracilis, 344, 370, 417–421, 423, 426, 427, 429–431, 434, 440, 441, 462, 473, 482, 486, 487
helicis major, 775
helicis minor, 775
hyoglossus, 151, 233, 576, 585–587, 615, 625, 722, 727, 735, 737, 739–741, 744, 745, 756
iliacus, 243, 321, 333, 335, 339, 340, 345, 359, 376, 419, 431
iliococcygeus, 365, 386
iliocostalis cervicis, 525
iliocostalis lumborum, 320, 525
iliocostalis thoracis, 524, 525
iliopsoas, 333, 335, 417–423, 431, 434
infraspinatus, 23, 24, 33–35, 51, 52, 59, 63, 523, 524
intercostal, 219, 230, 321
external 16, 151, 218, 219, 227, 233, 237, 238, 299, 300, 525, 526, 573, 592
innermost, 227
internal, 16, 148, 151, 218, 227, 233, 237, 238, 299, 300, 526, 573, 592
interfoveolar, 239, 243, 244, 340
internal oblique, 151, 233, 237–239, 241, 242, 244, 245, 248, 255, 320, 321, 332, 359, 374, 376, 431, 434, 522–526
interossei, dorsal, of foot, 469, 470, 500, 502, 516
of hand, 61, 79, 103–105, 108, 109, 111, 112, 114, 116, 121
palmar, 111, 114, 115, 121
plantar, 494, 495, 498, 499, 504, 516
interspinales, 526, 529
intertransversarii, 219, 526, 529, 594
ischiocavernosus 244–246, 344, 346, 354, 368, 370, 371
latissimus dorsi, 7, 12, 14, 16, 19–22, 49–52, 57, 61, 151, 231–233, 237, 238, 332, 521–525, 527
levator anguli oris, 576, 603, 606, 608, 614, 655–657, 748
levator ani, 312–314, 339, 344–346, 354, 355, 359, 365, 370, 371, 376, 382, 383, 387, 434
iliococcygeus part, 387
pubococcygeus part, 387
levator costae, 525
levator labii superioris, 576, 603, 604, 606, 608, 654, 657, 730, 748
levator labii superioris alaeque nasi, 576, 603, 604, 606, 608, 650, 654–657, 748
levator palpebrae superioris, 638, 652, 655–657, 664, 667, 668, 670, 672–676

Muscle(s), *continued*
levator scapulae, 23, 49, 51, 52, 61, 151, 233, 523–525, 527, 575–577, 592, 594, 608
levator veli palatini, 615, 647, 685, 690, 722, 735, 746–748, 750, 751, 753, 780, 793
levatores costarum, 525, 526
lingual, transverse, 736
longissimus, 525, 551
longissimus capitis, 151, 525–529, 584, 594, 647
longissimus cervicis, 525
longissimus thoracis, 320, 524, 525
longitudinal, of tongue, 737, 739, 741, 744, 745
inferior, 735
superior, 736
longus capitis, 151, 233, 593, 640, 647, 722, 747
longus colli, 151, 584, 591, 593, 594, 747
lumbrical, of foot, 494, 495, 516
of hand, 107–112, 114, 117, 121
masseter, 524, 576, 597, 603, 606, 608, 615, 622–624, 647, 727, 733, 737, 740
mentalis, 576, 603, 604, 606, 608, 730, 748
multifidus, 259, 320, 526, 528, 529
mylohyoid, 151, 233, 576, 580, 581, 583, 585–587, 597–600, 615, 625, 647, 693, 694, 722, 726–728, 733–736, 739, 740, 746–748
nasalis, 576, 603, 604, 608, 654, 748
nuchal, 574
oblique, external, 7, 12, 14, 19, 20, 61, 151, 152, 219, 231–233, 235–242, 253, 255, 259, 320, 321, 332, 334, 359, 374, 376, 521–525
aponeurosis of, 239, 240, 244
inferior, 14, 652, 656–658, 660, 664–667, 671, 673, 676, 677
internal, 151, 233, 237–239, 241, 242, 244, 245, 248, 255, 320, 321, 332, 359, 374, 376, 431, 434, 522–526
superior, 656, 657, 667, 668, 670, 672, 674–676
obliquus capitis, 526, 528, 529, 647
obturator externus, 359, 423, 429–431, 434
obturator internus, 243, 314, 345, 359, 364, 376, 387, 417, 427–431, 434, 523
occipitalis, 528, 606, 622, 635
occipitofrontalis, 527, 528, 576, 577, 603, 604, 606, 608, 654, 656, 657
belly of frontalis, 656, 657
ocular, 664–670, 672–677
of ear, 775, 776, 792
of larynx, 753, 764, 765
of orbit, 654, 655
of pharynx, 217, 228, 576, 582, 590, 597, 608, 741, 744, 745, 748, 749, 753–755
of soft palate, 753
of tongue, 736, 740
of uvula, 690, 731, 753
omohyoid, 20–22, 49–53, 61, 524, 575, 581, 582, 584–587, 592, 597, 608, 722, 735, 741, 748
belly of, inferior, 151, 233, 576, 579, 580, 583, 585
superior, 151, 152, 233, 576, 579, 583, 599, 600
opponens digiti minimi, of foot, 463, 498, 516
of hand, 107, 108, 114, 119, 121
opponens pollicis, 107, 108, 111, 112, 114, 119, 121
orbicularis oculi, 576, 603, 604, 606, 608, 650, 654, 655, 657, 658, 660
orbicularis oris, 576, 603, 604, 606, 608, 623, 690, 722, 730, 735, 737, 740, 748
orbitalis, 675
palatoglossus, 615, 731, 735, 737, 740, 741, 743–747
palatopharyngeus, 615, 690, 731, 737, 740, 743, 746, 747, 753, 755
palmaris brevis, 106, 121
palmaris longus, 50, 65–67, 99, 100, 120
papillary, 202, 209, 211, 212
anterior, 203, 204, 206, 207
posterior, 201, 203, 206, 207
septal, 201, 203
small, 203, 204
pectinate, 201–203
pectineus, 412, 417–421, 423, 431, 434
pectoralis major, 7, 12–16, 19–23, 35, 37, 49, 50, 52–54, 56, 57, 61, 63, 151, 152, 186, 231–233, 237–240, 576, 579–583, 585, 586, 592
pectoralis minor, 14–17, 20–22, 35, 49, 50, 53, 151, 152, 186, 233, 580–582
perineal, transverse, deep, 344, 354, 366, 368–370, 387, 434
superficial, 344, 366, 368–371, 387
peroneus brevis, 463–465, 467, 474, 475, 478, 479, 482, 483, 490, 491, 502, 505

Muscle(s), *continued*
peroneus longus, 412, 431, 463–467, 474, 475, 477, 479, 482, 483, 489, 490, 504–506
peroneus tertius, 467, 469, 502, 505
pharyngeal constrictor, inferior, 217, 228, 233, 576, 582, 588, 590, 597, 608, 735, 741, 744, 745, 748–750, 754, 756, 770
middle, 590, 615, 735, 741, 744, 745, 747–750, 754
superior, 615, 690, 731, 735, 737, 740, 744–747, 749, 750, 753–755
pharyngoepiglottic, 747, 753, 755
piriformis, 354, 355, 376, 417–420, 423, 427–431, 434, 523
plantaris, 424, 434, 440, 443, 463, 473, 474, 476, 478, 489
platysma, 12, 13, 575, 577, 581, 583, 591, 597, 599, 603, 604, 608, 622, 727, 739, 748
popliteus, 434, 443, 447, 448, 455, 456, 477–479, 483, 489
procerus, 603, 604, 606, 654, 655, 722, 748
pronator quadratus, 65, 66, 69, 70, 80, 101, 108, 111, 114, 119
pronator teres, 49, 50, 56, 61, 66–71, 80, 81, 99, 100
psoas major, 215, 230, 243, 285, 319–321, 328, 331–333, 335, 338, 340, 359, 419, 431, 522
psoas minor, 215, 333, 335
pterygoid, lateral, 614, 615, 624, 625, 647, 722, 750
medial, 614, 615, 624, 625, 647, 693–695, 722, 727, 734, 737, 740, 747, 750, 751, 753, 793
pubococcygeus, 365, 386
puborectalis, 386,
pyloric sphincter, 268, 287, 331
pyramidalis, 11, 237–240, 253, 255
quadratus femoris, 417, 428–431, 434, 482, 523
quadratus lumborum 219, 229, 230, 285, 321, 331–333, 335, 338–340, 431, 522
quadratus plantae, 495–498, 504, 515
quadriceps femoris, 417, 419
rectouterine, 379
rectus, inferior, 657, 664–667, 673, 675–677
lateral, 659, 664–670, 672, 675–677
medial, 657, 665–669, 674–677, 680
superior, 638, 657, 664–668, 670, 672–674, 676, 677
rectus abdominis, 11, 149, 151, 152, 230, 237–241, 243, 253, 255, 259, 300, 321, 331, 332, 335, 340, 354, 359, 374, 393
insertion of, 240
rectus capitis anterior, 151, 594, 647
rectus capitis, lateralis, 151, 529, 594, 647
rectus capitis posterior major, 526, 528, 529, 647
rectus capitis posterior minor, 526, 529, 647
rectus femoris, 412, 417–424, 431, 434, 647
rhomboid major, 23, 49, 51, 52, 523, 524, 527, 528
rhomboid minor, 23, 49, 51, 52, 523, 524, 527, 528
risorius, 575, 603, 604, 606, 730
rotatores, 526
sacrococcygeus, 365, 376
ventral, 386
sacrospinalis, 522
salpingopharyngeus, 615, 690, 746, 753, 755
sartorius, 412, 417–424, 431, 434, 440, 462, 482, 486, 487
scalene, 583, 584
scalenus anterior, 20, 61, 151, 152, 184, 186, 224, 233, 576, 580, 582, 587, 591–594, 608
scalenus medius, 61, 151, 233, 524, 529, 575, 576, 587, 591–594, 608
scalenus posterior, 61, 151, 233, 524–526, 529, 587, 593, 594, 608
semimembranosus, 413, 424, 426–431, 434, 440, 441, 443, 462, 473–475, 483, 486, 487, 523
semispinalis capitis, 233, 523–529, 576, 604, 608, 647
semispinalis cervicis, 525, 526, 528, 529
semispinalis thoracis, 525, 526
semitendinosus, 370, 426–428, 430, 431, 434, 441, 443, 462, 473, 475, 482, 486
serratus anterior, 7, 12, 14, 16, 19–21, 49–51, 61, 151, 231–233, 237–240, 255, 524
serratus posterior inferior, 332, 524, 525
serratus posterior superior, 533–535
soleus, 412, 434, 443, 462–464, 467, 473–479, 483, 489, 490
sphincter, pupillary, 680
pyloric, 268, 287, 331
urethral, 354, 366, 387
vaginal, 387

Muscle(s), *continued*
sphincter ani, external, 312, 313, 343–346, 354, 368, 370, 371, 376, 387
internal, 312, 343, 376
spinalis capitis, 525, 526
spinalis cervicis, 525
spinalis thoracis, 524, 525
splenius capitis, 61, 233, 523–529, 575, 576, 581, 594, 604, 608, 647
splenius cervicis, 523–525, 529
stapedius, 640, 695, 789, 793, 798
sternalis (variation), 151
sternocleidomastoid, 12, 13, 15, 18, 20, 50, 52, 61, 151, 152, 231, 233, 257, 523, 527, 574–577, 579–581, 583–586, 588, 591, 594, 603, 604, 606, 608, 622, 623, 722, 727
sternohyoid, 52, 61, 148, 151, 152, 233, 576, 579, 582–584, 586–588, 591, 592, 597, 599, 600, 608, 647, 722, 727, 735, 741, 748
sternothyroid, 148, 151, 152, 233, 576, 579, 582–584, 587, 591, 592, 608, 726, 735, 766
styloglossus, 587, 608, 625, 647, 722, 727, 735, 737, 740, 741, 744–748, 750, 756
stylohyoid, 151, 233, 576, 580, 583, 585, 587, 597–600, 608, 615, 623, 625, 647, 722, 727, 735, 740, 744, 745, 748, 750, 753, 788
stylopharyngeus, 647, 722, 735, 737, 740, 744, 745, 747–750, 753, 754, 770
subclavius, 16, 21, 35, 50–53, 151, 152, 184–186, 188, 227, 592
subcostales, 333
sublingual, 735
subscapularis, 16, 21, 22, 31, 35, 49, 50, 53, 55, 63, 151, 233
supinator, 50, 66–69, 72, 77–81, 99
supraspinatus, 21, 23, 24, 33–35, 49–53, 57, 63, 520, 524
temporalis, 608, 615, 623–625, 635, 647, 669
temporoparietalis, 576, 603, 604, 622, 623
tensor fasciae latae, 412, 413, 416–419
tensor tympani, 695, 780, 781, 791–793, 795, 796, 798, 801
semicanal for, 778, 796
tensor veli palatini, 615, 647, 694, 735, 746–748, 753, 755, 793
teres major, 16, 19, 21–24, 34, 49–53, 56–61, 63, 151, 233, 523, 524, 761
teres minor, 23, 24, 33–35, 51, 52, 59–61, 63, 523, 524
thyroarytenoid, 765, 770
thyrohyoid, 152, 233, 576, 583, 586–588, 592, 597, 608, 722, 735, 741, 744, 745, 748, 766
tibialis anterior, 412, 431, 462–467, 482, 489, 490, 504, 506
tibialis posterior, 434, 476–479, 483, 490, 504, 506
tragicus, 775
transverse lingual, 736
transverse perineal, deep, 344, 354, 366, 368–370, 387, 434
superficial, 344, 366, 368–371, 387
transversospinal, 526
transversus abdominis, 151, 238–243, 245, 253, 255, 259, 321, 332, 333, 335, 338, 340, 359, 376, 431, 522, 526
transversus thoracis, 148, 149, 152, 299
trapezius, 13, 21, 23, 50–52, 55, 61, 523, 524, 527, 528, 575–577, 580, 583–585, 592, 604, 608, 623
triangularis, 576
triceps, 7, 14, 37, 49, 51, 60, 61, 75, 76, 81, 82, 98, 524
lateral head, 23, 49, 51, 52, 57–61, 64, 75, 523
long head, 21–24, 29, 31, 33–35, 49–53, 55–61, 63, 64, 523
medial head, 49, 51–53, 55–59, 61, 64–66, 74, 75, 77, 81
vastus intermedius, 420, 423, 431, 434, 486
vastus lateralis, 412, 417–420, 423, 424, 426, 427, 429, 431, 434, 440, 444, 446, 463, 467, 486, 487
vastus medialis, 412, 417–421, 423, 431, 434, 440, 444–446, 462, 486, 487
vocalis, 766, 770
zygomaticus major, 576, 604, 606, 622, 623, 647, 654–657, 730, 748
zygomaticus minor, 576, 603, 604, 606, 654–657, 748
Muscular coat, of stomach, 268, 270
Musculotubal canal, 594
Myocardium, 201, 202, 206, 208, 209
Myometrium, 380

Nail. See *Finger nail*
Nares, 602, 726, 732
Nasal cavity, 613, 669, 684, 685, 691, 692, 723
  lateral wall of, 684, 685, 691, 692, 723
  nerves of, 694
Nasal vestibule, 685, 690, 691, 732
Nasolacrimal duct, 658, 660
  opening of, 685
Nasopharynx, 746, 747
  mucosa of, 746
  muscles in, 746
  openings of, 685
Neck, 574–588, 591–594
  arteries of, 577–582, 586
  features of, external, 574
  lower, nerves and vessels in deep part of, 588
  muscles of, 523, 525–527, 583
  nerves of, 577–582
  root of, 83, 152, 592–594
    superficial structures at, 592
  triangles of, 576
  veins of, 585, 586
Nerve(s), abducens, 638–642, 644, 670, 672, 674, 676, 788, 801
  accessory, 18, 257, 528, 566, 577, 579–582, 638–640, 642, 644, 754, 755, 788, 801
    spinal and bulbar roots, 801
  alveolar, inferior, 623–625, 693, 694, 696, 724, 737, 739
    superior, 624, 625, 695, 696, 725
  ampullar, anterior, 800, 804
    posterior, 801
  anococcygeal, 336, 346, 371
  auricular, great, 517, 527, 575, 577, 622
    posterior, 577, 623, 787, 794
  auriculotemporal, 607, 611, 622–625
  axillary, 18, 19, 22, 24, 56, 60, 63, 517
  buccal, 611, 622, 624, 625, 724, 725, 737
    branches of, 740
  calcaneal, 493, 515
  cardiac, cervical, inferior, 226
      superior, 188, 219, 582, 591, 754
      sympathetic, 588
    middle, 219, 220, 228, 582, 593, 754
    sympathetic origin, 185, 593
    vagal origin, 185
  caroticotympanic, 695, 793, 794
  carotid, internal, 640, 695, 754
  cervical, 228, 528, 581, 592
    openings for, 722
    rami, ventral, 580, 593
    rootlets of, 540, 639
    view of, posterior, 755
  chorda tympani, 625, 640, 693–695, 742, 747, 781, 755–789, 791–795, 801
  ciliary, long, 674
    short, 674
  cluneal, inferior, 336, 346, 371, 407, 425, 426, 517
    medial, 517, 521
    middle, 407, 425, 426
    superior, 407, 425, 517, 521
  coccygeal, 339, 364, 567
  cochlear, 801–804
  cranial, 639
  cutaneous, antebrachial, lateral, 38–41, 47, 54, 56, 58, 60, 98–101, 106, 517
      hiatus for, 47
      medial, 14, 18, 22, 38–41, 54, 63, 64, 98–100, 330
        anterior branch, 38
        ulnar branch, 38
      posterior, 39, 40, 42, 43, 48, 58–60, 98–100, 102, 112, 517
    anterior, 10, 11, 13, 20
      hiatus for, 48
    brachial, anterior, 14
      lateral, axillary branch, 24, 38, 39, 42, 58, 64
      medial, 14, 22, 38–43, 54, 64, 517
      posterior, 38, 39, 43, 58, 60, 64, 517
    cervical, transverse, 577
    crural, medial, 460, 468
    distribution of, in lower extremity, 405, 407
      in upper extremity, 38, 39
    dorsal, intermediate, 461
      lateral, 10, 460, 468
      medial, 461, 527
    femoral, anterior, 11, 337, 405, 407, 409, 414, 486, 487
      lateral, 317, 318, 320, 321, 336–338, 340, 359, 405, 407, 408, 414, 418, 425, 517
      medial, 409
Nerve(s), cutaneous, *continued*
      posterior, 336, 346, 371, 407, 408, 425, 426, 428, 460, 486, 517
    iliohypogastric, anterior, 245, 336
      lateral, 336, 405, 407
    intercostal, anterior, 232, 239, 240, 585
      lateral, 232
    interosseous, anterior, 70
    lateral, 10, 11, 13, 14, 18–20
      mammary branches, 19
    obturator, 405, 407, 414, 421, 422, 486, 487
    of foot, dorsal, 515
    of forearm, 67, 68, 70–72
    of hand, 112
    peroneal, superficial, 515
    plantar, lateral, 493, 516
      medial, 493, 496, 516
    radial, posterior, 42
    saphenous, medial, 489
    subcostal, anterior, 232, 246, 521
      lateral, 339
    sural, 468
      lateral, 397, 405, 428, 441, 460, 487, 489, 490, 516, 517
      medial, 426, 428, 460, 487, 489
    ulnar, 43, 106
      dorsal, 70, 76, 78
      palmar, 106
  digital, of foot, 468, 471, 493, 496, 516
    of hand, 38, 39, 102, 106, 110–112, 122–124
      dorsal, branch, 39, 112
      palmar branch, common, 38
        proper, 38, 39, 112, 123, 124
  dorsal, of clitoris, 245, 369, 371
    of penis, 246, 346, 347, 366
  esophageal, 220, 573
  ethmoid, anterior, 674, 676, 692, 693
    posterior, 674, 676
  facial, 225, 577, 605, 609, 622–625, 638–640, 644, 694, 695, 787–789, 791, 793, 794, 796, 798, 799, 801, 802, 804
    auricular branch, posterior, 580, 605, 625, 787, 794
    buccal branch, 605, 622, 654, 655, 657
    cervical branch, 579, 597, 599, 605, 622
    digastric branch, 623, 625, 788
    genu of, 795, 796
    mandibular branch, 599, 605, 622
    stylohyoid branch, 788
    temporal branch, 605, 622, 650, 654, 656
    zygomatic branch, 605, 622, 654–657
  femoral, 243, 317, 336–338, 340, 341, 359, 418, 421, 422
    branch to iliopsoas muscle, 321, 336, 338
    cutaneous nerve fields of, in leg and thigh, 409
    muscular branch, 421, 422, 486
  frontal, 517, 611, 640, 670
  genicular, 475
  genitofemoral, 11, 239, 317, 318, 320, 321, 336–338, 340, 359, 364, 383, 405
    femoral, branch, 321, 336–338, 340, 405, 408, 414
    genital branch, 243–246, 321, 336–338, 340, 347, 364, 408
  glossopharyngeal, 225, 565, 609, 638, 639, 642, 644, 692, 724, 737, 740, 742, 746, 747, 754, 755, 758, 788, 794, 801
    hiatus for passage of, 735
    lingual branch, 692, 758
    pharyngeal branch, 754, 793
    tonsillar branch, 692, 758
    tympanic branch, 607
  gluteal, inferior, 336, 428, 430
    superior, 451, 430
  gluteus maximus, 523
  hypoglossal, 565, 566, 580–582, 585, 586, 597, 600, 608, 625, 638–640, 642, 644, 727, 737, 739, 740, 746, 747, 754–756, 801
  iliohypogastric, 11, 218, 232, 239, 240, 244–246, 317, 318, 321, 331, 332, 336–338, 340, 359, 405, 408, 414, 425, 517, 521
  ilioinguinal, 11, 232, 239, 240, 244–246, 248, 317, 318, 321, 336–338, 340, 341, 347, 359, 408, 414
  incisive, 690
  infraorbital, 608, 611, 622–625, 650, 655–657, 659, 675, 676, 694–696, 793
    labial branches, superior, 622, 725
    nasal branches, external, 622
    palpebral, branch, inferior, 623, 654, 655
  infratrochlear, 611, 622–625, 650, 654–657, 674, 676
Nerve(s), infratrochlear, *continued*
    nasal branch, 654
    palpebral branch, 655
      inferior, 654
  intercostal, 11, 184, 185, 218, 220, 224, 226–230, 239, 240, 255, 299, 321, 359, 521, 526, 563, 573
    meningeal branch, 226
    ramus of, anterior primary, 219, 518
      posterior primary, 219, 518
    thoracic branch, 226
  intercostobrachial, 14, 18–20, 40, 582
  intermedius, 638, 788
  interosseous, anterior, 100
    posterior, 78, 100, 105
  jugular, 754
  labial, anterior, 244, 245
    posterior, 371
  lacrimal, 611, 640, 659, 670, 672, 674
    branches of, 653
    palpebral branches, 650, 654–656
  laryngeal, inferior, 758
    internal, 588, 756
    recurrent 176, 177, 185–187, 218, 228, 582, 586, 588, 591–593, 699
      esophageal, branches, 588, 758
      left, 176, 188, 189, 218–220, 224, 586, 588, 591, 592, 755
      inferior cardiac branch, 186
      right, 186, 218–220, 224, 755
    superior, 581, 592, 597, 600, 640, 737, 740, 741, 744–746, 748, 751, 753–755, 758, 770, 793
      external branch, 593, 754
      internal branch, 217, 593, 754, 758
  lingual, 586, 611, 624, 625, 693, 694, 696, 724, 727, 731, 737, 739, 740, 742, 746, 747, 756
    branches of, 740, 746, 747
      to tongue, 740
  lumbar, 320, 332, 339, 354
  mandibular, 625, 638, 639, 670, 676, 694, 696, 740, 801
    branches of, 625
  masseteric, 623–625, 740
  maxillary, 638–642, 670, 672, 676, 694–696, 788, 801
  median, 7, 14, 18, 22, 38, 41, 51, 54, 56, 63, 64, 67, 68, 70–72, 98–101, 106, 108, 110, 120, 121
    lateral cord contribution to, 14
    palmar branch, 38, 67, 101
  meningeal, 638
  mental, 611, 622–624, 696, 724
    labial branches, 740
  musculocutaneous, 14, 18, 22, 54–56, 63, 64
  mylohyoid, 580–582, 599, 600, 624, 625, 693, 694, 737, 739, 740
  nasal, external, 611, 622–625, 654–657, 690, 692, 694
  nasal septal, 692
  nasociliary, 610, 670, 672, 779, 793
    branch of, 793
  nasopalatine, 690, 692–694, 725
  obturator, 243, 336–338, 359, 364, 376, 383, 405, 408, 421, 422, 517
  occipital, greater, 517, 527, 528, 577, 579, 582, 622, 623
    lesser, 517, 527, 577, 579, 582, 622, 623
  oculomotor, 225, 639–642, 644, 670, 672, 674, 676, 788, 793, 901
    inferior branch, 674–676
    superior branch, 672, 674
  of neck, 588
  of orbit, 654, 655
  of palate, 692, 723, 725
  of pancreas, 279
  of penis, 248, 347
  of porta hepatis, 279
  of pterygoid canal, 640, 693–695, 746, 794
  of spermatic cord, 347
  of stapedius muscle, 640
  of stomach, 279
  of upper extremity, cutaneous, 38, 39
  olfactory, 690, 692
    lateral, 691
    medial, 690, 691
  ophthalmic, 638–642, 670, 672, 674, 676, 694–696, 788, 801
    frontal branch, 793
    lacrimal branch, 793
  optic 636–642, 644, 659, 664, 666–670, 672–677, 679, 793
  palatine, 611, 640, 693

Nerve(s), palatine, *continued*
        greater, 692–694, 725, 731, 747
        lesser, 692, 725, 731, 747
        posterior, 693
    pectoral, 15, 18, 582
    perineal, 346, 366, 371
    peripheral nerves, cutaneous fields of, in thigh, 408
    peroneal, common, 336, 346, 347, 405, 426, 428, 441, 463, 465, 486, 487, 489
        cutaneous fields of, in leg and foot, 411
        deep, 405, 411, 465, 468, 471, 490, 491
            lateral cutaneous branch, 411
        superficial, 405, 411, 461, 465, 468, 490, 491, 515
    petrosal, deep, 640, 693, 695
    greater, 638, 640, 642, 645, 693–695, 788, 793–796, 798, 801, 802
        lesser, 607, 638, 642, 645, 694, 695, 793, 794, 801
    pharyngeal, 747
    phrenic, 20, 152, 175, 176, 184, 189, 220, 228, 229, 318, 573, 580–582, 586, 592, 593, 596
        accessory, 592
        left, 176, 177, 185–188, 208, 209, 218, 230
        right, 18, 177, 184, 186–188, 208, 209, 230
    plantar, lateral, 407, 410, 493, 496, 497, 515
        medial, 407, 410, 493, 496, 497, 515
    pterygoid, lateral, 625
        medial, 694, 793
    pterygopalatine, 640, 694, 695, 793
    pudendal, 336, 339, 345, 346, 359, 364, 366, 371
    radial, 18, 22, 38, 51, 54, 56, 58, 60, 63, 64, 67, 68, 70–72, 98, 112, 517
        deep branch, 67, 68, 70, 72, 76, 78, 81, 99
        superficial branch, 38, 39, 41, 43, 67, 68, 70, 72, 76, 78, 99–102, 112
    rectal, inferior 336, 346, 364, 371
    saccular, 800, 803, 804
    sacral, 320, 339, 343, 364
    saphenous, 374, 407, 409, 421, 422, 460, 468, 486, 487, 489–491, 517
        infrapatellar branch, 407, 414, 460, 487
    scapular, dorsal, 18, 527, 528, 592
    sciatic, 336, 428, 430, 486, 523
    scrotal, anterior, 246
        posterior, 346
    septal, 690
        nasal branch, 692
    spinal, 10, 518, 564, 568
        cervical, 527, 565
        lateral branch, 10, 518
        lumbar, 565, 567
        medial branch, 10, 518
        meningeal branches, 10, 572, 573
        ramus of, anterior, 10, 571, 572
            anterior primary, 573
            dorsal, 10
            posterior, 10, 517
            posterior primary, 518, 563, 571–573
        ramus communicans, 10, 518
        root of, dorsal, 10, 563
                dorsal root, ganglion, 10
            ventral, 10, 571, 573
        sacral, 354, 565
        thoracic, 320, 527, 563, 565
    splanchnic, greater, 10, 184, 185, 218–220, 224, 226–230, 299, 300, 320, 331, 328, 573
            root of, 10
        lesser, 218, 219, 224, 226, 229, 230, 299, 300, 320, 338
        lumbar, 318, 320, 332
    stapedial, 789, 793
    subclavian, 185
    subcostal, 218–220, 224, 226, 239, 240, 244, 246, 255, 332, 337–339
    sublingual, 740, 746, 747, 756
    submandibular, 740
    suboccipital, 528
    subscapular, 22, 24
    supraclavicular, 11, 13, 517, 577
        posterior, 40, 42
    supraorbital, 622–624, 650, 654–657
        ramus of, lateral, 650, 656, 657, 670, 672
            medial, 650, 656, 657, 670, 672
            superior palpebral, 654, 655
    suprascapular, 18, 22, 24 582
    supratrochlear, 611, 622–625, 650, 654–657, 670, 672
    sural, 405, 407, 410, 411, 460, 490, 491, 515, 517
        cutaneous nerve fields, of, in leg and foot, 410, 411

Nerve(s), *continued*
    temporal, deep, 624, 625
    thoracic, 14, 226, 228, 300, 338, 339, 405, 517, 527
        cutaneous branch of, lateral, 14
        intercostal branch, 226
        long, 14, 18–20, 582
        rami of, posterior, 517, 527, 528
        ventral root of T-5, T-6, T-7, T-8, T-10, 226
    thoracodorsal, 18–20, 22, 582
    thyrohyoid, 580, 597, 600
    tibial, 336, 426, 428, 441, 475, 476, 479, 486, 487, 489–491
        calcaneal branch, 410
        cutaneous nerve fields of, in leg, 410
    to coccygeus muscle, 336
    to digastric muscle, 793
    to gemelli muscles, 336
    to levator ani muscle, 336
    to levator scapulae muscle, 528
    to nasalis muscle, 656
    to obturator internus muscle, 336
    to omohyoid muscle, 592
    to pectineus muscle, 418
    to pterygoid muscle, medial, 747, 793
    to quadratus femoris muscle, 336
    to subclavius muscle, 592
    to tensor tympani muscle, 694, 793
    to tensor veli palatini muscle, 694, 747, 793
    to teres minor muscle, 24
    trigeminal, 610, 637–639, 644, 670, 672, 674, 676, 693, 695, 696, 788, 801
        alveolar branches, inferior, 611
            superior, 611
        mandibular branches, 611
        root of, motor, 639, 642, 694, 695, 793, 801
            sensory, 642, 694, 801
        semilunar ganglion of, 639
    trochlear, 638–642, 670, 672, 674, 801
    tympanic, 640, 695, 793–796, 798, 799
    ulnar, 14, 18, 22, 41, 51, 54, 56, 58, 60, 63, 64, 67, 68, 70, 71, 76, 78, 98–102, 106, 108, 110, 111, 120, 517
        dorsal branch, 39, 67, 101
        palmar branch, 38, 67
    utricular, 800, 803
    utriculoampullar, 803
    vagus, 20, 184, 224, 225, 228, 565, 581, 582, 591, 593, 608, 609, 625, 638, 639, 642, 644, 695, 737, 740, 754, 755, 788, 794, 801
        auricular branch, 622–625, 640, 787, 789, 793, 794
        bronchial branches, 218, 220, 224, 228, 229
        cardiac branches, 186, 188, 219, 220, 228, 582, 592, 593, 754, 755
            inferior, 592
            middle, 592
            superior, 592
        esophageal branches, 218, 220, 229, 230
        left, 176, 177, 185–188, 208, 209, 218, 220, 224, 318, 573, 586, 588, 591, 592, 755
        pharyngeal branch, 754, 793
        recurrent laryngeal branch, 177, 184, 208, 219, 220, 593
        right, 186, 187, 189, 218–220, 224, 228, 229, 573, 580, 582, 586, 592, 755
        superior ganglion of, 755
        tracheal branches, 228, 593
    vertebral, sympathetic, 226, 593
    vestibular, 801–804
    vestibulocochlear, 638, 639, 642, 644, 694, 788, 800
    zygomatic, 664, 659, 676, 695
    zygomaticofacial, 622, 623, 654–657
        cutaneous branch of, 650
    zygomaticotemporal, 611, 622, 654
Nervus intermedius, 639, 695, 801
Network, arterial, calcaneal, 479, 496
        dorsal carpal, 62, 105
        genicular, 422, 465
        malleolar, 465, 471
        nasal septal, 690
        patellar, 414
        trochanteric, 428
    venous, of dorsal foot, 460
        of dorsal hand, 102
Nipple, 3, 4, 7–9, 19, 136, 172, 232, 399
Node, atrioventricular, 210, 213
    sinoatrial, 213
Nodule of semilunar valve, 212
Nose, conchae of, 601, 684, 685, 688, 691–693, 732, 751
    external, bones of, 682

Nose, external, *continued*
        cartilages of, 655–657, 682, 683, 690, 726
    septum of, 601, 669, 683, 690, 692, 726, 751, 753
    vestibule of, 685, 690, 691, 732
Nostrils, 602, 726, 732
Notch, angular, of stomach, 269, 273
    cardiac, 153, 170
    interarytenoid, 751, 755, 758, 767, 772
    intertragic, 773
    jugular, 136, 257
    mandibular, 700
    mastoid, 646, 777, 787
    of Rivinus, 786
    sciatic, greater, 389, 390, 401
        lesser, 389, 390
    supraorbital, 652, 689
    thyroid, superior, 136, 257, 757, 761, 763
    trochlear, of ulna, 9, 87, 88, 93, 133
    tympanic, 785, 786

Occipital condyle, 610, 646, 748
Occipital, crest, internal, 645
Occipital protuberance, external, 523, 528, 574, 646, 647
    internal, 645
Odontoid process. See *Dens*
Olecranon, 43, 51, 57–59, 61, 71, 74, 75, 77, 78, 88, 92, 98
Olfactory bulb, 639, 644, 690, 691
Olfactory stria, 639
Olfactory tract, 639, 644, 690, 691
Olive, 639
Omentum, greater, 256, 259–261, 265, 266, 270, 273, 275, 277, 279, 296, 299–302, 305, 307, 310, 331, 332, 343, 399
    lesser, 111, 261, 264, 273, 300
Opening(s). *See also Orifice*
    adductor canal, opening of, 440
    atrioventricular, 202, 204, 206, 212–215
    bulbourethral gland, opening of, 361
    coronary sinus, opening of, 180, 191–193, 203, 208, 213–215
    ejaculatory duct, opening of, 356, 361
    frontal sinus, opening into, 685, 686
    lymphatic duct, right, opening of, 587
    nasolacrimal duct, opening of, 685
    nasopharynx, opening of, 685
    pancreatic duct, opening of, 284
    paranasal air sinuses, opening of, 685, 686
    paraurethral ducts, opening of, 367
    preputial, 343
    prostatic ductule, opening of, 356
    sphenoethmoidal sinus, opening of, 685
    sublingual gland, opening of, 737
    tympanic, for auditory tube, 778, 786, 791, 795, 798
    ureter, opening of, 356, 357, 361, 377
    urethral, external, 343, 351, 361, 367, 368, 370–372, 376, 377, 388
        internal, 343, 356, 357, 361
    uterine, 377, 380
    vaginal, 367, 368, 370, 372, 388
    vestibular gland, greater, opening of, 367, 368, 370
Optic canal, 645, 662, 669, 670, 676
Optic chiasma, 639, 644, 668
Optic disc, 679, 680
Ora serrata, 669, 680
Oral cavity, 694, 723–725, 729, 732, 737–740, 746, 747
    mucosa of, 603, 694, 699, 734, 735, 737, 739, 744, 746, 747
    musculature of, 735
    vessels and nerves of, 740, 746, 747
Orbit, 653–657
    apex of, 675
    dissection of, anterior, 654–657
    eyeball, exposure of, 657
    muscles of, superficial, 654
    vessels and nerves of, 654–656
        superficial, 654
Orbital cavity, bony, 661–663
    wall of, lateral, 663
        medial, 662
Orbital septum 650–652, 655–657, 673
Orifice, cardiac, of stomach, 268, 299, 316
    of appendix, 306
    See also *Opening(s)*
Oro-naso-pharyngeal lymphatic ring, 752
Oropharynx, 732, 746, 747
    muscles in, 746
Osseous spiral lamina, of cochlea, 800

Ossicles, of middle ear, 792, 797
Ovary (Ovaries), 310, 316, 318, 372, 374, 377–382, 399
  developing, 377
  stroma of, 360

Pad, distal digital, 37
Palate, 725
  arteries of, 692
  folds of, muscular, 685, 731
    transverse, 701
  glands of, 690, 701, 722, 731, 746, 747
  hard, 685, 690, 701, 702, 721, 722, 726, 729, 732, 747
  mucosa of, 735, 747
  nerves of, 692, 747
  soft, 685, 701, 726, 729, 732, 747, 751
  vessels of, 747
Palatine process, of maxilla, 631, 683, 684, 702
Palatine tonsil. See *Tonsil, palatine*
Palatoglossal arch, 729, 731, 732, 735, 737, 743
Palatopharyngeal arch, 726, 729, 731, 732, 735, 737, 751, 755
Palm, of hand, 1, 37–39, 106, 107, 520, See also *Hand*
  arteries of, 62, 106, 110, 111
    superficial, 106
  creases of, 37
  eminence of, hypothenar, 37
    thenar, 37
  monticuli of, 37
  muscles of, 37, 107, 114, 115
  nerves of, 38, 39, 106, 110, 111
    superficial, 106
  surface anatomy of, 37
  synovial sheaths of, 107
  tendons of, 107, 112
Palmar arch, deep, 62, 111, 113, 182
  superficial, 62, 110, 111, 113, 182, 183
Palpebra(e). See *Eyelid(s)*
Palpebral commissure, lateral, 648, 677
  medial, 677
Pancreas, 215, 261, 265, 266, 279, 285, 288–293, 300, 303, 316, 331, 332
  body of, 275, 277, 286, 302, 305
  duodenum and, 285
  head of, 25, 258, 272, 285, 296, 310, 316, 317, 332
  lymphatic vessels of, 279
  nerves of, 279
  tail of, 258, 272, 277, 285, 287, 305, 317, 331
  unciate process of, 286
Pancreatic duct. See *Duct, pancreatic*
Pancreatic duct system, 287
Papilla(e), conical, 743
  duodenal, 268, 283, 284, 287–293
    greater, 332
  filiform, 743
  foliate, 732, 743, 746
  fungiform, 743
  incisive, 701
  interdental, 740
  lacrimal, 649, 658, 677
    inferior, 658
  optic, 681
  renal, 325, 327, 328
  vallate, 692, 732, 743, 751, 753, 755, 758
Papillary process, of caudate lobe of liver, 305
Paradidymis, 363
Paroophoron, 372
Pars flaccida, of tympanic membrane, 766, 782, 786
Pars iridica retinae, 680
Pars optica retinae, 680
Pars tensa, of tympanic membrane, 782, 784, 786
Pecten, of pubis, 335, 389, 393, 417, 419, 420
Pedicles, of vertebrae, 553
    lumbar, 555
    thoracic, 547
  of vertebral arches, 557, 573
Peduncle, cerebral, 639
Pelvis, brim of, 359
  female, 316, 345, 377, 378, 385–388, 393, 396, 398
    ligaments of, 394
    outlet of, 394
    skeleton of, 388, 393
  lesser (female), 385
  lymph nodes and vessels of, 383
  male, 321, 343, 345, 359, 395, 397
  minor, 316, 321, 398
    diagrammatic section through, 385
  muscular floor of, 365, 386, 387
  nerves of, 340
Pelvis, *continued*
  radiograph of, 401
  renal, 285, 320, 323, 325–328, 331, 332, 363, 372
  section through, diagrammatic, 385
  skeleton of, 389–392
  vessels of, 340
Penis, 1, 12, 242, 348–351
  bulb of, 343, 352, 359, 361
  capsule of, fibrous, 348
  corpus cavernosum, 248, 348, 352, 355, 363
  corpus spongiosum, 248, 343, 344, 352, 363
  erectile bodies of, 352
  ligaments of, fundiform, 12, 231–233, 236, 343
    suspensory, 12, 232, 234, 239, 246, 248, 347
  nerves of, 347
  prepuce of, 246, 343, 349, 351, 361
  vessels of, 347
Pericardial cavity, 159, 215, 299
Pericardium, 170, 175, 176, 184–186, 188, 189, 191, 195, 208, 209, 212, 214, 215, 229, 256, 257, 399
  aortic recess of, 179, 188, 194, 214
  diaphragmatic, 170, 189, 214, 573
  over pulmonary surface of heart, 170
  parietal, 152, 159, 177, 189, 194, 204
  visceral, 159, 204
Perichorioidal space, 680
Perimetrium, 380
Perineum, central tendinous point of, 344, 367, 370, 376, 387
  female, 370
  male, 345, 346
Periorbita, 664, 667, 675
Periosteum, 606, 608, 664, 675
Peripheral nerves of thigh, cutaneous nerve fields of, 408
Perirenal fat, 332
Perithyroid space, 591
Peritoneal cavity, 215, 262, 299, 300, 331, 332
Peritoneal fold, over hepatic artery, 300
Peritoneal reflections, 264, 312, 353, 354, 373
Peritoneum, 255, 259, 268, 280, 316, 317, 331, 335, 340, 345, 352, 354, 379
  diaphragmatic, 321, 331
  inferior reflection of, over pelvic organs, 312, 353, 354
  parietal, 253, 259, 262A, 299, 300, 316, 321, 331, 332, 340, 359, 377
  rectouterine pouch, 181, 264, 305, 313, 316, 321, 343, 352–354, 373, 374, 377, 379, 383
  retovesical pouch, 305, 313, 321, 343, 352–354
  vesicouterine pouch, 264, 316, 373, 374, 378
  visceral, 299, 300, 331, 332
Pes anserinus, 420, 423, 444, 462, 489
Phalanx (Phalanges). See *Bone(s), phalanx (phalanges)*
Pharyngeal cavity, 691, 726, 732, 751, 770
Pharyngeal, recess, 685, 691, 726, 732, 751
Pharyngoepiglottic fold, 726, 746, 751, 753, 755
Pharynx, 258, 692, 723, 729, 731, 740, 744–747, 750, 751, 760
  junction with esophagus, 735
  muscular wall of, 747
  vessels and nerves of, 747, 754
Pia mater, 571, 633
Piriform recess, 751, 753, 755, 758, 770
Pituitary gland, 636–639, 641, 642, 685
  infundibulum of, 636
  lobes of, anterior (adenohypophysis), 636
    intermediate, 636
    posterior (neurohypophysis), 636
  sagittal section through, 636
Placenta, 181, 399
  septum of, 181
  sinus of, marginal, 181
Plate(s), cribriform of ethmoid bone, 645, 683, 684, 686, 687, 689
  perpendicular, of ethmoid bone, 613, 683, 687–689
    of palatine bone, 684
  pterygoid, medial, hamulus of, 747
  tarsal, 651, 652, 656
Pleura, 136, 155, 156, 164, 165, 170, 219, 257, 300, 320
  cervical, 227, 592
  cupula of, 153–155, 170, 215, 219, 227, 258, 593
  costal, 154, 170, 175, 184, 185, 215, 219, 220, 227, 256, 299, 320, 331, 573, 593
  diaphragmatic, 152, 154, 175, 177, 185, 186, 188, 189, 208, 209, 212, 215, 220, 227, 256, 299, 331, 333, 573
  margin of, anterior, 170
Pleura, margin of, *continued*
    inferior, 170, 320
  mediastinal, 154, 170, 175–177, 184, 185, 188, 189, 212, 220, 227, 255, 256, 258
  parietal, 136, 152, 157, 158, 170, 227, 258, 300
  recess of, costodiaphragmatic, 150, 152, 154–156, 170, 175, 186, 215, 219, 220, 230, 257, 258, 273, 299, 300
    costomediastinal, 152, 153, 158, 170, 189, 299
    phrenicomediastinal, 300
  vertebral, 219
  visceral, 157, 159, 215, 299
Pleural, cavity, 152, 157, 158, 299, 300, 331
Plexus, abdominal aortic, 226
  aortic, 185, 219, 224, 229, 339, 573
    ramus to, 227
  basilar, 642
  brachial, 19–22, 184, 185, 187, 188, 227, 228, 576, 577, 579–583, 586, 588, 596, 608
  cardiac, 184–186, 188, 189, 208
  carotid, 152, 638, 640, 693, 754, 755, 793, 794
  celiac, 318, 339
  cervical, 577, 579
  choroid, 639, 644
  coccygeal, 336, 337
  dental, inferior, 696
  esophageal, 184, 185, 224
  gastric, 224, 318
  hypogastric, 318, 340
  iliac, 339
  lumbar, 320
  lumbosacral, 336, 337
  marginal, of cornea, 678
  mesenteric, 296, 318, 339
  parotid, 605
  pelvic, 225
  pharyngeal, 754
  pterygoid, 620, 793
  pudendal, 336, 339, 364
  pulmonary, 188, 219, 220, 229
  renal, 339
  sacral, 336, 337, 339, 364
  tympanic, 793, 794
    branches to cochlear and vestibular windows, 793
  venous, biliary, 331
    pampiniform 248, 254, 255, 347
    prostatic, 359
    thyroid, 152, 586, 591
    vaginal, 382
    vertebral, 227, 573
      internal, 226
    vesical, 240, 355, 382
Plica fimbriata, 735, 738, 741, 756
Pole(s), of brain, frontal, 635, 639
    orbital gyri, 639
    orbital sulci, 639
    occipital, 635, 639
    temporal, 635, 639
  of kidney, 302, 305, 316, 328
Pons, 566, 639
Porta hepatis, 262B, 316
  lymphatic vessels of, 279
  nerves of, 279
Pouch, of Douglas, 264, 373
  rectouterine, 181, 264, 305, 313, 316, 321, 343, 352–354, 373, 374, 377, 379, 383
  rectovesical, 305, 313, 321, 343, 352–354
  vesicouterine, 264, 316, 373, 374, 378
Premaxilla, 631
Premolar teeth, 699, 701, 705, 710, 711, 714, 715, 717, 718
Prepuce, of clitoris, 244, 367, 368, 370, 388
  of penis, 246, 343, 349, 351, 361
Processus vaginalis, 253, 255
Prominence, laryngeal, 574–576, 585, 586, 589, 599, 600
  of facial canal, 795, 798, 799
  of lateral semicircular canal, 792, 795
Promontory, sacral, 303, 319, 321, 335, 355, 393, 546
  of tympanic cavity, 778, 781, 784, 785, 789, 790, 795–799
    subiculum of, 778, 799
Prostate. See *Gland, prostate*
Protuberance, mental, 598, 610
  occipital, external, 523, 528, 574, 646, 647
    internal, 645
Pterygoid canal, 663
  nerve and artery of, 746
Pterygoid hamulus. See *Hamulus, pterygoid*
Pterygoid plate, 702, 722, 748

Pterygoid plate, *continued*
hamulus of, 747
Pterygoid process, of sphenoid bone, 640, 646, 647, 684, 746
lamina of, lateral, 735
medial, 746, 747
hamulus of, 746
Pubis, body of, 386, 391, 392
pecten of, 335, 389, 393, 417, 419, 420
ramus of, inferior, 242, 352, 369, 389–392, 401, 430
superior, 242, 352, 369, 390–392, 401
tubercle of, 244
Pulp, splenic, 278
Pulp cavity, of tooth, 703, 720, 721
Punctum lacrimale, 649, 653, 657, 658
superior, 658
Pupil, 648, 665
Pyloric antrum, 268–270, 331
Pylorus, 136, 224, 257–259, 268, 270, 273, 277, 287, 305
Pyramid(s), of medulla oblongata, 566, 639
decussation of, 566, 639
renal, 325, 327, 331, 332
Pyramidal, eminence, tympanic, 778, 785, 795, 796, 799

Quadrangular space, 23, 51, 53, 60, 523

Radiograph of knee joint, 457
of large intestine, 311
Ramus (Rami), of ischium, 366, 369, 389–392
of mandible, 601, 608, 613, 697, 700, 722, 728, 737, 740, 754
of pubis, inferior ramus, 242, 352, 369, 389–392, 401, 430
superior ramus, 242, 352, 369, 390–392, 401
of spinal nerve, anterior ramus, 10, 571, 572
dorsal ramus, 10
posterior ramus, 10, 517
primary ramus, anterior, 573
posterior, 518, 563, 571, 573
of supraorbital nerve, lateral, palpebral ramus, 650, 656, 657, 670, 672
medial palpebral ramus, 650, 656, 657, 670, 672
superior palpebral ramus, 654, 655
of thoracic nerve, posterior rami, 517, 527, 528
of ulnar and median nerves, cutaneous ramus, 38
to aortic plexus, 227
Rami communicantes, 10, 185, 219, 224, 227, 320, 339, 518, 571, 573, 593, 654
Raphé, mylohyoid, 583, 598, 600
of deep transverse perineal muscles, 366
of scrotum, 248, 252, 344
palatine, 701, 729, 731
pharyngeal, 590, 750
pterygomandibular, 615, 722, 731, 735, 737, 740, 746–748
tendinous, of gastrocnemius muscle, 472
Recess, costodiaphragmatic, 150, 152, 154–156, 170, 175, 186, 215, 219, 220, 230, 257, 258, 273, 299, 300
costomediastinal, 152, 153, 158, 170, 185, 299
duodenal, inferior, 302, 305, 316
superior, 302, 305, 316
epitympanic, 780–782, 786, 791, 801
hypotympanic, 782, 792
ileocecal, 301, 302
intersigmoid, 316
mediastinovertebral, 219, 227
omental bursa, inferior, 275
splenic, 275, 300, 316
superior, 275, 316
pericardial, of aorta, 170, 188, 194, 214
of pulmonary trunk, 188, 194
pharyngeal, 685, 691, 726, 732, 751
phrenicomediastinal, 300
piriform, 751, 753, 755, 758, 770
retrocecal, 305
sphenoethmoidal, 685
subpopliteal, 678
superior, of tympanic membrane, 781
trigeminal, 778
Rectum, 257, 258, 264, 272, 303, 305, 311–314, 316, 317, 319, 335, 338, 339, 343, 345, 353, 354, 365, 373, 376, 378, 382, 383, 385
ampulla of, 312, 314, 343, 377
longitudinal coat of, 312, 313
Reflections, peritoneal, 264, 373
Region, acromial, 520
anal, 520
Region, *continued*
antebrachial, anterior, 1, 520
lateral, 37
medial, 37
posterior, 1, 520
atlanto-occipital, median section of, 539, 540
auricular, 602
axillary, 1, 37
brachial, anterior, 1
lateral, 37
medial, 37
posterior, 1, 520
buccal, 520, 602
calcaneal, 1, 520
clavicular, 37
crural, anterior, 1, 520
posterior, 1, 520
cubital, anterior, 1, 37
lateral, 37
deltoid, 1, 37, 520
digital, palmar, 37
elbow, 520
epigastric, 1
femoral, anterior, 1
posterior, 520
frontal, 1, 520, 602
gluteal, 520
hyoid, 520
hypochondriac, 1, 520
infraclavicular, 1, 37
infraorbital, 602
infrascapular, 520
infratemporal, 624, 625
inguinal, 1, 244–246
interscapular, 520
knee, anterior, 1, 444, 445
posterior, 520
laryngeal, 602
lumbar, 520
lateral, 520
mental, 1, 602
nasal, 1, 602
neck, anterior, 1, 602
lateral, 1, 602
posterior, 520
occipital, 520, 602
of body, anterior, view, 1
posterior view, 520
oral, 1, 602
orbital, 1, 520, 602
parietal, 1, 520, 602
parotidomasseteric, 520, 602
pectoral, 1, 37, 520
prevertebral, 594
pubic, 1
sacral, 520
scapular, 520
sternal, 1
sternocleidomastoid, 1, 520, 602
submental, 602
suboccipital, 528
suprascapular, 520, 602
temporal, 1, 520, 602
thoracic, lateral, 37
trochanteric, 1
umbilical, 1
vertebral, 520
zygomatic, 520, 602
Renal calyx, 325, 327, 331, 332
Renal cortex, 325, 327, 331
Renal fascia, 332
Renal hilum, 323, 329, 332
Renal pelvis, 271, 285, 304, 323, 325–328, 331, 332, 363, 372
Renal pyramid, 325, 327, 331, 332
Renal sinus, 332
Reticularis, of adrenal cortex, 324
Retina, 669
ciliary part, 669, 680
neural part, 669, 679
pigmented layer of, 679, 680
vessels of, 678, 679, 681
Retinaculum, extensor, of foot, 462–470, 505, 507
of hand, 61, 65, 74–76, 78, 101, 103, 105, 120
flexor, of foot, 462, 474–478, 491, 493, 496, 507
of hand, 107, 108, 110, 111, 114, 120
patellar, 431, 445, 462, 463
lateral, 444, 446
medial, 444, 446
peroneal, inferior, 463, 467, 469, 505
superior, 463, 467, 474–478, 491, 505, 513
Retroesophageal space, 726
Retromolar triangle, 700
Ridge(s), for insertion of pronator teres muscle, 90
longitudinal of stomach, 268
of nail matrix, 126–128
Rima glottidis, 766–768, 771, 772
Rima vestibuli, of larynx, 766
Ring, femoral, 383
fibrocartilaginous, of tympanic membrane, 629, 631, 781, 784–787, 791
inguinal, deep, 242, 243, 253, 340, 364
superficial, 231, 232, 235, 236, 244, 248, 255, 347, 388
lymphatic, oro-naso-pharyngeal, 752
tendinous, common, of orbit, 640, 664, 668, 675
umbilical, 181, 240, 243, 255, 256, 261
Roof, of tympanic cavity, 782
Root(s), dorsal, of spinal nerve, 10, 563
motor, of trigeminal nerve, 639, 642, 694, 695, 793, 801
of finger nail, 126–128
of splanchnic nerve, greater, 10
of mesentery, 264, 295, 301, 302, 310, 316
of mesocolon, sigmoid, 316
transverse, 301, 302, 305, 316
of tongue, 743, 745, 751, 753, 755, 769
of tooth, 703, 720, 721
of tympanic cavity, 787, 789
sensory, of trigeminal nerve, 642, 694, 801
ventral, of spinal nerve, 10, 571, 573
Root canal, 703, 721
Rugae, vaginal, 380

Sac, endolymphatic, 802, 805
lacrimal, 652, 656–658, 660
fornix of, 658
Sacculus, 800, 803, 805
Salivary glands, 733. See also *Gland, parotid; Gland, sublingual; Gland, submandibular*
Salpingopalatine fold, 685
Salpingopharyngeal fold, 726, 732, 751, 755
Scala tympani, 640, 800, 805
Scala vestibuli, 640, 800, 805
Scalp, 633, 635
dermis of, 633
Sclera, 669, 679, 680
cribriform area of, 679, 680
Scrotum, 242, 248, 252, 343, 344
dartos layer of, 255, 343, 344, 354
septum of, 248, 252
skin of, 248, 252, 255
Sella turcica, 613, 636, 685, 691
Semicanal, for tensor tympani muscle, 778, 796
of auditory tube, 796
Semicircular canal, common crus of, 801
Seminal vesicle, 258, 352–355, 357–360, 362, 363
Septum (Septa), dural, 801
interalveolar, 720, 721
interatrial, 180, 201, 202
intermuscular, lower limb, 464, 474, 486, 487, 489, 490
upper limb, 7, 47–49, 51–53, 55, 57–59, 61, 64–67, 69, 71, 74, 75, 77, 81, 98–101
interventricular, 201, 202, 204, 206, 207, 209, 211, 212
membranous part, 212
muscular part, 212
intra-alveolar, 720, 721
nasal, 601, 669, 683, 690, 692, 726, 751, 753
cartilaginous, 655–657, 682, 683, 690, 726
of musculotubal canal, 778, 785, 791, 796, 798
of penis, 348–350
of glans penis, 350
of testis, 250–252, 343
of tongue, 736, 739, 741
orbital, 650–652, 655–657, 673
placental, 181
primum, 180
secundum, 180
scrotal, 248, 252
Septum canalis, musculotubarii, 778, 785, 791, 796, 798
Sheath(s), carotid, 584
dural, 793
femoral, 337
vascular components, of, 335
of dura mater, for hypoglossal nerve roots, 726
for spinal nerve roots, 726
of rectus abdominis muscle, 61, 149, 300, 331, 332, 359
anterior layer, 7, 12–14, 16, 151, 231, 237–241, 253, 255

Sheath(s), *continued*
posterior layer, 151, 238–243, 255, 340
optic nerve, 669, 673, 675, 676, 680
synovial. See *Tendon sheath, (synovial)*
tendon. See *Tendon sheath (synovial)*
vaginal, of styloid process, 646
Shoulder joint, See *Joint, shoulder*
Sinus(es), air, paranasal, openings of, 685, 686
anal, 312, 343
aortic, left, 207
basilar, 638, 642
carotid, 588
cavernous, 587, 620, 638, 640–642, 801
confluence of, 620, 637
coronary, 180, 191–193, 213–215
orifice of, 180, 191, 202, 203, 208
valve of, 180, 191, 202, 203, 208, 214
costomediastinal, 159
dural, 637
frontal, 635, 683, 685, 686, 690, 691, 722
opening into, 685, 686
intercavernous, 638, 642
lactiferous, 8
marginal, of placenta, 181
maxillary, 660, 663, 664, 667, 695, 793
occipital, 637
of epididymis, 249, 252, 253
of finger nail, 126
pericardial, oblique, 189
transverse, 189, 190, 214
petrosal, 587, 620, 637, 638, 642
superior, 801
prostatic, 356
renal, 327, 332
sagittal, inferior, 620, 637, 638
superior, 620, 633–638, 645
at confluence of sinuses, 634
opening of cerebral veins into, 634
septum, 208
sigmoid, 536, 587, 620, 637, 638, 640, 645, 787–790, 801
sphenoethmoidal, opening of, 685
sphenoid, 613, 636, 641, 663, 669, 683–686, 690, 691, 693, 722, 726, 732, 747, 793
sphenoparietal, 637, 638
straight, 620, 637, 638
transverse, 587, 788
of pharynx, 754
tympanic, 778, 796, 799
Sinus septum, 208
Sinus venarum, 192, 195
Sinus venosus, of sclera, 678, 680
Skin, 9, 348, 518
of anus, 312
of labium majus, 244
of scrotum, 252
Skull, 601, 610, 633
at birth, 629–631
base of, 645–647
brachycephalic, 627
dolichocephalic, 628
Small intestine, 302, 343
mesentery of, root of, 264
Socket(s), for teeth, 598, 663, 697, 720, 721
of hip joint, 401, 437
Sole, of foot, 492–498, 520, Also see *Foot, sole of*
arteries of, 493, 496, 497
muscles of, 494, 495, 498
nerves of, 493, 496, 497
plantar aponeurosis, 493
Soleal line, of tibia, 448, 484
Space(s), calcaneal subtendinous, 491
cavernous, of penis, 361
epidural, 343, 571, 726
esophagotracheal, 591, 726, 769
infraglottic, 766, 769
interfascial, dorsal, of foot, 516
intervaginal, 673, 680
parapharyngeal, 737
perichorioidal, 680
perithyroid, 591
prevertebral, 591
prevesical, 240, 343
quadrangular, 23, 51, 53, 60, 523
retroesophageal, 726
retrogastric, 305
retropharyngeal, 740, 747
retropubic, 253, 255, 343, 353, 373, 385
subarachnoid, 343, 571–573, 633
subdural, 572, 573
suprasternal, 214
triangular, 21, 23, 51, 53, 523
Spermatic cord, 12, 231, 232, 236–239, 242, 246, 247, 251, 341, 347, 354, 355, 416
fascia of, 239
nerves of, 347
vessels of, 347
Sphincter, of Oddi, 288–293
urethral, 376
Spinal column. See *Vertebral column*
Spinal cord, 219, 518, 563, 565, 566, 570, 573, 639, 644,
cervical part, 563, 566
fissure of, anterior median, 566, 567, 568, 570, 572
gray matter, 518
lumbar part, 563, 566
meninges of, 570, 571, 633
sulcus of, anterolateral, 566, 568
posterior median, 570
thoracic part, 563, 566
with nerve roots, 568
within vertebral, canal, 230, 565
Spine(s), genioglossal, 698, 700, 728
iliac, anterior inferior, 387, 389, 390, 393, 401
anterior superior, 12, 233, 235–237, 253, 257, 387, 389, 390, 392, 393, 395, 398, 401, 412, 417, 420
posterior inferior, 258, 389, 390, 401
posterior superior, 258, 387, 389, 390, 394, 523, 526
ischial, 258, 314, 389–391, 393, 401
anterior inferior, 387
mental, 698, 700
nasal, anterior, 601, 610, 683, 684
posterior, 646, 702
of helix, 774
of scapula, 23, 24, 26, 30, 32, 33, 36, 51, 61, 258, 523
of sphenoid bone, 613, 646
palatine, 702
suprameatal, of temporal bone, 777
tympanic, greater, 785, 786
lesser, 785, 786
Spleen, 136, 257, 258, 261, 265, 272, 273, 275, 277, 278, 299, 300, 310, 331
extremity of, anterior, 256, 274, 276, 300, 316
posterior, 274, 276, 300
fibrous capsule of, 278
hilum of, 274, 277, 278
impression on, gastric, 274, 278, 305, 316
renal, 274, 278
margin of, inferior, 274, 276
superior, 274, 276, 305, 316
serous coat of, 278
surface of, diaphragmatic, 276, 278
visceral, 274
Squama, of frontal bone, 610
of occipital bone, 630
of temporal bone, 631
Stomach, 136, 215, 224, 257, 258, 261, 265, 269, 270, 272, 273, 279, 294, 299, 310, 330, 332
angular notch of, 269, 273
bed of, structures of, 277
body of, 259, 268–270, 273, 275
cardiac orifice of, 268, 299, 316
cardiac part, 136, 257, 258, 268, 270, 273, 305, 316, 317, 321, 338
coat of, muscular, 268, 270
curvature of, greater, 111, 256, 259, 268, 269, 273, 275, 294
lesser, 111, 257, 268, 269, 273, 275, 294
fundus of, 136, 268–270, 273, 294
interior, 268
lymphatic vessels of, 279
mucosa of, 299, 332
nerves of, 279
pylorus of, 136, 224, 257–259
serous layer of, 215, 268
wall of, 300
layers of, 215, 268, 299, 331, 332
Stria, olfactory, 639
Styloid process, of temporal bone, 735, 740, 790
Subarachnoid space, 343, 571–573, 633
Subdural space, 572, 573
Subiculum, promontory, of temporal bone, 778, 799
Substance, perforated, anterior, 639
Sulcus (Sulci), bicipital, medial, 49
carotid, 645
cerebral, 635
coronary, 188, 192, 195
cubital, 73
cuboid, 503
esophageal, on lung, 164
Sulcus (Sulci), *continued*
for corpus spongiosum penis, 352
for flexor hallucis longus tendon, 513
for peroneus longus endon, 513, 514
for petrosal nerve, greater, 645
lesser, 645
for petrosal sinus, inferior, 645
superior, 645
for popliteus muscle, 448
for sigmoid sinus, 536, 645
for splenic artery, on pancreas, 286
for sagittal sinus, superior, 645
for temporal artery, middle, 777
for transverse sinus, 645
for trigeminal ganglion, 642
for vertebral artery, 534
genitofemoral, 388, 393
gluteal, 388, 413
infraorbital, 661, 662
inguinal, 393
intertubercular, of humerus, 84, 85
interventricular, anterior, 176, 188, 208
posterior, 192, 195
lingual, median, 743
of aortic arch, on lung, 164
of azygos vein, on lung, 165
of spinal cord, anterolateral, 566, 568
posterior median, 570
posterolateral, 570
of splenic artery, on pancreas, 286
of subclavian artery, on 1st, rib, 143, 164, 165
of subclavian vein, on 1st rib, 143
of superior vena cava, on lung, 165
olfactory, 639
orbital, of frontal pole, 639
palatine, 702
palpebral, superior, 603, 648
paracolic, 260, 301, 302, 305
paraglenoid, 390
promontory, of temporal bone, 778, 799
pubic, 393
temporal, inferior, 639
tympanic, 778, 781, 784
Sulcus terminalis, of right atrium, 188, 192, 195
of tongue, 743, 753
of ulnar nerve, on humerus, 83
Supraorbital margin, 601, 667
Supraorbital notch, 652, 689
Supraventricular crest, of right atrium, 204, 209
Sustentaculum tali, 503, 509, 513
Suture, coronal, 601, 610, 626, 628–630, 635
frontal, 630
frontoethmoid, 645
frontolacrimal 601
frontomaxillary, 601, 661, 682
frontonasal, 601, 682, 683
frontozygomatic, 610
incisive, 646, 702
infraorbital, 661
internasal, 601
lacrimoethmoidal, 662
lacrimomaxillary, 610
lambdoidal, 610, 626, 628, 629, 635, 646
nasofrontal, 682
nasomaxillary, 601, 610, 682
occipital, transverse, 629, 631
occipitomastoid, 610, 645, 646
palatine, median, 646, 702
transverse, 646, 683, 702
parietomastoid, 610
sagittal, 626, 628, 630, 632, 633
sphenofrontal, 601, 610, 635, 645, 663
sphenoparietal, 601
sphenosquamosal, 610, 645, 735
sphenozygomatic, 601, 610, 659
squamosal, 610
vomeromaxillary, 683
zygomaticomaxillary, 601, 661
zygomaticotemporal, 610
Sympathetic trunk ganglia, 10, 184, 219, 224, 226, 299, 793
Symphysis pubis, 235, 240, 242, 253, 311, 333, 342, 343, 353, 364–366, 373, 387, 388, 393, 395, 396, 399, 401, 439
Synchondrosis, petro-occipital, 647, 750, 751, 753
spheno-occipital, 645
sternocostal, 147
sternomanubrial, 147
Syndesmosis, arycorniculate, 764
tibiofibular, distal, 480, 491
Synovial, tendon sheath(s). See *Tendon sheaths (synovial)*

Taenia coli, 259, 260
Taenia libera, 260, 277, 296, 301, 302, 306, 307, 332
Taenia, mesocolica, 277, 307, 332
Taenia omentalis, 259, 273, 277, 301, 305, 307, 332
Talus, head of, 501, 503
  process of, lateral, (malleolar), 501
  trochlea of, 501
Tarsal plates, 651, 652, 656
Tarsus, inferior, of eyelid, 651, 652, 673, 677
  superior, of eyelid, 651, 652, 655–657, 673
Taste pathways, 742
Tegmen tympani, 782, 787, 789, 792, 802
Temoral line, inferior, of parietal bone, 626, 627, 777
  superior, of parietal bone, 626, 627
Tendinous arc, for origin of levator ani muscle, 365, 376, 386
Tendinous intersections of rectus abdominis muscle, 151, 231
Tendon(s), Achilles, 473, 505, 507
  calcaneal, 413, 462, 463, 467, 472–479, 491, 502, 505, 507, 509
  central, of diaphragm, 184, 185, 215, 216, 230, 333–335
  conjoint, 242
  cricoesophageal, 753
  extensor, of dorsal, wrist, 103
  flexor, common, of forearm, 50, 80
  of abductor hallucis muscle, 507
  of abductor pollicis longus muscle, 61, 65, 75, 76, 79, 101, 105, 107, 108, 112
  of adductor magnus muscle, 423, 440, 447, 487
  of ankle region, 505, 506
  of biceps brachii muscle, 21, 23, 29–31, 34, 35, 49, 53, 55, 59, 63, 66, 67, 69, 70, 91, 98, 99, 151
  of biceps femoris muscle, 426, 430, 440, 441, 467, 472, 478
  of brachialis muscle, 55, 66, 99
  of brachioradialis muscle, 47, 65–70, 75, 101, 114
  of digastric muscle, intermediate, 598–600
  of extensor carpi radialis brevis muscle, 74, 75, 77, 79, 101, 112
  of extensor carpi radialis longus muscle, 74, 75, 77, 79, 100, 101, 112
  of extensor carpi ulnaris muscle, 74, 76–79, 101, 120
  of extensor digiti minimi muscle, 76, 77, 101, 120, 121
  of extensor digitorum brevis muscle, 470
  of extensor digitorum longus muscle, 462–464, 467, 469, 505, 515, 516
  of extensor digitorum muscle, 61, 77, 79, 103, 109, 112, 116, 120, 121
  of extensor hallucis brevis muscle, 467, 469, 471, 516
  of extensor hallucis longus muscle, 462–464, 469, 471, 491, 515, 516
  of extensor indicis muscle, 79, 120, 121
  of extensor pollicis brevis muscle, 61, 66, 75, 77–79, 101, 121
  of extensor pollicis longus muscle, 61, 77–79, 101, 105, 112, 121
  of flexor carpi radialis, muscle, 37, 65, 66, 68–70, 101, 108, 110, 111, 120
  of flexor carpi ulnaris muscle, 65, 66, 68, 70, 108, 114, 120
  of flexor, digitorum brevis, muscle, 495, 496, 498, 516
  of flexor digitorum longus, muscle, 462, 477, 478, 483, 491, 495, 498, 515, 516,
  of flexor digitorum profundus, muscle, 69, 70, 101, 108, 109, 114, 117, 120–124
  of flexor digitorum superficialis, muscle, 69, 70, 101, 108, 109, 114, 117, 120, 121, 123
  of flexor hallucis longus, muscle, 462, 477–479, 483, 515, 516
  of flexor pollicis longus, muscle, 69, 101, 111, 114, 121
  of foot, dorsal, 469. See also names of specific muscles
  of gastrocnemius muscle, 462, 474, 490
  of gracilis muscle, 423, 462, 489
  of hand, 112. See also names of specific muscles
  of iliopsoas muscle, 429, 430
  of latissimus dorsi muscle, 21, 49, 53, 55, 63, 625
  of lumbrical, muscles, of foot, 498
    of hand, 116
  of oblique muscle, inferior, 671
    superior, 652, 665–668, 671, 674, 676, 677
  of obturator internus, 387, 429
  of palm, 107. See also names of specific muscles
Tendon(s), *continued*
  of palmar, interossei muscles, 116
  of palmaris longus muscle, 37, 65, 66, 68–70, 101, 108
  of pectoralis major muscle, 23, 59
  of peroneus brevis muscle, 467, 469, 470, 483, 505, 508, 511, 514, 515
  of peroneus longus muscle, 467, 483, 491, 495, 498, 506, 514, 515
  of peroneus tertius muscle, 463–465, 467, 469, 470, 505, 516
  of piriformis muscle, 359
  of plantaris muscle, 473–475, 479
  of popliteus muscle, 448, 454
  of pronator teres, muscle, 77, 79
  of psoas minor muscle, 317, 335, 340
  of quadriceps femoris muscle, 444, 445, 455, 458
  of rectus muscle, inferior, 673
    lateral, 669–671
    medial, 669, 671
    superior, 671
  of rectus abdominis muscle, 151, 239
  of rectus femoris muscle, 417, 419, 435–437, 439, 446, 463
  of sartorius muscle, 420, 423, 489
  of semimembranosus muscle, 427, 429, 430, 434, 440, 447, 448, 462, 473, 477, 489
  of semitendinosus muscle, 423, 427, 429, 430, 440, 462, 473, 487, 489
  of stapedius muscle, 792, 795, 796
  of sternocleidomastoid muscle, 583, 647
  of subscapularis muscle, 34, 49, 63, 151
  of supraspinatus muscle, 51, 151
  of temporalis muscle, 737, 740
  of tensor tympani muscle, 781, 785, 791, 793, 795, 796, 801
  of tensor veli palatini muscle, 722, 731, 753
  of teres major muscle, 49, 55
  of tibialis anterior muscle, 462, 464, 466, 467, 469–471, 491, 506, 509, 514, 515
  of tibialis posterior muscle, 462, 474–479, 483, 491, 498, 509, 514, 515
  of trapezius muscle, 51, 647
  of triceps brachii muscle, 51, 57, 59, 74, 75, 77, 82
  of vastus intermedius muscle, 446
  of wrist, 103, 107. Also see names of specific muscles
Tendon sheath(s) (synovial), of abductor pollicis longus muscle, 112, 114
  of digital tendon, of hand, 123, 124
  of extensor carpi radialis brevis muscle, 112
  of extensor carpi radialis longus muscle, 112
  of extensor digitorum muscle, 112
  of extensor digitorum longus muscle, 462, 463, 505, 515
  of extensor hallucis longus muscle, 462, 507
  of extensor indicis muscle, 112
  of extensor pollicis brevis muscle, 112, 121
  of extensor pollicis longus muscle, 112, 121
  of flexor carpi radialis muscle, 37, 108, 114
  of flexor digitorum longus muscle, 462, 507
  of flexor hallucis longus muscle, 462, 492, 507
  of flexor pollicis longus muscle, 107, 108, 114
  of foot, 462, 463, 497, 505, 507, 515. See also names of specific muscles
  of hand, 103, 107–109, 114, 121, 123, 124. See also names of specific muscles
    of digital tendons, 123, 124
  of long biceps tendon, 31, 49, 55, 151
  of peroneus brevis muscle, 463, 505
  of peroneus longus muscle, 463, 495, 505, 515
  of tibialis anterior muscle, 462, 463, 507
  of tibialis posterior muscle, 462, 507
  of wrist, 103, 107, 108, 112, 114
Tenon's capsule, 669, 677
Tentorium cerebelli, 637, 638, 685, 722
Testis, 242, 247–253, 319, 353, 363
  appendix of, 247–249, 253, 363
  developing, 363
  ligament of, scrotal, 253, 255
  lobules of, 250–252, 343
  septa of, 250–252, 343
Thigh, anterior, 1, 337, 407, 414–424
    muscles of, 417–420, 423, 424
  cross section of, 486, 487
  fascia of, 416
  nerve fields of, cutaneous, 408, 409
    peripheral, 408
  posterior, 425–429, 520
    muscles of, 426–429
  surface anatomy of, 412, 413
Thoracic inlet, 592
Thoracic wall, anterior, 148, 149, 151, 231, 232, 257
  posterior, 218, 221, 226, 227, 258
    nerves of, 226, 227
    sympathetic trunks of, 226, 227
    vessels of, 226, 227
Thorax, area of superficial cardiac dullness in, 170
  aspect of, lateral, 14
  lymphatics of, 3, 5, 6, 187
  position of heart within, 170
  radiograph of, 150
  section of, cross, 157–159
    frontal, 215
    sagittal, 214
  sites for ascultation of heart, 171
  skeleton of, 137, 140, 153–156, 172
    natural contour of, 146
  vessels of, great, 586
  viscera of, 152, 257
Thymus gland, 152, 153, 175, 177, 184, 185, 256, 726
  left lobe, 175
  right lobe, 175
Thyroid ansa, 226
Thyroid cartilage. See *Cartilage, thyroid*
Thyroid gland, See *Gland(s), thyroid*
Thyroid notch, superior, 136, 257, 757, 761, 763
Toe(s), 425–437
  phalanges, 501, 503
  See also *Bone(s), phalanx (phalanges); Foot*
Tongue, 621, 625, 724, 732, 737–739, 743, 746, 756
  apex of, 741, 743, 744
    artery and vein of, 740
  dorsum of, 692, 729, 731, 732, 740, 743, 753, 758
  foramen cecum of, 685, 726, 736
  frenulum of, 733, 735, 737, 740
  lymphatic drainage of, 621
  membrane of, mucous, 741, 758
  muscles of, extrinsic, 741, 744, 745
    intrinsic, 736
    longitudinal, inferior, 735, 740
      superior, 736, 740
  root of, 743, 745, 751, 753, 755, 769
  septum of, 736, 739, 741
  surface of, inferior, 738
  vessels and nerves at base of tongue, 746
Tonsil(s), cerebellar, 639
  lingual, 732, 736, 743, 753
  palatine, 692, 726, 729, 731, 732, 735, 737, 743–746, 751–573, 755, 758
  pharyngeal, 685, 690, 691, 722, 726, 732, 746, 751–753, 755
Tooth (Teeth), 692–722, 729–731, 734
  bicuspid, 699
  canine, 699, 701, 705, 708–711, 713, 714, 717
  crown of, 703, 720, 721
  cusp of, 704
  deciduous, 707–710
  incisor, 699, 701, 705, 708–710, 712, 713, 716, 717, 720
    permanent, lower, 709, 710, 716, 717, 720
      upper, 710, 712, 713, 720
  molar, 699, 701, 704, 705, 708, 709, 711, 715, 718, 719
  neck of, 703
  nerves of, 696
  permanent, 705, 706, 709, 710, 712, 713, 716, 717, 720
  premolar, 699, 701, 705, 710, 711, 714, 715, 717, 718
  pulp cavity of, 703, 720, 721
  roots of, 703, 720, 721
  section of, longitudinal, 703
  sockets for, 598, 663, 697, 720, 721
  surfaces of, 705
Torus, of levator veli palatini muscle, 726, 751
Torus tubarius, 685, 722, 726, 732, 751, 755
Trabecula(e), of corpus cavernosum penis, 361
  septomarginal (moderator band), 202–204, 209
  splenic, 278
Trabeculae carnae, 201, 206, 211, 212
Trachea, 136, 151, 168, 177, 186, 188, 189, 214, 216, 219, 224, 257, 258, 582, 588–590, 592, 726, 744, 750, 755, 756, 758, 760, 766, 768, 769
  bifurcation of, 157, 168, 214–216, 768
  glands of, 591
  muscosa of, 591
  wall of, membranous, 591, 726, 759, 765
Tract, iliotibial, 296, 416, 424, 430, 444, 463, 482, 486, 487
  olfactory, 639, 644, 690, 691
Tractus solitarius, nucleus of, 742
Tragus, 773, 774

Triangle(s), carotid, 574, 602
deltopectoral, 1, 12, 15, 231, 575, 583
femoral, 1, 418
lumbar, 320, 521–524
of neck, 576, 579
retromolar, 700
sternocostal, 175
submandibular, 520, 597, 602, 727
suboccipital, 528, 529
Trigone, fibrous, of heart, 191
of bladder, 356, 357, 361
inferior, without pleura, of heart, 170
retromolar, of mandible, 737, 740
superior, without pleura, of heart, 170
Trochanter, greater, 333, 394-396, 401, 427, 429–433, 435, 436
lesser, 333, 401, 423, 429–433, 435, 436, 439
Trochlea, of humerus, 82–84, 86, 96–98
of orbit, 656, 657
of superior oblique muscle, 667, 676
of talus, 501
Trochlear notch, of ulna, 9, 87, 88, 93, 133
Trunk, brachiocephalic, 182, 188, 208, 214, 219, 593, 594, 726, 755
celiac, 181, 182, 216, 218, 226, 265, 266, 271, 272, 285, 317, 318, 321, 335, 338, 339, 341
costocervical, 218, 226, 569, 596
lumbosacral nerve, 336–339, 364
lymphatic, bronchomediastinal, 219, 220
anterior, 592
femoral, 384
iliac, 384
intestinal, 223, 296, 318, 320, 332, 341
jugular, 223, 587, 592, 593
lumbar, 223, 318, 320, 340, 341, 384
mammary, 593
preaortic, 384
subclavian, 223, 587, 592, 593
of body, dermatomes of, 2, 519
superficial vessels and nerves of, 11, 232
pulmonary, 136, 150, 172, 176, 181, 188, 190–192, 194, 204, 208, 209, 212
sympathetic, 10, 185, 218–220, 224, 226, 229, 230, 299, 300, 320, 332, 338–340, 573, 593, 740, 754, 755
in posterior thoracic wall, 226, 227
lumbar, 332, 338
thyrocervical, 17, 20, 182, 569, 581, 582, 592, 596, 755
spinal branches of, 569
vagus nerve, 229
Tube, auditory, 615, 694, 695, 779, 780, 791–793, 795, 798, 801
cartilage of, 647, 690, 693, 746, 747, 753, 780
mucosa of, 793
openings of, 685, 691, 726, 732, 746, 751–753, 778
semicanal of, 796
tympanic opening for, 778, 786, 791, 795, 798
view of, lateral, 793
uterine, 372, 379, 380, 382, 399
ampulla of, 310, 316, 318, 374, 375, 377–380
developing, 372
fimbria of, 374, 377–380
infundibulum of, 372, 377–379, 381, 382, 399
isthmus of, 316, 318, 374, 375, 377–380, 383
openings of, 375, 377, 380
Tuber cinereum, 639
Tuber omentale, 275, 277, 300
Tubercle, accessory, of upper 1st molar (Carabelli's), 704
adductor, 457
anterior, of atlas, 530, 531, 594
of cervical vertebrae, 541
articular, of temporal bone, 614, 646, 735, 777
auricular, of Darwin, 773
carotid, 151, 574, 593, 594
conoid, 138
corniculate, 726, 751, 753, 755, 765, 767–769
cuneiform 726, 751, 755, 765, 767, 769
epiglottic, 764, 767, 772
greater, of humerus, 30, 32, 36, 83–85
infraglenoid, 6, 26, 28
intercondylar, of tibia, 442, 444, 446, 452, 457, 459, 464, 466, 467, 480, 481
intervenous, of atrium, 202
jugular, 645
lesser, of humerus, 84
mental, 574, 598, 697
obturator, anterior, 389
posterior, 389, 390
of rectus muscle, lateral, origin of, 663

Tubercle, *continued*
of rib, 143, 551, 554
pharyngeal, of occipital bone, 646, 751
posterior, of atlas, 523, 526, 530, 531, 540
of cervical vertebrae, 541
pubic, 244, 333, 369, 386–389, 393, 397, 398, 417
scalene, of 1st rib, 143
serratus anterior, of 2nd rib, 143
supraglenoid, 28
thyroid, inferior, 735, 744, 745
Tuberculum sellae, 645
Tuberosity, calcaneal, 413, 467, 473, 474, 477, 492, 494, 495, 503, 509, 511, 513, 514
deltoid, of humerus, 84
gluteal, 436
ischial, 25, 258, 333, 344, 369, 370, 387–391, 393, 394, 401, 427, 523
maxillary, 720
muscular, of inferior angle of scapula, 26
of distal phalanx, 118, 132
of 1st metatarsal bone, 503
of 5th metatarsal bone, 501, 503, 510, 511, 513
of ilium, 390
of navicular bone, 513
of radius, 89, 96, 97
of scaphoid, 135
of tibia, 442, 444, 452, 464, 466, 467, 480, 481
of ulna, 87, 88
pterygoid, of mandible, 614, 730
of sphenoid bone, 698, 700
sacral, 560, 561
Tunica albuginea, of testis, 250, 251
Tunica vaginalis testis, 247, 248
cavity of, 253, 343, 353
parietal layer, 247–249, 252, 253, 255, 353
visceral layer, 248, 252, 253, 255, 343, 353
Tympanic air cells, 778
Tympanic antrum, 801
Tympanic cavity, 695, 778, 779, 781, 782, 785, 790, 791, 793, 795, 796, 798, 800–802
roof of, 695, 782, 792
wall of, lateral, 791
medial, 787, 789, 790, 793, 795, 796, 798
window of, cochlear, 799
vestibular, 799
Tympanic crest, 735
Tympanic membrane, 783, 784, 786, 787, 791, 795
secondary, 799
Tympanic notch, 785, 786
Tympanic plexus, 793, 794
Tympanic sinus, 778, 796, 799

Umbilical cord, 178, 255, 399
Umbilical fold, lateral, 256, 259, 321, 340, 359
medial, 256, 259, 321, 340, 352, 359
median, 243, 256, 320, 321, 340, 352, 359
Umbilicus, 12, 231, 232, 240, 242, 255, 257, 368, 393, 399
Umbo, 781, 784, 787, 789
Uncus, 639
Unguis, 125
Urachus, 240, 243, 255, 256, 261, 319, 321, 352, 356, 358, 363, 372, 377, 399
Ureter, 178, 243, 301, 319, 320, 325–328, 338, 363, 372
left, 258, 271, 285, 302, 316–318, 320, 321, 332, 340, 341, 352, 354, 363, 376, 379, 382, 383
orifices of, 343, 356, 357, 361
right, 258, 271, 285, 304, 305, 317, 318, 321, 322, 340, 352, 355, 358, 359, 363, 374, 377–379, 382, 383
Urethra, 248, 349, 362, 369, 372
female, 368, 369, 371, 372, 376, 377
male, 343, 348–353, 355–357, 361–363, 366
crest of, 356, 361
navicular fossa of, 343, 350, 361
penile, 343, 348, 361
prostatic, 356, 357, 361
membranous, 352, 357, 361, 365, 366
opening of, 343, 351, 361, 367, 368, 370–372, 376, 377, 388
Urinary bladder. See *Bladder, urinary*
Uterine cavity, 377, 380
Uterine tube. See *Tube, uterine*
Uterus, 178, 264, 310, 342, 372–375, 378–380, 382, 383, 385, 399
body of, 316, 318, 379–381, 399
cervical canal of, 377, 380
external os of, 181
fundus of, 318, 378, 380, 381
isthmus of, 375, 379, 380
supravaginal part, 379

Uterus, *continued*
surface of, anterior (vesical), 378
intestinal, 379
Utricle, prostatic, 343, 356, 361, 363
Utriculus, of inner ear, 800, 803, 805
Uvula, of bladder, 356, 361
of soft palate, 615, 690, 692, 693, 729, 731, 732, 751, 753, 755

Vagina, 181, 369, 372, 380–382
fornix of, 373, 377
surface of, posterior, 379
vestibule of, 367, 368, 388
Vallate papillae, 751, 753, 755
Vallecula, of epiglottis, 685, 736, 743, 758, 769
Valve(s), aortic, 171, 172, 191, 207, 212
left semilunar cusp, 191, 207, 212
posterior semilunar cusp, 191, 207, 212
right semilunar cusp, 191, 207, 212
nodule of, 212
atrioventricular, 171, 172, 191, 201
left (bicuspid, mitral), 171, 172, 191, 201, 211, 212, 215
anterior cusp, 191, 212
posterior cusp, 191, 201, 212
right (tricuspid), 171, 172, 180, 191, 201–205, 208, 214
dorsal cusp, 191, 201–205, 208, 214
septal cusp, 191, 201–204, 208, 214
ventral cusp, 191, 202–205, 208, 214
bicuspid. See *Valve(s), atrioventricular, left*
cross section of valve, in vein, 46
ileocecal, 306
mitral, See *Valve(s), atrioventricular, left*
of coronary sinus, 180, 191, 202, 203, 208, 214
of foramen ovale, 206, 208
of Houston, 312
of inferior vena cava, 180, 202, 208, 214
of navicular fossa, 361
pulmonary, 171, 172, 204
anterior semilunar cusp, 191, 204, 209
left semilunar cusp, 191, 204, 209
right semilunar cusp, 191, 204, 209
tricuspid. See *Valve(s), atrioventricular, right*
Vas deferens, 352, 357–360, 362, 363
ampulla of, 340, 352, 357–359, 362
Vasa recti, 332
Vein(s), alveolar, inferior, 740
angular, 587, 620, 654–657
antebrachial, median, 40, 41, 100
appendicular, 272
auricular, posterior, 527, 528, 577, 579, 587, 787
axillary, 14, 19, 20, 63, 185, 187, 188, 227, 581, 582, 585
azygos, 158, 159, 184, 188, 195, 214, 215, 218–221, 226, 229, 230, 299, 331, 573
arch of, 184
basilic, 3, 14, 40, 41, 43, 47, 54, 63, 64, 98–100, 102
hiatus for, 47
basivertebral, 214, 572
brachial, 19, 21, 22, 54, 98
brachiocephalic, left, 20, 152, 172, 175, 177, 178, 185, 186, 188, 189, 193, 214, 223, 580–582, 586, 588, 592–594, 726, 755
right, 152, 172, 175, 177, 178, 184, 186, 188, 189, 195, 208, 223, 586, 587, 592, 755
bronchial, 184, 188
cardiac, anterior, 190
great, 190, 191, 193–195, 204
middle, 191–193, 195
small, 191
cecal, anterior, 296
cephalic, 3, 7, 11–14, 20, 40–43, 47, 54, 63, 64, 98–102, 231, 576, 579–582, 585, 586
cerebral, 634
great, 637, 638
inferior, 637
middle, 793
superior, 637
cervical, deep, 528, 593, 620
median, 587
superficial, 11, 585–587, 592
transverse, 585–587
circumflex, humeral, anterior, 54
posterior, 19, 42
iliac, deep, 239, 245, 317, 338, 376
superficial, 11, 232, 239, 246, 416
scapular, 14, 19
dorsal, 527
colic, 320
left, 272, 297, 303

Vein(s), colic, *continued*
middle, 272, 277, 285, 296, 297
right, 272, 285, 297
collateral, ulnar, inferior, 54
conjunctival, 678
coronary (of stomach), 285
left (of heart), 190
cremasteric, 239, 246, 347
cross section of, with valves shown, 46
cubital, median, 40, 41, 82, 98, 99
cystic, 265, 272
digital, dorsal, of foot, 468, 516
diploic, 632, 633
frontal, 620, 633
occipital, 620, 632
temporal, anterior, 620, 632
posterior, 620, 632
dorsalis pedis, 515
emissary, 620, 633
epigastric, inferior, 11, 149, 239, 243, 245, 256, 297, 317, 318, 321, 338, 340, 352, 376, 383
superficial, 232, 239, 245, 246, 414
superior, 11, 255
esophageal, 220, 573
facial, 574, 579, 580, 585–587, 597, 599, 600, 620, 622, 623, 727, 737
deep, 623
transverse, 574
femoral, 335, 414, 416, 418, 421, 422, 486
anterior, 414
deep, 422
gastric, left, 265, 270, 272, 273, 277, 300, 317
right, 270, 272
short, 270, 272
gastroduodenal, 285
gastroepiploic, 273
left, 265, 272
right, 266, 270, 272, 296, 316, 332
genicular, 460
hemiazygos, 219, 300, 573
accessory, 185, 219
hepatic, 183, 215, 230, 280, 299, 300, 316–318, 321, 340, 573
humeral circumflex. See *Vein(s), circumflex, humeral*
hypoglossal, 586, 587, 600, 620, 727, 737, 739
ileal, 272
ileocolic, 272, 296
iliac, common, 316, 317, 321, 338, 341, 355, 364, 376, 382
external, 243, 317, 338, 352, 354, 359, 376, 377, 383, 418
internal, 321, 354, 355, 382
iliac circumflex. See *Vein(s), circumflex iliac*
iliolumbar, 320, 321, 338
intercapitular, 102
intercostal, 184, 185, 218, 227, 229, 230, 239, 255, 299, 359, 526, 573
spinal branches, 226, 572
superior, left, 176, 185, 188, 208
jejunal, 272, 297
jugular, anterior, 575, 585–587, 591
external, 527, 575, 579–582, 585–587, 592, 597, 620, 727
internal, 152, 172, 178, 186, 576, 579–588, 591–593, 608, 620, 621, 640, 740, 754, 755, 793
bulb of, 576, 586, 754, 782
labial, anterior, of vagina, 244, 245
inferior, of face, 620
posterior, of vagina, 244, 245
superior, of face, 620
lacrimal, 638
laryngeal, superior, 587, 741, 744, 748, 755, 770
lingual, 756
deep, 737, 756
lumbar, 317, 321, 354
ascending, 218–221, 338, 340
mammary, external, 11
marginal, lateral, of foot, 468
right, of heart, 190
masseteric, 740
maxillary, 620
meningeal, middle, 620, 635
frontal branch, 635
orbital branch, 635
parietal branch, 635
mesenteric, inferior, 272, 285, 296, 303, 316, 317, 320, 331, 332, 340
superior, 266, 272, 285, 286, 296, 297, 303, 316, 317, 331, 332
metatarsal, dorsal, 516
plantar, 516

Vein(s), *continued*
nasofrontal, 587, 620, 638
oblique, of left atrium, 195
obturator, 340, 359, 376, 383
occipital, 527, 528, 577, 579, 585, 586, 620
mastoid branch, 528
of clitoris, dorsal, 245, 369, 376
superficial, 244
of glans penis, 351
of neck, 585
of penile bulb, 366
of penis, dorsal, 248, 343, 347, 348, 351, 355, 366
superficial, 246, 343, 347
of tongue, apex of, 740
of vertebral column, 573
of vestibular bulb, 371
ophthalmic, 620, 638
ovarian, 326
left, 318, 323
right, 285, 318, 383
pancreaticoduodenal, 272
paraumbilical, 11, 240, 243
perforating, 486
pericardiophrenic, 175, 177, 187, 573
pharyngeal, 620, 754
phrenic, inferior, 243, 321, 341
superior, 230
plantar, lateral, 515
medial, 515
popliteal, 426, 428, 441, 475, 476, 487, 489
portal, 178, 181, 183, 265, 266, 270, 272, 285, 286, 299, 300, 317
branches of, 299
right, 300
formation of, 266
prepyloric, 265, 272
pudendal, external, 245, 246, 414, 416
internal, 346, 428
superficial, 244
pulmonary, 184, 188, 213, 228
left, 158, 159, 185, 187, 189, 192, 195, 215, 220
right, 158, 165, 187, 189, 192, 195, 220, 227, 229
rectal, inferior, 312
middle, 383
superior, 272, 303, 382, 383
renal, 285, 318, 323, 325, 326, 331
left, 178, 272, 321, 331
right, 321, 322, 331, 332
retromandibular, 579–581, 585–587, 620, 622, 623, 727, 737, 740
sacral, lateral, 354
middle, 317, 321, 338, 383
saphenous, accessory, 486
great, 11, 232, 341, 412, 414, 416, 418, 422, 425, 460, 461, 468, 486, 487, 489–491, 515
small, 425, 426, 428, 441, 460, 468, 472, 475, 487, 489–491, 515, 516
scapular circumflex. See *Vein(s), circumflex, scapular*
sigmoid, 272, 303, 317
small intestinal, 285
spermatic, 285
spinal, anterior, 226, 573
splenic, 266, 272, 274, 278, 285, 286, 300, 303, 317, 331
subclavian, 177, 184, 580, 587, 592, 593
left, 172, 223, 586, 588, 755
right, 18, 172, 186, 223, 586, 594, 755
subcostal, 218
sublingual, 586, 727, 747, 756
submental, 580, 585, 587, 599, 600, 620, 623, 727
subscapular, 19
superficial, of arm, 40
of foot, dorsal, 468
of forearm, 41
of hand, dorsal, 102
of gluteal region, 425
of leg, posterior, 460
of limb, upper, 42, 43
of thigh, posterior, 425
of trunk, ventral, 11
suprarenal, 300, 318, 321–324, 326
central, 300, 324
left, 321
right, 322
suprascapular, 587, 592
supratrochlear, 620
temporal, middle, 623
superficial, 587, 604, 620
testicular, 243, 254, 320, 326, 338, 340
left, 317, 321, 323, 332
right, 317

Vein(s), *continued*
thoracic, internal, 11, 20, 149, 152, 175, 177, 187–189, 243, 585, 586, 593, 594
perforating branches of, 13
lateral, 3, 19, 20
superficial, 232
thoracoacromial, 15, 42, 585
thoracodorsal, 19, 20
thoracoepigastric, 3, 7, 11, 13, 14, 19, 20, 232, 582
thymic, 175, 188, 189, 586
superior, 592
thyroid, inferior, 152, 188, 582, 585, 586, 592, 593
middle, 586, 592
superior, 152, 580, 582, 585, 586, 588, 592, 594, 597, 600, 620, 726, 766
thyroid ima, 175, 187, 189, 588, 726
tibial, anterior, 466, 490, 491
posterior, 475, 490, 491
ulnar, 108
ulnar collateral. See *Vein(s), collateral, ulnar*
umbilical, 178, 181, 232, 255, 256, 261, 262
uterine, 11, 181, 383
valves in, cross section of, 46
ventricular, posterior (of heart), 192, 193, 195
vertebral, 194, 528, 582, 591–593
internal, 226
vesical, inferior, 243, 355
superior, 255, 340, 355
vorticose, 638, 678
zygomatico-orbital, 656
Vena cava, inferior, 150, 178, 180, 181, 189, 192, 195, 202, 215, 220, 221, , 229, 230, 258, 262, 265, 266, 272, 277, 280, 282, 285, 299, 300, 305, 316-321, 331, 332, 338, 340, 341, 573
superior, 136, 150, 157, 172, 176–178, 180, 181, 184, 186, 188–190, 192, 194, 195, 202, 208, 209, 212, 213, 219, 223, 228, 257, 586, 755
Vena caval aperture, in diaphragm, 299
Vena caval foramen, 229, 333–335
Venous mesocardium, 195
Venous network, of dorsal foot, 460
of dorsal hand, 102
Ventricle, of heart, left, 136, 150, 158, 159, 188, 192, 194, 195, 206, 207, 215, 257
right, 136, 158, 159, 188, 192, 194, 195, 201–204, 214, 257
of larynx, 685, 726, 736, 766, 769
Vermis, cerebellar, 639
Vertebra(e), body of, 157, 215, 228, 230, 298, 518, 522, 532, 533, 535, 536, 538, 540–543, 551, 554–556, 573
cross section of, containing spinal cord, 262, 518, 571, 572
sagittal section of, showing costovertebral articulations, 551
cervical, 25, 137, 530–536, 539–546, 571
foramen of, transverse, 532, 541
vertebral, 532, 541, 542
process of, articular, 541, 542
spinous, 219, 523, 541, 542
transverse, 258, 529, 542
1st. See *Atlas.*
2nd. See *Axis.*
3rd, 540, 543
4th, 543
5th, 541, 543
6th, 137, 140, 543
7th, 25, 523, 542, 543, 565, 591
coccygeal, 258, 388, 544–546, 561, 562
horn of, 561, 562
process of, transverse, 562
lumbar, 137, 223, 311, 335, 342, 393, 395, 397, 544–546, 555, 556, 558
body of, 230, 522, 555–558
cross section of, 522
foramen of, vertebral, 555
lamina of, 555
pedicle of, 555
process of, accessory, 555, 558
articular, inferior, 331, 556, 558
superior, 331, 555, 556
costal, 331, 555, 556
mammillary, 555, 558
spinous, 219, 230, 331, 332, 393, 522, 523, 555
transverse, 219, 332, 333, 393, 401, 555, 556, 558
surface of, articular inferior, 556
1st, 25, 137, 140, 230, 331, 334, 523, 558, 565
2nd, 558
4th, 311
5th, 417, 523, 565

Vertebra(e), *continued*
sacral, 354, 393, 544–546, 557, 559–561
thoracic, 159, 215, 544-549, 552–554, 558, 572, 573, 594
body, 547–549, 554, 558
section of, transverse, 554
facet of, articular, 573
costal, 547, 548, 558
foramen of, vertebral, 547, 554
lamina of, 219, 300, 549
pedicle of, 547, 553
process of, accessory, 558
articular, 547–549, 558
spinous, 214, 320, 518, 547–549, 554, 558
transverse, 219, 547–549, 554, 594
6th, 547, 548
10th, 549, 558
11th, 558
12th, 558
Vertebra prominens, 523, 543, 546, 574
Vertebral column, 214, 228, 543–546, 551, 573
Vertex, corneal, 680
Vesicle, seminal, 258, 352–355, 357–360, 362, 363
Vesicouterine pouch, 264
Vessels. See also *Artery (Arteries); Vein(s)*
abdominal, anterior, 239
posterior, 338
brachial, 21
central, of retina, 676, 678, 679, 681
circumflex, iliac, deep, 355
superficial, 414
colic, left, 316, 320
middle, 301, 302
coronary, 190, 191
duodenal, 332
epigastric, inferior, 149, 240, 355, 377
superior, 149, 240
femoral, 419, 421
gastric, short, 305, 316
gastroepiploic, left, 266, 270
right, 265, 270
great, of thorax, 194, 195, 586
iliac, common, 374
external, 382
internal, 376
iliac circumflex. See *Vein(s), circumflex, iliac*
infratrochlear, 657
inguinal, superficial, 244
intercostal, 240, 299
anterior, 149
jejunal, 303

Vessels, *continued*
labial, 244
laryngeal, superior, 597, 751, 753, 784
lymphatic, of pancreas, 279
of porta hepatis, 279
of stomach, 279
meningeal, middle, 635, 637
mesenteric, inferior, 303
superior, 297
musculophrenic, 149
obturator, right, 355
of abdominal wall, anterior, 239, 240, 243, 244, 246
of abdomen, inferior, 340
of arm, 54, 62
of axilla, 20
of back, upper, 528
of cranial cavity, base of, 638, 643, 644
of elbow, 73
of eyeball, 678
of face, deep, 623
superficial, 622
of forearm, 62, 73
of gluteal region, 428
of head, superficial, 623
of leg, posterior, 476
of neck, 588
of orbit, 654, 655
of pelvis, 340
of penis, 246, 347
of popliteal fossal, 441
of spermatic cord, 347
of suboccipital region, 528
of thigh,. posterior, 428
of thoracic wall, posterior, 226, 227
of upper extremity, 62
of ureter, 340
ovarian, 316, 377–380, 382
palatine, greater, 747
lesser, 692, 747
pericardiacophrenic, 185, 186, 189
perineal, 366
pharyngeal, 747, 754
phrenic, inferior, 318
plantar, 497
pudendal, external, 347
internal, 345, 355, 359, 366, 371, 376, 382
rectal, inferior, 355, 382
middle, 355, 382
superior, 355
renal, 320, 338
scrotal, 246

Vessels, *continued*
splenic, 266, 277, 278, 300, 316
sublingual, 740
submental, 599
sural, 475
testicular, 250, 252, 253, 320
thoracic, internal, 152, 585
lateral, 582
thoracoacromial, 580
thyroid, inferior, 751
tibial, posterior, 474, 475
tympanic, 799
uterine, left, 382
Vestibular bulb, 368, 371, 372, 382, 388
Vestibular cecum, 800
Vestibular window, 778, 790, 799, 805
Vestibule, of ear, 640
of larynx, 767, 769
of mouth, 690, 729, 732, 736, 737
of nose, 685, 690, 691, 732
of omental bursa, 151, 275, 305
of vagina, 367, 368, 388
Vibrissae, 722
Vinculum, 108, 109
Vocal fold, 726, 766–770
Vocal process, of arytenoid cartilage, 768, 770–772
Vomer, 601, 613, 631, 646, 683, 751

Window, cochlear, 640, 799, 805
branches of tympanic plexus to, 793
vestibular, 790, 799, 805
branches of tympanic plexus to, 793
Wing(s), of crista galli, 645
of sphenoid bone, 601, 614, 629, 635, 642, 645, 646, 663, 728
Wrist, arteries of, 105
bones of, 118, 119, 132, 133
joints of, 129, 131–134
ligaments of, 129, 131
roentenogram of, 133
synovial sheaths, of, 107
tendons of, 103, 107

Xiphoid process, of sternum, 141, 142, 145, 176, 186, 188, 230, 255, 299, 399

Zona orbicularis, of hip, 438
Zygomatic arch, 606, 608, 614, 615, 624, 625, 635, 646, 678
Zygomatic process, of frontal bone, 671, 689
of temporal bone, 612, 777, 789

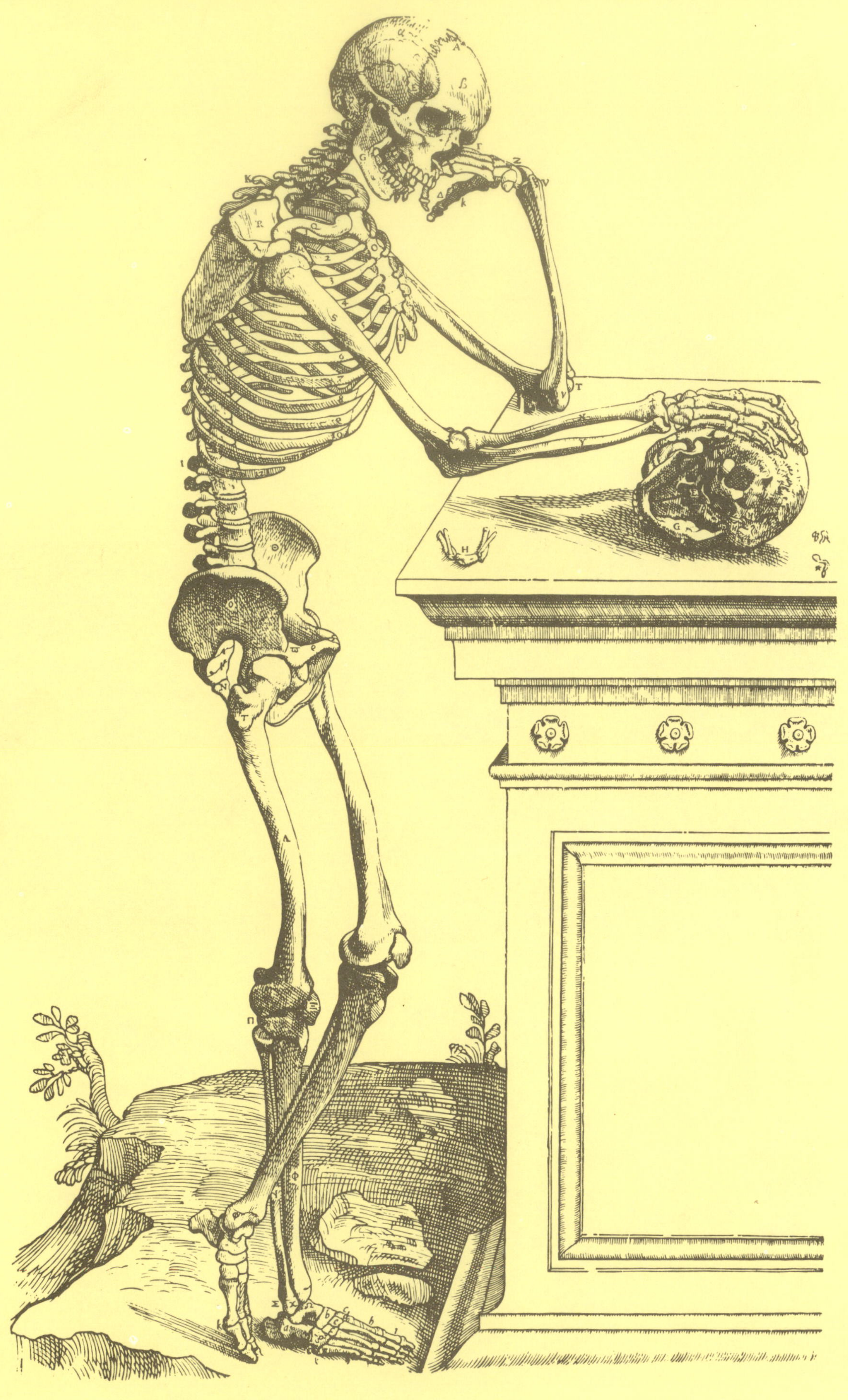